TABLE WINES

TABLE WINES

THE TECHNOLOGY OF THEIR PRODUCTION

M. A. AMERINE
& M. A. JOSLYN

SECOND EDITION

University of California Press
Berkeley, Los Angeles, London

University of California Press
Berkeley and Los Angeles, California
University of California Press, Ltd.
London, England
Copyright © 1951, 1970 by The Regents of the University of California
Second Edition, 1970
ISBN: 0-520-01657-2
Library of Congress Catalog Card Number: 69-12471

5 6 7 8 9

PREFACE

TO THE FIRST EDITION

The food industries, of which the wine industry is a part, have two important problems: reducing costs and maintaining or increasing the quality of their product. These are frequently interrelated and may involve complex economic problems.

The food industries are highly competitive; and the rising costs of operation, particularly for labor, and of materials, both processing and raw, have introduced serious economic problems. The success of any individual plant is dependent either on the development of specialty noncompetitive items which can be produced by virtue of localized availability of particular raw materials or through the ingenuity of the products-development staff, or on the more efficient production of standard items. Improvements in labor management through application of time and motion studies and improvements in facilities for and conditions of work and more efficient mechanization in harvesting, transportation, and handling of raw materials are necessary. This is particularly true of the wine industry, where the better selection of machinery for operation, improvement in plant layout, and balancing of unit operations are important.

Mechanization affects not only the economy of production but also the quality of the product, and particular care should be taken to make such changes as will both improve quality and reduce operating costs. Too often inefficient practices become so entrenched that producers fail to recognize that better (and even cheaper) ones are available. On the other hand, in an attempt to reduce handling costs, the product may be treated by processes which are inimical to quality. Finally, there are those who have maintained quality, but at such expense that the price is excessive.

The wine industry shares these problems. In this book we have tried to indicate the most desirable practices from the standpoint of cost and quality. But general recommendations are not easy to make because both high-quality and ordinary-quality wines are being produced.

The California industry already produces some excellent wines that can stand proudly on their own merits. It could develop more such wines,

v

and even finer ones. Only by producing fine native wines, consistent in quality and flavor from year to year and marketed under varietal or other distinctive names, will the industry fully exploit its potential market among discriminating consumers.

Of course the mass production and marketing is necessarily in ordinary-quality wines. But improvement in their quality would also benefit the industry. Many potential consumers might be converted into actual ones if the wines available at low or moderate cost were more generally fresh and fruity in flavor, as such wines should be. Improvement is often possible at little increased cost, sometimes even with decreased cost, by better choice of methods and greater care in all steps of wine production, from plant layout to bottling and labeling.

This book is intended to help producers, technicians, and chemists in the wine industry find ways to lower costs and improve quality of both fine and ordinary table wines. It has been organized in accordance with suggestions from wine makers, in the hope of making it a convenient handbook for them. The two introductory chapters point out some of the fields where improvements are needed, summarize the requirements of state and federal regulations, and report economic trends in the industry. Chapters 3 through 16 describe winery equipment and processes—both those now used in commercial practice in this state and those that have been suggested, or that are used in other wine-producing countries with similar problems. One chapter is devoted to each of the main types of table wines, and the special processes needed for it.

Because wine making is based on a variable raw material—grapes—and uses a complicated biological process—fermentation—it cannot at present be completely standardized or reduced to rules of thumb. Making fine wine is an art. Mastering such an art requires not only experience but also a thorough understanding of its principles. These principles are discussed in the final section of the book (chapters 17 through 20). We wish to emphasize that this has not been placed last because we consider it of lesser importance. On the contrary, we believe that progress in winery practice can only be developed on the basis of sound knowledge of grape and wine composition and of the principles of fermentation. We hope that wine makers who wish to improve techniques or have special difficulties to solve will find some useful suggestions in this section. Those who wish to do further reading will find a list of general references in the appendix, as well as a list of the literature cited in the text.

This book has developed out of two previous publications of the University of California: Agricultural Extension Service Circular 88, *Elements of Wine Making*, by M. A. Joslyn and W. V. Cruess; and our own Agricultural Experiment Station Bulletin 639, *Commercial Production of Table Wines.* Both of these publications, issued in 1934 and 1940, respectively, are now out of print. While material from these publications has been freely drawn upon, much new information has been added. Like the earlier publications, the book is based on wine investigations since Repeal in the University of California divisions of Food Technology and Viticulture; on the authors' observations of the wine industry, both in this country and abroad; and on our study of the literature.

The Wine Institute has kindly furnished many of the photographs reproduced here.

We wish to acknowledge the assistance of several of our colleagues. Professor S. W. Shear contributed most of the data and material for the chapter on economic trends. Professor George Marsh contributed the section on copper analyses. A number of staff members in the divisions of Viticulture and Food Technology have read the manuscript and made helpful suggestions. We are also most grateful for the critical reading of the manuscript by Mrs. Mary Rubo and Miss Lucy Lawrence, editors in the College of Agriculture. Needless to say, however, the authors alone are responsible for any errors in the text. Suggestions for improvement of the book would be gratefully received.

M.A.A. AND M.A.J.

Davis and Berkeley
June 1, 1950

PREFACE

The eighteen years since the first edition of this text have brought many advances in the technology of table wine production. The application of microbiology and biochemistry to wine production is now in full flower. This has been greatly facilitated by improved techniques for determination of small amounts of material: paper, thin-layer, and gas-liquid chromatography, enzymatic procedures, neutron activation procedures, and atomic absorption spectrophotometry.

It is clear that we should soon have a complete picture of the chemical components of wines which influence their color, taste, odor, and quality. It is not yet so clear how we can correlate this vast amount of information with the actual color, taste, and other characteristics of wines as perceived by the consumer. This is surely one goal of enologists for the last third of the twentieth century. From correlations of composition and sensory response we should be able to predict consumer response. To accomplish this, much study in the new field of the psychometrics of consumer reaction will be necessary.

Kielhöfer (1966) has defined four major problems of the German wine industry as sweetness, acidity, sulfurous acid, and stability. The most important future problems, he believed, are involved in more exact determination of the role of wine components: flavor materials, polyphenols and nitrogenous compounds and their changes in wine. One should hope, he says, to be able to detect falsification, define quality more precisely by analytical and subjective procedures, and to obtain more information on the health values of wine.

As Amerine (1955) has noted:

> The art of winemaking, like that of brewing, is undergoing a rapid metamorphosis in the 20th century. As more and more of the technical details influencing quality are brought under the control of the wine maker the art of winemaking will gradually become the science of winemaking.
>
> A variety of reasons may be given for this transformation. The scientific skills necessary for converting the industry have only been developed within the past 100 years and are by no means fully understood today. Even where the process variables

were recognized the necessary instrumentation for their automatic measurement and control were not available. However, the great progress in the antibiotic industry has shown how far the fermentation industries can be brought under strict control. A second reason is our desire, and need, for uniformity. The intuitive type of control produced some very fine products but also some very poor ones. Efficiency is another reason for these developments. The cost of the raw material and the labor overhead have gradually become so great that we must control their efficient utilization. Competition alone requires efficient controlled operation.

Of the three primary factors influencing the final quality of a wine we normally consider variety, fermentation procedure, and aging in a decreasing order of importance. Actually no fine wine can result when any of these three are neglected, so that in studying controlled fermentation there is no intention of minimizing the importance of the other two. Furthermore, there is an interrelation between the three which complicates their separate study. For example, a white grape requires a different fermentation procedure and method of aging than a red. A high acid must will ferment better and with less care than a low acid must and the wine will be more stable—other factors being equal.

This point of view was summarized by Silox and Lee (1948) as follows:

"Although fermentation is one of the oldest processes and has important economic significance due to its widespread use in industry, it must still be classified as an art rather than a science. This is in part due to the complex nature of fermentation variables that affect the evaluation of results. Secondly, although fermentation involves difficult interrelated problems in biology and engineering, men in these two fields have worked independently until recent years."

The application of the principles of bio-engineering to the wine industry has been slow in this country and even less exciting in Europe. No one doubts that automation and mechanization are desirable from the point of view of standardization and quality, cf. Amerine (1965a).

The problems of operating large wineries in the hot interior valley of California, particularly for the production of table wines, have been emphasized by La Rosa (1963). He noted the critical importance of control of harvesting and fermentations (especially cooling), new and simpler methods for developing stability, reduction of sugar and color pigment losses, purity of water, waste disposal, and quality control.

The text has been revised, amplified, and rearranged for this edition. The materials on principles (formerly chapters 17 through 20) have been incorporated into appropriate earlier chapters. As near as possible the current literature to December 31, 1966, has been checked, but considerable 1967–1968 material is included.

We are again indebted to the Wine Institute and our colleagues for their help. We are especially grateful that Dr. S. W. Shear came out of retirement to revise chapter 3. We also acknowledge the assistance of Professor L. L. Sammet and Mrs. Miriam A. Revzan. Mr. W. Allmendinger of the Wine Institute has been very helpful with statistical problems, and Mr. Hugh Cook of the same office supplied much useful information. Mr. Richard R. Lazarus, before his retirement from the Wine Institute, was of great help on many legal problems. Professors R. E. Kunkee, H. W. Berg, George Marsh, and H. J. Phaff have been especially helpful in reading certain chapters. Mr. C. S. Ough has been most generous in critically reading parts of the manuscript and checking a variety of points. We should add that none of the uncorrected errors are due to our collaborators. We hope our readers will let us know of such errors.

We acknowledge also the very kind cooperation of a number of wineries and government agencies in supplying photographs and tabular material. These are specifically acknowledged in the legends. Our thanks also to Mr. Arnold A. Bayard of Philadelphia for a generous gift to help in the final preparation of the manuscript.

M.A.A. AND M.A.J.

Davis and Berkeley
November 14, 1967

CONTENTS

CONTENTS

APPENDICES

LIST OF TABLES

xv

LIST OF ILLUSTRATIONS

INTRODUCTION

CHAPTER ONE

WINE TYPES OF THE WORLD

In this chapter only a broad general survey of wine types and the technological factors influencing their production is attempted. The continents and countries are listed roughly in the order of their importance in wine production. California wines are discussed in chapter 2.

EUROPE . . .

. . . is the main wine area of the world

There were nearly 25 million acres of grapes in the world in 1966. According to Protin (1966), the primary recent increases in acreage in recent years have been in Bulgaria, Czechoslovakia, and Latin America. World production has increased, on an acreage basis, from 534 gallons for 1946–1950 to 603 for 1951–1960, 689 for 1961–1965, and 745 for 1965. The overall increase is equivalent to only one to two tons per acre. Of course, total acreage was used and much of the fruit actually went for fresh fruit or raisins. Grapes are grown only in temperate zones (fig. 1).

Protin (1966) has also reported a slight but progressive decline in the amount of wine available for aging. At the end of the first year it amounted

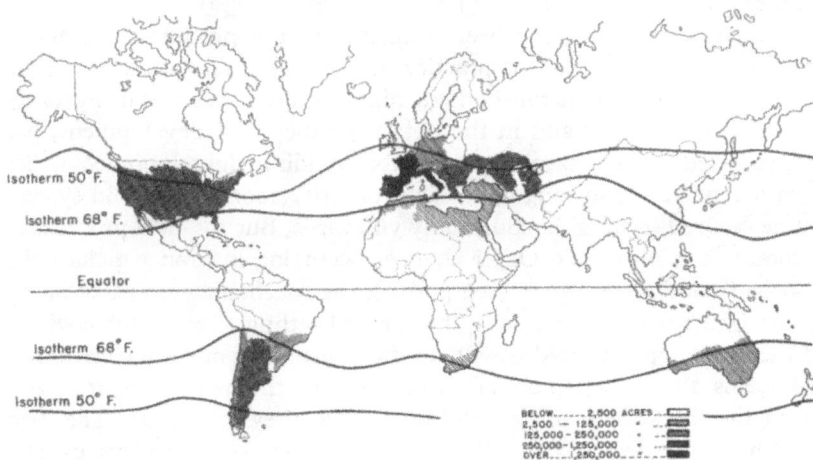

Figure 1. World distribution of grapes.

in 1962 to 46 per cent of the production; in 1963, to 47; in 1964, to 40; and in 1965, to 42. It is clear that wines are being brought to the market earlier. Protin (1966) gives tentative data on the relative cost of new wines (with an error for the United States of possibly ten fold). Nevertheless, costs appear to be remarkably similar for various countries at about 50 cents per gallon.

France

In any discussion of world wine types France must come first, not only because of the volume of wine produced but also because of the diversity of types and the distinctiveness of the wines produced. The widespread use of French wine type names by other countries is testimony to this. In 1965 France had over 3 million acres of grapes and produced over 1.7 billion gallons of wine. The main high-quality wine types produced in France, roughly from north to south, are the Alsatian varietals, Champagne, Chablis, white and red Burgundy, Beaujolais, Rhône, and the white and red Bordeaux.

Information on the legal restriction on labeling of French wines are given in chapter 2. France has extensive laws delimiting the viticultural districts and even the vineyards. These laws, *appellations d'origine*, not only set the exact boundaries of the district but may specify the varieties which may be grown, and the maximum production and minimum alcohol content required if the wines are to be labeled with the district or vineyard name. Wines labeled *appellation contrôlée* have the most prestige. Slightly less restrictive are the regulations on *vins délimités de qualité supérieur* (usually abbreviated V.S.Q.S.). (See also p. 7.) For an excellent history of the genesis of these laws see Kuhnholtz-Lordat (1963), Lichine (1967), Quittanson (1965), and Semichon (1930).

In recent years there has been a change in the production of many red wines of France. The old practice of fairly long fermentation on the skins, which resulted in full-bodied, high-tannin wines requiring fairly long aging in the cask and in the bottle for their full development, has been replaced by a shorter fermentation yielding lighter-bodied wines which are not aged so long and reach maturity sooner. The old system may be favored in the finer Burgundy vineyards, but the new procedures predominate in Bordeaux. Other changes occurring in France include the use of *maceration carbonique* (see p. 610), the mechanization of crushing and the fermentation operations, the earlier bottling of white table wines, and the more sophisticated use of clarification and stabilizing techniques.

Alsace is the only region of France where the wines are regularly labeled for the variety of grapes from which they are made. The best wines are made from White Riesling, Traminer, or nowadays usually

4

Gewürztraminer, and Pinot gris (used to produce a wine labeled Tokay d'Alsace—this appellation appears to be in clear conflict with the Hungarian Tokay appellation). Sylvaner is widely planted, but its wines are less distinctive. Other varieties are grown but their wines are seldom exported. District names from specified delimited districts are used on some labels.

In Alsace, according to Dietrich (1966), the one common practice is a long cool fermentation. Another important factor influencing quality is close control of the date of harvesting of each variety. Sugar content, total acidity, and the condition of the fruit (amount of mold, etc.) are taken into consideration. The vintage is generally between October 1 to 20. Sylvaner, Pinot blanc, Chasselas doré, and Muscat blanc are harvested first, and Traminer, Pinot gris, and White Riesling later. Moldy fruit may be kept separate. The grapes are ordinarily pressed without destemming and sometimes with a minimum of crushing. The musts from moldy fruit are usually, and preferably, settled. That of sound fruit is not settled; in fact, it produces better wine if not settled, according to Dietrich (1966). Nevertheless, both settling and centrifugation are practiced. The musts are sulfured, but this is kept to a minimum if a subsequent malo-lactic fermentation is desired, as is essential in cool years. Sugaring of the musts to ensure adequate alcohol, *chaptalization*, is used in certain years under limits fixed by a committee of experts. Wines made by sugaring are not entitled to use *Grand Cru, Grand Vin*, or *Grand Réserve* on the label. The tendency in recent years has been to harvest late to avoid *chaptalization*. Surprisingly, for such a cool region, addition of 0.2 gr per liter of tartaric acid before fermentation is recommended. Citric acid at the legal limit of 0.05 gr per liter does not give comparable sensory results. Dietrich (1966) found no difference in the volatile acidity of Traminer wines made without addition of acids compared to those to which citric or tartaric acid was added prior to fermentation. He also reported that grapes grown on calcareous soils had more sugar and titratable acidity that those grown on non-calcareous soils. The increase in sugar is probably a temperature effect. It is difficult to explain the higher acidity.

The delimited region of Champagne produces mainly white sparkling wines, with a small amount of rosé. The bottle fermentation process used is described in chapter 17. The cool climate necessitates growing the early-ripening Pinot noir in order to have adequate sugar for a balanced wine. Chardonnay, a white variety, has less sugar. Its lighter-colored and fruity wines are useful for blending. The usual blend contains about one-fourth to one-third Chardonnay. In the warmer years some wholly Chardonnay wine, labeled *blanc de blancs*, is produced. Connoisseurs are divided in their opinion as to the quality of these all-white wines. Not all parts of the Champagne region produce wines of equal value to make sparkling wine. The prices paid for grapes from the different areas reflect this (see Simon, 1962).

The actual production of Champagne is carried out by a limited number of firms. All bottle a blended nonvintage wine. The driest is labeled brut, and sweeter types sec and demi-sec. In many years a vintage wine is also produced. This will consist of wines, usually from the best districts. Since the quality of bottle-fermented wine depends partly on the length of time it is aged on the yeast in the bottle (p. 671), the French regulations provide that Champagnes must remain in the bottle at least nine months before disgorging and sale. The most notable recent change in Champagne production has been the use of concrete and lined steel containers for the fermentation, transportation, and storage of the new wines. This has resulted in lower costs of production and fresher wines. It may also have removed the finer differences in *cuvées* of the producers. There has been great improvement in recent years in the technological control in making the wine.

Sales and production of Champagne have increased markedly in recent years. Wines of the vintages of 1952, 1953, 1955, 1959, 1961, 1962, 1964, and 1966 are the best of recent years. Popular discussions of Champagne are given by Simon (1962) and Forbes (1967). For an excellent technological discussion see Chappaz (1951).

Chablis is a small delimited region, but the type name is copied by wineries from Spain to California. The imitations bear little resemblance to the true Chablis, which is made from Chardonnay growing on a chalky soil. The cool climate limits ripening; so the wines are generally rather low in alcohol and pH and high in total acidity. The reputation of this wine to serve with the fish course is thus justified. The best recent years are 1955, 1959, 1961, 1962, 1963, 1964, and 1966.

The true Burgundy wine district stretches from Dijon to just south of Beaune, barely thirty miles. It is called the Côte d'Or because of the quality and value of its wines. Most of these are red, produced from Pinot noir and its cultivars. In the more northern Côte de Nuits area the wines are especially fine: alcoholic and of full red color. The southern Côte de Beaune red wines are no less characteristic but slightly less distinguished. The cool climate requires supplemental addition of sugar to the musts in most years—sometimes to excess, in our opinion. Nevertheless, when made from fully ripened grapes the red wines of Burgundy are among the most distinguished produced in the world. (Fig. 2 shows the winery at Clos Vougeot in Burgundy.)

There is a limited production of white wines from the varieties Char-

Figure 2.   Clos Vougeot and its surrounding vineyard.

donnay and Aligoté. The former are by far the better in quality. They have, at best, a rich flavor and an inimitable aroma and bouquet. In recent years more sulfur dioxide than formerly has been used by some producers.

The finest red Burgundy wines of recent vintages, according to Lichine (1963) and others, are those of 1945, 1947, 1949, 1952, 1955, 1959, 1961, and 1962. The vintages of 1964 and 1966 have also produced good wines. The vintages of 1947, 1949, 1952, 1953, 1955, 1958, 1959, 1960, and 1961 were very good for whites, according to Lichine. There are also reports of good white Burgundies in 1962, 1963, 1964, and 1966.

Just south of the Côte d'Or are the southern Burgundy vineyards of the Côte Chalonnaise (red and white wines), Mâconnaise (also white and red wine), and Beaujolais (mainly red). Some are made from Pinot noir (e.g., Mercurey and Givry in the Côte Chalonnaise) and from Chardonnay (at Pouilly-Fuissé in Mâconnais), but the revealing grape, especially in Beaujolais, is Gamay, which produces early-maturing red wines with a spicy, fruity character that no other wine has. Some of these wines carry village names, and there is a great variation in quality. The recent vintages of special quality are 1961, 1962, 1964, and 1966. Over half of the wines are entitled to an *appellation contrôlée*.

According to P. André (1966), the traditional procedure for the vintage in Beaujolais is to harvest directly into 90-liter (about 26-gallon) containers (*bennes*). During the filling the grapes are pressed down by hand so that about 50 per cent of the fruit is crushed. At the wineries these are dumped directly into 50-hectoliter (1,320-gallon) open tanks without destemming or further crushing. The free-run juice is drawn off when the specific gravity is 1.01–1.02 (2.5°–5.5° Brix). The marc is

7

then pressed, first very lightly, to crush the still uncrushed fruit. This first-pressed juice is much appreciated for its high aroma (locally it is called *paradis*). The full pressure is then applied; this juice is more astringent and less sweet. The free-run, *paradis*, and final press are combined and fermented to dryness. André (1966) compared the wines made in this way with those made from fully crushed grapes. He notes that the acidity of the must is higher in those prepared in the traditional manner. In his small-scale test, 50 liters were free-run, 13 *paradis*, and 16 press. In the procedure using crushed grapes, 61 liters were free-run and 12 press. While the yield with two pressings is slightly less, this is compensated for by the higher quality. The sensory evaluation strongly favored the wines produced in the traditional manner. They were also slightly higher in alcohol and higher in fixed acidity. One wonders whether the quality difference was not due to the slower rate of fermentation at the cooler temperature.

The Jura is no longer an important wine-producing district. At Château Chalon and a few other places a film yeast wine is produced. Francophiles find it like a Spanish fino type sherry, but the wines are tart and of only moderate alcohol.

Wines are produced all along or near the Rhône from Vienne to Avignon. The four most important districts are Côte Rôtie, Hermitage, Châteauneuf-du-Pape, and Tavel, but there are a number of other interesting areas: Condrieu, Cornas, Saint-Péray, Crozes-Hermitage, Château-Grillet, Gigondas, and Lirac. The climate is more moderate here than in Burgundy and there is less difference between vintages. In fact, it can be too warm for the highest quality at Châtauneuf-du-Pape (e.g., as in 1959). Many of the vineyards in the first two regions are planted on steep slopes, and the soil at Châteauneuf-du-Pape is very rocky. The best recent vintages have been 1961, 1962, 1964, and 1966. Wines from the whole district are called Côtes-du-Rhône.

Côte Rôtie (less than 100 acres) is a red wine produced mainly from Syrah (80 per cent), but with white grapes of Viognier (20 per cent) mixed in. In spite of this the wines have a full color and age very well.

Hermitage (less than 1,000 acres) is made as a white wine (from Roussanne and Marsanne) and as a red (from Syrah but often with some of the above-mentioned white grapes added). The white is a very full-bodied wine in the best years, and ages very well. The reds are fine sturdy wines that can be profitably aged for many years.

Near Avignon in the southern Rhône is Châteauneuf-du-Pape (less than 6,000 acres). The district is planted to a number of varieties, mainly Grenache with a little Syrah and several other varieties. (Cinsault is favored by some growers as a quality variety.) The wines have a moderate red color and mature somewhat faster than the northern Rhônes. A small amount of white wine is made from Clairette and Grenache blanc.

Opposite Avignon a rosé, Tavel, is produced from mainly Grenache (up to 65 per cent), but with some other varieties (Cinsault, Mourvèdre, etc.) mixed in to modify the orange-pink color. It is rather alcoholic and should be drunk when young. For a critical discussion of Rhône wines see Hallgarten (1965). Less than 20 per cent of the Rhône wines are entitled to an *appellation contrôlée*.

The longest river in France is the Loire. As it turns west it supports at least five viticultural areas: Pouilly-sur-Loire and Sancerre, Vouvray, Angers and Saumur, Chinon and Bourgueil, and Muscadet. Most of the wines are white, but they vary widely in alcohol, sugar content, and quality.

The wines of Pouilly-sur-Loire (often sold as Pouilly-Fumé or Pouilly-Blanc Fumé) and of Sancerre are made from several varieties: mainly from Sauvignon blanc and Chasselas doré. The first variety is often called Blanc Fumé. It is apparently the same as the Sauvignon blanc of Bordeaux, but the wines do not have the aromatic character of the wines of that variety grown in Graves. The wines are generally dry, tart, and often low in alcohol.

Vouvray is the product of the variety Chenin blanc. The wines may be dry, slightly sweet, sweet and still, or slightly sweet and gassy. The sweeter wines are produced from botrytised grapes. In especially favorable years wines of a true natural sweet character and exceptional quality are produced. Some of the wine is used for producing sparkling wines. The best recent years are 1961, 1962, 1964, and 1966. The wines are fruity and mature rapidly, but they cannot, on the whole, be considered high-quality wines. The wines of Saumur-Angers are similar in all respects. Some of these are made sweet for the American market.

The wines from Bourgueil and the Chinon in Touraine are produced from Cabernet franc. They are often very light in color, almost a rosé. Though praised by some experts, they are often not aged sufficiently and can be thin and unbalanced. They require a malo-lactic fermentation which appears to be difficult to control.

Near the end of the Loire are the white wines called Muscadet, produced from the variety Melon. The product is fruity, like all the Loire wines, and not, even at its best, of exceptional quality. In spite of the name, these wines have no distinguishable muscat character.

The large Bordeaux district includes the following main areas: Médoc, Graves, Sauternes, Saint-Emilion, and Pomerol. Among the secondary regions are Entre-Deux-Meres (including Primières Côtes-de-Bordeaux, Sainte-Crois-du-Mont), Côtes-de-Blaye, Côtes de Bourg, Côtes de

Fronsac, Lalande-de-Pomerol, Montagne-Saint-Emilion, Néac, and others. In contrast to the other regions of France, there are often fairly large vineyard holdings and the practice of using vineyard names is well established (usually as "château").

The Médoc extends westward from Bordeaux along the Gironde. The main wine-producing communes are Margaux, Moulis, Saint-Julien, Pauillac, and Saint-Estèphe. Soils of varying exposure, composition, temperature, and depth are planted to a number of red varieties. These include Cabernet Sauvignon, Cabernet franc, Merlot, Malbec, Petit Verdot, Carmenère, and Saint-Macaire. It is apparent that the main quality factor is the variety, particularly the percentage of Cabernet Sauvignon or Cabernet franc. The well-drained soils may be a further favorable factor. Sequin (1966) considers that the physical conditions of Médoc soils provide for a regular supply of moisture and that this is an important factor in the quality of the wines produced. However, the soils are well drained and dry out in the fall. No comparative data are given. As Sequin himself admits, some of the best vineyards have as many as three types of soils.

Formerly the wines were fermented on the skins for one to several weeks. Recently the period on the skins has been reduced in order to produce earlier-maturing wines. Whether this will reduce the aging potential of the wines, and hence their eventual quality, is not yet known, but apparently it will.

The classification of 1855, made by the Bordeaux wine merchants, reflected the quality of the wines as of that date. The better vineyards were classified into five growths based on the quality of the wines sold before 1855. Because of changes in the size and location of the vineyards, as well as changes in ownership, the classification can no longer be valid, if it ever was. Furthermore, different varieties were planted after phylloxera, and, of course, the wine makers and wine making procedures are now different. Finally, even in the best years, some grapes were picked too soon or too late. In 1964 some of the most famous vineyards were picked after the heavy rains and were of lesser quality than those harvested before the rains. Because of the world-wide increase in demand for fine wines since World War II, the so-called first growths are in such heavy demand that they often are overpriced. Unfortunately, the demand has been so great that some of the more famous châteaux have château-bottled their wine even in years of ordinary quality. Some very poor quality château-bottled 1963's, for example, have reached the American market. The Médocs are full-flavored but light in body, often with only 10.5 to 11.5 per cent alcohol. However, the wines age for many years, thanks to their good tannin content. The best recent years have been 1955,

1959, 1961, 1962, 1964 (some), and 1966. Wines of 1953, 1952, 1949, 1947, and 1945 are still usually in good condition. In fact, the 1945's seem not yet to be mature.

Just to the east of Bordeaux lies Graves, a district known primarily for its white wines, although it actually produces more red than white. The reds are softer and perhaps less colored than the Médocs and with a few exceptions are not château-bottled. The whites are produced from Sémillon (about one-third) and Sauvignon blanc (about two-thirds). They are again full-flavored and full-colored. Usually they are dry, or nearly so. However, many of them are heavily sulfited to prevent yeast growth and darkening of color. Even this is not always successful.

Figure 3. Château d'Yquem and its surrounding vineyard.

Still further east is Sauternes. The wines are produced from botrytised grapes (p. 641) of Sémillon and Sauvignon blanc and are quite sweet, often cloyingly so. But at their best they are very luscious wines. Barsac is a part of the Sauternes district and some wines are so labeled. They are usually but not always less sweet than wines labeled Sauternes. The best wines are château-bottled. Recent vintages of high quality include 1955, 1959, 1961, 1962, and 1966. The famous Château d'Yquem and its surrounding vineyard are shown in figure 3.

Of the vineyards east and north of Bordeaux, the two important districts are Saint-Emilion and Pomerol. The climate here is a little more equitable than in the Médoc, but grapes do not get overripe. In fact, as in the Médoc, sugaring of the must is used in the cooler years. Similar varieties are planted, with more Cabernet franc than in the Graves or Médoc. Again the best wines come from certain châteaux. The location of the vineyards and the percentage of the varieties in the vineyard determine the quality of the wine produced. The better châteaux have a tradition of producing high-quality wines, handed down from some astute former owners, which guides the viticultural and enological practices. These actually may be more responsible for the quality than vineyard location and varietal assortment. However, the fact that no high-quality wine has been produced in many areas would seem to indicate that in these areas at least the environment is less favorable.

The wines of Saint-Emilion and Pomerol are sometimes called the "Burgundies of Bordeaux." We do not find this to have much meaning. Possibly it refers to their slightly fuller (more alcoholic and less acid?) character. They taste and smell very much like other Bordeaux wines. The best years are as in the Médoc. As in other regions, even in the best years some wines are of less satisfactory quality and in the poorer years a lucky havesting and vinification may produce a respectable wine.

This by no means exhausts the wine-producing districts of France. In fact, about 40 per cent of French wine is produced in the *départements* of the Midi: Aube, Gard, and Hérault. Most of this is low-alcohol *vin ordinaire*. There are locally appreciated wines in the Pyrénées-Orientales, Tarn, Dordogne, and elsewhere. For a general discussion of French wines see references in Appendix A. Galet (1956–1964) gives the best data on the varieties of grapes used in France.

Italy

Wine making in Italy may have preceded the Greeks—at least in a primitive way by the Etruscans. However, the main impetus was from the Greek colonies, particularly in Sicily. By the second century B.C. wine production was important in quantity and quality, both for local and export trade. This appears to be due to the demand and to increasing ability to store and ship wines without undue loss.

Among the best vineyards (of many) of Roman Italy were those of Campania and Latium. The Falernian of Campania was especially noted. Its modern Italian counterpart is not known. Others from these districts were Capuan and Cauline, Guaran (from between Cumae and Naples), Pompeian, Fromian, and Labican. Younger (1966) emphasizes that Tuscan wines were unappreciated or unwanted, possibly because mainly muscat varieties were used. The Ligurian Lunensan was appreciated, as

was the Rhaetic and Pucine from near Venice. Younger (1966) and others emphasize the wide distribution of Italian wines throughout the Roman Empire. To protect the home industry Rome tried unsuccessfully to limit grape growing in Gaul and elsewhere. For a detailed study of the history of Italian wines see Marescalchi and Dalmasso (1931–1937).

In the period 600 to 1400 there was a decline in the acreage and number of vineyards, but wine making continued. Younger (1966) believes that the Church played only a modest role in the survival of the grape and wine industry. However, the very large number of vineyards which were started by monasteries indicates the interest of the Church in wine making—for an essential part of the Mass, as a table beverage, and as an article of trade. Trade in wines continued and greatly expanded during the Renaissance.

At present, Italian wine production averages about 1.7 billion gallons per year, and in recent years has sometimes exceeded French production. There are nearly 6 million acres of grapes, but the figure is only approximate, for, with 66 per cent of the vines being grown in mixed plantings with other crops, conversion to acres of vines is difficult (Cosmo, 1966). These mixed plantings produce the least satisfactory grapes for wine production because of overproduction where vines are grown on trees or in arbors, and the difficulty of fungal control in such plantings. Italian wine production has increased remarkably in this century. Mixed plantings have gradually decreased, and specialized vineyards have increased. Perini (1966) indicates that about half the proprietors have less than twelve acres. In 1963 there were 554 coöperative wineries in Italy, and the number appears to be increasing. Per capita consumption has recently increased (95 liters in 1958 to 115.6 in 1963).

Grapes are grown in a wide range of climatic conditions from the very warm Mediterranean climate of southern Sicily to the cool, continental climate of the Alps. A wide range of soil conditions exists. In some areas grapes are grown on slopes as steep as any of those of the Rhine.

A bewildering number of varieties of grapes are grown in Italy. Some are known under a number of local names and are not easily identifiable. Certain regions produce table grapes as well as wine grapes, but these represent only a small fraction of the wine grape production and an insignificant amount would be crushed. However, many high-quality varieties are known; if they are not more widely grown it may be because of lack of demand for high-quality wines.

There are thousands of small vineyards in Italy, most of which make their own wine, although many of them produce only a small amount of wine. The small vineyardist–wine maker does not necessarily mean poor-quality wine, but there are too few trained wine makers among the

small growers. Proni and Pallavicini (1962) consider that the main problem of Piedmont viticulture is the necessity for larger vineyards to reduce cost of operation and hence to raise the standard of living. Reorganization of vineyards into larger units was strongly recommended.

The long history of political subdivision in Italy seems to have restricted movement of wine and hence reduced critical evaluation. Wine-making procedures vary widely from vineyard to vineyard in efficiency, sanitation, and the desire and knowledge of how to produce high-quality wine. A hopeful note is the number of modern coöperative wineries with modern equipment.

The most favorable factor is the long tradition of wine drinking. Recently the demands of the European Common Market have led to more precise definition of many types of wine and to delimitation of the regions where they can be produced. (See Cosmo, 1966; Capone, 1963; and Bruni, 1964.) The main stumbling block is still "Chianti." The traditional Chianti *classico* district presumably should be the only one producing chianti. However, demand for chianti is large and much wine from nearby and even remote districts finds its way to the market labeled as chianti.

Since 1963 Italy has begun to develop controlled geographical appellations on a nation-wide scale. As of 1966 the committee working on this subject had approved thirty-two appellations and reserved judgment on nineteen. Among those finally approved are Barolo and Barbaresco (Piemonte), Frascati and Est! Est!! Est!!! de Montefiascone (Lazio), Trebbiano, Sangiovese, and Merlot de Aprilia (Emilia-Romagna), Brunello di Montalcino, Vernaccia de San Gimignano, and Bianco di Pitigliano (Toscana), and Ischia Bianco and Rosso and Bianco Superiore (Campania). Also approved were Asti Spumante and Moscato d'Asti Spumanti (Piemonte), Chianti, Vino Nobile di Montepulciano, and Elba Bianco and Rosso (Toscana), and Gutturnio dei Colli Piacentini and Albana di Romagna (Emilia-Romagna).

In the final stages of approval are Gattinara (Piemonte), Lugana Rosso (Lombardia), and Conero and Verdicchio di Matelica (Marche). Still under study and definition are Sangiovese di Romagna (Emilia-Romagna), Soave, Valpolicella, and Bardolino (Veneto), Carema, Caluso Passito, and Erbaluce di Caluso (Piemonte), Verdicchio dei Castelli di Jesi (Marche), Franciacorta Pinot and Rosso, Riviera del Garda Rosso, Riviera del Garda Chiaretto, Bianco and Rosso della Parrina, and Tocai di Lugana (Lombardia), Franciatorta Bianco and Bianchello del Metanro (Marche), and Vernaccia di Serrapetrona. As with the French appellations of origin, each district is carefully defined and varieties and production techniques are specified. Many of the Italian appellations have variety

names in conjunction with the region. This list does not include many other Italian varietal types which presumably will eventually be defined. (For further information see Anon., 1966 *h*.)

Greater attention might be given by Italian enologists to encourage production of quality wines and to letting the connoisseur know which are the best wines. This would apply also to the export market.

Because of these favorable and unfavorable factors, it is more difficult to find a truly distinguished wine in Italy than in France or Germany.

The most French-looking district is Piedmont, where, in fact, French influence is more noticeable. Vines are generally planted in proper vineyards and well cared for. The wines are often sold with a varietal name: Barbera, Fresia, Grignolino, Nebbiolo, and so on. Apparently a number of cultivars of each variety are cultivated, for a wide range in color is found in the bottled wine. This may, of course, be due also to blending to different standards. Barbaresco, Barolo, and Gattinara are regional types made mainly from Nebbiolo. According to Proni and Pallavicini (1962), the production of various red grapes in Piedmont is: Barbera (3 million hectoliters), Fresia (800,000), Dolcetto (750,000), Nebbiolo (70,000), Bonarda (43,000), and Grignolino (32,000). The production of whites is: Cortese (145,000) and Muscat blanc (325,000). Total production is about 161.6 million gallons. It is surprising that most of this is still made by small proprietors.

The climate of Piedmont is moderately cool and the acidity of the musts is fairly high, especially in the Barbera. A malo-lactic fermentation is therefore highly desirable. Unfortunately some wine is bottled before the malo-lactic fermentation is completed. Gassy (*frizzante*) wines thus find their way to the market. This does not appear to hinder Italian appreciation of Piedmont wines, but it does deter consumers elsewhere.

The wines of Lombardy are seldom exported, although Valtellina is well known. Nebbiolo is the revealing variety, but, judging from the color of the wines produced, it may be a different cultivar than that used in Piedmont. The wines of Lake Garda are also occasionally exported. Lugana is a well-known wine.

The Italian Tyrol (Trentino-Aldo Adige), in our opinion, produces some of the best Italian wines. The standards of enology are usually impeccable, and a variety of fine red and white table wines are produced. Some of these are sold with a district name, Santa Maddalena, for example, or a district plus a variety name. Among the varietal wines easily found on the American market are Gewürztraminer, Italian Riesling (Wälsch-riesling), Sylvaner, Merlot, and Cabernet. This use of varietal names suggests that variety is the single most important factor in the quality of the wines of the district.

From near Verona come three well-known wines: the reds Bardolina and Valpolicella and the white Soave. The Bardolina and Valpolicella are light-colored and light-bodied reds made from a mixture of three or four varieties, including Negrara, Molinara, Rondinella, and Rossignola. Like most Italian wines they mature early and do not, except occasionally, improve by long bottle-aging. Soave is mainly made from Garganega and Saint Emilion (here called Trebbiano; in the Midi of France it is called Ugni blanc). The generally neutral character and lack of real quality of the wine is probably due to the latter variety. (It also accounts for the poor quality of most white chiantis.)

Tuscany's most famous wine is chianti. Since 1932 there have been a number of delimited districts: Montalbano, Rufina, Colli Fiorentine, Colli Senesi, Colli Aretini, and Colline Pisane. Recently other small districts have been added. Chianti is produced throughout Tuscany, but importation of wine from other districts is not unknown. The best is made from 70 per cent Sangioveto (syn. Sangiovese) and Cannaiolo nero (plus Colorino, Saint Emilion, and other varieties) grapes. Some of it is made by the usual red wine technique, aged in the wood and in bottles, and, when shipped with a vintage date by a reputable producer and carrying the black rooster seal on the neck, can be very good. The prefix "Vecchio" can be used if the wine is two or more years old. "Riserva" is used for wines three or more years old. The best is bottled for aging in Bordeaux-shaped bottles rather than in the difficult-to-bin straw-wrapped *fiasco*. The best chiantis have a minimum of 12.5 per cent alcohol, but lesser-quality wines can be sold at 11.5 per cent.

Some chianti is produced by the *governo* process which is almost unique to Tuscany. The red wine is made in the usual way from Sangioveto. The Cannaiolo nero grapes are placed in narrow boxes and stacked from the time of harvest until December or even later. They become shriveled during this time. They are then crushed and added to the already fermented Sangioveto wine. The slow secondary fermentation that ensues may last until March. This second fermentation is more efficient in utilization of sugar. It undoubtedly improves the color and increases the tannin and alcohol content. It may leave residual sugar, result in slightly gassy wines, and can yield wines with very special flavors which may be of bacterial origin. The process seems an anachronism in modern wine making and is difficult to control. From near Siena, Montepulciano (also called Nobile di Montepulciano) is made (again mainly from Sangioveto). It is sometimes very good.

Orvieto from Umbria is a rather neutral and uninteresting white table wine which is sometimes dry and sometimes sweet. A mixture of varieties (Procanico, Grechetta, Verdello, and Drupeggio) is used.

It has been delimited since 1931 and a *consorzio* has protected its interests since 1934. The modern wines do not seem worthy of its ancient fame.

Est! Est!! Est!!! (or the Vino di Montefiascone) is from the province of Viterbo (Lazio). The main varieties used are Rossetta, Trebbiano, and Malvasia toscana. Its medieval fame may have been due to the use of muscat grapes. Both dry and sweet wines are produced.

Castelli Romani (including Frascati) comes from a delimited district just south of Rome. The varieties are Bellone, Malvasia di Candia, Malvasia gentile, Buonvino, and the ubiquitous Trebbiano. It even has a *consorzio*. In spite of the great fame of Frascati since Roman times (the exact location of the ancient wine is not known), the wines are of variable quality and many are carbonated.

Falerno farther south and Capri, Ischia and Lacrima Christi del Vesuvio (or Lachryma Christi) near Naples are well-known wines. Falerno bianco is made from the variety Falanghina; Falerno rosso from Aglianico, Piede di Palumbo, and other varieties; Capri bianco from Greco, Fiano, Biancolella, and others; Capri rosso and rosato from Tintora, Guarnaccia, Piede di Palumbo, and others; Lacrima Christi (or Vesuvio bianco) from Greco di Torre, Coda di Volpe, Biancolella, and others; and Lacrima Cristi rosso or rosato (also called Lacrima di Somma) from Aglianico, Piede di Palumbo, Soricella, Palombina, and others. We confess that it has not been our good fortune to find these wines outstanding in any way. The varietal background is not uniform for any of them, and they often reach the market in poor condition.

In Sicily the table wines from grapes grown on the upper slopes of Etna can be agreeable, at least when drunk locally. An Etna bianco is produced from a number of white grapes. The Etna rosso is made from Nerello mascalese, Nocera, and other varieties. We prefer the red. In western Sicily the Corvo di Casteldaccia is a well-blended and generally well-stabilized table wine. It is generally available in Italy and is safe if not great. The white is made principally from Catarratto and Inzolia. There is a red made from Perricone, Catanese, and other varieties. It has been compared to the reds of Bordeaux—why we do not know.

In contrast to the average quality of the table wines of Italy are the large number of sweet wines, some not fortified but fermented to 15 or more per cent alcohol, and some fortified. Many of these are well aged and have real character.

When the grapes are allowed to dry somewhat, many *passito* wines are produced. The muscats of Asti and elsewhere, many Sardinian wines, Caluso, the Aleaticos from Tuscany, the Malvasias of several areas, particularly of the islands just north of Sicily, and muscatels of Siracusa,

Zucco, Noto, and elsewhere in Sicily, and even the muscatel of Pantelleria are all interesting sweet wines which may be classified as table or dessert wines depending on the alcohol content. Frequently at the place of production one will find a fine well-aged wine.

Throughout northern Italy a large number of *vino santo* wines are produced. They are often made from partially dried grapes and fermented under pressure for one or two years. Many are quite light in color.

The two internationally famous dessert wines of Italy are Marsala and vermouth. Both are classical "manufactured" wines, and owe their distinctiveness and quality to astute blending. The production of these wines is discussed in full by Joslyn and Amerine (1964).

For recent discussions of Italian wines see Bruni (1964), Capone (1963), Cosmo (1966), and Layton (1961).

Spain

More acres are planted to grapes in Spain than in any other country—about 4 million in 1965. But production per acre is so low that in 1964 and 1965 it averaged only 677 million gallons—far less than half that of France or Italy. Sherry and Málaga are known throughout the world as distinctive Spanish dessert wines, but far more table wines than dessert wines are made in Spain. The only important bottle-wine export is Rioja. However, much bulk wine is exported for blending, particularly from Alicante, Valencia, and Tarragona.

The Rioja district near Haro and Logroño has been delimited and the district has an appellation of origin. This delimitation has been facilitated by the limited number of companies which produce Rioja. Both red and white wines are produced, but the red is by far the best and the only one exported in quantity. The whites are rather neutral and some are high in sulfur dioxide.

Several varieties of grapes are grown for the reds, mainly Mazuela, Grenache (Garnacha here), and Graciano, Tempranillo, with the proportions used varying between wineries. The fermentation is conducted in some wineries in a rather rare fashion (the *maceration carbonique* occasionally employed in Beaujolais and elsewhere in France, see pp. 7–8). The uncrushed grapes are placed in open or closed tanks, where their respiration creates an air-free condition. This results in death of the skin cells, especially of the grapes in the lower part of the tank. These cells then lose their semipermeability and the color and flavor are easily extracted. The weight of the grapes gradually crushes the fruit and fermentation starts at the bottom of the tank. The process is slow and not efficient, and is apparently not used by the more modern producers, who crush the grapes and ferment in open tanks in the usual way.

Riojas are mainly aged in 50-gallon barrels. Most but not all of the important producers sell young, aged, and well-aged wine. The vintage dates sometimes represent the date the original barrel was filled. Since then it has been diluted by younger wine by a process of fractional blending so that the average age of the wine is considerably younger than the date printed on the label. However, this is no longer true of wines exported to this country, since the certificate of origin must state the actual age of the wine. Riojas do age well in the bottle. Most companies market a range of quality wines, usually based on age.

Some white wine is made from the Grenache blanc, Malvasia, Viura, and other varieties, and some is even used to produce a sound but unexciting sparkling wine. The table wines are often noticeably sweet and, like Graves, high in sulfur dioxide.

Red wine is produced in Navarre in northern Spain, but it is rather rough and thin compared to Rioja. The same is true of wines from Aragon to the east, those of Castile to the south, and those of Galicia to the west. Galician wines are often gassy and of low alcohol content.

Just south of Barcelona is an important white wine district near Villafranca del Panadés. It has a *denominación de origen.* A variety of wines from low-alcohol dry whites to 16 per cent sweet reds are produced. A considerable amount of this wine is made into sparkling wine. We would like it better if it were not often so sweet. Recently there has been renewed interest in upgrading the quality of the wines, particularly by planting better varieties. Alella is another Catalonian wine with a delimited area and a protected label. Both reds and whites are produced— the former from Sumoll, Grenache, and Tempranillo and the latter from Grenache blanc and Xarel-lo. Again the wines are of varying sweetness and alcohol content.

Much rather alcoholic red wine is made near Valencia. Some is exported in bulk. Little is bottled or worth bottling. Some of these wines are dry and low in alcohol, and others are sweet and fortified. The same is true of the table wines, both red and white, of La Mancha, Spain's largest wine-producing district. Valdepeñas is the best-known name, but it is not a high-quality wine. Although there is a red grape called Valdepeñas, the name refers to a region and is more often white than red (and the reds are more pink than red). The white is most often made from the variety Airen and the red from Cencibel or Grenache. These wines could be much improved if they were better finished (e.g., filtered clear). It must be admitted, however, that they are sold very cheaply, for a few pennies a glass.

There are many locally famous dessert wines in Spain: Priorato, Málaga, Montilla (and Moriles), and Sherry. See Joslyn and Amerine (1964) for a discussion of these types. For a popular discussion of Spanish wines see Rainbird (1966). Since 1935 per capita wine consumption has ranged from 57 to 64 liters per year. At the same time beer consumption has risen from 3.2 to 26 liters (Hidalgo, 1967).

Portugal

Portugal has over 800,000 acres of grapes and produced nearly 400 million gallons of wine in 1965. The two famous wines of this country are, of course, port and Madeira. (See Joslyn and Amerine, 1964). Actually, far more table wine is made than dessert wine. The most famous table wines are the Vinhos Verdes, Dão, Colares, and Bucelas, which are protected by a *designação de origem*.

Vinhos Verdes are produced in a delimited district in the Minho district north of Oporto, primarily at Monção, Basto, Brego, Lima, Amarante, and Penofiel. Grapes are grown mainly on trees and hence, even with pruning, tend to overproduce. The musts are therefore low in sugar (about 16 per cent) and high in acid (over 1 per cent). The new wines are bottled fairly young and undergo a malo-lactic fermentation both before and after bottling. The result is that the wines are gassy, thin in taste, and often still too acid for wine drinkers accustomed to traditional table wines. Nevertheless, they are popular in Portugal. The crackling or gassy rosé wines from Portugal on the American market are usually not Vinhos Verdes, but carbonated wines of normal composition.

The Dão region, near Vizeu, is a small strictly delimited area, but it produces some of the best red table wines of Portugal, especially when properly bottle-aged for several years. A mixture of early-ripening varieties are planted, including Tourigo do Dão, Tinta Pinheira, and Arinto. The latter is white and is blended with the two red varieties or used alone to produce a white wine. The red is more important than the white in both volume and quality. A number of table wines are produced in the Bairrada region (near Mealhada) and the whites are often made into bottle-fermented sparkling wines, called *espumante* (*not* champagne). While these sparkling wines are well-made, some are finished a little too sweet.

West and north of Lisbon there are more than 4,000 acres of grapes in the Colares district. The wines are mainly red. The primary variety is Ramisco and the region is delimited. Because of their location they are high in chloride. The wines are rather tannic when young but can be aged.

Bucellas is a white, often sweet table wine produced in a small region north of Lisbon from the variety Arinto. Modern ampelographers do not believe that this variety is a Riesling or Riesling type. In our opinion the wines suffer from the same defect as many Graves—too high a total and free sulfur dioxide. Other wines that are not yet delimited, usually of less constant quality, are Evel, Periquita, Grandjó, Pinhel, Palmela, Arealva, and Mateus. The last is a rosé, which in this country is noticeably sweet.

Millions of gallons of other table wines are produced each year in Portugal. Thanks to the intelligent work of the Junta Nacional do Vinho there has been a marked improvement in quality over the last twenty years. The chief handicap has been the mediocre quality of the grape varieties grown. For a recent discussion see Allen (1963).

U.S.S.R.

The Soviet Union is rapidly becoming one of the foremost viticultural regions. Production of grapes in 1950 amounted to 830,000 tons, in 1958 to 1,912,000, and in 1964 to 2,898,000. Since this was produced on 1,814,000 *bearing* acres, production is less than two tons per acre. However (Anon., 1966d), the ratio of nonbearing to bearing has been declining since 1959 and should stabilize in the near future. (For a view of the viticultural regions see fig. 4.) The 1964 acreages and ratio of nonbearing to bearing areas for the various grape-producing districts were as follows:

Figure 4. Hillside vineyard in the Crimea in the Soviet Union. (Courtesy Embassy USSR, Washington, D.C.)

TABLE WINES

Republic	Total area (000 acres)	Ratio of nonbearing to bearing Area (per cent)
Ukraine	850	27
Moldavia	593	42
R.S.F.S.R.	425	58
Georgia	253	44
Azerbaidzhan	185	108
Uzbekistan	121	63
Armenia	89	56
Kadzhikistan	40	82
Kazakhstan	37	85
Turkmenistan	22	47
Kirgistan	12	71
Total	2,627	45

Of the total acreage 2,293,000 acres are government-owned and 334,000 are privately owned.

Wine production has expanded rapidly with the increasing acreage. The following data (Anon., 1966d) are illustrative (in 1,000 gallons):

1958	163,327	1962	258,313
1959	176,726	1963	313,196
1960	205,218	1964	335,712
1961	223,895		

In spite of the increasing production, the Soviet Union is still a net importer of raisins, wines, and brandies. In 1964 imports amounted to 39,800 tons of raisins, 280,300 gallons of wine, and 619,800 gallons of brandy. Of the brandy and wine 53.5 per cent was from Bulgaria, 32.4 per cent from Rumania, and 11.9 per cent from Hungary.

A wide variety of wine types are produced in the Soviet Union. Most of them are red and white table wines. A number are very sweet and of only moderate alcohol, 15 to 16 per cent. A few are of the more conventional port, muscatel, and sherry types. Sparkling wines are popular and widely produced, and there is a flourishing brandy industry. These differences in wine types and quality reflect the wide climatic differences of the country, the large number of grape varieties used, and, perhaps, some traditions reflecting western European practices or local practices. For example, the French variety Aligoté is widely used for white table wines, as in France. However, it is used also as a sparkling wine stock.

The Ukraine is by far the largest viticultural region. It lies in a cool region subject to low winter temperatures. Many of the vines must be protected from freezing by covering. This may mean plowing furrows alongside the vine in the early winter, pushing the vine down into the furrows, and covering them. This weakening process probably accounts in part for the relatively low yield in certain areas. No less than six climatic zones exist in the Ukraine. The best viticultural area of the

Ukraine is the Crimea, followed by Odessa. A wide range of native and imported varieties are grown: Rkatsiteli, Saperavi, Aligoté, Chasselas doré, Wälschriesling, Sémillon, Cabernet Sauvignon, and others.

Moldavia is a very ancient viticultural region, most of which was not incorporated into the Soviet Union until after World War II. Greek and Roman colonies planted vines there before the Christian Era. Until phylloxera devastated the vineyards at the turn of the century, many native and European varieties were planted; thereafter direct producers were widely planted, but gave wines of low quality. (Direct producers are crosses of *V. vinifera* and of a phylloxera-resistant species.) Amerine (1963*b*) noted that nearly two-thirds of the vines were still direct producers in the early 1960's. Modernization of vineyards will gradually replace these with better varieties. Aligoté, Fetească, Sauvignon blanc, White Riesling, Sémillon, Rkatsiteli, Traminer, Muscat blanc, and Pinot gris were recommended for whites and Cabernet Sauvignon, Merlot, Malbec, and Saperavi for reds. Table wines predominate, but many types are produced. In recent years the number of small wineries has decreased and larger, more modern fermenting and processing plants have been constructed.

Georgia is one of the oldest viticultural regions of the world and was especially important in the early Christian period (Beridze, 1965). Kakhetie is perhaps the most famous of its numerous subregions. About three-fourths of the grapes are native varieties: Rkatsiteli, Tsolikoouri, Mtsvane, Saperavi, Tchinouri, and Gorouli Mtsvane. The most important European varieties are Aligoté, Chardonnay, Cabernet Sauvignon, Pinot noir, and several muscats. A few direct producers are still grown. Georgian white table wines are especially praised. Some of the red table wines are slightly sweet. Some wine is still produced in the traditional fashion by fermenting on the skins in amphora buried in the ground. Georgian sparkling wines are produced mainly by the tank process, and are popular throughout the Soviet Union; about 20 per cent of the total Soviet sparkling wine production is from Georgia.

The third largest producer is the White Russian area, R.S.F.S.R., which includes the region north of the Caucasus and along the Don River near Rostov. Since winter temperatures are low, about 80 per cent of the vines require covering. Irrigation is common in the summer. There are plantings of European varieties: Aligoté, White Riesling, Sylvaner, Pinot gris, Clairette (blanche?), Pinot noir, and Cabernet Sauvignon; and of native varieties: Rkatsiteli, Saperavi, Assyl-Kara, Narma, Aly Terski, Guliabi, Agadai, Kaitagui, and others. There is a large production of sweet wines of moderate alcohol. The products of the sparkling wine plant at Krasnodar (Abraou-Durso) are very well known throughout the Soviet Union. A number of muscat varieties are grown for sweet, table, and dessert wines.

Azerbaidzhan is an increasingly important viticultural region west of the Caspian Sea. Again both native and European varieties are grown. The variety Matras produces a good red table wine. Kidogna is used for a high-tannin red wine. A number of dessert wines are also produced.

Armenia is a very ancient viticultural region, perhaps dating from the Neolithic period. About one-fourth of the grapes grown are of table varieties. Several viticultural zones are planted. In spite of its southern location, winter temperatures are frequently very low and at certain elevations the summers are cool. Some rather low-sugar musts are reported. In spite of this, only 20 per cent of the wines are table wines. The usual varieties, Aligoté, Rkatsiteli, and Saperavi, are grown, plus some Chardonnay and Pinot noir and many native varieties. Armenian dessert wines are especially praised.

Turkmenistan is a small viticultural region of generally high temperature producing mainly table grapes and dessert wines. Uzbekistan is also warm and produces many table grapes and dessert wines; but also makes table wines, both still and sparkling. Wines are also made in Kazakhstan, Tadzhikistan, and Kirgistan.

Russian wines would be improved if every region did not feel it had to produce a wide range of wine types, some of which the climate of the region is not suited to produce. The white table wines are generally better than the reds, and the sweet types are better than the dry.

For data on Soviet sparkling wines see Frolov-Bagreev (1948). Frolov-Bagreev and Agabal'iants (1951) and Nilov and Skurikhin (1967) have published texts on the biochemistry of musts and wines.

Rumania

Rumania is another very ancient viticultural country, certainly dating to Greek colonies. At the turn of the century phylloxera caused serious damage to the vineyards. Since World War II there has been much replanting and expansion. As of 1965 nearly 750,000 acres had been planted, of which about two-thirds were in production. Over 170 million gallons of wine were produced. These new plantings have emphasized fine varieties; for this reason alone great improvement in the quality of Rumanian wines can be expected. An impressive ampelography has been published (Constaninescu et al., 1959–1966), new experiment stations have been established, and several training centers and schools inaugurated. The climate is especially favorable, with adequate heat summation but a long moderate autumn so that even the late-ripening varieties mature. In some areas rather low winter temperatures occur. The soils vary from sandy to calcareous and schistose.

According to Cosmo (1964) and Teodorescu (1966), the most important native varietes for ordinary white table wines are Galbenă, Mustoasă de Măderat, Creață, Plăvaie, Crîmpoşie, Frăncuşa, and Iordană. For fine white table wines they recommend Fetească albă, Grasă de Cotnari, Tămîioasă românească (Muscat blanc); for red table wines, Băbească neagră and Fetească neagră. The most important non-Rumanian varieties are Wälschriesling (here as elsewhere called Italian Riesling), Pinot gris, Furmint, Sauvignon blanc, Aligoté, Chardonnay, Traminer, Sylvaner, and Muscat Ottonel for the white; and Cabernet Sauvignon, Merlot, and Pinot noir for the reds.

The white table wine regions and their most important varieties are Odobeşti (and Panciu and Huşi)—Galbenă, Wälschriesling, and Furmint; Drăgăşani—Crîmpoşie, Iordană, Wälschriesling, Sauvignon blanc, and Tămîioasă românească; Teremia—Creață, Majarca and Steinschiller; Iassy (and Dealul, Bujor-prut, Coteşti, and Dealul Mare)—Fetească albă, Pinot gris, and Wälschriesling; and Stafăneşti-Arges—Sauvignon blanc and Fetească regală. The best white table wines are reported to be from Transylvania (Alba Julia, Tîrnave, Aiud, Blaj and Simleul). These are produced from Fetească albă, Fetească regală, Traminer, Furmint, and Pinot gris (often a dessert type). Many of the white table wines contain 0.6 to 1 per cent reducing sugar, according to Cosmo (1964).

The best red table wine regions and the varieties used are Nicoreşti—Băbească; Dealul Mare (including Tohani, Urlați, Valea, and Călugărească)—Cabernet Sauvignon, Malbec, Pinot noir and Merlot; Halînga (and Corcova, Segarcea, Iancu Jianu)—Pinot noir; Banat (at Miniş)—Cadarcă and Cabernet Sauvignon; and Recaş—Pinot noir. According to Teodorescu (1966), some of the best sweet table wines are produced at Cotnari in the north of Moldavia from its traditional mixture of varieties: Grasă, Fetească, Francuşa, and Tămîioasă românească; at Murfatlar in Dobrudja near the Black Sea, from Chardonnay and Pinot gris; at Alba Julia and Lechința Teaca, from Fetească albă and Pinot gris; at Tîrnave, from Traminer, Pinot gris, and Fetească albă; at Pietroasa, from Grasă; and at Drăgăşani, from Sauvignon blanc. Prisnea (1964) also considered the wines of Cotnari to be among the finest produced in the country.

Muscatel wines are produced in several regions: Pietroasa, Stefăneşti-Argeş, Drăgăşani, and Severin (from Tămîioasă românească) and in Transylvania at Tîrnave, Alba Julia, and Iassy (from Muscat Ottonel). Many of these are only moderately high in alcohol and are more like sweet table than dessert wines.

Sparkling wine production has increased from less than half a million bottles in 1959 to nearly four million in 1965. Centers of production are in Transylvania and Moldavia.

To process the rapidly increasing production of all types of wine many new and modern plants have been built, especially at Valea Călugărească and Tohani (region of Ploeşti), at Murfatlar (region of Dobrudja) and at Jidvei (region of Mures). Modern stabilizing and bottling plants are located at Iassy, Bacău, Focşani, Ploeşti, Constanţa, Craiova and Timişoara.

One of the many encouraging aspects of the Rumanian wine industry is the emphasis on planting high-quality varieties and improving the wine making, aging and stabilization. The quality of the wines reflects this.

Jugoslavia

There are over 600,000 acres of vines in the country, almost all of which are planted to wine grape varieties. Production is about 150 million gallons; hence the yield per acre is rather low. The export market, mainly to central European countries, amounts to about 12 million gallons a year. Consumption averages 25 liters (6.6 gallons), which is less than before World War II, according to Radenkovič (1962), who notes the tendency of wine prices to fluctuate. Protin (1966) reported an increase of 41.4 per cent in the wholesale value of wine between 1964 and 1965. (A devaluation of the dinar in July, 1965, may partially account for the change.)

Jugoslavia has a variety of climatic conditions—from the cool continental climate of Slovenia to the dry Mediterranean climate of the islands and coast of Dalmatia. During its long viticultural history (vines date from the Greek period or earlier) it has been subjected to an extraordinary variety of external influences: Bulgarian and Russian from the east, Turkish and Greek from the south, Italian and French from the west, and Austrian and Hungarian from the north—not to speak of the German and Rumanian agricultural enclaves within the country itself.

Its long history and foreign influences have given the country an extraordinary diversity of varieties and ampelographical problems. Thus in the north, in Slovenia, are planted varieties from France, Merlot, and Sémillon; from Italy, Barbera and Refško (Refosco); from Austria, Traminac (Traminer) and Veltlenač (Veltliner); and from Hungary, Kadaraka and Furmint. The Wälschriesling (Italian Riesling) is widely planted here, as in Austria and Hungary, but may have originated in Jugoslavia. Some of these varieties are planted also in Croatia, Bosnia and Hercegovina, Serbia, and Macedonia, but the main varieties in the southern republics are native or are from Bulgaria—Plavina, Prokupats, Žilavka, Blatina, Plavac Mali, and others.

Since World War II Jugoslavia has built more than fifty modern wineries with excellent equipment (fig. 5). These new wineries have markedly raised the standards of the wine industry. The problem of defining and controlling the types of wines produced in given regions has barely begun except for Dingač, a high-alcohol Dalmatian red wine produced in a small delimited region from the variety Plavac Mali. The high alcohol, 15 or more percent, of its wines makes it difficult for western tastes to appreciate. The nomenclature of Jugoslavian wines has not yet been standardized. Varietal, regional, and generic wines are all produced.

Figure 5. Modern winery in Jugoslavia. (Courtesy V. Žanko.)

Slovenia produces primarily varietal-named white wines: Silvanac, Rizling, Sauvignon (or Sovinjon), Traminac, and some red, Merlot.

Croatia produces mainly white table wines in Croatia proper. Among the best-known wines are Plješivica (a red from the variety Portuguese), Slavonia (mainly whites from Wälschriesling, Traminer, Sylvaner, Sauvignon blanc, etc.), and Zagorje (also whites). In Istria some reds are produced from Refosco, Barbera, Merlot, Cabernet Sauvignon, and Sémillon, and are so named. Some of these are rather low in alcohol and have undergone a malo-lactic fermentation.

Dalmatia experienced an extraordinary prosperity between 1870 and 1894. Most of its high-alcohol wines were exported to France for blending purposes. The replanting in France and the appearance of phylloxera in Dalmatia led to a great reduction in demand, a subsequent removal of vines, and eventually the migration of thousands of Dalmatians to Australia, New Zealand, and the United States. However, grapes are still grown along the coast and on many of the islands. Among the more than 250 native varieties are Plavac Mali, Vugava, Maraština, Pašip, Grk, and Kujundjuša. Besides Dingač, Postup is a well-known red; Opolo varies from rosé to red; and Vugava, Maraština, and Grk are white. Prošek is a sweet wine of 16 or more per cent alcohol, originally produced from partially dried grapes. Nowadays some appears to be made with grape concentrate. The same type is produced in Italy.

Bosnia and Hercegovina are still about one-third Moslem and produce only a few wines. The most famous is Žilavka, made from the variety of that name from the vineyards near Mostar. Blatina, from the same general region, has a deep red color and appreciable tannin. Macedonia likewise has few wines. The most important regions are Tikveš and Ohrid. Kavadarka and Vranats are popular native varieties. Montenegro has a red wine, Crmničko.

Serbia has a variety of vineyard areas and wines. Župa produces a dark red wine of the same name. A rosé or light red wine from Prokupats is sold as such. Krajina is a rather dark red wine produced on the right bank of the Danube. (There is also a white Krajina from Bagrina.) Other regions are Šumadija, Metohia (a red from Prokupats), Vlasotinačko, Venčac-Oplenac, Smederevo (Smederevka is the white wine). Plovdina (or Plavina) is a Serbian wine sold with a varietal label. Kosmet, Vlasotinci, Niš, and Prokuplje are regions in southern Serbia. In northern Serbia are Subotica-Horgoš (from Kadarka, Čoka, and Merlot), Vršac (white), Banat (white), and Fruška Gora (white, rosé, and red.)

Opinions differ in Jugoslavia on the quality of their wines. Some of this may be based on local pride and competition. Serbians seldom like Slovenian wines and Slovenians find some Dalmatian wines too high in alcohol. At present the wines reach the market well stabilized. Practically no bottle-aged reds are to be found. In general, the whites appeal to us more than the reds and the Croatian and Slovenian whites most of all. However, we have tasted a number of Serbian reds that had great potentialities.

Germany

Grapes were introduced into Germany by the Romans, and all the present grape-growing regions were planted by the sixteenth or seventeenth century. (For a detailed history see Bassermann-Jordan 1923.) Some cellars continue to use casks (fig. 6) and traditional methods but many employ the finest modern equipment.

Figure 6. Oak casks in underground cellar in Germany. Note growth of mold on the walls, probably of *Cladosporium cellare*. (Courtesy Press- und Informationsamt der Bundesregierung, Bonn, Germany.)

Germany, both West and East, has vineyards in spite of the low heat accumulation of the region. In 1965 there were over 165,000 acres of grapes in West Germany and production was over 130 million gallons, down from 190 million in 1964, an all-time high. Even in the warmest years in vineyards with excellent drainage and exposure to the sun the sugar content is low (except when *Botrytis cinerea* grows, or in areas where the grapes freeze on the vine and the sweet unfrozen solution is pressed out).

Becker and Lorenz (1966) noted that average production in Germany has increased from about 300 gallons per acre in 1900–1909 to 480 in 1930–1939, to 715 in 1950–1959, and to 1,008 in 1960–1964.

German wines can be discussed from the point of view of varieties used, climatic zones, or regions. The last will be used here, starting from vineyards in the south and proceeding to the more northern vineyards. For the varietal percentages of the various districts see table 1.

Becker and Lorenz (1966) indicate some changes in the percentages of the different varieties planted in 1925 and in 1960: White Riesling, 22.8 and 26.3; Sylvaner, 38.8 and 34.5; Müller-Thurgau, 0 and 14.2;

TABLE 1
ACREAGE AND PERCENTAGE OF VARIOUS VARIETIES IN GERMAN WINE DISTRICTS

Rheinpfalz[a] 42,434	Rheinhessen 38,412	Mosel-Saar-Ruwer 23,334.8	Baden 21,156	Württemberg 15,439	Rheingau 7,951	Nahe 7,221	Franken 897	Mittelrhein 2,501	Ahr 1,315
Sylvaner 50	Sylvaner 60	White Riesling 86	Müller-Thurgau 21	Trollinger 29	White Riesling 70	Sylvaner 48	Sylvaner 60	White Riesling 84	Portugieser 39
Portugieser 18	Müller-Thurgau 21	Elbling 8	Chasselas doré 16	White Riesling 21	Sylvaner 14	White Riesling 31	Müller-Thurgau 25	Müller-Thurgau 6	Pinot noir 27
White Riesling 13	White Riesling 7	Müller-Thurgau 4	Pinot noir 10	Sylvaner 18	Müller-Thurgau 13	Müller-Thurgau 18	White Riesling 6	Sylvaner 5	White Riesling 16
Müller-Thurgau 13	Portugieser 9		Rulander 10	Portugieser 9					Müller-Thurgau 14
			Sylvaner 8	Limberger 7					
			Riesling 7						

[a] Bearing acreage in 1965.
Source of data: Anon. (1967d).

Elbling, 7.2 and 1.5; Chasselas doré, 4.2 and 1.9; Portuguese, 5.8 and 8.2; Trollinger, 4.4 and 8.0; Pinot noir, 2.9 and 1.8; and others, 13.9 and 8.6.

The heat summation throughout the German grape-growing districts is usually insufficient to ensure complete maturation. In a really cool year, such as 1955, 1956, or 1965, virtually all the crush must be sugared. Correction of excessive must acidity is often necessary in such years, frequently with calcium carbonate or by using ion-exchange resins. Even in a very warm season, such as 1959, some musts will require sugaring, but few will need to have their must acidity reduced. The finished wines, however, in most years and most regions may still have too much acidity. The malo-lactic fermentation to correct wine-fixed acidity is therefore very widely induced (pp. 496–505).

The local protagonists of vineyards on the shores of Lake Constance and in Baden have made some overenthusiastic claims. The quality of the main variety, Chasselas doré, is at best minimal. However, individual growers do well with the traditional German varieties, White Riesling and Sylvaner, and they produce some excellent wines in good years.

Farther down (north) the Rhine there is the Pfalz or Palatinate, more or less an extension of Alsace. The vineyards face the east and hence get the morning sun. They are supposed to be the "biggest" (i.e., most alcoholic) German wines, but this is not always true. They do have less total acid than most other German wines, and at best a florid and full aroma. Some Pfalz wines are very good and when they are good they are superlative. The main varieties are Sylvaner, White Riesling, Portugieser, and Müller-Thurgau. But here it is not variety but method of harvesting that establishes the quality. Since this is typical of Germany, it should be discussed further.

We have already mentioned the importance of *Botrytis cinerea* for Sauternes and the Loire. The process is no less important for Germany, but here is has some special nuances. The botrytised grapes are picked late (*Spätlese*, possibly sweet), especially late (*Auslese*, always sweet), or very late from selected berries (*Beerenauslese* or *Trockenbeerenauslese* wines are always sweet-tasting—and, because of the small yield, very expensive).

Thus a Palatinate *Beerenauslese* may be very fine compared to a similar Moselle or Rheingau wine. The skill of the vintner and the vagaries of the local climate limit how fine a wine he can make. The best of the Palatinate do very well indeed. Why do not the rest of the Palatinate wines measure up to the standards of the Rheingau or Moselle? Perhaps they do, but they seldom achieve the superiority of many fine Moselles and

Rheingaus. Probably this is because the climate-variety-exposure relationship is less favorable, but we know of no critical experiment.

The Rheinhessen, extending north and west from the Palatinate, has mainly Sylvaner, with the same general eastern exposure, but (is it soil or climate or both?) it does not rate as high in quality as its neighbors to the north or south. The best wines of the best years are very fine whereas the *Weinstube* wines of the Rheinhessen are frequently oversulfured, thin, and sugared.

The two main high-quality regions of Germany are the Rheingau and the Moselle (of the latter we shall note two secondary offshoots). The Rheingau extends along the slopes of the Rhine barely 25 miles from Wiesbaden to Rüdesheim. Yet on these slopes are produced some of the finest German wines—usually White Riesling. However, 14 per cent is Sylvaner and 13 per cent Müller-Thurgau. In the best years and when they are suitably harvested *Spätlese, Auslese, Beerenauslese,* and even *Trockenbeerenauslese* wines are produced. But even the standard wines are finer than their southern counterparts. Why? Exposure is a patent answer, but there may be and undoubtedly are other parameters to the regional quality complex.

Figure 7. Hillside vineyards in Ahr district of Germany. (Courtesy Presse- und Informationsamt der Bundesregierung, Bonn, Germany.)

The Rheingau wines are slightly higher in alcohol and less tart than the wines of the Moselle. We know of no critical experiments, but our impression is that the Rheingau wines may have a little more color. But differences in processing, particularly in the amount of sulfur dioxide used, may and frequently do make it difficult to generalize.

White Riesling is the dominant variety in both regions, espècially in the best vineyards, though some are planted to Elbling and Müller-Thurgau. Near the Moselle are the wines of the Saar and of the Ruwer. Both are often of superior quality. The Ahr (fig. 7) is cooler, yet produces light red wines of moderate quality from Pinot noir.

In Franken, in northern Bavaria, Sylvaner is the predominant variety. The soil is chalky, which is said to give the wines a special flavor, but we know of no experimental evidence. There are suggestions from the work of Benda and Wolf (1965) that yeast flora may have an important influence on such nuances of flavor.

The best recent years in Rhine and Moselle were 1959, 1962, 1964, and 1966. Between 1950–1951 and 1962–1963 still wine consumption increased 109 percent in Germany, according to Becker and Lorenz (1966). During the same period consumption of sparkling wines increased nearly a thousandfold.

Greece

Wine was an important item of export by the ancient Greeks. In spite of their classical reputation, ancient Greek wines could be very poor, as Younger (1966) noted. Among the popular classical wines were Chain, Lesbian, Thasian, Lampsacus, Issan, Naspercenian, Bibline, and probably Magnesian and Pramnian. Pramnian, produced on the island of Icaria, in the Sporades, was too dry for some. Wines made from partially dried grapes were common and sweetness was considered desirable, possibly to cover acetic odors. Some wines were not considered drinkable until they were old—up to six or more years, according to Younger (1966). It is doubtful that the old wines were really better. Perhaps they were the wines that resisted spoilage best and thus had been kept for aging. Younger believes, correctly we think, "that most of the wine that was drunk in ancient Greece was drunk within the year of its vintaging." Logothétis (1965) has traced the origin of Malvasia to the Peloponnesian city of Monemvasia in 1214. From the fourteenth century it was produced in Crete, and its culture spread throughout the Greek islands and later to Italy, Spain, France, and Portugal.

The long domination of Greece by the Turks did nothing to help the wine industry. Another negative factor is the large amount of grapes planted for currants. The variety produces excellent currants but is too low in acid. The overplanting was in response to the heavy demand for dried grapes in France in the 1890's owing to the grape shortage there which resulted from phylloxera.

Wine is made in all parts of the mainland, and on many of the islands (fig. 8). About 550,000 acres are planted to grapes (54 per cent to wine grapes) and annual production is over 100 million gallons, about equally of red and white. Production per acre is small even taking into account those used for table purposes and raisins, but the industry is very important to the total economy (Davidis, 1956).

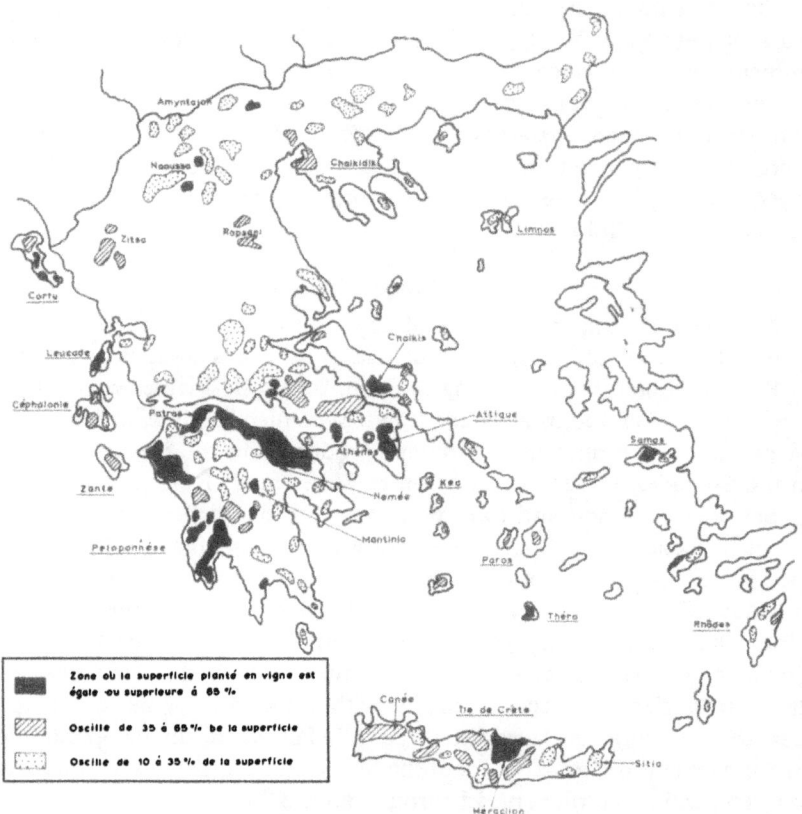

Figure 8. Distribution of vineyards in Greece. Within solid black areas vineyards represent 65 per cent or more of planted land, within shaded areas, 35 to 65 per cent, and within dotted areas, 10 to 35 per cent. (Courtesy Professor U. Davidis.)

In all parts of Greece retsina is produced; in fact, it amounts to about one-third of the production. This is ordinary red or white wine to which 1 to 3 per cent dried powdered resin has been added. The resulting turpentine odor is repellent to non-Greeks, but, surprisingly, a taste for it can be cultivated and foreigners living in Greece often consume it.

On Peloponnesos, the principal grape-producing region, besides retsina the leading types are Demestica (red and white), Mavrodaphne (port-type), Fhileri (rosé), and Mantinias (sweet). White Mantinias is mostly made from the varieties Moschofilero and Asproudes. In Thessaly a rosé (called "Pelure d'onion") and a red are produced from the local variety Agiorgitiko.

Attica is the driest and hottest region of Greece. The variety Sevatiano, however, thrives and produces a well-balanced white table wine which ages well.

A variety of table wines are produced in Macedonia and Thrace. Naoussa, one of the best Greek table wines, is a product of Xynomavron. Another Macedonian red is Amyntaion. Other popular varieties are Cinsault, Chalkidikis, and Liminés (an ancient type). On the island of Lemnos the latter variety produces a good red.

Wine is produced on most of the islands, particularly Crete. Maroussos is a dry red, but various sweet wines are also made. Roméïkon, Kotsifali, Manilani, and Liatiko are native varieties. Corfu has Brothoko, which may be red, pink, or white, and dry or sweet. A white table wine, Santa Mavra de Leucade, and a red, Ropa de Corfu, are also produced. Rhodes and Mytilene produce muscatels, red Malvazia, and Cantia as well as various other dry and sweet whites and reds. Whites are often made from Athiri and reds from Amorgiano. A red wine having the appellation of origin Rhodes is produced from this variety.

On Samos perhaps the most famous Greek muscatel is made, apparently, from Muscat blanc (and certainly not from Muscat of Alexandria); there is also Fokianos, which may be dry or sweet. Aegina, Paros, and Hydra produce red table wines; Chios, a variety of dry wines; and Kefalonia has Rabola, a white dessert wine. Germany, Holland, Belgium, and Sweden are the main export markets. There much of the wine is used for blending.

Bulgaria

In Bulgaria likewise the vine has been cultivated for many centuries. About 450,000 acres are planted to grapes and the acreage is increasing. Before World War II it was most famous for its fine grapes. In 1963 exports amounted to 170,000 tons. Recently wine making has been encouraged by the government and some fine wine grape varieties have been planted. Wine production was about 75 million gallons in 1964 and over 90 million gallons in 1965. New and modern wineries have been constructed. Both still and sparkling table wines and dessert wines are produced, and a rather high percentage is exported (about 3 million gallons of a total production of about 70 million gallons in 1963). A brandy called pliska is produced. For an excellent description of the varieties of grapes used in Bulgaria see Constantinscu *et al.* (1959–1966). Kondarev (1965) also gives a description of Bulgarian vineyards, as does Pancsev (1964).

Hungary

The Hungarian grape and wine industry is of very ancient origin, dating from the eighth to the sixth century B.C., according to Németh (1964), who notes the numerous relics of Roman wine making that exist in the country. The Hungarian grape acreage is now being expanded and amounts to about 600,000 acres. Production, as in Austria, is highly variable. In 1964 it amounted to almost 150 million gallons, but in 1965 to less than 50 million gallons. The objectives of the current expansion are to provide more table grapes for export, to increase wine production, and to plant better varieties in certain regions in order to improve quality.

Németh (1964) made a detailed study of 115 varieties. He considers Aligoté, Alicante Bouschet, Gamay teinturier, Gyöngyfehér, Királyleányka, Merlot, Muscat Bouschet, and Úrréti the best. However, other varieties gave high-quality wines, but better selections are needed: Ezerjó, Furmint, Hárslevelü, Kadarka, Kékfrankos, Médoc noir, Veltliner rouge, and Wälschriesling. Németh recommended importation of new clones of Cabernet franc, Cabernet Sauvignon, Chardonnay, Kövérszőlő, Pinot gris, Pinot noir, and White Riesling. Inspection shows that a number of other varieties produced quality wines: Sauvignon blanc and Traminer, for example. It is difficult to understand the high scores given to Alicante Bouschet and Kadarka. Kocsis (1958–1962) lists six new varieties recommended for Hungary. According to Fejes (1964), new and larger wineries are to be built and efforts are to be made to improve the quality and quantity of the wines produced by small vineyardists. Mechanization of vineyard operations and planting of better varieties are also taking place in Hungary, according to Katona (1966). His report predicts that the most important wines of the future in Hungary will have 9 to 11 per cent alcohol, retain considerable carbon dioxide, be only slightly aromatic, and have a red or greenish color. This, he adds, does not mean that high-quality wines will not also be produced, but one wonders.

There are seventeen primary wine-producing areas in Hungary (fig. 9).

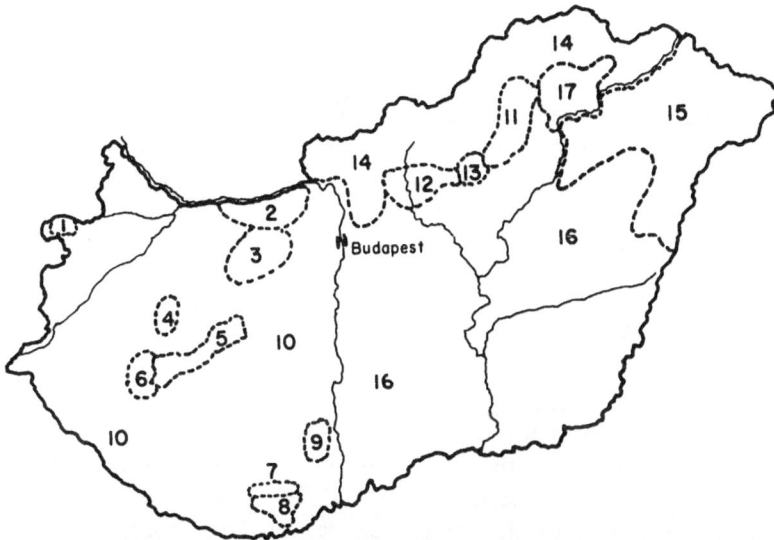

Figure 9. Hungarian viticultural districts: 1, Sopron; 2, Neszmély; 3, Mór; 4, Somlö; 5, Badacsony-Balatonfüred-Csopak; 6, Balatonmellek; 7, Mecsek; 8, Villany; 9, Szekszárd; 10, Dúnantúl; 11, Eger; 12, Gyöngyös-Visonta; 13, Debrö; 14, Felföld; 15, Nyjírség; 16, Alföld; 17, Tokaj-Hegyalja.

The region near Lake Balaton is considered climatically very favorable for wine grapes. The region has about 20,000 areas divided into four zones: Badacsony (the important varieties planted are Kéknyelű, Szürkebarat, Sylvaner, and Wälschriesling and the wines often have 13 or more per cent alcohol), Balatonfüred-Csopak (70 per cent of the grapes are Wälschriesling, the rest Furmint and Sylvaner, often with high alcohol), Balaton-Mellik (with similar varieties), and the south side of Lake Balaton (more noted for table grapes). In Balaton (or Balaton-Mellik) there are over 7,000 acres of grapes. The predominant varieties are Wälschriesling, Mézesfehér, Ezerjó, and various Sylvaners.

Another well-known region is Eger (from which comes Egri Bikavér, a red wine made with the native variety Kadarka). Eger has nearly 8,500 acres of grapes. The best and most prevalent varieties are Nagyburgundi (Pinot noir?), Kékfrankos, Kadarka, Médoc noir, Cabernet (franc? Sauvignon?), Wälschriesling, and Leányka.

The sandy soils of Alföld (nearly 250,000 acres) are planted mainly to white varieties: Wälschriesling, Ezerjó (very characteristic), Kövidinka, Mézesfehér, Sárfehér, and Slankamenka. Some wine from Kadarka is also produced.

37

In Mátra Alján there were nearly 34,000 acres of grapes in 1966. The dominant varieties were Hárslevelü, Wälschriesling, Leányka (Leánka), Kadarka, and Mézesfehér.

Tokaj (or Tokay in English) is certainly the most famous wine of the country. It is of ancient origin and was always expensive to produce. It comes from northeastern Hungary from the region known as Tokaj-Hegyalja, which now has about 12,000 acres of grapes. The important types are Tokajiaszu, Tokajiszamorodni, and Tokajipecsenyebor (literally wine for the roast). The traditional variety is Furmint, but other varieties, particularly Hárslevelü, are planted. The grapes are not harvested until the end of October, when some natural drying and some botrytis invasion has occurred. A detailed study of the soils of the Tokay district was made by Hidvéghy (1950). There was wide variation in the surface and subsoils. However, the best soils seemed to be those with the highest stone content (up to 75 to 80 per cent). This appears to be a drainage-temperature factor.

Wines made solely from the juice obtained without mechanical pressing from selected shriveled berries were designated as *exzencia*. Berries attacked and shriveled by *Botrytis cinerea* are sorted and stored separately in wooden vats until the end of the harvest season. The juice pressed out of the grapes by their own weight, which usually has a sugar content of 50 to 60 per cent, constitutes the base of the "essence." These wines were very sweet, of low alcohol (5 to 7 per cent), and expensive. We have seen none of them on the market since World War II. Wines made from pressed grapes, called *szamorodni*, are usually dry and moderately high in alcohol, but have an oxidized flavor.

Most of the Tokay now exported is labeled *aszu*. Aszu wines, according to Ásvány (1967a), are usually prepared by crushing the shriveled berries and adding these crushed grapes to must or wine. To a wine cask, containing 136–140 liters of must or wine, from 2 to 6 puttony (28–30 liters) of crushed shriveled berries are added. At present, instead of standardization based on the number of containers of ground shriveled berries added to normal must, the wine is standardized to a particular range of sugar and sugar-free extract content. The present regulations governing production of "Aszu of Tokay" specify the residual sugar and extract content and require that the wine have a distinct Tokay character. Both must and new wine, more rarely old wine, are added to the ground shriveled berries and the mixture is allowed to stand for 12 to 48 hours. The free-run must is separated, the pomace pressed, and the pressed must added to the free-run. The resulting mixture is fermented at 8° to 12°C. Depending on the grade (number of puttony), the Aszu wines are matured for at least four years or longer in wooden casks of about 200-liter capacity. This aging is required to develop the characteristic taste and bouquet of the Aszu wine. The sugar and extract content of different grades of Aszu are as follows:

Puttonyos	Sugar, g/l	Sugar-free extract, g/l
2	30	25
3	60	30
4	90	35
5	120	40

In recent years Aszu wines found on the American market have been amber in color and some have contained hydroxymethylfurfural. This may indicate use of a certain amount of grape concentrate to sweeten the wine, contrary to the Hungarian regulations.

A wide variety of fruity sweet white wines of the *Trockenbeerenauslese* type are produced. These are as pleasant in taste as the better German wines of this type, though perhaps not as flowery and distinctive in aroma.

Austria

Austria has been a grape-growing country since before Roman times and a wine-drinking country since, according to Schmidt (1965). In the Middle Ages monasteries cultivated many vineyards, but wine production and consumption decreased in the seventeenth century owing to a variety of factors: the competition of beer and coffee, changes in climate, the Turkish wars, politics, and so on. In the eighteenth and nineteenth centuries Austrian viticulture improved rapidly. The Blue Portuguese variety was introduced in 1772 and Pinot noir in 1800. Sparkling wines became popular in the second half of the nineteenth century.

Just why wine consumption dropped again in the nineteenth century is not known. In Vienna in 1815 consumption was 87 liters per capita per year; in 1870, 39 liters. The climate is cool and most of the vines are white: in 1961, 87.6 per cent white, 8.8 per cent red, and 3.6 per cent from direct producers. There are now over 100,000 acres planted to grapes. As late as the sixteenth century it seems to have been three times as large. A part of the decrease was due to phylloxera, which spread from 0.8 per cent of the vineyards in 1880 to 94.1 per cent in 1912. Since 1963 direct producers have been forbidden, but there were still about 2,500 acres planted in 1961!

Production in 1961 averaged just over 400 gallons per acre (for the total acreage). Increasing use of the "high" culture system of training vines is reducing costs and frost damage and increasing yields. Production varies according to climate, but averages about 50 million gallons. It was

74 million in 1964, about 37 million in 1965, and 38.3 million in 1966. According to Schmidt (1965) wine production was probably little larger than this in the sixteenth century.

In the southern Steirmark a few slightly sweet low-alcohol, white wines are produced, mainly for local consumption. The easternmost region, Burgenland, produces red and white wines, neither of outstanding quality. It produces about one-third of the Austrian wine and has been increasing in production.

The best and most important Austrian wines come from vineyards along the Danube from areas such as Dürnstein, Loiben, Langenlois, Krems, Traismauer, Rohrendorf, and Furth. Gewürztraminer, Müller-Thurgau, Sylvaner, Wälschriesling, Green and Red Veltliner, Rotgipfler, White Riesling, Zierfandler, Muscat Ottonel, Neuburger, Rulander, Blue Franken, Blue and Green Portuguese, Pinot noir, Chardonnay, and other varieties are planted (Schratt, 1966). Many other varieties, formerly widely planted, such as Honigler, Slankamenka, and Greazer (Banater Riesling), are seldom found today. The wines are often sold with the name of the variety on the label. Wines from late-harvested grapes of Weisser Burgunder etc., are labeled *Spätlese* or, when appropriate, *Auslese*. These do not appear to have as much botrytis character as their German counterparts and, according to Füger (1957), are very variable in alcohol and sugar content. Schmidt (1965) considers the wines from Wachau to be the best, but local experts might differ. Near and in Vienna there are a few vineyards, chiefly at Klosterneuburg, the home of one of the oldest viticultural and enological experiment stations in Europe.

A number of interesting wines are produced in villages just south of Vienna—Baden, Gumpoldskirchen, Pfaffstätten, Traiskirchen, Bad Vöslau, and others—from the varieties listed above. Vienna has been a center for sparkling white and red wine production for many years.

In the Austrian wine industry a large percentage of the wines are sold in bulk in wine restaurants, where they are consumed before, during, and after meals (46 per cent in 1939, more since). See Bundesverband der Weinbautreibenden Oesterreichs (1963) for photographic evidence.

Both viticulture and enology are conducted at a high level in Austria. About half the planting are trained to high (4 to 6 feet) wires. The importance of cool fermentation, the malo-lactic fermentation, and proper stabilization are well recognized. The Austrian wines are thus almost always of standard or better quality. Some of the best wines are exported.

Switzerland

Although Switzerland produces far more white wine than red, the demand for red is much greater. Peyer (1964) notes that between 1951 and

1960 the average annual consumption of red wine amounted to slightly over 30 million gallons, of which about 24 million gallons were imported. Of the 4.5 million gallons of white wine consumed, less than one-fourth

Figure 10. Hillside vineyards in western Switzerland showing harvested grapes being emptied into tanks for transportation to winery. (Courtesy Dr. H. Rentschler, Wädenswil.)

was imported. Gallay (1966) reported a steady increase in consumption of red and white wines since 1948 in Switzerland; however, per capita consumption has increased only slightly with population increase. Grape-juice consumption has tripled since 1950, but still represents only about 5 per cent of wine consumption.

The total grape acreage has decreased in this century and is now less than 30,000 acres. However, production per acre has increased. Cool climatic conditions limit the areas where grapes can be grown and the varieties that can be planted. In the western part of the country, 22,000 acres are planted, mainly white (81 per cent). Chasselas doré (there called Fendant) is the primary variety, although in the warmer parts some other *Vitis vinifera* varieties and some direct-producing hybrids are grown. The Chasselas doré is an early-ripening variety which seldom attains high sugar. Sugaring of musts is therefore essential. Its whites are rather neutral in flavor and often accidentally (from the malo-lactic fermentation) or intentionally (from a slight carbonation during bottling) somewhat gassy. The gassiness, however, adds to the interest of these neutral low-acid wines. The regions around Lake Neuchâtel and near Lausanne are best known for their white wines (fig. 10). They are often sold under the name of a village (e.g., Aigle, Dézaley, etc.). Gamay is a popular red grape in Valais.

In eastern Switzerland, 4,000 acres, 97 per cent of the grapes, are Pinot noir (here called Clevner). In some years because of botrytis on the grapes they are pressed as soon as possible to reduce the polyphenoloxidase content. Even so, browning of the color may occur. In the best years the wines have a moderate Pinot noir character and some distinction.

In southern Tessin, 4,000 acres of grapes are grown. Merlot is the popular red, but the wines are seldom exported.

Czechoslovakia

This country had over 65,000 acres of grapes in 1965, and 72,000 are projected for 1980 (up to 132,000 eventually). Of the twelve viticultural regions eight are in Slovakia, three in Moravia, and one in Bohemia. The vineyards near Melnik, north of Prague, are perhaps the best known. Most of the wines are white, produced from one or more of twenty-two white and five red varieties. Among the important varieties are White Riesling, Ezerjó, Traminer, Neuburger, Chardonnay, Pinot noir, Rulander, Müller-Thurgau, Wälschriesling, Sylvaner, Veltliner, Sauvignon (blanc?), and Leányka (see Valdner, 1967). Czechoslovakia has a very active program for the breeding and selection of grapes. This is necessary, of course, in regions of limited heat summation. The new varieties, according to Vereš (1966), are to replace the acreage of direct producers, with high-quality varieties.

The wines are of moderate alcohol, and sugaring is as common as in Germany. The wines are often sold in cask, as in Austria, but some are bottled. Consumption has more than doubled in the last ten years.

In the Tokay region of Czechloslovakia, Polakovič and Vereš (1966) consider the well-drained soils to be a major factor in quality. They believe that the rocky nature of the soils leads to easy warming up of the soil during the day. They also note that the best exposures are to the southeast and southwest. The long dry autumn is another quality factor and this, with the dry soil, leads to the shriveling of the fruit—an essential for production of high-quality wines of this type.

Albania

Little has been written about Albanian wines. In 1964 there were nearly 29,000 acres, of which only 18,000 were in production. Production was just over 800,000 gallons; almost a third of this was exported, nearly all to the Soviet Union.

Luxembourg

Only about 2,000 acres of vineyards exist in this small country. The region is really an extension of the German Moselle. The wines, however, are very low in alcohol and rather tart and have not achieved the reputation of the German Moselles.

SOUTH AMERICA...

... has a great potential for
grape and wine production.

No wines were produced in South America before the Spanish Conquest. Vines from Mexico were introduced into Peru in the early sixteenth century and spread to the more temperate Chile and Argentina. The establishment of vines in Brazil may have been directly from Europe. There were three strong impetuses: the need of wines for sacramental purposes, the desire for wines for table purposes, and the development of an export market.

Argentina

The low-quality Criolla-type varieties were imported into Argentina from Chile about 1557. But a wide variety of European varieties are now planted: Malbec and Merlot for reds, and Pedro Ximenes, Saint Emilion (Trebbiano), Riesling, and Sauvignon blanc for whites.

The negative quality factors are the slightly less favorable climatic condition (than in Chile), the very heavy yield per acre, and the lack of demand for high-quality wine.

Argentina had about 650,000 acres in grapes in 1965, almost all in full production. Argentina also produces table grapes, some of which are exported. Raisin production is small. The most important aspect of the Argentine wine industry is its large-scale production of ordinary wines and its relatively high consumption—over 22 gallons per capita in 1965. All types of wines are produced—dessert, table, sparkling, and vermouths. There is a large production of very low-alcohol wines. The large production and consumption of ordinary table wines reflect the influence of the important Italian immigrant population. While the climate of Mendoza and San Juan are moderately cool, the tendency to overcrop reduces the potential quality of the wines.

Argentina has only a small export market, owing in part to the lack of well-stabilized wines, properly bottled and labeled. Some of the Argentinean wines imported into California have been of minimal quality. For a discussion of current practices see Oreglia (1964).

Chile

Vines were first planted in Chile in 1548. The Chilean wine industry seems to have followed the same pattern as the Peruvian industry until the mid-nineteenth century, with use of the Criolla-type grape. Two factors favored the Chilean industry: cooler climatic regions for growing wine grapes, and the immigration of technically sophisticated immigrants from France in the mid-nineteeth century. Some of this immigration was due to the massive invasion of phylloxera into French vineyards, the economic distress of whole regions, and the subsequent migration of much of the population. Varieties such as Sémillon, Sauvignon blanc, Merlot, Malbec, and Cabernet Sauvignon were then imported and made possible the production of higher-quality wines.

In the twentieth century the major consumption of wine has been in the village at the hacienda level. The evidence is that the excessive use of wine in Chile has led to public health problems.

In 1965 Chile had over 270,000 acres in grapes. There is a sizable production of table grapes and even of some raisins, but most of the grapes are used for wine. In 1964 and 1965 this resulted in an average production of 112 million gallons. Consumption in 1964 was about 14 gallons per capita.

Chile has a number of regions with a very favorable climate for grape growing. Many varieties capable of producing quality wine have been planted. Some of the Criolla-type varieties, imported from Peru (and originally from Mexico), are still grown. Some of these are used for production of sweet wines. Technology and critical consumer demand have lagged and production has not reached its potential level. The large acreage of the Merlot variety favors production of a good red wine, and, in fact, the best Chilean wines have been well-aged red table wines—often named after the vineyard (fig. 11). Much Cabernet Sauvignon and Malbec and a little Pinot noir are planted also.

The whites, mainly from Sémillon, are often over-aged in the wood and are of lesser quality. Other white varieties planted include Sauvignon blanc and Folle blanche. The stabilization of the wines, particularly of the whites, could be improved. Contrary to the usual report, Chile is not a large exporter of wine. According to Protin (1966), in 1965 only 1.6 million gallons were exported. Some of the wine imported into the United States has not been of high quality. The words *Reservado* or *Gran Vino* on the label should indicate a superior quality, since these appellations are controlled, to some extent, by the government. Although many wines are sold under estate or varietal labels, there is, particularly for export, a regrettable use of European generic names such as chablis, borgona, and rhin.

Figure 11. Oak casks for aging table wines in Chile. (Courtesy Dean Ruy Barbosa.)

Brazil

The third largest producing country of South America is Brazil, which produced nearly 35 million gallons in 1964. The grape acreage is over 165,000, of which a considerable amount is planted to table grapes. The main vineyard areas are in São Paulo and Rio Grande do Sul—both in the south, with warm and humid climates.

The most serious negative factors for quality are the warm humid climate, the large acreage of varieties other than *V. vinifera*, and the lack of demand for high-quality wines. Until the late nineteenth century there was a large importation of wine from Portugal. For economic and political reasons this importation is now negligible. Per capita consumption of wine in Brazil was only about two-thirds of a gallon in 1964. This reflects the difficulty of developing a large grape production under unfavorable climatic conditions.

The positive factors favoring the development of the Brazilian wine industry are the interest in planting better varieties of grapes and the greatly improved technology which has been sponsored by the government.

Uruguay

The fourth largest acreage of grapes in South America is in Uruguay, with about 45,000 acres, much of this near Montevideo. Production is over 22 million gallons and consumption about 7 gallons per capita. Because of private enterprise some wines of quality are produced. European varieties such as Tannat (the most appreciated), Sauvignon blanc, and Pinot noir are planted. Otherwise the industry resembled that of its neighbor Argentina in producing a large volume of cheap wine.

Peru

Pastor (1967) believes that the cool Humboldt current has a moderating influence on the climate of southern Peru. The first plantings of grapes were made near Cuzco about 1550. This was not a fortunate choice of a region for the first plantings of *V. vinifera* on this continent, as its climate is very hot and dry. The original vines were seedlings of the Criolla- or Mission-type grape from Mexico, called "Quebranta" in Peru. As in California, these low-acid, late-maturing varieties were found to produce wines of poor quality. A large percentage of Peruvian wines were distilled in the nineteenth century. Because of the demand for alcohol in California during the Gold Rush there was a good market for brandy, and the distilled product of Peru, called pisco, filled this need.

Today there are only about 20,000 acres of grapes in the country, and

most of the wine (nearly 7 million gallons in 1965) appears to be consumed by agricultural workers or is distilled for pisco brandy. Protin
(1966) reported per capita consumption of only 1 liter in 1965. The use of
wine for distillation seems sensible because most of the wines are of
nominal quality. This is due not only to the climate but also to the poor
varieties planted, the primitive fermentation technology, and lack of a
critical or substantial home or export consumer market, which reduce the
incentive to improve either the varietal complement of Peruvian vineyards
or production processes.

There is little evidence of a permanent quality wine industry in
Peru. For local cultural reasons, a demand for high-alcohol distilled
products may justify the production of grapes. Modern technology and
better varieties could (and occasionally does) result in better wines.

Other countries

There is a small acreage of vines in Bolivia (nearly 6,000), Columbia
(less than 1,000), and Venezuela. The latter country produces some wine
from imported grape concentrate and imports over 3.5 million gallons of
wine annually.

AFRICA

Climate limits vines to
the north and south.

Algeria

Viticulture in Algeria dates from Roman times or earlier. Under
Moslem domination, table grape production was encouraged, but the
wine industry remained small. Starting in the mid-nineteenth century
under French influence, the wine industry developed so rapidly that by
1940 the annual production was 370 million gallons. Most of this was
exported to France. World War II greatly discouraged the industry, and
the departure of the French in 1952 further reduced production. The
decreasing French market has also been a factor. The main center of
production is near Oran, where about 70 per cent of the acreage is
located.

In 1964 there were about 960,000 acres in vines, with a production of
over 275 million gallons of wine in 1965. Of this, 240 million gallons were
exported, mainly to France, in 1964, and 220 million gallons in 1965.
The prospects are that exports to France will continue to decrease.
Only about 9 million gallons were consumed in Algeria in 1964.

47

The varieties planted in Algeria have obviously been greatly influenced by those grown in the south of France. This has been very unfortunate, because the climate of Algeria is very different from that of the Midi. Thus Alicante Bouschet, Carignane, Grenache, Cinsault, Mourvèdre, Clairette, and Mersegnera are the predominant varieties planted. These are either large producers or are best adapted to warm climates. Since the climatic conditions vary widely, the quality of Algerian wines could be improved by planting better varieties in the cooler regions.

However, the most serious effect on the quality of Algerian wines has been not the varieties but the methods of production. As long as Algerian wines were intended primarily as high-alcohol blending wines for the low-alcohol wines of the south of France, their inherent quality was not of great importance. Thus methods of rapid fermentation and movement to the French market were used. There is ample evidence that higher-quality wines could be produced in Algeria if a demand for them could be developed. One positive quality factor is the modern wineries, many of concrete construction, which were built before World War II. With adequate temperature control these could be used for producing better-quality wines.

Morocco

Morocco is also an ancient grape-growing area. In this century, under French influence, the acreage expanded to supply cheap blending wine for France. At present 180,000 acres are in grapes, and 1965 production of wine was over 91 million gallons. In 1965 exports dropped to only 40 million gallons; so overproduction is a major problem and will probably continue, depending on French wine import policy.

Much of the acreage is in the relatively mild climatic zone east of Rabat. The wineries are modern and quite similar to those of Algeria. Besides native varieties the ordinary standard French varieties are grown. The quality is very moderate, though we have tasted some pleasant red wines produced near Meknes. Again the largely Moslem population restricts home consumption—only 6.5 million gallons in 1965. For a recent report on the wine industry in Morocco see Joppien (1960).

Tunisia

There are over 120,000 acres in grapes, producing about 45 million gallons of wine. In 1964 most of this was exported, but exports dropped sharply as the French market decreased and probably will continue to decrease. Local consumption averages about 7 million gallons. The modern Tunisian grape and wine industry was primarily an adjunct of the

48

French demand for wines for blending. Most of the wine was fairly high-alcohol table wine produced from ordinary varieties such as Carignane and Alicante Bouschet. These and the warm climatic conditions plus the low consumption of wine in a Moslem country did not favor the development of a quality wine industry. The future does not appear promising except for low-priced wine for export.

Egypt

Undoubtedly Egypt is the oldest grape-growing country in Africa. Grapes were grown and wines made as early as 2500 B.C. However, the industry was a small one and use of wine appears to have been restricted mostly to the wealthy. Today there are nearly 25,000 acres of grapes. About 1.2 million gallons of wine were produced in 1963, of which nearly half was exported. Both white and red table wines are produced. As might be expected from a warm climate, the wines tend to be low in acid and flat in taste. Excessive storage on the lees seems to be a problem. The samples seen have had local names.

South Africa

This country represents an entirely different viticultural tradition from that of the north African countries. Vines from Europe were first planted by the early Dutch settlers just over three hundred years ago. The famous Groot Constantia vineyard was planted to a muscat-flavored variety in 1684 and its wine achieved a considerable reputation in Europe in the eighteenth and early nineteenth centuries. The reason for the decline in its fame thereafter is not known. However, muscat varieties are notoriously low in acidity and high in pH. If alcohol-tolerant acid-reducing bacteria became established, the consequent reduction in quality may have caused the decline in the popularity of the wine. The abolition of preferential tariffs for South African wines in Great Britain by the Gladstone government in 1861 and the invasion of phylloxera about 1885 were serious blows to the South African grape and wine industry.

The modern South African grape and wine industry is oriented toward home consumption and decreasingly to the export market. This market is mainly to Great Britain, Germany, and Canada. Acreage is increasing and in 1965 was over 190,000 acres. There is an important table grape industry; much of this is for export. Wine production in 1965 reached a record 118 million gallons of which nearly 5 million were exported. A large percentage of the wine produced (more than half) is distilled. Per capita consumption is more than twice that of the United States, over 8 liters per year (about 40 million gallons), but consumption of table wines has increased much more rapidly than that of dessert wines since about 1955.

A number of climatic regions are planted to vines. In the cool coastal belt of Stellenbosch, Paarl, Malmsebury, Ceres, Tulbagh, and the Constantia Valley, European varieties such as Cinsault (here called Hermitage), Steen (a high-acid bud sport), Palomino (here called White French), and Green are the important varieties, although Riesling, Cabernet Sauvignon, Clairette blanche, and other varieties are grown. The Little Karoo is a warmer district, more like region IV or V (p. 72) in California. It stretches from beyond the Drakenstein range to the Swartberg Mountains and the district of Worcester, Robertson, Montagu, Oudtshoorn, and Ladysmith. Besides Palomino and Cinsault, Muscat of Alexandria (here called Hanepoot), Thompson Seedless, and Muscat blanc are grown here. Some of the better varieties also are grown: Riesling, Cabernet Sauvignon, Petite Sirah, Gamay, and Pinotage (a cross).

Dry white table wines are produced mainly in the Stellenbosch, Paarl, and Tulbagh areas; red table types are produced in the Stellenbosch, Paarl, and Cape areas (Anon, 1967c).

A variety of types of wines are produced—many obviously of German influence. Pressure fermentations are common and some low-carbonated wine, called Perlwine, is made. The Riesling-type wines are often slightly sweet and a little heavy. Of the red table wines some are bottled for aging and often develop well. A few are rather heavy-bodied and alcoholic.

Generally the wineries are quite modern, and the K.W.V. (Ko-operative Wynbouwers Verenigung, a semigovernmental agency) has helped stabilize the industry by setting up strict quality standards at fixed prices. They have also encouraged the production of flor sherry and brandy. (See Anon., 1967c, for further details.)

South Africa is not plagued by the imitative generic names for its table wines as are Australia and the United States. Sparkling wines are Fonkelwyn, and many of the bottled wines have estate (or winery), vineyard, or varietal names.

NORTH AMERICA . . .

. . . produces wines from native as well as from
European grapes.

Eastern United States

Grapes are native to many of the states east of the Rocky Mountains. The earliest explorers of our eastern shores reported many wild varieties. Unfortunately, the clusters and berries were small, the sugar content was low and the acidity high, and they had a pronounced and, to the Europeans, strange flavor. Early in the Colonial period and thereafter, European varieties were imported, but all of them failed. The vines survived for a few years, but eventually succumbed to winterkilling, owing to the very low temperatures, to phylloxera, to virus or fungus diseases, or to insects. (See Hymans, 1965, for a perceptive analysis.)

Not until the beginning of the nineteenth century were extensive efforts made to domesticate the native species. Some of the early workers with vines may inadvertently have produced crosses with adjacent experimental *V. vinifera* vines, notably the variety Alexander, which Jefferson recommended. From whatever origin a number of truly indigenous varieties were developed: Agawan, Catawba, Concord (now by far the most popular), Delaware, Dutchess, Elvira, Fredonia, Herbemont, Isabella, Ives, Niagara, Vergennes, and over a thousand others. All are characterized by a pronounced (*labrusca* or foxy) aroma (which, however, varies markedly in intensity and quality from variety to variety). Until recently this was attributed to methyl anthranilate, and the odor of the fruit does resemble that of this ester. Just why the aroma is called "foxy" is not known.

The most important grape-growing state east of the Rocky Mountains is New York, with 33,211 acres in 1966, of which 26,901 were planted to Concords. The Concord variety is widely planted in the Chautauqua district west of Buffalo on Lake Erie for table and juice purposes. Catawba and Delaware are the next most popular varieties in New York. A few grapes are grown along the Hudson for wine. The main wine-producing area is in the Finger Lakes district in upper New York. Vines are planted on the slopes leading to these narrow lakes: Canandaigua, Keuka, and so on. The moderating influence of these deep lakes, especially on the winter temperature, permits survival of grapes.

United States law permits simultaneous addition of sugar and water providing the volume is not increased more than 35 per cent. Otherwise the methods of wine production are much the same as in California, except that some red wine is made by the hot-press method. The crushed red grapes, usually Concord, are heated to 60° to 65.5°C (140° to 150°F) and then pressed, cooled, and fermented. The strong flavor of *labrusca* varieties appears to withstand the deleterious effects of heat much better than *vinifera* varieties.

In recent years there have been renewed attempts to grow *vinifera* varieties in the eastern United States. In 1966 the New York Crop Reporting Service reported 73 acres of *vinifera* varieties in the Finger Lakes area. The quality of some of the wines has been excellent. Only time will tell whether these experiments will be economically feasible.

The extent of winterkilling and the speed of recovery are major problems. There are about 1,000 acres of direct producers (mainly Seibel 5279) in the state of New York.

A variety of types of wine are produced in New York and other states east of the Rocky Mountains. A few are named after the variety of grape from which they are produced. Most of the wines, however, have such generic names as burgundy, champagne, claret, and sauterne. These seem particularly inappropriate for the foxy-flavored wines. California wines are sometimes blended in to reduce costs and decrease the special flavor. As long as the blend contains 75 per cent New York wine it may still be labeled "New York". Some very sweet, low-alcohol kosher-type wines are produced in the eastern United States. Concord grape concentrate and inexpensive California red wine have found their way into some of these products. A few of these wines are now being produced at a lower sugar content.

A few wines are made in Ohio, Michigan, Illinois, Maryland, Missouri, Arkansas, and elsewhere, mainly from *labrusca* varieties or from direct producers. Ohio wines have had some reputation since Nicholas Longworth was so successful with Catawba just before the Civil War. The Ohio Valley was then called the Rhineland of America. In 1847 Longworth produced a "Sparkling Catawba" which achieved a considerable reputation, whether for quality or uniqueness is not known. The vineyards of Missouri were also well known in the nineteenth century, and Cook's sparkling wine was produced there soon after Longworth's Ohio sparkling wine. This brand is still made in California. Maryland has been particularly successful with direct-producer hybrids; some of their wines have an almost completely *vinifera*-like character.

There is some use of *V. rotundifolia* (muscadine) grapes for wine in the South Atlantic states. Most of these wineries appear to survive because of local pride and local demand. There surely must be a place for an industry that has survived this long. More attention to production and stabilization would produce better-quality wines. Some amelioration of the strong flavor seems desirable, through the use of less Concord or muscadine.

In summary: the eastern United States is suited mainly to the production of labrusca-flavored varieties. This is sometimes modified by blending with California wine and may be further changed by the use of direct-producer hybrids. Eastern wines have proved popular in sparkling wines.

Canada

Grapes are grown in Canada mainly on the Niagara peninsula of Ontario. This is possible because of the moderating influence of the adjacent lakes Erie and Ontario. Most of the vines are *V. labrusca* (Concord, Catawba, Niagara, Delaware, etc.). However, a number of direct-producing varieties have been planted. These are mainly hybrids of varieties of *V. vinifera* and *V. labrusca*. Since most of the hybridization was done in France they are often called French hybrids. The original purpose of the crossing was to combine the phylloxera-resistance of American species with the quality of European varieties. The product did not have the foxy flavor of pure *labrusca* varieties. Nowadays the selections seem to be made primarily for their productivity, flavor, and disease- and cold-resistance. Phylloxera-resistance would be desirable, but is not essential.

There is a small acreage in the Okanagan Valley in British Columbia. The climate there is quite cool, but persistent attempts to grow *V. vinifera* varieties are being made. The success of these vineyards is still not clearly established.

Since World War II total acreage in Canada has increased to nearly 23,000. Wine production was about 12 million gallons in 1964. Very liberal sugaring and watering regulations make possible the rather high gallon per ton yields. Canada imports over 4 million gallons of wine a year. Consumption was 2.3 liters per capita in 1963. As in the United States, European generic appellations are widely used.

Mexico

Vines were planted in Mexico at Parras north of Mexico City in the sixteenth century. They were apparently seedlings of seeds from Spain, but their origin is not known; Hymans (1965) believes that some European cuttings were grafted on native vines. These vines were later planted in the states of Chihuahua, Coahuila, and Durango, and by 1697 in Baja California, primarily by the missionaries who settled there. The predominant seedling, called Criolla, was soon planted in Peru, Chile, and Argentina, and somewhat later in California. There it was and still is called Mission. It is unfortunate that this variety should have been so widely planted, for it ripens very late, has low color, and very low total acidity. We have seen Mission musts in California with a pH of over 4.0—a very unfavorable medium for the production of a table wine of high quality.

A further handicap for the Mexican wine industry was the very warm climate and the lack of a wine-drinking population. In recent years there has been great interest in growing better varieties of grapes and in improving the quality of the wines. In 1965 there were 32,400 acres planted to vines. Wine production in 1964 amounted to just over 3 million gallons, some of which is apparently distilled, for only 1.8 million gallons were consumed (per capita consumption is estimated to be only 0.2 liter).

Since 1945 Mexico has nearly tripled its acreage in grapes. The principal vineyard areas are: the Laguna region around Torreón, the states of Durango and Coahuila; Parras and Saltillo, Coahuila; the Aguascalientes region, Aguascalientes; the San Juan del Rio region, Queretaro; the Ciudad Delicias area, Chihuahua; and northern Baja California. The most important of these, with about 12,000 acres, is Laguna, roughly 500 miles south of El Paso, Texas. In this very warm region the conditions are similar to those of Bakersfield (region V). About 80 per cent of its grapes are table varieties, but a large part of the crop is delivered to the wineries for wine production.

There is no information on the percentage of the fruit used for table purposes, but Mexico City is an important market. Of the grapes that are fermented, about 75 per cent are used for the production of brandy. At the present time there are about 26 wineries in Mexico. Several of these are in Mexico City, to which the grapes have to be shipped.

At one time there was considerable use of sugar, water, and other sophisticants by the Mexican wine industry. These deleterious practices now seem to have been brought under control. Nevertheless, with the exception of an occasional table wine, only the dessert wines seem to have real quality, particularly some of the muscatels. With better varieties and the modernisation of the wineries and technology suited to the warm climatic conditions, the quality of Mexican wines could be improved.

OCEANIA . . .

. . . is the newest viticultural region.

Australia

Vines are not native to Australia. The first European varieties were planted near Sydney in 1790, and wine was first shipped to Great Britain in 1822. An intelligent viticulturist, James Busby, dominated the industry after 1824. In 1830 he imported 20,000 cuttings and, reputedly, 570 varieties from Europe.

Today Australian vine acreage is increasing; it was nearly 135,000 in 1965. South Australia has the largest plantings, followed by New South Wales, Victoria, Western Australia, and Queensland. The climatic conditions vary from mild and humid in the Hunter River Valley to cool to warm and dry in the south, where irrigation is often required. A limited range of grape varieties is grown: Syrah, Grenache, Doradillo, and Pedro Ximenes. Until recently this has been restricted by the tight embargo on the importation of new varieties into phylloxera-free South Australia. This has now been overcome and new plantings of varieties producing high-quality wines are being made.

Wine production was nearly 50 million gallons in 1964, but much of this is distilled. The export market is still critically important to the industry. In 1965 about 2.5 million gallons were exported. If half the wine produced is distilled (Boehm, 1966) the export market accounts for 10 per cent of the finished wine. A wide range of wine types are produced, mainly under generic names of European origin such as champagne, port, and burgundy. However, an increasing group has varietal or specially coined names. Wine consumption averages 5.5 liters per capita—compared to 3.7 for the United States.

The most important positive factors for quality have been the pride of production, particularly of the old established companies (fig. 12), some of which were founded more than a hundred years ago; the admirable technological developments; and the keen interest of many consumer groups in better and aged wines. The negative factors are the lack of a large acreage of high-quality varieties and the warm climatic conditions of some regions. (Both of these apply to California also.).

Figure 12. Australian winery and vineyards. (Courtesy B. Seppelt and Sons Pty., Ltd., Seppetsfield, South Australia.)

Moreover, Australia is one of the world's greatest beer-drinking countries. For descriptions of the Australian industry see Webb (1959) and Simon (1967).

New Zealand

Although New Zealand has a rather cool climate for grape growing, strenuous efforts are being made to develop a grape and wine industry. Less than 1,500 acres are planted to vines, and 1964 production was just under 2 million gallons. Sugaring because of the poor ripening of the grapes is common and essential. Some of the wines appear to have been made from diluted and sugared musts (as in the eastern United States), but without the *labrusca* flavor as here.

ASIA
produces few wines.

Japan

There have been vineyards in Japan for several centuries. The Kofu and Jaraku varieties are believed to be from China or are sports of some early-introduced variety. Grapes are grown in several areas, but chiefly near Kofu, on Hokkaido, and near Osaka. The climate is cool, and only early-maturing varieties of *V. vinifera* ripen and often do not fully mature. Some growers are cultivating *V. vinifera* varieties in order to improve their wines and a number of *V. labrusca* varieties and hybrids have been planted also.

In 1964 Japan had over 52,000 acres of vines. Production of wine in the same year amounted to 12 million gallons. However, only 3.5 million were truly wine as defined in this country. The rest was about 20 per cent wine plus alcohol, acids, sugar, and other materials. Wine making has been modernized in a number of wineries (fig. 13), and there appears to be a demand for better wines. Some of the table wines are very creditable. The white table wines are sometimes sugared to produce a rather high alcohol content. However, they are sound wines and well finished. See also Amerine (1964).

China

China has vineyards in several areas and produces a number of wines and brandies. (See Cosmo, 1957.) Taiwan also has some vineyards, but a recent Taiwan Pinot blanc was more of a sherry than a table wine.

Figure 13. Crushing grapes in Japan. (Courtesy Suntory Ltd., Osaka, Japan.)

Cyprus

Wine has been produced here for many centuries. The crusaders who settled here helped to maintain the vineyards during the Middle Ages. At present over 90,000 acres are planted to grapes. Most of the holdings are very small and wine-making procedures tend to be primitive. Production is about 11 million gallons. Much of this is exported—over 4 million gallons in 1964. Dessert wines are important for the export trade, particularly the Commanderia—a sort of tawny port made from partially raisined grapes. The most important variety is the red grape Mavro (over three-fourths of the acreage). The Xinisteri is a white variety accounting for 20 per cent of the acreage. Improvement of varietal plantings and modernization of production techniques are essential steps for improving Cypriot wines.

Israel

There have been vineyards in the Holy Land since Biblical times (Goor, 1966), but wine making was greatly repressed during the long period of Turkish domination. The modern period began in the 1870's when an agricultural school was started. Baron Rothschild imported varieties from France and planted a large vineyard and erected a winery in the 1880's. (Viteles, 1928; Goor, 1966.) At present the most important wineries are at Rishon-le Zion and Zichron Jacob.

De Leon (1959) discusses the French and Italian influences on Israeli viticulture and some Arabic influences. He notes the survival of several native varieties and presence of virus and other diseases.

Climatic conditions are not favorable for the vines for producing table wines in Israel because of the high temperature. In some areas two crops a year are produced. Generally, therefore, the grapes are low in acid. The most important varieties, according to Ough (1965), are Carignane (39 per cent), Alicante Grenache (Grenache?) (37), Muscat of Alexandria (11), Sémillon (6), Clairette (blanche? also called Boorbanlenc) (5), and Alicante Bouschet (1.5). Grenache is sometimes used to produce white musts.

In 1965 Israel had nearly 25,000 acres, of which 22,000 were in production. About 10 million gallons are produced each year; in 1965, according to Protin (1966), nearly half a million gallons were exported. Much of this is sent to the United States. Per capita consumption is 4.75 liters per year, somewhat greater than in this country.

Current practices in wine making in Israel are described by Rappaport (1958), Rosenstein (1964), Lichtblau (1965) and Ough (1965). In 1962 about 41,000 tons of grapes were harvested, of which 75 per cent were wine and 25 per cent table varieties. Production includes table, dessert, sparkling, and flavored wines. The dessert wines, because of the favorable climate, appear to be the best. Of the wines consumed in Israel about 60 per cent were dry, 23 per cent were light sweet, and the balance were sweet dessert wines. The red varieties account for two-thirds of the crop.

Israel is rapidly modernizing its vineyards and wineries and is attempting to raise the standards of production. In the future, area names may be used in Israel: Aichron-Ya'akov (syn. Zichron-Jacob), Negev, Gedera, Saidoren-Gezner, Judean Hills (syn. Beit Shemesh), and Upper Galilee. Israel no longer uses European appellations such as burgundy or chablis.

Turkey

There are at least five important viticultural regions: Thrace and Marmara, the Aegean Sea and southwestern Asia Minor, central Asia Minor, and southern and southeastern Asia Minor. Grapes have surely been cultivated here since 5,000 B.C. or earlier. Total acreage in 1962 amounted to nearly 1.9 million acres. In the east and northeast of Asia Minor most of the 12 million gallons of wines are produced in wineries controlled by the government. Turkey exported about 1.3 million gallons in 1964. Among the best-known red wine types are Yakut Damlasi and Buzbag; red and white types include Kalup, Doluca, Trakya, Kavaklideres, and Kalecik; white types are Guzelbag, Cankaya Yildizi, Doluca, Sanver (muscat), Tekirdag, Narbag (sweet), and Misbag (muscat and sweet). (See Alleweldt, 1965, for further information.)

Lebanon

Lebanon has 60,000 acres of grapes from which just over one million gallons of wine are produced annually. Consumption is only 1.74 liters per capita, since the country is part Moslem. Even so this is more than three times as great as that of Syria. The industry is still rather primitive.

Syria

Nearly 170,000 gallons are produced annually in Syria. Per capita consumption, however, is only 0.45 liter per year. The warm climatic conditions and the low consumption do not favor large-scale wine production.

Jordan

There are about 42,000 acres of grapes in Jordan, many planted to table grapes. Annual wine production was just under 400,000 gallons in 1959, according to Protin (1966). No information is available on the effect of the hostilities of 1967 on the industry.

Iran

Grapes have been grown in Iran since Neolithic times. At present there are about 180,000 acres of vines. Since it is a Moslem country, wine production amounts to only 100,000 gallons per year. Most of the wines are of moderate quality and are often sweet. A legendary quality has long been attributed to Shiraz wine, but at present it is of mediocre quality. One of us recently visited wineries in Iran and found the new wineries progressive in outlook and some of the wines noteworthy.

Other countries

There are a few vineyards in Afghanistan and Pakistan. Formerly some wine was made near Quetta. Most of the grapes are used for raisins or table purposes. India has about 20,000 acres of vines, primarily in the southern part of the country. Very little wine is produced, but there has been some recent interest in producing wines for tourist use.

CHAPTER TWO

CALIFORNIA GRAPES, WINE TYPES, AND LEGAL RESTRICTIONS

WINES AND TYPES OF WINE

*Standard wines and fine wines require different
procedures in fermenting and aging.*

Wine is the product of the partial or complete fermentation of the juice of grapes. Table wines, those containing not over 14 per cent alcohol, are served with various appropriate courses at meals; they are referred to also as "light," "natural," "dinner," and often as "dry" wines. But the terms "dry" for wines containing less than 14 per cent alcohol and "sweet" for wines containing more than 14 per cent alcohol are confusing: the so-called "dry"-type wines are sometimes sweet to the taste and the "sweet"-type wines sometimes dry. These terms have therefore been rejected for legal use and are being discouraged by the California industry. The term "table wine" includes sweet wines of not over 14 per cent alcohol; but it may not be appropriate to classify some of the very sweet blended types as table wines. Most of these are usually produced by fermentation procedures. Sparkling wines, such as champagne,[1] are also considered to be "table wines."

Wines containing over 14 per cent alcohol, produced by the addition of brandy or grape spirits (after partial or complete fermentation) for the purpose of preserving some natural grape sugar in the wine, are now called "dessert" and "appetizer" wines. These present very different problems to the wine maker and are not discussed here. (See Joslyn and Amerine, 1941a, 1941b, 1964.)

----

[1] The system of capitalizing wine type names used in this book deserves explanation. Wine types named after a specific variety of grape are capitalized: Chardonnay, Pinot blanc, Zinfandel, etc. Generic types of wine are not capitalized: claret, hock, chianti, etc.—even though the name has a geographical origin—burgundy, rhine, sauterne, etc. However, when these names are used for wines of the region of that name, they are capitalized: Burgundy (France), Rhine (Germany), Sauternes (France), etc.

Methods are given in this book for the domestic production of both standard, or ordinary, table wines (basically clean, sound wines of average quality) and fine table wines (those of choice, premium, or fancy quality).

Ordinary wine, known in France as *vin ordinaire*, supplies most of the consumption in the great wine-producing countries, where it is made by simple, natural, and inexpensive methods; it is usually aged for only a short time and most of it is consumed within a year after fermentation. As more and more of this wine is bottled, increasingly complex stabilization procedures are being used. Methods of producing ordinary wines in California under present conditions have been fairly well worked out, but there remains the problem of improving the quality (color, flavor and bouquet) in order to stimulate demand.

The fine wines, which make up only a small part of world production, are made with great care from selected grapes, and are sometimes aged for considerable periods both in the cask and in the bottle; they are the high-priced wines upon which a reputation for quality in wine making is built. The varieties, equipment, and methods of handling recommended for ordinary wines are usually not desirable or satisfactory for the production of fine wines. The desired quality may be obtained by proper fermentation of the better varieties of wine grapes and correct aging of the wine. With such wines, even a slight change in the method of aging or finishing may have a marked influence upon the quality of the finished product.

An understanding of the principles discussed in chapters 4 to 11 is essential to the wine maker who wishes to improve present practices or to create distinctive types whether of standard or fine wines.

IMPROVING CALIFORNIA WINES

*Demand could be stimulated by developing more
fine varietal or other native wines and by
standardizing the types of ordinary wines.*

The unusually favorable soil and climatic conditions of California offer many opportunities for the profitable development of unique and characteristic native wines. Only a few such wines have been produced thus far. Varietal wines—those named for the variety from which they are primarily produced and from which they take their distinctive character, such as Cabernet Sauvignon, Sauvignon blanc, and Zinfandel—have won increasing acceptance and are a step in the right direction; but the varietal character must not be obliterated by unwise treatment and blending. The success of varietal labeling is confirmed by the rapidly increasing importation of varietal-labeled wines, particularly from France.

So far as we can determine, there are no French laws dealing specifically with varietal labeling. The well-known French *appellation d'origine* (p. 4) limits the production in a certain district to certain varieties. For example, at least six varieties are approved for production of a wine entitled to the *appellation d'origine* Bordeaux. Since Cabernet Sauvignon is only one of the varieties which can be used, there is no legal guarantee that any Cabernet Sauvignon has been used in an imported wine carrying the varietal label "Cabernet" or "Cabernet Sauvignon" even though it is entitled to the *appellation d'origine* "Bordeaux." In view of the recent plethora of cheap varietal wines from France and elsewhere it would be desirable for the California industry to insist on some guarantees of composition—at least equal to the 51 per cent American requirement (p. 91 f). Although French regulations specify that no untruthful information may be used on the label, we are skeptical about the rigorous enforcement of the veracity of the varietal origin of some.

This applies not only to varietal-labeled wine from France but also to those of other countries. We have, for example, seen a Spanish Zinfandel on the American market, although this variety is not known in Spain! In view of the extremely mixed plantings in some parts of Italy we are skeptical also in regard to the composition of varietal-labeled Italian wines, for we have seen Italian Grignolinos that vary in color from pink to deep red. There may be, and undoubtedly are, a number of Grignolino wines of varying color, but such variation in commercial exports seems excessive.

However, the industry has neglected many opportunities in this field and has catered to the trade demand for foreign wine types, such as burgundy, chablis, and rhine. These are legally known as generic types. (See also Amerine and Cruess, 1960; Amerine *et al.*, 1967.)

HISTORY . . .

. . . Grapes first planted in
southern California.

The first European grapes in California were imported from Baja California—apparently the Criolla variety of Mexico which came to be known as Mission. This late-ripening, low-purple-colored, low-acid, high-producing variety dominated vineyards in all parts of the state until the 1860's.

Other importations were apparently made in the 1840's by Louis Vignes in Los Angeles and by Agostin Haraszthy in San Diego, San

Francisco, and Sonoma. Until the 1860's more grapes were grown in southern than in northern California, but by the late 1870's there were widespread plantings of numerous European varieties in all parts of the state. Many of the early wineries had cellars for better temperature control (fig. 14).

Figure 14. Entrance to hillside wine cellar in California. (Courtesy Wine Institute, San Francisco.)

Starting in 1880 and continuing to 1893 the University of California conducted far-reaching experiments on wine grape varieties (Amerine, 1962a). This research was greatly expanded and refined by the post-Prohibition research of Amerine and Winkler (1944, 1963a). (For further discussions of the history of the California wine industry see Carosso, 1951; Peninou and Greenleaf, 1954, 1967; Amerine, 1969.) Huber (1967) has recommended a program of industry control of grape quality to prevent unsuitable grapes from reaching the winery. He noted especially the need for grower education.

RECOMMENDED GRAPE VARIETIES

To obtain wines of quality, the grapes
must have the proper composition and character for the
type of wine that is to be produced.

The raw material for all table wine is sound, mature, fresh grapes (or, with certain types of sweet table wines, partially dried grapes). Many varieties without distinctive flavor produce palatable and well-balanced if undistinguished wines. But the best wines are made from grape varieties that impart a recognizable and distinctive flavor.

Different grapes are needed for red and for white wines, for dry and for sweet table wines. Even for ordinary wines, and especially for fine wines, the grapes must be suitable not only in color and flavor but also in chemical composition, especially in the balance between sugar and acid in the must. The composition and even the color and flavor of the grapes depend on the climate and on their maturity, as well as on the variety. The maturity recommended for dry red, rosé, dry white, and sweet table wines is discussed in chapters 12 to 15.

INFLUENCE OF ENVIRONMENT

Seasonal and regional differences, owing primarily to temperature,
markedly influence the composition of the grapes.

When mean winter temperatures fall below $-3.6°C$ (30°F), Prescott (1965) noted that special precautions (usually covering) must be taken to protect vines of *Vitis vinifera*. This greatly limits the extension of vines in many areas of the world.

Furthermore, even if protected, grapes should not be subject to winter temperatures much below $-18°C$ (0°F). Prescott notes also that to bring grapes to maturity the length of the season above $10°C$ (50°F) must be about six months and the temperature summation during this period should be at least 1,600 to 1,800 degree-days.

While temperature summation is a controlling factor, solar radiation and other factors must be taken into account. In calculating degree-days the difference between the average daily temperature and 50°F is computed and these are summated for the growing season or for some part of the season. On this basis Winkler (1962) distinguished five climatic zones in California, from the coldest, I (less than 2,500 degree-days), II (2,501 to 3,000), III (3,001–3,500), IV (3,501–4,000), to the warmest,

TABLE 2

INFLUENCE OF REGION ON COMPOSITION OF GRAPES PICKED AT APPROXIMATELY SAME STATE OF MATURITY

Variety	Brix (Degrees)			Total acid (Per cent)			pH		
	Fresno[a]	Davis[b]	Bonny Doon[c]	Fresno	Davis	Bonny Doon	Fresno	Davis	Bonny Doon
1937									
Alicante Bouschet	—	18.8	18.9	—	0.78	1.20	—	3.47	3.10
Burger	—	17.8	17.6	—	0.55	0.81	—	3.46	3.15
Cabernet Sauvignon	22.9	22.4	20.7	0.65	0.67	1.10	3.48	3.63	3.41
Carignane	21.8	21.8	—	0.62	0.70	—	3.67	3.58	—
Sauvignon vert	23.3	—	20.3	0.50	—	0.67	3.81	—	3.24
Sémillon	—	19.0	18.0	—	0.67	0.97	—	3.45	3.10
Zinfandel	—	22.4	21.3	—	0.61	0.86	—	3.58	3.28
1938									
Alicante Bouschet	—	19.9	19.1	—	0.63	1.29	—	3.63	2.95
Cabernet Sauvignon	22.0	20.7	—	0.46	0.68	—	3.58	3.53	—
Sauvignon vert	24.2	24.3	22.4	0.44	0.57	0.57	3.67	3.81	3.28
Sémillon	21.0	18.1	34.1[d]	0.41	0.55	0.69	3.42	3.36	3.07
Zinfandel	21.0	22.8	24.7[d]	0.55	0.55	1.16	3.51	3.58	3.14

[a] Fresno—4,680 degree-days of temperature above 50°F during growing season.
[b] Davis—3,618 degree-days of temperature above 50°F during growing season.
[c] Bonny Doon—2,400 degree-days of temperature above 50°F during growing season.
[d] Figures are high due to late harvesting and small crop on vines.
Source of data: Amerine and Joslyn, 1940.

V (4,001 or more). Kraus (1966) emphasizes the close inverse correlation between heat summation and time to ripen that exists in northern or cool countries.

Koblet and Zwicky (1965), on the basis of twelve years' data, have shown that the correlation between seasonal temperature summation and sugar content is greater than the correlation with the temperature summation of any particular month or with hours of sunshine. The negative correlation with acidity followed the same pattern, except that there was a high negative correlation between September heat summation and total acidity. This agrees with Winkler's (1958) observation in California that high temperatures late in the season results in especially low-acid musts.

The grape-growing regions of the interior valleys of California are approximately 55 to 60 per cent hotter during the growing season than those of the coastal counties. (See footnotes to table 2 for the method of calculation.) In the warmer districts, grapes mature much earlier and, at the same degree of sugar, have less total acid, a higher pH, and less color.

Table 2 indicates the differences in acidity and pH of certain varieties in the Fresno, Davis, and Bonny Doon (near Santa Cruz) districts for the 1937 and 1938 seasons, harvested at approximately the same degree Brix. The average summation of temperature at Davis is only 77 per cent that of Fresno, and Bonny Doon receives barely 50 per cent as much effective heat during the ripening season as does Fresno. Not all the samples could be harvested at the same stage of maturity. During the later and more rapid stages of maturation, the total acidity drops at remarkably varying rates with time, for different varieties, and in different climatic regions. The total acidity drops about 0.004 to 0.012 per cent per day from a Brix of about 20 to 23 degrees. Or, for each degree rise in Brix, the total acid decreases 0.03 to 0.10 per cent. The figures indicate significant differences in the total acid and pH of the grapes in these districts. The total acid in the grapes from Bonny Doon is almost twice that of the same varieties from Fresno.

Table 3 indicates the marked differences between seasons. Not only does the picking date for the same degree Brix come earlier in the hot season but the acids and colors are lower. Even though the samples in 1936 were picked considerably earlier, the Brix degrees were slightly higher. This slightly higher degree Brix will not, however, account for the much lower average total acid content in 1936. (See further data in Winkler and Amerine, 1937.) Table 4 indicates that varieties differ markedly in their sugar/acid ratio even when grown in the same region and harvested at nearly the same sugar content.

TABLE 3

EFFECT OF SEASON AND VARIETY ON COMPOSITION OF MUST AND WINE

Variety	Average date collected		Must		Wine					
			Brix		Alcohol		Total acid		Color intensity[a]	
	1935	1936	1935 (degrees)	1936 (degrees)	1935 (per cent)[b]	1936 (per cent)[b]	1935 (per cent)	1936 (per cent)	1935	1936
Alicante Bouschet	Oct. 1	Sept. 23	21.7	22.9	10.9	12.3	0.61	0.57	64	63
Burger	Sept. 30	Sept. 19	19.8	19.5	9.6	11.0	0.66	0.55	—	—
Carignane	Sept. 30	Sept. 19	22.3	23.3	10.9	12.4	0.70	0.50	27	22
Palomino	Sept. 24	Sept. 30	21.5	22.9	11.0	12.5	0.38	0.36	—	—
Petite Sirah	Sept. 28	Sept. 22	23.2	25.6	11.3	13.9	0.67	0.56	77	59
Zinfandel	Oct. 1	Sept. 19	24.7	23.9	12.8	13.5	0.71	0.58	31	19
Average 6 varieties	Sept. 29	Sept. 19	22.2	23.1	11.1	12.6	0.62	0.52	50	41
Average of 240 samples[c]	Sept. 30	Sept. 21	22.1	23.0	11.0	12.2	0.61	0.50	39	33

[a] These figures represent the relative intensity of color expressed on an arbitrary scale; the higher the figure, the greater is the concentration of pigment.
[b] Per cent by volume.
[c] Represents about 40 varieties from several districts of California.
Source of data: Winkler and Amerine (1937).

TABLE 4

BRIX, TOTAL ACID, AND pH OF COMMON WINE-GRAPE VARIETIES

All grown in the same region during the 1938 season

Variety	Date picked	Brix (degrees)	Total acid as tartaric (per cent)	pH	Ratio Brix to total acid
Alicante Bouschet	October 18	21.9	0.60	3.57	36.5
Barbera	October 17	21.5	0.95	3.20	22.6
Cabernet Sauvignon	October 17	20.7	0.68	3.53	30.5
Carignane	October 3	21.0	0.47	3.69	44.6
Chasselas doré	September 22	20.5	0.44	—	46.6
Fresia	October 17	20.7	0.75	3.55	27.6
Mission	October 10	24.0	0.45	3.98	53.4
Muscat of Alexandria	October 17	20.9	0.43	3.55	48.6
Nebbiolo	September 28	22.4	0.81	3.16	27.6
Palomino	October 10	23.7	0.37	3.79	63.0
Petite Sirah	September 21	22.3	0.50	3.46	44.6
Pinot noir	September 14	22.8	0.55	3.86	41.5
Sylvaner	August 31	21.0	0.69	3.54	30.4
White Riesling	October 3	22.1	0.51	3.47	43.4
Zinfandel	September 21	22.8	0.55	3.58	41.5

The large and varying differences of climate on wine grape composition in Europe are indicated in the following tabulation of Wejnar (1965). The warmest seasons were 1963 and 1964; the coolest, 1962. There were low malic acids in the 1963 and 1964 musts compared to those of 1962.

Item	Müller-Thurgau	Chasselas doré	Weisser Burgunder	Sylvaner	Fröhl Sylvaner	White Riesling
Date of harvest:						
1962	10/3	10/3	10/12	10/12	10/20	10/20
1963	9/17	9/18	9/26	9/26	9/27	9/27
1964	9/17	9/18	9/17	9/25	9/24	9/25
Sugar:						
1962	72	56	60	55	62	62
1963	70	60	72	66	66	69
1964	70	77	82	78	74	82
Total acid:						
1962	1.22	1.33	1.67	1.62	1.64	1.76
1963	0.80	0.94	0.68	0.79	0.84	0.98
1964	0.60	0.58	0.75	0.64	0.65	1.00
Malic acid:						
1962	0.57	0.73	0.77	0.92	1.02	0.78
1963	0.28	0.35	0.37	0.36	0.44	0.38
1964	0.21	0.22	0.33	0.29	0.27	0.34
Tartaric acid:						
1962	0.61	0.65	0.80	0.74	0.67	1.02
1963	0.75	0.82	0.55	0.66	0.67	0.84
1964	0.59	0.62	0.71	0.77	0.71	0.96

Wejnar indicates that average daily temperatures as well as number of sunny days are important.

In a later study Wejnar (1967) showed that although 1965 was slightly cooler than 1966 the musts of 1965 had lower total and lower malic acid contents and higher sugar than those of 1966. He attributes this to the lower precipitation in 1965 (10.8 mm during October 1965 versus 91.2 mm during October 1966). However, the 1965 wines were harvested over a month later than those of 1966. This late harvesting would seem to be the main reason for the lower malic acid content. Similar results with late harvesting have been obtained in California.

The influence of regions on the color of grapes is indicated in table 5. In a moderately cool coastal valley such as Napa, a variety may be suitable for making a well-colored red table wine, whereas at Fresno the same variety would not have sufficient color or acid. The generally higher pH at Fresno (see table 2) would indicate a must which was less satisfactory for dry-wine fermentation; and the color would be less attractive at the higher pH. Varieties such as the Grenache make satisfactory red wines in the cooler districts, but in the San Joaquin Valley area they are frequently used for distilling material, or pressed for use as white musts for dessert wines. They do not make good rosé wines when grown in these hot regions. (For a further discussion of the factors influencing color, see Winkler and Amerine, 1938.)

In general, the warmer the season or the district, the higher the sugar content and pH on a given date, and the less the total acid and color. Other factors being equal, the cool districts are the most likely to give naturally well-balanced musts for dry table wines. According to the

TABLE 5

INFLUENCE OF REGIONAL CONDITIONS ON COLOR OF GRAPES

	Regions and averaged color value[a]				
Variety	Delano, Fresno	Lodi, Guasti, Davis	Livermore Valley, Asti, Ukiah	Napa Valley, Santa Clara Valley	South Sonoma County, Santa Cruz Mts.
Alicante Bouschet	74	85	92	143	235
Carignane	45	49	57	83	100
Mataro	8	14	20	55	65
Petite Sirah	70	80	89	143	200
Zinfandel	27	44	52	62	200

[a] These figures represent the relative intensity of color expressed on an arbitrary scale; the higher the figure, the greater is the concentration of pigment.
Source of data: Winkler and Amerine (1938).

experience of most wine makers, the best quality of such wines is produced in districts where the musts are naturally well-balanced. But even in districts where the musts are deficient in acidity, wineries can produce satisfactory standard table wines by using one of the varieties, such as Barbera, which retains its high acid when grown in a warm climate; by picking high-acid, second-crop grapes at the same time as the low-acid, first-crop; by picking earlier; by picking a high-acid, low-sugar variety at the same time as the low-acid, high-sugar variety and combining the musts; or by adding acid to the must. However, not all of these practices are suited to the production of the highest-quality wines.

Factors other than temperature which influence the composition of grapes are hours of sunshine and cloudiness, maximum and minimum night and day temperatures (as distinct from mean daily temperatures), relative temperature of various months, humidity, and so on. Kliewer and Schultz (1964) reported greater incorporation of $C^{14}O_2$ into organic acids and sugars under full sunlight than under 20 to 30 per cent of full sunlight. Low sunlight increased the labeled carbon dioxide in the organic acids compared to full sunlight, and decreased the amount in the sugars. However, the acids differed from each other: more labeled carbon dioxide was found in malic acid under shaded conditions and more in tartaric acid under full sunlight. They showed also that reduced sunlight delayed maturity and resulted in less heat accumulation and more total acidity at harvest. They noted that varieties seemed to differ in the heat accumulation required for maturation. The critical importance of heat summation on the accumulation of sugars and the titratable acidity of wine grapes was confirmed by Wejnar (1965). Malic acid, especially, is reduced, through respiration, by higher temperatures, and by a large number of sunny days in May, June, and July.

Winkler (1962) has reviewed the influence of viticultural practices on vine growth and fruit composition. Overcropping results in delayed maturity, lower sugar and acidity, and higher pH in musts. In California irrigation had little influence on composition unless moisture deficiencies developed. Branas (1967) reported that in some vineyards in the south of France irrigation had no effect, in others it delayed maturity, and in still others it hastened it. He attributed this to differences in soil conditions. The report of Hendrickson and Veihmeyer (1950) gives more and better information. They concluded that irrigation is without effect on grape maturation or composition if the soil moisture throughout the root zone is kept above the permanent wilting percentage. The problem of soil temperature was not a limiting factor in their experiments. In Europe, where heat summation is a limiting factor, it is conceivable and indeed probable that well-drained soils where the soil moisture content is low will be warmer and thus result in earlier maturation, more sugar, less acidity, and better-quality wines. It seems probable that it is not irrigation but soil temperature which influences must composition. If irrigation were injurious, most of the wines of Europe would be poor, for summer rainfall occurs to some extent throughout the northern European vineyard areas during the growing season.

Grapes and climate

The adaptation of specific varieties to the different climatic regions of California continues to be an important study of the California Agricultural Experiment Station. Information on origin, cultural characteristics, and type of the important wine-grape varieties will be found in other publications of the College of Agriculture (Amerine and Winkler, 1943, 1963*a*, 1963*b*).

On the basis of a five-year study in the Napa Valley, Martini (1966) concluded that only 2 per cent of the grapes had mold infection at harvest; of the 2 per cent, 90 per cent were infected by *Botrytis cinerea* and about 8 per cent by mildew (*Uncinula necator*). Other molds identified were *Aspergillus* sp., *Cladosporium* sp., and *Penicillium* sp. Mildew was a factor in only one season, but mold infection remained rather constant during harvesting except when more than an inch of rain fell during the harvest season. Botrytis does not constitute a problem, since it is beneficent itself unless secondary infection occurs. This may occur under Napa Valley conditions but is normally not serious.

Table 6 summarizes the recommendations on grapes for the various types of wine in the five grape-growing regions[2] of California (fig. 15).

---

[2] The regional classification is based on the amount of heat received during the growing season. (See Amerine and Winkler, 1943, 1944, 1963*a*.). Region I is the coolest area where grapes are grown. Typical vineyard areas in this zone include parts of Mendocino, Napa, and Sonoma counties, the mountain areas of Santa Clara County, part of Santa Cruz County, and Mission San Jose in Alameda County. San Juan Bautista in San Benito County and Gonzales in Monterey County are also in this climatic zone. Region II, the most important table wine district, comprises the larger part of the Napa and Santa Clara valleys, and includes warmer areas in Sonoma, Napa, Santa Clara, San Benito, and Monterey counties. Region III is moderately warm, and covers the southern end of Mendocino County, the warmer parts of Sonoma and Napa counties, Livermore Valley, King City in Monterey County, Templeton in San Luis Obispo County, and Alpine in San Diego County. Region IV includes the central part of the Sacramento–San Joaquin Valley from Livingston in Merced County to Davis in Yolo County, Cordelia in Solano County, the Cucamonga region in San Bernardino County, and Escondido in San Diego County. Region V is the warmest district where grapes are produced in California. The San Joaquin Valley from Merced south is the most important grape-growing district in this region. Some of the desert areas and the region north of Sacramento are in region V. See pp. 65, 67 for degree-days for each zone.

TABLE 6

RECOMMENDED VARIETIES, WINE-GRAPE REGIONS, AND WINE TYPES

Variety	Productivity[a]	Type of wine	Region of adaptation and quality of product				
			Region I	Region II	Region III	Region IV	Region V
WHITE VARIETIES							
Chardonnay	Low	Varietal	Excellent	Good	Good?	No	No
Chenin blanc	Moderate	Varietal, dry or sweet	Possible	Standard	Standard	Standard	No
Emerald Riesling	High	Varietal, dry or sweet	No	No	Good	Good	Standard
Folle blanche	Moderate	Varietal, generic, or sparkling	No	Standard	Standard?	No	No
French Colombard	High	Varietal, generic, or sparkling	Possible	Standard	Standard	Standard	Standard
Gewürztraminer	Low	Varietal	Good	Good?	No	Standard	No
Peverella	High	Varietal, generic, or sparkling	No	No	Standard?	No	No?
Pinot blanc	Low	Varietal	Good	Good	Possible	No	No
Red Veltliner	Moderate	Generic	No	Standard	Standard	No	No
Sauvignon blanc	Moderate	Varietal or sweet table in III	Good	Excellent	Good	Possible	No
Sémillon	High	Generic, varietal, or sweet table in III	Possible	Standard	Good	No	No
Sylvaner	Moderate	Varietal	Possible	Good	Standard	Good	No
White Riesling	Low	Varietal	Excellent	Good	No	No	No
RED VARIETIES							
Aleatico	Moderate	Sweet table	No	No	No	Standard	Standard
Barbera	Moderate	Varietal	No	No	Good	Good	Possible
Cabernet Sauvignon	Low	Varietal	Excellent	Excellent	Good	No	No
Carignane	High	Generic	No	Possible	Standard	Standard	Possible
Gamay	Moderate	Pink	Possible	Standard	Standard	No	No
Gamay Beaujolais	Moderate	Varietal or pink	Good	Standard	No	No	No
Grenache	High	Pink varietal	Good	Good	Standard	Standard	No
Grignolino	Moderate	Pink varietal	No	No	Possible	Possible	No
Petite Sirah	Moderate	Generic	Possible	Standard	Standard?	Standard	No
Pinot noir	Low	Varietal or pink	Excellent	Good	No	No	No
Refosco	Moderate	Generic	Possible	Standard	Standard	No	Standard
Ruby Cabernet	Moderate	Varietal	No	Possible	Good	Good	Standard
Zinfandel	Moderate	Varietal or pink	Good	Good	No	Possible	No

a Productivity varies with the clone used, soil and climatic factors, severity of pruning, and other factors. In general low means below 5 tons per acre, moderate 6 to 10 tons, and high above 10 tons.

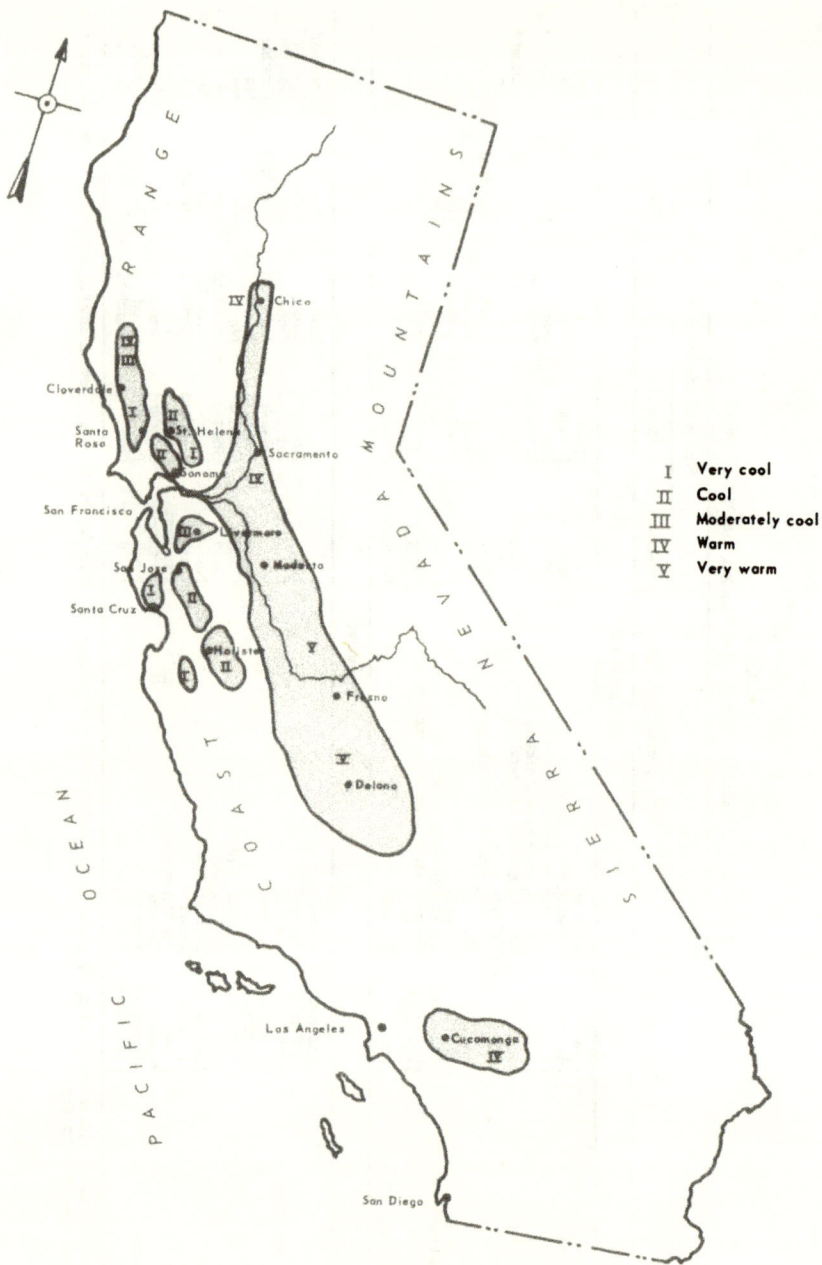

Figure 15. Climatic zones of California.

Most of the high-quality varieties for dry wines are best suited to regions I and II. Region III seems best adapted for dry and sweet white table wines, and regions IV and V for standard table wines. Vineyards in the cooler regions are shown in figures 16 and 17.

In most districts there is a shortage of the varieties of wine grapes best suited to climatic and soil conditions and to the types of wine being made. New plantings of the varieties of wine grapes recognized for their high quality are rather limited (table 21, p. 138); such red wine grapes as

Figure 16. Hillside vineyard scene in San Benito County. (Courtesy Almadén Vineyards, Inc.)

Carignane and Alicante Bouschet, which usually produce only ordinary wine, predominate. Any improvement in this regard depends upon the grape growers, of course. But the wine maker could probably do much to encourage the production of finer and better-adapted varieties in his district: he could insist on the best grapes for his higher-quality wines and compensate the grower adequately for the low-producing quality varieties. Unless the grower is so paid, he cannot be expected to plant the low-yielding, finer varieties.

In summary: the main factors influencing the quality of California wines are the grape variety and the regional conditions under which it is grown. This does not mean that poor processing and neglect during aging cannot spoil wine of the best grape variety. It does mean that distinctive and quality wines are not easily produced from nondescript material.

Figure 17. New vineyard scene in the Napa Valley. (Courtesy Beaulieu Vineyard.)

For identifying grape varieties the ampelographer uses time of maturity, vegetative and fruiting characteristics, and in special cases, the shape, size, and other characteristics of the seed. The following indicates the range of characteristics (adapted from Bioletti, 1938):

I. Time of maturity: very early, early mid-season, mid-season, late, and very late.
II. Fruit: color of skin and pulp, texture, composition, bloom, brush, torus, adherence of berry, method of attachment, etc.
III. Seed: number, size, shape of body, beak and keel, etc.
IV. Cluster: number, shape, size, framework.
V. Leaf: form, sinuses, dentation, color, surface, texture, etc.

Grapes mature in early June in the Imperial Valley for very early-ripening varieties and in late October in region I (if then) for late-ripening varieties.

76

Grapes for varietal wines

For high-quality wines deriving their flavor and color from a single variety, the following grapes are recommended, provided they are grown in regions to which they are suited (see also table 6):

For white varietal wines:

Chardonnay, I, II
Chenin blanc,ᵃ I, II, III
Emerald Riesling, III, IV, V
French Colombard, I, II, III, IV, V
Flora, I, II
Gewürztraminer, I, II

Pinot blanc, I, II, III
Red Veltliner, II, III
Sauvignon blanc, I, II, III
Sémillon, I, II, III
Sylvaner, I, II, III
White Riesling, I, II

For red varietal wines:

Barbera, III, IV, V
Cabernet Sauvignon, I, II, III
Carignane, II, III, IV, V?
Gamay, II, III
Gamay Beaujolais,ᵇ I, II
Grenache, I, II, III

Petite Sirah, I, II, III
Pinot noir, I, II
Refosco, I, II, III
Ruby Cabernet, II, III, IV, V
Zinfandel, I, II, III, IV?

Blending these varieties with each other or with other varieties reduces their distinctiveness but is sometimes necessary to produce a wine having the desirable alcohol-to-acid ratio, and may sometimes produce new and distinctive types. By blending varieties of grapes we believe that new and recognizable types or sub-types of wine will be developed. Figure 18 illustrates several important wine grape varieties of California.

A number of other varieties with distinctive and attractive varietal flavors or other desirable characteristics are not now used to any extent in producing such wines in California. These include, among the whites, Peverella, Aligoté, and Melon; and among the reds, Fresia, Nebbiolo, Sangioveto, Grignolino, Tannat, Merlot, and Valdepeñas.

Petite Sirah is grown on a large acreage in California. When properly aged it makes wine of characteristic flavor which is frequently too heavy and alcoholic for straight varietal wine; when diluted too much by blending it does not maintain its characteristic flavor. It does not equal in quality the varieties listed above, but is used because of its availability, high color content, and body.

Grapes for generic wines

Specific recommendations for nonvarietal types of wine cannot be made until these wines are better standardized and more information is available on the adaptation of varieties to regions. Varieties that produce

---

ᵃ The Chenin blanc (Pineau blanc de la Loire) should not be confused with the Pinot blanc, nor the Red Veltliner with the Red Traminer.

ᵇ The Gamay Beaujolais should not be confused with the variety called simply Gamay in the Napa Valley, which has quite different vine and fruit characteristics.

satisfactory table wines in the cooler districts frequently cannot do so in the hot districts, and vice versa. The nonvarietal wines such as claret, burgundy, chablis, and sauterne are usually made from whatever grapes the wine maker has available. Most of these varieties continue to be used simply because they are productive.

Use of better varieties for these wines would improve their quality. New plantings intended for generic wines should be made only in suitable locations with the best of the highly productive varieties. The most important of these varieties (see table 6) include Chenin blanc, French Colombard, and Sémillon, for the whites; and Carignane, Petite Sirah, and Refosco for the reds. These, properly made and aged, should produce satisfactory wines of the California chablis, rhine, claret, and burgundy types (if production of these must continue). Other varieties that are less satisfactory but sometimes available are the Burger, Charbono, Green Hungarian, Gray Riesling, Mataro, Saint Macaire, Sauvignon vert, and Saint Emilion (Ugni blanc).

Rarely if ever advisable for these types of wines, even when grown in the coolest areas, are the Aramon, Alicante Bouschet, Blue Portuguese, Grand noir, Mission, Palomino, and Pinot St. George, although the Aramon may occasionally be useful for pink-wine production. We have tasted pleasant and even distinctive wines made from Pinot St. George, but they are not a substitute for wines made from Pinot noir, although it must be admitted that some California Pinot noirs on the market leave much to be desired in varietal character and quality.

For ordinary dry white table wines, varieties with a varietal flavor are sometimes useful. Too many are made from Thompson Seedless. The wines of this variety are very neutral, which is precisely the defect of many of these wines on the market.

The sweet table wines—California sauterne and California sweet sauterne—require high sugar, medium body, moderately low acid, and good flavor, all of which are obtainable from the Sémillon when it is grown in moderately warm districts. Because its flavor is too pronounced and its acidity too low, the Sauvignon vert is useful only for blending, but should not be a major constituent of the blend. Sauvignon blanc is a much more desirable variety for blending with the Sémillon. Palomino, Burger, Folle blanche, Saint Emilion, and Green Hungarian are undesirable for these types of wine.

The deficiencies of many of the varieties mentioned, so far as their adaptability for producing ordinary-quality table wines is concerned, can be corrected by judicious blending, preferably of the grapes at the time of crushing; the composition of the blend for each type of wine should be kept as uniform as possible from year to year.

Much of the ordinary wine in California is sold under generic names such as burgundy and chablis. With these wines the absence of sharply defined, well-understood, and easily enforceable standards has resulted in confusing arrays of widely differing California wines under the existing type labels. Standardization of wine types is necessary: at present there is too great a variation in both composition and flavor of the same type of wine produced at different wineries. With the hope of promoting such standardization, some suggestions on the preferred usage of various common type names have been made (p. 90 f). The familiarity of the public with these generic type names and their great economic importance to the California industry should stimulate efforts toward standardization.

We feel that little will be accomplished by trying to standardize the present generic types. The industry would be wise to abandon their use, develop new names and standardize them, and thus promote a market for new types. This is especially true of wines which are made wholly or in part from *Vitis labrusca* grapes. Because of their distinctive aroma these wines have no resemblance to their European counterparts. The California generic wines are usually better in this regard.

Wines that do not fit under either varietal names or standardized generic names could be sold under proprietary names or new names unique for California wines, or be labeled simply as red or white table wines.

LABELING

The nomenclature of California wine types needs clarification and standardization to prevent overlapping and confusion.

The nomenclature for California wines has never been standardized. Many of the type names which are used commercially originated abroad. The advisability of using such foreign names as "burgundy" and "sauterne" for California wines has often been questioned (Amerine and Cruess, 1960; Amerine *et al.*, 1967; Berg and Webb, 1955; Cruess, 1947). The development of distinctive names which would be characteristic of our wines should stimulate the production of native wines of more standardized composition and higher quality, because such types could be defined and controlled on the basis of actual potentialities of our grapes and production methods and not by reference to conditions abroad. Several type names already widely used—angelica, Cabernet, Pinot blanc, Pinot noir, Chardonnay, Sauvignon blanc, White Riesling, Zinfandel—are not derived from any foreign geographical appellation, but originated in

79

California or are taken from names of grape varieties used here. A number of wineries have given proprietary names to their wines. These are encouraging developments, but there is much room for improvement. The legal restrictions on vineyard, vintage, and varietal labeling and estate bottling vary greatly from country to country. Vineyard appellations generally mean what they say, but there are exceptions. When a German vineyard name is used, only three-fourths of the wine need come from grapes of the vineyard named; however, the rest must come from vineyards of the same district. Both in France and Germany, where the same vineyard may be owned by more than one proprietor, some movement of wines may occur without infringing on the right to label the wine *Original-Abfüllung* in Germany or *mise du domaine* in France. In California, in contrast, "Estate-bottling" means that the wine was made exclusively from grapes grown by the producer. The term "Produced and bottled by" requires that only 75 per cent actually come from the vineyard or district. Thus New York burgundy need be made only 75 per cent from New York grapes. On the other hand, California law requires that wines labeled "California" be 100 per cent of California grapes. "Made by" simply indicates that at least 10 per cent of the wine was produced by the shipper.

On vintage labeling, Champagnes must be 100 per cent of the year named. In the United States the vintage must be of the year indicated. Enforcement in France, Italy, and Spain is probably less severe than here. Many blended Spanish wines have been shipped with the oldest year of the blend being used on the label, even though it may constitute only a small fraction of the wine. Recently, with tritium dating being used by the Alcohol and Tobacco Tax Division, this practice has nearly ceased for imports into the United States, but it continues within Spain (p. 19).

According to Lichine (1963), there are differences in the degree to which regulations are enforced in France. In Bordeaux the château-bottled wines are authentically labeled with the vintage marked. In Burgundy, however, wines are occasionally mislabeled with old vintages. It is best to buy estate bottlings of a reputable grower or wines from a reliable shipper. (See Capus, 1947, for a history of the French regulations.)

The strongest, clearest, and most straightforward objection to the use of European geographic names for non-European wines is probably that of Wilkinson (1918), who argued for sincerity in the merchandising of Australian wines. He considered the question of probable fraud in the use of geographical names and reviewed the pertinent provisions of usage, legislation, and treaties. Wilkinson notes the early attempts of Australian wine makers to avoid European geographical appellations: "... let me mention a curious anomaly: while it takes on the continent of Europe widely divergent regions, both climatically and geographically, separated by hundreds and even thousands of miles, to produce Champagne, Claret, Burgundy, Chablis, Sauterne, Hock, Port, Sherry, Tokay, etc., the spectacle is not uncommon in Australia of one small vineyard of under 100 acres in total extent trying, in all seriousness, to reproduce most, if not all, of the great European wines mentioned."

Wilkinson may have reduced the effectiveness of his arguments by suggesting coined names such as Burgalia, Chabalia, and Champalia, which were obviously intended to "suggest" Australian burgundy, chablis, and champagne. Nevertheless, his arguments are still persuasive:

Nothing short of the complete abandonment of European geographical wine names will meet the Australian needs and save us from international discredit, however irksome the short transition period may be. This reform in the nomenclature must be carried out if we desire to assure the future prosperity of our Australian wine industry; for by such a step only will our wines ultimately achieve the distinction in the world's commerce which their inherent good qualities merit. We cannot practice the petty deception any longer of designating our wines as Champagne, Burgundy, Sauterne, Chablis, Port, Sherry, etc. They cannot truthfully be so named unless produced in France, Portugal, or Spain. Being Australian, they must bear only Australian names. Need we feel ashamed to name them so that the world may know them as Australian ?

A similar point of view has been expressed by Amerine and Joslyn (1940), Cruess (1947), Joslyn and Amerine (1964), Amerine and Cruess (1960), and Amerine *et al.* (1967).

The nomenclatural difficulties of the California wine industry are by no means unique. The official position of the South African industry has been clearly expressed (Anon., 1967c) as follows:

In several countries designations such as "sherry", "port," and "champagne" are accepted as generic names, reflecting the nature of the wine or the particular process of vinification. On the other hand in some countries, notably in western Europe, these designations, as well as a host of others, are regarded as geographical names, or appellations of origin, and therefore as the names of wines produced exclusively in prescribed areas of production within the countries concerned.

Attempts by the latter countries to obtain broad international acceptance of their views are resisted both in producing and in consuming countries. In the first instance, referring to South Africa's position only, the history of wine production at the Cape is of particular significance.

It is too often forgotten that the people who founded the South African wine industry were Europeans who brought with them the traditions, customs, and tastes of Europe. They imported their vines from Europe and (because they were European immigrants and not aborigines) they made European wines. They did this not only to satisfy their own European taste, but also to satisfy the palates of their export market in Europe. Today, after 300 years, the South African wine industry is wholly orientated to the production of the traditional, classical wine types which may, because of soil and climate, be different in character from their European prototypes, but which, unmistakably, belong to the same class.

Clearly any suggestion that the South African wine industry produces traditional wine types such as port or sherry, in a cynical desire to exploit the popularity enjoyed by these wine names in the world markets is therefore without foundation.

Secondly, sherries and ports from the Cape wine cellars are long established exports to the United Kingdom, North America and elsewhere. Indeed, in some export markets South African sherries and ports today rank with those of almost any other supplying country both in quality and in quantity consumed.

This confirms the attitude of the South African wine industry that even should traditional wine type designations such as "sherry" have had a geographical connotation at some stage, the designation has clearly, in many countries, through long-standing usage and consumer acceptance, become generic.

Turning to the difficulties confronting South African exports, reference has already been made to those countries in which the various classical wine type names are regarded as strictly geographical names. South Africa is completely denied the use of these names in those countries. In addition, some of the countries concerned have managed to obtain protection for selected names through bilateral agreements with other countries. In this manner France concluded an agreement with South Africa in terms of which a prescribed list of names are not used on any South African wines nor permitted on any imported wines other than French. To make matters more difficult from the marketing view-point, some of the Cape's competitors outside the historical vineyards of Europe have no such agreement with France and use the descriptive names freely. Similarly, in the Netherlands, the terms "sherry," "malaga," "port" and some others may be used only in relation to Spanish sherry and malaga, Portuguese port, etc.

In the United Kingdom South Africa may use the words "sherry" and "port," but only with qualification. The word "type" stigmatises the port even before it reaches the consumer. "Sherry" may, however, be sold as such provided the name is prefixed by the country of origin, only Spain being exempt from this requirement. The phrase "South African Sherry" naturally facilitates product-identification by the consumer.

Joslyn and Amerine (1964) reviewed the international agreements on protection of geographical appellations:

European wine producers and distributors, however, have zealously attempted to guard the right to exclusive use of generic regional wine type-names. Action to protect the names of wines identified by their place of origin has been taken at the following times and places: Paris, March 20, 1883; Madrid, April 14, 1891; Brussels, December 14, 1900; Washington, June 2, 1911; The Hague, November 6, 1925; London, June 2, 1934; Lisbon 31, 1958. And the Office International du Vin, under the agreement of November 29, 1924, by which it was set up, explicitly concerns itself with appellations of origin, and has opposed the use of generic regional type-names in countries other than those of origin.

The protection of appellations of origin in international trade is discussed in some detail by Anon. (1927), Hitier (1923), Queuille (1927). The early international agreements were enforced not only in trade practices, but also at expositions where wines were displayed and judged. Several American wines exhibited at the Paris Exposition of 1900 were disqualified because of the use of foreign type-names (Wiley, 1903, 1919). In international trade, European wine producers were given exclusive rights to appellations of origin by the law of May 6, 1919 (Anon., 1927) and the law of January 1,

1930 (Semichon, 1930; Anon., 1933; and Capus, 1935). These restrictions were renewed and strengthened with formation of the European Economic Community. The International Madrid Agreement of April 14, 1891, and its subsequent revisions, designed to prevent repression of indications of false or fallacious claims of origin, has been replaced by the Lisbon Agreement of 1958. This latter enactment applies to all countries except those specifically noted in the Madrid Agreement as revised in London or the Hague, and has been signed by representatives of nine countries, who have also approved regulations for its execution. The final ratification date stipulated for other countries was May 1, 1963. . . .

Certainly it would be of considerable value to evolve a system of naming to which objection cannot be made. But in view of the large commerical interests involved, it is clear that any changes must originate with the wine industry itself. At this time, any ideas on the subject can only be expressed as suggestions.

The type names follow the present commercial usage, with certain recommendations for improvement. The terms "dry red table wine" and "dry white table wine" often adequately describe the wine, and their use should be encouraged. In selecting a type name and label, the producer should be guided by state and federal regulations (see p. 101) and particularly by the character and composition of the wine. There should be a real difference, recognizable to the wine maker, between the various types he bottles. To bottle claret and burgundy from the same tank is to confuse the public and eventually to injure sales. There is too much variation among wineries in their use of type names also. Some of these differences arise solely because of a lack of standards acceptable to the whole industry. Others, however, are due to the variable raw material available and to divergent fermentation and aging practices.

It is sometimes thought that European wines and pre-Prohibition California wines were of uniform chemical composition and that by conforming to certain chemical standards a distinct type of wine might be produced. The analyses in table 6 (p. 73) of various types of European wines and some pre-Prohibition California wines indicate that this is not correct. Very wide variations exist among all the important chemical constituents. Amerine (1947) and Ough (1964a, 1966c) have shown how great this variability is for commercial California wines produced since Repeal. These results are summarized in tables 7, 8, 9, and 10. Uniformity of type will probably appear first in the varietal wines because of better selection of the raw material and greater standardization in the methods of producing these wines. However, there is no reason why an individual winery cannot standardize the composition and quality of its generic wines. (See also p. 565.)

The wine type specifications of the Wine Institute (1968a) are of little value for generic types because of their very general nature. We present them here, with slight changes in wording and spelling, mainly because of

TABLE 7

Comparison of Composition of Types of California Table Wines before Prohibition and Similar European Types

Wine type	Number of samples	Total acid as tartaric		Volatile acid as acetic		Alcohol	
		Range (grams per 100 cc)	Average (grams per 100 cc)	Range (grams per 100 cc)	Average (grams per 100 cc)	Range (per cent vol.)	Average (per cent vol.)
California chablis[a,b]	7	0.49-0.67	0.57	—	0.104[c]	10.6-12.8	12.1
Chablis (French)[d]	9	0.34-0.74	0.60	0.038-0.112	0.069	9.4-13.4	11.4
California Riesling[a,b]	24	0.47-0.73	0.58	0.093-0.174[c]	0.134[c]	10.5-14.5	12.7
Riesling (Rhine)[e]	68	0.36-1.62	0.77	0.020-0.080	0.050	5.9-13.1	10.1
Riesling (Moselle)[e]	187	0.46-1.46	0.77	0.020-0.140	0.050	5.4-12.1	9.2
California sauterne[a]	9	0.42-0.63	0.54	—	—	11.6-14.7	12.8
Sauternes[e]	11	0.65-1.28	0.81	0.066-0.159	0.114	9.8-15.4	12.9
Sauternes[f]	16	0.38-0.96	0.74	0.051-0.095	0.068	12.8-15.8	14.0
California claret[a,b]	20	0.60-0.82	0.67	0.088-0.139[c]	0.115[c]	11.5-14.1	12.6
Claret (Médoc)[e]	40	0.38-0.68	0.58	0.092-0.106*	0.098*	8.4-11.6	10.2
Bordeaux[f]	378	0.49-1.04	0.66	—	—	9.3-13.5	11.4
California burgundy[a,b]	20	0.58-0.78	0.69	—	0.109[c]	9.7-14.1	12.6
Burgundy (French)[e,g,h]	15	0.50-1.30	0.75	0.072-0.160	0.096	7.1-15.8	11.4
Burgundy (French)[f]	31	0.48-1.05	0.63	0.041-0.115	0.077	10.5-14.8	13.1
California red chianti[a,b]	2	—	0.71	—	0.118	—	11.9
Chianti (Italian)[j]	29	0.56-0.59	0.65	0.049-0.101	0.071	11.1-14.1	12.5
Chianti (Italian), white[f]	252	0.50-0.97	0.67	0.042-0.192	0.077	9.5-13.6	11.7
Chianti (Italian), red[f]	985	0.51-1.03	0.70	0.042-0.132	0.082	10.0-14.5	11.6
Sparkling wine, French[k,l]	7	0.53-0.85	0.67	0.045-0.073	0.058	12.7-14.5	13.6**
Sparkling wine, German[k]	4	0.61-0.72	0.65	—	—	10.9-13.6	12.3

TABLE 7 (Continued)

Wine type	Number of samples	Extract Range (grams per 100 cc)	Extract Average (grams per 100 cc)	Sugar Range (grams per 100 cc)	Sugar Average (grams per 100 cc)	Tannin and coloring matter Range (grams per 100 cc)	Tannin and coloring matter Average (grams per 100 cc)	Total sulfur dioxide Range (parts per million)	Total sulfur dioxide Average (parts per million)
California chablis[a,b]	7	1.8– 2.2	2.1	0.10– 0.45	0.21	0.03–0.06	0.04	—	—
Chablis (French)[d]	9	1.8– 2.9	2.2	0.05– 0.86	0.24	—	—	—	—
California Riesling[a,b]	24	1.6– 2.6	2.0	0.50– 0.63	0.14	0.02–0.08[c]	0.04[c]	—	—
Riesling (Rhine)[e]	68	2.5– 5.3	2.9	0.01– 0.83	0.23	—	—	0–255	50
Riesling (Moselle)[e]	187	1.6– 4.6	2.3	0.04– 1.11	0.20	—	—	—	—
California sauterne[a]	9	1.9– 4.0	2.9	0.17– 1.75	0.83	0.03– 0.06	0.05	—	—
Sauternes[e]	11	2.3–12.8	5.1	0.65– 8.27	3.33	—	—	133–468	235
Sauternes[f]	16	2.3–19.6	8.1	0.55–14.07	5.23	—	—	—	—
California claret[a,b]	20	2.1– 3.3	2.7	0.04– 0.63	0.16	0.15–0.29	0.21	—	—
Claret (Médoc)[e]	40	2.0– 3.9	2.4	0.11– 0.84	0.23	—	—	—	—
Bordeaux[f]	378	1.8– 3.0	2.3†	0.08– 1.57	—	—	—	—	—
California burgundy[a,b]	20	2.1– 3.5	2.8	0.03– 0.42	0.14	0.03–0.33	0.21	—	—
Burgundy (French)[e,g,h]	15	1.6– 2.9[i]	2.1[i]	0.15– 0.23	0.18	—	—	—	—
Burgundy (French)[f]	31	2.1– 3.2	2.8†	0.13– 0.62	0.24	—	—	—	—
California red chianti[a,b]	2	—	2.6	—	0.15	—	0.26	—	—
Chianti (Italian)[j]	29	2.2– 3.1	2.5	—	—	—	0.20	—	—
Chianti (Italian), white[f]	252	1.6– 2.3	1.9†	—	—	0.09–0.30	—	8–361	108
Chianti (Italian), red[f]	985	1.8– 3.3	2.4‡	—	—	—	—	—	—
Sparkling wine, French[k,l]	7	2.9–19.2	11.1¶	0.4–17.3	0.94§	0.01–0.05	0.03	—	—
Sparkling wine, German[k]	4	13.6–15.7	14.6	11.4–13.5	12.4	—	—	—	—

* Five samples only.
† Extract minus sugar.
** Exceptionally high. Recent results indicate 12 is a better average.
‡ Two samples only.
§ One sample omitted.

Sources of data:
a Bigelow (1900).   d Filaudeau (1912b).   g Filaudeau (1912a).   j Anon. (1932).
b Wiley (1903).   e König (1903).   h Filaudeau (1916).   k Vogt (1945).
c Wiley only.   f Ribéreau-Gayon and Peynaud (1958).   i Curtel (1912).   l Pacottet and Guittonneau (1930).

TABLE 8

COMPOSITION OF CALIFORNIA WHITE TABLE WINE SINCE PROHIBITION

Type	Number of samples	Total acid as tartaric gr/100 cc		Volatile acid as acetic gr/100 cc		pH		Alcohol per cent/vol.		Extract gr/100 cc		Reducing sugar gr/100 cc	
		Range	Av.	Range	Av.	Range	Av.	Range	Av.	Range	Av.	Range	Av.
Dry sauterne													
Prewar	94	0.33-0.71	0.55	0.039-0.178	0.091	3.12-3.73	3.46	10.7-14.1	12.3	1.8-5.9	3.0	2.19-5.3	0.83
Postwar	22	0.33-0.71	0.59	0.020-0.138	0.052	3.22-3.90	3.49	11.2-12.7	11.8	2.0-2.5	2.3[a]	0.10-2.0	0.61
White chianti													
Prewar	19	0.49-0.69	0.57	0.048-0.112	0.086	3.29-3.88	3.58	10.9-13.0	12.0	2.2-3.3	2.6	0.23-0.91	0.39
Postwarxx	16	0.45-0.82	0.61	0.03-0.05	0.039	—	—	10.7-13.3	11.8	—	—	0.1-0.2	0.11
Chablis													
Prewar	57	0.42-0.73	0.57	0.045-0.176	0.086	3.13-3.78	3.40	10.3-13.0	12.1	1.7-3.0	2.4	0.17-0.84	0.32
Postwar	28	0.46-0.78	0.63	0.03-0.149	0.057	2.84-3.80	3.51	10.4-13.5	11.7	2.0-2.8	2.3[a]	0.1-0.3	0.17
Riesling													
Prewar	65	0.36-0.69	0.56	0.040-0.178	0.089	3.03-3.68	3.40	9.2-14.2	12.1	1.8-2.9	2.5	0.13-0.56	0.32
Postwar	31	0.45-0.77	0.61	0.020-0.135	0.052	3.10-3.62	3.41	10.2-12.9	11.8	2.0-2.6	2.3[a]	0.1-1.0	0.21

TABLE 9

COMPOSITION OF SWEET CALIFORNIA WHITE AND RED TABLE WINE SINCE PROHIBITION

Type	Number of samples	Total acid as tartaric gr/100 cc		Volatile acid as acetic gr/100 cc		pH		Alcohol per cent/vol.		Extract gr/100 cc		Reducing sugar gr/100 cc	
		Range	Av.	Range	Av.	Range	Av.	Range	Av.	Range	Av.	Range	Av.
Sauterne													
Prewar	56	0.39-0.96	0.60	0.052-0.160	0.093	3.10-3.89	3.43	10.2-14.6	12.5	2.1-8.3	5.3	0.14-5.6	3.10
Postwar	47	0.39-0.76	0.57	0.02-0.120	0.062	3.19-3.71	3.44[a]	10.5-13.6	12.1	1.9-5.2	2.6[a]	0.11-4.58	1.0
Château													
Prewar	39	0.39-0.71	0.59	0.054-0.160	0.099	3.03-3.72	3.42	10.2-14.0	12.1	2.2-7.9	5.7	0.16-5.60	3.65
Postwar	16	0.44-0.72	0.60	0.030-0.115	0.067	3.03-3.82	3.52	10.9-13.2	12.2	2.6-6.7	4.6[a]	0.20-6.00	2.98
Mellow red													
Postwar[b]	7	0.49-0.65	0.56	0.03-0.05	0.043	3.58-3.69	3.63	12.0-13.0	12.4	—	—	0.6-3.5	1.5

TABLE 8 (Concluded)

COMPOSITION OF CALIFORNIA WHITE TABLE WINE SINCE PROHIBITION

Type	Tannin and coloring matter gr/100 cc		Total sulfur dioxide mg/liter		Color		Acetaldehyde mg/liter		Iron mg/liter		Copper mg/liter	
	Range	Av.	Range	Av.	Range	Av.	Range	Av.	Range	Av.	Range	Av.
Dry sauterne												
Prewar	0.02–0.13	0.05	67–514	190	6–39	15	21–242	78	2.0–10.2	5.1	—	—
Postwar	0.02–0.05	0.03	66–314	151	9–22	14[a]	60–89	69[a]	1.0–13.0	4.2	0.0–0.96	0.24
White chianti												
Prewar	0.03–0.08	0.05	51–268	166	14–20	17	26–89	61	4.0–6.8	6.1	—	—
Postwar[b]	—	—	125–170	150	—	—	—	—	2.5–9.0	5.1	0.0–0.4	0.08
Chablis												
Prewar	0.02–0.08	0.05	13–360	147	9–25	17	14–170	69	2.0–9.1	5.3	—	—
Postwar	0.02–0.04	0.03	95–312	180	8–18	13[a]	39–145	99[a]	1.0–7.5	3.9	0.0–1.3	0.33
Riesling												
Prewar	0.03–0.09	0.05	45–260	182	9–26	18	37–143	75	1.0–9.0	4.4	—	—
Postwar	0.01–0.04	0.03[a]	57–211	132	10–29	17[a]	31–132	75[a]	0.0–9.0	4.6[a]	0.0–1.25	0.25

[a] Incomplete analyses.

Sources of data: Amerine and Joslyn (1951); and Ough (1964a, 1966c).

TABLE 9 (Concluded)

COMPOSITION OF SWEET CALIFORNIA WHITE AND RED TABLE WINE SINCE PROHIBITION

Type	Tannin and coloring matter gr/100 cc		Total sulfur dioxide mg/liter		Color		Acetaldehyde mg/liter		Iron mg/liter		Copper mg/liter	
	Range	Av.	Range	Av.	Range	Av.	Range	Av.	Range	Av.	Range	Av.
Sauterne												
Prewar	0.03–0.11	0.05	77–520	280	7–30	15	15–198	63	1.0–14.0	6.1	—	—
Postwar	0.03–0.07	0.03[a]	13–454	160	5–27	14[a]	27–198	88[a]	2.0–6.0	3.8[a]	0.0–0.43	0.10[a]
Château												
Prewar	0.03–0.11	0.06	230–394	318	8–25	15	48–155	89	2.0–20.0	6.0	—	—
Postwar	0.02–0.05	0.03[a]	76–590	206	5–33	17[a]	81–292	128[a]	1.8–10.7	5.1	0.0–2.2	0.41
Mellow red												
Postwar[b]	—	—	66–130	106	—	—	—	—	2.0–8.4	5.0	0.03–0.2	0.09

[a] Incomplete analyses.

Sources of data: Amerine and Joslyn (1951); and Ough (1964a, 1966c).

TABLE 10

COMPOSITION OF COMMERCIAL CALIFORNIA SPARKLING ROSÉ AND RED TABLE WINES SINCE PROHIBITION

Type	Number of samples	Total acid as Tartaric, gr/100 cc		Volatile acid as acetic, gr/100 cc		pH		Alcohol per cent/vol.		Extract gr/100 cc		Reducing sugar gr/100 cc	
		Range	Av.	Range	Av.	Range	Av.	Range	Av.	Range	Av.	Range	Av.
Champagne													
Postwar													
Bulk	9	0.51–0.76	0.63	0.040–0.091	0.056	3.10–3.50	3.28	11.3–13.0	12.1	2.7–4.9	3.8a	0.13–0.73	0.66a
Bottle	5	0.66–0.82	0.73	0.03–0.05	0.044	—	—	12.2–12.9	12.5	—	—	1.5–2.2	1.84
Sparkling Burgundy													
Postwar	12	0.40–0.77	0.65	0.042–0.118	0.066	3.12–3.73	3.44	7.4–14.2	11.8	3.5–5.9	4.7a	1.03–2.83	2.20
Pink Champagne													
Postwar	9	0.50–0.79	0.62	0.030–0.104	0.059	3.17–3.55	3.34	7.4–13.9	11.9	3.5–5.2	4.1a	1.3–2.4	1.8
Rosé													
Postwar b	18	0.49–0.74	0.58	0.02–0.05	0.034	3.31–3.70	3.45	11.2–12.9	12.0	—	—	0.11–3.6	1.16
Barbera													
Prewar	12	0.52–0.69	0.62	0.04–0.104	0.064	3.32–3.75	3.57	10.3–13.0	12.1	2.4–3.3	2.9	0.22–0.46	0.29
Postwar	6	0.52–0.84	0.65	0.052–0.084	0.068	3.12–3.80	3.55	11.3–13.6	12.5	2.4–3.4	2.9a	0.09–0.30	0.17
Burgundy													
Prewar	88	0.46–0.88	0.65	0.044–0.176	0.095	3.29–3.97	3.60	10.8–14.1	12.6	2.1–4.4	2.9	0.17–1.90	0.47
Postwar	69	0.54–0.81	0.65	0.02–0.137	0.071	3.22–3.80	3.54a	10.9–13.6	12.3	2.5–3.8	3.0a	0.06–0.85	0.23a
Cabernet													
Prewar	36	0.48–0.82	0.64	0.048–0.226	0.108	3.05–3.64	3.47	11.1–13.8	12.5	2.4–3.8	2.9	0.20–0.76	0.39
Postwar	21	0.53–0.82	0.65	0.042–0.134	0.071	3.33–3.80	3.53a	11.5–13.8	12.6	2.3–3.4	2.9a	0.06–0.33	0.15a
Chianti													
Prewar	24	0.56–0.79	0.66	0.042–0.136	0.098	3.38–3.68	3.57	9.3–13.5	12.2	2.3–3.4	2.9	0.23–1.13	0.54
Postwar	9	0.46–0.73	0.62	0.030–0.076	0.056	3.12–3.81	3.52	11.3–13.5	12.3	2.4–3.0	2.8a	0.08–0.20	0.12
Claret													
Prewar	58	0.45–0.79	0.64	0.052–0.156	0.087	3.18–3.69	3.49	11.3–16.0	12.4b	2.3–7.3	2.9a	0.17–5.00	0.45b
Postwar	46	0.49–0.86	0.64	0.030–0.115	0.071	3.10–3.80	3.52a	10.7–13.9	12.2	2.5–6.3	3.1a	0.06–3.28	0.28a
Zinfandel													
Prewar	64	0.45–0.87	0.64	0.020–0.156	0.087	3.18–3.89	3.52	11.5–14.3	12.7	2.3–3.4	2.8	0.19–1.37	0.40
Postwar	25	0.52–0.83	0.64	0.035–0.108	0.066	3.27–3.93	3.54	11.4–14.0	12.6	2.4–3.1	2.8a	0.07–0.61	0.20

TABLE 10 (Concluded)

Composition of Commercial California Sparkling Rosé and Red Table Wines since Prohibition

Type	Tannin and coloring matter gr/100 cc		Total sulfur dioxide mg/liter		Color		Acetaldehyde mg/liter		Iron mg/liter		Copper mg/liter	
	Range	Av.	Range	Av.	Range	Av.	Range	Av.	Range	Av.	Range	Av.
Champagne												
Postwar												
Bulk	0.01–0.04	0.03	54–228	156	7–23	15[a]	84–166	124	6.5–16.7	8.6	0.0–1.5	0.19
Bottle	—	—	153–175	162	—	—	—	—	4.0–4.0	4.0	0.0–0.1	0.04
Sparkling Burgundy												
Postwar	0.09–0.17	0.13[a]	0–177	94	87–213	141[a]	21–89	47[a]	1.6–5.0	3.3	0.0–0.5	0.25
Pink Champagne												
Postwar	0.03–0.14	0.06	16–169	109	12–76	43[a]	9–82	58[a]	1.0–7.0	3.0	0.0–0.6	0.27
Rosé												
Postwar[b]	0.05–0.06	0.05	95–237	147	—	—	—	—	1.0–8.0	3.6	0.0–0.2	0.07
Barbera												
Prewar	0.13–0.18	0.19	0–110	28	174–454	276	10–59	30	5.6–10.0	8.5	—	—
Postwar	0.15–0.18	0.16[a]	63–132	96	204–286	239[a]	11–49	24[a]	0.8–8.8	4.3	0.0–3.9	0.77
Burgundy												
Prewar	0.13–0.37	0.21	10–93	52	89–625	266	7–494	75	0.0?–18	7.1	—	—
Postwar	0.11–0.26	0.18	0–256	81[a]	71–400	233[a]	12–133	43[a]	0.5–20	4.4[a]	0.0–3.6	0.34[a]
Cabernet												
Prewar	0.15–0.38	0.23[a]	24–143	63	103–476	282	18–54	34	1.0–16.7	6.5	0.0–8.58	0.24[b]
Postwar	0.15–0.36	0.28[a]	0–170	87[a]	105–476	233[a]	18–79	42[a]	1.1–10	4.1	—	—
Chianti												
Prewar	0.14–0.32	0.21	0–140	51	121–500	227	11–33	21	0.0?–9.3	3.4	0.0–0.57	0.24
Postwar	0.09–0.16	0.13[a]	95–239	152	141–196	163[a]	22–85	53[a]	0.7–5.0	2.4	—	—
Claret												
Prewar	0.12–0.35	0.20	0–120	51	105–410	215	17–56	30	3.0–9.0	5.6	—	—
Postwar	0.09–0.25	0.17[a]	19–170	101[a]	77–370	176[a]	28–118	58[a]	0.5–10.0	3.4[a]	0.0–0.91	0.30[a]
Zinfandel												
Prewar	0.10–0.38	0.19	0–249	62	107–654	238	9–95	29	2.0–29.2	6.6	—	—
Postwar	0.09–0.33	0.23[a]	0–218	70	125–1,000	303[a]	5–103	40	1.1–19.2	4.4	0.0–1.25	0.46

a Incomplete analyses.
b Sample with highest alcohol, extract, and reducing sugar omitted.
Sources of data: Amerine and Joslyn (1951); Ough (1964a, 1966c); Valaer (1950).

their historical interest, and add comments on how they might be improved. There seems little chance of making them more specific because of wide and deeply ingrained differences of opinion in the California industry.

Dry red wines

The California burgundy is a type name which has been utilized for dry red table wines of variable color and alcohol content. The Burgundy wines of France are produced from the Pinot noir with or without mixture with the Gamay varieties. In California the Petite Sirah has frequently been used for this type of wine. If this practice were uniformly followed and if the Petite Sirah were not used for any other type of wine, its varietal flavor would constitute a good point of differentiation for California burgundy. The Wine Institute early recommended that preference be given by judges at fairs to wines with the varietal flavor and aroma of Pinot noir. This recommendation was later dropped because it seemed impractical in view of the limited acreage of Pinot noir in California (p. 154) and its preferred utilization for producing a varietal wine. A total acidity of at least 0.55 per cent and an extract content of at least 2.3 are highly desirable.

The analyses of table 10 (p. 88) show that there has been considerable overlapping in the composition of the types California burgundy and California claret, especially with regard to what is usually considered a good point of distinction, namely color. Amerine et al. (1967) suggested that if California burgundy and California claret are to be produced it be done on the primary basis of alcohol content. California claret would then be a dry red table wine of about 11.5 per cent alcohol and at least 0.65 per cent total acidity, and California burgundy a wine of 12.5 per cent alcohol, or more, and not over 0.65 per cent total acidity. Nightingale (1961), considered this a pertinent suggestion, but so far it has not been accepted by the wine industry. The result is that California burgundies and clarets are almost indistinguishable from each other or have varying standards from winery to winery.

The Wine Institute (1968a) specifications for claret and burgundy are: "The dry table wine called California claret should be light to medium red in color, tart, of light or medium body." "The dry table wine type called California burgundy should be medium to deep red in color and full-bodied. It should have balance and softness on the palate derived from proper aging." In general, however, California claret has come to be applied to red wines of no predominant or characteristic flavor or aroma, which contain less color and extract than California burgundy. Originally "claret" was used for the red wines of Bordeaux, which are derived mostly from the Cabernet Sauvignon variety. In this state, wines so produced usually receive the name Cabernet. Clarets with a Cabernet flavor and aroma are obviously superior to those without such flavor and aroma, but, as with Pinot noir, the acreage is limited and the available Cabernet grapes are used solely for producing the varietal wine. Use of some Zinfandel as a means of giving distinctiveness and character to our clarets has been suggested, but the blend should not be so strong in Zinfandel flavor as to be confused with wines labeled Zinfandel. A minimum total acidity of 0.60 per cent and an extract content of between 2.2 and 2.8 are indicated for the better-quality clarets.

California Cabernet is a distinctive type of wine which is produced from at least 51 per cent of the Cabernet Sauvignon variety and has the predominant aroma and flavor of that grape. As table 10 indicates, it is a moderately well-colored wine of good extract and alcohol. The better Cabernet wines are heavy-bodied, have a full ruby-red color, and a total acidity of not less than 0.60 per cent. The Wine Institute (1968a) specification for Cabernet is: "A dry red table wine produced of Cabernet Sauvignon grapes. In this type the heavier-bodied and darker-colored wines should be on an equal basis with the lighter-bodied and lighter-colored wines. It should be well-balanced and aged, both in wood and bottle. It should have a distinctive bottle bouquet." California Cabernet is likely to be of lesser quality than California Cabernet Sauvignon.

Other dry red table wines which take their name from a grape variety are the California Barbera and Zinfandel. These likewise must derive at least 51 per cent of their volume and their predominant aroma and flavor from the respective varieties.

The Barbera grape should produce a full-bodied, very tart wine, since it has the highest total acid of any of the common red-wine grape varieties. A total acidity of at least 0.65 per cent is indicated and more is desirable. Unfortunately, some California Barberas on the market have undergone considerable malo-lactic fermentation and are not high in total acidity. Some of these are very good aged red table wines, but we find it difficult to identify them as "Barbera" wines. The Wine Institute (1968a) specification for Barbera is: "A dry red table wine produced of Barbera grapes. In this type the heavier-bodied and darker-colored wines should be on an equal with the lighter-bodied and lighter-colored wines. In addition to the pronounced varietal aroma and flavor of the Barbera grape, these wines should possess a distinguishable high total acidity."

The Zinfandel is apparently of unique California usage; it is easily recognized by its distinctive berry-like aroma and flavor. It is generally most pleasing when consumed young, although some of its best wines improve with bottle-aging. It should not possess a residuum of sugar.

The better Zinfandels have at least 0.55 per cent acidity and an extract of from 2.2 to 2.8. The Wine Institute (1968a) specification for Zinfandel is: "A dry red table wine produced of Zinfandel grapes. In this type the heavier-bodied and darker-colored wines should be on an equal basis with the lighter-bodied and lighter-colored wines. It should be of moderate aging, well-balanced, and should have the fruity aroma characteristic of the variety." Apparently two different types of California Zinfandel are envisaged by this specification—one heavy-bodied and dark-colored and the other light-bodied and low in color.

California red chianti is a type name used for fairly heavy, moderately astringent red wines. It was formerly marketed in a raffia-covered bottle. The red chianti of Italy owes its character mainly to the variety Sangioveto, which has only occasionally been used here. California red chianti, in contrast to California claret and burgundy, could be a piquant fruity wine and preferably should have a distinct flavor of the Sangioveto or other Italian varieties since these varieties are not used in varietal types at present. A minimum acidity of 0.55 per cent and an extract of 2.3 are desirable. The Wine Institute (1968a) specification for chianti is: "A dry red table wine type of medium red color, of medium tartness, fruity and full-bodied. It should be moderately aged and well-balanced." This has no clear meaning for distinguishing this type from other generic types.

A number of varietal types found a market before and just after the war, including Carignane, Charbono, Durif, Gamay (see footnote, p. 77), Green Hungarian, Malvoisie, Mourastel, Petite Sirah, Pinot noir, and Ruby Cabernet. Each should possess the distinct aroma of the grape variety from which it is produced and named. The Wine Institute (1968a) specifications for Pinot noir, red or black Pinot, and Gamay are: Pinot noir is "a dry red table wine type produced of Pinot noir grapes. In this type the heavier-bodied and darker-colored wines should be on an equal basis with the lighter-bodied and lighter-colored wines. One of the main distinguishable characteristics is the relatively soft and smooth flavor. It should be well-balanced and well-aged, both in wood and bottle. The varietal aroma and flavor of the Pinot noir grape should be pronounced." For the red or black Pinot: "A dry red table wine type produced of red Pinot grapes, such as Pinot St. George, Pinot Meunier, and Pinot Pernand. The varietal aroma and flavor of the Pinot grape should be pronounced. In this type the heavier-bodied and darker-colored wines should be on an equal basis with the lighter-bodied and lighter-colored wines." (We doubt that Pinot St. George can meet this specification.) For Gamay: "A dry red table wine produced of the Gamay grape. In this type the heavier-bodied and darker-colored wines should be on an equal basis with the lighter-bodied and lighter-colored wines. The varietal aroma and flavor of the Gamay grapes should be pronounced." (We doubt that Gamay, as grown in California, has a recognizable varietal aroma.) Since Carignane, Charbono, Malvoisie, and Mourastel seldom possess a distinguishable varietal character, their use as type labels is not recommended. The distinction between Durif and Petite Sirah is not clear to us. However, we have recently tasted some Petite Sirah wine which seemed to justify a varietal label. Ruby Cabernet clearly merits a place as a varietal type. It deserves better control of the malo-lactic fermentation than it has sometimes received.

The Wine Institute (1968a) specification for California red table wine is as follows:

It is recommended that any red table wine containing not over 1.5 per cent reducing sugar, not specifically named in the foregoing recommendations be automatically judged in the class known as California Red Table Wine. It is recommended that wines with varietal or generic names entered in this class be grouped together by name and that such groups be judged in an appropriate sequence. These wines should be judged on their general elements of quality rather than characteristics of any particular variety of grape unless they have a varietal label in which case they will be judged on the appropriate varietal character. In this type the heavier-bodied and darker-colored wines should be on an equal basis with the lighter-bodied and lighter-colored wines. An amber color is objectionable.

For California sweet red table wine: "The wines in this class should have the same characteristics as California Red Table Wine with the exception that the reducing sugar should be above 1.5 per cent." (It is unfortunate that more California wines are not labeled "red table wine" or "sweet red table wine" instead of California "claret" or "burgundy.")

Pink or *rosé* wines, which have found an increasing place in the industry, should be fruity, light-bodied, tart (preferably at least 0.60 per cent in total acidity), with alcohol content in the lower range—11 to 12 per cent. Although Grignolino is a variety of low color content, wines on the market with this name are frequently as deep in color as a claret or burgundy as a result of blending. Since the color of the grape is too orange for commercial acceptance, some blending is indicated; but the wine should be kept light in color, have a relatively high total acidity (at least 0.60 per cent), and, of course, have the varietal flavor and aroma of the Grignolino grape. This is true of the type planted in the variety collection of the University of California at Davis. We understand there may be darker-colored clones.

Gamay and Grenache rosés are also being produced from these varieties as are several pink wines labeled simply as rosé. The Grenache seems to be the best if it is grown in a cool region and harvested early. Some wineries produce both a dry and a sweet rosé, but the discriminating consumer finds it difficult to distinguish the two from the label. We believe that the public would welcome information as to the sugar content of the wine, when above 1 per cent, on the label. The Wine Institute (1968a) specification for rosé is: "The dry table wine called rosé should be fruity, light, and tart. Color should be pink with or without an orange modifying tint, and should not have an amber tint. A noticeable Muscat aroma is undesirable in rosé." The Wine Institute (1968a) specification for sweet rosé is: "This rosé should have the same characteristics as rosé with the exception that the reducing sugar should be above 0.5 per cent. A noticeable Muscat aroma is undesirable in sweet rosé." (Many California rosé wines have more than 0.5 per cent reducing sugar without being labeled "sweet rosé.") The Wine Institute (1968a) specification of Grignolino is: "A dry red table wine produced of the Grignolino grape. The varietal flavor of the Grignolino grapes should be pronounced. This wine type should be medium-bodied. The color varies from orange-red to medium-red. It should be aged sufficiently to reduce the natural astringency to a palatable level." (We have not found Grignolino wines to have such a distinct varietal flavor.)

Dry white wines

Under California standards (table 11), white table wines under 14 per cent alcohol should have a total fixed acidity, calculated as tartaric acid, of not less than 0.3 grams per 100 cc, the extract not less than 1.7 grams per 100 cc, the alcoholic content not less than 10.0 per cent and not more than 14.0 per cent by volume, and the volatile acidity, calculated as acetic acid, not in excess of 0.11 grams per 100 cc after sulfur dioxide correction. (The regulations specify cc, not ml.)

According to the Wine Institute specifications (1968a), all white table wines mentioned in these specifications should evidence proper maturation prior to bottling, but should not have a woody or oaky taste. They should be free of excessive oxidation, excess sulfur dioxide, and excess gassiness (evolution of bubbles at time of serving), and should not be cloudy.

California chablis is used for light, straw-colored wines of dry, fruity character and medium tartness. In France this type is produced almost entirely from Chardonnay grapes. Wines made from Pinot blanc or Chardonnay grapes are considered best here as well, but because of the shortage of these varieties and the demand for the grapes to produce varietal types, it is a rare California chablis that is produced from them. Not too much alcohol—11 to 12 per cent—is desired in this type, and total acidity should not be less than 0.65 per cent. The Wine Institute (1968a) specification for chablis is: "The white table wine called California

TABLE 11

LEGAL LIMITS FOR CERTAIN CONSTITUENTS IN TABLE WINES
Set by California State Department of Public Health and by
Federal Regulations[a]

Authority	Type	Alcohol range[b] (per cent)	Volatile acid, maximum[c] (gr/100 cc)	Fixed acid, minimum (gr/100 cc)	Extract, minimum (gr/100 cc)	Sulfur dioxide, maximum (mg/liter)
California	White	10.0–14.0	0.110	0.30	1.7	350
California	Red	10.5–14.0	0.120	0.40	1.8	350
Federal	White	not over 14	0.120	—	—	350
Federal	Red	not over 14	0.140	—	—	350

[a] Alcohol as per cent by volume, volatile acid as acetic acid, fixed acid as tartaric acid, and sulfur dioxide as sulfur dioxide.
[b] The general practice of government authorities is to permit a 0.5 per cent tolerance although there seems to be no provision for such a tolerance in the regulations.
[c] Exclusive of sulfur dioxide.
Sources of data: California State Department of Public Health, Bureau of Food and Drug Inspection, 1946; and U.S. Internal Revenue Service (1961b).

chablis should be light-to-medium straw color. It should be light to medium-bodied, of medium acidity, fruity, well-balanced, and have a good bottle bouquet." (See comments on California sauterne below.)

Dry wines made from, and labeled, Chardonnay and Pinot blanc have become increasingly important in the California industry. They should be a pale to a light-gold color, of medium body, and of not less than 0.60 per cent total acidity. Naturally they must possess the appropriate characteristic varietal flavor and aroma. California White Pinot wines (made from Chenin blanc) should conform to the same general requirements. The varietal characteristics of the Pinot blanc and Chenin blanc are much less pronounced than those of the Chardonnay. We favor the abandonment of the confusing White Pinot label. There is a difference of opinion in the industry as to the most desirable sugar content of California Chenin blanc. We hold no brief for dry or sweet Chenin blanc, but believe that the customer should be able to distinguish between them from the information on the label.

The Wine Institute (1968a) specification for White Pinot is: "A dry white table wine produced from Chenin blanc (White Pinot, Pinot de la Loire) grapes. The flavor and aroma of the Chenin blanc grape (White Pinot, Pineau de la Loire) should be pronounced. It should be tart, light-bodied, and have a good bottle bouquet." The specification for California Pinot blanc is: "A dry white table wine produced from Pinot blanc grapes. The flavor and aroma of the Pinot blanc grape should be pronounced. It should be moderately tart, medium-bodied, and have a good bottle bouquet." The specification for California Chardonnay is: "A dry white table wine produced from Chardonnay grapes possessing pronounced amount of varietal aroma and flavor of the Chardonnay grape. It should be medium to full-bodied with medium acidity, pale to light-golden in color, and have a good bottle bouquet." (This type is often labeled Pinot Chardonnay.)

There has been considerable overlapping in the composition of California hock and moselle. It is very difficult to suggest distinctive differences between these types that would hold uniformly true in the industry. Since most of these type names have been replaced with California rhine, they will not be considered further here and they are not now in use in California so far as we know.

California rhine wine should have a pale to light-gold color and be dry and tart (not less than 0.60 per cent total acidity) for the highest quality. To differentiate it from California chablis, the presence of at least some Riesling flavor is indicated. Since this seems difficult in view of the shortage of White Riesling, Amerine et al. (1967) suggested that the distinction between California chablis and dry sauterne be based on alcohol and acid content and that the type California rhine be abandoned. Nightingale (1961) recognized the merit of the suggestion, but it has not found favor with the California industry. It is unfortunate that the industry has offered no suggestion as to how these types might be made reasonably distinguishable to the American consumer. The Wine Institute (1968a) specification for California rhine or Riesling is: "The white table wine called California rhine wine or California Riesling should be pale to medium straw in color, it should be medium-bodied, of medium acidity to tart, fresh and fruity, and have a good bottle bouquet." (See comments on California sauterne below.)

California Riesling is a type name used for dry white wines made from at least 51 per cent Riesling grapes and deriving their predominant flavor and aroma from these varieties. The greatest successes have been achieved with the White Riesling (commonly called Johannisberger Riesling in California) and the Sylvaner (Franken Riesling), which are the most distinctive. The Wine Institute (1968a) specification for California White (Johannisberger) Riesling is: "A dry white table wine produced from White (Johannisberger) Riesling grapes. It should be light straw in color preferably with a slight greenish tinge. It should have a distinguishable varietal aroma and flavor, be fruity, medium to full-bodied and rather tart. It should have a good bottle bouquet." The specification for California Sylvaner (incorrectly known as Franken Riesling) is: "A dry white table wine produced from Sylvaner grapes. It should be light straw in color, preferably with a slight greenish tinge. It should have a distinguishable varietal aroma and flavor, be fruity, medium-bodied, and of medium acidity. It should have a good bottle bouquet."

Producers should avoid using the Riesling label alone on wines of the Wälschriesling, Kleinberger Riesling, and particularly the Gray Riesling; for, although these grape varieties produce mild, pleasant wines that sometimes fit the designation "California rhine" or the specific varietal name, they may disappoint those who desire Riesling wines. The desirable characteristics of distinctive flavor and aroma for Riesling wines are not ordinarily obtained from grapes grown in warm regions. Use of muscat-flavored wines for blending with Riesling wines is not recommended. Pale- to light-gold-colored wines, with at least 0.60 per cent acidity and only a medium body and alcohol, are preferred. The Wine Institute (1968a) specification for California Gray Riesling is: "A dry white table wine produced from Gray Riesling grapes. It should have a distinguishable varietal aroma and flavor, be medium-bodied and of medium acidity. It should have a good bottle bouquet."

Other types of wine which take the name of the variety from which they are predominantly derived and which appear to be distinctive and satisfactory for use when the grapes are grown in the proper region are Sémillon, Sauvignon blanc, and Gewürztraminer. Sémillon wines should have a light-gold color, a full body, and be dry, with a moderate total acidity—about 0.65 per cent. The Sauvignon blanc has a highly individual aroma and a unique aromatic flavor. It should be a heavy-bodied wine of medium alcohol content and a total acidity of not less than 0.60 per cent. The Wine Institute (1968a) specification for California Gewürztraminer is: "This wine should be straw-to-light-golden in color. It should have an easily distinguishable varietal aroma and flavor which is spicy and almost muscat-like. It should be of medium body and of medium acidity and have a good bottle bouquet." (This should not be labeled Traminer.)

Emerald Riesling has a slight muscat flavor and currently is marketed as a slightly sweet type. Its natural high total acidity would indicate that it should be tried also as a dry, high-acid, spicy-flavored type.

The industry has recommended that wines labeled Ugni blanc (Saint Emilion or Trebbiano) and Gutedel (Chasselas doré), being somewhat neutral, be judged at expositions with California chablis and California rhine wines. This would seem to indicate that the industry feels that these names should not ordinarily be used on labels. At least some wines made from Folle blanche grapes seem worthy of a varietal label. The Wine Institute (1968a) specification for California Folle blanche is: "A dry white table wine produced from the Folle blanche grapes. It should be light in color and body and tart, fruity and fresh." The specification for California French Colombard is: "A dry white table wine produced from the French Colombard grapes. This wine should be pale to light-golden in color, medium-bodied, fruity and of medium acidity." We do not find that Green Hungarian makes a white table wine with a distinguishable varietal flavor.

California white chianti is a dry white wine which formerly was bottled in the same kind of flask as California red chianti. In Tuscany (Italy) it is produced from Saint Emilion (also called Ugni blanc or Trebbiano) and Malvasia grapes. It probably could be differentiated from the other nonvarietal dry white wine types by a slightly higher astringency produced by the method of fermentation, and by inclusion of a slight muscat flavor. It is not, however, a wine type of high quality, and its production in California would seem to be superfluous.

California dry sauterne is another confusing type; the sauterne appellation might better be reserved for a sweet table wine type, as indicated below. The Wine Institute (1968a) specification for California sauterne is: "The table wine type called California sauterne should be straw to light-golden color, full-bodied without noticeable high acidity and contain not over 1.5 per cent reducing sugar. It should have balance and softness on the palate and a good bottle bouquet." (The difference between this and California chablis and white chianti and rhine is minimal.)

The Wine Institute specification for California dry white table wine is:

It is recommended that any dry white table wine not specifically named in these specifications be automatically judged in a class to be known as California Dry White Table Wine. It is recommended that wines with varietal or generic names entered in this class be grouped together by name and that such groups be judged in an appropriate sequence. The wines should be judged on their general elements of quality rather than characteristics of any particular variety of grapes unless they have a varietal label in which case they will be judged on the appropriate varietal character. The color may range from pale straw to golden; they should be well-balanced on the palate. The reducing sugar should be less than 1.0 per cent.

Sweet table wines

Two sweet white table wines are widely produced in California— California sauterne and California sweet sauterne (or haut sauterne or chateau type). As indicated in table 8 (p. 86), these may be differentiated on the percentage of sugar they contain. California dry sauterne (table 7) is really a dry white table wine type and should contain less than 0.25 per cent reducing sugar; California sauterne should have from 1 to 3 per cent reducing sugar; and the California sweet sauterne from 3 to 6 or more per cent reducing sugar. The Wine Institute (1968a) specification for California medium sauterne is: "The wine in this class should have the same characteristics as California Sauterne with the exception that the reducing sugar should not be above 3.0 per cent. A slight Muscat flavor and aroma is permissible." The better-quality wines of this type in the Sauternes district of France as well as in California have been made mainly from Sémillon grapes with a lesser amount of Sauvignon blanc. These wines should have a light-gold color and at least 0.50 per cent total acidity. The only wine produced wholly from botrytised grapes in California was the Premier Sémillon of the Cresta Blanca Wine Company (see p. 649). This was a wine of outstanding quality and it is hoped that its production will be resumed. The Wine Institute (1968a) specification for California sweet Sauvignon blanc is: "The wine in this class should have the same characteristics as California Sauvignon blanc with the exception that the reducing sugar should be above 1.5 per cent." The specification for California Sweet Sémillon is: "The wine in this class should have the same characteristics as California Sémillon with the exception that the reducing sugar should be above 1.5 per cent."

A pink-colored sweet table wine called Aleatico, made from Aleatico grapes, was formerly produced in California. It should contain over 2 per cent reducing sugar and have a spicy and aromatic muscat flavor. The Aleatico character should not be masked by an excessive sulfur dioxide content. In order to differentiate it from a dessert wine of the same name, the label "Aleatico table wine" has been recommended.

Another type of muscat-flavored table wine is frequently called California light muscat. Varieties such as Muscat blanc (Muscat Canelli) and Malvasia bianca could be used more successfully than the extensively planted Muscat of Alexandria if they were available. The best-balanced and most pleasing of these wines contain from 5 to 10 per cent of residual sugar and a total acid content of at least 0.55 per cent. (See also Amerine, 1947.) The Wine Institute (1968a) specification for California light sweet muscat is: "A white table wine produced of muscat-flavored grapes. The table wine called light sweet Muscat should have unmistakable muscat aroma and flavor. It should have more than 7 per cent reducing sugar. It should be medium to full-bodied and well-balanced. It should be light-golden to golden in color." The Wine Institute (1968a) specification for California sweet white table wine is: "The wine in this class should have the same characteristics as California White Table Wine with the exception that the reducing sugar should be 1.0 per cent or over."

Sparkling and carbonated wines

The present legal requirements for California sparkling wines are given on page 655. They have been the subject of much controversy

99

in the industry because of the desire to call all wines produced in closed containers "California champagne." Regardless of the merits of this proposal (see also Amerine and Monaghan, 1950), the better-quality california sparkling white wines should have at least 0.65 per cent total acidity, a light-straw color, and be free of excessive oxidation and any sulfur dioxide odor. No doubt wines made partially or wholly from the various Pinot varieties or blends of these varieties are of better quality, but because their acidity is frequently too low in California they have been combined with Folle blanche, French Colombard, and similar high-acid varieties, which have proved useful for blending. It is customary to indicate on the label the degree of sweetness and, for the better-quality sparkling wines, the vintage. The terms "brut," "dry" (or "sec"), and "demi-doux" indicate increasing amounts of sugar (*dosage*) in the wine. The practice of making the "dry" types very sweet is to be deplored. The Wine Institute (1968a) specification for California champagne is: "California champagne should be pale to straw color. It should have a good acidity and body. It should be fresh, fruity, well-balanced, and show a distinctive bottle bouquet." It is recommended that the California champagne be rated in three groups according to the percentage of reducing sugar: Group 1, 0.0–1.5 per cent; Group 2, 1.5–3.0 per cent; Group 3, over 3.0 per cent.

Other types of sparkling wine are California pink champagne, California sparkling burgundy (sometimes called California champagne rouge or red champagne), and California sparkling muscat. The pink champagne should be a light-pink color with no amber, and should have a fruity flavor, which usually requires at least 0.65 per cent acidity. It usually contains 1.5 to 4.0 per cent reducing sugar. Sparkling muscat should have a pronounced muscat flavor and aroma, be of a light-gold color, and have at least 4 per cent reducing sugar and over 0.60 per cent total acidity. The Wine Institute (1968a) specification for California pink champagne is: "The color of this wine type should be true pink. It should be fruity, fresh, light-bodied, tart and well-balanced. The reducing sugar should be 1.5 per cent or over." The specification for California champagne rouge and sparkling burgundy is: "On this type, the heavier-bodied and darker-colored wines should be on an equal basis with the lighter-bodied and ligher-colored wines. It should have a good acidity and be semi-sweet to sweet. Reducing sugar should be 1.5 per cent or over. It should be fruity, smooth and well-balanced and show distinctive bottle bouquet." Again there is a suggestion that two different types of California sparkling burgundy exist. The specification for California sparkling muscat is: "California sparkling muscat should be pale to straw color, have good acidity and body. It should be fresh, fruity, well-balanced and have the unmistakable flavor and aroma of Muscat grapes. The reducing sugar should be 4 per cent or over."

Other type names

Individual wineries may desire to produce distinctive types of wine either from special varieties or by unique processes, and to label them with type names other than those mentioned above. In using such proprietary and varietal names, the producer should consider not only securing approval from governmental authorities but also whether the type represents a desirable contribution to California wine nomenclature, and, if named after a variety, whether the grapes used are correctly identified and the wine has a unique and recognizable character. Undue multiplication of wine type names is confusing to the consumer and should be avoided.

The character of the varieties used for producing varietal types of wine changes from season to season and from district to district. To guide the consumer, sufficient information on place of production and, if possible, the vintage should be given for wines which carry varietal names or for wines of high quality which will improve by aging in the bottle.

FEDERAL AND STATE LAWS

The regulations of the various state and federal agencies concerning the production, names, and composition of wines are complex and are changing constantly.

Those interested in bonding winery premises, securing the necessary permits, and producing wines should consult the Wine Institute, 717 Market Street, San Francisco, 94103, or the following government agencies: Basic Permit and Trade Practice Division of the Alcohol and Tobacco Tax Division of the Internal Revenue Service, United States Treasury Department; and the Department of Public Health, the Board of Equalization, and the Department of Agriculture of the State of California. An outline of the pertinent restrictions is presented here.

Definitions

The California Administrative Code defines "wine" as follows (in part):

"*California Wine*" (or "*California grape wine*").—"California wine" is any grape wine produced in the State of California, and is the product of the normal alcoholic fermentation of the juice of sound, ripe grapes (including pure condensed grape must), with or without added grape brandy or other spirits derived from grapes or grape products, and containing not to exceed 21 per cent alcohol by volume, but without any other addition or abstraction whatsoever except such as may occur in normal cellar treatment; provided, that there may be added before, during or after fermentation pure condensed grape must; and

Provided further, that no sugar or material containing sugar, other than pure condensed grape must, and no water in excess of the minimum amount necessary to facilitate normal fermentation, may be used in the production or cellar treatment of any grape wine in the State of California. (California State Department of Public Health, 1946.)

The Federal Alcohol Administration gives the following definition:

Section 21, Class 1. Grape wine.—(a) "Grape wine" is wine produced by the normal alcoholic fermentation of the juice of sound, ripe grapes (including restored or unrestored pure condensed grape must), with or without the addition, after fermentation, of pure condensed grape must, and with or without added grape brandy or alcohol, but without other addition or abstraction except as may occur in cellar treatment . . . (U.S. Internal Revenue Service, 1961b.)

Although various government agencies have defined wine for regulatory and taxation purposes in slightly different fashion, all indicate that it is essentially a fermented alcoholic beverage produced from grapes. (See also pp. 104 and 105.) Addition of alcohol, from whatever source, to table wines is prohibited in many countries, but detection of fortification is difficult. Garoglio and Boddi-Giannardi (1967) found a caramel-type reaction in wine distillates and proposed its use to detect addition of brandy to wine. However, old wood-aged wines also gave a reaction. Nevertheless, since such compounds are undoubtedly formed when wines are heated at boiling temperatures, a more specific test would be useful. This appears to be more practical than the use of ratios of components, such as Rebelein's (1958, 1962, 1964) (p. 475). Siegel et al. (1965) found that these are influenced by degree of ripeness, yeast strain, and other factors. Patschky (1967) used an extract number and a variety of ratios to detect fortification. The results were significant for the latter only when there was a low alcohol yield, and for the former when combined with determination of fermentation by-products.

Numerous types of wine are produced from the many varieties of the several *Vitis* species, but normally, in California, only varieties of *V. vinifera* are used. The juice of grapes varies in character, composition, and suitability for wine making, not only with the different varieties but also for the same variety grown under different environmental conditions and treated in different ways before, during, and after fermentation.

Fermented beverages from other fruits, if called "wine" must be preceded by the name of the fruit from which they are produced: orange wine, peach wine, and so on. They are not considered in this publication.

Federal regulations

The most important restrictions on the composition and character of wines are found in Regulations No. 4 of the Federal Alcohol Administration (U.S. Internal Revenue Service, 1961*a,b*) regarding labeling and advertising. These regulations define wines and limit their composition as follows: for natural red wine, a maximum volatile acidity (calculated as acetic acid and exclusive of sulfur dioxide) of 0.14 gram per 100 cc; for other wines, a maximum of 0.12 gram per 100 cc. Wines named for a specific variety of grape must derive from it at least 51 per cent of their volume from grapes grown in the district and must be fermented in and conform to the regulations of the district for which they are named. Generic and semigeneric designations for wines are also defined. Vermouth and sake are examples of generic names. Wines with a semigeneric name such as burgundy, claret, chablis, champagne, chianti, moselle, rhine wine (syn. hock), and sauterne must have the actual district of origin stated in conjuction with the type on the label (e.g., California claret), and the wine must conform to the standard of identity for the type as set up in the regulations or, if there is no such standard, to the generally accepted trade understanding of such class or type.

Labeling requirements include brand name, type, name and address, blends, alcohol content, and contents. The alcohol content need be only within ± 1.5 per cent of that stated on the label. (But this does not mean that the alcohol content may be above or below that permitted by state or federal regulations.) The year of production of bottled wine can be specified only if certain conditions are met. A certificate of label approval must be obtained for wine domestically bottled or packed (with certain exceptions for wine for export).

Advertising statements or claims which are false, misleading to the consumer, disparaging of a competitor's products, and the like are prohibited by the federal regulations. The wine industry itself has adopted certain rules about advertising.

Finished wines, without special labeling, should not have over 350 parts per million of sulfur dioxide. The federal regulations also specify that filtration, pasteurization, refrigeration, and other treatments should be used to the minimum extent necessary to stabilize the wine. See also U.S. Laws (1963, 1965).

The regulations of the Food and Drug Administration define "wine" and specify limits for the content of certain constituents, such as acids, sugar, sodium chloride, potassium sulfate, and sugar-free solids (U.S. Department of Agriculture, Food and Drug Administration, 1939). The present federal minimum limit for sugar-free solids is far lower than that encountered in California.

103

The regulations of the Alcohol and Tobacco Tax Division of the Internal Revenue Service are framed to guarantee revenue protection and thus include certain limitations in processing methods on bonded winery premises (U.S. Internal Revenue Service, 1961a). Many operations which are permitted on bonded winery premises are prohibited on wholesale or retail premises. Practices which jeopardize the revenue or would be otherwise objectionable from a supervisory or control aspect are not permitted. These regulations also have to do with the bonding, construction, and equipment of wineries. All commercial wine makers must become familiar with these provisions, particularly those having to do with forms 701 and 702, which cover the operations of the winery and must be filed monthly. The federal regulations are so complex and lengthy, with many confusing statements and administrative interpretations, that the wine industry has recommended a comprehensive law to simplify them.

State regulations

The Department of Public Health of the State of California has established minimum standards for wines in order to keep poor, unsound wines from being offered to the public, to protect the wine industry, and to induce better wine making. Those relating to table wines are summarized in table 11. Besides meeting the requirements for chemical composition, the wines must have the desired composition and color for the type and possess the clean vinous taste and aroma of normal wines.

To correct natural deficiencies in grape musts and wines and to ensure the soundness, purity, and palatability of the finished wine, normal cellar treatment is permitted. This includes the addition of neutral potassium tartrate or calcium carbonate to reduce excess total acidity (rarely necessary in California); addition of tartaric acid produced from grapes, or of commercial malic or citric acids, to correct deficient acidity; addition of sugar for the *dosage* of carbonated and sparkling wines; blending processes; addition before, during, or after fermentation of pure condensed grape must; and treatments, where desirable or necessary, to improve the quality of the finished product. Prohibited practices are the use of water in excess of the minimum amount necessary to facilitate normal crushing; use of monochloracetic acid; production of wine, for consumption, from raisins, dried grapes, or residues and water; addition of synthetic materials; and others.

Table wine (including light wine, light grape wine, light red wine, light white wine, and natural wine) is defined as wine containing not over 14 per cent alcohol. Red wine is defined as wine which contains the red coloring matter of the stems, juice, or pulp; white wine does not contain this red coloring matter. California wine must derive 100 per cent of its volume from grapes grown, fermented, and finished in California.

Vintage wine means wine produced wholly from grapes that were gathered and their juice fermented in the calendar year and in the viticultural area given on the label or in the advertisement of the wine.

Sparkling wine (including sparkling grape wine, sparkling red wine, and sparkling white wine) is defined as wine made effervescent solely with carbon dioxide from a secondary fermentation in a closed container, tank, or bottle. Champagne is restricted to the product of a secondary fermentation in glass containers of not greater than one-gallon capacity, which has the taste, aroma, and other characteristics attributed to champagne. Other sparkling wines with these characteristics, but which have been produced by a secondary fermentation in containers of over one-gallon capacity, may be designated as "sparkling wine—champagne style," "sparkling wine—champagne type," "American (or California, etc.) champagne, bulk process," or "American (or California, etc.) champagne, natural fermentation in bulk."

Carbonated wine (including carbonated grape wine, carbonated red wine, and carbonated white wine) is wine made effervescent by a process other than a secondary fermentation in a closed container, tank, or bottle.

Varietal names for wines are authorized if the wine derives its predominant taste, aroma, and other characteristics, and at least 51 per cent of its volume, from the given variety. However, where the consumer is not familiar with the variety as a type name, an explanatory statement must appear on the label along with the variety name.

Geographical appellations of origin are permitted if at least 75 per cent of the volume is derived from fruit grown and fermented in the place or region indicated, and the wine has been produced and finished within that place or region. It must conform to the laws and regulations of such a place or region governing composition, method of production, and labeling for wines for consumption within the given place or region. Cellar treatment and blending of wines from a given place or region are permitted outside the original location provided such treatment or blending would be proper in the original region.

Under the California law, any wine labeled "California Central Coast Counties Dry Wine" must be produced 100 per cent from grapes grown in any one or more of the central coast counties (Sonoma, Napa, Mendocino, Lake, Santa Clara, Santa Cruz, Alameda, San Benito, Solano, San Luis Obispo, Contra Costa, Monterey, and Marin).

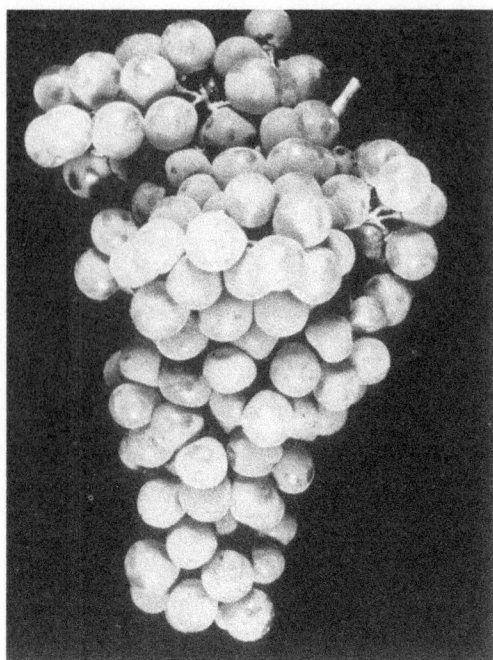
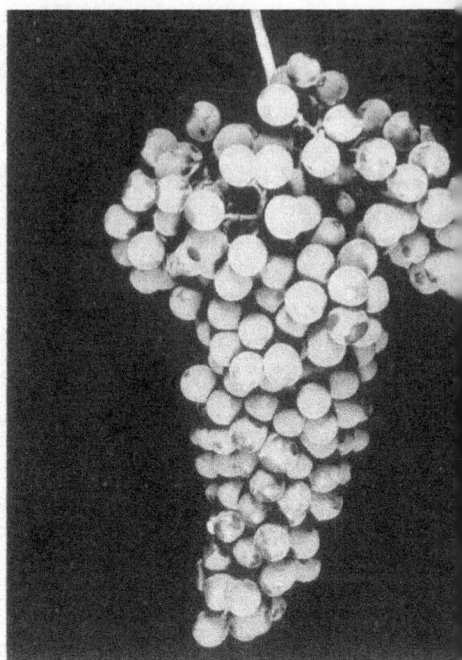

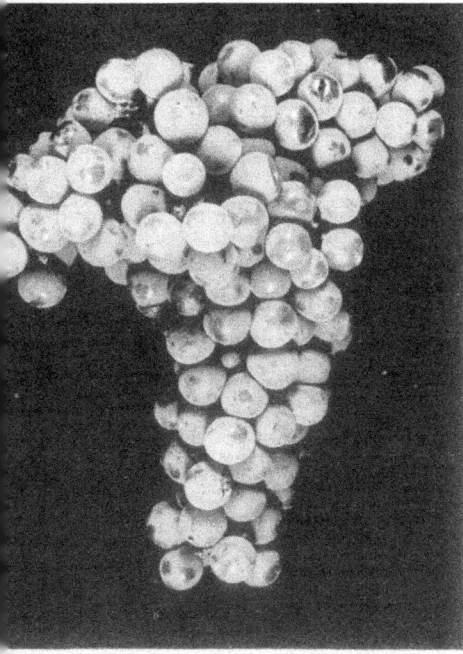

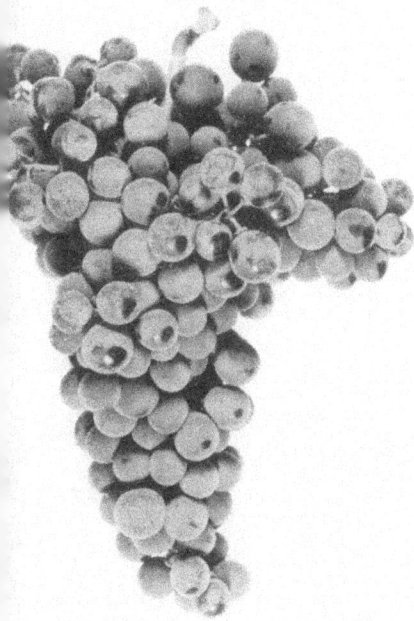

Figure 18. Important wine grape varieties of California. (Note: Illustrations not to the same relative scale.) Top row, left to right: Sémillon, Emerald Riesling, White Riesling, French Colombard. Bottom row, left to right: Zinfandel, Petite Sirah, Carignane, Cabernet Sauvignon.

PERMITTED ADDITIVES

There is increasing interest in international trade in the materials which may be added to wines, their quality and quantity, and the residues which remain in the wine. The Office International de la Vigne et du Vin (1964) has published a list of such materials together with data on their composition. This list is constantly being upgraded. An excellent list of permissible materials which may be added to wines (and other foods) in Germany was given by Petritschek-Schillinger (1967). Among the permitted materials which are not commonly used or mentioned in this country are agar, asbestos, cellulose, egg white, potassium hexacyanoferrate (limit 20 mg per kg), charcoal, oxygen, and isinglass. According to this list, sorbic acid and sorbates are not permitted in German wines. Cerutti (1963) reported a similar list of additives permitted in Italy.

The materials permitted in this country and their amounts are as follows (this list omits some common additives such as sulfur dioxide):

Material	Use	Reference or limitation[a]
Acetic acid	To correct natural deficiencies and to perfect wine	Not to exceed 0.4 gallon of pure 100 per cent acetic acid per 1,000 gallons and not more than 0.140 gr per 100 cc of acetic acid in the finished red wine (or 0.120 for white).—Rev. Rul. 66-155; Sec. 240.364, 26 CFR
Afferin Cufex	To remove trace metals from wine	No insoluble or soluble residue in excess of one part per million may remain in finished wine, and basic character of wine may not be changed by such treatment.—Sec. 240.524, 26 CFR[b]
Antifoam "A" Antifoam AF emulsion	To reduce foam in fermenters	Sec. 240.524, 26 CFR
Atmos 300	To reduce foam in musts and wines	Not over 25 p.p.m.—Rev. Rul. 63-60; Secs. 240.524, 240.1051, 26 CFR
Antifoam "C"	To reduce foam in musts and wines	Silicone content of wine not to exceed 10 p.p.m.—Rev. Rul. 63-110; Sec. 240.524, 26 CFR
Polyoxyethylene-40-monostearate and silicone dioxide, and defoaming agents composed of sorbic acid, carboxy methyl cellulose, dimethyl polysiloxane, polyethylene-40 monostearate, and sorbitan monostearate	To reduce foam in musts and wines	Must meet requirements of FDA. Not over 0.15 lb per 1,000 gallons of 100 per cent active agent. Agents 30 per cent active, not over 0.5 lb per 1,000 gallons of wine. Silicone dioxide must be completely removed by filtration.—Rev. Rul. 65-183, 64-266; Sec. 240.524, 26 CFR

Material	Use	Reference or limitation*
Ascorbic acid Isoascorbic acid	To prevent darkening of color and deterioration in flavor of wines and wine materials, and overoxidation of vermouth and other wines	May be added to fruit, grapes, berries, and other materials used in wine production, to juice of such materials, or to wine, within limitations which do not alter class or type of wine. Its use need not be shown on labels.
Calcium carbonate	To reduce excess natural acids in high-acid wine	Natural or fixed acids may not be reduced below 5 parts per 1,000.—Sec. 250.524, 26 CFR

Clarifying agents:

Material	Use	Reference or limitation*
AMA special gelatine solution	To clarify and stabilize wine	Sec. 240.524, 26 CFR
Bentonite (Wyoming Clay)	To clarify wine	Sec. 240.524, 26 CFR
Bentonite slurry	To clarify wine	1 lb of bentonite to not more than 2 gallons of water. Total quantity of water not to exceed 1 per cent of volume of wine.—Rev. Rul. 59-150; Secs. 240.524, 240.1051, 26 CFR
Bentonite, activated carbon, and copper sulfate	To clarify and stabilize wine	Addition of copper must not exceed 0.5 p.p.m.; residual copper must not exceed 0.2 p.p.m.—Rev. Rul. 67-133; Sec. 240.1051, 26 CFR
Copper sulfate	To clarify and stabilize wine	Not more than 0.5 mg per liter may be added nor may residue be more than 0.2 mg per liter.—Rev. Rul. 66-229; Sec. 240.524, 26 CFR
Diatomaceous earths or similar products (which do not contain active chemical ingredients or have not been so treated that they may alter character of wine)		
Fumaric acid	To correct natural deficiencies in fruit and berry wines	Not to exceed 25 lbs per 1,000 gallons of wine.—Rev. Rul. 63-243; Secs. 240.524, 240.404, 26 CFR
Fumaric acid	To correct natural deficiencies and to stabilize grape wine	Not to exceed 25 lbs per 1,000 gallons of wine.—Rev. Rul. 63-61; Secs. 240.364, 240.524, 26 CFR
Fulgar (aluminum silicate and albumen)	To clarify wine	Not more than 6.6 lbs per 1,000 gallons of wine.—Rev. Rul. 60-335; Sec. 240.524, 26 CFR
Locust bean gum, carragheenan, alginate, bentonite, agar agar, and diatomaceous earth	To clarify and stabilize wine	Not more than 2 lbs per 1,000 gallons of wine.—Rev. Rul. 61-199; Sec. 240.524, 26 CFR

Material	Use	Reference or limitation[a]
Clarifying agents: (continued)		
Polyvinylpolypyr-rolidone (PVPP)	To clarify and stabilize wine	Not more than 6 lbs per 1,000 gallons of wine. Must be completely removed by filtration.—Rev. Rul. 66-173; Sec. 240.524, 26 CFR, and FDA Reg. Sec. 121.1110, 21 CFR
Promine-D	To clarify and stabilize wine	Not more than 1.5 lbs per 1,000 gallons of wine. Water used must not exceed 0.5 per cent of volume of wine treated.—Rev. Rul. 60-334; Sec. 240.524, 26 CFR
Protovac PV-7916 Sparkaloid No. 1 Sparkaloid No. 2	To clarify and stabilize wine	Not more than 2 lbs per 1,000 gallons of wine
Tansul clay No. 7, 710, 711	To clarify and stabilize wine	Not more than 10 lbs per 1,000 gallons of wine.—Rev. Rul. 64-134; Sec. 240.524, 26 CFR
Combustion product gas	For counterpressure in champagne production	CO_2 content of gas must not exceed 1 per cent.—Rev. Rul. 65-198; Sec. 240.524, 26 CFR, and FDA Reg. Sec. 121.1060, 21 CFR
Compressed air	To aerate sherry wine during baking process	Not to cause changes other than those occurring during usual storage in wooden cooperage over a period of time.—Rev. Rul. 59-189 (I.R.B. 1959-21, 20)
"DEPC," Diethyl-pyrocarbonate	As a sterilizing or preserving agent	No longer permitted
Diammonium phosphate	Yeast food in fermentation of wines	Not more than 10 lbs per 1,000 gallons of wine.—Rev. Rul. 61-83; Sec. 240.524, 26 CFR
Glycine (amino acid)	Yeast food in fermentation of wines	Not more than 2 lbs per 1,000 gallons of wine
Granular cork	To treat wines stored in redwood and concrete tanks	Not more than 10 lbs per 1,000 gallons of wine
Hydrogen peroxide	To facilitate secondary fermentation in champagne production	Not more than 3 p.p.m.—Rev. Rul. 63-231; Secs. 240.524, 240.1051, 240.1052, 26 CFR
Mineral oil	On surface of wine in storage tanks to prevent access of air to wine	No oil may remain in finished wine when marketed
Oak chips (uncharred and untreated)	To treat wines	Rev. Rul. 61-126; Sec. 240.524, 26 CFR
Oak chips (charred)	To treat Spanish Type Blending Sherry	Finished product after addition of oak chips must have flavor and color of Spanish Type Blending Sherry commonly obtained by storage of sherry wine in properly treated used charred oak whisky barrels
Oak chip sawdust	To treat wines	Rev. Rul. 61-126; Sec. 240.524, 26 CFR
Pectolytic enzymes:		Ind. Circ. 60-3, March 1, 1960
Pectinol 59L	To clarify wine	Not more than 0.3 lb per 1,000 gallons of wine

Material	Use	Reference or limitation[a]
Klerzyme 200	To clarify wine	Not more than 0.5 lb per 1,000 gallons of wine.—Rev. Rul. 64-37; Sec. 240.524, 26 CFR
Klerzyme 100	To clarify wine	Not more than 1 lb per 1,000 gallons of wine.—Rev. Rul. 66-287; Sec. 240.524, 26 CFR
Klerzyme liquid	To clarify wine	Not more than 0.5 lb per 1,000 gallons of wine.—Rev. Rul. 64-37; Sec. 240.524, 26 CFR, and IRS Sec. 5382
Pectinol 100-D	To clarify wine	Not over 10 lbs per 1,000 gallons of wine.—Rev. Rul. 62-63; Sec. 240.524, 26 CFR
Pectinol 41-P, 42E	To clarify wine	Not more than 1 lb per 1,000 gallons of wine.—Rev. Rul. 62-47; Secs. 240.524, 240.1051, 26 CFR
Pectinase 2LM	To clarify wine	Not more than 5 lbs per 1,000 gallons of wine.—Rev. Rul. 64-323; Sec. 240.524, 26 CFR
Pectinase Concentrate Pectinol A, M, O Pectinase Regular Pectinase D Regular	To clarify wine	Not more than 10 lbs per 1,000 gallons of wine
Pectinase LM	To clarify wine	Not more than 10 lbs per 1,000 gallons of wine.—Rev. Rul. 60-298; Sec. 240.524, 26 CFR
Pectolase Spark-L	To clarify wine	Not more than 6 lbs per 1,000 gallons of wine.—Rev. Rul. 65-85; Sec. 240.524, 26 CFR
Sodium bisulfite	As a sterilizing or preserving agent	Sec. 240.523, 26 CFR
Sodium carbonate	To reduce excess natural acidity	May be used within limitations of Sec. 240.523 provided natural or fixed acids are not reduced below 5 parts per thousand
Sorbic acid Potassium salt of sorbic acid Sodium salt of sorbic acid	As a sterilizing and preservative agent and to inhibit mold growth and secondary fermentations	Not more than 0.1 per cent of sorbic acid or salts thereof may be used in wine or in materials for production of wine
Takamine cellulose	To clarify must or wine	Not more than 5 lbs per 1,000 gallons of wine.—Rev. Rul. 63-111; Sec. 240.524, 26 CFR
Uni-Loid Type 43B (agar-agar and Supercel)	To clarify and stabilize wine	Not over 2 lbs per 1,000 gallons.—Rev. Rul. 66-356; Sec. 240.524, 26 CFR
Veltol	To stabilize and smooth wines	Not over 250 p.p.m.—Rev. Rul. 62-145; Sec. 240.524, 26 CFR
Vitagen gas	To remove air and reduce oxidation in still wines	Rev. Rul. 61-200; Secs. 240.524, 240.531, 26 CFR, IRS, 5041-(a) (1954) and Food Additives (Added) Reg. 21 CFR 121
Vitagen gas	For counterpressure in champagne production	Rev. Rul. 63-70; Sec. 240.524, 26 CFR. The gas must be substantially free of CO_2

Material	Use	Reference or limitation[a]
Yeastex 61	Yeast food in fermentation of wines	Not more than 2 lbs per 1,000 gallons of wine.—Rev. Rul. 62-120 and 61-51; Section 240.524, 26 CFR

Anion exchange resins:

Amberlite IR-45 Duolite A-7 (Tartex 180) Duolite A-30 (Tartex 181) SAF Anion Exchanger	To reduce natural acidity of wine	May be used in a continuous column process for reducing natural acidity of wine provided resins are essentially in hydroxyl (OH) state: fixed acids are not reduced below 5 parts per 1,000, inorganic anion content of wine is not increased more than 10 mg per liter (10 parts per million) calculated as chlorine, sulfur, phosphorus, etc.; treatment does not remove color in excess of that normally contained in the wine; and original character of the wine is not altered
Duolite A-6	Treatment of white wines	Restrictions (above) imposed on use of other approved anion exchange resins applicable
Amberlite XE-168	Treatment of white wines	Color must not be reduced below 0.6 Lovibond unit in a $\frac{1}{2}$ in. cell.—Rev. Rul. 63-166; Secs. 240.524, 240.1051, 240.1052, 26 CFR

Cation exchange resins:

Amberlite IR-120H Amberlite IR-120 Duolite C-3 Na (Tartex 160) Duolite C-20 Na (Tartex 161) Duolite C-25 Na (Tartex 162) Permutit Q Permutit Q Spec 157 SAF Cation Exchanger	To stabilize wines To exchange sodium ions for undesirable metallic ions	May be used in continuous-column process. Resin must be essentially in sodium state so that certain metallic elements in the wine will be replaced with sodium ions. After regeneration and before reuse, the resin must again be essentially in sodium state. Over-all change in pH of wine before and after treatment shall not be greater than 0.2 pH units. Wine-water eluate from column during "sweetening on" process or any wine contained in wash water during "sweetening off" process should be discarded or used solely for distilling material
Duolite C-3 (Tartex 160)	To stabilize table wines	May be used in a batch process employing no more than 4 lbs per 100 gallons of wine
Duolite C-20, C-20L	To remove metallic substances from wine	Resin is regenerated with mineral acid. Regenerated column must be washed with ion-free water until pH of effluent and influent water is the same. Inorganic anions may not be increased more than 10 mg per liter. Potassium content may not be reduced below 300 p.p.m. and pH not below 3.0.—Rev. Rul. 63-69, 65-246, 65-278; Secs. 240.524, 240.1051, 26 CFR

Material	Use	Reference or limitation[a]
Non-ionic and conditioned resins:		
Duolite S-30 (Tartex 260) After special conditioning— Duolite A-7 Duolite A-30 Amberlite IRA-401 Amberlite IRA-400 Amberlite IRC-50 Amberlite XE-89	For removing excessive oxidized color, foreign flavors and odors, and for stabilizing wines by removal of heavy molecules and proteinaceous material	May be used in continuous-column process after suitable conditioning and/or regeneration. Conditioning agents and regenerants consisting of fruit acids common to the wine, and inorganic acids and bases may be employed provided conditioned or regenerated resin is rinsed with ion-free water until pH of effluent water is same as that of influent water. Water equal in quality to water obtained by distillation will be considered to be ion-free water. Tartaric acid may not be used as a conditioning agent when resin is to be used in treating wines other than grape. Inorganic anions or cations may not be increased more than 10 mg per liter; color may not be removed in excess of that normally contained in the wine; pH of the wine may not be changed more than 0.2 pH units nor acidity more than 0.5 gm per liter; original character of the wine may not be altered.—Rev. Rul. 62-29; Sec. 240.1051, 26 CFR

[a] Except as noted this list is based on Revenue Ruling 58-461 (IRB 1958-37, 51) amplified by later Revenue Rulings as indicated. It was also published in Federal-Revenue Rulings by Commerce Clearing House, No. 167-58, June 24, 1959, 30472 (p. 30, 344-30, 346). CFR is Code of Federal Regulations.

[b] This section reads: "In the process of filtering, clarifying, or purifying wine on bonded wine cellar premises, materials, methods, and equipment may be used to remove cloudiness, precipitation, and undesirable odors and flavors, but the addition of any substance foreign to wine which changes the character of the wine, or the abstraction of ingredients which will change its character, to the extent inconsistent with good commercial practice, is not permitted on bonded wine cellar premises." See U.S. Internal Revenue Service (1961a).

CHAPTER THREE

STATISTICAL TRENDS IN THE WINE INDUSTRY

INTERNATIONAL ACREAGE
and production continues to increase.

At the turn of the century grape acreage was down in most European countries owing to the invasion of phylloxera in the last third of the nineteenth century. Only parts of Greece, California, Australia, and a few other regions escaped phylloxera, which caused great economic losses in abandoned vineyards. As a direct result of phylloxera the varieties of grapes planted in many areas changed. Many vineyards were replanted with American varieties such as Noah and Clinton, which produce poor-quality wines but are more or less resistant to phylloxera and mildew. These still constitute a problem in Moldavia and other regions. Phylloxera had one beneficial effect. When vines were replanted in rows instead of helter-skelter as formerly, mechanical cultivation became possible. Grape acreage increased rapidly early in the twentieth century. However, the population was also increasing and, though there were temporary surpluses, the wine could usually be easily disposed of.

World War I had a serious impact on grape and wine production in France and Germany, and to a lesser extent in Austria and Jugoslavia. But this soon led to the expansion of the 1920's and the serious over-production during the depression of the 1930's. These had more disastrous effects on wine prices than on many other agricultural products. Nevertheless, governmental control programs effected some stabilization of prices by World War II. After World War II, acreage and production expanded rapidly throughout the world.

After World War II

Acreage reached about 25,000,000 in 1962, where it has remained essentially constant. At present losses are occurring, or may be expected to occur, in France, Hungary, Italy, North Africa, Spain, and possibly Jugoslavia. Increases have recently taken place in Argentina, Austria, Bulgaria, Chile, Czechoslovakia, Germany, Greece, and South Africa. The European Common Market envisaged free trade by the end of 1968. The effect of cheap Italian wines (owing to low costs of production) on higher German wine prices (owing to greater production costs) is not known, but it could be profound.

Vineyard acreage in the Soviet Union, after many years of increase, showed a slight decrease between 1964 and 1965 and a gain in 1966. Nevertheless, long-range Soviet agricultural projections still call for a further 50 per cent or more increase in acreage. There is considerable nonbearing acreage in many countries. With modern methods of vineyard management, production should expand rapidly whether or not new plantings are made. However, unlike most orchard crops, grape acreages tend to be very flexible, since they can be brought into production so rapidly. For a general discussion of recent changes in world production of grapes and wine see Commonwealth Economic Committee (1961). See also the occasional Mementos of the Office International de la Vigne et du Vin (1965).

The 1965, 1966, and 1967 total acreage in various countries was as follows (thousands of acres):

	1965	1966	1967
Spain	4,032 (91.6)[a]	3,969 (91.5)[a]	3,912 (94.6)[a]
Italy	4,001 (98.0)	3,945 (97.9)	3,903 (98.0)
France	3,328 (93.6)	3,325 (93.6)	3,283 (94.2)
USSR	2,502 (73.3)	2,538 (73.7)	2,544
Turkey	1,880	1,992	1,992
Algeria	864	864	766 (98.1)
Portugal	828 (98.9)	831 (99.8)	835 (98.8)
Rumania (1964)	726	726	801
Jugoslavia	686 (91.3)	668 (93.1)	617
Argentina	650 (99.8)	675 (98.8)	683 (98.8)
United States	579[b]	616 (est.)	616 (est.)
California	486 (94.6)	490 (94.2)	487 (94.2)
Hungary	592 (81.7)	586 (81.8)	575 (76.5)
Greece	551 (94.9)	552 (94.2)	552 (94.2)
Bulgaria	458 (82.0)	466 (87.4)	481 (85.7)
Chile	272	272	262
Germany	199 (82.8)	201 (82.7)	201 (82.9)
South Africa	190	190	190
Iran (est.)	180	180	180
Morocco	180 (92.0)	180 (92.0)	180
Syria	168	168	168
Brazil	166	166	166
Australia	135 (90.4)	136 (90.8)	135 (91.7)
Tunisia	122 (98.0)	122	122
Austria	109 (78.3)	109 (78.3)	110 (87.7)
Cyprus (1962)	92	92	92
Czechoslovakia	66 (75.4)	69 (74.2)	70 (69.1)
Lebanon	36 (95.3)	37 (94.5)	43 (86.0)
Japan	54 (78.2)	55 (81.5)	55 (est.)
Uruguay	45	45	45
Jordan	44	44	44
Mexico	32	35	35
Albania (1964)	29	29	29
Switzerland	28	29	29
Israel	24 (87.3)	23 (86.6)	23 (88.4)
Egypt	24	24	24
Peru	20	20	20

[a] Percentage in production.

[b] Approximate 1964 acreage. United States average data are difficult to obtain because information is only partially available, and for many states only on a number-of-vine basis.

Source of data: Protin (1966, 1967, 1968).

Production

Wine production has been increasing regularly since World War II, from about 5.6 billion gallons annually for the period 1953–1957 to 6.8 for the period 1961–1965. Production was 7.4 billion gallons in 1964, 9.3 billion in 1965, 7.2 billion in 1966, and 7.4 billion in 1967. This increase in production is due less to expanding world grape acreage than to greater production per acre. The 1965, 1966, and 1967 production was as follows (in millions of gallons):

	1965	1966	1967
Italy	1,842 (25.1)[a]	1,720 (23.9)	1,981 (26.6)
France	1,757 (23.9)	1,609 (22.4)	1,717 (23.1)
Spain	727 (9.9)	832 (11.6)	598 (8.0)
Argentina	482 (6.6)	579 (8.1)	735 (9.9)
USSR	510 (6.8)	483 (6.7)	450 (6.0)
Portugal	394 (5.4)	240 (3.3)	267 (3.5)
Algeria	317 (4.3)	237 (3.3)	166 (2.2)
United States	219 (3.0)	202 (2.8)	182 (2.5)
California	188 (2.6)	158 (2.2)	168 (2.3)
Rumania	172 (2.3)	172 (2.4)	172 (2.3)[h]
Jugoslavia	136 (1.9)	150 (2.1)	140 (1.9)
Germany	133 (1.8)	127 (1.8)	160 (2.2)
South Africa	118 (1.6)	110 (1.5)	114 (1.5)
Greece	102 (1.4)	102 (1.4)	102 (1.0)[i]
Chile	96 (1.3)	125 (1.7)	129 (1.7)
Bulgaria	91 (1.2)	94 (1.3)	73 (1.0)[b]
Morocco	91 (1.2)	57 (0.8)	29 (0.4)[b]
Hungary	65 (0.9)[b]	88 (1.2)	126 (1.7)
Tunisia	44 (0.6)	33 (0.5)	24 (0.3)[b]
Austria	37 (0.5)[b]	38 (0.5)[b]	68 (0.9)
Brazil	33 (0.5)[c]	33 (0.5)[c]	33 (0.4)[c]
Switzerland	24 (0.3)	21 (0.3)	22 (0.3)
Australia	22 (0.3)	24 (0.3)	24 (0.3)[i]
Uruguay (1962)	22 (0.3)[d]	22 (0.3)[d]	22 (0.3)
Japan[d]	12 (0.2)[e]	12 (0.2)[e]	12 (0.2)[e]
Canada	12 (0.2)[f]	12 (0.2)[f]	12 (0.2)[f]
Turkey	12 (0.2)	11 (0.2)	11 (0.2)
Cyprus	11 (0.1)	11 (0.2)	11 (0.2)[h]
Israel	10 (0.1)	10 (0.1)	10 (0.1)
Czechoslovakia	7 (0.1)	9 (0.1)	19 (0.3)
Peru	7 (0.1)	7 (0.1)	7 (0.1)
Mexico (1964)	3 (0.04)	3 (0.04)	3 (0.04)
Luxembourg	3 (0.04)	3 (0.05)	3 (0.04)

[a] Figures in parentheses are percentage of total world production.
[b] Year of very small production.
[c] 1964 data.
[d] 1962 data.
[e] Only about one-third is strictly "wine"; the rest represents blends of wine, alcohol, sugar, etc. The 1966 figure is from 1965.
[f] 1963 data.
[g] 1965 data.
[h] Estimate.
[i] 1966 data.
Source of data: Protin (1966, 1967, 1968) except for United States and California, where data of Wine Institute were used.

Of the ten highest producing countries seven are in Europe. Of total world production about 79.1 per cent is produced in Europe, 8.8 in South Africa and 7.7 per cent in Africa, 3.0 in North America, 0.6 in Asia, and 0.6 in Oceania. The United States, with only about 2.8 per cent of the world's grape acreage, produces 6 per cent of the grapes.

Export and import

In 1965, exports of wine from one country to another amounted to 680 million gallons, and in 1966 to 741 million gallons. The major importing countries were France, Germany, Switzerland, Portuguese territories, Great Britain, Belgium and Luxembourg, Russia, East Germany, French franc countries, and the United States. The main exporting countries are Algeria, France, Italy, Portugal, Spain, Morocco, Bulgaria, Hungary, Tunisia, Greece, Jugoslavia, and Rumania. The 1965 and 1966 export and import data by countries are summarized in tables 12a and 12b.

TABLE 12a

IMPORTATION OF WINES BY COUNTRIES, 1965 AND 1966

(000 gallons)

	1965		1966	
Country	Gallons	Per cent of total	Gallons	Per cent of total
France	248,128	37.1	264,172	38.7
Germany	112,725	16.9	141,768	20.8
Switzerland	41,368	6.2	41,403	6.1
Portuguese Territories	36,392	5.4	42,669	6.3
Great Britain	32,435	4.9	32,770	4.8
Belgium & Luxembourg	26,610	4.0	25,577	3.7
USSR (1963)	20,803	3.1	b	
East Germany (1963)	18,636	2.8	b	
French franc countries*	17,424	2.6	18,484	2.7
United States	16,275	2.4	b	
Czechoslovakia	11,703	1.7	4,933	0.7
Holland	10,843	1.6	12,054	1.57
Sweden	8,894	1.3	8,955	1.3
Austria	6,777	1.0	12,600	1.8
Poland (1963)	4,422	0.7	b	
Canada	4,179	0.6	5,059	0.7
Denmark	3,640	0.5	b	
Italy	1,901	0.3	2,370	0.3
Finland	1,662	0.2	1,759	0.3
Norway	1,438	0.2	1,636	0.2
Ireland	1,220	0.2	1,034	0.2
Venezuela	575	0.1	576	0.1
New Zealand	282	0.04	282	0.04

* Includes Algeria, Madagascar, Morocco, and French colonies, or former French colonies, in Africa and the Pacific.
b No data.
Source of data: Protin (1966, 1967).

TABLE 12b

(000 gallons)

Country	1965		1966	
	Gallons	Per cent of total	Gallons	Per cent of total
Algeria	219,120	32.7	224,400	30.3
France	101,636	14.9	101,680	13.7
Italy	69,383	10.4	67,814	9.1
Portugal	64,824	9.7	73,174	9.9
Spain	59,454	8.9	66,123	8.9
Morocco	38,785	5.8	53,845	7.3
Bulgaria	29,251	4.4	37,286	5.0
Hungary	18,718	2.8	16,192	2.3
Tunisia	17,160	2.6	41,693	5.6
Greece	12,538	1.9	12,215	1.6
Jugoslavia	11,690	1.7	10,571	1.4
Rumania	11,193	1.7	11,193	1.5
Germany	5,347	0.8	5,006	0.7
South Africa (1964)	5,066	0.8	4,817	0.6
Cyprus (1963)	4,134	0.6	a	
Australia (1963)	2,398	0.4	a	
Chile	1,646	0.2	1,309	0.2
Luxembourg	1,554	0.2	1,542	0.2
Turkey	1,279	0.2	a	
Egypt (1963)	478	0.07	487	0.7
Israel	426	0.06	421	0.6
Austria	385	0.06	a	
Belgium	1,521	0.2	2,115	0.3
United States	286	0.04	286	0.04

a No data.
Source of data: Protin (1966, 1967).

Stocks of wine

Most of the wine of the world is drunk within the year it is produced. Protin (1966, 1967) reported that world consumption in 1965 was 6.0846 billion gallons and in 1966 it was 6.2023. Total stocks were only 3.1273 billion gallons in 1965 and 3.2694 in 1966. More than half of the stocks were held in France, Italy, and the United States. Other countries with appreciable stocks are Germany, Argentina, Spain, Algeria, USSR, Portugal, South Africa, Tunisia, Morocco, and Australia. Stocks as percentage of total annual consumption are given in table 13. These figures are somewhat deceiving, since they are based only on total consumption. However, they do reveal the relatively minor stocks in several countries. Since these countries are not large importers this means that a good deal of the wine is consumed before it is one year old. The relatively large stocks of the United States are also noteworthy.

TABLE 13

STOCKS OF WINE AND PERCENTAGE OF TOTAL CONSUMPTION BY COUNTRIES
(000 gallons)

	1965		1966	
	Stocks	Stocks as related to total consumption	Stocks	Stocks as related to total consumption
Country	Gallons	Per cent	Gallons	Per cent
France	985,195	63.8	1,042,562	67.3
Italy	514,800	34.8	607,200	37.8
United States	262,114	138.2	262,114	137.2
Germany	186,252	81.7	181,764	75.1
Argentina	180,809	35.7	162,990	33.7
Spain (1964)	116,213	21.8	116,213	24.8
Algeria (1964)	111,139	1,169.4	a	
USSR	110,880	32.3	110,880	32.3
Portugal	93,318	36.0	121,677	51.1
South Africa	59,727	150.7	56,638	139.6
Tunisia	39,946	665.2	48,405	523.9
Morocco	54,342	918.9	54,342	1,629.8
Australia	53,687	325.0	a	
Switzerland	43,840	73.9	44,919	73.3
Brazil (1963)	43,349	105.2 (of 1962)	a	
Austria	42,240	74.2	44,880	77.3
Chile	39,775	35.9 (of 1963)	a	
Hungary	32,216	75.1	25,784	46.9
Bulgaria	27,772	78.2 (of 1964)	28,248	64.6
Greece	18,480	21.5	21,120	24.3
Czechoslovakia	17,517	81.7	15,319	71.9
Belgium	11,231	48.6	12,099	51.4
Great Britain	10,131	30.5 (of 1964)	a	

ᵃ No data.
Source of data: Protin (1967).

Consumption

The per capita consumption of wine has declined in many countries since 1900. The main wine-drinking countries are still France, Italy, and Portugal. The 1966 per capita consumption was as follows (in liters): France (120), Italy (115), Portugal (98), Argentina (80.2), Spain (60), Chile (54, in 1964), Switzerland (39.1), Greece (38.7), Luxembourg (33), Austria (30), Rumania (29, in 1964), Uruguay (26, in 1963), Jugoslavia (25, in 1964), Bulgaria (20), Hungary (20; was 31 in 1964), Germany (15.4), Belgium (9.4), South Africa (8.2), USSR (5.8), Czechoslovakia (5.7), Australia (5.5, in 1964), Israel (4.5), Algeria (4, est.), East Germany (3.8, in 1963), United States (3.7), and Holland (3.6).

TRENDS IN THE CALIFORNIA WINE INDUSTRY[1]

The trend in national per capita consumption of wine has been markedly upward for over fifty years, even during Prohibition. California now produces over three-fourths of the national total and crushes over half of its big grape crop.

The gross wine production and percentage of the total for the ten most important states in 1966 were as follows: California, 157.968 million gallons, 81.5 per cent; New York, 17.480 million, 9.0 per cent; Illinois, 4.396 million, 2.3 per cent; New Jersey, 4.236 million, 2.2 per cent; Michigan, 2.177 million, 1.1 per cent; Washington, 1.925 million, 1.0 per cent; Virginia, 1.842 million, 0.9 per cent; Georgia, 1.237 million, 0.6 per cent; Arkansas, 0.877 million, 0.5 per cent; Ohio, 0.858 million, 0.4 per cent. All other states produced only 0.909 million, or less than 0.5 per cent of the total. (Data from Allmendinger, 1968.) In 1965 California production was 85.9 per cent of the United States total and in 1964 83.9 per cent.

PRE-PROHIBITION PERIOD

Table wines were most important up to 1918.

The commercial wine industry of California began in Los Angeles in the middle of the nineteenth century and grew until the passage of the Eighteenth Amendment in 1918 restricted trade in wine (Leggett, 1939). According to Bioletti (1915), California wine production amounted to about 150,000 gallons in 1857. By the early 'eighties it had reached an average of about 10 million gallons, and from 1890 to 1892 it averaged about 17.4 million gallons a year. Of this, 15.2 million gallons were table wines (Shear and Pearce, 1934). The industry grew rapidly during the next two decades; production for 1909–1913 reached a pre-Prohibition peak of about 43.6 million gallons a year, or about 85 per cent of national

---

[1] This section was written and tables 14 to 21 and 24 to 29 were compiled by S. W. Shear, Associate Agricultural Economist Emeritus in the Experiment Station and on the Giannini Foundation of Agricultural Economics, University of California, Berkeley. For more comprehensive discussions of economic trends and outlook and basic historical statistical data for the California grape industry, see Mehren and Shear (1950); 80 tables compiled by Shear, included as a statistical compendium in Mehren (1950); Farrell and Blaich (1964); Farrell (1966); Allmendinger (1965, 1967); McColly (1967), and Anon (1966*f*). Current statistical data may be found in the Wine Institute Bulletin (1965, 1968*b*) in their Annual Wine Industry Statistical Survey (San Francisco, annual issues); Wines & Vines (San Francisco, annual statistical issues); and Federal-State Market News Service, Marketing California Grapes, Raisins, Wine (San Francisco, annual issues).

production. Since dessert-wine production rose more rapidly than that of table wines, the average table-wine production of 24.4 million gallons in the period 1909–1913 constituted only 56 per cent of the state's total production.

During 1909–1913, consumption of commercial wine in the United States averaged about 50 million gallons, or 0.52 gallon per capita per year—about 0.32 of table wine and 0.20 of dessert. Foreign wine then constituted about 7.4 million gallons, about 15 per cent of the total. Most of the imported wine was dry table wine, and more than half of the total national consumption consisted of dry table wine.

Eastern buyers of juice grapes would pay little or no premium for some of the varieties that were used before Prohibition to make the better-quality commercial wine in California. Many pre-Prohibition vineyards of soft-textured, thin-skinned, high-quality commercial wine grapes, particularly white varieties, were grafted or replanted to varieties that, although they made poorer wine, were heavier yielders and could be shipped to eastern markets with minimum wastage. For the same reason, most new plantings were in irrigated vineyards in the San Joaquin Valley where the juice-grape varieties then in demand yielded heavily at low costs of production.

HOMEMADE AND OTHER NONCOMMERCIAL WINE...

... very popular during Prohibition, has decreased greatly
since Repeal, and now constitutes a very small
part of table-wine consumption.

Prohibition reduced the consumption of commercial wines to small quantities for sacramental and medicinal purposes only. Foreign imports practically ceased, and commercial production declined during Prohibition, averaging considerably less than 5 million gallons a year during the period 1918–1932.

Before 1918 almost no homemade wine had been produced and consumed in the United States. Before Prohibition there was less incentive to produce wine in the home because ample supplies of cheap imported table wines were available. Prohibition resulted in large shipments of California "juice" grapes to eastern industrial centers where they were used for making lawful homemade and some illicit wine. Prohibition so stimulated homemade wine consumption that per capita wine consumption in the United States is estimated to have risen much above the peak of pre-Prohibition years, averaging about 0.8 gallon during 1925–1929 as compared with 0.52 gallon of commercial wine during 1909–1913.[2]

Homemade wine is mostly a natural unfortified table wine of low alcoholic content and hence competes with commercial table wines containing less than 14 per cent alcohol. An average of over 90 million gallons a year of noncommercial wine is estimated to have been made from California grapes and consumed in the United States from 1927 to 1929, the peak years of wine consumption during Prohibition. Since Repeal in 1933, increasingly large supplies of reasonably priced tax-paid commercial wine have been the chief cause of the big decrease in homemade wine production and consumption. After Repeal, consumption of noncommercial wine made from California grapes averaged about 33 million gallons, or 60 per cent of a 55-million-gallon total of all still table wine during the period 1935–1939. Since then it has declined steadily, averaging only about 11 million gallons a year for 1962–1964, or less than 14 per cent of total consumption of all table wine of about 80 million gallons a year. (No estimates of the very small volume of noncommercial wine made from eastern grapes are available.) Per capita consumption of homemade wine in the United States has decreased steadily since Repeal, dropping from about 0.25 gallon, or 32 per cent of the 1935–1939 average, to only 0.06 gallon, or 6 per cent of the 1962–1964 average.

One factor tending to restrict commercial wine consumption in the United States and to encourage the continuation of some home wine making has been the prohibition of retail bulk sales in every state but Louisiana. In every other state the consumer can buy only at retail in the original sealed container; usually the container must be glass, not over 1 gallon in size in many states and not over 4.9 gallons in most of the others. High taxes and license fees and restrictive and discriminatory regulations still curtail consumption of California commercial wines in some states, although conditions have been much improved in a number of states since 1950.

PRODUCTION AND CONSUMPTION SINCE REPEAL...

... have increased dramatically.

Overall production and per capita consumption of commercial wine in the United States have increased steadily since Repeal, aside from the market

---

[2] Shear has made very approximate preliminary estimates of the tonnage of grapes used in the United States for the production of legal and illegal wine and brandy during the Prohibition period, 1919–1932, and the approximate gallonage produced and consumed, published in a monograph by the U.S. Tariff Commission (1939, table 180).

decline since 1956 in regular dessert wine. Imports and consumption of foreign wines have increased even faster than domestically produced commercial wine. Much of the decrease in consumption of regular dessert wine has been offset by steady increases in vermouth and other aperitif "special natural wines" over 14 per cent in alcoholic content. Commercial table wines, both still and sparkling, have increased fairly steadily since Repeal and more rapidly since 1958 (tables 14 and 15).

Table 14 and figure 19 show that the national consumption of sparkling table wine, although relatively small in total, has increased at a much more

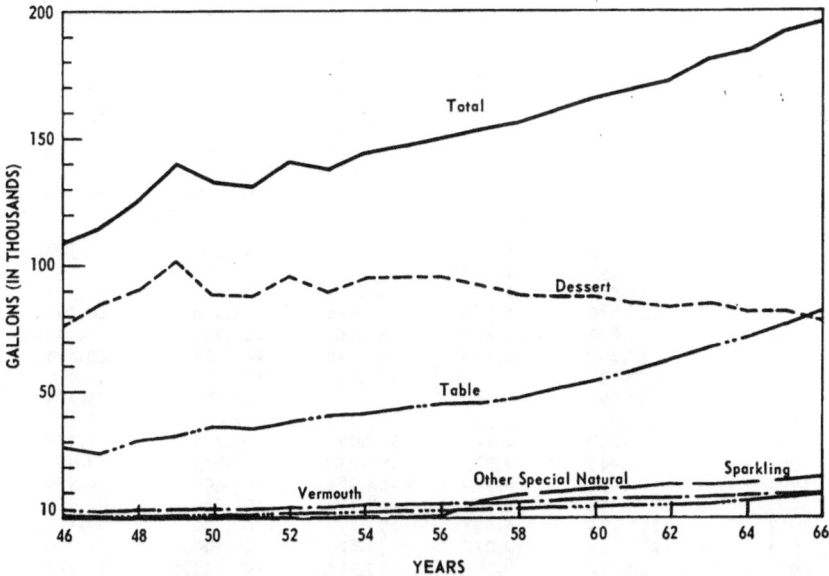

Figure 19. Commercially produced wine entering distribution channels (includes imports).

rapid rate than have still table and dessert wines. Sparkling wine consumption jumped from less than 1 million gallons in the period 1935–1939 to about 10 million gallons in 1966, or about 5 per cent of all wine. Total consumption of commercial still wines has risen 170 per cent since Repeal, increasing from an average of about 68 million gallons for 1935–1939 to over 186 million gallons in 1966 (or a 0.95 gallon per capita). Over half of this increase in still wine of 118 million gallons was in table wine 14 per cent or less in alcoholic content, which rose from 0.17 gallon per capita, or about 22 million gallons, to 0.42 gallon per capita, or 82 million gallons in 1966, an increase of 270 per cent.

TABLE 14

APPARENT CONSUMPTION OF COMMERCIAL STILL AND SPARKLING WINES IN THE
UNITED STATES: TAX-PAID WITHDRAWALS OF UNITED STATES AND IMPORTED COM-
MERCIAL STILL AND SPARKLING WINES

			Still wine, commercial, tax-paid withdrawals		
	Total, still and sparkling[a]	Sparkling wine	Total[b]	Dessert, over 14 per cent alcohol[c]	Table, not over 14 per cent alcohol
Years beginning July 1	1	2	3	4	5
			1,000 wine gallons		
Averages					
1909–1913	d	d	49,445	19,196	30,247
1935–1939	68,392	867	67,525	45,399	22,126
1940–1944	101,204	1,185	100,019	67,196	32,823
1945–1949	121,314	1,674	119,640	89,584	30,056
1950–1954	137,651	1,935	135,716	96,986	38,730
1955–1959	153,771	3,260	150,511	103,861	46,650
1960–1964	175,004	5,365	169,639	106,806	62,833
Annual					
1950	132,655	1,852	130,803	94,303	36,500
1951	131,355	1,723	129,632	93,511	36,121
1952	141,516	1,907	139,609	100,836	38,773
1953	138,086	1,964	136,122	95,526	40,596
1954	144,642	2,228	142,414	100,752	41,662
1955	147,400	2,562	144,838	101,165	43,673
1956	150,138	2,882	147,256	102,270	44,986
1957	153,994	3,089	150,905	105,182	45,723
1958	156,029	3,546	152,483	104,987	47,496
1959	161,296	4,223	157,073	105,702	51,371
1960	166,245	4,315	161,930	107,843	54,087
1961	169,674	4,710	164,964	106,408	58,556
1962	173,351	4,951	168,400	106,086	62,314
1963	181,104	5,831	175,273	108,136	67,137
1964	184,647	7,018	177,629	106,560	71,069
1965	192,540	8,185	184,355	107,533	76,822
1966	196,257	9,797	186,460	104,253	82,207

ᵃ Column 2 plus column 3; totals differ slightly because of rounding.
ᵇ Column 4 plus column 5; totals differ slightly because of rounding.
ᶜ Dessert includes vermouth, domestic and imported, and also sake imports.
ᵈ The very small consumption of sparkling wine is included in still wine totals 1909–1913.
Sources: Compiled by S. W. Shear, Giannini Foundation of Agricultural Economics, largely from official reports. Cols. 3–5: Tax-paid withdrawals of wine produced in the United States from U.S. Bureau of Internal Revenue annual statistical reports, plus imports for consumption, tax paid, and, before 1953, very small imports, tax free, from U.S. Department of Commerce reports.

United States consumption of all wine over 14 per cent in alcoholic content increased gradually after Repeal until it reached 0.64 gallon per capita, or a total of about 100 million gallons in 1952 (tables 14 and 15). Since then, however, the rate of increase has slowed down. Consumption

TABLE 15

APPARENT PER CAPITA CONSUMPTION OF COMMERCIAL STILL AND SPARKLING WINES
IN THE UNITED STATES: TAX-PAID WITHDRAWALS OF UNITED STATES AND IMPORTED
COMMERCIAL STILL AND SPARKLING WINES

			Still wine, commercial, tax-paid withdrawals		
	Total, still and sparkling[a]	Sparkling wine	Total[b]	Dessert, over 14 per cent alcohol[c]	Table, not over 14 per cent alcohol
Years beginning July 1	1	2	3	4	5
			Wine gallons per capita		
Averages					
1909–1913	[d]	[d]	0.52	0.20	0.32
1935–1939	0.53	0.01	0.52	0.35	0.17
1940–1944	0.76	0.01	0.75	0.50	0.25
1945–1949	0.84	0.01	0.83	0.62	0.21
1950–1954	0.87	0.01	0.86	0.62	0.24
1955–1959	0.89	0.02	0.87	0.60	0.27
1960–1964	0.93	0.03	0.90	0.57	0.34
Annual					
1950	0.87	0.01	0.86	0.62	0.24
1951	0.84	0.01	0.83	0.60	0.23
1952	0.89	0.01	0.88	0.64	0.25
1953	0.86	0.01	0.85	0.60	0.25
1954	0.89	0.01	0.87	0.62	0.25
1955	0.89	0.02	0.87	0.61	0.26
1956	0.90	0.02	0.88	0.61	0.27
1957	0.89	0.02	0.87	0.61	0.26
1958	0.89	0.02	0.87	0.60	0.27
1959	0.90	0.02	0.88	0.59	0.29
1960	0.91	0.02	0.89	0.59	0.30
1961	0.92	0.03	0.89	0.58	0.32
1962	0.93	0.03	0.90	0.57	0.33
1963	0.95	0.03	0.92	0.57	0.35
1964	0.96	0.04	0.92	0.55	0.37
1965	0.99	0.04	0.95	0.55	0.39
1966	1.00	0.05	0.95	0.53	0.42

[a] Column 2 plus column 3; totals differ slightly because of rounding.
[b] Column 4 plus column 5; totals differ slightly because of rounding.
[c] Dessert includes domestic and imported vermouth and also sake imports.
[d] The very small consumption of sparkling wine is included in still wine totals, 1909–1913.
Source: Computed by S. W. Shear, Giannini Foundation of Agricultural Economics, from data in preceding table by dividing by December population of continental United States (excluding armed forces overseas) as reported by U.S. Bureau of Census.

in 1965 was only 0.53 gallon per capita, totaling about 104 million gallons, 135 per cent greater than during 1935–1939. Until 1956 nearly all the increase was in traditional dessert wines, consumption of which reached a peak of about 95.6 million gallons in 1956 (table 16 and fig. 19). Thereafter, however, it dropped to 78.5 million gallons by 1966, a decrease of 17.1 million gallons. This decrease was more than offset by a marked increase of 19 million gallons in the consumption of vermouth and other "special natural wines," mostly the latter.[3] In this decade, vermouth consumption rose from about 5.5 million gallons to 9.5 million, an increase of 4 million gallons. Meantime, consumption of other "special natural" flavored wines rose from less than 1 million gallons in 1956 to 16 million in 1966, an increase of 15 million gallons. Table 16 shows that the rapid rate of growth in consumption of the popular other "special natural" wines over 14 per cent in alcohol content slowed down markedly after 1964, increasing less than 500,000 gallons from 1965 to 1966.

Before World War II, California wineries probably bottled somewhat less than half of the wine they produced themselves and sold the balance to wholesale bottlers. After 1941, wartime conditions of supply, demand, and price in the United States were so favorable to bottled wines that wineries themselves were bottling most of their output by the end of the war, either in California or at their own eastern bottling plants.

IMPORTED WINE

*Imports since Repeal have been increasing
and are nearly all bottled rather than bulk.*

During the period 1909–1913, before Prohibition, an average of about 7.5 million gallons of wine a year was imported, constituting nearly 15 per cent of United States total wine consumption. After Repeal, before World War II, nearly 4 million gallons a year were imported (table 17 and fig. 20), constituting about 4 per cent of national consumption of all wines—sparkling and still, commercial and homemade—and between 5 and 6 per cent of the commercially produced wine. During the war, total imports first fell; then, in 1943, under the stimulus of high prices, they rose both absolutely and relative to total United States wine consumption. This is illustrated in figure 20.

---

[3] Internal Revenue Service statistics on "special natural wines" over 14 per cent in alcohol content includes vermouth and, beginning in 1948, a separate category designated as "other special natural wines" flavored with herbs and fruit flavors. A very large majority of "other special natural wines" are made in California, where a minimum alcohol content of 20 per cent is required by the state. Federal regulations require a minimum alcohol content of only 14 per cent, and some states require as little as 15 per cent. In 1965 California produced about 95 per cent of the United States total consumption of "other special natural wines" and 34 per cent of vermouth consumption, of which 47 per cent of the total was imported.

TABLE 16
COMMERCIALLY PRODUCED WINE ENTERING DISTRIBUTION CHANNELS IN THE UNITED STATES, BY TYPE OF WINE, CROP YEARS 1946–1966[a]

Crop year beginning July 1	Table Total (1,000 gallons)	Per cent of total	Dessert Total (1,000 gallons)	Per cent of total	Vermouth Total (1,000 gallons)	Per cent of total	Other special natural Total (1,000 gallons)	Per cent of total	Sparkling Total (1,000 gallons)	Per cent of total	Total[b] (1,000 gallons)	Per cent of total
	1	2	3	4	5	6	7	8	9	10	11	12
1946	28,746	26.2	76,023	69.2	3,117	2.8	—	—	1,951	1.8	109,837	100.0
1947	25,939	22.5	85,424	74.2	2,452	2.1	—	—	1,255	1.1	115,070	100.0
1948	30,132	24.0	90,631	72.1	3,163	2.5	204	0.2	1,494	1.2	125,624	100.0
1949	33,661	24.0	101,344	72.3	3,458	2.5	220	0.1	1,498	1.1	140,181	100.0
1950	36,536	27.5	89,773	67.7	4,257	3.2	235	0.2	1,853	1.4	132,654	100.0
1951	36,182	27.6	88,721	67.7	4,269	3.2	218	0.2	1,723	1.3	131,093	100.0
1952	38,816	27.4	95,909	67.8	4,649	3.3	200	0.1	1,908	1.3	141,482	100.0
1953	40,643	29.4	90,325	65.4	4,936	3.6	218	0.2	1,963	1.4	138,085	100.0
1954	41,716	28.8	95,090	65.7	5,285	3.7	323	0.2	2,228	1.5	144,642	100.0
1955	43,728	29.7	95,228	64.6	5,503	3.7	379	0.3	2,562	1.7	147,400	100.0
1956	45,054	30.0	95,644	63.7	5,626	3.7	932	0.6	2,882	1.9	150,138	100.0
1957	45,797	29.7	92,913	60.3	5,933	3.9	6,266	4.1	3,089	2.0	153,998	100.0
1958	47,598	30.5	88,928	57.0	6,546	4.2	9,411	6.0	3,546	2.3	156,029	100.0
1959	51,479	31.9	87,965	54.5	6,809	4.2	10,820	6.7	4,224	2.6	161,297	100.0
1960	54,203	32.6	87,982	52.9	7,387	4.4	12,357	7.4	4,315	2.6	166,244	100.0
1961	58,675	34.6	85,664	50.5	7,896	4.7	12,729	7.5	4,709	2.8	169,673	100.0
1962	62,443	36.0	84,183	48.6	7,805	4.5	13,969	8.1	4,951	2.9	173,351	100.0
1963	67,276	37.1	85,705	47.3	8,597	4.7	13,694	7.6	5,831	3.2	181,103	100.0
1964	71,227	38.6	82,779	44.8	8,947	4.8	14,676	7.9	7,019	3.8	184,648	100.0
1965	76,970	40.0	82,605	42.9	9,200	4.8	15,579	8.1	8,186	4.3	192,540	100.0
1966[c]	82,380	42.0	78,534	40.0	9,538	4.9	16,007	8.2	9,798	5.0	196,257	100.0

[a] Includes tax-paid withdrawals of U.S. produced wine and imports for consumption of foreign wine.
[b] Sum of components is not equal to total in all cases as a result of rounding individual figures.
[c] Preliminary.
Sources: Prepared by Wine Institute from reports of Internal Revenue Service, U.S. Treasury Department, and Bureau of Census, U.S. Department of Commerce. See latest data in Wine Institute, *Annual Wine Industry, Statistical Reports*, Part IV.

When wartime restrictions on crushing raisin grapes were removed in 1945, California wine production increased greatly, prices fell, and wine imports dropped much below the wartime peak. After the war, through 1949, annual import totals averaged about 3.5 million gallons a year, or about 3 per cent of national consumption. After 1949 there was a marked improvement in conditions favoring the United States as a market for

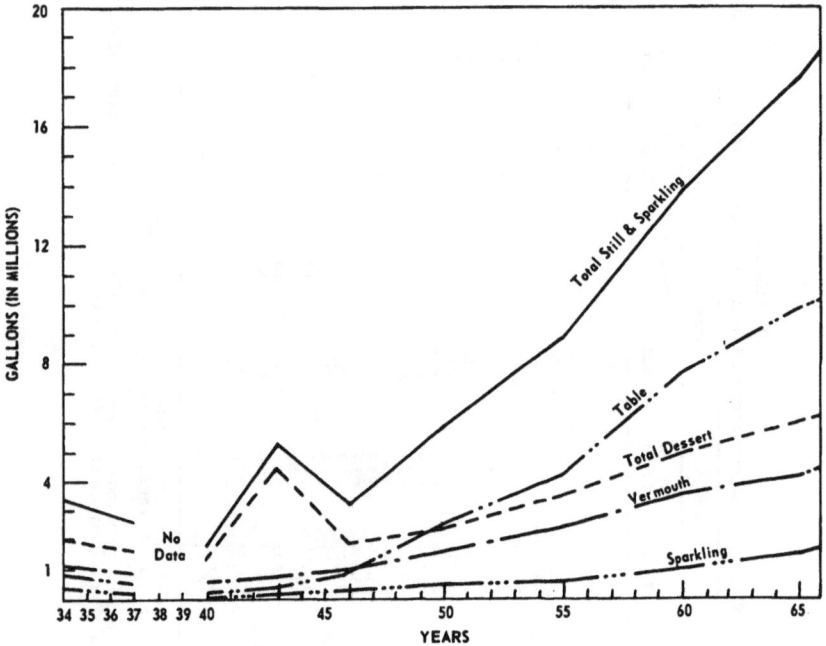

Figure 20. Importation of wines of various types into the United States.

foreign wine. Imports rose to an average of 5.8 million gallons for 1950–1954, rapidly increasing to 8.8 million gallons a year for 1955–1959, 13.8 million gallons for 1960–1964, to a peak of over 18 million gallons in 1966.

The proportion of United States consumption of all wine accounted for by imports from foreign countries rose steadily from 4 per cent in 1950 to over 9 per cent in 1964 (table 18). However, the proportion of consumption of the different categories of wine contributed by imports differs greatly and is more significant from the competitive point of view. Sparkling wine imports, although small in total, have increased steadily but not as rapidly as domestic production; the percentage of consumption imported fell from over 30 per cent for 1950–1954 to less than 20 per cent in 1966. Imports of vermouth, small but increasing steadily, accounted for about 47 per cent of national consumption of this kind of wine in 1966. Of the large total of national consumption, imports of traditional dessert wines constituted only about 2 per cent of the dessert total, while imports of table wine rose to almost 14 per cent. However, import percentages do not reveal the fact that many imports, particularly of table wines, are relatively high-priced bottled wines that compete directly with the small but significant volume of fine premium wines produced in California.

In peacetime, Italy and France have held either first or second place as the source of United States wine imports in total and also of vermouth and sparkling and still table wines. France has long supplied a big majority of our sparkling wine imports. As sources of still table wine imports, France and Italy have vied for first place, accounting for about 70 per cent of total imports of such wine in recent years, with Germany about 15 per cent, and smaller quantities from Chile and Israel. Spain and Portugal, as sources of a very large majority of dessert-wine imports, have long held third and fourth place in total wine imports and, during World War II, first and second place, when wine shipments from France and Italy nearly ceased.

Plentiful supplies and low prices of wines in the United States tend to discourage imports from abroad, and high prices and consumer purchasing power to attract them. The war-curtailed supply of imported wines available in the United States in 1940–1942 gave California wines the opportunity of becoming more widely known in this country. A number of domestic wineries producing table and dessert wines of high quality were able to get consumer acceptance and approval of their products more rapidly because many of the well-known European brands or types were not available. Some of these wineries appear to have maintained and even expanded this market, and a number of new ones producing quality wines have been established.

Analyses by Farrell and Blaich (1964) indicate that available supplies of western European wine for export to the United States and other non-European countries will likely continue to be high during the next few years. Further substantial increases in wine imports into the United States might result in a decrease in California prices if wine production in the United States approximates the total growth in demand for wine in this country.

TABLE 17

GRAPES: UNITED STATES WINE IMPORTS FOR CONSUMPTION BY KINDS

WINE GALLONS

Year beginning July 1	Still and sparkling, total	Sparkling, all	Still wines		Dessert, over 14 per cent and not over 24 per cent alcohol					
			Total	Table, 14 per cent or less	Total	Vermouth	Except vermouth			
							Total	Sake[a]	N.e.s.[b]	Other
	1	2	3	4	5	6	7	8	9	10
Averages										
1934–1938	3,408,380	445,187	2,963,193	894,344	2,068,849	1,125,365	943,484	139,887	2,603	800,995
1940–1942	1,728,104	151,748	1,576,356	268,417	1,307,939	563,981	743,958	27,881	2,153	713,924
1943–1945	5,361,462	165,012	5,196,450	426,802	4,769,648	714,207	4,055,441	41	1,733	4,053,667
1946–1949	3,265,678	374,274	2,891,404	967,564	1,923,840	1,038,340	885,500	10,902	27,339	847,259
1950–1954	5,833,727	616,721	5,217,006	2,637,231	2,579,775	1,712,410	867,365	44,300	111,645	711,420
1955–1959	8,820,962	799,138	8,021,824	4,380,781	3,641,043	2,575,609	1,065,434	81,286	225,968	758,179
1960–1964	13,836,303	1,066,811	12,769,491	7,711,002	5,058,490	3,775,050	1,283,440	132,439	395,479	755,521
Annual										
1950	5,303,609	663,178	4,640,431	2,182,010	2,458,421	1,584,234	874,187	36,264	49,641	788,282
1951	5,197,370	598,245	4,599,125	2,293,383	2,305,742	1,524,740	781,002	40,327	82,917	657,758
1952	5,613,738	568,053	5,065,685	2,596,624	2,469,061	1,634,197	834,864	43,411	102,579	688,874
1953	6,231,794	586,741	5,645,053	2,904,181	2,740,872	1,841,111	899,761	47,237	142,531	709,993
1954	6,802,123	667,387	6,134,736	3,209,958	2,924,778	1,977,768	947,010	54,258	180,559	711,193
1955	7,559,409	728,850	6,830,559	3,540,998	3,289,561	2,283,385	1,001,176	55,056	208,052	738,068
1956	8,095,642	759,071	7,336,571	4,064,419	3,272,152	2,201,545	1,070,607	67,270	227,323	776,014
1957	8,586,761	772,268	7,814,493	4,252,995	3,561,498	2,503,991	1,057,507	73,284	219,602	764,621
1958	9,416,057	808,157	8,607,900	4,645,601	3,962,299	2,873,267	1,089,032	102,502	239,578	746,952
1959	10,446,943	927,345	9,519,598	5,399,804	4,119,704	3,010,857	1,108,847	108,320	235,285	765,242
1960	11,166,790	913,163	10,253,627	5,806,864	4,446,763	3,241,417	1,205,346	115,727	287,692	801,927
1961	13,399,571	976,618	12,422,953	7,422,295	5,000,658	3,657,456	1,343,202	118,425	381,561	843,210
1962	13,607,704	1,027,953	12,579,751	7,422,977	5,156,774	3,662,899	1,493,875	129,622	379,581	984,672
1963	14,979,606	1,129,125	13,850,481	8,101,585	5,748,896	4,074,567	1,674,329	149,123	443,518	1,090,688
1964	16,027,843	1,287,198	14,740,645	8,801,288	5,939,357	4,238,911	1,700,446	158,299	485,045	1,057,102
1965	17,576,463	1,564,821	16,011,642	9,882,606	6,129,036	4,348,637	1,780,399	148,773	459,296	1,172,330
1966	18,414,820	1,812,246	16,602,574	10,229,182	6,373,392	4,502,962	1,870,430	172,486	510,131	1,187,813

[a] For years 1934–35, sake total assumed as total wine imports from Japan, as sake imports 1936 to date were all or nearly all imported from Japan (except from Hong Kong and China in 1936) and total imports from Japan were nearly all sake.

[b] 1934–1935 imports from only Hong Kong, China, Poland, and Danzig, the chief sources of this category of wine imports since 1935.

Sources: Assembled by Giannini Foundation of Agricultural Economics largely from compilations by Wine Institute, based on official monthly records of U.S. Bureau of Census, which publishes calendar-year and preliminary monthly import data by country of origin in FT 110. For most recent data see Wine Institute, *Annual Wine Industry Statistical Survey*, Part IV (San Francisco, annual issues). Cols. 4, 5, 7, 9, and 10: 1934–35 estimated by S. W. Shear.

TABLE 18

COMMERCIALLY PRODUCED STILL AND SPARKLING WINE ENTERING DISTRIBUTION CHANNELS IN THE UNITED STATES, BY ORIGIN, 1950–1966

Year beginning July 1	United States produced wine				United States		Foreign wine		Total, all wine
	California[a]		Other states[a]						
	Total	Per cent of total	Total	Per cent of total	Total	Per cent of total	Total	Per cent of total	Total
	1,000 wine gallons		1,000 wine gallons		1,000 wine gallons		1,000 wine gallons		1,000 wine gallons
	1	2	3	4	5	6	7	8	9
Averages									
1950–1954	114,457	83.2	17,300	12.6	131,757	95.8	5,834	4.2	137,591
1955–1959	123,767	80.5	21,184	13.8	144,951	94.3	8,821	5.7	153,772
1960–1964	134,817	77.1	26,351	15.0	161,168	92.1	13,836	7.9	175,006
Annual									
1950	109,977	82.9	17,374	13.1	127,351	96.0	5,304	4.0	132,655
1951	109,463	83.5	16,433	12.5	125,896	96.0	5,197	4.0	131,093
1952	119,116	84.2	16,731	11.8	135,847	96.0	5,634	4.0	141,481
1953	114,782	83.1	17,071	12.4	131,853	95.5	6,232	4.5	138,085
1954	118,950	82.2	18,890	13.1	137,840	95.3	6,802	4.7	144,642
1955	120,649	81.9	19,192	13.0	139,841	94.9	7,559	5.1	147,400
1956	121,083	80.6	20,959	14.0	142,042	94.6	8,096	5.4	150,138
1957	125,298	81.4	20,111	13.0	145,409	94.4	8,589	5.6	153,998
1958	124,930	80.1	21,683	13.9	146,613	94.4	9,416	6.0	156,029
1959	126,877	78.7	23,973	14.8	150,850	94.0	10,447	6.5	161,297
1960	131,740	79.3	23,337	14.0	155,077	93.3	11,167	6.7	166,244
1961	129,307	76.2	26,966	15.9	156,273	92.1	13,400	7.9	169,673
1962	133,803	77.2	25,941	15.0	159,744	92.2	13,606	7.8	173,360
1963	138,348	76.4	27,776	15.3	166,124	91.7	14,980	8.3	181,104
1964	140,887	76.3	27,733	15.0	168,620	91.3	16,028	8.7	184,648
1965	143,997	74.8	30,966	16.1	174,963	90.9	17,576	9.1	192,540
1966	147,173	75.0	30,699	15.6	177,842	90.6	18,415	9.4	196,257

[a] California wine excludes small quantity of California wine exported. Breakdown of U.S. exports, Col. 5, into California and other states is subject to small revision which will be given in revised series to be published by Wine Institute in 1968.

Source: Compiled by Giannini Foundation of Agricultural Economics from Wine Institute, *Annual Wine Industry Statistical Survey*, Part IV (San Francisco, annual issues).

WINE PRODUCTION BY DISTRICTS

*The San Joaquin Valley produces most of the wine
and brandy. Premium table wines are mainly a
product of the central coast counties.*

The most recent detailed statistics on California commercial crush of
grapes and gross wine production are those for 1965 (table 19). The 1966
crush by individual varieties is given in table 20, summarized by districts.
(See also fig. 21.) Table 21 is based on a detailed county survey by the
California Crop and Livestock Reporting Service. That survey summarized bearing acreage for about 150 varieties of wine grapes, by districts, for 1964. Table 21 gives acreage data for eleven individual white
varieties of a total of fifty-two whites reported by individual name and for
fifteen dark varieties of a total of ninety-three of these varieties. The data
on bearing acreage for various districts are not necessarily indicative of
the tonnage contributed by each.

The yield per acre for any given variety is much heavier in the interior
valley than in the central coast and southern California counties. Average
yields per bearing acre for the whole state in 1966 were 5.3 tons for wine
varieties, compared with 5.8 for table varieties and 8.6 for raisin varieties.
Table and raisin varieties are nearly all grown under irrigation in the
San Joaquin Valley, which also accounts for 53 per cent of the state's
acreage of wine varieties. Wine varieties grown in the San Joaquin Valley
are the heavier-producing varieties with 1964 yields averaging about 3.5
times the 2.2-ton average for southern California and the coast counties.

Comparison of columns 5 and 6 of table 21 shows that in 1964 about
43 per cent of the acreage of dark varieties of wine grapes and 62 per cent
of the white varieties were in the low-yielding table wine producing areas
of the central coast and southern California. In this area the acreage of
every white variety except Burger substantially exceeded that in the rest
of the state. Eight dark varieties in the interior valley greatly exceeded
the acreage in the table wine area, and only three varieties had a substantially smaller acreage.

In 1965 roughly one-third of the state's gross table wine production
was in southern California and the central coast counties (table 19).
The San Joaquin Valley in 1964 produced about 80 per cent of the
tonnage of true wine grape varieties grown in the state; in 1966, about
70 per cent, as well as very nearly all the raisin and table grape varieties
normally crushed for wine and brandy. Almost all the state's dessert
wine and brandy is made from the high-yielding grapes of this area and a
majority of the table wine. The concentration of grape acreage in the four

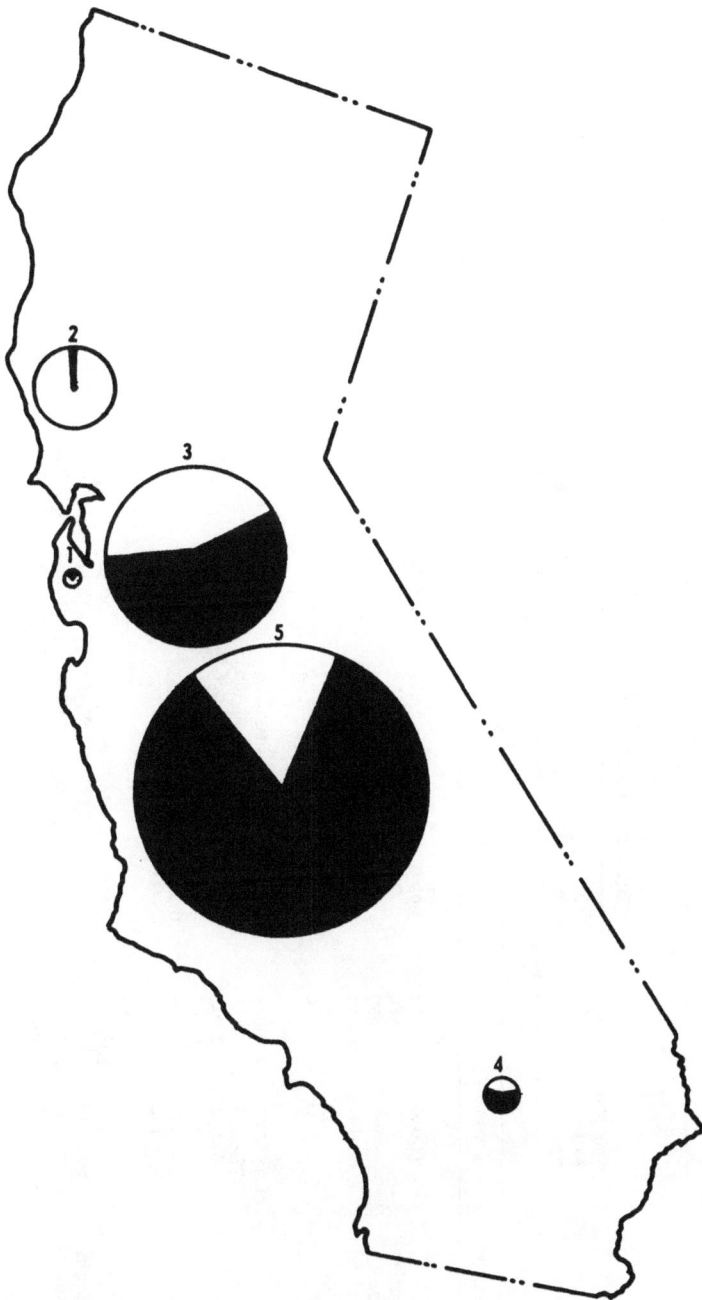

Figure 21. Relative production of table wines (light sector) and dessert wines (dark sector) in: 1, south coast counties; 2, north coast counties; 3, Sacramento and northern San Joaquin; 4, central and southern San Joaquin; 5, southern California.

TABLE 19

CALIFORNIA COMMERCIAL GRAPE CRUSH, STORAGE COOPERAGE, GROSS WINE PRODUCTION AND INVENTORIES, 1965

District and county[a]	Commercial grape crush[b]	Storage cooperage[c]	Gross wine production[d] (1,000 gallons)			Gross wine inventories[d] (1,000 gallons)			Active bonded[e]	
	Tons		Table wine[f]	Dessert[i]	Total[g]	Table wine[h]	Dessert[i]	Total	Wineries and cellars	Fruit distilleries[j]
	1	2	3	4	5	6	7	8	9	10
District 1										
total	74,090	20,472	3,690	4,307	7,997	5,047	4,429	9,476	38	8
Los Angeles	389	584	74	0	74	368	149	517	5	0
Riverside	3,671	677	115	0	115	24	76	322	5	1
San Bernardino	69,895	19,051	3,477	4,307	7,784	4,363	4,165	8,528	23	7
San Diego	135	160	23	0	23	68	38	106	2	0
Other	0	0	0	0	0	2	1	3	3	0
District 2										
total	1,351,344	200,949	23,593	81,452	105,045	25,424	88,494	113,918	35	28
Fresno	770,530	123,836	12,644	43,935	56,569	12,210	50,350	62,560	19	17
Kern	236,534	26,044	1,619	15,287	16,906	2,334	15,094	17,428	4	3
Madera	171,619	26,395	3,036	11,781	14,817	3,731	12,751	16,482	3	2
Tulare	167,552	23,203	6,210	10,043	16,253	6,947	9,871	16,818	4	5
Other	5,109	1,471	84	416	500	202	428	630	5	1
District 3										
total	482,603	108,463	30,736	24,257	55,003	36,183	33,389	69,572	36	15
San Joaquin	363,062	61,817	21,900	16,664	38,564	18,781	20,087	38,868	25	13
Stanislaus	118,906	42,711	8,551	7,603	16,154	16,650	12,708	29,358	4	2
Other	635	3,935	285	0	285	753	594	1,347	7	0
District 4										
total	104,659	40,636	20,673	169	20,842	27,437	1,327	28,764	61	4
Mendocino	4,499	3,592	849	0	849	1,099	2	1,101	2	0
Napa	39,625	16,018	8,060	106	8,166	11,285	784	12,069	30	1
Sonoma	60,027	20,776	11,670	63	11,733	14,857	540	15,397	27	3
Other	508	250	95	0	95	196	1	197	2	0

District and county[a]	Commercial grape crush[b]	Storage cooperage[c]	Gross wine production[d]			Gross wine inventories[d]			Wineries and cellars[e]	Fruit distilleries[j]
	1	2	3	4	5	6	7	8	9	10
	Tons		Table wine[f]	Dessert[f]	Total[g]	Table wine[h]	Dessert[i]	Total		
			1,000 gallons			1,000 gallons				
District 5 total	42,272	17,143	6,273	938	7,211	10,548	3,037	13,585	58	3
Alameda	5,394	3,854	978	3	981	2,221	607	2,828	10	0
Contra Costa	156	135	28	0	28	88	0	88	6	0
San Francisco	0	85	0	0	0	50	12	62	1	0
Santa Clara	20,208	10,333	2,691	556	3,247	5,729	1,779	7,508	29	2
Other	16,514	2,735	2,576	379	2,955	2,460	638	3,098	12	1
State total	2,054,968	387,663	84,965	111,133	196,098	104,639	130,676	235,315	228	58

[a] Some of these district boundaries for 1965 differ from those for 1966 given in Table 23. County cooperage, production, and inventory figures may not add to exact district and state totals, as figures are rounded to nearest thousands.

[b] July 1–December 31. Includes fresh grapes crushed for wine, brandy, grape juice, and grape concentrate.

[c] As of December 31. Includes fermenters and other large containers usable for storage.

[d] As of December 31, without allowances for losses, removals of standard wine for distillation, increases resulting from amelioration, etc. Includes a small amount of fruit and berry wines.

[e] As of July 1, 1967. Includes some that did not operate during year beginning July 1, 1966, but does not include three experimental wineries operated by colleges and universities.

[f] Table wine with 14 per cent or less alcohol; includes dessert wine stock, 14–24 per cent alcohol.

[g] Includes wine used in the production of vermouth, other natural wines, and sparkling wine. For production of these wines, on a calendar year basis, see page 4 of source cited.

[h] Includes dessert wine stock and table wine lees.

[i] Includes dessert wine lees.

[j] Plants producing distilled spirits, including those producing but not storing.

[k] Not adjusted to Treasury Department figures which vary fractionally in some instances. For vermouth, other special natural wines, and sparkling wine inventories, see page 4 of source cited.

Source: Wine Institute, Twenty-Ninth Annual Wine Industry Statistical Survey, Part I (San Francisco, 1965), p. 9.

TABLE 20

Commercial Crush of Fresh Grapes in California, by District Where Crushed and Variety, July 1 to December 31, 1966[a]

Type and variety	Southern California[b] Total	Per cent of total	Southern San Joaquin[c] Total	Per cent of total	Central San Joaquin[d] Total	Per cent of total	Escalon-Modesto[e] Total	Per cent of total
	Tons		Tons		Tons		Tons	
	1	2	3	4	5	6	7	8
Wine varieties								
Alicante Bouschet	1,293	2.1	708	0.3	5,511	0.7	1,202	0.6
Carignane	2,152	3.5	16,114	7.2	39,080	5.1	40,405	19.4
Grenache	6,495	10.5	4,800	2.1	38,802	5.1	38,874	18.7
Mission	14,207	22.9	620	0.3	12,032	1.6	15,739	7.6
Palomino	6,617	10.7	5,729	2.6	21,791	2.9	11,213	5.4
Petite Sirah	0	0	1,771	0.8	323	1	164	0.1
Zinfandel	6,302	10.2	257	0.1	739	0.1	7,905	3.8
Other white	7,392	11.9	16,419	7.3	8,528	1.1	10,102	4.8
Other dark	11,782	19.0	16,941	7.6	25,572	3.4	16,472	7.9
Total wine[f]	58,457	94.3	63,359	28.3	152,378	20.0	142,076	68.2
Table varieties								
Emperor	0	0	31,522	14.1	42,182	5.5	0	0
Tokay	17	1	1,590	0.7	609	0.1	30,704	14.7
Malaga	126	0.2	8,709	3.9	22,568	3.0	500	0.2
Other table	1,505	2.4	31,077	13.9	23,894	3.1	267	0.1
Total table	1,648	2.7	72,898	32.5	89,253	11.7	31,471	15.1
Raisin varieties								
Thompson Seedless	1,339	2.2	77,407	34.6	486,461	63.8	34,451	16.5
Muscat	469	0.8	9,847	4.4	29,546	3.9	380	0.2
Other raisin	60	1	520	0.2	5,369	0.7	0	0
Total raisin	1,868	3.0	87,774	39.2	521,376	68.3	34,831	16.7
All varieties	61,973	100.0	224,031	100.0	763,007	100.0	208,378	100.0

[a] Crushed at bonded wineries and distilled spirits plants. Does not include small quantities crushed for juice and concentrate at other processing plants.
[b] Includes Los Angeles, Riverside, San Bernardino, San Diego, and Santa Barbara counties.
[c] Includes Kern, San Luis Obispo, and portion of Tulare County south of City of Tulare.

Type and variety	Lodi-Sacramento[f]		North coast north Bay[g]		North coast south Bay[h]		All districts	
	Tons	Per cent of total	Tons	Per cent of total	Tons	Per cent of total	Total	Per cent of total
	9	10	11	12	13	14	15	16
	Tons		Tons		Tons		Tons	
Wine varieties								
Alicante Bouschet	1,179	0.6	1,349	1.4	53	0.1	11,295	0.7
Carignane	29,067	15.1	18,143	18.7	1,487	3.8	146,448	9.2
Grenache	7,596	4.0	1,876	1.9	4,040	10.4	102,483	6.5
Mission	3,518	1.8	228	0.2	87	0.2	46,431	2.9
Palomino	2,265	1.2	2,154	2.2	1,263	3.3	51,032	3.2
Petite Sirah	845	0.4	10,154	10.5	498	1.3	13,755	0.9
Zinfandel	21,467	11.2	17,845	18.4	1,900	4.9	56,415	3.6
Other white	3,159	1.6	28,204	29.1	11,032	28.4	84,836	5.4
Other dark	6,169	3.2	15,585	16.1	4,984	12.8	97,505	6.2
Total wine[j]	75,406	39.2	96,677	99.8	26,347	67.9	614,700	38.8
Table varieties								
Emperor	90	0.1	0	0	1,206	3.1	75,000	4.7
Tokay	100,217	52.2	0	0	1,047	2.7	134,184	8.5
Malaga	282	0.1	0	0	0	0	32,185	2.0
Other table	1,227	0.6	12	[i]	27	0.1	58,009	3.7
Total table	101,816	53.0	12	[i]	2,280	5.9	299,378	18.9
Raisin varieties								
Thompson Seedless	14,474	7.5	41	[i]	9,952	25.6	624,125	39.4
Muscat	426	0.2	117	0.1	232	0.6	41,017	2.6
Other raisin	0	0	0	0	0	0	5,949	0.4
Total raisin	14,900	7.8	158	0.2	10,184	26.2	671,091	42.3
All varieties	192,122	100.0	96,847	100.0	38,811	100.0	1,585,169	100.0

[f] Includes Amador, Butte, El Dorado, Placer, Sacramento, and portion of San Joaquin County north of Highway 4.

[g] Includes Marin, Mendocino, Napa, Solano, and Sonoma counties.

[h] Includes Alameda, Contra Costa, Monterey, San Benito, San Francisco, San Mateo, Santa Clara, and Santa Cruz counties.

[i] Less than 0.05 per cent.

[j] Includes following quantities of wine varieties for which variety breakdowns not reported: southern California, 2,217 tons; Lodi-Sacramento, 141 tons; north coast-north Bay, 1,139 tons; North coast-south Bay, 1,003 tons; and all districts, 4,500 tons.

Sources: Prepared by Wine Institute from confidential reports obtained from California wineries and distilled spirits plants. Wine Institute, *Annual Wine Industry, Statistical Report*, Part I (San Francisco, October 31, 1967), table 2.

TABLE 21
California Bearing Acreage of Wine Varieties of Grapes, by Chief Varieties, by Districts, 1964

| | | Districts | | | | | | | | |
| | | Interior valley | | | | Other | | | Central coast | | |
Class and variety	State total	Sacramento	Central	San Joaquin	Total	Total	Southern California	Total	North Bay	South Bay	Other counties[a]
	1	2	3	4	5	6	7	8	9	10	11
					ACRES						
All varieties[b]	454,922	670	66,187	314,838	381,695	72,228	36,646	35,582	26,142	9,440	999
Wine varieties[c]	120,060	625	33,021	29,382	63,028	56,281	21,174	35,107	25,974	9,133	751
Dark, all	99,027	614	30,371	24,228	55,213	43,167	18,096	25,071	19,131	5,940	647
White, all	21,033	11	2,650	5,154	7,815	13,114	3,078	10,036	6,843	3,193	104
Palomino[d]	8,165	10	1,693	2,913	4,616	3,514	1,767	1,747	1,441	306	35
Burger	2,413	—[e]	591	652	1,243	1,166	697	469	382	87	4
French Colombard	2,183	—	168	617	785	1,396	—	1,396	1,218	178	2
Riesling and Sylvaner[d]	1,527	—	43	22	65	1,426	65	1,361	574	787	36
Sémillon	1,284	—	55	191	246	1,038	—	1,038	385	653	—
Sauvignon blanc[d]	1,213	—	23	73	96	1,099	—	1,099	918	181	18
Colombard[f]	1,056	—	70	24	94	957	34	923	664	259	5
Pinot blanc[d]	944	—	—	13	13	931	—	931	538	393	—
Pedro Ximenes	787	—	—	280	280	507	507	—	—	—	—
Green Hungarian	378	—	4	46	50	328	—	328	287	41	—
Other[d]	1,083	1	3	323	327	752	8	744	436	308	4
Dark, all	99,027	614	30,371	24,228	55,213	43,167	18,096	25,071	19,131	5,940	647
Above 9 per cent non-bearing, all	46,542	49	16,986	14,079	31,114	15,320	4,879	10,441	7,533	2,908	108
Carignane	24,198	33	10,953	5,716	16,702	7,469	1,327	6,142	5,101	1,041	27
Grenache	13,701	16	5,198	5,398	10,612	3,055	2,687	368	66	302	34
Valdepeñas	1,106	—	555	493	1,048	58	—	58	58	—	—
Cabernet Sauvignon[d]	834	—	—	—	—	821	—	821	593	228	13
Gamay	868	—	—	—	—	868	—	868	769	99	
Pinot noir[d]	716					714		714	707		

| Class and variety | State total | Interior valley | | | | Other | | Central coast | | | Other counties[a] |
		Sacramento	Central	San Joaquin	Total	Total	Southern California	Total	North Bay	South Bay	
	1	2	3	4	5	6	7	8	9	10	11
Below 9 per cent non-bearing, all	52,485	565	13,385	10,149	24,099	27,847	13,217	14,630	11,598	3,032	539
Zinfandel	22,004	223	7,287	644	8,154	13,426	5,858	7,568	5,670	1,898	424
Alicante Bouschet	9,112	19	2,570	4,328	6,917	2,172	1,220	952	847	105	23
Mission	8,581	76	2,617	1,598	4,291	4,218	4,040	178	38	140	72
Petite Sirah[d]	4,506	—	310	321	631	3,869	—	3,869	3,567	302	6
Mataro	2,649	247	46	56	349	2,296	1,765	531	37	494	4
Salvador	2,250	—	391	1,681	2,072	178	169	9	7	2	4
Malvoise[d]	1,120	—	98	758	856	254	156	98	67	31	10
Feher Szagos[g]	746	—	34	712	746	—	—	—	—	—	—
Grand noir[d]	440	—	10	7	17	423	—	423	381	42	—
Burgundy[d]	1,077	—	22	44	66	1,011	9	1,002	984	18	—

[a] The very small acreage in "other counties" in column 11 for which individual county data by variety are not reported separately in the source cited are not included in district totals. The names of these "other" minor counties excluded from district totals are listed in the source cited, in footnote 3 for 12 raisin counties, footnote 4 for 5 table grape counties, and footnote 9 for 21 wine counties. Counties included in each district except as noted above are as follows (see also Table 23): *north Bay:* Mendocino, Lake, Sonoma, Napa, Solano, and Marin; *south Bay:* San Francisco, Contra Costa, Alameda, Santa Clara, Monterey, San Benito, San Mateo, Santa Cruz, and San Luis Obispo; *southern California:* Los Angeles, San Bernardino, Riverside, San Diego, Orange, and Ventura; *San Joaquin Valley:* Fresno, Kern, Madera, Tulare, Merced, and Kings; *central valley:* Sacramento, San Joaquin, Stanislaus, Amador, Calaveras, and Tuolumne. The interior valley and the central valley include the small portions of the Sacramento Valley counties north of Amador, Sacramento, and Solano counties.

[b] All varieties, including table and raisin varieties (see detail in Table 20).

[c] Grapes classified as wine varieties by California Crop and Livestock Reporting Service are those grown and used almost exclusively for making wine, brandy, and juice.

[d] For name of 21 counties for which data are not given for wine varieties individually, see footnote 9, page 5, in source cited. For synonyms and names of similar varieties included in specified category, see detailed footnotes in source cited, page 5. In addition to varieties listed individually in this table, 42 other white varieties are included in total and 77 other dark varieties, a total of about 150.

[e] Dashes indicate either zero or very small acreage not reported separately by individual county in source cited.

[f] Probably Sauvignon vert.

[g] Erroneously listed as a dark variety.

Source: Compiled by S. W. Shear, Giannini Foundation of Agricultural Economics, from California Crop and Livestock Reporting Service, *California Grape Acreage by Varieties and Principal Counties, as of 1964* (Sacramento, 1965), 5 pp.

TABLE 22
California Bearing Grape Acreage of Raisin and Table Varieties by Chief Varieties, by Districts, 1964

		Districts									
		Interior valley				Other					
									Central coast		
Class and variety	State total	Sacramento	Central	San Joaquin	Total	Total	Southern California	Total	North Bay	South Bay	Other counties[a]
	1	2	3	4	5	6	7	8	9	10	11
							ACRES				
All varieties[b]	454,922	670	66,187	314,838	381,695	72,228	36,646	35,582	26,142	9,440	999
Raisin varieties[c]	251,489	5	9,355	232,012	241,372	10,011	9,766	245	108	137	106
Muscat	19,953	5[d]	207	18,784	18,996	923	691	232	96	136	34
Other, all	231,536	—	9,148	213,228	222,376	9,088	9,075	13	12	1	72
Thompson Seedless	228,180	—	9,148	209,944	219,092	9,016	9,003	13	12	1	72
Sultana	1,534	—	—	1,462	1,462	72	72	—	—	—	—
Currants	1,813	—	—	1,813	1,813	—	—	—	—	—	—
Table varieties[c]	83,373	40	23,811	53,444	77,295	5,936	5,706	230	60	170	142
Emperor	29,905	—	151	29,748	29,899	6	6	—	—	—	—
Ribier	7,474	—	171	7,202	7,373	92	62	30	30	—	9
Almeria	2,047	—	11	2,036	2,047	—	—	—	—	—	—
Calmeria	1,157	—	—	1,157	1,157	—	—	—	—	—	—
Tokay	23,272	27	22,757	322	23,106	148	113	35	9	26	18
Cardinal	4,077	—	154	2,445	2,599	1,476	1,476	—	—	—	2
Perlette	3,317	—	2	652	654	2,663	2,663	—	—	—	—
Malaga	4,999	8	308	4,493	4,809	180	175	5	—	5	10
Red Malaga	2,774	—	3	2,591	2,594	180	180	—	—	—	—
Italia	1,432	—	2	1,428	1,430	2	2	—	—	—	—
Rose of Peru	297	—	66	—	66	219	214	5	2	3	12
Concord	156	—	3	6	9	76	61	15	6	9	71
Other	2,466	5	183	1,364	1,552	894	754	140	13	127	20

^a The very small acreage in "other counties" in column 11 for which individual county data by variety are not reported separately in source cited are not included in district totals. Names of these "other" minor counties excluded from district totals are listed in source cited, in footnote 3 for twelve raisin counties, footnote 4 for fifteen table grape counties, and footnote 9 for twenty-one wine counties.

Counties included in each district with exclusion of minor counties noted above are as follows (see also Table 23): *north Bay*: Mendocino, Lake, Sonoma, Napa, Solano, and Marin; *south Bay*: San Francisco, Contra Costa, Alameda, Santa Clara, Monterey, San Benito, San Mateo, Santa Cruz, and San Luis Obispo; *southern California*: Los Angeles, San Bernardino, Riverside, San Diego, Orange, and Ventura; *San Joaquin Valley* (designated as southern San Joaquin Valley in Table 23): Fresno, Kern, Madera, Tulare, Merced, and Kings; *central valley* (designated as northern San Joaquin Valley in Table 23): Sacramento, San Joaquin, Stanislaus, Amador, Calaveras, and Tuolumne. Interior valley and central valley include small portions of Sacramento Valley counties north of Amador, Sacramento, and Solano counties.

^b All varieties, including wine varieties (see varietal details in following table).

^c Varieties as classified by California Crop and Livestock Reporting Service; raisin varieties are those well adapted for making raisins, although Muscats and Thompson Seedless are also used for making wine, brandy, and juice, and Thompson Seedless for fresh table consumption and canning. Table varieties are those well adapted for fresh table consumption but not for making raisins, although most varieties are also used in part for making wine, brandy, and juice. In addition to 12 varieties for which data are given individually in table, source cited lists in footnotes 35 individual varieties lumped as "Other."

^d Dashes indicate either zero or very small acreage not reported separately by individual county in source cited.

Source: Compiled by S. W. Shear, Giannini Foundation of Agricultural Economics, from California Crop and Livestock Reporting Service, *California Grape Acreage by Varieties and Principal Counties, as of 1964* (Sacramento, 1965). 5 pp.

heavy-yielding counties of Madera, Fresno, Tulare, and Kern is highly significant for future grape production and utilization in the state, as is the big increase in percentage of Thompson Seedless of the total bearing acreage (table 22) from 36 per cent in 1946 to 51 per cent in 1966. See also Henderson *et al.* (1965) for data on the grape acreage in California.

VARIETAL COMPOSITION OF THE CRUSH

Table and raisin grapes have been crushed in increasing amounts by California wineries. Since Repeal in 1933 there has been a great increase in the tonnage and the proportion of the California grape crop utilized for wine and brandy and a small increase in the proportion used for canning, offset by a substantial decrease in the relative importance of drying and fresh table market outlets (fig. 22). In 1965 about one-third of the state crop of grapes was utilized for raisins, 14 per cent for fresh table consumption, and nearly 2 per cent for canning. Utilization for commercial and noncommercial production of wine and brandy rose from about 1,000,000 tons a year during 1935–1939 to 1,650,000 tons during 1960–1964, or from about 44 per cent of the total grape crop to 55 per cent. All this increase was in the commercial crush. Shipments of juice grapes for noncommercial or homemade wine decreased from about 210,000 tons during 1935–1939 to only 105,000 tons during 1960–1964.

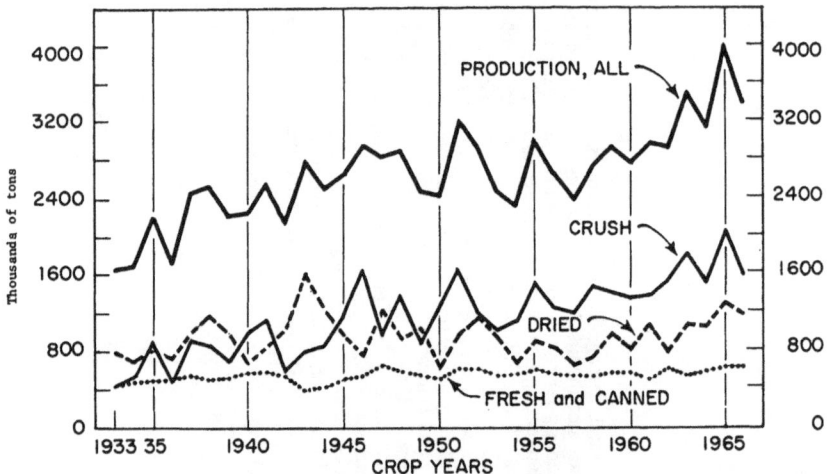

Figure 22. Production and utilization of California grapes, 1933–1966. (Much of the marked increase in California grape production in recent years has been utilized by the commercial crush for wine, brandy, and other juice products.) Source: Basic data from table 30.

The commercial crush of fresh grapes within the state, utilized for wine, brandy, and other juice, rose from 773,000 tons a year for 1935–1939 to 1,520,000 tons for 1960–1964, accounting for 50 per cent of the state's total production of grapes. During the latter period the equivalent of 30,000 fresh tons a year of dried waste raisins and packing-house culls were utilized by the state's wineries and distilleries.

Within a decade, state production of raisin grape varieties rose about 300,000 tons to an average of about 3 million tons for 1960–1964 and a peak of almost 4 million tons in 1965. Nearly all this increase was in Thompson Seedless raisin grapes and was crushed by the wineries of the state. During 1960–1964, dried raisin grapes utilized about 970,000 tons of grapes a year, accounting for about 50 per cent of the production of these varieties and over 30 per cent of all grape production.

Production of table and wine grape varieties did not increase in the past decade, but the proportion used for crushing increased while that for fresh shipments declined correspondingly. As a result, during the period 1960–1964 wineries utilized 55 per cent of the crop of table grape varieties and 90 per cent of the crop of wine varieties. Of the average commercial crush of 1,522,000 tons a year during those five years, about 709,000 tons (47 per cent) were raisin varieties, 512,000 tons (34 per cent) were wine varieties, and 301,000 tons (20 per cent) were table varieties.

Most of the increasing proportion of California production of wine and brandy has been made from Thompson Seedless grapes and table varieties, nearly all grown in the interior valley (tables 22 and 23), that are generally conceded not to make as desirable wine as the true wine varieties and muscats. Large quantities of seedless raisin grapes and of Tokays, Malagas, and Emperors are crushed for this purpose in the San Joaquin Valley, and they have constituted a larger proportion of the state crush for wine and brandy since World War II than before. The Thompson Seedless crush increased more than that of any other variety since Repeal and, proportionately, probably more than the muscat crush.

The state's commercial crush of all grapes in 1964 and in 1966 was a little greater than the 1960–1964 average of 1,522,000 tons, and the proportion of raisin, table, and wine varieties was about the same. Tokays accounted for a little over 7 per cent of the total crush in 1964 and Emperors nearly 5 per cent. Nearly 41 per cent of the total was seedless grapes, and 5 per cent was muscats.[4] Although much of the crush of table and raisin varieties is used for making beverage brandy and neutral

---

[4] For additional detail on the 1964 crush by counties and for individual wine varieties, see Wine Institute (1965).

TABLE 23

CALIFORNIA GRAPE ACREAGE TOTAL, BEARING AND NONBEARING, BY VARIETAL CLASSES, BY DISTRICT AND COUNTY, 1966

District and county	Wine varieties				Raisin varieties				Table varieties			
	All ages	Bearing	Nonbearing		All ages	Bearing	Nonbearing		All ages	Bearing	Nonbearing	
	Acres	Acres	Acres	Per cent[a]	Acres	Acres	Acres	Per cent[a]	Acres	Acres	Acres	Per cent[a]
	1	2	3	4	5	6	7	8	9	10	11	12
State total[b]	142,020	126,200	15,820	11.1	261,700	253,380	8,320	3.2	86,170	82,660	4,110	4.8
Interior valley total[c]	77,770	63,370	7,534	12.1	254,830	246,870	7,960	3.1	79,950	76,450	3,500	4.4
Sacramento total	550	550	0	0	0	0	0	0	40	40	0	0
Butte	0	0	0	0	0	0	0	0	0	0	0	0
Colusa	0	0	0	0	0	0	0	0	0	0	0	0
El Dorado	0	0	0	0	0	0	0	0	0	0	0	0
Glenn	0	0	0	0	0	0	0	0	0	0	0	0
Nevada	0	0	0	0	0	0	0	0	0	0	0	0
Placer	430	430	0	0	0	0	0	0	30	30	0	0
Shasta	0	0	0	0	0	0	0	0	0	0	0	0
Siskiyou	0	0	0	0	0	0	0	0	0	0	0	0
Sutter	0	0	0	0	0	0	0	0	0	0	0	0
Tehama	0	0	0	0	0	19	0	0	0	0	0	0
Yolo	120	120	0	0	0	0	0	0	10	10	0	0
Yuba	0	0	0	0	0	0	0	0	0	0	0	0
Northern San Joaquin Valley total	36,050	33,260	2,790	7.7	8,980	8,930	50	0.6	23,830	23,410	420	1.8
Amador	570	530	40	7.0	30	30	0	0	0	0	0	0
Calaveras	110	110	0	0	0	0	0	0	0	0	0	0
Sacramento	100	60	40	40.0	0	0	0	0	120	120	0	0
San Joaquin	24,200	22,380	1,820	7.5	690	680	10	1.4	23,300	22,880	420	1.8
Stanislaus	11,070	10,180	890	8.0	8,260	8,220	40	15.4	410	410	0	0
Tuolumne	0	0	0	0	0	0	0	0	0	0	0	0
Southern San Joaquin Valley total	41,170	34,560	6,610	16.1	245,850	237,940	7,910	3.2	56,080	53,000	3,080	5.5
Fresno	12,610	11,090	1,520	12.1	149,380	145,100	4,280	2.9	12,360	12,000	360	2.9
Inyo	0	0	0	0	0	0	0	0	0	0	0	0
Kern	5,990	4,990	1,000	16.7	21,910	21,110	800	3.7	11,660	10,440	1,220	10.5
Kings	720	720	0	0	3,380	3,380	0	0	10	10	0	0
Madera	7,210	5,540	1,670	23.2	27,490	25,360	2,130	7.7	500	420	80	16.0
Merced	7,960	6,160	1,800	22.6	8,590	8,530	60	0.7	660	660	0	0
Tulare	6,680	6,060	620	9.3	35,100	34,460	640	1.8	30,890	29,470	1,420	4.6

District and county	Wine varieties				Raisin varieties				Table varieties			
	All ages Acres	Bearing Acres	Nonbearing Acres	Nonbearing Per cent[a]	All ages Acres	Bearing Acres	Nonbearing Acres	Nonbearing Per cent[a]	All ages Acres	Bearing Acres	Nonbearing Acres	Nonbearing Per cent[a]
	1	2	3	4	5	6	7	8	9	10	11	12
Central coast total	43,760	37,560	6,200	14.2	160	160	0	0	200	200	0	0
North Bay total	31,300	27,450	3,850	12.3	20	20	0	0	30	30	0	0
Humboldt	120	100	0	0	0	0	0	0	0	0	0	0
Lake	0	0	20	16.7	0	0	0	0	0	0	0	0
Marin	0	0	0	0	0	0	0	0	0	0	0	0
Mendocino	5,700	4,700	1,000	17.5	10	10	0	0	0	0	0	0
Napa	12,030	10,450	1,580	13.1	10	10	0	0	20	20	0	0
Solano	720	640	80	11.1	0	0	0	0	0	0	0	0
Sonoma	12,730	11,560	1,170	9.2	0	0	0	0	10	10	0	0
South Bay total	12,460	10,110	2,350	18.9	140	140	0	0	170	170	0	0
Alameda	2,000	1,840	160	8.0	10	10	0	0	10	10	0	0
Contra Costa	1,400	1,400	0	0	10	10	0	0	40	40	0	0
Monterey	1,090	1,070	20	1.8	0	0	0	0	0	0	0	0
San Benito	4,300	2,240	2,060	47.9	0	0	0	0	0	0	0	0
San Luis Obispo	550	530	20	3.6	0	0	0	0	0	0	0	0
San Mateo	0	0	0	0	0	0	0	0	0	0	0	0
Santa Clara	3,030	2,940	90	3.0	120	120	0	0	110	110	0	0
Santa Cruz	90	90	0	0	0	0	0	0	10	10	0	0
Southern California total	20,160	20,100	60	0.3	6,620	6,260	360	5.4	5,940	5,330	610	10.3
Imperial	0	0	0	0	0	0	0	0	0	0	0	0
Los Angeles	0	0	0	0	0	0	0	0	0	0	0	0
Orange	0	0	0	0	0	0	0	0	0	0	0	0
Riverside	1,300	1,270	30	2.3	5,670	5,310	360	6.3	4,830	4,230	600	12.4
San Bernardino	18,350	18,320	30	0.6	500	500	0	0	610	600	10	1.6
San Diego	510	510	0	0	450	450	0	0	500	500	0	0
Santa Barbara	0	0	0	0	0	0	0	0	0	0	0	0
Ventura	0	0	0	0	0	0	0	0	0	0	0	0

[a] Percentage of all ages of varietal class to nearest 10th per cent, that is, 0.0.
[b] Includes small amounts not listed by individual counties.
[c] Interior valley total is total of three districts—Sacramento, northern San Joaquin, and southern San Joaquin Valley. In table grape acreage, by variety, northern San Joaquin is designated as "central," and southern San Joaquin as "San Joaquin."

Source: Compiled by S. W. Shear, Giannini Foundation of Agricultural Economics, from California Crop and Livestock Reporting Service, *California Fruit and Nut Acreage, Bearing and Nonbearing as of 1966*, Sacramento, June, 1967, p. 8.

brandy for fortifying dessert wines, a considerable quantity is used for making concentrate and wines.

If the production trends indicated elsewhere prevail during the next few years, the proportion of normal production of table and wine varieties crushed is likely to change little on the average and, of raisin varieties, to increase a little above the 37 per cent of recent years. Because of the upward trend of raisin grape production and crush these varieties will probably contribute a somewhat larger proportion of the crush by wineries and distillers. Table and seedless raisin varieties will probably continue to constitute at least two-thirds of the grapes used in making wine and brandy in California and will therefore be the source of most of the production and marketing problems of the wine industry. The Thompson Seedless will still be the leading grape in the state in total tonnage produced, dried, and crushed, and annual variations in its production and utilization will therefore be a most important supply factor affecting the wine industry of the state. In the past, variations in the tonnage of this one variety utilized for drying usually have affected the total commercial winery crush more than fluctuations in its total production.

The Wine Institute (1965) gives a breakdown for nine individual varieties of wine grapes for the 1966 crush of 614,700 tons (table 20). The corresponding tonnage data for 1964, which are similar to 1966, show the relative importance of the chief varieties of wine grapes in percentage of the total crush of wine varieties as follows: all whites, 22 (premium whites, 5; other whites, 17); all blacks, 78 (Carignane, 28; Grenache and Mission together, 27; Zinfandel, 8; Alicante Bouschet, 2; premium blacks, 5; all other blacks, 11 per cent).

THE TREND IN WINERY NUMBERS AND SIZE...

... has been toward fewer and larger wineries in California.

During Prohibition the number of individual bonded wineries in California was less than 10. These produced sacramental wines. After Repeal the number of bonded wine-making premises in the United States and California producing still wine reached a maximum in 1934 in California with 804 and in other states in 1937 with 659 (table 24). Since 1935 the number has steadily decreased. By 1949 there were only 428 in California and 451 in other states, and by 1965 this had decreased to 204 in California and 191 in other states. Most of the wineries in other states are very small, for they make only about 15 per cent of the wine produced in the United States. The number of bonded wine premises and tax-paid wine-bottling houses authorized to operate for recent years is given in table 25.

TABLE 24

BONDED WINE PREMISES IN OPERATION IN CALIFORNIA, OTHER STATES, AND THE
UNITED STATES: CROP YEARS 1934 TO 1965[a]

Year beginning July 1	California	Other states	United States	Year beginning July 1	California	Other states	United States
1933	542	302	845	1950	408	428	836
1934	804	404	1,208	1951	401	393	794
1935	800	557	1,357	1952	386	364	750
1936	697	613	1,310	1953	381	342	723
1937	618	659	1,277	1954	360	319	679
1938	581	658	1,239	1955	317	297	614
1939	538	650	1,188	1956	296	272	568
1940	526	644	1,170	1957	278	265	543
1941	502	624	1,126	1958	271	260	531
1942	485	618	1,103	1959	256	244	500
1943	451	571	1,022	1960	249	240	489
1944	465	548	1,013	1961	241	230	471
1945	468	525	993	1962	233	217	450
1946	445	502	947	1963	230	207	437
1947	443	489	932	1964	226	198	424
1948	430	415	905	1965	204	191	395
1949	428	451	879				

[a] Data are the number operated during any part of the year. Specific designations of bonded premises included have changed in some cases during period shown because of changes in Internal Revenue Service regulations, but all premises handling wine in bond are included.

Sources: Prepared by Wine Institute from reports of Internal Revenue Service, U.S. Treasury Department, Washington, D.C.

TABLE 25

BONDED WINE PREMISES AND TAXPAID WINE-BOTTLING HOUSES AUTHORIZED
TO OPERATE ON JUNE 30 IN CALIFORNIA, OTHER STATES, AND THE UNITED STATES,
1955 TO 1966[a]

Year	Bonded wine premises[b]			Taxpaid wine-bottling houses		
	California	Other states	United States	California	Other states	United States
1955	322	299	621	21	90	111
1956	303	277	580	28	108	136
1957	288	258	546	25	100	125
1958	273	260	533	26	95	121
1959	263	247	410	28	91	119
1960	254	241	495	28	89	117
1961	246	228	474	26	87	113
1962	240	218	458	26	80	106
1963	237	211	448	26	77	103
1964	231	207	438	27	72	99
1965	233	202	435	25	68	93
1966	231	195	426	26	59	85

[a] Authorized to operate under federal laws and regulations.
[b] Includes premises designated as bonded wineries or bonded wine cellars.
Sources: Prepared by Wine Institute from reports of Internal Revenue Service, U.S. Treasury Department, Washington, D.C.

The data in table 26 show the changes in the numbers of winery premises of different sizes in California between 1941, 1947, and 1965. From 1941 to 1965 there was a decrease of 55 per cent in the number of smaller wineries, for the most part those making table wines and having combined storage and fermenter capacity of less than a million gallons. The productive capacity of wineries with large outputs—the great bulk of whose output is dessert wine—increased even more than the number of such wineries. From 1947 to 1965 the number of the largest wineries with storage capacity of over 5 million gallons increased 29 per cent. Nearly all the increase in facilities has been in the central San Joaquin Valley. Of the eighteen wineries in California with a storage capacity in 1965 of over 5 million gallons, eleven had 5 to 10 million gallons and seven had over 10 million gallons. Table 19 shows 228 bonded winery premises in 1964, when average storage capacity was as follows: the state, 1.7 million gallons; the San Joaquin Valley—Stanislaus through Tulare counties, 4.4 million gallons; southern California, 520,000 gallons; coast counties north of San Francisco Bay, 640,000 gallons; and south of the Bay, 300,000 gallons.

The larger firms producing and marketing wine have several winery premises in different parts of the state. Several of these firms have built

TABLE 26

FREQUENCY DISTRIBUTION OF SIZE OF CALIFORNIA BONDED WINERY PREMISES, DECEMBER 31, 1941, 1947, AND 1965

Storage capacity[a] (gallons)	Number of wineries			Percentage change from 1947
	In 1941	In 1947	In 1965	
Less than 10,000	201	87	17	−80
10,000–25,000	22	32	13	−59
25,000–50,000	28	32	20	−38
50,000–100,000	28	30	21	−30
100,000–250,000	54	59	27	−54
250,000–500,000	28	43	23	−47
500,000–1,000,000	20	25	18	−28
1,000,000–2,000,000	31	28	25	−11
2,000,000–5,000,000	22	33	30	−9
Over 5,000,000	9	14	18[b]	+29
Total	443	383	212[c]	−45

[a] Including fermenters usable as storage as of December 31.
[b] Of eighteen wineries with storage capacity in 1965 of over 5,000,000, eleven had 5,000,000–10,000,000, and seven had over 10,000,000.
[c] Table 20 shows 228 bonded winery premises in 1965. Average storage capacity for state was 1,700,000 gallons; for San Joaquin Valley, Stanislaus, through Tulare counties, 4,400,000; southern California, 520,000; coast counties north of San Francisco Bay, 640,000; and south Bay, 300,000 gallons.
Source: Compiled by Wine Institute, San Francisco, from unpublished detailed reports.

or purchased additional wineries since 1941; so their wine-making and storage capacity has increased relatively much more than the number and size of large individual winery premises accounted for in table 26. The number and the capacity of wineries producing table wines of outstanding quality, particularly of varietal wines, changed relatively little from prewar to postwar years. However, a number of premium quality wineries have planted extensive new vineyards and have increased the winery capacity.

The Wine Institute estimated in 1948 that approximately 75 per cent of California wineries, including coöperatives, grew at least part of the

TABLE 27

AVERAGE RETURNS PER TON TO CALIFORNIA GROWERS FOR
GRAPES CRUSHED, BY VARIETAL CLASSES, 1947–1967

Crop year	Average returns for grapes crushed (dollars per ton, fresh weight)			
	Raisin[a]	Table[a]	Wine[a]	Total
Averages				
1947–1949	28.23	23.53	32.83	28.97
1950–1954	36.96	30.62	44.70	38.44
1955–1959	43.08	34.76	45.90	42.40
1960–1964	38.68	35.72	63.90	46.68
Annual				
1947	29.40	25.70	33.00	29.80
1948	30.00	22.40	35.70	30.30
1949	25.30	22.50	29.80	26.80
1950	59.50	50.10	70.80	61.00
1951	29.20	21.40	40.30	30.90
1952	23.80	18.60	30.50	25.60
1953	33.80	31.10	40.50	36.20
1954	38.50	31.90	41.40	38.50
1955	29.90	23.50	35.60	30.10
1956	39.30	30.00	42.90	39.20
1957	52.30	44.50	53.00	50.90
1958	52.10	39.50	50.20	48.90
1959	41.80	36.30	47.80	42.90
1960	36.10	30.00	54.60	40.70
1961	38.40	35.60	76.80	49.60
1962	44.10	41.90	63.40	51.00
1963	33.50	30.20	52.40	39.80
1964	41.30	40.90	72.30	52.30
1965	29.50	24.50	48.30	34.90
1966	27.50	26.80	55.40	38.30
1967	39.60	34.50	61.90	47.80

[a] See table 28 for chief varieties of grapes included in each varietal class.

Sources: Compiled by S. W. Shear, Giannini Foundation of Agricultural Economics. California Crop and Livestock Reporting Service, *California Fruit and Nut Crops, 1909–1955*, Special Publ. 261, Sacramento, 1956, pp. 36–46. *Idem, California Fruit and Nut Crops, 1949–1961*, pp. 15–20, and annual issues thereafter (*California Fruit and Nut Statistics*), Sacramento, 1965. *Idem,* "California Fruits, 1965 Grape Crushing Price" (Corrected), February 15, 1967, table 1, and February 19, 1968 (Mimeo.)

grapes they crushed. The grapes grown by commercial wineries and by grower members of the coöperatives are roughly estimated to have accounted for about 70 per cent of the total tonnage crushed in California

TABLE 28

GRAPES FOR CRUSHING: AVERAGE GROWER RETURNS, DOLLARS PER TON, DELIVERED BASIS FOR PURCHASES IN CALIFORNIA DURING 1966 SEASON, BY AREAS WHERE GROWN

Type and variety	North coast	Lodi region[a]	Modesto region[b]	Other San Joaquin Valley[c]	Southern California	1966 state average	1967 state average
Raisin grapes							
Thompson Seedless	d	o	27.70	27.40	d	27.40	39.80
Sultana	o	d	o	27.50	d	27.50	39.40
Muscat	o	o	d	30.30	39.80	30.40	37.00
All raisin grapes	d	d	27.70	27.50	39.80	27.50	39.60
Table grapes							
Emperor	o	d	o	22.10	o	22.30	24.00
Tokay	o	29.90	d	d	d	29.90	39.40
White Malaga	o	o	28.50	29.60	d	29.50	36.00
Other table	d	d	d	23.80	d	23.90	29.00
All table grapes	d	29.90	28.80	24.70	26.10	26.80	34.50
Wine grapes, black							
Alicante Bouschet	81.60	40.80	d	29.90	31.80	42.20	47.40
Carignane	89.30	55.00	50.70	39.90	33.60	50.10	56.60
Grenache	85.70	d	43.00	38.40	32.70	40.60	47.80
Mataro	93.50	d	o	d	33.10	38.10	54.00
Mission	72.80	d	37.50	37.90	32.50	36.40	43.10
Petite Sirah	100.00	d	o	d	o	92.30	99.30
Royalty	o	o	d	42.00	o	42.70	54.10
Rubired	o	o	d	44.00	d	44.30	54.00
Salvador	d	d	45.30	42.80	d	44.10	55.40
Zinfandel	100.30	65.30	62.10	42.60	32.90	70.40	77.60
Other black	158.40	d	62.20	40.50	32.70	84.00	81.50
All black grapes	110.60	60.50	49.80	39.40	32.80	52.70	58.90
Wine grapes, white							
Burger	90.30	d	d	d	28.30	52.40	56.70
Palomino	79.10	33.50	35.10	34.40	26.60	38.00	53.10
Other white	129.10	d	d	40.10	d	91.10	96.20
All white grapes	120.00	45.30	51.60	36.30	26.70	66.80	75.20
All wine grapes	114.30	59.20	51.10	38.90	31.90	55.40	61.90
All varieties	114.20	41.20	45.10	29.50	31.80	38.30	47.80

[a] "Lodi region" includes Sacramento County and San Joaquin County north of Eight Mile Road.
[b] "Modesto region" includes San Joaquin Valley south of Eight Mile Road and north of Merced River including Escalon area.
[c] "Other San Joaquin Valley" includes San Joaquin Valley south of Merced River and north of Tehachapi Mountains.
[d] Included in "state average" but not shown, to avoid disclosure of individual operations.
[e] No data available.
[f] Includes Royalty and Rubired.
Source: California Crop and Livestock Reporting Service, *California Fruits, Grape Crushing Price*, February 15, 1967, table 1 and February 19, 1968, table 1 (Mimeo).

in recent years. Hence returns to growers per ton for grapes bought outright by wineries in the state, as estimated by the California Crop and Livestock Reporting Service (tables 27 and 28), reflect actual returns from only about 30 per cent of the total tonnage of grapes crushed, rather than returns from all the grapes crushed for making wine and brandy and other crush products.

GRAPE ACREAGE, YIELDS AND PRODUCTION TRENDS...

... are upward.

When Prohibition was repealed in 1933, the California grape industry was near the bottom of a depression. Acreage, yields, production, and prices were far below those of the prosperous 1920's. After 1936 they began to increase. Bearing acreage increased only slightly from about 437,000 acres in 1936 to about 497,000 acres in 1950 (table 29). However, with larger yields per acre production increased more rapidly, particularly after 1940; under the stimulus of high wartime prices (table 28) production rose almost one-third from 1934–1938 to 1945–1948, when it averaged nearly 2.8 million tons (table 30).

Table 29,[5] from a report by Allmendinger (1965), shows the course of California bearing grape acreage from 1946 to 1967 as determined by new plantings and removal of old acreage. Heavy removals from 1946 to 1957 exceeded new plantings enough to reduce bearing acreage to a low of 399,000 in 1957. The bearing grape acreage of the state had risen to a little over 462,000 by 1967, nearly 35,000 acres less than the postwar high of nearly 497,000 acres in 1949. About 49 per cent of the 1967 bearing acres reached bearing age after 1949. With about 5 per cent of the total acreage in 1967 not yet in bearing, the bearing acreage will increase slightly in the next three years unless the removal rate of recent years is increased.

Since 1957 the productive capacity of the California grape industry has increased much faster than bearing acreage because of a marked upward trend in yield per acre (table 31). The average yield per bearing acre of California grapes has increased at a fairly consistent rate of about one ton per acre every twelve years and normally may continue to do so for the next few years as a result of improved cultural practices and shifts to higher-yielding varieties and districts.

Allmendinger (1965) estimated normal yields per bearing acre of California grapes for 1965 as follows: raisin varieties, 8.4 tons; table

---

[5] Updated through 1966.

TABLE 29

CALIFORNIA GRAPES, ALL VARIETIES: ACREAGE

Crop year	Bearing	Non-bearing	Total	Reaching bearing age	Removals since previous year	Plantings 1946–1966
	1	2	3	4	5	6
Average						
1946–						
1949	492,769	52,188	544,957	16,894	17,710	13,924
1950–						
1954	468,065	19,767	487,832	9,770	19,992	6,585
1955–						
1959	409,246	37,951	447,197	8,340	15,624	15,784
1960–						
1964	438,324	40,438	478,762	16,154	6,989	9,998
Annual						
1946	490,790	59,792	550,582	7,783	16,892	19,586
1947	489,052	63,555	552,607	15,072	16,810	18,835
1948	494,498	49,298	543,796	25,134	19,688	10,877
1949	496,636	36,109	532,745	19,586	17,448	6,397
1950	496,364	23,664	520,028	18,835	19,107	6,390
1951	480,834	19,136	499,970	10,877	26,407	6,349
1952	463,510	18,183	481,693	6,397	23,721	5,444
1953	454,096	17,665	471,761	6,390	15,804	5,872
1954	445,522	20,187	465,709	6,349	14,923	8,871
1955	423,774	25,888	449,662	5,444	27,192	11,145
1956	407,507	30,384	437,891	5,872	22,139	10,368
1957	399,000	35,102	434,102	8,871	17,378	13,589
1958	406,847	40,975	447,822	11,145	3,298	17,018
1959	409,101	57,407	466,508	10,368	8,114	26,800
1960	416,180	59,601	475,781	13,589	6,510	15,783
1961	424,876	50,161	475,037	17,018	8,322	7,578
1962	444,146	36,096	480,242	26,800	7,530	12,735
1963	451,496	29,702	481,198	15,783	8,433	9,389
1964	454,922	26,628	481,550	7,578	4,152	4,504
1965	459,850	31,530	491,380	12,735	7,057	17,637
1966	461,640	28,250	489,890	9,389	7,599	6,109
1967	462,500[a]					

[a] Preliminary estimate.
Sources: Fermenting Material Processors Advisory Board, *California Grape Industry: Some Production and Marketing Trends and Prospects*, by W. Allmendinger (1965), table 2; and California Crop and Livestock Reporting Service, *Acreage Estimates, California Fruit and Nut Crops* (Sacramento, annual issues). Col. 2: Three-year sum of column 6 for year shown plus column 6 for last two prior years. Col. 3: Column 1 plus column 2. Col. 4: Column 6 three years earlier. Col. 5: Column 1 for preceding year plus column 4 for year shown less column 1 for year shown. Col. 6: Data shown represent largest acreage planted in year shown which is reported as standing in any year.

varieties, 6.7 tons; wine varieties, 5.2 tons—an over-all average of 7.2 tons for all varieties. This compares with an overall average yield of almost 7.7 tons per bearing acre for 1963–1966, based upon revised production estimates (which include dried raisins at actual drying ratios which are higher than 4 to 1, the ratio previously used).

TABLE 30

TRENDS IN UNITED STATES GRAPE PRODUCTION AND CALIFORNIA UTILIZATION, 1935–1967

	1935–1939	1945–1949	1950–1954	1955–1959	1960–1964	1965	1966	1967
			1,000 fresh tons					
Production (of value)[a]								
United States	2,444	2,941	2,893	2,982	3,283	4,313	3,734	3,007
Other states	220	181	211	237	222	338	334	337
California	2,224	2,760	2,682	2,745	3,061	3,975	3,400	2,670
Raisin varieties	1,270	1,592	1,512	1,632	1,945	2,575	2,175	1,620
Table varieties	392	572	582	540	544	650	560	425
Wine varieties	562	596	588	573	572	750	665	625
Utilization (sales)[a]								
California	2,221	2,757	2,679	2,742	3,059	3,973	3,398	2,658
Fresh, all	502	538	532	521	525	566	560	402
Raisin varieties	136	160	166	201	223	245	249	192
Table varieties	206	274	285	246	243	266	261	156
Wine varieties	160	104	81	74	59	55	50	54
Canned (raisin)	6	21	26	35	46	55	62	56
Dried (raisin)[b]	940	986	871	821	966	1,297	1,190	760
Crush, all	773	1,212	1,251	1,365	1,522	2,055	1,586	1,450
Raisin varieties	194	427	450	574	709	979	675	611
Table varieties	180	296	297	293	301	382	297	269
Wine varieties	399	489	504	498	512	694	614	570

[a] Preliminary estimates for 1967 include some unofficial estimates for utilization.
[b] In recent years, 2,000–3,000 tons of dried grapes other than raisin varieties are included.
Source: Compiled by S. W. Shear, Giannini Foundation of Agricultural Economics, University of California, Berkeley, from latest reports of California Crop and Livestock Reporting Service and U.S. Department of Agriculture, *Agricultural Statistics.*

TABLE 31

TRENDS IN CALIFORNIA GRAPE YIELDS PER BEARING ACRE, 1924–1967

	Averages						Annual		
	1924–1928	1934–1938	1945–1948	1950–1954	1955–1959	1960–1964	1965	1966	1967[a]
	1	2	3	4	5	6	7	8	9
Varietal class				Fresh tons per acre					
All varieties	3.8	4.4	5.7	5.7	6.7	7.0	8.6	7.4	5.9
Raisin varieties[b]	4.4	5.1	6.6	6.6	7.8	8.2	10.2	8.6	6.7
Table varieties	3.8	4.7	7.2	6.6	6.9	6.6	7.8	6.8	5.2
Wine varieties	2.8	3.3	3.6	3.9	4.6	4.8	6.1	5.3	4.9

[a] Preliminary estimates.
[b] Through 1958, total raisin variety grape production includes dried raisin production computed to fresh weight by using a drying ratio of 4 to 1; from 1959 to date, includes actual fresh weight of raisin variety grapes dried as raisins.
Source: Computed by S. W. Shear, Giannini Foundation of Agricultural Economics, from latest revised estimates of production and bearing acreage from most recent reports of California Crop and Livestock Reporting Service.

California grape production averaged about 3,340,000 tons a year during the period 1963–1967. Allmendinger (1965) estimated that a normal crop for 1967 would have been over 3,400,000 tons, or about 100,000 tons more than the average harvested during 1963–1967. He estimated that a continuation of the recently prevailing trend in normal yields would result in an annual increase in the productive capacity of existing bearing acreage of about 37,000 tons a year. Actual yields per acre, however, may vary considerably from year to year, fluctuating with weather conditions, pest infestation, and cultural care. In 1965, with unusually good weather, the average of 8.6 tons per acre was much greater than an estimated normal of 7.2 tons. In 1967, when weather conditions were poor, it was only 5.9 tons.

The percentage of nonbearing acreage of vines to those of all ages (table 23), by varieties, districts, and counties in 1966, gives some indication of where to expect the most increase in bearing acreage and production in the next few years. For the state as a whole the highest proportion of nonbearing acreage was for wine varieties—approximately 11 per cent—compared with slightly over 3 per cent for raisin and almost 5 per cent for table varieties. The coast counties south of the Bay had nearly 19 per cent of their wine grape acreage in nonbearing vines, the largest proportion of young vines in any district in the state. The southern San Joaquin Valley had about 16 per cent, with over 23 per cent for Madera County, the highest proportion of nonbearing acreage for any individual county.

The proportion of nonbearing acreage for most white wine varieties in 1964 (table 23) exceeded the state average of 9 per cent for all wine varieties. Approximately 56 per cent of the acreage of dark wine varieties in the San Joaquin Valley exceeded the state average for nonbearing vines compared with only 35 per cent in the central coast counties and southern California.

In 1966 the largest percentages of total acreage of the variety as nonbearing in California vineyards were Cabernet Sauvignon (48), French Colombard (25), Gamay (77), Pinot blanc (69), Pinot noir (70), and Royalty (66). However, the actual nonbearing acreages standing in 1966 for the ten largest were Thompson Seedless (8,147), Carignane (2,916), French Colombard (1,482), Zinfandel (1,258), Emperor (1,131), Calmeria (815), Pinot noir (732), Cabernet Sauvignon (1,212), and Royalty (595). This indicates the continued large planting of table and raisin grapes in the state, but shows also that wine grape varieties have the largest percentage increases in tonnage in prospect. See also Winkler (1964) for information on high-quality wine grape acreage in the coast counties of California.

INCREASING GRAPE PRODUCTION...

... is outstripping commercial marketings.

The continued upward trend in California grape production and the commercial crush for wine and brandy has resulted mainly from the substantial increase in Thompson Seedless raisin grape production, most of which had to be crushed because of the slower increase in other uses of raisin grapes. Also, a significant but smaller increase in production of wine varieties has added to the increase in the commercial crush. Table varieties, however, have shown almost no increase either in production or commercial crush.

Figure 22 shows that changes in the pattern of grape utilization in the state reflect the larger annual variations and the major trend in total grape production; most of the increase in production from larger crops tends to be absorbed by the commercial crush for wine, brandy, and other juice products. Variations in the quantities of grapes utilized for shipment to fresh markets and for canning are small, and have resulted mostly from a gradual upward trend in the demand for canned fruit cocktail. The tonnage of raisins dried has tended to parallel changes in the raisin grape crop, both in annual fluctuations and the long-time trend. The parallelism between changes in total production and the commercial crush, however, is more marked than for utilization of dried raisins. Generally, the larger the crop, the larger the tonnage and the proportion of total production that is crushed by wineries and distilleries. Utilization other than the crush for all other purposes combined has not increased as rapidly as production. Hence the big increases in grape production in recent years have usually been absorbed mostly by the commercial crush of wineries and distilleries. In turn, most of the surplus production in excess of total marketings in regular trade channels as fresh grapes and grape products has cumulated in the inventory of wine and brandy.

Commercial marketings of California fresh grapes and grape products moved into domestic and foreign markets have increased steadily but slowly since 1960, reaching the equivalent of about 3,100,000 tons of fresh grapes in each of the marketing years 1965–66 and 1966–67, of which one-half was as wine and brandy (fig. 23 and table 32). During this decade, California grape production and the inventory of grape products both increased much more rapidly than regular commercial marketings. The inventory of California grape products, mostly in wine and brandy, in the summer of 1967 was the equivalent of almost 2,900,000 tons of grapes, compared with normal production of about 3,400,000 tons and annual marketings of almost 3,200,000 tons. However, actual production in 1967 was more than 25 per cent below a normal crop of 3,400,000 tons. Normal crops may be even larger during the next few years, and will continue to exceed the most optimistic expectations for regular commercial marketings of all California grapes, which may continue to increase only gradually above the maximum to date of nearly 3,200,000 tons reached in 1967. Excessive inventories of wines and brandy are likely to be maintained unless remedial action is instituted by the industry. A majority of any increase in marketings of California grapes will continue to be in the form of wine and brandy, which increased a total of only about 200,000 tons equivalent of fresh grapes in the seven years from

TABLE 32

CALIFORNIA GRAPES: TOTAL COMMERCIAL SUPPLY AND MARKETINGS, CALCULATED NORMAL MARKETINGS AND SUPPLY, AND MARKETINGS RATIO, CROP YEARS 1960–1967

Crop year	Incoming inventory[a]			Grape production	Total commercial supply	Commercial marketings[b,c]			
	Grape crush products	Raisins	Total			Quantity		Calculated normal index number 1960 = 100	Supply-marketings ratio[e]
						Actual	Calculated normal[d]		
	1	2	3	4	5	6	7	8	9
				1,000 tons, fresh-weight basis					
1960	1,628.0	123.0	1,751.0	2,766.0	4,517.0	2,846.8	2,812.5	100.0	163.6
1961	1,628.5	116.2	1,744.7	2,695.0[f]	4,439.7	2,850.8	2,864.7	101.9	157.9
1962	1,393.3[g]	114.0	1,507.3	2,724.0[f]	4,231.3	2,887.5	2,916.9	103.7	147.7
1963	1,335.9[g]	102.3	1,438.2	3,439.0[h]	4,877.2	2,949.5	2,969.1	105.6	167.2
1964	1,725.8	110.2	1,836.0	3,145.0	4,981.0	3,007.8	3,021.3	107.4	167.8
1965	1,817.0	208.8	2,025.8	3,975.0	6,000.8	3,140.0	3,073.5	109.3	198.6
1966	2,348.0	304.2[i]	2,652.2	3,400.0	5,952.2	3,101.4	3,125.7	111.1	196.9
1967	2,519.9	351.7[i]	2,871.6	2,670.0[j]	5,541.6[j]	—	3,177.9	113.0	176.7[j]

a July 1 for grape crush products and September 1 for raisins.
b For grape crush products, marketings are shipments into distribution channels and losses, excluding interprocessor shipments in California and bulk beverage brandy shipments out of state for storage and aging; for raisins, they are shipments by California processors, excluding interprocessor shipments in California; and, for fresh grapes and canning, they are utilization.
c July 1 to June 30 for grape crush products and September 1 to August 31 for raisins.
d Calculated normal marketings for any year is least-squares trend-line value for the year as computed from actual marketings for 1960–1966 inclusive.
e Percentage of total supply (col. 5) is of calculated normal marketings (col. 7) during prior year.
f Excludes grapes crushed to produce products held, set aside to meet requirement of Federal Grapes for Crushing Marketing Order.
g Excludes products held, set aside to meet requirement of Federal Grapes for Crushing Marketing Order.
h Excludes rain-damaged raisins lost in field during drying.
i Excludes surplus raisins sold to U.S. Department of Agriculture but not delivered.
j Preliminary.

Sources: W. Allmendinger, "California Grape Industry—Economic Situation and Outlook" (San Francisco, California, June 9, 1967), table 1; prepared by Wine Institute from reports of California Crop and Livestock Reporting Service; Internal Revenue Service, U.S. Treasury Department; and Raisin Administrative Committee. Revised H-16-67.

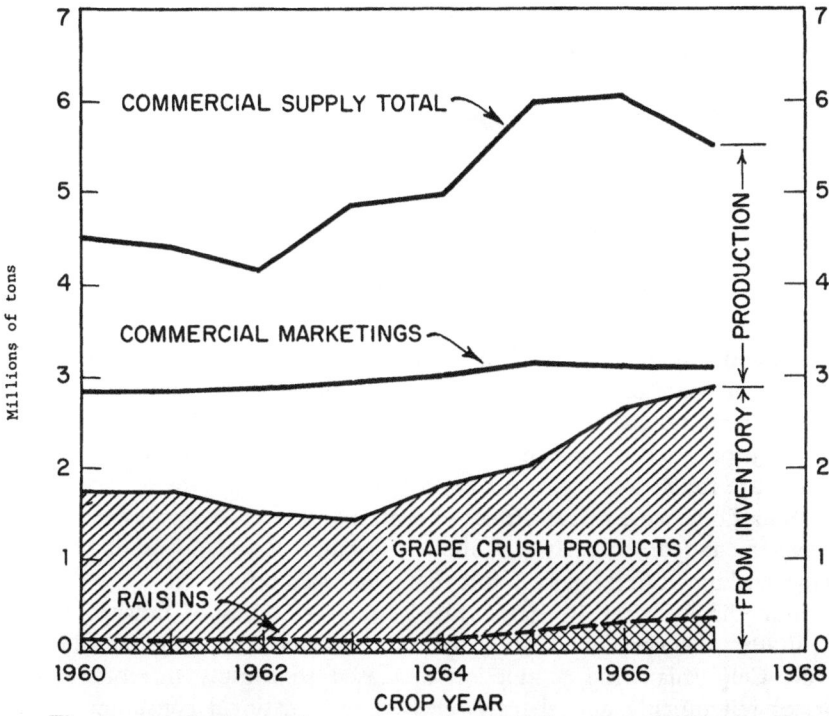

Figure 23. Commercial supply and marketings of California grapes, 1960–1967. (Production and commercial supply of California grapes has increased greatly in recent years and much faster than regular commercial marketings of all grapes and grape products.) Source: Basic data from table 32.

1960 to 1967. Table wines may be expected to account for a significant part of any such increase in wine marketings. California production of these wines has increased steadily since World War II doubling output during the decade through 1966.

The marked increase in the state's grape production and commercial supply of grapes and grape products since 1958 (figs. 22, 23) reveals that the commercial supply has increased greatly since 1960 and much faster than regular commercial marketings of all grapes and grape products. Total supply in 1966 was almost double the quantity of grape products moved into consumption that year. Much of this marked increase in available supplies reflects the large increase in the inventory of wine and brandy and other crush products. This cumulative increase in inventory has resulted from grape crops far in excess of commercial marketings moving into regular consumption channels. The excess of commercial

157

supply would have been greater if the program of the Grape Crush Administrative Committee had not sharply curtailed surplus supplies of dessert wine from the 1961 and 1962 grape crops. The drop in the commercial supply of grapes in 1967 resulted entirely from the exceptionally small 1967 grape crop—more than 25 per cent below normal expectations.

WINE AND GRAPE PRICES...

. . . have fluctuated greatly since Repeal.

Indexes of wholesale prices of standard bottled California wine f.o.b. California wineries, calculated by the Bureau of Labor Statistics of the Department of Labor for the calendar years 1947–1965, are given in table 33 for still red table and dessert wines. Fluctuations of these bottled prices are much less from month to month and from year to year than for bulk f.o.b. wine prices. Prices of dessert and red table wines generally move up and down relatively about the same. Retail prices of bottled wine are known to be more stable than wholesale prices.[6]

Immediately after Repeal, United States supplies of finished wine naturally were very small and prices of California wine very high. However, California wine production increased so rapidly thereafter that prices fell quickly and sharply, even though national consumption of wine rose nearly as fast as production for several years. From 1935 to 1941, f.o.b. bulk prices of standard California table wines were mostly 20–30 cents a gallon; of dessert wines, 30–40 cents; while grower returns from sales of wine grape varieties to wineries averaged only about $15 a ton for the period 1934–1940 (table 27). Thereafter, however, under the stimulus of wartime demand and governmental curtailment of production of alcoholic beverages (including wine), prices rose rapidly. Growers' returns for wine grape varieties averaged $88 a ton for 1943–1946. Wine prices reached a peak in 1945 when the Office of Price Administration raised maximum prices for bottled wine to $1.42 a gallon for dessert wine, 88 cents for red table wine, and $1.01 for white table wine. With the resumption of peacetime conditions of production, consumption, and demand, prices for wine collapsed in 1946–47. By the time the 1947 crop was being crushed, prices had fallen to 45–50 cents a gallon for dessert wines and 40–45 cents for table wine. Prices continued to decline, reaching a low point of about 35 cents a gallon for table wines and about

---

[6] For a more detailed summary of prices, demand, and consumption for California wine, brandy, raisins, and fresh table grapes, see Farrell (1966).

TABLE 33

INDEXES OF WHOLESALE PRICES FOR CALIFORNIA WINE, 1947–1967

	Red table and dessert[a]	Red table	Dessert
	1	2	3
Calendar year	Per cent of 1957–1959 as 100[b]		
Averages			
1947–1949	107.1	87.1	114.8
1950–1954	88.8	84.3	90.1
1955–1959	96.2	95.3	96.4
1960–1964	107.1	108.1	106.8
Annual			
1947	114.6	90.8	124.0
1948	104.8	85.4	112.4
1949	101.9	85.2	108.1
1950	96.0	86.3	99.5
1951	97.0	90.0	99.3
1952	82.7	82.3	82.2
1953	82.5	81.6	82.2
1954	85.8	81.5	87.1
1955	90.3	87.6	91.1
1956	90.5	88.7	91.0
1957	94.5	92.5	95.1
1958	102.8	103.8	102.5
1959	102.8	103.8	102.5
1960	102.9	103.8	102.6
1961	103.6	106.1	102.4
1962	110.5	110.9	110.5
1963	110.2	110.4	110.4
1964	108.3	109.5	107.9
1965	111.4	112.8	110.7
1966	109.9	111.4	109.4
1967	111.6	113.8	111.6

[a] Weighted averages.
[b] Based on prices f.o.b. California winery, in cases of twelve fifths.
Sources: *Wine Institute Bulletin*, No. 871 (San Francisco, 1957), p. 12, No. 1380 (1966), p. 5, and No. 1420 (1967), p. 7.

40 cents for dessert wines in the fall of 1949.[7] Grower returns averaged only about $33 a ton from 1947 to 1949, even though the market for grapes was supported by substantial government purchases of surplus raisins for overseas relief.

---

[7] The California Federal-State Market News Service has reported weekly f.o.b. winery prices for bulk wine since 1947, summarized in its annual issues of Marketing California Grapes, Raisins, Wine. The Wine Institute (1957, 1968b) has plotted simple averages of these weekly prices, and annual inventory shipment ratios between the percentage of the December 31 inventory of California wine shipments and all markets during the calendar year ending December 31, for table and dessert wines in graphs in its annual wine industry statistical reports for 1947–1957 and 1952–1967.

With an abnormally low inventory of California wine in the fall of 1950, grower returns jumped to an average of $60 a ton for wine varieties; during the winter, table wine prices rose to over 60 cents and dessert wine to over 85 cents a gallon. During the rest of the decade, beginning in 1950, grower returns for the wine grape crush showed a slight upward trend, fluctuating around $45 a ton. Bulk prices of table wines were fairly stable, fluctuating from 35 to 40 cents a gallon; dessert wines, 40 to 50 cents.

From 1956 to 1965, wine supplies relative to consumption were lower than during the previous decade, and prices higher; grower returns for the crush of wine varieties ranged from $50 to $70 a ton, averaging $64 during 1960–1964. Bulk prices of table wine from 1957 to 1965, with the exception of the 1961–1962 and 1962–1963 seasons, fluctuated near a fairly stable level of 50 cents, but declined gradually to about 45 cents in 1966–1967 and 1967–1968. The range in dessert wine prices, however, was much wider—from 50 to 75 cents in 1961–1962 and 1962–1963, when surplus supplies of dessert wine were sharply curtailed by the program of the Grape Crush Administrative Committee. The bumper crop of all grapes in 1965 depressed grower returns for the grape crush of wine varieties to $48 a ton and other varieties to about $27. Dessert wine prices were as low as 45 cents in the spring of 1966, recovering to a level of about 55 cents for the 1966–1967 and 1967–1968 marketing seasons. Table wine prices declined during 1966 and 1967, selling mostly at less than 45 cents a gallon f.o.b. California wineries.

There is a demonstrable and substantial interdependence among the raisin, table, and wine grape sectors at the producer level, fairly strong and direct between the raisin and table grape sectors, but weaker between either of those two sectors and the wine grape sector. During the decade 1947–1957 the upward trend in prices of the crush for each varietal class of grapes was due primarily to the comparatively static production of all grapes in conjunction with the rising demand for wine, particularly for table wine. Since 1957, however, there has been significant divergence in these relationships: the price of raisin and table varieties crushed has trended downward and the price of the crush of wine varieties has trended upward. These changes are probably attributable to the relative changes in the production of each varietal class and the relative changes in demand for grape products which were favorable to wine, particularly table wine.

Although there has been a close correlation among the several areas in year-to-year movement of prices for wine grapes crushed in very recent years the price differential for the north coast has tended to increase over prices in other areas. Comparative returns per ton to growers in 1965 for cash purchases by wineries for different varieties of grapes for each of the five chief areas of the state (table 28) show how very large and highly significant the differences in prices of different varieties of wine grapes within any one area and of the same variety in other areas have become. These differences reflect basic economic factors in supply and demand, particularly in the kind, quality, and quantity of the different wines produced. The highest prices are always paid for wine grapes grown in the north coast counties, primarily because they generally make the highest quality of table wine produced in the state; moreover, the yields per acre in this area are low and costs of production per ton high compared to those in the Central Valley. Grape growers in the north coast area would not grow wine grapes unless they received much higher prices than are paid for the same varieties of grapes grown in other areas where corresponding yields are much greater and costs per ton much less. Grapes that make premium quality table wine can be grown only in limited areas, and yields per acre are generally low in these areas. High yields and low costs per acre can be obtained only where soils are rich, summers are hot, and irrigation is available as in the Central Valley. Grapes grown in these high-yielding areas, however, do not produce as high a quality of table wine with the varieties currently available, although many of them do make acceptable standard and good quality dessert and flavored wines and brandy. The lower price paid for the crush of raisin and table varieties, used mostly for making dessert wine and brandy, largely accounts for the fact that normal market prices for these products are not nearly as much higher per gallon than for table wine, even with their much higher alcohol content.

Table 28 reveals that in 1966, with fairly representative production and prices, wineries paid growers in the north coast area much higher prices for every one of the twelve varieties listed. Prices were unusually high in 1967 because production was abnormally low. Prices for grapes grown in the Lodi area were the next highest, although substantially lower than for the north coast counties. The Modesto area ranked third, with varietal prices significantly higher than those paid in the southern San Joaquin area, where prices were the lowest for any area of the state, with the exception of a few white wine varieties for which southern California growers received the lowest prices.

The extreme differences in winery prices to growers are strikingly revealed by the relationship between the high prices received in the north coast counties and the low returns in the southern San Joaquin Valley. Returns per ton for the chief varieties of wine grapes in the north coast ranged from twice to three and a half times those in the southern San Joaquin Valley. White varieties in the north coast area brought top prices averaging $109 a ton, or more than three times the corresponding returns for the southern San Joaquin Valley; black varieties averaged $100, or slightly less than three times as much. Comparative returns per ton for different varieties, considered in the light of the yields per acre of individual growers, should be helpful in judging their relative profitability.

Wine and brandy are the most important grape products for which the United States per capita consumption and demand have increased significantly in recent years or seem likely to do so in the foreseeable future. Prospective increase in demand for wine, particularly table wine, is much more favorable than for fresh table grapes or raisins. Per capita consumption and the demand for California fresh table grapes and raisins have trended downward since 1933, and there is little reason to believe that they will improve. There appears to be little likelihood of significant improvement in export markets for any important California grape products. Aside from increasing national population, the economic outlook for the whole California grape industry, therefore, seems to depend largely on expansion of consumption and demand for wine and brandy in the United States. This is also the conclusion of Huber (1967) and Kasimatis et al. (1967). Huber recommends planting of better varieties and quality determination at the time of harvesting (to prevent overcropping and other undesirable grower practices). He believes that quality improvement is very important if California is to retain or increase its share of the American market.

Since the above summary was prepared, new viticultural areas in a number of California regions have been developed, particularly in San Benito and Monterey counties. In addition, there has been widespread planting of table wine grape varieties in the traditional coastal valleys as well as in the interior valleys. Table wine production has obviously assumed much greater importance than was envisaged when this text was written.

THE WINERY AND ITS EQUIPMENT

CHAPTER FOUR

WINERY DESIGN, CONSTRUCTION, AND EQUIPMENT

The winery should be of functional design, attractive in appearance, properly located, and constructed to facilitate operation and cleaning. The equipment should be selected for function, sanitation, resistance to corrosion, ease of maintenance, and assembled to permit straight-line operation.

Surprisingly little information on the architectural and engineering factors involved in the construction of winery facilities has been published. Most of the available data are descriptive and are not reported in sufficient detail. The construction and operation of wineries in the older reference texts, such as the classical German handbook by Babo and Mach (1922–1927), are incompletely described. Hengst (1963) pointed out that the architectural, engineering, and technological factors in construction and operation of wineries in Germany were neglected until quite recently. Even in the second edition of the two-volume treatise on enology by Ribéreau-Gayon and Peynaud (1964–1966) French equipment is described only briefly (vol. 1, chap. 15; vol. 2, chap. 22) and without construction details or operating instructions. Useful information on winery equipment, however, is reported in the catalogues of firms in Italy and France that produce winery machinery. Until recently the Italian equipment has been widely used in California, particularly the Garolla crusher and stemmer, which have now been largely replaced by locally constructed machinery.

A survey of wine-making equipment, with line drawings, was made by Manfredi and Ropa (1962). Some useful information is given also by Tarantola *et al.* (1954) and particularly by Troost (1961), who includes a list of suppliers as do Benvegnin *et al.* (1951) and Wines & Vines (1967). The basic unit operations and processes are described briefly by Joslyn and Amerine (1964). Henwood (1957) stressed the savings possible through improved engineering design of wineries. Troost (1966*a*) also pointed out the economic advantages of better engineering of processing operations and included some detailed information, especially on German equipment.

Principles

Surprisingly, a recent report by Rimskiĭ (1966) from the Soviet Union emphasized the importance of industrial aesthetics in winery construction for improved quality, greater labor efficiency, and development of an industrial culture. Attractive surroundings and good working conditions are stressed. The report of Raspino (1966) is an example of the increasing attention paid to labor costs and comfort. He considers the need for a separate bottling room, properly arranged and with adequate ventilation, heat, and light, and for adequate storage space for empty and filled bottles.

In contrast to the dearth of published reports on the wine industry, much more data are available on the construction and operation of breweries. Fehrmann and Sonntag (1962) discuss the machinery used in beer production from handling of grains to storage of finished product. Herz (1967), in a critical survey of the scientific and technological literature on brewing, listed the important reference works and periodical sources. For a similar review of the literature of enology, see Appendices A, B, and C.

Although continuous production of fermentable media and continuous controlled fermentation of organic acids, glutamic acid, and other amino acids, beer, bread, vinegar, and whisky has been developed (Rose, 1961, 1965; Rainbow and Rose, 1963), wine is still made by batch process operations. Continuous production of champagne has been reported in Russia (Amerine, 1963b), but difficulties have been encountered in reproducing those results here by Kunkee and Ough (1966). Holzberg et al. (1967) reported inhibition by alcohol and possible depletion of an essential nutrient from grape juice as limiting factors in engineering a continuous fermentation process. Continuous separation of juice from skins, seeds, and stems of grapes, while developed for *Vitis labrusca* (Concord) grapes, has not been applied to *V. vinifera* grapes (Celmer, 1961). Continuous pressing of fruit and grape juices and newer developments in production of juices from berries, grapes, and deciduous fruits are discussed by Koch (1956a, 1956b), Lüthi (1958, 1959a, 1959b), and Swindells and Robbins (1966). One restriction on applying continuous automated fermentation procedures to the wine industry is the limited period of time the raw material, grapes, is available for fermentation.

The unit operations and processes used in wine making are not reported in sufficient detail. Flow sheets are shown in relatively few publications (Cruess and Havighorst, 1948; Lee, 1950; Joslyn and Turbovsky, 1954; Ziemba, 1954). While these flow sheets indicate the major operations and the sequence in which they are performed, they do not include material and energy balances on the basis of which a definitive engineering evaluation could be made. There is not enough information to be useful in design and operation. The limited surveys of operations used in California wineries indicate wide variability. Berg (1949, 1951, 1953a, 1953b) surveyed most of the larger California wineries and examined their production methods. In 1949 he described several new techniques for crusher feed regulation, pumping-over operations, improved fining devices, wine tank handling operations, and wine heating for hot bottling. In 1953 he briefly reviewed the current fermentation practices, which have since been changed. Most of his attention was paid to current stabilization practices and wine stabilization factors. He reported in 1951 that none of the twenty-four wineries producing generic white table wines agreed on a finishing procedure, either as to type of operations, the sequence in which they were applied, or the method and frequency of application. This was true also of the twenty-three wineries producing generic red table wines. There was better agreement among the wineries producing varietal wines, particularly the reds. In view of the variation in finishing operations, both then and now, it is surprising that Amerine et al. (1967) should claim that finishing operations were more similar than crushing and fermenting operations. They recommended the separation of the latter from the wine aging and finishing operations.

This separation has been justified on the basis that crushing and fermenting should be carried out close to the source of grapes and that aging and finishing require more close technical control and hence are more economically carried out in large central wineries. Winkler (1935, 1962) emphasized the importance of reducing the interval between harvesting and fermentation, particularly in the warmer regions of California. He recommended that wine grapes be crushed and pumped into the fermenter within an hour or two of harvesting in warmer regions, and that hauling be limited to the viticultural region in which the grapes are produced. This would make it possible to deliver grapes to the wineries in more nearly sound condition than is usual with long hauls.

In the older established viticultural regions, particularly when grapes were picked by hand into small containers and transported on foot or by horse-drawn cart to the winery, it was customary to locate the winery close to the vineyard. Today, with rapid transport by truck and the possibilities of refrigeration in transit, it is not so essential for crushing and fermenting facilities to be near the vineyards. Wines from Concord grapes are produced economically at a large winery in Chicago from single-strength Concord grape juice shipped in tank trucks at $-2.2°C$ (28°F), with overhead ultraviolet germicidal lamps, from New York, Ohio, and Michigan. California wineries import frozen Concord grapes and berries from Washington and Oregon for fermentation. It is conceivable that mechanical harvesting coupled with automatic crushing and sulfiting may be applied to produce musts which can be shipped in open gondola trucks or closed tank trucks under refrigeration to a centralized winery. Economical and technological problems still remain to be solved, however. But if tuna can be caught as far away as Australia, frozen, and transported to canneries in Los Angeles, this should be possible.

The main and probably long-range justification for the separation of crushing and fermenting operations from aging and finishing operations and the location of the former close to the vineyards lies in the disposal of pomace from the fermenter. Waste disposal, both liquid and solid, is an important limiting factor in location of crushing and fermenting facilities. Stems from the crusher-stemmer and pomace from the fermenter are still disposed of largely on land. Economic conversion of these wastes into usable products remains an unsolved problem. While proximity of crushing and fermenting facilities to the vineyard may reduce cost and time of transportation, it increases the hazard of contamination of grapes with *Drosophila melanogaster* and similar fruit flies. Berg *et al.* (1958) and Berg (1959a) called attention to contamination of grapes delivered to California wineries with fruit flies, fruit fly excreta, and moldy and otherwise damaged berries.

The separation of crushing and fermentation from other operations was assumed to be the usual cellar practice in the older federal regulations (U.S. Internal Revenue Service 1961a, 1961b). Current regulations provide for bonded wine cellars (which may be designated bonded wineries under certain circumstances and which handle all phases of wine making from crushing through bottling). Bonded warehouses may also be installed to store bottled wine. In contrast, the finishing of standard wines has long been carried out in France and particularly in South Africa in central wineries. This practice is being established in California by larger wineries which contract for new wines to be produced by smaller wineries. The growers' coöperative winery in South Africa, the K.W.V., operates a central finishing cellar to which the individual winegrowers ship new wine made in their own premises (Anon., 1967c), whereas in California grower coöperatives usually transport their grapes to the winery. Amerine (1963b) has noted the centralization of finishing operations in the Soviet Union; similar practices are now standard in Jugoslavia, Bulgaria, Rumania, and Hungary, and to a lesser extent in Italy and Spain.

LOCATION

Selection of a site requires careful consideration.

Besides the need for proximity to producing vineyards to facilitate transport, two other considerations were involved traditionally. One was concerned with limitations of material transfer. Before the development of the must pump, near the turn of the twentieth century, it was necessary to locate the winery on a hillside so that grapes could be brought up an incline to the crusher and stemmer at the top level of the winery (or could be mechanically transported to the top level). The must was then transferred by gravity flow from a receiving sump or tank to fermenters on the floor below and from the fermenters to storage and aging cellars at a still lower level. Because of lack of proper insulation, air conditioning, and refrigeration, wines were stored for aging in cellars or in natural or artificially created caverns and tunnels. Modern hauling, conveying, and pumping facilities and modern mechanical refrigeration have eliminated these limitations. Aging in smaller oak containers, barrels, and ovals, however, is still preferred for finishing the finer types of table wines. This too is subject to change as we learn more about the rate and extent of oxygen absorption required for a particular type of wine. It is a reflection on the state of the art that oxidation-reduction phenomena in wine are still only partly defined both as to the types of changes and the extent to which they occur.

In selecting the site for a winery, consideration should be given to the types of operation to be performed; the facilities required to meet the planned production; the availability of grapes and other supplies required for the rate of processing; proximity to highways and roads for trucking and to spur tracks and railroads for rail transport; availability of manpower, both skilled and unskilled; availability of potable water for use in wine making and for steam generation, and of cooling water for primary cooling of musts during fermentation; availability of economic fuel and electric power; and availability of facilities to dispose of winery wastes without creating a nuisance.

Location near centers of population has advantages and disadvantages. The possibility of attracting visitors has been exploited both here and abroad. Convenient and attractive tasting rooms with adjoining retail facilities have proved to be important assets for advertising and marketing. Town populations may, however, expand until they surround a winery that was originally at the outskirts, and thus create problems. The direction of prevailing winds must be considered in order to avoid the creation of nuisances to surrounding settlements and to avoid contamination of wines by undesirable odors from manufacturing plants.

The site selected for the winery should be large enough to accommodate the main building, the office, the laboratory, tasting and display buildings, amenities for workers, boiler and machinery buildings, crusher and stemmer sheds, roadways to platform scales, testing facilities, unloading conveyors, and roadways from loading platform to trucks and rail cars used for shipping wines. Traffic control should be planned to avoid traffic jams. The traffic route for receiving grapes, supplies for control of fermentation and amelioration of musts and wines, and other raw materials should be separated from that involved in receiving, bottling, and casing supplies, and in shipment of wines. Adequate roadways and parking facilities for visitors should be provided. The land area should be adequate not only for the planned production but also for future expansion.

The site should be a level, well-drained area with soil structure adequate to support the dead load of buildings. Soils that would absorb run-off or rain water and slide must be avoided. The compression load the soil would be required to bear will depend on size, material of construction, and number of fermentation and storage tanks to be installed.

Information on these and other factors governing selection of site and winery location is given in Joslyn and Amerine (1964) and Amerine *et al.* (1967). General information on food processing, plant location, and plant design is summarized by Heid and Shipley (1961).

MATERIALS OF CONSTRUCTION

*Insulation, air conditioning, and ventilation are
important and when properly selected
will aid quality control and improve efficiency.*

The construction materials should provide sufficient insulation to help maintain the required temperature. The structural details, design of floors, walls, and roof, will vary with the layout of facilities and the degree of departmentalization. The crushing facilities are usually separated from the winery proper, and consist of unloading units, open conveyors to crusher-stemmers, and closed pipe lines from must sump to the fermenting or juice-separating tanks.

The construction of the covered walled-in sections, particularly the aging and finishing departments, bottling and bottle-storage facilities, where uniform temperatures are to be maintained, will depend upon the environmental conditions. In the cooler northern coastal valleys, concrete, stone, or brick walls may suffice, and cool temperatures may be maintained by opening vents to admit cold night air and closing the winery premises during the day. In the hot interior area more attention must be paid to insulation and air conditioning.

The problems of wine making in hot climates were surveyed and discussed by Bioletti (1905) on the basis of his visits to Algeria and elsewhere, and in more detail by Brémond (1957), La Rosa (1963), Ough and Amerine (1963b) and Rosenstein (1964).

Ready ingress of must and other supplies into the winery and convenient shipment of wine via truck or rail should be provided. Single-story, straight-line production is preferable to multistory construction for ease of operations. Lighting and ventilation requirements vary with operations. High ceilings and adequate window or vent space are needed in the fermenting room to remove gaseous fermentation products and for cooling. The bottling room should be well lighted and ventilated and easily cleaned and sanitized. Forced ventilation to remove carbon dioxide from the fermenting room or from large indoor fermenting vats is desirable. If carbon dioxide is used to blanket wines stored for finishing operations in stainless steel tanks, care must be taken to remove it from the room. The window and door area should be properly designed for adequate lighting and convenience of entrance and exit. To avoid heating during the day and to prevent contamination with fruit flies, lighting should not be excessive. Air blankets are preferable to screens over openings into the fermenting area and new wine storage areas.

The construction of the winery should be earthquake-proof, with adequate concrete foundations. If the soil area is likely to slide when wet, a floating slab support may be used. The coöperation of a reliable architect, a qualified structural engineer, and the wine technologist in deciding on essential operations and the design of facilities suited for these operations should be secured. This is important not only in the construction of new plants but also in redesign of existing plants for more efficient operation or expansion of production facilities.

The floor should be properly reinforced concrete covered with a smooth waterproof cement finish resistant to wine. The final cement coat on floors and over drain surfaces should be covered with an adherent, inert, flexible coating. The design of cement floors for food plants is discussed in general by Rader (1946), and the use of acid-proof brick or tile set with sulfur cement by Sheppard (1947). The properties and uses of the newer synthetic corrosion-resistant coatings are discussed by Ploderl and Weyman (1957). Various types of epoxy and other synthetic plastics for coating concrete surfaces are available.

The floors should be strong enough to bear the compression and vibration load required. Throughout the winery, particularly in the fermenting room, they should be well sloped for drainage so they can be easily washed and kept clean. There should be no low areas where water or wine may accumulate. Drains should have smooth well-rounded bottoms to prevent accumulation of materials, should be covered with a removable iron grating, and should be numerous, large enough, and with sufficient slope for maximum drainage flow. Drains are particularly useful underneath the discharge valves of tanks. There should be no corners where refuse and pomace can accumulate—no areas which cannot be reached by a stream of water.

Fermenting and aging and storage tanks, vats, or casks should be set on fairly high concrete piers to facilitate cleaning and inspection and repair of bottom surfaces. Racks for storage of barrels for aging smaller lots should be accessible for handling by fork-lift trucks, and facilities for bottle storage in suitable racks should be provided.

*Arrangement of facilities should
fit the needs of the plant.*

The main facilities required for the production of table wines are: (1) a grape-receiving area including platform scales and testing facilities; (2) crushing facilities including unloading area, and conveyors to crusher-stemmer and stem receiving and hauling facilities; (3) a fermenting room for receiving and fermenting red grape must and for receiving white grape must, separating free-run juice and pumping this into closed fermenters; (4) a storage room where fermentation is completed, wines are racked, filtered, fined, aged, and stabilized; (5) the bottling, bottle-storage, and casing room; and (6) the shipping department. Facilities should be provided also for a laboratory for quality control of grapes, fermentation, new wines, and finished wines, and for other necessary research and development. Separate facilities should be provided for boiler and refrigerator compressor, water-cooling tower, and machine shop for repair, maintenance, and construction operations. All working areas generating off-odors or subject to possible contamination should be well separated from the wine storage area. Adequate washroom, toilet, and eating facilities for employees must be provided. These are regulated by federal and state labor acts. A floor plan of a small winery in the Napa valley is shown in figure 24.

The arrangement of these units will vary with the size of the winery, the types of wine produced, and the degree of automation and mechanization. Crushing and fermenting facilities for red and for white wines are separated when the scale of operation justifies it. This separation reduces possible contamination of white musts and wines with red, and provides for more efficient arrangement of equipment, which differs in some respects for these operations.

Facilities should be planned so that the grapes, must, wine, stems, and pomace travel by the most direct and economical route. The unloading, conveying, and pumping lines should be operated with the minimum of interference with each other. The conveyors and pumps should be of suitable type, of sufficient capacity to meet all peak needs, distributed so as to facilitate all transfer of materials, and assembled so that they can be thoroughly and easily cleaned. Where possible, in-line cleaning operations should be planned. Overhead pipes and passageways or platforms should be avoided.

In larger wineries steel fermenter and storage tanks are installed outside the winery buildings, and only the finishing and bottling operations take place indoors. These steel tanks are usually lined with an adherent inert plastic to protect the must or wine against contamination with iron and are insulated outside with aluminum foil. The economic and technological factors involved in selection of these wine storage and processing tanks are discussed by Yamada (1959).

Yamada (1959) noted that as the number of wineries in California has been decreasing the storage capacity has been increasing (see p. 148).

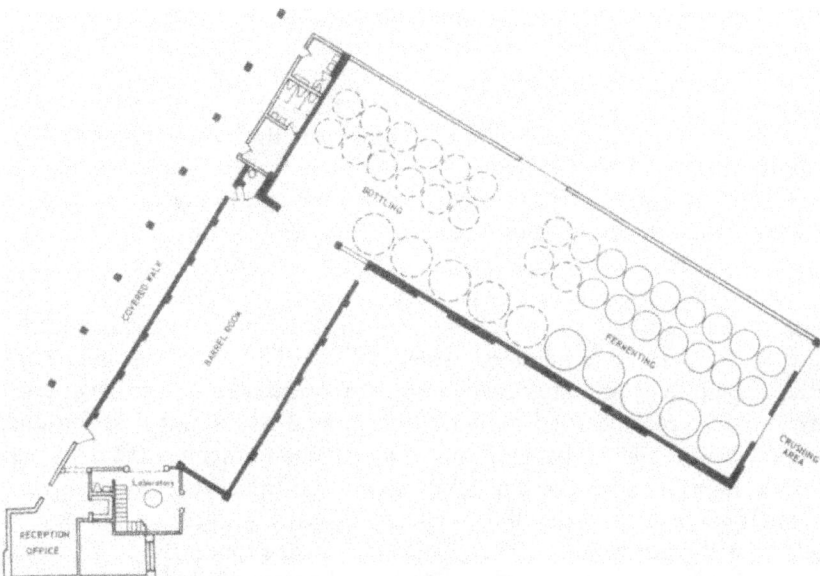

Figure 24. Floor plan of the Robert Mondavi Winery at Oakville, California (reception wing not shown).

173

This means that our wineries have become much larger, and a shift from wooden or concrete storage containers to steel containers has become possible. Among the advantages of steel tanks he noted their easier insulation, lack of calcium pickup (as from concrete), possibility of large sizes and lower costs, ease of sterilization and storage in an empty condition, lower insurance costs and slower depreciation, and greatly reduced wine losses and cleaning costs.

Improved engineering layout for separation and transfer of material has been described by R. Rossi, Jr. (1962), for large-scale processing of standard wines. This includes use of receiving tanks designed to facilitate separation of free-run juice from skins, seeds, and pulps; screen conveyors to remove pomace and feed it into a continuous expeller presser; and conveying of pressed pomace to disintegrators and into scalping screen separators where it is washed to leach out remaining sugar. While this layout facilitates handling and ensures economic and complete sugar recovery, its use is limited mainly to the production of standard dessert wines.

SIZE . . .

. . . should be adequate for the production planned
and for possible future expansion.

The appropriate size for the winery depends upon the needs of the proprietor. Neither a large nor a small winery can guarantee quality unless the wine maker knows and controls the factors that affect quality, such as selection, harvesting, and transport of the grapes and the conditions of fermenting and aging.

The fermenting facilities should be large enough to meet the needs of a fermenting season lasting from six to sixteen weeks. The size and number of fermenters should be sufficient to provide for peak-load operation of crushers and presses. The actual fermenting capacity varies with the quantity of grapes received, the total fermenting season, and the rate of turnover of fermenters. Joslyn and Turbovsky (1954) have suggested a simple relation to calculate actual fermenter capacity. If varieties are to be segregated for production of varietal wines and free-run and press wines are to be separated, more and smaller fermenters are required. The storage and aging capacity will depend on the average aging period used. If the average time is three years, a yearly production of 100,000 gallons will require over 300,000 gallons of storage space. But if the wines are aged only a year, the storage cellar need be only a little larger than the space required for the average yearly production. Joslyn and Amerine

(1964) reported that in California bonded wineries producing table wines the ratio of storage to fermenting capacity ranges from 2.00 to 9.28, with an average of 4.11 for fifty-three wineries of 30,000–2,000,000 gallons storage capacity. There is no sharp division between storage and fermenter areas; closed storage vats may be used for fermentation and open fermenters may be closed over for storage when necessary.

SERVICE
Water, steam and electricity are needed.

A high-pressure cold water supply, with many readily accessible outlets throughout the winery, is essential. Water is used in washing and rinsing tanks, conveyors, and floors, and in primary cooling of musts during fermentation. Water cooling is more economical than mechanical refrigeration, but is limited by availability of cool well water or of economic water cooling facilities. Insulated steam and hot water outlets will make it easier to clean and heat tanks, pipes, and equipment. The hot water may be reticulated as such or obtained by specially designed steam and water mixing valves.

Although the water supply in most California areas is pure enough for winery use, this is not true everywhere. Elaborate and costly water-purifying systems may need to be installed. The quality of water for efficient boiler operation should also be considered The sources, quality, use, and treatment of water in food processing are discussed by Joslyn (1963a) and Matz (1965). Steam production and reticulation in food processing are discussed by Joslyn (1963b), and steam generation and use in wineries by Fetter (1966b).

Adequate lighting, particularly in the bottling area and the laboratory, is necessary. Diffuse daylight is preferable in the laboratory, but suitable artificial lighting in the bottling area and other working areas will suffice. Numerous protected electric outlets for current of various power loads are a great convenience in winery operations. Nonsparking waterproof connections should be made available where alcohol vapors may be present or where water might splash onto and into a motor or connection. The application of electricity for motor drives, for regulating, recording, and controlling temperature, pressure, and flow, for inspection, and for lighting in food processing operations is described by Dodds (1963).

EQUIPMENT

All materials with which the musts and wines come into contact should be resistant to corrosion; should add no undesirable flavors, odors, or metallic ions to the musts or wine; and should be easily cleaned and sterilized.

The historical development of stainless steels and the factors influencing selection and use of stainless steels in wine making has been discussed recently by Fetter (1968). In general, the 18-8 chrome-nickel steels are satisfactory except where contact with sulfur dioxide and chlorine antiseptics is involved. Fetter recommends a high surface polish to reduce adherence of argols and facilitate cleaning. Troost (1967b) described the vertical and horizontal stainless steel pressure tanks used to ferment and store white and red wine in Germany and the properties of the stainless steels used in their fabrication.

A detailed study of Soviet-produced plastics in the wine industry, made by Tiurin et al. (1964), revealed that most plastics, such as polyethylenes, were suitable for wine, but detailed sensory data were not given.

Recently lightweight fiberglass tanks have been successfully used in this country for fermentation and storage of wine. Fiberglass tanks have the advantage of being self-insulated, flexurally strong, and corrosion-resistant.

METALLIC CONTAMINATION...

*... should be guarded against to protect
both wine and equipment.*

Much information is now available in the United States and foreign publications describing wine-making equipment. Hengst (1963) described equipment and installation and discussed factors influencing selection of particular types. Descombes and Crespy (1956), De Rosa (1965b), Dehore and Dellenbach (1966), and Troost (1966a, 1966b, 1966c) discussed new materials available for construction or coating of wine receptacles, including fermenting and storage vats, transportation tanks, and glass containers. The increasing use of metal tanks makes it essential to prevent metal contamination and undesirable changes in appearance and odor by the application of suitable inert coatings. The older coatings for cement and steel tank surfaces were compared by Cruess et al. (1937), who found that asphalt-based paints were best. These however, like the microcrystalline wax coatings, have a tendency to peel. They have been largely replaced by plastic paints. A wide variety of synthetic plastics are now available for coating cement and metal surfaces. Synthetic plastic coatings are available as prepared water emulsions for direct application to clean surfaces, as separate monomers and catalysts which can be applied at room temperature, and as coatings which must be heated or baked to obtain full protection.

Epoxy resins contain epoxide and hydroxyl, which react with polyfunctional reagents, such as phenolic resins or polyamines, to yield cured resin systems of good chemical and water resistance. Curing is done by heating and treatment with acids. Ploderl and Weyman (1957) have shown that it is possible to modify the formulation of epoxy resins, their solvents, and the curing process to make them suitable for a wide range of uses. Chlorosulfonated polyethylene coatings were shown to have excellent resistance to chemical attack. Sigalin (1965) reported data on epoxy resins produced in Poland for protective coatings for steel and concrete containers for musts and wines.

Rubber hose constructed of laminated reinforced fabric and, more recently, plastic pipe (Stowe 1957), as well as permanently installed coated steel, stainless steel or glass lines, are used for conveying must and wine. They should be of the proper size and be fitted with corrosion-resistant quick-opening sanitary connections.

It is imperative that the must and particularly the finished wine be protected from metallic contamination during handling and storage. In the small winery these requirements can be met fairly well by wooden presses, concrete or wooden vats, casks, corrosion-resistant crushers and buckets, rubber hoses, and glass bottles. The use of metal crushers, stemmers, and must lines frequently leads to iron and copper pickup, and there is appreciable calcium pickup from new concrete tanks and poorly prepared filter pads. Since these cause various disorders in the wine, such as clouding, off-flavors, and off-aromas, they must be avoided (p. 794 f).

Unfortunately, the simplest solution—the use of noncorrodible metals—is costly. Since the danger of metal contamination is not equal at all stages or during all processes, corrosion-resistant metals or nonmetal materials can be used on surfaces where it is most likely to occur. Must or wine varies in corrosiveness according to its acidity, pH, aeration, temperature, sulfur dioxide content, and rate of flow. The volume of wine that passes over or through a given metal or concrete surface is important. The acidity and pH of musts and wines vary from region to region, and this, of course, influences corrosion. Metal contamination is more of a problem with high-acid, low-pH wines from the cooler regions, all other conditions (such as sulfur dioxide control) being equal.

Since conditions that affect corrosion vary from plant to plant, no general recommendations can be made on what to use for each piece of equipment. But information about the places where corrosion is most severe and about the behavior of metals and metal alloys may be helpful in making a choice.

Corrosion-resistant metals

Extensive tests with various kinds of metals and alloys were made by Searle *et al.* (1934) and by Mrak *et al.* (1937*b*) soon after repeal of the Eighteenth Amendment. Data on newer metals and alloys are not available. In the tests of Mrak *et al.* (1937*b*) stainless steels (18-8-3 alloys), Inconel, Aluminum-alloy 76, Durimetl, and certain metalized metals withstood corrosion satisfactorily. Relatively resistant metals were aluminum, copper, Aluminum-alloy 12, Cupro-nickel, Everdur, Waukesha metal, and acid, aluminum, manganese, and phosphorus bronzes. In some instances the absence of corrosion may have been due to deposition of tartrates, which prevented contact between the metal and the liquid. Durimetl N2795 and metalized metals 110, 120, and 130 had very irregular surfaces and showed this type of deposition. Less satisfactory metals tested included Aluminum-alloy 55, Aluminum 86, chrome plate, Monel metal, and silver. Least satisfactory were cast iron, machine brass, nickel, nickel plate, Niresist, steel, tin, and tin plate.

In these tests, deep pitting of the metals was not observed; but under severe winery conditions pitting of aluminum alloys has been noted. Aluminum, although it has the advantage of high heat conductivity and high structural strength per unit weight, has the disadvantage of being subject to pitting corrosion (Gilroy and Champion, 1948). Aziz and Goddard (1952) investigated the pitting corrosion characteristics of aluminum and have recommended alloying with magnesium and/or manganese to reduce this. At present, aluminum is used only for picking buckets or tubs.

The selection and proper finishing, fabrication, and specification of stainless steels to achieve sanitation and corrosion resistance in food processing are discussed by Cunningham (1947). Of the many stainless steels available to the food industries, experience indicates that the chromium nickel nonmagnetic steels No. 304 and 316 are best. Even these will pit and corrode when improperly machined and made. Welding failure is the most common cause of trouble with stainless-steel equipment. Recent improvements in design and fabrication of stainless steel tanks are discussed by Peters (1965). Troost and Fetter (1966*a*) emphasized the proper surface polishing of stainless steel.

The proper passivation of stainless steels before they are used is important. The passivation should be preceded by thorough cleaning with a slightly alkaline solution followed by treatment with 5 per cent oxalic acid. Tufanov *et al.* (1966) investigated corrodibility of Russian stainless steels and changes in color and taste of wines in contact with them. They reported that the rate of corrosion increased for the first five days and then slightly decreased. This was ascribed to the presence of protective oxides on the surface. The effects on wine quality with the wines studied were insignificant. Different stainless steels suitable for short exposure, for prolonged exposure, and for construction of presses and pumps were recommended. See also Terrile (1965).

Effects of metals on wines

In the tests of Mrak *et al.* (1937*a*, 1937*b*), the metals showing the least effect on the wines were Durion, Durimetl, Inconel, Allegheny C (a stainless steel), aluminum-bronze, and Monel metal. Thoroughly cleaned copper filings had no detectable influence on the appearance of red and white table wines after four weeks' storage; but uncleaned copper turnings rapidly caused turbidity. Hydrogen sulfide was formed in the wines when cast iron, steel, tin, and some of the aluminum alloys were added to the wine. Aluminum and manganese-bronze tended to bleach the red wine. Usually changes in appearance of the wine occur at much lower concentrations than to changes in flavor or odor.

These workers also tested metallic salts, such as might be formed when metals and must or wine are in contact. When salts of aluminum, calcium, chromium, cupric copper, ferrous and ferric iron, lead, nickel, silver, stannous and stannic tins, and zinc were added to red and white table wines, changes occurred in appearance, odor, and taste. The salts of all the metals except nickel, chromium, zinc, and stannous tin, in both wines, and ferrous iron in the red table wine, caused the wines to become turbid. The amounts of the metals used were 2, 5, 10, 25, and 50 parts per million, and the wines were stored up to 280 days. All the salts except those of calcium, and ferrous iron and tin, in the white wine, or aluminum, copper, or ferrous iron, in the red wine, eventually caused the wines to have an aldehyde odor.

Factors affecting corrosion

Only one kind of metal or metal alloy should be used on a single piece of equipment, at least on surfaces that come into contact with wine. The use of two or more metals that touch each other increases corrosion and pitting from the electric potentials that are set up where they join.

The relative hardness of metal surfaces should be considered, particularly in the construction of pumps and mixers. Industrial hard chrome plating, because it gives a particularly hard and inert surface, has been used with advantage in other food industries (Scull, 1949). Polished, smooth hard chrome surfaces applied directly on iron or steel as well as on copper were tested in the laboratory by Joslyn (1950) and Joslyn and Lukton (1953) and were found to be resistant to wine. E. A. Rossi, Jr. (1951), however, reported that, while hard chrome plating is more satisfactory for wines than unplated bronze, it does not appear to be as satisfactory as stainless steel. He found that where there was metal-to-metal contact, where the parts were difficult to plate, where mechanical wear occurred, or where wine and air were in contact with the equipment, hard chrome plating did not prevent corrosion and was not as satisfactory as stainless steel.

New alloys should be thoroughly tested in the plant under all potential conditions of operation (temperature, type of wine, and so on) before being used for expensive equipment. Manufacturers of winery equipment should make such tests before using the new alloys.

If the metals are tested in places where the must and the wine will touch them, the metal best adapted to withstand corrosion at those points may be selected. Corrosion appears to be most severe at the crusher; hence the most corrosion-resistant metals are needed for crushers. Filter chargers (prefilter mixing tanks) are also very easily corroded because the wine is agitated and aerated in them.

White musts and wines are more corrosive than red ones. Corrosion is generally but not always greater if the must or wine contains sulfur dioxide.

Adequate cleaning of metals surfaces before and after use is very important, even with corrosion-resistant metals, if corrosion is to be controlled. Poorly welded equipment may show excessive corrosion.

CRUSHERS AND STEMMERS

The combined crusher-stemmer is now generally used in California.

Unloading, crushing, and stemming operations are now usually located outside the winery proper, preferably but not necessarily under roof, to facilitate delivery of grapes and disposal of stems, for ease of cleaning, and to avoid introducing fruit flies into the winery. When the daily volume handled is small, the grapes can be dumped directly into the crusher, but this is rarely done at present. Separate crushers for red and for white grapes are preferred. The crusher, stemmer, must pump, and must line should be constructed of corrosion-resistant metals. Metal contamination by grapes and must during conveyance to and from the crusher-stemmer and to the fermenter can be minimized by coating the inner surfaces of conveyors with acid-resistant plastic. Most of the older crushers are made of steel, but the later models are usually of bronze. Since sulfiting by injection of measured volumes of liquid sulfur dioxide at the discharge end of the must pump is now common practice, particular care should be taken to avoid metallic contamination. Most of the iron dissolved from the steel surfaces of the crusher and stemmer is precipitated during fermentation and storage, but dissolved copper may be harmful to wine quality and stability.

Unloading platforms

Grapes may be delivered to the winery in wooden field boxes in which they are packed, in one-ton sheet-metal tanks handled by fork-lift trucks, in two-ton detachable tanks set on narrow two-wheel trailers, or in three-to five-ton mounted tanks or gondolas (fig. 25), mounted on semitractors or four-wheel tractors (Winkler, 1962). The trucks of grapes in boxes or in small tanks and the gondolas are weighed upon arrival at the winery, usually on platform scales, sampled, inspected, and unloaded at specially constructed unloading platforms or receiving tanks connected by conveyor to the crusher.

Figure 25. Gondola truck being unloaded at winery. (Courtesy Almadén Vineyards, Inc.)

Two types of unloading platform for boxed grapes are in common use. In one type the conveyor leading to the crusher operates at about the level of the bed of the delivery truck so that the boxes of fruit can easily be dumped directly into the conveyor. Unloading may be done from one or both sides of the conveyor. The platform is usually of wood, is difficult to clean, and may leak. Furthermore, no means of cleaning the boxes before returning them to the truck is usually provided.

In the second type of unloading shed (fig. 25) the conveyor is at the ground level with a sloping concrete platform leading to it from one or both sides. The slope of the platform should be sufficient so that the grapes can easily be pushed into the conveyor.

When grapes are delivered in gondola trucks, mechanical unloading (fig. 25) has largely replaced the older unloading by hand. This varies from crane dumping of smaller tanks to larger tanks mechanically tilted by hoist or specially constructed fork-lift truck. After unloading, the grapes are fed into the crusher by a screw conveyor or by hydraulic rams.

Crushers

Several types of crusher are available: the roller crusher, the Garolla crusher and stemmer, the centrifugal crusher (fig. 26), and the disintegrator (fig. 27). In California the Garolla crusher predominates, and the disintegrator is used mainly to facilitate recovery of sugar from pomace and stems and for production of must for distilling from grapes. The disintegrator, however, is useful in producing better-balanced musts from San Joaquin Valley destemmed red grapes of low tannin content. Wines from disintegrated unstemmed grapes usually are objectionably contaminated with stem extractives and are cloudier and more difficult to clarify.

The design of crushers is now being investigated as more information on the engineering properties of grapes becomes available. Zhdanovich *et al.* (1967*a*) pointed out that very little basic information is available on the rational construction of crusher-stemmers. Data are needed on the influence of berry composition, size and turgor, cluster size, tenacity of attachment of the berry to the stem, degree of oxidizability of the crushed grape juice, degree of brittleness or dryness of the stems, and other factors. The fact that all these variables are changing during the harvesting period needs to be taken into account. Emel'ianov *et al.* (1966) correctly noted that construction of efficient crushers depends on gaining better information on the resistance of grape berries to crushing. Their apparatus was similar to that of Briza (1955), which was used by Amerine (1962*b*). He showed that resistance to crushing decreased during ripening.

Figure 26. Vertical crusher-stemmer. Grapes enter at top; capacity is about 30 tons per hour. (Courtesy Budde & Westermann Machinery Corp.)

Table grapes were much more resistant to crushing than wine grapes. Manfredi and Ropa (1962) described the available equipment for crushing, stemming, and pressing, and for separating juice or wine from pomace.

Roller-crusher

The earlier roller type has two grooved metal rollers which crush and tear the grapes between them while revolving in opposite directions at the same or different speeds. A difference in speeds helps to tear the skins.

183

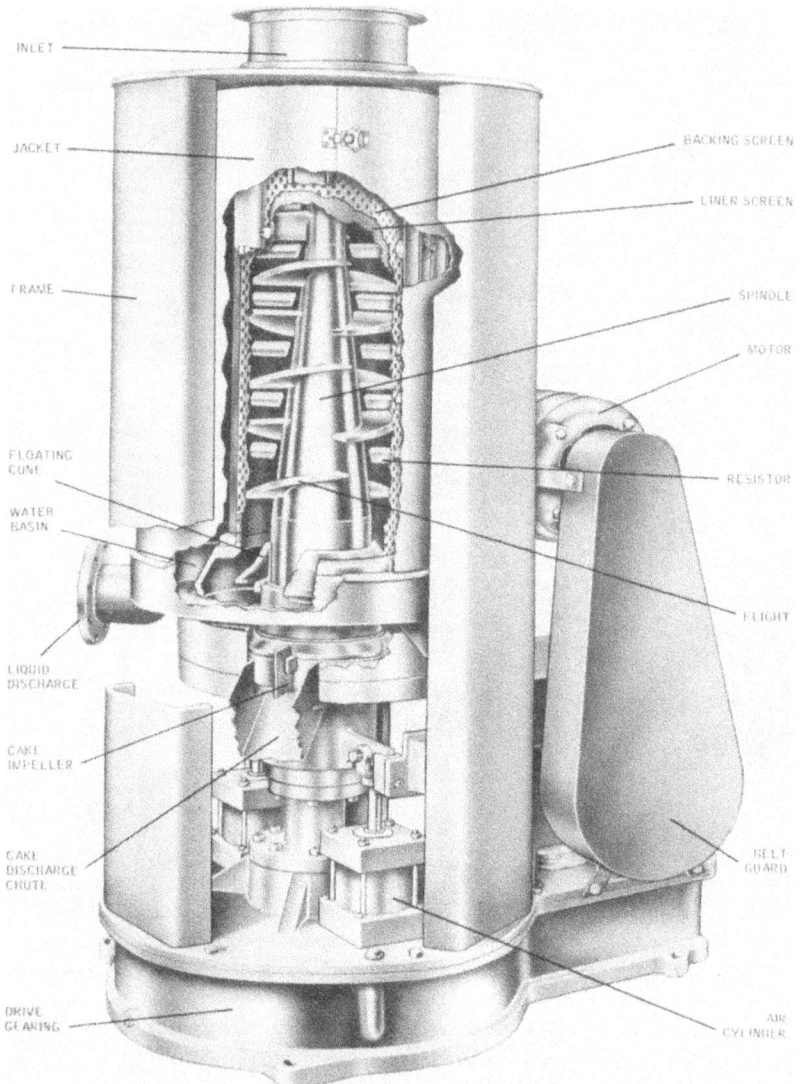

Figure 27. Cutaway of a Pressmaster, a vertical screw press used for preparing clear grape juice. (Courtesy Jones Division, Beloit Corp.)

In some machines the two rollers have fairly deep flutings that fit into each other, and both rollers operate at the same speed. The rollers must be set far enough apart so that the seeds are not broken, but close enough to crush all the fruit. The distance between them should be adjusted according to the variety and maturity of the grapes and the season. Sometimes the grapes are only crushed by this system, but usually the crushed grapes and stems drop down into one end of a perforated cylinder, where they are separated from each other by revolving metal paddles. The stems are ejected at one end, and the must collects in a sump, from which it is pumped.

Crusher-stemmer

Other types of crushers utilize only very rapidly moving metal paddles, set in a more slowly revolving cylinder, for crushing the grapes. In the crusher-stemmer, the most important type used in California, the grapes are knocked against the sides of a revolving cylinder, crushed, and separated from the stems. The must falls through the holes in the cylinder, and most of the stems continue out through the open end. Very thin-skinned, juicy varieties are crushed easily; thick-skinned, overripe, tough ones, with more difficulty. Controlling the speed of revolution of the paddles helps to secure a higher percentage of crushed fruit.

A crusher constructed of an upright cylinder with rapidly revolving paddles between fixed horizontal plates is sometimes used in Europe. The grapes are fed onto the top plate and are thrown centrifugally against the fixed sides of the cylinder. The crushed grapes fall from the sides and the process is repeated on the lower level. With a suitable speed of rotation, good crushing is achieved.

MUST AND WINE LINES
Corrosion resistance and cleanability are needed.

Some wineries have permanently installed corrosion-resistant or suitably lined pipes from the crusher to the fermenters (fig. 28), from the fermentation vats and tanks to the sumps, from the fermenting room to the storage room, and between the tanks. These should be self-draining and arranged for easy cleaning with steam, water, or antiseptic solutions. As far as possible, they should be free of tees and elbows where liquid or trash can accumulate.

Glass pipe is used satisfactorily in many wineries for the transfer of wine. This is a good development, particularly for handling hot sweet table wines that contain a good deal of sulfur dioxide.

Stainless steel or enamel- or glass-lined steel is preferable to iron or copper. The inside of iron pipe may be coated with protective enamels or baked-on paints or may be hard chrome-plated. Several resistant

Figure. 28.   Redwood fermenting vats and must line.

linings, free from objectionable flavors and tastes, are available for coating pipes. They should, however, be applied carefully, and the lining should be examined periodically for cracks or other imperfections. Individual tests should be made under winery conditions, especially as to the durability of the paint, its protective properties, and its possible effect on the wine.

Rubber hoses are used for much of the transfer of wines. Only the best-quality, smooth-lined hoses should be used. These should be thoroughly conditioned before use by running dilute citric acid through them. After use they should be disinfected and kept dry. Lüthi and Hochstrasser (1950) have shown that rubber hoses lose their smooth interior surface with age and become a source of microbial infection.

CONTAINERS

Selection of material depends on use; proper preparation is essential.

Types of containers

Vats are straight-sided open containers and may be constructed of oak, redwood (fig. 28), or concrete (fig. 29); similarly constructed containers

Fig. 29. Concrete fermenting vats.

with a top are called "tanks" (fig. 30). Actually, wooden vats and tanks may have a slight bulge or may be smaller at the top than at the bottom, but the essential feature of their construction is that the staves are not bent or coopered. They are constructed in varying sizes with capacities of over 1,000 gallons. Redwood is used more commonly than oak because the latter is much more expensive.

Barrels, casks, and pipes have a bulge in the center: they are made with staves larger in the middle than at the ends, for strength. They are usually made of white oak, the preferred material for wine storage containers. Barrels (fig. 31) are constructed in varying capacities from 3 to 50 gallons; the smaller ones are known as "kegs." Containers of the 100- to 1,000-gallon size are known as "butts," "pipes," "puncheons," "ovals" (fig. 32), or "casks," depending on their shape and country or district of origin. Oak casks are occasionally made in sizes up to about 5,000 gallons, but wooden cooperage larger than this is usually made in the form of a tank.

Figure 30. A puncheon, oak tanks, and ovals.

Vats are used for fermenting red musts; tanks for fermenting both red and white musts and also for storage. Large fermenters (over 2,000 gallons) are not desirable for the better wines because they involve difficulties in controlling both the quality of the fruit used and the temperature and rate of fermentation. But for standard-quality table wines, vats and tanks (fig. 33) up to or over 50,000 gallons' capacity are satisfactory if enough cooling is available.

Figure 31. Oak barrels for aging of table wine. Note staggered arrangement to permit cleaning underneath barrels. (Courtesy Beaulieu Vineyard.)

Wooden cooperage is used mostly for storage, although oak casks are sometimes used for fermenting both red and white musts, and puncheons are often used as white-wine fermenters.

New containers, whether of oak, redwood, or concrete, must be treated before wine is placed in them. They should first be checked for leaks by filling them with water. Methods for conditioning are discussed on later pages. Both open and closed fermenters are used in California, but for sanitary reasons the latter are preferred. To keep the floating cap moist, liquid is pumped from the bottom of the fermenter and released on the cap.

Special tanks for draining off the free-run juice are now common (fig. 34). These usually have false bottoms or steeply sloping floors to facilitate drainage. Most of the new installations are of stainless steel. Figure 35

Figure 32.   Oak ovals and tanks, lined metal tanks, and plate-and-frame filter. (Courtesy Paul Masson Vineyards.)

Figure 33.   Redwood tanks in foreground and oak barrels in rear for table wines. (Courtesy Paul Masson Vineyards.)

shows a pressure installation in Australia and figure 36 shows a pressure one in South Africa. Valuïko *et al.* (1967) showed that application of a vacuum increased the amount of free-run. It also reduced the amount of oxidation and resulted in musts of lower tannin and pigment content. There was a slight increase in the amount of suspended material and a detectable loss of aromatic components. The construction of a vacuum separator is complicated and was used only for small-scale operation.

Wines are stored in oak, redwood, concrete, lined iron, or steel and stainless steel containers. The latter are especially used in blending finished wine and holding it prior to bottling so as to obtain wine of more uniform quality. The aged polish-filtered wines, after removal of dissolved and occluded oxygen by nitrogen, may be held under a blanket of carbon dioxide or nitrogen in stainless steel tanks (Cant, 1960; Mondavi and Robe, 1961).

Proper shapes for wine tanks were discussed by Buïko (1966), who recommended square or rectangular rather than round tanks. Haushofer (1966) compared the space occupied by cylindrical metal tanks when standing or when horizontally stacked and the interface wine/deposit area. He calculated that standing tanks occupy a smaller cellar volume but result in larger waste of cellar space.

Figure 34.   Settling tanks in California winery. (Courtesy Beaulieu Vineyard.)

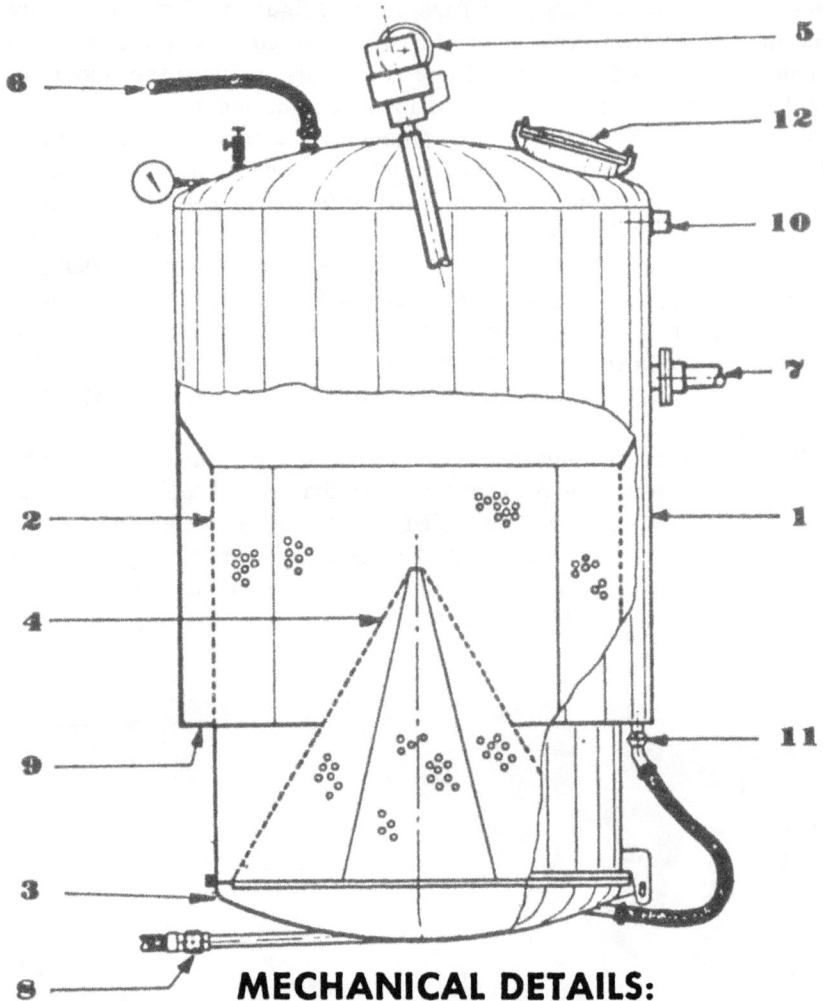

MECHANICAL DETAILS:

Figure 35. Mechanical details of a grape extraction unit used in South Africa: 1, 12 × 7 butt; 2, internal removal perforated cylindrical screens; 3, domed hinged bottom door; 4, conical detachable screens; 5, electrically operating door opening; 6, CO_2 gas inlet line; 7, crushed must inlet line; 8, juice outlet line; 9, side juice collecting ring; 10, must level float switch; 11, juice delivery ball float valve; 12, manhole. (From Berg, 1967.)

Concrete fermenters

Most new fermenters in California are of lined iron or stainless steel. Concrete fermenters (fig. 29) take less room than wooden ones and are

Figure 36. M.A.C. separator from Australia for separating juice under carbon dioxide pressure. (Courtesy Dr. J. C. M. Fornachon.)

easy to keep in good condition between vintages. They can be neatly arranged in rows. They are permanent and need little upkeep. But if the inside surface does not stay smooth, they may become hard to clean, and undesirable microörganisms may multiply in the crevices. They should at least be smooth-finished. Frequently they are specially lined (see also p. 196). One disadvantage is the low heat loss.

Closed concrete tanks were most often constructed for fermenting red wines. These, which are common in Algeria, have the advantage of complete submersion of the cap in either liquid or carbon dioxide, and can be used also for storing ordinary bulk wines at the end of the vintage season. All types of concrete containers should be properly designed and constructed. Adequate reinforcing is needed to prevent cracking.

Corsetti (1966) reports that large wooden tanks are being replaced in Italy by reinforced concrete tanks. Slow-setting, fast-hardening, high-resistance, or aluminous cements are recommended. Sizes of 15,000 to 30,000 gallons are mentioned—smaller than would be considered economical in California. In constructing concrete tanks Corsetti rightly emphasizes the need for a proper gravel bed under the foundation, for rounded internal corners, and for smooth internal walls (preferably treated with fluorsilicates or glass-lined). He recommends filling the tanks with a 2 to 3 per cent solution of quicklime during setting. To remove alkaline materials he first fills the new tank with 10 per cent sulfuric acid and then with 10 per cent tartaric acid. Before wine is placed in the tanks they should be thoroughly washed with water.

Redwood fermenters

Redwood is still a widely used material for fermenting and storage vats. Wooden vats require a smaller initial outlay than do concrete or insulated steel tanks. When properly erected and cared for they last for many years. The rate of heat loss from them is usually greater than from concrete; so the temperature is somewhat more easily controlled. Heat radiates more rapidly into the air around them because of their arrangement, wall thickness, shape, and larger ratio of surface to volume.

Skilled help should be used in erecting wooden vats. If a porous stave is found when the vat is filled, it should be replaced. The center hoops should never be tightened as much as the upper and lower ones. If a leak appears when the vat is being filled, the vat should not be filled very far above the leak until the opening has had a chance to swell shut; otherwise the weight of the liquid above it may prevent swelling from closing the leak.

Open fermenters should never be filled so that they overrun at the height of the fermentation. At least 20 per cent unfilled space must usually be left. Since a ton of grapes occupies about 220 gallons, with the 20 per cent margin about 250 to 275 gallons of vat space are needed for each ton of grapes.

Fermenter floor drains

It will be easier to separate wine from pomace and to withdraw wine from redwood, concrete, or steel fermenters if the floor of the fermenter is steeply sloped to a central or side drain. Formerly a slatted false bottom was installed, but these easily become plugged with pomace. Manholes at back as well as at front of tanks facilitate pomace removal. In larger installations fermented pomace is removed by being washed out with water or distilling material from a lower manhole into a pomace conveyor. Or pomace may be removed from the fermenter by a heavy-duty centrifugal pump. A trough at one side of the fermenter is convenient for withdrawing wine or fermenting must. In many wineries no attempt is made to screen the juice from the tank. Instead, special screens with mechanical self-cleaners are installed over the drainage sumps. These screens should be of corrosion-resistant metal. Vibrating separating screens as well as stationary ones are used.

Storage containers

The problem of the storage cellar is to keep the wine out of direct contact with air and to permit it to age normally. Slow and limited oxidation occurs by oxygen diffusing through the pores of wood or being absorbed from the headspace. Oxidation in large redwood vats, in closed concrete tanks, and in lined iron or stainless containers is limited and aging is slower. These containers, however, are useful in providing more permanent storage with minimum loss of wine by evaporation or aeration. The redwood tank is satisfactory for all except the highest-quality wines, for which oak puncheons and ovals (fig. 32) are recommended. Redwood is seldom used for making small-sized cooperage. Red wines may be kept in properly treated barrels without loss of quality; indeed, for best aging they should not be placed in very large containers. (See pp. 510 ff.)

Concrete tanks with various types of linings have been used for ordinary dry wines. No completely satisfactory conditioning material for concrete storage tanks has yet been found. The acid treatment outlined below is probably the best for routine use. Lined steel tanks or stainless steel tanks are more inert.

Concrete containers have the advantage of taking less space. However, since the wine in them has less contact with the air than it has through wooden casks, there is not only less evaporation but also less aging. If the concrete tanks have not been properly lined or treated, the wines may show a considerable amount of calcium pickup, which may eventually lead to the precipitation of calcium tartrate from the wine; or the acidity may be reduced so that the wines taste flat and vapid. Unlined tanks should therefore not be used for the better-quality table wines or for aging, and particularly not for wines containing much free sulfur dioxide. When properly lined, however, they may be used for red table wines for brief periods.

New containers, whether of oak, redwood, concrete, or steel, should first be checked for leaks by filling with water. Then, before wine is placed in them, they must be conditioned.

Conditioning concrete containers

New concrete containers should first be sanded off, washed, and dried. Some claim that no further treatment is necessary, but most wineries now precondition their concrete tanks.

Treatment with tartaric acid is the most practical and desirable method of conditioning new concrete containers for standard wines. One method is to paint or spray them with a strong (5 to 10 per cent) solution of tartaric acid. After this has dried, a second and even a third coat are similarly applied. Fermenting and storing distilling material in them is also useful for final conditioning. Or they may be filled with a dilute (1 per cent) solution of tartaric acid or a strong (20 per cent) solution of oxalic acid for about 24 hours. They should then be thoroughly washed with water. See also Corsetti's recommendation, p. 194.

Coating the fermenting tanks, and particularly the sumps and drains that are subject to wear, with various paints, plastics, and baked-on enamels has also been tried. Such materials should be free of lead and linseed oil, must not flake, and should not permit the wine to pick up too much calcium. Cruess *et al.* (1937) found Bass-Hueter enamel useful, although it did permit calcium pickup. Pioneer Flintkote asphalt emulsion gave good results in their tests. Amercoat, Corrosite, Tygon paints, and others have frequently proved satisfactory for wine-storage tanks over a fairly short period of time. But these and other coatings have not proved resistant to blistering and flaking in fermenters.

More recently, microcrystalline wax, a petroleum by-product, has been used for both fermenters and tanks. It is satisfactory when properly applied to the cleaned dry surface with a special blowtorch, which leaves a smooth firm surface. When so applied, the wax penetrates a short distance into the concrete and thus resists flaking. It is easily washed; its high melting points, about 76.7°C (170°F), permits normal cleaning and sterilizing. However, it gradually erodes and must be replaced.

Glass linings are sometimes used in Europe. These must be carefully applied and when applied as plates must be cemented together. They must also be used carefully so that none of the glass plates is broken, thus allowing the wine to get behind all the other plates. They are expensive and are not recommended.

Conditioning wooden containers

Before new wooden containers are used, they are treated with a hot alkaline solution to remove much of the soluble, bitter, oak- or wood-flavored constituents that would otherwise taint the wine. Wooden containers usually cannot be used without removal of excess extractives, tannins, and other water- and alcohol-soluble constituents. The chemistry of wood extractives is described by Hillis (1962), and the extractives present in redwood are discussed by Anderson (1961). Singleton and Draper (1961) reported on the rate and extent of extraction of aqueous alcohol solubles from wood chips.

Small wooden containers may be completely filled with a hot 0.5 to 1 per cent solution of soda ash or sal soda. The solution is left in the container for several days and then the container is washed with a dilute acid solution, or sulfur is burned in it, or it is filled with distilling material and finally washed several times with plain water. Some prefer to use hot water as well as the hot soda solution for conditioning. Clean distilling material of 10 to 12 per cent alcohol makes an admirable liquid to place in the new container to complete the conditioning.

The finest table wines, particularly whites, should never be placed in new cooperage, even when it has been treated as just outlined. Rather, it should be further conditioned by storing lesser-quality red or white wine in it for a year or two.

Large cooperage is more difficult to get into condition, but the lower amount of surface per unit of volume makes its curing less critical than that of smaller cooperage. Use of live steam is never recommended, for the tank or cask may easily warp out of shape and the surface may be damaged. One simple method is to fill the tank to a depth of two feet with water, add 1 to 2 per cent soda ash or sal soda, and heat with steam to 71.2° to 82.2°C (170°–180°F). This solution is then pumped onto the sides of the tank, or a sprinkler is arranged at the top so that the sides are treated with the material for three to six hours. The tank is then washed and the treatment repeated. Care should be taken that it is not continued long enough to injure the smooth surface. A similar result can be obtained by filling the tank with a dilute solution (0.5 per cent) of sulfuric acid and allowing it to set for about two days. The acid may be used for other tanks, but should not be allowed to stand in contact with the wood of any single tank for more than two or three days. Or a 0.5 per cent solution of unslaked lime may be placed in the tank for several days. The lime should be stirred up once or twice a day. Afterward the tank is washed as described for smaller containers.

The outside of wooden cooperage should be treated, usually by spraying with linseed oil. The cask-borer (*Scobicia declivis*) attacks mainly oak containers, but is occasionally found in redwood tanks that contain wine or fermenting must. Since it works mainly in the light, containers near doors or windows should be watched. Those in a dark place are less likely to be attacked. A strong hot solution of alum applied to the outside of containers or a 3 per cent rotenone solution followed, when dry, by a linseed-oil spray, will give adequate protection. Or rotenone and linseed oil may be mixed and applied together.

To prevent rusting, hoops should be painted or galvanized. After the hoops are painted and the tanks are dry, a thin layer of linseed oil should be applied by brush or spray gun.

POMACE ELEVATORS AND CONVEYORS . . .

. . . save labor and time.

Formerly, the pomace left in the fermenting vat was shoveled over the sides and into a basket-type press. Aside from the inconvenience of working in a hot tank filled with carbon dioxide, this practice was insanitary and required much hand labor. This procedure was replaced first by electrically operated elevators used for raising the pomace from the floor of the fermenting vat. One man in rubber boots moves the elevator and pushes the pomace into it. The elevator may dump directly into a basket press or into a conveyor. In the small winery, where movable basket presses on trucks are used, the pomace is almost as conveniently transferred directly from the fermenting vats to the presses. At present the fermented pomace is either pumped out with a centrifugal pump lowered by hoist into the fermenter or washed out of the fermenter.

Pomace conveyors operate in a concrete or metal trough. Small wooden crosspieces, called cleats, on a continuous chain carry the pomace along. The most satisfactory arrangement seems to be a concrete trough at the level of the bottom of the tank. To prevent erosion of the floor of the concrete trough, a layer of heavy glass may be laid on it.

PRESSES AND DRAINAGE

Wineries requiring large volumes of distilling material and using only the free-run liquid for wine usually prefer to wash the pomace from the fermenters. After the free-run must or wine has been drained off, the manhole at the side and bottom of the tank is opened. The floor of the tank is constructed at a slope of 10 to 30 per cent (fig. 34) and a conveyor runs along the base of the tank just below the manhole. One man, using a hose, washes the pomace from the tank into the conveyor. Instead of fresh water, the wash water from other tanks may be used. To partially separate the liquid from the pomace, stainless-steel stationary or vibrating screens with a mechanical means of keeping them clean are placed at the end of the conveyor.

Wineries throughout the world have developed drainage systems to increase the amount of free-run white juice and to decrease the volume of material that needs to be pressed (figs. 34, 35, 36). To obtain information on increasing juice yields by draining, Zhdanovich et al. (1967b) studied the effect of depth of grapes on percentage of juice extracted. A static gravitational procedure was used. The optimum pulp layer for yield of free-run juice was 450 to 500 mm. Increasing the pulp thickness to 1400 mm resulted in a 15 per cent reduction of juice yield. Increasing the pulp thickness from 200 mm to 1400 mm did not increase the amount of suspended material or the tannin content of the juice. Reducing the pulp layer from 200 to 100 mm increased the amount of suspended solids in the juice by 10 to 19 per cent. Increasing the number of perforations on the draining surface from 4 to 36 per cent did not, surprisingly, increase the amount of free-run juice or of suspended solids. However, increasing the size of the holes did increase the amount of suspended solids. Round perforations of 5 mm or larger allowed some seeds and skins to pass. This type of research should be extended to other varieties at various stages of maturation and from different regions.

Presses

After separating the free-run wine from the pomace in the red fermenters, or of free-run juice from white grapes, the residual pomace may be pressed to obtain pure wine, or refermented, or its residual alcohol and sugars may be washed out in several ways, depending on the size and type of other operations in the winery. Until recently, basket presses were preferred for both unfermented and fermented pomace. When properly operated they give a good yield of clear juice or wine, but they are slow, expensive to operate, and difficult to keep clean, and are now being replaced both in California and abroad with Willmes (figs. 37, 38), Vaslin (fig. 39), and similar types of cylindrical presses. Manfredi and Ropa (1962), in a discussion of presses for separation and recovery of wine, divide the processes into continuous and discontinuous operation cycles, either mechanical or hydraulic. They discuss the continuous-screw press (both horizontal or vertical) and combination mechanical and hydraulic presses. While they consider the Willmes type of pneumatic press useful, they point out its limitations and recommend the continuous mechanical press designed by Colin. Continuous-screw presses are easier to operate but produce cloudy wines.

In the design of presses the results of Molnar and Mercz (1964) should be studied. They showed that pressure losses during pressing varied with thickness and moisture content of the pomace, friction in the pomace and

Figure 37. Willmes press. (Courtesy H. C. Stollenwerk, Inc., Egg Harbor City, New Jersey.)

in the press basket, variation between varieties and the shape of the press basket, and relation of pressure to different sides of the press. The specific pressure is 70 per cent lower on the passive side of the press.

Humeau (1966) compared the effect of three types of presses, a horizontal type and two continuous types, on the yield and quality of juice obtained. The horizontal press yielded 85.4 per cent of must, but it contained stem juices. With continuous presses, removal of stems before pressing gave higher yields, but the tannin content was smaller. Variety of grape, degree of ripeness, and their composition as well as the type of press used affected yield and quality of the must.

Palieri (1938) compared the hydraulic and continuous presses as to yield and composition of the wine. Wines from the continuous press were higher in tannin (0.033 versus 0.021 per cent), and wine yield was 10 to 12 per cent greater. There was also a saving of about 12 per cent in cost of operation. The use of continuous presses often results in wines of abnormal composition, according to Cappelleri et al. (1968)—particularly a methanol content of over the Italian legal limit.

Konlechner and Haushofer (1962) tested the Austrian "Garnier" horizontal press, which is very similar to the French horizontal Vaslin press (fig. 39). Chains connect the ends of the press so that after the first pressing the press cake can be broken up by turning. There was higher

Figure 38. Filling Willmes press in South Africa. (Courtesy Co-operative Winegrowers Association of South Africa, Ltd.)

ash, extract, iron, and tannin in the second and third pressings than in the first, but less titratable acidity.

Coffelt and Berg (1965b) and Coffelt (1965) recently reported on the new type of press, called Serpentine, which expresses juice from either crushed or uncrushed grapes by passing them between two perforated belts over a series of pulleys. It is claimed to have the advantages of high capacity and continuous processing. The prototype press has a capacity of 5 tons per hour of crushed fruit, and somewhat less with whole fruit. It appears to have great flexibility of layout, to be light in weight, to have low power requirements, and could be constructed in a number of configurations and sizes. The average yield of liquids from grapes was

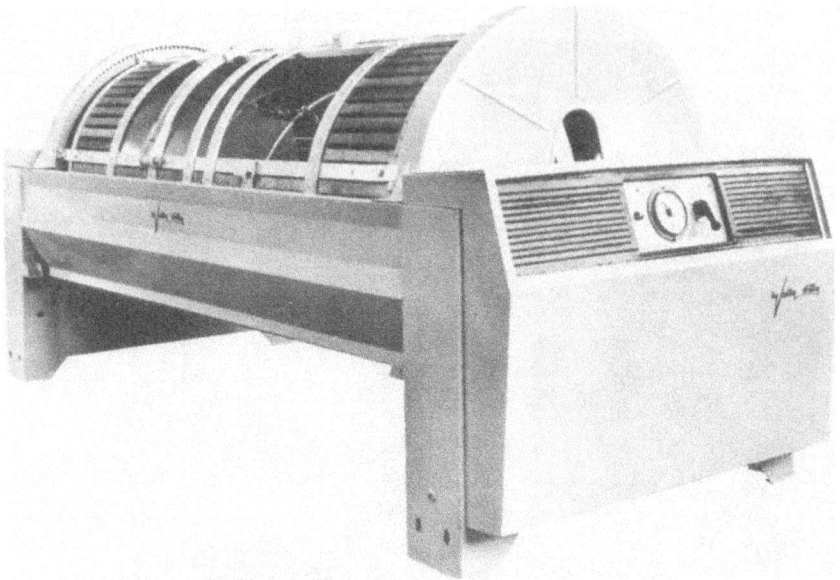

Figure 39. Vaslin-type grape press for batch operation. Handles up to 12 tons. (Courtesy Budde & Westermann Machinery Corp.)

about 9 per cent over that obtained with the conventional basket-press method. The first commercial model has a 12-inch-wide perforated plastic belt and nine pressing stations. It has been successfully tried with apple pomace, but the capacity is low because of flooding. A new dejuicing screen is now under construction.

At a recent panel discussion the various methods of pressing and separating juice of white wines were considered by the Technical Advisory Committee of the Wine Institute (Anon. 1965b).

Basket presses

Basket presses should be made of hardwood, preferably well-seasoned oak. The metal bands and the rivets and metal basin around the base should be of corrosion-resistant metal. Basket presses are always round in California, but square ones are sometimes used elsewhere. The simplest are hand-operated, with a central screw for applying the pressure. This system is slow and requires too much manual labor.

Mechanically operated types are used nowadays (fig. 40). The pressure may be obtained by direct gear action, using a central screw from motors. More generally, rams are hydraulically operated, using oil or water pressure, and the whole press is raised against a stationary head. The

Figure 40.   Hydraulic basket press.

pressure is sometimes from above. In the best types the loss of energy due to friction is only about 10 to 15 per cent as compared with 50 per cent or more in the screw type. They may be horizontally installed (fig. 41).

Basket presses may be either permanently located (in which case the pomace must be brought to them) or the ram or screw may be in one position with a movable press basket. The hydraulically operated press is permanently installed, but the press basket itself is usually arranged on wheels or tracks so that it can be moved for loading and unloading. Some

Figure 41. Large horizontal presses in Germany. (Courtesy Bucher-Guyer Maschinfabrik.)

wineries have more than one basket for each press; one can then be unloaded and reloaded while the other is being pressed.

A pressure gauge should be used to measure the rate of application and the amount of pressure. Gendron (1967b) reports that when the stems were pressed increased amounts of arginine and alanine were released and that these inhibited alcoholic fermentation. There were only traces of arginine in the free-run juice, but successive pressings contained 175, 450, 625, 750, and 937 mg per liter. For alanine the free-run contained 100 mg per liter and successive pressings 150, 250, 260, 312, and 374. It is possible that some other material inhibiting fermentation was also expressed.

The wooden parts of basket presses tend to acetify during the crushing season. To prevent this, they should be steamed occasionally and allowed to dry in the sun.

Continuous-screw presses

Continuous-screw presses are the most common type of continuous presses commercially used in California. They should be constructed of corrosion-resistant metal. They are very difficult to operate with fresh pomace, and require frequent adjustment. Much of the juice coming from the press, particularly that from near the outlet, where the pressure is highest, is filled with ground-up organic matter. Although the total yield of juice may be higher, the quantity of lees after settling, or from the first racking, is very large and the resulting wines may be difficult to clear. It is possible to operate a continuous press so that the clear juice, which is pressed out first, is kept separate from the thick, murky material that is expressed last. The latter material is higher in minerals, color, extract, and tannin, and lower in total and fixed acidity and sugar than the juice pressed first.

For fermented pomace, however, continuous presses present no difficulties in operation and give good yields. By adjusting the spacing at the outlet they can be operated at greater or lesser pressure. Some European types control the pressure by adjusting weights over the exit. If the press wine is to be used as commercial wine, the press should not be operated as tightly as if for distilling material. When operated for maximum yields, a considerable amount of tannin and bitter material is also extracted; these wines are more difficult to clarify and have an unpleasant flavor. If high pressures are used to obtain maximum yield, it is better not to use the press wine except for distilling material. Wineries with a distillery may, however, wash the pomace, grind and referment it for direct distillation, or extract the fermentable sugars. For information on countercurrent extraction of pomace, see Berg and Guymon (1951), Berg et al. (1968), and Coffelt et al. (1965).

Rack-and-cloth presses

In the rack-and-cloth press, a cloth is wrapped around thin layers of grapes, and a thin wooden rack is placed between each layer. The whole stack is placed under a hydraulic press, and high pressure is applied. Unless the cloths are thoroughly cleaned and sterilized after use, they may become a source of contamination. The process of loading the press is, moreover, slow. This is used mostly in the smaller wineries producing Concord wines.

To remove fermented pomace from tanks, pomace elevators are often used (Konlechner and Haushofer, 1956a). However, most fermenters are arranged with sloping floors for gravity removal of pomace.

PUMPS . . .

. . . should be selected for particular use.

Pumps for obtaining and maintaining fluid flow in food processing are discussed by Peterson (1964). The pumps used in the German wine industry are described by Hengst (1963) and Fetter (1966a). The general aspects of fluid flow are clearly but briefly presented by Caddell (1959).

Pumps are used in wineries for transferring musts, lees, and wine. The wine may be pumped rapidly, as in moving from one tank to another; at a moderate rate, as in filtering or cooling; or slowly, as in bottling. They must have the required capacity and pressure at the maximum suction head and at the maximum delivery head. They must be capable of handling fluids of varying viscosity and density and should give the required suction head. If the suction inlet is above the liquid level the pump ought to be self-priming, or at least easily modified to make priming quick and easy. Different types of pumps are thus required for different winery operations.

Other factors to consider in purchasing a pump are convenience and

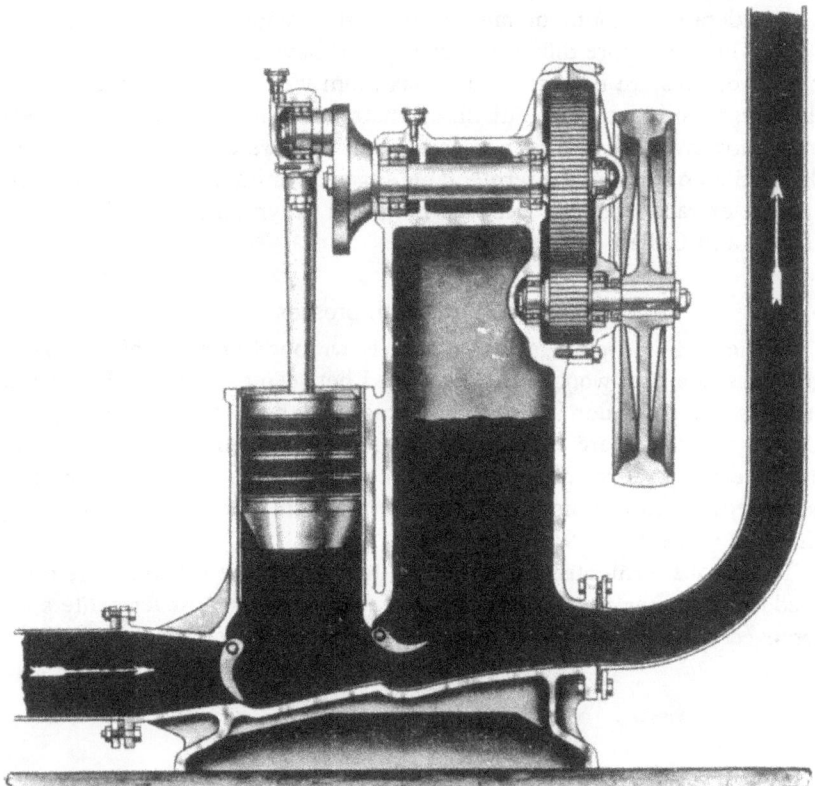

Figure 42. Piston-type pump for pumping crushed grapes. (Courtesy C. Coq & Cie.)

cheapness of installation, cost of operation and maintenance, size and available space, resistance to corrosion and to mechanical injury, strength for withstanding the maximum load even when improperly operated, easy and positive means of lubrication, simplicity for dismantling and cleaning, safety of operation, and high mechanical efficiency. Most pumps sold for winery use are constructed from corrosion-resistant metal, but the purchaser would do well to make certain of this.

The preferred pump for handling musts is the centrifugal type. It has the advantage over the piston (ram) type (fig. 42) in that it does not continue operation against a line pressure. The disadvantage of the centrifugal pump is that it does not "feed" well on crushed grapes of low liquid content.

For pumping lees the piston type pump is used. Vertical piston pumps are considered better than horizontal. In ordinary moving of wines from one tank to another, centrifugal pumps are generally satisfactory. They are available in a variety of sizes to fit the needs of large or small operations. Some, however, are of low efficiency, have poor suction, and pump too much air.

Three types of pumps are used in filtering: ordinary centrifugal, turbine, and positive-displacement gear (including the worm and bump types). These have usually been direct-drive pumps, but the varispeed types are much preferred, particularly for filtration operations. Of the three the positive-displacement type gives the best control of filtration pressure. Since the worm type is free of pulsation it is perhaps the best for low-pressure filters.

In vacuum-type automatic bottling machines, a positive-displacement pump is used for returning wine. The stainless-steel impeller bump type or the rubber impeller "Jabsco" type have both proved satisfactory.

COOLING EQUIPMENT...

... cannot be neglected by large or small wineries.

The role of refrigeration in cooling fermenting musts and in stabilization and clarification of wines is well established. Jordan (1911) early stressed the need for cool fermentation in producing high-quality dry wines. Heat transfer factors in winery refrigeration are discussed by Schreffler (1952). Skofis (1960) discussed refrigeration and other in-plant production problems, and reported (1953) on the role of refrigeration in stabilization and clarification of wines. Marsh and Guymon (1964) discussed the engineering and technological aspects of refrigeration in wine making. The cooler of Bioletti (1906a) is an early example of cooling equipment in California.

Equipment must be provided for adequate temperature control during fermentation and storage. Marsh and Guymon (1964) calculated that the minimum cooling requirements for fermentation are 150,000 B.T.U. per 1,000 gallons, with a safe figure of 250,000 for load calculations. The larger the size of the fermenters, the greater the capacity of the cooling equipment needed per ton of grapes.

California wineries use either cooling coils in the fermenters, or external coolers, or both. The refrigerant may be cold well water, ice water, cold brine, or some other suitable refrigerant. In emergencies, cooling has been accomplished by adding ice directly to the fermenter, although this dilutes the must and may be illegal.

The use of a properly designed cooling coil mounted within the fermenter will result in a more nearly uniform temperature of fermentation, lower the cooling load, and reduce oxidation. Coils must be properly placed and large enough for adequate temperature control.

More positive results are obtained by passing the must through tubular heat exchangers (fig. 43) in which it is cooled by indirect contact with a suitable refrigerant. Except with small fermenters in cool regions, extra

Figure 43. Tubular heat exchangers for cooling musts. (Courtesy Off. Mecc. Padovan, Conegliano, Italy.)

cooling during fermentation is essential to produce sound table wines, particularly red ones. Heat exchangers are used also to chill wines after fermentation to remove tartrates. The cooled wine is stored in insulated tanks or in tanks located in insulated rooms.

Cooling coils in fermenters

Having cooling coils for cold water, brine, or other refrigerants in all the tanks is more convenient but also more expensive than having only one or two sumps with such coils. Cooling coils in the fermenter should be one to two feet away from the walls, constructed of corrosion-resistant metal, and periodically examined for leaks. The cooling coils are usually mounted one foot from the walls, and centrally with respect to the height of the wall. Soft-drawn, two-inch copper tubes mounted vertically at ten- to twelve-inch centers are commonly used in concrete fermenters, but stainless-steel tubes would be better. At least one linear foot of piping for each 100 gallons of fermenter capacity should be provided.

In the pre-Prohibition era this use of cooling coils in red-wine fermenters was not satisfactory because the central coils and their supporting frames, as installed in wooden fermenters, interfered with punching down the cap and did not control the temperature of the cap. Their present successful use in concrete fermenters is due to improved design and the substitution of pumping over for punching down in managing the cap.

Heat exchangers

Any cooling system should be constructed of corrosion-resistant metals. To make it easier to pump through the heat exchanger, the must is drawn off from the fermenter into a sump passing it through a screen to separate the skins and seeds. The refrigerant may be spread over the coils through which the must is pumped, or conveyed in outer tubes past inner tubes of must.

The heat exchangers most commonly used by wineries producing table wines are single-pass, water-cooled, counterflow, shell, and multi-tube coolers (fig. 43). They consist of several 20-foot lengths of 3- or 4-inch galvanized iron pipe, inside of which are mounted from three to seven $\frac{3}{4}$- or 1-inch, thin-walled copper tubes joined together by a system of return bends designed to produce separated water and wine flow. The cooling surface varies from 121 to 228 square feet in the various types of units.

Plate coolers (fig. 44), which are widely used, have the advantage of flexibility of operation, high rate of heat transfer, small size, and corrosion-

Figure 44. Three-stage unit for wine pasteurization and hot bottling. (Courtesy Off. Mecc. Padovan, Conegliano, Italy.)

resistant metal construction. Their chief disadvantage is that the must has to be screened carefully to keep seeds and skin particles from plugging the channels between the plates. Portable plate coolers which can be inserted directly into the fermenter have also been used successfully.

For cooling wine to low temperatures to remove excess tartrates (see p. 518), either the usual countercurrent, tubular heat exchanger or the plate type may be used.

The rate of heat transfer in shell and tube coolers was found by Marsh and Guymon (1964) to vary from 111 to 390 B.T.U. per hour per square foot of cooling area per degree of mean temperature difference. The most economical use of cooling water results when the volumes of wine and water flowing through the unit are approximately equal.

Heat exchangers can be used for warming wine when this is needed. Manfredi and Ropa (1962) include descriptions of refrigerators and pasteurizers in their section on machinery for the conservation, clarification, and stabilization of wine.

EBRO cooler

Banolas (1948) developed an apparatus for carrying out fermentation at a constant temperature and recovering the alcohol vapor and volatile esters from the carbon dioxide gas evolved in fermentation. His patented "EBRO" cooler consists of an external vertical-type shell-and-tube heat exchanger attached to a closed fermenter. The must is allowed to flow by gravity into a small receiving tank below the fermenter, which serves also to receive the cooled, alcohol-enriched must. The must is pumped from this constant-level receiving tank into the cooler, at the top of which the fermentation gases are allowed to enter by a pipe leading from the top of the fermenter. The alcohol vapor and volatile esters in the gas are mixed with and absorbed by the must during its passage through the cooler. The cooled must falls into a constant-level cooler receiving tank connected with the fermenter receiving tank, and the carbon dioxide gas is vented into the air through a vertical pipe at the top of the cooler receiving tank. The system is so arranged that the must can be pumped

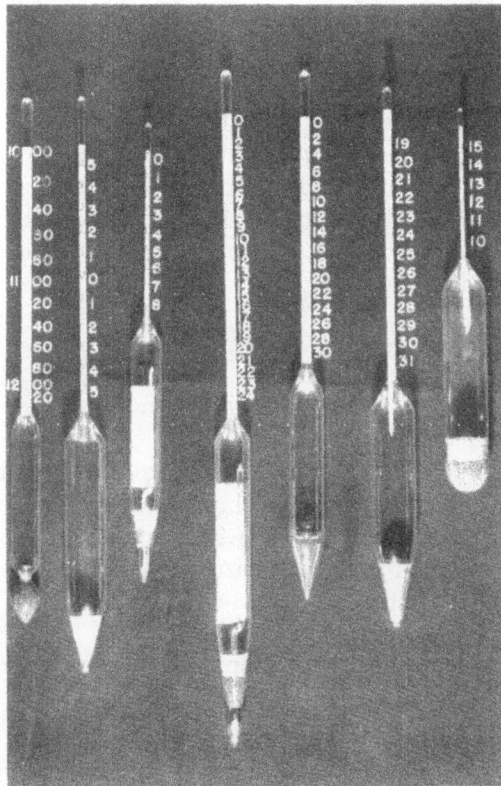

Figure 45 (*Left*).   Long-stemmed thermometer—approximately 3 feet long.
Figure 46 (*Right*.)   Hydrometers used in wineries, left to right: Specific gravity; five Balling or Brix types ($-5°$ to $5°$, used for table wines, $0°$ to $8°$, used for extract determination, $0°$ to $24°$, $0°$ to $30°$, and $19°$ to $31°$, all three used for fermenting musts and grape juice); 10 to 15 per cent alcohol. Two of the hydrometers have enclosed thermometers for temperature correction.

from the bottom of the fermenter receiving tank into the top of the fermenter, or from the fermenter through the cooler and then back into the fermenter. The cooler tubes are fitted with hoods having spiral slots to increase the rate of heat transfer and to improve contact between the fermenter gas and the must. A refrigerant at 0°C (32°F) is circulated countercurrent around the tubes and shell in the tube cooler. So far as we know it has not been used in California.

Thermometers

Thermometers are used in the fermenters, on heat exchangers, in the storage room, and in the laboratory. Different types of thermometers are needed for these various places. In the fermenters a long-stemmed thermometer is used (fig. 45). If the fermenter is very large, a recording type of thermometer may be installed. On heat exchangers, thermometers of special design are used. These usually screw directly into the apparatus. A simple maximum-minimum thermometer or a recording thermometer may be used to indicate the temperature of the storage rooms. In the laboratory, various types are used, depending on the needs of the analyst.

Hydrometers (see p. 599) are used to follow the course of the fermentation (fig. 46).

PASTEURIZERS . . .

. . . should be easily cleaned and corrosion-resistant.

Many wineries use pasteurizers to control bacterial diseases of wine or to prevent spoilage of sound wines under unfavorable conditions. The role of pasteurization in sterilization and clarification of wine is discussed by Pilone (1953). Pasteurization has been largely replaced by germproof filtration, either of the asbestos pad or of the more recently introduced membrane filter type (Mulvany, 1966).

Pasteurizers should be made of corrosion-resistant metal. Copper and brass are particularly undesirable in contact with hot sulfited wines. Pasteurizing such wines in copper equipment almost always results in enough metal pickup to cause copper cloudiness.

Pasteurizers should be so designed that dead pockets are avoided and all the wine is brought to the proper temperature and held there for the desired period of time. Plate pasteurizers will accomplish this purpose more efficiently than tubular heat exchangers. Care should be taken to see that indicating or recording thermometers are properly placed, especially in tunnel pasteurizers (fig. 47).

Figure 47. "Atlantico"—tunnel pasteurizer for bottled wines; 48°C (118.4 °F) for 50 minutes. (Courtesy Off. Mecc. Padovan, Conegliano, Italy.)

FILTERS . . .

. . . are widely used to clarify wines.

Several types of filters are used in wineries. These vary from the plate-and-frame or vacuum filters used for rough filtration to fine or polish filtration for bottling. The general principles of filtration are discussed in the German text of Kufferath (1954) and in the American texts of Dickey and Bryden (1946) and Dickey (1961).

Recent engineering data on design and use of filtration equipment is summarized by Smith and Giesse (1961), Tiller and Huang (1961), Van Note and Weems (1961), and selection of filter fabrics by Ehlers (1961). Recent data on the prevention of clogging of filter media were discussed by Kehat et al. (1967). Filtration in the brewing industry is described in some detail by Kutter (1959) and Kutter and Hirt (1959). Selection, arrangement, and use of filtration equipment in the winery are considered by Peters (1953) and Fessler (1966). Holden (1953) reported on polish filtration of wine with plate-and-frame filter presses. Hoffman and Berg (1967) discussed the use of vacuum filters in wineries.

Four or more types of filters are commonly used in California wineries:

plate-and-frame, screen or leaf, asbestos or fiber-pad screen, and membrane. The first three types are commonly used for filtering rather cloudy material or for handling large volumes of wine. The latter two are used for finishing or polishing filtration prior to bottling. (For their use see pp. 547 to 562.) Pulp and candle filters are no longer used in California, but membrane filters (fig. 48) are finding increased use in this state.

Figure 48. Taking sample from Millipore-filtered wine with hypodermic needle to test for sterility. (Courtesy Beaulieu Vineyard.)

Filters should be made of easily cleaned, corrosion-resistant material and should be easy to dismantle and service. They should always be left dry after use. When a filter is set up, it should be thoroughly washed before wine is pumped in. The first wine is usually cloudy and should not be sent to the receiving tank. In screen or leaf filters, circulation is, of course, required.

Sakthivadivel and Irmay (1966) critically reviewed the theories proposed on filtration of microscopic colloidal suspensions. While the laws governing seepage of liquids through porous media have been known for over a century (first enunciated by Darcy in 1856), most of the investigations concerning the transport of fines through porous media were initiated only about 30 years ago. The theory of flow of liquids through a porous matrix (filter) was critically discussed by Irmay (1958). During the flow and simultaneous transfer of fines many of the physical and physico-chemical properties of the porous matrix (filter) are modified, particularly its porosity, density, and hydraulic conductivity. There is very little data on the nature and extent of this modification by the colloids of wine. In the design of sand fillers the important parameters are the equivalent diameter of matrix grains (dm) and of the fines (df). For spherical uniform particles the filter clogs when $dm/df < 5$ to 6 and the fines are washed out by laminar flow when $dm/df > 10$ to 12. The equivalent diameters in these relations refer to particles 15 per cent smaller and 85 per cent larger than the dimensions cited by Irmay (1958).

AUXILIARY EQUIPMENT

Self-desludging continuous centrifugers are used in separating wine from lees with maximum recovery, as described by Fessler (1953). The engineering design factors of centrifuging equipment are summarized by Smith (1961). Ion exchange columns are now being installed and widely used (McGarvey et al., 1958). Various types of improved sugar recovery systems are in use (Coffelt et al., 1965), and a variety of new bottle-filling and -closing machines (see pp. 576–579).

CHAPTER FIVE

SANITATION[1]

Adequate sanitation is required by state and federal agencies. It pays because it lessens losses of wine, leads to the production of more stable wines, and helps to maintain consumer interest. Federal law now defines a food as adulterated "if it has been prepared, packed, or held under sanitary conditions whereby it may have become contaminated with filth, or whereby it may have been rendered injurious to health" (U.S. Department of Agriculture, Food and Drug Administration, 1939). The Federal Food and Drug Act of June 25, 1938, is now generally known as the Federal Food, Drug, and Cosmetic Act. It has been strengthened by the factory-inspection amendment of 1953, which makes refusal to permit inspection an offense subject to federal prosecution.

Every winery, no matter how small, should set up a regular program of sanitation. This program is best supervised by a qualified sanitarian who is also thoroughly familiar with the technology of wine making. Some member of the winery staff who understands the technical problems that may arise could also serve as sanitation supervisor after he has had training in basic sanitation procedures. The sanitation supervisor or sanitarian, in addition to technical training in wine making and sanitation, should have the ability (1) to set up a sanitation program keyed to the winery where he is employed; (2) to make thorough and regular sanitation surveys (a check list may be helpful in keeping data on the progress of the sanitation program); (3) to evaluate sanitation surveys and make recommendations to management; (4) to communicate with the winery employees and act as their leader in matters of sanitation; (5) to encourage the employees to form good work habits and follow necessary rules; and (6) to comprehend both state and federal food and drug laws.

Management should support the winery's sanitation program by giving leadership status to the supervisor or sanitarian in all matters pertaining to sanitation. They should give due consideration to recommendations made by the sanitarian in regard to repairs and new equipment. General procedures for setting up a sanitary plan are given in a manual published by

---

[1] This chapter has been prepared in consultation with Mr. Almond D. Davison, formerly Sanitarian of the Wine Institute. We are grateful to Mr. Davison for his generous help.

the University of California (1946). For sanitation programs in California see Bell (1967) and Davison (1961, 1963).

The basic principles are given in the Wine Institute Sanitation Guide for Wineries,[2] Davison (1961, 1963):

1. Keep the winery clean and free of refuse at all times—inside and outside.
2. Inspect premises and equipment regularly.
3. Get rid of harmful bacteria, insects, rodents.
4. Use plenty of water, cleaning and sterilizing aids.
5. Keep equipment in good repair at all times, particularly any that comes into contact with the wine.

Compliance with these principles will increase plant efficiency, protect the quality of the wine, reduce deterioration of equipment, lessen spoilage, satisfy the aesthetic values of consumers, and comply with government regulations.

CLEANING AND STERILIZING AGENTS . . .

. . . should be selected for a particular purpose.

Sterilization is defined as the complete destruction of all life. It does not imply cleanliness. Fecal matter can be sterilized. Sanitization implies cleaning and sufficient antimicrobial treatment to kill microbes down to a level consistent with the demands of public health.

Detergents

Materials which, when added to water, increase its ability to clean surfaces are called "detergents." Water or steam under pressure and, of course, scrubbing or scraping have a cleansing effect, but chemical detergents are often needed. The ideal detergent wets surfaces readily, is easy to rinse off, softens water, saponifies fats, emulsifies or deflocculates materials, has good dissolving, buffering, and neutralizing power, is not corrosive, and has some sterilizing action. No single detergent is equally good in all these qualities. The agent must be selected according to the nature of the deposit to be removed, the material in the surface to be cleaned, the method of cleaning, the chemical nature of the water to be used, and bacteriological needs. The number of different detergents used in the winery should be kept to a minimum. Usually the general-purpose types are employed to avoid confusion.

Ideal detergents should, according to Jennings (1961), have good wetting properties, since they lower the surface tension so that one gets

---

[2] This was first issued in 1946 (Wine Institute, 1946), revised in 1957, and revised and brought up to date in 1961–1963 by Davison. A revision of this guide is now being prepared.

closer contact to the surface. The process involves displacing the foreign material from the surface, dispersing it in the cleaning media, and removing it before it can redeposit. Adequate alkalinity is needed to saponify fats, to cause molecular degradation in proteins (known as peptization), and to remove tartrates and argols. Alkali solutions stronger than 3 per cent will etch glass. Their corrosiveness can be minimized by adding sodium metasilicate.

Sequestering and chelating agents are used to tie up calcium and magnesium and render them unavailable for precipitation. But at high temperatures they revert to simpler phosphates and can result in calcium phosphate precipitation. Rinsing properties are also important. They are usually complex phosphates or silicates, or contain organic additives.

Alkaline detergents, such as soda ash (commercial sodium carbonate) and caustic soda (commercial sodium hydroxide), are commonly used in the winery; but for most equipment, less caustic chemicals are preferable, such as silicates (sodium metasilicate, sodium orthosilicate, or liquid silicate), phosphates (such as trisodium phosphate, sodium hexametaphosphate), or even sodium sesquicarbonates, aluminium acetate, sodium sulfate, sulfonic acid derivatives, and sulfated esters.

Acid detergents are used only for difficult scales in metal pipes, since they are too destructive for wood or concrete. Greenfield (1967) makes the useful point that all sanitizers exert their maximum influence on clean surfaces.

Stainless steel and glass enamel are much easier to clean than wood or concrete.

Greenfield (1967) notes that modern in-place cleaning systems using permanently installed sprays, automatic solution preparation, and automatic control programing usually achieve better results than manual operations.

Caron (1964) recommends a wide variety of detergents, mineral salts, bases, acids, esters of fatty acids, and sequestrants for bottle washing. Many of these are useful solely for special conditions, but some may find use in the wine industry. Caron's text should be studied carefully by those with special bottling problems, especially the washing techniques for used bottles. See also Upperton (1965).

Sterilizing agents

Although the detergents mentioned above have some sterilizing action, more powerful agents are normally used for sterilizing. Hypochlorite solutions are widely used in the wine industry for this purpose. Unfortunately, this and other common sterilizing agents have little cleansing effect. Greenfield (1965) found hypochlorite-bromide combinations consistently more effective than hypochlorites alone for destruction of yeasts. Chlorinated detergents, however, cleaned epoxy-lined fermenters more effectively than nonchlorinated cleaners. Some recent developments are chlorinated isocyanurates, which provide good solubility and have stability in the dry form, and chlorinated hydantoins, which are also stable but are of more limited solubility.

New materials combining detergents and sterilizing agents are coming into use. These are usually sold under trade names such as Sterichlor, Hayamon, Metso, Roccal, and Emulsept.

Useful time-temperature concentration relationships for germicidal effectiveness of sanitizers as given by Greenfield (1967) are shown in table 34.

TABLE 34

GERMICIDAL EFFECTIVENESS OF VARIOUS SANITIZERS

Product or treatment	Active agent concentration (mg/liter)	Temperature °C	Temperature °F	Minimum time at temperature (Minutes)
Hypochlorite ⎫ Chlorinated isocyanurate[a] ⎬ Chlorinated hydantoin[b] ⎭	50	23.9	75	1
Chloramine T[c]	200	23.9	75	1
Chloramine[d]	100	23.9	75	1
Iodine	12.5	23.9	75	1
Hypochromite-hypochlorite	12.5	23.9	75	1
Quaternary ammonium	200	23.9	75	0.5
Acid surfactants	234	23.9	75	1
Caustic[e]	21,600	54.4	130	5
Hot water	—	76.7	170	5
Hot water	—	76.7	170	15
Steam	—	93.3	200	5
Hot air	—	82.2	180	20

[a] pH 9.5 or less.
[b] pH 7.0 or less.
[c] pH 7.2.
[d] pH 6.8.
[e] For bottle washing.
Source of data: Greenfield (1967).

Use of potentially toxic chemicals in wines is never justified. There is considerable doubt whether such chemicals should be used in vineyards or in wineries unless the chemical is completely destroyed or dissipated before the wine is made. Lehman (1950) has summarized the toxicological reasons why certain chemicals cannot be tolerated. Concerning the detergents (wetting agents) Lehman says: "The quaternary ammonium compounds and alkyl dimethyl benzyl ammonium chloride (Roccal) . . . are bactericidal in their action and . . . are toxic and have no place in foods." The anionic wetting agents, such as sodium sulfate and the alkyl aryl sulfonates (Noccanal), are also toxic but can be removed by adequate rinsing.

All detergents or sterilizing agents must be removed from containers or equipment by a thorough washing with water before introducing wine. Hypochlorites are particularly harmful to the color and flavor of wines and, if followed by sulfur dioxide, the chlorine is reduced to chloride ion. Quaternary ammonium compounds, if used at all on surfaces that come in contact with wines, must be removed with special thoroughness. A summary of the various chemicals and their uses is given on pages 221 and 222.

CLEANING THE WINERY
Year-round sanitation is necessary.

Much of the difficulty in keeping winery floors clean is caused by poorly arranged equipment that creates corners and crevices that cannot be easily cleaned. Equipment that has not been designed with sanitation in mind is usually difficult to keep clean because of inaccessible pockets or areas.

A combination of dry and wet cleaning is best. The heavy-duty wet-dry vacuum cleaner has been found to be an excellent tool for cleaning. Floors should be cleaned regularly by dry or wet methods depending upon the material to be removed. Wine on the floor, especially that which has spoiled, should be washed away immediately and the area should be scrubbed, washed with lime or strong hypochlorite solution, and thoroughly rinsed off with water. The floors and the outside of the wood cooperage should periodically be thoroughly washed and then disinfected with a dilute hypochlorite solution. With the use of dry cleaning wherever possible the humidity of the winery can be kept lower than where only wet methods are used. The high humidity in the still air of the winery cellar is very conducive to mold growth on the outside of wooden cooperage. The top of the tank, overhead platforms, and ramps can be vacuumed, cleaned, or washed off, taking precautions that no water gets into the wine.

Improper care of equipment is perhaps the commonest source of contamination. Crushers, must pumps and lines, presses, filters, hoses, pipes, and tank cars are all difficult to clean completely. Even such apparently simple equipment as buckets, shovels, wine thieves, and hydrometer cylinders are not easily cleaned. A mere washing with water is rarely sufficient. Under the warm climatic conditions of California, any wine or must left in the equipment spoils rapidly and becomes a source of contamination for the whole winery.

After use, the equipment should be dismantled as completely as possible, thoroughly washed with water, and with a detergent if necessary, sterilized with wet steam, boiling water, or dilute solutions of hypochlorite (or another sterilizing agent, if the material is adversely affected by hypochlorite), and then rinsed with clean water and drained. Circulating the cleaning solutions is recommended, but at certain places in the equipment hand scrubbing with a brush may be necessary.

Hoses, after thorough cleaning and rinsing, should be placed on sloping

---

MATERIAL TO USE FOR CLEANING AND DISINFECTING

For cleaning wooden or concrete containers:
1. Materials containing sodium carbonate, metasilicates, and phosphates, usually sold under trade names
2. Trisodium phosphate and other complex phosphates such as hexametaphosphate, tetraphosphate, pyrophosphate and tripolyphosphate.
3. Soda ash, or sal soda

For cleaning metal equipment:
4. Any of the three above
5. Caustic soda (injures wood and concrete)
6. Carbon abrasive
7. Citric acid (to remove scale and oxides; if acids are used to remove mineral incrustations one must consider their corrosiveness)

For neutralizing, in the form of solution or spray:
8. Lime (slaked)
9. Soda ash, or sal soda

For sterilizing:
10. Sodium or calcium hypochlorite (bromine- and iodine-containing sanitizers are now available. Obviously, no residue may enter the wine)
11. Combinations of sodium or calcium hypochlorite and alkaline materials sold under trade names
12. Chlorinated trisodium phosphate
13. Chloramines, especially the new faster-acting types
14. Combinations of chloramines and alkaline materials sold under trade names
15. Chlorinated hydantoins
16. Sulfur dioxide in the form of liquid sulfur dioxide or metabisulfite
17. Sulfur wicks or pots, burned inside closed wooden cooperage (now seldom used in California)

Any of these cleaning, neutralizing, or sterilizing agents must be thoroughly rinsed off.

FOR WALLS, FLOORS, AND OUTSIDE SURFACES OF EQUIPMENT

For cleaning:
18. Same as nos. 1 to 3

For sterilizing:
19. Chlorine compounds mentioned under nos. 10 and 11
20. Sulfur dioxide
21. Quaternary ammonium compounds, usually sold under trade names; must be used so that no trace of the compound can get into the wine. Their tendency to cling to treated surfaces is an advantage when used on walls and floors, but a disadvantage on equipment that may come in contact with the wine.
22. Acid-anionic surfactant sanitizers. These liquid formulations consist of anionic surfactants blended with moderately strong inorganic and organic acids, and are reported to be more effective than quaternary ammonium compounds.

FOR OUTDOOR PURPOSES

For cleaning and disinfecting wood or concrete platforms or crushing equipment, or for spraying on grounds and pomace piles:
23. Lime (unslaked)
24. Chloride of lime
25. Sodium bisulfite

For disinfecting and insect control, usually in the form of a spray:
26. Chlorinated benzenes (do not use near processing equipment or cooperage or anywhere inside the winery)
27. Diazinon
28. Chloridane
29. DDT (do not use near processing equipment or cooperage or anywhere inside the winery)
30. Sulfur dioxide
31. Pyrethrum
32. Vapona

The latter materials, alone or in combination, are also sold under various trade names. Since some are toxic they must be used only as directed.

Source: Adapted from Davison (1961, 1963) and other sources.

racks instead of on the floor so that they will drain and dry. Both ends of the hose are screened to keep out fruit flies. Rubber loses its elasticity with time. Periodic soaking overnight in an alkaline solution will help to maintain the elasticity of the hoses.

Filter cloths should be washed in a washing machine with hot soapy water, thoroughly rinsed, and dried. Filter candles should be boiled and may need to be soaked in hot acid occasionally in order to remove organic deposits. (As previously indicated, these are rarely used in California.)

Thorough cleaning and sterilizing are especially needed for equipment that has been on contact with spoiled or contaminated wine. Every surface with which wine comes in contact can be considered a source of contamination and should be treated as such both before and after use.

Care of the bottling room is particularly important, for it is the final possible source of bacterial or metallic contamination. It is also usually the first point of inspection by public health agencies. The room should be well-lighted and ventilated, with easily cleaned walls and floors. Many spoilage organisms are present in a winery (p. 328; see also Engle, 1950).

Care of crushers

Stemmers and crushers should be carefully checked for mechanical defects before the vintage season. A breakdown during the crushing season can cause serious inconvenience. The must lines, must pumps, and other equipment used in these operations should be examined and placed in working order. Many crusher-stemmers could be repaired to eliminate pockets or crevices where stems, skins, or seeds collect. Parts of equipment that come in contact with grape materials or wine should be painted with a good lacquer or acid-proof varnish. Some of the recent epoxy resins have been approved for use in food containers. However, the Food and Drug Administration does not actually approve paints or coatings, but finds no objections to them if none of their component parts migrate into the must or wine.

The epoxy resin paints have been successfully used in California for coating the interior of gondola trucks, for receiving hoppers, and in wine storage tanks. Mammut, which has been used for many years in the beer industry, is melted by heating and applied to fairly dry surfaces, including wood. Tanks lined with Mammut should not be heated above 120°F or the Mammut will leech out.

The chemical action of sanitizers or equipment varies with the material used. Greenfield (1967) rates stainless steel the most resistant and wood the least. Epoxy plastics, phenolic plastics, and glass are usually satisfactory unless extreme exposure to acid or alkali or extremes of pH are encountered.

During the season, the conveyors, crushers, and must lines should be kept clean. They should not be allowed to stand with must in them for more than an hour or two. After use (usually twice a day) they are washed out and properly drained, and before re-use are thoroughly flushed with water.

Disposal of pomace

After pressing, the pomace must be disposed of as rapidly as possible. Under no circumstances should it stand in or close to the fermentation room, for it rapidly acetifies, and the fruit flies (*Drosophila*) quickly carry acetic acid bacteria (*Acetobacter*) from the pomace pile to clean fermenting vats. (See p. 228.) The present practice is to scatter the fermented pomace as a thin layer in the fields, where it dries quickly and does not become a serious breeding ground for fruit flies.

CARE OF CONTAINERS

Constant attention is needed to prevent spoilage.

Cleaning used cooperage

Used cooperage should be properly conditioned to avoid losses in both quality and volume of wine. After emptying the containers, Berg (1948) recommends washing them thoroughly with water and spraying with a hot (120°F) 20 per cent solution of "Winery Special"[3] (90 per cent soda ash and 10 per cent caustic soda). Then wash with hot water and spray with a solution of a chlorine compound containing 400 parts per million of available chlorine.[4] Again wash thoroughly with cold water, drain, and dry, using dry-wet vacuum. Burn a sulfur wick in the tank (2 ounces per 1,000 gallons) or add an equivalent amount of sulfur dioxide. Rewash the tank before use. Cooperage should be carefully inspected visually and by smelling before being filled.

Wines may increase 0.01 per cent or more in volatile acidity after being placed in improperly cleaned vinegar-sour tanks. A hot dilute soda-ash solution effectively neutralizes any vinegar present in used barrels and, if followed by wet-steaming and washing, usually leaves the barrels in good condition. The use of hot soda-ash at too high a concentration for too long a time, however, causes deterioration of the wood. About once a year the outside surface of wooden containers can be washed down with a dilute cleaning solution followed by water.

Mold on the surface of the tanks may be a problem. Quaternary ammonium compounds, propylene glycol, redwood bark extract, Cunilin[5] (copper-8-quinolinolate), and dilute chlorine solutions have been used. None are 100 per cent effective. O'Brien (1960) notes that Cunilin has no flavor or odor, and claims that it minimizes stave shrinkage. It does not build up a gummy deposit on the surface. Control of humidity and

---

[3] Not to be confused with proprietary solutions of somewhat different composition and use.

[4] The number of ounces, p, of chlorine compound required to produce a sanitizing solution of a given strength may be calculated from the formula $p = 128xv/10,000a$, where x is the number of gallons of sanitizing solution to be produced, v the desired concentration in parts per million, and a the percentage of available chlorine in the compound or solution.

spillage are also helpful. Low humidity and a good circulation of air are especially recommended. To clean the outside of dirty tanks, Berg (1948) recommends V.E.X.,[5] 4 to 12 ounces per gallon at $71.1°$ to $82.2°C$ ($160°$ to $180°F$). This is applied at the top of the staves to remove grime, varnish, and linseed oil. The surface should then be treated as for new containers (p. 197). Mold formation on casks can be prevented by applying a thin layer of Wolmanit M (a patented Austrian antimold preparation), according to Haushofer and Rethaller (1966).

Removing tartrate deposits

During storage, wines deposit cream of tartar and other substances. Some of this impure potassium acid tartrate—argols—sticks to the sides of the containers, though much of it is found in the lees. It is necessary to remove this in order to recover the tartrates (see p. 518) and also to smooth the inner surface, which becomes very rough. Traditionally the argols are removed by scraping, but this is laborious and may injure the wood. Tartrates can be removed from concrete containers by a jet of steam directed toward the wall at an acute angle. Most wineries now install a circular sprayhead inside the tank. Even with cold water, most of the tartrates will be removed in 24 to 48 hours. For soaking, Berg (1948) suggests 80 pounds each of soda ash and caustic soda and 3 pounds of Oronite D-40[5] per 1,000 gallons of water at $48.9°C$ ($120°F$). This solution can be used on several tanks, but should be removed for tartrate recovery when it reaches 12 to 15 per cent of tartrates.

Treating moldy cooperage

Casks that leak because of cracked staves should be recoopered. Recoopering consists of scraping the staves, replacing the leaky or moldy ones, and sometimes building a new head. Although recoopering is expensive, it prevents the use of contaminated containers and frequently saves the winery much trouble later from leakage and spoilage. A single moldy stave may contaminate a whole tank of wine and damage its quality irreparably.

If the winery chooses to treat the bad cooperage itself instead of recoopering, the inside should be carefully examined for molds, cracks, and other defects. Any growths should be scraped off, as they can rarely be washed off completely. One or two treatments with hot alkaline solution should be tried first. A hypochlorite solution (containing from 500 to 1,000 parts per million of available chlorine) should then be sprayed over

---

[5] Trade designation of a proprietary product.

the inner walls of the container, or be filled into it and be left for several hours. This should be followed by several washings with hot water. If the container still does not smell clean, a second and shorter application of hypochlorite solution may be needed, or a combined hypochlorite and steam treatment. All traces of hypochlorite must be removed before use. (See also Berg, 1948.)

For very badly spoiled cooperage, several preliminary treatments with dilute hydrogen peroxide, a 1 per cent solution of potassium permanganate, or neutral oils may be necessary; but recoopering is generally best. It is questionable whether recoopering of badly spoiled, thin-staved cooperage is worth while. Prevention is the wisest course, for cooperage will not become moldy if it has been properly rinsed and sterilized after use.

Inert and impervious plastic or composition wax linings have proved useful in temporary conditioning of old or badly leaking tanks. Mammut has been successfully used as an inside coating. Placite, a two-component epoxy-type resin, has also found favor with the wine industry. Several other products are still under test. Outside sealants (e.g., epoxite) are reported to withstand up to two tons per square inch of hydrostatic pressure. One advantage of the outside sealants is that they are more likely to be approved for use since they do not come in contact with the wine. Another advantage is that outside sealants are more easily applied.

Removing red pigment

Sometimes it is necessary to use a red-wine storage tank for white wines. Berg (1948) suggests the following procedure: after washing, spray five or six times at two-hour intervals with a hot (48.9°C, 120°F) mixture of 30 pounds of soda ash, 24 pounds of Winery Special, and 15 pounds of Sterichlor, or other chlorine preparation, in 50 gallons of water. Brush the walls and wash until the water comes out clean. Then spray with a 1 per cent solution of citric acid. Rewash and, if the walls are not free of red pigment, fill the tank with water and add 6 pounds of sulfur dioxide per 1,000 gallons. After two or three days the tank should be ready for use. This program may be modified according to the amount of red color in the tanks. Small cooperage can be more conveniently filled with the alkaline solution than sprayed with it.

Storing empty containers

Concrete tanks are best left open when not in use. But they should be kept dry and before re-use must be carefully inspected and cleaned.

Storing empty wooden containers is a difficult problem. The hot, low-humidity conditions of California quickly dry out empty containers, which then tend to fall apart. Open wooden fermenters are sometimes painted with a lime paste when not in use, but this is hard to remove. The best procedure is to clean the fermenters thoroughly with alkaline solutions followed by chlorine solution. They are then filled with water, and about 12 to 15 pounds of unslaked lime per 1,000 gallons is added. One winery has been successful in adding 15 pounds of pebble lime per 1,000 gallons. The crust that forms on the surface prevents the propagation of insects.

If tanks, casks, or barrels are to be kept empty, the hoops should be tightened from time to time, and the staves not allowed to fall apart. To prevent drying out they should be washed weekly. If sulfur wicks are burned in the barrels or if sulfur dioxide is introduced, they should remain in a sanitary condition; but resterilizing once a month is a useful precaution. Care must be taken not to use too much sulfur dioxide, for it is absorbed by the wood and may be difficult to remove even with several washings of water. If the excess is not eliminated, too much may be absorbed from the wood by the wine (Wanner, 1938). Addition of excess sulfur dioxide is more likely when liquid sulfur dioxide, rather than sulfur wicks, is used.

Some wineries fill their wooden tanks and cooperage with various preparations, such as a dilute hypochlorite solution (250 parts per million). This rapidly loses its strength, however. Saturated limewater, though satisfactory for short intervals, has only a limited period of effectiveness in preventing contamination. A dilute solution of sulfuric acid and potassium metabisulfite (half a pound of concentrated sulfuric acid and one pound of metabisulfite per 1,000 gallons) will sometimes keep the containers in good condition; but the containers should be inspected occasionally to make sure that evaporation has not occurred, which would permit the drying out of the top of the barrel or tank. Fluosilicic acid and other compounds containing fluorine must not be used in the winery.

PEST CONTROL

Animals should not be allowed in wineries.

Every effort should be made to control the entrance of rodents to winery premises. All new construction should be rodentproof, and old buildings should be inspected to determine whether places of entry and nesting spots can be eliminated. Sources of food should be removed and a persistent program of trapping should be carried out. Storer (1949) gives methods of rodentproofing buildings and trapping. A common rodenticide currently used is Warfarin[6] or similar products containing hydroxy coumarin. Cats and dogs should not be allowed on winery premises, even for the desirable objective of rodent control. Birds are also to be excluded, as their droppings are as objectionable as those of rodents. Screening is the only practicable method.

The immediate area of all winery buildings should be cleared and kept free of discarded materials, equipment, or weeds. Control of rodents in areas around the building should also be maintained, rather than waiting to destroy them after they enter the building. Bait stations should be set up at strategic points outside the buildings. The feed or bait in the bait station boxes must, of course, be kept fresh. Rodent control is most effective just before and during the cold rainy season. Bait stations should, therefore, be established before the rainy season begins.

Government agencies have frowned on the use of DDT preparations to control flies and insects on account of the danger that some of the material may get into the wine. The common fruit or vinegar fly (*Drosophila melanogaster*) is a means of spreading infection about the winery. Fruit flies breed in pomace piles, waste organic matter near the winery, areas on and under leaking wood storage, and pools from leaky tanks. Their life cycle is short (about six days); so control must be frequent as well as thorough if the breeding cycle is to be broken. Mechanical cleanup and chlorine solutions should be used. The *Drosophila* have been found to travel up to six miles in 24 hours. Wineries located in vineyards can probably not avoid them completely, but at least the breeding places in and around the winery should be eliminated. Recent tests by Yerington (1958, 1963, 1964, 1967) show DDVP (dimethyl-dichloro-vinyl-phosphate, dichlorvos for short, or Vapona, its trade name) to be an excellent insecticide for killing *Drosophila* inside or outside the winery. Because it is toxic it should be used only by trained personnel and strictly according to directions. Deodorized kerosene is the preferred carrier. The authorization for winery use specifies that no more than one gallon of 1 per cent dichlorvos be used for 64,000 cubic feet. Machinery that comes in contact with the spray should be covered or washed after spraying.

Pressurization of the winery and air-screen barriers have been used to keep fruit flies out of the winery (Davison, 1959; Overby, 1959), but pressurization cannot be used if there are too many openings. Screening not only helps keep the flies out but also increases resistance to loss of pressure. Pressurization helps to reduce dangerous carbon dioxide accumulation inside the winery. Air current over open doors can also be used.

---

[6] Trade designation of a proprietary product.

The air velocity should be about seven to ten miles per hour. Since air currents do not prevent entrance of fruit beetles along the floor, an insecticide might be placed on door sills to prevent their entry.

HEALTH OF WORKERS . . .

. . . cannot be neglected by any winery.

Although no occupational diseases seem to be associated with winery workers, temporary conditions that might constitute a health or accident hazard should be prevented. Delaunay (1925) notes that the humidity of certain French caves and their lack of aeration might be undesirable. Russell *et al.* (1939), in a survey of the industrial hazards of California wineries, recommend that prospective employees be given a physical examination, with annual physical checkups thereafter. Men should be placed in work for which they are physically fitted. Chronic alcoholics, epileptics, and persons suffering from other nervous disorders should not be assigned to areas where they might be exposed to alcohol vapors. The need for sufficient ventilation and lighting is stressed. Adequate sanitary facilities and an educational program to encourage good personal-hygiene habits are emphasized. Waterproof boots and clothing are recommended for certain workers.

Where there is danger of exposure to alcohol vapors or carbon dioxide, two men should always work together, one inside the tank and the other outside. The man working inside should have a rope around his waist with one end outside the tank. Forced-draft ventilation should be used to provide adequate air and to keep the carbon dioxide content, the temperature, and the humidity within acceptable limits. The danger from excessive concentrations of carbon dioxide needs to be constantly re-emphasized. The accumulation of carbon dioxide is a major problem in winery fermenting rooms, particularly where wines are fermented in closed refrigerated rooms. Concentrations of carbon dioxide of below 1 per cent are considered safe. Increasing amounts reduce the efficiency of the worker, and, at about 4 per cent, prolonged work becomes untenable. Since carbon dioxide is heavier than air it tends·to settle to the floor. It is best, therefore, to withdraw air at the floor level and introduce fresh air from the ceiling.

A safety director, an active safety educational program, first-aid cabinets, and the proper types of gas masks for ammonia or the refrigerant being used, should be provided. Cold rooms should use a nontoxic and non-inflammable refrigerant gas. Extension-cord lights should be enclosed in approved vaporproof glass. The wiring should be of the four-wire grounded type. Splashproof motors are recommended, and covered electrical outlets for portable pump plugs should be installed.

The possible toxic effects of sulfur dioxide in the winery have not always been adequately guarded against. The atmospheric concentration of this gas should not exceed 10 parts per million. Workers who must be exposed to greater concentrations should wear approved eye and respiratory protection. The acute and chronic effects of sulfur dioxide and the recommended treatment and precautionary measures are summarized in a publication of the American Petroleum Institute (1948).

COMPOSITION AND HANDLING
OF GRAPES

CHAPTER SIX

COMPOSITION OF GRAPES

To obtain wines of quality, the grapes must have the proper composition and character for the types of wines to be produced.

PHYSICAL COMPOSITION AT MATURITY

Stems constitute 2 to 6 per cent of the grape.
The must is composed of 0 to 5 per cent seeds,
5 to 12 per cent skins, and 80 to 90 per cent juice.
Yields of juice show wide variations.

The stems (rachis and pedicels) on which the berries are borne constitute from 2 to 6 per cent of the total weight at maturity, with an average of about 3 per cent. The fleshy pericarp (known as pulp, which is surrounded by skin and in which the seeds are embedded) composes the greater portion of the fruit itself. From 83 to 90 per cent of the crushed, stemmed grapes (must)[1] is juice. The skins account for 5 to 12 per cent of the weight of the must and the seeds from 0 to 5 per cent. A few thick-skinned, small-berried, or heavy-seeded varieties may exceed these limits. The approximate composition of the different parts of a grape cluster are as follows (in percentages):

	Weight	Water	Acids and salts	Cellulose and lignin	Tannin
Stems	2–6	60–80	2–9	6–8	1–4
Berry	95–97				
Skin	5–12	70–80			1–2
Pulp	85–87	60–85	0.2–1.2		
Seeds	0–5	30–40		12–13	2–5

Source of data: Verona and Florenzano (1956).

Marsais (1941), on the basis of a large number of wine-grape varieties, found an average of 80.2 per cent clear juice and 19.8 per cent pomace in unstemmed grapes. This pomace consisted of 15.1 per cent stems (3 per cent of the total weight), 41.55 per cent skins (8.2 per cent of total weight),

---

[1] Although "must" sometimes refers to the freshly crushed grapes, it usually denotes the free-run unfermented juice from the tank or press.

22.0 per cent seeds (4.4 per cent of total weight), and 21.3 per cent solids (4.2 per cent of total weight). The clear juice, constituting 80.2 per cent of the whole grapes, yielded 72.6 per cent wine based on the original total weight. For a ton of grapes these figures indicate 1,491.4 pounds of wine, 336 pounds of stemmed pomace, and 60 pounds of stems. If the specific gravity of the wine is assumed to be 1, this results in an average yield of 178 gallons per ton. Commercial operations in California frequently show yields as high as 190 to 195 gallons per ton. If no raisins were present in the crushed fruit and if no water was used to facilitate crushing, we believe the yield would be closer to 178 gallons per ton. It is, of course, possible to show yields of new wine higher than this if stems, pulpy material, skins, and other solids pass into the fermenters. Some of the newer grinding, squeezing, rolling, and continuous pressing procedures do lead to such conditions. (See also p. 205.)

Stems

The stems are high in tannin (1 to 4 per cent), minerals (2.0 to 2.5 per cent), and acids (0.5 to 3.0 per cent as acid tartrate). According to Girard and Lindet (1898), ether-extractable resinous materials (called phlobaphenes by them) account for about half of the total tannin content. The bitter taste of the stems is attributed to such compounds by Ventre (1930). The percentage of moisture in the stems decreases from over 75 per cent to about 65 per cent during ripening, but may drop to only 40 or even 30 per cent in overripe grapes where the stems become very desiccated. Only about 1 per cent sugar is found in the stems (Amerine and Bailey, 1959; Amerine and Root, 1960). However, the stems, as separated by the crusher, are covered with sweet juice. They also contain a varying percentage of raisins, depending on the variety and maturity of the grapes and the type of crusher-stemmer used. The actual sugar content of a sample of stems may therefore be much greater than 1 per cent.

The ratio of stem to berry weight differs with variety, season, and maturity. With a given variety, this ratio will be much lower in a dry, hot season or district than in a cool season or district. Table 35 gives typical data on the changes that occur in the weight of the stems, skins, pulp, and seeds during ripening. Because of the possibility of extracting a bitter taste from the stems, they are usually removed before fermentation. However, to facilitate pressing, the stems may be placed in the press basket. Some believe that red grapes low in tannin, grown in regions as warm as IV or V in California, may profit by fermentation with some of the stems. No persuasive data have yet been presented to show that this is true.

TABLE 35

CHANGES IN WEIGHT OF SKINS, PULP, SEEDS, AND STEMS OF PETITE VERDOT GRAPES
DURING MATURATION

	July 20		August 7		August 23		September 6		September 25	
	Grams	Per cent[a]	Grams	Per cent[a]	Grams	Per cent[a]	Grams	Per cent[a]	Grams	Per cent[a]
Weight per 100 berries										
Skins	5.9	13.7	6.4	10.5	6.6	9.2	7.7	9.8	11.0	8.1
Pulp	31.6	73.5	47.9	78.5	58.3	81.0	64.4	81.8	118.5	87.8
Seeds	5.5	12.8	6.7	11.0	7.1	9.8	6.6	8.4	5.5	4.1
Average weight per berry	0.43	—	0.61	—	0.72	—	0.79	—	1.35	—
Average weight of stems of average cluster	2.5	—	2.3	—	3.1	—	2.7	—	2.1	—

[a] Percentage of total weight.
Source of data: Laborde (1907).

Skin

In the American grape, specifically Concord, Winton and Winton (1935) report the cross-section of the pericarp to consist of (1) an epicarp (polygonal cells, with thick outer walls and porous radial and inner walls), (2) a hypoderm (also polygonal cells, with porous walls, increasing in size inward), (3) a mesocarp (large, thin-walled, rounded cells, showing sugar crystals in the alcohol-fixed material, with occasional fibrovascular bundles and reticulated bundles), and (4) an endocarp (polygonal cells). In the black and red American varieties the coloring matter is dissolved in the epicarp and the hypoderm.

Winton and Winton (1935) state that there is no fundamental difference in the structure of American grapes, just given, and that of *V. vinifera*. The toughness of the skin of American grapes is primarily due to the thickness of the outer wall of the epicarp, but this varies greatly between varieties and species. Bonnet (1903) reported the distances through the center of the outer walls of Concord to be 7.0 to 8.1 μ and only 3.0 to 5.4 for Tokay, 4.0 for a muscat variety, and 2.7 for a seedless variety. De Villiers (1926) reported reducing sugars throughout the fruit, except around the primary and secondary vascular strands (fig. 49). This is interesting in view of Amerine and Root's (1960) report of nonreducing sugars in the stem up to the brush. (See also Amerine and Bailey, 1959.) Apparently the hydrolysis of sucrose takes place between the vascular system and the parenchyma cells of the flesh. In the ripe fruit, tannin (as detected by the ferric chloride test) was confined to the vicinity of the vascular tissues and in the subepidermal layers. Except in a few red-juice varieties the pigments are confined to the epicarp.

Winkler and Williams (1936) showed that the growth stages of a grape berry can be divided into three distinct periods: I, rapid growth; II, depressed growth; and III, final swell. Nakagawa and Nanjo (1965, 1966) confirmed this. They also studied the relative rates of cell division and enlargement in the different parts of the berry. Cell division in the epidermal layer continued for 33 days after full bloom. The length of period II was different for different varieties. In period III the increase in thickness of the outer wall tissue was rapid and was related to berry enlargement.

The skin is covered with a thin waxlike layer known as the bloom, which normally constitutes from 1 to 2 per cent of the total weight of the skin. Some varieties, such as Perruno and Palomino, have little bloom; others—Meunier, for example—have a very thick bloom. Markley et al. (1938) found the following in the pomace of V. labrusca: free oleanolic acid; linoleic, oleic, palmitic, stearic, and higher fatty acids; glycerin; the hydrocarbons nonacosane and hentriacontane; sitosterol; and fractions representing mixtures of primary alcohols of the series C_{22} to C_{28}. The bloom provides a protective covering for the berry. Apparently yeast cells stick to the bloom during the maturation of the grapes.

Further data on the surface waxes of grapes were given by Radler (1965a, 1965b). The main constituent, about two-thirds, is oleanolic acid, a triterpene. The amount of total wax was 0.09–0.11 mg per square centimeter of surface; this is fairly constant during growth and maturation of the grape. Of the alcohols about 40 per cent is C_{26} with C_{24} and C_{28} next most prominent, though alcohols from C_{14} to C_{34} are present. Radler (1968) also noted that Isabella (V. labrusca) has about 10 times more paraffins than V. vinifera varieties, particularly of C_{29} and C_{31}. These compounds are often odorous and so will contibute to the aroma of wine. They may also have some retarding effect on the activity of yeasts and bacteria.

The outer layers of the fruit (mainly the skin) seem to contain the greater portion of the aroma, coloring, and flavoring constituents. As the grapes pass the stage of full maturity some migration of these constituents from the cells in and near the skins into the inner cell tissues occurs. The skins of red grapes are high in tannin (3.0 to 6.5 per cent). The skin-to-pulp ratio is greatest with small-berried varieties. Thus, with an equal concentration of color or flavor per square inch of skin, the juice—and thus the wine—of a small-berried variety will have the greater concentration of color and flavor. This is one reason why the large-berried table grape varieties, such as Malaga, Tokay, Ribier, and Emperor do not produce good wines.

The toughness of the skin differs markedly among varieties. This is important, as it partially determines resistance to handling injury and to crushing. In the data of Amerine (1962b) the breaking weight decreased as the fruit ripened. Tokay and Muscat of Alexandria were much tougher-skinned than Cabernet Sauvignon or Sémillon. Howard (1905) and Alwood (1914) noted that many needle-like crystals were attached to the fibrovascular bundles and in the cells just under the skin—undoubtedly of potassium acid tartrate.

Seeds

The seeds are highest in tannin content (5 to 8 per cent) and contain important amounts of oils and resinous material. The phenolic compounds of the seeds of different varieties tended to resemble each other in the proportions present, according to Singleton et al. (1966). Epicatechin gallate is present in the seeds, a result not always previously found. Oil constitutes from 10 to 20 per cent of the weight of the seeds. This oil consists of about 10 per cent solid fatty acids (palmitic and stearic) and 80 per cent liquid fatty acids (oleic, linoleic, and others), with small quantities of other materials.

Flanzy and Flanzy (1959) reported a fairly constant amount of linoleic acid in grape-seed oil (65 to 70 per cent of the fatty acids of the oil). Similar results were obtained by Prévot and Cabeza (1962). Silvestre (1953) reported tocopherols in grape-seed oil. Flanzy and Dubois (1964) reported 0.05 to 0.07 per cent in the oil of seeds of Carignane and Mourastel and 0.026 per cent in a commerical oil. According to Dubois (1964a), they are the principal natural polyphenolic antioxidants of grape-seed oil. There is a marked decrease in these tocopherols at temperatures of 65°C (149°F) and 88°C (188.4°F). A synergistic effect on the antioxidant property of tocopherols seems to be exercised by phospholipids (Dubois, 1964b).

Fruit zones

Fig. 49 shows the structure of a grape berry. It is customary to distinguish three zones within the fruit: I, near the skin; II, the intermediate region; and III, near the seeds. The composition of the three zones is not at all similar. Usually the intermediate, but sometimes the peripheral, region has the highest sugar content and lowest acidity, and that near the seeds has the highest acid and lowest sugar. The following data from Benvegnin et al. (1951) for Chasselas doré and other Swiss varieties are illustrative:

	Zone I		Zone II		Zone III	
Constituent	Chasselas doré	Other	Chasselas doré	Other	Chasselas doré	Other
Sugar, per cent	19.3	18.0	19.1	18.7	18.7	16.6
Acid, per cent as tartaric	0.29	0.45	0.74	0.87	0.90	1.38

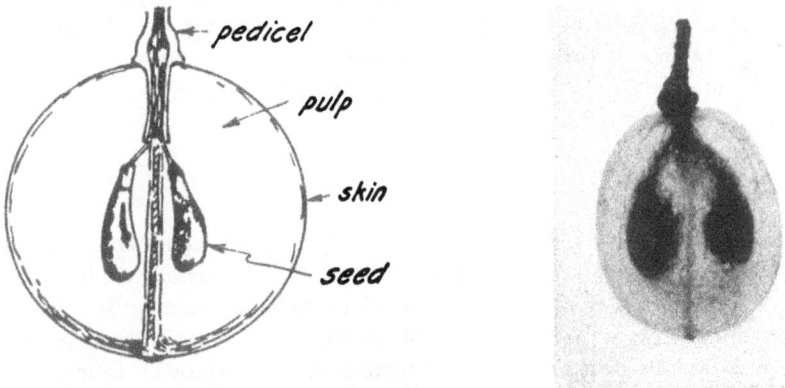

Figure 49. Left, diagrammatic sketch of berry showing principal parts; right, photograph of xylene-cleared berry with skin removed and vascular system and seeds exposed.

It is primarily the uneven composition of the berry which makes the free-run and press juice vary in composition. With ripe but *not overripe* grapes, the free-run juice, mainly from zone II and to a lesser extent from zones I and III, has the highest percentage of sugar. But with green grapes the first-press juice may be higher in sugar than the free-run. This is probably the result of unequal ripening of the pulp, and reflects a ripening of the grape from the exterior to the interior. Ventre (1930) reported that with overripe grapes the free-run juice is less sweet than the press. This is probably because sugar from the shriveled or raisined fruit is represented in the press juice to a greater extent than in the free-run. This condition is very common in California. The problem is complicated by the fact that the free-run juice always includes some material from zone I, and perhaps more from III, as well as that from II. Thus the composition of the free-run juice, compared with that of the press, varies with the maturity of the fruit and the variety. Also, for a given variety, composition varies with the season and the region.

Juice yield

The yield of juice from grapes depends upon various factors, including the variety, stage of maturity, condition and size of fruit, the season, the region of production, the method of crushing, the size and effectiveness of the crushers, fermenters, and pressers, the temperature, and length of time on the skins. Table 36 is therefore only indicative; the actual yield obtained will depend on a number of factors. Since many wineries do not use a press, the actual yield of juice may be somewhat less, but the

TABLE 36
Yields of Stems, Pomace, Juice, and Wine from a Ton of Grapes

Source	Stems (from crusher)	Pomace		White wine			Red wine		
		Unfermented	Fermented	Fresh juice	New	After racking	Free run	From press	Total
		Pounds		Gallons					
Bioletti (1914)									
Maximum	64.6	513.0	392.1	183.4	—	—	143.0	41.1	179.5
Minimum	16.2	265.0	252.0	160.4	—	—	127.8	34.7	170.9
Average	27.2	391.0	345.2	168.7	151.0	145.5	136.4	38.4	174.9
Marsais (1941)									
Average	60.0	337.6	—	176.8	—	—	—	—	—
Benvegnin et al. (1951)									
Average	134.0	266.8	—	176.9	174.8	167.7	—	—	—
Amerine (1949b)									
Maximum	89.2	203.7	—	189.7	169.9	—	—	—	—
Minimum	31.6	180.5	—	164.7	156.5	—	—	—	—
Average	72.6	192.9	—	177.6	164.4	—	—	—	—

alcohol recovery obtained by washing the pomace free of sugar may be larger than if the grapes were pressed. Based on present information, fresh-juice yields of 160 to 190 gallons per ton may be expected. The yield of new white wine will be 5 to 10 per cent less, or about 150 to 175 gallons per ton. Red wine yields are usually about 15 to 20 gallons per ton higher than those of white wine. Only general averages can be given because of the many variables listed above.

CHEMICAL COMPOSITION AT MATURITY

To the wine maker, sugars and acids are the most important constituents of the must.

The range in percentages of the more important components of the must is given in the data below, taken mainly from Ventre (1931) and von der Heide and Schmitthenner (1922). Water constitutes 70 to 85 per cent of the weight of the must, depending on the maturity, seasonal and regional conditions, and the variety.

Constituent	Range (per cent)	Comment
Water	70–85	
Extract	15–30	
Carbohydrates		
Sugars	12–27	Glucose and fructose
Pectins	0.01–0.10	Includes gums, etc.
Pentosans	0.01–0.05	Small amounts of pentoses also present
Inositol	0.02–0.08	Larger amounts in high-acid musts
Acids, total	0.3–1.5	pH range 2.9 to 3.9
Malic	0.1–0.8	Varies with variety, region, and season
Tartaric	0.2–1.0	Mainly potassium acid tartrate
Citric	0.01–0.05	
Tannin	0.0–0.2	
Nitrogen	0.01–0.20	In proteins, amino acids, ammonia, etc.
Ash	0.2–0.6	

The same authorities give the following ranges in amounts of the various components of the ash in the must:

Ash constituent	Range (gr per liter)	Comment
Iron	0.001–0.030	Contact of must with iron surfaces increases this
Potassium	0.400–2.00	Increased by addition of potassium salts; varies markedly with maturity and growing condition
Calcium	0.040–0.150	Contact with concrete tanks increases this
Magnesium	0.050–0.200	
Aluminum	0.001–0.040	Contact with certain filter aids increases this
Sodium	0.050–0.200	
Manganese	0.000–0.050	Highest values in varieties other than *V. vinifera*
Chloride	0.030–0.150	High in grapes grown near the sea
Phosphate	0.100–0.420	
Sulfate	0.028–0.330	

Sugars

The following sugars were found in the fruit of *V. vinifera* by Kliewer (1966): stachyose, raffinose, melibiose, maltose, sucrose, galactose, glucose, and fructose. The two most important sugars are *d*-glucose and *d*-fructose, in approximately equal amounts. Fructose is considerably sweeter than glucose. During maturation the fructose to glucose ratio increases from 0.5 to about 1.0. Contrary to expectations, the glucose to fructose ratio of green grapes was little different from that of ripe grapes in the study of Weger (1965c). However, Amerine and Thoukis (1958) and Kliewer (1967b) reported the expected decrease, but found rather wide differences in the ratios of different varieties at maturity. Kliewer (1967b) noted lower glucose to fructose ratios in overripe grapes. Kliewer found certain varieties—Chardonnay, Pinot blanc, and Green Hungarian, for example—that could be classed as high-fructose varieties, and others —Chenin blanc, Emerald Riesling, and Zinfandel, among others—that were high-glucose varieties. The glucose to fructose ratio tended to be lower in a warm season than in two cool seasons. He postulated that differences in the ratio could occur as a result of enzymatic conversion of glucose to fructose or preferential loss of glucose via the pentose cycle. Surprisingly, the fructose to glucose ratio of six *V. vinifera* varieties studied by Lott and Barrett (1967) varied from 1.06 to 1.40, i.e., greater fructose than glucose.

From 12 to 27 per cent or more of the weight of the must, when the grapes are harvested, consists of these two sugars. Musts from overripe grapes, particularly of late-ripening varieties, have a fructose to glucose ratio above 1, and drying or cooking does not change the ratio. The data of Szabó and Rakcsányi (1937) in table 37 are typical.

Amerine and Bailey (1959) showed that there is more sucrose and starch in the main stem than in the lateral branches of the fruit cluster. More reducing sugar was found in the lateral branches, particularly in the brush. They reported 1 to 2 pounds of fermentable sugar in the stems per ton of grapes. In other varieties Amerine and Root (1960) obtained 2.3 to 4.8 pounds of fermentable sugar per ton in the stems. Small amounts of pentose have been noted in the European literature, but Kliewer (1965) was unable to detect them in California grapes. He found raffinose (0.01 to 0.32 per cent) melibiose, stachyose, and maltose, but only 0.019 to 0.18 per cent sucrose in musts of *V. vinifera*.

Lott and Barrett (1967) confirmed that significant amounts of sucrose are found only in clones derived from *V. labrusca*. The amounts in six *V. vinifera* varieties varied from 0.17 to 0.60.

Malya (1965) studied the increase in reductones during ripening of grapes—a process which he believes plays a role in the development of

TABLE 37

CHANGES IN TOTAL SUGAR, GLUCOSE, FRUCTOSE, AND THE FRUCTOSE-GLUCOSE RATIO DURING MATURATION

Date	Chasselas doré				Furmint			
	Total sugar (per cent)	Glucose (per cent)	Fructose (per cent)	ratio	Total sugar (per cent)	Glucose (per cent)	Fructose (per cent)	ratio
8–22	11.3	6.1	5.2	0.85	—	—	—	—
8–27	12.9	7.3	5.6	0.77	2.7	1.8	0.9	0.50
9–11	18.1	9.0	9.1	1.01	14.4	7.2	7.2	1.00
9–24	15.2	7.5	7.7	1.03	15.2	7.4	7.8	1.05
10–9	19.2	9.2	10.0	1.09	19.5	9.3	10.2	1.10
10–29	22.7	10.4	12.3	1.18	22.3	9.9	12.4	1.25
11–22	20.0	8.8	11.2	1.27	20.3	8.9	11.4	1.28

Source of data: Szabó and Rakcsányi (1937).

the odor of the resulting wine. Settling musts with sulfur dioxide reduced the reductone content. Varieties could be classified into three groups, depending on the behavior of reductone during ripening. Varieties which show an increase during ripening were Traminer and Chardonnay; a variety where the changes are nominal was Furmint. Among those in which the reductone content drops only to 5 to 15 mg per liter during ripening were Wälschriesling, Kővidinka, Izsáki, Sárfehér, and Pirosszlanka. Varieties of the first group profit more by must sulfiting. Surprisingly, wines made from botrytised musts had a very low reductone content (6 to 8 mg per liter), while those from musts of moldy grapes (other than botrytis) were high in reductone (45 to 55).

Acids

Several organic acids are present in grape juice. Those occurring in sufficient quantities to be of primary interest to the wine maker are *l*-citric, *l*-malic, and *d*-tartaric. The latter two account for over 90 per cent of the total acid constituents of the juice. The following organic acids were identified in grape juice by Kliewer (1966): malic, tartaric, citric, isocitric, ascorbic, *cis-aconitic*, oxalic, glycolic, glyoxylic, succinic, lactic, *glutaric*, *fumaric*, *pyrrolidone carboxylic*, α-ketoglutaric, pyruvic, oxalacetic, galacturonic, glucuronic, shikimic, quinic, chlorogenic, and caffeic. The acids italicized were newly established by him. Clauss *et al.* (1966) isolated mucic acid from musts. The solubility product of calcium mucate is given as 1.4×10^{-7} $(mol/liter)^2$.

In California the titratable acidity of the juice of mature grapes varies from about 0.30 to 1.20 per cent, calculated as tartaric acid. The acids are formed mainly in the leaves and translocated to the fruits during maturation. (See also Bobadilla and Navarro, 1949; for recent data see Kliewer *et al.*, 1967*a*.)

Tartaric acid does not seem to be formed from malic acid. Apparently it is a by-product of photosynthesis, with carbon atoms 2 and 3 of tartaric acid the same as atoms 3 and 4 of glucose. So far as is known, tartaric acid is not a part of the Krebs cycle. In studies with radioactive compounds Drawert and Steffan (1966) showed that malic and tartaric acid are formed by different metabolic patterns. Little tartaric acid was formed from labeled glucose or glutamic acid, whereas much malic acid was produced. Tartaric was respired only in the dark, while malic acid produced glutamic acid, acetate, and sugar. Sugar formation from malic acid increased in the light. These studies indicate that these organic acids are not part of a stationary pool but part of the energy reserve of green grapes and are produced in the berries during ripening. G. Ribér-eau-Gayon and Lefebvre (1967) reported maximum conversion of glucose to tartaric acid at 20°C (68°F) compared to higher or lower temperatures and more tartaric acid formed at low light intensity compared to high. They consider malic acid to be a by-product of photosynthesis and tartaric acid to be derived from glucide degradation.

Recent work by G. Ribéreau-Gayon (1968) does not confirm some of the conclusions of Drawert and Steffan. He found that malic acid is formed principally by direct fixation of carbon dioxide by enolpyruvic acid. This mechanism was particularly active in green berries. Some malic acid is synthesized from citric acid during transport from the roots. The loss of malic acid during maturation is due not only to its complete degradation but also to conversion of some to glucose. Only growing organs (leaves and green fruit) synthesize tartaric acid. He also found it to be formed from glucose but by a separation between atoms 4 and 5.

Few varieties or climatic conditions in California yield grapes with musts containing over 1.0 per cent acid, and the average is probably about 0.5. This is much lower than the average for French and German grapes, but resembles those reported for Algeria and Spain. The titratable acidity also varies with season and variety.

Kliewer *et al.* (1967*b*) studied artificial shading, which gave 21 to 30 per cent of normal solar radiation, from the beginning of ripening (*veraison*) to maturity, as a method of delaying maturity and improving the sugar-acid balance. This resulted in delays of one to five weeks in maturity and higher total acidity and malate content. If shading could be done economically it would extend the area in California for the production of premium-quality grapes for table wines. See also Kliewer (1964) and Schultz and Lider (1964).

There is a marked difference, among varieties, in the ratio of tartaric to malic acid. Amerine and Winkler (1942) reported some varieties with a relatively high percentage of their total acid as tartaric acid; others were relatively rich in malic acid, though seldom over 50 per cent. Ferré (1928a) found only 10 to 30 per cent of the acidity present as free malic acid under relatively cool French conditions. In Burgundy, at least, the tartrate content constitutes a much higher percentage in warm years (70 to 90 per cent) compared to cool years (50 per cent). (See Ferré, 1925.) Bremond (1937a) reported higher percentages of malic acid in grapes from the cool hill regions of Algeria that in those from the plains. Peynaud (1939a) found marked difference in the malic acid content in five important varieties of the Bordeaux region, and suggested that the practice of blending two or three varieties of grapes has developed in that region in an effort to maintain the malic acid content as nearly constant as possible in years of widely different climatic conditions. The malic acid content (and consequently the total titratable acidity) is believed to be lower in warm years, but Amerine (1951) was unable to prove this. Kieffer (1949), for example, found that the tartaric acid to malic acid relation was normally 3 to 1, but in cool years only 2 to 4 or 2 to 3. He also found that some varieties respired malic acid much easier than others. See also Peynaud and Maurié (1953a, 1958).

Amerine and Winkler (1942) found that the acidity caused by tartrates amounted to 25 to 40 per cent of the total early in the season, but that late in the season it varied from 45 to 81 per cent, and the varieties had characteristic differences. Thompson Seedless, for example, is a high-tartrate variety when it reaches maturity; Carignane and Ribier may be classified as low-tartrate varieties. One would expect the latter varieties to be more susceptible to bacterial spoilage than the former, other factors being equal.

Kliewer et al. (1967a) also reported wide variations in the tartrates and malates in different varieties as well as the percentage of titratable acidity due to malates. They were able to divide grape varieties into four groups based on the tartrate to malate ratio. Typical high malate varieties were Carignane, Chenin blanc, Malbec, and Pinot noir. Examples of moderately high malate varieties were Aleatico, Chardonnay, Grenache, and Petite Sirah. Among the intermediate malate varieties were Cabernet Sauvignon, Gamay, Gewürztraminer, and Sémillon. Low malate varieties included Chasselas doré, Palomino, Thompson Seedless, and White Riesling. Their observations thus in general corroborate those previously reported by Amerine (1956) and Peynaud (1948).

Information on the tartrate, malate, glucose, and fructose content of a number of *Vitis* species are given by Kliewer (1967a).

During ripening there is a gradual decrease in the total titratable acidity accompanied by a slow but steady increase in pH.[2] There is no direct correlation between titratable acidity and pH, undoubtedly because of the widely varying buffer capacity[3] of the must. Brémond (1937a) found a general correlation between the percentage of free tartaric acid and the pH. Figure 50 shows that for a given variety there is a simultaneous and inverse change in pH and the total acidity. In general, grapes with the highest titratable acidity also have the lowest pH.

Wejnar (1968) from a comparative analysis of 38 German wines for total acidity, tartaric acid, malic acid, lactic acid, pH, ash and alkalinity of ash found a close correlation between pH and tartaric acid content (correlation coefficient 0.785). The pH was found to be regulated largely by tartaric acid content (correlation coefficient 0.789) particularly by the

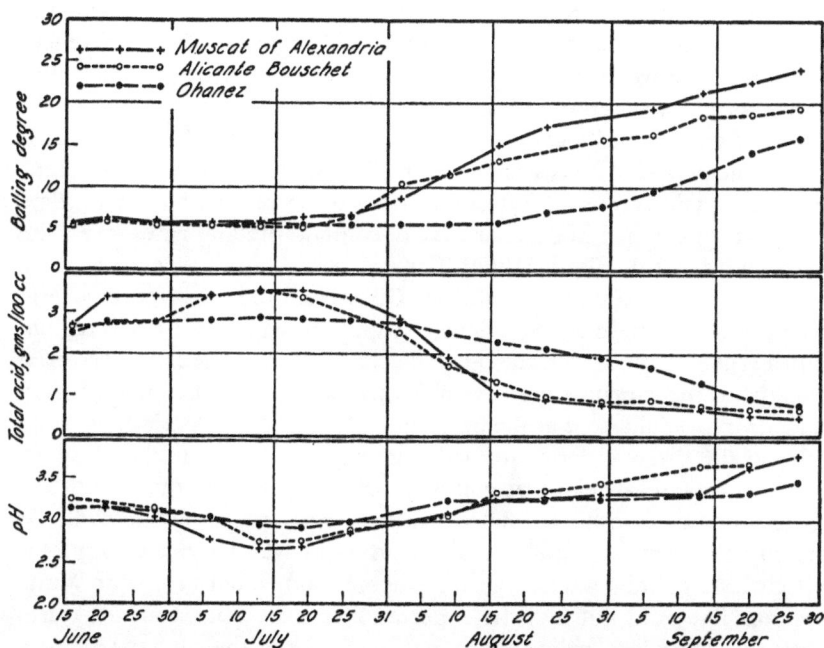

Figure 50. Seasonal changes in total acid, pH, and degree Balling of Alicante Bouschet, Muscat of Alexandria, and Ohanez grapes at Davis.

---

[2] The pH is a measure of the active acidity as distinguished from the total or titratable acidity. It is approximately an inverse function of the concentration of hydrogen ions supplied by the dissociation of the acids. The lower the pH, the greater the effective or active acidity.

[3] The buffer capacity is a measure of the resistance of a solution to change in its pH when acids or bases are added.

ratio of tartaric acid to potassium content ($\rho = 0.914$) and tartaric acid to alkalinity of ash ($\rho = 0.933$). Since the alkalinity of ash is a measure of potassium acid bitartrate, it appears that the buffer capacity of wine is determined largely by relative concentration of the acid salt.

The total content of both tartrate and malate decreases during ripening, especially the malate, which, in certain varieties and during hot weather, is reduced to a very low figure. Thus the titratable acidity which is due to tartaric acid and acid tartrate increases during ripening. The percentage of free acid decreases and the pH increases. The acid salt of tartaric acid increases during ripening, both per berry and in percentage by volume. If the free tartaric acid content of the must is 0.2 per cent or more, the resulting wine seldom contains excessive calcium (Anon., 1966g).

Peynaud and Maurié (1953a, 1958) substantiated the rapid decrease in malic acid during the ripening period. They found variations in the tartrate to malate ratio for the same varieties in years of similar climatic conditions. Peynaud and Maurié (1953a, 1958) noted marked differences in the ratio between seasons. For four German varieties, Liebert and Wartenberg (1965) reported losses of 47 to 63 per cent of the malic acid between Sept. 12 and Oct. 27, and 29 to 50 per cent of the tartaric acid in the same period. The largest loss of malic acid was in Müller-Thurgau; for tartaric acid it was in Sylvaner. It is not correct that a wine grape variety is ripe when the tartaric acid content predominates over the malic acid content. Thus Müller-Thurgau was "ripe" when the malic to tartaric ratio was 0.41 to 0.58, but Chasselas doré was ripe at a ratio of 1.15 to 0.57. These were obviously grapes grown in a very cool region. Ţirdea (1964) showed that the tartaric acid content of the fruit decreased throughout the ripening period in Rumania. In contrast, there was a sharp drop in malic acid at the beginning of the ripening period. Surprisingly, in the three varieties studied, the malic acid constituted 78 to 80 per cent of the total acidity at maturity. Obviously not all the parameters controlling the tartrate to malate ratio of ripe grapes are known. At maturity the tartaric acid to malic acid ratio was very low (1:7) for Wälschriesling, moderate (1:3) for Aligoté, and equal (1:1) for White Riesling. They did not find the expected low malic acid in all the warm years; there was less malic acid in mature grapes of the cool 1954 season than in the warm seasons of 1952, 1953 and 1955.

DL-Malic acid has been recommended for addition to foods. Addition of DL-malic acid to musts or wines could easily be detected by paper chromatography after an induced malo-lactic acid fermentation using *Leuconostoc mesenteroides*, according to Pilnik and Faddegon (1967). Only D(+) malic acid, which does not naturally occur in grape musts or wines, would remain.

In contrast to the changes during maturation in malates and tartrates in the pulp, Peynaud and Maurié·(1953a) reported increases in these components in the skins and stems. The increases in malate was generally greater than that of tartrate. In fact, the composition of the stems and skins most resembles that of the leaves.

The acids are not evenly distributed throughout the flesh of the grape. At maturity the zone near the skin has the lowest titratable acidity, the intermediate has more, and that around the seeds has the highest. This is nearly the opposite of the distribution of the sugars. (See p. 237.) In green grapes, however, the distribution of acids may be different, for the peripheral zone sometimes has the higher titratable acidity.

In the flesh of mature grapes the distribution of potassium acid tartrate (cream of tartar) differs from that of the total titratable acidity in that more occurs in the zone near the skins than in the interior (von der Heide and Schmitthenner, 1922). In ripe grapes cream of tartar crystals are found in and near the skin (Howard, 1905; Alwood, 1914), indicating that these cells are more than saturated with respect to this substance.

The unequal distribution of free acid and acid salt in the different zones at different stages of maturity probably accounts for Françot's (1945) finding that the first press (of whole grapes) in Champagne is lower in titratable acidity *and in pH* than are the second and third presses. (See also table 65, p. 659.)

The tartaric acid in grapes and wines is dextrorotatory. On heating, this can be converted into the optically inactive racemic acid. The calcium salts of the various isomers of tartaric acid vary in solubility. Cambitzi (1947) gives the solubility of these, in grams per 100 ml, as: *dextro-* calcium tartrate, 0.023; *levo-* calcium tartrate, 0.025; and racemic calcium tartrate, 0.003. The differences between the solubilities of the various optical isomers of potassium hydrogen tartrate are not so large. Cambitzi called attention to the autoracemization which the natural dextrorotatory tartaric acid of wine undergoes to produce the optically inactive acid whose calcium salt is sparingly soluble and forms crystalline deposits in old wines.

The solubility of potassium acid tartrate at various temperatures has been reported by Benvegnin *et al.* (1951) as follows:

Temperature °F	°C	Water gr/liter	Solubility in 10.5 per cent alcohol gr/liter
77	25	5.4	3.72
68	20	4.9	3.05
59	15	4.4	2.53
50	10	4.1	2.12
41	5	3.2	1.75

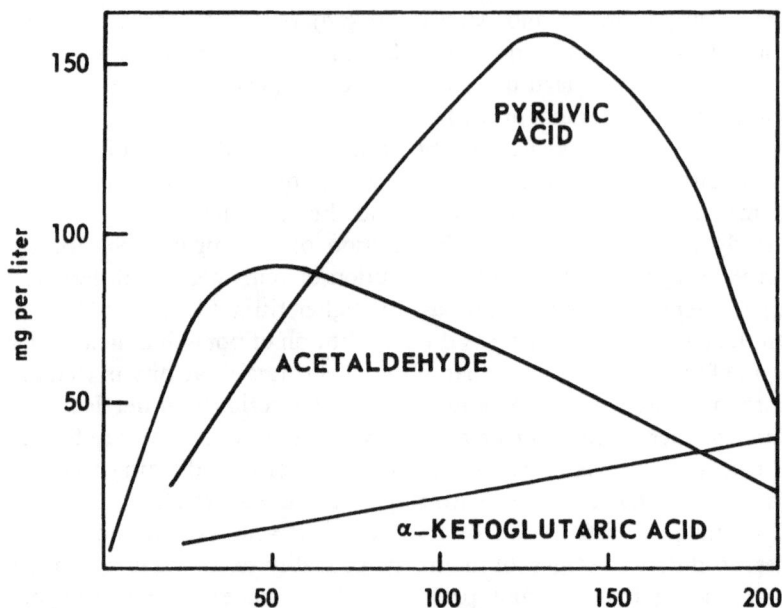

Figure 51. Formation of pyruvic acid, acetaldehyde, and α-ketoglutaric acid during fermentation.

The effect of pressing on the composition of musts and of their resulting wines is very complex. Carles et al. (1963) reported a decrease in tartrate but an increase in malate in the musts during pressing. The potassium and phosphate contents increased during pressing. Those of calcium, magnesium, iron, and copper at first decreased and then increased; so did the pH. These results were from one variety with two types of hydraulic presses in one region.

Deibner (1964a) reported rather variable results on the α-ketoglutaric acid content of musts during ripening (whether expressed on a volume, weight, or per berry basis). On a volume basis it was higher early in the ripening period and fell to about 3.6 mg per liter at maturity. There is a marked increase during fermentation. Storage of the wine on the lees resulted in greater losses than when not on the lees. Deibner (1964a) also noted no pyruvic acid at the beginning of maturation, sudden appearance amounting to about 5 mg per liter, and a slow decrease until full maturation. During fermentation there is a marked increase. Storage with or without the lees eventually resulted in wines with little or no pyruvic acid. Lafon-Lafourcade and Peynaud (1966) report sound grapes to contain about 35 mg per liter of pyruvic acid and 22 mg of α-ketoglutaric acid. Most of the grapes attacked by *Botrytis cinerea* contain larger amounts. See figure 51.

Hennig and Lay (1965) reported 0.03 to 0.04 gr per liter of oxalic acid in three German musts.

Both galacturonic and glucuronic acids have long been considered to be present in small amounts in musts and in larger amounts in musts or wines of grapes attacked by *B. cinerea*. Rentschler and Tanner (1955) reported little glucuronic acid in normal or botrytised musts but considerable galacturonic acid. Blouin and Peynaud (1963a) reported both present, but Dimotaki-Kourakou (1964a) was unable to detect any. Georgeakopoulos *et al.* (1963), however, consider galacturonic acid to be a normal component of wines, derived from enzymatic hydrolysis of pectin. They were unable to isolate glucuronic acid from either musts or wines, even from wines made from botrytised grapes.

In the cloudy material of musts from sound and moldy grapes Schormüller and Clauss (1967) reported succinic acid and L-α-pyrrolidone-α-carbonic acid. In the moldy turbidity they also found 2-ketogluconic acid and a C_7 polyvalent acid. Enzymatic oxidation of galacturonic acid is the source of mucic acid, according to Schormüller *et al.* (1967), which is found primarily in botrytised grapes. (See also Clauss *et al.*, 1966.)

Sugar to acid ratio

The ratio of sugar content (usually expressed as degree Brix) to acidity varies markedly among varieties in a given region and for the same variety grown in different regions. Amerine and Winkler (1941a) have shown that California grapes may be divided into three classes based on this ratio. At Davis, varieties with Brix-acid ratios of below 28.6, 31.4, and 34.3, at 20°, 22°, and 24° Brix, respecively, can be classified as typical table wine grape varieties. A second group of varieties had ratios below 28.6 and 31.4, at 20° and 22° Brix, but exceeded 34.3 at 24°. The third group exceeded the ratios given at all three degrees Brix. The latter varieties are typical dessert wine grapes. The Brix-acid ratio is probably a more rational method of classifying the utility of grape varieties than is either Brix or percentage of total acidity alone.

Maltabar (1959) proposed a coefficient of maturity $K = P \times S/100$ where P is the weight of 100 berries and S the percentage of sugar in the must. For Sylvaner he gives the following data:

Date	Weight (gr.) 100 berries	Sugar (per cent)	K
Aug. 10	160.0	13.2	21.2
Aug. 22	175.0	16.4	28.7
Aug. 31	181.0	17.2	31.3
Sept. 3	181.0	17.9	32.4
Sept. 12	173.7	18.4	32.0

Kozenko (1962) proposed the maturation formula $K = K_0/(1 + at)$ where K is the titratable acidity as grams of tartaric per liter at time t, K_0 the titratable acidity at t_0 and a a constant. An approximation of a is given from $(K_0 - K)/Kt$. Ungarian (1953) proposed another maturity index, $P_m - (Ar \times 100)/(24 - 0.755S)$, where Ar is the grams of titratable acidity per liter and S is the percentage of sugar. Nikandrova (1959) gave the following maximum limits for P_m for different types of wines: sparkling, 92–118; table wines, 81–91; red table wines, 76–80; natural sweet wines, 81–100; and fortified dessert wines, less than 76. Ungarian's index is obviously an acid to sugar ratio. Maltabar's ratio has the advantage of reflecting the decrease in weight that occurs late in the season. The Kozenko index is based only on titratable acidity and would be unduly influenced by climatic conditions.

In some varieties Deibner et al. (1965b) did not find that the maximum sugar coincided with the attainment of maximum weight. Furthermore, the calculated acidity and sugar proposed by Soviet viticulturists were applicable to only one of four varieties.

Nitrogenous Compounds

Nitrogenous compounds of various kinds are found in musts. Colby (1896) found that the total organic nitrogen, expressed as protein, ranged from 0.01 to 0.20 per cent in musts of several varieties from different parts of California. Similar results were later obtained by Niehaus (1938) in South Africa. Drought-injured grapes and nitrogen-fertilized grapes had about the same nitrogen content as normal fruit. Muth and Malsch (1934), in two German musts, found most of the organic nitrogen present as amino acids, peptides, and purines, with only traces of proteins. In a filtered 1932 Pinot noir must they found (as mg per liter) a total nitrogen, by Kjeldahl, of 616.4; protein nitrogen, 28.0; amino acid nitrogen, 226.2; and humin nitrogen, 7.8. They thus accounted for 609.5 of the 616.4 mg present.

Murolo (1967) reported 190 to 395 mg per liter of total nitrogen in thirty-two Italian musts. Of this 63.2 to 78.2 per cent was assimilable by yeasts. There was a small variation between yeast strain and season in the percentage assimilable.

During maturation the organic nitrogen in the fruit steadily increases, including amino acids and proteins, according to Ferenczi, (1966a, b), while ammonia decreases. At maturity there is a wide variation in nitrogen content for the same variety—nearly 100 per cent between years and 50 per cent between vineyards (Cordonnier, 1966).

The large range in the total nitrogen content mg per liter is indicated by the figures of Zinchenko (1964a):

Traminer rosé	403–799
Furmint	274–563
Leánka (Leányka)	262–550
Wälschriesling	206–471

The ratio of mg of nitrogen per liter to percentage of sugar also varied:

Traminer rosé	17.8–40.2
Furmint	15.2–33.1
Leánka (Leányka)	11.6–30.5
Wälschriesling	12.0–25.9

Since the solid parts of the grapes are richer in nitrogen than the pulp, red wines generally are higher in nitrogen than whites. The press wine is also higher in total nitrogen and in proteins than the free-run. Heating musts with the skins also results in wines with higher nitrogen. Bentonite treatment of musts reduces the must nitrogen content, as does filtration. Hennig (1943) determined seven nitrogen fractions in German musts. In fresh musts of the 1941 and 1942 vintages he found the following (as mg per liter):

Nitrogen fraction	1941	1942
Total nitrogen	1,075	748
Protein nitrogen	40	18
Ammonia nitrogen	112	15
Phosphotungstic precipitable nitrogen	349	231
Amino acid nitrogen	408	319
Humin nitrogen	20	17
Amide nitrogen	34	16

This confirms the earlier findings of the insignificant amount of protein present, but the total nitrogen content is rather high. In eleven California-grown grape juices Bayly and Berg (1967) reported 20 to 260 mg per liter of protein (average 121).

Genevois and Ribéreau-Gayon (1936) reported 300 to 660 mg per liter of organic nitrogen. The most important organic fractions are those of phosphotungstic and amino acid nitrogen. The former includes the tri- and tetrapeptides, diamino acids such as arginine and lysine, and heterocyclic acids such as histidine and proline, as well as any purines present. The pyrimidines and purines in this group, and the appreciable quantities of amino acids found, are believed by these workers to be of importance to bacterial growth. The amino acid fraction includes the dipeptides. The humin nitrogen fraction includes tyrosine and tryptophan; the amide nitrogen fraction contains asparagine and glutamine. The nitrogen compounds found in the residue are of unknown composition and amount to 10 to 20 per cent of the total nitrogen.

Koch (1963) and Koch and Sajak (1959) have summarized their investigations on the proteins of grapes. They found soluble proteins in grapes, the nature of which differed somewhat between varieties. The amounts increased during ripening and more appeared to be formed in warm seasons. (See also Koch and Schwahn, 1958.) Short-time heating and bentonite fining gave protein-free grape juice. The White Riesling and Müller-Thurgau varieties apparently contain glucoprotein with some eighteen amino acids. Glucosamine has been noted as a part of the glucoprotein in musts by Koch and Geiss (1955), Weis et al. (1960), and others.

A quantitative volumetric procedure for the determination of the hexosamines and total amino acids was given by Deibner and Bayonove (1965). They reported 8.7 to 29.2 mg per liter of nitrogen from hexosamines in thirteen musts and wines and 9.9 to 53.55 from amino acids. Deibner (1964a, b) has reported a small increase in hexosamines on a per berry or per weight basis during ripening. The total amino acid content increased on all three bases.

Earlier workers found the following amino acids: alanine, valine, tyrosine, leucine, proline, serine, threonine, glutamic, and aspartic acids, and possibly phenylalanine. (See Castor and Archer, 1956.)

Kliewer et al. (1966) (see also Nassar and Kliewer, 1966) identified thirty-two amino acids in various species of Vitis: α-alanine, β-alanine, arginine, asparagine, lysine, methionine, glycine, valine, cysteine, threonine, isoleucine, phenylalanine, proline, histidine, serine, γ-aminobutyric acid, aspartic acid, tryptophan, tyrosine, glutamic acid, leucine, cystine, cysteic acid, citrulline, norvaline, glutamine, pipecolic acid, pyrrolidone carboxylic acid, hydroxyproline, homoserine, α-aminobutyric acid, norleucine, and eight unidentified compounds that reacted positively with ninhydrin. (Compounds in italics were isolated for the first time.) Hydroxyproline and an unidentified component were primarily present in only five species, which suggests that they may be related. There were quantitative and qualitative differences between species of some of the other amino acids, but none that appeared to warrant their use in classification of species.

During ripening Nassar and Kliewer (1966) reported large increases in arginine, proline, glutamic acid, aspartic acid, histidine, phenylalanine, tryptophan, lysine, valine, threonine, leucine, isoleucine, serine, tyrosine, cysteic acid, pipecolic acid, and α-alanine. Total amino acid content increased from 708 μ moles per 100 gr of fresh weight on August 18 to 1400 on September 28 and 1757 on November 1. In five Portuguese musts Silva and Petinga (1962) found considerable variability in the number of amino acids present. One must had only eleven amino acids, another had eighteen. Similar reports of the extreme variability in the total nitrogen and individual amino acid content between varieties and musts from different regions is found in the work of Navara *et al.* (1966*a*) from Czechoslovakia. The amount of some of the amino acids (e.g., proline, cystine, and glutathione) was not related to the variety of grape, the season, or the region. However, the wines of the variety Bouvier grown at one location had a high glutamic acid content (90 to 140 mg per liter), but at another vineyard it amounted to only 0 to 25 mg per liter. They consider that alanine may be the most important amino acid for quality. In general, they consider the free amino acids to be an important quality factor; but convincing sensory data and correlations are not available. Total amine content varied from 130 to 1420 mg per liter.

During grape maturation, there is a significant increase in proline, serine, and threonine, and a decrease in arginine, ammonia, and amine nitrogen in musts of Cabernet Sauvignon and Merlot grapes, according to Lafon-Lafourcade and Guimberteau (1962). As grapes ripen, the forms of nitrogen which are less easily utilized by yeasts tend to increase. This may explain why musts of overripe grapes sometimes ferment slowly.

Flanzy and Poux (1965*b*) reported a much higher content of proline in musts of 1962, a warm year, than in 1963, a cool year. Most of the other amino acids were present in larger amounts in 1963 compared to 1962. However, the total was greater in the 1962 must. The percentage of utilization of amino acids during fermentation was generally greater in 1963 (except for arginine and threonine, which were utilized about the same in both years). Ough (1968) reported 50 to 80 per cent of the proline in the juice and 5 to 20 per cent in the pulp. The proline content increases with maturity, both in amount and as per cent of total nitrogen. From a number of varieties and sources he reported 304 to 4,600 mg per liter of proline (average of samples 742). In 48 wine samples the average was 869. He notes the Cabernet group of varieties to be high in proline.

The wide variation in qualitative and quantitative data on the amino acid contents of musts and wines is attributed by Bidan and André (1958) to be due to differences in the methods employed, the heterogeneity of samples (particularly as to the age and treatment of wines), differences in methods of fermenting the wines and the presence or absence of malo-lactic fermentation.

The amine nitrogen of musts, especially the arginine, phenylalanine, histidine, valine, and glutamic acid, are largely assimilated. Lysine, proline, and glycine are much less easily assimilated by yeasts.

The rapid decrease in ammonia in grapes during ripening, particularly during the latter stages, is a well-established phenomenon. Ammonia is of some significance in alcoholic fermentations (see p. 427). The analyses of the 1941 and 1942 wines mentioned above show 15 and 112 mg per liter, respectively, which are within the range of 0 to 200 reported by Genevois and Ribéreau-Gayon (1936, 1947).

A summary of the qualitative amino acid composition of musts and wines is given in table 38. For recent information on the amino acid content of a wide variety of musts and wines, see Ough and Bustos (1969).

TABLE 38

AMINO ACID COMPOSITION OF MUSTS AND WINES

	Musts			Wines					Sparkling wines		
	1	2	3	4	1	5	6	3	7	8	9
α-Alanine	+	+	+	+	+	+	+	+	?	+	+
γ-Aminobutyric acid	+	+	+	+	+	+	−	+	−	+	+
Arginine	+	+	+	+	+	+	−	+	−	−	+
Aspartic acid	+	−	+	+	+	+	+	+	−	+	+
Citrulline	−	−	−	−	−	+	−	−	−	−	−
Cystine	−	−	−	+	−	+	−	−	?	+	+
Glutamic acid	+	+	+	+	+	+	+	+	?	+	+
Glutamine	−	−	−	+	−	−	−	−	−	−	−
Glycine	+	+	+	+	+	+	−	+	?	+	+
Histidine	+	+	−	−	+	+	−	+	?	+	+
Isoleucine	+	+	−	−	+	+	−	+	−	+	+
Leucine	+	+	−	−	+	+	−	+	+	+	+
Lysine	+	−	−	−	+	+	−	−	−	−	+
Methionine	+	+	−	+	+	+	−	−	?	+	−
Ornithine	+	−	−	−	+	−	−	−	−	−	−
Phenylalanine	+	+	−	+	+	+	−	+	+	+	+
Proline	+	−	+	+	+	+	+	+	+	+	+
Serine	+	−	+	+	+	+	+	+	−	+	+
Threonine	+	+	+	+	+	+	+	+	−	+	+
Tryptophan	−	−	−	−	−	−	−	−	+	+	+
Tyrosine	+	−	−	+	+	+	−	+	?	−	−
Valine	+	+	+	+	+	+	+	+	+	+	+

1, Koch and Bretthauer (1957); 2, Hennig (1957); 3, Lüthi and Vetsch (1953); 4, Prillinger (1957); 5, Hennig (1955); 6, Sissakian and Besinger (1953); 7, Schanderl (1959); 8, Sissakian et al. (1961); 9, Bergner and Wagner (1965).

Grapes attacked by cochylis (an insect pest) and various kinds of mold are reported to be high in total nitrogen. The must from the second and third pressings is higher in ammonia and total nitrogen than is the free-run juice. Traces of nitrates and lecithins are reported in musts. The nitrogen content is undoubtedly of significance in the nutrition of yeasts, the production of higher alcohols during fermentation, the clarification of the wine, and probably certain types of bacterial spoilage, but more information is needed on the quantities of the nitrogen fractions present.

A summary of the nitrogen fractions of musts and wines is given in table 39.

TABLE 39

NITROGEN CONTENT OF MUSTS AND WINES
(mg per liter)

	Musts			Wines		
Fraction	Minimum	Maximum	Average	Minimum	Maximum	Average
Total	98	1,130	390 (28)[a]	70	781	350 (164)[a]
Ammonia	0	146	44 (25)	0	143	14 (83)
Amine	15	182	75 (22)	8	348	73 (158)
Amide	—	—	—	1	40	4 (55)
Polypeptide	38	132	81 (5)	28	273	148 (7)
Protein	28	97	51 (5)	1	125	42 (23)
Hexoseamine	18	29	23 (2)	8	29	14 (12)
Nucleic	—	—	—	—	—	23 (2)
Vitamin	—	—	—	—	—	0.16

[a] Figures in parentheses refer to number of samples upon which the average is based. Source of data: Cordonnier (1966).

Pigments

Normally, the pulp of mature grapes is very light yellow, and it is the skins which contain the important coloring materials of both red and white grapes. However, the Alicante Bouschet and a few other varieties have a red-colored pulp and, very late in the season, red-skinned grapes may have some color in the peripheral region of the pulp. The amount of color in grapes varies markedly with the variety, maturity, seasonal conditions, amount of crop, and other factors. (See also p. 70.)

Immature grapes contain chlorophyll, which gradually diminishes during ripening. A few varieties, when grown under cool climatic conditions, retain their green color to maturity, and occasionally a wine with a slight greenish tint is produced. Traces of carotene and xanthophyll are present also, mainly in the outer layers of the skin, according to von der Heide and Schmitthenner (1922).

In both red and white wines, the yellow pigments quercetin (a flavonol) and its glycoside, quercitrin (3-glycoside), have been reported. Peyrot (1934), however, found that the absorption spectra of white wines resembled cyanin, a reduction product of quercetin, more than it resembled either quercetin or quercitrin. However, Williams and Wender (1952) isolated in a pure form from grapes of three varieties of *V. vinifera* quercetin (3,3,4,5,7-pentahydroxyflavone) and isoquercitrin (quercetin-3-glucoside). Owing to its insolubility, quercetin is found in wine only in traces, but quercitrin is present in measurable quantities—1 to 30 mg per liter. Oxidation of these pigments apparently produces brownish-colored compounds. Genevois and Ribéreau-Gayon (1947) suggest that the browning of grape juice is characteristic of a polyphenol of this type. As the pH is raised toward 7, the color of these pigments become very yellow. For a review of the pigments of white grapes and their behavior during fermentation see Cantarelli and Peri (1964). Weinges (1964) reported catechin to be the most widely occurring phenolic in grapes and other fruits.

P. Ribéreau-Gayon (1964b, c) has reported kaempferin-3-monogluco-side (5, 10), quercetin-3-monoglucoside (50, 55), myricetin-3-monogluco-side (15, 0), and quercetin-3-monoglucuronoside (30, 35) in grapes. The proportion of each in red and white grapes, respectively, is given in parentheses. Since very little of these are present in white wines, Hennig and Burkhardt (1960b) and Ribéreau-Gayon concluded that flavonols do not contribute to the color of white wines.

Nuzubidse and Gulbani (1964) reported the three most important flavonols of grapes to be quercitrin, quercetin, and kaempferol. They are present in especially large amounts in the leaves. Different varieties were reported to have characteristic differences in flavonols. More data would be welcome, though this result appears to be reasonable. Cantarelli (1966) also reported myricetin. Using a new technique for isolation of flavonols Bourzeix (1967a) reported 37 to 97 mg per liter (as rutin) in four red wines and 44 to 46 in two whites.

The four grape leucoanthocyanins (proanthocyanidins) isolated from the skins and seeds of Pinot blanc grapes by Joslyn and Dittmar (1967a, 1967b) yielded catechin and epicatechin on heating with dilute acid. On heating with butanol-hydrochloric acid reagent, cyanidin was formed. They are, therefore, apparently compounds of catechin and leucocyanidin-like materials. Glucose was found in the acid hydrolyzate. At least one of the skin proanthocyanidins is a diglucoside. Seeds of the same grapes yielded (+)-catechin, (−)-epicatechin, and a small amount of gallo-catechin. No glucosides were found in the proanthocyanidin material from seeds.

Cantarelli (1966) noted that most of the leucoanthocyan of the berry is in the seed but that only 1 to 4 per cent of this is found in the wine. More than half of the leucoanthocyanins originally present are lost during fermentation. To reduce the leucoanthocyanin content poly-amides (nylon) are most effective, activated carbon less, and albumen, casein, and gelatin least. Peri (1967) investigated the conversion of two leucoanthocyanins, isolated from the seeds of white Trebbiano grapes, on treatment with hot butanol-2N hydrochloric acid solution. This conversion was found to occur in two stages, first the formation of catechin derivatives in absence of oxygen and then cyanidin formation which is oxygen dependent.

Naito (1966) reported that with certain varieties, such as Delaware, there was nearly as much leucoanthocyanin in the skin as anthocyanin. Two other varieties had little or no leucoanthocyanins in the skin.

Mareca Cortés and Gonzalez (1964, 1965a, b) found colorless flavonoids in red grapes before the red color begins to form. These pigments brown in the presence of air. After fermentation there is a gradual decrease in both anthocyanin and flavonoid pigments. They distinguished three regions of importance in the spectrophotometric absorption curves: (1) the typical anthocyanin maximum at 520 mμ, which is characteristic of extracts of the coloring matter of the skins and of wines during and just after fermentation; (2) the maximum between 400 and 500 mμ, usually at about 450, which is characteristic of mature red wines; and (3) a maximum at about 370 mμ, which is found in very old red wines and in white wines. It is not clear how the polarographic curves of Mareca Cortés and Miguel (1965) contribute to our knowledge of the flavonoid pigments. They claim that the polarographic curves are highly reproducible.

The color of red grapes was identified by Willstätter and Zollinger (1915, 1916) as being mainly the water-soluble monoglucoside, malvin (oenin). Its tint is believed to be affected by the presence of tannin, by flavones and other co-pigments (e.g., alkaloids), by iron and other metals, and particularly by the acidity and pH. Tannin gives it a markedly bluer tint; and a violet coloration may be caused by the addition of minute amounts of iron. Other metals that form complex combinations also influence its color. The anthocyanins are amphoteric: acids increase the red tint and alkalies the blue tint. This change in color from red to blue cannot be used as an acid-base indicator because it is not sharp enough, and the color change occurs at too low a pH—from 4 to 5.8; Casale (1930) gives 5.4 to 5.8 as the usual range. Furthermore, there is much difference in tint and rate of color change among different varieties as the end point is approached.

Differences in the shade of color produced by naturally occurring anthocyanins have been ascribed to variations in the relative and total concentration of the acid-stable flavylium salts caused by changes in acidity, or to formation of anthocyanin complexes with organic substances (co-pigments) or with metals. Jurd and Asen (1966) discount the widely claimed co-pigmentation effect since they found that substances such as quercitrin, chlorogenic acid, and methyl gallate have no effect on the color, spectra, or stability of cyanidin 3-glucoside in aqueous solution at

TABLE 40
ANTHOCYANIN PIGMENTS IN VARIOUS SPECIES OF VITIS

	rotundi-folia	riparia	rupestris	labrusca	arizonica	ber-landieri	mon-ticola	cordi-folia	rubra	lince-cumii	aesti-valis	coriacea	amu-rensis	vinifera	
														A	B
Total no.	5	15	12	11	11	10	6	11	12	17	9	7	5	5	9
Cyanidin															
monoglucoside		2		5	8	8	3	10	20	29	31	58		20	3
acylated monoglucoside										7					
diglucoside	9	5	2		1	2			1	2		4			
acylated diglucoside										5					
Peonidin															
monoglucoside				10	14	16	5	11	5	7	11	6	13	45	15
acylated monoglucoside[a]				3	1	1			1			4	15		2
acylated monoglucoside[a]															2
diglucoside	6	2	8	1	10			2	1	3					
acylated diglucoside										2		4			
Delphinidin															
monoglucoside		14	9	21	13	23	36	15	30	17	31	20		6	12
acylated monoglucoside		1	3							6	4				
diglucoside	38	12	34						1	1					
acylated diglucoside		2	6							3					
Petunidin															
monoglucoside		10	3	15	10	20	26	18	20	8	10	4	5	9	12
acylated monoglucoside		1	22	1	1	1		2	2						
diglucoside	29	17	2							1					
acylated diglucoside		3								3					
Malvidin															
monoglucoside		6	2	34	29	25	27	30	16	4	6		27	20	36
acylated monoglucoside[a]		2		7	3	2	3	4			2				9
acylated monoglucoside[a]								1							9
diglucoside	18	21	8	1	10	2		5	1	1			40		
acylated diglucoside		2	1	2				2	2	1	2				

a The difference in structure between the two acylated pigments is not known.
A = Muscat Hamburg; B = all other varieties tested.
Source of data: P. Ribéreau-Gayon (1964b, 1964c).

pH 3–6.5. In acetate buffer solutions (pH 5.45) containing aluminum salts, however, quercitrin and chlorogenic acid form highly colored coordinate complexes with the anthocyanin. The formation of these co-pigment aluminum-anthocyanin complexes depends on pH and type of organic acids present. They do not form in citrate buffers because of the preferential chelation of aluminum cations by citrate.

The concentration of the anthocyanin and its state of aggregation also affect the color. The latter is determined in part by the pH of the cell sap and the presence or absence of polysaccharide protective colloids, such as pentosans. Malvin (oenin), on hydrolysis, yields glucose and the aglycon malvidin (oenidin). Malvidin is 3,5,7,4'tetrahydroxy-3',5'-dimethoxy-2-phenyl benzopyrilium chloride. The important structural feature of this compound is that two of three adjacent hydroxyl groups are methylated. The anthocyanidin of the American (*Vitis labrusca*) varieties of grapes has but one substituted hydroxyl group and will thus give a positive color test with ferric chloride, whereas that of the European (*V. vinifera*) varieties will not. Ribéreau-Gayon (1933*a*), however, reports a much lower methoxy content in crude pigment preparations from aged Bordeaux wines, which were obviously made from *V. vinifera* grapes, than in young wines. The anthocyanidin may be present in combination with one or two molecules of glucose and thus may exist as a monoglucoside or a diglucoside. The stability of the anthocyanin pigments, particularly their resistance to oxidative change, is believed to be related to the extent of methylation and glucosidal bond formation. Tannin and sulfur dioxide have a marked protective effect on anthocyanin pigments, but aldehydes may cause loss in color by combination with anthocyanins (Joslyn and Comar, 1941). The presence of large amounts of sulfur dioxide cause bleaching of the color, but if an excess is not added or if the high sulfur dioxide concentration is not maintained too long, the color will return to its original intensity when the sulfur dioxide is removed. In addition to the colored anthocyanins in grapes, a colorless precursor, the leucoanthocyanin, is present.

The actual anthocyanin pigments found in different species of *Vitis* are given in table 40. In *V. vinifera* malvidin is the principal pigment. P. Ribéreau-Gayon (1964*b*, 1964*c*) found no aglycones of these pigments. The presence of diglucosides is a unique feature of American species of *Vitis*. He does note that diglucosides were reported by Soviet enologists in Muscat Hamburg, and this discrepancy remains to be explained.

Ribéreau-Gayon and P. Ribéreau-Gayon (1958) reported six to seventeen colored anthocyanins in twelve species of grapes as follows (in per cent):

Species	Total no. constituents	Monogly-sides	Diglyco-sides	Unidentified	Constituents carried on side ring 2 OH	3 OH
V. riparia	14	32	57	11	9	80
V. rupestris	12	14	74	12	10	78
V. lincenumii	17	65	8	27	41	32
V. aestivalis	9	87	9	4	48	48
V. coriacea	7	88	8	4	72	24
V. labrusca	10	84	4	12	16	72
V. arizonica	12	74	21	5	33	62
V. berlandieri	9	93	5	2	26	72
V. rubra	12	91	7	2	27	71
V. monticola	6	97	—	3	8	89
V. cordifolia	11	84	9	7	23	70
V. vinifera	8	77	—	23	18	59

Note the abundance of diglucosides in *V. riparia* and *V. rupestris*. Three species have a rather high percentage of two hydroxyl groups on the side ring (cyanidin and peonidin): *V. lincenumii, V. aestivalis,* and *V. coriacea.*

Easy preparation of the paper chromatograms with young wines is common. Later, little coloring material moves from its origin. This is attributed to transformation of some of the soluble anthocyans into a colloidal state. Van Wyk and Venter (1964) found only traces of malvin in South Africa *V. vinifera* varieties.

The main conclusion of P. Ribéreau-Gayon (1959) was confirmed by Somers (1966*a*, 1966*b*): no 3,5-diglucosides of any of the common anthocyanidins were identified in a variety of *Vitis vinifera* (Shiraz, apparently similar to the California variety Petite Sirah). He reported malvidin-3-glucoside to constitute 48 per cent of the pigment of this variety, malvidin-3-glucoside-*p*-coumaric acid (1 and 2) to amount to 3 and 15 per cent, petunidin-3-glucoside to 10 per cent, peonidin-3-glucoside to 8 per cent, delphinidin-3-glucoside to 6 per cent, peonidin-3-glucoside-*p*-coumaric acid (2) to 5 per cent, and malvidin-3-glucoside-caffeic acid to 3 per cent. The other pigments were detected in very limited amounts: delphinidin-3-glucoside-*p*-coumaric acid (1 and 2), delphinidin-3-glucoside-caffeic acid, petunidin-3-glucoside-*p*-coumaric acid (1 and 2), petunidin-3-glucoside-caffeic acid, peonidin-3-glucoside-*p*-coumaric acid (1), and peonidin-3-glucoside-caffeic acid. Somers did not detect cyanidin-3-glucoside in this variety of *V. vinifera.* The main point of attachment of the acyl substituent in the main group of *p*-coumaryl derivatives is the glucose moiety.

In the South African variety Barlinka, Koeppen and Basson (1966) found malvidin to be the chief pigment. Mono-*p*-coumaryl malvidin and peonidin were about half as concentrated; only traces of petunidin and delphinidin were found. Iakivchuk (1966) reported that in Chasselas rosé and Muscat Hamburg 90 to 92 per cent of the anthocyan pigments were present as glucosides and the rest as the aglucones. He found 70 to 75 per cent of the former was malvidin, 15 to 17 per cent cyanidin and 10 to 15 per cent delphinidin. Cappellari *et al.* (1966) reported considerable amounts of the diglucosides of malvidin, and peonidin in several varieties of *V. vinifera*. They also reported small amounts of diglucosides of cyanidin in a few varieties.

About twenty-one possibly different pigments were isolated from grapes by Zamorani and Pifferi (1964). Of these eleven were not identified. They showed that Merlot (a pure *V. vinifera*) had little or no anthocyanin diglucoside but that hybrids with American species (such as Baco and Clinton) had seven or eight diglucosides with especially large amounts of the diglucosides of malvidin, delphinidin, and petunidin. In fact, of the total anthocyans, 47.4 per cent were diglucosides for Clinton and 55.5 per cent for Baco. For recent data on the pigments of *V. vinifera* and its hybrids see Liuni *et al.* (1965).

Traces of the diglucoside malvidin were found in the *V. vinifera* varieties Pinotage, Hermitage, Gamay noir, and Souzão by Van Wyk and Venter (1964), but not in their wines. In Alicante Bouschet its presence was uncertain. It was absent in six other *V. vinifera* varieties. Bentonite fining or pasteurization did not affect the amount of malvidin present, and activated carbon removed only a very small amount. In Jacquez, a hybrid, as much as 0.5 gm per liter of malvin is present.

In 200 varieties of *V. vinifera* Getow and Petkow (1966) found malvidin (3,5-diglucoside malvin) in thirteen. Three of these, Khyndogny, Tchkaveri, and Matrassa, are important commercially in some areas of the Soviet Union. It thus appears more and more difficult to claim that diglucosides are *always* absent from varieties of *V. vinifera* unless the techniques used are exceptionally clumsy. The anthocyanins identified by Colagrande and Grandi (1960) and the relative amounts present were as follows:

	Barbera	Isabella
Delphinidin-3-galactoside	−	−
Delphinidin-3-glucoside	+ +	+
Petunidin-3-fructoside	+ +	+
Cyanidin-3-fructoside	+	±
Malvidin-3-galactoside	−	−
Malvidin-3-glucoside	+ + +	+ + +
Malvidin-3,5-diglucoside	−	+ ±
Peonidin-3-fructoside	−	−
Peonidin-3-glucoside	+	+ +
Peonidin-3,5-diglucoside	−	±
Acylated pigments of malvidin	−	+ +
Acylated pigments of peonidin	+	+ ±

Pigment patterns for a number of *V. vinifera* varieties are given by Rankine *et al.* (1958). They varied markedly from each other.

In contrast, Cappelleri (1965a, 1965b) found the diglucoside of malvidin in two genuine *V. vinifera* wines, one of Negrara and the other of Negronza. The results were the same by two methods in two years. No diglucoside was found in 107 other varieties. They noted caffeic and *p*-coumaric acids as constituents of the acylated pigments of grapes. Malic acid appeared to be the acylating agent of one variety and tartaric acid of two other varieties. One assumes that no contamination occurred in sampling.

From the biogenetic point of view, Somers postulates two enzyme systems controlling acylation with *p*-coumaric acid because of the presence of two different sets of *p*-coumaryl derivatives of the same anthocyanins.

Somers suggests that the failure of Somaatmadja and Powers (1963) to find peonidin-3-glucoside in Cabernet Sauvignon may be due to the use of neutral lead acetate in preparing the samples. Neutral lead acetate precipitates this compound scarcely at all. Using gel filtration, Somers (1966a, 1966b, 1967) claimed to have separated the anthocyanin pigments from the condensed tannins. This fraction was red in color, suggesting that anthocyanidin or anthocyanin pigments are incorporated in the tannin structure. This is contrary to the views of P. Ribéreau-Gayon (1964b, 1964c), who supposed they were yellow-brown in color and composed principally of leucoanthocyanidin pigments. The red color of wines thus appears, according to Somers, to be due to monomeric anthocyanin pigments and condensed flavonoid pigments, i.e., tannins. Mild acid hydrolysis produced cyanidin (mainly) and delphinidin, petunidin, malvidin, peonidin, and glucose (8 per cent). This indicates that only weak chemical bonds, if any, are involved between the main leucoanthocyanidin matrix and the anthocyanins. He suggested a ratio of one anthocyan molecule to about four leucoanthocyan moieties. A minimum molecular weight of 2,000 was suggested, but a range of molecular weights is believed to exist, from 2,000 to 5,000, with some up to 50,000.

Somers (1968) used gel column filtration to separate various phenolics. The pigments in red grapes consisted not only of anthocyanins and acylated anthocyanins but also of tannin pigments—in most cases the latter predominating. The pigment pattern of six red wine grapes showed distinct differences. There were also regional differences for the same variety. Owing to better extraction of anthocyanins compared to tannin pigments the pigment pattern of new wines is not the same as that of the grapes but they are similar. During aging there is a progressive loss of monomeric pigments so that the relative proportion of tannin pigments increases with age.

Reversible decoloration of anthocyanins occurs, according to Jurd

(1964), (1) by raising and lowering the pH, (2) by adding and removing bisulfite ion, (3) by adding and removing materials such as hydroxylamine, hydrazine, etc., which react with a carbonyl group, (4) by bacterial fermentation of citric acid, and (5) at the height of alcoholic fermentation. The effect of bisulfite is due (Jurd, 1964) to a reaction between the 2-carbonium ion and the bisulfite ion to form a neutral, colorless chromen-2 (or 4)-sulfonic acid. Jurd (1964) does not feel that the decoloration is due to reduction to leucoanthocyanins because of the observed difficulty of obtaining colored forms from leucoanthocyanins. He suggests a reaction of the flavylium-2-carbonium ion with some transient inter-mediate anion that is involved both in bacterial fermentation of citric acid and in alcoholic fermentation of hexose sugars. The destructive effect of B. cinerea on flavonols and anthocyanins is well known. Bolcato et al. (1964) showed that this consisted of modifications of the pigments so that a number of the characteristic spots on paper chromatography disappeared

Confirmation of higher pigment content and in some instances more tannin in wines fermented with sulfur dioxide (25 to 300 mg per liter) is found in the report of Valuĭko (1965b). There was little difference in pigment extraction between 25 and 150 mg per liter of sulfur dioxide. However, between 150 and 300 the increase was proportional to the sulfur dioxide content. Valuĭko recommended the use of small amounts of sulfur dioxide to prevent pigment loss during aging. The data on p. 400 also indicate how sulphur dioxide and titratable acidity favor color extraction.

An excellent summary of the phenolic compounds of musts and wines has been given by P. Ribéreau-Gayon (1964b, 1964c, 1968). Table 41 summarizes the structure of the phenolic compounds found in musts and wines. A summary of our knowledge of grape anthocyanins to 1964 has been given by Webb (1964). While paper chromatographic techniques have been of great value, he notes correctly that the isolated and purified pigment obtained may differ from that actually present in the grape skin or in wine. The path for biosynthesis of anthocyanins suggested by Durmishidze (1959) was: catechin → cyanidin → quercetin → gallo-catechin → delphinidin → malvidin (oenidin). For a discussion of methods for determining anthocyanins and tannins in wine see P. Ribéreau-Gayon and Nedeltchev (1965) and P. Ribéreau-Gayon (1968).

Breider and Wolf (1966) noted various abnormalities in chickens fed hybrid wines made from hybrids of V. vinifera and American species compared to those fed wines made only from V. vinifera varieties. The degree of toxicity of the hybrids was directly correlated to their Perono-spora resistance, a characteristic of American species. The cause of the

TABLE 41
THE PHENOLIC COMPOUNDS OF MUSTS AND WINES

Compounds identified	Amount		Nature of compounds
	Reds	White	

Benzoic acids

R = R' = H p-hydroxybenzoic acid
R = OH, R' = H protocatechuic acid
R = OCH₃, R' = H vanillic acid
R = R' = OH gallic acid
R = R' = OCH₃ syringic acid

Amount — Reds: 50–100, White: 1–5; Nature of compounds: esters

R = H salicylic acid
R = OH gentisic acid

Cinnamic acids

R = H p-coumaric acid
R = OH caffeic acid
R = OCH₃ ferulic acid

Amount — Reds: 50–100, White: 2–10; Nature of compounds: esters with anthocyans and tartaric acid

Flavanols

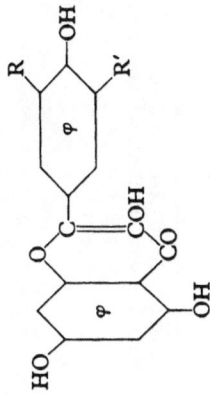

R = R′ = H kaempferin
R = OH, R′ = H quercetin
R = R′ = OH myricetin

15

0

2 or 3 glucosides and 1 glucuronoside in musts; 3 aglycones in wines

Anthocyanidins

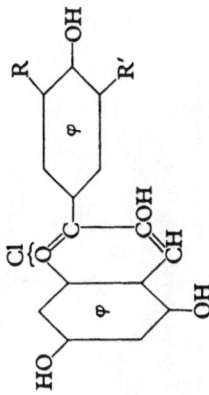

R = OH, R′ = H cyanidin
R = OCH₃, R′ = H peonidin
R = R′ = OH delphinidin
R = OCH₃, R′ = OH petunidin
R = R′ = OCH₃ malvidin

20–500

0

glucosides and acylated glucosides (to p-coumaric acid); varies for different species of *Vitis*

Tannin-flavan-3-ols

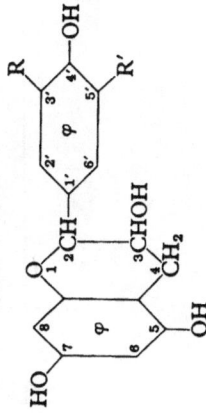

R = OH, R′ = H catechin
R = R′ = OH gallocatechin

1,500–5,000
50–100

0
0

0–100 polymers of flavones[a]

TABLE 4I (continued)

Compounds identified	Amount		Nature of compounds
	Reds	White	

Tannin-flavan-3,4-diols

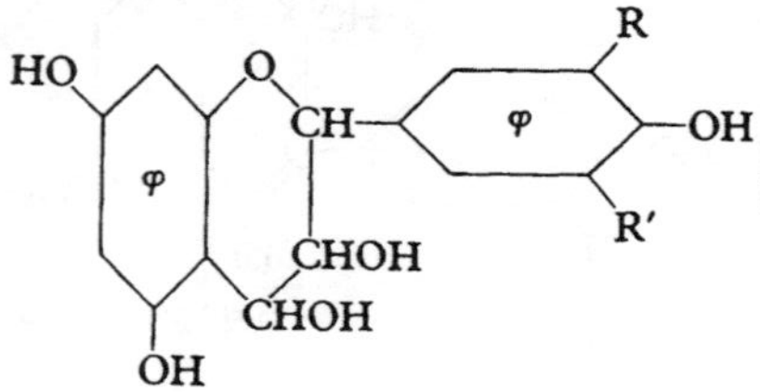

R = OH, R′ = H leucoanthocyanidin
R = R′ = OH leucodelphinidin

traces   o

[a] Principally of 3,4-flavan diols; as monomers, small amounts of flavans are found in red wine.
Source of data: P. Ribéreau-Gayon (1964b, 1964c).

toxicity is not known, but it may be due to calcium or other mineral deficiency. No such evidence has been reported in this country, where nearly 15 per cent of the wine produced is from native varieties or hybrids containing diglucosides. No evidence of toxicity has been reported from other regions where wines of *V. labrusca* varieties or of direct-producer varieties are widely planted, as in Moldavia or parts of the Midi of France. See also Bosticco (1966).

A concentrated powdered pigment extract, called Enocianina, is well known in Italy. The studies of Bosticco (1966) showed that when only 0.05 per cent was mixed with chicken feed it reduced growth and food utilization but had no effect on skin color. In view of numerous reports on the toxicity of anthocyanins to bacteria, he suggests that the effects are due to changes in the intestinal flora.

The opposite problem, adulteration by *V. vinifera* juice in Concord (*V. labrusca*) juice, is a problem in the United States. Fitelson (1967) noted that on paper chromatography three major spots are formed by the anthocyanidins. The two lower spots contain the three pigments found in Concord grapes in the largest amounts; the upper spot contains malvidin and peonidin, which are present in small amounts in Concord juice but form the major polyphenolic in California juice. Thus the addition of California juice to Concord juice greatly increases the intensity of the color of the top spot.

The controversy over the presence of diglucoside pigments in wines has to be seen in the perspective of the development of the Common Market in Europe. Free flow of wines into Germany would prove a distinct disadvantage to the German industry with its higher costs of production. The German authorities may have introduced the presence of diglucosides as a means of reducing import competition. This is based on tentative evidence that varieties containing diglucosides are toxic.

The present German restrictions constitute a barrier to importation of American wines into Germany or into any countries where the diglucoside pigment prohibition exists. This is an almost absolute barrier against wines from the eastern United States. It will also prevent wines made of or blended with wines made from Rubired or Royalty (and presumably Salvador) from being exported to Germany. Rubired and Royalty are complex hybrids developed by the California Agricultural Experiment Station.

The report of Smith and Luh (1965) indicates that Rubired has a pigment complex different from that of *V. vinifera* varieties, having a predominant diglucoside pattern. The decreasing order of pigments was: malvidin-3,5-diglucoside, peonidin-3,5-diglucoside, malvidin-3-

monoglucoside, peonidin-3-monoglucoside, delphinidin-3-monogluco-side, petunidin-3-monoglucoside, an unidentified pigment (3b), petu-nidin-3,5-diglucoside, malvidin-3,5-diglucoside acylated with p-cou-maric acid, another unidentified pigment (1b), malvidin-3-monoglucoside acylated with p-coumaric acid, and delphinidin-3,5-diglucoside. Chen and Luh (1967) showed that Royalty grapes contain several diglucosides and thus appears to contain non-*Vitis vinifera* parentage. The pigments isolated (in decreasing order of concentration) were malvidin-3,5-diglu-coside, *ditto* acylated with p-coumaric acid, malvidin-3-monoglucoside, peonidin-3,5-diglucoside, malvidin-3,5-diglucoside acylated with caffeic acid, peonidin-3-monoglucoside, *ditto* acylated with p-coumaric acid, and peonidin-3,5-diglucoside acylated with p-coumaric acid. Traces of the 3-glucosides of cyanidin, delphinidin, and petunidin and of a 3,5-di-glucoside of petunidin were noted. For information on the pigments of other varieties in California see Albach *et al.* (1959) and Bockain *et al.* (1955). The presence of diglucoside pigments is worthy of attention (possibly due to an error in identification ?).

The report of P. Ribéreau-Gayon (1963) that pure *V. vinifera* grapes contain little or no diglucosides, particularly of malvidin, while varieties of crosses with various American species do, has led to a number of European studies on the precise determination of diglucosides. The official French procedure is that of Jaulmes and Ney (1960). Further precision appears to be possible with the procedure of Deibner and Bourzeix (1964). The Bouschet hybrids have been shown to contain diglucosides by Deibner and Bourzeix (1960, 1964). The results with Royalty and Rubired are thus easier to understand. These authors (1965a, 1965b) have also reviewed the procedures for detection of di-glucosides.

No diglucosides of malvidin were found in the varieties of *V. vinifera* examined by Biol and Michel (1961, 1962). However, twenty-nine of the thirty-seven direct-producer hybrids tested contained the diglucoside. Among the hybrids without the diglucoside were Burdin 7 and 705, Landot 244 and 4411, Seibel 5, 455, 10,878, 11,803, and 14,596, and Seyve-Villard 23,353. Biol and Foulonneau (1961) did find the digluco-side of peonidin in *V. vinifera* varieties and also some acylated diglucoside of malvin.

Rice (1965a) identified the mono- and diglucosides of delphinidin, petunidin, cyanidin, malvidin, and peonidin in Concord grapes. Several acyl mono- and diglucosides also appear to be present. Ingalske *et al.* (1963) reported a similar pigment content in Concord grapes to the *V. labrusca* fruit used by P. Ribéreau-Gayon (1959). However, they found fourteen instead of eleven pigments; cyanidin-3-monoglucoside was present in relatively larger amounts, and malvidin-3-glucoside was not the predominant pigment. Further, more of the pigments were acylated. They attribute this to differences in climate and to the use of methanolic hydrochloric acid extraction, a better solvent for acylated pigments than hydrochloric acid alone which was used by P. Ribéreau-Gayon. Acetylated anthocyanins were reported recently by Webb.

To detect addition of wine of direct producers to wine of *V. vinifera* varieties, Gentilini and Cappelleri (1966) recommended use of a nitrite reaction which results in a lemon-yellow fluorescence in ultraviolet light. They found this much more rapid than chromatographic techniques.

Pifferi and Zamorani (1964) showed that fermentation of Merlot with high amounts of sulfur dioxide resulted in a high level of diglucoside pigments. Refermentation of a high diglucoside (Baco) variety with low diglucoside grapes did not reduce the diglucoside content sufficiently to prevent detection. Visintini-Romanin (1967) identified the oxidation product of malvin as malvon. He believes this to be connected with the oxidation of ascorbic acid.

Masquelier and Jensen (1953) and Powers *et al.* (1960) found that the anthocyanidins have bactericidal activity but not their aglycones. Masquelier and Delaunay (1965) presented evidence that several of the phenolic acids had bactericidal properties to *Staphylococcus* spp., especially *p*-coumaric, ferulic, caffeic, and gallic acids. Protocatechuic and chlorogenic acids have less bactericidal activity. See also p. 307.

Brown (1940) failed to find malvidin in muscadine (*Vitis rotundifolia*) grapes, but identified a diglucoside which he called muscadin that appears to be similar to petunidin. The fact that the color of wines of different varieties is of markedly different stability during aging (Amerine and Winkler, 1947) suggests the presence of a pigment complex of variable composition.

Tannin

Tannin occurs chiefly in the skins, stems, and seeds. Very little is found in the free-run juice, but prolonged contact of the juice with the skins, stems, or seeds, or fermentation in the presence of these compounds, leads to dissolving of tannin materials.

Singleton *et al.* (1966) noted the extreme variability of paper chromatograms of the phenolic compounds of red wines. They undertook "mapping" studies of the phenolic compounds of various parts of the grape berry during ripening. Only Calzin and one of its parents, Refosco, showed any large amounts of phenolic material in the juice. The juice phenolics of the other varieties tended to be blue-fluorescing substances which resemble chlorogenic acid and its analogues. During ripening, the tannin content of the skins increases at about the same rate as the color. Benvegnin et al. (1951) report the quantities present in the seeds, stems, and skins.

	Tannin (per cent)		Tannin (kg per 100 kg of grapes)	
Parts of cluster	Chasselas doré	Pinot noir	Chasselas doré	Pinot noir
Seeds	5.2	6.4	0.17	0.26
Stems	3.2	3.1	0.12	0.11
Skins	0.6	1.7	0.05	0.10

White musts contain only 0.01 to 0.03 per cent tannin.

The naturally occurring tannins were separated into three classes by Nierenstein (1934) and Russell (1935). More recently two main groups of vegetable tannins have been recognized, the hydrolyzable and the condensed tannins (Haslam, 1966). The hydrolyzable tannins occur as galloyl esters of glucose (gallotannins), as galloyl esters of other sugars, or as galloyl esters of quinic acid. Ellagitannins are hydrolyzable tannins which differ from gallotannins by the formation of ellagic acid on acid hydrolysis. The structure of hydrolyzable tannins, particularly the gallotannins, is now fairly well established (Haslam, 1967). The condensed tannins are not as well defined. They may be polymers of catechin, polymers of flavan-3,4-diols, or compounds of catechin with leucoantho-cyanins (Harborne, 1967). They were termed originally condensed tannins because on hydrolysis with acid they were converted largely into red or brown amorphoric compounds designated as phlobaphenes. The term now is restricted to naturally occurring phenolic polymers, insoluble in water but soluble in neutral organic solvents, Zavarin et al. (1963, 1965), Zavarin and Snajberk (1965).

According to Herrmann (1963), grapes contain the following poly-phenolic compounds (other than the anthocyan pigments): chlorogenic acid, isochlorogenic acid, neochlorogenic acid, p-coumaric acid ester; flavandiols and proanthocyanins: (+)-catechin, (−)-epicatechin, (+)-gal-locatechin, (−)-epigallocatechin, epicatechin gallate; flavonols and flavones: isoquercitrin, quercitrin, myricitrin camphoryl glycoside; and other phenolic compounds: gallic acid, ellagic acid, and procatechuic acid.

Rakcsányi (1964) showed that the tannins of grapes belong to the group of condensed tannins. Hot water extraction was used to remove 60 to 65 per cent of the tannins from the skins. The purity of the tannins thus extracted was poor. It is well known that the monomeric forms of the catechins (flavan-3-ols) and leucoanthocyanidins (3,4-flavanols) do not give the typical reactions of true tannins. P. Ribéreau-Gayon and Stonestreet (1966) therefore consider them to be preliminary stages of the condensed tannins. They found d-catechin, l-catechin, and dl-gallocatechin in Bordeaux red wines. They also reported but did not identify three other flavonoids. The leucoanthocyanidin polymers were short-chained (2 to 8 mols), but gave the normal tannin reactions.

Pifferi (1966) isolated the following ethyl acetate-soluble polyphenolic compounds from grapes: catechins: d-catechin, l-epicatechin, gallocatechin, and gallocatechin gallate; flavone: quercetin; flavone-glycosides: rutin and quercitrin; phenolic acids: gallic, protocatechins, γ-resorsylic, quinic, and the cis- and trans- isomers of p-coumaric, caffeic, and chlorogenic acids; depsides: ellagic and m-digallic acids; and coumarins: esculetin and methyl esculetin.

Jurics (1967) reported 10.1 mg per liter of D(−)-catechin and 3.97 of L(−)-epicatechin in grape musts.

According to Nègre (1942–43), the coloring materials are dissolved mainly during the early stages of fermentation, and the tannins in the later stages. The free-run wine has about one-half the tannin content of the wine pressed from the residue.

The results of Singleton and Draper (1964) indicate that grape seeds contribute significantly to the tannin content of red wines. They showed that complete extraction of the tannins of the seed could contribute 0.2 to 0.4 per cent tannin to the wine. Half or less actually appears in red wines in the normal process of fermentation.

Singleton (1966b) reported that during maturation, total phenolics increased on a per berry basis (but decreased on a weight basis) until about 30 days before harvest, when they leveled off. Using the Folin-Denis-Pro procedure and calculating the phenolics as gallic acid, the results were as follows (mg per gr of berry and mg per berry at maturity): Aligoté, 3.24 and 5.12; Calzin, 5.42 and 8.29; Catawba, 4.06 and 8.56; Delaware, 4.36 and 5.57; Emerald Riesling, 5.32 and 9.20; French Colombard, 2.53 and 4.53; Grenache, 2.92 and 3.56; Muscat of Alexandria, 2.84 and 11.69; Petite Sirah, 3.80 and 5.96; Pinot blanc, 6.06 and 8.09; Sauvignon blanc, 2.25 and 5.02; and Sémillon, 2.43 and 5.80.

The presence of shikimic acid in grapes may indicate that it is involved in some synthesis. Quinic acid is reported in musts. Chlorogenic and caffeic acids were found in musts by Henning and Burkhardt (1958), but not by Weurman and de Rooij (1958) or P. Ribéreau-Gayon (1964b,

1964c). Chlorogenic acid is derived from one mol of caffeic acid and one of quinic. Their formulas are:

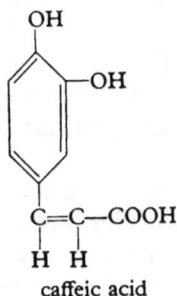

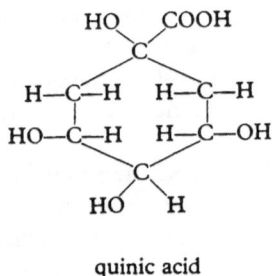

caffeic acid   quinic acid

Weurman and de Rooij (1958) did find chlorogenic acid isomers in grapes (possibly *neo*-chlorogenic acid and *p*-coumarylquinic acids). See Williams and Wender (1952). Tanner and Rentschler (1956) found chlorogenic acid *only* when the stems of Chasselas doré were crushed and fermented with the fruit. The stems of two other varieties, however, contained no chlorogenic acid. They therefore believe that neither this acid nor quinic acid is present in grape juice or wine. In Cabernet Sauvignon leaves, grape skins, and wine P. Ribéreau-Gayon (1965) was unable to find chlorogenic acid. He believes that previous workers confused caffeic acid with chlorogenic. He was, however, able to identify monocaffeyltartaric and monoparacoumaryl tartaric acids in grapes and wines.

Both gallic and protocatechuic acids are present in small amounts in wines, especially the former. Burkhardt (1965a) found *p*-coumarylquinic acid (or its calcium salt) in musts. For recent information on the separation of several components of isochlorogenic acid see Corse *et al.* (1965). See p. 450 for further information on tannins.

The work of Durmishidze (1967) and others have clearly established the polyphenols as one of the principal oxidation-reduction systems of musts. The auto-oxidation of phenols is accelerated in the presence of ions such as Cu, Fe, Mn, Ni, and Co. The oxidation of (+)-catechin is four times more rapid in the presence of cuprous ions than in their absence; it is eight times faster in the presence of ferrous ion than in its absence. Ferrous ion also greatly accelerates the decarboxylation of malic, tartaric, aspartic, and glutamic acids. The polyphenols act as inhibitors for dehydrogenases or oxidative phosphorylation. Also (+)-catechin has an effect on the decarboxylation of different organic acids.

Mineral constitutents

The ash content of grapes amounts to 0.2 to 0.6 per cent of the fresh weight. Potassium, sodium, calcium, and iron as carbonates or oxides, phosphates, sulfates, and chlorides account for most of the ash.

Hennig and Villforth (1938) found 14.9 to 37.9 mg of sodium per liter in 98 German wines; Reichard (1936) reported 5.2 to 18.5 mg in 70 Pflaz wines. Wines from grapes grown near the sea or from saline soils are always higher in sodium and chloride. Jouret and Poux (1961c) reported about 4 times as much sodium and 5 to 6 times as much chloride in wines produced from vines grown on saline soils compared to those on nonsaline soils. Surprisingly, there is more potassium but a lower alkalinity of the ash. Garoglio and Stella (1965) reported a trace of organic chloride in musts of botrytised grapes (0.5 mg per liter).

The potassium content as mg per liter changes very little during ripening, according to the results of Puissant (1960). He found that botrytised grapes are higher than normal in potassium, calcium, magnesium, and ammonia. During ripening, the calcium and magnesium contents, as mg per liter, decrease, calcium by about 50 per cent and magnesium by less. Potassium (calculated as the oxide) alone accounts for 50 to 70 per cent. Lasserre (1933) found 120 to 130 mg per liter of calcium and 13 to 90 of magnesium in grapes. The ratio of magnesium to calcium varied from 0.4 to 1.1 in the grapes, but the ratio was from 3.0 to 4.0 in red Bordeaux wines and from 2.0 to 2.6 in white Bordeaux wines. The calcium content of wine, limited by the solubility of calcium tartrate, seldom exceeds 100 mg per liter.

Fresh must contains 1 to 30 parts per million of iron present as free ferrous ions and, to a lesser extent, as ferrous complexes and ferric ions. The latter are bound mostly as complexes of tartaric, citric, and malic acids.

Duquy et al. (1955) found that soils high in phosphorus often produced wines with higher iron content. Much of the iron appeared to have been dissolved from soil adhering to the fruit. The enrichment was due not to simple dissolving of iron from the soil but to reductive biological processes during fermentation. This was confirmed by Flanzy and Deibner (1956), who also found iron pickup from crushers and presses. It was emphasized, too, by Cordonnier (1953) and Nègre and Cordonnier (1953).

The ratio of ferric ions (mainly in complex form) to ferrous ions (mainly free) was found by Ribéreau-Gayon (1933a, 1947) to depend on the state of oxidation of the wine. Marsh and Nobusada (1938) give methods of analysis and data on the distribution between free ferrous, free ferric, and bound ferric iron in California wine. Great care must be exercized in these determinations to prevent aeration of the wine. Their results, in mg per liter, were as follows:

Number of Samples	Fe^{++}	Fe^{+++}	Fe (bound)
2	5	1	0
4	3	1	3
7	1	1	11
20	5	0	0
53	4	0	2
59	4	0	1

Usually less than 1 mg of copper per liter is found in grape juice, and much of the copper content of the must is absorbed by the yeast during fermentation. In thirty-one Bordeaux wines which had been made without contact with copper, Lherme (1931–32) found only seven to have more than 0.5 mg per liter. The inherent copper of California musts should be as low or lower, because copper sprays are not used here. Amerine (unpublished data on wines of the 1948 vintage) found a range of 0.16 to 0.39 mg per liter (average 0.25) in the copper content of forty-six new commercial wines of the 1948 vintage from wineries in all parts of the state—wines whose contact with copper during vinification was not specially controlled. Under conditions of minimum exposure to copper-bearing metals, California wines contain less than 0.3 parts per million of copper. In new wines produced at Davis and crushed in a Garolla-type crusher with limited contact with copper, the following quantities were found:

Type of wine	Number of samples	Copper (mg per liter) Maximum	Minimum	Average
White table	39	0.43	0.04	0.12
Red table	33	0.28	0.04	0.09

Surprisingly high copper contents were reported in twenty commercial Soviet grape juices by Gasiuk et al. (1966): 2.2 to 5.2 mg per liter (average 4.1). Washing the grapes reduced the copper content by only 12 to 39 per cent. Pressing reduced the copper content of musts by 0 to 58 per cent (Gasiuk, 1965).

Aluminum and manganese, from a trace to as much as 40 to 50 mg per liter, are reported in grapes and wines, but usually only 1 to 2 mg are present. Wines made from *Vitis labrusca* or its hybrids are reported by Flanzy and Thérond (1939) to contain about three times as much manganese as *V. vinifera* grapes. Their thirty-seven samples of the hybrids and *V. labrusca* averaged 3.32 mg. Wurziger (1954) found no manganese in 23 per cent of the samples of German wines; 39.2 per cent contained 1 to 40 mg per 100 gr of ash; 28.4 per cent 41 to 100; 4.0 over 100 but less than 1,000; and 5.4 per cent over 1,000. Von der Heide and Hennig (1933) reported zinc in musts in amounts of from

9 to 19 mg per liter, with 1 to 6 being the most common range. From traces to as much as 30 mg per liter of boron, as $B(OH)_3$, are reported. Boron fertilization markedly delays grape maturity (in fact, prevented satisfactory ripening) in France, according to Decau and Lamazou-Betbeder (1964). In California the lead and arsenic content of the musts is exceedingly low, since sprays containing these elements are rarely used in this state. Musts contain 0.025 to 0.4 mg per liter of lead with most samples containing 0.1 to 0.25, according to Jaulmes et al. (1960). They found that there was usually but not always a decrease in lead during fermentation.

Bertrand and Bertrand (1949) found 2 to 9 mg of rubidium per kg of dry material in French grapes, averaging 0.46 mg per liter in French white wines, and 1.15 mg per liter in red wines. Some of the cadmium present in musts is lost during fermentation in the lees, according to Eschnauer (1965a).

Of the inorganic anions, phosphates and sulfates predominate. Soil conditions markedly influence the amounts of the constituents present in musts. Some of the sulfate of the wine is derived from the sulfur dioxide added to the musts. (See also p. 394.) Chlorides are likewise present in variable amounts, depending on soil conditions. Grapes grown on sandy soils near the ocean, and the wines made from them, are frequently very high in chloride. (See p. 460.) Traces of bromide, iodine, and even fluoride are normally found in musts and wines. Hennig and Villforth (1938) found 0.064 to 0.543 mg of fluorine per liter in ninty-eight German wines. But they note that higher amounts were reported by earlier workers. They found 0.25 to 0.30 mg of iodine per liter in musts, 0.10 to 0.20 mg in normal dry wines, and 0.4 to 0.6 mg in sweet table (*Trockenbeerenauslese*) wines.

There have been various reports that the rootstock might influence the composition (and hence the quality) of the fruit of the scion. Bénard et al. (1963) used nine stocks for Grenache. The analytical differences on the musts and wines were small, except that musts of fruits on 161–49C and 5BB seemed to be less in ash, alkalinity of ash and potassium, and higher in iron. We are not convinced of the statistical significance of the results. If a factor is to be developed to predict fermentability it will be very complex (Amerine and Kunkee, 1965).

Vitamins

A number of vitamins have been reported in grapes. Ascorbic acid (vitamin C) is present in small but measurable quantities in fresh grapes. During ripening, Ournac and Poux (1966) reported an early increase in the total amount of ascorbic acid in the berries followed by a decrease

Variety	Ascorbic acid (mg per 100 gr of grapes)	Source
Green (?)	1.2, 2.6	
Muscadine	5.5, 1.9	Floyd and Fraps
Scuppernong	7.2, 7.2, 7.2	(1939)
Small pink (?)	2.3	
Madeleine Celine	3.4, 4.9	
Mission	4.7	
Folle blanche	7.1, 7.8, 7.5	Amerine and
Muscat of Alexandria	3.3, 3.8	Sternberg[a]
Burger	9.3, 8.4	
Pearl of Csaba	5.6	
Corbeau	3.8	
Gamay noir	3.8	
Malbec	3.8	
Sauvignon blanc	3.8	Schatzlein and
Refosco	2.0 to 3.8	Fox-Timmling
White Riesling	2.0 to 3.8	(1940)
Sémillon	2.0 to 3.8	
Furmint	2.0	
Sylvaner	2.0	
Traminer	2.0	
31 varieties	1.05 to 18.3	Venezia (1938)
20 varieties	7.1 to 9.7	Muradov (1966)
10 white varieties, 1941	1.26 to 15.39, av. 8.32	Venezia (1944)
10 white varieties, 1942	1.84 to 6.24, av. 4.51	

[a] Unpublished data, 1941.

which lasted until after the fruit started to ripen (until after the *veraison*); it then increased until the fruit ripened and afterward decreased. They were not able to correlate these changes with changes in sugar or acidity. When expressed on the basis of percentage by weight, an early increase in ascorbic acid was followed by a long decrease. The skin area contains a greater percentage of ascorbic acid than the juice, but the total amount in the juice is greater than that of the skin. This vitamin accounts in part for the low oxidation-reduction potential of fresh grape juice. Storage of sulfited grape juice on the skins for thirty days resulted in juices of high ascorbic acid and tannin. No data on the alcohol content of the juice thus prepared was given. The research should be verified for other regions and varieties.

Muradov (1966) reported that pasteurization resulted in the loss of 35.5 to 70.1 per cent of the ascorbic acid. Most if not all of the ascorbic acid is lost during crushing and fermentation. Reports of several earlier authors indicated that fresh grapes contain from 1 to 18 mg of ascorbic acid per 100 gr. In Canada Zubeckis (1964) reported that fresh grapes contained 1.1 to 11.7 mg per 100 ml of juice of ascorbic acid, except for the Veerport variety, which had 18.5 to 33.8 mg per 100 ml. There was a slight increase during ripening. Pasteurized grape juice retained about one-third the original ascorbic acid. More recently, using improved techniques, Ournac (1965) reported 20 to 60 mg per liter of ascorbic acid in deep-colored musts. Zimmerman *et al.* (1940) showed that treatment of musts with pectin-splitting enzymes reduced the ascorbic acid content about one-third and that filtration caused a loss of almost two-thirds.

The B vitamins have been studied by Morgan *et al.* (1939) and by Perlman and Morgan (1945) in this country. The amounts of B complex vitamins in Bordeaux grapes have been summarized by Peynaud and Lafourcade (1958). The thiamine content of grapes, both per liter of juice and per berry, increases during maturation, especially during the final stages of ripening, according to Ournac and Flanzy (1957). About 75 per cent is found in the expressed juice and the remainder in the solid part of the skin and pulp. For the varieties Aramon, Carignane, and Alicante Bouschet the amounts were 0.170, 0.200, and 0.295 mg per kg of grapes. Thiamine is present in amounts of from 300 to 500 μg per liter of grape juice. In 10 musts 0.375 mg per liter was reported by Peynaud and Lafourcade (1958). Schanderl (1959) found 0.12 to 0.27 mg per liter in fresh grape juice and 0.03 to 0.13 in commerical grape juice. Mathews (1958) detected traces to 0.25 mg per liter in four commercial Swiss grape juices. The vitamin content of 20 grape musts was 0.27 to 0.50 mg per liter for thiamine and 1.06 to 1.83 for riboflavin, according to Muradov (1966). Genevois and Flavier (1939) found somewhat larger amounts of B vitamins in the wine than in the must.

Riboflavin is present in both grapes in amounts of up to 1.45 mg per kg but usually about 0.40 mg or less. In grapes it increases slowly until near maturation, when it decreases. If the juice is exposed to the light, riboflavin rapidly disappears. Added riboflavin is retained if unsulfited grape juice is stored in dark-colored bottles.

Pyridoxine (B_6) is present in grapes up to 1.81 mg per kg. Added pyridoxine is retained well in grape juice. In Bordeaux grapes the pyridoxine content reached a maximum (0.4 to 0.6 mg per liter) about two weeks before maturing and then declined, according to Peynaud and Lafourcade (1957). In seventeen musts of Bordeaux they reported 0.16 to 0.53 mg per liter (average 0.32) of free pyridoxine and an average of 0.42 of the total. These values compare favorably with Castor's (1953a) 0.47 mg per liter, but are much lower than the 0.60 to 1.06 mg. per liter reported by Morgan *et al.* (1939). In *V. vinifera* and certain other species Radler (1957) reported pyridoxine contents of 0.3 to 2.9 mg per liter. The amounts of pyridoxine, pantothenic acid, and nicotinic acid present were positively correlated. Perlman and Morgan (1945) found 0.83 and 1.82 mg per kg of pyridoxine in two grape juices. Radler (1957) found 1.8 to 8.8 mg per liter of nicotinic acid in various musts.

Nicotinic acid does not increase on a per liter basis but does when expressed on a per berry basis.

Pantothenic acid was found in two grape samples (89 and 150 μg per 100 gr) by Perlman and Morgan (1945). Radler (1957) gave values of 0.3 to 3.4 mg per liter in several musts. During maturation the pantothenic acid content of grapes increases to 0.50 to 1.38 mg per liter (average 0.80). Smith and Olmo (1944) reported significantly higher amounts in tetraploid varieties than in diploid varieties. Interspecific *V. labrusca* × *V. vinifera* hybrids were higher in pantothenic acid than were *V. vinifera* × *V. vinifera* hybrids. (See also Peynaud and Lafourcade, 1955a.) These and other data on the vitamin (other than ascorbic acid) content of musts and wines are summarized below.

For musts Lafon-Lafourcade and Peynaud (1958) showed a marked increase during maturation of *p*-aminobenzoic acid and a slight increase in pterolglutamic acid and choline. Musts of Bordeaux grapes contained 0.015 to 0.092 mg per liter (average 0.47) of *p*-aminobenzoic acid, 0.0009 to 0.0015 mg (average 0.0012) of pterolglutamic acid, and 0.24 to 0.039 mg per liter (average 0.033) of choline. They state, however, that the extremes found in a large number of determinations were 0.015 to 0.093, 0.0009 to 0.0018, and 0.013 to 0.039, respectively, and they

Vitamin	Amount in must (ug per 100 gr)	Amount in wines (ug per 100 gr)	Source
Thiamine	30–57[a]	0–12	Morgan et al. (1939)
	49	0–24	Perlman and Morgan (1945)
	—	15–30	Genevois and Flavier (1939)
	31.6	7.4	Scheurer (1944)
	—	0.8–8.6[b]	Cailleau and Chevillard (1949)
	60[c]	—	Watt and Merrill (1963)
Riboflavin	23[a]	27–50	Morgan et al. (1939)
	40–145[b]	25–112[b]	Perlman and Morgan (1945)
	—	6–22[d]	Perlman and Morgan (1945)
	—	5–40[a]	Genevois and Flavier (1939)
	14.1	9.3	Scheurer (1944)
	—	7.5	Schön et al. (1939)
	—	8.0–45.0[a]	Cailleau and Chevillard (1949)
	40[c]	—	Watt and Merrill (1963)
Pyridoxine	83, 182	66, 70, 72	Perlman and Morgan (1945)
Pantothenic acid	89, 150	40–110	Perlman and Morgan (1945)
	—	7–45[d]	Perlman and Morgan (1945)
	—	20–120[a]	Cailleau and Chevillard (1949)
Nicotinic acid	—	65–210[a]	Cailleau and Chevillard (1949)
	200[c]	—	Watt and Merrill (1963)

[a] Per 100 ml.
[b] In experimental samples after one month's storage.
[c] Per 100 gr of edible portion.
[d] In commercial wines.

give averages of 0.051, 0.0012, and 0.026, respectively. These compare with Castor's (1953a) values of 38 µg per liter for p-aminobenzoic acid and 40 µg per liter for choline.

Ough and Kunkee (1968) reported 1.35 to 6.8 micrograms per liter of biotin in 33 red grapes (average 2.85) and 0.6 to 2.55 in 34 white grapes (average 1.47). They did not find any regional effect. Biotin increased markedly in Cabernet Sauvignon during maturation. Biotin content is positively correlated with rate of fermentation but is, in itself, not directly a significant factor in determining the fermentation rate of grape juice.

The presence of inositol in grapes and wines has been known for many years. During ripening inositol increases by a factor of about two or more, according to Peynaud and Lafourcade (1955b). Some differences between varieties and seasons were noted. More was found in high acid musts. It passes into the wine unchanged. Castor (1953a) and Peynaud and Lafourcade (1955b) showed that musts contain about 1,000 times as much inositol as required for yeast growth (up to 400 mg per liter). Pasteurization, fining, ion exchange, filtration, addition of sulfur dioxide and other treatments do not reduce the amount of inositol. On heating it is reported to yield furfural. It is not fermented by yeasts. In Bordeaux musts Peynaud and Lafourcade (1958) reported 1 to 2 µg per liter of folic acid. They reported biotin decreased to 2.6 µg per liter during ripening of 10 Bordeaux musts.

In summary: grapes are comparatively rich in thiamine and mesoinositol. They have fair amounts of nicotinic and pantothenic acids. They are normal in pyridoxine and biotin but very poor in riboflavin.

Traces of vitamin A and total carotenoids, 0.5 µg per 100 mg, were found in wine by Schön et al. (1939). Scheurer (1944), however, reports 7.15 µg of "provitamin A" in 100 gr of grapes. Watt and Merrill (1963) give a vitamin A volume of 80 international units per 100 gr of the edible portion of grapes, but Zimmerman and Malsch (1938) found no carotene in grapes. In twenty grape musts Muradov (1966) reported 0.37 to 1.30 mg per liter of vitamin A. Pasteurization resulted in the loss of 16 to 17 per cent.

Flavones having the chemical and physiological properties of rutin and similar compounds of the so-called "vitamin P" group occur in grapes and wine. Grapes have been reported to be one of the richest sources of "vitamin P" by Lavollay and Sevestre (1944) and by De Eds (1949). De Eds found a high physiological activity in preparations obtained from pomace.

Enzymes

The enzyme complex of grapes has been very little studied. Oxidizing enzymes (probably phenolase and peroxidase), invertase, tannase, and pectic enzymes have been reported. Delp (quoted by Genevois and Ribéreau-Gayon, 1947) found an oxidizing enzyme in certain German wines; and Hussein and Cruess (1940a, b) reported peroxidase in California grapes. Poux (1966a, b) has reported that the mechanism for the oxidation is not completely known, but it apparently proceeds through at least three stages leading to the formation of colored compounds by polymerization.

Pallavicini (1967) reported that the enzymes of grapes (phenolase, phosphatase, proteinase, and sucrase) are found in the skins and to a lesser extent in the pulp. However, in Muscat Hamburg and Raboso the phenolase was primarily in the pulp. Centrifuging or settling the musts removed most of the enzymes.

The amounts of peroxidase, polyphenoloxidase, and catalase in early- and late-ripening varieties were reported by Golodriga and Pu-Chao (1963). Catalase was particularly high in early-ripening varieties. Polyphenoloxidase was present in largest amounts in late-ripening varieties.

Ivanov (1967) confirmed that polyphenoloxidase is the main oxidizing system of grapes, with the greatest activity in the skins in the varieties tested. Cassignard (1966) found the skins, seeds, and stems to contain the most enzyme. He considers this to be the prime reason why the pomace should not be pressed too long or hard. It also explains why settling reduces the polyphenoloxidase content of musts. Ivanov (1967) showed that varieties differed in the total amount of enzyme activity (more in a muscat variety than in Riesling) and in the relative amounts in the pulp compared to the skins (more in Riesling and Aligoté). The Bordeaux varieties tested by Cassignard (1966) had about the same level, except Merlot blanc, which had about twice as much as the other varieties. Poux's (1966a, b) studies of Languedoc varieties, showed that Carignane blanc was low in polyphenoloxidase, Terret blanc higher, and Grenache blanc and Maccabeo the highest. He used a modification of Ponting and Joslyn's (1948) procedure for the determination. The enzyme content was not related to color content, as, for example, in Grenache blanc, Grenache rosé, and Grenache noir the blanc had the most and the rosé the least. Pressing greatly increased the polyphenoloxidase content, and not in the same proportion for each variety. Surprisingly, the musts of botrytised fruit had less polyphenoloxidase than sound fruit in Cassignard's experiments but not in Poux's. Differences in the molds present may account for some of the differences reported and varying procedures for others.

Grapes which have been attacked by *Botrytis cinerea* normally contain rather large amounts of an oxidizing enzyme which rapidly leads to the oxidation of phenolic compounds in the must. The enzyme is apparently secreted into the grape from the mycelia of the fungus. The danger of this enzyme is that it may remain in the wine after fermentation in such quantities as to cause cloudiness even after the wine is bottled, or to cause darkening when the wine is exposed to the air. Rentschler (1950) has identified the enzyme causing darkening in the 1950 Swiss wines as a phenoloxidase derived from *B. cinerea*. Different grape varieties differ in their degree of browning not only because of the presence of varying amounts of polyphenoloxidase but also because of their varying amounts of suitable substrate material. Some varieties seem to have more of this substrate material in the stems. Ivanov (1966, 1967) and Cassignard (1966) reported higher polyphenoloxidase activity at the beginning of ripening. (See also Deibner and Rifaï, 1963). The nature of the polyphenoloxidases and methods for their determination were discussed by Poux (1966a, 1966b). They are proteins with a molecular weight of about 100,000 and contain 0.25 per cent copper (or 4 atoms of copper per molecule).

Heating to 70° to 75°C (158° to 167°F) for thirty minutes completely inactivated the polyphenoloxidase. Heating at 55°C (121°F) partially deactivated the enzyme (Poux, 1967). Very slow heat inactivation of polyphenoloxidase occurs when musts are heated at temperatures of 60°C (140°F), according to Demeaux and Bidan (1967). At 70°C (158°F) more than 99 per cent inactivation occurs in about 3 minutes. Inactivation is more effective at pH 3 than at 4. Guinot and Ménoret (1965) using a tubular heat exchanger found that peroxidases and polyphenoloxidases were all destroyed at 70°C (158°F) or above. Heating to 80°C (176°F) or higher markedly increased the phenolic content of the must. Small amounts of sulfur dioxide, 10 to 50 mg per liter, also inhibit its activity. The quantity of sulfur dioxide used to prevent the reaction of dissolved oxygen and the enzyme needs to be calculated on the basis of the amount of the enzyme present. Cassignard admits that the method of determination could be improved. Poux (1967) found that 0.4 to 1 gr per liter of bentonite adsorbed the polyphenoloxidase. Ivanov (1967) found sulfur dioxide to be a better inactivator than bentonite, which confirms practical observations. However, Cassignard (1966) recommended bentonite for fining musts during settling. Musts containing oxidizing enzymes of the catechin oxidase type should be treated with bentonite and polyacrylamide (to speed bentonite sedimentation), according to Tagunkov (1966). Iliescu et al. (1965) recommended fining musts with bentonite and using 180 mg per liter of sulfur dioxide in the preparation of sweet table wines. The treatment appears to be for removal of enzymes. Poux (1966a, 1966b) showed that the polyphenoloxidase content decreases during fermentation, but Cassignard (1966) did not find this.

Pantanelli (1912, 1915) found a protein-splitting enzyme in musts but not in wines. Baglioni *et al.* (1935–37) found the juice of fresh grapes to have proteolytic activity. An aldehydomutase acting on acetaldehyde to produce ethyl alcohol and acetic acid is reported by Ribéreau-Gayon (1943). He also found a yeast-secreted reductase which in young wines may reduce the aldehyde content. Catalase is present. Invertase is present in green grapes, increasing slightly with maturity (Casale and Garina-Canina, 1935–37). Sucrase has been found in musts and wines. Since young wines had more sucrase than their musts, some of the enzyme apparently is derived from the yeast.

Using Sephadex G-50, Bolcato *et al.* (1965) separated invertase, an acid phosphatase, a phenolase, two proteases, and a peroxidase from musts and wines. Pallavicini (1966) separated four enzymes from wine using Sephadex G-50: phenolase, acid phosphatase, protease, and sucrase were present in all, phenolase in 15, proteinase in 3, and acid phosphatase in 2.

An enzyme which demethoxylated pectins was reported by Fellenberg (1914). Semichon and Flanzy (1926) also found pectic enzymes. Three types of pectic enzymes may be distinguished: (1) those splitting off methyl ester groups from the polygalacturonic acid chain (sometimes called pectin methyl esterases [PME], pectin pectyl-hydrolases [I.U.B. (1) No. 3.1.1.11], or formerly pectases or pectin esterases), (2) those hydrolyzing the α-1,4-glycosidic linkages in pectin polyuronides (often called polygalacturonases [PG], polygalacturonide glucanohydrolases [I.U.B. No. 3.2.1.15], or formerly pectinases, pectin polygalacturonases, or pectin acid depolymerase), and (3) those splitting the α-1,4-glycosidic bonds by a transelimination mechanism (also called pectate transeliminase [PTE], pectate lyase, poly-α-1,4-D-galacturonide lyase [I.U.B. No. 4.2.99.3], or polygalacturonic acid transeliminase). For a critical discussion of recent methods of determining pectic enzymes see Vas *et al.* (1967).

Pectic substances

The pectin materials hydrolyze as follows:

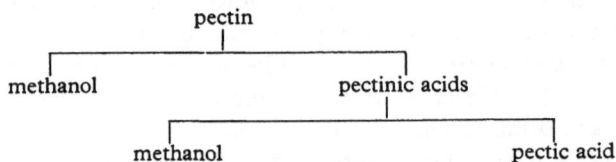

```
                        pectin
                          |
        ┌─────────────────┴─────────────────┐
    methanol                           pectinic acids
                                            |
                              ┌─────────────┴─────────────┐
                          methanol                   pectic acid
```

The variety of results obtained for pectin, pectinic acids, and pectic acid is due primarily to differences in the methods employed. Flanzy and Loisel (1958) criticize alcohol-precipitation procedures at low pH because of possible demethoxylation of pectin. They reported the following for four musts and their wines (mg per liter):

	Total	Pectin[a]	Pectic materials Pectinic acid	Pectic acid	Methanol
Musts	389–846	452–913	292–674	88–150	58–89
Wines	198–522	251–594	130–403	42–98	63–93

[a] Pectin = pectinic acid + pectic acid + methanol.

Pectic substances occur in the grape, increasing as it matures. *Vitis labrusca* varieties are reported to contain as much as 0.6 per cent of pectin and other acid-alcohol precipitable materials, with variable but generally lower amounts in *V. vinifera* grapes. Some available data are summarized in the table below.

Species and authority	Locality	Pectin content (per cent) Min.	Max.	Av.
V. vinifera:				
Heide and Schmitthenner (1922)	Germany	0.11	0.33	—
Marsh and Pitman (1930)	California	0.03	0.17	0.12
Besone (1940)	Davis, Calif.	0.02	0.21	0.13
Ventre (1930)	France	0.05	0.11	0.08
		0.14	0.39	0.26
Françot and Geoffroy (1951)	France	0.03	0.07	0.05[a]
		0.05	0.08	0.06[b]
		0.04	0.53	0.20[c]
Garina-Canina (1928)	Italy	0.08	0.39	—
V. labrusca:				
Willaman and Kertez (1931)	New York	—	0.60	—
Besone (1940)	Davis, Calif.	0.11	0.29	0.18

[a] First press.   [b] Second press.   [c] Last press.

Heat extracts a considerable quantity of soluble pectic materials. In the grape-juice industry the presence of colloidal pectic materials is undesirable, and pectin-hydrolyzing enzymes are commonly used to facilitate clarification. In musts and young wines such enzymes have been recommended as an aid to clarification (see p. 308). European enologists frequently mention the beneficial influence of pectin on the smoothness of wines (Semichon and Flanzy, 1926). Since most of the pectic materials are hydrolyzed during fermentation by naturally occurring pectic enzymes and some are precipitated by the alcohol formed during fermentation, it is difficult to see any direct connection. However, hydrolysis products of the pectic materials might favorably influence quality. Sadjadi (1936) believed that he had identified monogalacturonic acid as one of the components of grape pectin.

Datunashvili *et al.* (1967) reported that higher pH in red musts increased hydrolysis of pectins but not in white. They found that at 45°C (113°F) 90 per cent of the pectins were hydrolyzed in three hours, while at 20°C (68°F) only 22 per cent were hydrolyzed. At a pH of 3.53 the hydrolysis of pectins was complete in 24 hours.

Françot and Geoffroy (1951) separated the pectins from the gums. During fermentation, 30 to 90 per cent of the pectins were precipitated, but there was little change in the percentage of gums. The nature of the gums and their influence on clarification should be investigated.

Odorous constituents

Windisch (1906) has reviewed the early observations on the odorous constituents in *V. vinifera* grapes. Contrary to the usual opinion, these are not entirely localized in the cells of the skin. Vanillin was identified in the seeds and in red wines as early as 1883. The presence of terpenes in brandy was long considered as evidence of their presence in the fruit, but proof of this was lacking until recently. Sale and Wilson (1926) reported a volatile ester and volatile acid in grape juice, but did not identify them. The muscat aroma is, however, distinct in both grapes and wine. Some varieties have a slight aroma in the grapes and produce a distinctly flavored wine—for example, Cabernet Sauvignon, Sauvignon vert, and Sémillon. Other varieties such as Sauvignon blanc, Zinfandel, and Fresia have an even less distinct odor or flavor in the grape, but produce quite recognizable wines. The odorous constituents seem to be localized in the skins and to be formed in, or translocated into, the fruit during the later stages of ripening. Theron and Niehaus (1938) state that much of the odorous material is lost under hot South African climatic conditions.

The skins are known to be richer in aromatic materials than the pulp. The accumulation or development of the odorous compounds in the grapes during ripening has not been systematically studied. Ethyl alcohol constitutes the most important part of the volatile materials. Procedures for distillation of the volatile materials into chromic acid can therefore give us little assistance. The results of Deibner *et al.* (1965a) show that oxidizable volatile materials increase until the sugar content reaches its maximum and thereafter decrease.

The most pronounced odor present in grapes is the foxy aroma identified with different varieties of *V. labrusca* and related American species. This has been studied by Power (1921) and Power and Chestnut (1921, 1923) who identified it as methyl anthranilate. They investigated authentic juices of many varieties of *V. labrusca* and its hybrids as well as six varieties of *V. vinifera* from California. Not all *V. labrusca* hybrids contained the ester. Juice of Catawba, a *V. labrusca* × *V. vinifera* hybrid, gave no test; but other such hybrids—Niagara, Brighton, and others— gave a strong or moderate test. Muscadine grapes, *V. rotundifolia*, failed to give the test, as did the Burger, Sauvignon vert, Muscat (of Alexandria?), Alicante Bouschet, Petite Sirah, and Zinfandel juices from California. Sale and Wilson (1926) reported a methyl anthranilate content of 0.0 to 3.8 mg per liter in musts of grapes other than *V. vinifera*. Scott (1923) reported 0.80 to 1.49 mg per liter in commercial Concord juice and 0.08 and 0.40 mg per liter in commercial Catawba juices. Since Power and Chestnut failed to get a test with authentic Catawba juice, the commercial Catawba juices were probably mixtures. If so, some other odorous constituent must be present, for fresh Catawba grapes do have a very distinct "eastern" or "foxy" odor. In contrast to the earlier results, Brunelle *et al.* (1965) found little methyl or ethyl anthranilate in pure grape extracts. They believe that the small amounts detected in the official A.O.A.C. procedure are due to the presence of interfering substances. Imitation grape flavors, however, contained considerable methyl anthranilate and some ethyl anthranilate. (The quantities given seem too high to us.) They used gas-liquid chromatography to determine methyl and ethyl anthranilates. Similar results for total anthranilates were given by ultraviolet spectrophotometry (characteristic peak at 335 mμ). Mattick *et al.* (1963) recovered 98 ± 1.7 per cent of added methyl anthranilate by separating it by gas chromatography with detection by an electron affinity detector. This appears to be more rapid and foolproof than the A.O.A.C. procedures or the Shaulis and Robinson (1953) modification.

Stern *et al.* (1967) reported that esters represent the largest percentage of the components in a Concord grape essence, with ethyl acetate and methyl anthranilate predominating. A relatively large amount of ethyl crotonate was found. Methyl butyrate, ethyl isobutyrate, ethyl 2-hexenoate, ethyl benzoate and methyl benzyl acetate, not previously reported in Concord grapes, were also identified. An ethyl alkylthioester appears to be present. They note that these esters have a pleasant odor and probably contribute to the over-all pleasant aroma of Concord grapes.

The volatile essence of *V. rotundifolia* grapes (the predominant native variety of the south Atlantic states) was analyzed by Kepner and Webb (1956). No nitrogen- or sulfur-containing compounds were found in the essence. Since *V. labrusca* varieties contain esters of anthranilic acid as their principal odorous principle, Kepner and Webb suggested that freedom from nitrogen in the volatile essence would be a good way of differentiating non-*labrusca* wines from *labrusca*. The most significant compound in the *V. rotundifolia* volatile essence appeared to be α-phenethyl alcohol, a compound with a roselike odor. However, this compound now seems to be of general occurrence in wines (Usseglio-Tomasset, 1967a).

Haagen-Smit et al. (1949), investigating the volatile oil of fresh Zinfandel grapes, identified ethanol (244), acetaldehyde (1.80), acetic acid (0.0053), n-butyric acid (0.003), n-caproic acid (0.0015), glyoxylic acid (0.118), n-butylphthalate (2.250), leaf aldehyde (0.327), sulfur (0.004), acetylmethylcarbinol (0.013), waxy substances (0.024), and carbonyl compound (0.025). (Figures in parentheses are grams per 1,000 kg of fruit.) (See also pp. 477–481.)

The components and relative amounts of the compounds isolated from a volatile extract of Sauvignon blanc were reported by Chaudhary et al. (1964) as follows (T = trace, L = large, M = medium, S = small): ethanol (L), 1-propanol (S), 2-methyl-1-propanol (S), 1-butanol (T), 3-methyl-1-butanol (M), 2-methyl-1-butanol (M), 1-pentanol (T), 1-hexanol (L), 2-phenethyl alcohol (L), ethyl formate (T), ethyl acetate (S), ethyl propionate (T), isoamyl acetate (M), act-amyl acetate (S), n-amyl acetate (T), n-hexyl acetate (L), ethyl caproate (L), ethyl heptanoate (S), ethyl caprylate (T), ethyl caprate (M), isoamyl caprylate (S), hexyl heptanoate (S), hexyl caprylate (L), 2-hexenal (L). The following compounds were probably present: methyl acetate, n-propyl acetate, n-propyl propionate, isobutyl propionate, ethyl valerate, n-amyl propionate, isoamyl butyrate, isobutyl caproate, hexyl valerate, isobutyrate ester, and 3-hexenoate ester. See Chaudhary et al. (1968) for analyses of botrytised and non-botrytised Sauvignon blanc fruit.

The composition of the aromatic constituents of muscat grapes has been studied by Cordonnier (1956), Webb and Kepner (1957), and Stevens et al. (1966). Cordonnier identified geraniol, terpineol, limonene, and linaloöl. Webb and Kepner identified thirteen compounds and several C_4 to C_{12} carboxylic acid esters. Owing to the use of more advanced technology (capillary gas-liquid chromatography, GLC and rapid scan mass spectrometry), the report of Stevens et al. is more complete. They reported some 78 peaks, of which sixty were identified. The compounds of greatest importance to the "muscat" aroma appeared to be linaloöl and geraniol, which have rather sweet floral odors. Other terpene alcohols present in smaller amounts were nerol, α-terpineol, and citronellol. Also present were two cyclic ethers and two tetrahydropyrans which they believe were derived from linaloöl by air or enzymatic oxidation with subsequent rearrangement. The other compound present in abundance was 1-hexanol (also reported by Webb and Kepner).

Cordonnier (1956), Stevens et al. (1966), and Usseglio-Tomasset et al. (1966b) have identified linaloöl as the compound responsible for the muscaty odor in Muscat blanc. It has also been identified in White Riesling. In the essential oil of Salvia sclarea linalyl acetate was the principal constituent. In an extract of the seeds of *Coriandrum sativum* linaloöl was present. At least 80 per cent of the essential oil of *S. sclarea* is linalyl acetate, according to Matta and Astegiano (1967). Since neither muscat nor neutral-flavored wines contain linalyl acetate, its presence in wines is adequate proof of sophistication. Using gas and thin-layer chromatography, Usseglio-Tomasset (1967*b*) and Usseglio-Tomasset *et al.* (1966) identified linaloöl as the most important component of aromatic wines.

Webb *et al.* (1966), using a gas chromatographic technique, identified linaloöl in eight muscat-flavored varieties, although in varying amounts. 1-Hexanal and *trans*-hex-2-enal were also present, again in varying quantities. The corresponding alcohols, 1-hexanol and *trans*-hex-2-en-1-ol, were present in relatively large quantities in all except one variety, and there does not seem to be any inverse correlation between the levels of the corresponding aldehydes and alcohols. They could not correlate muscat aroma and linaloöl content. They also found the furan compounds and pyran linaloöl cyclization compounds, also reported by Stevens *et al.* (1966). A number of lower- and higher-boiling materials (than linaloöl) were found, but none had a characteristic muscat aroma. They conclude that "this particular aroma results from blending in various proportions of the materials investigated in this research."

Webb (1968) has summarized the present information on the muscat aroma. He considers linaloöl and geraniol as the two principal components but indicates that other components play an important role in the odor of certain varieties and suggests that compounds, yet to be identified, may be present in small amounts and could possess an intense and characteristic muscat odor.

For eight varieties he has summarized the relative concentration of the lower boiling point components as follows (L = large, M = average, S = small, T = trace,):

	Aleatico	Early Muscat	Malvasia bianca	Muscat of Alexandria
1-Hexanal	S	L	L	M
1-Butanol	S	S	S	S
2-Methyl-1-butanol 3-Methyl-1-butanol	S	S	S	S
Trans-hex-2-enal	S	L	M	L
Hexyl acetate	T	T	T	T
1-Hexanol	L	L	L	L
Trans-hex-2-en-1-ol	M	M	M	M
Furan I	T	S	T	T
Furan II	T	S	T	T
Linaloöl	S	M	S	S

	Muscat Hamburg	Orange Muscat	P-20-59[a]	Q26-39[a]
1-Hexanal	M	S	M	S
1-Butanol	S	S	T	S
2-Methyl-1-butanol ⎫ 3-Methyl-1-butanol ⎭	S	M	T	S
Trans-hex-2-enal	M	T	S	T
Hexyl acetate	T	S	T	T
1-Hexanol	L	L	M	L
Trans-hex-2-en-1-ol	M	L	S	M
Furan I	T	S	S	T
Furan II	T	S	S	T
Linaloöl	S	L	L	M

[a] Unnamed muscat-flavored hybrid.

Drawert and Rapp (1966) considered the biogenesis of 1-hexanol to proceed as follows: linolenic acid $(CH_3CH_2CH:CHCH_2CH:CHCH_2-CH:CH(CH_2)_7COOH) \rightarrow$ 2-hexen-1-al $(CH_3CH_2CH_2CH:CHCHO) \rightarrow$ 1-hexanol $(CH_3(CH_2)_4CHOH)$. They note that varietal flavor components are derived from ripe grapes. With *V. vinifera* varieties these develop fully only in ripe grapes. High-quality wines thus can be produced only from ripe grapes. It should therefore be possible to identify wines made from unripe grapes by addition of sugar to the must.

During maturation of White Riesling grapes most of the volatile components remained remarkably constant (Drawert and Rapp, 1966). Among those showing marked increases during ripening were *n*-amyl acetate, 1-butanol, ethyl caproate, *n*-butyl butyrate, and 1-hexanol. Most surprising were the decreases in heptanal, 2-hexenal (1), *cis*-3-hexen-1-ol (1) and nonanol, decanal, octanol, and nonyl acetic acid ester, ethyl caprinate, and decanol (2).

Including their own results, Drawert and Rapp (1966) reported the following compounds present in grape musts: methanol, ethanol, 1-propanol, 2-methyl-1-propanol, 1-butanol, 3-methyl-1-butanol, 2-methyl-1-butanol, 1-pentanol, 1-hexanol, 1-heptanol (1), *1-octanol*, *1-decanol*, *cis*-3-hexen-1-ol, 1-phenethyl alcohol*, acetaldehyde, 1-butanal, 1-pentanal, 1-hexanal, 1-heptanal, *1-decanal*, 2-hexen-1-al, *lauryl aldehyde*, acetone*, 2-butanone*, 2-pentanone*, formic acid, ethyl formate, acetic acid, methyl acetate*, ethyl acetate, *n*-propyl acetate*, isobutyl acetate, *n*-butyl acetate, isoamyl acetate*, *act*-amyl acetate*, *n*-amyl acetate, *n*-hexyl acetate, *n*-heptyl acetate, *n*-octyl acetate, *n-nonyl acetate*, propionic acid*, ethyl propionate, isobutyl propionate*, *n*-amyl propionate*, butyric acid*, ethyl butyrate, *n*-butyl butyrate, isoamyl butyrate*, valeric acid*, ethyl valerate*, caproic acid*, ethyl caproate*, isobutyl caproate*, ethyl enanthate*, *n*-hexyl enanthate*, ethyl caprylate, isoamyl caprylate*, *n*-hexyl caprylate*, caprinic acid*, ethyl caprinate, ethyl laurate*, *n*-butyl phthalate*, methyl anthranilate,ˋ geraniol*, terpineol*, limonene*, and linaloöl. (Italicized compounds were not finally identified. Compounds with an asterisk were not found in Drawert and Rapp's White Riesling must.)

Drawert and Rapp (1966) and Van Wyk et al. (1967) reported on the flavor components of White Riesling grapes grown in Germany and California. Their lists were similar except that ethyl acetate, linaloöl, benzyl alcohol, trans-hex-2-en-1-ol, and a number of volatile acids were found in the latter but not in the former. The presence of trans-hex-2-en-1-ol is attributed by both to enzymatic action on linolenic acid when grapes were crushed, ground, or pulped in the presence of air. The green odor of the freshly crushed grapes is attributed by Van Wyk et al. to both these compounds and possibly to 1-hexanol, 3-methyl-1-butanol, and 2-methyl-1-butanol. The fruity note they attribute to 2-phenethyl alcohol, linaloöl, ethyl acetate, and isoamyl acetate. In White Riesling wines they isolated the following volatile organic acids: acetic, n-butyric, n-caproic, n-caprylic, n-capric, 9-decenoic, succinic, and ethyl acid succinate. They consider n-caproic and n-caprylic to be significant to the odor of White Riesling. Traces of formic, propionic, isobutyric, 2-methyl-butyric, isovaleric, lactic, 2-hydroxyisocaproic, n-pelargonic, and malic were reported. Malic and succinic are not normally considered volatile acids. The most common neutral components were ethanol, 1-propanol, 2-methyl-1-propanol, 2-methyl-1-butanol, 3-methyl-1-butanol, 1-hexanol, levo-2,3-butanediol and 2-phenethyl alcohol, ethyl acetate, isoamyl acetate, ethyl n-caproate, ethyl n-caprylate, n-hexyl acetate, 1,3-propanediol monoacetate, and 2-phenethyl acetate. Smaller amounts of a number of other alcohols and esters were reported. They also found γ-butyrolactone, N-ethylacetamide, diethyl acetal, and acetaldehyde. No single component had the typical White Riesling odor. They conclude that it results from blends of the identified or still-to-be identified components. They note that a number of the compounds isolated can have little direct effect on the White Riesling odor since they are essentially odorless.

De Francesco (1967) presented excellent chromatographs of the odorous constituents of a number of specific varieties. No correlation of height of peaks and odor sensitivity is presented. For data on the biogenesis of aroma compounds see Heimann (1965). Webb and Ough (1962) concluded that added essence stripped from grapes and added to wines did not produce a desirable flavor. They believe that "fermentation alters some of the grape aroma components to produce substances of more specific aroma intensity." See U.S. Laws (1963) for regulations on addition of flavor concentrates. For analysis of Cabernet Sauvignon aroma, see Webb et al. (1964b).

Chaudhary et al. (1968) isolated and analyzed the essence from wines prepared from normal Sauvignon blanc grapes and those from similar grapes treated with *Botrytis cinerea*. Gas chromatography did not reveal any significant difference in either the acidic or neutral components of the volatile essence. Prillinger and Madner (1968) found differences in the gas chromatograms of normal red wines and of wines prepared from pomace, lees, or raisins. Prillinger *et al.* (1968) showed that oxidized grape juice contained hexanal, isovaleraldehyde, and 1-propanol.

ADDENDUM

Since the introduction of paper partition chromatography by Bate-Smith in 1948 for the separation and identification of anthocyanins and related flavanoid pigments (J. B. Harborne. The chromatographic identification of anthocyanin pigments. J. Chromat. *1*, 473-488, 1958), acid solvents have been used for extraction from plant tissue and for development of paper chromatograms and elution from them. The solvents widely used for extraction are methanol or water acidified with hydrochloric acid. The solvent systems used for development are butanol-acetic acid-water, butanol-hydrochloric acid, isopropanol-hydrochloric acid, water-hydrochloric acid, water-acetic acid-hydrochloric acid and less frequently water-acetic acid. Recently (at the Wine Institute, Technical Advisory Committee meeting in June, 1969) A. D. Webb reported that the use of hydrochloric acid in extracting and developing solvent systems resulted in the hydrolysis of acetylated anthocyanins. By substituting a weaker acid (acetic acid), he demonstrated the existence of acetylated anthocyanins in some grape varieties.

In a later report (R. A. Fong, R. E. Kepner and A. D. Webb: Acetic-acid-acylated anthocyanin pigments in the grape skins of a number of varieties of Vitis vinifera. Am. J. Enol. Viticult. *22*:155, 1971) the anthocyanin skin pigments of 44 grape varieties were investigated. Of those 29 contained acetic-acid-acylated pigments. Clones related to Pinot noir did not contain acylated pigments.

CHAPTER SEVEN

FROM GRAPES TO
THE FERMENTER
TO NEW WINE

Producing sound wines demands close control of every step from grapes to storing. This chapter gives the steps, from handling the grapes through fermentation, so far as they are the same for all types of table wines. Fermentation, finishing and aging, and special processes for different types of table wines are discussed in later chapters.

HARVESTING AND TRANSPORTING GRAPES

Grapes should be picked when they are properly mature
for the type of wine to be made. Maturity can
be tested by the use of a hydrometer or refractometer.

The finest table wines are produced when the winery has a constant source of high-quality varieties of wine grapes from year to year. When a limited number of such varieties are received year after year, the wine maker becomes familiar with their characteristics as to fermentation, aging, and finishing and thus learns how to handle each to its best advantage. Two of the problems of many small wineries that are attempting to produce fine wines is that they do not have a constant source of high-quality wine grapes from year to year or they attempt to handle too many varieties.

Scheduling

Since a winery cannot crush its whole vintage at once, it must co-operate with its vineyard superintendent or with the growers from whom it purchases grapes to arrange picking schedules. The varieties that ripen earlier, as determined by repeated field tests, must, of course, be harvested first. Both the grower and the winery should be interested in making maturity tests. Samples should be collected from each variety for each vineyard or, where the grapes from different parts of the vineyard do not ripen at the same time, from different parts of the vineyard. The tests should begin in August and be continued until harvesting is well advanced and the order of picking all varieties and vineyards is definitely established.

The idea that the vintage should start on the same date every year is erroneous even for climatic conditions as uniform as those in California. In the warm seasons, most varieties can be harvested two to four weeks earlier than usual; in 1935 and 1948, cool, late-maturing seasons necessitated a delay of at least four weeks. Normally, however, it is better to start picking some of the crop too early rather than to pick a large percentage of the crop too ripe.

Another and better solution of the problem of a proper harvesting schedule is to increase the size of the fermenting room so that a larger percentage of the crop may be handled at one time. Separate fermenting rooms for red and white table wines are particularly desirable. Another reason for early harvesting is the control of fruit flies. (See chap. 5.)

Measuring maturity

The relative proportions of sugars and acids in the pH of the must largely determine the maturity of the grape and its suitability for various types of wines. The most obvious changes during maturation are the gradual increases in the sugar content and in the pH of the fruit; at the same time, the titratable acidity, together with the total malate and to a lesser extent the total tartrate content, rapidly decreases. In red varieties the amount of anthocyanin color pigment gradually increases, and in white varieties the greenish color fades.

The proper stage for picking depends on the use for which the grapes are intended. Dessert and appetizer wines and natural sweet table wines require grapes high in sugar and relatively low in acid. Wines of this composition are obtained from grapes grown in the warmer districts or left on the vines until late in the season, or of varieties which ripen early or are naturally high in sugar. Grapes for dry table wines should be picked when the sugar content of the fruit is barely enough to produce the proper alcohol content. The fruit must not remain on the vines until the total acid has become abnormally low. Undue drying out of the fruit while still on the vines should therefore be avoided. Table wines made from overmature grapes may not ferment out dry and will surely be of poorer quality that those made from properly matured grapes.

Measuring maturity of grapes for table wines involves determining the percentage of total acid, the pH, and the sugar content. Different varieties have widely different total acid contents when their sugar reaches the same level. This condition, more than any other, determines the suitability of the several varieties for the different types of wine.

Table 4 (p. 69) shows the differences in composition between some of the common varieties of grapes, all grown in the same vineyard and picked as nearly as possible at the same sugar content. In countries where the minimum alcohol content is not important, the total acidity might be a partial measure of maturity. In California, however, where the wines must have a minimum alcohol content of 10.0 or 10.5 per cent, the total acid differs too much from variety to variety to be useful by itself; so the sugar content also must be measured.

For this reason the Brix-acid ratio represents a more rational basis for determining field maturity and grape utilization. Musts with a Brix of approximately 22 and Brix-acid ratio of 30 or less are particularly desirable for production of dry table wines.

The pH must be considered also. Musts of relatively high total acidity may have a pH which is too high for the best fermentation and taste for dry table wines. The must pH for such wines should not exceed 3.5: the optimum may be as low as 3.2 for white musts of the Riesling type and 3.35 for the lighter Cabernets.

La Rosa (1963) noted the need for frequent and continued testing of the grapes in order to harvest them at the proper maturity, harvesting them early in the morning rather than in the afternoon, cooling must and wine during fermentation, and cool storage during aging and finishing. Earlier (1955) he had proposed pH rather than a Brix-acid ratio as an index of maturity for a given variety in a certain location. La Rosa and Nielson (1956) pointed out the serious effect of delay in harvesting on the quality of grapes grown in the Central Valley of California.

In California musts a fairly uniform percentage of the soluble solids of the mature fruit is sugar—over 90 per cent. The soluble-solids content of the juice, usually measured by a Brix (or Balling) hydrometer, gives a good estimate of the amount of sugar present. The usual commercial Brix hydrometers (fig. 46, p. 211) are of variable accuracy. The chemical determination of sugar is more accurate but much less simple. The refractometer may also be used to determine the soluble-solids content of the juice. Relatively inexpensive hand refractometers are now available. Arnold (1957) confirmed the value of the refractometer for determination of approximate sugar content. Since the sugar values determined did not agree with the true values, a correction table and a temperature correction were developed. In this country the refractometer is usually checked against a standard sugar solution and adjusted accordingly. Winkler (1932) found that the Brix and refractometer readings agreed very closely. One advantage of the refractometer is that it requires only a few drops of juice. However, these few drops must be representative of the whole sample of fruit being tested.

The Brix hydrometer reading is in terms of grams per 100 gr of juice and should not be directly compared with the chemical results, which are naturally higher, being expressed as grams per 100 ml. (See table 73 and pp. 744 for the conversion from grams per 100 gr to grams per 100 ml.)

The sample used for determining the soluble-solids content, whether taken from fresh grapes in the vineyard during ripening or from a load of fruit being delivered at the winery, must be representative of all the grapes in the vineyard or in the load.

Amerine and Roessler (1958a, 1958b,) and Roessler and Amerine (1958) reported on methods of field sampling. They recommended berry or cluster sampling in preference to whole vine sampling. For berry sampling they recommended 200 to 500 berries. For cluster sampling about one-tenth of the vines had to be sampled in order to reduce the standard error of the sample mean to ±0.25 per cent. Similar results were obtained by Rankine et al. (1962), except that they were able to use smaller samples, presumably because of less field variability. They found that from 100 to 200 berries were adequate in berry sampling. For cluster sampling they required single clusters from 25 random vines out of 200 to reduce the standard error to ±0.25 units. They also found that the variability was greater the higher the acidity. Irrigated grapes were more variable than nonirrigated when the sugar was less than 20 per cent. Amerine and Roessler (1963) reiterate their faith in the reliability of berry sampling; see Rankine et al. (1962) for a similar conclusion.

Baker et al. (1965a) also showed that sequential measurements on grape juice of refractometer reading, pH, and total acidity gave non-normal distributions. Vineyard heterogeneity was considered to be the cause. The variations were more subtle than could be handled by the usual statistical trial designs. They used a computer program of finite Fournier series to analyze the data.

Other measures of maturity have been proposed (Deibner et al., 1965b). The only one of promise appears to be that based on weight per 100 berries times the percentage of sugar divided by 100. For other measures of maturity see p. 249 f.

The actual maturity test is made by crushing the berries in a press, in a screw-type crusher, or with the hands, and straining the juice into a cylinder or centrifuging it until it is clear. The Brix hydrometer is then floated in the liquid, and a reading is made at the bottom of the miniscus. Hydrometers used for this purpose are shown in figure 46 (p. 211). With the refractometer a few drops are placed on the prism. With both the hydrometer and the refractometer a temperature correction may be needed. (See Amerine, 1965b.)

Since grapes are often paid for on the basis of the Brix test in California, care must be exercised in its determination. A California law (Rogers *et al.*, 1940) regulates this when the grower is to be paid on the basis of the sugar content. Accuracy depends on a valid sample, which is dependent on a uniform and complete extraction of the sugar from a representative sample of grapes, a clean cylinder, a reliable hydrometer, proper correction of the reading for any difference from the calibrated temperature, and freedom from extraneous matter and air bubbles in the strained juice. With overripe fruit it is difficult to obtain a sufficient and complete extraction of the sugar (particularly if partly dried berries are present). A screw-type crusher is very desirable in such cases, but if many raisins are present they will not be proportionately represented in the juice.

Picking

Grapes should be delivered to the crusher in a cool condition and as soon as possible after picking. This is best done by picking grapes very early in the morning and moving them immediately to the crushers (fig. 52). If such a practice is not feasible, those picked in the afternoon can be allowed to cool overnight in unstacked boxes in the vineyard; but

Figure 52. Harvesting grapes in Japan. (Courtesy Suntory Ltd., Osaka, Japan.)

Figure 53. Gondola being unloaded from truck. (Courtesy Wine Institute.)

their condition will be less satisfactory, because of the delay, than if they were crushed immediately and the musts properly cooled. Ideally the grapes should reach the winery in their original condition, and this is best accomplished by reducing the delay between picking and crushing to the shortest possible time. In California this is done mainly in small, (fig. 53) or large (fig. 54) gondolas or occasionally in boxes, (fig. 55).

Most picking in California is done with knives. Although these may be somewhat faster and less tiring than shears, they do not allow the picker to remove rotten berries and other undesirable material from the cluster; and they frequently slash the fruit so that the juice escapes, which may allow fermentation or growth of spoilage organisms before the grapes reach the crusher. Knives are satisfactory if carefully used, as long as the fruit is in good condition and is delivered to the winery

Figure 54. Two-ton gondolas being unloaded. (Courtesy Paul Masson Vineyards.)

immediately. Shears are advisable if the grapes are to be used for the best-quality wines or if the clusters are in poor condition, since they permit trimming out the undesirable parts of the clusters without unduly cutting the fruit.

Selective picking and careful handling should bring the grapes to the wine maker uninjured. Moldy clusters and rotten fruit obviously should not be harvested, since they tend to lower the quality and soundness of the wines. The careful hand-sorting of the fruit to remove diseased berries,

Figure 55.   Unloading boxes at winery. (Courtesy Wine Institute.)

which is practiced in certain European districts, has as part of its object the removal of the undesirable microflora present in such berries.

A few varieties have second-crop grapes. If the first-crop grapes are overripe, second-crop fruit is useful in raising the acidity and lowering the sugar content. In this case they should be picked along with the overripe grapes. But in a cool season, when the sugar content of the first crop is already too low, they should probably be left on the vines.

The boxes must be as clean as possible. During the picking season they should be washed and sterilized occasionally with live steam, especially after being used to transport the grapes over long distances, or late in the season after rains when the boxes are likely to be very muddy and the grapes are not so clean. Dirty boxes are a source of contamination. If boxes used for red grapes are later used for white grapes, there may be an undesirable color pickup. Boxes are generally being replaced by picking buckets which are emptied into small gondola trailers (fig. 56). The gondola is usually coated with a resistant epoxy paint and should be washed out between loads.

Gondola trucks (fig. 57) are less desirable for thin-skinned, juicy varieties and must not be used for storage of grapes for more than a few hours. The weight of the fruit will crush the grapes at the bottom of the cluster and fermentation may start.

Aluminum picking containers are successfully used in Germany,

Figure 56.   Harvesting grapes into gondolas. (Courtesy Almadén Vineyards, Inc.)

according to Eschnauer (1963b). They must be properly constructed and lined. Of the small aluminum pickup by the grapes, over 70 per cent is lost during fermentation. However, Eschnauer (1963a, 1964) cautions against undue aluminum pickup, since undesirable changes in odor of the wine are caused by amounts of over 10 mg per liter.

Mechanical harvesting

With the increasing cost of field labor it has been obvious for some time that cheaper methods of harvesting grapes would need to be developed. The first steps were to harvest directly into gondola trucks or into 50-gallon drums which could be picked up by machinery and emptied. These reduced costs somewhat, but the major cost of hand harvesting the fruit remained.

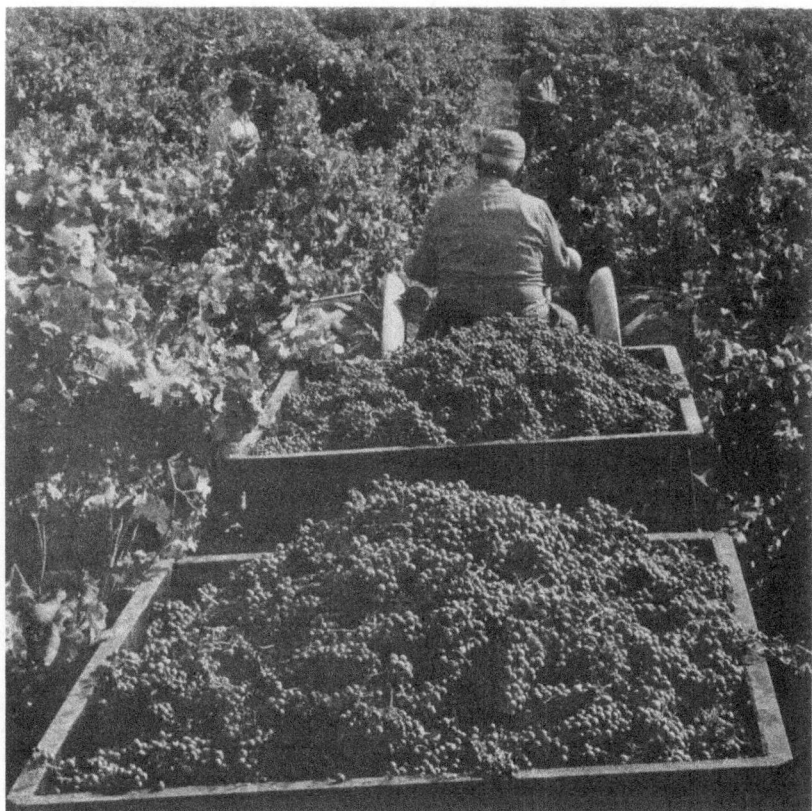

Figure 57.   Gondolas in vineyard. (Courtesy Paul Masson Vineyards.)

Winkler *et al.* (1957) and Lamouria *et al.* (1961) showed that grapes could be trained on an L-shaped trellis so that the fruit could hang free. A mowing machine arrangement would then cut the grapes and they would be conveyed by conveyor belts to the gondola truck. This system had

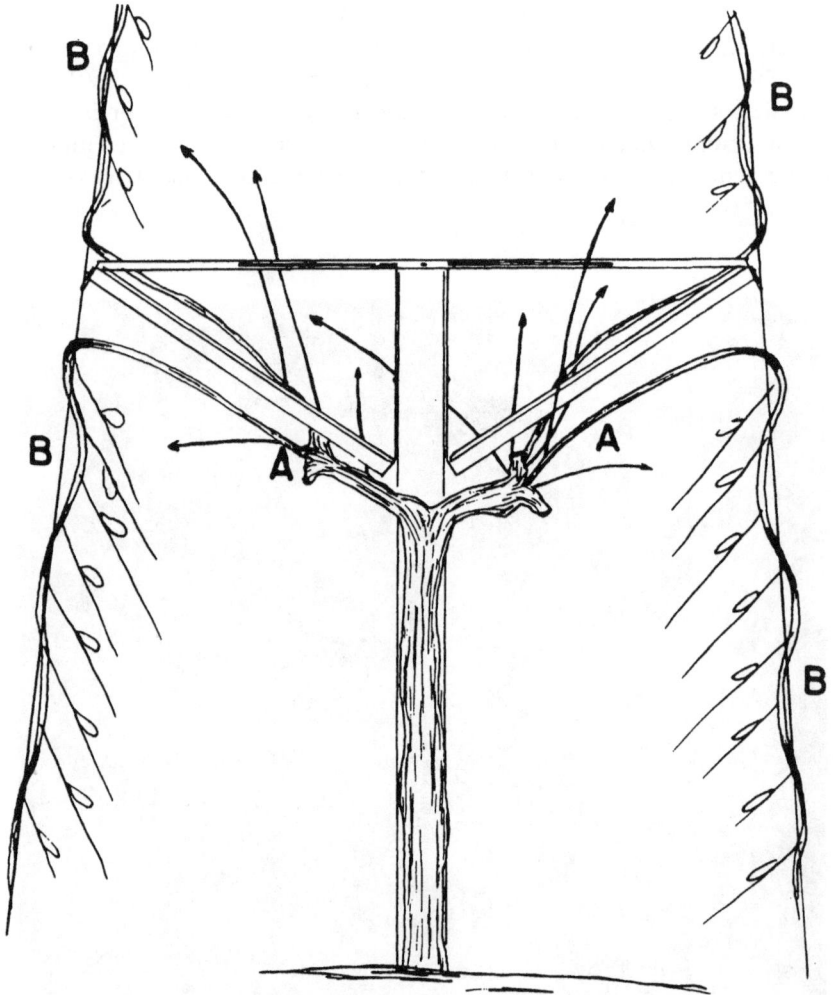

Figure 58. Duplex training system, mature vine. (A) Replacement zone to furnish fruiting canes, the selected shoots remaining after spring deshooting and deflorating indicated with arrows. (B) Fruit-bearing zone on wires, the pendent shoots with flower clusters arising from the fruiting canes.

disadvantages: the cost of training and maintaining the vines in the correct shape, and the difficulty of finding suitable varieties with stems long enough so that they could be cut off without injuring the fruit. Even under the best conditions 10 to 15 per cent of the fruit remained unharvested.

For Concord-type grapes in New York, where the fruit is easily separated from the vine, a shaking process has been sufficient to remove the fruit from the vine. Shaulis *et al.* (1960, 1964) and Shepardson *et al.* (1962) have published drawings of this type of equipment.

A similar type of equipment has been recommended for California by Olmo and Studer (1967) and Studer and Olmo (1967). The method of vine training and the impacter mechanism are shown in figs. 58 and 59, from Olmo *et al.* (1968). They were able to get a large percentage of the fruit clusters to drop off Thompson Seedles vines by vibration. The vines are cane-pruned to ensure that the fruit arises from fruiting canes on the laterals. The apparatus has been tested for two seasons and appears especially useful for Thompson Seedless, the most widely planted variety in California, but many leaves also fall when the vine is shaken, and rejection of low-quality fruit is difficult or impossible. It is also necessary to deshoot or debud the central area of the vine, which is not accessible to the vibration machine. The crop must be confined to the lateral canes. The pilot equipment was capable of harvesting one acre per hour. The varieties tested differed appreciably in separation of clusters, from 5 to 95 per cent.

A vacuum crusher-harvester has also been devised for use in California (Osteras, 1966). Recently several new or improved machines for mechanical harvesting have been developed. The Chisholm-Ryder appears to be most commonly used in California.

CRUSHING AND STEMMING . . .
. . . *must be complete.*

Blending before crushing

Every effort should be made to sort the varieties and qualities of grapes into their proper types or blends before crushing. Good grapes should not be mixed with less sound ones.

The best time to blend the grapes used for making certain nonvarietal types of wines is during crushing. Blending the grapes before fermentation does not give as close control over the composition as when wines are blended, and may not always eliminate the need for blending wines; but it is useful in making minor color corrections or in balancing a

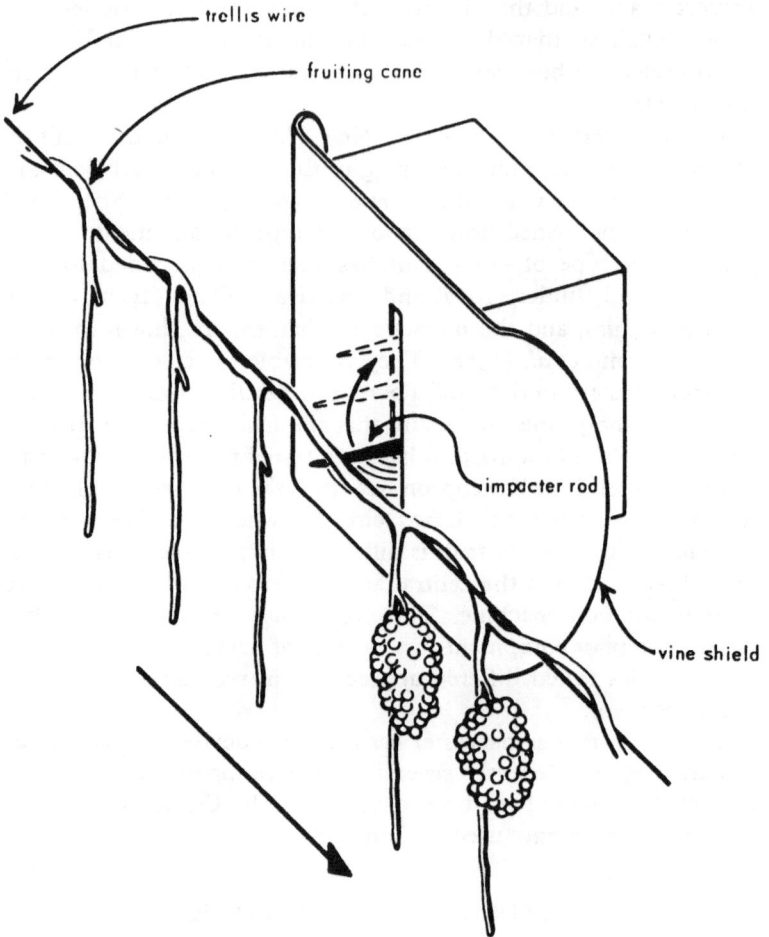

Figure 59. Action of impacter mechanism.

must for proper fermentation. Usually the blend is more stable and more likely to remain brilliant if the grapes ferment together than if the wines are mixed.

Adjusting the crusher

The equipment for crushing and stemming has been described in chapter 4. The crushing should be as complete as possible. Both under- and overripe grapes are more difficult to crush than turgid, mature grapes. When the greener grapes are not crushed, the must is of lower acidity than it should be. When the shriveled, overripe grapes are not crushed, the potential alcohol yield of the grapes is not realized because the high sugar content of these grapes cannot be completely extracted and fermented before pressing and drawing off. Crushers should therefore be adjusted according to the condition of the grapes to secure, as nearly as possible, 100 per cent crushing.

When the grapes are well crushed, the fermentation starts sooner, is more regular and complete, and the wines are of better color than if they are not crushed or are poorly crushed. Crushing also aerates the must, which aids in the multiplication of the yeasts. (For the effect of aeration on yeast growth, see pp. 420-422.)

When grapes are crushed, particularly those of certain white varieties, the tissues exposed to the air often turn brown. This is due to the oxidation of phenolic compounds in the tissues by polyphenoloxidases.

Stemming

Stemming is practiced throughout California. Machines which crush and stem the grapes at the same time are now used everywhere. The crusher should not be operated so that too many of the stems are broken up and get into the must, or so that the seeds are broken. However, the usual California crusher-stemmer does break up an undue amount of stems, especially of varieties with brittle stems. This may reduce the quality of the resulting red wines. In Europe the stems are sometimes left with the crushed grapes, but the wines are said to require longer aging to reach maturity. No conclusive evidence on their value has been reported; since early maturation is considered desirable in this state, the stems should probably be removed before fermentation. Leaving the stems in to facilitate pressing of white musts is another matter. We know of no evidence that it is an undesirable practice if the must is pressed immediately after crushing.

Durmishidze (1965) called attention to the complex biochemical reactions which occur in crushed grapes or during pressing. He added ^{14}C-labeled malic acid, aspartic acid, and alanine to crushed grapes. Tyrosine, threonine, and alanine and citric, glycolic, succinic, and fumaric acids were formed from malic acid. Durmishidze believes that this is an indication of an active fumarate cycle enzyme system in crushed grapes. Alanine formed cystine, lysine, arginine, aspartic and glutamic acids, serine, glycine, threonine, methionine, tryptophan, and leucine, and glycolic, malic, succinic, and fumaric acids. Aspartic acid was converted into lysine, histidine, threonine, alanine, proline, methionine, tryptophan, and leucine. If the labeled compounds were incubated on grape-skin preparations, a larger number of amino acids resulted (and in higher concentrations) than without the skins. Obviously, the fact that the products formed take part in metabolic processes adds to their complexity. Presence of oxidation products of catechin inhibited the decarboxylation of keto- and oxy- acids in the pulp.

Noncrushing

There have been many reports of placing uncrushed grapes in tanks and allowing them to ferment. The theory appears to be that the fermentation of the free-run juice will produce carbon dioxide in sufficient quantities to asphyxiate the cells of the uncrushed grapes, resulting in greater color and flavor release. Bénard and Jouret (1963) reported fermentation times of three to five days for crushed grapes and eight to eleven days for uncrushed. The higher fermentation temperatures 4° to 10°C (7 to 18°F) of the former are noteworthy. The alcohol and aldehyde contents of the wines from the free-run juice of the crushed grapes were higher than from the press. The opposite was true of the free-run and press of the uncrushed grapes. As expected, the wines of uncrushed grapes had lower color but matured earlier. Unfortunately, no critical sensory tests or statistical analyses of the results were made. Lacking such results the claims of the Narbonne school of enologists must be considered applicable only to their varieties and conditions. However, see page 609 for successful use of a similar system in Beaujolais.

Peynaud and Guimberteau (1962a) showed that three types of reactions occur when grapes are placed under carbon dioxide or nitrogen for several days: (1) alcoholic fermentation, (2) reduction in malic acid, and (3) internal movement of constituents into solution, particularly of nitrogenous compounds, polyphenols, and aroma materials. There is also some hydrolysis of pectins with liberation of methanol and an increase in free amino acids. For further information on noncrushing see chapter 13 on red wine production.

CONTROL OF MICROBIAL GROWTH

Successful wine making depends on the
skillful control of the various agents and
conditions of fermentation.

Undesirable yeasts and bacteria must be controlled. The appearance, flavor, and aroma of the wine, its soundness and freedom from disorder or diseases, and its resistance to bacterial attack during storage will depend mainly upon the kind of microbial growth that occurs in the fermenting mass and upon the character of the fermentation. The growth and activity of the desired organisms—wine yeasts—must be furthered, whereas that of the undesirable organisms—wild yeasts and bacteria—must be hindered. Fortunately, microörganisms differ in their nutritional requirements, their tolerance for acid and alcohol, their resistance to antiseptic agents, such as sulfur dioxide, and their response to temperature.

The most serious spoilage in musts is caused by the unchecked activity of acetic and lactic acid bacteria, and in wines by the activity of lactic acid bacteria. The susceptibility of wine to bacterial attack varies with the pH, alcohol content, presence of substances liberated by decomposing (autolyzing) yeast cells, and degree of exhaustion of nutrients, such as phosphates and nitrogenous compounds. Wines from musts of moderate to high acid content (low pH), fermented rapidly and continuously at low temperatures, are most resistant to bacterial attack and are of better quality. For propagating the desired alcoholic fermentation, true wine yeasts should predominate over those of other microörganisms; and the environmental conditions (composition of must, temperature, aeration) should be made favorable to their growth and activity. See page 404 for amounts of sulfur dioxide to add.

BALANCING THE MUST

Acids are most important in California.

For best results, the sugar, acid, and tannin content of the grape juice should be properly balanced. This balance is best obtained by selecting suitable varieties and by harvesting at the proper stage. Unfortunately, such practices are not always followed, especially in California, where the grapes are often harvested with too much sugar for dry table wines rather than too little, as in Europe, where it is often necessary to ameliorate the must by adding sugar.

Acid

The acid content of the must greatly affects the course of fermentation and the resistance of the wine to spoilage. At the optimum acidity, flavor formation occurs to a higher degree, the growth of undesirable organisms is checked more completely, and the wine obtained is more resistant to bacterial attack.

What this optimum acidity is for California conditions has not been determined. In France an acidity (expressed as tartaric acid) of 0.6 gr per 100 cc of must is considered satisfactory for sweet table wines, and at least 0.7 to 0.9 gr for most of the dry white and red table wines. The acidity required increases with the temperature of fermentation. Under Portuguese conditions the acidity of the must is adjusted by addition of up to 10 pounds of tartaric acid per 1000 gallons, depending on type of must, initial acidity, and fermentation temperature (Pato, 1932). Similar if less drastic corrections would improve the properties of ordinary California dry table wines.

The acid deficit in the must is made up, preferably, by the addition of tartaric acid, since an appreciable amount of the added tartaric acid is later removed from wine as cream of tartar; this lowers the titratable acidity and diminishes the acid taste of the wine without greatly changing the pH. Because of this loss of acid, however, a greater amount of tartaric is required than of acids such as citric. Citric acid is not recommended for addition before fermentation because some may be converted to acetic acid during fermentation. Pato (1932) recommends 3.3, 2.1, and 1.3 lbs of tartaric acid per 1,000 gallons for musts of 0.5, 0.6, and 0.7 per cent total acidity (as tartaric) respectively. This was for grapes of 25°C (77°F). If the temperature was 26° to 30°C (78.8° to 96.0°F) he added 5.0, 2.9, and 1.8 lbs., respectively. A more rational system of correction was devised by M. A. da S. Pato (1967). He calculated the amount of tartaric acid to add to musts of pH 3.4 to 4.0 to bring the must pH to 3.3 to 3.9. For example, to lower the must pH from 3.6 to 3.4 required 9.5 lbs. of tartaric acid per 1,000 gallons—considerably more than recommended above.

According to Fornachon (1938, 1943), the addition of tartaric acid to the must ensures that the primary fermentation will take place under conditions favorable for the healthy growth of yeast cells, and that fermentation will proceed more steadily in the acidified musts, so that control of temperature is facilitated. Acidification also aids in preventing the clouding caused by metallic impurities such as iron and copper salts.

Fornachon has shown that the susceptibility of the wine to bacterial spoilage is governed almost entirely by the acidity of the must and of the resulting wine. He suggests that the pH of the must and wine should be adjusted to below 3.6 provided the taste of the resulting wine is not made too acid. An even lower pH, 3.3–3.4, is usually desirable. To correct food acidity DL-malic has been recommended. This addition is easily detected (see p. 780).

Excess acidity seldom occurs in California musts. Ion-exchange or Kielhofer and Würdig's (1963a) double salt precipitation of excess tartrate can be used on such musts.

The titratable acidity is a better indication of the degree of acid taste than is pH, but other factors, such as sugar content, also affect it. (See

p. 422.) It is better to add acid to the must than to the wine, for the maintenance of a low pH during fermentation inhibits autolysis of yeast cells, aids in the extraction of the pigments from the skins of red grapes, and results in a better-flavored wine.

Tannin

In the fermentation of red wine the addition of tannin is not normally necessary, since the skins and seeds of grapes form a ready source of tannin.

In certain regions, particularly Algeria, tannin is added to the free-run white grape must; this practice is common in Australia also. The amount usually added is $\frac{2}{3}$ to 1 ounce per 100 gallons. Specially prepared tannins for wine use, the so-called enological tannins obtained by alcohol extraction of grape seeds and pomace, are preferred to the gallotannic acids obtained from quebracho bark. In California, tannin deficiencies in white wines are customarily made up by the less satisfactory procedure of adding tannin to the wine rather than to the musts.

Turbvosky et al. (1934) did not find that the addition of tannin prevented the development of undesirable organisms in musts and wines. However, Fornachon (1943) has shown that the addition of normal amounts (1 gr per liter) of commercial enological tannins definitely retards the growth of spoilage bacteria in dessert wines. The action of tannin in checking bacterial growth appears to be that of effect on the bacteria rather than precipitation of essential nutrients or accessory growth substances. The most extensive report on the effect of various polyphenolic compounds on fermentation is that of Šikovec (1966). Chlorogenic and isochlorogenic acids stimulated fermentation, whereas caffeic, ellagic, and gallic acids inhibited it. High-tannin wines also developed more involution yeast forms. Yeasts adsorb polyphenols and partly assimilate them. Yeasts vary markedly in their sensitivity to various polyphenols. (See also p. 271 for the extraction of tannin from seeds.)

Other substances

Deficiencies in other constituents necessary to successful alcoholic fermentation are rare in California grapes. There is usually sufficient potassium, phosphate, and assimilable nitrogenous material to support the growth and activity of yeast. The addition of yeast foods, such as urea and ammonium phosphate, is rarely necessary and sometimes harmful; the phosphates, particularly, contribute to white casse. (See p. 793.)

Occasionally, however, musts are encountered which are difficult to ferment dry and which respond to the addition of ammonium salts. Since ammonium ion appears to play a role in promoting fermentation, deficiencies in nitrogenous substances, if they occur, are best made up by the addition of ammonium salts. Ribéreau-Gayon and Peynaud (1950) have shown that ammonium salts are particularly useful with cold musts.

Cantarelli (1958) examined a number of products as possible activators of the rate of fermentation. Especially effective was lees, which accelerated the rate of fermentation without increasing yeast development. Nitrogenous compounds such as ammonium, amide, and amino forms were also very useful. For simple nitrogenous compounds there was a direct relation between the nitrogen content of must and that of yeast. The relationship was logarithmic for complex nitrogen compounds. Between yeast nitrogen content and speed of fermentation it was direct.

Wucherpfennig and Franke (1967) reported that grape juices treated with 3 gr of bentonite per liter did not contain heat-unstable protein (as indicated by remaining clear in an accelerated heat test). The wines, however, contained more protein than the must. The source of this protein seemed to be from the yeast, and the proteins appeared in the wine eleven to fifteen days after the start of the fermentation. These also appeared to be proteins of low molecular weight. Nevertheless, they confirm that bentonite-treated musts, because they are low in protein, produce protein-stable wines. Bayly and Berg (1967) believed that the apparent increase is due not to proteins but to dialyzable peptides. The Diemair and Maier (1962) procedure previously used records these as proteins because it is based on the biuret color reaction, which gives a positive test for peptides.

Commercial pectolytic enzyme preparations are sometimes helpful in fruit and grape juice clarification and in clarifying wines, but the results usually cannot be correlated with the pectin methyl esterase (PE) or polygalacturonase (PG, whether of the endo- or exo-type) content of the enzyme, according to Joslyn et al. (1952), Berg (1959b), and Marteau et al. (1963). See also Datunashvili et al. (1967). Marteau et al. found no direct correlation between speed of clarification and amount of methanol produced. Natural clarification of grape juices seems to be limited by their low endo-PG activity. The commercial preparations, in contrast, appear to have a large excess of endo-PG activity, which apparently limits their efficiency.

The study of Marteau et al. (1961) shows clearly that the demethoxylation of pectins occurs rapidly in musts owing to PE, and that the level of methanol depends on the level of pectinic acids in the must. Naturally occurring PE can effect the demethoxylation, but added pectolytic enzymes act more rapidly, although the final level of methanol is the same. This natural PE activity explains the low pectin content of the wines.

A rational explanation of the clarifying properties of the pectolytic enzymes has been given by Marteau (1963). He notes, correctly, that commercial preparations vary widely in the relative activity of their pectolytic enzymes. Endopolygalacturonases attack the glucosidic bonds at random and thus result in a rapid decrease in viscosity. The exopolyglacturonases attack only the terminal glucosidic bonds. They decrease viscosity only slightly. Marteau (1963) notes that commercial preparations contain varying amounts of oxidases which result in browning. He showed that the low pectin methyl esterase of some of these preparations limits their activity. The high endopolygalacturonase activity of commercial preparations may also limit the activity of pectin methyl esterase.

Ough and Amerine (1960) reported that 12.5 pounds of bentonite per 1,000 gallons of must did increase clarity and filter speed and lightened the color, but they did not recommend its use.

Hennig (1956) added commercial enzymes with esterifying and proteolytic activity to fermenting musts. The former gave a slight increase in the volatile ester content of the resulting wines. The latter had little effect either alone or in combination.

It has been reported that giberellin-treated grapes resist browning when crushed (Winter, 1967). This may be of some importance in the San Joaquin Valley, where grapes are often harvested late and the polyphenoloxidase content may be high. (See p. 280.) Under field conditions giberellin is believed to promote internal browning of the berries.

FERMENTING AND PRESSING

Conversion of sugar to alcohol, carbon dioxide,
and other products occurs during fermentation.
Pressing is done to separate skins from juice.

Fermentation is the basis of all types of wines. The grapes provide the sugar and other substances required for the growth of the yeasts. The yeasts provide a complicated enzyme system by which sugar is finally converted to carbon dioxide, alcohol, glycerin, and other products. Although the principles are much the same, the details differ greatly among red, pink, and white wines. For the practical operations of fermentation, see the chapters on each of these types of wine.

White, pink, and red wines likewise differ from each other as to time of pressing. Grapes for white wines are usually pressed before fermentation; those of pink wines, after a brief fermentation (one to three days). Grapes for red wines are usually fermented on the skins and pressed, in four to fourteen days, i.e., after fermentation is partially or almost completed.

The effect of pressing on the composition and quality of musts was studied by Erczhegyi and Mercz (1966). The pressed musts (15 per cent of the total) were higher in total nitrogen, proteins, extract, reductones, ash, tannins, iron, titratable acidity, pH, and malic acid. The pressed musts had less tartaric acid and tended to be brown in color. For sparkling wines, especially, the authors recommended using only the free-run.

The tannin content of fermenting musts is proportional to the length of time that the juice is in contact with the skins and seeds, according to Berg and Akiyoshi (1956a). However, maximum color density is usually reached when only 3 to 6 per cent alcohol has formed. However, after nine months of aging, the color of wines fermented on the skins was directly proportional to the length of time on the skins. Thus the wine maker must determine whether to press early and obtain a wine of lower tannin but with less color or press later and have higher tannin and color. Berg and Akiyoshi used McAdam's ΔC unit, which is proportional to the minimum perceptible difference between two colors as judged by the average observer, as a measure of the time to press. McAdam units are calculated from trichromatic coefficients from transmission data. They suggested that the proper time to press would be when ΔC first fell below 10 compared to a commercial one-year-old California red wine. They also noted increases in color after three to nine months of aging. They indicated that colorless sulfur dioxide pigment complexes, which were formed during fermentation, hydrolyze, thus resulting in a gain in color.

Berti (1965) considers that immediate separation of juice and pomace is especially important in California. Gravity juice separators, usually holding tanks with perforated or slotted bottoms, have the disadvantage of allowing the juice to be in contact with the pomace for relatively long periods of time. The product, however, is reasonably clear. Juice separators operating with drag or cleat or other mechanically operated screens give almost immediate separation of juice and solids, but the resulting juice is relatively cloudy and the wine has a large quantity of lees. Vibrating screen juice separators also operate almost instantaneously but aerate the juice to a froth. Some use screw or semiscrew presses which work like a screw press except that they operate at low pressure.

Coffelt and Berg (1965a) developed equipment for heat-treating whole grapes with steam under pressure. Because of the high temperatures obtainable, only a few seconds of heating were required to give juices which, after partial fermentation and fortification, produced ports of superior color and quality equal to or better than those produced by more conventional procedures.

STUCK WINES . . .
. . . *should be treated promptly.*

Stuck wines can be refermented by aerating and mixing with fresh grapes or actively fermenting must, or sometimes by heating. To obtain sound dry wines, fermentation should continue until all the fermentable sugar has been used up. A stuck wine is one in which the alcoholic fermentation has stopped while much unfermented but fermentable sugar still remains. Stuck wines often spoil rapidly in storage because of the development of lactic acid bacteria unless sulfur dioxide has been used. The high temperatures that often cause the sticking favor the development of these and other wine-disease bacteria. Red wines stick more often than white.

Sticking may be caused by too high (or occasionally by too low) a temperature in the fermenting vat, the presence of too much sugar, infection of the must with acetic acid bacteria or wild yeasts, or a combination of these conditions.

In southern France (Roussillon) Brugirard *et al.* (1965) reported that stuck fermentations are common. They are most often caused by high alcohol, excessive temperature, and insufficient aeration of the must. Less often it may result from use of too much sulfur dioxide or the presence of fungicide sprays in the must. They made the useful point that stuck fermentations must be cared for quickly, for after about ten days there is a rapid rise in the volatile acidity.

Prompt measures should be taken to re-ferment stuck wines before they spoil. The best method varies according to the cause of the sticking. Moderate aeration to invigorate the yeast is recommended in all attempts to re-ferment stuck wines. Adding small amounts of ammonium phosphate, urea, or potassium phosphate is sometimes useful.

The most widely used method of handling wines stuck because of high fermenting temperatures is to pump a portion of the wine into another vat and add a large (10 per cent) starter of actively fermenting must. When this is in full fermentation more of the stuck wine is added; if the fermentation continues satisfactorily, the rest is gradually added.

Another successful method, useful for sticking caused by high temperatures or high sugar content, consists of adding small amounts of the stuck wine to freshly crushed grapes or to vats in active fermentation. If the sticking is caused by too high a sugar content, grapes or musts of low Brix must be used to dilute the wine so that after fermentation the alcohol content will not exceed about 13 per cent by volume. If fermentation has become very slow but has not stopped entirely, 20 to 50 per cent by volume of actively fermenting must can be mixed with the stuck wine.

Cold weather during the late autumn or early winter may result in sticking because of lowered temperature. To complete the fermentation it may then be necessary to warm the wine by using a tubular heat exchanger such as a pasteurizer. The wine should be warmed to about 23.9°C (75°F).

Acetified musts

Musts spoiled by acetification are very rare or nonexistent nowadays in California. They cannot be converted into a marketable wine product. They are probably best pasteurized and then added in small amounts to rapidly fermenting musts. Some of the acetic acid will be utilized during the fermentation. If the spoilage is due primarily to *Acetobacter*, the must may be sold to vinegar producers. Lactic-sour musts or wines do not produce acceptable vinegar. (See pp. 783 and 785.)

RECORDS . . .
. . . are kept by the better wine makers.

The maintenance of a good set of records, although often difficult in the rush of the vintage season, is of great value. An easily referred-to record of the Brix degree and temperature can be written directly on the vat; but there should also be a permanent record of the source and type of grapes in the vat, the original total acid and pH, the amount of sulfur dioxide and pure yeast used, number and times of pumping over, the Brix degree and temperature at various stages of the fermentation, and special instructions for crushing, pressing, and so on.

This information will later prove useful in interpreting the success or failure of the fermentation. It will thus enable the wine maker to eliminate undesirable practices and standardize successful ones. The record card should remain with the wine so that information on the analyses, tasting, and finishing of the wine can be added later.

COMPLETING FERMENTATION . . .
. . . requires care to prevent sticking or oxidation.

When the white or red wine in the fermenting vat is drawn off, it will contain a small amount of fermentable sugar. In making sound dry wines the fermentation must not cease until practically all this sugar has disappeared and prompt and complete clarification has been obtained. To complete the fermentation the new wine is placed in closed storage tanks, preferably tanks equipped with fermentation bungs (traps which permit escape of gas and prevent entry of air). A temperature of about 15.5° to 21.1°C (60° to 70°F) is maintained. If the wine still contains over 0.2 or 0.3 per cent sugar a week after being placed in these storage containers, it should be well aerated by pumping over. In cold weather it may have to be warmed to 21.1° to 23.9°C (70° to 75°F) to complete its fermentation.

FERMENTATION, COMPOSITION, AND FINISHING

CHAPTER EIGHT

PROPAGATION OF
FERMENTATION

Successful wine making depends on the control of the various agents and conditions of fermentation. A knowledge of the biochemistry of fermentation is necessary.

YEASTS . . .
. . . are necessary for fermentation.

The word "fermentation" has undergone numerous changes in meaning. Originally it meant merely a gentle bubbling or boiling condition, and it was in this sense that the term was first applied to the reactions occurring during the production of wine and other alcoholic beverages. After Pasteur's research the word became more closely associated with microörganisms or their enzymes. It is now recognized that these transformations are brought about through the combined activity of a number of enzymes present in (and secreted by) the cells of living microörganisms. Usually the chemical transformations in which oxygen is present are termed "respiration," and those occurring in the absence of oxygen, or in which oxygen is not involved, are termed "fermentation."

Microörganisms

In alcoholic fermentation the chief products are alcohol and carbon dioxide, produced essentially in equimolecular proportions. Although alcoholic fermentations may be accomplished by various microörganisms —yeasts, several of the lower fungi, and some bacteria—the yeasts have the power of fermenting sugars more efficiently and producing the largest amounts of alcohol per cell and per unit weight of sugar fermented. Some species or strains of yeast are more alcohol-tolerant, and will ferment sugars more rapidly, more completely, and at higher concentrations, to produce beverages of higher alcoholic content than will other microörganisms. Yeasts differ widely in their ability to ferment sugars, in the completeness of this fermentation, and in the character of the by-products produced. In the fermentation of grape musts, the only species of yeasts used are those which can ferment the juice until it is practically free of sugar, and which produce desirable by-products. These yeasts are usually strains of *Saccharomyces cerevisiae* var. *ellipsoideus* (fig. 60, *A*), further classified on the basis of origin, fermentation characteristics (such as agglomeration), or minor morphological differences.

Under natural conditions, no single variety of yeast but a mixture of several varieties differing in alcohol-, ester-, and extract-forming powers, together with certain types of acid-tolerant bacteria, will enter into the fermentation. When these are present in proper proportions, under suitable conditions, a mixed fermentation ensues that is responsible for an occasional unique wine obtained by natural fermentation.

Because of lack of knowledge about the relation of the associative and competitive activities of mixed cultures on quality of wine fermentations, because of the difficulty of obtaining the special flora required, and because of the danger of spoilage in natural fermentations, only selected varieties of wine yeast have normally been used in the California wine industry. These yeasts have been chosen primarily on the basis of alcohol formation rather than flavor formation, and are usually imported from select enological regions abroad. It is now known, however, that flavor formation by yeasts usually varies inversely with their alcohol-forming powers, and that ester formation in wines may occur through the direct agency of yeasts and bacteria. The application of this knowledge to California conditions is yet in its infancy. Because of the danger of spoiling wines, mixed cultures are not recommended at present. (See also p. 334 *et seq.*)

Classification of yeasts

The designation of yeasts according to their industrial behavior, source, and growth requirements, which has received attention in the literature, is not scientifically valid (Mrak and Phaff, 1948). The classification proposed by Stelling-Dekker (1931), Lodder (1934), Diddens and Lodder (1942), Lodder and Kreger-Van Rij (1952), and Kreger-Van Rij (1964) is now widely accepted by workers in this field. On the basis of this classification, the true wine yeasts are differentiated from the bottom-fermenting or lager beer yeasts (*Sacch. carlsbergensis*) on the basis of cell shape and inability to ferment melibiose, and are most often designated as *Sacch. cerevisiae* var. *ellipsoideus* (fig. 60). However, the ellipsoidal cell shape is not currently considered as important as it once was. (See also Peynaud, 1957.) In the eastern European literature wine yeast is frequently called *Sacch. vini* or *Sacch. ellipsoideus*.

The classification of Kudriavtsev (1960) is used in the Soviet Union and elsewhere. His book is particularly valuable for its introduction to the Russian literature. He indicates that 300 or more strains of the true wine yeast have been isolated. As to exact differences between them there is little data. He makes the interesting point that some of the strains of true wine yeast that are distributed in Europe are in reality strains of *Sacch. oviformis* (e.g., Bordeaux 1893, Massandra III, Riesling A, etc.). For a general discussion of yeasts see Cook (1958) and Reiff *et al.* (1960–1962).

A very useful survey of the species of yeasts found in grapes and wines was made by Galzy (1956). This survey is especially valuable for data on the synonyms used. Galzy admitted the following yeasts as being for some reason, or at some time or place, of significance on grapes or in wines: *Brettanomyces bruxellensis* Kufferath et Van Laer;[1] *B. lambicus* Kufferath et Van Laer; *Candida brumptii* Langeron et Guerra; *C. guilliermondii* (Castelli) Langeron et Guerra; *C. krusei* (Castelli) Berkhout; *C. mycoderma* (Reess) Lodder et Kreger-Van Rij; *C. pulcherrima* (Lindner) Windisch, D;[2] *C. rugosa* (Anderson) Diddens et Lodder; *C. tropicalis* (Castelli) Berkhout; *Cryptococcus albidus* (Saito) Skinner; *Debaryomyces kloeckeri* (Guilliermond et Péju; *Hanseniaspora valbyensis* Klöcker; *Hansenula anomala* (Hansen) H. et P. Sydow, D; *H. saturnus* (Klöcker) H. et P. Sydow; *Kloeckera africana* (Klöcker) Janke, D; *Kl. apiculata* (Reess) Janke, D; *Kl. corticis* (Klöcker) Janke; *Kl. jensenii* (Klöcker) Janke, D; *Kl. magna* (De' Rossi) Janke; *Pichia farinosa* (Lindner) Hansen; *P. fermentans* Lodder, D; *P. membranaefaciens* Hansen, D; *Rhodotorula aurantiaca* (Saito) Lodder; *Rhod. glutinis* (Fres.) Harrison; *Rhod. mucilaginosa* (Jorg.) Harrison; *Rhod. rubra* (Demme) Lodder; *Saccharomyces acidifaciens* (Nickerson) Lodder et Kreger-Van Rij, D; *Sacch. bailii* (Lindner) Guilliermond; *Sacch. bayanus* Saccardo, D; *Sacch. bisporus* (Naganishi) Lodder et Kreger-Van Rij; *Sacch. carlsbergenesis* Hansen, D; *Sacch. cerevisiae* Hansen var. *ellipsoideus* (Hansen) Dekker, D; *Sacch. chevalieri* Guilliermond, D; *Sacch. delbrueckii* Lindner var. *mongolicus* (Saito) Lodder et Kreger-Van Rij; *Sacch. exiguus* (Reess) Hansen; *Sacch. fermentati* (Saito) Lodder et Kreger-Van Rij; *Sacch. florentinus* (Castelli) Lodder et Kreger-Van Rij, D; *Sacch. fructuum* Lodder et Kreger-Van Rij, D; *Sacch. heterogenicus* Osterwalder, D; *Sacch. italicus* Castelli; *Sacch. oviformis* Osterwalder, D; *Sacch. pastorianus* Hansen; *Sacch. rosei* (Guilliermond) Lodder et Kreger-Van Rij; *Sacch. rouxii* Boutroux; *Sacch. rouxii* Boutroux var. *polymorphus* (Kroemer et Kumbholz) Lodder et Kreger-Van Rij, D; *Sacch. steineri* Lodder et Kreger-Van Rij, D; *Sacch. uvarum* Beijerinck, D; *Sacch. veronae* Verona;

---

[1] The name of the investigator or investigators credited with first description of the species. The name in parentheses is that of the investigator who first isolated and described the yeast.

[2] Those marked with "D" were found by Domercq (1956) in Bordeaux musts or wines.

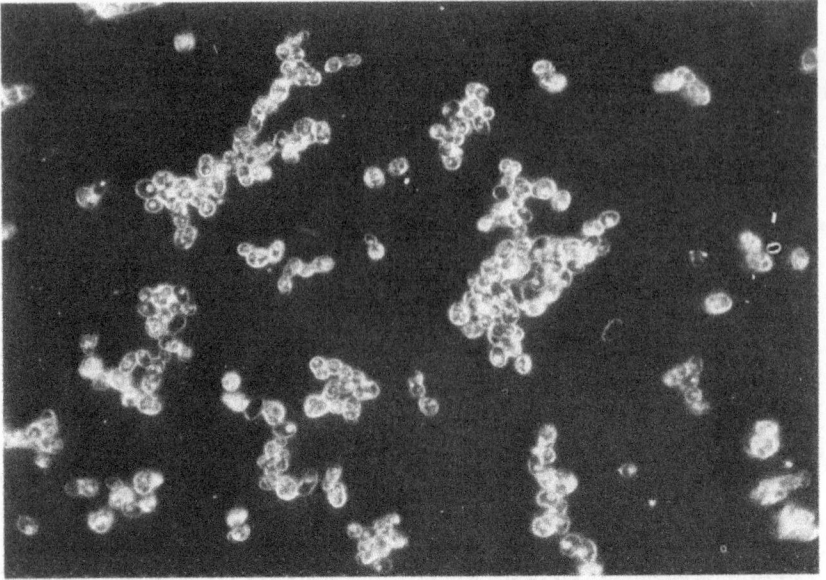

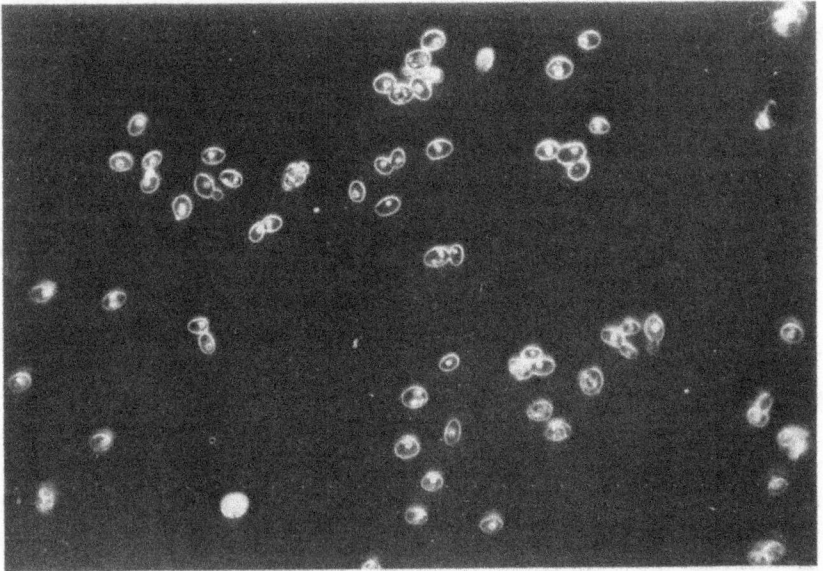

Figure 60. Yeast cells of *Saccharomyces cerevisiae* var. *ellipsoideus*: (above) in various stages of development and (below) during alcoholic fermentation. (Courtesy Seitz-Werke GmbH, Bad Krueznach, Germany. See p. 4 of their Informationen No. 29, 1967.)

Sacch. willianus Saccardo; *Saccharomycodes bisporus* Castelli; *Saccharomycodes ludwigii* Hansen, D; *Schizosaccharomyces octosporus* Beijerinck; *Schiz. pombe* Lindner; *Torulopsis bacillaris* (Kroemer et Krumbholz) Lodder, D; *Trichosporon pullulans* (Lindner) Diddens et Lodder.

Domercq (1956) also reported *Brettanomyces vini; Debaryomyces hansenii; Rhodoturula vini; Sacch. delbrueckii* Lindner; *Sacch. elegans* Lodder et Kreger-Van Rij; and *Torulopsis famata* (Harrison) Lodder et Kreger-Van Rij. The Czechoslovak Collections of Microörganisms (1964) lists also *Candida vinaria* Ohara, Nonomura et Yunome; *C. zeylanoides* (Castelli) Langeron et Guerra; *Debaryomyces globosus* Klöcker; *Sacch. coreanus* Saito; and *Sacch. globosus* Osterwalder (same as *delbrueckii*). The species involved in film formation has been variously attributed to *Sacch. beticus, Sacch. cheresiensis, Sacch. fermentati,* and *Sacch. oviformis,* among others. Kudriavtsev (1960) favors *Sacch. oviformis.* He also recognizes *Sacch. chodati* Steiner and *Sacch. prostoserdovi* Kudriavtsev as isolates from grapes and wines. He further classified a new genus and species, *Issatchenkia orientalis* Kudriavtsev, which he isolated from fruit and grape juices. Kudriavtsev accepts *Debaryomyces dekkeri* Mrak *et al.* and *D. kursanovi* Kudriavtsev, the first isolated from California grapes and the latter from Armenian fermenting musts, as well as *D. vini* Zimmermann (from diseased wines). Finally, he accepts *Schiz. acidodevoratus* Chalenko.

Verona and Florenzano (1956) reviewed the theory that *Torulopsis* probably represents the imperfect stage of species of *Saccharomyces* and *Debaryomyces, Mycoderma* the imperfect state of *Pichia, Kloeckera* or *Kloeckeraspora* and *Hanseniaspora, Candida* of species of *Saccharomyces* and *Hansenula, Rhodotorula* or *Sporobolomyces, Cryptococcus* of *Bullera,* and so on. Thus *Candida pseudotropicalis* may be related to *Sacch. fragilis, C. macedoniensis* to *Sacch. marxianus,* and *C. pelliculosa* to *Hansenula javanica.*

There are many known strains of wine yeasts. Where their desirable characteristics can be maintained by proper culturing and storage, supplementary industrial classification may be useful. The environmental and genetic conditions governing variation are now being investigated actively in several laboratories.

The yeast population of the white wine producing region of Cortese in Piedmont was studied by Malan and Cano Marotta (1959). The initial must population was 44 per cent *Kl. apiculata,* 40 per cent *C. pulcherrima,* and 12 per cent *Sacch. rosei.* In nonsulfited musts *Sacch. rosei* carried on much of the fermentation. During the early stages of fermentation of sulfited musts these were succeeded by *Sacch. chevalieri, Sacch. uvarum,* and mainly by *Sacch. cerevisiae* var. *ellipsoideus.* In some instances of high alcohol content *Sacch. italicus* and *Sacch. oviformis* were present. Cells of *Kl. apiculata* survived fermentation. For some wines *Sacch. uvarum* appeared to be the yeast associated with the final stages of fermentation.

Minárik and Kocková-Kratochvílová (1966) suggested that the usual criteria for alcohol production, acid formation, and residual sugars were of little value in the classification of strains of *Sacch. cerevisiae* var. *ellipsoideus* and related species. This seems obvious when one considers the wide range of strains and types which can be successfully used in fermentation.

Variations are known to occur in the amounts of by-products, such as succinic acid, glycerin, higher alcohols, and esters, produced by different strains of wine yeasts under similar conditions. (See Wahab *et al.*, 1949; Fornachon, 1950.) Differences in fermentation rate, fermentation efficiencies, flocculation or aggregation, and requirements for accessory growth factors are also recognized. Furthermore, important differences exists in the ability of different strains to utilize organic acids. Castan (1927) found three strains of wine yeast that utilized the following amounts of organic acids from a yeast-extract medium in 60 to 83 days at 22°C:

| | Percentage of acid utilized by | | |
Acid added[a]	*Malvoisie flétrie 36*	*Montibeux 29*	*Cortaillod 15*
Acetic	2.67	3.74	28.50
Succinic	2.75	4.58	1.20
Lactic	0.00	22.00	83.50
Malic	7.20	4.63	42.8
Tartaric	0.00	3.40	29.3
Citric	12.46	2.45	50.0

[a] Initial acid content in range of 0.5 to 1.0 per cent.

During fermentation an increase in total fixed acidity (0.1 to 0.4 gr per 100 ml) is often noted. A small amount of acetic acid is formed. Thoukis *et al.* (1965) demonstrated that the main increase is due to two organic acids: succinic (90 per cent) and lactic (10 per cent). This assumes that no malo-lactic fermentation occurred. Succinic acid is apparently formed mainly from the glyoxylate cycle and not from glutamic acid. The increase was not influenced by three commercial California yeast strains. The fermentations appear to have been anaerobic. Ribéreau-Gayon and Peynaud (1946) showed that the major increase occurs in the early stages of the fermentation. This was confirmed by Thoukis and co-workers. Thoukis *et al.* (1965) also showed that the pH did not significantly influence the amount of acids formed, but the percentage increase is greater in musts of high pH (low initial total acidity). They believed that the value of the succinic acid to the quality of wines has not been fully appreciated. Not only is it valuable for its own acid taste, with its salty-bitter character, but also because it may form esters.

Using a variety of species of yeasts, Iñigo Leal and Bravo Abad (1963) reported significant differences in the percentage of fixed acidity of new wines. *Candida pulcherrima* appears to form fumaric acid. They also found important differences in the rate of formation and utilization of volatile acids during fermentation by the different species of yeasts. As gas chromatographic procedures are more regularly used we may expect more information in this field.

New information on the formation of malic, tartaric, and succinic acids by different yeasts has been given by Drawert *et al.* (1965a, 1965c). They conclude that malic and tartaric acids were regular products of yeast growth and that succinic acid usually is. Glucose was the source.

Besides the early work of Castan (1927), previously reported (p. 322), Peynaud (1939–40) showed that yeasts metabolized from 10 to 24 per cent of the *l*-malic acid of musts. Rankine (1966a) found *Saccharomyces* yeasts that utilized up to 43 per cent. The amount metabolized was inversely proportional to the pH of the must. He confirmed the complete utilization of *l*-malic acid by added *Schiz. malidevorans* and no pH-dependence was observed. While this yeast was not overgrown by *Saccharomyces* sp. as *Schiz. pombe* is overrun, the strain of *Schiz. malidevorans* employed in his study produced hydrogen sulfide in excessive amounts. Rankine suggests that yeasts may be selected for their minimum or maximum utilization of *l*-malic acid for use on musts of low or high malic acid, respectively. (See pp. 305–307.)

Peynaud *et al.* (1967a, 1967b) showed that species of yeast formed primarily D(−) lactic acid with a small percentage of L(+). This amounted to not over 10 per cent of the L(+) isomer for *Saccharomyces* and up to 20 per cent for *Pichia* sp., *Hansenula* sp., and others. The amount of L(+) formed was about the same in musts as in synthetic media and under aerobic or anaerobic conditions. Likewise, pH, temperature of fermentation, amount of sugar fermented, or presence or absence of vitamins did not influence the amount of L(+) formed. Eleven of sixteen cultures of *Sacch. veronae*, on the contrary, formed large amounts of L(+), sometimes exclusively, and the other five produced moderate amounts of L(+). This was more evident on synthetic media under anaerobic conditions. This specific characteristic of *Sacch. veronae* can, they believe, be used as a diagnostic test for the species.

In Australia Rankine (1953) showed that the ethyl alcohol production was high in yeasts isolated from old compared to new viticultural regions. This one would expect by adaptation. There were differences in volatile acid and glycerin production at different temperatures, but the volumes used were small and oxidation effects may have interfered. One important result of Rankine's study was to show that acetaldehyde production was higher for two yeast strains, particularly at 15°C (59°F), than from other strains.

Yeasts occurring naturally on grapes

Very probably, in the districts of Europe where wine making has been practiced for centuries, wine yeasts especially adapted to bringing out the best qualities of the variety of grape grown in those districts are found on the grapes in predominating numbers during the later stages of ripening. The composition of the musts may also exert some selective effect. At any rate, in normal years, sound wines can be made in those regions merely by a light sulfiting of the musts without the addition of yeast. In many of the wine-making districts, when yeast starters are used—as in years of unfavorable weather conditions when the natural microflora is defective—they are prepared from the yeasts indigenous to the region.

In general the types of yeasts found on grapes and in wines throughout the world are remarkably similar. There are, however, distinct differences in the proportion of each yeast in different regions. The yeasts found under almost all conditions are *Sacch. cerevisiae* var. *ellipsoideus* (often incorrectly labeled *Sacch. ellipsoideus* or *Sacch. vini*) and *Kl. apiculata*. Climate seems to be one factor influencing the flora. For example, in Italy Castelli (1954) has shown that *Kl. apiculata* predominates over *Sacch. cerevisiae* var. *ellipsoideus* in the cooler northern region, about equals it in the warmer central area, whereas in the hot southern area *Kl. apiculata* is replaced by *Hanseniaspora* sp. as the predominant yeast on the fruit. Domercq (1957) believes that this may be due to adaptation of the yeasts to different levels of sugar, but temperature affects more than the sugar level of the fruit. Castelli and Del Giudice (1955) found more sporogenous types at the lower elevation of Mount Etna in Sicily, while asporogenous types predominated at the higher levels.

Several investigations of the yeasts occurring naturally on grapes and present in the new wine have been made, but most of these have not been complete. The yeasts described in the older publications were often poorly defined and usually their identity is uncertain. The origin of the yeasts (presumably the soil in which they may survive during winter or other periods of unfavorable conditions), the mode of transfer to the grapes in the vineyard (presumably by bees or other insects), and the type and succession of flora on the grapes and in the wine are still largely unknown.

Among the early complete investigations of the yeasts isolated from grapes, must, and wine were those of Ciferri and Verona (1941), Castelli (1938, 1939a, 1939b, 1947a, 1947b, 1947c, 1948a, 1948b), and Florenzano (1949) for Italian conditions, and Mrak and McClung (1938) for California conditions. Some of these studies were limited to the description of the genera and species isolated, give no quantitative data on relative numbers present, and do not describe their effect on the quality of the resulting wine. Castelli, however, gives some interesting data on the relative importance of the yeasts found on the grapes of several regions of Italy. In Sicilian musts Castelli and Del Giudice (1955) reported that 100 per cent of the musts contained *Sacch. cerevisiae* var. *ellipsoideus*, 75 per cent *Sacch. chevalieri*, 67 per cent *Sacch. oviformis*, 58 per cent *Hanseniaspora guilliermondii*, 50 *Kl. apiculata* and *Sacch. italicus*, 29 *Sacch. carlsbergensis*, 25 *Sacch. rosei*, 16 *Sacch. elegans*, 8 *Sacch. unisporus*, *Sacch. fructuum*, *Sacch. rouxi*, *Sacch. marxianus*, and *Candida pulcherrima*, and 4 *Sacch. fermentati*, *Sacch. heterogenicus*, *Sacch. bisporus*, and *C. utilis*. The high percentage of musts with *Sacch. oviformis* and *Sacch. chevalieri* is noteworthy. They also reported local differences in yeast distribution. Most of the *Sacch. carlsbergensis* was found in the western part of the island. Of 335 cultures, only 46 were asporogenous and all but four of these were *Kloeckera* sp. At the beginning of the fermentations *Kloeckera* sp. and *Hanseniaspora* sp. predominated.

Stevič (1963) confirmed the usual result that few or no yeasts occur on green grapes (see, in addition, p. 330). Contrary to some reports, very few yeasts were found on ripe grapes. Stevič believes that bees and wasps are important vectors of yeasts, and suggests that bees might be used for spreading desirable yeasts throughout a viticultural area.

Italian and Bordeaux fermentations generally start with nonspore-forming yeasts. At Bordeaux fermentation starts with *Kl. apiculata* for the red grapes and *Kl. apiculata* and *T. bacillaris* for the white grapes. These are rapidly succeeded by spore-forming yeasts such as *Saccharomyces* sp. and *Sacch. rosei* so that soon no nonspore formers remain. *Sacch. oviformis* predominated in fermenting high-sugar musts. It was found in only about 2 per cent of red and 8 per cent of white musts. Domercq (1956) considers *Sacch. oviformis* to be particularly important in Bordeaux wines, especially as a cause of spoilage of sweet table wines. *Saccharomycodes ludwigii* and *Sacch. acidifaciens* were rarely found in Bordeaux musts, but are common in white wines, where they appear to be one cause of spoilage (cloudiness).

Brèchot et al. (1962) found species of *Kloeckera* in only 13.4 per cent of the fermenting musts of Beaujolais. In contrast, they were found in 58.2 per cent of the Bordeaux wines in Domercq's (1956) study and the 76.4 per cent in Italian wines reported by Castelli (1954). *Sacch. cerevisiae* var. *ellipsoideus* was the dominant yeast, followed by *Sacch. steineri*, in the study of Bréchot *et al.* In contrast to Domercq's (1956) studies, *Sacch. oviformis* was rare in Beaujolais. The microflora of Beaujolais was relatively rich in *Hansenula* sp., *Brettanomyces* sp., *Candida* sp., *Endomyces* sp., *Rhodotorula* sp., and *Torulopsis* sp.

Domercq (1956) reported *Torulopsis bacillaris* only from botrytised grapes. *Sacch. oviformis* was found in less than 2 per cent of the red musts but in nearly 8 per cent of the whites. *Sacch. heterogenicus, Sacch. acidifaciens, Sacch. elegans,* and *Saccharomycodes ludwigii* were not found on red grapes, but were present in whites.

Peynaud and Domercq (1953) were unable to find any distinctive difference between the flora of the finest Bordeaux château and those of lesser-quality vineyards. They stated: "Dans l'état actuel de nos connaissances, il est difficile de dire si ces levures impriment au vin un caractère particulier, et si leur intervention peut, pour certaines, être considérée comme un facteur de qualité." They reported that the fermentation by-products and sensory quality were essentially the same in various species of *Saccharomyces*. The other yeasts, *Kloeckera* sp., *Pichia* sp., *Hansenula* sp., and so on, ferment only a small amount of the sugar and this slowly. They are, however, often found in finished wines. (See pp. 788, 790.)

The yeasts of grapes, musts, and wines were studied by Minárik *et al.* (1960) and Minárik (1964*a*, 1964*b*). In musts *Sacch. cerevisiae* var. *ellipsoideus* occurred in 96.8 per cent, *Sacch. oviformis* and *Sacch. uvarum* in 20.8, *Sacch. rosei* in 12.5 *Sacch. pastorianus* in 9.3, *Sacch. bayanus, Sacch. carlsbergensis,* and *Hansenula anomala* in 5.2 each, *Sacch. heterogenicus, Sacch. veronae, Sacch. acidifaciens,* and *Pichia fermentans* in 3.1 each, *Sacch. chevalieri, Sacch. exiguus, Sacch. steineri, Sacch. italicus,* and *Sacch. elegans* in 2.0 each, and *Sacch. willianus* and *Sacch. fructuum* in 1.0 each. Whereas no musts failed to have at least one of the sporogenous yeasts, only 72.9 per cent had asporogenous yeasts. In musts the asporogenous yeast *Kl. apiculata* was found in 64.5 per cent, *C. pulcherrima* in 28.1, *Torulopsis stellata* (probably *T. bacillaris*) in 3.1, *T. bacillaris* in 2.0, and *C. mycoderma* and *Kl. africana* in 1.0 each.

In wines at the end of the fermentation 98 per cent contained sporogenous yeasts of which *Sacch. cerevisiae* var. *ellipsoideus* amounted to 80.1 per cent, *Sacch. oviformis* to 5.5 per cent, *Sacch. uvarum* to 3.1 per cent, *Sacch. acidifaciens* to 2.4 per cent, and *Sacch. pastorianus* to 2.1 per cent. Of the other thirteen yeasts listed above, seven were present at less than 1 per cent and six were absent. Asporogenous yeasts were found in only 1.8 per cent of the wines at the end of the fermentation. In 1.5 per cent of the cultures this was *Kl. apiculata* and in 0.3 per cent it was

C. pulcherrima. The other four asporogenous yeasts listed were absent. *Sacch. acidifaciens* and *T. bacillaris* are fructophiles, i.e., they ferment fructose faster than glucose. *Hansenula anomala* and *C. pulcherrima* ferment the two hexose sugars with about equal rapidity. All the others are glucophiles. The authors believe that more research of this type should be carried on, since it is likely to be especially fruitful in stabilizing wines and in producing new types of wines.

In Czechoslovakia, Minárik and Nagyova (1966*a*) reported qualitative differences between various regions in the yeast microflora. In one region *Sacch. carlsbergensis* was relatively abundant in musts and *Sacch. chevalieri* in the wines. In general, *Kl. apiculata* and *C. pulcherrima* characterized the initial fermentation, followed by *Sacch. cerevisiae* var. *ellipsoideus* and *Sacch. oviformis.* In new wines the microflora was characterized by *Sacch. cerevisiae* var. *ellipsoideus* (35 to 47 per cent), *Sacch. oviformis* (13 to 17), *Sacch. chevalieri* (4 to 11), *C. mycoderma* (21 to 28), and *C. zeylanoides* (2). They stress the importance of winery sanitation to prevent contamination by film yeasts, and particularly insisted on the need for maintaining strictly anaerobic conditions.

In Uruguay Cano Marotta and Bracho de Kalamar (1962–1964) reported *Sacch. cerevisiae* var. *ellipsoideus*, *Sacch. fructuum*, *Sacch. carlsbergensis*, *Sacch. chevalieri*, *Sacch. rosei*, *Sacch. oviformis*, *Pichia membranaefaciens*, *Candida mycoderma*, *C. krusei*, and *Kl. apiculata* on the grapes in one region. In another only *Sacch. cerevisiae* var. *ellipsoideus*, *Saach. carlsbergensis*, and *Kl. apiculata* were found—probably, they believe because it is a region where grapes have been grown for only ten years. Their cultures were very sensitive to high temperatures.

A number of studies on the yeast flora of South African grapes and wines have been made: Van der Walt and Van Kerken (1958, 1960), Van Zyl and Du Plessis (1961), and Van Zyl (1962). The notable part of these reports is the widespread occurrence of *Brettanomyces* in that region: *B. intermedius* Krumbholtz et Tauschanoff and *B. schanderlii*. Van Zyl and Du Plessis (1961) reported that on vines and ripening grapes the most important species were *Kl. apiculata*, *Rhodotorula glutinis*, and *C. krusei* in that order of incidence. Very few *Saccharomyces* were found. In grape musts, however, *Sacch. cerevisiae* var. *ellipsoideus* and *Sacch. oviformis* predominated. The near absence of *Kl. apiculata* in fermenting musts they attribute to the general use of sulfur dioxide in South Africa. The rare occurrence of *Sacch. rosei* is also attributed to this practice. Species of *Schizosaccharomyces*, *Hanseniaspora*, *Pichia*, *Candida*, and *Kloeckera* were also found. Although *Sacch. cerevisiae* var. *ellipsoideus* and *Sacch. oviformis* occurred in wines, they were remnants from the alcoholic fermentation. The main spoilage organisms, besides the *Brettanomyces* species indicated above, were *Sacch. acidifaciens, Saccharomycodes ludwigii, Pichia membranaefaciens,* and *P. fermentans.*

The microflora of California grapes and wine has not been studied in detail and as systematically as it deserves. Holm (1908) described incompletely several yeasts from samples of California grapes and concluded that yeasts found on grapes produced in regions remote from wineries form low amounts of alcohol and produce films, turbidity, and unpleasant flavors and tastes. Cruess (1918) described species of six genera of yeasts obtained by him from California grapes. The true wine yeast was never abundant on grapes he examined, and was usually outnumbered many thousands of times by injurious microörganisms. Furthermore, as Cruess and others have found, most strains of wine yeast present on California grapes ferment certain musts only incompletely, give a lower yield of alcohol, and, while fermenting, form a finely divided cloud throughout the liquid. After settling, when the fermentation is complete, such yeast is easily disturbed and clouds the wine.

Mrak and McClung (1938) found a wide variety of yeast genera on California grapes and grape products. In the 241 cultures they described, the most common genera were *Saccharomyces, Candida, Torulopsis,* and *Kloeckera.* The ratio of *Saccharomyces* to *Zygosaccharomyces*[3] (diploid to haploid) isolates was 7 to 1. Eight genera of sporulating yeasts were found.

The fruit fly *Drosophila melanogaster* is extraordinarily sensitive to differences in yeast strain (Wolf and Reuther, 1959). Benda and Wolf (1965) were thus able to isolate two strains of *Sacch. cerevisiae* var. *ellipsoideus.* One appeared to be a slow-growing haploid and the other a diploid. The haploid strain, surprisingly, had a higher fermenting power. There were slight differences in the vitamin requirements of the two strains. The haploid strain produced more glycerin but less volatile acid than the diploid strain. This type of research appears very profitable. To differentiate strains of *Schiz. pombe* from those of *Schiz. malidevorans,* Wolf and Benda (1965, 1966) successfully used *D. melanogaster* to distinguish these yeasts.

The winery has a microflora of its own, according to Peynaud and Domercq (1959a), as the following data indicate:

Sampling location	Species and number
Outside of tanks	*Sacch. oviformis* (6), *Sacch. cerevisiae* var. *ellipsoideus* (3), *C. mycoderma* (13), *Pichia* sp. (1)
At bungs	*Sacch. elegans* (7), *Sacch. acidifaciens* (2) *C. mycoderma* (2)
Bottling equipment	*Sacch. oviformis* (4), *Sacch. acidifaciens* (4), *C. mycoderma* (10), *Brettanomyces* sp. (5)
Floors of cellars	*Sacch. cerevisiae* var. *ellipsoideus* (1), *Pichia* sp. (7), *C. mycoderma* (7)

[3] Now *Saccharomyces.*

A natural or spontaneous fermentation caused by the yeast carried into the vat on the skins of the grapes occurs in at least two stages. In the first stage a wild yeast, *Kl. apiculata* (formerly incorrectly called *Sacch. apiculatus*), is most numerous. This yeast ceases to ferment when the must contains about 4 per cent alcohol; it then gives way to the true wine yeast, *Sacch. cerevisiae* var. *ellipsoideus*, which completes the second stage.

During the early stages of fermentation of Czechoslovakian musts Minárik (1965) found that *Kl. apiculata* and *C. pulcherrima* constituted 75 per cent of the yeast flora (the latter being especially notable in red wine fermentations). At the end of the fermentation 94 per cent consisted of *Sacch. cerevisiae* var. *ellipsoideus*, *Sacch. oviformis*, *Sacch. carlsbergensis*, and *Sacch. chevalieri*. The latter two were believed to cause the clouding of some wines. The main surface yeast was *C. mycoderma*, but *C. zeylanoides* and *Pichia* sp. were also found. Minárik and Rágala (1966) reported much more *Sacch. cerevisiae* var. *ellipsoideus* in Czechoslovakian vineyards in a warm year (1959) than in a cool year (1958), but slightly less *Kl. apiculata*. During the early stages of fermentation *Kl. apiculata* was the dominant species in New Zealand, according to Parle and di Menna (1966). Thereafter *Sacch. cerevisiae* var. *ellipsoideus* dominated. Other species reported were *Sacch. chevalieri* and *Sacch. rosei*.

Habala and Švejcar (1965) classified the yeasts involved in spontaneous alcoholic fermentation into four chronological groups: *Kl. apiculata* and *C. pulcherrima* followed by *Sacch. cerevisiae* var. *ellipsoideus* and occasionally *Sacch. oviformis*. Later the main fermentation may also include *Sacch. uvarum* and *Sacch. pastorianus*. *Sacch. oviformis* is especially typical of the after-fermentation. Film yeasts, *C. mycoderma* and *C. zeylanoides*, may develop after the main fermentation, and *T. bacillaris* may develop later.

The general opinion on the contribution of *Kl. apiculata* given by Domercq (1957) is that this yeast disappears soon after the start of the fermentation and probably adds little to the quality of the wine. Further research is certainly necessary. Preliminary data of Kunkee and Amerine (1965) indicate that "successive" culture of yeast strains might prove profitable, with *Kl. apiculata* followed by a *Saccharomyces* sp., but many problems remain, particularly in preventing early contamination and over-growth by *Saccharomyces* sp.

Brettanomyces sp. do not seem to be of any value to wines. Schanderl and Draczynski (1952) found the species in German sparkling wines; Barret *et al.* (1955) found it in deposits in Arbois wines; Galzy and Rioux (1955) showed that *B. bruxellensis* was associated with a film yeast spoilage of Midi wines. See page 327 for its occurrence in South Africa, where it causes undesirable turbidity.

From Brazilian grape isolates Kreger-Van Rij (1964) identified *Endomycopsis vini,* a new species. In contrast to the usual results, Shimatani and Nagata (1967) reported *Penicillium* sp. and *Aspergillus* sp. in Japanese vineyards during the growing season. During the harvest season *Pullularia pullulans* was found throughout the vineyard, and from damaged grapes *Kl. apiculata, C. mycoderma, C. krusei, P. membranaefaciens, P. fermentans, Torulopsis famata,* and *Cryptococcus laurentii* (as well as two unidentified species). No *Saccharomyces* were found in the vineyard. During the crushing season, in wineries and in wines, they reported *P. membranaefaciens, C. mycoderma, C. krusei, C. guilliermondii, C. parapsilosis, C. pulcherrima, Sacch. oviformis, Sacch. mellis,* and *H. anomala.* About 80 per cent of the isolates were the first three.

During the fermentation of Jerez de la Frontera musts, Iñigo Leal *et al.* (1963) found thirteen species, but *Sacch. cerevisiae* var. *ellipsoideus, Sacch. italicus,* and *Sacch. mangini* (probably *Sacch. chevalieri*) were the yeasts primarily responsible for alcoholic fermentation. *Sacch. oviformis* was found in only one of twenty musts in contrast to its appearance in half the musts from Montilla (a district to the east). *Saccharomycodes ludwigii* and *Sacch. delbrueckii* were reported for the first time in Spanish musts. *Candida utilis* also appeared frequently. It is notable that flor films appeared within ten to twelve days of the completion of alcoholic fermentation. Schanderl (1959) reported that film formation under aerobic conditions is characteristic of a number of yeasts. Such rapid film development does not normally occur in other wine regions.

Schizosaccharomyces pombe has a normal haploid vegetative phase with only a transitory diploid state, whereas *Sacch. cerevisiae, Sacch. carlsbergensis, Sacch. chevalieri,* and *Sacch. italicus* are diploid in the vegetative phase. As Mortimer and Hawthorne (1966) point out, chromosome maps of *Sacch. cerevisiae* have progressed so rapidly that they now exceed 300 centimorgans and contain about 100 genes. About fifty of the genetic loci are identified with particular enzyme steps.

It is important to realize that many of these species are found in an occasional must and some have been isolated only once or twice. They cannot, therefore, be considered of great importance to the grape or wine industry.

A summary of the occurrence of yeasts is given in table 42.

TABLE 42

Yeasts Reported on Grapes and in Wines [1]

Genus and species	Notes and where reported
Anthoblastomyces	
campinensis	in Brazilian grape flowers
cryptococcoides	in Brazilian grape flowers
saccharophileas	in Brazilian grape flowers
Brettanomyces	
bruxellensis	in musts and wines
bruxellensis var. *lentus*	differs in rate of growth
bruxellensis var. non-membranaefaciens	in a must
claussenii	
custersii	in musts and wines
intermedius[b]	in South African wines
italicus (see *T. bacillaris*)	
lambicus	in musts and wines
patavinus	in musts and wines
schanderlii[b]	in wines
vini[d]	in bottled wines
Candida	
albicans	
boidini	
brumptii	in wines, rare
guilliermondii[a]	in a dry table wine and in winery
guilliermondii var. membranaefaciens	from must
ingens	
intermedia var. *ethanophila*	
krusei	common in musts and in wines
lipolytica	from fermenting must
melinii	in musts and wines, but rare in Greek and S. African musts
mycoderma	(incorrectly *Mycoderma vini*) common; the perfect form of *Pichia membranaefaciens*
parapsilosis	in a cloudy wine, in winery, and in musts and wines
pelliculosa	from must
pulcherrima	common in musts and wines; found on grapes
reukafii	from a white Sicilian wine
rugosa[c]	found in a California wine; rare
scottii[f]	in Czechoslovakia; on green grapes from New Zealand
solani	in Greek musts
sorbosa	
stellatoidea	in S. African musts; on grape flowers
tropicalis	isolated from a film on grape juice; in musts and wine
utilus	in Spanish musts
vinaria[e]	in Czechoslovakia
zeylanoides[e]	reported in wines
Cryptococcus	
albidus	in wine, rare
diffluens[a]	found on green grapes in New Zealand and in Sardinian must
laurentii	in wine, including a fortified wine, but rare
luteolus[a]	reported in a table wine

Genus and species	Notes and where reported

Debaryomyces
  dekkeri
  globosus[e, 2]   on grapes and in grape juice
  hansenii[d]   in must and wine
  kloeckeri[c]   very rare; reported once in California
  kursanovi   in German musts
  nicotianae
  subglobosus   from wine
  vini   in a spoiled wine

Endomycopsis
  vini   isolated in Brazil

Hanseniaspora
  apuliensis   in Italian musts; in wine (Galzy rejects the species)
  guilliermondii   now *valbyensis*
  uvarum
  valbyensis   in wine, especially in warm countries
  vineae   from vineyard soil

Hansenula
  anomala   common; the perfect stage of *Candida pelliculosa*
  saturnus   in wine made from late-harvested grapes
  schneggi   in wine
  suaveolens   in wine
  subpelliculosa   in musts

Issatchenkia
  orientalis

Kloeckera
  africana   fairly common on grapes and in musts and wines
  apiculata   very common but produces only 4 per cent alcohol
  corticis   in musts and wines; rare
  jensenii   in wines in Sardinia
  magna   in musts in Italy and Spain

Pichia
  alcoholphila   (probably same as *membranaefaciens*)
  farinosa[c]   in wine
  fermentans   on grapes and in wine; uncommon
  membranaefaciens   relatively common in musts at start of fermentation or
    in films
  silvestris   from wine

Rhodotorula
  aurantiaca[c]   very rare
  glutinus   rarely reported on grapes, musts, and wines
  minuta[f]   from green grapes in New Zealand
  mucilaginosa   in dry and sweet table wines; in Greek musts
  pallida[c]   from southern Italy
  rubra[c]   from musts
  vini[d]   from wine, rare

Saccharomyces
  acidifaciens   in spoiled wines and on grapes, resistant to sulfur
    dioxide (a fructophile)
  bailii[a, c]   in South African and Italian dry table wine (formerly
    Zygosaccharomyces)
  bayanus   on grapes and in musts and wines; common but low
    frequency
  bisporus (possibly *Sacch.*
    oviformis)   on grapes; low alcohol yield
  capsensis   from wine

Genus and species	Notes and where reported
Saccharomyces (cont.)	
carlsbergensis	on grapes and less often in wine (used for lager beer fermentation)
cerevisiae	probably includes the variety
cerevisiae var. ellipsoideus	the classical wine yeast and possibly the most widely distributed
chevalieri	(synonymous with *Sacch. mangini*) in many wines and occasionally in grapes
corneanus[e, 2]	rare, from grapes
delbrueckii[d]	in must and wine (formerly *Torulaspora*)
delbrueckii var. mongolicus	in German musts and Sicilian wine
elegans[d]	in sweet table wine (a fructophile); in French, Italian, and Spanish wines
eupagicus	
exiguus	originally found in pressed yeast, later in grape juice and wine
fermentati	believed the same as *S. beticus* (formerly *Torulaspora*); common in musts
florentinus	in musts and wines
fructuum	in musts and in grape juice
globosus[e, 2] (same as *Sacch. delbrueckii*)	on grapes and grape juice and wine, rare
heterogenicus	on grapes, in musts, and in grape juice, but rarely
italicus	on grapes and in grape juice, especially from warm climates
kluyveri	in Greek musts and wines from must
marxianus	sporogenous form of *Candida macedoniensis*; from must
mellis[g]	in Japanese wines and wineries; originally from honey (formerly *Zygosaccharomyces*)
microellipsoideus	in musts and wines
montuliensis	
oviformis	on grapes, in musts, grape juice, and wines; common
pastorianus	on grapes and frequently in Loire fermentations
prostoserdovi	
rosei	common on grapes and in wines (formerly *Torulaspora*)
rouxii	originally *Zygosaccharomyces*; from over-ripe grapes, musts, and wines
rouxii var. polymorphus	in German musts
steineri (same as *Sacch. italicus*)	in wine and on grapes
transvalensis	in Greek musts
unisporus	
uvarum	in musts and less frequently in wines
vanudenii	
veronae	in *Drosophila* and in musts and wines
willianus	found more in beer than musts
Saccharomycodes	
bisporus	in Italian wine, probably a variety of the following
ludwigii	from grape juice and wine; common; resistant to sulfur dioxide
Schizosaccharomyces	
acidodevoratus	
octosporus	originally found in a must from sulfited currants; from raisins
pombe	ferments malic acid; from musts, muté and wine

Genus and species	Notes and where reported
Schizosaccharomyces (cont.) pombe var. liquefaciens versatilis	originally isolated from home-canned grape juice; from raisins
Sphaerulina intermixta	
Torulopsis anomala bacillaris	in Czechoslovakian musts and wines (synonymous with T. stellata) (a fructophile) in musts and wine
behrendi	in Greek musts
burgeffiana	from grapes (Benda, 1962)
candida[a]	in sweet table wine
cantarelli	from grape musts
capsuligenus	from a winery culture
colliculosa	in must and wine
domercquii	
famata	in musts and wines
glabrata	in musts and wines
inconspicua	in a S. African winery; on musts and wines
molischiana	from Israeli must
pulcherrima	
vanzylii	from a refrigerated cellar floor
versatilis	in Czechoslovakian musts and wines
Trichosporon fermentans[c] hellenicum	in musts or on diseased grapes in Greek musts
intermedium	in musts
pullulans	in musts
veronae	in musts

[1] Verona and Florenzano list eight other yeasts for which evidence is not conclusive.
[2] These species not recognized by Lodder and Kreger-Van Rij (1952).
Source of data: Verona and Florenzano (1956) and Castelli (1965), except [a] from Van Zyl and Du Plessis (1961); [b] Van der Walt and Van Kerken (1958); [c] Galzy (1956); [d] Domercq (1956); [e] Czechoslovak Collections of Micro-organisms (1964); [f] Parle and di Menna (1966); [g] Shimatani and Nagata (1967).

USE OF PURE YEASTS . . .

*. . . is required because of lack of desirable yeasts
and presence of undesirable yeasts and bacteria
in California vineyards.*

The use of selected pure cultures of yeasts is common in wine making, particularly in conjunction with the control of fermentation by the use of sulfur dioxide, in order to render conditions favorable to the growth of the desired organisms and unfavorable to the growth of others. This practice, when properly used, results in a fermentation that begins promptly, proceeds regularly, and goes to completion in a relatively short time. A more complete utilization of the sugar occurs, which assures a better preservation of the product and an increased yield of alcohol.

Authorities agree that the fermentation should be conducted so as to favor the wine yeasts; but not all agree on the proper method of accomplishing this result. Some enologists have maintained that fine wines could be obtained by inoculating ordinary musts with yeasts derived from the better wine districts. Although a particular strain of wine yeast may have some inherent flavor-producing quality, it is as idle to think that a fine Burgundy wine can be made by fermenting Alicante Bouschet must with Burgundy yeasts as to think that a high-quality Sauternes can be made from Burger or Concord grapes. However, beer top yeasts or ale yeasts (commonly called *Sacch. cerevisiae* strains) certainly give a cereal or yeasty flavor to grape musts, whereas wine yeasts (*Sacch. cerevisiae* var. *ellipsoideus* strains, but see below) produce a fruity or vinous flavor even in sweet wort. The various strains of *Sacch. cerevisiae* var. *ellipsoideus* differ in the amount and type of desirable vinous flavors they produce, in their fermentation rate, fermentation efficiency, rate of settling, foaming tendency, and so forth. Windisch (1906) has reviewed the early literature and concludes that one cannot change the varietal character of a must by using a special yeast. He suggests, however, that fruit wines do seem to have more of a wine character when wine yeasts are used, and that the wine yeasts of a certain region may give a somewhat similar character when used to ferment musts of other regions. But he believes that the value of a pure yeast is in producing clean fermentation. Fornachon (1950) also indicates that too much should not be expected of pure yeasts, and that a pure culture alone will not produce a high-quality wine from poor grapes.

Use of single strains

The relative desirability of conducting fermentations with a single strain of selected wine yeast (*Sacch. cerevisiae* var. *ellipsoideus*) in comparison with fermentations conducted by mixed flora has long interested European enologists (Ventre, 1935; Renaud, 1939–40). Renaud believed that wines of quality can be made only by relying upon the natural mixed flora of the grapes. He feels that the use of selected yeasts tends to produce wines of uniform flavor without those subtle and valuable differences so much appreciated by the connoisseur. However, ordinary wines are considerably improved if their fermentation is carried out with selected yeasts of good quality. (See also p. 318.)

The role of yeasts in the production of aromatic compounds is emphasized by the work of Suomalainen and Nykänen (1964). They fermented an artificial medium and identified the odor components by gas chromatography, after separation. There was a marked similarity in the compounds identified and those found in brandy and whiskey: 2-methyl-1-propanol, 3-methyl-1-butanol, 2-phenethyl alcohol, ethyl esters of acetic, caproic, caprylic, capric, lauric, and palmitic acids, isoamyl caprylate, isoamyl caproate, isoamyl caprate, isoamyl acetate, hexanol, and an unidentified compound, probably an ester of some fatty acid of high molecular weight.

The possibility exists that differences in wine quality of adjacent blocks of grapes may be due to different strains of yeast. (See p. 328.) Webb and Kepner (1961), for example, found three strains of *Sacch. cerevisiae* var. *ellipsoideus* with varying ability to produce higher alcohols (data in weight per cent based on these four alcohols only):

Strain	1-Propanol	2-Methyl-1-propanol	2-Methyl-1-butanol	3-Methyl-1-butanol
Burgundy	18.2	12.4	12.0	57.4
Jerez	20.2	8.4	4.5	66.9
Montrachet	2.6	2.7	16.5	78.2

Van Zyl *et al.* (1963) reported different strains of yeast producing more or less of ethyl caprylate.

A large (8-fold) influence of yeast strain on hydrogen sulfide formation was emphasized by Rankine (1963). Not only do strains of wine yeast (like beer yeasts) differ remarkably in their ability to produce hydrogen sulfide, but Zambonelli (1964a) was able to show that hybridization of positive and negative strains was possible. The F_1 progeny were positive producers of hydrogen sulfide, but he did not obtain (1964b) any clear indication of the quantitative aspects of the genetic inheritance (see also 1965a, 1965b, 1965c). He found that 96 of 100 strains produce amounts of hydrogen sulfide varying from traces to 200 γ per 50 ml.

Cantarelli (1964) found more sulfide in wines fermented with sulfur dioxide using *Sacch. cerevisiae* var. *ellipsoideus* and *Sacch. rosei* than in those fermented without sulfur dioxide. However, with *Sacch. pastorianus* and *Sacch. oviformis* the opposite was true. Of these yeasts the most sulfide was formed by *S. oviformis* in the absence of sulfur dioxide and the least sulfide with the same yeast in the presence of sulfur dioxide. However, the differences were not sufficient, in Cantarelli's opinion, to justify selecting one yeast over another.

To reduce sulfide problems Cantarelli recommended fermenting with the least possible free sulfur, to prevent too rapid a fermentation, to aerate the new wine, to prevent contact with iron, aluminum, or zinc, and adding sulfur dioxide to the new wine. He also suggests the use of silver metal (illegal in this country) and of activated charcoal. Ion-exchange resins may also prove useful. Use of selected yeasts was recommended by

Rankine (1963). For data on the biochemistry of hydrogen sulfide formation see Colloque sur la Biochimie du Soufre (1956).

Fornachon (1950) is inclined to the use of single strains of yeast as starters, but notes that they must be carefully prepared and properly used for best results. Sisani (1948), however, in making the sweet table wine of Tuscany (*vino santo*), claimed that, although the alcohol yield of *Sacch. cerevisiae* var. *ellipsoideus* was superior, a mixed culture gave a better flavor. The yeast flora of the mixed culture was mainly *Sacch. ellipsoideus* and *Sacch. bayanus*, but *Sacch. chevalieri* and *Sacch. exiguus* (Reess) Hansen were almost as common as the latter. (See also Mayer and Pause, 1965b.)

Rankine and Lloyd (1963) showed that when 2 per cent of a pure yeast culture was used the added yeasts dominated the fermentation in the presence or absence of skins. However, the dominance was significantly more marked in the absence of skins. They reported "significantly better" wines made with a starter than without, but no detailed sensory data were presented.

Rankine (1968a) summarized data obtained in Australia over a period of twenty years on the importance of the variety and strain of yeasts used on the composition and quality of wines produced. Over twenty individual cultures of yeast were investigated, both in laboratory and pilot winery fermentations. The physical aspects of fermentation investigated included cell counts, flocculation and clarification, and oxidation-reduction potential. The physiological and metabolic studies included resistance to sulfur dioxide and other inhibitors, and metabolism including ethanol production, production of higher alcohols, decomposition of malic acid, production of hydrogen sulfide, and formation of sulfur dioxide binding compounds. Dominance of yeasts in mixed cultures and influences of yeasts on flavor and aroma were studied on both pilot winery and large laboratory scales. Rankine emphasizes that in addition to the more obvious advantages of the use of selected yeasts (rapid and predictable onset of fermentation, its evenness and completeness and absence of undesirable aromas and flavors) this can also influence flavor directly and control reduction in acidity during fermentation.

In contrast to the generally reported higher yield of alcohol with *Sacch. oviformis* in Europe, Ferreira (1959) found better yields with *Sacch. cerevisiae* var. *ellipsoideus* (Montrachet strain) at 30°C (86°F). A Burgundy strain was about equal in alcohol production to that of *Sacch. oviformis*, *Sacch. fragilis*, and *Schiz. pombe*. Least satisfactory was *Sacch. marxianus*. The volatile acidity production of the last three yeasts was higher, especially that of *Sacch. marxianus*. The slowest fermenter was *Sacch. fragilis*. At the higher temperature (30°C) the reduction in alcohol yield was especially large (about 50 per cent) for *Sacch. oviformis, Sacch. fragilis*, and *Sacch. marxianus*. We cannot explain the low result with *Sacch. fragilis*, which is generally considered a thermophile. It is of interest that *Sacch. oviformis* was not found by Tarantola and Gandini (1966) to have a higher alcohol-forming property than *Sacch. cerevisiae* var. *ellipsoideus*.

Fructose is much sweeter than glucose. During alcoholic fermentation glucose normally ferments faster than fructose. Several yeast strains are known which ferment fructose faster (see Gottschalk, 1946; Koch and Bretthauer, 1960). The glucose-fructose ratio can be increased by sugaring with partially fermented wine, sucrose, or grape juice. This gives the wine more extract and a less sweet taste. Use of selected yeasts could result in sweeter wines at the same total sugar content.

Yeasts fermenting glucose more rapidly than fructose were identified by Peynaud and Domercq (1955) as most species of *Saccharomyces*, *Saccharomycodes ludwigii, Brettanomyces* sp., *Kl. africana*, and *P. fermentans*. Those fermenting fructose faster were *Sacch. elegans, Sacch. acidifaciens, Sacch. rouxii*, and *Torulopsis bacillaris*. *Kl. apiculata* fermented the two sugars at about equal velocity. It is probable that various strains of a given species may have somewhat different characteristics.

Mestre and Jané (1946) found that pure yeast cultures produced more alcohol and fermented the sugars faster and more completely than did the native yeasts on the grapes. However, the pure yeasts did not appreciably convey the characteristic flavor of their region of origin. Again, insufficient sensory data were presented.

Florenzano (1949) found differences in the microflora of fermenting musts, refermenting musts (by the Tuscan *governo* system, p. 16), and new wine lees. *Sacch. cerevisiae* var. *ellipsoideus, Kl. apiculata, C. pulcherrima, Sacch. rosei*, and *Kl. magna* were the predominant organisms in the fermenting musts; but Florenzano believes that the less prevalent organisms (of which he isolated many) may also be active agents in natural fermentation. In the refermentation of the new wines by the *governo* technique used in Tuscany the following organisms predominated: *C. pulcherrima, Sacch. cerevisiae* var. *ellipsoideus, Sacch. florentinus*, and *Torulopsis bacillaris*. In the wine lees the prevailing species isolated belonged to the genus *Pichia*. He concludes from this that not only are the mixed cultures important but that changes in the prevailing species during fermentation may also be significant. From the lack of sensory data presented we cannot judge this.

Minárik (1965, 1966) concluded that mixed cultures of *Sacch. cerevisiae* var. *ellipsoideus, Sacch. oviformis*, and *Sacch. carlsbergensis* with *Sacch. rosei* gave the best results. However, *Sacch. oviformis* tended to dominate in mixed cultures, and neither the chemical nor the sensory data are statistically convincing. Masuda *et al.* (1964) fermented cider with mixtures of *Sacch. cerevisiae* and *Torulopsis bacillaris*. The results were no better than with *Sacch. cerevisiae* alone. Fermentations of *Sacch. cerevisiae* and *Sacch. rosei* or *Kl. magna* resulted in cider of lesser quality.

While it is likely that the proper succession of fermentation flora may be involved in the production of fine wines, there is little conclusive evidence on the subject. The possibility of controlling flavor by the use of species of yeasts other than *Sacch. cerevisiae* attracted the attention of Castelli (1942). He pointed out that *Sacch. rosei*, which has been found to be widely distributed in Italian wines, can ferment must with a resulting production of over 10 per cent of alcohol, small amounts of volatile acids, and no acetylmethylcarbinol, and may be more desirable than strains of *Sacch. cerevisiae* var. *ellipsoideus*. In a systematic study of strains of *Kl. apiculata* Tarantola (1946) found them undesirable for the production of wine in northern Italy, mainly because of their low alcohol, and high volatile acid- and aldehyde-producing properties.

Interest in the possible use of mixed yeast cultures has been stimulated by the discovery of the ability of *Schiz. pombe* to ferment malic acid. Dittrich (1963*a*, 1963*b*), Ribéreau-Gayon and Peynaud (1963), and Peynaud and Sudraud (1964) have attempted to utilize this ability of *Schiz. pombe*. This would obviously be a useful procedure where musts have excessively high malic acid, but it is a slow fermenter having a high temperature requirement. It was found to be difficult to prevent other yeasts from overgrowing *Schiz. pombe*. Pilot-plant use of *Schiz. pombe* to reduce excessive malic acid was successfully accomplished by Rzędowski and Rzędowska (1960) and Benda and Schmitt (1966). The latter reported no difficulty in reduction of the acidity if the natural yeast flora was first reduced. However, the wines were found to be uncharacteristic of the variety and to have an atypical flavor. Wines made by fermentation at 20°C (68°F) scored better than those fermented at 10°C (50°F). They suggest further studies with selected strains of the yeast. Usseglio-Tomasset (1966) reported that this yeast produced a pink-colored pigment (maximum absorption at 488 μ.)

Gandini and Tarditi (1966), in a review of the history of these investigations, recommended *Sacch. rosei* plus *Sacch. cerevisiae* var. *ellipsoideus* (or plus *Sacch. oviformis*) and *Schiz. pombe* plus *Sacch. cerevisiae* var. *ellipsoideus* (or plus *Sacch. oviformis*). The weakness of their results is that no statistically verifiable sensory evaluation of the quality of the wines was presented, although the treated wines were said to be "disarmonico." A systematic investigation of these fields in California would be highly desirable.

The results of Peynaud and Sudraud (1964) illustrate some of the problems of mixed cultures. They reported that musts had to be highly sulfited (100 to 150 mg per liter of free) and inoculated with at least 5 per cent *Schiz. pombe* in order to get adequate growth of this yeast. Peynaud and Sudraud question whether *Sacch. cerevisiae* var. *ellipsoideus* may not have the property of destroying or lysing the cells of other species. Mayer (1965) cautioned against the use of *Schiz. pombe* for reduction of malic acid because of its high optimum temperature and the possibility of undesirable changes in odor and taste. He noted that only under anaerobic conditions is ethanol a product. Benda and Schmitt (1966) obtained undesirable wines when *Schiz. pombe* was used as an inoculum, either alone or before addition of *Sacch. cerevisiae* var. *ellipsoideus*. Excessive loss of malic acid and absence of varietal characteristics in the aroma were noted. Samples fermented at 20°C (68°F) were better than those fermented at 11°C (51.8°F). Wienhaus (1967) reported that *Schiz. pombe* var. *liquefaciens* fermented malic acid better in the presence of sugar. Ethanol and carbon dioxide were the two products. For practical winery use a high temperature (25°C; 77°F) was recommended. Wienhaus adds 20 to 30 per cent grape must to a wine with the yeast culture. Up to 1 per cent malic acid can be fermented. He noted that the legal status of using such a process has not yet been determined in Germany.

Adams (1966) reported that mixtures of *Sacch. cerevisiae* var. *ellipsoideus, Sacch. florentinus, Sacch. steineri,* and *Torulopsis* sp. were being used in Canadian wineries. (See also Crowther, 1951–52.)

Use of starters

If sufficient active starter of the proper pure wine yeast is added, the undesirable organisms will be so outnumbered that they cannot act upon the must. They may be further inactivated if sulfur dioxide is added before the pure yeast. The must is not usually pasteurized before fermentation, although the practice has merit in cases of excessive contamination. If sulfur dioxide is used, the must should stand for several hours before the yeast starter is put in. White must should be thoroughly aerated after settling, either before or after the addition of an adequate starter.

According to the investigations of Bioletti and Cruess (1912), the starter is at its maximum activity when the Brix degree of the must in which it is grown has been reduced about one-half. Its efficiency does not greatly diminish until nearly all the sugar has disappeared.

Burgundy and champagne strains of wine yeast used in California are considered best for starters because they form a heavy, granular, and compact sediment toward the end of the fermentation and produce satisfactory fermentations. Wines made with them, if the musts were sound and well-balanced, clear rapidly. Yeasts are generally supplied to the winery as slant cultures on nutrient agar, in cotton-stoppered test tubes, or in bottles. For propagation at the winery, the bottle culture is better because of its larger size. Once a satisfactory pure culture is obtained, the winery should maintain slant cultures from year to year.

Where rapid fermentation to dryness is desired, *Sacch. oviformis* is probably the yeast of choice. It (or *Sacch. acidifaciens*) should not be used where sweet table wines are being produced. It is useful also in restarting "stuck" wines. Peynaud and Domercq (1959a) praised the flinty odor of wines fermented with it.

Propagating the yeast

The yeast starter is increased in volume before use, preferably by growing in a suitable pure-culture system. One system is to increase the volume of the yeast culture gradually, with precautions against infection, by successive transfers from the original culture to quantities of sterile must increasing in size. The half-pint culture, for example, may be transferred in succession to 1 gallon, 5 gallons, 50 gallons, and 500 gallons of sterile must. The transfers should be made when the rate of fermentation is most rapid. The smaller quantities of must are bulk-pasteurized and the larger (over 50 gallons) are sulfited before use.

A better procedure is to use an apparatus that propagates pure yeast. It consists essentially of closed tanks (fig. 61) equipped with steam and cooling coils and air distributors. One tank is often set above the other (fig. 62). Grape juice is introduced into the upper tank, pasteurized, cooled, aerated, and then dropped into several gallons of active pure yeast starter in the lower tank. When the must in the lower tank is in active fermentation all but a few gallons of it are withdrawn, and fresh, sterile must is then reintroduced into the culture vessel from the upper tank. See De Soto (1955) for the procedure used in a large California winery.

Some years ago Castor (1953b) noted that compressed yeasts might have technical and economic advantages, and pointed out that the use of such yeast was already twenty to thirty years old. He noted that true wine yeasts could be produced as compressed yeast if they had a strong respiratory mechanism. He also showed that the rate of fermentation was markedly increased with large inocula. His data did not indicate any sensory differences between normal and compressed yeast fermentations. He used 5 to 120 million yeast cells per ml for white musts and 0.3 to 60 gr per gallon for red musts. With the larger inoculum, final yeast counts in the new wine were 350 to 600 million cells per ml.

341

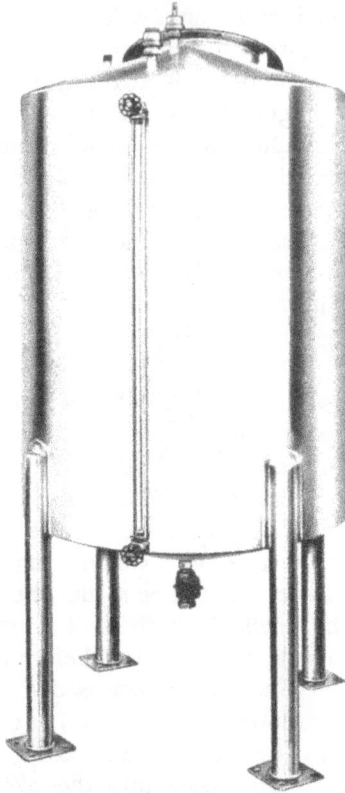

Figure 61. Stainless steel yeast culture tank. (Courtesy Valley Foundry and Machine Works, Inc., Fresno.)

Recently pressed yeast has become commercially available in this country for winery use. The requisite amount is added directly to the fermenting tank while it is being filled. Thoukis *et al.* (1963) described the large-scale commercial production and use of *Sacch. cerevisiae* var. *ellipsoideus*. The wet compressed yeast cake was found to be suitable for use in wineries. By means of mass pitching little grape sugar was used for growth of yeast. The rapid establishment of anaerobic conditions permitted lower levels of sulfur dioxide in the musts. The yeast cakes were much simpler to use than the traditional winery procedures. Goldman (1963b) recommended the use of wet compressed yeast cake for sparkling wine production because the yeast population for the tirage bottling could be more accurately calculated and controlled.

Saller and Stefani (1962) compared wine yeasts and pressed bakers'

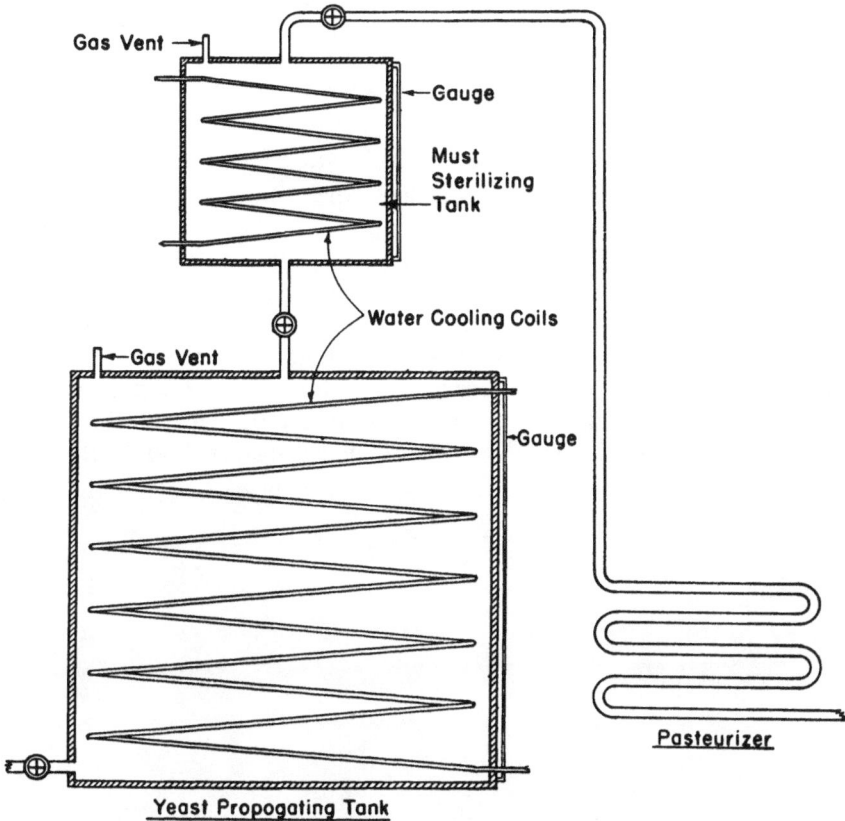

Figure 62. A tank for propagating pure-yeast starter.

yeast. The pressed yeast sometimes became inactive before the end of the fermentation. However, they were attempting to ferment to over 16 per cent alcohol. They also attributed an off-taste to pressed yeast, but no controlled sensory data were offered. Adams (1966) recommended frozen ($-29°C$; $-20°F$) yeasts. These could be stored several years at this temperature without total loss of viability or of essential fermenting characteristics. Active dry yeast is now commercially available.

Inoculating the must

It is best to inoculate each fermenting vat with a fresh yeast starter, because transfer from vat to vat is sure to cause contamination with undesirable yeasts sooner or later. As little as 1 per cent by volume of active starter is enough when the juice is clean and is fairly free from undesirable yeasts and bacteria; rarely is it necessary to add over 3 per cent. Too much starter may produce rapid and violent fermentation during which heat may be generated too rapidly to permit easy control of the temperature, even by good cooling systems.

Simultaneous alcoholic and malo-lactic fermentations

It was noted by Webb and Ingraham (1960), Peynaud and Domercq (1959b), and others that the malo-lactic fermentation could take place at the same time as the alcoholic fermentation. There have been a number of experimental trials in California, primarily with red wines. Not all the variables have yet been studied, but the procedure does not seem to present any serious problems. The main difficulty is to complete the fermentation rapidly and to separate the wine from the lees promptly thereafter. Careful day-to-day attention should be paid to the malic acid content (p. 496) as well as to the reducing sugar, pH, and titratable acidity.

CHAPTER NINE

ALCOHOLIC FERMENTATION, ANTISEPTIC AGENTS, AND CHEMICAL PRESERVATIVES

The transformation of organic substrates, such as carbohydrates, into products characteristic of the particular microörganism and of the effective environmental conditions is generally called "fermentation."

STAGES IN FERMENTATION...
... are many and complicated.

The process of alcoholic fermentation is less simple than is indicated by the Gay-Lussac equation (see p. 348). Alcoholic fermentation occurs in a series of well-defined stages, involving the formation of several important intermediates and the interaction of several enzyme systems (Axelrod, 1967; Baldwin, 1967; Hollmann, 1964). The glucose molecule passes anaerobically through twelve stable intermediary stages before alcohol and carbon dioxide appear as the primary final products. At least three, possibly eight, dissociable organic coenzymes, twenty or more enzyme proteins, and several inorganic catalysts (ammonium, potassium, manganese, magnesium, and copper ions) must be provided by the yeast cell (Meyerhof, 1945; Neuberg, 1946). (For early information see Harden, 1932.) The most important of these transformations are shown in table 43.

Carbohydrate metabolism was subdivided into three phases by Krebs and Kornberg (see Mahler and Cordes, 1966). In phase I of catabolism the sugars are mobilized for their subsequent transformations by conversion to phosphorylated hexoses. This entails either simple phosphorylation of oligo- or polysaccharides, or hydrolysis of oligosaccharides followed by phosphorylation of the sugars produced to hexose phosphates. No useful energy is provided by reactions of phase I. In phase II the hexose phosphates produced in phase I are incompletely degraded with the liberation of about one-third of the free energy potentially available by their complete combustion. In phase II, 2- and 3-carbon compounds are generated. In phase III these intermediates are either completely

TABLE 43

STAGES IN FERMENTATION

Reaction	Substrate	Product	Enzyme	Co-enzyme	Metals
1	Glucose	Glucopyranose-6-phosphate	Hexokinase	ATP → ADP	Mg^{++}
2	Glucopyranose-6-phosphate	Fructofuranose-6-phosphate	Phosphohexoisomerase	—	—
3	Fructofuranose-6-phosphate	Fructofuranose-1,6-diphosphate	Phosphofructokinase	ATP → ADP	Zn^{++}, Co^{++}, Fe^{++} or Ca^{++}
4	Fructofuranose-1,6-diphosphate	Phosphodihydroxyacetone and d-3-phosphoglyceraldehyde	Aldolase	—	—
5	Phosphodihydroxy-acetone	3-phosphoglyceraldehyde	Phosphotrioseisomerase	—	—
6	3-phosphoglycer-aldehyde + inorganic phosphate	1,3-diphosphoglyceric acid	Glyceraldehyde-3-phosphate dehydrogenase	NAD → NADH	—
7	1,3-diphosphoglyceric acid	3-phosphoglyceric acid	Phosphoglyceric kinase	ADP → ATP	Mg^{++}
8	3-phosphoglyceric acid	2-phosphoglyceric acid	Phosphoglyceromutase	—	Mg^{++}
9	2-phosphoglyceric acid	2-phosphoenol pyruvic acid + water	Enolase	—	Mg^{++}
10	2-phosphoenol pyruvic acid	Pyruvic acid	Pyruvic kinase	ADP → ATP	Mg^{++}, K^{+}
11	Pyruvic acid	Acetaldehyde and carbon dioxide	Carboxylase (pyruvate decarboxylase)	TPP	Mg^{++}
11a	Pyruvic acid	Lactic acid	Lactic dehydrogenase	NADH + H⁺ → NAD	—
12	Acetaldehyde	Ethyl alcohol	Alcohol dehydrogenase	NADH + H⁺ → NAD	—

346

oxidized by aerobic respiration to CO_2 and H_2O by the citric acid and certain ancillary processes or converted into definite end products such as alcohol by anaerobic fermentation. (See also figure 63.)

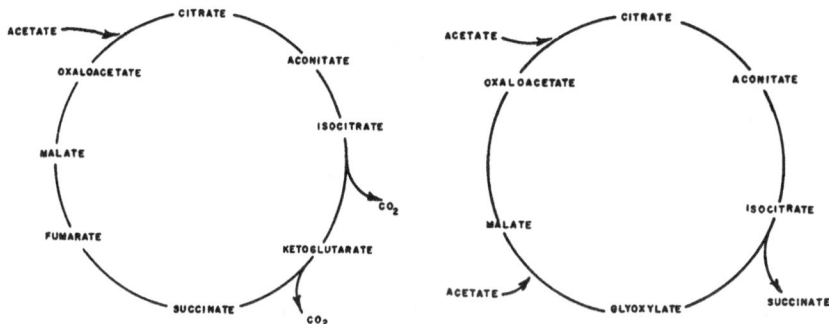

Figure 63.   Citric acid and glyoxylic acid cycles.

In essence, the processes of energy metabolism of the yeast are determined by the availability of oxygen to the cell. Oxygen deficiency leads to the accumulation of high-energy substances such as ethanol and higher alcohols. According to Nemec and Drobnica (1963) the pentose cycle constitutes an alternative form of hexose metabolism and this explains the formation of the essential pentose cell constituents.

Adenosine triphosphate, active in inter- and intramolecular phosphate transfer, is a compound of the purine base, 6-aminopurine or adenine, the sugar β-d-ribofuranose, and esterified phosphoric acid. Cozymase I (coenzyme I) is adenine-nicotinic amide dinucleotide, a compound of adenosine monophosphate with nicotinic amide mononucleotide; in the latter the base is nicotinic acid amide and the pentose sugar d-ribofuranose. Cocarboxylase is thiamine pyrophosphate (the diphosphate of vitamin B_1). The coenzymes and the corresponding enzyme proteins must be present in proper balance and at optimum activity for the conduct of a continuous and complete fermentation.

Nicotinamide-adenine dinucleotide (NAD) plays a central role in anaerobic and aerobic metabolism of yeasts. According to Suomalainen (1963), under aerobic growth conditions yeasts do not need more NAD than the yeast normally contains. Addition of nicotinic acid to growing yeasts does increase the NAD content. Under anaerobic conditions yeast synthesizes a smaller amount of its nicotinic acid than aerobically grown yeast. Thus nicotinic acid acts as a growth factor under anaerobic but not aerobic conditions.

347

Alcoholic fermentation is essentially a process of a series of reversible inter- and intramolecular oxidation-reductions, phosphorylations, and an irreversible decarboxylation. In the induction period the hexose diphosphate undergoes an intramolecular oxidation-reduction (dismutation) into two triose fractions: the ketose part, dihydroxyacetone phosphate, is reduced to l-glycerophosphate, while the glyceraldehyde phosphate is oxidized to 3-phosphoglyceric acid. After pyruvic acid is formed, the reduction step is the hydrogenation of acetaldehyde formed by decarboxylation of pyruvic acid.

Oesper (1968) reported an interesting historical account of the development of our knowledge of the consecutive reactions involved in glycolysis. From 1920 to 1939 all reactions except one, the oxidation of glyceraldehyde-3-phosphate by aldolase, remained unknown. The sequence of events occurring in this reaction required more than ten years to clear up. This delay was caused partly by a misleading statement and partly because the concept that enzymes actually participate in the reactions which they catalyze was not clearly stated and experimentally proven until 1947.

Alcohol yields

The quantity of alcohol and carbon dioxide formed and the kind and concentration of by-products vary with the strain of the yeast and the composition, temperature, and extent of aeration of the medium. It is customary to summarize the important over-all changes occurring during alcoholic fermentation by the Gay-Lussac equation:

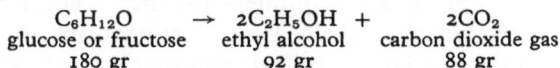

$$C_6H_{12}O \rightarrow 2C_2H_5OH + 2CO_2$$

glucose or fructose	ethyl alcohol	carbon dioxide gas
180 gr	92 gr	88 gr

Theoretically, according to the equation given above, the weight of alcohol produced should be 51.1 per cent of the weight of the sugar consumed. In practice the yield of alcohol by weight is generally considered to be 47 per cent; the remainder of the sugar is converted into other products and assimilated by the yeast in growth under fermenting conditions (Niehaus, 1937; Trauth and Bässler, 1936). See also table 44.

Judging from analyses, made by G. E. Colby (1896), of California wines in the pre-Prohibition era, the alcohol content by volume in wine is approximately equal to 57.7 per cent of the Brix degree of the must. This value is somewhat too high, however, at least under commercial conditions when the free-run juice is a truly representative sample of the juice; it is then nearer to 55 per cent by volume of the degree Brix.[1] The free-

---

[1] The Brix is expressed as gr per 100 gr. For a Brix of 23°, this means 25.2 of soluble solids per 100 ml.

run juice does not indicate the yield of alcohol obtainable when large quantities of raisins are present and the fermentation is conducted in the presence of the semidried fruit. Even higher alcohol yields than Colby reported are obtained in such cases, frequently as high as 65 per cent of the Brix reading. This is substantiated by the data of Winkler and Amerine (1937) for the 1936 vintage.

There are few accurate data on the quantitative yields of alcohol and carbon dioxide in alcoholic fermentations conducted under experimental conditions that exclude losses of alcohol by entrainment and evaporation. Gvaladze (1936) reported alcohol yields varying from 47.86 to 48.12 per cent and carbon dioxide yields varying from 47.02 and 47.68 per cent of the weight of sugar fermented, compared with the biologically unattainable yields of 51.1 per cent of alcohol and 48.9 per cent of carbon dioxide (table 44). French enologists commonly assume that 1 per cent alcohol by volume is obtained per 17 grams of sugar fermented per liter. Aside from changes in weight and volume during fermentation, a solution containing 170 grams of sugar per liter, on this basis, would yield a wine containing 10 per cent alcohol by volume in comparison with a theoretical yield of 11.8 per cent by volume if all the sugar were converted to alcohol and carbon dioxide.

TABLE 44

PRODUCTS OF ALCOHOLIC FERMENTATION, THEORETICAL AND ACTUAL, AND WITH THREE YEAST TYPES

	Percentage of fermentable sugar transformed					
Product	Theo-retical	In industrial fermenta-tions[a]	Pasteur data	With champagne wine yeast	With Rkatsateli wine yeast	With Steinberg wine yeast
Alcohol	51.1	48.4	48.4	47.86	48.22	48.07
Carbon dioxide	48.9	46.5	46.6	47.02	47.61	47.68
Acetaldehyde	0.0	0.00–0.08	—	0.01	0.04	0.02
Acetic acid	0.0	0.05–0.25	—	0.61	0.50	0.65
2,3-Butylene glycol	0.0	—	—	0.06	0.09	0.10
Glycerin	0.0	2.5–3.6	3.2	2.99	2.61	2.75
Lactic acid	0.0	0.0–0.2	—	0.40	0.28	0.40
Succinic acid	0.0	0.5–0.7	0.6	0.020–0.045	—	0.015–0.053
Furfural	0.0	trace	—	—	—	—
Fusel oil (higher alcohols)	0.0	0.05–0.35	—	—	—	—
Yeast (dry weight)	0.0	—	1.2	0.55	0.55	0.57

[a] Calculated on basis of Gay-Lussac's equation.
Sources of data: For industrial fermentations, Rahn (1932); for Pasteur data, Hewitt (1928); for type of wine yeast, Gvaladze (1936).

According to Tiurina (1960), Soviet enologists believe that 0.60 ml of alcohol are normally formed per gr of sugar fermented. This is equivalent to 1 ml of alcohol per 1.667 gr of sugar fermented. In open red fermentations the yield was only 0.55 to 0.57 ml per gr of sugar fermented, or, on the average, 1 ml of alcohol per 1.786 gr of sugar fermented. Tiurina reported that the alcohol yield per gr of sugar fermented was less than the theoretical yield in the early stages of the fermentation and greater in the later stages. This she attributed to the formation of intermediate by-products in the early stages and their later final conversion to alcohol and carbon dioxide.

Banolas (1948) calculated, on partial-pressure considerations, that the loss of alcohol by entrainment with carbon dioxide gas at 33°C (91.4°F) would be equivalent to a decrease of 1.65 per cent alcohol by volume. He assumes that a must containing 180 gr of glucose per liter would theoretically yield 92 gr or 115 ml of alcohol, or 11.5 per cent alcohol by volume (on the basis of 1 per cent alcohol per 15.65 gr of sugar). If this fermentation is conducted at an average temperature of 33°C (91.4°F) the partial pressures over the resulting wine would be as follows: alcohol vapor pressure, 94.78 mm of mercury; water vapor pressure, 37.40 mm; and carbon dioxide pressure, 627.87 mm. If the carbon dioxide gas is given off slowly and carries away the alcohol vapor and water vapor as they are formed, then 88 gr of carbon dioxide gas should carry and give up 13.2 gr of alcohol by volume. In practice, under these conditions, the alcohol yield obtained is 1 per cent by volume per 18 gr of sugar fermented per liter, corresponding to a theoretical loss of 1.5 per cent alcohol by volume.

This is in agreement with the results obtained by Mathieu and Mathieu (1938), who reported that carbon dioxide gas, on liberation, entrains alcohol and volatile aromatic compounds; the higher the temperature of the wine, the greater was this loss. According to them, 1 liter of must containing 180 gr of sugar liberates (at 35°C; 95°F) about 50 liters of carbon dioxide, which at this temperature entrains 0.25 gr of alcohol, equivalent to a loss of 1.56 per cent alcohol. Recently, however, Stradelli (1951) has calculated that the loss of alcohol with carbon dioxide is negligible and that the losses obtained during fermentation are due to other causes. For practical data and tables see Marsh (1958).

Warkentin and Nury (1963) note the varying reports of alcohol losses during fermentation that have been reported in the literature. The high losses reported by Banolas (1948) they attribute to using the vapor pressures of pure compounds in the calculations. The low results of Stradelli (1951), they believe, is due to the assumption that the alcohol-water-sugar system follows Raoult's law. The loss of alcohol by entrainment during alcoholic fermentation of laboratory-scale fermentations was shown by Zimmermann et al. (1964) to increase with the temperature of fermentation, the alcohol level of the wine being fermented, agitation of the fermenting liquid, and the presence of the pomace cap. The rate of loss was at a maximum during the middle of the fermentation, i.e., during the period of maximum fermentation. The evaporation and entrainment losses were 0.65 per cent at 26.5°C (79.7°F) for grape juice with an initial

Brix of 21° and 0.84 per cent for musts. This result is not in conflict with the 0.83 per cent loss reported by Warkentin and Nury (1963) with a higher Brix juice. In plant-scale operations the maximum loss was toward the end of the fermentation and did not differ between open and closed fermenters. The losses in plant operations were 0.7 per cent for a juice of 16° Brix at 25.5°C (77.9°F). See Dietrich (1954) for methods of reducing alcohol loss.

FACTORS INFLUENCING ALCOHOL YIELD . . .
. . . are chemical and biological.

The concentration of alcohol produced from grape juice by a given strain of yeast, the degree to which the sugar content is used up, the efficiency of fermentation, and the nature and concentration of by-products are influenced by temperature, extent of aeration, sugar concentration, acidity, strain and activity of yeast, and by other factors.

Temperature
Temperature is extremely important; the lower the temperature, the higher the yields of alcohol in fermentation, not only because of more complete fermentation (assuming the fermentation is completed) but also because of the reduction in loss of alcohol by evaporation and entrainment by escaping carbon dioxide gas. (See pp. 367-373 for a discussion of temperature as it affects alcoholic fermentation.)

Yields from various grapes
Jacobs and Newton (1938) found the yield of alcohol per ton of eastern grapes having an average fermentable sugar content of 11.5 per cent and a high sugar content of 15.0 per cent to be 15.1 gallons of 99.5 per cent alcohol and 19.7 gallons of 99.5 per cent alcohol, respectively.

Considerably higher yields are obtainable with California grapes. Berti (1949), on the assumption that the average loss in skins, seeds, and stems is 10 per cent and the nonfermentable sugar content is 2.2 per cent, calculates that one ton of grapes testing 23° Brix would yield 51.8 proof gallons.[2] Marsh (unpublished data), on the assumptions that degree Brix minus 3.0 is equal to the percentage of fermentable sugar, that the average pomace yield is 13 per cent, and that the fermentation is 92 per cent efficient (yield of alcohol by weight is 0.47 of sugar fermented), calculated that the ratio of proof gallons per ton to degree Brix is 2.16 at 24° Brix, corresponding to a yield of 51.8 proof gallons. Amerine (unpublished

---

[2] A proof gallon is a wine gallon (231 cubic inches) of 100 proof (50 per cent) alcohol.

data), on the basis of seven commercial fermentations during the 1948 season, found a yield of 17.5 to 25 gallons of alcohol per ton, depending on the maturity of the grapes. These grapes, like most of those of the 1948 season, were low in sugar.

Insufficient data are available for accurate determination of the yield of alcohol under all existing commercial conditions in California. Such knowledge is desirable to estimate production efficiency and to develop uniform systems of reporting inventories. The accurate determination of the yield of alcohol under commercial conditions is complicated by the difficulty of obtaining representative samples, by the presence of raisins or uncrushed grapes, by the use of sugar by yeasts and bacteria, by losses through entrainment, and other factors.

The kinetics of laboratory anaerobic batch and continuous grape juice fermentations were studied by Holzberg et al. (1967). Equations which predicted operation were derived for both exponential growth and stationary phases. The data indicated that alcohol completely inhibited fermentation at about 8.5 per cent alcohol—a very low figure. Studies of this sort with normal musts and without the excessive stirring (350 rpm) might reveal useful information on the various factors influencing alcohol yield.

During fermentation increasing amounts of pyruvic acid appear in the early stages of fermentation and less in the later stages (see fig. 51) which is similar to acetaldehyde. On the other hand, α-ketoglutaric acid increases throughout the fermentation. Lafon-Lafourcade and Peynaud (1966) showed that certain species of yeast (*Saccharomycodes ludwigii* and *Schizosaccharomyces pombe*) are especially ketogenic. More of both acids is found at increasing pH's—from 3 to 4 especially. More is also found during the fermentation of musts of moldy grapes than those of sound grapes. Addition of glycerin, aspartic acid, cystine, isoleucine, proline, threonine, or tyrosine greatly stimulated production of the acids. Most of the α-ketoglutaric acid is derived from glutamic acid. Lack of thiamine markedly increased their formation. This would be expected since thiamine is a constituent of cocarboxylase.

CHANGES IN BY-PRODUCTS ...
... due to various factors.

Blouin and Peynaud (1963b) noted, in a large number of wines, 11 to 460 mg per liter of pyruvic acid (average 71) and 2 to 346 mg per liter of α-ketoglutaric acid (average 80). α-Ketoglutaric acid increased throughout fermentation. There was considerable variation in the maximum and final pyruvic acid level produced by different yeasts and in the final level

352

of α-ketoglutaric acid. *Schizosaccharomyces pombe* produced especially high amounts of both compounds. Pyruvic acid formation was much greater at pH 4 than at any lower pH. There was a wide variation in the maximum amounts produced when ammonium sulfate was used as the nitrogen source. Addition of glutamic acid especially favored α-ketoglutaric acid production, but addition of ammonium sulfate and aspartic acid resulted in very little increase in α-ketoglutaric acid. Presence of glycine resulted in high pyruvic acid production. The most important factor in pyruvic acid production is thiamine. Production of pyruvic acid (up to 0.1 per cent) occurred in fermentation of musts from botrytised grapes. They consider these compounds to be especially important because of their ability to fix sulfur dioxide (see p. 386).

Deibner and Cabibel-Hughes (1965) reported 16.8 to 58.4 mg per liter of pyruvic acid in sweet, slightly fortified wines. They also found 25.6 to 45.6 mg per liter of α-ketoglutaric acid. Heating in the absence of air caused some degradation, more for α-ketoglutaric than for pyruvic. Fermenting with the skins resulted in more of these keto acids than fermentation of the juice alone.

Effect of aldehydes

When acetaldehyde is kept from serving as the hydrogen acceptor (as in the presence of added alkali or of acetaldehyde-binding agents such as bisulfite or dimedone), the normal course of fermentation is modified. Under such conditions, increased quantities of glycerin and acetic acid accumulate in the medium.

Genevois (1950) proposed in 1936, that acetaldehyde can serve as precursor of acetic acid, acetoin, and succinic acid. He pointed out that this hypothesis was substantiated by Peynaud's demonstration that, by the progressive addition of acetaldehyde to the fermentation medium, the quantities of acetic acid, succinic acid, and 2,3-butylene glycol could be increased. Conversely, by the addition of dimedone, which binds the acetaldehyde, the proportion of these three substances in the final product could be decreased.

L. André (1966) treated wines with hydroxylamine to obtain the oximes of the aldehydes and ketones, extracted these with ether, and placed them directly in the gas chromatograph. He found acetaldehyde, furfural, acetoin, and γ-butyrolactone. This included wines from Château Chalon and Madeira which have high acetaldehyde contents. This is in contrast to earlier work by Bayer (1957), who reported, in addition to acetaldehyde, propionaldehyde, butyraldehyde, valeraldehyde, hexenal, and traces of other higher aldehydes. Mecke *et al.* (1960) found, in addition, iso-butyraldehyde, isovaleraldehyde, and cinnamaldehyde; Rodopulo and

Egerov (1965) reported formaldehyde and enanthaldehyde. Some of these were found as acetals also. Webb and Kepner (1962) and Webb et al. (1964a), however, found only acetaldehyde and a nonidentified compound (later identified by Webb, (1965) as act-amyl lactate). The recent publication of Ronkainen et al. (1967) makes it clear that the following carbonyl compounds are normal products of alcoholic fermentation: acetaldehyde, propionaldehyde, isobutyraldehyde, butyraldehyde, and isovaleraldehyde (or 2-methylbutyraldehyde). Other normal fermentation products are pyruvic and 2-oxoglutaric acids and, to a lesser extent, 2-oxobutyric, 2-oxoisovaleric, 2-oxoisocaproic, and a trace of 2-oxomethylvaleric acid.

The first cell-free conversion of α-ketobutyrate to propionaldehyde and propanol appears to be that of Kunkee et al. (1965a). They also showed that the reported conversion of α-ketobutyrate to 2-methyl-1-butanol (act-amyl alcohol) appears to be in error.

Effect of pentoses

Fermentation mechanisms involve pentoses also. The first step is isomerization of the pentose to the corresponding ketose. The end products are lactic acid or acetic acid, and sometimes mannitol (from fructose).

Horecker (1963) noted the widespread distribution of pentoses in plants (either as free sugars, as pentosans, or in nucleic acids). The important pentoses are D-xylose and L-arabinose in polysaccharides and D-ribose and 2-deoxy-D-ribose in nucleic acids. The most important 5-carbon ketose is L-xylulose. Since nucleic acid is needed universally, the latter two are synthesized by all (or nearly all) cells. Indeed, as Horecker pointed out, most cells have more than one metabolic pathway from hexose to pentose: (1) the oxidative pathway of D-glucose-6-phosphate to D-ribose-5-phosphate (this is TPNH-dependent), or (2) the non-oxidative pentose phosphate pathway from fructose 6-phosphate. This pathway may also involve sedoheptulose 7-phosphate from ribulose 5-phosphate. Reductive pathways of anaerobic pentose metabolism are also known: L-arabinose to α-ketoglutaric acid, D-arabinose to pyruvic acid, and D-xylose to L-arabitol.

OTHER BY-PRODUCTS

Factors influencing higher alcohol formation

Windisch (1906) noted that the production of higher alcohols in wine was due not only to transformations of amino acids but might also arise from bacterial fermentation of carbohydrate. In spontaneous fermentations of musts, he got two to three times as much higher alcohol as he did when musts were fermented with pure yeasts. The conversion of amino acids to alcohols by yeast was first described by Ehrlich (1907–1912) as occurring by deamination and decarboxylation followed by reduction to yield an alcohol containing one carbon atom less than the original amino acid. The mechanism of this conversion is discussed in detail by Thorne (1950) and Harris (1958). The decomposition of amino acids by yeast during alcoholic fermentations may occur also by the Strickland reaction in which one amino acid is oxidized (serves as a hydrogen donor) and another is reduced (serves as a hydrogen acceptor). More recently the mechanism of formation of alcohols from amino acids was shown to occur through a feedback inhibition of one of the normal sequences of utilization (see p. 346; see also Ingraham and Guymon, 1960). Nordström and Carlsson (1965) investigated yeast growth in relation to formation of fusel alcohols. Äyräpää (1967b) used ^{14}C-labeled valine and leucine to investigate the mechanism of formation of higher alcohols.

The source of alcohols and their corresponding aldehydes and α-keto and amino acids is summarized in table 45. (See also Kunkee et al., 1965a.)

The rate of formation of higher alcohol has been variously reported. Castor and Guymon (1952) and Villforth and Schmid (1954) reported greater formation toward the end of the fermentation. Peynaud and Guimberteau (1962b), however, found greater formation in the early stages of fermentation, particularly of 3-methyl-1-butanol. 2-Methyl-1-propanol was formed throughout fermentation. Veselov and Gracheva (1963) noted that the higher alcohols are formed in the course of yeast cell multiplication, that is, in the first stage of fermentation, but Lewis (1964) believed the opposite. We agree with the view that when amino acid supply is low these exhausted amino acids must be synthesized through other pathways and this gives rise to formation of higher alcohols. This would normally occur in the later stages of fermentation (fig. 64). Castor and Guymon (1952) pointed out that the amino acids isoleucine, leucine, and valine disappeared in the early stages of alcoholic fermentation, while the higher alcohols appeared throughout the course of fermentation. Apparently the Ehrlich mechanism did not apply. Using radioactive tracer techniques, Thoukis (1958) showed that very little of the tagged leucine or isovaleraldehyde was found in 3-methyl-1-butanol.

Smaller amounts of higher alcohols were formed in musts with the following nitrogen sources: α-alanine, ammonium sulfate, arginine, asparagine, aspartic acid, glutamic acid, glutamine, ornithine, phenylalanine, tryptophan, or tyrosine. Larger amounts of higher alcohols were formed when the following nitrogen sources were supplied: asparagine and ammonium sulfate (together), cysteine, cystine, glycine, histidine, methionine, proline, serine, or threonine. A very high content of higher

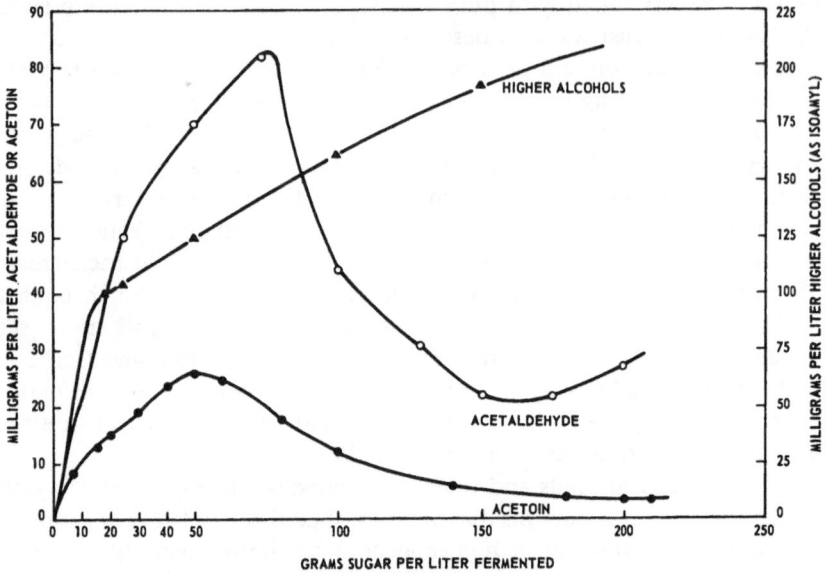

GRAMS SUGAR PER LITER FERMENTED

Figure 64. Formation of acetaldehyde, acetoin, and higher alcohols during alcoholic fermentation.

alcohols resulted when the nitrogen source was isoleucine, leucine, nor-leucine, or valine. When excess leucine was supplied, more 3-methyl-1-butanol was formed than predicted by the Ehrlich mechanism at low concentrations and less at higher concentrations. When valine was supplied in excess, less 2-methyl-1-propanol than predicted was produced at all levels. Only about one-sixth of the higher alcohols appear to be formed by the Ehrlich mechanism, that is, by deamination of the appropriate amino acid.

Very variable amounts of higher alcohols are formed by different yeasts and there is considerable variation in higher alcohol formation by the same yeast at different times. More was formed by *Sacch. cerevisiae* var. *ellipsoideus*, *Sacch. oviformis*, *Sacch. acidifaciens*, and *Sacch. chevalieri*; less by *Sacch. elegans*, *Sacch. rosei*, and *Torulopsis bacillaris*; and much less by *Kloeckera apiculata* or *Brettanomyces schanderlii*.

Rankine (1967a) found variations in production of higher alcohols by different yeasts as follows: 1-propanol, 13 to 106 mg per liter; 2-methyl-1-propanol, 9 to 37; and 3-methyl-1-butanol and 2-methyl-1-butanol, 115 to 262. Increasing the temperature from 15°C (59°F) to 25°C (77°F) produced an average of 24 per cent more 3-methyl-1-butanol plus 2-methyl-1-butanol, 39 per cent more 2-methyl-1-propanol, and 17 per cent less 1-propanol.

TABLE 45

SOURCE OF ALCOHOLS IN ALCOHOLIC FERMENTATION

Alcohol[a]	Aldehyde found	Keto acid found	Corresponding amino acid
Ethanol	Acetaldehyde	Pyruvic	Alanine
1-Propanol	n-Propioanaldehyde	α-Ketobutyric	α-Aminobutyric
2-Propanol	—	—	—
Isopropanol			
1-Butanol	n-Butyraldehyde	α-Ketovaleric	Norvaline
2-Methyl-1-propanol	Isobutyraldehyde	α-Ketoisovaleric	Valine
Isobutanol			
2-Butanol	—	—	—
sec-Butanol			
—		α-Keto-γ-methiobutyric	Methionine
2-Methyl-2-propanol			
tert-Butanol			
1-Pentanol	Valeraldehyde	α-Ketocaproic	Norleucine
n-Amyl			
3-Methyl-1-butanol	Isovaleraldehyde	α-Ketoisocaproic	Leucine
Isoamyl			
2-Methyl-1-butanol	act-Valeraldehyde	α-Keto-β-methylvaleric	Isoleucine
act-Amyl			
1-Hexanol	n-Hexanal	—	—
n-Hexanol			
1-Heptanol	n-Heptanal	—	—
n-Heptanol			
—		Oxalacetic	Aspartic
—		α-Ketoglutaric	Glutamic
2-Phenylethanol	β-Phenylacetaldehyde	β-Phenylpyruvic	Phenylalanine
β-Phenethyl			
2-(4-Hydroxyphenyl)-ethanol	p-Hydroxyphenylacetaldehyde	p-Hydroxyphenylpyruvic	Tyrosine
Tyrosol	Indolacetaldehyde	Indolpyruvic	Tryptophan
Tryptophol			
2-(3-Indole)-ethanol			
1,2-Ethanediol	glyoxal	—	Serine
glycol			

[a] In this text we have followed the IUPAC nomenclature for the higher alcohols. The more common names are listed in column 2. For a discussion of the rules see Hodgman, C.D. Handbook of Chemistry and Physics. Cleveland, The Chemical Rubber Publ. Co. Source of data: Suomalainen (1965), Ingraham (1966), and Suomalainen et al. (1968).

Peynaud and Guimberteau (1962b) generally found a higher level of higher alcohols in wines fermented under anaerobic conditions. Guymon *et al.* (1961) and Crowell and Guymon (1963) reported an increased production in agitated aerobic fermentations. Veselov and Gracheva (1963) showed that their formation depends on the intensity of metabolism of yeasts and the duration of fermentation. It is not dependent on the nitrogenous composition of the medium and is not proportional to the amount of carbohydrates fermented.

More of the higher alcohols are formed at pH 4.5 than at 5.0 or 4.0, 3.5 or 3.0. More was formed at 20°C (68°F) than at 15°C (59°F) or 25°C (77°F), 30°C (86°F) or 35°C (95°F). However, Veselov and Gracheva (1963) reported that at the same degree of utilization of carbohydrates less of the higher alcohols were formed at higher temperatures than at lower (8°C; 46.4°F). The amount formed at different levels of sugar depends on the nitrogen source, but the amount is usually proportional to the amount of sugar fermented. Growth factors such as thiamine, biotin, and panthothenic acid had very complex effects on higher alcohol formation.

Yoshizawa (1965) reported greater yields of higher alcohols at pH 5 (compared to lower pH) and with aeration. Using a washed yeast technique, little 2-methyl-1-propanol and 3-methyl-1-butanol were formed when pyruvic acid was not added to the media. Addition of leucine stimulated 3-methyl-1-butanol formation, and decreased the production of acetolactic acid, diacetyl, acetoin, and 2-methyl-1-propanol. Addition of α-ketoisovaleric acid increased the production of both 3-methyl-1-butanol and 2-methyl-1-propanol. Yoshizawa showed that α-ketocaproic acid was an intermediate product. Yoshizawa (1966) has also reported relatively more 3-methyl-1-butanol (and 2-methyl-1-butanol) than 2-methyl-1-propanol in wine than in whiskey (except American) or brandy. This he attributed to the higher levels of leucine and valine in wort compared to must. In amino acid-rich media the amount of higher alcohol formed is largely regulated by the Ehrlich mechanism. In amino acid-poor media the keto acids undergo decarboxylation to form the corresponding aldehyde from which the higher alcohol is formed. The composition of the higher alcohol in this case depends more on the ability of the yeast to synthesize amino acids. Also he states that 3-methyl-1-butanol is formed more easily than 2-methyl-1-propanol through the synthetic pathway. This would explain the higher 3-methyl-1-butanol/ 2-methyl-1-propanol (isoamyl/isobutanol) ratios in wines.

It is noteworthy that valine, leucine, isoleucine, tryptophan, and tyrosine begin to accumulate in the fermenting solution simultaneously with higher alcohol formation, according to Suomalainen and Ronkainen (1963).

The nitrogen source and the presence or absence of vitamins have a profound effect on the formation of higher alcohols, as the data in table 46 show. There obviously are varying mechanisms for the formation of higher alcohols. The presence of pantothenic acid appears especially critical, possibly as a coenzyme in some mechanisms.

Ribéreau-Gayon and Peynaud (1964, 1966), found different species of *Saccharomyces* to produce 37 to 300 mg per liter of 3-methyl-butanol and 7 to 105 of 2-methyl-1-propanol. Addition of 200 mg per liter of leucine increased higher alcohol formation with *Sacch. cerivisiae* var. *ellipsoideus*, *Sacch. oviformis*, *Sacch. chevaleri*, *Sacch. elegans*, and *Brettanomyces schanderlii*, but had little effect with *Sacch. acidifaciens*, *Sacch. rosei*, *Torulopsis bacillaris*, or *Kl. apiculata*.

In general, more higher alcohols are formed by fermentations carried out in the absence of air, since anaerobic conditions favor the formation of 3-methyl-1-butanol alcohol more than 2-methyl-1-propanol. With certain yeasts more than three times as much are formed. The pH effect is not large in the range of normal grape musts. Lesser amounts of higher alcohols are produced at pH 2.6, more at 4.5, and less at 5.0. The temperature optimum for both 3-methyl-1-butanol and 2-methyl-1-propanol formation is 20°C (68°F). Ough *et al.* (1966), however, found less 1-propanol produced at 20°C (68°F). Nordström (1966) reported optimum production of total higher alcohols at 25°C (77°F).

Äyräpää (1967a) reported that about 13 per cent of the leucine of yeast was transformed to 3-methyl-1-butanol alcohol during fermentation. He calculated that at least 10 per cent of the 3-methyl-1-butanol alcohol formed was derived from the protein of the yeast. In general Äyräpää (1967b) believes that the fraction of amino acid (valine, leucine, isoleucine, phenylalanine) transformed to higher alcohols is inversely correlated to the total nitrogen level of the substrate.

Based on careful gas chromatographic studies, Usseglio-Tomasset (1964) was unable to find 2-propanol, 2-methyl-2-propanol, 3-methyl-2-butanol, 2-pentanol, 3-pentanol, or 2-hexanol in Italian fused oils. Always present were 1-butanol, 2-butanol, and 1-pentanol. The earlier reports of some authors on the presence of secondary alcohols in fusel oil may have been due to the presence of esters and failure to actually identify the alcohols from the chromatography column. No relation between formation of 1-butanol and 1-hexanol was noted.

2-Phenethyl alcohol is formed parallel with ethanol. It is found up to 20 to 25 mg per liter in the absence of 2-phenylalanine (apparently because some yeasts can synthesize this precursor). Various yeasts produce different amounts of 2-phenethyl alcohol. *Schiz. pombe* was especially active. There was usually much more than 25 mg per liter in 24 Italian wines, 23.5–137.5 (average 47.6).

BY-PRODUCTS OTHER THAN ALCOHOLS

. . . are important to the quality of the product.

Products other than alcohol

Acetaldehyde, glycerin, 2,3-butylene glycol, lactic acid, succinic acid, and acetic acid are constant products of the alcoholic fermentation (Joslyn, 1940; Genevois, 1950). The acetaldehyde that accumulates at the end of the fermentation represents mostly the acetaldehyde that is not reduced to alcohol or is derived from autolysis of yeast (figs. 51, 64).

Carbonyls produced by alcoholic fermentation include: formaldehyde, acetaldehyde, propionaldehyde, butyraldehyde, isobutyraldehyde, valeraldehyde, isovaleraldehyde, 2-methylbutyraldehyde, hexanal, furfural, acetone, 2-butanone, and 2,3-butylene glycol (2,3-butanedione).

It is believed that the 2,3-butylene glycol is obtained by the condensation of two molecules of acetaldehyde, through the agency of the enzyme carboligase, to acetoin, and its reduction by other enzymes to the diol. The glycerin, which normally accumulates in the initial stages of the fermentation, is formed by the reduction of dihydroxyacetone phosphate by glycerophosphate, and the hydrolysis of the latter by phosphatase to glycerin and free phosphate. Formation of glycerin, acetic acid, and fixed acids during alcoholic fermentation is shown in figure 65.

Gancedo *et al.* (1968) investigated the pathways of utilization and production of glycerin by *Candida utilis* and *Sacch. cerevisiae.* Glycerol

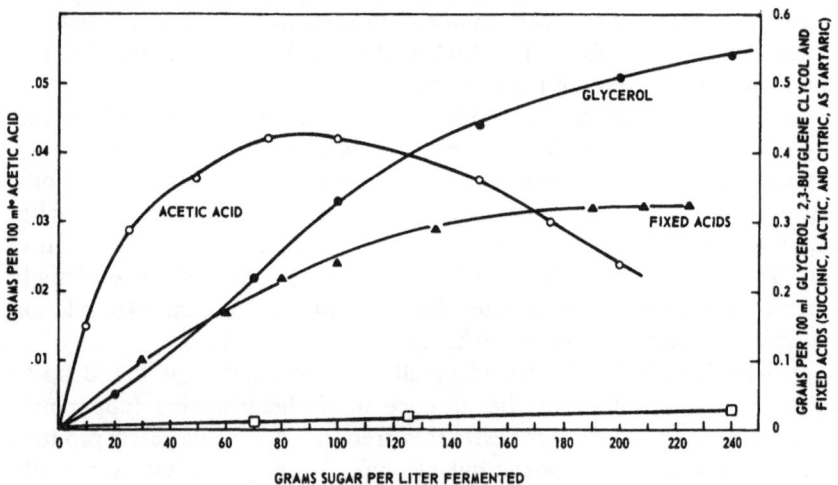

Figure 65. Formation of acetic acid, glycerin, and fixed acids during alcoholic fermentation.

was found to be produced by reduction of triose phosphate to α-glycerin phosphate by an NADH dependent enzyme. This is then hydrolyzed by an α-glycerophosphatase specific for the L form. The concentration of this phosphatase in yeasts is higher when they are grown on hexoses than when grown on non-sugar carbon sources.

Brockmann and Stier (1948) reported that at 30°C (86°F) the yield of glycerin after 40 to 50 per cent of the initial sugar was fermented varied from 3.7 to 4.2 gr per 100 gr of glucose utilized, and later decreased to between 3.0 and 3.5 gr. At 37°C (98.6°F) to 40°C (104°F) the yield of glycerin is considerably lower than at 30°C (86°F), being 2.7 to 3.2 gr per 100 gr of sugar when approximately half the sugar has been utilized. By the time 85 to 90 per cent of the sugar was fermented, the yield of glycerin amounted to only 1.8 to 2.3 gr per 100 gr of glucose utilized. This decrease in glycerin production with increase in temperature was accompanied by marked decrease in phosphatase activity at the higher temperature.

Lactic acid, which accumulates at a relatively constant rate (Hohl and Joslyn, 1941a) throughout the period of sugar utilization, is probably formed by reduction of pyruvic acid.

Succinic acid is now believed to be formed primarily by a condensation of pyruvic acid and carbon dioxide to form oxalacetic acid, which is reduced to succinic acid, and not, as formerly believed, by conversion of glutamic acid, derived from autolysis of yeast cells, to succinic acid (Kleinzeller, 1941). Kleinzeller demonstrated that the quantities of succinic acid formed in alcoholic fermentation were far in excess of those that could be obtained if all the glutamic acid present in the yeast cells were converted into succinic acid. Using C^{14} uniformly labeled malic acid, Mayer et al. (1964) found that it was mainly converted to succinic acid with a smaller production of lactic acid. Different yeasts produced different ratios of the two. They could not demonstrate alcohol formation from malic acid, which contradicts the results of Drawert et al. (1967) obtained by an apparently foolproof procedure. Drawert et al. had reported considerable metabolism of malic acid by Saccharomyces—about 34 per cent to ethanol, 31.7 per cent to 2-methyl-1-propanol, and 30.35 per cent to 3-methyl-1-butanol and 2-methyl-1-butanol.

Acetic acid, which accumulates only in the initial stages of the fermentation, is formed by the dismutation of acetaldehyde (oxidation of one molecule of acetaldehyde to acetic acid and the simultaneous reduction of another to alcohol) (Joslyn and Dunn, 1938; Peynaud, 1939–40; Ribéreau-Gayon and Peynaud, 1946). Since the accumulation of alcohol inhibits the activity of the oxidizing enzyme and stimulates the activity of the enzyme capable of reducing acetic acid to alcohol, the final acetic acid content will depend on the balance between dismutation of acetaldehyde and reduction of acetic acid.

Traces of formic acid (Hohl and Joslyn, 1941b) and citric acid (Ribéreau-Gayon and Peynaud, 1946) are produced during alcoholic fermentation. Carles et al. (1966) reported relatively large amounts of dimethylglyceric acid in wines. Usseglio-Tomasset (1967d) reported that wines contain about the same amount of formic acid as the musts from which they were fermented. During fermentation there is at first a slight increase and then a decrease. New dry and sweet table wines had about the same amount, but sweet wines show an increase during aging or during heat treatment. In 24 dry wines he found 0–8.4 mg per liter (average 1.7) and in 31 sweet wines (some aged) 0.67 to 8.7 (average 2.9). There was no correlation between the volatile acidity and the formic acid content.

According to Dubois and Jouret (1965), the volatile acids of some Carignane wines stored in bottles and in tanks were as follows (mg per liter):

| Acid | Bottle | | | Tank |
	1962	1963	1964	1964
Formic	2.4	6.3	1.5	4.1
Propionic	0.6	1.3	1.5	1.5
Isobutyric	1.05	1.4	1.8	0.8
n-Butyric	0.6	0.7	0.7	0.45
Isovaleric	0.4	0.7	0.6	0.02
Caproic	1.4	1.15	0.80	0.35
Acetic (total)	525.	660.	470.	450.

Red wines from moldy grapes fermented without sulfur dioxide had much higher formic acid contents—up to 40 to 50 mg per liter. In addition, small quantities of other by-products are formed—acetal, apparently arising from the biological condensation of alcohol and acetaldehyde, and esters, chiefly ethyl acetate. Other substances, such as the higher alcohols, accumulate as a result of the action of yeast on various precursors.

The amount of α-ketoglutaric acid formed during fermentation was found by Rankine (1968b) to vary with yeast strain (10-fold differences in a single must), musts (significantly yeast-must interaction), temperature (60 per cent as much at 15°C (59°F) as at 25°C (77°F)), pH (twice as much at 4.2 as at 3.0), amount of nitrogen (less with added ammonium sulfate), and L-glutamic acid (more when added). The reduced α-ketoglutaric acid when ammonium sulfate was added is interpreted as indicating use of endogenous α-ketoglutaric acid (formed from carbohydrate) in order to form glutamic acid. The mean values of the keto acids for 10 strains varied from 9 to 117 mg per liter (average 53). Rankine recommends use of low-α-ketoglutaric-acid producing yeast, low temperature fermentation and low pH. Less sulfur dioxide in the must also reduced α-ketoglutaric acid production.

The extraordinary lability of the by-products of alcoholic fermentation was demonstrated by Durmishidze (1966) using C^{14} labeled compounds. Acetic acid, acetaldehyde, and ethanol are rapidly assimilated by yeasts; glycerin and lactic and succinic acids, more slowly. Acetic acid was converted into acetaldehyde, ethanol, 2,3-butylene glycol, and glycolic and lactic acids. Lactic acid was converted to ethanol, acetic and glycolic acids, 2,3-butylene glycol, and glycerin. Glycerin was converted to acetaldehyde, ethanol, glycolic and lactic acids, and 2,3-butylene glycol. Succinic acid was converted to acetaldehyde, ethanol, acetic, fumaric, and malic acids, 2,3-butylene glycol, and glycerin. These results can be explained by the intervention of the tricarboxylic and glyoxylic acid cycles with the glycolytic cycle. See also p. 303 for similar results in crushed grapes on amino acids.

Acetic, butyric, and propionic acids, in contrast, have a marked inhibiting effect on alcoholic fermentation. Acetic acid, chiefly a product of bacterial fermentation, has a strong retarding influence. This is apparent at 0.2 per cent and increases with larger amounts until, at 0.5 to 1.0 per cent, all yeast activity stops. The toxicity of acetic acid toward yeast becomes greater with increase in temperature (Porchet, 1935).

Carbon dioxide may be utilized by the yeast in the synthesis of compounds such as succinic acid in the presence of pyruvic acid formed from sugar, but it exerts a slight restrictive action on growth and possibly on fermentation. Brandt (1945), in an investigation of the role of carbon dioxide in fermentation, found that under anaerobic conditions carbon dioxide has no influence on cell metabolism, rate of absorption, fermentation of glucose, or pH. Under aerobic conditions, however, carbon dioxide has a distinctly inhibitory effect upon growth of yeast, but activates endogenous respiration and absorption of glucose, and produces a change in pH. The relation of these differences to certain winery practices in Europe where the fermentations are conducted under a slight carbon dioxide pressure (see Garina-Canina, 1948, and p. 609) have not been studied in sufficient detail.

The mechanism of formation of ethyl acetate by *Hansenula* and other aerobic yeasts was investigated by Bedford (1942), Gray (1949), Davies *et al.* (1951), Peel (1951), and Tabachnik and Joslyn (1953*a*, 1953*b*). The data indicate that ethyl acetate is not formed through esterification by esterase of acetic acid and ethyl alcohol, but by other means. Peel (1951) and Tabachnik and Joslyn (1953*b*) demonstrated that the acetate part of ethyl acetate is formed aerobically within the yeast cell and is not derived from the acetic acid of the medium. These investigators are in agreement that ethyl acetate is the only ester formed during the fermentation of glucose by aerobic yeasts, and that it arises from alcohol oxidation. Oxygen, pH, and alcohol concentration as well as temperature influence ester accumulation. The esterase present in the yeast is not involved in ester formation, but affects ester accumulation as a result of hydrolysis of ethyl acetate formed in the oxidation of ethyl alcohol. Peynaud (1956b) pointed out that esters are formed also during fermentation under anaerobic conditions.

Laurema and Erikama (1968) recently investigated the effect of aeration, pH, and concentration of ethanol, acetaldehyde, and acetic acid on the production of ethyl acetate by washed *Hansenula anomala* cells. Yeast cultivated under aerobic conditions produced ethyl acetate from ethanol only in presence of added acetic acid. The maximum yield of ethyl acetate obtained was dependent on the concentration of undissociated acetic acid. They suggested that acetyl CoA formed from extracellular acetic acid was used for the ester formation as previously indicated by Nordström (1963a).

The most extensive and important recent contributions to our knowledge of ester formation by fermentative yeasts under anaerobic conditions were the studies of Nordström (1961, 1962a and b, 1963a–c, 1964a–e, 1965a–c). Nordström (1965a), in summarizing his investigations, pointed out that ethyl acetate was formed during the active alcoholic fermentation, and increased as alcohol formation increased, but was not formed at the later stages of fermentation when the fermentable sugar content was exhausted. The percentage of the added acid that was esterified increased with the number of carbon atoms from o for C_3 acids to 12 for C_6 to C_8 and then decreased to 10 for C_9 acids. The percentage of the alcohol added that was esterified increased from 1.3 per cent for C_4 alcohol to 20 per cent for C_7 alcohol. Yeast growth was closely related to ethyl acetate formation even under anaerobic conditions. Depression of yeast growth by deficiency in potassium, phosphate, magnesium ion, pantothenic acid, and other growth factor content resulted in decreased yield of ethyl acetate. Nordström found that coenzyme A (Co A) was involved in esterification during anaerobic fermentation. He reported that the nonenzymic esterification was 1,000 times too slow to account for ethyl acetate formed during anaerobic fermentation. The possibility that the reaction

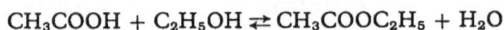

$$CH_3COOH + C_2H_5OH \rightleftharpoons CH_3COOC_2H_5 + H_2O$$

could be speeded up by esterase during alcoholic fermentation was ruled out by the fact that increase in ethyl acetate formation did not occur on addition of acetic acid to the fermentation medium (Nordström, 1961). High amounts of ethyl acetate are formed during aerobic growth of *Hansenula* and *Pichia* species (Peynaud, 1956b). The concentration of other esters formed may be several times the equilibrium value of the reaction above. Nordström (1962b, 1963a, 1965b, 1965c) obtained evidence that coenzyme A was involved in the esterification reaction and that ethyl acetate was formed by reaction of acetyl coenzyme A with alcohol, as follows:

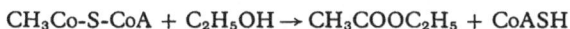

$$CH_3Co\text{-}S\text{-}CoA + C_2H_5OH \rightarrow CH_3COOC_2H_5 + CoASH$$

Peynaud (1956b) investigated the ability of various species of yeast to form ethyl acetate during fermentation of must. He reported that the film-forming aerobic yeasts, Hansenula and Pichia, produced the highest concentration of ethyl acetate, followed by Kloeckera apiculata, then S. ludwigii, and the group Candida pulcherrima, Kl. africana, Kl. magna, and Brettanomyces sp. The lowest levels of ethyl acetate were obtained by fermentative yeasts, Saccharomyces sp., Torulaspora sp. (now Sacch. sp.), and Torulopsis sp. In the absence of air the ethyl acetate produced varied from 416 mg per liter for Hansenula anomala to 23 for Sacch. cerevisiae var. ellipsoideus. In the presence of air these levels were increased to 879 to 34, respectively.

All other conditions being equal, Peynaud (1937) reported that the organic acids can be arranged in the following order of decreasing velocity of esterification: succinic, malic, lactic, tartaric, citric, propionic, and butyric. The hydrogen ions are the most active catalysts of esterification; the proportion of esters formed at 100°C, for example, is doubled by a decrease of one unit in pH value. The polyhydroxy acids can form both neutral and acid esters; the proportion of neutral esters is higher, the lower the pH, but in wines of pH 2.6 to 3.8 they are present in small amounts.

Other compounds

Hydrogen sulfide, methyl mercaptan, and ethyl mercaptan are foul-smelling compounds often found in wines. Ricketts and Coutts (1951) demonstrated that hydrogen sulfide was liberated during fermentation with six of the eight yeasts studied. Macher (1952) showed that the amount produced differed markedly between yeasts. Rankine (1963) has also shown differences in the tendency of different yeasts to produce hydrogen sulfide, and recommends selection of special yeasts that produce low sulfide. (See also p. 336.)

Rankine (1963) was unable to identify sulfate or sulfur-containing amino acids as a source of hydrogen sulfide. More hydrogen sulfide is apparently produced at lower must pH, higher temperatures, and greater depth of fermenter, according to Rankine.

Eschnauer and Tölg (1966) reported a maximum of 1.5 mg of hydrogen sulfide per liter during fermentation of normal sulfur-free musts, with most wines containing about 0.4 mg per liter. In four wines no sulfide was found after the first racking. However, wines with strong original sulfide odors may contain several mg per liter and a considerable amount may persist after racking. The detection threshold for hydrogen sulfide in wines is given as 1 mg per liter by Rankine (1963). Addition of sufficient sulfur dioxide to provide free sulfur dioxide removed 97 per cent of 10 mg per liter of hydrogen sulfide in five days.

Rankine (1963) reports ethanethiol, presumably formed by reaction of acetaldehyde and hydrogen sulfide, as one of the objectionable odors of some wines. Prompt removal of hydrogen sulfide from young wines is recommended as a means of preventing its formation.

Thoukis and Stern (1962) showed that the presence of even 1 mg per liter of free sulfur on grape musts would result in the formation of detectable amounts of hydrogen sulfide during alcoholic fermentation. This amount of hydrogen sulfide was easily removed by aeration. When as much as 5 mg per liter of free sulfur was present, sufficient hydrogen sulfide was formed to make its removal by aeration difficult. In their studies sulfur dioxide was a minor source of hydrogen sulfide. Accumulation of hydrogen sulfide is worse with cold anaerobic fermentations. The primary source of free sulfur is from vineyard spraying for mildew. Therefore, late spraying for mildew should be avoided.

Sulfur dioxide can interact with hydrogen sulfide. The first step in the reaction of hydrogen sulfide and sulfur dioxide, according to Albertson and McReynolds (1943), is the formation of a compound with a hydrogen bridge between hydrogen sulfide and the solvent. This compound then hydrolyzes to release hydrosulfide ions. The net result is oxidation of sulfite to sulfate and reduction of hydrogen sulfide to sulfur.

Cantarelli (1964) followed the formation of hydrogen sulfide and mercaptans during fermentation and aging with the following results (mmg per liter):

Days	Hydrogen sulfide	Mercaptan
0	3	0
2	41	10
6	126	7
9	480	27
13	398	86
31	300	72
43	260	120
62	210	230
93[a]	73	120
180	21	150
270	30	162

[a] Date of first racking.

Fermentation balancing equations

The quantity of some of these products, calculated on the basis of percentage of fermentable sugar transformed, is shown in table 44. Since the nature and proportions of these by-products influence the flavor of wine, several investigations have been made of the conditions for their accumulation and the balance between them (Gvaladze, 1936; Genevois et al., 1946, 1948a, 1948b). (See pp. 472-476.)

FACTORS INFLUENCING FERMENTATION
The products of alcoholic fermentation are influenced by many variables.

The chemical composition of the wine is influenced directly, but in a most complex manner, by that of the original must and by conditions of fermentation. Products of alcoholic fermentation other than alcohol and carbon dioxide contribute to the flavors and aroma acquired by the wine during fermentation. Their formation depends on the strain of yeast present, the nature of the must, the temperature, the aeration, and other conditions (see table 47).

The acidity and the tannin content of the must influence the course of fermentation as reflected in differences in amounts and kinds of by-products accumulated. At lower acidities, more acetaldehyde, glycerin, and volatile and fixed acids, but fewer aromatic principles, are formed. The extent of aeration influences the formation of fixed acids, such as succinic, and volatile acids and aldehydes, which increase in amount with increase in aeration.

The effect of sugars and other constituents of the grapes on the production of aromatic principles during fermentation is poorly understood. Peynaud (1939b) believes that the dicarboxylic amino acids are related to the quantity of bouquet produced during fermentation, but correlation of analytical and sensory data is needed.

Effect of temperature

Alcohol yield and the rate, amount, and proportions of the products of fermentation are affected by temperature. The influence of temperature on germination of spores on growth and on spore formation has been summarized in table 47.

Temperature is among the most important factors in flavor formation. According to most enologists, more bouquet is formed in a wine by a long, slow fermentation at a low temperature than by a short, rapid fermentation at higher temperatures. In cool fermentations the yeast apparently produces more esters and other aromatic bodies. This

367

TABLE 46

Formation of Higher Alcohols with Different Nitrogen Sources and Vitamins
(mg per liter)

Nitrogen source	+ All vitamins	− All vitamins	Minus thiamine	Minus biotin	Minus pantothenate	Minus nicotinamide	Minus pyridoxine	Minus meso-inositol	Minus p-amino-benzoate
Ammonium sulfate	47	242	88	100	122	30	40	60	57
Glutamic acid	47	82	52	40	15	40	52	40	40
Alanine	37	102	35	65	150	150	67	70	150
Phenylalanine	75	115	90	115	55	65	100	65	82
Glycine	172	257	235	172	265	177	177	172	177
Proline	150	277	175	237	140	165	187	140	145
Serine	155	132	132	147	125	140	85	132	140
Leucine	430	240	400	340	242	242	430	410	440
Valine	305	375	315	320	270	295	325	—	320

Source of data: Ribéreau-Gayon and Peynaud (1964-1966).

TABLE 47

EFFECT OF TEMPERATURE (°C) ON YEAST ACTIVITY

Saccharomyces species	Germination		Growth			Sporulation		
	Maximum	Minimum	Maximum	Optimum	Minimum	Maximum	Optimum	Minimum
cerevisiae	40	1–3	33–34	20–22	6–7	35–37	30	2–11
ellipsoideus	40–41	0.5	33–34	33–34	6–7	30.5–32.5	25	4–7.5
pastorianus	34	0.5	26–28	26–28	3–5	29–31.5	27.5	0.5–4
intermedius	40	0.5	26–28	26–28	3–5	27–29	25	0.5–4
validus (i.e. williamus)	39–40	0.5	26–28	26–28	3–5	27–29	25	4–8.5
turbidans	40	0.5	36–38	33–34	3–5	33–35	29	4–8

Source of data: Verona and Florenzano (1956).

statement is true of other fermented fruit products as well (e.g., cider). However, the possibility of odors arising from autolyzing yeasts in slow fermentations has not been adequately considered. See Uchimoto and Cruess, 1952.

Hohl and Cruess (1936) found that the production of alcohol in sterile grape juice inoculated with champagne yeast was markedly affected by temperature when a siruped method of fermentation was used. (Sirup was added at intervals and fermentation allowed to go until it stuck.) Their data, for 22° Brix grape juice in 500-ml bottles, are as follows:

Temperature of incubation °C	°F	Period of fermentation (days)	Number of sirup additions	Alcohol (per cent)
7.0	44.6	126	4	16.45
10.0	50.0	41	4	16.4
16.0	60.8	29	4	16.65
20.0–22.0	68–71.6	24	3	16.5
25.0	77.0	13	3	13.9
28.0	82.4	19	3	13.3
31.0	87.8	19	3	12.2
34.0	93.2	13	1	8.6
37.0	98.6	13	1	6.35

On rate and products

Temperature affects the rate of fermentation and the nature and amounts of by-products formed, but information on these factors is incomplete. Such experimental data as are available were obtained for the most part in fermentations carried out in small containers in which alcohol losses would be expected to be greater than under commercial conditions, particularly in closed fermenters. The most widely quoted data are those obtained by Müller-Thurgau (1884)—cited with slightly variant figures by Ribéreau-Gayon and Maurié (1938) and von der Heide and Schmitthenner (1922)—which are as follows (unfortunately the percentage of residual sugar is not given):

Sugar content (per cent by weight)	Fermentation temperature °C	°F	Duration of fermentation (days)	Alcohol content at end of fermentation (per cent)
12.7	37	96.8	13	4.21
	27	80.6	21	6.94
	18	64.4	33	6.95
	9	48.2	65	7.00
21.75	37	96.8	17.5	4.81
	27	80.6	24	9.37
	18	64.4	46	10.94
	9	48.2	100	11.76
30.3	37	96.8	14	5.17
	27	80.6	24	7.63
	18	64.4	34	9.05
	9	48.2	100	9.82

Roos and Chabert (1897) found somewhat similar results for grape musts containing 19.0 per cent sugar:

Fermentation temperature		Alcohol content	Residual sugar
°C	°F	(per cent)	(per cent)
25	77	10.9	0.25
30	86	10.9	0.20
35	95	9.6	0.23
40	104	9.3	0.27

Castelli (1941) investigated the effect of temperature on the fermentative activity of twenty cultures of yeast isolated from Italian musts. The yeasts were inoculated into 250 ml of white grape juice in heat-sterilized, cotton-stoppered, 300-ml cylindrical flasks and stored at various temperatures. The juice had a sugar content of 25.75 per cent and an acidity, as tartaric, of 1.22 per cent. After about forty days he analyzed the resulting wines for alcohol, total acidity, volatile acidity, and acetoin (acetylmethylcarbinol). His data for ten strains of *Saccharomyces cerevisiae* var. *ellipsoideus* are given in table 48. For these yeast strains the optimum temperature for alcoholic fermentation seems to be about 15.5°C (60°F), with considerably lower alcohol yields obtained at temperatures above or below that figure. The volatile acidity, which is higher than that found under California conditions, is lower with the lower fermentation temperature. Acetoin, when it was formed, was always present in higher concentrations at the higher temperatures.

Guymon and Castor (unpublished data) fermented French Colombard and Carignane musts at different stages of maturity at several temperatures. These fermentations were conducted in 2-gallon carboys containing one gallon of must. Three strains of yeast—Jerez, distillers', and Montrachet —were used. The average values obtained were as follows:

Variety	Initial Balling	Average alcohol content (per cent by volume)		
		At 12.2°C (64°F)	At 21.7°C (71°F)	At 33.9°C (93°F)
French Colombard	15.6	7.68	7.75	7.71
	20.4	11.31	11.16	11.11
	25.8	14.72	14.39	14.15
Carignane	20	11.50	11.34	11.07
	25.7	14.72[a]	15.22	11.77[b]

[a] One sample stuck; average without this, 15.19.
[b] All samples stuck, residual sugar.

Ough (1964b) reported a linear relationship between temperature and rate of fermentation. A multiple regression prediction equation based on degree Brix, pH, and ammonia content was calculated. Later, Ough

TABLE 48

EFFECT OF TEMPERATURE ON ALCOHOL AND ACID CONTENT OF WHITE GRAPE JUICE FERMENTED BY SACCHAROMYCES ELLIPSOIDEUS STRAINS

Strain, locality, date of isolation	5°–6.1°C (41°–43°F)			10°–12.2°C (50°–54°F)			15°–17.2°C (59°–63°F)		
	Alcohol (per cent)	Total acidity (per cent)	Volatile acidity (per cent)	Alcohol (per cent)	Total acidity (per cent)	Volatile acidity (per cent)	Alcohol (per cent)	Total acidity (per cent)	Volatile acidity (per cent)
1 Umbria, 1912	6.00	1.035	0.046	11.26	1.1	0.063	12.06	1.2	0.111
6 Colli Romani, 1939	12.28	1.115	0.022	12.20	1.16	0.054	12.48	1.2	0.051
7 Sier. M, I, 1932	9.22	1.095	0.060	13.26	1.05	0.091	12.48	1.125	0.098
20 Umbria, 1933	11.45	1.105	0.038	14.00	1.11	0.050	14.40	1.2	0.060
24 Umbria, 1933	6.16	1.095	0.074	11.72	1.05	0.090	14.30	1.03	0.091
26 Colli Romani, 1939	7.62	1.170	0.055	11.56	1.055	0.070	13.80	1.03	0.085
18 Chianti, 1936	10.66	1.055	0.061	11.20	1.07	0.091	13.50	1.1	0.108
31 Chianti, 1936	10.50	1.060	0.072	13.58	1.015	0.065	13.26	1.17	0.108
40 Var. *umbra*, Umbria, 1933	7.02	1.195	0.086	11.62	1.12	0.087	14.30	1.12	0.090
364 Var. *major*, Chianti, 1936	11.30	1.150	0.068	12.30	1.2	0.082	13.58	1.1	0.091

Strain, locality, date of isolation	25°C (77°F)			30°C (86°F)			37°C (98.6°F)		
	Alcohol (per cent)	Total acidity (per cent)	Volatile acidity (per cent)	Alcohol (per cent)	Total acidity (per cent)	Volatile acidity (per cent)	Alcohol (per cent)	Total acidity (per cent)	Volatile acidity (per cent)
1 Umbria, 1912	12.32	1.125	0.120	7.70	1.18	0.125	7.70	1.18	0.130
6 Colli Romani, 1939	9.00	1.125	0.054	5.48	1.13	0.097	3.50	1.13	0.098
7 Sier. M, I, 1932	11.86	1.2	0.086	5.02	1.18	0.113	4.20	1.085	0.143
20 Umbria, 1933	12.80	1.125	0.061	5.30	1.1	0.086	5.36	1.1	0.080
24 Umbria, 1933	12.40	1.05	0.091	5.10	1.025	0.108	2.70	1.06	0.109
26 Colli Romani, 1939	12.32	1.105	0.122	4.80	1.2	0.128	3.08	1.2	0.127
18 Chianti, 1936	10.76	1.08	0.126	6.00	1.17	0.100	5.30	1.2	0.113
31 Chianti, 1936	11.40	1.075	0.104	4.40	1.11	0.105	5.08	1.06	0.113
40 Var. *umbra*, Umbria, 1933	12.20	1.1	0.117	7.56	1.2	0.099	3.64	1.2	0.144
364 Var. *major*, Chianti, 1936	9.90	1.075	0.084	5.00	1.1	0.095	4.56	1.125	0.116

Source of data: Castelli (1941).

(1966a) showed that at constant nitrogen content maximum yeast cell population was inversely related to initial degree Brix. Both increased initial alcohol content and higher fermentation temperatures decreased yeast vitality (Ough, 1966b).

Ough and Amerine (1966) studied fermentation rate as influenced by ammonia and total nitrogen content. The previous prediction equations were more successful at low fermentation temperatures than at high.

The effect of fermentation temperature on higher alcohol formation was investigated by Ough et al. (1966). On the average, 47 per cent more 2-methyl-1-butanol and 3-methyl-1-butanol were formed at 21° or 26.7°C (70° or 80°F) and 1-propanol least. 2-Methyl-1-propanol formation seemed fairly independent of temperature. Ough and Amerine (1967a) found that the amount of levo- and meso-2,3-butylene glycol increased with increasing temperature from 10° to 21° to 33°C (50° to 70° to 91°F). Acetoin also increased greatly at the highest temperature. In contrast, total volatile esters, acetaldehyde, 2-methyl-1-butanol and 3-methyl-1-butanol alcohols were higher at the middle temperature. Acetic acid was lowest at the middle temperature.

Relation to yeast strain

The optimum temperature, as far as rate of fermentation is concerned, for most varieties of wine yeast is about 29.4°C (85°F), although for cold-tolerant yeasts it is considerably lower. Castan (1927) found that the optimum temperatures for three strains of yeast were 30°, 26.1°, and 25°C (86°, 79°, and 77°F), and that one yeast continued activity at 38.3°C (101°F). At temperatures below 21.1°C (70°F) the rate of fermentation is slow; growth and activity of many wine yeasts practically cease at 15.5°C (50°F).

From 82 fermenting musts Mayer and Pause (1965b) isolated a variety of yeasts and determined their fermentation velocity at temperatures of 5° to 20°C (41° to 68°F): production of volatile acids and hydrogen sulfide was less at lower alcohol contents. They confirmed Rankine's (1963) result that hydrogen sulfide formation markedly varies with the strain of the yeast. Malic acid was, surprisingly, retained less well in pure yeast cultures than in spontaneous fermentations.

Porchet (1938) reported that some varieties of yeast can cause alcoholic fermentation at temperatures below 0°C (32°F). She isolated two varieties of Sacch. cerevisiae var. ellipsoideus from wort stored at −3.0°C (26.6°F), at which temperature they brought about a spontaneous fermentation. She showed that resistance to cold was a characteristic racial property of these strains and not merely a temporary adaptation to low temperature. Wine yeast can be acclimatized to low temperatures by proper exposure and selection (Osterwalder, 1934*b*, 1941). Berand and Millet (1949) found that yeasts grown at 7.0°C (44.6°F) produced more alcohol per cell than did those grown at 25°C (77°F). They noted that after acclimatization to the lower temperature the yeasts retained their greater alcohol-producing properties for several fermentations at higher temperatures.

Above 29.4°C (85°F), yeast vigor decreases with increase in temperatures, and fermentation, especially in musts of high sugar content, generally ceases—the wine "sticks." This stuck wine is liable to attack by disease-producing bacteria and usually will not re-ferment unless special measures are taken, even though the temperature is reduced to normal (see p. 311).

The alcohol tolerance of yeast decreases with increase in temperature. The amount of alcohol tolerated by the enzyme systems is also less at higher temperatures. Thus Müller-Thurgau (1884) found that fermentation stopped with one yeast at about 36.1°C (97°F) with 3.8 per cent alcohol by weight; at about 32.8°C (81°F) with 7.5 per cent; at about 17.8°C (64°F) with 8.8 per cent; and at about 8.9°C (48°F) with 9.5 per cent. The higher alcohols (e.g., 1-propanol or 1-pentanol) exert a greater inhibitive action compared to ethanol. Some of those produced by wild yeasts may be involved in the sticking of wine (see p. 311).

In commercial production

In industrial trials in California, fermentations were successfully carried out at temperatures of 10°C (50°F). Tchelistcheff (1948) found that white musts fermented between 7.2°C (45°F) and 15.5°C (60°F), with a finishing temperature of 18.9°C (66°F) to 20.0°C (68°F), produce wines with greater freshness and fruitiness of flavor, lower volatile acidity, higher glycerin content, and a relatively smaller amount of lees than do musts fermented at temperatures of from 25.5°C (78°F) to 31.1°C (88°F). The wines fermented at low temperatures had a better bacteriological stability, a higher production of alcohol, and a better precipitation of tartrates and iron. They were slightly higher in pH than the normally fermented wines. He noticed an excessive yeastiness in some of the cold-fermented wines, and the cost of production was greater.

Cruess (1948) notes that cooling of red musts to a low temperature, 7.2°C (45°F), before fermentation has been tried as a means of preventing excessive temperature rise. However, this makes color extraction difficult and the quality of the product always suffers.

Heat generated during fermentation

Considerable heat is generated during alcoholic fermentation and, unless dissipated, will cause a rise in the temperature of the must. The amount liberated in the fermentation has been calculated from the difference of heats of combustion for the fermented material and for the products formed. This computation is not very accurate: errors occur in combustion-heat determinations; corrections must be made for the heats of solutions of products and for escaping gases; the concentration of reacting substances continuously decreases while that of the products increases; and the fermentation occurs in successive stages of decomposition of sugar. The heat evolved from 180 gr of sugar consumed in the reaction of $C_6H_{12}O_6 = 2C_2H_5OH + 2CO_2$ (see p. 348) has been calculated by Rahn (1932) to be 26.0 calories (kg calories) when the sugar, alcohol, and carbon dioxide are in their standard state; other reported calculations vary from 22 to 33. Genevois (1936) has reported 28.0 calories when the reacting substances and products are in dilute solution. Winzler and Baumberger (1938) give a calculated value of 22.5 for glucose at a concentration of $1 \times 10^{-4}M$, alcohol at $2 \times 10^{-3}M$, and carbon dioxide gas at a pressure of 0.0003 atmospheres.

The heat evolved in actual fermentations has been measured by a number of investigators. Of the earlier measurements, those of Bouffard (1895) are most consistent; he found the heat of fermentation of grape juice to vary from 23.4 to 23.7 calories per 180 gr of sugar fermented. Rubner (1904, 1913) reported 24.0 and, after correcting for the formation of by-products (such as glycerin and succinic acid) during the alcoholic fermentation, calculated a value of 24.055, which agrees closely with his experimentally measured value. Although some European enologists consider the heat liberated in fermentation to be about 27 calories, the majority accept Bouffard's early measurement of 23.5 calories.

On the basis of 23.5 calories per 180 gr of sugar, a must containing 22 per cent of sugar would rise about 28.9°C (52°F) if all the heat developed by fermentation were prevented from escaping. That is, for each Brix degree of sugar in the must, sufficient heat is generated during fermentation to raise the temperature approximately 1.3°C (2.34°F). If its initial temperature were 15.5°C (60°F) it would reach 37.8°C (100°F) and stick while still containing 5 per cent sugar. In practice, however, these temperatures are not reached because heat is lost by radiation from the open surface of the fermenter to the surroundings, by conduction through the walls of the fermenter, and by loss in the evolved carbon dioxide gas. The temperature of the surrounding air affects the percentage of heat lost.

Heat dissipated

Large volumes of carbon dioxide gas are given off during fermentation. Thus 1,000 gallons of must of 22 per cent sugar content, occupying about 134 cubic feet, will generate, at 25.5°C (80°F), about 9,380 cubic feet of gas. The heat content of this volume of gas is 210 B.T.U. per degree Fahrenheit, or approximately one-fifth that formed in fermentation (1,040 B.T.U.). Actually, the proportion of the heat liberated in fermentation which is absorbed and eliminated by carbon dioxide gas may be higher, since the gas evolved entrains moisture vapor. Mathieu and Mathieu (1938) reported that 1 liter of must containing 180 gr of sugar will liberate at 35°C (95°F) about 50 liters of gas carrying about one-fifth of the heat liberated in fermentation.

CONTROL OF TEMPERATURE
Cooling and control of temperature during fermentation are necessary to obtain wines of fine quality and good stability.

Need for cooling

In the opinion of most enologists, better bouquet and aroma are formed by a long, gradual fermentation at low temperatures than by a short, quick fermentation at higher temperatures (see Jordan, 1911). In cool fermentations the yeast apparently produces more esters and other aromatic bodies. This statement holds true for other fermented fruit products also (e.g., cider). At lower temperatures the yield of alcohol is greater, partly because of lower losses of alcohol by evaporation and by entrainment in the escaping carbon dioxide gas, and partly because of the more efficient transformation of fermentable sugar into alcohol. The wine obtained by cool fermentation is more easily cleared and is less susceptible to bacterial attack and spoilage.

At high temperatures the yeast causes reactions unfavorable to the quality. A wine fermented at 21.1°C (70°F) is smoother and fresher, with a more desirable bouquet, than one fermented at 32.2° to 35°C (90° to 95°F). At the latter temperature range the wine, if fermented on the skins, has more color, tannin, and body because of the greater solubility of those substances at the higher temperature (Bioletti, 1905). But at high temperatures there is danger of "sticking" while considerable amounts of sugar remain in the wine. Excessively high temperatures during fermentation encourage the growth of wine-disease bacteria, and these result in an undesirably high production of volatile acids and off-flavors.

The temperature to which the fermenting grapes or must will rise is determined by the temperature at time of crushing, plus the rise in temperature from the heat generated by fermentation, and minus the heat lost during fermentation by radiation, conduction, and gas liberation. The warmer the grapes and the more sugar they contain, therefore, the higher the rise in temperature. Temperature will rise less with a smaller fermenting mass, cooler air, and slower fermentation. Cooling the grapes or must before fermentation, fermenting in vats with a large radiating surface per unit volume, and cooling the fermenting must itself are the means by which dangerously high temperatures may be prevented. (See pp. 207–209.) Even in cool coastal regions where small fermentation vats are used, cooling during fermentation is necessary to control the temperature, particularly early in the vintage season. Where large fermenting vats of 3,000–10,000-gallon capacity are used for fermenting dry table wines, the fermenting must has to be cooled artificially, even if the grapes are cool when crushed (fig. 66).

Figure 66. Stainless steel fermenting tanks with refrigerated cooling jackets. (Courtesy Valley Foundry and Machine Works, Inc., Fresno.)

Saller (1957) recommended settling of white musts, inoculating with cold-tolerant pure yeast cultures, and fermenting at 4°C (39.2°F) as a means of preventing growth and competition of undesirable yeasts. Saller (1958) summarized the advantages of cold fermentations for producing white table wines as follows: more alcohol, less volatile acidity, more aromatic substances, lower reducing material required to reach the same redox potential—thus requiring less sulfur dioxide, earlier maturation for bottling, greater tartrate stability, and a fruitier flavor. Possible disadvantages are lower glycerin and extract. (Fig. 67 is a typical example of a controlled fermentation.) Saller also developed an automatic regulator which lowered the temperature if the amount of carbon dioxide produced increased too rapidly.

The heat lost during the full course of the average fermentation, conducted in open vats not exceeding 3,000-gallon capacity, is about 50 per cent of that generated by fermentation. The exact amount of heat lost depends on the size and shape of the fermenter, the material out of which it is constructed, and the temperature of the surrounding air. Much of the available data on heat losses were obtained for wood fermenters during the

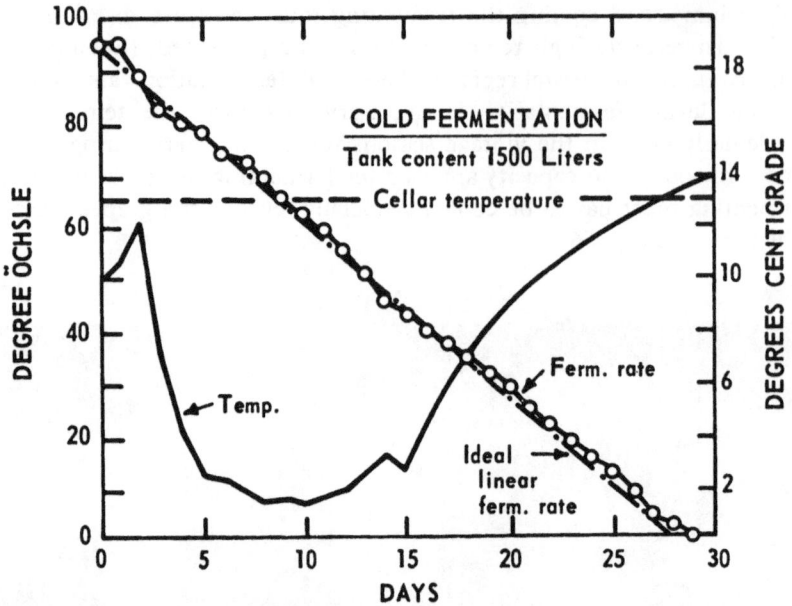

Figure 67. Controlled fermentation, linear decrease in specific gravity. Source of data: Saller (1958).

pre-Prohibition period, and few data are available for present conditions, particularly for the concrete fermenters now commonly used, in which heat loss is limited mostly to the open surface. During the initial stages of fermentation (first-half stage), the heat loss is, on the average, about 33 per cent. In very hot weather and in larger vats (up to 10,000-gallon capacity) the over-all loss may be less than 33 per cent.

Amount of cooling needed

The amount of cooling necessary has been calculated by Bioletti (1960a) as follows:

Let

S = Brix degree of must (approximate sugar content),
T = temperature of contents of vat in degrees Fahrenheit,
M = maximum temperature desired in degrees Fahrenheit, and
C = number of degrees Fahrenheit necessary to remove by cooling.

Then

$$C = 1.17° S + T - M.$$

Suppose it is desired to ferment out a must of 24° Brix, initially at 70°F, so that 80°F will not be exceeded during fermentation. Then

$S = 24$, $T = 70$, $M = 80$, and

$C = (1.17° × 24) + 70° − 80° = 28° + 70° − 80° = 18°F$.

or, in °C,

$C = (0.65 × 24) + 21.1°C − 26.7 = 10.0°C$.

Every gallon in the vat must be cooled 18°F below the initial temperature 21.1°C (70°F) in order not to exceed a maximum of 26.7°C (80°F). This estimate is based on the assumption that heat equivalent to half of the actual temperature rise of 2.34° per °F of sugar fermented is dissipated into the surroundings. With large fermenters, however, this loss is somewhat less in practice, and about 1.5 to 1.8°F of heat per degree of sugar fermented must be eliminated by cooling, instead of 1.17°.

COOLING . . .
. . . is required for most fermentations.

Cooling the must before and during fermentation is necessary in California to obtain wines of high quality and stability. Fermentations at high temperatures are always undesirable. The amount of cooling required depends on the initial temperature and sugar content of the grapes, the external temperatures during fermentation, the size and type of fermenters, and the type of wine. For the cooling requirements for different types, see chapters 13 to 17 and page 207 f.

When to cool
Cooling must not be too extreme and should not be carried out during the later stages of fermentation, or fermentation may be too prolonged. It is better to cool the must initially, if necessary, to a point some 5.5°C (10°F) below the maximum temperature desired, and then to cool again at the height of the fermentation, when the Brix degree has decreased to approximately half its initial value and the yeast cells are more numerous and more active. At this stage, cooling will have less retarding effect upon fermentation than if done earlier or later. The temperature of the must at this stage should be reduced about 0.83°C (1.5°F) for each degree Brix so that the final temperature will not exceed the maximum desirable. Addition of sulfur dioxide to musts before or during the fermentation does not ordinarily control the rate of fermentation enough to make cooling unnecessary.

A uniform low fermentation temperature is best for white table wines and can sometimes be obtained by means of cooling coils (in the fermenter) through which cold water or other refrigerant is passed. Another method is to pass the wine through tubular or plate heat exchangers (see p. 209). Because such exchangers clog easily, only screened juice should be used in them. It is possible to secure a linear rate of fermentation by controlling the temperature (fig. 67).

USE OF SULFUR DIOXIDE . . .
. . . is of ancient origin and is universal.

Sulfur dioxide, as the fumes of burning elemental sulfur, has been used in wine making since antiquity to disinfect containers, control fermentation, and preserve wine. It was probably known to the early Egyptians and Romans, and the enological literature of the eighteenth century contains numerous references to its use. It has been and is used in all wine-making regions of the world. It is indispensable in the production of wines in hot countries as attested to by the early observations reported by Bioletti (1905) and more recently by Marcilla Arrazola (1946), Dugast (1929), and others. The historical basis of its use is discussed by Bioletti (1912), Ventre (1931), Brémond (1937b), Joslyn and Braverman (1954), Blouin (1966), Ribéreau-Gayon and Peynaud (1964–1966), and Amerine et al. (1967).

The available literature on chemistry and technology as well as toxicology of sulfur dioxide was recently reviewed in the article by Heydenreich (1967) and in the more comprehensive text of Schroeter (1966).

In spite of its ancient origin, only recently has the mechanism of its antimicrobial action and the chemistry of its use in preventing undesirable changes in color and flavor been elucidated. If properly used it is beneficial to the quality of even the finest wines, particularly white table wines. When used under optimal conditions in minimum quantities it leads to the production of sound wines free from undesirable microörganisms. It will prevent microbial spoilage and protect wines against excessive oxidation during storage and aging. In excessive amounts it detracts from the aroma and bouquet, interferes with natural aging, and may cause undesirable odors and undesirable changes in clarity and color of wines. In spite of the wide dissemination of information on the proper use of sulfur dioxide and sulfites in California, there is still a wide variation in levels of sulfur dioxide in marketed wines, and occasionally excessive quantities are used. Joslyn (1954a, 1954b) expressed concern about this variability in sulfur dioxide content of California wines, and Kielhöfer (1966) considers it to be a major problem in the German wine industry.

The utility of sulfur dioxide and its acid salts is due to the unique properties of its aqueous solutions. They combine antimicrobial with antioxidative activity. Sulfurous acid has selective bactericidal and fungicidal action upon which depends its use in controlling fermentation of musts and its ability to prevent microbial spoilage of wine. Just as salt is used to inhibit undesirable microbial activity in pickling by lactic acid fermentation, so sulfur dioxide is used in wine making by promoting desirable alcoholic fermentation and inhibiting activity of undesirable yeasts and bacteria. In wine making, unlike brewing, sterilization by sulfiting instead of pasteurization by heating, prior to the addition of a starter of desirable pure wine yeast, is used to control fermentation. Only in the production of wines from *V. labrusca* (Concord) grapes or from various berries is heat treatment used for both color extraction and sterilization before fermentation. Sulfur dioxide also functions as an antioxidant in inhibiting enzyme-catalyzed oxidative discoloration and nonenzymic browning during fermentation, aging, and storage of wine. No other known preservative combines the desirable antimicrobial and antioxidative activity of sulfur dioxide.

The utility of sulfur dioxide and its compounds, however, is limited by its ready combination with carbonyl compounds (aldehydes, sugars, etc.), unsaturated aliphatic compounds, proteins, and other compounds, and its oxidizability by oxygen. In the combined form it is much less toxic and less effective as an antioxidant, and in its oxidized form as sulfate it is inert. The presence of sulfite-binding compounds which compete for the added sulfite with microörganisms to be destroyed will reduce the efficacy of sulfite as a preservative. While the technique of break-point chlorination has been well established in the application of chlorine as a sterilizing agent in the food industries, similar techniques for evaluating and overcoming the competitive effect of organic matter for sulfurous acid have not been developed. In break-point chlorination the chlorine demand of the organic matter present is satisfied and a minimum level of free chlorine available for reaction with microbial cells is maintained (Mercer and Somers, 1957). The nearest approach to this is the emphasis of Moreau and Vinet (1928) on the determination of the conditions in sweet wines that reduce the effective concentration of added sulfite. Moreau and Vinet (1928, 1937) proposed the determination of two indices, T and R, as measures of the ability of wine to combine with sulfurous acid. Subsequently Benevegnin et al. (1951) and Benevegnin and Michael (1952) modified this procedure. Rankine (1966c), however, cautioned against reliance on the total sulfur dioxide content as a measure of the free sulfur dioxide level. Blouin (1963, 1966) stressed the variable proportion of sulfur dioxide bound to other constituents of wine: pyruvic, α-ketoglutaric, and galacturonic acids, and so forth.

The distinction between "free" sulfur dioxide, capable of directly and rapidly reducing iodine, and "combined" sulfur dioxide, that not oxidizable by iodine, has been recognized since 1892, when Ripper introduced his procedure for differentiating them. It is only recently, however, that the chemical transformations of sulfur dioxide in wine have been investigated and elucidated in detail. The most notable additions to our knowledge of these transformations are the thesis of Blouin (1966) and the report of Rankine (1966c). In most of the viticultural regions of the world, maximum limits for both total and free sulfur dioxide are set, but in California and the United States only the total sulfur dioxide content is regulated. Heydenreich (1967) has summarized the limits for total and free sulfur dioxide for thirty-nine countries in a wide variety of foods, including wines. Amerine (1965b) also gives information for a number of countries. Because of concern about the possible toxicity of sulphur dioxide some countries have reduced, or are discussing reducing, the maximum limit to 200 mg per liter.

The chemical and physical properties of sulfur dioxide were reviewed by Prillinger (1963), who emphasized the fact that sulfur dioxide acts not only as a reducing agent but also as an enzyme inhibitor. Schroeter's (1966) monograph on sulfur dioxide discusses in some detail its preparation, properties, reactions, analyses, and applications. It is recommended for its information on sulfur dioxide and its compounds.

Sulfur dioxide and its compounds

Sulfur dioxide (SO_2), except for common salt (sodium chloride) and the nitrates and nitrites, is the main inorganic preservative used in processing of foods as such or as its sulfites. Over eight million tons are produced annually for conversion into sulfuric acid: one million tons are generated for use in production of sulfited pulp, and about 300,000 tons are used in industrial and agricultural applications, according to Jacobs (1951) and Schallis and Macaluso (1954). Sulfur dioxide is a readily condensible, colorless gas. It is produced commercially by burning sulfur in air, by roasting of pyrites and other metal sulfides, or by reduction of gypsum with coke. Sulfur dioxide was originally obtained for use in wine making by burning sulfur in the form of a wick or candle, but this is no longer commonly practiced. Liquid sulfur dioxide, the gas liquefied under pressure and stored in heavy-walled steel cylinders, is now being used widely. Sulfites, bisulfites, and metabisulfites of alkali metals, such as potassium or sodium, are used where smaller quantities of wines are to be sulfited. Their utility depends on their conversion in the presence of acids, such as the organic acids of musts or wine, into available sulfurous acid.

Sulfur dioxide gas is readily absorbed in water, but Whitney and Vivian (1949) and Whitney et al. (1953) reported that both gas and liquid film resistances are involved in this solution. (See also Johnstone and Leppla, 1934.) While hydration of sulfur dioxide was recently shown to be one of the most rapid hydrolytic reactions known, modern evidence indicates that aqueous solutions of sulfur dioxide contain no detectable molecules of sulfurous acid (H_2SO_3). The existence of the compound sulfurous acid was questioned as early as 1938. On the basis of ultraviolet, infrared, Raman, fluorescent, microwave, and mass spectra and on recent electron spin resonance and X-ray diffraction data summarized by Schroeter (1966), it is now recognized that sulfurous acid *per se* does not exist. The reactions occurring in aqueous solution of sulfur dioxide are as follows:

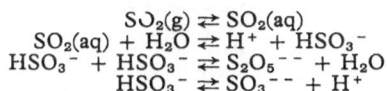

$$SO_2(g) \rightleftarrows SO_2(aq)$$
$$SO_2(aq) + H_2O \rightleftarrows H^+ + HSO_3^-$$
$$HSO_3^- + HSO_3^- \rightleftarrows S_2O_5^{--} + H_2O$$
$$HSO_3^- \rightleftarrows SO_3^{--} + H^+$$

Although the un-ionized sulfur species in aqueous solution of sulfur dioxide is almost entirely uncombined sulfur dioxide molecules, the term "sulfurous acid" is used to mean an aqueous solution of sulfur dioxide ($H_2O + SO_2$). The ionization constants of "sulfurous acid" have been reported in the literature since 1904. The range in values reported is shown in table 49, adapted from the Gmelin Institute (1960). In 1911,

TABLE 49

RANGE IN REPORTED VALUES FOR DISSOCIATION CONSTANTS
OF SULFUROUS ACID

Dissociation constant		Temperature			
$K_1 \times 10^2$	$K_2 \times 10^7$	°C	°F	Method	Authority
1.6	—	20	68	Cryoscopic	Drucker (1904)
1.74	—	25	77	Conductivity	Kerp and Bauer (1907c)
>3<1	—	0–50	32–122	Cond., Cryo.	Lindner (1912)
—	50	25	77	Conductivity	Jellinek (1911)
—	1.0	15	59	Neutralization	Kolthoff (1920)
1.2	—	25	77	Conductivity	Sherrill and Noyes (1926)
—	5.0	15	59	Potentiometric	Drucker (1927)
2.84–0.23	—	5–90	41–194	Conductivity	Campbell and Maass (1930)
1.73	—	25	77	Conductivity	Campbell and Maass (1930)
3.074–1.408	—	0–25	32–77	Conductivity	Morgan and Maass (1931)
3.25	—	25	77	Conductivity	Morgan and Maass (1931)

383

TABLE 49 (*cont.*)

RANGE IN REPORTED VALUES FOR DISSOCIATION CONSTANTS
OF SULFUROUS ACID

Dissociation constant		Temperature			
$K_1 \times 10^2$	$K_2 \times 10^7$	°C	°F	Method	Authority
0.2	0.182	14	57.2	Antimony electrode	Britton and Robinson (1931)
—	0.013	room		Tungsten electrode	Britton and Dodd (1931)
1.66	1.0	18	64.4	Glass electrode	Britton and Robinson (1931, 1932)
1.5	—	25	77	Solubility	Davis (1932)
1.3	—	25	77	Conductivity	Johnstone and Leppla (1934)
—	1.0	20	68	Glass electrode	Rumpf (1933)
—	−2.51	room		pH measurement	Bäckström (1934)
1.27	0.62	25	77	Glass electrode	Yui (1940)
1.72	0.62	25	77	Potentiometric	Tartar and Garretson (1941)
1.66	0.72	room		pH measurement	Airola (1941)
1.70	1.0	room		Neutralization	Stenger and Kolthoff (1947)
1.25	0.56	27	77	Energy of formation	Latimer (1952)

Source of data: Gmelin Institut für Anorganische Chemie (1960).

Jellenek, using three different methods, obtained widely diverse values for the second ionization constant (7×10^{-5}, 5.8×10^{-6}, and 3.2×10^{-7}) and gave as his final value the mean of these (5×10^{-6}). For many years this value was reported in handbook tables and cited in reference texts. It was used by Vas and Ingram (1949), Josiyn and Braverman (1954), and Amerine *et al.* (1967). The correct values for the ionization constant (as given by Tartar and Garretson in 1941, and cited by Schroeter, 1966) are $K_1 = 1.72 \times 10^{-2}$ and $K_2 = 6.24 \times 10^{-8}$. Prakke and Stiasny (1933), using the values $K_1 = 1.7 \times 10^{-2}$ and $K_2 = 1 \times 10^{-7}$, calculate the concentration of various ionic species in aqueous solution at various pH values as shown in table 50. This indicates that at pH values below 1.77, sulfur dioxide predominates; at pH levels between 1.77 and 5.0, HSO_3^- predominates and SO_3^{--} is not present in appreciable amounts until the pH is 6 or above. These considerations apply also to wine, but in wines the alcohol, sugar, and other organic compounds and salts influence the ionization constants. Some data are available on the effect of ionic strength, ethanol content, and temperature on dissociation of sulfurous acid, but the true ionization constants in musts and in wines are not known. Most of the data on the effect of pH on efficacy of sulfurous acid are based on approximations assuming that conditions are

TABLE 50

RELATIVE CONCENTRATION OF VARIOUS MOLECULAR AND IONIC
SPECIES IN AQUEOUS SOLUTIONS OF SULFUR DIOXIDE AT
VARIOUS pH LEVELS

pH	Per cent H_2SO_3	Per cent HSO_3^-	Per cent SO_3^{--}
0.0	98.3	1.7	—
1.0	85.5	14.5	—
1.77	50.0	50.0	—
3.0	5.5	94.5	—
3.3	0.2	99.8	—
4.0	—	100.0	—
5.0	—	99.0	1.0
6.0	—	80.9	19.1
7.0	—	50.0	50.0
8.0	—	9.0	91.0
9.0	—	1.0	99.0
10.0	—	0.1	99.9

Source of data: Prakke and Stiasny (1933).

similar to those in aqueous solutions of sulfur dioxide. Thus even recently Schelhorn (1951b, 1953) reported undissociated sulfurous acid to be 100 to 500 times as lethal as benzoic acid on the basis of calculating the effective concentration of sulfur dioxide in solutions of sodium sulfite at pH 4.5 and 6.0. Moreau and Vinet (1937) had used similar considerations to evaluate the antiseptic power of sulfur dioxide. They reported that at pH 2.8, in aqueous solution, about 10 per cent of sulfurous acid was in the "free" condition, whereas at 3.8 only 1 per cent was free. They reported that a decrease in total acidity of an acid must containing 260 mg of free sulfur dioxide per liter of 0.3 per cent (expressed as tartaric) would decrease antiseptic power by 37.6 per cent. Nègre and Dugal (1967) recommend no addition of sulfur dioxide for musts with a pH below 3.1. Even up to pH 3.3 they state that only minimal amounts be used. The lower the alcohol content and titratable acidity and the higher the pH, the more sulfur dioxide required.

Combination of sulfur dioxide and bisulfites with constituents of must and wine

The combination of sulfur dioxide and sulfites with various constituents of must, wine, and other food products has been known for over seventy-five years. The early literature is reviewed by Kerp (1904a, 1904b, 1904c), Monier-Williams (1927), Joslyn and Braverman (1954), Gehman and Osman (1954), Kielhöfer and Würdig (1960f, 1961c), Rankine (1966c), Blouin (1966), and Schroeter (1966). It is now well established that

385

sulfurous acid and the acid sulfites can combine with the carbonyl groups of aldehydes, of aldose sugars, of glyoxylic, pyruvic, α-ketoglutaric, or galacturonic acids, of certain unsaturated compounds, and of certain phenolic compounds, particularly caffeic and p-coumaric acid (Hennig and Burkhardt, 1960a, 1960b). Kerp called attention to the fact that in wines the sulfite is bound almost entirely by acetaldehyde and that only in strongly sulfited wines is the excess sulfur dioxide bound by the sugars. In sulfited musts and strongly sulfited sweet wines, aldose sugars, particularly glucose, bind the sulfite. Most of the published data on the rate and extent of combination of sulfite with carbonyl and other compounds are based on inference from data on decrease in rate of reduction of iodine solution by sulfite solutions containing various added chemicals. Only rarely has the actual sulfite derivative been isolated and investigated. This was done early for formaldehyde (Suter, 1944) and more recently for glucose and arabinose by Braverman (1953) and for mannose and galactose by Braverman and Kopelman (1961). Rehm et al. (1965) prepared and used aldehyde, sugar, and keto acid sulfonates in their investigations of the antimicrobial properties of these compounds.

The bisulfite addition products of compounds containing reactive carbonyl groups are α-hydroxy sulfonic acids (Suter, 1944; Schroeter, 1966). The acetaldehyde-bisulfite compound is more correctly designated as acetaldehyde-α-hydroxy sulfonate. The extent of combination is less and the rate of binding is slower the lower the pH. The relative amount of combined bisulfite is given by the dissociation constants for the particular compound sulfonated. Kolthoff and Stenger (1942) published dissociation constants of aldehyde sulfonic acids at 25°C (77°F) shown in table 51. The lower the value for K, the less is the compound dissociated. Thus acetaldehyde in 0.1 N solution is dissociated to the extent of only 0.45 per cent, while glucose under the same conditions is dissociated

TABLE 51

DISSOCIATION CONSTANTS OF ALDEHYDE SULFONIC ACIDS AT 25°C

Aldehyde	K
Formaldehyde	1.2×10^{-7}
Acetaldehyde	2.5×10^{-6}
Benzaldehyde	1.0×10^{-4}
Acetone	$3.5-4.0 \times 10^{-3}$
Furfural	7.2×10^{-4}
Chloral	3.5×10^{-2}
Arabinose	3.5×10^{-2}
Glucose	2.2×10^{-1}

Source of data: Kolthoff and Stenger (1942).

to 74.61 per cent. The dissociation constant is known to increase with temperature. For acetaldehyde it is five times as great at 37.5°C (99.5°F) as at 25°C (77°F). For further information see Kerp (1903) and Kerp and Bauer (1904, 1907a, b).

The dissociation constants and the percentage of sulfur dioxide fixed at 20°C (68°F) at pH 3–4 with 50 mg per liter of free sulfur dioxide in cider were determined for various compounds by Burroughs and Whiting (1960) as follows:

Compound	Dissociation constant	Percentage combined
Acetaldehyde	1.5×10^{-6}	100
Pyruvic acid	4.0×10^{-4}	66
α-Ketoglutaric acid	8.8×10^{-4}	47
L-Xylosone	2.1×10^{-3}	27
Monogalacturonic acid	3.0×10^{-2}	2.5
Trigalacturonic acid	3.7×10^{-2}	2.1
Xylose	6.9×10^{-2}	1.1
Glucose	6.4×10^{-1}	0.12

Kielhöfer (1958) obtained similar results in wine.

The power of several carbonyl compounds to combine with sulfur dioxide was given by Rankine (1966c) as follows:

	Normal range (mg per liter)	pK	SO_2 at 50 mg per liter free Total bound (per cent)	Bound (mg per liter)
Acetaldehyde	20–100	1.5×10^{-6}	100	29–145
Pyruvic acid	0–100	4.0×10^{-4}	66	0–48
α-Ketoglutaric acid	15–40	8.8×10^{-4}	47	0–8
Glucose	0–10 (per cent)	6.4×10^{-1}	0.1	0–50

Fructose has even less affinity for sulfur dioxide than glucose. Arabinose has a higher affinity for sulfur dioxide than glucose, but very little is present in wines.

More extensive data on the combination of various compounds with bisulfite were presented by Blouin (1966). The dissociation constants he obtained are shown in table 52. At a level of free sulfur dioxide of 20 mg per liter, substances with a K value of 0.003×10^{-3} or less will completely bind the sulfur dioxide; substances with a K value of 31×10^{-3} or higher will bind less than 1 per cent. The compounds in the first category (e.g., acetaldehyde) are those most likely to play an important part in binding sulfur dioxide.

Blouin (1966) observed an interesting effect of metals on the fixation of sulfur dioxide by tartaric acid solutions in the presence and absence of various sugars when exposed to air. At pH 3.3 and a storage period of 26 days the amount of fixed sulfur dioxide with tartaric acid present increased from 3 mg without metals to 42 mg with 20 mg of iron and 2 mg

TABLE 52

DISSOCIATION CONSTANTS OF SULFITE ADDITION OF VARIOUS SUBSTANCES
AT 20°C (68°F)

Substance	$K \times 10^{-3}$	Substance	$K \times 10^{-3}$
Aldehydes and ketones		Sugars	
Glyceric aldehyde	0.05	Arabinose	40.0
Glycolic aldehyde	0.02	Xylose	150.0
Propanal	0.01	Ribose	30.0
1-Butanal	0.01	Glucose	900.0
3–methyl–1–propanal	0.01	Mannose	80.0
3–methanyl–1–butanal	0.01	Galactose	165.0
Hexanal	0.01	Rhamnose	165.0
Decanal	0.01	Fucose	210.0
Acetone	3.8	Fructose	15,000.0
Dihydroxyacetone	1.5	Fructose-1,6-diphosphate	300.0
Triose phosphate	0.15	Sorbose	110.0
Carbonyl acids		Maltose	800.0
Glyoxylic	0.008	Cellobiose	550.0
Glucuronic	50.0	Lactose	500.0
Galacturonic	17.0	Sucrose	5400.0
Glucurone	2.5	Raffinose	1800.0
Dioxy tartaric	200.0		
Dioxyfumaric	0.2		
Oxalacetic	0.2		
α-Ketoglutaric	0.5		
Pyruvic	0.3		
Levulinic	15.0		

Source of data: Blouin (1966).

of copper per liter; in the presence of glucose from 25 without metals to 155 with metals; in the presence of fructose from 4 to 175; and in the presence of arabinose from 22 to 175.

Considerable information is available on changes in free sulfur dioxide content of musts and wines under experimental conditions, but the detailed changes in sulfur dioxide during fermentation, storage, and aging under commercial conditions have not been reported. Mills and Wiegand (1942) recorded data on changes in sulfur dioxide content of representative wines stored in full half-pint bottles. Ough et al. (1960) reported changes in total and free sulfur dioxide content of white wines stored in bottles at various temperatures. Neither of these papers included complete data on the chemical composition of wines tested and the nature of the bound sulfur dioxide. Peynaud and Lafourcade (1952) gave data on free and total sulfur dioxide content of white Bordeaux wines and also on total aldehyde and total sugar content on the assumption that the excess of combined sulfur dioxide over that required to bind the acetaldehyde is bound with sugar. They calculated the amounts bound by sugar. As shown in table 53, from 16.8 per cent to 73.5 per cent of the

TABLE 53

State of Sulfur Dioxide in White Bordeaux Wines

Wine	Sugar (gr/liter)	SO₂ free (mg/liter)	SO₂ total (mg/liter)	SO₂ combined (mg/liter)	Acetaldehyde (mg/liter)	SO₂ combined to acetaldehyde (mg/liter)[a]	SO₂ combined to sugar (mg/liter)[b]	SO₂ combined to sugar (per cent)
Barsac 1933	56	68	404	336	123	179	157	46.8
Graves 1936	20	48	164	116	33	48	68	58.6
Entre-Deux-Mers 1937	40	80	416	336	140	204	132	39.3
Sauternes 1938	62	96	336	240	64	93	147	61.3
Graves 1946	3	92	218	122	86	102	20	16.8
Loupiac 1945	48	76	292	216	51	77	139	64.4
Barsac 1943	74	104	515	411	74	109	302	73.5
Preignac 1943	82	56	224	168	44	64	104	61.9

[a] Calculated on basis that 44 mg of acetaldehyde combine with 64 mg of sulfur dioxide.
[b] Calculated by difference from total combined SO₂.
Source of data: Peynaud and Lafourcade (1952).

Type of wine	Grape variety	pH	Total Acidity (gr tartaric per 100 ml)	Total Acetaldehyde (mg liter)	Reducing Sugar (g/100 ml)	Free SO₂ (mg liter)
Rosé (Davis)	Pinot noir	3.72	0.48	8	0.20	16
Rosé (Davis)	Pinot noir	3.58	0.63	37	0.19	26
Rosé (Davis)	Pinot noir	3.71	0.48	7	0.12	13
Rosé (Davis)	Pinot noir	3.54	0.63	33	0.18	19
Rosé (Davis)	Pinot noir	3.72	0.47	14	0.11	7
Rosé (Davis)	Pinot noir	3.57	0.60	37	0.14	12
Dry White (Davis)	Sauvignon blanc	3.29	0.77	58	0.14	25
Dry White (Davis)	Sauvignon blanc	3.18	0.80	35	0.09	24
Dry White (Davis)	Sauvignon blanc	3.16	0.80	42	0.11	19
Dry White (Oakville)	White Riesling	3.02	0.67	38	1.55	15
Dry White (Oakville)	White Riesling	3.01	0.72	37	1.59	16
Dry White (Oakville)	White Riesling	2.98	0.75	37	0.60	14
Dry White (Oakville)	White Riesling	3.02	0.70	38	0.90	16
Natural sweet (Davis)	Sémillon	3.74	0.52	139	3.60	785
Natural sweet (Davis)	Sémillon	3.74	0.61	144	9.90	525
Natural sweet (Davis)	Sémillon	3.58	0.53	89	3.00	395
Natural sweet (Davis)	Sémillon	3.58	0.55	139	8.40	190

total bound sulfur dioxide is bound by sugar. They did not give the actual glucose content of these wines and did not actually determine the concentration of glucose sulfonic acid present.

Similar data for California wines are shown in table 54. These wines were analyzed by Ough (1959b) in January, 1959, one year after fermentation. The Sémillon wines were purposely oversulfited. It is surprising that all the bound sulfur dioxide even in the oversulfited wines could be accounted for by acetaldehyde: only in five wines was there a surplus of bound sulfur dioxide over that which could combine with acetaldehyde present.

The forms of sulfur dioxide present in several Australian wines and a must were calculated by Rankine (1966c) as follows:

Compound	Riesling	Riesling	Sauternes	Dessert	Must
Reducing sugar (per cent)	0.2	2.2	5.0	8.0	25.0
Glucose (per cent)	0.05	0.5	1.5	3.0	12.5
Arabinose (per cent)	0.07	0.09	0.07	0.07	0.06
Aldehyde (mg per liter)	30.0	30.0	50.0	64.0	—
pH	3.2	3.4	3.4	3.7	3.8
Total SO₂ (mg per liter)	100.0	150.0	180.0	130.0	100.0
Free SO₂ (mg per liter)	20.0	30.0	50.0	15.0	40.0
Undissociated H₂SO₃ (mg per liter)	0.7	0.7	1.2	0.2	0.4
Combined SO₂ (mg per liter)	80.0	120.0	130.0	115.0	60.0
SO₂ combined to aldehyde (mg per liter)	43.0	43.0	73.0	93.0	—
SO₂ combined to sugar (mg per liter)	1.5	4.0	11.0	4.5	40.0
Residual combined SO₂ (mg per liter)	36.0	73.0	46.0	17.5	20.0

rtain California Wines

Total SO₂ ng liter)	Bound SO₂ mg/liter	Bound SO₂ millimols per liter	Acetaldehyde mg/liter	Acetaldehyde millimols per liter	Sugar gr/100 ml	Sugar millimols per liter	Bound SO₂ To acetaldehyde per cent	Bound SO₂ To sugar per cent
29	13	0.206	8	0.182	0.20	11.1	89.7	10.3
65	39	0.609	37	0.841	0.19	10.6	100.0	0.0
48	35	0.563	7	0.159	0.12	6.67	28.3	71.7
83	64	1.000	33	0.750	0.18	10.0	75.0	25.0
76	69	1.077	14	0.318	0.11	6.11	29.5	70.5
50	38	0.594	37	0.841	0.14	7.78	100.0	0.0
80	55	0.860	58	1.318	0.14	7.78	100.0	0.0
53	29	0.453	35	0.795	0.09	5.00	100.0	0.0
57	38	0.594	42	0.955	0.11	6.11	100.0	0.0
55	40	0.625	38	0.855	1.55	86.1	100.0	0.0
64	48	0.750	37	0.841	1.59	88.3	100.0	0.0
68	54	0.843	37	0.841	0.60	33.35	99.7	0.3
64	48	0.750	38	0.855	0.90	50.0	100.0	0.0
845	60	0.937	139	3.160	3.60	200.0	100.0	0.0
660	135	2.110	144	3.275	9.90	551.0	100.0	0.0
405	10	0.156	89	2.025	3.01	166.0	100.0	0.0
255	65	1.015	139	3.160	8.40	466.5	100.0	0.0

Rankine (1966c) showed that the wines and the must he examined did not have sufficient glucose and acetaldehyde to account for all the bound sulfur dioxide. Kielhöfer (1963b) previously reported that the glucose present in wines combines with but a small proportion of the bound sulfur dioxide. He used the term "Rest-SO₂" for the sulfur dioxide bound to compounds other than acetaldehyde or glucose. In some wines this is greater than the amount combined to sugars; and in a few wines made of botrytised grapes, for example, it exceeds that combined with acetaldehyde (Kielhöfer and Würdig, 1960b). Blouin (1963) reported that in sweet table wines of Bordeaux from 40 to 80 per cent of the total sulfur dioxide seems to be combined to these undefined components. Blouin (1966) reported that the most important of these compounds are glucuronic, galacturonic, pyruvic, and α-ketoglutaric acids.

Blouin (1966) reported that at least 20 per cent of the combined sulfur dioxide is still not defined. He calculated the level of combined sulfur dioxide necessary to reach a given level of free sulfur dioxide (called C-L 50 for 50 mg per liter of free sulfur dioxide). For Bordeaux wines of less than 0.3 per cent reducing sugar this is fixed at C-L 15 and for sweet table wines at C-L 100. (In California we usually feel that these limits are too high.) Blouin also calculated the amount of combined sulfur dioxide not fixed by acetaldehyde and glucose; this is called R-L 50, where 50 is the mg per liter of fixed sulfur dioxide not so combined.

For nonmoldy grapes there is generally a decrease in C-L during maturation, that is, the amount of combined sulfur dioxide required to produce 15 mg per liter of free sulfur dioxide decreases. The decreases was less for Sémillon and Sauvignon blanc than for Muscadelle. When the grapes were not free of mold the opposite occurred; the amount of bound sulfur dioxide required to produce a given level of free *increased* during ripening. The increase was greater for Sémillon than for Saint Émilion. Kielhöfer and Würdig (1959) had previously investigated the increased binding of sulfur dioxide in must and wine made from moldy grapes.

The main cause of the increase is the action of acetic acid bacteria, although some of the increase results from *Botrytis* and *Aspergillus*. The conditions of vinification also influence the C-L 50 and R-L 50 values. Maceration of grape tissues increases combination; increase in level of added sulfite also increases C-L 50 and R-L 50. C-L 50 values increase during fermentation, more rapidly under anaerobic than aerobic conditions, and reach a maximum when 120 gr of sugar per liter have been fermented.

Blouin found, in three samples of must, that 55 to 80 per cent of the combined sulfur dioxide was not bound by glucose, and 14 to 70 per cent was not bound by the keto acids (pyruvic, α-ketoglutaric, glyoxylic, or uronic). In seven samples of dry white wines, from 26 to 79 per cent of the combined sulfur dioxide was not bound by acetaldehyde plus glucose, and from 11 to 72 per cent was not bound by acetaldehyde plus glucose plus keto acids. In eight samples of sweet white wine from 45 to 78 per cent of the combined sulfur dioxide was not bound by acetaldehyde plus glucose, and from 19 to 65 per cent was not bound by these and keto acids. The as yet undefined components of combined sulfur dioxide are thus appreciable. On the other hand, Weeks (1969) accounted for 92 to 102 per cent of the bound to acetaldehyde, pyruvic acid, and α-ketoglutaric acid.

Blouin reported no sulfur dioxide bound by glucose in dry wine, from 26 to 163 mg per liter bound by acetaldehyde, 3 to 43 by pyruvic acid, 15 to 43 by α-ketoglutaric acid, 1 to 7 by glyoxylic acid, and 1 to 4 by uronic acids. In sweet wines, 58 to 151 mg per liter were bound by acetaldehyde, 2 to 15 by glucose, 41 to 136 by pyruvic acid, 13 to 50 by α-ketoglutaric acid, none by glyoxylic acid, and only 1 to 5 by uronic acids. In dry wines and in sweet wines, acetaldehyde and pyruvic acids predominate among constituents binding sulfur dioxide. Both Kielhöfer (1963b) and Blouin (1963, 1966) stress the need to add sulfur dioxide after fermentation in sufficient quantities to combine completely with the acetaldehyde present to obtain its most beneficial effects. Hennig (1943) also had reported that sulfur dioxide binds aldehyde, but he believed that some free aldehyde was important to the development of bouquet. More recent data on the binding of sulfur dioxide by acetaldehyde and the factors affecting it are reported by Diemair *et al.* (1960a) and Kielhöfer and Würdig (1960a, b, c, e). Since we lack complete data on all the factors influencing binding of sulfur dioxide we recommend empirical tests on each wine.

The addition of sulfur dioxide to must is known to result in increase in production of aldehydes and glycerin. It is not known whether this accumulation, which under certain conditions is quite high, is due to the direct formation of acetaldehyde sulfonate, as demonstrated by Neuberg and Reinfurth (1918), at pH values close to neutrality, or to interference by sulfur dioxide in the normal reduction of acetaldehyde to alcohol.

The importance of the sulfurous acid bound to pyruvic and α-keto-glutaric acids was emphasized by Peynaud and Lafon-Lafourcade (1966). They showed that pyruvic acid was formed in the first part of the fermentation and then gradually decreased. When thiamine was added during fermentation the pyruvic acid decreased rapidly. The α-ketoglutaric acid increased slowly throughout the fermentation.

To reduce the by-products which fix sulfurous acid, Peynaud and Lafon-Lafourcade (1966) recommended (especially for musts from moldy grapes): low temperature fermentation, anaerobic fermentation, addition of ammonium salts, and use of a nonketogenic strain of yeast. They note that use of 0.5 mg per liter of thiamine one to two days *before* addition of sulfurous acid results in lower levels of these acids and hence in less fixation of sulfurous acid; but the use of thiamine is not authorized in France. Removal of these acids might be possible, by a malo-lactic fermentation, but in sweet table wines the risk of a lactic fermentation of sugar must be taken into account.

Mestre Artigas and Campllonch y Romen (1942) reported accumulation of as much as 360 mg per liter of acetaldehyde in wines produced from heavily sulfited musts; Joslyn and Comar (1938) found over 1,000 mg per liter of acetaldehyde in a wine made from oversulfited must. Lafourcade (1955) reported the following increases in acetaldehyde content of the wine with increase in level of added sulfur dioxide (mg per liter):

SO$_2$ added	SO$_2$ in wine	Acetaldehyde in wine
0	0	24
56	32	48
112	79	76
168	118	101
224	150	131

Blouin (1966) also found that addition of sulfur dioxide during fermentation increased the sulfite-binding capacity of wine apparently by increasing acetaldehyde content. The accumulation of acetaldehyde is undesirable, as it may later influence color stability, particularly of red wines, and aging. Care should be taken to produce and maintain wine at a low aldehyde level. High pH and high temperature during fermentation and exposure to air during racking, finishing, and storage should be avoided.

Zang (1963) maintained that sulfur dioxide could be demonstrated in musts and wines that had not been treated with sulfur dioxide. The tests of Rankine (1963) and of Amerine (1965c) failed to substantiate this fully. However, Zang and Franzen (1966), using peroxide oxidation and barium precipitation, showed that sulfur dioxide is formed during alcoholic fermentation. The possibility of other oxidizable volatile sulfur compounds being formed should also be considered. Zang and Franzen (1967) emphasized how variable the amounts of sulfur dioxide produced might be (0 to 128 mg per liter); the largest amount was formed in a filtered and pasteurized must. Franzen (1968) reported sulfur dioxide production of 7 to 128 mg per liter (average 69) in twenty German wine fermentations. Some yeast strains produce more sulfur dioxide than others. Loss of sulfur dioxide during storage varied from 4 to 390 mg. There was no relation between the total sulfur dioxide present at the beginning of storage and that remaining after storage. Care should, therefore, be exercised in not adding too much sulfur dioxide in the hope that the excess will be lost.

Würdig and Schlotter (1967) reported an increase in sulfur dioxide during alcoholic fermentation. At the same time there was a decrease in sulfate, and 32 to 86 per cent (average 55) of the sulfate lost was recovered as sulfur dioxide. From 13 to 114 mg per liter of sulfur dioxide (average 47) were produced in twenty fermentations. Somewhat more sulfur dioxide was formed in fermentations in 500-liter casks compared to 5-liter casks. The amount of sulfur dioxide formed increased with the sugar content of the must, but the amount of sulfate present is the limiting factor. They note that the first report of this reduction is that of Haas et al. (cited by Kerp, 1904c) from data obtained in the years 1889 to 1902. Schanderl (1959) also reported reduction of sulfate to sulfite, as did Diemair et al. (1960a, 1960c). The strain of Sacch. cerevisiae var. ellipsoideus used by Weeks (1969) produced no sulfur dioxide. A strain of Sacch. oviformis produced 35 to 76 mg. per liter.

The free sulfur dioxide content (including dissolved sulfur dioxide, sulfurous acid, and bisulfite ion) is the product which exerts the desired effects in wine making and wine preservation. Its utility is based upon its reducing or antioxidative properties, its selective antiseptic effect, and its clarifying, dissolving, and acidifying influence. Kielhöfer (1960a, 1960b) and Kielhöfer and Würdig (1959, 1960a, 1960b) summarize the available information on these effects of sulfur dioxide. The levels of free sulfur dioxide recommended for Australian wines at various pH's by Rankine

(1966c), however, are too high for some California wines. He recommended 20, 40, and 70 mg per liter for white table wines with pH's of below 3.3, 3.3 to 3.6, and above 3.6.

Antioxidative and enzyme inhibitory action

Sulfur dioxide tends to protect the wine from excessive oxidation by inhibiting enzymic and nonenzymic oxidative discoloration and also by reducing the available molecular oxygen. It is known to be effective in inhibiting nonenzymatic browning. The mechanism of this effect was reviewed by Joslyn and Braverman (1954), Gehman and Osman (1954), and Reynolds (1963, 1965). Inhibition of browning by inhibiting the initial step in the reaction between D-glucose and other carbonyl compounds with amino acids, the so-called Maillard reaction, was proposed to explain the temporary inhibition by bisulfite of browning in model systems (glucose-glycine) and certain foods. It was proposed that bisulfite addition compounds of the carbonyl compounds were formed and thus prevented interaction between carbonyl groups and amino groups. Another explanation offered was that the added bisulfite oxidized reducing sugars including D-glucose to the corresponding gluconic acids and was itself reduced to sulfur. Inhibition of the browning reaction by oxidation of glucose to gluconic acid by bisulfite would result in the accumulation of sulfur. While the formation of sulfur has been noted in model systems under some conditions (Reynolds, 1963), its formation in sulfited fruit products has not been reported. Bolin et al. (1964) took special precautions to search for its formation during storage of sulfited apricots, but failed to find it. They obtained a good correlation between increase in sulfate content and decrease in sulfur dioxide content of dried fruit exposed to light or protected from it. They found that exposure to light increased the rate of loss of sulfur dioxide. A third theory proposed that bisulfite reduced the intermediates formed in the reaction mixture and thus inhibited formation of melanoidins while itself being oxidized. Song and Chichester (1967a) examined these proposals and pointed out that they did not correspond to the actual kinetics of the reaction between glucose and glycine. They reported that the bisulfites inhibited the reaction step prior to the steady-state browning step, and suggested that this occurred by attack on intermediates of the Maillard reaction by a reactive form of bisulfite, such as a free radical, rather than by the original bisulfite molecule itself. Song and Chichester (1967b) presented additional kinetic data supporting the hypothesis of inhibition of the Maillard reaction via free radicals formed from bisulfite. Whether similar reactions occur in fruit products is not known, although there is some evidence of free radical formation during browning. Their theory, however, explains the fact that sulfur dioxide will not give lasting protection of wine against nonenzymatic oxidation because of oxidation occurring during aging. While the inhibition by sulfur dioxide of browning by polyphenoloxidase has long been known (e.g., Diemair et al., 1960b), the mechanism of this inhibition was not completely understood until recently. Embs and Markakis (1965) demonstrated that in model systems at pH 6.5 sulfite prevented oxidation of catechin by mushroom polyphenoloxidase by combining with the enzymatically produced o-quinones and stopping their condensation to melanins. However, pre-incubation of polyphenoloxidase with sulfite gradually reduced the ability of the enzyme to cause browning. Markakis and Embs (1966) found that ascorbic acid, by preventing the reaction of sulfites with quinones, facilitated the sulfite inhibition of the enzyme. In turn, this inhibition resulted in a decreased over-all destruction of ascorbic acid. Sulfite initially did not protect the ascorbic acid from oxidation by the quinones formed enzymatically. Sulfite, however, gradually decreased both the diphenolase and monophenolase activity of mushroom phenoloxidase and thereby diminished the over-all destruction of ascorbic acid.

In addition to inhibiting activity of polyphenoloxidase both indirectly by reacting with the intermediate, o-quinone, and directly by inactivating the enzyme itself, bisulfite is known to inhibit other enzymes. Wyss (1948) early proposed that microbial inhibition by sulfite was due to disruption of alcoholic fermentation either by combination with acetaldehyde or by reduction of the S-S linkage essential for enzymatic activity. This explanation was accepted by Schelhorn (1951a). For its use in preventing browning of musts see Lepădatu et al. (1963).

Antiseptic and preservative effect

The antiseptic effect of sulfur dioxide and sulfites was long known to depend on the free, undissociated sulfurous acid or molecular sulfur dioxide, and to be less or absent in combined sulfites or neutral sulfites. Müller-Thurgau and Osterwalder (1914) were the first to observe that sulfurous acid and its salts were effective as preservatives only in acid media. This was confirmed by Cruess et al. (1931), Cruess and Irish (1932), Rahn and Conn (1944), Vas and Ingram (1949), and Schelhorn (1951a, 1951b, 1953). It is now known that even slight changes in pH in the region of 3–5 markedly affect the proportion of undissociated sulfurous acid present and consequently its antiseptic efficacy (p. 384).

The antiseptic power of sulfur dioxide is known to decrease on combination with aldehydes, sugars, or other constituents of must and wine. This was observed by Ripper in 1892 and subsequently confirmed by Hailer (1911). Bioletti and Cruess (1912) reported that free sulfur dioxide has more than thirty times the disinfecting power of bound sulfur dioxide and is sixty times as effective as bound sulfur dioxide in inhibiting fermentation. Yeasts transferred to fresh must from water suspensions containing 350 mg per liter of free sulfur dioxide did not ferment, whereas those from grape must containing 240 mg per liter of total sulfur dioxide of which only 227 mg per liter were free did ferment. Ingram (1948) proved that the growth of a *Zygosaccharomyces* (now *Saccharomyces*) species of yeast was related to the concentration of free sulfur dioxide and not to the concentration of glucose bound sulfur dioxide. An excellent summary of the effects of sulfur dioxide in wines and musts has been presented by Kielhöfer (1963b). He showed that the antiseptic effect is characterized by free sulfur dioxide, that is, by undissociated sulfurous acid. Its effect is less in the presence of yeasts.

Fornachon (1963a) reported evidence that bound sulfur dioxide was inhibitory to growth of lactic acid bacteria. He found that *Lactobacillus hilgardii* and *Leuconostoc mesenteroides* rapidly attack acetaldehyde. In a media containing sulfur dioxide and acetaldehyde this resulted in the release of sufficient sulfur dioxide (from the bound sulfur dioxide) to inhibit growth. *Lactobacillus arabinosus* consumed much less acetaldehyde. Thus it may be difficult to start a malo-lactic fermentation in wines containing a high level of bound sulfur dioxide. Sulfur dioxide not only increases the time before the malo-lactic acid fermentation starts, but also increases the duration of this fermentation.

In a laboratory-scale continuous fermentation of musts of a relatively high pH (3.5 to 3.6), Flanzy et al. (1966) reported growth of lactic acid bacteria and formation of volatile acidity. If sulfur dioxide was periodically added, growth of bacteria and volatile acid formation were controlled, but the fermentation rate was reduced and acetaldehyde formation increased.

Sulfur dioxide is known to have a selective antiseptic effect. Cruess (1911) and Bioletti and Cruess (1912) reported that acetic acid bacteria, lactic acid bacteria, and many varieties of molds are more sensitive to sulfur dioxide than yeasts are. Among the yeasts, the more strongly aerobic species are generally more sensitive than the more fermentative species.

Several yeast species extremely resistant to sulfur dioxide have been found. Krömer and Heinrich (1922) observed one in an oversulfited must, and Osterwalder (1924) reported on the occurrence of a *Schizosaccharomyces* resistant to sulfur dioxide. Highly-resistant yeasts were observed by Osterwalder (1934a), Schanderl (1952), and Domercq (1956, 1957). The yeasts resistant to sulfur dioxide may tolerate 700 mg per liter and more of sulfur dioxide. Some strains of *Mucor* and other molds resistant to sulfur dioxide have been observed also. Dubaquié and Débordes (1935) showed that antiseptics had more effect on strains of yeast which preferentially fermented glucose than on fructose-fermenting Sauternes strains. Schanderl and Staudenmayer (1964) point out that *Brettanomyces*, in contrast to other yeasts, are much less affected by sulfur dioxide.

The antiseptic action of sulfur dioxide varies with stage of growth and development, microbial population, temperature, pH, and composition (particularly acetaldehyde, sugar, and alcohol content). Sulfur dioxide is only a temporary preservative when used in moderate amounts because of loss of preservative value on oxidation to sulfate, on volatilization, and on combination with carbonyl compounds.

Certain microörganisms, particularly the fermentative yeasts, can be adapted to sulfur dioxide. This adaptation was demonstrated and investigated very early by Porchet (1931) and more recently by Scardovi (1951, 1952, 1953). Scardovi was able to obtain, by adaptation and subsequent plating, variants having from ten to twelve times the resistance to sulfur dioxide of the parent strain. The resistance of the adapted yeasts was found to be permanent and highly specific. The glutathione present in the yeast cell occurred only in the oxidized form during the initial stages of growth in presence of sulfur dioxide. The higher the sulfur dioxide resistance the higher was the glutathione content of the cells. A difference was found between the acetate oxidation and glucose-fermentation ability of the original strain and the strain resistant to sulfur dioxide. The sensitivity of acetate oxidation and of glucose fermentation to sulfur dioxide were completely independent characters, but were related to sensitivity of growth to sulfur dioxide.

In spite of its long use, the mechanism of the antimicrobial action of sulfur dioxide (or sulfurous acid) was not known until recently. Rehm and Wittman (1962) reported on factors affecting the antimicrobial activity of sulfurous acid; Rehm and Wittman (1963) investigated the effect of dissociated and undissociated sulfurous acid on various microörganisms; Rehm *et al.* (1964) investigated the dissociation and antimicrobial activity of several sulfonates; and Wallnöfer and Rehm (1965) reported on the nature of the antimicrobial activity of sulfur dioxide. In confirmation of the early investigations of Fornachon (1963a), Rehm and Wittman (1962) found that some sulfonates (e.g., benzaldehyde sulfonate) have little antimicrobial action, but others (such as salicylaldehyde and glucose sulfonate) have considerable. Carbonyl compounds like pyruvate, α-ketoglutaric, and acetaldehyde, however, form sulfonates with high antimicrobial activity. These sulfonates markedly decreased growth and respiration of yeast. The inhibition of respiration by free sulfurous acid was greater than that by sulfonates, but the sulfonates of acetaldehyde, pyruvate, and acetone had a significant inhibiting effect, while glucose sulfonate had only slight inhibiting effect.

Sulfur dioxide is known to be a strong inhibiting agent on enzymes containing SH-groups and is a primary inhibitor of nicotinamide adenine dinucleotide (NAD). Rehm (1964) and Wallnöfer and Rehm (1965) observed that 3-phosphoglycerate was not fermented by sulfite-inhibited cultures of yeast and of E. coli. The yeast ethanol dehydrogenase, however, was not inhibited by small amounts of sulfur dioxide. Investigation of the action of sulfurous acid on the metabolism of fermenting and respiring cells of Saccharomyces cerevisiae and E. coli proved that NAD-dependent steps in the metabolism of carbohydrates were inhibited by sulfurous acid. This was believed to be due to the formation of an addition compound between NAD and bisulfite.

Wines made from sulfited musts keep much better than those obtained by natural fermentations. Sulfur dioxide is particularly useful in preventing the development of lactic acid bacteria during storage and thus preventing spoilage. The amount to be added to the wine, however, and the concentration at which it is to be maintained should be as small as possible in order to avoid undue hindrance to normal aging.

Very small amounts of sulfur dioxide (usually equivalent to 5 ounces of potassium metabisulfite per ton of grapes) suffice to prevent growth of molds and wild yeasts. As Cruess (1912) and others have shown, 100 mg per liter of sulfur dioxide (6 ounces of potassium metabisulfite per ton) will eliminate over 99.9 per cent of the active cells of microörganisms from normal musts. By properly timing the sulfiting and the addition of wine-yeast starter, the full effect of the maximum amount of free sulfur dioxide is exerted on the injurious organisms, while the wine yeast is exposed to the minimum amount of free sulfur dioxide. Furthermore, the wine yeasts can adapt themselves to sulfur dioxide and become comparatively resistant to it. Porchet (1931) and Scardovi (1951, 1953) have found this adaptation to be quite marked.

The antiseptic action of sulfur dioxide toward microörganisms, particularly yeasts, varies with their stage of development and the numbers present. It is greater toward the resting, or sporulating, yeasts and more effective the lower the numbers present. Yeast in full activity is more resistant to sulfur dioxide because of the rapid fixation of the latter by the aldehydes formed in fermentation, the mechanical entrainment of sulfur dioxide gas by carbon dioxide gas, and the natural increase in resistance of the cell.

The lower the temperature of fermentation, the lower the concentration of sulfur dioxide required to prevent the development of undesirable microörganisms. Sulfur dioxide, according to Skavysh (1965), is the best cure for wines infected by lactic bacteria when it is added at a level of 70 to 80 mg per liter.

Clarification

In sufficiently high amounts, sulfur dioxide acts as an acid in causing a rapid and complete clarification of must through neutralization of the negative charge on the colloids present in suspension. In the quantities usually added, its use for this purpose is limited to the clearing of white must before fermentation. The clearing is mainly mechanical: the sulfur dioxide merely inhibits the fermentation long enough for the skins, seeds, particles of pulp, and other debris to settle out. The naturally occurring pectic enzymes may assist in this clarification. It is possible that the activation of pectin esterase by bisulfite may be involved in the increased clarification of sulfited musts. Edwards and Joslyn (1952) observed that bisulfite appreciably increased the activity of orange pectin esterase. About 60 to 75 mg per liter of sulfur dioxide—approximately half a pound to 1,000 gallons—are used.

Dissolving action

Sulfurous acid may be considered as a fairly strong acid, able to cause the solution of certain substances. Thus sulfiting increases the fixed acidity, the extract, and the alkalinity of the ash because of its solvent effect on cream of tartar. In small amounts it results in a higher color in red wine because of its solvent action on the coloring matters in the skin, although, if excessive amounts are used, its bleaching action will mask this solvent action. The following data on the color values in new wines made with varying acidity and sulfur dioxide (Amerine and Joslyn, 1951) are illustrative:

Sample no.	Total acid (per cent)	Sulfur dioxide (mg per liter)	Color index[a]
I	0.64	50	78
2	0.66	250	125
3	0.95[b]	50	128
4	0.95[b]	250	182

[a] Color intensity increases as figures increase.
[b] Tartaric acid added before fermentation.

Berg and Akiyoshi (1962) confirmed the greater extraction of color in the presence of sulfur dioxide. They also showed greater retention of color of wines fermented with sulfur dioxide.

Pifferi and Zamorani (1964) found that fermentation of Merlot grapes with 100 mg per liter of sulfur dioxide helped dissolve diglucoside pigments and that in the resulting wines as much as 6.6 per cent of the anthocyans were diglucosides. Harvalia (1965) reported an increase in color with 80 mg per liter of sulfur dioxide at 3 to 6 per cent alcohol compared to no sulfur dioxide. A small increase in color was found when sulfur dioxide was added to musts from red grapes without the skins at 80 mg per liter of sulfur dioxide, and a large increase at 400 mg per liter. He attributed this to the formation of anthocyans from colorless pseudobases. A shift in the colorless to colored equilibrium certainly occurs. More work on the nature of the pseudobases is obviously needed. There are many empirical reports on increases in color of red wines during aging.

Acidifying action

A constant increase in total acidity of wines obtained from sulfited musts is among the most characteristic effects of the use of sulfur dioxide in vinification. This acidification results from the dissolving and antiseptic powers of sulfur dioxide, by which cream of tartar is converted into soluble acid salts while the development of microörganisms that destroy fixed acids is prevented. Part of the sulfurous acid present is converted to sulfuric acid by oxidation. The acidifying action of the sulfur dioxide itself is small, however—only 0.082 per cent of acid, as tartaric— when the maximum permissible amount of sulfur dioxide (350 parts per million) is used. In practice the actual increase in fixed acid amounts to 0.05 to 0.10 per cent. Repeated additions of sulfur dioxide over long periods of time will certainly approach or exceed these limits.

Bleaching and discoloration

The discoloration effect of sulfur dioxide on the red pigments (on low acid wines) is a commonly observed fact. If acetaldehyde is added to remove the free sulfur dioxide the reaction is reversible and the color reappears. This occurs with the glucosides but not with the glycones.

Various theories have been proposed for the decolorizing properties of sulfite for anthocyanins. Some consider it due to the reducing properties of sulfite; others, to formation of additive compounds with substances possessing active carbonyl groups. Since the process is reversible at low pH reduction apparently does not occur. P. Ribéreau-Gayon (1959) suggested that the principal decoloration reaction is the formation of a chalcone-bisulfite addition compound:

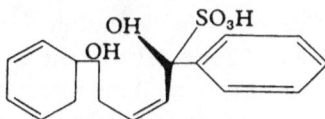

$$\text{OH} \quad \text{SO}_3\text{H}$$

On the basis of studies on the spectrum of various anthocyan pigments at various pH's and sulfite concentration, Jurd (1964) believes that sulfite decoloration of anthocyan compounds is dependent on carbonium, bisulfite, and hydrogen ion concentrations. He considers that the reactions are:

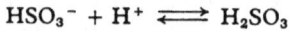

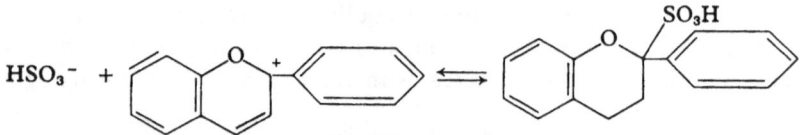

$$HSO_3^- + H^+ \rightleftharpoons H_2SO_3$$

The slow and incomplete reaction at a low pH is due to the decrease in bisulfite ion concentration. The product is colorless chromen-2(or 4)-sulfonic acid (R—SO$_3$H), similar in structure and properties to an anthocyanin carbinol base (R—OH).

It is now considered that acid solutions of the anthocyanins exist as equilibrium mixtures of a colored oxonium or carbonium ion and a colorless carbinol or pseudobase. When the pH is near 7 or higher, colored quinoidal or anhydrobases are formed when the compound contains an ionizable phenolic group. If the sugar molecule at position 3 is replaced by a hydrogen, the resulting anthocyanidin is less stable, and in acid solutions is rapidly changed to colorless carbinol bases. Keto forms of these bases may slowly change to their corresponding enol forms, with oxidative degradation occurring simultaneously. Timberlake and Bridle (1967a) found that the anthocyanin 3,5-diglucosides were stronger acids than the 3-glucosides. Light sometimes had a pronounced effect on the stability of the compounds.

One disadvantage of the use of anthocyans as coloring agents is their bleaching by sulfur dioxide. Timberlake and Bridle (1967b) consider that sulfur dioxide reacts with the flavylium or anthocyan cation to form a colorless chromen-2(or 4)-sulfonic base—the whole ratio of the reactants being 1 to 1. The 3,5-diglucosides form stronger complexes with sulfur dioxide than the 3-glucosides. The anthocyanins formed weaker complexes than expected, and they suggested that anthocyan–sulfur dioxide complex formation is not due solely to combination with the HSO$_3^-$ ion. Flavylium salts vary widely in the extent of their combination with sulfur dioxide.

Control of fermentation

According to Bioletti and Cruess (1912), the use of sulfur dioxide in fermentation, especially when it is combined with pure yeast, has several advantages. A more nearly perfect fermentation and sounder wines are obtained. The volatile acidity is uniformly lower in wines from sulfited musts than in those from untreated musts. This fact is clearly indicated

TABLE 55

EFFECT OF SULFUR DIOXIDE IN PREVENTING HIGH VOLATILE ACIDITY IN WINES

	Method of fermentation			Number of samples	Percentage of samples containing viable lactic acid bacteria	Composition of wine (per cent)			
Year	Metabi-sulfite added	Cooling	Pure yeast			Alcohol	Volatile acid	Total acid	Sugar
1913	No	No	No	101	100	11.5	0.118	0.66	0.49
	Yes	No	No	6	0	12.6	0.048	0.50	0.42
	Yes	No	Yes	67	0	12.1	0.066	0.57	0.21
1914	No	No	No	68	80	12.1	0.088	0.61	0.20
	Yes	No	Yes	56	6	12.1	0.052	0.62	0.24
1934	No	No	No	81	81	12.7	0.173	0.80	0.52
	Yes	No	No	64	20	11.6	0.064	0.71	0.21
	Yes	No	Yes	21	80	13.4	0.087	0.62	0.36
	Yes	Yes	Yes	69	14	12.4	0.060	0.50	0.17
1935	No	No	No	51	—	12.9	0.137	0.51	0.57
	Yes	No	No	98	—	12.5	0.043	0.66	0.15

Source of data: Cruess (1935a, 1935b).

in experiments reported by Cruess (1935a, 1935b), and summarized in table 55. Sulfur dioxide was added in the form of potassium metabisulfite. The fixed acidity is protected by the use of sulfur dioxide, and the sulfited wines consequently show a higher total acidity than do the untreated ones. The use of sulfur dioxide or metabisulfite, by ensuring a purer fermentation, increases the yield of alcohol, often by as much as 1 per cent but usually by only 0.2 to 0.3 per cent. The extract content of dry wines is higher, and the color of red wines more desirable in tint.

Dosages

An excessive amount of sulfur dioxide must be avoided: it detracts from the flavor of the wine, interferes with the natural aging, and leads to the production of undesirable turbidities and deposits when copper salts are present in the wine. Over a long period of time it may also lead to the formation of excessive quantities of sulfate (see p. 456). Furthermore, the dissolving action of sulfurous acid on metal and cement surfaces is undoubtedly considerable. Only the minimum quantity necessary for proper control of the fermentation should therefore be used. *When proper sanitary precautions are used in picking, crushing, fermenting, and storage, sulfur dioxide need not be used in objectionable amounts.*

The amount of sulfur dioxide to be added to a given must to control the fermentation depends on the degree of ripeness and soundness of the grapes, the temperature of the grapes and must, and the weather conditions at the time of crushing. Moldy grapes, warm grapes, and overripe grapes rich in sugar, high in pH, and low in acid require more sulfur dioxide than cool, sound grapes of moderate sugar, low pH, and high acid content. In Europe the amount usually added in hot regions is 100 to 200 mg per liter; in cool regions, less than 100. There is some indication that certain red-juice varieties (e.g., Alicante Bouschet) should have more than others. The quantity of sulfur dioxide to be added under various conditions of ordinary commercial operations in California is indicated in table 56.

TABLE 56

AMOUNT OF SULFUR DIOXIDE TO BE ADDED UNDER VARIOUS CONDITIONS

Condition and temperature of grapes	Liquid sulfur dioxide		6 per cent sulfurous acid solution		Potassium metabisulfite	
	Ounces per 1,000 gals. of must	Ounces per ton of grapes	Gals. per 1,000 gals. of must	Pints per ton of grapes	Ounces per 1,000 gals. of must	Ounces per ton of grapes
Clean, sound, cool, underripe	10	2	1¼	2	20	3¼
Sound, cool, optimum maturity	15	2½	2	3	31	5
Moldy, bruised, hot, overripe, low in acid	36	6	3¼	4¼	56	9

Even though the musts have been sulfited, sulfur dioxide is often added to wine to prevent bacterial spoilage and to decrease oxidation. To avoid adding too much, the total and free sulfur dioxide content of the wine should be determined first. If 100 to 150 mg per liter of sulfur dioxide have been used on the must before fermentation, not more than half to three-fourths of a pound of sulfur dioxide should be needed per 1,000 gallons of wine; this amounts to 60 to 90 mg per liter. This, together with the residual sulfur dioxide, is usually high enough to check the development of undesirable bacteria and maintain the desired reducing conditions in the wine. Quinn (1940) recommended frequent small additions of sulfur dioxide rather than an occasional large addition to take advantage of the antiseptic power of free sulfur dioxide. Peynaud and Lafourcade (1952) cautioned against the use of high amounts of sulfur dioxide either in settling must or addition prior to or during fermentation because under these conditions acetaldehyde accumulates and a larger percentage of the sulfur dioxide added is fixed. If possible, the addition of sulfur dioxide should be avoided in the presence of yeast. In adding sulfur dioxide to new wines they recommend a large initial addition to cause death of most of the undesirable yeasts rather than several small additions. Total sulfur dioxide from 75 to 150 mg per liter and free sulfur dioxide contents of 10 to 30 mg per liter are usually maintained in California table wines during storage.

SOURCES OF SULFUR DIOXIDE
Liquid sulfur dioxide is now the preferred source.

Sulfur wicks

The oldest way of obtaining sulfur dioxide is by burning sulfur. Sulfur in the form of a wick or candle is first burned in the cask or tank, and the sulfur dioxide is absorbed by the must or wine as the vessel is filled. This method is used at present only in disinfecting empty cooperage and, to a limited extent, in lightly sulfuring wine in tanks or casks during racking, since the sulfur dioxide produced by the burning wick is present in the cask. Some is also absorbed by the wood and later released to the wine.

One serious objection to this method is that the sulfur which falls or sublimes into the cask may be reduced by yeast to hydrogen sulfide (p. 366). Another objection is that it is difficult to measure the quantities being added. It is, however, the cheapest source of sulfur dioxide for small-scale operations. Burning of sulfur also exhausts the oxygen in the casks.

Liquid sulfur dioxide

Liquid sulfur dioxide, the gas liquefied under pressure and held in heavy-walled steel cylinders, is now widely used because it is easy to handle. Sulfur dioxide gas exists as a liquid at $21.1°C$ ($70°F$) under an absolute pressure of 50 pounds per square inch. Under these conditions it has a density of 86 pounds per cubic foot and a specific gravity of about 1.4. Its advantages are purity and relative cheapness.

By means of any one of several measuring devices for dispensing the sulfur dioxide, the necessary amount of the gas from the cylinders can be introduced directly into the must or wine. The available methods of measuring liquid sulfur dioxide by weighing, by volume, and by metering have been described by Willson et al. (1943a, 1943b). In the method formerly employed, a pressure vessel, usually of nine-pound capacity so that it could be easily transported, was fitted with a gauge glass calibrated in volumes equivalent to a pound. Sulfur dioxide is now usually weighed from cylinders or, preferably in large-scale operations, metered. Several metering devices are now available. The articles by Willson and others discuss the properties of liquid sulfur dioxide, the precautions to be followed in its use, and methods of using it in the food industries. For data on the solubility of sulfur dioxide see Johnstone and Leppla (1934).

Another way of using liquid sulfur dioxide is to prepare a solution of sulfurous acid from it. For musts low in total acid, this form (or direct addition of liquid sulfur dioxide) is preferable to metabisulfites. Only fresh solutions should be used, for they lose strength in storage. A stock solution of sulfurous acid, usually containing 6 per cent, is made by adding a weighed amount of gas to cold water. Willson (1943a) gives the specific gravities as follows:

Concentration of sulfur dioxide (per cent)	Specific gravity		
	At 15°C (59°F)	At 20°C (68°F)	At 30°C (86°F)
1.0	1.004	1.003	1.000
2.0	1.009	1.008	1.005
3.0	1.014	1.013	1.010
4.0	1.020	1.018	1.014
5.0	1.025	1.023	1.019
6.0	1.030	1.028	1.024
7.0	1.035	1.032	1.028
8.0	1.040	1.037	....

The solution is prepared by letting the gas bubble through water chilled with ice to below 4.4°C (40°F), usually in a glass or stainless steel container. The gas must be added slowly to prevent undue losses. To weigh the gas used, the cylinder is placed on a platform scale and weighed before and after the gas is drawn off into the ice water. At 4.4°C (40°F) a saturated solution will contain 20 per cent sulfur dioxide, or 1.67 pounds of sulfur dioxide per gallon. At 20°C (68°F) a saturated solution will contain 11.29 per cent sulfur dioxide; and at 100°F, only 5.41 per cent. Usually a 6 per cent solution is made; this will contain approximately 8 ounces of the liquid sulfur dioxide per gallon of water. The solubility of sulfur dioxide in water and in alcohol is given by Fabre (1929):

Temperature		Grams per liter dissolved in:	
°C	°F	Water	Alcohol
0	32	—	95.10
8	46.4	16.80	—
10	50	—	55.10
12	53.9	14.20	—
15	59	—	41.75
20	68	10.40	33.05
24	75.2	9.20	—
30	86	8.30	—

Sulfites, bisulfites, and metabisulfites

Sulfites, bisulfites, and metabisulfites of alkali metals, such as potassium or sodium, can be used because when dissolved in an acid solution—must, for example—they are readily converted into available sulfurous acid. In this conversion they neutralize an equivalent amount of the acid in the must and hence are less desirable than liquid sulfur dioxide or a 6 per cent solution of it. Sodium bisulfite is a cheaper source, but is usually not so reliable in composition or so free from contamination by heavy metals as is potassium metabisulfite. The latter, which is most commonly used, provides a higher percentage of available sulfur dioxide than some of the other sulfites of potassium: it yields 57.6 per cent sulfur dioxide, although only 50 per cent is usually considered to be available under practical conditions. The crystalline or the powdered salt is to be dissolved in water at the rate of 8 ounces to a gallon; it should be completely dissolved and thoroughly mixed before it is added to the must. Grinding the salt to a fine powder before adding it to the water aids dissolving. One gallon of this solution is enough for a ton of grapes (see table 56, p. 404). The solution may be added slowly to the crushing sump or to the stream of must as it discharges from the delivery pipe of the must line. Since the metabisulfite, even in powdered form, decreases in strength on standing, only freshly prepared solutions from the fresh dry salt are acceptable.

Distributing sulfur dioxide

When any form of sulfur dioxide is used, the amount must be measured accurately, applied as soon as possible after crushing the grapes, and distributed quickly and evenly throughout the mass. Any method that accomplishes all these requirements is satisfactory. Too little attention is often given to the even distribution of sulfur dioxide in the crushed grapes. Lepădatu et al. (1963) found the free-run excessively high in sulfur dioxide, sometimes leading to mercaptan formation.

Sulfur dioxide should be metered into the must line with a proportioning device. When using liquid sulfurous acid solutions, one can add the amount needed per ton of grapes slowly and regularly to the vat being filled or to the crushing sump by a controlled dropping device, such as a large bottle fitted with a siphon. Another possible method is to distribute the sulfurous acid solution uniformly over the surface of the must at regular intervals—for example, after each 1,000 gallons of crushed must have collected in the fermenting tank. Or the amount of solution required (see table 56) could be added after the entire quantity of must has collected in the fermenter; then the whole mass could be thoroughly pumped over where possible, or punched down, at the completion of filling. This is the least desirable procedure.

Nakayama et al. (1966) obtained a constant sulfur dioxide content during storage by adding Amberlite IR-45 saturated with bisulfite. Watanabe et al. (1966) used Dowex A-1 in the tartrate form for removing excess sulfur dioxide. Dowex A-1 was superior to Amberlite IRA-93 for this purpose.

For data on the fermentation of desulfited musts see Flanzy and Ournac (1963).

Toxicity

There is voluminous literature on the possible toxic effects of drinking wines containing sulfur dioxide (Monier-Williams, 1927; Nichols and Cruess, 1932; Nichols, 1934; Schroeter, 1966; Heydenreich, 1967). Lanteaume et al. (1965) were unable to find any effect on growth.

There is a long controversy as to the toxicity of sulfites. The most recent study, that of Causert et al. (1964), did not show any anatomical or physiological problems caused by ingestion of sulfite, but there was a significant decrease in excretion of thiamine and an increase in urinary excretion of calcium. They recommend caution in consumption of wines approaching the 450 mg per liter limit. Hötzel et al. (1966) concluded that free and total sulfur dioxide in wine had equal toxicity. Rats with a deficiency of all B vitamins were more susceptible to sulfur dioxide toxicity than those not deficient or deficient only in thiamine.

OTHER ANTISEPTICS
Some are of limited use; others are proscribed.

Sorbic acid and sorbates

Control of yeast growth in sweet table wines was obtained at 80 mg per liter of sorbic acid, and 30 mg per liter of total sulfur dioxide by Ough and Ingraham (1960). Control of refermentation was obviously better at $11.7°C$ ($53°F$) than at $27.7°C$ ($72°F$). The average threshold for sorbic acid was 135 mg per liter, but some members of the panel were very sensitive to sorbic acid. They were able to detect amounts as low as 50 mg per liter. They concluded that its use in high-quality wines was contraindicated. Sorbic acid was reported to have little effect on yeast growth below 300 mg per liter by Terčelj and Adamič (1965). This is, of course, in the range where serious sensory effects are noted. Lambion (1963) rejected sorbic acid because of its inadequate antiseptic action.

Peynaud (1963) found sorbic acid in amounts up to 200 mg per liter in the presence of 30 to 40 mg per liter of free sulfur dioxide to be useful in the conservation of sweet Bordeaux wines. At a pH of 3.1, 150 mg per liter was sufficient, but at pH 3.5 more than 200 mg per liter was required. The percentage of undissociated sorbic acid is 98 at pH 3.1 and 94 at pH 3.5. The poor odor associated with the use of sorbic acid appears to be due to the formation of crotonic aldehyde ($CH_3CH = CHCHO$) and unsaturated compounds of the type $—CH = CO—C = O$. Geraniol may be the undesirable compound. They recommend use of freshly prepared pure sorbic acid, storage of treated wines in the absence of oxygen, and sufficient sulfur dioxide to prevent a malo-lactic fermentation.

The sorbic acid should be introduced into the wine slowly to prevent its precipitation by the acids of the wine. While sorbic acid may reduce the amount of sulfur dioxide required for stability of sweet table wines, the possibility of development of off-odors remains.

The use of sorbic acid or sorbates in nineteen countries was reviewed by Lück and Neu (1965). On two Moselle and two Rheingau wines containing 270 mg per liter of potassium sorbate (the equivalent of 200 mg of sorbic acid), using the triangular test (p. 409), they were unable to demonstrate any difference between treated and untreated samples. We question whether the panel was large enough and it may not, therefore, have included any tasters sensitive to sorbic acid.

Tacit recognition that sorbic acid is detrimental to the quality of wines is found in the experiments of Minárik and Nagyová (1966b), who do not recommend it for wines of low extract or "dans les fins vins de bouquet." This seems to modify Minárik's (1962) earlier recommendation of 50 to 200 mg per liter. They do note that use of sorbic acid for sweet low-alcohol wines reduces the quantity of other antiseptics needed. It may indicate a synergistic effect of sorbic acid.

Ásvány (1963), in contrast, found wines treated with sorbic acid superior in quality to untreated, even after four years' storage. This, he believed, was due to lack of undesirable odors from a secondary fermentation. He recommended 200 mg per liter of potassium sorbate for wines of 11 per cent and 120 to 150 mg for wines of 13 to 15 per cent. More sorbic acid was required for high pH than for low pH wines.

At present it is our opinion that sorbic acid should seldom be used for table wines, and then with extreme caution.

Diethyl ester of pyrocarbonic acid (DEPC)*

One of the most interesting preservatives introduced recently is the diethyl ester of pyrocarbonic acid. This is a compound of two naturally occurring harmless components of wine, carbon dioxide and ethyl alcohol. This preservative decomposes in wine fairly rapidly and almost completely into carbon dioxide and ethanol: $(C_2H_5OCO)_2 + H_2O \rightarrow 2C_2H_5OH + 2CO_2$. A summary on the antiseptic values of and possible toxicity of DEPC was given by Pauli (1967). Its decomposition in dry wines in the absence of carbon dioxide was believed to be fairly rapid and complete by Merzhanian (1952) and Parfent'ev and Kovalenko (1951). According to Pauli and Genth (1966), it is only slightly soluble in water (0.6 per cent) but readily soluble in alcohol. At room temperature it is almost completely hydrolyzed in twelve hours even at 10 atm pressure. The

---

* Discovery of small amounts of urethane (a carcinogenic) in DEPC-treated beverages has led to its prohibition in this country.

hydrolysis is more rapid at higher temperatures and pH's. It readily forms compounds with primary and secondary amines, as, for example:

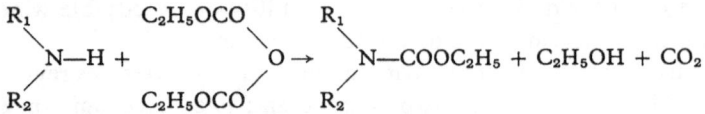

$$\begin{array}{c} R_1 \\ \diagdown \\ N{-}H \; + \\ \diagup \\ R_2 \end{array} \qquad \begin{array}{c} C_2H_5OCO \\ \diagdown \\ O \rightarrow \\ \diagup \\ C_2H_5OCO \end{array} \qquad \begin{array}{c} R_1 \\ \diagdown \\ N{-}COOC_2H_5 \; + \; C_2H_5OH \; + \; CO_2 \\ \diagup \\ R_2 \end{array}$$

The hydrolysis of DEPC occurs in two stages, as discussed by Genth (1964). In the first stage, DEPC is converted into an unstable intermediate, diethyl carbonate. This is then hydrolyzed into ethyl alcohol and carbon dioxide. Diethyl carbonate apparently is the active intermediate involved in reaction with amino acids, intracellular proteins, and other amines. The reaction with intracellular proteins is considered to be the basis of the lethal action of DEPC.

As Thoukis *et al.* (1962) indicate, if pure DEPC is added to new wine or to water it leaves an odor similar to that of a new wine or a water solution of diethyl carbonate. They demonstrated that, while all the DEPC decomposed in a water solution in 120 hours, only 97 to 98 per cent was lost from a sauterne wine. If the pH of the wine was reduced to 1.0 the remaining DEPC decomposed within one week. Apparently a complex of DEPC is formed in wine. This compound does not appear to be an amino acid addition compound, or a reaction product with organic acids, sugars, ethanol, or higher alcohols. Tannic acid does react with DEPC to give a complex which was hydrolyzable at pH 1.0. They suggested that DEPC might react with ammonium ions to yield ethyl carbamate.

To clarify the reactions of DEPC in solutions, Thoukis *et al.* (1962) and Duhm *et al.* (1966) used a C^{14} labelled carbonyl group in the DEPC:

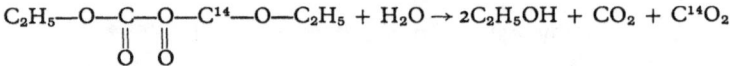

$$C_2H_5{-}O{-}\underset{\underset{O}{\|}}{C}{-}O{-}C^{14}{-}\underset{\underset{O}{\|}}{O}{-}C_2H_5 \; + \; H_2O \rightarrow 2C_2H_5OH \; + \; CO_2 \; + \; C^{14}O_2$$

They showed that, when 100 mg per liter was added, activity in wines was reduced to 5.4 to 6.5 per cent in four or five hours. DEPC apparently reacts with all compounds having a free hydroxyl group:

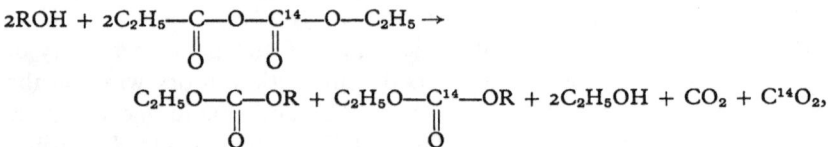

$$2ROH \; + \; 2C_2H_5{-}\underset{\underset{O}{\|}}{C}{-}O{-}C^{14}{-}\underset{\underset{O}{\|}}{O}{-}C_2H_5 \rightarrow$$

$$C_2H_5O{-}\underset{\underset{O}{\|}}{C}{-}OR \; + \; C_2H_5O{-}\underset{\underset{O}{\|}}{C^{14}}{-}OR \; + \; 2C_2H_5OH \; + \; CO_2 \; + \; C^{14}O_2,$$

or with a secondary amino group:

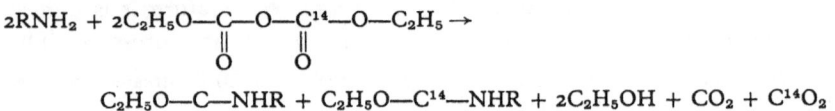

$$2RNH_2 \; + \; 2C_2H_5O{-}\underset{\underset{O}{\|}}{C}{-}O{-}C^{14}{-}\underset{\underset{O}{\|}}{O}{-}C_2H_5 \rightarrow$$

$$C_2H_5O{-}\underset{\underset{O}{\|}}{C}{-}NHR \; + \; C_2H_5O{-}C^{14}{-}NHR \; + \; 2C_2H_5OH \; + \; CO_2 \; + \; C^{14}O_2$$

Amino acids, proteins, and tannins account for most of the side reactions in wines. In model studies with pure compounds, residual activity varied from 0.1 per cent or less for malic, tartaric, and citric acids. It was only 0.3 to 0.4 per cent for glucose and fructose. For a variety of amino acids it amounted to 1.2 to 4.5 per cent. The extent of the DEPC side reactions depends mainly on the pH value of the beverage. If the pH is lowered, the exchange of DEPC-C^{14} produces an increased residue of activity. With increasing pH, on the contrary, the extent of the side reactions is considerably increased and the activity is thus reduced. The constituents of the beverage are involved to different degrees in these pH-dependent reactions. Mehlitz et al. (1967) have shown that DEPC affects not only catalase activity but that of alcohol dehydrogenase and of trypsin. In the latter two, 50 per cent of the activity is reduced by 50 to 80 mg per liter of DEPC, whereas 130 to 580 mg per liter reduces their activity 90 per cent.

The toxic effect seems to be due to the reaction of DEPC with the amine and sulfhydryl groups of cellular enzymes. DEPC is more toxic to yeast, particularly film-forming yeast like Candida spp., and fermentative bacteria than to spore-forming bacteria and molds. Hennig (1959) suggested that its antimicrobial activity is related to inhibition of fermentation enzymes. Mehlitz and Treptow (1966), however, observed that in cold sterilized fruit juices the naturally occurring enzymes, particularly those involved in formation of aroma in grape juice, were not deactivated. They found that a number of enzymes (β-amylase, glucose oxidase, catalase, invertase, peroxidase, polyphenoloxidase, and pectinase) were resistant to DEPC. Catalase was particularly resistant and was completely inactivated only at a level of 255 mg DEPC per liter or higher. Mehlitz et al. (1967) observed that yeast ethanol dehydrogenase and trypsin were quite sensitive to DEPC. The activity of both enzymes decreased rapidly with increase in concentration of added DEPC. The activity of the ethanol dehydrogenase was reduced markedly at a level of 110 mg per liter and completely prevented at 550. Trypsin was more resistant, and required at least 600 mg per liter to reduce its activity to a low level and 1,000 mg per liter to inactivate completely.

Garoglio and Stella (1964) reported complete nontoxicity of DEPC added to beer, wines, and carbonated beverages if consumption was twenty hours or more after use of DEPC. They recommended its use. There was some reduction of efficiency of DEPC in the presence of tannin.

If it were not for these side reactions DEPC would be a perfect preservative, a chemical equivalent of pasteurization, for it would sterilize the wine, destroy the spoilage organism, and then disappear without leaving traces of its presence (since ethanol and carbon dioxide in small amounts are natural components of wine and harmless). Because of its decomposition DEPC is only a temporary preservative and is used only at the time the wine is bottled. The possible toxicity of DEPC depends on the probable reactions of its by-product diethyl carbonate and the toxicity, if any, of the products of these reactions.

Although Paulus and Lorke (1967) reported that DEPC carboxylated amino acids, gallic acid, chlorogenic acid, lactic acid, ascorbic acid, and catechin, none of the products was toxic either orally or peritoneally. Lang et al. (1967) reported that most of the products were enzymatically hydrolyzed. That of ascorbic acid was an exception. Its half life in water solution was 5 to 10 days. However, it did not seem to be resorbed. The authors conclude that no toxicological problem should arise from the use of DEPC in fruits, juices, or wines.

Merzhanian (1952), Parfent'ev and Kovalenko (1951, 1952), Kozenko (1952, 1957), and Frolov-Bagreev (1952) all reported the presence of DEPC in wine. Parfent'ev and Kovalenko discussed the possible methods by which carbon dioxide could be bound in organic combinations in sparkling wines and concluded that diethyl ester of pyrocarbonic acid was the most likely. They claimed to have isolated DEPC from sparkling wine. Kozenko (1952, 1957) reported that DEPC was present in wine but not in must, and that the highest concentration occurred in sparkling wines. Both Merzhanian (1952) and Parfent'ev and Kovalenko (1951, 1952) detected esterified carbon dioxide by extracting must or wine with ethyl ether and then hydrolyzing the ester in the ether extract by either alkali or acid treatment. The former was preferred, as it was faster and more complete. The ester was determined from measurement of the carbon dioxide evolved directly in the acid treatment, or after acidification in the alkali treatment. This procedure did not differentiate between various possible esters of carbonic acid. The Soviet enologists attributed the ability of bottle-fermented sparkling wines to resist refermentation on addition of sweetened brandy and its superior keeping quality after opening to the sterilizing action of naturally occurring DEPC. Prillinger and Horwatitsch (1964), however, were not able to detect DEPC in sparkling wines up to 10 atm. of pressure. Kielhöfer and Würdig (1963c) could not detect diethyl carbonate, an intermediate in the acid-catalyzed hydrolysis of DEPC, in any sparkling wines to which DEPC was not added. They always found it in wines to which it was added, and thus concluded that DEPC is not formed naturally.

Pauli and Genth (1966), from the Bayer factory, consider the Russian claims on the natural occurrence of DEPC to be unsubstantiated. However, the Russian enologists described the properties of DEPC long before it was synthesized at the Bayer factory. The existence of diethyl carbonates other than DEPC was recognized by Parfent'ev and Kovalenko (1951), and particularly by Konzenko (1957).

DEPC was first recommended as a preservative for German sweet table wines by Hennig (1959). Ough and Ingraham (1961) were the first to recommend it as a cold-sterilizing agent for California wines. Joslyn (1961b) pointed out that the commercially available DEPC from different sources varied in flavor, that off-flavor formed on storing DEPC in open bottles, and that this could be inhibited by addition of sulfite. Pauli and Genth (1966) blame the foreign flavor of early DEPC samples on the presence of impurities. They note that people vary in their sensitivity to amounts of DEPC from 100 to 500 mg per liter. Canadian tasters rejected as undesirable wines to which 200 mg per liter of DEPC had been added, according to Adams (1965b). This certainly appeared to be so four to seven days after addition, but two months after addition the results did not appear to be significant. Adams also reported that a minimum of 200 mg per liter of DEPC was necessary to inhibit yeast growth in grape juices—unfortunately tested at pH 4.0 or above. See also Adams (1965a).

Pauli and Genth (1966) and Pauli (1967) give complete reviews of the properties, method of action, and analytical determination of DEPC. According to them, it was legal for wine (1966) only in Brazil, Canada, Germany, Israel, and the United States. It is legal also for fruit juices in Argentina, Brazil, Bulgaria, Chile, Czechoslovakia, Greece, South Africa, Sweden, and Tunisia. Rentschler (1965b) and Mayer and Lüthi (1960) used it in Switzerland. Minárik and Laho (1962) recommended 150 to 200 mg per liter of DEPC for sweet table wines in Czechoslovakia. See Treptow and Gierschener (1968) for a favorable report on the use of DEPC.

Hennig (1959, 1963) has summarized his experiments on the use of DEPC for German sweet table wines. He recommends 50 to 100 mg per liter with 10 to 15 mg per liter of free sulfur dioxide. In Germany the amounts which can be added depend on the alcohol content of the wine:

Alcohol (per cent)	DEPC (ml per 100 liter)
7.5	18
8.5	16
10.0	14.5
12.0	11.5
15.0	10.0

Hennig (1963) believed that 0.5 gr per liter of DEPC would be needed before sensory detection of diethyl carbonate would be obvious. The amount may be considerably less.

Ivanov et al. (1966) recommended 75 mg per liter of DEPC for wine stabilization and found no undesirable flavor effects up to 150 mg per liter. The wines contained *Sacch. oviformis, Brettanomyces vini* or lactic acid bacteria, and the test lasted 75 days. But Tonchev *et al.* (1967) reported that 100 mg per liter or more of DEPC gave an unpleasant flavor to grape juice. Only 25 mg per liter was required to control yeast growth, but 300 and 500 mg per liter were necessary for lactic and acetic acid bacteria, respectively.

DEPC has been recommended as a preservative for fruit juices and sweet wine. When used for fruit juices Pauli and Genth (1966) recommend 40 to 80 mg per liter, but apple juices may require up to 200 mg per liter. The advantage of DEPC is that the juices can be cold-filled and thus retain greater freshness than hot-filled. Pauli (1967) has summarized the practical information on the amounts to use. Even at 30 mg per liter many yeasts are destroyed, and at up to 200 mg per liter no sensory differences were noted. The suggested limit of the World Health Organization (WHO) and Food and Agriculture Organization (FAO) is 300. Since sulfur dioxide has a synergistic effect, 100 seems adequate for wines containing normal amounts of sulfur dioxide and yeasts. Hennig (1963) correctly recommended that DEPC be used just before bottling, that the corks be sterilized, and that there be immediate dissolution of the DEPC in the wine.

Three methods of adding DEPC to wines were tested by Kielhöfer and Würdig (1964): direct addition to the wine and very rapid stirring to mix, pumping with the membrane Orlita, and a special Orlita-Sta. pump. The best results were with the first. Troost (1968) also recommended direct and continuous addition of DEPC using a special pump. Genth (1965) strongly recommends a metering pump rather than stirring the DEPC into the wine. DEPC must be stored in a cool dark place and be used within the time limit stated on the label. The manufacturer's precautions about handling DEPC should be heeded.

From 5.9 to 8.3 mg per liter of diethyl carbonate are formed when 100 mg per liter of DEPC are added. Diethyl carbonate, found as a by-product, gives an odor to the wine. Carbon dioxide, nitrogen, helium, and oxygen did not remove it, but storage in the presence of air did. Since untreated wines do not contain diethyl carbonate, its presence is proof that DEPC has been used. Pure DEPC does not form diethyl carbonate during storage, but Kielhöfer and Würdig (1963c) and Prillinger (1964), using a gas chromatographic technique, reported 3 to 6 mg of diethyl carbonate in DEPC.

Würdig (Anon., 1966a) reports that about 5 per cent of the DEPC added is found as diethyl carbonate. If a 10 mg limit of diethyl carbonate is set, as in Germany, then no more than 200 mg per liter of DEPC may be used. It seems that the primary use of DEPC will be at the time of bottling of sweet table wines. Only protein-stable wines of low yeast count (less than 500 per ml) should be used.

DEC or diethyl carbonate can be determined by ether extraction and hydrolysis of the extract as originally done by Parfent'ev and Kovalenko (1951, 1952) and Merzhanian (1952). This procedure was improved by several investigators; the most recent contribution was that of Garschagen (1967). To detect diethyl carbonate GLC has been used.

Kielhöfer (1963a) noted that different yeast strains varied in resistance to DEPC, but 25 mg per liter was normally sufficient to prevent growth. For *Saccharomyces acidifaciens*, which is more resistant, Kielhöfer reports that 40 to 50 mg per liter is necessary. He agrees with Ough and Ingraham (1961) that DEPC is effective within ten to thirty minutes at 25 to 100 mg per liter. DEPC is effective at 50 mg per liter, and 100 mg is adequate for stabilizing sweet table wines, according to Terčelj and Adamič (1965). Garoglio and Stella (1964) reported 0.00017 per cent as the lethal dose of DEPC for *Sacch. cerevisiae* when the yeast population was 6×10^5 per ml. For greater effectiveness DEPC should always be used in closed tanks. Its lethal action depends on the type and number of microörganisms. For a number of yeasts and bacteria the sterilizing concentrations are given in table 57.

TABLE 57

EFFECT OF DIETHYLPYROCARBONATE

Microörganisms	Number	Sterilizing concentration (mg per liter)
Saccharomyces cerevisiae	5×10^6	30–80
var. *ellipsoideus* Burgundy	5×10^4	100
var. *ellipsoideus* Champagne	6×10^5	100
Sacch. oviformis	5×10^4	50
Sacch. pastorianus	4×10^2	100
Sacch. acidifaciens	5×10^2	200
Pichia farinosa	4×10^5	100
P. membranefaciens	5×10^2	25
Torula utilis	1×10^5	250
Candida mycoderma	5×10^2	200
Acetobacter pasteurianum	5×10^2	80
Bacterium aceticum	5×10^4	500
Lactobacillus arabinosus	1×10^6	250
Micrococcus annulatus	4.8×10^2	30
Bacterium coli	1×10^5	400
Staphylococcus aureus	1×10^3	70–100
Botrytis cinerea	1×10^3	100
Aspergillus niger	1×10^4	150
Penicillium glaucum	3×10^2	250

Source of data: Pauli (1967)

ASCORBIC ACID
Proposed as a substitute antioxidant.

In chemical preservation of fruit juices the antimicrobial and antioxidative properties of sulfur dioxide have been partially replaced by other substances to allow the use of lower quantities of sulfur dioxide. Sodium benzoate has been used as the primary antimicrobial agent, with only the quantities of sulfur dioxide required to reduce oxygen present in the juice and in the headspace of the container added. Mixtures of sodium benzoate and sulfurous acid, at first prohibited in Great Britain, are now allowed. The antioxidative properties of sulfur dioxide have been replaced or supplemented by ascorbic acid. These replacements sometimes proved beneficial, particularly when the effectiveness of the mixture was increased by synergistic action.

The replacement of the antioxidative effect of sulfur dioxide by ascorbic acid was proposed for fruit wine and sparkling wines. Its use to prevent nonenzymatic oxidation of wine was investigated in the Soviet-Union, Germany, and Switzerland from 1956 on. As pointed out by Ribéreau-Gayon and Peynaud (1964–1966), in spite of many investigations in the United States and abroad, the efficacy of ascorbic acid is still in dispute, owing to different methods of application in wines and to incomplete knowledge concerning its mode of action and transformations in wines.

Ascorbic acid has been reported to be effective in prevention of nonenzymatic oxidations by some investigators and to be harmful or of questionable value by others. Since 1956 its use has been allowed by the Internal Revenue Service to prevent deterioration of flavor and darkening of color in apple wine, and in 1957 it was allowed in the production of vermouth and other wines. The use of isoascorbic acid, a cheaper synthetically produced isomer (now called erythorbic acid), was allowed in 1958. Its use in Switzerland was investigated by Rentschler and Tanner (1958) and Rentschler (1960). Its use in Germany was investigated by Kielhöfer (1956, 1958, 1959, 1960a), Kielhöfer and Würdig (1960b), Diemair *et al.* (1960), and Haushofer and Rethaller (1965). Its use in Spain was investigated by Mareca Cortés *et al.* (1959); in Italy by Santagostino and Sapetti (1963); in Japan by Kushida *et al.* (1964, 1965); and in France by Maurel *et al.* (1964). Fessler (1961) reported on the effect of ascorbic acid and erythorbic acid on residual oxygen content, discoloration, and flavor of California bottled wines. The red wines he used had a total sulfur dioxide content of 150 to 165 mg per liter, the rosé from 155 to 165 and the white from 240 to 270. At these levels of sulfur dioxide the addition of from half a pound to two pounds per 1,000 gallons increased the objectionably high "sulfury" taste and aroma, and increased the tendency to development of hydrogen sulfide. The oxygen content at the time of bottling, however, was reduced by addition of these antioxidants, more rapidly by erythorbic acid than by ascorbic acid.

Kielhöfer (1956) was one of the first to point out that ascorbic acid does not solve the problem of finding a substitute for sulfur dioxide. Ascorbic acid in wines has no antiseptic or sterilizing properties for either yeasts or bacteria, and does not possess anti-enzymatic effects. Prillinger (1963) recommended ascorbic acid as an antioxidation agent only when enzymatic oxidation is not a factor and when, after combination with acetaldehyde, free sulfur dioxide is still present. Kushida *et al.* (1964) reported that the decolorizing effect of 100 mg per liter of sulfurous acid was about equal to 200 mg per liter of ascorbic or erythorbic acid. The sensory quality of wines containing sulfurous acid and either one of the other acids was superior to wines containing only an organic acid or no additive.

Addition of ascorbic acid up to 100 mg per liter was authorized in France on October 22, 1962, according to Maurel *et al.* (1964). A 10 per cent tolerance was admitted. While ascorbic acid is legal in this country, it is rarely used because it frequently leads to problems of browning.

The mechanism of oxidation of ascorbic acid in wines was studied by Santagostino and Sapetti (1963). They favored direct oxidation and autocatalytic reactions which included the following:

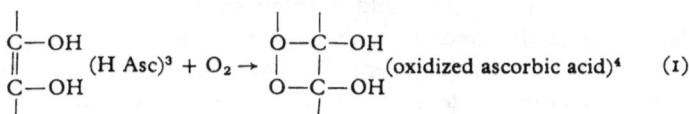

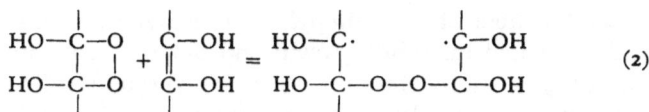

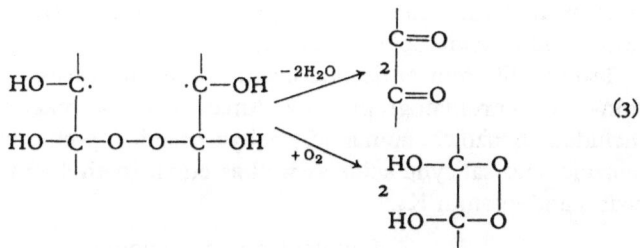

$$
\begin{array}{l}
\overset{|}{C}-OH \\
\overset{\|}{C}-OH \\
\overset{|}{}
\end{array}
(H\ Asc)^3 + O_2 \rightarrow
\begin{array}{l}
\overset{|}{O}-\overset{|}{C}-OH \\
\overset{|}{O}-\overset{|}{C}-OH \\
\overset{|}{}
\end{array}
(\text{oxidized ascorbic acid})^4 \qquad (1)
$$

$$
\begin{array}{l}
\overset{|}{HO-C-O} \\
\overset{|}{HO-C-O} \\
\overset{|}{}
\end{array}
+
\begin{array}{l}
\overset{|}{C}-OH \\
\overset{\|}{C}-OH \\
\overset{|}{}
\end{array}
=
\begin{array}{l}
HO-\overset{|}{C}\cdot \\
HO-\overset{|}{C}-O-O-\overset{|}{C}-OH \\
\end{array}
\cdot\overset{|}{C}-OH \qquad (2)
$$

$$
\begin{array}{l}
HO-\overset{|}{C}\cdot \qquad \cdot\overset{|}{C}-OH \\
HO-\overset{|}{C}-O-O-\overset{|}{C}-OH
\end{array}
\xrightarrow[+O_2]{-2H_2O}
\begin{array}{l}
2\begin{array}{l}\overset{|}{C}=O \\ \overset{|}{C}=O \\ \overset{|}{}\end{array} \\
2\begin{array}{l}HO-\overset{|}{C}-O \\ HO-\overset{|}{C}-O \\ \overset{|}{}\end{array}
\end{array}
\qquad (3)
$$

---

[3] Ascorbic acid.
[4] Dehydroascorbic acid.

Addition of ascorbic acid must be made in amounts of 50 or more mg per liter, according to Haushofer and Rethaller (1965), since smaller amounts are completely destroyed after the wine is bottled. Whether ascorbic acid should or should not be added depends on the composition of the wines. They do not recommend ascorbic acid for late-harvested Austrian wines of relatively high alcohol. Some fresh, fruity wines (of Veltliner, Müller-Thurgau, and similar varieties) profited by addition of ascorbic acid if they had been well-matured. Other slower-maturing wines (of Sylvaner, Wälschriesling, etc.) could profitably be handled with added ascorbic acid only after long storage; with varieties which require long aging (White Riesling) ascorbic acid should be avoided. Use of ascorbic acid to inhibit browning reactions was tested in model studies using potato or apple phenolase with pyrocatechin by Duden and Siddiqui (1966). Even large amounts of ascorbic acid (1,000 mg per 100 gr) did not change oxygen uptake. However, under such conditions enzyme activity ceased in the course of the reaction.

Other antiseptic agents

As a partial substitute for sulfur dioxide, Yang and Orser (1962) recommended vitamin K_5. They found that 10 mg per liter of the vitamin and 100 mg per liter of sulfur dioxide gave good protection to sweet table wines. The sensory threshold is about 50 mg per liter. There was some darkening of the color of white treated wines. The material is not yet approved for use by the federal Food and Drug Administration.

As a possible supplement to sulfur dioxide, Lafourcade (1955) believed that actidione offered possibilities. So far this has not proven correct.

Combinations of the n-propyl and n-heptyl esters of p-hydroxybenzoic acid prevented microbial growth and did not result in an off-flavor in Rice's (1968) study. However, the n-heptyl ester resulted in haziness. He believed 100 mg per liter of the n-propyl or 60 mg per liter of the n-butyl ester would be sufficient and not present off-flavors. However, the samples tested had considerable free sulfur dioxide.

Twelve different antiseptic agents added to musts were identified by thin-layer chromatography by Amati and Formaglini (1965). These included benzoic, cinnamic, dehydroacetic, p-chlorobenzoic, p-oxybenzoic, and salicylic acids as well as ethyl, methyl and other p-oxybenzoates and vitamin K_5.

Fungicides and pesticides

The increasing use of chemical agents to control growth of fungi and pests has been of concern to viticulturalists (as to their effect on vine physiology) and enologists (as to their effect on fermentation and development of bouquet during aging). Gärtel (1967) in reviewing this problem, warns against the use of such agents late in the season. Several of them can subsequently reduce the rate of fermentation. Unless special precautions are taken, aerobic conditions and growth of undesirable microörganisms (bacteria or yeast) may then occur.

Castor et al. (1957) noted possible delays in fermentation owing to captan[5] residues on grapes. Delays of twenty to forty hours with residues of up to 2 mg per liter of captan were reported. Once fermentation started, it appeared to be normal. Cantarelli et al. (1964) did not find that zineb[6] influenced the rate of fermentation up to 500 mg per liter. In fact, it may actually speed up fermentation. Some hydrogen sulfide, carbon bisulfide, and mercaptans were produced, but they were not harmful. Most of the spray residue is recovered unchanged in the lees. They also showed that zineb did not have a selective effect favoring development of yeasts that produce volatile sulfur compounds. Thus the volatile sulfur compounds are due to normal yeast metabolism of the sulfur in the zineb molecule. Some of the organic fungicides tested by Minárik and Rágala (1966) had a selective influence on the yeast flora. Captan, for example, stimulated the development of Torulopsis bacillaris, which is important in the final stages of fermentation, but generally inhibits other yeasts. Zineb inhibited Candida pulcherrima but favored growth of Sacch. oviformis.

The effects of certain fungicides and pesticides on fermentation and fermentation residues by Sacch. cerevisiae, Sacch. oviformis, and Hansenula anomala were investigated by Bidan (1966). The fungicides tested were zineb, mezineb, captan, phaltane,[7] difoltan,[8] and dinitrophenylcrotonate. Of these the most active in inhibiting yeast growth were captan, phaltane, and difoltan. The fungicides tested included HCH,[9] DDT,[10] malathion,[11] and several commercial preparations. None had any effect on yeast growth at 10 mg per liter. Even at 100 mg per liter none prevented fermentation.

Moţoc and Dimitriu (1963–1964) found little fungicidal effect of penicillin, streptomycin, eritomycin, aureocycline, or chloromycetin on Sacch. cerevisiae var. ellipsoideus. The tetracycline antibiotics inhibited Candida mycoderma (incorrectly cited as Mycoderma vini). Acetic acid bacteria were strongly inhibited by aureocycline and at the beginning of their growth only by penicillin and chloromycetin.

---

[5] N -trichloromethylmercapto-4-cyclohexene-1,2-dicarboximide.
[6] Zinc ethylenebisdithiocarbamate.
[7] N-trichloromethylthiophthalimide.
[8] N-(1,1,2,2-tetrachloroethylsulfenyl)-cis-Δ-4-cyclohexene-1,2-dicarboximide.
[9] 1,2,3,4,5,6-Hexachloro-cyclohexane.
[10] 2,2-Bis (p-chlorophenyl)-1,1,1-trichloroethane.
[11] O,O-dimethyl S-1,2,-di(ethoxycarbamyl)ethyl phosphorodithioate.

Monochloracetic acid, bromchloracetic acid, salicyclic acid, benzoic acid, ethylene oxide, among others, have been tried and abandoned. Either they failed to prevent bacterial growth or legal restrictions on their use were imposed because of their toxicity. Regardless of their merits or their toxicity, they are subject to abuse when used in wines because, when added in excessive quantities, they do not reveal their presence by smell or taste as sulfur dioxide does.

EFFECT OF AERATION
Its influence on yeast growth is not simple.

Yeasts require oxygen for their maximum development, although most of them can live and multiply for a limited time in its absence. Usually not more than two or three generations are formed in the complete absence of oxygen. It has been demonstrated that certain strains of yeast can grow continuously in the absence of oxygen (Chinn, 1950; Ephrussi and Slonimski, 1950; Neilson and Joslyn, 1951). Under such conditions the yeasts are deficient in cytochrome, cytochrome oxidase, and succinic acid dehydrogenase. In the absence of oxygen, however, they exert their greatest power of alcoholic fermentation. The fermentative enzymes secreted by the yeast require no oxygen for their activity. Higher amounts of alcohol are formed in the absence of air. In fermentations conducted in the presence of air, the efficiency of alcohol production is reduced, not only because oxygen interferes with the alcoholic fermentation by the enzymes, but also because of increased active respiration of sugar by yeast. The greater loss of alcohol by evaporation or entrainment, in open fermentations and under conditions of aeration, is also a factor.

In wine making, therefore, the desirable procedure is first to promote the multiplication and vigor of the yeast by growing it with abundant oxygen and then to conduct the alcoholic fermentation with a limited supply of oxygen. These conditions are brought about automatically in the usual methods of wine making. The crushing and stemming of the grapes thoroughly aerate the must. The yeast multiplies vigorously in this aerated nutrient solution until it has consumed most of the dissolved oxygen. Then alcoholic fermentation ensues. If the yeast is weakened before the wine is dry, it may usually be invigorated by aeration, as by pumping over. This aeration mixes the yeast thoroughly with the fermenting liquid; removes carbon dioxide, which has a retarding influence on fermentation; and supplies the oxygen that favors multiplication of the yeast. Excessive aeration during fermentation results in a flat, oxidized wine of poor flavor.

Ribéreau-Gayon *et al.* (1951) showed that aeration has its maximum effect on the speed of fermentation during the initial stages of fermentation. They ascribe the inefficiency of aeration during the later stages of fermentation to the presence of alcohol, which interferes with the assimilation of the necessary nitrogen. In red-wine fermentations, therefore, they recommend no aeration after the second day.

The role of oxygen in fermentation has been summarized by Paronetto (1966):

	Time on skins (hours)				
	0	1	3	6	12
Nonsulfited					
Redox potential, v	0.321	0.390	0.416	0.436	0.455
Oxygen, mg per liter	—	3.3	5.2	6.3	7.3
Tannin, ml 0.1 N KMnO₄	5.9	5.4	4.8	4.2	3.6
Quinone, mg per liter	—	0.6	4.5	6.7	10.5
Ascorbic acid, mg per liter	12.5	2.5	0.8	0.5	0.3
Sulfited, 100 mg per liter					
Redox potential, v	0.321	0.346	0.356	0.376	0.386
Oxygen, mg per liter	—	0.7	1.0	1.3	1.5
Tannin, ml 0.1 N KMnO₄	5.9	5.8	5.8	5.7	5.7
Quinone, mg per liter	—	0	0	0	0.5
Ascorbic acid, mg per liter	12.5	10.2	9.9	9.5	9.3

Not only do nonsulfited musts have a higher redox potential while on the skins, but they form more quinones and retain their higher potential during fermentation. Paronetto distinguishes direct oxidation by oxygen from enzymatic and catalytic oxidation.

Schanderl (1948a, 1948b, 1950–51) and Lichev et al. (1966) showed clearly that the oxidation reduction potential decreases during fermentation. In the latter study it was reported that Bulgarian white grape must had higher oxidation-reduction potential than red and oxidized more intensely. White wine thus requires more sulfur dioxide for its protective effect in reducing dissolved oxygen.

Jaulmes (1967) has reviewed the present practices of using carbon dioxide during handling of the must and wine. He reported that its use is legal in most European countries except France. He noted that non-aerated musts can be heated with much less injury to their flavor than aerated musts. The main advantages for handling wines under carbon dioxide are that it prevents access of air and subsequent oxidation and acetification and also preserves the sulfur dioxide content. Jaulmes recommended handling white musts from moldy grapes (which are very susceptible to browning in the presence of oxygen) under carbon dioxide to prevent browning. Use of carbon dioxide in handling white wines is especially common in Switzerland. Capt and Hammel (1953) have shown that the optimum carbon dioxide content is 1.13 to 1.42 gr per liter. For red wines Jaulmes (1967) recommended use of a mixture of nitrogen and carbon dioxide. Old red wines contain 0.1 to 0.5 gr per liter of carbon dioxide. The undesirable gasiness is perceptible at 0.7. A wine of 10 per cent alcohol in contact with 25 per cent carbon dioxide will contain only 0.37 gr per liter of carbon dioxide, which is less than can be detected.

EFFECT OF SUGAR CONCENTRATION
Rate and extent of fermentation are affected.

Sugar above a certain concentration will retard fermentation; and, if its concentration is high enough, the activity of the yeast may be entirely inhibited. The optimum sugar concentration for maximum alcohol production is about 28 per cent. If the initial sugar concentration is very high (30° Brix or over), sufficient alcohol and other by-products may be formed to arrest fermentation before the sugar is entirely consumed. The paralyzing influence of high sugar content is indicated by the following data of Benvegnin *et al.* (1951):

Sugar (per cent)	Alcohol (per cent after two months)
37	8.6
42	6.3
47	5.9
55	3.4
75	0.0

Vogt (1945) reports similar results in German musts of high sugar content:

Sugar (per cent)	Alcohol (per cent)
26.9	14.3
33.2	12.8
37.9	11.6
45.9	9.5

All strains of *Sacch. cerevisiae,* including the wine yeasts, are characterized by low sugar tolerance. Thus Gray (1945) reported the percentage of sugar utilized by different strains of distillers' yeast to decrease from over 99.2, at about 5 gr of glucose per 100 ml of media, to from 33 to 57 at about 30 gr of glucose per 100 ml. The inhibition of sugar utilization, which varied in degree with the strains studied, was due in part to plasmolysis, which occurred in sugar solutions of higher concentrations. Gray (1946) found it possible to increase the glucose tolerance of distillers' yeast by daily transfer to a medium of high glucose content. Such acclimatization, however, resulted in decreased alcohol tolerance. This is of interest in view of the fact that the naturally sugar-tolerant or osmophilic yeasts which can multiply in solutions of high osmotic pressure, such as honey and sirups, are not noted for their fermentative activity and usually do not tolerate much alcohol.

The studies of Ough (1964b) again show that sugar has an inhibitory effect on fermentation rate. He reported ammonia content, pH, and temperature as the other variables. The rate of fermentation from 20° to 0° Brix at 21.1°C (70°F) was 0.117 + 0.00085 NH_3 + 0.1066 pH − 0.0169° Brix. This equation accounted for about 80 per cent of variation in the fermentation rate of white grape juice. He also reported (1966a) that lower pH's (3.0 versus 3.5 and 4.0) retarded fermentation and that the retardation was greater at higher temperatures—32.8°C (91°F)—than at lower temperatures—21.1° or 10°C (70° or 50°F). Finally (1966b) he correlated the effects of initial alcohol content, pH, and temperature on fermentation rate. The greatest inhibitory effect of alcohol on fermentation rate was at the higher temperature and lower pH.

EFFECT OF ALCOHOL CONCENTRATION
Alcohol tolerance of yeasts varies.

Although 16 per cent alcohol by volume is the maximum produced by wine yeast in ordinary wine making, under certain conditions as much as 21 per cent alcohol can be obtained. The rate of fermentation decreases markedly with increase in alcohol content; and, under practical conditions, fermentations often stick at 13 to 15 per cent by volume, especially when a selected yeast is not used. In making dry table wines it is usually desirable to adjust the must to not over 23° Brix before fermentation so that the wine, when dry, will contain not more than 13 per cent alcohol by volume. Better white table wines are frequently produced with a Brix degree of not over 21° or 22°.

The wine yeasts vary remarkably in their ability to withstand alcohol as well as in their fermentation capacity. Castan (1927) found that different yeasts produced different quantities of alcohol when grown in a sterile, sweetened must of 31.71 per cent sugar, as follows:

Strain of yeast	Alcohol (per cent) by volume)	Strain of yeast	Alcohol (per cent) by volume)
Portici 3	11.9	Pully 59	15.9
No. 24	11.1	Dôle 22	16.4
Chablis 13	12.4	Argle 6	16.5
Dizaley 75	13.4	Pully 20	16.7
Pully 21	14.9	Montibeux 29	16.9
Montibeux 1	15.7	Cortaillod 15	17.0
Pully 8	15.5	Malvoisie Fletrie 36	17.2

More data of this type with greater repetition, statistical analysis of the results, and more careful control of the prefermentation treatment of the yeast cultures would be desirable.

Hohl and Cruess (1936) and Hohl (1938) were able to increase the alcohol content of fermented grape must to as high as 19 per cent by gradual addition of concentrate during fermentation. Jerez (Spanish sherry) flor yeast was found to be more alcohol-tolerant than other strains by Hohl and Cruess (1939). Strains of the Jerez flor yielded over 19 per cent alcohol by siruped fermentation and as high as 18 per cent by normal fermentation of a 30° Brix solution. We have observed alcohol production of 16 to 17 per cent in which no special precautions were taken with musts of high sugar content, especially when considerable amounts of raisined berries were present.

Gray (1941), in his studies of the alcohol tolerance of distillers' yeasts, found marked variation in the ability of various strains of yeast to utilize sugar in the presence of alcohol. Yeasts of extremely low alcohol tolerance utilized not over 77 per cent of the sugar in the presence of 4.8 per cent alcohol and were largely contaminants, such as *Hansenula anomala*. At the other extreme were the strains of *Sacch. mellis* isolated by Phaff and Douglas (1944) from a spoiled dessert wine of 21 per cent alcohol. They believed that this yeast was able to develop in the wine, although they could not grow pure cultures of the yeast either in a media containing over 12 per cent alcohol or in dessert wine.

EFFECT OF POLYPHENOLS

Šikovec (1966) found considerable variation in the effect of various polyphenolic compounds on the course of alcoholic fermentation. Chlorogenic and isochlorogenic acids stimulated fermentation, whereas gallic, ellagic, and caffeic acids inhibited it. With *Saccharomycodes ludwigii* (a sulfite-tolerant yeast) and 6 per cent alcohol the results were different: gallic and ellagic at first inhibited then accelerated fermentation. At 11 per cent alcohol all the polyohenols inhibited yeast growth at first, but only chlorogenic acid accelerated it at the end of the fermentation. (See p. 307.)

During aging, chlorogenic acid is hydrolyzed, producing caffeic acid. This may be one reason why bottle-fermented sparkling wines sometimes ferment slowly. Another reason is that wines high in tannin often lead to the development of yeast involution forms and, at high alcohol contents, to the death of such forms. Possibly this is the origin of the practice of adding tannin to sparkling wine *cuvées*. Šikovec also noted that some strains of yeast were more inhibited by polyphenols than others. Yeast cells adsorb polyphenols and may sometimes partly assimilate them.

The alkyl esters of gallic acid have a microbicide effect on *S. cerevisiae*. Apparently these compounds are absorbed until a lethal concentration is reached. With rapid fermentation this concentration is seldom reached. Denaturation of cell proteins is believed to be the primary cause of death.

Dittrich and Kerner (1966) also give data on the repressive effect of the alkyl ester of gallic acid on fermentation.

YEAST NUTRITION

Carbon, nitrogen, and minerals are required, but in special forms.

The rate of alcoholic fermentation depends also on the amount of yeast present. Since the number of yeasts initially present in the must is low, conditions must be made favorable for their growth and activity, which depend on sufficient food, moisture, and air supply, and on correct temperature. Yeast requires a source of carbon, a source of nitrogen, certain minerals (both the macronutrient and the micronutrient), and several accessory growth factors (Joslyn, 1951). The general nutritive requirements of *Sacch. cerevisiae* are known fairly well, but the particular requirements of the wine yeast strains have not been investigated thoroughly. Nutrients such as potassium salts, phosphates, other mineral salts, and accessory growth factors are rarely limiting in grape musts, but nitrogenous constituents, particularly ammonium salts, occasionally are.

Sugars

Wine yeasts will ordinarily utilize the sugars—glucose, fructose, mannose, sucrose, and maltose—with equal readiness. Most strains will ferment glucose faster than fructose (Dubrunfaut, 1947; Hopkins and Roberts, 1935); but some, the so-called Sauternes yeasts, will ferment fructose faster (Gayon and Dubourg, 1890; Dubourg, 1897; Harden, 1932; Gottschalk, 1946). Domercq (1956, 1957) identifies *Torulopsis bacillaris, Sacch. acidifaciens,* and *Sacch. elegans* as fructose-fermenting yeasts. (See also p. 338.)

Szabó and Rakcsányi (1937) have shown that in Hungarian musts of 17 to 20 per cent total sugar, glucose is fermented preferentially; at sugar contents of from 20 to 25 per cent, the rate of fermentation of the two sugars was about equal; from 25 to 30 per cent total sugar favored the fermentation of fructose (see table 58). This was substantiated by their analyses of sweet table wines, where those of low sugar content were found to have two to six times as much fructose as glucose, whereas in the very sweetest, *essenz* types, there was more glucose than fructose. This is illustrated by the data in table 59.

According to Gottschalk (1946), the relative rate of fermentation of a sugar by a given strain of yeast depends on the concentration of the sugar, its affinity for hexokinase, and its rate of passage through the cell wall.

TABLE 58

CHANGES IN RATIO OF GLUCOSE TO FRUCTOSE DURING FERMENTATION OF
MUSTS OF DIFFERENT TOTAL SUGAR CONTENT

Original must			Original must + 7.6 per cent sugar			Original must + 19.2 per cent sugar		
Total sugar (per cent)	Glucose (per cent)	G/F[a] ratio	Total sugar (per cent)	Glucose (per cent)	G/F[a] ratio	Total sugar (per cent)	Glucose (per cent)	G/F[a] ratio
17.9	8.5	0.91	25.5	13.2	1.08	37.1	18.8	1.02
17.3	8.2	0.90	23.9	12.8	1.16	35.3	18.5	1.10
14.0	6.7	0.92	21.3	12.3	1.37	32.8	17.9	1.20
10.4	4.3	0.71	17.9	9.5	1.13	31.7	17.7	1.26
6.4	2.0	0.45	15.5	8.2	1.13	29.8	16.8	1.29
1.2	0.02	0.01	13.9	7.2	1.09	27.4	15.9	1.38
—	—	—	11.2	5.8	0.98	26.1	15.0	1.35
—	—	—	6.2	2.6	0.70	21.9	12.3	1.30
—	—	—	3.7	1.0	0.37	19.6	10.7	1.20
—	—	—	2.4	0.6	0.30	16.2	8.4	1.08
—	—	—	1.1	0.3	0.30	—	—	—

[a]Glucose/fructose.
Source of data: Szabó and Rakcsányi (1937).

TABLE 59

RATIO OF GLUCOSE TO FRUCTOSE IN SWEET TABLE WINES

Type of wine	Alcohol (per cent)	Total sugar (per cent)	Glucose (per cent)	Fructose (per cent)	G/F[a] ratio
Tokajer Szamorodni	13.5	1.6	0.5	1.1	0.45
Mórer Gutedel	13.6	4.6	1.1	3.5	0.31
Tokajer Aszu	13.2	6.7	2.7	4.0	0.67
Tokajer Aszu	12.0	9.5	3.6	5.9	0.61
Tokajer Aszu	11.2	18.8	8.3	10.5	0.79
Tokajer Aszu	9.0	25.2	11.9	13.3	0.89
Essenz, 1888	8.2	23.3	16.6	6.7	2.52
Essenz, 1890	7.9	35.1	21.7	13.4	1.62
Essenz, 1906	4.9	42.6	27.4	15.2	1.80

[a] Glucose/fructose.
Source of data: Szabó and Rakcsányi (1937).

He believes that the Sauternes strain of yeast ferments faster than glucose because its cell walls exhibit a selective permeability for fructose. When the yeast cells were frozen and ground, they fermented glucose faster than fructose. All strains have the enzymes necessary for the conversion of both disaccharides, sucrose and maltose, into their component sugars—glucose and fructose, and glucose, respectively. More research is needed in this area.

Nitrogenous Compounds

Ammonium phosphate, urea, and the amino acids asparagine, aspartic acid, glutamic acid, leucine, and arginine are good sources of nitrogen for growth. Ammonium salts and urea are of particular importance. It has been shown that the ability of yeast to utilize these is dependent on the type and amount of accessory growth factors or bios constituents (Schulz et al., 1940); and that ammonium salts and biotin influence the fermentative ability of yeast (Winzler et al., 1944). The stimulation of fermentative activity by ammonium salts is well recognized (Thorne, 1946; Tressler et al., 1941). Biotin is an essential growth-promoter for yeast.

Tarantola (1955) found that the usual wine yeasts during alcoholic fermentation assimilated 45.7 to 48.5 per cent of the total nitrogen present in musts. During fermentation ammonia nitrogen is rapidly and completely assimilated. According to Cordonnier (1966), it is transformed to organic nitrogen and is found as free amino acids in the wine. The total amine content can be increased by four times by adding up to 250 mg per liter of ammonia. After a few months traces of ammonia (usually 10 to 20 mg per liter) will be found in the wine. This has been attributed to lactic acid bacteria. However, red wines aged in the bottle for many years may also have appreciable ammonia.

Significant changes take place in the nitrogenous fractions of must. Hennig (1943) found a rapid decrease in total nitrogen, protein nitrogen, ammonia, phosphotungstic precipitable nitrogen, amino acids, and humin nitrogen. Niehaus (1938) found a loss of 50.7 to 58.5 per cent of the total nitrogen during fermentation. He reported that the greater part of this loss took place within 12 to 48 hours after fermentation had started. Different strains of yeast removed varying quantities of nitrogen, but the differences were not of practical value, in his opinion. Aeration during fermentation increased the growth of yeast and the total quantity of nitrogen removed from the must, but the loss per unit of yeast cells was greater in the unaerated must. As the temperature of fermentation is increased from 15° to 30°C (59° to 86°F), the loss of nitrogen decreases.

Peynaud and Lafon-Lafourcade (1962) have confirmed that yeasts can assimilate all the amino acids except lysine. However, assimilation and nutritive value, as measured by weight of yeasts formed, are not always the same. They reported a variable excretion of amino acids from yeast cells (with no autolysis). The amino acids supplied have a clear influence on the by-products of alcoholic fermentation. Deibner (1964a, 1964b) also noted a decrease in hexosamines during fermentation.

After fermentation there is a slow increase in several nitrogenous fractions, mainly owing to autolysis of the yeast, especially of amino acids and to a lesser extent of hexosamines. This phenomenon was reported as early as 1898 by Laborde, according to Genevois and Ribéreau-Gayon (1936). They also reported differences among the wine yeasts in the ability to utilize all or only part of the ammonium nitrogen of the must.

The release of nitrogen by the yeast was believed by Schanderl (1942) to account for bacterial growth in certain German sparkling wines in bottles. Niehaus (1938) noted significant increases—10 to 50 per cent—in total nitrogen content after storage on the lees for 56 days; the largest increase was in the dry white table wines. Nitrogenous compounds, particularly nucleotides, may be released into must during the early stages of fermentation, Delisle and Phaff (1961), Lewis and Phaff (1963), and Lee and Lewis (1968a, b).

The isoelectric point of the proteins of different varieties in Koch's (1963) study varied from 3.3 to 4.0.[12] Where the pH of the wine is near to the isolectric point of the protein (as in warm years) greater precipitation will occur. In some instances the malo-lactic fermentation causes the same result. This may occur also during blending. Fermentation, of course, results in precipitation of proteins, more of Koch's fraction II than I, and there is a progressive reduction of proteins during aging. The protein can be removed by ultrafiltration or by heat. However, heating removes most of fraction I and less of II. In years of high protein content, heating may leave troublesome amounts of protein in solution. Bentonite fining, however, removes most of fractions I and II.

Cordonnier (1966) divides the nitrogen fractions into a nondialyzable group (proteins, nucleic acids, and nitrogen associated with polysaccharides or tannins) and a dialyzable group (peptides, amino acids, amides, purine bases, pyrimidines, and vitamins). The former group have molecular weights of over 10,000. They are important in the prefermentation phases of vinification and in the physicochemical stability of wines. The latter group was important to fermentation and in the biological stability of wines.

Koch's test for successful removal of protein was to add 5 per cent of saturated ammonium sulfate to a sample of the wine and heat at 45°C (113°F) for nine hours. The sample is then placed in water at 0°C (32°F) for fifteen minutes. If there is no precipitation the wine can be considered protein-stable.

The nitrogen source also influences the production of various by-products, as shown in tables 60 and 61.

Maximum glycerin and acetic acid are produced only with ammonia or a mixture of amino acids (Ribéreau-Gayon and Peynaud, 1964–1966). Ammonia was effective in increasing the production of 2,3-butylene glycol and acetaldehyde but not of succinic acid. Arginine especially increased

---

[12] Koch and Sajak (1963) reported isoelectric points of 3.3 to 3.7 for the proteins of six varieties. The proteins contained nineteen amino acids and seven sugars and glucosamines.

TABLE 60

BY-PRODUCTS OF ALCOHOLIC FERMENTATION ON DIFFERENT NITROGEN SOURCES
(mol per liter)

Source of nitrogen	Glycerin	Acetic acid	Succinic acid	2,3-Butylene glycol	Acetaldehyde
Aspartic acid[a]	29	3.0	3.7	1.6	3.0
Arginine	40	5.3	3.8	4.2	5.7
Cysteine	43	5.2	5.5	1.8	2.7
Histidine	31	4.2	4.0	0.9	1.2
Methionine	39	3.0	6.2	1.8	0.6
Tyrosine	36	2.8	4.0	1.4	2.2
Valine	35	6.2	3.8	1.8	2.2
Ammonia	50	10.3	3.5	5.0	4.6
Mixture of amino acids	57	10.2	6.2	2.0	2.8
Ammonia + mixture	57	13.7	4.0	5.6	1.8

[a] The results with threonine, leucine, isoleucine, and phenylalanine were essentially the same as those with aspartic acid.
Source of data: Ribéreau-Gayon and Peynaud (1964–1966).

production of 2,3-butylene glycol and aldehyde. For a study of the nitrogen metabolism of yeasts see Poux et al. (1964).

Minerals

Potassium, calcium, and magnesium salts, as well as inorganic phosphates and sulfates, are required for growth and activity in relatively large amounts; iron, manganese, copper, zinc, cobalt, and iodine are required in smaller amounts (Joslyn, 1941). It is now known that potassium as well as ammonium functions in fermentation. For a general discussion of inorganic and organic fermentation activators see Cantarelli (1957).

Growth factors

The growth-promoting nutrilites, or growth accessory factors, required to complete the medium are inositol, thiamine, biotin, pantothenic acid, pyridoxine, and possibly nicotinic acid and p-aminobenzoic acid (Williams, 1941; Peterson and Peterson, 1945). Of these factors, inositol is required in relatively high concentration, thiamine and pyridoxine in smaller amounts (from 1/10 to 1/1000 as much), pantothenic acid in even smaller amounts, and biotin in traces. Thiamine and biotin, particularly the latter, stimulate fermentation as well as growth.

Addition of thiamine (0.5 mg per liter) to musts generally reduced the amounts of pyruvic and α-ketoglutaric acid in the finished wine, according to Lafon-Lafourcade et al. (1967). This in turn reduced the level of fixed

TABLE 61

CLASSIFICATION OF AMINO ACIDS ACCORDING TO ASSIMILABILITY, NUTRITIVE VALUE, AMOUNTS IN YEASTS, AND EXCRETION AND LIBERATION BY AUTOLYSIS

Speed of assimilation	Individual assimilation	Assimilation in mixtures	Yeast Production	Consumption of yeast	Amino acids secreted	Amino acids liberated by autolysis
Complete, immediate Glutamic acid Aspartic acid Serine Threonine Lysine Arginine	**90–100%** Leucine Isoleucine Methionine Cysteine Glutamic acid Tryptophan	**70–100%** Arginine Tryptophan Isoleucine Cysteine	**1.0–2.0 gr/liter** Leucine Methionine Serine Glutamic acid Valine Tryptophan Proline	**> 10%** Glutamic acid Lysine Valine	Glutamic acid Proline Methionine Arginine Cysteine Glycine Serine Valine Threonine Histidine	Glutamic acid Proline Lysine Serine Asparagine Leucine Valine Isoleucine Alanine Phenylalanine Threonine Glycine Tyrosine Cysteine Histidine Arginine Methionine Tryptophan
Rapid, progressive Valine Methionine Leucine Isoleucine Histidine	**50–90%** Asparagine Phenylalanine Valine Threonine	**50–70%** Methionine Histidine Valine Asparagine Threonine Tyrosine Phenylalanine	**0.5–1.0 gr/liter** Tyrosine Isoleucine Phenylalanine Asparagine Glycine	**5–10%** Threonine Serine Isoleucine Leucine Asparagine Glycine Proline		
Slow Glycine Phenylalanine Tyrosine Tryptophan Alanine	**16–50%** Tyrosine Arginine Serine Proline Histidine	**> 50** Serine Lysine Glycine Glutamic acid Leucine	**> 0.5 gr/liter** Histidine Threonine Arginine Cysteine	**< 5%** Alanine Arginine Phenylalanine Tyrosine Methionine Histidine Cysteine Tryptophan		
None Proline	Lysine	Proline	Lysine			

Source of data: Ribéreau-Gayon and Peynaud (1964–1966).

sulfur dioxide. Less volatile acidity was produced. They conclude that these acids are often but not always the cause of high fixed sulfur dioxide. They especially recommended use of thiamine in the production of sweet table wines from botrytised grapes where high fixed sulfur dioxide and excessive volatile acidity are often a problem.

Ournac and Flanzy (1957) believed (see also Dupuy and Flanzy, 1954) that the amount of thiamine present was slightly less than that needed for complete fermentation of the sugar and that the rate of fermentation could be increased by addition of thiamine. Ribéreau-Gayon and Peynaud (1952) and Giudice (1955) found some benefit from added thiamine. However, other experiments have not shown any increase in the rate or extent of fermentation from added thiamine. In fact, Knuchel et al. (1954) found that too much thiamine retarded fermentation. For information on the amounts present in musts, see page 277.

There has been little study of the effect of the fermentation and stabilizing practices on the B-complex vitamin content of the resulting wines. The report of Bur'ian et al. (1964) is, therefore, interesting. They tried four fermentation procedures. Floating or submerged cap fermentations or in a special apparatus resulted in essentially equal B-content wines. Fermentations without the skins resulted in reductions of 33 per cent in inositol, 55 per cent in pantothenic acid, 60 per cent in thiamine, and 49 per cent in nicotinic acid.

Martakov and Kolesnikov (1967) reported that yeast cells release β-fructofuranosidase into the media when the enzyme is absent or inactivated in the media. Yeast cells will also absorb the enzyme when it is present. During the growth of yeast there is absorption of several of the growth factors from the media, particularly thiamine, nicotinic acid, and pantothenic acid, which accumulate within the yeast cells.

The autolysis of yeasts is an important problem in enology because of the development of desirable and undesirable microörganisms and off-odors. Datunashvili (1964) showed that increasing the temperature from 8° to 48°C (44.4° to 118.4°F) greatly increased yeast autolysis, as did increasing the alcohol content from 10 to 12 per cent. The optimum pH for yeast autolysis is 5.5, far above that encountered in normal winery practice. Under anaerobic conditions autolysis was more rapid than under aerobic conditions. The most rapid autolysis was during the first two months of storage and was much more rapid than under aerobic conditions. Moreover, autolysis during this same period was more rapid and was much more complete at 18° to 25°C (64.4° to 77°F) compared to 0° to 10°C (32° to 50°F). See also Belitzer (1934) and, for early data, Nilsson and Alm (1936), Iwanoff (1913, 1921a, 1921b, 1921c) and Navassart (1910, 1911).

EFFECT OF ORGANIC ACIDS...

... may be favorable or unfavorable.

The natural acids of grapes in musts have little inhibiting effect on wine yeasts, since these acids may, under certain conditions, be utilized as carbon sources. Yeasts will grow in solutions of tartaric, malic, and citric acids when the acid concentration is higher than that of the usual must.

The natural grape acids, in fact, by discouraging growth of competing organisms more sensitive to them, exert an indirect favorable action. At low acidities the growth of undesirable microörganisms is greater; hence, to stimulate a sound fermentation, fruit acids such as tartaric, citric, and malic are sometimes added to a must low in acidity (see p. 306). When de-ionized juice was treated with various acids, citric acid inhibited fermentation more than tartaric or sulfuric. Malic acid showed a lightly repressive effect, according to Ough and Kunkee (1967). Cf. p. 303.

Moţoc and Dimitriu (1966) note that a number of common esters are toxic to yeast growth and alcoholic fermentation—the toxicity increasing with molecular weight.

CHAPTER TEN

COMPOSITION OF TABLE WINES

The chemical constituents of wines are useful in differentiating the various types, in determining soundness (freedom from products of spoilage or to show compounds which might cause spoilage or cloudiness), in indicating quality in order to conform to legal requirements, and in measuring the amounts of components that are desirable or undesirable in the diet.

SIGNIFICANCE OF CONSTITUENTS

The most important are alcohol, acids, and sugars.

Alcohol content

Under California conditions the alcohol content of table wines does not differ markedly from type to type, since the climate is warm enough to mature practically all varieties of grapes in all seasons and regions. To comply with California regulations, white and red table wines must have from 10.0 to 10.5 and 14.0 per cent alcohol, respectively. The lighter types of dry white table wines should have an alcohol content somewhat lower than the red table wines; we suggest 11.5 versus 12.5.

Methanol

In fermentation of synthetic media, Bertrand and Silverstein (1950) found only traces of methanol. They believe that earlier reports of production of methanol in alcoholic fermentation were due to insufficient nutrients in the culture media used. Flanzy and Bouzigues (1959) reported no change in the methanol content of white musts during fermentation, but there was an increase with reds. It is not derived from glycine, but more likely results from hydrolysis of naturally occurring pectins. This is substantiated by the fact that the methanol content is higher (1) when pectolytic enzymes are added to the musts, (2) in wines made by fermentation on the skins, and (3) in wines made from macerated grapes compared to those from nonmacerated grapes (Flanzy and Loisel, 1958). Flanzy and Bouzigues (1959) found more methanol is wines made from musts stored on the skins at 0°C (32°F) for thirteen to twenty-two days compared to those not so stored. This seems reasonable, since the pectin methyl esterase is present mostly in the skins.

There are reports that wines made from *Vitis labrusca* varieties or from their hybrids are higher in methanol than those from *V. vinifera* varieties. We have not been able to substantiate this, but it may occur if the reported characteristic ester of *V. labrusca* grapes, methyl anthranilate, is hydrolyzed, or if their pectin content is especially high. Since fruit wines are usually high in methanol, the methanol content may be useful as an indication of admixture of fruit and grape wines. In the amounts normally present, methanol has little sensory significance. However, methyl esters are often aromatic.

Marescalchi (1966) reports that the most recent Italian law limits the methanol content of white wines to 0.20 ml per 100 ml of absolute alcohol and red wines to 0.25 ml. He recommended that this be increased to 0.30 and 0.35 ml, since many wines exceed these limits. For further information on the methanol content of table wines see Amerine (1954), Ribéreau-Gayon and Peynaud (1958), and Guimãraes (1965). The amounts of methanol varies from 0 to 635 mg per liter, with an average of about 100. Larger amounts are found in red wines than in white (about 2 to 1). Feduchy Mariño *et al.* (1964) have summarized previous research and their own on methanol in musts and wines as follows (mg per liter).

Source	No.	Type	Minimum	Maximum	Average
France	?	must	58	89	—
France	?	wine	63	93	—
France	21	wine	38	188	—
France	?	*V. vinifera*	—	—	100
France	?	hybrids	162	215	—
Spain	220	wine	39	624	145

The methanol content of twenty-two white Italian table wines of Verdicchio varied from 0.046 to 0.114 gr per liter (average 0.083), according to Pallotta and Donati (1965). In forty red and rosé Italian table wines Manelli and Mancini (1966) reported a minimum of 59.2 mg per liter of methanol and a maximum of 245 (average 104.2). In twenty-seven white table wines the range was 32 to 146 (average 60.6). The alcohol content varied from 9.4 to 19.8, but only four of the sixty-seven wines had more than 14 per cent. In 159 Italian wines of ordinary quality from the provinces of Modena and Reggio Emilia (including those of the first pressing) Mascolo (1966*a*) reported a maximum of 0.77 mg per liter of methanol and a minimum of 0.11 (average 0.32). In terms of ml per 100 ml of absolute alcohol the maximum was 0.90, the minimum 0.11, and the average 0.36. About two-thirds of these wines exceeded the Italian legal limit for methanol. In contrast, in 160 high-quality and free-run wines the maximum ml per liter was 0.37, the minimum 0.05, and the average 0.18, or, in terms of absolute alcohol, per 100 ml 0.36 maximum, 0.19 minimum, and 0.19 average. Only about one-sixth of these wines were above the legal limit. In a small (23) group of second-press and spoiled wines, more than 60 per cent were above the legal limit. Since the alcohol content averages about 10 per cent, the percentage of wines above the legal limit is about the same by either method of calculation. He recommended (1966b) that the limit be expressed as ml per liter of wine. He notes that the grapes in this region are strongly attacked by fungus diseases in August and September, and presumably are high in pectin-hydrolyzing enzymes. He recommended a suspension of the regulations for wines of the first group (high quality, etc.) but not on the ordinary wines (*vini posti al consumo diretto*). Moţoc *et al.* (1966) confirmed that the natural pectin esterase and added pecto-lytic enzymes markedly increase the amount of methanol formed. They recommend inactivation of the natural enzymes by heating the musts to 80°C (176°F) before the pectolytic enzymes are added.

Higher alcohols (fusel oil constituents)

Usseglio-Tomasset (1964), Guymon (1964), and Morgan (1965) used gas chromatographic techniques to identify the higher alcohols of various alcoholic beverages. Absent (or present in trace amounts), according to Usseglio-Tomasset, were 2-propanol, 2-methyl-2-propanol, 3-methyl-2-butanol, 2-pentanol, 3-pentanol, and 2-hexanol. Always present were 1-propanol, 1-butanol, 2-butanol, 2-methyl-1-propanol, 2-methyl-1-butanol, 3-methyl-1-butanol, 1-pentanol, and 1-hexanol. There was a significant correlation between the formation of 1-butanol and 1-hexanol. Data on the higher alcohol content of wines are tabulated below:

			Mg of higher alcohols per:			
			100 ml wine		100 ml ethanol	
Author	Type of wine	Number of wines	Range	Average	Range	Average
Cioffi (1948)	white	12	44–137	87	395–1230	729
Cioffi (1948)	red	22	14–172	77	86–1300	592
Pettigiani (1943)	?	24	15–30	—	100–300	—
Windisch (1906)	white	14	6–68	35	153–611	353
Pallotta and Donati (1965)	white	22	22–46	33	—	—
Guymon and Heitz (1952)	white	120	16–37	25	131–308	208
Guymon and Heitz (1952)	red	130	14–42	29	142–344	235
Peynaud and Guimberteau (1958)	red	32	29–55	39	245–487	337
Peynaud and Guimberteau (1958)	white, rosé	25	22–44	31	167–326	245

The distribution of higher alcohols in two French and two Finnish wines was as follows, according to Sihto *et al.* (1964), in percentage of total higher alcohols:

	Bordeaux White	Finnish White	Beaujolais	Finnish Red
3-Methyl-1-butanol	58	82	64	75
2-Methyl-1-butanol	15	12	15	12
1-Pentanol	0	1	+	+
1-Butanol	1	0.5	1	1
2-Methyl-1-propanol	23	4	16	11
1-Propanol	3	0.5	4	1

Obviously 3-methyl-1-butanol and 2-methyl-1-butanol predominate.

Peynaud and Guimberteau (1958), in eight red Bordeaux wines found a 3-methyl-1-butanol alcohol content of 205 to 356 mg per liter (average 267); 2-methyl-1-propanol, 66 to 200 (average 100). In eight white wines the 3-methyl-1-butanol ranged from 110 to 234 mg per liter (average 164) and the 2-methyl-1-propanol from 42 to 120 (average 77). In red wines about one-fourth of the higher alcohol was 2-methyl-1-propanol, in white one-third. In general red wines have more higher alcohols than whites.

The aroma compounds formed by vinous fermentation were reviewed by Webb (1967). He listed the alcohols: 1-propanol, 2-methyl-1-propanol, 1-butanol, 2-methyl-1-butanol, 3-methyl-1-butanol, 1-pentanol, 1-hexanol, 2-phenethyl, (−)-2,3-butanediol, *meso*-2,3-butanediol, 3-methyl-1-pentanol, 4-methyl-1-pentanol, and 3-hydroxy-2-butanone acetoin. Webb believes that the higher alcohols are recognizable in some California Cabernet Sauvignon and Zinfandel wines. To control the higher alcohol level of wines, and hence their flavor, he suggested changing the amino acid balance or using special yeasts which cannot synthesize certain amino acids. Both are possible, but no commercial applications seem to have been made to date.

Tyrosol 2-(4-hydroxyphenyl)-ethanol in the amount of 15 to 45 mg per liter was reported in Bordeaux red and white wines by Ribéreau-Gayon and Sapis (1965). It is one of the few phenolic compounds present in about equal amounts in red and white wines. It is not found in musts. They also detected 0 to 0.8 mg per liter of tryptophol, but rarely in white wines. In the same wines they found 10 to 75 mg per liter of phenethyl alcohol and 0 to 5 mg per liter of γ-butyrolactone.

Nykänen *et al.* (1966) found 8 to 43 mg per liter (average 18) of tyrosol in six table wines and 0.7 to 3.1 mg per liter (average 1.4) of tryptophol. These are less than the 2-phenethyl alcohol content of 7 to 50 mg per liter reported by Äyräpää (1962) in wines of France, Hungary, and South Africa.

All these compounds were formed during fermentation, some at the expense of the corresponding amino acids, but mainly by other pathways (see p. 354).

Sugar content

According to previous California regulations, table wines should not have a Brix reading of over 14° in the presence of their normal alcohol, but this requirement was repealed in 1949 in order to permit the production of very sweet table wines.

The sugar content of the wine is a good indication of the completeness of fermentation and of its sweetness. Glycerin, however, also tastes sweet—somewhat sweeter than glucose, according to Cameron (1944). However, the amount present is barely detectable. The quantity of reducing sugar present is useful in differentiating, both analytically and by taste tests, certain types of sweet table wines, such as the California sauterne types (p. 98). Wines with less than 0.2 per cent sugar are "dry," since this amount of sugar is not perceptible. It is also low enough to prevent refermentation by normal wine yeasts, for most of the residual sugar is nonfermentable. However, Tarantola (1948) found it possible to reduce the residual sugar content by dealcoholizing the wine and refermenting; the reducing values of twenty Italian wines was decreased from 0.24 per cent to 0.09 per cent calculated as invert sugar. A figure of 0.1 per cent would thus appear to be closer to the true nonfermentable reducing value. However, in practice wines with 0.2 per cent, or even higher, are biologically stable.

The residual sugars of wines are still incompletely known. Navara et al. (1966b) reported glucose, fructose, maltose, melibiose, lactose, sucrose, galactose, and unidentified sugars which they believed to be maltotrioses and pentoses.

Rice et al. (1968) reported the occurrence of raffinose, lactose, maltose, sucrose, galactose, glucose, arabinose, xylose, ribose, and rhamnose as residual sugars in several *V. lubrusca* varieties of grapes.

A number of pentoses are reported in wines: arabinose, xylose, and methylpentose. It is probable that these are formed mainly during fermentation, though some may be derived from hydrolysis of lignin. Esau (1967) considers that they are formed from precursors present in the grapes. In 89 Italian wines, Tarantola (1950) found the percentage of pentoses to be: below 0.05, 3; 0.051 to 0.100, 44; 0.101 to 0.150, 33; 0.151 to 0.199, 9. He reported the pentose content of red wines to be higher than that of whites, but the procedure used may have given high values. Esau and Amerine (1964) tentatively identified D-glycero-D-*manno*-octulose (8 carbon atoms) and *manno*-heptulose and *altro*-heptulose (7

carbon atoms) in a California red wine. Later (1966) they noted pre-
sumptive evidence of an additional heptulose and octulose. Their data on
the residual sugars in wines are given in table 62.

TABLE 62

RESIDUAL SUGARS IN WINE
(mg per 100 ml)

| Sugars | Grape variety | |
	Cabernet Sauvignon	Pinot blanc
Raffinose	0.4	0.3
Lactose	3.0	3.5
Maltose	1.7	1.3
Sucrose	5.2	2.3
Galactose	3.2	3.5
Glucose	10.2	8.8
Fructose	1.5	1.0
Arabinose	4.6	2.1
Xylose	2.2	2.2
Ribose	2.3	3.3
Rhamnose	0.8	1.0
Total	35.1	29.3
D-*glycero*-D-*manno* octulose	1.5	1.0
manno-heptulose	2.0	1.1
altro-heptulose	2.2	1.4
unknown-heptulose	0.8	0.5
Total	6.5	4.0

Source of data: Esau and Amerine (1966).

Possibly little or none of the sucrose present in grapes remains un-
hydrolyzed in wines. However, some wines may be sweetened before
shipment. A sensitive procedure for detecting added sucrose is given by
Guimberteau and Peynaud (1965). Since sucrose is rapidly hydrolyzed
in wines, the test must be carried out a few days after addition of the
sucrose if it is to detect sophistication.

Paul (1963) defines reductones as constituents of wine which have
reducing properties with respect to iodine. In musts, 10 to 32 mg per
liter was calculated as ascorbic acid. Addition of 50 to 100 mg per liter of
sulfur dioxide to musts increases the reductone content because in the
absence of sulfur dioxide the natural must reductones are oxidized. In
the presence of sulfur dioxide the reductones apparently are protected
(in a hydroxyl·form) and thus are more easily extracted from the skins
and pulp. Finely ground musts have a higher reductone content. Paul

believes that high reductone content leads to high sensory quality. Possibly for white wines of the Riesling type this is true.

Wucherpfennig and Lay (1967) showed that a slow but significant increase in hydroxymethylfurfural (2.0 mg per liter) content occurred when wines of 0.4 to 3.4 per cent reducing sugar were heated at 50°C (122°F) for 120 hours. When heated for the same period at 70°C (158°F), much higher amounts of HMF (20 mg per liter) were formed. Addition of 3.0 per cent invert sugar increased the HMF content 3.2 mg per liter at the lower temperature and 74.4 at the higher. Reducing the pH (from 3.35 to 3.05, 2.92, or 2.36) increased the HMF produced. Addition of amino acids before heating had little effect on HMF production. Most important, they showed a higher HMF content in old German white wines (reducing sugar content of 0.11 to 2.4 per cent). A 1939 wine with 0.36 per cent sugar had 8.2 mg per liter of HMF; a 1940 wine with 0.78 per cent sugar, 10.0 mg per liter. Young wines of the vintages of 1955 to 1960 had 0.8 to 2.4 mg per liter. These wines had been stored in bottles at cellar temperatures. The threshold of pure HMF is 100 to 200 mg per liter, according to Wucherpfennig and Lay (1967).

Extract content

The extract content (soluble, nonsugar solids) distinguishes the heavy- and light-bodied types. Wines having an extract below 2 per cent are very light or thin on the palate compared with wines having over 3 per cent. Before comparing the real extract contents of two different wines, the sugar content should be subtracted from the apparent extract or total solids content. The extract as determined by a Brix or Balling hydrometer will be in grams per 100 gr. To convert to grams per 100 ml, multiply by the specific gravity. Full-bodied red table wines may have an extract content of about 2.5 per 100 ml; dry white table wines have an average extract content of about 2.0.

The present federal limits on sugar-free extract content call for a minimum of 1.7 for white wines and 1.8 for red table wines. California wines normally meet these minimums. However, the fermentation of low-sugar grapes and their subsequent fortification may yield wines of very low extract. Also, some eastern wines made with the 35 per cent dilution with water and sugar may be too low in extract. The minimum legal total extract (as gr per 100 ml) for Austrian wines, according to Weger (1966) is: white, 1.5; rosé, 1.6, red, 1.7. The minimum may also be defined as the extract-minus sugar. The legal minimum then is 1.45, 1.55, and 1.65, respectively. For Switzerland the limit for total extract is higher: 1.5, 1.7, and 1.8.

Volatile acid content

The volatile acid content, largely acetic, is a good indication of soundness. Wines with high volatile acid content usually taste vinegary. The present California regulations provide that white table wines should not exceed 0.11 gr of volatile acid per 100 cc, calculated as acetic acid and exclusive of sulfur dioxide. The limit for red table wines is 0.12. The federal limits are 0.12 and 0.14, respectively. With rare exceptions, these limits keep the objectionable, vinegary wines from moving into trade, and the industry has little difficulty in producing wines which meet these limits.

Peynaud (1936, 1937) and others have shown, however, that the ethyl acetate formed by acetic acid bacteria, rather than the acetic acid itself, usually produces the objectionable, sharp, vinegary, or acescent odor of such wines. Since acetic acid and ethyl acetate are most often produced simultaneously by bacterial action, measurement of the percentage of acetic acid is usually but not always a measure of the quantity of the objectionable ethyl acetate present. If the volatile neutral ester content exceeds about 180 mg per liter, calculated as ethyl acetate, the wine will have a distinct spoiled or acescent character. (See also p. 446.) Convenient methods have been devised for separate determinations of the undesirable volatile products of wines, as aids to identification of the nature, type, and objectionable quality of the spoilage.

A recent federal ruling approved the use of 0.5 gallon of acetic acid per 1,000 gallons of wine. This, which amounts to about 0.04 per cent acetic acid in the final product, appears to be a retrogressive proposal.

The following fatty acids were identified in an Algerian red wine by Soumalainen (1965): acetic, propionic, isobutyric, butyric, isovaleric, valeric, α-methyl-n-valeric, isocaproic, caproic, caprylic, pelargonic, and capric. Webb (1967) reported in addition the following organic acids (not all volatile) produced by alcoholic fermentation: formic, 2-methyl-butyric, enanthic, lauric, myristic, palmitic, crotonic, 9-decenoic, succinic, lactic, 2-hydroxyisocaproic, and phenylacetic as well as ethyl acid malate and ethyl acid succinate.

Usseglio-Tomasset (1967c) reported varying portions of acids produced by various species of *Saccharomyces:* acetic, propionic, isobutyric, butyric, isovaleric, capronic, caprylic, pelargonic, caproic and seven unidentified. *Sacch. elegans* was an especially notable producer of propionic, *Kloeckera apiculata* of capronic, *Sacch. cerevisiae* var. *ellipsoideus* of caprylic, and *Sacch. italicus* of one of the unidentified acids. Acetic acid was the primary acid but there was no correlation between the per cent acetic and of its homologues.

Usseglio-Tomasset (1967d) reported that wines contain about the same amount of formic acid as the musts from which they were fermented.

During fermentation there is at first a slight increase and then a decrease. New dry and sweet table wines had about the same amount but sweet wines show an increase during aging or during heat treatment. In 24 dry wines he found 0 to 8.4 mg per liter (average 1.7) and in 31 sweet wines (some aged) 0.7 to 8.7 mg per liter (average 2.9). There was no correlation between the volatile acid and the formic acid contents.

Lactic acid has been known for a long time as a normal product of alcoholic fermentation. Peynaud et al. (1967a, 1967b) estimate that the normal level is 0.2 per cent but that occasionally it may be more. Pallotta and Donati (1965) in 22 wines of Verdicchio reported 0.02 to 0.255 per cent. More is formed at pH 6.0 than at pH 2.8, and more is formed when more sugar is fermented. Temperature of fermentation did not seem to influence the amount of lactic acid formed. Lack of vitamins favored lactic acid production.

Fixed acid content

Nègre and Dugal (1967) have shown that the fixed acidity (the total less the volatile[1]) of new wine in the spring after the vintage is decreased when the initial fixed acidity was high, but when the initial fixed acidity was low that of the new wine increased. The fixed acid content is helpful in differentiating certain types. Dry table wines of high fixed acidity (over 0.60 per cent) taste fresh and tart, whereas those of low acidity (below 0.40 per cent) are flat and insipid. Sweet table wines should not be so high in fixed acidity as dry table wines, since they will then have an undesirable, sweet-sour taste. Aged dry red and white table wines also are better balanced with only moderate acid content. The very low fixed acidity of some dry California table wines is a serious defect. For the changes in fixed acids during fermentation see Münz (1965b) and page 360.

Ventre (1931) showed that it is the tartaric acid content of wines that makes them appear more tart. He also noted that the higher the alcohol content the less noticeable is the acidity. He believed that tartaric acid accounts for the harsh acid taste, while malic acid gives the fresh fruity taste. Bobadilla and Novarro (1949) reported that tartaric acid resulted in a flavor of lower quality than other acids. They rated lactic and succinic acids as most important for body and flavor, citric and acetic acids important as indices of soundness, and malic acid least desirable from a sensory point of view. Neither they nor Ventre actually cite complete sensory data. Pilnik (1964), in comparing various organic acids for

---

[1] Calculated as tartaric.

sourness, found that the concentration of undissociated acid was more important than pH in determining sourness of citric, lactic, and tartaric acids. Amerine *et al.* (1965*b*) determined the relative sourness of citric, lactic, malic, and tartaric acids and found that both pH and percentage of titratable acidity were important in determining sourness. The order of sourness was malic > tartaric > citric > lactic. The factors influencing the relative sourness of the acids and the over-all sour taste are not yet fully known.

The wide variation in the acid composition of wines of a given region is illustrated by the data of Pallotta and Donati (1965). In twenty-two white table wines of Verdicchio they reported malic acid to vary from 0.045 to 0.295 gr per 100 ml, tartaric from 0.180 to 0.510, citric from 0.004 to 0.065, and succinic from 0.045 to 0.124. Similar data could be cited for many areas.

Succinic acid, a normal by-product of alcoholic fermentation (p. 361), occurs in all wines either as the free acid, as ethyl acid succinate, or as diethyl succinate. Thoukis *et al.* (1965) reported that the predominant nonvolatile organic acid formed during fermentation was succinic acid and that it was accompanied by lesser amounts of lactic acid. They reported formation of 0.1 to 0.4 gr per 100 ml of nonvolatile organic acid, of which 90 per cent was succinic acid in experimental fermentations of diluted grape concentrate. This is higher than the range usually given: 0.05 to 0.15. Succinic acid is formed in larger amounts in apple than in berry fruit juice fermentations. More recently Caputi, Jr. (1967) reported 0.075 to 0.087 gr of succinic per 100 ml in grape wines containing 0.158 to 0.250 malic acid. These wines had a total acidity of 0.56 to 0.72 gr per 100 ml.

Citric acid is slowly decarboxylated during aging. Citramalic acid is one product. Citramalic (or α-methylmalic) acid was also shown to be formed during alcoholic fermentation by Dimotaki-Kourakou (1962). Its presence was confirmed by Vitagliano and Mura (1966). In twenty-three Italian wines they reported 0.052 to 0.187 gr per liter (average 0.118). In the same wines the citric acid content ranged from 0.002 to 0.171 gr per liter (average 0.054). The results of Vitagliano and Mura (1966) show a certain parallelism between alcohol and citramalic acid contents. Addition of sulfur dioxide, up to 350 mg per liter, did not influence its formation in their tests. However, lowering the pH resulted in a marked increase in its formation (0.065 gr per liter at 4.06, 0.081 at 3.64, 0.118 at 3.35, 0.222 at 3.18, and 0.282 at 2.68). More glycerin, 2,3-butylene glycol, and acetaldehyde were formed at the lower pH values. Citramalic acid may originate from a condensation of acetic and pyruvic acids:

$$
\underset{CH_3}{\overset{COOH}{|}} + \underset{\underset{CH_3}{|}}{\overset{COOH}{\overset{|}{C}}}\!\!-\!\!O \longrightarrow \underset{\underset{COOH}{|}}{\overset{\overset{COOH}{|}}{\underset{C(OH)CH_3}{\overset{CH_2}{|}}}}
$$

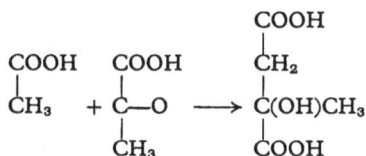

It might also originate in the deamination of glutamic acid. The first mechanism, which clearly associates its formation with fermentation, appeals to us. Sai and Amaka (1967) reported greater production of (−)-citramalic acid by respiratory-deficient yeast mutants.

Castino (1968) isolated an acid which he believes is dihydroxyisovaleric. From about 50 to 150 mg per liter were reported. Since it is easily oxidized in acetone it probably gives the excess results for citramalic acid.

A small amount of oxalic acid is found in wines, particularly those which have been exposed to aerobic conditions. In twenty German table wines Hennig and Lay (1965) reported 0.0 to 0.06 gr per liter.

Small amounts of pyruvic and α-ketoglutaric acids are reported in various types of wines by Egorov and Borissova (1955, 1956, 1957), Markmann and El'gort (1962), and Blouin and Peynaud (1963b). With one exception the amounts are less than 100 mg per liter. Suomalainen and Ronkainen (1963) identified pyruvic, α-ketoglutaric, α-ketobutyric, α-ketoisovaleric and α-ketoisocaproic acids in wines. Rankine (1965a), in sixty-seven Australian wines, reported 1 to 128 mg per liter (average 29) of pyruvic acid. Variety, region, and year did not affect this. Red table wines which had undergone a malo-lactic fermentation were lower in pyruvic acid. Yeast strain and fermentation temperature did have an effect—more at 30°C (86°F) than at 15°C (59°F). Rankine (1967b) reported consistent differences in amount of pyruvic acid produced by *Saccharomyces cerevisiae* and *Sacch. cerevisiae* var. *ellipsoideus* compared to *Sacch. oviformis* and *Sacch. carlsbergensis*. Lafon-Lafourcade and Peynaud (1965, 1966) consider pH, the media, the nitrogen sources, and thiamine deficiency to be the most important factors in pyruvic acid production. The release of pyruvic acid during alcoholic fermentation indicates that alcoholic fermentation occurs under a partial thiamine deficiency. The α-ketoglutaric acid is derived mainly from glutamic acid.

In ten musts and twenty white Bordeaux wines Blouin and Peynaud (1963b) reported 20 to 70 mg per liter of pyruvic acid and 15 to 40 mg per liter of α-ketoglutaric acid. Recent data of Deibner and Cabibel-Hugues (1966) indicate 11.6 to 142.4 mg per liter of pyruvic acid (average

63.6) in fourteen wines and 0 to 70.4 mg per liter of α-ketoglutaric acid (average 29.2). Pyruvic acid appeared to be higher in rosé wines made with destemmed musts. Many more results are needed on the normal levels of these acids in wines.

The greater bound sulfur dioxide content of high pH wines is due to their greater pyruvic acid content, according to Peynaud and Lafon-Lafourcade (1966) and Rankine (1962). Pyruvic acid and sulfur dioxide react to form pyruvate sulfonic acid: $K = 4.0 \times 10^{-4}$. Rankine (1965a) showed that at 30 mg per liter of free sulfur dioxide 53 per cent was bound by pyruvic acid. At the mean value of 29 mg per liter of pyruvic acid, 11 mg per liter of sulfur dioxide were bound to it.

Baraud et al. (1966) reported the following α-ketonic acids in thirteen red Bordeaux wines: α-ketoglutaric, 3–10 mg per liter; pyruvic, 8–25; ketobutyric, 0.7–1.8; ketoisovaleric, 0.5–2.2; ketoisocaproic, 0.9–2.0; and levulinic 0.8–2.0. In fourteen white Bordeaux wines the amounts were 15–36, 60–170, 1.5–5.0, 0.9–2.7, 1.0–3.2, and 0.5–1.8, respectively. Obviously the white wines are much higher in α-ketonic acids. They also reported appreciable differences in these acids between wines fermented by different yeasts. Aerobic fermentations favored their production.

Soumalainen and Keränen (1967) note p-hydroxyphenylpyruvic acid as a constant product of anaerobic fermentation with lesser amounts of α-keto-β-methylvaleric, α-ketoisovaleric, α-ketoisocaproic, β-phenyl-pyruvic, α-ketobutyric, oxalacetic, and α-keto-γ-methiobutyric.

Blouin and Peynaud (1963a) report 1 to 5 milliequivalents of galacturonic and glucuronic acids in normal musts and wines. They are significant in that they bind some of the sulfur dioxide: 1 milliequivalent of galacturonic binds 5 to 6 mg of sulfur dioxide while the same amount of glucuronic binds 2 to 3 mg. Dimotaki-Kourakou and Sotiropoulos (1967) reported 0.00 to 1.12 gr per liter of galacturonic acid in eleven Greek wines (average 0.55). From the alcohol-precipitatable material (gums) Corrao (1957) identified from the hydrolyzed solution galacturonic acid, galactose, mannose, arabinose, rhamnose, and a trace of xylose. The constant presence of glucuronic and galacturonic acids in musts and wines helps to explain, according to Blouin and Peynaud (1963a), the deficit of anions in the acid balance, the high dextrorotatory condition of certain wines, and some of the previously unknown compounds that combine with sulfur dioxide.

Baraud (1951, 1953) was unable to find dihydroxymaleic acid in wines. This is surprising in view of the Soviet reports on its presence in wines. He did report wine to contain an unknown degradation product of glucides which gave a reaction with tartrazine. For data on the acid balance see Kozenko and Romanova (1962).

pH value

The pH is of interest to the wine maker because it is related to the resistance of the wine to bacterial spoilage, to its tendency toward ferric casse, to the percentage of sulfur dioxide present in the free condition, to the tint—particularly that of red wines—and to the acid taste. The pH of California table wines ranges from about 3 to over 4; however, table wines with less than 3.4 have better resistance to spoilage, taste fresher and fruitier, and have a better shade of color. The pH of young German table wines made in cool years is sometimes below 3, but this is usually increased by a malo-lactic fermentation, by use of calcium carbonate, or by ion exchange.

The effect of pH on the color of the anthocyanin pigments is well known. As the pH increases there is an increasing formation of a colorless pseudobase. For cyanidin-3-5-diglucoside, P. Ribéreau-Gayon (1964b, 1964c) showed that the optical density at pH 3.9 is only one-tenth that at pH 2.9.

Esters

The significance of esters to the aroma and bouquet is still a subject of controversy. The earlier concept that only ethyl acetate was important, and this chiefly as a spoilage product, is certainly incorrect. While ethyl acetate at relatively high concentrations is a factor in acetic spoilage, at lower concentrations it may be a pleasant part of the odor of wines. Other esters which seem to be important are ethyl laurate, ethyl propionate, ethyl butanoate, amyl acetate, pentyl acetate, and hexyl acetate.

Ribéreau-Gayon and Peynaud (1964–1966; vol. 2, pp. 239–259) reviewed the literature on ester formation in wine from the original investigations of Berthelot in 1863 to the more recent work of Peynaud, but did not include the contributions of Nordström (1962 to 1965c).

Esters are usually not present in appreciable concentration in grapes, except for methyl anthranilate in Concord grapes, but appreciable amounts of esters are generally present in wines. They are formed during alcoholic fermentation by yeast and particularly by acetic and lactic acid bacteria, and their concentration slowly increases on storage of wine in casks and bottles. Among the volatile esters identified in wine by Nordström (1965c) were ethyl acetate, isopentyl acetate, and ethyl caprylate. While previously the separation and identification of volatile constituents in grapes and wine were restricted to those present in larger concentrations (Haagen Smit et al. 1949), today, by the application of capillary gas chromatography combined with rapid-scan mass spectroscopy, this separation has been facilitated (Stevens et al., 1965, 1966, 1967; Bayer, 1966). Jakob and Bachmann (1964) applied thin-layer chromatography to the isolation of flavor-active (including ester) components of wine. The data obtained with older large-tube gas chromatography were reported by Webb and Kepner (1957). The present status of sensory evaluation of flavor and advances in fruit flavor chemistry is summarized in Committee on Foods (1957), Little (1958), Kuehner (1964), Hornstein (1966), and Schultz et al. (1967). By newer techniques, Stevens et al. (1966, 1967) reported that, although the bulk of the essential oils of both the muscat and Grenache grapes was composed of alcohols, a fair number of esters occur (twelve in muscat oil and four in Grenache oil). The esters form a small percentage of muscat essence but a higher percentage of Grenache oil. In both varieties ethyl acetate was present in smaller amounts than other esters.

Peynaud (1956b) reported that the level of ethyl acetate in wines with the characteristic odor of acescence is 180 to 200 mg per liter. The threshold concentration at which esters may be detected in wines varies with the nature of the ester. Esters of higher alcohol (e.g., isopentyl acetate) may be detected at a concentration of 0.01 mg per liter or less.

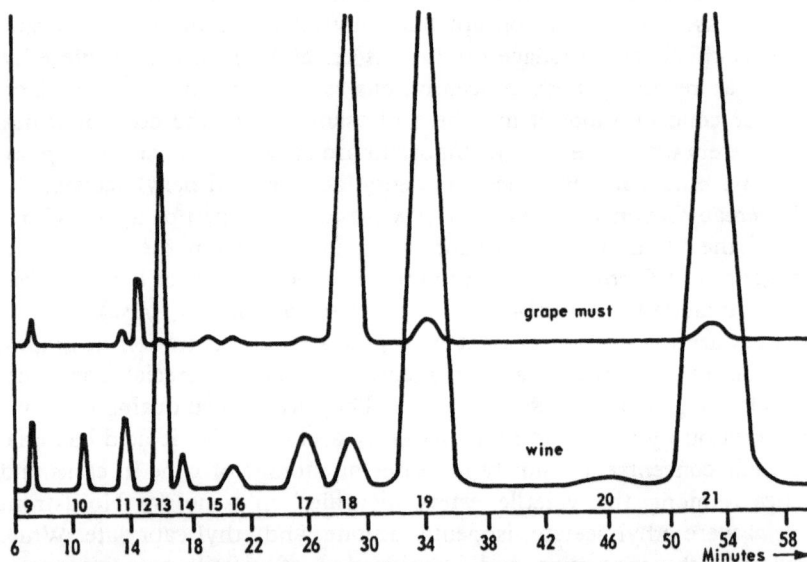

Figure 68. Gas chromatograms of a must and its wine. Numbers in chromatogram are: 9, ethanol, ethyl acetate, and 2-propanol; 10, 1-propanol and isopropyl acetate; 11, 2-butanol and ethyl propionate; 12, n-propyl acetate; 13, isobutanol and n-valeraldehyde; 14, ethyl isobutyrate; 15, unidentified; 16, 1-butanol and isobutyl acetate; 17, unidentified; 18, n-butyl acetate; 19, isoamyl alcohol; 20, 1-pentanol; 21, isoamyl acetate.

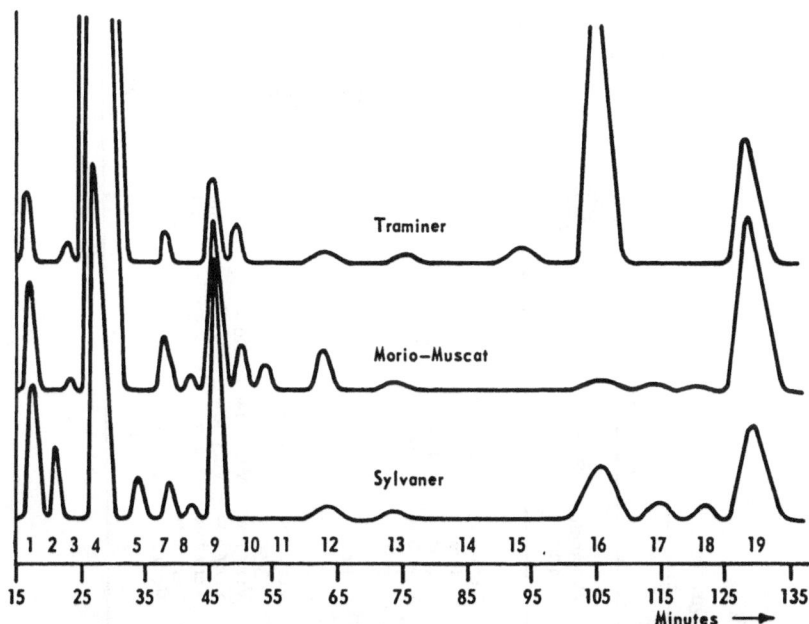

Figure 69. Gas chromatograms of musts of three varieties. Numbers in chromatogram are: 1, ethyl caproate; 2, isoamyl butyrate; 3, 2-heptanol; 4, hexanol and ethyl enanthate; 5, 6, and 7, unidentified; 8, 1-heptanol; 9, ethyl caprylate; 10 and 11, unidentified; 12, 1-octanol; 13, ethyl perlargonate; 14 to 18, unidentified; 19, ethyl caprinate.

Bayer (1966) recently reported data on ester content in various wines obtained by gas chromatography. The values he found for esters of different acids are shown in table 63.

Figure 68 shows how the gas chromatograms of a grape must differ from that of the wine made from it. Note particularly the appearance of higher alcohols in the wine and the disappearance of n-butyl acetate. Figure 69 shows how the gas chromatograms of the wines of three varieties differ from each other. Note especially the similarities and differences between Traminer and Morio-Muskat and, in turn, their differences from Sylvaner.

The number of esters in wines as summarized by Webb (1967) was thirty-two: ethyl formate, acetate, caproate, heptanoate, caprylate, pelargonate, lactate, 3-hydroxybutyrate, caprate, 9-decanoate, undecanoate, laurate, and 2-hydroxyisocaproate; diethyl malate, succinate, and phthalate; ethyl acid malate, acid succinate, and acid tartrate; isoamyl acetate,

447

TABLE 63
ESTER CONTENT IN DIFFERENT WINES

Wine	Formic acid	Acetic acid	Propionic acid	Butyric acid	Valeric acid	Aliphatic acids C_6-C_{10}	Aromatic	Cinnamic
			Ester (μmol/liter) of					
1953 Morio-Muskat, Geilweilerhof	11.01	10.26	Very low	Very low	Very low			18.20
1955 Morio-Muskat, Geilweilerhof	17.55	76.00	low	low				8.02
1949 Serrig, Staatsdomäne	15.4	34.5	low	low	20.4	4.8	80.0	
1953 Serrig, Staatsdomäne	11.3	26.4	Very low	low	28.0	3.1	65.0	
1955 Riesling, Natur, Geilweilerhof	11.95	24.70	low	low	low	7.30		
1955 Riesling, verbessert, Geilweilerhof	9.41	32.30	low	low	low	8.44		
1934 Riesling Nr. 34094, Bad Kreuznach	12.17	14.44	2.12	Very low	7.60	2.51	5.55	
1934 Riesling Nr. 34095, Bad Kreuznach	21.65	23.56	Very low	low	10.49	2.28	9.77	
1934 Riesling Nr. 34096, Bad Kreuznach	9.86	13.68	Very low	Very low	6.46	3.80	11.10	
1934 Riesling Nr. 34097, Bad Kreuznach	low	31.92	3.50		low	3.2	9.50	
1937 Riesling Nr. 37091, Bad Kreuznach	9.26	36.48	Very low		low	4.2	28.58	
1937 Riesling Nr. 37095, Bad Kreuznach	9.26	9.66	Very low			1.15	25.31	
1943 Riesling Nr. 43064, Bad Kreuznach	9.11	8.44	Very low	Very low	low	3.61	9.40	
1944 Riesling Nr. 44055, Bad Kreuznach	4.70	15.50	Very low	Very low	low	3.04		
1946 Riesling Nr. 46022, Bad Kreuznach	8.96	6.35	Very low	low	1.56	3.42	4.29	
1955 Portugieser, Siebeldingen (Red)	low	33.44	Very low	8.63	3.57	21.28		
1955 Müller-Thurgau, Geilweilerhof	8.21	26.60		Very low		4.37	8.51	
1955 Sylvaner, Geilweilerhof	6.27	114.76		2.20	6.76	8.25	53.28	

Source of data: Bayer (1966).

caproate, caprylate, and caprate; n-hexyl acetate; 1,3-propanediol monoacetate; 2-phenethyl acetate; n-propyl acetate; isobutyl acetate; act-amyl acetate; ($-$)-2,3-butanediol monoacetate; dimethyl phthalate; and γ-butyrolactone.

In summary: esters could be formed by direct enzymatic action during fermentation, they could be present in the original raw material, and they could result from hydrogen ion catalyzed esterification reactions during aging. Unfortunately, there is little available information on the thresholds of the esters or of their effects on each other, either as to intensity or quality. Therefore, reports of the amounts present do not necessarily indicate their relative importance to the aroma or bouquet of the wine.

Webb *et al.* (1966) have isolated the acid esters of tartaric and malic acids. They also reported (1964b) ethyl acid succinate in Cabernet Sauvignon wine; and (1967) ethyl acid tartrate and an isomer of ethyl acid malate with a free carboxylic acid group adjacent to the carbinol group in a dessert wine. They suggested that the malic acid ester was produced enzymatically during the primary alcoholic fermentation—possibly by a hydrogen ion catalyzed reaction of ethanol with malic acid to produce two parts of the isomer indicated above and one part of the normal isomer. They suggested, therefore, that the increase in esters during aging is due mainly to increases in ethyl acid succinate and ethyl acid tartrate. Ribéreau-Gayon and Peynaud (1936) reported that the acid esters accumulate owing to nonbiological changes on aging. Esterification is discussed further on page 445. P. Ribéreau-Gayon (1964d) has also identified cinnamic acid esters of tartaric acid in grapes.

Aldehydes

The accumulation of acetaldehyde during fermentation was investigated by Fornachon (1953), Paul (1958), and others (see p. 356). The acetaldehyde content is an indication of the degree of aeration of table wines. In dry white table wines the presence of over 100 parts per million usually indicates an aerated or oxidized wine, and is surely the maximum permissible amount. The acetaldehyde content of red wines is much lower than that of white wines—usually below 50 mg per liter—because of the presence of tannins and anthocyans. Acetaldehyde formed by oxidation or added to red wine was shown by Joslyn and Comar (1941) to combine with the anthocyanin pigments. Mareca Cortés and Salcedo (1957) confirmed the combination of acetaldehyde with polyphenolic compounds present in red wine. Valuĭko (1965a) also confirmed this combination of acetaldehyde with tannins and anthocyanins. He reported

449

that acetaldehyde reacted rapidly with tannins; over 50 per cent of tannins present were precipitated by aldehydes within a few days. Reaction with anthocyanin pigments was slower but more complete. When the concentration of acetaldehyde was sufficiently high, the wines became decolorized upon prolonged storage. However, in ports Singleton *et al.* (1964*a*) reported better retention of the red color when high aldehyde spirits were used for fortifications.

Sherries are, of course, much higher in aldehydes owing to their method of production. During storage of Riesling and Cabernet wines for three years, formation of isobutyraldehyde, isovaleraldehyde, enanthaldehyde, and caprylaldehyde was reported by Valuĭko (1965*a*). In young wines there was a minimum of free aldehydes and a maximum of bound. The reverse was true in old wines. While the total aldehyde content decreased with time, the percentage as free increased. This should be investigated further. A priori, we would expect the opposite. The free aldehydes react with ethyl alcohol to form acetals. This reaction is catalyzed by the slightly acid medium of wine. Bayer (1966) reported that the acetaldehyde content of wines increases with age. In Riesling wines he reported 3 mg per liter in one-year-old wine, increasing to 45 mg per liter in eight-year-old wines. In the same wine, acetal content increased from 0.5 mg per liter in young wines to 24 mg per liter in old wines. See also p. 511. '

Tannins and related compounds

The tannins improve the body of wine, stabilize color, and are of importance in fining. Their flavor is considered desirable in small quantities in red table wines, but undesirable in white table wines. The tannins also produce an astringent taste; a tannin content of over 0.25 per cent is indicative of a young red wine. The darker-colored red wines of high extract content usually contain more tannin than do the lighter-colored and lighter-bodied ones, although a few varieties (e.g., Grignolino) are high in tannin but low in coloring matter. Young wines are higher in tannin content than old wines, and one of the purposes of aging is to permit the astringency of the wines to be reduced by the oxidation and precipitation of excess tannin. Wines of quality do not contain sufficient tannin to give an objectionable astringent taste. The usual range of tannin content is 0.01 to 0.04 per cent for white table wines, and 0.10 to 0.20 for red table wines. Wines containing over 0.20 per cent are usually undesirably astringent and puckery, but the method currently used for determining the tannin content is too nonspecific to make any general recommendation. The type of tannin present, the age and composition of the wine, and the presence of nontannin, permanganate-reducing matter all influence the palatability of a wine at a given tannin content.

P. Ribéreau-Gayon and Stonestreet (1966) reported recently that the Folin-Denis reagent-reducing phenolics present in wine do not change appreciably with aging but that marked decreases in vanillin reactive phenolics and in leucoanthocyanin occur. This was confirmed by Joslyn and Little (1968). P. Ribéreau-Gayon (1968) indicated that oxidative condensation involving the B phenolic ring of flavonoids was responsible for the decrease in vanillin reactive phenolic content. Somers (1968), however, believes that condensation between anthocyanins and tannins is involved in aging.

Among the phenolic compounds isolated from wines by Hennig and Burkhardt (1957, 1960a, 1960b) were d-catechin, l-epicatechin, l-epigallo-catechin, gallic acid, protocatechuic acid, and ellagic acid. They also reported the polyphenols: chlorogenic, isochlorogenic, cis- and trans-caffeic acids, and probably caffeic acid lactone (esculetin). Among the phenolic compounds were cis- and trans-coumaric acid and probably coumaric acid lactone (umbelliferone). Quinic acid is a component part of chlorogenic acid. It is reported in young and old wines. Shikimic acid may be present in young wines. Rebelein (1965a) reported 0.002 to 0.033 gr per 100 ml of catechin (average 0.007) in sixty-three German white wines. In Spätlese wines the amounts were less, 0.002–0.006 (average 0.004). In thirty-three non-German white wines the range was 0.002–0.032 (average 0.008). In thirty-nine German and non-German reds the range was 0.045–0.79 (average 0.215). In four rosés he found 0.054–0.137 (average 0.075).

Gallic acid tended to disappear during aging. Old wines seldom contain condensable tannins. Pomace and pomace wines are high in free gallic and ellagic acids. Burkhardt (1965a) reported p-coumarylquinic acid (or its calcium salt) in wines. It does not lead to browning.

Durmishidze (1955, 1959) reported similar data on occurrence of catechin, gallocatechin, and epicatechin gallate in Russian grapes and wine. Nilov and Skurikhin (1967), on the basis of various Soviet studies, consider 58 per cent of the tannins of red wines to be catechins of which 15 per cent are (−) gallocatechin, 2 per cent (±) gallocatechin, 5 per cent (±) catechin, 15 per cent (+) catechin, and 21 per cent unidentified.

Nilov and Skurikhin (1967) report that the sensory effect of tannin increases up to a certain degree of oxidation and then decreases. They also believe that tannins give "fullness" and a special "tone" to red wines. Application of this to the aging of red wines would be very desirable. Nondialyzable tannins were found to be tasteless. In young wines as much as 75 per cent of the tannins are dialyzable, whereas in old red wines this is greatly reduced. This is due to polymerization of tannins and their conversion to colloidal forms. The concept was first proposed by Ribéreau-Gayon (1933a). Nilov and Skurikhin considers tannins to be important components of the oxidation-reduction system during aging.

Herrman (1959) stressed the role of catechins in wines, both chemical and sensory. Weinges (1964) pointed out that the most commonly occurring flavonoid in fruit is d-catechin.

Šikovec (1966), recently reported that (see p. 424) polyphenolic components of grapes and must influence growth, respiration, and fermentation of yeast. Refermentation particularly is sensitive to polyphenolics.

P. Ribéreau-Gayon (1964b, 1964c, 1968) classified the phenolic compounds of grapes and wine into phenolic acids, flavanosides, chromonins, and tannins, or, more specifically: benzoic acids (seven, which exist as derivatives of an unknown nature but with an ester-type linkage), cinnamic acids (three, which exist as esters of tartaric acid and anthocyanins), flavanols (three, which occur as monoglucosides and monoglucuronosides), anthocyanidins (five, which are present as mono- and diglucosides, sometimes acylated with cinnamic acid), and tannins, which result mainly from polymerization of flavan-3,4-diols (leucoanthocyanidins). He reported red wines to be higher in phenolic acids than white. While esters of benzoic and cinnamic acids occurred in grapes, only the free acids were present in wines. Syringic and coumaric acids occurred in largest amounts; vanillic and caffeic acids in smaller amounts. The glucosides kaempferol, quercetin, and myricetin were found in grape skins, but only their aglycones in wine. The anthocyans were present as the monoglucosides of the corresponding anthocyanidins, but in a few grape species diglucosides occurred.

P. Ribéreau-Gayon (1964b, 1964c) characterized the astringent phenolics of grapes and wine as being either polymers of flavan-3-ols (catechins) or as polymers of flavan-3,4-diols. Apparently condensation products of flavan-3-ols and flavan-3,4-diols may occur.

Stocker (1967) recently summarized the information on the synthesis and chemistry of flavonoid tannins, with particular reference to leucocyanidins isolated by Freudenberg and Weinges. He gives data on the conversion of these into cyanidin in yields of 22 to 50 per cent of theoretical. He reported on a new synthesis of leucoanthocyanin by reduction of dihydroquercetin after acetification with $LiAlH_4$ and on the chromatographic separation of condensed leucocyanidins on columns of polyamide or dextran gel and thin-layer chromatography or kieselguhr and cellulose.

A variety of methods have been used for the separation identification and determination of the individual phenolics of wine and the investigation of changes in these during aging. They ranged from the early fractional precipitation procedures, such as those used by Nègre (1942–1943), to the use of polyamide resins like nylon for selective adsorption of leucoanthocyanins, to Sephadex gel separation by Wucherpfennig and Franke (1964) and by Somers (1966a, 1966b). Somers separated phenolics in dry red wine by precipitation with basic lead acetate solution, deleaded the phenolics, and separated them on Sephadex gel using aqueous acidified ethanol solution for application and elution. The astringent, gelatin-precipitating tannins isolated in this manner from red wine was found to be a condensed flavonoid containing anthocyanin pigment. He reported that the purified wine tannins contained one anthocyanin to four leucoanthocyanin moieties. Most of the tannins in a 1964 wine were reported to have an average molecular weight of 2,000. In a 1959 wine, most of the condensed polyphenols were in the molecular weight range of 2,000 to 5,000, but some polyphenolics had molecular weights up to 50,000. Between five and fifteen flavonoid units apparently are involved in the tannin structure.

Somers also separated the phenolics in the acidified aqueous acetone extract of the skins of Shiraz berries. The tannin fraction obtained amounted to 42 per cent by weight and accounted for 12 per cent of color. The astringent reddish-brown tannins obtained were similar to those isolated from wine. These tannins are not the products of fermentation but occur naturally in the grape. However, some may be artifacts from the method of preparation.

A fractionation of the tannins present in wines was made by Diemair and Polster (1967). The main fraction was brown, a lesser fraction was blue, and others ranged from yellowish-green to red. None of these fractions reacted with vanillin-hydrochloric acid, i.e., were not hydroxymethylfurfural. The molecular weight of the brown fraction was 5,000 to 10,000 and of the blue fraction was 1,000 to 5,000.

Chlorogenic acid, generally present in table wines, decreases with age. Caffeic acid is formed later in the maturation process. Diemair and Polster believed that chlorogenic and caffeic acids caused the browning in some young wines, since polyphenoloxidase converts these acids into brown-colored quinoid-type compounds. Since sulfur dioxide inactivates polyphenoloxidase, this explains why these acids are always found in young wines that contain adequate sulfur dioxide. Masquellier and Ricci (1964) reported 50 to 150 mg per liter of chlorogenic acid in wines. Only 3 to 20 mg per liter of caffeic acid were reported. They also mention p-coumarylquinic, and feruloylquinic acids.

Weurman and de Rooij (1958) and P. Ribéreau-Gayon (1964b) were not able to confirm the presence of chlorogenic acid (the ester of caffeic acid and quinic acid) in wines. They also failed to find esculetin and umbellifin, which had been reported by Hennig and Burkhardt (1958).

J. A. Rossi, Jr., and Singleton (1966a) report that oxygen absorption of phenolics in acid solution was slow. Maximum oxygen absorption was obtained under alkaline conditions. Surprisingly, phenolics of white wines took up more oxygen per unit of phenol than did those of red wines. Since total phenolics are poorly correlated with the tendency of white wines to brown, J. A. Rossi, Jr., and Singleton (1966b) note that there must be differences in the relative amounts of specific fractions of phenolics. They used seed phenolics which were fractionated into catechin, leucoanthocyan and condensed tannin fractions. These fractions contain at least three, five and two different major phenolic substances, respectively. The catechin fraction appeared to account for most of the browning.

Rebelein (1965a, 1965b, 1965c) found wines with low catechin and low methanol (normal), others with high catechin and high methanol (evidence of pomace wines), some with high catechin and low methanol (evidence of use of red grapes pressed early), and others with low catechin and high methanol (high methanol fortifying spirits). Where both catchin and methanol were low and there was a red color, artificial coloring may be suspected.

Pifferi (1966) identified d-catechin, l-epicatechin, gallocatechin, gallocatechin gallate, quercetin, rutin, quercitrin, and the following phenolic acids: gallic, protocatechuic, γ-recocilic, quinic, and some cis and trans isomers of p-coumaric, caffeic, and chlorogenic acids. He also reported ellagic and m-digallic acids and esculetin and methyl esculetin.

For information on the colored compounds in grapes and wines, see pp. 255–269.

P. Ribéreau-Gayon and Stonestreet (1964) consider that the aging of red wines gradually results in anthocyanin pigments passing to a colloidal state and precipitating. This allows the yellow-brown color of the tannins to be seen. This yellow-brown color increases with aging as the tannins condense. They reported that the percentage of leucoanthocyans increased with the age of red wines as follows: 1962, 0.15; 1961, 0.19; 1960, 0.15; 1958, 0.23; 1953, 0.23; 1947, 0.37; 1938, 0.27; and 1921, 0.44. The tannin content as determined by permanganate oxidation tended to increase in the same wines: 34, 37, 34, 48, 48, 64, 54, and 68, respectively. Whether changes in vinification practices influenced these results is not known. For further information on the role of tannins in color, see p. 453.

Robinson et al. (1966) studied the changes during aging in absorption of the mono- and diglucosides of malvidin, petunidin, delphinidin, peonidin, and cyanidin. The diglucosides were more stable to decolorization but were more prone to browning. Of the pigments, malvidin and peonidin are the most stable and delphinidin the least.

Masquelier and Laparra (1967) reported cinnamic constituents of red wines to be choleric (bilious) to rats, and they believe that wine may have a similar effect on man. Data with human subjects, however, are needed.

Glycerin

Wines having considerable glycerin are heavier-bodied and smoother than those containing little. Because a smooth texture is desirable in sweet table wines, it is generally considered that they should be fairly high in glycerin. The usual range of glycerin for California white table wines is from 0.7 to 1.1 per cent (about 0.9 average). Red table wines average a little higher—about 1.0 per cent. Sweet table wines in California have no more (or less) glycerin than dry table wines. Glycerin has a distinct sweet taste, but in the range at which it is present in wine its sweetness is barely detectable and it is questionable if it contributes to the "feel" of sweet table wines.

Sulfur dioxide

The sulfur dioxide content is of importance because of the regulatory limitation of 350 parts per million total. (The limitation on free sulfur dioxide has been removed from California and federal regulations because it is difficult to define precisely and because, during storage, many factors influence the proportion present as free sulfur dioxide and this leads to uncertain and variable results.) The determination of free sulfur dioxide, nevertheless, is of value in winery control work and as a measure of the degree of objectionableness of the odor. Dry white California table wines range from 50 to 250 mg per liter in total sulfur dioxide; the better wines have less than 100. Sweet table wines range from 100 mg per liter to over the 350 limit, but they should not exceed 200, and the objectionable free sulfur dioxide odor should be kept to a minimum. Although the presence of 50 to 75 mg per liter of total sulfur dioxide in red table wines is not objectionable, some California red table wines have over 100, and in these the odor of sulfur dioxide is sometimes noticeable and objectionable. In critical sensory evaluation, judges apparently react unfavorably to fixed sulfur dioxide in dry red table wines. (See also pp. 385 ff.)

The sulfate content, as potassium sulfate, averages below 1 gr per liter in California wines, although an occasional wine exceeds the federal limit of 2 gr. Newly fermented wines made without addition of sulfur dioxide or calcium sulfate contain below 0.5 gr per liter of sulfate, calculated as potassium sulfate. Sulfur dioxide, upon oxidation, is converted into sulfate. The complete oxidation of all the sulfur dioxide present in a wine containing, initially, 100 parts per million of sulfur dioxide would raise the sulfate content only 0.025 per cent as potassium sulfate. Since calcium sulfate is not added to table wine musts in California, it is not a source of sulfate. The presence of sulfate in amounts over 1 gr, then, indicates the oxidation of successive small additions of sulfur dioxide over a period of time.

A detailed study of thirty-four white Bordeaux table wines by Peynaud *et al.* (1964) showed that sulfate was the most important factor in stability other than potassium and tartrate. Other significant correlations between mineral content and wine stability are noted. The authors conclude that more data is needed.

Mineral constituents

The direct and indirect influences of iron and copper on taste have been noted by numerous workers. Belavoine (1950) believes that a limit of about 10 mg of iron per liter is indicated on sensory grounds. Corrao and Gattuso (1964a) also fix this as the level above which clouding is likely to occur. In eighty-five white Sicilian wines the iron content varied from 0.5 to 36.3 mg per liter (average 12.3). In seventy rosé and red wines the range was 1.0 to 62.0 (average 13.2). The range and average are much greater than would be expected in California wines. In forty Sicilian wines Corrao (1963a) reported total iron contents of 2.5 to 36.2 mg per liter; however, twenty-seven of the forty were between 5.0 and 20.0 mg per liter, eight below 5, and five above 20. The generally high values would indicate appreciable winery contamination during handling and aging. Even smaller quantities of copper are reported to affect the taste adversely.

During fermentation large amounts of the must copper and iron are adsorbed on or incorporated into the yeast cell. Thoukis and Amerine (1956) reported that 40.9 to 89 per cent of the copper was removed, and 47.5 to 70.0 per cent of the iron. Since their experiments were made in small containers, the conditions may not have been as anaerobic as under winery conditions. Incorporation of inorganic iron into yeast cells occurred in the early stages of fermentation followed by excretion of soluble organic iron, according to Shijo and Tanaka (1965). Thereafter the iron content gradually decreased. Addition of amino acids did not affect the amounts of organic iron formed. The soluble iron content of red wines was twice that of white wines. Adding up to 15 per cent ethanol to a must resulted in a rapid decrease in soluble iron content for four days; thereafter no change was observed.

Ribéreau-Gayon (1933a, 1933b) showed that wine held in the absence of air does not contain ferric ions, that only after exposure to air will ferric ions be formed and then almost completely as complexes with acids. When the oxidation is prolonged, colloidal hazes and precipitates of ferric compounds will form. When musts are fermented in small containers exposed to air, and when the resulting wines are exposed to air, considerable amounts of iron are precipitated, particularly during the oxidation following fermentation. Byrne et al. (1937) and Mrak and Fessler (1938), among others, found this to be true. Some of the iron is lost with the yeast cells. Byrne et al. found that the iron content of the California grapes ranged from 1.5 to 23 parts per million, depending on variety and degree of contamination with dust or soil. Quantities of iron are dissolved by musts during passage through iron crushers and must lines. A pickup of from 1.0 to 48.5 parts per million was reported by Mrak and Fessler (1938). During fermentation in small glass containers, the iron content decreased from an average of 9 to an average of 2 parts per million. Under commercial conditions, decrease in iron content is smaller; the iron content at the end of fermentation is usually about the same as that of the uncontaminated grapes.

Cappelleri (1968) showed that 40 laboratory samples had less than 5.5 mg per liter of iron (average 2.6). In 56 samples from small wineries the maximum was 13.2 (average 5.1). However, in 142 samples from large industrial wineries the maximum was 89 (average 10.7).

Frolov-Bagreev and Andreevskaia (1950) suggested that manganese and molybdenum may influence the taste and quality of wine; they further suggest that these and other metals, such as vanadium, titanium, boron, and radium, may affect flavor either by poisoning of yeast enzyme systems or by changing the morphology or physiology of the yeasts. We do not find the sensory evidence convincing.

In 842 Hungarian wines (503 white, 138 red, and 201 direct producers) Tuzson (1964) reported 98.2 per cent with less than 3.5 mg per liter of manganese. The direct producer wines had slightly more manganese than those of Vitis vinifera—1.75 to 1.51 mg per liter. (See also Flanzy and Thérond, 1939.) Certain regions produced wines of significantly higher manganese content than others. Varietal differences are reported, but these may be related to regional, i.e. climatic, differences.

The manganese content of direct producer wines was about double that of V. vinifera wines in the study of Carrante and Perniola (1967) on the basis of either mg per liter or percentage of ash. The Seibel hybrids were low in manganese, but four of the six Seyve Villard hybrids and a Ravat hybrid were very high in manganese. Corrao (1963b) investigated fifty-nine Sicilian white and red wines for manganese. The amounts ranged from 0.25 to 2.20 mg per liter, with 42.4 per cent of the samples in the range 0.61 to 0.90; 23.7 per cent from 0.91 to 1.20; 16.9 from 1.21 to 1.50; and 3.4 each below 0.30 or above 0.150. This amounted to 20.1 to 50.0 mg per 100 gr of ash. There were no noticeable differences in the manganese content of the white or rosé and red wines.

The limit of 0.2 mg per liter of lead in foods in Great Britain has resulted in much interest in the normal lead content of foods and how it can be reduced. Jaulmes et al. (1960) reported that for French wines 29 per cent exceeded the 0.2 mg limit and 0.6 would be a more practical maximum. The sources are lead capsules, unlined cement tanks, lead-based paints, rubber hoses, lead-containing metals in pumps, filters, fillers, faucets, and gaskets, and even lead in filter pads, bentonite, and glass. Of course, no lead sprays should be used in the vineyard. Similar findings were reported by De Almeida (1947), Rankine (1957), and Brémond and Roubert (1958). The studies on lead content of wine may be summarized as follows (mg per liter):

		Number	Minimum	Maximum	Average
Bolotov (1939)	Russia	44	0.24	0.90	—
De Almeida (1947)	Portugal	54	T	2.6	—
Rankine (1955)	Australia	55	0.04	0.86	0.23
Brémond and Roubert (1958)	Algeria	67	0.02	1.78	0.25
Tarantola and Libero (1958)	Italy	29	0.08	0.66	0.17
Jaulmes et al. (1960)	France	477	0.05	> 1.00	0.18

In 25 European table wines Ney (1965) reported 0.129 to 0.254 mg per liter of lithium. Amati and Rastelli (1967) reported 0.001 to 0.092 mg per liter of lithium in 112 Italian wines. Eschnauer (1967) found only traces of lithium in German wines. Little antimony occurs in grapes or wine. In 88 wines of various countries Eschnauer (1966a) reported only one to have detectable antimony, 0.021 mg per liter. However, contact of wine with rubber tubing containing antimony can lead to higher amounts. In sixteen German wines Eschnauer (1966b) found 0.0013 to 0.0041 mg per liter (average 0.0031) of cadmium. In 52 French wines Mestres and Martin (1964) reported the cobalt content to vary from 0.02 to 0.3 mg per liter. The average in seventeen red wines was 0.23, in 13 rosés 0.13, and 22 whites 0.11. Very little nickel is found in wines. Eschnauer (1965b) reported none in 6 German wines and only 0.0006 to 0.00011 mg per liter in seven others.

Aluminum is seldom used in wineries, as it leads to cloudiness. For data on aluminum see Bryan (1948). Eschnauer (1964, 1965b) found 0.51 to 0.93 (average 0.8) mg per liter of aluminum in 18 German wines. At 10 mg per liter (due to contamination) undesirable changes in clarity, color, and odor occurred.

Recent changes in Italian laws have made it necessary to control the sodium content. Weger (1967) noted that appreciable amounts of sodium were picked up by treatment with sodium bentonite. Ion exchange treatment, however, is the most important source of excess sodium.

The sodium content should not exceed 100 mg per liter, according to the Office International de la Vigne et du Vin, and if possible should be lowered to 60 (Anon., 1967b, 1968). However, in 101 Australian white table wines Rankine (1965b) found 11 to 275 mg per liter of sodium (average 80); in 58 ted table wines the range was 9 to 275 (average 110). In 146 California table and dessert wines Amerine and Kishaba (1952) reported 26 to 400 mg per liter (average 85). In 155 California table wines Lucia and Hunt (1957) found 10 to 172 mg per liter (average 55). In 101 Australian white table wines Rankine (1965b) reported 0.390 to 1.520 gr per liter of potassium (average 0.880); in 58 red table wines the range was 1.050 to 2.340 (average 1.650). (See also Amerine, 1958.) In 24 Italian wines Garoglio and Salati (1967) reported potassium contents of 0.453 to 1.260 gr per liter (average 0.809); the sodium contents were 10 to 150 mg per liter (average 56). The potassium to sodium ratios (by weight) varied from 4.65 to 92 per liter (average 24.69 per liter).

The calcium content of 101 Australian white table wines ranged from 0.038 to 0.110 gr per liter (average 0.068), according to Rankine (1965b); in 58 red table wines the range was 0.040 to 0.110 (average 0.062). The magnesium content of 101 Australian white table wines ranged from 0.056 to 0.115 gr per liter (average 0.083), according to Rankine (1965b); in 58 red table wines it was 0.069 to 0.180 (average 0.125). For data on the calcium and magnesium content see Lasserre (1933). Jouret and Bénard (1967) reported the zinc content of normal French wines to range from 0.11 to 3.85 mg per liter (average 0.87). A limit of 5 mg appears reasonable. Zinc-containing insecticides may have to be avoided.

Eschnauer (1961, 1962, 1967) estimates that there have been 500 reports on the trace element content of musts and wines from 1900 to 1965. His review of trace elements in wine (mainly German) was as follows (mg per liter):

Metal	1953 red	General limit or range	Usual average
Chlorine	—	20–80	50
Iron	5.9	5–25[a]	10[a]
Boron	4.5	3.4–7.8	7
Silicon	4.4	0.1–5	1
Manganese	3.3	0–2	0.05
Zinc	2.6	0.1–3	0.9[a]
Aluminum	0.83	0.51–0.93	0.7
Copper	0.78	about 1	0.4[a]
Rubidium	0.67	about 1	—

[a] Estimate.

Metal	1953 red	General limit or range	Usual average
Fluorine	0.18	0.06–0.49	0.3
Vanadium	0.11	0.06–0.26	0.1
Iodine	0.09	0.10–0.60	0.4
Bromine	—	T–1	0.005
Titanium	0.07	0.04–0.23	0.1
Cobalt	0.007	0.0005–0.021	0.005
Strontium	0.007	0.001–0.25	0.05
Arsenic	0.0035	trace–0.02	0.005
Lead	0.002	trace–0.05	0.01
Cadmium	0.002	trace–0.01	0.005
Molybdenum	0.0008	0.001–0.1	0.005
Barium	0.0001	0.0001–0.05	0.01
Chromium	0.0001	trace–0.0005	0.00001
Nickel	0.0001	0.00005–0.001	0.0005
Thallium	0.00032	0.00006–0.0007	0.0005
Lithium	+ ?	1	—
Silver	+ ?	trace–0.0005	0.0001
Bismuth	—	—	
Tin	+	trace–0.7	0.1
Antimony	—	0–0.02	—
Uranium	—	0–0000	—

In general, Eschnauer's figures are lower than those to be expected in non-sugared wines, such as Italian, Spanish or Californian wines.

Guyot and Balatre (1966), on the basis of their own and other studies, reported that the total bromide content of wines should be less than 1 mg per liter. Corigliano and Pasquale (1965) reported 0.2 to 3.4 mg per liter (average 1.4) of fluorine in 36 Italian table wines.

On the basis of extensive analysis of authentic wines, Cabinis (1962) reported that when the chloride content was less than 60 mg per liter the bromide content was less than 0.6 mg per liter. When the chloride content was 60 to 80 mg per liter the bromide could be as high as 1 mg per liter; for more than 80 mg per liter of chloride, up to 3 mg per liter of bromide were reported. Since no organic bromine compounds are found naturally in wines, detection of addition of bromine-containing organic agents is simple: extraction with ether. Even should organic bromine compounds hydrolyze (as they do), the chloride-bromine relationship can still be used as an indication of sophistication.

Jouret and Bénard (1965) have demonstrated wide variations in chloride-sodium and potassium-sodium ratios in wines from different rootstocks. Cultural and climatic conditions were also important.

The requirements of multiplying yeasts for phosphates are not known. Apparently, once the yeast cells are formed phosphates are normally adequate so that the alcoholic fermentation cycle can begin. Thereafter the phosphate can be used over and over. As expected, Archer and Castor (1956) showed that phosphate uptake continued as long as yeast multiplication was occurring and ceased thereafter. Some phosphate was returned to the media in the latter stages of fermentation. In fermentations at 11.1°C (52°F) and 21.1°C (70°F) phosphate uptake per 10^4 cells averaged 0.00146 and 0.00158 mg. The uptake was remarkably independent of fermentation conditions (addition of phosphate or ammonium ion). Prolonged storage on the lees results in yeast autolysis and an increase first of orthophosphate and later of meta- and pyrophosphate, according to Bourdet and Hérard (1958). Corrao and Gattuso (1964a) reported 80 to 682 mg per liter (average 312) of phosphate (as P_2O_5) in eighty-five Sicilian white wines and 130 to 640 (average 324) in seventy rosé and red wines.

The present suggested international limit of sulfate, as K_2SO_4, is 1.5 gr per liter, (Anon., 1966a). Actually, this report indicated that virtually all table wines have 0.70 or less of sulfate. Old sauternes and sherries are the exceptions. See also pp. 455–456.

The minimum legal ash content of Austrian wines, according to Weger (1966) is (as gr per 100 ml): white, 0.13; rosé, 0.14; and red, 0.16. The Swiss minimums are 0.16, 0.18, and 0.20, respectively.

In fifty-four table wines of the province of Trento (north of Venice), Margheri and Rigotti (1964) reported the following data on mineral content:

	Minimum	Maximum	Average
Ash (gr per liter)	1.824	3.232	2.583
Alkalinity of ash (ml 0.1 N NaOH)	12.60	32.00	19.74
Potassium (gr per liter)	0.590	1.190	0.880
Sodium ,,	0.008	0.092	0.038
Calcium ,,	0.056	0.424	0.153
Magnesium ,,	0.028	0.164	0.104
Iron (mg per liter)	1.20	16.80	8.03
Sulfate (SO_4, gr per liter)	0.252	0.920	0.508
Phosphate (PO_4, gr per liter)	0.135	0.392	0.236
Chloride (Cl, mg per liter)	2.0	14.0	6.4
K/Na ratio	4.03	73.75	34.87

Very good balance of the anions and cations was obtained by Rebelein (1965d).

Nitrogenous constituents

Rebelein (1967b) reported that 200 German table wines had about twice the nitrate content of 65 non-German table wines (7.2 mg per liter as N_2O_5 vs. 3.5). Young wines contain more nitrogen than old. Whites contain more than rosés or reds. Cordonnier (1966) believes that reactions of proteins with condensed tannins accounts for the difference. Fining, filtration, and pasteurization also result in lower protein contents. C. Flanzy et al. (1964) confirmed that new wines have less total nitrogen than the musts from which they were derived. They also showed that when the wine was left in contact with the lees for one to three months

461

the nitrogen content of the wine increased, reaching a maximum in about two months. Most (87 per cent) of the increase was in amine nitrogen and the rest in amide and protein nitrogen. During this period yeast autolysis is obviously occurring.

The amount of different forms of nitrogen reported in wines depends on the methods used for their determination. Amine nitrogen by formol titration is low because it does not include proline, the most abundant amino acid of wines. The microbiological procedure for amino acids often gives high results because the microörganisms may utilize some small-size peptides in their nutrition. Only the free amino acids will be detected by chromatography. This is shown by the following data of Lafon-Lafourcade and Peynaud (1961) (mg N per liter):

	Bordeaux white	*Still champagne*
Total N	190	319
Ammoniacal N	5.5	11.2
Protein N	50	60
Amine N (by formol)	15	59
Amine N (by chromatography)	61	61
Amine N (by microbiology)	118	189
Total polypeptides	73	180
Polypeptides (by microbiology)	57	127
Polypeptides not determined	16	53

Krug (1967) reviewed the physico-chemical properties of proteins as they affect wine clouding. The effect of tin and flavonols on protein clouding of white wines was emphasized. In red wines iron is the metal that causes most of the clouding. Krug recommends a heating test (three to five days) as the best measure of protein stability, but notes that a rapid test may be made by adding 5 ml of 1 per cent tannin. If excessive protein is present, clouding occurs in a few minutes. For the warm test he recommends regular clear-glass bottles, heating at 80° to 90°C (176° to 194°F) for two to three hours, holding overnight at 40° to 60°C (104°to 140°F), and leaving in a warm, lighted room for four to five days.

C. Flanzy et al. (1964) classified the amino acids in four groups:

I. Used during fermentation and not restored by autolysis: arginine, phenylalanine, and histidine.
II. Used during fermentation and restored by autolysis: proline.
III. Used during fermentation and present in increased quantities after autolysis: glutamic acid, aspartic acid, leucine, isoleucine, valine, serine, lysine, tyrosine, and tryptophan.
IV. Increased by fermentation and by autolysis: cystine, methionine, and glycine.

The results are somewhat complicated by the fact that all the wines underwent a malo-lactic fermentation during storage. Sensory examination showed improvement in the quality up to a month on the lees, but no statistical analysis of the sensory data is presented. The amino acids, through secondary reactions, probably are important factors in the quality of wines.

A comprehensive summary of the nitrogenous components of wines has been given by Ferenczi (1966b). It is significant that this survey was based on reports of several wine-producing countries. He emphasized the importance of nitrogenous compounds to the clarity and quality of the wine. Ferenczi noted the effect of climate (more protein in warm years), time of harvest, method of vinification, and other factors on the types of nitrogenous compounds in the final wine. Settling the must reduced the nitrogen content from 10 to 15 per cent. The changes during fermentation are very complicated. Total nitrogen decreased from 30 to 80 per cent, according to Ferenczi. Some amino acids are almost completely utilized by yeasts: methionine and valine, for example, are almost completely assimilated within 24 hours. Others, arginine and proline, remain essentially unchanged. During fermentation it was noted that 60 per cent of the proteins were lost at 15°C (59°F), whereas at 40°C (105°F) the loss was only 35 per cent. We believe that this type of research should be extended. Ferenczi accepts the Soviet concept that there are two protein fractions in wines: one composed of sixteen or seventeen amino acids and relatively unstable; the other composed of the same amino acids (minus rhamnose). This fraction is stable. The concept appears attractive. Ferenczi (1966a) noted also that in European countries the problems of protein instability occur widely.

The total nitrogen of a number of Greek wines varies from 50 to 550 mg per liter with averages of 141 to 330, according to Danilatos and Sotiropoulos (1968). In contrast to other areas the wines of the warmest district of Greece were lowest in protein nitrogen. Tarantola (1955) reported total nitrogen contents of 55 to 255 mg per liter in nine Italian wines. Of this 51 to 67 per cent was assimilable by yeasts. In 2,000 Hungarian wines Ferenczi (1966a) reported 100 to 1800 mg per liter of total nitrogen and 7 to 160 mg per liter of protein nitrogen. The wines of the high-quality regions in Hungary generally had higher nitrogen contents. Wines contain only 20 to 70 per cent as much total nitrogen as musts and 2 to 7 per cent as much protein nitrogen. Bentonite fining removed 10 to 130 mg per liter of total nitrogen and 3 to 35 mg per liter of protein nitrogen.

In European wines Beridze and Sir'iladze (1963) reported 95 to 193 mg per liter of total nitrogen. Of this, 21 to 29 per cent was amine nitrogen. With the variety Mtsvane it was 30 to 40 per cent. Amino acid nitrogen was from 44 to 50 per cent of the total (50 to 60 with Mtsvane). The most important amino acids were proline, alanine, and glutamic acid.

In Georgian table wines total nitrogen varied from 208 to 269 mg per liter; in natural sweet wines, from 70 to 124. In the former, amine nitrogen amounted to 31 to 49 per cent; in the latter, 18 to 43 per cent. Amino acid nitrogen was 43 to 62 per cent and 52 to 86 per cent of the total respectively. In dessert wines total nitrogen was 106 to 405 mg per liter. Nineteen amino acids were identified in Georgian wines. In fourteen California wines Bayly and Berg (1967) reported 30 to 275 mg per liter of protein (average 116). Total protein content did not correlate well with heat stability. These values are higher than those given for Russian wines by Avakiants (1967) and Begunsova and Linetskaia (1967), who reported only 5 to 70 mg per liter.

A number of studies on the nature of the proteins in German table wines have been made: Koch et al. (1956), Koch and Bretthauer (1957), Koch (1957, 1963), Koch and Sajak (1959, 1963), Diemair et al. (1961a, 1961b, 1961c). These show higher proteins in the wines of warmer seasons. The varieties differed considerably in the amounts present. The isoelectric point was very close to the pH of the wine. One reason for protein instability is the change in pH which occurs during the malo-lactic fermentation. As many as four fractions were identified. Fermentation, heating, and bentonite treatment removed more of one fraction than another. Similar results were obtained in California by Moretti and Berg (1965). One of the four protein fractions that they isolated was especially heat-labile and had a lower isoelectric point. The ratio of this fraction to the total was believed to be an index of protein stability.

Protein clouding is due not only to precipitation of thermolabile proteins but also to formation of protein-tannin complexes in the presence of traces of heavy metals (especially tin), according to Krug (1968). Tin-tannin complexes are white and therefore do not show up. However, the tin-tannin reaction is very sensitive and occurs with very small amounts of tin.

Kielhöfer and Aumann (1955) reported tin-protein precipitates and hydrogen sulfide in wines containing tin and high protein and sulfur dioxide.

A more sophisticated study of the nitrogenous compounds of wines is that of Terčelj (1965). He confirms the spectacular decrease in amino acids during fermentation, noting, however, the wide variation between amino acids. Filtration also results in a very large reduction in amino acids, again varying widely between the various amino acids. He considers this a practical means of stabilizing sweet table wines.

When ammonia was added to musts there was a spectacular increase

(10 to 30 times) in certain amino acids in the resulting wines (C. Flanzy et al., 1964). This was especially marked for arginine, histidine, lysine, valine, tyrosine, isoleucine, methionine, aspartic acid, and alanine, and less marked for glutamic acid, tryptophan, cystine, serine, and proline. There were, however, marked differences in the increases for certain amino acids between the musts of the two varieties studied and with those of a later study (C. Flanzy and Poux, 1965b). Aeration during fermentation influenced these changes differently for various amino acids—naturally being less favorable for amino acids such as glutamic acid and aspartic acid, which are involved in the respiratory cycle. When ammonium tartrate was the source of the ammonia the resulting wines were lighter in color. Addition of ammonia usually improved the quality, and C. Flanzy and Poux (1965a) confirmed this for wines of another season. No statistical analysis of the sensory data is given.

During aging there is an increase in nitrogen prior to the first racking, owing to yeast autolysis. Thereafter the nitrogen content diminishes. Ferenczi (1966a) noted that this is not a simple decrease, for temperature, tannin content, metal content, and other factors are involved.

Lafon-Lafourcade and Peynaud (1961) showed clearly that proline is the most abundant amino acid of grapes and wines, followed by threonine, lysine, glutamic acid, and serine. Proline and threonine constitute about 70 per cent of the total amino acid content. The total amine nitrogen calculated as nitrogen per liter for the four years 1955 to 1959 was 207, 126, 144, and 115, respectively. The years 1955 and 1959 were the two warmest seasons. The mg per liter of proline for these years was 700, 560, 560, and 395, but for threonine was 148, 138, 283, and 205. Considerable differences in total amino acids for different varieties were reported.

Dimotakis (1958) also showed that proline was the most abundant amino acid in Greek wines, followed by alanine, glutamic acid, aspartic acid, histidine, leucine, serine, glycine, α-aminobutyric acid, hydroxyproline, cystine, valine, methionine, threonine, arginine, norvaline, and tryptophan. Of the eleven wines tested, only proline was found in all. Lüthi and Vetsch (1952) reported alanine to be especially important in the nutrition of lactic acid bacteria.

Hennig (1955) seems to be the only investigator who has reported the amino acid citrulline in wines, and Koch and Bretthauer (1957) the only ones to find ornithine. The first report of oxyproline, norvaline, and α-aminobutyric acid in wines is evidently that of Dimotakis (1956). Cases (1959) identified ethanolamine in wine.

The malo-lactic fermentation of white wines generally resulted (Lafon-Lafourcade and Peynaud, 1961) in slight changes in the amino acid content of wines: for red wines there were slight decreases in arginine, α-alanine, phenylalanine, and serine, and increases in glutamic acid, methionine, threonine (sometimes), and tryptophan. Clearly there are complicated bacterial effects on amino acids.

During prolonged storage of wine on the lees, its nitrogenous content increases by absorption of the products of yeast autolysis. According to Bourdet and Hérard (1958), there is an increase in free amino acids. If the contact of lees and wine is brief, peptides increase. The qualitative nature of the amino acids did not change much on storage. There was an accumulation of arginine, lysine, histidine, and methionine. In two young wines 90 to 91 per cent of the nitrogen was present as free amino acids and 6 to 9 per cent as amides. Ammonia and nonprotein nitrogen amounted to only 0.4 to 3.3 per cent.

The influence of heating the must to 80°C (176°F) and then pressing, as against fermenting in open and closed tanks for four days, is shown by the following data (mg per liter) of Lafon-Lafourcade and Peynaud (1961):

	Closed tank	Open tank	Heated juice
α-Alanine	10	10	10
Aspartic acid	18	12	0
Arginine	0	0	0
Cystine	5	5	2
Glutamic acid	35	41	40
Glycine	25	23	0
Histidine	11	3.6	3.8
Isoleucine	12.5	9	9
Leucine	1.0	1.0	0
Lysine	14	0	0
Methionine	1.5	1.5	0
Phenylalanine	2.0	2.0	2.0
Proline	500	500	500
Serine	50	46	40
Threonine	126	240	62
Tryptophan	7	7	4.1
Tyrosine	0	0	0
Valine	19	20	19
Total as N per liter	102	111	83

The disappearance of some amino acids and the reduction in others as a result of heating is noteworthy. The increase in threonine in the open tank is also large. Feuillat and Bergeret (1967a) showed that bentonite added to musts for 24 hours to remove polyphenoloxidases did not result in large losses of any of the amino acids in the new wine (compared to new wines to which no bentonite was added). In fact, there was a large increase in tyrosine and α-aminobutyric acid. However, fermenting in the presence of bentonite resulted in loss of about half of the α-aminobutyric acid.

Using gas chromatography, Glonina and Avakiants (1966) reported the following amounts of amino acids in wines (mg per liter):

	Sparkling	Dry white
Proline	264.0	770
Alanine	72.2	70
Glutamic acid	42.2	315
Leucine	26.5⎫	28
Isoleucine	11.6⎭	
Tyrosine	36.5	37
Arginine	34.8	91
Lysine	23.4	37
Threonine	19.4	32
Aspartic acid	16.6	72
Histidine	15.6	13
Glycine	14.8	16
Valine	9.8	16
Serine	9.3	11
Phenylalanine	8.2	15
α-Aminobutyric acid	8.7	5
Oxyproline	7.1	—
Methionine	6.4	49
Cysteine	5.3	47

Tarantola (1954) observed marked increases in valine, leucine, and tyrosine in wines stored on the lees. This was not observed by Bidan and André (1958). The study of Bourdet and Hérard (1958) showed markedly large increases in aspartic acid, glutamic acid, lysine, α-alanine, glycine, serine, threonine, valine, leucine, isoleucine, γ-aminobutyric acid, phenylalanine, tyrosine, asparagine, and glutamine in red wines stored seven days on the lees (in small containers stirred once each day), compared to new wine separated before the seven days' storage. There was, however, a decrease in arginine, proline, histidine, and cystine and little change in methionine and α-aminobutyric acid.

In fourteen Portuguese wines Silva and Petinga (1962) reported considerable variation in the number and amounts of the amino acids present. One wine had twenty-eight amino acids; another had only nine. Among the rare amino acids found were α-aminobutyric acid in one wine (compared to γ-aminobutyric acid in ten), β-alanine in two (α-alanine in thirteen), cysteine in two (and cystine in six), hydroxyleucine in two (isoleucine and leucine in fourteen), methionine sulfide in one, methionine sulfoxide in one, norvaline in one, ornithine in two, oxyproline in two (compared to proline in fourteen), and histamine in three.

In Soviet wines Nilov and Skurikhin (1967) reported amino nitrogen contents of 10 to 500 mg per liter, with red wines usually much higher than whites. Nilov and Ogorodnik (1965) reported reactions between amino acids (150 to 200 mg per liter) and glucose, or arabinose (0.1 to 0.5 gr per liter). Arabinose was the more reactive. The reaction was faster at higher pH's and temperature. Without sugars very little deamination occurred in 60 to 70 days at 54°C (131.8°F) and pH 3. But in the

467

presence of sugars over half of the glutamic acid was deaminated. Agabal'iants and Glonina (1966) could not find a connection between the amino acid content and the oxidizability of wines. Oxidative deamination does not, therefore, seem to be important to the quality of the wine.

Nucleotides and nucleosides identified in wine by Terčelj (1965) included guanine, adenine, cytosine, hypoxanthine, xanthine, thymine uracils, cytidine, thymidine, adenoside, uridine, guanosine, GMP, CMP, AMP, TMP, and IMP.

Methyl-, ethyl-, propyl-, isopropyl-, butyl-, isobutyl-, amyl-, isoamyl-, and β-phenylethylamines and pyrrolidine as well as other unidentified amines were identified in wines by Drawert (1965). There were some quantitative differences, depending on the sources of the wine, but their significance to quality was not obvious. These compounds may account for certain of the pharmocological properties attributed to wines.

Mayer and Pause (1966) isolated six indol compounds from seventy wines as follows:

Compound and Formula	White No.	White mg/liter	Red No.	Red mg/liter
(β-Hydroxyethyl)-indole $C_6H_4NHCH=CCH_2CH_2OH_3$	12	1–7	7	1–5
5-Methoxyindolyl-2-carbonic acid $CH_3OC_6H_3NHC(COOH)=CH$	4	1–2	0	—
β-(3-Indolyl)-acrylic acid $C_6H_4NHCH=CCH=CHCOOH$	8	2–10	7	1–10
3-Indolyl-acetic acid $C_6H_4NHCH=CCH_2COOH$	23	1–5	15	1–5
Tryptophan $C_6H_4NHCH=CCH_2CH(NH_2)COOH$	25	2–10	11	1–6
5-Hydroxytryptophan $HOC_6H_3NHCH=CCH_2CH(NH_2)COOH$	4	1–3	1	1

Except for tryptophan, these appear to be products of alcoholic fermentation and are present mainly in new wines. Mayer and Pause were able to isolate histamine from only one wine (ca 5 mg per liter). Some of these compounds caused headaches when added to wines. They suggest that this may be the reason why young wines are more toxic than old wines, at least to some sensitive consumers.

Recent reports show that histamine is present in wines, more in reds than in whites. Histamine is known to have a toxic effect on the liver, and Marquardt and Werringloer (1965) demonstrated that this is much more marked in alcoholic solutions than in aqueous solutions. Since histamine is not a product of alcoholic fermentations but is produced by bacterial fermentations, the authors recommend control of bacterial activity to reduce histamine formation. The maximum amount reported from 100 wines of Europe, North Africa, and Chile was 22 μg per ml; however, Werringloer (1966) reported 3 mg per liter of histamine in a white German wine made from Chasselas doré. For earlier reports on the presence of histamine see the reports of Tarantola (1954) and Carles et al. (1958). Ough (1971. J. Agr. Food Chem. *19*:241) found an average of only 1.8 mg per liter in 253 California table wines.

In both red and white wines Mayer and Pause (1966) found that there was no apparent relation between the histamine content and the presence of indole derivatives. (See p. 468.)

Foam constituents

Newly fermented wines frequently foam excessively, which is undesirable. Sparkling wines, in contrast to beer, should not have a persistent foam. Foaminess is particularly objectionable during bottling. Wines made from muscat varieties seem to foam more than those made from other grapes. Amerine et al. (1942) summarized the available information on foaming of wines. Removal of foam from a wine decreases its tendency to subsequent foaming. This may be the basis for the practice of filling the fermenters so full that they foam over during fermentation. The saponins were not found to be a factor in foaminess in wines. Fermentations at 0°, 11.8°, and 32.8°C (32°, 55°, and 91°F) gave wines with less foaminess than those fermented at 22.2°C 72°F). There was little difference in foaminess among wines fermented with three different yeasts, with or without sulfur dioxide. Wines that were fermented on the skins foamed more than those made of the same grapes but fermented off the skins. The difference in tannin content was insufficient to account for the difference in foaminess. The possibility is that prefermentation cultivation of the yeast on sterile must may influence degree of foaming.

Vitamins

Ascorbic acid decreases steadily in fermenting wines and little is found in wines. However, Ournac (1965) reported 3 to 9 mg per liter in some dark-colored wines. Baraud (1951, 1953), however, doubted that ascorbic acid was present in the wines he studied, for he was unable to obtain a color reaction with a tetrazine.

Thiamine is lost during fermentation owing to its incorporation into cocarboxylase and its destruction by sulfur dioxide; the amount retained is less, the higher the free sulfur dioxide content. Filtration with bentonite removed or destroyed much of the vitamin. Perlman and Morgan (1945) found 0.001 to 0.12 mg per liter in new wines. When thiamine was added, 85 to 100 per cent was retained after three months' storage at 45°F (7.2°C). Genevois and Flavier (1939) reported 0.15 to 0.3 mg per liter. Diffusion of thiamine from the lees into the wine was observed by Ournac and Flanzy (1967). Different yeasts varied in the amount of thiamine released. By this means they hoped to restore the thiamine removed by alcoholic fermentation.

The Soviet enologists report a vitamin P component, probably mainly flavones. Muradov (1966) found 0.23 to 0.37 per cent in twenty Russian musts. Pasteurization resulted in the loss of only 1.7 per cent. See also Lavollay and Sevestre (1944) and Ournac (1953).

Riboflavin is present in larger amounts in wines than in musts, but less than 0.4 mg per liter. Schön et al. (1939) reported 0.075 mg per liter in Portuguese red wines. Riboflavin rapidly disappears when wines are exposed to light, as in clear glass bottles. In dark glass 68 to 96 per cent was retained in the experiments of Perlman and Morgan (1945). Sulfur dioxide exercised some protective effect on its retention. Filtration with bentonite reduced the riboflavin content about half.

Pyridoxine is present in most if not all wines. Perlman and Morgan (1945) reported 0.66 to 0.72 mg per liter in three California wines. They found it was well retained when added to wines. Peynaud and Lafourcade (1957) investigated the free and total pyridoxine content of Bordeaux wines. In thirty-one white wines the free pyridoxine ranged from 0.12 to 0.67 mg per liter (average 0.31) and the total from 0.22 to 0.82 (average 0.44). In fifty-eight reds the free was 0.13 to 0.68 (average 0.35) and the total from 0.25 to 0.78 (average 0.47). Obviously most of the pyridoxine is present in the free condition, and red and white wines contain similar amounts. Aging in the wood, fining with gelatin, isinglass, casein, and white of egg, or albumen, did not greatly affect the pyridoxine content, but fining with charcoal or bentonite or treatment with oxygen destroyed 15 to 40 per cent of it. Tiurina et al. (1967) reported pyridoxine contents of 0.56 to 1.0 mg per liter in five Soviet red wines.

Nicotinic acid was reported in amounts of 0.6 to 2.1 mg per liter in French wines by Cailleau and Chevillard (1949). The nicotinic acid content of five Soviet red wines varied from 0.38 to 1.24 mg per liter in the report of Tiurina et al. (1967).

Pantothenic acid was found in two wines at 0.40 and 1.1 mg per kg by Perlman and Morgan (1945). In commercial wines the amounts found were less, 0.07 to 0.45. Cailleau and Chevillard (1949) reported 0.2 to 1.2 mg per liter. In twenty-five Bordeaux white wines Lafon-Lafourcade and Peynaud (1958) showed 0.55 to 1.25 mg per liter (average 0.81) as calcium pantothenate and in forty-five red wines 0.47 to 1.87 (average 0.98). Only a little is lost during fermentation unless bacterial growth occurs. Added pantothenic acid is well retained in wines. The pantothenic acid of five Soviet red wines varied from 1.89 to 3.78 mg per liter, according to Tiurina et al. (1967).

In fifteen Bordeaux white wines Lafon-Lafourcade and Peynaud (1958) found 0.015 to 0.133 mg per liter (average 0.069) of *p*-amino-benzoic acid. They also detected small amounts of pterolglutamic acid, 0.0004 to 0.005 mg per liter (average 0.0024). The amounts of these two vitamins in red wines were similar. They also found 0.019 to 0.027 mg per liter of choline (average 0.021) in twelve Bordeaux white wines. Red wines averaged 0.029. For these three vitamins, wines are as high or higher than the musts from which they were derived.

In forty-one white Spanish wines Cabezudo *et al.* (1963) reported 0.002 to 0.013 mg per liter of biotin compared to 0.002 to 0.018 in forty-six red wines. Red wines aged in wood tended to be higher in biotin. The biotin content of five red Soviet wines varied from 0.005 to 0.0058 mg per liter (Tiurina *et al.*, 1967).

Peynaud and Lafourcade (1955*b*) found 0.00 to 0.59 gr per liter of inositol in fifty-seven red Bordeaux wines. In thirty-one white Bordeaux the range was 0.22 to 0.73 gr per liter (average 0.33). Also, using a micro-biological technique, Tiurina *et al.* reported 0.293 to 0.439 gr per liter of inositol in five red wines. The low amounts in some wines is attributed to microbial activity.

The harmful effects of wines made from crosses of *Vitis vinifera* and *V. riparia* on chickens were attributed by Breider *et al.* (1965) to vitamin deficiencies (due to inactivation of the vitamins) and to some unknown toxic substance. The toxicity appeared to be greater in the presence of alcohol. None of the harmful effects could, however, be explained by the alcohol content itself. The possibility of calcium deficiency might also explain the differences. For further data on the components of wine as related to human diet see pages 481 to 493.

Miscellaneous constituents

The colloidal content of new wines, according to Usseglio-Tomasset (1963), results from must components (90 per cent) and materials contributed by the yeasts. The colloids of the pulp are 70 per cent arabans, galactans, and pectins; 80 per cent in the skins. The inactivation of pectolytic enzymes during fermentation accounts for the relatively high pectin content of new red wines. See also Tarantola and Usseglio-Tomasset (1963).

Peynaud and Lafon (1951) found diacetyl (up to 3 mg per liter) in all the red wines examined. They suggested its origin from yeast fermentation, as did Charpentié *et al.* (1951). However, Radler (1962*a*) found more acetoin and diacetyl in wines which had undergone a malo-lactic fermentation; and Fornachon and Lloyd (1965) demonstrated that wines which have undergone a malo-lactic fermentation contain significantly more diacetyl and acetoin than wines which have not undergone such a fermentation. In Australia, *Leuconostoc mesenteroides* was the principal organism and pyruvic and citric acids the main substrates. The homo-fermentative rods *Lactobacillus brevis* and *Lact. hilgardii* did not produce diacetyl or acetoin under normal conditions. Since wines which have not undergone detectable bacterial activity frequently contain small amounts of diacetyl, Fornachon and Lloyd (1965) suggest that it is formed in small amounts by yeasts during normal alcoholic fermentation. They noted that the sensory significance of diacetyl in various wines is not known. Amounts of 4 mg per liter or more in table wines seemed to them excessive. Diacetyl is present in normal wines at an average of 0.2 mg per liter, according to Dittrich and Kerner (1964). Above 0.89 mg per liter wines acquired a sour-milk odor. One wine containing 4.3 mg per liter was reported. More data on the diacetyl threshold in wines is needed. Guymon and Crowell (1965) showed that most of the acetoin and diacetyl formed in the early stages of fermentation was later lost.

Acetoin is present in normal wines at from 3.0 to 31.8 mg per liter (Dittrich and Kerner, 1964); but one wine had as much as 56.2 mg per liter and Ribéreau-Gayon and Peynaud (1958) reported one with 84.0.

RATIOS BETWEEN COMPONENTS

The ratios between certain components may be used to detect dilution of wine with water, fortification with alcohol, or sweetening with must or concentrate.

The ratios between the concentrations of certain components or between components determined in several ways are commonly used abroad to detect adulteration of or abnormalities in wines. Several are used to detect diluting with water or fortification with alcohol; one is used to detect sweetening with must or concentrate. These ratios have not been adapted to California conditions, but should prove more useful than actual composition because, for a particular natural wine, they vary over a narrower range. (See Fabre, 1929.) In order to use them here, the ranges of the ratios that occur in natural California wines must be known, and standard ratios would have to be worked out.

Recent Italian legislation, according to Weger (1966), fixes maximum content for certain metals, sulfates, methanol, and so forth. It fixes minimum limits on extract-minus-sugar, the ratio of extract-minus-sugar to ash (a minimum of eight to thirteen suggested) and the ratio of alkalinity of ash to ash (a minimum of eight to twelve suggested). In some northern European wines the ratio of extract-minus-sugar to ash was below the suggested limit. A minimum alcohol to ash ratio of eight was suggested. About half of the northern Italian wines tested failed to meet this limit. As tests for the genuineness of Italian wines Vettori (1966) recommends alcohol (actual and potential), extract, ash, alkalinity of the ash, glycerin, and the following ratios: extract to ash, alkalinity of the ash to ash, alcohol × 8 to reduced extract, and glycerin to alcohol. For special purposes he also uses the sucrose, caramel, glucose, and fructose contents, polarimetric deviation, methanol, sorbite, mannite, ethyl acetate, acetaldehyde, acetoin, and various minerals, the tartrate, malate, citrate, succinate, and lactate contents, the titratable and volatile acidity, and nitrogen and polyphenolic compounds. He discusses the various enological rules as they apply to Italy. The need for large amounts of analytical data before establishing limits on the chemical constituents is emphasized by the work of Garoglio and Salati (1967). They found that some wines had less than the minimum limit of ash, ratio of sugar-free extract to ash, and potassium (435 mg per liter). The sodium limit (100 mg per liter) was not exceeded, but bentonite treatment did increase the sodium content. More data on normal California wines are needed before limits can be established.

Dilution or fortification

Ratios between alcohol, acid, extract, glycerin, or potassium contents are used to determine if a wine is a normal or natural one—that is, undiluted with water or unfortified with alcohol. In these ratios the acid is expressed as grams of sulfuric acid *per liter*. Multiply the percentage of acid as tartaric by 6.5 to convert to grams of sulfuric per liter. The volatile acid, expressed as the percentage of acetic acid, multiplied by 8.1 gives volatile acid as grams of sulfuric acid per liter. Total acid as grams of sulfuric acid per liter minus volatile acid as grams of sulfuric per liter gives fixed acid calculated as sulfuric acid. (Unless otherwise stated, all other total-acid contents given in this book are expressed as grams of tartaric acid per 100 ml of grape juice or wine.)

a) Gautier sum: This is the sum of alcohol by volume + fixed acid + $\frac{1}{10}$ volatile acid (as sulfuric). It is customary to subtract from this figure added acid, added alcohol, free tartaric, above 0.5 gr for reds or above 1.0 gr for whites, and 0.2 per gr of potassium sulfate over 1 gr per liter. This sum is characteristic of the initial sugar content and acid content of the grapes, and usually varies from 13 to 17 for undiluted French wines. Below 12.5 there is a presumption of watering.

b) Halphen ratio: fixed acid × 0.70/alcohol content. Added acids must be subtracted before using this ratio. A maximum volatile acidity of 0.086 per cent (as acetic) is used. For lesser volatile acidities than 0.086, change the 0.7 to the appropriate figure calculated as sulfuric per liter.

The ratio found is compared with normal ratios for wines of the same type and alcohol content; if less than that given in Halphen's curve, there is a presumption of watering. The Gautier sum and Halphen ratio apply only to sound dry table wines.

c) Blarez ratio: alcohol content/fixed acidity. Maximum ratios for French wines vary from 2.0 to 5.8, according to type and origin of the wine.

d) Extract ratio: alcohol by weight/reduced extract. Maximum is about 6.5 for dry white table wines and 4.5 for dry red table wines. Extract minus sugar (less 0.1) minus potassium sulfate over 0.1 per cent, minus free tartaric (calculated as sulfuric), over 0.5 gr for reds or 1.0 for whites, minus any added soluble solids equals the reduced extract.

e) Roos ratio: Gautier sum/extract ratio, or a/d. This is usually over 3.2 for red wines and 2.4 for white wines. There is a strong presumption of watering when these minimums are not reached.

f) Tartrate index: total tartaric acid/total potassium. In this ratio, both numerator and denominator are expressed as grams of potassium acid tartrate per liter. The ratio is below 1 for unwatered wines–even for those of low alcohol made in cool years and to which the other ratios may not apply.

g) Glycerin ratio: alcohol content/glycerin content. When the alcohol and glycerin are expressed in percentage by weight, normal California wines should not have ratios above 18.2. For French wines the maximum value is 14.2. In sixty-eight California white table wines, Amerine and Webb (1943) found a range of alcohol to glycerin ratios of 5.6 to 12.9, (average 10.2). In fifty-six red table wines they found a range of 5.0 to 11.1 (average 9.3). Peynaud (1948) has criticized their method for sweet wines, and it is possible that their lowest values are too low. Their averages, however, are only slightly below those reported in the literature (see data below). Hickinbotham (1948) found an average of 14.2 in white Australian table wines and 15.6 in red table wines. In the literature, this alcohol-to-glycerin relation is usually expressed as the weight of glycerin per 100 parts of alcohol, or the reciprocal of that given above.

Analyst and source	Number of analyses	Alcohol-glycerin-ratio in wines, by weight	
		Range	Average
Ferré (from Peynaud, 1948)	16	10.4–15.5	11.9
Fresenius (from Peynaud, 1948)	1,000	12.6–16.7	11.9
Laborde (from Peynaud, 1948)	34	10.3–16.1	11.9
Peynaud (1948)	48	10.2–15.4	12.2
Hickinbotham (1948), white wines	—	—	14.2
Hickinbotham (1948), red wines	—	—	15.6
Amerine and Webb (1943), white wines	68	5.6–12.9	10.2
Amerine and Webb (1943), red wines	56	5.0–11.1	9.3

h) Rebelein ratio: 2,3-butylene glycol/glycerin. This ratio is also supposed to detect fortification of table wines. See Garina-Canina (1953).

The principal objection to these rules, particularly to the first four, which have been widely used, is that the composition of wines is so variable that the ratios are only roughly applicable, and then only to a given region. The acidity, for example, is subject to additions or subtractions which markedly influence most of the ratios. High-acid varieties subjected to a malo-lactic fermentation may give misleading results.

A more natural system has been suggested by Genevois *et al.* (1948*b*) and Genevois (1950), based on studies of the by-products of alcoholic fermentation. They have found these products to be produced in young wines in a fairly constant ratio to each other. This interrelation can be expressed very closely by the equation $5s + 2a + b + 2m + h = 0.9g$, where a is acetic acid, b 2,3-butylene glycol, s succinic acid, m acetoin, h acetaldehyde, and g the glycerin content, each expressed as mols per liter. The wine yeasts, however, differ markedly in their ability to form acetic acid, succinic acid, 2,3-butylene glycol, and glycerin. The ratio of acetic acid to succinic acid varied from 0.45 to 2.0; the ratio of 2,3-butylene glycol to glycerin, from 49 to 95 (Peynaud and Ribéreau-Gayon, 1947).

One deficiency of this equation is the lack of data on the amount of cell substance produced (Genevois, 1961; Nordström, 1966). Nordström has noted also that not all the by-products are formed from acetaldyhyde, as proposed by Genevois. A considerable part of the carbon atoms are derived from carbon dioxide. Some of the coefficients need, therefore, to be changed. While the equation for the by-products of alcoholic fermentation of Ribéreau-Gayon and his colleagues (based on the hypothesis of Genevois, 1950) obviously fits some fermentations, the exceptions are many and critical. We doubt if any meaningful equations can be established without exact specification of biological and environmental conditions during fermentation. Indeed, Peynaud and Domercq (1955) give balance sheets for wines produced by various species of yeasts fermenting under aerobic and anaerobic conditions.

In order to detect sugaring, watering, or fortification, Rebelein (1957*a*, 1957*b*, 1962, 1964) proposed a K value,[2] an REZ limit of 61, a $\sqrt{f}$ minimum of 0.89, and a H limit of 68. The values depend on lengthy calculations which in turn depend on accurate analytical values. Rebelein (1964) gave a more detailed defense of these limits. The problem is an important one in Europe. It is potentially a problem in the eastern

---

[2] Calculated from the ratio glycerin × 2,3-butylene glycol/ethanol[3], where all are in gr per liter. Normal wines are supposed to have K values of about 8×10^{-6}.

United States where sugaring and watering are used, although in this country the government's control of the winery's basic permit prevents gross abuses.

Siegel et al. (1965) reported that sixteen yeast strains produced wines with Rebelein K values (p. 475) of 3.20 to 4.81 × 10^{-6}, averaging 3.63; in another must with fifteen strains, 7.36 to 12.29 × 10^{-7}, averaging 10.07. Addition of 2,3-butylene glycol or glycerin did not affect the amounts of these compounds formed, but did affect the K value. Until there is a better understanding of the factors influencing the K value it should not be used as a means of detecting sophistication.

Corrao and Gattuso (1964b) reported normal glycerin content in Sicilian white wines. The 2,3-butylene glycol content was slightly higher than normal. Rebelein's index varied from 2.4 to 14.0; the geometrical mean was slightly above 5. They conclude that Rebelein's index is not a good measure of genuineness for Sicilian wines. This is based on the premise that German wines have the same by-product composition as those from the fermentation of Sicilian musts. This is not an easily acceptable assumption.

The formulae of Ribéreau-Gayon and Peynaud or of Rebelein (1957a) are valid if various yeast strains produce the same relative amounts of by-products. Siegel et al. (1965) in one study found a low but relatively constant Rebelein K value; with another must there was considerable variation. Patschky (1967) has examined a number of ratios designed to detect addition of alcohol to table wines. Some seem rather complicated and one was valid only in combination with determination of other fermentation by-products. (See also p. 475.)

Several investigators have attempted to characterize wines on the basis of their content of volatile components detectable by gas liquid chromatography. Prillinger et al. (1967), and Prillinger and Madner (1968) focused attention on the ratio of peak quantities of ethyl ester of lactic acid to hexanol as an index of identity. This is 0.5 for both red and white wines but is higher in pomace wines, lees wine, raisin wines, and artificial wines.

Added must or concentrate

The proportions of glucose and fructose are used to determine whether a natural sweet wine has derived its sugar from the original must or has been made by adding must or concentrate to a dry wine. The Blarez ratio, concentration of copper-reducing material/optical rotation, is an index characteristic of the proportion of glucose and fructose present and is usually below 4 for natural sweet wines.

SUBJECTIVE FACTORS . . .

. . . are difficult to measure.

Unfortunately, several of the constituents that characterize the various wine types cannot be estimated easily by chemical analysis. The aroma- and bouquet-contributing substances, at present measured subjectively by tasting, are present in small amounts. Probably, as better methods are discovered and new, distinctive components are differentiated, more rigid standards for the different types of wines, based on the quantitative determination of these substances, will be developed.

The most obvious and distinctive flavoring constituents are those that give the wine its varietal character—for example, muscat, Zinfandel, or Cabernet. Some of these aromatic principles are readily detectable, while others are faint. The problem of their chemical detection and quantitative measurement is complicated not only by the minute quan- tities present but also by their susceptibility to oxidation and by the subsequent changes they undergo in aging. (See p. 446 f.)

Hennig and Villforth (1942) made an extensive study of the odorous constituents of a 1938 Müller-Thurgau wine. They identified the following aldehydes, ketones, and related compounds: formaldehyde, acetalde- hyde, propionaldehyde, cinnamaldehyde, vanillin, acetone, acetyl- methylcarbinol, and acetal. Caproaldehyde, and higher members of this series, benzaldehyde, methyl ethyl ketone, and furfural, were not identi- fied with certainty but probably occur. The remaining odorous con- stituents apparently were mainly esters, since little free alcohol or free acid was present in the pentane extract which contained all the odorous constituents. Because of the high boiling point of some of the esters and the danger of decomposition by fractional distillation, their component parts, alcohols and acids, were identified after saponification. Alcohols found included methyl, ethyl, isopropyl, isobutyl, isoamyl, and α- terpineol. Probably also present were *n*-propyl, *n*-heptyl, and *sec*-nonyl (2-nonanol). The eleven acids identified were formic, acetic, propionic, *n*-butyric, caproic, caprylic, capric, and lauric. Probably also present were isobutyric, isovaleric, and enanthic acids. Hennig and Villforth, on the basis of these studies, consider the esters to be the most important in quantity and quality of the odorous constituents of the wine studied. However, many important problems remain, particularly as to the source, nature, and relative importance of these odorous materials. Legent-Fournès (1944) has summarized much of the early information.

Considerable progress has been made in our knowledge of the chemistry of flavors by the application of newer methods of separation and identi- fication. Gas chromatography has been applied to the examination of the volatile essences of grapes and wines (Bayer, 1966; Lemperle and Mecke, 1965; and others). Bayer (1966) reported close correlation between quality and volatile compounds in flavor extracts of wine. The more important of these compounds have been isolated and characterized by gas chromatography.

Prillinger et al. (1967) hoped to be able to characterize different wines on the basis of differences in their gas chromatographic patterns. This assumes that the height of a gas chromatographic peak is proportional to the quality and intensity of its odor. This is surely unproved. It also assumes that no synergistic or masking effects occur between odorous constituents. The data in figure 70 illustrate why this should be utilized with caution for quality evaluation.

Webb (1968b) presented evidence on the presence of 9 gamma lactones in wines of which six were identified: gamma-butyrolactone, pantoyl lactone (2-hydroxy-3,e-dimethyl-4-hydroxybutyric acid lactone), 4,5-dihydroxycaproic acid gamma lactone (2 isomers) and 4-carboethoxy-4-aminobutyric gamma lactam. Their significance to wine odor has not been determined.

The processes occurring during aging of wine have been examined in some detail. More information is needed, however, on the nonvolatile flavor components and the interrelation between these and the volatile constituents. Modern techniques have yielded more information in this field than the older and more tedious techniques of flavor chemistry, as illustrated in the extensive investigation of Hennig and Villforth (1942) and Haagen Smit et al. (1949).

The detection and determination of the actual components of volatile and relatively nonvolatile flavor and odor components must be accompanied by investigation of the odor and taste thresholds of these components, individually and in combination. Investigations in these areas, combining subjective organoleptic evaluation with objective determination of chemical composition, have been initiated (Amerine et al., 1959b, 1965b). Webb (1967) supplies information on the biogenesis of the flavor compounds in wines and their possible importance to the character and quality of wines. More remains to be done, however, before odor and flavor and the factors influencing these sensory properties can be objectively evaluated. Progress is being made in the elucidation of the chemistry and physiology of flavors, but the tactile properties which contribute to mouth, feel, or body still remain to be defined. The physiology of the sensation of astringency, which is so important in sensory evaluation of wines, is poorly defined and neither satisfactory subjective nor objective tests for its evaluation are available (Joslyn and Goldstein, 1964). The analytical procedures based on determination of gross

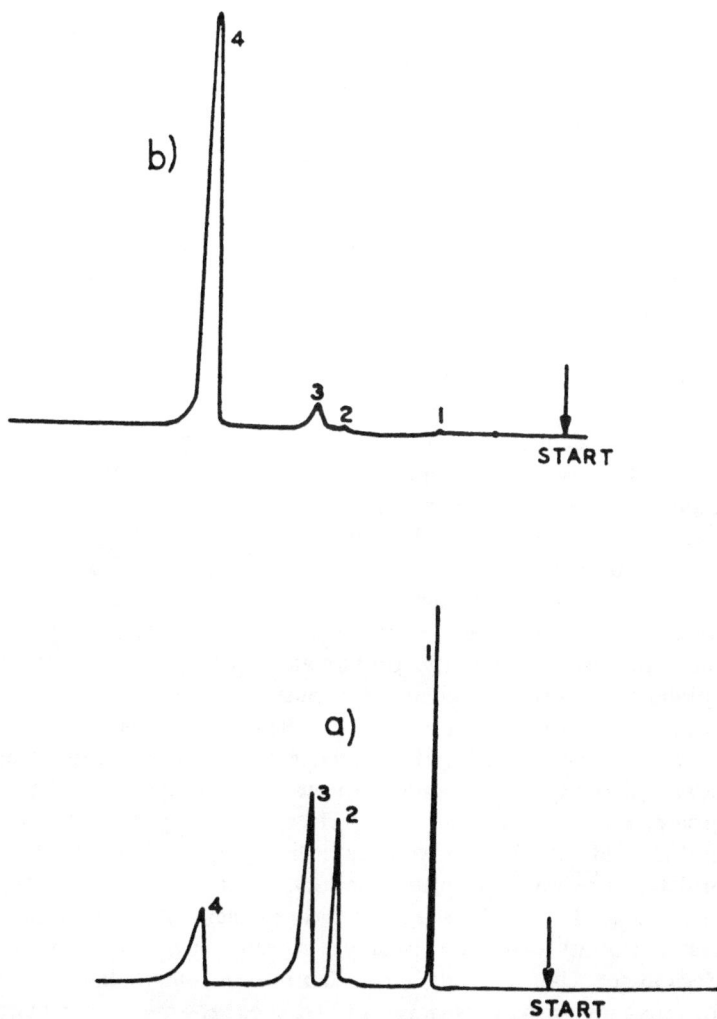

Figure 70. Comparison of response of a flame ionization detector (*a*) and human response to odor (*b*) for: 1, ethanol; 2, methyl acetate; 3, citral; and 4, β-ionone. (From Bayer, 1966.)

chemical composition are not sufficient and cannot be substituted for sensory evaluation (Baker and Amerine, 1953). Le Magnen (1966) pointed out that human olfactory sensitivity is 1,000 to 10,000 times that of present-day gas chromatography equipment. Boidron and P. Ribéreau-Gayon (1967) consider the most important volatile constituents in wines probably are acetates and lactates, *n*-butanol, *n*-hexanol, and 3-methyl-2-butanol. They also report the ratios of diethyl succinate to γ-butyrolactone and of 3-methyl-1-butanol to 2-methyl-1-butanol are important characteristics. They note correctly that with increasingly sensitive equipment other important compounds will surely be isolated.

Several hundred compounds account for the odor of wines, according to Drawert and Rapp (1966). In wines they consider the following 121 odorous materials to be established: (identification of compounds followed by an asterisk was considered inadequate): methanol, ethanol, 1-propanol, 2-propanol, 1-butanol, 2-butanol, 2-methyl-1-propanol, 2-methyl-1-butanol, 3-methyl-1-butanol, 3-methyl-2-butanol, 2-methyl-2-propanol, 1-pentanol, 2-pentanol, 3-pentanol, 1-hexanol, 2-hexanol, 1-heptanol, 2-heptanol, 1-octanol, 2-octanol, 1-nonanol, 2-nonanol, 1-decanol, 2-decanol*, 1-undecanol*, 2-lauryl alcohol*, 2-phenethyl alcohol, formaldehyde, acetaldehyde, propionaldehyde, 1-butanal*, 2-butanal*, 1-pentanal, 2-pentanal, 1-hexanal, 2-hexen-1-al*, 1-heptanal, 1-octanal, 1-nonanal, 1-decanal, 1-laurylaldehyde*, benzaldehyde*, cinnamaldehyde*, vanillin*, furfural, α-terpineol, linaloöl, α-ionone*, citral*, acetone, acetoin, methyl ethyl ketone, diacetyl, formic acid, ethyl formate, acetic acid, methyl acetate*, ethyl acetate, isopropyl acetate*, isobutyl acetate, *n*-hexyl acetate, *n*-heptyl acetate, β-phenylethyl acetate, propionic acid, ethyl propionate, *n*-propyl propionate, isobutyl propionate, *n*-butyric acid, ethyl *n*-butyrate, isoamyl *n*-butyrate, isopropyl *n*-butyrate*, isobutyric acid, ethyl isobutyrate, isobutyl isobutyrate, *n*-valeric acid, ethyl *n*-valerate, isovaleric acid, ethyl isovalerate, *act*-amyl isovalerate, caproic acid, ethyl caproate, isoamyl caproate, *act*-amyl caproate, isobutyl caproate, β-phenylethyl caproate, enanthic acid, ethyl enanthate, caprylic acid, methyl caprylate, ethyl caprylate, *n*-propyl caprylate, isobutyl caprylate, isoamyl caprylate, pelargonic acid*, ethyl pelargonate, capric acid, methyl caprate, ethyl caprate, isobutyl caprate, isoamyl caprate, *act*-amyl caprate, ethyl undecanoate*, lauric acid, ethyl laurate, *act*-amyl laurate, myristic acid, ethyl myristinate, isoamyl myristinate, *act*-amyl myristinate, ethyl pentadecanoate, ethyl palmitate*, methyl anthranilate, ethyl benzoate, ethyl cinnamate, and ethyl salicylate*.

Hexanol (hexyl alcohol) is identifiable as an odor at 0.001 mg per liter, according to Drawert *et al.* (1965*b*). It is reported to have a woody odor. A small amount seems to be formed during yeast fermentation, but the main source is by reduction of 2-hexen-1-al. This compound is normally present in grapes, the amount varying with variety and maturity.

The quantity of oxidizable volatile materials has been used to determine the quality of grape and fruit juices (Koch and Breker, 1953;

Deibner and Rifaï, 1958). Since this measures mainly ethanol and acetaldehyde it does not, in our opinion, give useful information on the quality of the juices.

Esterification

The bouquet is due in part to the harmonious blending of the large numbers of esters, produced from the various alcohols and acids, with other aromatic constituents. See also p. 445 f.

Peynaud (1937) reported that the formation of esters by chemical reactions during aging is about as great as the formation of esters by the esterases present in yeast and bacteria. Chemical esterification, however, yields mainly acid esters, and biological esterification yields mainly neutral esters (Ribéreau-Gayon and Peynaud, 1936).

The total content of esters in wine is determined by its composition and age; it varied, in French wines examined by Peynaud, from about 176 to 264 mg per liter in new wines, to 792 to 880 mg in old wines, calculated as ethyl acetate. Amerine (1944) determined only the neutral esters in commercial California wines, and found the total neutral esters (mg per liter), calculated as ethyl acetate, to be:

Wine	Number of samples	Total	Volatile
Burgundy	19	356	145
Zinfandel	19	300	113
Sauterne	10	137	65
Dry sauterne	21	174	120

While the early enologists (Rocques, 1899) believed the esters to have the most desirable odors of wine, the work of Peynaud (1936) indicates that for table wines the volatile neutral ester, ethyl acetate, is generally an undesirable aromatic constituent. (See also Legent-Fournès, 1944.)

WINE IN THE NORMAL DIET[3]

. . . offers certain advantages.

The ethanol in wines, as in beer and distilled spirits, is the major constituent which provides the euphoric, analgesic, vasodilating, relaxing, sedating, and even intoxicating properties of these beverages. These physiological effects, however, may be significantly modified by the other components which accompany the ethanol, and which occur in strikingly different concentrations in the various beverages. Many studies

[3] This section was written by the late Professor Emeritus Agnes Fay Morgan. For a more complete text see Proc. Congrès Med. Intern. Bordeaux, 25–27 Sept. 1961: 74–100 and Wines & Vines 49 (8):22–24. 1968.

have been aimed at the identification of substances which favorably or unfavorably influence the acceptability of the beverages as well as their physiological effect.

The great variety of chemical constituents in wines is due to differences in the composition of the soil and the climate of the vineyards; the variety of the grapevine; the kind and amount of insecticide used; the maturity of the grapes at harvest; the processes used in fermentation, including the amount and kind of yeast employed: the absence or presence of stems and skins, the amount of sulfur dioxide employed, the use of various clarification and filtration procedures, and the application of baking, racking, aging, and other operations well known to enologists. These factors are mentioned here in order to emphasize the difficulty of interpreting the results of physiological research involving a wine of unspecified type and mode of production. Even when the wine under investigation is well described and samples are carefully prepared, replication of results is often difficult, undoubtedly because of variation in composition.

The main questions concerning the metabolic value of wine in the daily diet are these: (1) What is the concentration range in wines of such recognized nutrients as amino acids, inorganic elements, vitamins, and carbohydrates? (2) How are these nutrients affected by grape variety, vinification process, and other factors? (3) Are these nutrients in wines fully utilized in the body, and are the differences in the concentrations in various wines of physiological significance? (4) Is the ethanol of wine indistinguishable in its physiological effects from the ethanol in other beverages? Much work has been done on each of these problems, but no conclusive answers are available. The reason for our persisting uncertainty, especially where physiological effects are concerned, may be tied in some way to differences in the concentrations of non-ethanol constituents found in every wine tested, even in those of supposedly identical origin.

Inorganic elements

Traces of many metallic ions as well as of anions, such as the halogens, sulfites, sulfates, carbonates, and phosphates, have been found in musts and wines. The most troublesome of the metals to the wine maker are iron and copper, often the cause of cloudiness. Iron, however, may also be particularly important to the nutritionist; most wines contain 5 to 30 mg per liter of iron, generally in the ferrous form and thus physiologically available. Calcium in natural wines, not treated with calcium sulfate, varies from 50 to 200 mg per liter; this amount may have some nutritional importance, since the daily calcium requirement of adults may be as low as 200 to 400 mg. The possible significance of the magnesium content of wines and other alcoholic beverages has been suggested in research of Gottlieb *et al.* (1959).

The sodium and potassium values of wines have often been determined, in recent years, chiefly by means of the flame photometer. The metabolic significance of these metals has been recognized in connection with dietary restrictions in control of edema associated with heart, liver, and kidney ailments. Lucia and Hunt (1957), who analyzed 155 red and white California table wines and 104 dessert wines containing less than 10 mg of sodium per 100 ml, and relatively large amounts of potassium, concluded that these should be desirable additions to the often monotonous low-sodium diets prescribed by many physicians. (See also p. 459.) For a recent report on the possible health values of minerals (potassium, magnesium, calcium, sodium, iron, manganese, zinc, phosphorous, sulfur, chloride, boron, silicon) and trace elements (aluminum, fluoride, copper, vanadium, iodine, cobalt and molybdemum) see Kliewe and Eschnauer (1967).

Vitamins

Relatively little research has been recorded on these nutrients in alcoholic beverages, even though vitamin deficiencies are frequently involved in the nutritional disturbances of the excessive drinker.

Wine grapes are usually rather poorly endowed with ascorbic acid, from 1.2 to 18.3 mg per liter of juice, and the musts are likely to lose much of this substance in the course of vinification. Analyses reported from Russia, France, Germany, Portugal, Italy, and California all indicate variable and rather small amounts occurring naturally in grapes and musts, but only 2 to 5 per cent are retained in young wines. Ascorbic acid added to wines is said to be lost rapidly, with only 7.5 per cent left after two months.

In grapes, as in all foods of vegetable origin, carotene is the pigment from which vitamin A is formed in the animal body. Carotenoid pigments are notably lacking in grapes and consequently in the musts and wines.

B-Vitamins

The group of vitamins now generally discussed together as B-vitamins is represented fairly well in grapes, musts, and wines. Their presence may be considered a significant characteristic distinguishing wines from distilled spirits and even from beer. Most of the significant studies of wines as vitamin carriers have been concentrated on these substances. Not only the original endowment from the grape but also the additions and subtractions occurring during juice extraction and yeast multiplication must be taken into account. (See p. 469.)

Only five members of this group—thiamine, riboflavin, pantothenic acid, pyridoxine, and nicotinic acid—may be considered indispensable in the diet. There are several others, including biotin, folic acid, vitamin B_{12}, and possibly inositol and p-aminobenzoic acid, which have vitamin-like functions and occur in foods, but may be more or less adequately supplied through intestinal biosynthesis.

Fresh grapes contain measurable amounts of thiamine which increase during the maturation process, and may reach 500 μg or more per liter of juice. One study on California grape juices and wines (Morgan et al., 1939) showed that fresh grape juices contained 120 to 570 μg of thiamine per liter, but less than half these amounts remained after six months of frozen storage. Wines from these grapes contained 50 to 120 μg after one to twelve months storage, or only one-third to one-fourth the amount present in the fresh musts.

Sulfiting rapidly destroys thiamine, and most of it is removed during clarifying with bentonite clay. In a study by Perlman and Morgan (1945) in which both rat growth and thiochrome methods were used for assay, it was again noted that thiamine is lost during fermentation and after sulfite treatment, in proportion to the amount of free sulfur dioxide present. The addition of 1 mg of thiamine to each liter of grape juice or wine before vinification resulted in 85 to 100 per cent retention of the vitamin after three months' storage. When the extra vitamin was added after vinification, the wine retained 69 per cent. Since it is well established that yeasts use much thiamine during multiplication, the low content of thiamine in wines must be expected even without sulfiting or clarifying.

The analyses by Hall et al. (1956) of California grapes, musts, and wines indicate a fair amount of thiamine present in fresh grapes, averaging 410 μg per kg in three white varieties and 490 in four red varieties. Concentrations of 340 and 500 μg per liter, respectively, were found in the corresponding musts, and only 10 and 20 μg per liter in the corresponding dessert wines. Again it is obvious that some loss occurs in must production, and almost complete loss during vinification.

The amount of riboflavin in grapes and grape juices is relatively small. It has often been noted that the yeast cells provide extra riboflavin rather than consume it. In wines the riboflavin content does not seem to be affected by normal sulfiting and aging, but it is influenced by light. One of the California studies of Perlman and Morgan (1945) involved the addition of 1 mg per liter of riboflavin to juice or wine, and observation of its retention during storage in the light and in the dark, and at refrigerator or room temperatures. Only 5 to 35 per cent of the riboflavin was retained in the juices and wines bottled in clear glass, but those kept in dark glass retained 68 to 95 per cent of the vitamin at room temperature and slightly more at refrigerator temperature. Even after four years at room temperature, substantial amounts of the riboflavin were found in the juices and wines in dark bottles (fig. 71). Both excessive sulfite treatment and total omission of sulfur dioxide were found to produce excessive loss of the vitamin.

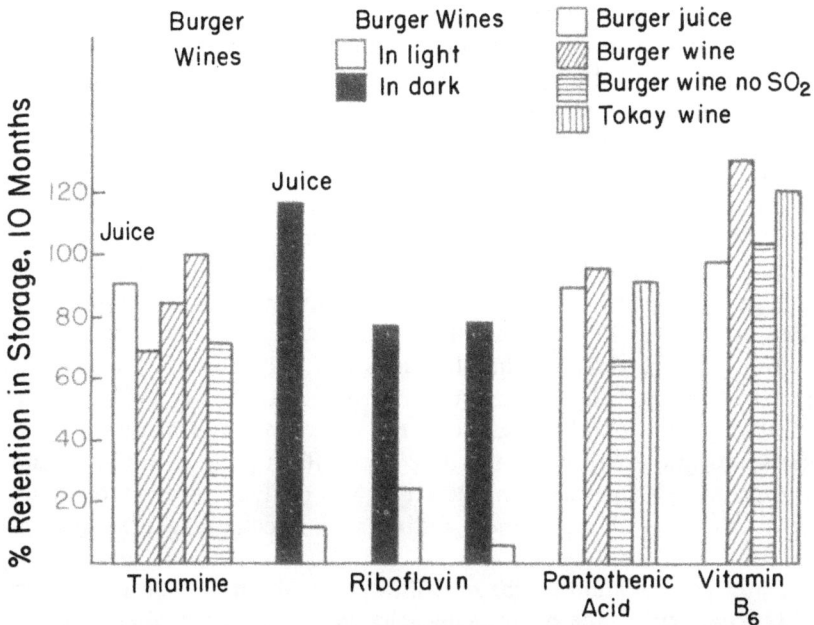

Figure 71. Effect of storage in four B-vitamins in grape juices and wines and the riboflavin content as affected by light.

Hall *et al.* (1956) determined the riboflavin content of several varieties of red and white grapes, and their respective musts and wines. These red and white grapes were found to contain only 240 and 220 μg per kg, respectively, with about half this concentration in the musts and slightly less than half in most of the wines. Exceptional were the sweet red wines, which contained an average of 240 μg per liter, about the same amount as in the grapes. This higher endowment of sweet red wines, which has been noted also for thiamine, nicotinic acid, and vitamin B_6, was ascribed to the shorter period of fermentation, the presence of the grape skins in the crush, and the protection from light afforded by the pigments.

Although vitamin B_6, including pyridoxine, pyridoxal, and pyridoxamine, occurs commonly in grapes and wines, it has been assayed less frequently than have thiamine and riboflavin. This vitamin is affected by claryifying and fining but not by aging. Perlman and Morgan (1945) found that pyridoxine added to grape juices and wines in the amount of 1 mg per liter was retained during eighteen months of storage at room temperature; in fact, increases in the content of this vitamin were often seen.

Hall et al. (1956) reported that 24 and 33 per cent of the pyridoxine were lost in must production for white and for red grapes, respectively, 43 to 72 per cent in the vinification of the dry white and red wines, but only 24 per cent in the vinification of the sweet red wines. It seems obvious that the grape and its products are well endowed with vitamin B_6, but much of this valuable substance is lost by present wine-making processes.

Relatively few studies have been conducted on the presence of pantothenic acid in grapes, musts, and wines. Perlman and Morgan (1945) found concentrations of 1.50 and 0.89 mg per liter in Burger and Tokay grape juice, respectively, and 0.4 to 1.10 in the wines made from them. When 10 mg per liter of pantothenic acid was added, from 66 to 100 per cent was retained during fermentation and storage. No adverse effect from sulfur dioxide treatment, fermentation, or aging was seen. Hall et al. (1956) found 0.68 and 0.85 mg per kg in four red and six white varieties of grapes, respectively, 0.63 and 0.76 in their musts, 0.37 and 0.12 in the dry white table and dessert wines made from them, 0.29 and 0.37 in dry red and rosé wines, and 0.24 in sweet red wines (figs. 72 and 73).

During fermentation of these California wines, it was observed that 41 to 80 per cent of the pantothenic acid disappeared, probably owing in part to the yeast varieties used and to other differences in the treatment of the musts and wines. Thus, as for thiamine, losses of pantothenic acid are small during must extraction but large during fermentation. An investigation of the factors causing these losses seems worth while.

Approximately 1.80 and 2.07 mg per liter of niacin have been found in California white and red musts, respectively, 0.41 to 0.96 in the dry white and red wines made from them, and 2.05 in the sweet red dessert wines. About 60 per cent of the vitamin is thus lost during fermentation in production of the dry wines, but none in production of the sweet red wines.

The antianemia vitamins, folic acid and vitamin B_{12}, are generally considered to be made available in part from intestinal bacterial activity, although their presence in many foods has been well established. Grapes contain 4 to 10 μg per 100 gr of folic acid, but only one-third to two-thirds of this amount remains in the musts and in the wines (Hall et al.,

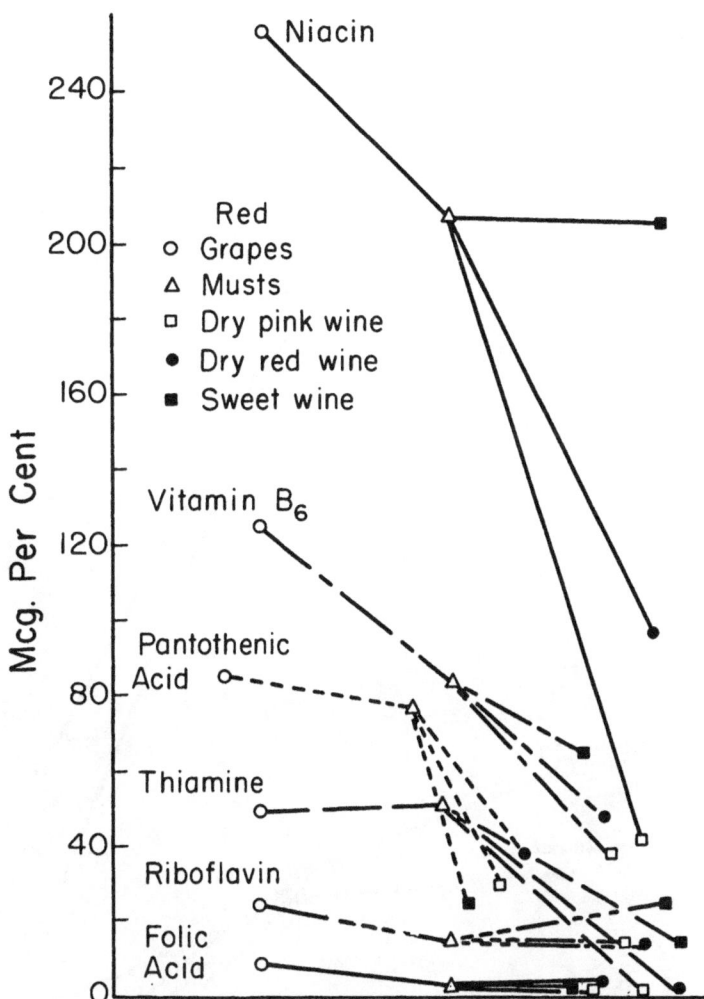

Figure 72. Changes in B-vitamins during production of various types of red wines (mcg per 100 gr).

1956). The amount of folic acid in sweet red wines has been found to be higher than that in any other wines tested. Folic acid is sensitive to light, to clarifying compounds, and to sulfite.

Vitamin B_{12} is not usually found in foods of vegetable origin, but some B_{12}-like activity has been seen in red and white grapes, musts, and wines, with concentrations ranging from 9 to 25 μg per kg or per liter. The

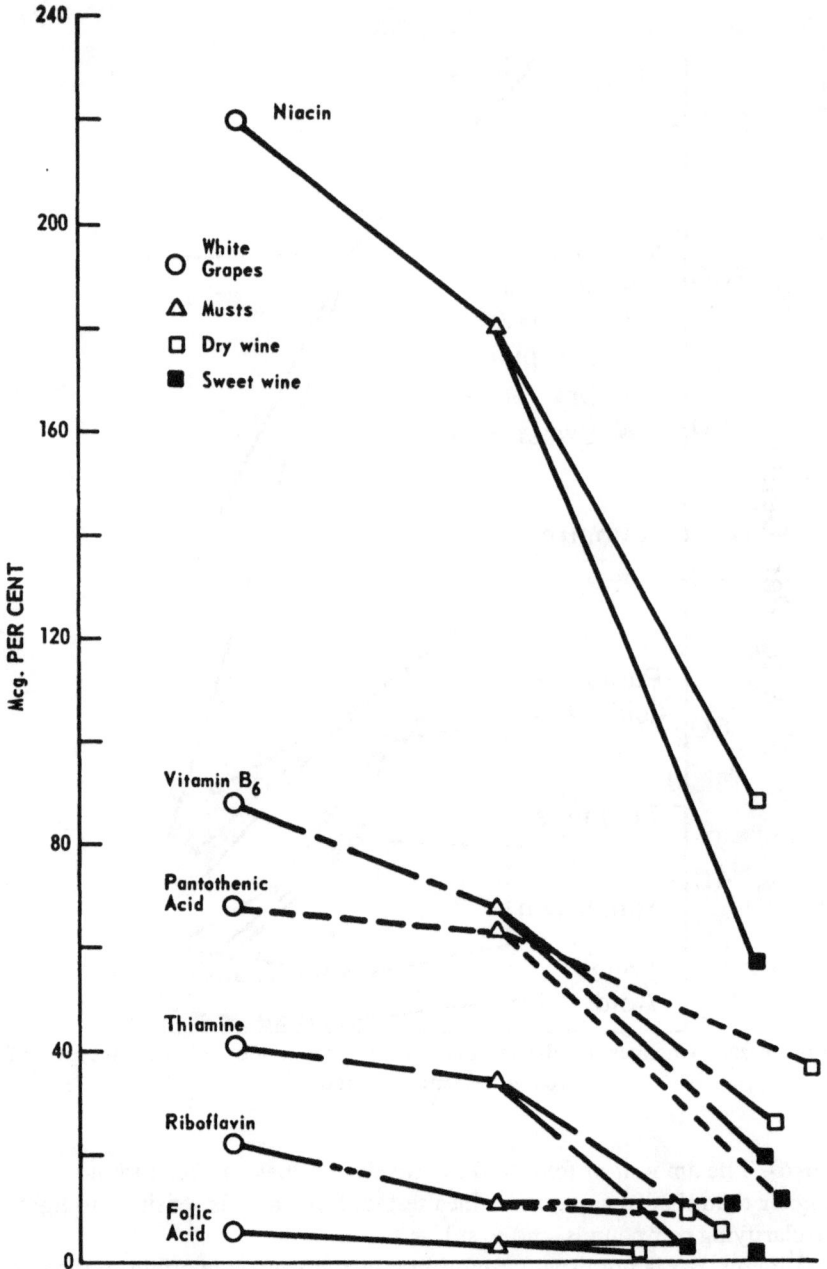

Figure 73. Changes in vitamins during production of various types of white wines (mcg per 100 gr).

red wines tested had three to five times as much of this activity as did the white wines.

In summary, the analysis of the six B vitamins in must production from red and white grapes, and in production of dry white, rosé, and red wines and of sweet red wines, as reported by Hall *et al.* (1956), indicates relatively complete loss of thiamine in vinification, large losses of ribo-flavin and folic acid in must production but conservation or increase during fermentation, and the superior conservation of all such vitamins except pantothenic acid in the sweet red wines (fig. 72). The same differ-ence was not found in the production of white sweet wines (fig. 73). If wines were to be fortified by added vitamins, then thiamine and panto-thenic acid should be considered as the most important possibilities. These, along with other B vitamins, are remarkably stable in wines under proper storage conditions. For the possible vitamin P content of wines see Lavollay and Sevestre (1944).

Utilization of alcohol calories

Another aspect of the B-vitamin endowment of wines is the contribution of these vitamins to the utilization of the calories in the beverages. For such caloric use, all the wines are inadequate as to thiamine, but they provide 13 to 18 per cent of the needed riboflavin, 59 to 94 per cent of the pyridoxine, 2 to 5 per cent of the pantothenic acid, and 5 to 14 per cent of the nicotinic acid.

The extent of utilization of alcohol calories has often been considered. Some investigators have indicated that only 65 per cent of these calories are available for rat growth; others have assumed that 100 per cent of the calories are available. Morgan *et al.* (1957) concluded that only 75 per cent could be utilized by growing rats and hamsters, since the growth rate of young alcohol- or wine-fed animals was found to correspond to the provision of only about 5.25 calories from each gram of alcohol, rather than to the theoretical value of 7 calories per gram (fig. 74).

When rats or other animals are offered no choice between water or an alcohol solution, but are given only one or the other *ad libitum* as the sole source of fluid, those on alcohol show a significantly lower rate of growth. This has been attributed by some observers to a direct growth-inhibiting effect of the alcohol. When such alcohol solutions are the only fluids provided, however, the intake of dry diet is reduced immediately, undoubtedly because of the inhibiting effect of alcohol on fluid intake and the reduction of total water intake.

The importance of this phenomenon was demonstrated by Morgan *et al.* (1957), who found that when the water intake of control water-fed rats is restricted to the same amount of water taken by animals receiving

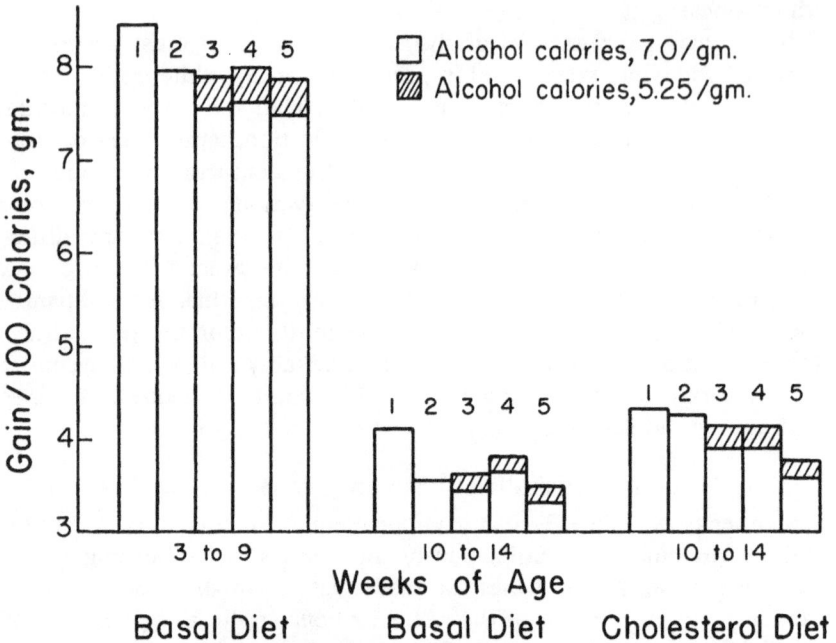

Figure 74. Gain in weight per 100 calories of food: 1, unrestricted water intake; 2, water restricted to that taken by alcohol- and wine-fed rats; 3, 15 per cent alcohol; 4, 15 per cent alcohol dry red wine; 5, 15 per cent alcohol dry white wine. For the 10–14 week period same without and with cholesterol added. Shaded part of columns indicates values when utilization of three-fourths of alcohol calories is assumed; unshaded, values when 100 per cent utilization is assumed.

alcohol solutions or wine, the food intake and growth rates of the two groups are the same.

Under such conditions the restriction of water may act as a stress factor, with a resultant increase in ACTH production, reduced adrenal ascorbic acid and cholesterol levels, increased nitrogen excretion, increased blood sugar levels, and increased total liver lipids and liver cholesterol. Most of these changes have been noted in rats and guinea pigs subjected to intoxicating doses of alcohol. Morgan and coworkers found some of these effects in rats given alcohol solutions and wine, as well as in rats on restricted water intake.

The theoretical amount of carbohydrates (sugars), calories, and sodium (mg) derived from servings of 4 ounces of table wines was summarized as follows by Leake and Silverman (1966): dry red, 0.3, 95, and 10;

rosé, 1.3, 94, and 19; dry white, 0.4, 90, and 14; sweet white, 4.9, 110, and 12; sparkling wine, 1.8, 99, and 8. Kosher-type wines give as much as 131 calories per 4-ounce serving and dessert wines 156 to 196. In contrast, the calorie content of beer amounts to 162 to 211 per 12-ounce serving and of whisky or brandy to 107 calories per 1.5-ounce serving of 86° proof. Distilled spirits contain very little sodium (from 1.4 to 18 mg per liter) or potassium (20 to 28 mg per liter). Data on the iron and calcium content are not given by Leake and Silverman (1966), but are known to be low. (See p. 456 f.)

Cholesterol metabolism and use of wine

Certain of these effects of alcohol use or of water restriction may have clinical significance. For example, there has been an impression among some medical men that the use of alcohol tends to decrease the occurrence and severity of atherosclerosis, perhaps by influencing liver metabolism and cholesterol levels. In contrast, there is statistical evidence pointing to an increase in liver damage, especially increased cirrhosis, among alcoholics. Much research, both experimental and clinical, has been conducted on these important and possibly related problems. It has been shown experimentally, however, that liver cirrhosis is not necessarily the direct result of alcohol ingestion, since the same type of liver damage is produced on the same basal diet when excessive sugar is given instead of excessive alcohol (Best et al., 1949). Some investigators have suggested that the liver damage results from lack of sufficient lipotropic factors in the diet; while others have attributed most of the pathological changes, especially those noted in alcoholics, to vitamin deficiencies due to the low intake of solid food.

There is evidence, nonetheless, that alcohol may influence the liver. Since blood β-lipoproteins and cholesterol originate in the liver, and these blood constituents are generally believed to be significantly related to the development of atherosclerosis, it has seemed desirable to examine the effects of alcohol on cholesterol levels. In particular, it has seemed necessary to study the effects of cholesterol-rich diets, already known to produce arteriosclerotic lesions in many species, in the presence or absence of alcohol and of specific alcoholic beverages such as whisky, beer, or wine.

In comparing the effects of water, alcohol solutions, and wines, we therefore fed young rats and hamsters either normal or cholesterol-rich diets with (1) water ad libitum, (2) water restricted to the amount taken by the alcohol- and wine-fed animals, (3) a 15 per cent alcohol solution, (4) a dry red wine with 15 per cent alcohol content, or (5) a dry white wine with 15 per cent alcohol content.

The growth of the alcohol and wine-fed groups was about the same as that of the water-restricted rats on the basal diet (fig. 75). However, the cholesterol-rich diets produced large increases in the levels of liver fat, liver cholesterol, adrenal cholesterol, and serum cholesterol (in the hamsters), as shown in figure 75. These increases were significantly *less* in the wine-fed groups than in those given water *ad libitum*, restricted water, or comparable solutions of alcohol (Morgan *et al.*, 1957).

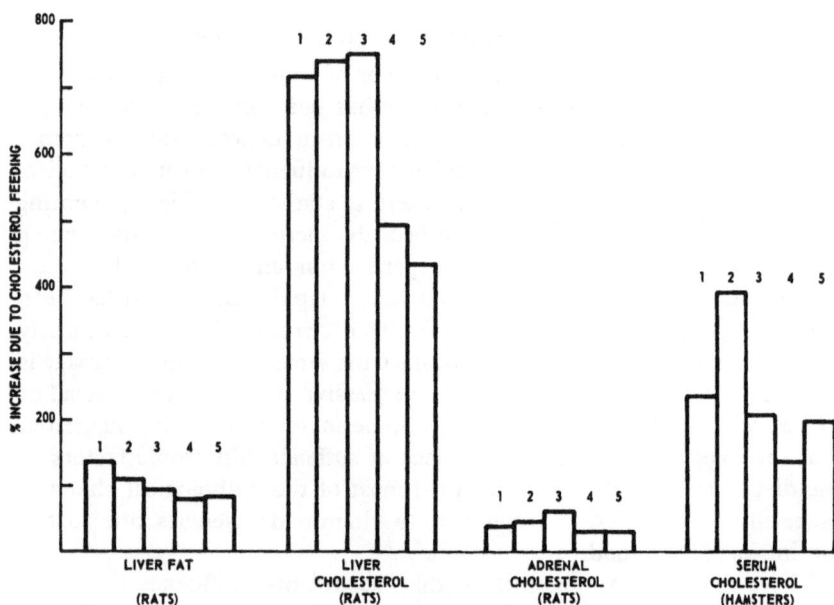

Figure 75. Increases in liver fat and in liver, adrenal, and serum cholesterol on various diets. Basal diet *ad libitum* for all: 1, water *ad libitum*; 2, water restricted to amount taken by ethanol group; 3, 15 per cent ethanol; 4, 15 per cent alcohol dry red wine; 5, 15 per cent alcohol dry white wine.

Whatever the clinical importance of these findings, it is obvious that alcohol solutions and wines of the same alcohol content do not necessarily have the same physiological effects. The differences may be presumed to come from inhibiting or stimulating compounds in the wines which are yet to be identified. For recent articles on the use of wine in therapy see Lucia (1954, 1963a, 1963b), Leake and Silverman (1966, 1967), Eylaud (1960), and Funk and Prescott (1967). For a perceptive review of the possible nutritional values of wine and of the possible deleterious effects of processing see Flanzy (1961) and Flanzy *et al.* (1955).

Masquelier (1956) considers flavones and leucoanthocyanins to have vitamin P activity. Cider had an especially high content of the latter and Bordeaux white of the former. Beer had less of both. Masquelier and Laparra (1967) have reported a choleric effect from cinnamic constituents. See Diemair and Diemair (1966) for data on the presence in wines and possible toxicity of histamine. (See also pp. 468–469.)

Masquelier (1959) considered that bactericidal action of wines is due to a combination of chemical and physical factors. He reported that caffeic acid and flavonoid pigments do not play an essential role. Since pigment extractions made directly from grapes show relatively little bactericidal action, the bactericidal compound (or compounds) present apparently occurs in the grape as inactive precursor (or precursors). Malvidin apparently had the highest bactericidal activity. During aging the anthocyanins are slowly hydrolyzed and the aglycones liberated. The aglycones are eventually degraded and precipitated. Masquelier reported maximum bactericidal activity during the first ten years of aging and decreasing activity thereafter.

Somaatmadja et al. (1965) identified a leucocyanidin and a leuco-delphinidin in grapes which inhibited reproduction of Staphylococcus aureus, Lactobacillus casei, and Escherichia coli.

CHAPTER ELEVEN

DEVELOPMENT AND STABILIZATION OF WINE

While partly fermented newly made ciders and new wines are consumed during the season they are made (the tradition of serving "heuriger" wines is well established in Austria, France, Germany, and Switzerland), only clear, properly aged wines are widely accepted and esteemed. Settling, racking, fining, filtering, and other stabilization operations are used to promote development and to age and clarify the wine before marketing.

Wine is transferred from fermenting vats into storage tanks and vats to complete the alcoholic fermentation, to undergo a secondary fermentation (the malo-lactic fermentation, traditionally called the "secondary" fermentation), and to undergo aging and stabilization. The existing demand, both in California and elsewhere, is for a brilliantly clear and well-stabilized wine. The presence of sediment or development of haze during marketing and chilling for table use is considered by American consumers as evidence of incipient spoilage or instability. This is not always true, for some of the more delicately flavored, less stable old wines and even young wines which often contain considerable sediment are still sound and of acceptable and sometimes excellent flavor.

Newly fermented wine is usually cloudy and may have a pronounced yeasty odor. The purpose of aging is to facilitate the natural clarification of the wine and to bring about certain desirable changes in color, odor, and taste. While wine is being aged in the wood, settling of suspended matter and racking (the transfer of wine from one container to another) help both in the clarification and in aeration of the wine.

Mercz et al. (1963) have made a study of the size of particles in musts and wines. In musts and new wines before the first racking the particles could be classified into three groups: 0.7 to 2.1 μ, 2.1 to 6.2 μ, and over 6.2 μ. The first two predominate in musts and new wines, but during full fermentation most particles are yeasts and over 6.2 μ. Within two or three months and with normal racking the particles fall in three groups: 0.13 to 1.3 μ, 1.3 to 4.2 μ, and over 4.2 μ, with most in the last two groups.

Should natural clarification not occur, some artificial method or inducement for clarification must be used. The more common method is to clarify with an organic or inorganic substance—a process commonly called fining. Filtering is also widely used; and pasteurizing, heat stabilizing, and chilling all have their place. Some of these processes significantly accelerate or modify the chemical changes which normally take place during aging.

SETTLING AND RACKING

*Transfer of wine from one container to another constitutes
an important phase of clarification and aging.*

After fermentation nearly ceases, the yeast and suspended particles of skins and pulp settle rapidly and form a sediment known as the first or crude lees. When the fermentation has reached this stage, the new wine is drawn off (racked) from the lees to aid in clearing and to avoid extraction of undesirable flavors from the old or dead yeasts. Prompt removal of the yeast cells also improves the keeping quality of the wine by preventing release of cellular constituents such as amino acids, proteins, and vitamins from the yeast. The racked, partially cleared wine is stored in completely filled and sealed tanks which are kept full during storage. Racking is repeated regularly to aid in clearing. During racking, the wine loses the carbon dioxide with which it is charged and absorbs the oxygen necessary to aging. Particular care has to be taken in the first racking to prevent agitation leading to sudden release of dissolved and occluded carbon dioxide which would stir up the sediment and cloud the wine. White wines to be marketed with a small amount of carbon dioxide, the so-called *pétillant* wines, particularly when this is obtained by a secondary malo-lactic fermentation, have to be specially racked and handled under a blanket of carbon dioxide gas. No oxidation and consequently little aging will occur so long as the wine is charged with carbon dioxide. Overaeration during racking should be avoided, particularly with white wines, which mature best under reducing conditions. By varying the method of transfer and the amount of aeration, the oxidation can be controlled to suit the type and condition of the wine.

To increase aeration, allow the wine to run into a tub or sump and then pump it to the top of the tank being filled; or allow air to enter the hose with the wine and pump it to the bottom of the new tank. To reduce aeration, pump the wine directly from the bottom of the tank from just above the crude lees to the bottom of the new tank. Aeration can be further reduced by handling the wine under a blanket of carbon dioxide and racking into a tank where the air has been replaced with carbon dioxide; with this method the pumps should be selected and operated to prevent aeration.

Certain wines, particularly those of high acidity, are sometimes left on the lees to promote the malo-lactic fermentation, and the cellars are kept warm until the desired degree of deacidification has been achieved. The slight gassiness noted in some red wines early in the spring as the cellar temperature starts to rise may be due to organisms fermenting malic acid rather than to fermentation of residual sugar by yeast. Since their activity results in a decrease in acidity, such bacterial fermentations are usually not desirable under California conditions. Racking and the addition of sulfur dioxide in moderate amounts are usually adequate for control of bacterial spoilage, but periodic checks of the total and volatile acidity and the pH should be made.

MALO-LACTIC FERMENTATION

The "great" wines of the world are usually the product of
mixed rather than pure culture fermentation.
Secondary fermentations are common in the production of ale,
flor sherry, the bottle fermentation of sparkling wines,
and the biological reduction of excess acidity in some wines.

Bacterial fermentation of malic acid and its conversion into lactic acid has long been recognized, and is promoted in the cooler viticultural regions of the world where grapes mature with excessive amounts of malic acid. Malo-lactic acid fermentation is an established feature of wine making in Austria, France, Germany, Portugal (in at least one region), and Switzerland for both red and white wines. It also occurs to some extent in wines of many other countries.

The bacterial conversion of the excess malic acid in these wines into lactic acid and carbon dioxide has been known since 1890. The historical development of our knowledge of the bacteria involved and the chemistry of transformation is discussed in detail by Vaughn (1955), Schanderl (1959), Ribéreau-Gayon and Peynaud (1964–1966), Amerine et al. (1967), Kunkee (1967b), and Amerine and Kunkee (1968). The early contributions of Seifert (1901, 1903) in Austria and of Müller-Thurgau and Osterwalder (1913, 1917, 1919) in Switzerland were continued and expanded by Lüthi (1957a, 1957b). Lüthi (1957a) emphasized the fact that malo-lactic acid fermentation plays an important role in Swiss wine making, since it is necessary for the production of acceptable Swiss wines. This secondary fermentation was also recognized to occur in, and to account for the quality of, the great red wines of Bordeaux (Ribéreau-Gayon, 1936, 1946; Ribéreau-Gayon and Peynaud, 1938a, 1938b; Peynaud and Domercq, 1961). Ferré (1928a) noted the importance of this fermentation in the production of the best wines of Burgundy. Lüthi (1957a) did not consider the biological deacidification as always necessary, and considered it harmful to the quality of the sweet white wines of Bordelais whose acidity was regulated by the "noble rot" *Botrytis cinerea* (see p. 641). Charpentié (1954), however, pointed out the importance of a malo-lactic fermentation for white Bordeaux wines made from nonbotrytised grapes. Such wines have greater softness and mellowness without the high sulfur dioxide content of wines with residual sugar. But he emphasized that every wine constitutes a different problem as to the desirability or nondesirability of a malo-lactic fermentation.

Malo-lactic acid fermentation is now known to occur in Australian, Austrian, Canadian, Spanish, Jugoslavian, and other wines as well as those of the countries indicated above. Fornachon (1957, 1963a, 1963b, 1964) investigated various aspects of malo-lactic fermentation in Australian wines. In 400 samples of Australian wines which were collected in 1953–1955, Fornachon (1957) reported that 46 per cent of the dry table wines had undergone a malo-lactic fermentation within eight months. Only 4 per cent of the dessert wines had undergone the secondary fermentation. After reviewing the general problem of malo-lactic fermentation, Fornachon (1963b) stated that for Australian conditions species of *Leuconostoc* were the most suitable organisms. Adams (1964) reported very general occurrence of a malo-lactic fermentation in Canadian wines. However, he determined this on the basis of the presence of lactic acid and not on the reduction or disappearance of malic acid. In Rumania Alexiu *et al.* (1967) noted that some wines with low free sulfur dioxide did not undergo a malo-lactic fermentation.

For an early report of control of the process in California see Vaughn and Tchelistcheff, 1957. In the cooler regions and seasons in California, a reduction of acidity by means of a malo-lactic fermentation may be desirable; but since many of our wines are already too low in total acidity, this is not now generally recommended for California conditions. Suverkrop and Tchelistcheff (1949) observed malo-lactic fermentation in Napa Valley red wines, and believed that these wines improved as a result. Amerine (1950a) observed the formation of large quantities of lactic acid in wines during or immediately after the primary fermentation. During storage, formation of lactic acid was accompanied by the disappearance of malic acid. He stressed that such malo-lactic acid fermentation should be prevented in California wines of low total acidity.

In 144 California commercial wines, Ingraham and Cooke (1960) reported that over half had undergone a malo-lactic fermentation as evidenced by the disappearance of malic acid. The malo-lactic fermentation was much more common in red table wines (75 per cent) than in white table wines (32.1 per cent) and rosé (12.5 per cent). They believed that there was a causal relationship between the malo-lactic fermentation and the quality of red table wines. However, most of the high-quality wines tested were from the cooler coastal districts, and other quality factors were probably present. Kunkee et al. (1965b) also found that most commercial wines from southern California had undergone a malo-lactic fermentation. Most of the wines which had undergone this fermentation were high in acetoin (acetylmethylcarbinol). Six of the nine high acetoin wines had 4 mg per liter or more of diacetyl.

Bobadilla and Navarro (1949) reported on the destruction by respiration and changes in organic acids during fermentation and aging. They emphasized the importance of lactic and succinic acids in contributing to the body and flavor of wines. In Jerez wines malo-lactic acid fermentation begins in the last stages of the alcoholic fermentation and continues to the first racking. This reduces the initial malic acid content to one-third its initial level and improves the sensory quality of the wine. Malic acid is considered by Bobadilla and Navarro to be less desirable than lactic acid. They report lactic and succinic acids to be most important for body and flavor of Jerez wines, and certainly more important than citric or acetic acids.

Differences in the sensory characteristics of wines fermented with several strains of bacteria were demonstrated by Pilone and Kunkee (1965). However, the panel used could not consistently rank wines with and without a malo-lactic fermentation in the order of their quality.

The nature of the improvement in sensory quality by malo-lactic fermentation is not completely known. While the reduction in acidity and the formation of carbon dioxide are important contributions, other factors may be involved, such as maintenance of reducing conditions at later stages of yeast fermentation and storage, and absorption of certain constituents by the bacteria in the formation of secondary products. Certainly, if the activity of the bacteria were limited to a decrease in acidity, this would rarely be desirable under the usual California conditions.

For a discussion of the loss of color in red wines during the malo-lactic fermentation see Vetsche and Lüthi (1964).

Bacteriological aspects

The cocci associated with the malo-lactic fermentation were recognized as species of Micrococcus by Seifert (1903), Müller-Thurgau and Osterwalder (1913, 1917), and Arena (1936). Vaughn (1955) pointed out that, on the basis of the published information concerning their morphology and physiology, it is questionable whether they should be placed in the genus *Micrococcus*. Since they are homofermentative lactic acid-producing cocci which are catalase-negative, Vaughn (1955) favored their classification in the genus *Streptococcus* or *Pediococcus*. Lüthi (1957a), however, accepted the suggestion of Müller-Thurgau and Osterwalder that the causative organism be designated as *Bacterium gracile*.

Fornachon (1957) reported *Lactobacillus hilgardii* and *Lact. brevis* in Australian wines undergoing the malo-lactic fermentation. Later (1964) he identified *Leuconostoc mesenteroides*. In general, this strain tolerated a low pH better than species of *Lactobacillus*. Since these are the wines that most need a malo-lactic fermentation the importance of this bacteria is emphasized.

Lambion and Mekhi (1957) isolated forty-seven strains of lactic acid bacteria from twenty-six French wines. The rods appeared to be *Lactobacillus plantarum*. The rods provoking a malo-lactic fermentation in wines were identified as *Lact. plantarum* var. *gracile* (Müller-Thurgau) Orla-Jensen. The heterofermentative cocci were characterized best as *Leuconostoc mesenteroides*, which they proposed to call *Leuc. mesenteroides* var. *gracile* (Van Tieghem) Bidan. Homofermentative cocci may be *Streptococcus vini* Megula. Ozino (1967) found that the microflora during the malo-lactic fermentation is not exclusively composed of lactic acid bacteria. She found of 151 strains, 43 of *Acetobacter* (*A. roseus* and *A. pasteurianus*) and 108 of *Micrococcus* (*M. varians*).

The classification of the bacteria associated with decarboxylation of malic acid is still difficult and often confusing. As Fornachon (1943) has noted, it is normal for newly isolated cultures of the lactic acid bacteria to differ from the presently classified species. Barre and Galzy (1960) also found this to be true, and similar conclusions were reached by Bidan (1956) and Dupuy (1957c).

The possibility of using selected cultures of lactic acid bacteria was posed by Bidan (1956). The cultures were from northern (Alsace), central (Loire), and southern (Gaillac) regions of France. Most investigators using lactic acid bacteria have found that the cultures isolated were somewhat atypical of those previously identified. Thus Bidan's (1956) *Lactobacillus fermenti* had a different optimum temperature and utilized xylose, in contrast to the type culture. Bidan's *Lact. fermenti* resembled *Lact. hilgardii* and the *Bacterium intermedium* of Müller-Thurgau. Another culture resembled *Leuconostoc citrovorum*, but differed from it in several physiological characters. The classic *B. gracile* he proposed calling *Leuconostoc gracile*. Still another culture he found to be a *Pediococcus*, for which the name *P. vini* was proposed. This report of *Pediococcus* in wine clearly antedates the report of Gini and Vaughn (1962), but Vaughn's (1955) suggestion (see above) preceded Bidan's.

Bidan believed it to be the same as the *Micrococcus variococcus* of Müller-Thurgau and Osterwalder (1913, 1917) or the *M. malolacticus* of Seifert (1903).

Ingraham *et al.* (1960) identified three rods, *Lactobacillus hilgardii* (15 isolates), *Lact. brevis* (1), and *Lact. delbrueckii* (2), and two cocci, *Leuconostoc* sp. (10) and *Pediococcus* (?12) from California wines.

Probably no part of enology has been under more active study the past twenty years than that of malo-lactic fermentation, particularly in Europe. Peynaud and Domercq (1961) have summarized the Bordeaux studies. Especially important is their observation that the mesoinositol content is much less in wines which have undergone the malo-lactic fermentation.

A complete review of the malo-lactic fermentation in wines was recently presented by Radler (1962b, 1966). He reported the following homofermentative species (those with a temperature optimum of 37°–60° C; 98.6°–140°F): *Lactobacillus caucasicus*, *Lact. lactis*, *Lact. helveticus*, *Lact. acidophilus*, *Lact. bifidus*, *Lact. bulgaricus*, *Lact. thermophilus*, and *Lact. delbrueckii;* and (with a temperature optimum of 28°–32°C; 82.4-89.6°F): *Lact. casei*, *Lact. leichmannii*, and *Lact. plantarum*. Homofermentative cocci reported are *Pediococcus cerevisiae* and *P. acidi lactici*. Heterofermentative species include *Lact. pastorianus*, *Lact. buchneri*, *Lact. brevis*, *Lact. fermenti*, and possibly *Lact. trichodes* and *Lact. hilgardii*. Heterofermentative cocci are represented by *Leuconostoc mesenteroides*, *Leuc. dextranicum*, and *Leuc. citrovorum*.

Pilone and Kunkee (1966) found few differences between the composition of wines that had undergone a malo-lactic fermentation and those which had not (except for the malic to lactic conversion). The main differences were more volatile acidity and acetoin plus diacetyl and possibly diethyl succinate in the samples which had undergone this fermentation. They were unable to find significant differences between the malo-lactic strains of lactic acid bacteria used: *Lactobacillus delbrueckii*, *Lact. buchneri*, *Lact. brevis*, *Leuconostoc citrovorum*, and two strains of *Pediococcus cervisiae*.

Poittevin *et al.* (1963) isolated *Lactobacillus plantarum* var. *pentosus* and var. *arabinosus* and *Lact. buchneri* from Uruguayan wines. Since the *arbinosus* strain does not produce mannitol and attacks malic but not tartaric acid, it is preferred to the *pentosus* strain, which can utilize tartaric acid (one isolate did not attack tartaric acid). *Lact. buchneri* is not recommended, for it produces mannitol. They also have a fermentation index (volatile to total acid) of over 0.1. All the other strains had an index of 0.1 or less. None of the strains studied utilized glycerin.

All the strains of *Lact. plantarum* var. *pentosus* and *Lact. buchneri* isolated from Uruguayan table wines by DeCores *et al.* (1966) attacked this acid. Nor did any strain of the three utilize glycerin. *Lact. buchneri* strains also produced mannitol. The strain of preference for inducing a malo-lactic fermentation is therefore *Lact. plantarum* var. *arabinosus*. Wejnar (1966) confirms that *Bacterium gracile* is clearly a *Leuconostoc*. He also reported that *Micrococcus acidovorax* is clearly different from *M. variococcus* or *M. malolacticus*. *Bacterium mannitopoeum* he believes to be *Lactobacillus buchneri*. (Further data on the bacteria involved in malo-lactic fermentation are given on pp. 775–782.)

Fornachon and Lloyd (1965) reported that wines which have undergone a malo-lactic fermentation contain more diacetyl and acetoin than wines which have not. In Australia *Leuconostoc mesenteroides* is the organism chiefly responsible for their formation. The main substrates were pyruvic and citric acids. The presence of even a low percentage of glucose in-inhibited their formation. The usual organisms which account for the malo-lactic fermentation in Australian wines, *Lact. hilgardii* and *Lact. brevis*, produce diacetyl and acetoin only under conditions which are unlikely to occur in normal winery operations.

Growth requirements

Lüthi (1957*b*) reported that the presence of yeast cells was required for the growth of the malo-lactic acid bacteria and that they functioned by providing the bacteria with growth stimulants not present in wine. Peynaud and Domercq (1959*b*, 1960) reported that growth of lactic acid bacteria was much easier in fermenting grape juices than in wines. The report of R. B. Webb and Ingraham (1960) substantiated this for California conditions. See also Marques Gomes *et al.* (1956) for Portugal.

The susceptibility of wine to spoilage by lactic acid bacteria and to malo-lactic fermentation is influenced by the balance between the growth inhibiting properties of alcohol formed during fermentation, the initial depletion of microbial nutrients and growth factors present in grape juice, and the subsequent post fermentation liberation of some nutrients and growth factors. These opposing effects of yeasts on the growth of bacteria in wine have been discussed recently by Radler (1966) and Fornachon (1968). It is recognized that yeasts differ in their ability to absorb and retain nutrient and growth factors, but there is a difference of opinion as to whether this difference is significant in determining susceptibility of wine to support growth of lactic acid bacteria. Ribéreau-Gayon and Peynaud (1960–1961) reported 11 of 15 different species of yeast grown in grape juice inhibited subsequent growth of two lactic acid bacteria. Webb and Ingraham (1960) reported that the lactic acid bacteria used by them grew equally well with any of the five strains of wine yeast. More recently Fornachon (1968) reported data on the growth of two species of lactic acid bacteria in wine fermented with 9 strains of *Saccharomyces cerevisiae*, 1 of *Sacch. fructuum*, 1 brewery yeast, and 1 of *Sacch. carlsbergensis*, before and after infection with several spoilage yeasts, showing marked difference of these yeasts on bacterial growth. In some experimental wines bacterial growth was delayed or failed altogether. This difference was more pronounced in wines racked off the lees immediately after fermentation and was less in wines allowed to remain on the lees for three or more weeks.

The malo-lactic fermentation may require special catalysts for growth, but these are still unknown. Schanderl (1959) believes that there is a symbiosis between the bacteria and the yeasts. Fell (1961) was able to freeze-dry (lyophilize) mixtures of yeast and lactic acid bacteria. He used the samples to ferment grape musts successfully with a simultaneous malo-lactic fermentation. However, in plant experiments even wines which had no added bacteria underwent a malo-lactic fermentation. This is not surprising, since the winery equipment probably contained many lactic acid bacteria.

R. B. Webb (1962) has described the original isolation of pure strains and various experiments using them. In some experiments there was an undue rise in volatile acidity. He noted that it was difficult to induce the malo-lactic fermentation in old wines. Several yeast types also appeared to inhibit a malo-lactic fermentation He felt that some of his strains contributed desirable odors to Pinot noir wines.

Kunkee *et al.* (1964) were able to induce a malo-lactic fermentation at any stage of alcoholic fermentation. This was much more rapid with a strain of *Leuconostoc* than with one of *Lactobacillus*. Although there was a slight increase in volatile acidity in the treated wines, it did not adversely affect quality. Kunkee (1967a), using *Leuconostoc citrovorum*, showed fermentation was inhibited by added acid (0.69 to 0.91 per cent as tartaric) and by sulfur dioxide. It was accelerated by increasing the amount of inoculum, by fermenting on the skins, or by raising the pH. Similar results have been reported in Australia by Fornachon (1957). Above pH 3.7 about half of his samples underwent a malo-lactic fermentation even in the presence of 150 or more mg per liter of sulfur dioxide. For data on the malo-lactic fermentation under commercial conditions in Canada see Adams (1967); see Rice (1965b) for New York conditions.

Silva Babo (1963) found that the vitamin B complex was especially favorable to the growth of lactic acid bacteria, and considers that the favorable effect of adding lactic acid bacteria during the alcoholic fermentation was due to the better nutrition provided by the presence of yeasts. More data would be useful. The amino acid requirements of the lactic acid bacteria have been studied by a number of laboratories (e.g., Radler, 1958). The formation of the "malic" enzyme by *Lactobacillus arabinosus* (now *plantarum*) requires eleven amino acids, according to Bocks (1961): arginine, cysteine, histidine, isoleucine, leucine, lysine, methionine, phenylalanine, tryptophan, tyrosine, and valine. Also necessary were adenine sulfate, guanine, thymine, and uracil.

The amino acids in an enriched culture by *Leuconostoc* sp. or *Lactobacillus* sp. following fermentation appear to be very different from each other and from those in the original. For example, Peynaud and Domercq (1961) report the following (mg per liter):

	Before fermentation	After fermentation by *Leuconostoc sp.*	After fermentation by *Lactobacillus sp.*
α-Alanine	88	110	110
Arginine	150	41	41
Aspartic acid	110	20	19
Cystine	5	3	3
Glutamic acid	194	71	73
Glycine	102	36	35
Histidine	20	9	9
Isoleucine	130	29	35
Leucine	155	46	62
Lysine	63	0	0
Methionine	45	16	16
Phenylalanine	120	56	32
Proline	205	148	148
Serine	124	92	92
Threonine	275	130	130
Tryptophan	8	4	4
Tyrosine	106	35	29
Valine	132	40	47
Total N	756	621	627
Ammoniacal N	87	78	78

The malo-lactic acid fermentation occurs in the range of 10° to 37°C (50° to 98.6°F). Lüthi (1957a) reported that it would not occur in wines or ciders at temperatures below 8°C (46.4°F). Acidity is just as important as temperature. The optimal pH for a wine of 10 per cent alcohol was reported by Lüthi to be 3.3–3.4; the growth of bacteria is retarded at pH below 3.2. At the same pH tartaric acid inhibits growth of lactic acid bacteria much more than malic acid. The malo-lactic fermentation is accelerated at lower alcohol levels and decreased at higher alcohol concentration.

Chemistry

The main reactions in the malo-lactic fermentation, according to Korkes *et al.* (1950), are:

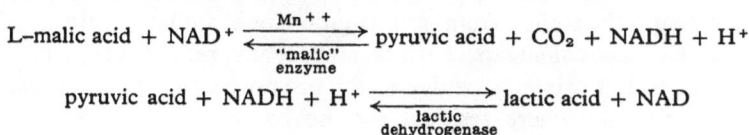

$$\text{L–malic acid} + \text{NAD}^+ \underset{\substack{\text{"malic"}\\\text{enzyme}}}{\overset{\text{Mn}^{++}}{\rightleftarrows}} \text{pyruvic acid} + CO_2 + \text{NADH} + H^+$$

$$\text{pyruvic acid} + \text{NADH} + H^+ \underset{\substack{\text{lactic}\\\text{dehydrogenase}}}{\rightleftarrows} \text{lactic acid} + \text{NAD}$$

The first reaction thus is:

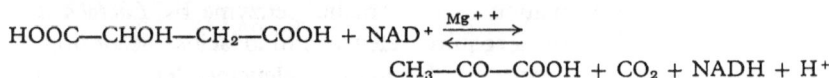

$$HOOC—CHOH—CH_2—COOH + NAD^+ \underset{\longleftarrow}{\overset{Mg^{++}}{\longrightarrow}}$$

$$CH_3—CO—COOH + CO_2 + NADH + H^+$$

Further data on the enzyme systems of lactic acid bacteria (*Lactobacillus plantarum*) were reported by Flesch and Holbach (1965). Three enzymes were studied: malic dehydrogenase (which catalyzes the reversible oxidation of malic acid to oxalacetic acid), "malic" enzyme (which catalyzes the reversible decarboxylation of malic acid to pyruvic acid), and oxalacetic decarboxylase (which catalyzes the reversible decarboxylation of oxalacetic acid to pyruvic acid). Flesch and Holbach (1965) reported that the latter two enzymes were distinctly different, based on Michaelis constants, *p*-chloromercuribenzoate inhibition, and the effect of avidin. Malic dehydrogenase is apparently not very active in wines, since its pH optimum is much too high. The conclusion to date is that malic acid is decarboxylated directly to pyruvic acid rather than being dehydrogenated to oxalacetic acid and then decarboxylated.

Peynaud et al. (1968) stress the fact that in malo-lactic acid fermentation L-lactic acid occurs, whereas in the corresponding fermentation of glucose, D-lactic acid or a mixture of isomers is formed. *Leuconostoc vinos* A., *Leuc. gracile*, and *Lactobacillus hilgardia* isolated from wines were used. If conversion of malic acid into lactic acid involved lactodehydrogenase and pyruvic acid intermediate, fermentation with *Leuconostoc* would give D-lactic acid while fermentation with *Lact. hilgardia* would give a mixture of the D and L isomers. In view of the above results, malo-lactic acid fermentation is a simple decarboxylation process without intermediates.

A good review of the malo-lactic problem was presented by Fell (1961), who noted that the problem of the source of energy for the malo-lactic fermentation has not been solved. (The reaction is endothermic.) The three possibilities are: from fermentation of hydrocarbon, from decomposition of malic acid to carbon dioxide, and from decomposition of nitrogenous materials (Kunkee, 1968; and Amerine and Kunkee, 1968). Fell (1961) emphasized that malic acid is not the energy source for growth of these bacteria. So fermentable carbohydrate is needed as a source of energy. Because the malo-lactic fermentation is endothermic, the bacteria must secure energy from some other source. Melamed (1962) believes that this comes from fermentation of residual sugars. In his review of the residual sugars of wines, sucrose was reported as absent in many studies, but this may be due to the insensitive procedures used. In some experiments where residual sucrose has been reported, 0.02 to

504

0.15 per cent, it may represent in part hydrolysis of glucosides. Melamed found no sucrose in twenty-seven French table wines.

Meyrath and Luthi (1969) reported that all strains of *Leuconostoc* isolated from Swiss wines or fruit juices ferment arabinose better than any of nine sugars tested. These strains apparently lack the ability to reduce the intermediate acetic acid to ethanol at a sufficiently high rate, a necessary step in glucose fermentation but not in pentose fermentation. For two strains found to ferment fructose as readily as arabinose, reduction of fructose to mannitol is favored over reduction of acetic acid to ethanol. Glucose is utilized, provided that small amounts of citric, pyruvic, or oxalacetic acid are added.

The amount of residual reducing sugars varies, of course, depending on the yeasts used, the conditions of fermentation, and the amount of sugar in the original fruit. For "dry" table wines it usually ranges from 0.08 to 0.30 per cent. Of this, 0.015 to 0.21 per cent is fructose. Melamed (1962) showed that wines which have undergone a malo-lactic fermentation were generally lower in reducing sugars than those which had not. The decrease was primarily in glucose, but fructose was attacked. The pentoses remained unchanged.

The pentoses likewise vary widely in "dry" table wines, from 0.03 to 0.20 per cent (average below 0.10). They consist primarily of arabinose and xylose and occasionally of rhamnose. Melamed (1962) reported 0.03 to 0.17 per cent (average 0.07) arabinose and 0.00 to 0.04 per cent (average 0.02) xylose. Rhamnose was found in only one wine, a wine which had undergone a malo-lactic fermentation.

YEAST AUTOLYSIS

The prompt removal of yeast cells from the new wine is desirable; this protects the wine from nitrogenous substances released both by excretions of living yeast cells and by autolysis of dead cells (Drews, 1936).

During initial growth, yeast cells excrete appreciable quantities of nucleotides and amino acids (both as free amino acids and polypeptides). In the presence of fermentable sugars the nonnucleotide nitrogenous constituents are reabsorbed, but the nucleotides are not, and their excretion continues during fermentation. The amount of nitrogenous excretion varies with the strain of yeast, the composition of the must or wine, and particularly with temperature (the higher the temperature the more extensive the excretion). The release of nitrogenous substances by living brewer's yeast cells has been investigated by Hartelius and others (Barton-Wright, 1949; Joslyn, 1956; Delisle and Phaff, 1961; Lewis and

Phaff, 1963; Lewis, 1964; and Lee and Lewis, 1968*a*, 1968*b*). Similar data for wine yeasts, however, are not available.

Nitrogenous excretions increase after death of the yeast cells when autolysis (self-digestion of cellular constituents) occurs. The changes in protein, nucleic acid, and amino acid content during autolysis have been studied in several species of yeast by Vosti and Joslyn (1954*a*, 1954*b*), Joslyn and Vosti (1955), and Joslyn (1955). Autolysis liberates strongly reducing enzymes and proteins, which not only inhibit oxidation of the wines and produce compounds unpleasant to the taste and smell, but also favor the growth of lactic acid bacteria and so render the wine more susceptible to bacterial spoilage.

Nutrient elements are absorbed by yeast during fermentation; if this absorption is so managed that it causes a depletion of some necessary element, the wine will be immune both to refermentation and to bacterial attack. Although the conditions favoring the most active and complete removal of nutrients by wine yeasts are not known, it has been demonstrated that yeast will remove thiamine, ammonium ion, and other constituents. Boulard (1926) and Vandecaveye (1928) have shown that successive generations of yeast can deplete the nutrients sufficiently to prevent refermentation. Ribéreau-Gayon and Maurié (1938) report that a must initially containing 86.7 mg of ammonia per liter and 300 mg of phosphoric acid, filtered four days after fermentation, contained 12.7 mg of ammonia and 50 mg of phosphoric acid (with an alcohol content of 10.3 per cent, and 3.5 per cent of sugar). After filtration this must fermented exceedingly slowly. The addition of 300 mg of phosphoric acid as disodium phosphate had no stimulating effect on fermentation, but the addition of 75 mg of ammonia as ammonium sulfate produced an immediate increase in fermentation rate, and the wine completed its fermentation in fourteen to fifteen days. Apparently, ammonium salts were limiting refermentation. This has been amply confirmed.

The bacteria that cause spoilage of wine require accessory growth substances which are absorbed from the grape juice by yeast cells, or which are formed by the yeast during growth. These substances, however, can be liberated into the wine by the autolysis of the yeast. Two of the most important factors influencing autolysis are pH and temperature; alcohol content and phosphate content are additional factors (the alcohol acts as a plasmolyzing agent). Although the conditions of autolysis of wine yeasts are not known, bakers' yeast, *Saccharomyces cerevisiae*, will autolyze most rapidly at pH 6 and at 45° to 50°C (113° to 122°F), according to Vosti and Joslyn (1954*a*, 1954*b*) and Virtanen (1955). Fornachon (1943) found that must fermented at a pH of 3.5, at 15°C (59°F), yielded wine which did not support growth of *Lactobacilli* even when the wine was left for six weeks on the gross or first lees. The same must at pH 4.0, fermented at 15°C (59°F), supported bacterial growth after three weeks of storage on the lees, and at 35°C (95°F) after one week. The average total nitrogen content of the wines after storage on crude lees for one to six weeks was as follows:

pH	Temperature		Total N
	°C	°F	mg per liter
3.5	15	59	127
	25	77	133
	35	95	145
4.0	15	59	142
	25	77	159
	35	95	168

Schanderl (1942) found that insufficient organic nitrogen for bacterial growth was present in German sparkling wines and the inorganic nitrogen was not assimilated by the bacteria. The yeasts, however, could utilize the inorganic nitrogen added to the wine, and produce organic nitrogen compounds which the bacteria utilized when the yeast cells released them by autolysis. Control of autolysis is therefore important in both still and sparkling wines. Bourdet and Hérard (1958) reported marked enrichment of inorganic and organic phosphorus in wines stored on the lees. This was especially true of a film yeast wine.

Low temperature, high acidity (low pH), and prompt removal of the yeast cells thus favor the stabilization of wine by avoiding autolysis of yeast cells. The possible favorable results of yeast autolysis in bottle-fermented sparkling wines are considered on pages 671–673.

CONTROL OF AERATION . . .

. . . is needed to prevent excessive oxidation.

During clarifying and aging, the wine must be protected against excessive aeration. To avoid darkening of color and loss of fruitiness, white table wines, particularly, should not be unduly exposed to air. Excessive amounts of air may be introduced by leaky pumps, improper racking, use of compressed air in mixing fining agents and wine, and improper transfer from one tank to another. Aeration during filtration (particularly finishing filtration) and bottling must be controlled. Under the usual California winery conditions, wine picks up excessive amounts of oxygen during all these operations. The oxygen absorbed is used up in the oxidation of alcohol, tannins, and other constituents which, after oxidation, contribute to discoloration, haze formation, and loss in flavor. This undesirable oxidation may be reduced by removal of dissolved oxygen, before its reduction, by stripping with nitrogen (Bayes, 1950). Stripping oxygen from wine with nitrogen has been used in California since 1950. (See Berg, 1950, Cant, 1960, and Air Reduction Co., 1961.) In addition, in-line sparging has been introduced. The effectiveness of any system of using an inert gas to remove dissolved oxygen before it has been reduced by wine constituents, as emphasized by Nightingale (1965), depends on the equipment and its use. Centrifugal pumps with poor packing can introduce more oxygen than could be removed by an in-line sparger or stripper. Hoses or glass tubes, if used, should have air-tight connections; otherwise, pumping wine through them will continually introduce oxygen. Bonetti (1965) stressed that air-tight tanks are needed to store table wines under a blanket of nitrogen. Wooden cooperage is too porous. Only lined or stainless steel tanks can be used.

The pickup of oxygen by wine during various operations has been reported for California wines. Wines pumped into top-filled tanks have an oxygen pickup of over 2.5 ml per liter as against less than 0.3 ml pickup when they are pumped into bottom-filled tanks. Filtration operations result in pickups of from 0.8 ml to about 5.0 ml depending on the method used. Bottling operations introduce from 0.5 ml to over 2.0 ml of oxygen per liter.

Prillinger (1965) recently reported information on the approximate amount of oxygen pickup during various winery operations (in mg per liter): racking under pressure, 0.2–0.3; racking through the bung, 0.4–0.6; one year in full 50-gallon barrels, 2.8–7.0; the same not filled, 32–38; bottom filling, 0.38–0.50; top filling of bottles, 1.35–1.39; filling down sides of bottle, 0.48–0.62; pouring from one bottle to another, 1.30–3.00; bottling sparkling wine under pressure (a) with compressed air in the wine, 0.6–4.0; (b) compressed air over the liquid, 5.9–15.7; ditto (a) with carbon dioxide in the liquid, 0.4–1.1; (b) with carbon dioxide over the liquid, 5.7–8.4, and full bottle open 24 hours, about 0.3. He recommended 50 to 100 mg per liter of ascorbic acid plus about 30 mg per liter of free sulfur dioxide.

Geiss (1966) reported that the amount of oxygen absorbed by the wine, either in hot or cold bottling, was insignificant. However, oxidation did occur owing to absorption of oxygen from the air space over the wine. He therefore recommended, according to the case: removal of air with carbon dioxide, or removal of air and introduction of carbon dioxide at the desired pressure, or filling the bottle to the desired level and removal of air with carbon dioxide from the neck of the bottle.

Cant (1960), however, reported that during bottling the greatest oxygen pickup occurred as the wine was discharged into the filling machine bowl. Under German conditions, Bielig (1966) reported absorption of 0.6 mg of oxygen per liter during filling at normal temperature.

The absorption of oxygen by wine will increase as the temperature decreases. According to Ribéreau-Gayon (1947), the maximum oxygen which table wines can absorb at 20°C (68°F) is 5.6 to 6.0 ml per liter. However, Wucherpfenning and Kleinknecht (1966a) reported 19 to 20 mg per liter of oxygen absorbed at 12°C (53.6°F), 10 to 11 at 20°C (68°F), and no oxygen absorption at 50°C (122°F).

A high oxygen level in stored wine, particularly white wine, is un- desirable. It is generally agreed that the oxygen content should be kept at a minimum during storage. This is not easily accomplished, since the wine is continually subjected to various treatments during processing in modern cellar operation—clarification, filtration, metal removal, stabilization. Nightingale (1965) recommended that the use of inert gas for removal of oxygen be carried out at the later stages of stabilization, where it is more effective and economical. Cant (1960) and Nightingale (1965) preferred stripping to sparging for removal of oxygen from wine in storage. The latter operation, however, is useful during bottling to prevent further pickup of oxygen. No published data seem to be available on the need for aeration of wines stored in very large, 200,000 gallon, tanks. See p. 366 for the need for aeration in new wines containing hydrogen sulfide.

CHANGES DURING AGING...

... profoundly affect the quality of the wine.

Effect of aging

Aging is a complex process of oxidation, reduction, and esterification that results in the formation of a desirable bouquet and the loss of the raw flavor of new wine. Separation and deposition of all or some of the excess cream of tartar occur during storage.

Sound wine, under suitable aging conditions, becomes mellow and smooth, loses its early harshness, and forms a bouquet which is more complex and delicate than the simple, fruity fragrance of a new wine. There is a decrease in acidity, caused chiefly by the precipitation of cream of tartar but also by the combination of the acids with the various alcohols, especially ethyl alcohol. The amount of alcohol in combination with acids increases; various aldehydes, acetals, and other oxygenated bodies are formed. A decrease in the tannin content is accompanied by formation of a deposit containing oxidized tannin and coloring matter. Red wines decrease in color and gradually become tawny or brown; white wines acquire an amber hue with age.

The Russian practice of designating changes that occur in the wines before bottling as "maturation" and those that occur in the bottle later as "aging" seems a rational one.

Cask-aging

Most wines should be oxidized to the proper degree by aging in the cask before bottling. If the wine is bottled too young, it may spoil or mature too slowly; if bottled when too old, it will lack fruitiness and be vapid and off-colored. The best wines result from a proper balance of cask- and bottle-aging. It is desirable to bottle choice wines, particularly white wines, relatively early in order to allow them to mature more slowly in the bottle. The present tendency is to ferment and store white wines in stainless steel tanks. A short aging period in oak, six months to a year, may precede bottling. Except for sweet table wines, most of the fine white table wines now reach the market within one to two years of the vintage. We feel that some of the wines, particularly in California and Germany, have been bottled too young—with noticeable sulfide or yeasty odor. The lighter types may be expected to reach their maximum quality after only six months to two years of bottle-aging; the heavier and sweeter types may continue to improve in the bottle for a number of years.

Because of the cost of handling in small cooperage, the tendency is to hold red wines in large containers of 5,000 to 50,000 gallons for a year and then, during normal stabilization, to store the finest wines in smaller oak cooperage (50 to 500 gallons) for one or two years before bottling.

For white, rosé, or red wines of ordinary varieties the whole process is speeded up. Wines that will not profit by cask- or bottle-aging ought to be finished rapidly and be bottled. (Bottle-aging is discussed on p. 512, and the changes that take place in wine during aging on p. 511 f.)

After most of the suspended material has settled out and the surplus carbon dioxide gas has been removed, the wine is slowly oxidized by the oxygen absorbed through the pores of the cask and by contact with the air during loss by evaporation and during racking. The oxygen is fixed by the wine soon after absorption, apparently by the reduced metallic ions present. These, in turn, after oxidation takes place, act upon other constituents; eventually the oxygen absorbed is transferred to certain oxidizable principles, notably the tannins and coloring matter. The action of oxygen is not completed at once; the preliminary effects are often undesirable but transitory; and the primary oxidized substances continue their action. Oxygenated bodies are formed from the alcohol and tannins; the result is a deposit of insoluble matter, brown or red in color. Excessive oxidation, however, causes decolorization and browning of red wines, the browning of white wines, and the formation of colloidal iron deposits (Ribéreau-Gayon, 1933a). The only satisfactory method of controlling oxidation in wines at present is through use of containers of desirable size (surface to volume ratio is limiting) and of appropriate porosity (that is, allowing a desirable rate of diffusion of oxygen from the air into the wine).

The rate of maturing varies with the kind of wine, the extent of aeration (especially during racking), the type of storage container, and the storage temperature. (See Singleton, 1962). The higher the content of tannin, residual sulfur dioxide, and other reducing substances, the slower the aging, and the longer the period of storage required. The more intense the aeration, the more rapid the aging. Small casks and frequent rackings increase the aeration and therefore the rapidity of aging. Large casks and low temperatures retard these changes. If the process is too rapid, the wine does not acquire its finest qualities, but becomes vapid; if it is too slow, the aging is unduly prolonged, and the wine is exposed longer to the possibility of injurious changes or contamination. The best quality for red wines generally results from the use of small casks (50 to 500 gallons) and relatively low temperatures (10° to 15°C; 50° to 59°F). Where such temperatures are not possible, larger casks must be used, at least for part of the aging. Oak casks are particularly desirable, not only because of their porosity but also because a *limited amount* of oak extractives improves the flavor of wines, particularly red table wines.

Oxidation

Although oxidation by molecular oxygen is recognized as an important factor in stabilization and development of wine, the actual quantity of oxygen required for a particular wine and the rate at which this is to be supplied are not known. Ribéreau-Gayon (1933a) early recommended that oxygen absorption be limited in rate to that which could be catalytically reduced by oxidizable constituents such as the flavonoid compounds and sulfurous acid. There is very little data on the total quantity of oxygen absorbed and reduced by wine during aging. Frolov-Bagreev and Agabal'iants (1951) reported that wine stored in small casks (250-liter) absorbed 40 ml of oxygen the first year and 30 ml the second year, but whether this was sufficient or excessive for aging was not reported.

When oxygen is introduced into a wine at too rapid a rate or in excessive amounts, the metallic ions can no longer serve as oxygen carriers, and direct oxidation results. This causes the wine to take on a decidedly rancio flavor, characterized by an accumulation of free acetaldehyde. The aldehyde formed results in the more rapid precipitation of anthocyan pigments of red wine, and leads to the accumulation of undesirable resins. Joslyn and Comar (1941) found a gradual increase in acetaldehyde content and a more rapid decrease in tannins which was lessened by the addition of tartaric acid. The rate of disappearance of aldehyde added to wine stored in absence of air was dependent upon the type of wine. Sulfur dioxide markedly reduced the rate of disappearance of added aldehyde by combining with it. The first effect of air or oxygen is to make the wine flat, unpleasant, and sometimes bitter; but if protected from too much air and stored for some time, the wine recovers and improves. This is the explanation of the "bottle sickness" of some newly bottled white wines.

Following fermentation, wines stored in wood and racked frequently show a rapid increase in the oxidation-potential, but in the absence of oxygen the potential may decrease. Wines stored in the bottle several years also show a decrease in potential. Joslyn (1949) showed that California red table wines generally had a lower potential while aging in the wood than white wines. A firm conclusion of Deibner (1957c, 1957e) was that low oxidation-reduction potentials do not always correspond to better quality, and that, in fact, mediocre wines might have low potentials. The entire history of a wine must therefore be followed to establish any correlation between oxidation-reduction potential and sensory quality.

Garino-Canina (1935) and Schanderl (1950–51) reported marked decreases in potential of wines exposed to direct sunlight. Deibner (1957c) noted potentials of up to 0.5 v when wines were aerated. But as Joslyn (1949) reported, the relation does not seem to be direct. In five of six wines Georgiev et al. (1950–51) reported lower potentials in wines stored about seven weeks in the bottle at 7° to 8°C (44.2° to 46.4°F) compared to those stored at 5° to 0°C (41° to 32°F). There is an especially large increase in potential at the time of the first racking, unless it is made in the absence of air. Gelatin or charcoal fining and pasteurization raised the potential.

Joslyn (1938) suggested that reduction plays an important role in bottle-aging of wine and reported data (1949) indicating lower oxidation-reduction potential in bottled wine. Krasinskii and Pryakhina (1946) also observed lowering of the oxidation-reduction potential in bottled wine during aging. The extent of this lowering depends on type of wine, the extent of oxidation prior to bottling, and its state of oxidation at time of bottling. The excessive reduction on bottle-aging which has been observed with some wines is undesirable because it leads to change in flavor. During bottle-aging of red wines Deibner and Mourgues (1964b) reported a slight but irregular decrease in the redox potential. They also noted considerable variation in the potential in different parts of the bottle.

Ribéreau-Gayon (1933a, 1952), as a result of his extensive investigations of oxidation and reduction in French wines, has concluded that the chief oxidizable constituents in wine are tannins, anthocyans, and sulfurous acid. He has shown that iron and copper salts are important as intermediary oxidants and catalysts of oxidation. Gatet and Genevois (1941) reported that reduced ascorbic acid exists in wine, and that iron salts catalyze the oxidation of tartaric acid into dihydroxymaleic acid and dihydroxytartaric acid. Geloso (1930–31), on the other hand, reported the existence of two oxidation-reduction systems in wine—one with a normal potential of $E_h = -0.115$ at pH 9 and 20°C (68°F) which reacts rapidly and gives up its hydrogen to various acceptors, including molecular oxygen, the other with a normal potential of $E_h = -0.160$ at pH 9 and 20°C (68°F) which reacts slowly and occurs in the reduced form only when it is not mixed with another system which can oxidize it. Geloso believes that these systems are identical with those that develop in a glucose solution stored out of contact with air; and he suggests that they represent an enolic form of some sugar derivative. Since they were done at pH 9 their application to normal wine aging is doubtful.

For other early studies on the oxidation-reduction potential see Garino-Canina (1935), Joslyn (1938), and Joslyn and Dunn (1938). Joslyn (1949) concluded that his data on the oxidation-reduction potential of California wines would favor Ribéreau-Gayon's hypothesis, but pointed out that more data are needed before the redox systems in wine can be characterized.

Nelson and Wheeler (1939) were unable to detect a reversible redox system and, therefore, do not accord great significance to the limiting potential. Schanderl (1948a, 1948b, 1950–51) found that the oxidation-reduction potential of normal table wines varies over fairly wide limits. He reported that sparkling wines and the lighter dry white wines have a lower optimum potential for their best quality than do the sweet table wines or dessert wines, and that the potential increases during aging. Joslyn (1949), however, investigated some table wines which became more reducing even when stored in small oak cooperage. Whether this reduction was due to addition of sulfur dioxide or to malo-lactic fermentation is not known. The studies of Ribéreau-Gayon and Gardrat (1957) were done at the normal pH of wines. Ascorbic acid does not seem to be a factor in determining the oxidation-reduction titration curves of wines. The minimum potential found by reduction was lower in old than in young wines. The polyphenolic constituents begin to be reduced at relatively high potentials, 0.3 to 0.4 v. However, if the polyphenols are first reduced they begin to be oxidized at relatively low potentials, 0.0 to 0.1 v. The polyphenolic compounds thus appear to be important in determining the level of the oxidation-reduction potential of wines. The inflection point for the curve for the potential of the skins of red wines is 0.26 v for wines of pH 3. For malvidin the normal potential at pH 2.5 is 0.3 v. Ribéreau-Gayon and Gardrat were careful not to claim that the polyphenolic compounds, the anthocyanins in particular, constitute the oxidation-reduction system of wine. But they do state that these may at least have an influence on such a system. It is maintained that the condensation of anthocyans during aging markedly modifies the oxidation-reduction potential of the wine. In studies on oxidation-reduction potentials it is important to distinguish those due to genuine reversible thermodynamic systems from those due to nonreversible systems.

The role of heavy metals as catalyzers of oxidation-reduction reactions has been fully explored, especially in France, the Soviet Union, and the United States. Whether dioxymaleic acid is the primary product of oxidation of tartaric acid, as envisaged by Rodopulo (1965), is not known. The suggested reaction is:

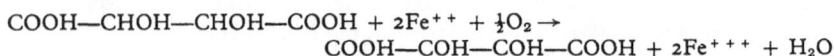

$$COOH—CHOH—CHOH—COOH + 2Fe^{++} + \tfrac{1}{2}O_2 \rightarrow$$
$$COOH—COH—COH—COOH + 2Fe^{+++} + H_2O$$

The other products postulated (diketosuccinic acid, *meso*-oxalic acid, glyoxylic acid, oxalic acid, the aldehyde of *meso*-oxalic acid, glycolic aldehyde, and glyoxal) may be present, but the route of their formation may not be as clear as stated by Soviet biochemists. Nevertheless, Rodopulo considers the dioxymaleic acid $\rightleftarrows$ diketosuccinic acid oxidation-reduction system to be important in the maturation of aging wines, helping to maintain reducing conditions. According to this, addition of dioxymaleic acid to wines should accelerate their aging, and this appears to occur, at least in sparkling wines, according to Rodopulo (1965). Similar but less rapid results were obtained by use of ascorbic acid. Wines kept in the bottle several years also contain dioxymaleic acid and have a low oxidation-reduction potential. Rodopulo believed that this aided in the development of the bouquet.

The Russian procedure for rapid maturation of table wines is based on holding wines at 40°C (104°F) with no more than 0.05 mg per liter of oxygen. This is said to lead to protein hydrolysis and increase in the free amino acid content. Oxidation-reduction systems such as cysteine $\rightleftarrows$ cystine and glutathione (reduced) $\rightleftarrows$ glutathione (oxidized) reduce the potential and hasten aging. Formation of tyrosol from tyrosine and 2-phenethyl alcohol from phenylalanine is also considered important. Similar results in aging white table wines rapidly by heating in bottles were obtained by Singleton et al. (1964b). They presented statistically significantly sensory results as to the differences between treated and untreated wines. However, no widespread commercial use seems to have taken place in California. Rodopulo (1965) proposed acceleration of aging of sparkling wine by heating the *cuvée* to 40°C (104°F) in the presence of a small amount of oxygen to secure maximum formation of dioxymaleic acid as well as protein hydrolysis. The new oxidation-reduction systems thus produced were believed to accelerate the maturing and aging of the wine after the second fermentation. No detailed statistical analysis of the sensory results was presented. More research is required to establish all the factors influencing maturation and aging. See p. 672 for the results of a similar process in California.

Rodopulo (1965) reiterated his belief that the reducing condition of musts is due to ascorbic acid *and* dioxymaleic acid as well as to unidentified compounds. After crushing, these are oxidized by quinones and the appropriate oxidases, and the musts lose their reducing property. The oxidation-reduction potential rapidly rises from 0.325 v to 0.454 v. The formation of quinones (quinoids ?) is due to the activity of polyphenoloxidase. Both the quinones and the reduced ascorbic acid eventually participate in oxidation-reduction reactions leading to the formation of colored compounds of unknown composition.

During fermentation, glutathione and cysteine are produced and ensure the reduction of the quinones. The oxidizing enzymes with heavy metal in the molecule are deactivated by glutathione and cysteine. Also some of the oxidases are absorbed by the yeasts. These new oxidation-reduction systems, cysteine ⇌ cystine and glutathione (reduced) ⇌ glutathione (oxidized), rapidly reduce the oxidation-reduction potential. However, if the wine is aerated the polyphenols will be oxidized to produce quinones, which rapidly dehydrogenate reducing compounds such as cysteine and glutathione. The potential then rises again—to as much as 0.45 v.

The system ascorbic acid ⇌ dehyroascorbic acid is nearly inactive in wines because dehydroascorbic acid is not reduced in the wine.

The redox potential of the Rumanian wines tested by Bălănescu (1963–64b) did not seem to be closely related to the quality of the wines, though common young wines usually had a higher potential than high-quality aged wines. There did seem to be some regional differences in redox potential of white wines, and less of red wines.

The oxidation-reduction potential of wines made from direct-producer hybrids appears to be lower than that of wines of pure *Vitis vinifera*, according to Deibner and Mourgues (1964b). The complicated nature of the redox system of wines has been emphasized by Ribéreau-Gayon (1963). Copper complexes are much more active as catalyzers of oxidation than are iron or iron complexes. Traces of sulfhydryl derivatives (e.g., glutathione) inhibit their activity. Kielhöfer (1960a) showed that ascorbic acid does not function as an antioxidant but rather as a catalyst for the oxidation of certain constituents of the wine, particularly of sulfurous acid. Without ascorbic acid the sulfurous acid content of a wine decreased from 128 to 106 mg per liter in 15 days; with ascorbic acid it decreased in the same time to 29 mg per liter. The redox potential attains very low values in the presence of ascorbic acid, sufficient to dissolve, at least partially, precipitated ferric complexes. For best results ascorbic acid should be added to wines just before they are to undergo aeration, as in racking or filtration. Ribéreau-Gayon (1963) notes that some wines treated with ascorbic acid are more susceptible to oxidation than untreated wines, probably because ascorbic acid catalyzes the oxidation of certain constituents that are not oxidized in its absence.

The changes in the oxidation-reduction potential during fermentation and processing have been summarized by Deibner (1957a), who reviews a large number of papers on this subject from the Soviet Union. During fermentation there is a rapid decrease in potential from about $+0.4$ v to 0.1 v (with considerable variation among musts).

Deibner believes that the decrease in potential is associated with the reduction of quinones. The decrease depends on the conditions of culture of the pure yeasts, according to Schanderl (1948a, 1948b). The redox potential of the new wine is higher when the yeasts were grown in the presence of air. Addition of ascorbic acid definitely modifies the potential. The changes in the oxidation-reduction potential during fermentation are due to complex factors: composition of media, method of fermentation, temperature and degree of aeration, presence of sulfur dioxide, ascorbic acid, and so forth. In sparkling wines there is also a reduction in potential during the secondary fermentation, whether in bottle or tank. After final dosage there was a small decrease during bottle-aging.

White wines are known to be more sensitive to browning and development of madeirized flavor on oxidation and usually are not aged as long as red wines before bottling. Valuĭko (1965b) recently reported that red wines do not accumulate aldehydes on oxidation as rapidly as white wines because of combination of aldyhydes as formed with tannins and pigments. He attributed the low susceptibility of red wines to madeirization to this factor.

The state of oxidation of a white wine may be rapidly and easily determined colorimetrically by measuring the rate and extent of reduction of a pigment which is colored in the oxidized state and colorless in the reduced state. Clark (1919–1956) investigated many such systems and recently (1960) summarized his own and others' contributions on reversible and irreversible oxidation-reduction potential of organic systems. Gray and Stone (1939a, 1939b) introduced the use of 2,6-dichlorophenolindophenol and measurement of the time (in seconds) to reduce the dye (decolorize) by 80 per cent. This measure, expressed as ITT value, was applied to wine by Koch (1955) and more recently by Wucherpfennig and Kleinknecht (1966a). The ITT value measures the concentration of reducing substances present (including free sulfur dioxide, reductones, etc.).

The ITT value, moreover, can be measured only in white wines, as the red pigments of rosé and red wines interfere. The potentiometric method of determining the oxidation-reduction potential of a wine is more difficult because it is poorly poised and subject to change by absorption of even small amounts of oxygen. Joslyn (1949) and Costa (1959) called attention to the difficulty of obtaining replicate values with the usual platinum-calomel cell electrode systems. The measurement of oxidation-reduction potentials in wines, however, has been improved by the introduction of modified pipette electrodes. These are described by Deibner (1953, 1966c) and Costa (1959). Mareca Cortés (1954), Deibner (1956a, 1956b, 1957a, 1957b), and others (see Ribéreau-Gayon and Peynaud, 1964–1966) have reported data on changes in oxidation-reduction during fermentation and aging of wines and its relation to sensory evaluation. Deibner (1956b) particularly has stressed that the oxidation-reduction potential is related to sensory qualities of wine, particularly to those of sweet table wines. More data are needed.

Alcohol content

During aging in large redwood or concrete tanks, little decrease in alcohol occurs. In puncheons and smaller wooden cooperage, both decreases and increases in alcohol content have been noted. Whether or not there is an increase or decrease depends on the relative humidity of the air. In small casks (3 gallons), Dicenty et al. (1935) investigated this phenomenon, with the results shown below. The explanation of these results is that diffusion and evaporation are two different physical phenomena which influence the movement of water and alcohol through the wood. Water diffuses through wood faster than alcohol does. At low humidities it is rapidly evaporated from the surface, more is lost, and the alcohol content is increased. But at high humidities it does not evaporate so rapidly from the surface as does the alcohol; hence there is a decrease in alcohol during storage. This assumes that the amount of alcohol vapor in the air is low, which is normally the case.

Average temperature, °C	Average relative humidity, per cent	Per cent alcohol Original	Final	Per cent change
14.0	94.8	15.09	14.55	−3.57
14.0	94.8	15.09	14.50	−3.90
14.6	94.0	15.07	14.51	−3.71
14.6	94.0	15.13	14.54	−3.89
20.0	74.7	15.11	15.09	−0.03
20.0	74.7	15.01	15.09	+0.33
34.2	57.2	14.98	15.20	+3.47
34.2	57.2	15.26	15.40	+0.90
34.2	37.4	15.13	15.50	+2.38
34.2	37.4	15.06	15.48	+2.71

Temperature during aging

Temperature markedly affects aging. The lower the temperature, the more speedily the yeasts and other microörganisms become inactive and settle out, and the more rapidly and completely excess cream of tartar is precipitated. A wine should therefore be kept cold for several weeks after fermentation, in order to throw the excess salts out of solution and to deposit the microörganisms. Many of the gums and proteins, in contrast, are eliminated more rapidly at higher temperatures, at which aging is more rapid. When separated from all the sediment, table wines develop best at an even temperature: 10° and 15.5°C (50° and 60°F).

Period of aging

Since many factors affect the length of time necessary for obtaining optimum quality by aging, the period of aging alone is not a sufficient sign of maturation. Thus wines stored in large containers may be nearly as young in flavor after several years' storage as they were initially, whereas wines in small casks in warm cellars may mature so fast that they become over-aged within a year. Wines improve with age only up to a certain point; beyond that they decrease in quality. (See also p. 584.) For use of ultrasonic treatment for rapid aging and its deleterious effects see Singleton and Draper (1963).

Singleton (1963) investigated the effects of cobalt-60 gamma irradiation on the quality and composition of dry red and dry white table wines. There was a marked decrease in redox potential, an increase in aldehyde and a decrease in color and spectral absorption. In general, sensory scores decreased with 100,000, 500,000, and 1,000,000 rads. Singleton suggested that the new flavors produced at the lower irradiation level might improve complexity of color and hence quality. He, therefore, recommended further experimentation.

TARTRATE REMOVAL AND STABILIZATION

New wines contain an excess of potassium bitartrate. This excess is removed by precipitation during storage and aging as cream of tartar. Potassium bitartrate crystals may occur in the grapes as harvested and precipitate in the crude lees after the initial settling and in the argols formed during aging. The formation and removal of excess potassium bitartrate is facilitated by chilling. Crystallization of excess potassium bitartrate is not instantaneous. Its rate depends on degree of supersaturation (which is increased by chilling) and on the concentration of nuclei or foci for crystallization. Tartrates have minimal value as a by-product in this country.

Von Weimarn (1925) reviewed the early investigations of his own and others on the conditions governing precipitation of solid substances from solution. The investigations published during the period 1906–1925 indicated that concentration of reacting solutions, solubility, and time were the most important variables affecting precipitation. The basic principles of formation of insoluble precipitates and the factors governing their properties have been critically reviewed by Walton (1967). Tammann (1925) emphasized the importance of formation of crystalliza-, tion centers, either by a limited number of molecules of the substance in supersaturated state or in undercooled melt after favorable collision or by foreign particles serving as loci or nuclei for crystallization. For excess potassium bitartrate to separate out it is necessary that conditions be made available favoring association of the component ions, their orientation into positions similar to those occupied in the crystal lattice, and crystallization and growth of crystals. Since not all these aspects have yet been defined for a particular wine, refrigeration to remove excess cream of tartar still depends more on empirical art than on science. Differences between wines, between old and new wines, and in wine components affecting crystallization are yet to be determined. Wines may also be stabilized against post-bottling tartrate deposition by ion-exchange treatment to reduce the potassium concentration. Chilling is preferred for finer table wines, and ion-exchange treatment is usually reserved for ordinary table wines and for dessert wines (fig. 76). Contamination of the wine with calcium salts from improperly prepared filter pads or concrete tanks and the contamination of wine with oxalate from impure citric acid or from improperly treated concrete vats should be avoided. In addition to deposits of crystalline potassium bitartrate, calcium tartrate and calcium oxalate deposits may occur. De Soto and Yamado (1963) reported from 50 to 90 per cent oxalate in deposits from California sherry.

The variables influencing tartrate stability have been studied in many laboratories. Among the factors which have been shown to be important are alcohol, acids, cations, anions, pH, and various complexing compounds. Data comparing the rate and extent of precipitation of potassium bitartrate from new unclarified wines with partially clarified aged wines still are not available. The empirical cold stability test has been widely used in the California wines industry. Warkentin (1950, 1951) showed that California practice varied from 24 to 240 hours at temperatures of −11.1° to 0°C (10° to 32°F) in containers of two ounces to one-fifth of a gallon. The samples were observed by a variety of procedures from casual visual to nephelometric. Such tests, if extended long enough, could provide the necessary information, but the winery rarely has the

Figure 76.   Refrigerated tanks in Italian winery. (Courtesy Padovan, Conegliano.)

time available. As De Soto and Yamada (1963) correctly point out, it is often difficult to interpret crystalline precipitation because of the presence of other interfering materials. Also, laboratory tests often fail to duplicate plant conditions and this may change the stability.

Berg and Keefer (1958, 1959), therefore, calculated the relative stability of potassium and calcium tartrates in wines of various alcohol contents under different conditions of time and temperature. They gave tables for using approximate concentration products (CP) to calculate tartrate stability: CP = (mol per liter K) × (mol per liter total tartrate) × (per cent acid tartrate) and CP = (mol per liter Ca) × (per cent total tartrate).

On the basis of extensive commercial tests De Soto and Yamada (1963) suggested the following maximum concentration product levels for potassium acid tartrate and calcium tartrate for various types of California wines:

Wine type	Potassium acid tartrate 10^6		Calcium tartrate 10^8	
	Highest level found	Suggested safe level	Highest level found	Suggested safe level
White table	18.5	16.5	230	200
Red table	34.7	30.0	590	400
Pale dry sherry	14.6	10.0	170	90
Muscatel	18.0	17.5	310	250
Port	23.3	20.0	410	275

Although these levels appear safe for most wines, not all the factors influencing tartrate stability are known. Furthermore, these levels were determined for one winery and presumably might be different for another winery under different plant and climatic conditions.

Koch and Schiller (1964) found little influence of pH on the rate of crystallization of potassium acid tartrate. The tartrate concentration had less effect on rate of crystallization than the potassium content. Reduced temperature, of course, speeded up the precipitation rate, but with a constant supersaturation the opposite was observed. Presence of magnesium, calcium, and ferrous iron speeded up crystallization; sodium slowed it down.

B. F. Pilone and Berg (1965) showed that the changes in potassium and tartrate content during storage or after ion-exchange or charcoal treatment could not be explained solely on the basis of a simple reaction between the two ions. In red wines resolubilization of potassium acid tartrate occurred. This they attributed to polyphenol-tartrate reactions. They also suggested potassium-colloidal pigment reactions. In white wines solubilization of tartrates was attributed to the binding power of proteins for tartaric acid. Red wines complex tartrate but not potassium according to Balakian and Berg (1968). This complexing, which differs greatly between wines, prevents much of the tartrates from precipitation. They reported that the tartrate-holding capacity of red wine was influenced more by pigments present in a grape-skin extract than by addition of various crude preparations of catechins, leucoanthocyanins, and condensed tannins. They concluded that the anthocyanin pigments complex tartrate but did not give any conclusive data. Anthocyanins were not added as such and the tartrate complex was not isolated or identified. Wucherpfennig and Ratzka (1967) consider the condensed polyphenols to be important.

Cantarelli (1963) has summarized the available information on inhibitors of potassium bitartrate precipitation. The best known of these is metatartaric acid, but Cantarelli shows that the polylactide of malic acid has approximately the same inhibiting effect. He also reported that 0.1 per cent of a 1 per cent solution of carboxy methyl cellulose (CMC 7 LP Hercules Power) had good inhibiting properties. Surprisingly, 0.2 per cent of a 2 per cent solution of purified apple pectin (Fluka) also inhibited bitartrate precipitation in a high degree, and sodium metaphosphate (2 per cent of a 20 per cent solution) was the most effective agent tested. It is unlikely that its use in wines in adequate quantities will be approved by government agencies. Wucherpfennig and Ratzka (1967) reported that addition of 1 gr per liter of tannin strongly inhibited precipitation of tartrates (even when gelatin, kieselguhr, or asbestos were subsequently added). In general, condensed polyphenols interfered with tartrate precipitation. When these are removed by filtration, further tartrate precipitation occurs. This suggests removal of the condensed polyphenols with polyamides (nylons), filtering, and then cold-stabilizing.

Metatartaric acid, a polylactide of tartaric acid, is produced by heating the latter at 170°C (338°F) for 120 or more hours. It has been approved for use in this country. Peynaud and Guimberteau (1961) recommended that the degree of esterification (or polymerization) of the product should be at least 30 per cent. On hydrolysis, only tartaric acid should be present. They reported that 50 to 100 mg per liter protected young wines from tartrate precipitation even when stored at low temperature for several months. The mechanism of the inhibiting action of metatartaric acid seems to be an interference in the formation and growth of potassium bitartrate or calcium tartrate crystals.

Benes and Krumphanzl (1964) reported that Hungarian wines could be stabilized by addition of metatartaric acid produced by heating tartaric acid to 160–170°C (320°–338°F) for one and a half hours. Chenard (1963) noted that the presence of activated charcoal or bentonite reduced the efficacy of metatartaric acid. He recommended that metatartaric acid which is more than 40 to 41 per cent polymerized should not be used, as it may cause clouding. When the iron content of red wines was high, products over 38 per cent polymerized also gave cloudiness. Since it is effective for periods of six months or less, it does not seem that it would be of general utility in this country and we do not recommend its use. See also Chenard (1963).

The effect of temperature on the solubility of calcium tartrates at various temperatures and percentages of alcohol was given by Berg and Keefer (1958) as follows (in milliequivalents):

Temperature		Alcohol (per cent)			
°C	°F	10	12	14	16
−4	24.8	0.30	0.25	0.21	0.18
0	32	0.34	0.29	0.24	0.21
5	41	0.40	0.34	0.29	0.24
10	50	0.47	0.40	0.33	0.28
15	59	0.56	0.47	0.40	0.34
20	68	0.66	0.55	0.47	0.39

The percentage of tartrates present as the bitartrate is at a maximum at about pH 3.6. Other factors being equal, maximum precipitation of potassium bitartrate will occur at this pH. Therefore, if the pH is less than 3.6 and is raised toward this figure (by a malo-lactic fermentation, etc.) there will be a greater tendency for potassium bitartrate to precipitate. Contrariwise, if the pH is above 3.6 and is lowered toward that figure there will be a tendency for bitartrates to precipitate.

Chilling to separate excess cream of tartar

Chilling wine close to the freezing point is an effective means of hastening elimination of excess cream of tartar from new wines and may be used in preparing ordinary table wines. According to data obtained by Marsh and Joslyn (1935) on the conditions affecting the rate of cream of tartar deposition, a mere cooling of the wine to a temperature just above freezing is not sufficient to separate appreciable amounts of cream of tartar. To remove a considerable portion, the wine must be stored at the cool temperature for a period that depends upon the nature of the wine and the temperature of storage. Thus, for dry white wine, storage for three days at 10°C (50°F), for one to three days at 5.5°C (40°F) and 0°C (32°F), and for less than one day at −3.9°C (25°F) reduced the cream of tartar content to that of the sample stored at room temperature for 225 days. The rate of precipitation of cream of tartar was fairly rapid in the initial period and then decreased, becoming very slow, after twelve days, at virtually all temperatures used, because of decrease in the extent of supersaturation. The decrease in rate of separation, however, appeared to be more rapid than would be accounted for by decrease in cream of tartar concentration. The rate of precipitation increased with decrease in temperature; thus 30 per cent of the cream of tartar present was removed after storage for five days at 10°C (50°F), for about three days at 0°C (32°F) and 5.5°C (40°F), and for two days at −3.9°C (25°F).

Dry table wines of 10 to 14 per cent alcohol congeal at temperatures of −5.5° to −3.3°C (22° to 26°F); wines containing sugar congeal at lower temperatures. It is the practice now to chill dry wines to about −3.3° to −2.2°C (26° to 28°F)—just above their freezing point—and store cold for about ten days to remove excess tartrates. The optimum conditions for detartrating wines of various ages and the age at which wine can best be chill-proofed are not known. Furthermore, some believe that high-quality wines should not be subjected to such extremes of temperature, particularly since they usually undergo a longer period of aging and thus become tartrate-stable. After chilling, the wine must be filtered before the temperature rises. A cold filter must be used to prevent redissolving of the precipitated tartrates.

For ordinary wines it is sometimes very desirable to preheat the wine to remove colloidal material before chilling. This seems to facilitate tartrate removal during the chilling process. Rapid cooling also seems to give larger crystals than slow cooling. Tartrate removal from wines frozen to a sludge is often less satisfactory than from wines cooled to a slightly higher temperature. The factors influencing speed of crystal formation and size of crystals need to be studied for a variety of types of wine.

Daghetta and Amelotti (1965) found that the best results for laboratory tartrate removal were obtained by precipitation at low temperature with agitation. However, only slightly less satisfactory results were found by adding potassium while boiling. The lowest results were obtained by precipitation at low temperatures without agitation.

Ion exchange

Use of ion-exchange resins in wineries for prevention of potassium acid tartrate precipitation is now common throughout the world, although it is still not permitted in France. When properly applied, little or no harmful effect on the quality of the wine need occur.

There are two main types of ion-exchange resins: cation and anion. In the former the resin or matrix contains labile ions capable of exchanging with cations in the wine. Usually the labile ion is sodium and it exchanges with potassium in the wine. There will be slight reduction in acidity if the resin is entirely in the sodium form. To prevent this or to increase the acidity the resin may be prepared in the mixed sodium and hydrogen form. If the acidity is too low it can be increased by using the hydrogen form. Anion-exchange resins may be employed when the wines are too high in acidity. Tartrate stability also occurs in this case, since part of the tartrate ions are replaced by hydroxyl ions.

If cation resins in the hydrogen form are used, the wine will be lower in pH; if in the sodium form, the wine will be appreciably enriched with sodium ions. Kielhöfer (1957) showed that ion-exchanged wines were of equal or superior quality *as far as reducing excess acidity is concerned* compared to those treated with calcium carbonate. However, Rankine (1965b) states that there may be some effect on the flavor of delicate white table wines and that some wine makers claim to be able to detect differences between treated and untreated wines. If new resins impart an off-flavor to the wine, the resin should be stored with some wine for a few days and then regenerated with warm hydrochloric acid prior to sodium chloride regeneration. Most instances of harmful effect can be traced to excessive use of the process.

Some of the iron and copper present in wine will be removed by cation-exchange resins. However, this is usually insufficient to remove excess iron or copper except from wines of very low pH (see Pato, 1959). For acid losses during ion-exchange see Villforth (1954).

The exchange capacity of different resins varies considerably. The manufacturer's specification should be noted. However, actual testing during operation is usually necessary.

Most applications of ion exchange are now by use of columns filled with the resin (fig. 77). The columns may be large or small depending on

Figure 77. Ion-exchange columns for California wines. (Courtesy Valley Foundry and Machine Works, Inc., Fresno.)

the volume of wine to be treated. In preparing a cation exchanger, the resin is placed in the sodium form by treating the column with a solution of sodium chloride. The column must then be washed with water until all the chloride, as determined by a silver nitrate test, is removed. Ion-exchanged water is used for this purpose. The wash water is then drained and wine is introduced from the bottom of the column until the column is full. Wine is then introduced from the top.

To determine the break-through point when a cation-exchange resin is used, the potassium content is periodically determined on the wine passing through the column. This is usually done with a flame photometer (p. 759). It is important not to overrun the column. If this occurs the wine will again be subject to tartrate precipitation. Such an overtreated wine is also high in sodium and cannot be rerun through a column. After the resin is exhausted, as determined by appearance of potassium, the column must be washed. This is usually accomplished by passing water up through the resin until the affluent is clear. The utilization of cation-exchangers in the sodium form may be seriously limited if the suggested limit of 100 mg per liter of sodium for wines in international commerce applies. Public health authorities may also object to low-potassium high sodium wines. It seems wise, therefore, to maintain the potassium content at 500 mg per liter or higher by blending treated and untreated wine.

If the column is not to be used for some time after use for wines containing sugar, the resin may be subject to mold growth. Rankine (1965b) recommends storage with sulfurous acid solution. This has the advantage of acting as a mild acid regenerant and removes metal ions bound to the resin.

After prolonged use the resin progressively loses its exchange capacity. This is due to absorption of proteins, tannins, and pectins, thus reducing the available exchange sites on the resin. Washing or sodium chloride or hydrochloric acid regeneration does not remove these. Soaking with hydrogen peroxide solution or treatment with 3 per cent sodium hypochlorite is usually effective. Rankine (1965b) does not recommend chromic acid because it is corrosive and difficult to handle.

Ion-exchanged wines are often difficult to re-ferment because of the removal of vitamins and growth factors by the treatment. Therefore, some wine makers prefer not to treat sparkling wine blends lest the secondary fermentation in the bottle or tank be too slow. Gerasimov and Kuleshova (1965) reported that a cation-exchange resin (sodium or hydrogen form) might reduce the pyridoxine content by 93 per cent. Bentonite lowered the pyridoxine by 40 to 79 per cent. Thiamine was completely removed from wines either by ion-exchange resins or by bentonite. The ion-exchange resins did not remove appreciable inositol or biotin, but did lower the pantothenic acid content: cation exchangers removed 58 to 81 per cent. Bentonite had little effect on free pantothenic acid content. In contrast, nicotinic acid was absorbed more by bentonite than by ion-exchange resins. In general, the cation-exchange resin in the hydrogen form removed more of the B vitamins (especially pantothenic and nicotinic acids) than in the sodium form.

With a column cation-exchanger Dehner (1965) did not find speed of passage, within limits, to have a marked effect on efficiency. For sparkling wines he recommended regeneration with a mixture of hydrochloric acid and sodium chloride, since this improved flavor (by keeping the pH lower). The best ratio has to be determined for each wine, for it will depend on the composition of the wine. He also noted the increase in potassium content which occurs if the capacity of the column is exceeded.

Ion-exchange treated wines frequently fine poorly with bentonite, according to Rankine and Emerson (1963). They recommend fining with sodium bentonite *prior* to ion-exchange treatment. If this is not possible, the bentonite fining should be accompanied by addition of gelatin.

The existing federal regulations (U.S. Internal Revenue Service, 1961a) allow the use of anion-exchange resins (Amberlite IR-45, Duolite A-7 (Tartex 180), Duolite A-30 (Tartex 181), and SAF) to reduce the natural acidity of wine; Duolite A-6 for the treatment of white wine; cation-exchange resins (Amberlite IR-120-H, Amberlite IR-120, Duolite C-3 Na (Tartex 160), Duolite C-20 Na (Tartex 161), Duolite C-25 Na (Tartex 162), Permutit Q, Permutit Q Spec. 157, and SAF Cation Exchanger) to stabilize wine by exchanging hydrogen or sodium ions for undesirable metallic ions. The anion-exchange resins are allowed in the hydroxyl state, but if they are used the inorganic anion content of the wine cannot be increased more than 10 mg per liter; the natural or fixed acids may not be reduced below 5 parts per 1,000; the treatment must not remove color in excess of that normally contained in wine; and the basic character of the wine must not be altered. The cation-exchange resins, with the exception of Amberlite IR-120-H, are allowed in the sodium state. The over-all change in the pH of the wine before and after treatment is limited to 0.2 pH unit. Nonionic and conditioning resins are allowed for removal of excessive oxidized color, foreign flavors and odors, and for stabilizing wines by removal of proteins and other macromolecular constituents. These include Duolite S-30 (Tartex 260), Duolite A-7, Duolite A-30, Amberlite IRA-401, and Amberlite IRA-401-S. At present Amberlite IRC-120 is widely used in California, but comparable products of other companies appear to be equally good. (See pp. 111–112.)

The effect of anion and cation exchangers in four white table wines is reported by T. D. Ionescu (1966) as follows:

	Cation treated (cretă)			Anion treated (ionit)			Not treated		
	Min.	Max.	Av.	Min.	Max.	Av.	Min.	Max.	Av.
Alcohol (per cent)	9.7	13.15	11.1	9.7	13.15	11.1	9.7	13.15	11.2
Titratable acidity (per cent)	0.69	0.76	0.74	0.665	0.72	0.70	0.875	1.05	0.93
Sulfur dioxide, total (mg per liter)	5.1	96	53	3.2	79	40	5.1	96	64
Tannin and coloring matter (gr per liter)	0.07	0.17	0.13	0.08	0.16	0.12	0.16	0.22	0.18
Sensory quality	8.1	8.35	8.3	8.1	8.35	8.2	8.1	8.3	8.2

Amano and Kagami (1966) removed appreciable copper from wines using ion-exchange resin Imac C-21 in the sodium cycle. They also presented equations showing the relation between height of exchange zone and linear velocity of wine at different initial copper ion concentrations.

T. D. Ionescu (1966) has suggested that ion exchangers be used to correct off-colors in wines and to recover tartrates from distillery slops. Their use in analytical procedures is well known (pp. 740 and 759). For their effect on the redox potential see Goranow et al. (1965). For use on grape juice see Popper and Nury (1964).

FINING

Although a sound wine often becomes brilliantly clear by natural settling, cloudiness may persist. Clarification can be aided best by fining, filtration, centrifuging, or sometimes by heating.

Fining hastens defecation, aging, and bottle maturity. Even the best wines are nearly always fined, at least once, immediately before being bottled. In fining, the small particles of suspended material are induced to coalesce and form larger particles which settle out by gravity and carry other suspended matter down with them. According to Nègre (1939), fining gives a more permanent clarification than filtering. He believes that filtering may sometimes remove substances which help to stabilize the wine and that after their removal the wine may cloud more easily than if not filtered. The possibility of contamination by micro-örganisms during filtration and their subsequent growth and clouding of the wines needs to be considered.

Choice of fining agent

The commonly used fining agents are gelatin, isinglass, egg white, casein, sodium caseinate, proprietary products, and calcium or sodium bentonite or their bentonite slurries. These cause flocculation and settling either by combining chemically with the colloids or by neutralizing the electrical charge of colloid particles. Gelatin and similar fining agents that combine with tannin decrease the tannin content of wine and cause a noticeable decrease in color. In certain light wines where loss of color is not desirable, egg albumen or isinglass should be substituted for casein or gelatin. Prior addition of tannin also will help prevent the bleaching of color by casein or gelatin and facilitate the rate of sedimentation. Ibarra and Cruess (1948) and O'Neal et al. (1951) discuss removal of excess color with casein.

The fining agents are usually dissolved in water and then thoroughly mixed with the wine, which is stored until the suspended matter settles out. It can then be racked and, if necessary, filtered.

The amount and type of fining agents used will depend upon the nature of the suspended matter and the type and composition of the wine. The rate of settling of the flocculant formed, the degree of clarity desired in the wine, and the volume and compactness of sediment formed are other factors that influence selection of fining agents. In fining table wines, particularly those of high quality, a low percentage of lees is desired. The percentage of lees in gelatin- or isinglass-fined wines is small, usually less than 2 per cent by volume; whereas that in bentonite-fined wines is much higher, ranging from 5 to 10 per cent. The compactness of the sediment is of importance when the wine is to be racked off the lees without filtering: the isinglass sediment is particularly light and fluffy and more care must be taken in racking isinglass-fined wines. The bulk wines are usually filtered through a filter press after fining so that loss of wine in the lees does not occur. There is, however, considerable work involved (racking, centrifuging, filtering, etc.), and the quality of the wine recovered is reduced.

Although enough of the nitrogenous fining agents must be added to achieve the desired degree of clarity, an excess should be avoided because of the danger that any surplus may support growth of undesirable microörganisms.

The United States Internal Revenue Service regulations (1961a) restrict the conditions of use of fining agents so that the removal of the cloudiness, precipitate, or undesirable odor and flavor will not change the character of the wine or abstract ingredients that will change its character. Specific requirements for certain authorized fining materials are as follows:

Aferrin (calcium phytate)—no insoluble or soluble residue in excess of one ppm (mg per liter)
Bentonite slurry—1 pound of bentonite in not more than 2 gallons of water. Total quantity of water not to exceed 1 per cent of volume of wine treated.
Cufex (ferrocyanide-bentonite complex)—no insoluble or soluble residue in excess of one ppm
Gum arabic—2 pounds per 1,000 gallons of wine
Protovac PV–7916, Sparkalloid No. 1 and No. 2 (polypyrrolidone)—2 pounds per 1,000 gallons of wine
Wine clarifier—2 pounds per 1,000 gallons of wine

When fining agents such as gelatin are added in excess, they may act as protective colloids and stabilize the colloids present in wine. Wines containing excessive amounts of gelatin may become permanently cloudy or may form a cloud after treatment. Ribéreau-Gayon and Peynaud (1934–35) studied the factors which influence the amount of fining agent required. They noted that in the absence of minerals (particularly traces of ferric ion) gelatin and tannin failed to precipitate. This may explain why certain wines fine better when aerated. They also found that the removal of tannins by gelatin was about ten times more active in red than in white wines. In some wines the pH was so low (2.8) that the fining was unsuccessful. They also reported very important influences of temperature. At too high a temperature, 25°C (77°F), gelatin gave a poor clarification. Simply lowering the temperature may help to clarify some wines. Most fining agents cause an increase in the oxidation-reduction potential.

One should always conduct small-scale fining tests in the laboratory before attempting to clarify the cask or tank of wine in the winery: to determine the efficiency of the fining agent, its effect on the flavor of the wine, and the quantity necessary to clarify adequately. Particular care should be taken to determine accurately the minimum quantity of fining agent required.

Principles of fining

Turbidity of wine may be due to presence of coarse suspensions of particles of grape tissue; to suspensions of yeast and bacteria; or to colloidally dispersed particles derived from grapes, formed during fermentation or secreted by microbial cells, or formed as a result of changes during storage and aging of wine. The suspended particles of grape tissue and most of the microbial cells usually settle out fairly completely, but the colloidally suspended matter may remain dispersed for some time. The colloidally dispersed matter has been classified into proteins derived from grape tissue or yeast, pectins and gums, and related hemicelluloses from grape tissue, glucosan from yeast, and metallocolloids formed by flocculation of insoluble oxidized salts of iron (ferric phosphate, ferric tannate) or reduced compounds of copper (cupric and cuprous sulfide) with proteins. Colloidal degradation products of polyphenolics may also be present. The chemical composition and macromolecular structure of these colloids is still not completely elucidated, but the conditions influencing their formation, stabilization or flocculation, and sedimentation are generally known.

The colloid chemistry of clarification of wine was investigated early in Germany, in the classical studies of Rüdiger and Mayr (1929), and in France by Ribéreau-Gayon and Peynaud (1964-1966). The colloid chemistry of interaction of proteins with tannins in wine was reviewed by Joslyn (1953) and recently by Singleton (1967). Joslyn and Goldstein (1964) discussed protein binding in relation to astringency of fruit phenolics.

The reaction between proteins like gelatin and tannins was investigated by leather chemists interested in better understanding of the tanning process (Gustavson, 1949, 1956); by the early brewing chemists (Trunkel, 1910; Hartong, 1929; Page, 1942); and more recently by Moeller (1957).

The combination of proteins with tannins has been investigated by plant biochemists, since this is a factor in limiting recovery of enzymes from tissues of plants rich in tannins. Tannins are known to precipitate proteins and inactivate enzymes in plant tissue on homogenization (Badran and Jones, 1965; Goldstein and Swain, 1965; Loomis and Battaile, 1966). The stoichiometric relations between the gelatin precipitated by tannin and tannins were investigated very early by Trunkel (1910) and later by Hartong (1929) and Page (1942). Kain (1967a) has reported on the weight of tannin precipitated per gram of different gelatins. This varied from 1.2 to 2.4 gr per gr, depending on the source of gelatin.

The combination between gelatin and tannin was early shown to be labile. Trunkel (1910) found that the tannin precipitated by gelatin could be recovered by treatment with alcohol. Subsequently this was shown to be due to the fact that the bond formed between phenolic compounds and proteins is a hydrogen bond which can be broken by addition of urea, dilute alkali, and aqueous organic solvents capable of hydrogen bonding, such as alcohols or acetone. Detannining can be obtained also by addition of caffeine, polyethylene glycol, polyvinylpyrrolidone, and nonionic and cationic detergents. Gustavson (1954) established the fact that hide powder, hydrated nylon, and insoluble polyvinylpyrrolidone (PVP) have nearly the same capacity for binding tannins. The peptide or amide linkage of proteins is involved in the formation of the reversibly hydrogen-bonded complex between tannins and proteins. With hydrolyzable tannins, like gallotannic acid, hydrogen bonding is very strong at pH 3 to 4, but decreases above pH 5. With flavonoid or condensed tannins, binding is almost independent of pH, below pH 7 to 8. With these the binding involves largely un-ionized phenolic hydroxyl groups; with gallotannins strong hydrogen bonds are formed by un-ionized carboxyl groups and weaker hydrogen bonds by un-ionized phenolic hydroxyl groups. Oxidation of phenols to quinones, which may occur before or after reaction with proteins, results in poly-merization and reaction to form covalent bonds with proteins. Oxidation followed by covalent condensation results in an irreversible combination of phenolic compounds with proteins.

Calderon *et al.* (1968) confirmed the now well-established fact that precipitation of proteins by gallotannic acid is pH dependent while that by condensed tannins is not. They reported less gelatin precipitated by tannic acid at pH 3.5 than at pH 5.0, while quebracho tannin precipi-tated about the same amounts at both pH levels. They reported that addition of 15 per cent ethanol decreased the percentage of tannic acid precipitated at pH 5.0 but had no effect on percentage of gelatin pre-cipitated. Alcohol had more effect on tannic acid than on quebracho tannin. Trunkel (1910) had previously reported alcohol to liberate tannin from gelatin-tannates.

In the early investigations of Trunkel (1910) and Rüdiger and Mayr (1929), the precipitation of tannins by added gelatin was considered to be a mutual flocculation of oppositely charged colloidal micelles. Atten-tion was focused on the charge, the distribution of the electrical charge as measured by the electrical double-layer potential, the size and shape of particles, and stabilization of suspension by absorbed water layer as well as electrical charge. This necessarily emphasized the effect of source and method of producing gelatin. The isoelectric point of gelatin obtained by acid hydrolyses of collagen-containing tissues (hide and bones) is about pH 8, whereas that obtained by alkali treatment is pH 4.7–5.0. Rüdiger and Mayr (1929) observed that in fruit wines having a pH ranging from 3.5 to 4.1 the haze particles were negatively charged. The gelatin used in wine fining caused a positive charge in these wines. Flocculation occurred when the positively charged gelatin micelles neutralized negatively charged haze particles, and rapid precipitation resulted. When there is an excess of either gelatin or haze the secondary particles assume the sign of the colloid in excess, and haze persists.

On the basis of the more recent evidence of secondary hydrogen bonds, pH and macromolecular structure are limiting conditions. The distance between the phenolic carboxyls or hydroxyls involved in hydrogen bonding and the N-substituted amides or peptide bonds limits the inter-action (Gustavson, 1956). The composition and molecular size of the phenolic compounds present, the composition and molecular size of protein added, and the pH, temperature, alcohol content, and other environmental factors influence the rate and extent of flocculation. The capacity of a protein-type fining agent would then depend primarily on the number of bonding sites for phenolic hydrogen. The selectivity of the protein-type fining agent would depend on the optimum type, placement, and geometry of the bonding sites. The following examples of phenol removal (weight of phenol removed/weight of fining agent added) by various fining agents are reported by J. A. Rossi, Jr., and Singleton (1966b):

Agent	Condensed tannin	Leuco-anthocyan	Catechin
Gelatin, 100 mg per liter	1.14	0.86	—
200 mg per liter	1.02	0.77	—
Casein, 150 mg per liter	0.79	0.80	—
Isinglass, 150 mg per liter	1.00	1.20	—
Nylon 66 column	0.03	0.10	0.08
Insoluble PVP column	0.02	0.25	0.19

As reported by Gustavson (1956) and others, the larger phenolic molecules should be preferentially removed because of the greater amount of "fit" between the phenol and the fining agent. On this basis it should be possible to produce synthetic resins with high selectivity for undesirable phenolics. Since anthocyanins are incorporated into tannins during aging, it would be possible to remove the tannins early in the aging and thus maintain color. However, color stability is reduced in the absence of tannins.

Fining with clays such as bentonite is based partly on mutual flocculation and partly on specific absorption. Clays such as kaolin and Spanish earth have been used for many years in Europe for wines which are difficult to clarify. Ribéreau-Gayon and Peynaud (1934–1935) used kaolin for wines overfined with gelatin. Bentonite, a montmorillonite clay mineral, has largely supplemented all other fining agents for bulk wines in California and elsewhere (Silica Products Co., 1930). The structure and composition of kaolins, montmorillonites, and related minerals is discussed in the older classical text on mineralogy by Dana and Ford (1932) and in the more modern reference texts of Brown (1961) and Olphen (1963). Kaolin is recognized essentially as a hydrated alumino-silicate present in all clay and shale deposits with an oxide ratio, $Al_2O_3:SiO_2:H_2O$, of $1:2:2$. The montmorillonites are composed of two-dimensional arrays of silicon-oxygen tetrahedra and two-dimensional arrays of aluminum or magnesium oxygen-hydroxyl octahedra superimposed as sheets. Montmorillonites are expanding three-layer clays in which the tetravalent silicon is sometimes replaced by trivalent aluminum and by divalent magnesium. Bentonite is a special mont-morillonite clay characterized by tremendous swelling and a high cation-exchange capacity. The main interlamellar cations are sodium and calcium. These cations are absorbed on the layer surface to compensate the nega-tive lattice charge resulting when an atom of higher valence (Al) is replaced by an atom of lower valence (Zn, Li, Mg). Clays have the property of absorbing organic anions at the edges of the particles and organic cations at the negative face surface by exchange absorption.

Bentonite was first introduced for the removal of colloids in honey by Lathrop and Paine (1931). This use was based on flocculation of the positively charged honey colloids by negatively charged clay particles. Subsequently it was applied to the clarification of vinegar by Saywell (1934a) and then to the clarification of wine (Saywell, 1934b). Bentonite and other clays, however, are known to react with and absorb proteins and similar colloids also. The absorption of enzymes on clays was ob-served very early by Dauwe (1905) and Rideal and Thomas (1922), and was investigated during the period 1922–1925 by Willstätter and his colleagues, and later by Willstätter (1927) and Waldschmidt-Leitz (1929). Alderton et al. (1945) reported that bentonite would absorb lysozyme from egg white and could be used for its isolation. McLaren (1954) called attention to the absorption of proteins on kaolinite and other clay minerals, and McLaren et al. (1958) investigated this in detail for kaolinite and montmorillonite clays. The mechanism of this effect is discussed also by Olphen (1963). Kean and Marsh (1956b) observed that bentonite fining removed proteins from wine and thus increased its tolerance to heavy metals. This was confirmed by Berg and Akiyoshi (1961), Nasledov (1963), Rankine and Emerson (1963), Ásvány (1965), Petró (1965), Terčelj (1966), and Scott (1967), and was investigated in detail by Kain (1967b).

The results of Tarantola (1966) confirm that protein stability results from bentonite fining. Bentonite treatment at low temperature—5°C (23°F) for 24 hours or at 0°C (32°F) for 12 days—was less satisfactory than at higher temperatures. He recommended 1 gr per liter for protein removal, in either the must or the wine. A good review of the use of bentonite as a fining agent was given by Ferenczi (1966c). In amounts of 0.30–0.60 gr 'per liter stability of Hungarian wines is assured. He re-ported that the reduction of protein nitrogen amounted to 4 to 11 mg per liter, depending on the bentonite fining. The polphenol content and color were reduced, but sodium and calcium, ash, and alkalinity of the ash were increased.

Hennig (1950) showed that appreciable chemical changes occur in wines treated by bentonite (at levels of 100 to 150 gr per 100 liters):

decreases in titratable acidity (0.005 to 0.03 per cent) and in total nitrogen (20 to 50 mg per liter) or increases in ash (20 to 300 mg per liter), calcium (5 to 50 mg per liter), magnesium (10 to 20 mg per liter), sodium (10 to 45 mg per liter), iron (0.2 to 1 mg per liter), and pH (0.02 to 0.2).

The Codex Oenologique International of the Office International de la Vigne et du Vin (1964) specifies that bentonites to be used in wines should not have any abnormal odor. When used at the rate of 2 gr per 100 ml of distilled water, they should have a pH of 10, should not lose more than 5 to 15 per cent in weight when heated at 105°C (221°F), should have no more than 20 mg per liter of lead in a 10 per cent acetic acid extract, or 4 mg per liter of arsenic, 400 mg per liter of iron, 60 milliequivalents of calcium plus magnesium in a citric acid extract, should give no qualitative test for magnesium or aluminum, and should have no more than 300 milliequivalents of base in 100 gr of bentonite.

Ribéreau-Gayon (1947) recommended using a small amount of activated carbon along with the bentonite to absorb any undesirable flavor. Bentonite is now available in a powder which can readily be made up into a smooth suspension and added directly to wine. About one-half to one pound is used per 100 gallons of wine. Excessively accurate control tests need not be carried out, since a small excess of added fining agent does not cause clouding. Settling is complete and rapid; the sediment is usually heavy and compact. Increasing the temperature to 48.8°C (120°F) has been shown to speed up materially the coagulation and settling of bentonite. When used with hot wine, one to two pounds per 1,000 gallons is sufficient.

Kain (1967a) tested nine commercial bentonites (only three of known origin and brand names not given). Besides the tests suggested above, he developed several new procedures: as an example of the variations encountered, sand content varied from 0.1 to 1.7 per cent, lead from 0.2 to 2.1 mg per liter, sodium from 0.02 to 1.6, calcium from 0.3 to 0.95, and swelling capacity from 9 to 49 ml per 2 gr. None of the samples passed all the tests, suggesting that closer control of bentonites for winery use needs to be established—at least in Austria, where his studies were made. Kain (1967b) reported data on the protein absorption of nine selected bentonites from eight kinds of white wine. Bentonites of high swelling capacity absorbed more protein than those of low swelling capacity. Bentonites were observed to absorb coloring matter from red wines; the anthocyanin monoglucosides were absorbed to a greater extent than the diglucosides.

Ásvány (1965) showed that addition of bentonite to fermenting musts reduced the total nitrogen content of the resulting wines 14 to 75 per cent. Addition of about 1.5 to 2 pounds per 1,000 gallons ensured protein stability in the wines. Petró (1965), however, found that bentonite fining of musts reduced the total nitrogen of the wines only about 20 per cent. This resulted in better-quality wines with a lesser tendency to oxidasic casse. She recommended bentonite treatment of musts (up to about 3.5 pounds per 1,000 gallons) and blue fining of wines to ensure protein stability (pp. 542 ff and 802).

Terčelj (1966) reported that clarification with gelatin and tannin, bentonite, or ferrocyanide, or heat treatment to 50°C (122°F) reduced the protein but not the free amino acid content. Treatment with ion-exchange resins reduced the free amino acids markedly, but had a lesser effect on the protein content.

The major possible advantages of bentonite fining are protein removal, prevention of copper cloudiness, reduction of the possibility of iron clouding, absorption of growth factors, absorption of oxidases and other materials, and mechanical clarification. The disadvantages are possible exchange of sodium, potassium, calcium, magnesium and other ions, absorption of red colors, removal of vitamins, and quantity of sediment.

According to Weger (1965b), bentonite can be used as needed in Italy, but German law limits it to 1.5 gm per liter and Austria to 2.0 (also requiring low sodium and iron in 10 per cent acid extracts of the bentonite). He called attention to the difference in the properties of calcium and sodium bentonites. Sodium bentonite has greater swelling properties and results in a larger volume of precipitate and less wine than with calcium bentonite. The amount of protein removed may differ by 60 per cent. They also appear to differ in degree of reduction of acidity and amount of ion exchange. The effects varied between wines. Therefore, Weger recommended clear labeling of bentonites and laboratory testing before use. Weger (1965b) preferred gelatin-bentonite as the general fining agent for red wines and albumen-tannin-gelatin-bentonite for whites. The use of gelatin with bentonite, however, would obscure the protein-removing properties of the latter.

The diversity of qualities and prices of bentonites was emphasized by Weger (1963b). He showed (1965a) that stability tests at different temperatures can give very different bentonite requirements. Better stability was achieved by fining at 50°C (122°F) or higher. Removal of precipitated color pigments then occurred.

Litschev et al. (1966) found the flocculation of sodium bentonite in wines to be colloidal and not to depend on the source of the bentonite. The swelling properties as a measure of fining effectiveness, they believe, have been overrated. The reaction of bentonite in the wine was found to be primarily one of ion exchange. Hydrogen bentonites remove iron and aluminum from wines. For convenience in handling, calcium bentonite appears best, as it is easier to add and to filter. The problem of using sodium bentonite instead of calcium bentonite is apparently not a critical problem in this country. In Germany, where there is a limit of 60 mg per liter of sodium in commercial wine, use of sodium bentonite would be contraindicated. Also, sodium bentonite may leave colloidal material in suspension. Troost (1966a) found that in German wines sodium bentonite resulted in a large, loose precipitate which was difficult to remove and resulted in wines of higher sodium content. Therefore, he recommended calcium bentonite for fining since it yields a heavier precipitate and does not result in sodium pickup. Ferenczi (1966c) notes that sodium bentonites give good flocculation of proteins, but produce a large flocculent deposit from which it is difficult to filter. Calcium bentonites do not remove as much protein and the precipitation is slower, but the precipitate is smaller and filtration is simpler.

Ferenczi (1966c) notes that enologists are unanimous in recommending bentonite as a method of removing protein. He noted reduction of total nitrogen of 10 to 130 mg per liter and of 3 to 35 mg of protein by bentonite fining. Pasteurization at 40°C (104°F) for eight to ten days gave good protein stability, but no data on the sensory quality of the treated wines were given.

Jakob (1965) reported an increase in the ash content of 30 to 310 mg per liter when 4 gr per liter of sodium or calcium bentonite were added to wines. Sodium bentonite has a swelling capacity of 4.6 to 8.3 times its weight in water, whereas calcium has a swelling capacity of 2.4 to 2.9 times. The protein absorption capacity of a bentonite is related to its swelling capacity, and all bentonites are more effective if preswelled in water or wine before use. Jakob favors an ion-exchange type of explanation of the effect of bentonite. Sodium pickup from sodium bentonite amounted to 1.7 to 3.5 gr per 100 gr of sodium bentonite; from calcium bentonite the calcium pickup was 0.7 to 1.5 gr per 100 gr of calcium bentonite.

In untreated Hungarian wines, Tuzson (1967) reports that the most common crystalline deposit was calcium tartrate followed by potassium acid tartrate. Potassium oxalate was seldom detected. Following bentonite treatment, calcium tartrate and potassium acid tartrate were reported in finished wines, whereas only the latter was found when no bentonite was used.

For degassing of wines, Geiss (1963a, 1963b) recommended treatment with bentonite, stirring, application of a vacuum, and pumping over. Wines of high viscosity which contain excess carbon dioxide sometimes appear cloudy when opened. He recommends EK-filtration (p. 560) for these wines.

Bentonite has replaced tannin in many wineries as the fining agent to follow gelatin.

Isinglass

The best organic fining agent is isinglass. This is added in preparation for bottling at the rate of 0.5 to 1.5 ounces per 100 gallons. The isinglass may be dissolved in water or wine, preferably cold, by soaking it overnight. It should then be ground or rubbed on a fine screen. A brilliantly clear wine results if the clarification is successful; but, as the isinglass sediment is light and fluffy, care should be taken to avoid disturbing it in racking. In using this claryifying agent, add it slowly to the wine, with vigorous stirring or pumping over to mix it in thoroughly. If necessary, tannin, approximately equal in weight to the isinglass, may be added several days before the isinglass, but if isinglass alone gives satisfactory clarification, tannin should be omitted.

Gelatin

In fining with gelatin, use only the purest edible grade, free from objectionable odor or taste. It should be dissolved in warm water to form a solution of about two ounces per gallon. From one-third to one ounce of gelatin is used per 100 gallons of wine. In white wines, about the same amount of tannin, depending on the wine, should be added to the wine several days before the gelatin. Tannin need not be added to red wines. When used on red wines, there is a loss of color and tannin that is proportional to the amount of gelatin used, as indicated by the following figures:

Gelatin added (mg per liter)	Color	Tannin (gr per 100 ml)
0	750	0.193
50	625	0.189
100	588	0.184
250	526	0.172

The figures for color represent the relative intensity expressed on an arbitrary scale: the higher the figure, the greater the concentration of pigments.

The gelatin is added slowly with stirring or pumping over; the wine is then allowed to settle and is racked from the sediment after several days. Liquid gelatin is also available, (Weger, 1963a).

Kain (1967b) reported significant differences in composition, tannin precipitation capacity, and nonprecipitable nitrogenous constituents in samples of different gelatin preparations. In seventeen commercial wine-clarifying gelatins, the pH of a 1 per cent aqueous solution varied from 5.2 to 7.0, the chromium content from 3 to 640 mg per liter, the iron content from 31 to 169 mg per liter, the zinc content from 0 to 600, and the nontannin precipitable nitrogen from 0.2 to 0.98 per cent of the total nitrogen. Kain warned about the introduction of off-odors, off-tastes, and chromium into wine by the use of impure grades of gelatin. The relative tannin-combining property should also be a check on each sample. In wines of high tannin content Klenk and Maurer (1967) noted abnormal tartrate crystals. To prevent this, removal of some of the tannin was recommended.

Casein

Specially prepared soluble, odorless, and tasteless casein preparations are available for fining wines. They are usually added at the rate of 4 ounces per 100 gallons of wine. Potassium caseinate, sodium caseinate, and edible casein are used. The latter must be solubilized by treating with ammonia or sodium bicarbonate before use. All casein preparations give best results if added after being dissolved in water made slightly alkaline with bicarbonate. Although casein preparations decolorize wine, they differ in their decolorizing action. O'Neal et al. (1951) found that the higher the formol titration value of the casein, the greater its decolorizing power. They recommended that casein fining be done at cellar temperature. Somewhat better results were obtained with a combination of casein and bentonite than with casein alone. They also recognized the possibility that casein may adversely influence the odor of certain wines, and advised that laboratory tests should be conducted on each wine to determine the appropriate amount to add.

Egg albumen

The whites of from four to eight fresh eggs or the equivalent in egg albumen (1 ounce of dried egg albumen) are used per 100 gallons of wine. Fresh egg whites are preferred by some, but they must be separated from the yokes in a sanitary manner. For this reason, egg albumen is often used. For either, a suspension in water is prepared.

Other clarifying agents

Among other clarifying agents, sodium alginate, polyvinylpyrrolidone (PVP), phytate, and nylon paste have been recommended. PVP was recommended for fining California red wines by La Rosa (1958) and Clemens and Martinelli (1958). In a comparative test between PVP and gelatin, Ough (1960b) reported that PVP removed more tannin than gelatin and that the treated wines were easier to filter. However, PVP also removed more color, the amount of sediment was greater, and in color stability tests there was a slight difference in favor of gelatin. In sensory tests, also, a slight preference for gelatin-treated wines was expressed. De Rosa and Biondo (1959) also reported PVP inferior to gelatin for white wines.

Sodium alginate has been recommended for some time for fining of wine. Geoffroy and Perin (1960) used 2 gr per hectoliter (about 0.2 pound per 1,000 gallons) for Champagne stocks. The resulting wines gave less masking (p. 670), and the quality was equal to or better than samples treated with bentonite or tannin-gelatin. The wines also were easier to filter. Deibner and Bénard (1958), however, were unable to show any removal of iron from wines with sodium alginate. However, the three wines they tested were *exceptionally* high in iron, 97 to 187 mg per liter.

Calcium phytate can eliminate the ferric iron of wines, and its use is authorized in many countries. If too much phytate is added some will remain in the wine. If the wine is subsequently aerated and ferric iron is formed, a precipitate will form. For this reason Léglise and Michel (1958) recommend careful laboratory tests to establish the exact amount of calcium phytate necessary to remove only the ferric iron. A safety margin of 4 gr per 100 gallons was allowed. They found bentonite unsatisfactory as a fining agent following phytate treatment, and preferred albumen (20 gr per 100 gallons).

Use of phytates for removal of iron, however, has not been popular because reprecipitation sometimes occurs. This happens when the wine is not aerated during treatment and is not left in contact with the phytate long enough to allow for reaction and precipitation. Storage for at least a week with frequent aeration is recommended (to ensure that all the iron is present in the ferric condition). Of course there should be a preliminary laboratory study to establish the proper amount of phytate.

As a means of removing iron and copper and maintaining the color, Maruyama and Kushida (1966) added EDTA (disodium calcium ethylenediaminetetraacetate) or ascorbic acid or both. Ascorbic acid alone reduced the color of white wines. Ascorbic acid with EDTA or EDTA alone increased the color of white wines. Surprisingly, ascorbic acid plus EDTA reduced the color of pomace wines. Joslyn and Lukton (1953) found EDTA effective in preventing turbidity due to iron or copper. It does not appear to be approved for use in wines in this country at present.

De Villiers (1961) and Cantarelli (1962) recommended use of nylon paste to improve the color of white wines, to remove browning precursors, to increase resistance to browning, and to reduce anthocyanogen and tannin content. Fuller and Berg (1965) recommended nylon in preference to casein because it gave greater protection against browning. They emphasized the necessity for laboratory trials in view of the finding of Caputi, Jr., and Peterson (1965) that nylon treatment sometimes increases the tendency to browning. Nitrogen stripping, carbon dioxide blanketing, and addition of sulfur dioxide all have their place in the handling of such wines. Singleton (1966a) pointed out the similarity in hydrogen-bonding capabilities of polyamides (nylon) and PVP compared to gelatin.

Sparkolloid (a proprietary product) is a refined polysaccaride fining material, according to Scott (1967). He recommends it for hard-to-clarify wines and notes its small lees yield. A drawback is that the material must be heated before use. Scott has compared several fining agents on a cloudy dry white wine (table 64). He noted that the character of the lees

TABLE 64

CLARIFICATION TESTS OF DRY WHITE WINES[a]

Fining agents	Visual observation		Turbidity (Nephlos units)		Filtration Time (seconds per 100 ml)
	Lees (mm)	Clarity (1-10)	Before filtration	After filtration	
Control	1	1	46	10	389
1 lb. Bentonite	4	3	28	6	295
3 lbs. Bentonite	8	3	24	8	213
1 lb. Casein	2	1 −	97	9	307
3 lbs. Casein	3	1 −	100 +	8	261
1 lb. Sparkolloid	2	3	26	5	★★[b]
2 lbs. Sparkolloid	3	2	15	6	310
4 oz. Gelatin	1	1 −	78	7	298
4 oz. Gelatin, 1 lb. Bentonite	4	5	13	5	297
1 lb. Cufex, 1 lb. Sparkolloid	2	3	14	3	277
1/2 lb. Casein, 2 lbs. Bentonite	7	5	13	4	288
3 lbs. MM-7 Clay	3	1 −	44	3	281

[a] Alcohol 11.5; reducing sugar 0.25; total acid 0.683; pH 3.34; iron 4.8 mg per liter; and copper, trace. The wine was heat-stable and cold (tartrate)-unstable.
[b] Pressure failure, no reading.
Source of data: Scott (1967).

varied from the slimy nonfilterable types to granular sediments from which the wine could be filtered.

A new commercial fining agent to supplement Sparkolloid has been developed by Scott (1969). It is called Cold Mix Sparkolloid and requires no heating before use. Both go into the wine as negatively charged macromolecules. Their advantage is in their compact lees. Scott (1969) developed a fibrillated asbestos which could be used as a clarifying agent prior to filtration. It increased filter pad life without reducing pad performance.

PROTEINS IN WINE

Of the colloids known to be present in wine,
polypeptides and proteins are of particular
importance in influencing stability.

Ribéreau-Gayon (1932) called attention to the importance of proteins as causative factors in the instability of wine. The proteins in wine were investigated by Kielhöfer (1942, 1948a, 1948b, 1949a, 1949b, 1951, 1954), Koch et al. (1956), Koch (1957, 1963), Koch and Bretthauer (1957), Koch and Schwahn (1958), and Diemair et al. (1961a, b). Proteins serve as nuclei about which insoluble copper, iron, and other salts deposit, and on denaturation by heat or cold may flocculate and form unsightly suspensions and deposits. Proteins were shown by Ribéreau-Gayon (1933b) to be involved in both ferric phosphate casse and cupric sulfide casse formation. This was confirmed by Kean and Marsh (1956a, 1956b) and Lukton and Joslyn (1956). Both of the latter investigations used the modified gravimetric estimation of protein by precipitation with 10 per cent trichloroacetic acid at 90°C (194°F), originally proposed by Hoch and Vallie (1953).

The "warm" test for excess protein in wines was devised by Kielhöfer (1948a, 1948b). The Koch and Sajak (1961) modification is now widely used: 95 ml of wine and 5 ml of cold saturated ammonium sulfate are mixed in a 100-ml Erlenmeyer flask. The flask is held seven hours at 55°C (131°F), and is then held in an ice bath for fifteen minutes. Cloudiness or a precipitate is an indication of excess protein. This is obviously a test for only the heat-unstable proteins. Troost and Fetter (1960) found that it did not always predict the proper amount of bentonite to use.

To detect "soluble" protein in musts and wines Werner and Hartmann (1966) used a quantitative colorimetric biuret reaction, which consists of treating 10 ml of wine with 10 ml of phosphomolybdic acid reagent[1] and shaking well. Wines containing soluble proteins become cloudy, and after 24 hours show a precipitate. For quantitative protein determination after 24 hours at room temperature, centrifuge at 3800 rpm for twenty minutes. Discard the supernatant liquid, add 10 ml of 96 per cent ethanol, and shake well. Centrifuge for ten minutes at 3800 rpm. Discard the supernatant liquid and add 2 ml of 2 N sodium hydroxide. Warm the centrifuge tube with the hand to dissolve the protein precipitate. Add four drops (0.12 ml) of 20 per cent copper sulfate solution and shake well. Bring to 10 ml with 2 N sodium hydroxide and shake again for two minutes.

---

[1] Prepared from 5 gr phosphomolybdic acid, 15 gr concentrated sulfuric acid, 5 gr sodium sulfate, and 0.25 gr glucose brought to a liter with distilled water.

Leave two hours for the biuret color to develop. Centrifuge for ten minutes at 3500 rpm and filter through hard filter paper. The color intensity was made at 546 mμ compared with a distilled water blank. The solution and filter paper were then washed into a Kjeldahl flask and the total nitrogen was determined. After several wines have been thus treated, a standard curve is drawn. A separate curve for musts is needed. Their test differentiated excess protein from fresh grape juice and low protein grape juice made by diluting grape concentrate. The advantage of their test is that it gives quantitative data which are useful in determining the appropriate type and amount of treatment. However, amount of protein and protein stability are not correlated.

A practical test for protein stability of wines was proposed by Berg and Akiyoshi (1961). To 10 ml of wine 1 ml of 55 per cent trichloroacetic acid was added and the tube was placed in boiling water for two minutes. The sample was left at room temperature and the degree of clouding was tested in a Coleman Model 9 Nepho-Colorimeter. The amount of haze determined in this fashion correlated very well with the amount of haze produced by storing 3-ounce screw-cap bottles at 48.8°C (120°F) for four days. Absence of haze or amorphous deposit classified the wine as protein-stable.

Danilatos and Sotiropoulos (1968) compared various tests for protein stability: tannin, trichloroacetic acid, heat plus ammonium sulfate, heat, and phosphomolybdic acid. The tannin test was most sensitive. In general Greek wines are low in protein and are protein-stable—8 to 25 mg per liter. The levels of protein present in wine as reported by various investigators, and summarized by Amerine (1954), varied from 0.015 to 0.143 per cent of nitrogen as protein. Joslyn and Lukton (1956) investigated wines varying from 0.015 to 0.110 per cent. Koch and Sajak (1959) reported that the protein levels varied considerably with method of precipitation: lower levels were found with the trichloroacetic acid procedure than with acetone, ammonium sulfate, sodium tungstate, or sodium sulfate-alcohol precipitants. They found that the proteins present in grape must were nonhomogeneous and could be separated electrophoretically into two or more fractions. The proteins in grape must and in wine (p. 464) are negatively charged polyelectroytes. Holden (1955) proposed a combined heat and cold treatment for the stabilization of wine. The actual protein level at which table or dessert wines are stable and the changes in protein content during production and processing of wines are still not known with sufficient accuracy to predict wine behavior.

Where the pH of the wine is near the isoelectric point of the wine proteins, greater precipitation will occur. The malo-lactic fermentation, by increasing the pH, sometimes results in protein clouding by bringing the pH of the wine nearer to that of the isoelectric point of the proteins. This may occur during blending also. Fermentation, of course, also results in precipitation of proteins, more of Koch's (1963) fraction II than I. There is also a progressive reduction of proteins during aging. They can be removed by ultrafiltration or by heat. However, heating removes more of fraction I and less of II. In years of high protein content, heating may leave troublesome amounts of protein in solution. Bentonite fining, however, removes most of fractions I and II. See also Kean and Marsh (1956a, 1956b) and Koch and Sajak (1959, 1963).

Böhringer and Dölle (1959) reported that the eight amino acids primarily involved in clouding are aspartic acid, isoleucine, leucine, methionine, phenylalanine, threonine, tyrosine, and valine. Wines containing aspartic acid and threonine may be stable; so the other six are most involved. Masuda and Muraki (1966) reported that immediate centrifuging to remove yeast cells after fermentation markedly reduced the amino acid content of the wine. Only proline was found in appreciable amounts in wines treated in this way. Holding wines on the lees one to four months resulted in increases in the amino acid content especially of α-alanine, arginine, aspartic and glutamic acids, glycine, histidine, leucine, and lysine. In fact, cystine, methionine, phenylalanine, and valine appeared only in wines that had been stored for four months with the yeast deposit.

FINING TO REMOVE IRON AND COPPER

Although metallic cations, particularly ferric and cupric ions, are necessary catalysts to promote desired oxidative changes, their presence in excess may cause hazing or formation of deposits under certain conditions. This clouding, because of the conditions under which it occurs, is referred to as "casse" in the enological literature. Iron clouds, particularly ferric phosphate casse, were common in the early post-Repeal period in California; copper casse was even more troublesome when iron and steel crushers, pumps, lines, and other equipment were replaced with copper and copper alloys. These conditions have been largely overcome by introduction of more corrosion-resistant metals, glass conduits, reduction in protein content by special fining, application of ion-exchange resins, and addition of permissible metal-chelating agents, such as citric acid. Methods for removal of excess copper and iron cations and prevention of copper and iron turbidities in wine are discussed by Joslyn et al. (1953) and Joslyn and Lukton (1953). (See also pp. 794–798.)

The removal of excess iron by the addition of potassium ferrocyanide was introduced over fifty years ago in Germany by W. Möslinger. It was investigated and described in the monographic treatises of Heide (1933) and Ribéreau-Gayon (1935a). This treatment, the so-called "blue-fining," is widely used in Germany, but is not permitted in the United States. It is apparently still illegal in Italy, Spain, and Portugal, although there seems to be some surreptitious use; it is legal, under strict control, in France (Cordonnier, 1961).

Blue-fining, unless carefully supervised, may lead to the liberation of cyanide. Both California and federal regulations limit the residual cyanide content, as determined by Hubach test, to 1 mg per liter (Hubach, 1948). When excess ferrocyanide is added and hydrocyanic acid is formed, some of the hydrocyanic acid is removed by absorption on the blue deposit.

Small amounts of hydrogen cyanide (free and fixed) occur in untreated wines. Jaulmes and Mestre (1962) fix the maximum at 0.25 mg per liter. In three wines fined with potassium ferrocyanide and poorly handled (blue deposit) there were 0.0006, 0.05, and 0.015 mg per liter of free hydrogen cyanide and 0.095, 0.55, and 0.62 mg per liter of total. They give a sensitive method for its determination. The more complicated technique of Deibner and Mourgues (1963) permits detection of 0.06 mg per liter, and recovery appears to be 90 per cent.

Another problem with blue-fining is that ferrous iron combines rapidly with ferrocyanide, while ferric iron is partly complexed with organic acids and reacts slowly with ferrocyanide. For this reason wines should be blue-fined in the absence of oxygen. It has even been suggested that ascorbic acid be added before blue-fining to ensure that as much as possible of the iron is present in the ferrous condition.

The complicated and not completely resolved chemistry of the reaction of ferrocyanide and iron has been studied by Bonastre (1959). He showed that the presence of an organic fining agent, isinglass, markedly reduced the efficiency of the ferrocyanide. The removal of iron is greater in dilute solutions even when the ratio of ferrocyanide to iron is constant. The reaction is slow, lasting up to seven days.

Tarantola (1963) recommends that at least two hours be allowed for the preliminary laboratory test to determine the quantity of ferrocyanide to add. Since filter paper may absorb some ferrocyanide, it is recommended that the laboratory samples be centrifuged. The ferrocyanide should be added and allowed to react before the fining agent. We also recommend that no attempt should be made to remove all the iron. It is necessary to remove only the excess iron that might cause cloudiness. Tarantola stated that in properly treated wine no free or combined hydrocyanic acid caused by the treatment will be found in measurable quantities.

When wines are stored on the ferrocyanide precipitate at $15°C$ ($59°F$) for a month, the free and total cyanide content of the wine remains less than 0.01 mg per liter according to Tarantola and Castino (1964). At $22°C$ ($77°F$), however, as much as 0.1 mg per liter will be found after a month's storage. If excess potassium ferrocyanide is used, much more cyanide is formed. Even so, with a greatly exaggerated excess, the amount of cyanide formed is only 1-60th to 1-120th of the lethal dose. See also Miconi (1962a).

Jaulmes et al. (1963–64) fined with ferrocyanide (20 mg per liter), bentonite (1 gr per liter or 8 pounds per 1,000 gallons), gelatin-tannin (100 mg per liter of each), albumen (1 gr per liter), and kaolin (3 gr per liter). There were large losses of vitamins: for cyanocobalamine, 25, 25, 13, 1, and 12 per cent, respectively; for nicotinamide, 16, 16, 16, 37, and 43 per cent; for riboflavin, 9, 37, 8, 0, and 0 per cent; for folic acid, 1, 36, 2, 14, and 53 per cent; for thiamine, 19, 65, 6, 13, and 22 per cent; and for biotin, 33, 64, 61, 58, and 48 per cent. The same treatments caused loss of amino acids as follows: tryptophan, 19, 23, 9, 10, and 16 per cent; methionine, 10, 14, 2, 10, and 7 per cent; lysine, 21, 13, 0.6, 13, and 16 per cent; and leucine, 12, 19, 12, 5, and 0.9 per cent. The authors conclude that bentonite causes the greatest losses in amino acids.

Several proprietary complexes of ferrocyanide were proposed for use in heavy metal removal (Joslyn and Lukton, 1953). Of these only the Fessler compound, now sold as Cufex, is permitted in the United States. The use of Fessler's compound (Fessler, 1952) was endorsed by Marsh (1952). Cufex is a buffered potassium ferrocyanide compound, according to Scott (1967).

Martini (1965) has proposed using racemic tartaric acid before refrigerating wines as a method of removing excess calcium. Crystallization was encouraged by pumping over. It should be used only on clear wines. The costs of racemic tartaric acid are not given.

PECTIC ENZYME TREATMENT

Fungal pectic enzymes have been introduced for the purpose of increasing the yield of free-run juice from must, of free-run wine from red wine, or for clarification of juice or wine. Their use for the clarification of apple and grape juice was established by the investigations of Kertesz in the United States and Mehlitz in Germany about forty years ago. Their use in wine making was proposed in 1941 by Besone and Cruess (1941). Commercial fungal pectic enzymes differ in concentration and type of pectic enzyme present and in type of diluent (diatomaceous earth, gelatin, sugar) used in standardizing their activity (Joslyn et al., 1952; Neubeck, 1959; Reed, 1966). Some twelve proprietary enzyme preparations are permitted by federal regulation; these may be added in amounts varying from 0.3 pound to 10 pounds per 1,000 gallons depending on the product used.

Pectic enzymes are still widely used in the production of Concord juice and wine in which the activity of the naturally occurring pectic enzymes is inhibited by the heat treatment used in color extraction and in which the added pectic enzyme promotes increase in yield of juice and improves the pressing properties of pomace by maceration of uncrushed berries and decomposition or precipitation of slime-producing pectins and gums. Under California conditions where crushing is more complete, where heat treatment for color extraction is not widely used, and where the sulfite added to must promotes activity of naturally ocurring pectic enzymes, particularly pectin esterase (Edwards and Joslyn, 1952), addition of pectic enzymes is limited. Pectic enzymes are used commercially only when the conditions warrant. In some varieties of grapes and under some conditions appreciable increases in yield and clarity of free-run juice do occur with use of pectic enzymes.

Berg (1959b) compared several commercially available pectic enzymes under controlled commercial conditions and found appreciable variability in results. The variability of results with pectic enzymes previously reported was due, according to Berg (1959b), to differences between the enzymes and differences between musts. In general, the enzymes increased juice yield, clarity, and rate of browning. In the production of Concord wines, where the activity of naturally occurring pectic enzymes is inhibited by tannins and heat destruction during color extraction, pectic enzymes are widely used. The variation in activity of normally occurring pectic enzymes in California grapes and consequently in the content of residual pectic substance in the must and wine still is not known.

FILTRATION

In the clarification of wine, filtration is usually a necessary supplement to settling, racking, and fining.

Very cloudy wines may be filtered to clear them rapidly; bulk filtration of this type of wine is used also in preparing ordinary bulk wine for shipment. However, all wines to be bottled are given a finishing or polishing filtration just before bottling. The type of filter used depends on the quality of wine and its condition. Close filtration is used rather extensively abroad, and recently has found favor in California for sterilizing wines by removing practically all microörganisms. For data on filtering grape juice see Harris (1964); for beer, Horne (1942). Kowala (1965) gives useful information that is applicable to the wine industry on filtering in an inert atmosphere.

Particularly in a finishing filtration, the wine must be protected from excessive aeration and from contamination with metallic impurities from the filter lines and the filter medium. Contamination of wine by calcium and iron salts absorbed from filter aids and filter materials was reported by Saywell (1935) and subsequently by several others. Special care is now taken to prepare filter aids (by calcining) and filter pads and asbestos (by acid washing) to prevent contamination.

To prevent color loss when this is an important factor, Feuillat and Bergeret (1966) recommended minimum aeration during filtration.

The rational use of filtration has been emphasized by Marsh (1949) and Geiss (1963b). Geiss recommends use of diatomaceous earths as filter aids—as free as possible of iron and calcium. Asbestos filters are considered the best. Their correct preparation is very important. The proper use of filters depends on keeping the pressure differential as low as possible, especially for germ-proof filtration. Geiss recommends low-pressure filtration so that no more than 50 liters per hour are filtered for 40 × 40 cm filter pads.

Factors influencing filtration

Filtration is essentially straining through a porous medium which allows liquids to pass but retains the solids. The rate of filtration depends on the size of the pores in the filter medium, the percentage and types of solids present, the rate of deposition of the solids, the resistance to flow of the filter medium, the flow characteristics of the liquid being filtered, and the thickness of the filter medium and cake. When filtering begins, the pores in the surface of the filtering medium are throttled by particles which settle on them; as filtration proceeds, the liquid has to escape through the throttled orifices and through a layer of particles of increasing thickness. In filtering out nonrigid solids, such as those in wine, pressure must be built up gradually and uniformly to minimize the effect of squeezing together the nonrigid solids under increased pressure.

The theory and practice of filtration are discussed by Dickey and Bryden (1946) and the more recent advances by Miller (1947–1949). The application of filter aids to wine filtration is described by Saywell (1935) and in recent circulars of the manufacturers of diatomaceous filter aids, such as the Johns-Manville Corporation and the Dicalite Corporation.

In general, the clarity of the filtered wine varies inversely with rate of filtration; a compromise must therefore be made to attain adequate clarity without unduly reducing the rate of filtration.

Figure 78.   Filter press in foreground; filter-aid mixing tank in background.

Bulk filtration

For bulk filtration, plate-and-frame filter presses are commonly used (figs. 78, 79, 80, 81); leaf filters are less common, and pulp filters are rarely used in California. In the plate-and-frame filter press, the wine is forced through the filter medium by a nonaerating pump, preferably of the positive-displacement gear type. The wine is forced, under pressure, through canvas sheets lined with a filter paper and precoated with diatomaceous earth. The individual filter compartments are formed by corrosion-resistant metal frames, and the canvas with its paper liner and precoat is supported by a metal plate (figs. 82, 83a). The filter-paper liner, which is now common, keeps particles of filter aid from going through the canvas, and materially assists in cleaning the canvas and the press. To minimize the sliming and clogging effect of the soft, amorphous particles which tend to pack together and clog the filter pores, the wine is continuously mixed with a filter aid. For best results the filter aid must be uniformly mixed with the wine and held in suspension by constant

549

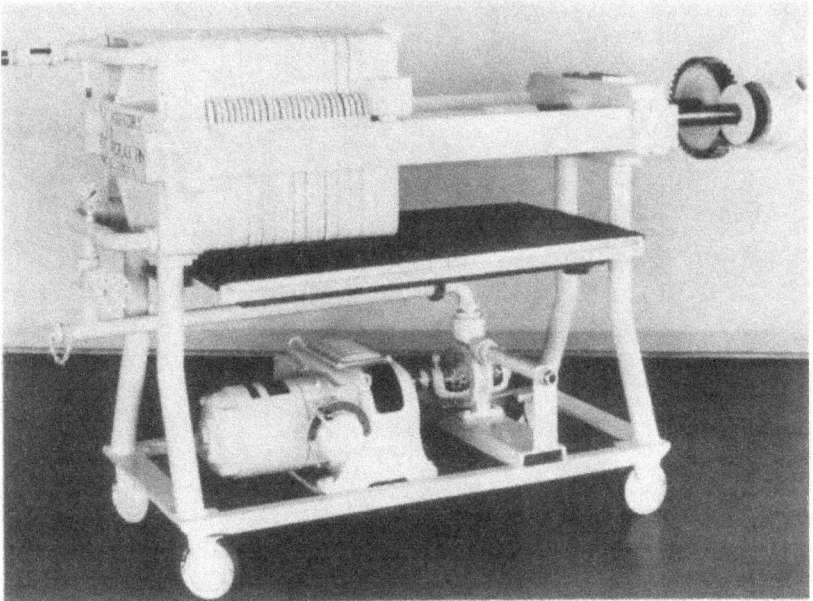

Figure 79. Stainless steel plate-and-frame filter, with varidrive motor, turbine type pump, and drip pan. (Courtesy Valley Foundry and Machine Works, Inc., Fresno.)

Figure 80. Filter press for lees filtration. (Courtesy Budde & Westermann Machinery Corp.)

Figure 81.   Stainless steel plate-and-frame filter showing screen plate and frame. (Courtesy Valley Foundry and Machine Works, Inc., Fresno.)

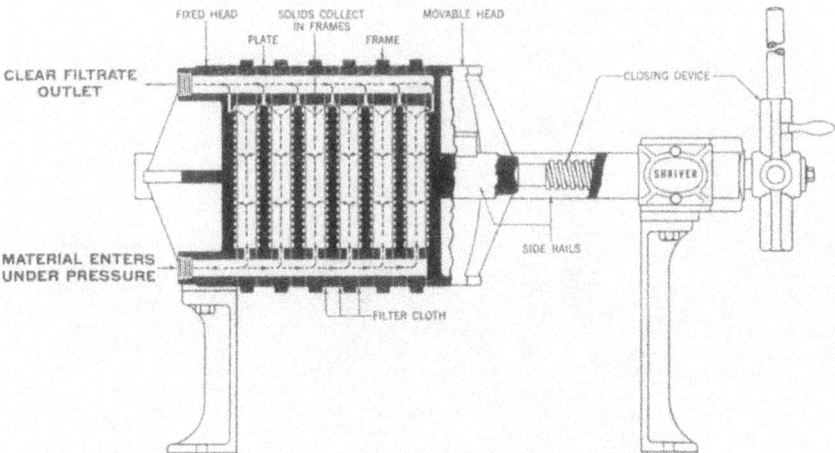

Figure 82.   Sectional view of a filter press showing arrangement of plates, filter cloth, and frames and indicating flow of material. (Courtesy T. Shriver & Company, Inc., Harrison, New Jersey.)

551

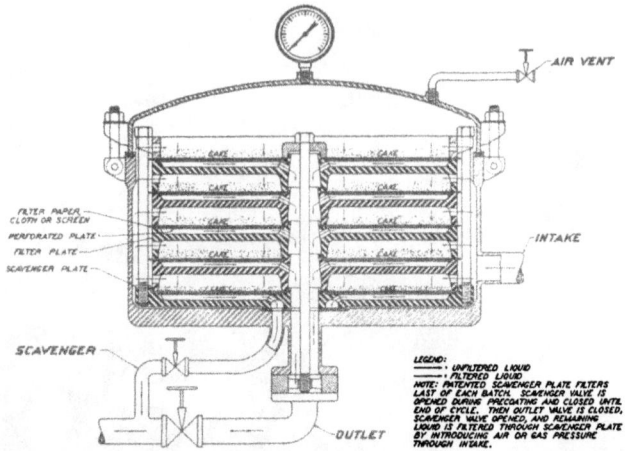

Figure 83*a*.   Cross section of a horizontal plate filter showing arrangement of plates, filter paper (or cloth or screen), and cake. (Courtesy Sparkler Manufacturing Co.)

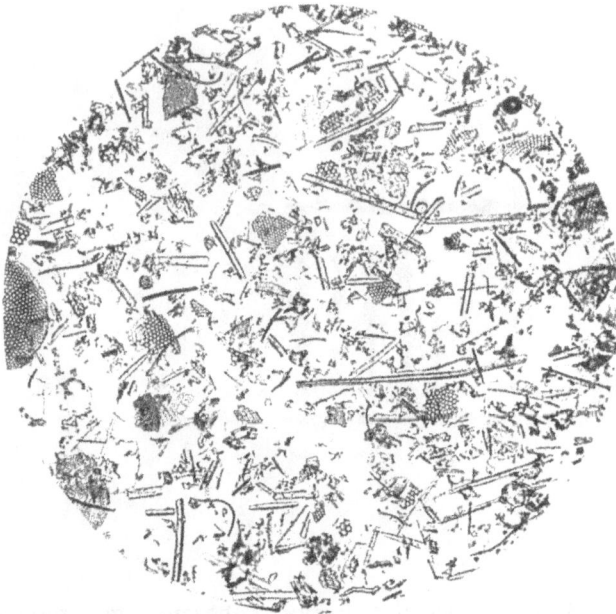

Figure 83*b*. Enlarged (500×) photograph of Hyflo Super-Cel. (Courtesy Johns-Manville Corp.)

552

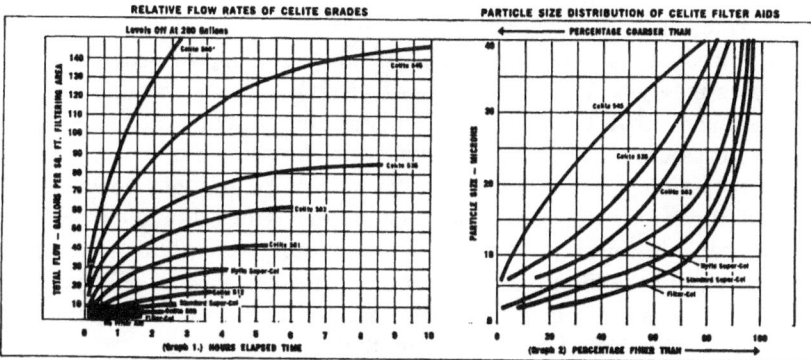

Figure 83c. Relative flow rates for different filter aids and their particle size distribution. (Courtesy Johns-Manville Corp.)

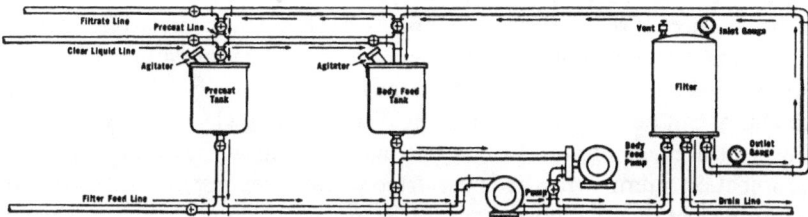

Figure 83d. Typical filtration system using filter aid. (Courtesy Johns-Manville Corp.)

agitation during filtration. The amount of filter aid necessary varies with its character (fig. 83b) and the type of wine. The more cloudy the wine, the larger should be the particle size of the filter aid (fig. 83c). An excess is to be strictly avoided. If all particles of the aid are not removed the wine may take on undesirable flavors. Filter aids as a possible source of calcium, magnesium, and aluminum should also be checked. One disadvantage of these filters is the high pressures necessarily used. The plate-and-frame filter has a low initial cost, occupies the smallest floor space, and is quite versatile, but it often leaks, and labor costs for discharging or cleaning have been relatively high. These are now constructed for more efficient operation, with recessed plates to reduce leaking, mechanical handling of plates to reduce labor, automatic sluicing, or even pressure operation.

In precoating filters, air must be kept out of the system. It may cause cavitation at the precoat pump and disrupt flow, or it may get in frames

553

in the back of a plate-and-frame filter and prevent precoat deposition. Later, when the pressure rises, wine will rise into this area. If flow is in at the bottom at the front of the filter and out at the top, this will not occur. Air vents may also be installed. Spatz (1967) suggests that the precoat tank be higher than the pump, with a large short suction pipe to provide positive pressure; that baffles be installed in the precoat tank to eliminate liquid vortexing, or that the filtrate return line be brought down to the level of the heel to eliminate aeration of the falling liquid. When the precoat tank is lower than the filter, siphoning can occur. This may partially empty the filter and hinder precoat formation. Siphoning can be prevented by installing a restrictive orifice in the return pipeline to maintain a pressure of 3 to 4 psi in the filter during precoating. See fig. 83d for a typical arrangement.

For addition of filter aids during filtrations either a dry or a wet type of feed system may be used. To prevent large particles from plugging valves or jamming the pump, the suction line should be six inches above the bottom. The suction pipeline opening should point in the same direction as the rotation of the slurry in the tank. The slurry tank should be drained and washed to prevent settled filter aid from accumulating above six inches. Body-feed pipelines should have a bypass line through which the slurry can be circulated when the filter is being cleaned or precoated. This will prevent filter aid from settling out, and will avoid overloading the injection pump when it is restarted. The most popular filter aids in California are Hyflo Super-Cel and Standard Super-Cel for rough and fine filtrations, respectively. Filter-Cel and Celite have also been used for polishing filtrations. One device gives control of filter aid by maintaining a uniform pressure drop across the filter (Anon., 1967a). A controlled-volume pump delivers the exact amount of filter aid with each stroke (fig. 84). It is possible to adjust the pump to maintain a constant rate of pressure drop (fig. 85), which has the same effect as decreasing the amount of filter aid used. Either way, it is the rate of differential-pressure increase which governs body-feed addition rate.

Filter cloths (also called septums) may not function efficiently when they are plugged with solid particles or when they sag into the grid work of the plate and impede wine flow. Perforated metal or plastic sheets can be installed between the plate and cloth or paper. Polypropylene cloths are said to function well, not to sag, and can be used for red or white wines alternatively with only a citric acid rinse.

Leaf-type filters consist of perforated hollow metal screens in a suitable housing (fig. 78). In this type, filter aid is deposited as a precoat on the outer surface of the leaf. Wine is forced through the filter aid into the center space, from which it is pumped. Pressure leaf filters are batch-

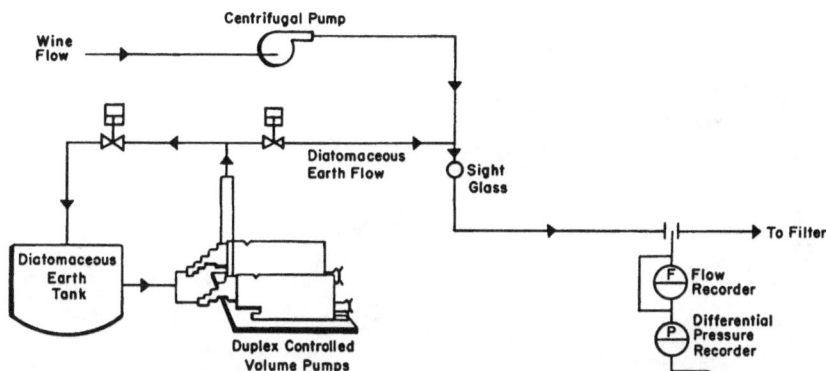

Figure 84. Schematic diagram of addition of filter aid to maintain a uniform pressure drop across the filter. (Courtesy Food Engineering.)

type filters: after a period of operation they must be cleaned. Two different types of pressure leaf filters are used by the wine industry: vertical-shell, vertical leaf; and horizontal shell, vertical leaf. In operation these are the same. They can be cleaned by a dry discharge system (where the accumulated solids and filter aid are mechanically removed as moist solids) or by a wet discharge system (where water is used to slurry the solids as they are washed from the filter leaves).

Trabert (1967) recommended automated control of filters, not only because it ensures uniform operation but also for the saving in labor costs. The controls can be simple switches or fully automatic operation. Fail-safe circuits are desirable to prevent improper operation through mechanical failure in the system.

In pulp or mass filters the filter medium—asbestos or paper fiber—is washed and applied to the filter. The main advantage is that only low pressures are required. The disadvantage is that time is required to prepare the filter for use, and if the pulp is reused it may prove a source of infection.

More recently continuous rotary vacuum filters have been introduced in California wineries. Hoffman and L. J. Berg (1967) studied the use of vacuum filters for table wines. They used a rotary drum precoat vacuum filter. They believe that it may have applications in the recovery of wine from lees and press wines or for producing clear grape juice. They correctly note that a major problem is aeration and stripping of volatile components. The former is not a problem with grape juices and can be minimized with wines by stripping the filtrate with an inert gas immediately after filtration. With red press wines the high tannin content

555

Figure 85. Adjusting micrometer dial on controlled-column pump to change amount of filter air injected. (Courtesy Food Engineering.)

might made the aeration beneficial. Their statement, "In all aspects, except varietal character, it is equal to the free-run product," indicates that the problem of loss of volatile constituents has not been solved. Vacuum filtration would appear to be useful for large wineries, at least for the recovery of wine from lees.

Polishing or finishing filtration

Asbestos- or fiber-pad filters and porous porcelain- or carbon-candle filters are used in final filtration just before bottling. The pads are held between corrosion-resistant metal frames in such a way that each pad acts as an individual filter. The pads are available in various porosities, but are costly and must be discarded after use. To reduce contamination with acid-soluble iron and calcium salts, the pads may be washed with a 1 per cent solution of tartaric acid before use. It is desirable to attach a silk bag filter to the outlet end of a pad filter to remove particles of fiber which may carry through. In the larger wineries, plate-and-frame filter presses (figs. 79–81) are used for finishing filtration, with asbestos fibers as filter aid and precoat. The filter aid should be of small particle size in this case.

The candle filter consists of several large, unglazed porcelain or porous carbon tubes closed at one end and housed in a closed metal chamber. The fine pores of the candle act as a filter. The disadvantage of these tubes is that they are difficult to clean and if not thoroughly sterilized may prove a source of contamination.

The wine to be filtered in either the pad or the candle filter must be nearly clear. The clarity of the filtered wine should be checked periodically during filtration, visually or preferably by some physical method. (See methods of measuring brilliance, p. 706.)

Membrane filters

The pad and diatomaceous earth filters are classified as depth filters because their filtration is by entrapment inside the filter. Screen filters filter at the top surface of the filter. For high particle loading with fine particles, depth filters are preferred. For further clarification, or final filtration with sterilization, a tighter depth or a screen filter may be used. See Mulvany (1966) and Aronson (1967) for information on their use.

Membrane or screen filters are thin membranes made from pure and biologically inert cellulose esters and similar polymers, penetrated with pores of exactly uniform size which occupy approximately 80 per cent of the filter volume. Membrane filters were developed by the German colloid chemist Zsigmondy in the first quarter of this century (Ferry, 1936), but were perfected during World War II and subsequently one of the early commercial applications of membrane filters was that of Goetz and Tsuneishe (1951) to water. Later Castor (1952) suggested that they might be of interest for wines. They are now widely used in many areas of biology, biochemistry, and pharmacology for sterilization of air and liquids, direct microbial counts, and sterilization filtration of wine as well as beer.

Membrane filters are made by several manufacturers (Millipore, Gelman and Schleicher, and Schnell). They contain 2 to 3 per cent of their dry weight as detergent, a fact not mentioned in the descriptive brochures of any of the manufacturers. Millipore filters contain Triton X-100 or similar detergent. The manufacturers state that the detergent is added to promote efficiency of filtration (Cahn, 1967) and to allow filters to be sterilized by autoclaving. Without the detergent the filters are unwettable, and excessive pressures are needed to affect filtrations.

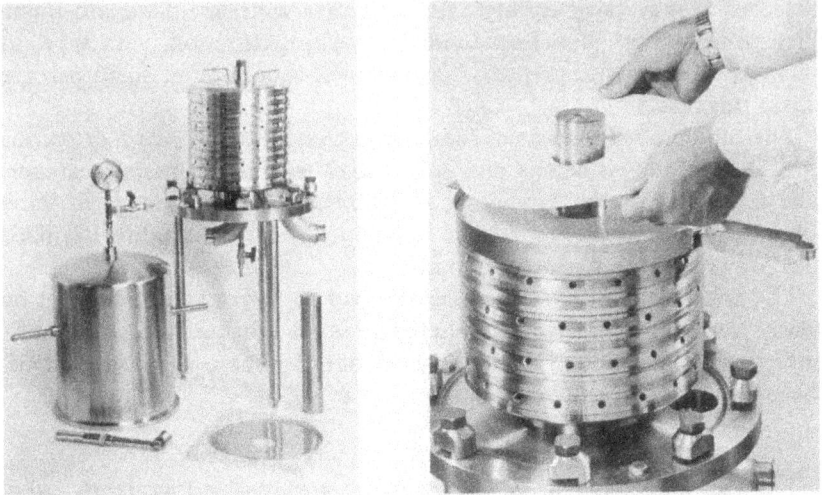

Figure 86. (*Left*) Multiple filter holders; (right) placing discs on assembly. After insertion of the discs the dome is fitted on and secured. (Courtesy Millipore Filter Corp.)

When distilled water or a saline solution is filtered the resulting filtrates develop a persistent foam. Glass-distilled water at 90° to 100°C (194° to 212°F) effectively removes most of the detergent in the filter.

The wetting agent and plasticizer added to Millipore membranes are nonionic, are present in low concentration, and are largely removed during the preliminary hot water sterilization of the filter units. For high-volume filtration and biological stability, Millipore recommends multiplate filter-holder tubes, with type DA, 0.65μ pore size filters. The flow rate through a 20-plate Multiplate Filter Holder of prefiltered wine is reported to be 4,000 gallons an hour and the total volume passed is 120,000 gallons up to a differential pressure of 80 psi. Typical installations are shown in figures 86, 87, and 88. Membrane filters are now widely employed in California wineries without adverse effects on quality when properly used.

Sterilization filtration

When the pore size of the filter medium is small enough and the filtration is carefully conducted at not too high a pressure, all microörganisms may be removed from the wine. If such closely filtered wine is filled into sterile bottles, under aseptic conditions, and is closed with sterile corks, it will not be subject to bacterial spoilage. Scrupulous care must be taken however, to avoid reinfection at bottling, and all the operations must be

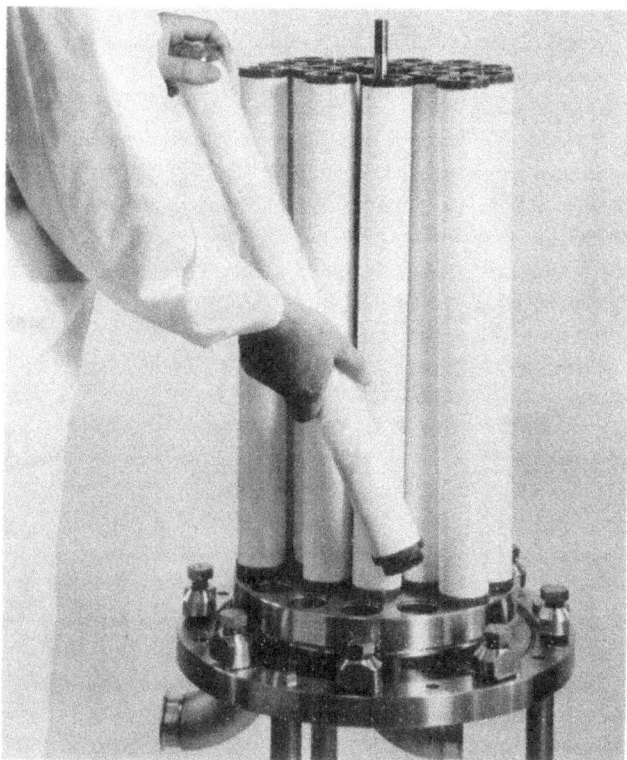

Figure 87.   Cartridges for Millipore multitube filter. (Courtesy Millipore Filter Corp.)

conducted in a clean room in the most sanitary way. The bottle may be sterilized by thorough washing and then rinsed with a solution of sulfurous acid and sterile water before use. Only corks of high quality, properly sterilized, will serve. Sterilization filtration, at best difficult, may be dangerous when improperly used because of reinfection during bottling. There is no reason why it should be impracticable in the average large winery for sweet table wines.

The sterilization filtration of fruit juices, wines, vinegars, and biologicals has been under investigation for some time both here and abroad. Carpenter *et al.* (1932) investigated the application of the Seitz EK asbestos filter pads to the sterilization of apple and grape juice. They point out that the removal of microörganisms by such pads was not merely sieving, since the average pore size of the pads, 11 to 55 μ, was several times as large as the longest diameter of the microörganisms which

Figure 88. Millipore multiplate or multitube filters. (Courtesy Millipore Filter Corp.)

could be removed. For example, the yeast they investigated varied from $1.7 \times 1.2 \mu$ for small-celled, to $1.2 \times 10 \mu$ for long yeast, and $3.5 \times 4 \mu$ for spherical yeast. The high absorption capacity of ultra-filters, the influence of electrical charge on separation, the deformability of microbial cells under pressure, and the heteroporosity of such filter media as unglazed porcelain filter candles have long been known. (See review by Ferry, 1936.) Carpenter et al. (1932) called attention to the effect of pump pressure on sterilization filtration; at pressures over ten pounds per square inch, for example, sterility could not be obtained in cider with EK pads. They pointed out also that filter disks of uniform porosity are difficult to manufacture. The variable porosity of filter pads is well recognized in the brewery industry, and the deposition of filter aid to cover up the larger holes or pores in such filter media is now common practice in that industry.

This knowledge was applied by Fessler (1949) and Fessler et al.

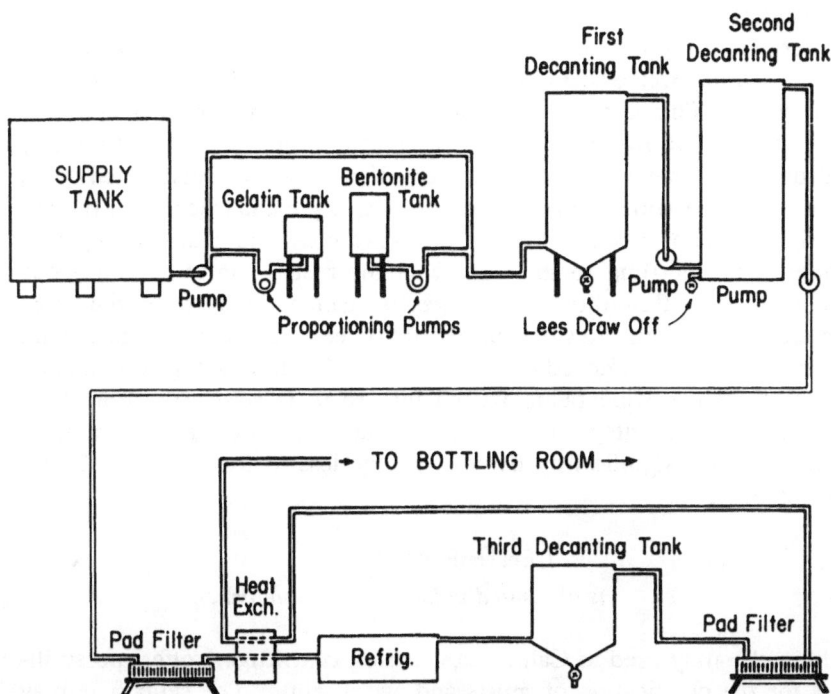

Figure 89. Flow diagram of a stabilizing system used in South Africa. Source: Berg (1967).

(1949) to the removal of microörganisms from California wines, particularly yeast, which could grow in the wines and produce turbidity. In this process the filter pads are properly precoated with filter aid to seal the larger pores, and the filter assembly, the bottling line, and bottles are sterilized by treatment with a quaternary ammonium germicide. The filtration is then conducted under pressure so that the yeasts that cause clouding are retained on the filter pads. As applied to wine, sterilization filtration does not and need not remove all viable microbial cells. All that is necessary is to remove the microörganisms which can develop in objectionable amounts in the bottled wine. This has apparently been accomplished in several industrial installations (Turbovsky, 1949). Great care should be exercised in removing *all* the quaternary ammonium solution from the filter, pump, bottling line, and bottles before introducing the wine.

Peynaud and Domercq (1959a) reported that sterile filtration of fine Bordeaux wines "seems to fatigue the wines." More comparative sensory data would be useful. Use of sterile filtration in Germany was reviewed by Schanderl (1965a).

Berg (1967) reported on a continuous wine-stabilizing setup which has been used for twelve years in South Africa. Figure 89 is a flow diagram of the system. This is a gelatin-bentonite clarifying system. The first pad filter is a 16 × 16-inch 40-plate filter into which aid is continuously introduced. Note that the wine from the insulated refrigerated tank, into which the wine is introduced at −5°C (23°F), is taken from the top of the tank and passed through a filter-pad filter and heat exchanger en route to the bottling room. Lees from the first decanter for a 125,000 gallon a week flow-through averages 90 gallons; that from the third decanter, 70 gallons. So small an amount of lees is deposited in the second decanter that it is removed only once a year. The filter setups last about a week, since the wine being passed into them is relatively clear. Berg indicates the advantages of the continuous system: (1) greatly reduced labor cost, (2) significantly decreased wine loss in lees, and (3) shorter processing time.

CENTRIFUGING . . .

. . . is of limited value for dry table wines.

Although rarely used in California wineries, centrifuging offers possibilities for the clarification of musts and sweet wines. The centrifuge may unduly aerate dry wines and in used mainly for musts and dessert or appetizer wines. It should be constructed of corrosion-resistant metal and be thoroughly drained and washed after use. In operation the centrifuge will remove both large and small particles according to the speed at which it is turning. When improperly operated there may be a slight loss of alcohol from the wine.

Centrifuges may be continuous or batch-operated. The continuous self-desludging centrifuge has been introduced in Europe and used in some wineries (Fessler and Nasledov, 1951). Its capacity is too small for large-scale winery production. It is useful, however, in clarification of wine lees and in recovery of wine from crude lees after the first racking. (See also Fessler, 1953.)

A detailed study of the use of the centrifuge for clarifying musts and wines was made by Mercz et al. (1963). They recommend holding the wine for two weeks after completion of fermentation before attempting centrifuging. At this period fining and filtration are very difficult. They noted average losses of 61 per cent of the carbon dioxide during centrifuging. They did not find any other unfavorable changes in composition. One advantage is the large loss of microörganisms, notably yeasts.

PASTEURIZING

Wines are pasteurized primarily to destroy injurious microörganisms capable of developing in them; but sound wines need not be pasteurized.

Pasteurizing is widely used to check the progress of bacterial diseases but sound wines are sometimes pasteurized to ensure their keeping under unfavorable conditions and to promote stability. The process does not, however, render the wine immune, and the treated product must be run into clean, sterilized casks and protected from reinfection.

If the wine is heated to a suitable temperature and held there for the proper period, the organisms that could develop will be destroyed. The higher the temperature, the shorter the necessary time of heating. The amount of heating needed will depend upon the extent and type of infection and the composition of the wine. The smaller the number of microörganisms and the higher the acid and alcohol content, the less heating required. Pasteurizing, when necessary, should be accomplished with minimum injury to the flavor of the wine. For table wines this is done best by heating the wine in the presence of as small an oxygen content as possible. The harmful effect on quality of the hydroxymethyl-fural produced by heating grape juice is discussed by Flanzy and Collon (1962b) and Kern (1964). Use of pasteurization to stabilize table wines was recommended by Sudraud (1963).

Some California wineries have occasionally been troubled by a yeast which is tolerant to sulfur dioxide. This yeast grows in bottled white wines and produces an undesirable cloud and sediment (p. 788). For this reason several wineries pasteurize and hot-bottle. This processing is being replaced by sterilization filtration which appears to promote better bottle-aging than hot filling. Bacterial spoilage can be controlled so readily by sulfur dioxide that wines spoiled by bacteria are rarely encountered.

Pasteurizing as a means of preserving wines particularly susceptible to bacterial diseases or to refermentation by yeasts may be accomplished in several ways. The wines may be filled cold into the bottles, sealed with a special closure, heated lying on their sides in water at 60°C (140°F) for about thirty minutes, and cooled. In another method the wine is flash-pasteurized in a suitable pasteurizer, cooled to a bottling temperature of 60°C (140°F), and filled at that temperature into steamed bottles which are then closed, turned over to sterilize the closure, and cooled. Since pasteurizing will usually precipitate some colloidal material, the wine should be bulk-pasteurized, cooled, and filtered before bottling.

Pasteurizing has been used along with fining to promote the more complete separation of suspended matter, but cloudy wine should be cleared before pasteurizing to avoid injuring its flavor.

HEAT-STABILIZING . . .

. . . is of value for some wines.

It is the practice in some of the larger wineries to heat-stabilize the wines by heating them to 60°C (140°F) in continuous heat exchangers, filling hot into tanks, and allowing them to remain warm for three days before filtering and cooling. This practice effectively pasteurizes most wines, but it is a rather severe treatment which many table wines cannot stand. Usually it leads to excessive oxidation and metal pickup, and should be considered only when all other methods of clarification have failed. Heating in the presence of bentonite has been tried for wines difficult to clear and is sometimes effective.

Wines of low pH, 3.0 or lower, are usually more heat-stable than wines of higher pH. See p. 519 for heat-stability tests.

For the early maturation of standard-quality table wines in Germany, Koch (1956a) recommended (1) pasteurization of the must, (2) fermentation in refrigerated rooms, (3) blending, (4) polish-filtration, (5) pasteurization at 68° to 75°C (154.4° to 167°F) and holding at this temperature for two minutes, (6) cooling to 4°C (21.8°F) and stirring slowly at first and more rapidly toward the end of five to seven days, and (7) filtering cold and bottling immediately.

CONCENTRATION

In some regions of France, owing either to the cool climate (Burgundy) or to heavy crops (the Midi), some means of increasing the alcohol content is needed. In Burgundy the addition of sugar to the musts is common. As early as 1910 Monti showed that partial freezing of the water and removal of the ice would result in higher-alcohol content of the residual liquid. Bacquet (1966) has reviewed the history of the process and indicated its present status in France. He estimates that several hundred installations have been made in France and Algeria over the past thirty years, with perhaps a hundred still being used in France. The advantage of the process is that there is no loss of volatile materials— even an increase. The disadvantage is the cost and the concurrent increase in acidity. A typical example of before-and-after composition was: alcohol, 9.0 to 10.3; total acidity, 0.83 to 0.89; and volatile acidity, 0.078 to 0.086. In red wines there is some darkening of the color and usually the pH is lowered. Wines of high initial acidity are likely to be too acid in taste following concentration.

CHAPTER TWELVE

FINISHING OPERATIONS

Care in bottling is essential. Bottle-aging greatly improves many red wines.

BLENDING

Blending is used to produce standardized wine of a certain type,
to accentuate a special flavor, or to balance the wine.

Where the character and composition of the grapes used in wine making are known and uniform, the blending, when necessary, is done when the grapes are crushed. More frequently, however, the wines must be blended also. Wines with excessive or deficient acidity, or with too little or too much alcohol, may sometimes be improved by blending. The improvements in quality and uniformity of wine that are possible by a careful blending of selected wines in the cellar are often overlooked.

In making blending tests, the taster must have the desired result in mind. Haphazard blending is of little value and may do damage to wines of desirable characteristic flavors. Furthermore, the final chemical composition of the product will make certain blends impossible from the stock at hand. Therefore, only wines that are basically suitable need be considered. Certain blends will also be impractical if the stock wine is too expensive because of its quality and age. After the best combination of usable wines is found by laboratory blends and tasting, a larger test blending should be very accurately measured. Several gallons of the blend may be mixed well in a glass demijohn and stored for a few days. This final testing should reveal any remaining deficiencies and permit more extensive chemical and sensory testing.

Blending of two components is commonly done by use of the equation $A/B = (m - b)/(a - m)$ where a and b are the percentages by volume of wines A and B, and m is the desired percentage in the blend. The gallonage of either A or B is fixed or the proportion of one is given as 1. For convenience, the wine of a percentage less than m is taken as b.

For example, suppose it is desired to produce a rosé wine of 2.0 per cent sugar using x gallons of A wine of 2.5 per cent sugar and 10,000 gallons of B wine of 0.2 per cent sugar. Then

$$\frac{x}{10,000} = \frac{2.0 - 0.2}{12.5 - 2.0} = 1,717 \text{ gallons.}$$

Or, if the proportions of A and B are required, then:

$$\frac{A}{B} = \frac{2.0 - 0.2}{12.5 - 2.0} = \frac{1.8}{10.5} = \frac{1}{5.83}$$

Thus blending one part of A to 5.83 parts of B will produce the desired percentage of sugar in the blend. This algebraic formula is usually used as the Pearson square, where:

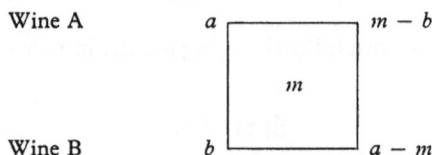

Wine A $\quad a$ $m - b$

m

Wine B $\quad b$ $a - m$

Suppose it is desired to know the proportions for blending A wine of 13.6 per cent alcohol with B wine of 11.3 per cent alcohol to produce a wine of 12.5 per cent alcohol:

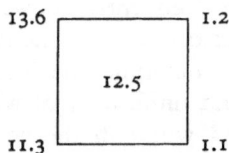

13.6   1.2

12.5

11.3   1.1

Thus 1.2 parts of wine A and 1.1 parts of B will produce the desired percentage. For further details of blending formulae see Joslyn (1961*a*) and Ambrosi and Flockemann (1961). When three or more components are to be blended the calculations can become complex. Costa (1968) constructed a simple analog computer for this type of calculation.

Even though all components of the blend may have been brilliantly clear before mixing, either clouding or precipitation often occurs after blending. Sometimes this change is due to the actual precipitation of tartrates or the formation of metal cloudiness (casse); at other times it results from overaeration and disappears on standing. If the wine should not remain clear, either in the wood or in test bottlings, it may have to be refined, filtered, or otherwise treated. The final blend should always be stored for several weeks before processing for bottling.

Actual blends must be made and tested—not putting so much of one wine in for flavor and so much of another for balance based on calculations. The reasons for this are outlined in the experiments of Singleton and Ough (1962). They found that blends usually scored higher than the mean of the scores of the base wines; seven of thirty-four blends scored higher than the highest score of two base wines tested separately. They attribute this partly to an increase in complexity of the blend. There may also be a dilution effect of an undesirable component. They suggest that the differences are perceived on a geometric rather than a linear scale. Whatever the reasons, it is clear that the quality of blends cannot be predicted from the composition or quality of either of the components. There is no reason to doubt that these effects occur also in blends of more than two components.

Solid-state control with digital-analogue circuitry for handling blends of 10,000 gph with great accuracy is now on the market (Anon., 1965a).

Although blending is a great advantage to the wine maker—for instance, in permitting him to standardize his types—it is not a cure-all. Bad wines, even in small amounts, may contaminate or dilute good ones, and the good wine may thus depreciate in value. The desirable characteristic flavors of certain wines may be diluted beyond recognition, or be masked by other, less desirable flavors. Varietal wines, therefore, usually should be blended with wines of compatible character. In blending, the consumer acceptance of wines of varying prices and quality must be kept in mind. Large-scale consumer-acceptance tests may be of value in determining the potential market. Blending of several distinctive high-quality wines to produce a uniform type may rob the consumer of a chance to make a selection between them and thus may actually reduce sales.

BOTTLING

Bottling protects the wine from the action of microörganisms and oxygen, permits secondary aging, and facilitates distribution.

The bottle should be well chosen for consumer acceptance and sales appeal. There has been too little investigation of the desirability of various shapes, colors, and sizes of bottles for different types of wines. While many different shapes have been used, it is possible that new shapes might achieve consumer acceptance or preference.

The wine used must be of suitable quality, age, and clarity. The actual operation of bottling, from filling to labeling, should be carefully supervised. Fine, sound wines are often spoiled during bottling by unnecessary or extreme aeration, by infection with bacteria, yeasts, or molds, by metal contamination, or by faulty closure. *Sanitation in bottling, as well as in wine making, cannot be too strongly stressed.* The more important steps in bottling still table wines are final clarification or filtration, choice and preparation of bottles, filling, closing, pasteurizing (for some wines), labeling, and casing.

Preparing wine for bottling

Only sound wines, properly aged, should be bottled. There is a tendency to bottle table wines, particularly the whites, rather young, both here and abroad. If such wine is properly mellowed and stabilized, early bottling need introduce no difficulty in respect to clarity and keeping quality. We have, however, noted some California white table wines which had been bottled so young that they retained a yeasty sulfide odor.

Just before bottling, the wine usually undergoes a final clarification or close filtration. (See p. 556 for details.) Some wine makers prefer to fine their best wines with isinglass to produce a crystal-clear brilliance. The cost of isinglass is one factor restricting its use. The final step before bottling, however, is usually a close filtration (fig. 87). This is often done through membrane filters (of the Millipore type) which filter out yeasts.

A test bottling should always be made before the actual bottling. The wine to be bottled is run through a small laboratory filter (see fig. 94) and into bottles of the same shape, color, and lot as will be used for the final bottling. Enough wine should be filtered to fill five or more bottles. After corking or capping, these are stored under the following conditions: at $-3.9°C$ ($25°F$), at room temperature in the light and in the dark, at $29.4°C$ ($85°F$) or $32.2°C$ ($90°F$), and at $60°C$ ($140°F$). The bottles should be inspected daily for about a week, but the most significant changes occur in two to four days; the sample at $60°C$ ($140°F$), need remain at this temperature for only 24 hours. All the bottles should be inspected in both direct and reflected light. If cloudiness does not develop under these conditions, or when the bottles are returned to room temperature, the wines will probably stay brilliant under commercial conditions for one, two, or many years, but only if the actual commercial operations of filtering, filling, and capping are the same as the trial operations. This is not always so. A poorly conducted polish filtration through an improperly cleaned filter may lead to copper contamination under plant conditions. Clouding might then occur after two to six months, even though the trail bottling showed that the wine was stable.

Bottles

Caron (1964) gives the nominal capacity of various types of bottles and half bottles in France as follows (in centiliters):

Type	Bottle	Half bottle
Rhine	70	35
Bordelaise, bourguignonne, Anjou, etc.	75	37.5
Champagne	75	37.5

Caron also gives detailed data on the dimensions of various types of bottles. See figure 90 for typical dimensions for sparkling wine bottles.

The standards of fill for containers for wine in this country are: 4.9 gallons, 3 gallons, 1 gallon, $\frac{4}{5}$ gallon, $\frac{1}{2}$ gallon, $\frac{2}{5}$ gallon, 1 quart, $\frac{4}{5}$ quart, 1 pint, $\frac{4}{5}$ pint, $\frac{1}{2}$ pint, 4 ounces, 3 ounces, and 2 ounces. For aperitif wines only a 15/16 quart container is legal. (See figs. 91 and 92.)

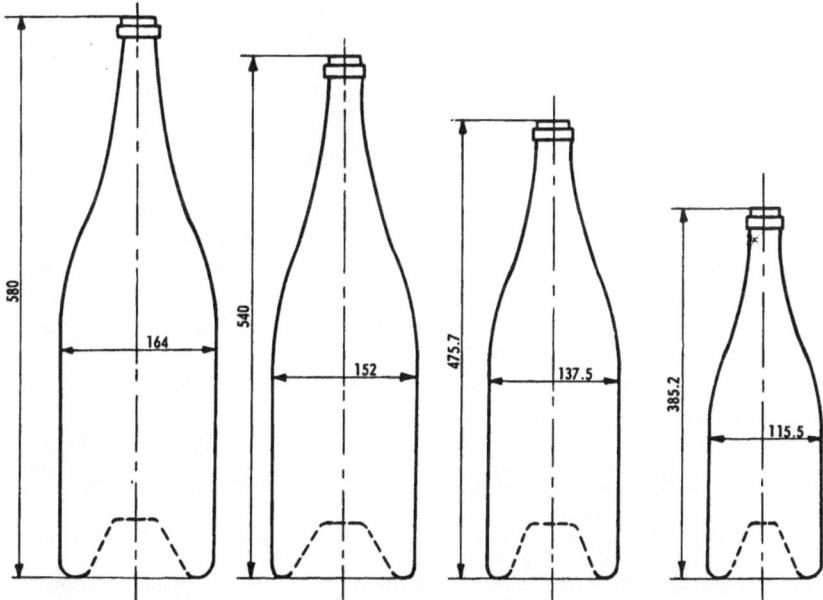

Figure 90. Quadruple magnum, triple magnum, double magnum, magnum (dimensions in mm).

The net contents of the container must be given. This need not be on the label. At present the net contents are commonly indicated by blowing the appropriate figures or letters in the bottle itself, usually at the bottom on opposite sides.

The tolerance on the fill is given in Regulations No. 4, U.S. Internal Revenue Service (1961b), as follows:

(1) Discrepancies due exclusively to errors in measuring which occur in filling conducted in compliance with good commercial practice.

(2) Discrepancies due exclusively to differences in the capacity of containers, resulting solely from unavoidable difficulties in manufacturing such containers so as to be of uniform capacity: *Provided*, That no greater tolerance shall be allowed in case of containers which, because of their design, cannot be made of approximately uniform capacity than is allowed in the case of containers which can be manufactured so as to be of approximately uniform capacity.

Figure 91. Bottles used for table wines. Left to right: Bordeaux-type, green, cork finish, used for Cabernet Sauvignon, Zinfandel, and claret; Bordeaux type, clear glass, screw cap finish, used for sauterne; burgundy, green, cork finish, used for Pinot noir, Chardonnay, chablis, and burgundy; German-type, brown, cork finish, used for Johannisberg Riesling and rhine; German-type, dark green, cork finish, used for Johannisberg Riesling and Moselle; fiasco, raffia-covered, used for chianti; and champagne, dark green, used for various sparkling wines.

(3) Discrepancies in measure due to differences in atmospheric conditions in various places and which unavoidably result from the ordinary and customary exposure of alcoholic beverages in containers to evaporation. The reasonableness of discrepancies under this paragraph shall be determined on the facts in each case.

(4) Unreasonable shortages in certain of the containers in any shipment shall not be compensated by overages in other containers in the same shipment.

However, in the new wine regulations (U. S. Internal Revenue Service, 1961a) the tolerance is given as follows:

Bottles must be filled as nearly as possible to conform to the amount shown on the label or blown in the bottle to be contained therein, but in no event may the amount of wine contained in any bottle, due to lack of uniformity of the bottles, vary more than two per cent from the amount stated to be contained therein; and further in such case there shall be substantially as many bottles overfilled as there are bottles underfilled for each lot of wine bottled.

The bottles must be carefully selected. They should be of suitable size and shape and should offer the best protection for the type of wine. They must also be strong and free of defects, which is particularly important when bottling is done mechanically. Modern bottle glass is apparently of sufficient insolubility and strength to meet industry needs. The chief complaints have been the spreading finish on the neck, which reduces the

Figure 92. Bottles and appropriate glasses for (left to right) Rhine or Riesling, Bordeaux or Cabernet Sauvignon (1/10), Sauternes or Sémillon, Bordeaux or Cabernet Sauvignon (magnum), and Burgundy or Pinot noir or Chardonnay.

tightness of the cork (especially in rhine-type bottles), and the poor color. Dillon (1958) notes that fractures during handling start at the surface. To prevent this, exterior coating of bottles may be used.

Bottle stability, according to Caron (1964), includes consideration of mechanical, temperature, and pressure factors. Mechanical stability includes a large number of factors concerned with automatic bottling, corking, labeling, and casing. Temperature effects are easily recognized, but the pressure problem is not one simply of total pressure. Caron (1964) suggests parameters of resistance to immediate pressure, to continuous pressure, and to temperature and pressure.

The glass should be of the proper hue as dictated by custom: clear white, greenish, or brown, according to the type, for white table wines; dark green for red table wines (fig. 91). There is ample evidence that greenish-brown bottles cut off the wave lengths of light that are most damaging to the keeping quality of wines. Caron (1964) indicates the ferrous oxide content of clear wine bottles as 0.02 to 0.08 per cent, and of Bordeaux type green bottles as 0.08 to 0.15 per cent. Champagne bottles contain up to 2 to 5 per cent. The desired tint may be obtained by including other metals. The highly prized amber tint is achieved by maintaining a reducing action for sulfur during fusion (fig. 93). Dillon (1958) demonstrates why amber-colored glass is preferred for wine: it filters out more of the ultraviolet rays that cause undesirable photochemical reactions in wines. Some manufacturers claim that the greenish-amber bottles are more expensive to produce. According to Leonhardt (1963), blue-green glass is produced by incorporating iron and copper; green, with chromium and uranium; blue, with cobalt; brown, with manganese and nickel; and reddish-brown, with manganese. When there is doubt as to the effect of light on the wine, it should be laboratory-filtered (fig. 94) and test-bottled.

The glass should be without flaws and air bubbles, uniform in thickness, and free of strains. Abnormalities are readily observable when the bottle is examined by polarized light. This is easily done by placing the bottle in

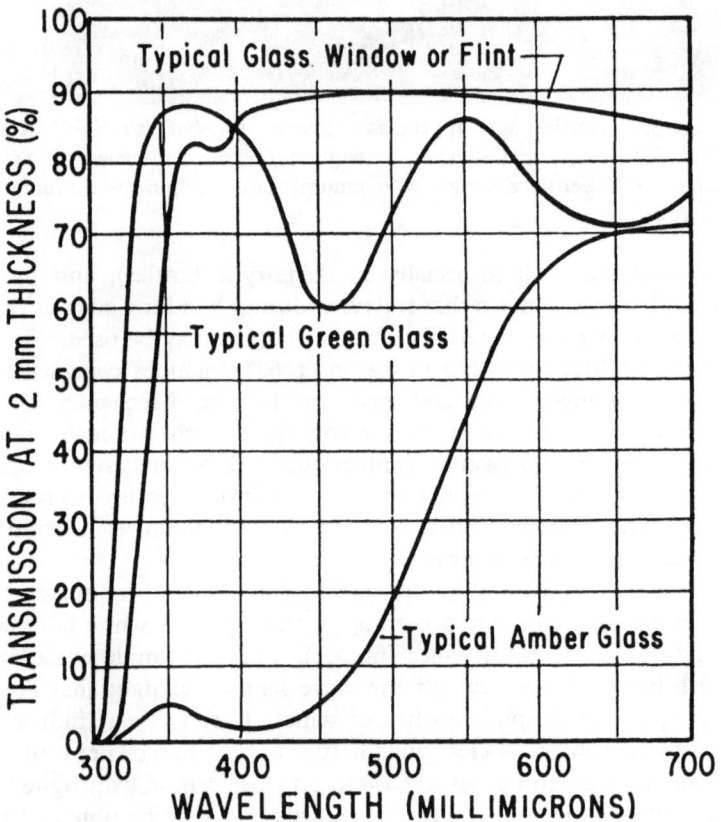

Figure 93. Effect of glass color on transmission of light of different wave lengths.

Figure 94.   Laboratory filter for filtration tests.

the path of a strong polarized light and observing it through spectacles which have polarizing lenses with the plane of polarization at an angle of 90° to the plane of the polarized light. The necks should be properly blown to suit the type of corkage or seal desired.

The bottles should be thoroughly cleaned and as nearly sterile as possible. Castor (1956) and Schanderl (1957) reported microörganisms in new bottles shipped directly from the factory, but none of these seemed dangerous to the wines. Haltar (1960) did find yeasts and molds in new bottles taken directly from the heat-annealing conveyor. He recommended stoppering the bottles before they leave the factory. Castor (1956) found no molds in new bottles delivered to the winery capped. After exposure he found some molds and bacteria. Bottles delivered uncapped contained molds and bacteria, and after exposure occasionally had viable yeasts. Some believe that new bottles can be cleaned by blowing air into them, and this is the usual practice (figs. 95, 96). Washing is still used, for new bottles can be cleaned best by thorough washing with water and sterilizing with steam.

Used bottles can be cleaned only by washing with a suitable detergent solution—soda ash, trisodium phosphate, or metaphosphate solution—and rinsing with clean, sterile water. Halter (1960) showed that, even after

Figure 95. Unscrambler and air cleaner for bottles. (Courtesy Biner-Ellison Manufacturing Co., Los Angeles.)

Figure 96.   Automatic air cleaner and filler.

fifteen to eighteen minutes in the antiseptic solution, many yeasts and molds remained in used bottles. For small installations for grape juice or wines, Halter (1959) concluded that it was preferable to use new bottles, and that these should be washed.

Used bottles are not employed in American wineries. Where they are used abroad their complete sterilization is a difficult problem, according to Thedden (1965). The sterility depends not only on the concentration of hydroxide used but also on the amount of contamination, the pressure, and the temperature. The length of time the alkali is used is also a problem; it depends to some extent on the number of bottles processed. The correct procedure for automatic bottle-washing machines is to clean the bottles thoroughly first and then sterilize them. Thedden recommends titration values, not pH, for determining the value of the alkali solution.

The manufacturer's directions for use of a detergent should be followed carefully. Bottles should be washed until free of all adhering foreign matter. No trace of chemical which would affect the stability, odor, or flavor of wine should be left in the bottle. Caron (1964) has given very useful data on washing used bottles. Equivalent sodium hydroxide detergent action was as follows:

Temperature (o°F)	109.9	119.8	129.7	139.6	149.5	159.4
Temperature (o°C)	43.3	48.8	54.3	59.8	65.3	70.8
Time (min.)			NaOH (per cent)			
1	11.8	7.9	5.3	3.5	2.4	1.6
3	6.4	4.3	2.9	1.9	1.3	0.9
5	4.8	3.2	2.16	1.4	1.0	0.6
7	4.0	2.7	1.8	1.2	0.8	0.5
9	3.5	2.3	1.6	1.0	0.7	0.5
11	3.1	2.1	1.4	0.9	0.6	0.4
13	2.8	1.9	1.3	0.8	0.6	0.4
15	2.6	1.7	1.2	0.8	0.5	0.3

The washed bottles should be allowed to drain until dry, but should be sterilized with steam and filled soon after drying so that they do not become dusty or contaminated. The practice of rinsing bottles with wine before filling is generally not desirable because of the possibility of spreading infection. Bottles should be inspected just before filling and again before labeling to make sure that they contain no foreign matter.

Cans

Prior to and just after World War II a number of large-scale experiments in canning wines were made in California. Failure to find a sufficiently inert lining and possibly bottling with too much oxygen led to clouding of the wines, pitting of the lining, and development of off-odors. So far as we know, no wines are being canned in this country at present. However, French table wine in cans began to appear on the American and European markets in 1966. Sudario (1966) indicated that the wines should be canned under vacuum or under an inert gas. Storage for no more than eight to twelve months was recommended. Since the notes on the sensory examination of these wines are very general, they are not convincing.

Plastic containers

Development of and approval for use in foods of a number of plastic containers has turned the attention of wine bottlers to their use for wine. *So far as we know*, none now has Food and Drug Administration approval. However, with the development of new stabilizers and more resistant materials, the plastic bottle may offer real advantages to the wine industry over glass bottles (e.g., lighter in weight and nonbreakable). They have recently been successfully tested in Europe for short-term storage.

Filling

Details of various methods of bottling (Haubs, 1966) include cold sterile filling with a counterpressure or vacuum filler, warm filling at 50° to 55°C (112° to 131°F), with or without a holding tank, or filling using diethyl ester of pyrocarbonic acid (DEPC) with a counterpressure or vacuum filler. In cold sterile or DEPC-filling, the wine may profitably be held in pressure tanks (about 3 atm). The latest German bottling technique, according to Troost (1966a), is to evacuate the flask, fill with carbon dioxide, and fill warm under pressure.

Bottling wines hot (50°C, 122°F) produced small amounts of hydroxymethylfurfural (HMF), according to Wucherpfennig and Lay (1967). Holding wines at this temperature for five days resulted in formation of 2 mg per liter; heating at 70°C (158°F) produced 20 mg. The HMF content of 1960 wines was only 0.81 mg per liter; that of 1939 wines was 8.2. The author concludes that HMF increased with age, but one would like to see analysis of the same wine over a period of time. The HMF contents of musts and new wines of the 1939 or 1960 vintage are not known. For a similar increase in HMF during storage of grape juice see Flanzy and Collon (1962a).

In filling the clean, dry, sterile bottles with the freshly clarified or

Figure 97.   Rotary bottom filler with thirty tubes. (Courtesy Biner-Ellison Machinery Co., Los Angeles.)

577

filtered wine, one should be careful to avoid infection, excessive aeration, and contamination with metallic impurities. The bottling room should be well lighted, well ventilated, and scrupulously clean. The fillers (fig. 97) should be thoroughly cleaned, sterilized, and rinsed with a little of the wine before use. Discard—do not re-use—the rinsing wine. The wine should not come into contact with any metal parts made of iron or other corrodible metals.

The saturation level of oxygen in wine is generally considered to be 6 to 7 ml per liter. Cant (1960) showed that wines being removed from cold stabilization approached oxygen saturation. He used a nitrogen stripping column which on a single pass reduced oxygen to 2 ml. An in-line nitrogen sparger was less effective than the column for high oxygen wines, but equally effective for low oxygen wines. Carbon dioxide was less effective. The cost of stripping was about 50 cents per 1,000 gallons.

During bottling the greatest oxygen pickup occurred as the wine was being discharged into the filling machine bowl. Purging the air from the bottles with carbon dioxide had contradictory effects on oxygen level, according to Cant (1960). However, several wineries purge the bottles before filling. Wucherpfennig and Kleinknecht (1965a, 1965c) recommended pre-evacuated bottles and bottling under carbon dioxide pressure. Even with hot filling, loss of carbon dioxide was not excessive when it was used as a counterpressure, according to the same authors. Vacuum filling did reduce the carbon dioxide content. Prillinger (1965) noted an oxygen pickup during the bottling of sparkling wine cuvées of 0.6 to 4.0 mg per liter when the empty flask was filled with air, and 5.9 to 15.7 when the air pressure continued on the filled bottle. When the empty flask was filled with carbon dioxide the oxygen pickup was 0.4 to 1.1; when the carbon dioxide continued on the filled bottle it amounted to 5.7 to 8.4. Geiss (1965) recommended purging with carbon dioxide and leaving carbon dioxide in the neck of the bottle. He believes that oxygen pickup from the air space between the cork and the wine is more important than from air contact during filling, and this seems logical.

Wucherpfennig and Kleinknecht (1965b) reported much lower color in cold-filled white wines compared to those bottled and stored for 24 hours at 80°C (140°F). However, when the cold-filled wine was bottled in the presence of air and the hot-bottled wine (50°C, 122°F) was bottled under carbon dioxide there was little difference in their absorption curves between 380 and 500 mμ. There were greater differences in color between hot-filled bottles than between cold-filled bottles. However, no specific changes in color pigments or polyphenolic compounds could be demonstrated in the two treatments.

Compared with the amount of oxygen absorbed during bottling, that present in the space above the liquid and below the closure is much more important:—up to eight times as much oxygen may be absorbed from this space as was absorbed during bottling. Filling the space with an inert gas would eliminate this source of oxygen.

Bielig (1966) showed that when bottling at normal temperatures an average of 0.6 mg per liter of oxygen will dissolve. This is irrespective of the system of bottling except when bottling under pressure when the oxygen absorption is 1.5 mg per liter (0.6 mg of oxygen are equivalent to 2.4 mg of sulfur dioxide). With high-temperature bottling the absorbed oxygen disappears more quickly. If the bottles are prefilled with nitrogen, the oxygen pickup will be only about 0.4 mg per liter. E. A. Rossi, Jr. (1965), recommended nitrogen as the counterpressure gas of choice.

Wucherpfennig and Kleinknecht (1966a, b) found no analytical difference between wines bottled at room temperature with air, at 50°C (122°F) with a vacuum and carbon dioxide as the counterpressure gas, and at the same temperature with the gas without a vacuum, and at the same temperature with a vacuum. The ITT value and free sulfur dioxide contents, for example, were not different for warm-filled wines than for germproof filtered wines. However, sensory evaluation indicated that bottling at 50°C (122°F) with a vacuum resulted in wines of lower quality. Next best was bottling at 50°C (122°F) with carbon dioxide but without a vacuum. Best was room temperature or 50°C (122°F) with a vacuum and carbon dioxide. Analysis of variance showed that the first procedure differed significantly from the others.

Jakob and Schrodt (1966) reported better quality of Riesling, Sylvaner, and Müller-Thurgau that were cold-filled compared to hot-filled. Flushing the bottles with carbon dioxide had no special beneficial effect for Sylvaner or Müller-Thurgau, but significantly improved the Riesling, particularly with hot filling.

In a filling cost analysis (Anon., 1967e) for German wines it was noted that cold-sterile filling or hot-filling had essentially the same cost per bottle or per hour. Filling with addition of DEPC was slightly more expensive.

Schanderl (1965a) reports that in a wine left in the filling machine over the weekend a growth of *Pulularia pullulans* occurred and then passed into the first 100 to 200 bottles when bottling was renewed.

Use a siphon or other proved fillers with spouts long enough to reach to the bottom of the bottle. Fill the bottles from the bottom up. Many of the automatic bottle fillers actuated by vacuum aerate the wines too much. Aeration during bottling is an important cause of subsequent darkening and haze formation with white table wines. Furthermore, too much air space should not be left in the neck of the bottle, for it permits undue oxidation of the wine. For red wines which are to be aged in the bottle for many years this space should be minimal, about one-fourth of an inch. The so-called "bottle sickness" of newly bottled wines is usually the result of too much aeration during bottling.

The retention of heat in hot-filled bottles is often neglected. Bottles filled at 60°C (140°F) and stored at 18°C (64.4°F) did not reach room temperature for 22 hours (Haushofer and Rethaller, 1964). Obviously, greater oxidation will occur when wines remain at higher temperatures for longer periods of time. Holding wines at higher temperatures is not recommended as a routine practice.

In the traditional practice, wine for bottling was drawn from the barrels in which it was aged and then given a final filtration, after which it was filled into bottles and closed with corks. This practice resulted in considerable variation from bottle to bottle, owing to differences in oxygen content at time of closure and to introduction of oxygen during corking. In order to reduce these variations, blending during and after aging was introduced and oxygen pickup was controlled by removing the oxygen present in the wine after final clarification or filtration by stripping with oxygen-free nitrogen (Air Reduction Co., 1961). Compressed gas containing 99.9 per cent nitrogen and free of gaseous products introduced from the lubricant used in the compressor was employed. The ordinary nitrogen compressed with oil-lubricated pumps and compressors is not satisfactory because it contains volatile products obtained from the lubricant. Water-lubricated compressors are preferred and these are used in the United States.

A specially constructed nonlubricated compressor, developed in Switzerland, is being used for compression of gases where lubricant cannot be used. The wine, after removal of oxygen by bubbling nitrogen gas through, is transferred into glass-lined tanks, where it is held under nitrogen. This practice allows large-scale bottle-aging of wines (Mondavi and Robe, 1961) and production of more uniform wines. However, long storage in partially filled tanks, even when held under pressure of an inert gas, is not recommended. Some bacterial spoilage has occurred, especially with low sulfur dioxide reds.

Among the inert gases available to remove dissolved and occluded oxygen and to blanket the wine to protect it against exposure to air are pure nitrogen, pure carbon dioxide, and a mixture of nitrogen and carbon dioxide. Nitrogen gas was the first to be allowed in wine making, and subsequently carbon dioxide. The latter is more soluble and must be used with caution to avoid excessive absorption. It is heavier than air, however, and is more useful in blanketing open surfaces such as the filler bowl.

A recent development is an inert gas generator which operates by controlled burning of the oxygen with natural gas, propane, or butane. The gas resulting from carefully controlled combustion is a mixture containing 9.8 per cent carbon dioxide, 1.2 to 3.0 per cent carbon monoxide, 83.2 to 86.3 per cent nitrogen, 1.8 to 2.9 per cent hydrogen, and 0.00 to 0.02 per cent oxygen. The resulting gas is compressed and stored for use at the plant as "Vitagen gas" (Cavanaugh et al., 1961).

The wine is transferred by nitrogen pressure through stainless steel, Pyrex glass, or Tygon plastic lines to the filler, but must be protected from oxygen pickup at the filler. The washed clear bottles are gassed out with nitrogen gas to replace the air present and are filled in an automatic bottom-filling unit. If corks are used, these are specially selected paraffined corks. During corking the bottled wine is vacuumized to avoid compression of gas in the headspace and is corked automatically.

Closures

The closure preferred by custom for dry table wines is the straight, untapered wine cork at least 1½ inches long. A 2-inch (or longer) cork is used for choice red wines to be stored for long periods. Successful closure depends on selecting corks of the highest quality and the right size, on softening and preparing them properly, on using wine bottles with the correct "corkage" (that is, very slightly tapered, with a neck free from grooves, bumps, or other imperfections), and on driving the cork into place properly. Freshly cut corks contain not only particles of cork dust adhering in the small pores or cavities, but also soluble gallic acid, iron gallate, nitrogenous and fatty matters, and other substances that impart a disagreeable corky taste. The corks, especially the poorer and more porous ones, are usually infected with resistant mold spores, yeasts, and bacteria; these may and occasionally do develop in the wine that seeps into the cork. Geiss (1962a, 1962b) found corks with at least seven annual rings to be best. If the corks are too porous their properties can be improved by impregnating with paraffin. Laminated corks (two or more pieces glued together) are commonly used for sparkling wines. They were first produced in 1895, according to Sharf and Lyon (1958).

When the cork contains a mold growth, the wine may acquire an off-taste during bottle-aging. This is the "corked" taste reported in the literature. The "corked" odor is attributed to the presence of *Aspergillus glaucus* in the cork by Sharf and Lyon (1958).

It may be desirable to test different samples of corks prior to use. Select five or ten at random from the lot to be tested. Cut the corks into small pieces and mix well. Weigh out about 5 gr of the pieces, place in 100 ml of 15 per cent alcohol, warm to 37.8°C (100°F), cover, and allow to sit in a warm place overnight. Filter through pure, washed cotton. If the extract has a moldy smell a more extensive examination of the corks should be made; their rejection may be necessary.

Loose cork dust and cork extractive must be removed before use, and the corks must be sterilized and softened. This is accomplished by soaking about an hour in a cold dilute o.o2 per cent solution of sulfur dioxide containing a little glycerin, and rinsing well in clear water. Corks should be well drained before use. An untreated dry cork sometimes will not work with semiautomatic corkers.

The industry has encountered difficulty in avoiding contamination of wine with the liquids pressed out of the cork during compression with automatic corking machines. For this reason, treatment of the cork with glycerin-sulfite solutions or other liquid sterilizers is frequently avoided. To do this the clean, sound cork is thoroughly shaken to remove cork dust and is then coated with paraffin. Paraffining is usually accomplished by tumbling the corks with 1 × 3 × 6-inch blocks of low melting point paraffin (58.9° to 61.1°C; 138° to 142°F) for about an hour at room temperature, or until the desired thickness of coating is secured.

The usual corking machine operates in two stages: first the cork is compressed to a small cylinder by the operation of horizontal movable jaws, or stationary jaws and plunger; next, the compressed cork is driven into the bottle by a vertical plunger. The best type of corker is one that compresses the cork from all sides by jaws that close uniformly like the iris of a camera lens. However, machines in which the cork is rolled as it is compressed from three sides by movable jaws are satisfactory. Stationary-jaw, horizontal-plunger type compression is less desirable because of the danger of breaking off pieces of the cork; but when chamfered corks are used, this danger is averted. If the cork is not dry and is compressed, drops of moisture ooze out from the bottom; these must be wiped off with a clean, dry cloth to prevent contamination of the wine. For large commercial operations this is, of course, impossible; air is blown across the bottom of the compressed cork to remove the liquid.

In corking, the air in the headspace of the bottle is compressed, and if the newly corked bottle is placed on its side, the wine will occasionally leak through the corks, particularly if they have been soaked or if the corking machine makes a crease on the side of the cork as it is driven into the bottle. Therefore, after corking, the bottled wine should be allowed to stand upright for a day or two until the air in the headspace diffuses out and the pressure in the bottle is equalized. The bottle is then placed on its side so that the neck is completely filled with wine. This seals the cork and prevents undesirable oxidation of the wine.

When stored in relatively warm cellars, especially in the light, uncapsuled corks may be attacked by cork borers. The primary cork borers (Götz, 1966) are: *Nemapogon cloacellus* (May to mid-September, especially June), *N. granellus* (April to October, especially June), *Oenophila v-flavum* (mid-June to October, especially August), *Dryadaula pactolia, Hofmannophila pseudospretella* (June to August), and *Endrosis lacteella*. (The dates refer to the main periods of activity of the insects.) To prevent cork-borer activity the corks may be chemically treated or insect-resistant capsules may be used. According to Götz (1966), one bottle storage room was sprayed with pyrethrum in a petroleum base (Flit). Release of sulfur dioxide has also proved effective.

Screw caps

Instead of corks, screw caps are frequently used as closure, particularly for the standard-quality table wines which sell at highly competitive prices and are not aged after bottling. These caps are available in a variety of types and patterns. Particular attention should be paid to the finish on the neck of the bottle. The cap should seat properly and seal the bottle completely. When wines are bottled hot and covered with a vacuum seal, the screw cap acts simply to protect the seal. Screw caps should be tested before use to make certain that the liner will not communicate off-odors or -tastes to the wines.

Aluminum foil screw caps or special two-piece stoppers of cork or plastic may be used. To prevent possible tampering and to obtain a hermetical seal, a special inside paper seal is applied at the top of the bottle before capping. The bottle may also be finished to prevent spilling of wine during pouring by coating the pouring surface with silicone.

For anything except short-time storage, polyethylene stoppers are unsuited for still or sparkling table wines. The reason seems to be the amount of oxygen entering around or through such corks. Stephan (1964) found the amounts of entering oxygen per month (in cc) as follows: natural pressed corks, 0.01; polyethylene stopper, (a) standard, 0.16; (b) coated with Disfan,[1] 0.03; and (c) coated with acid-free gelatin, 0.01.

Vodret (1965) bottled two white and three red table wines with cork and plastic and observed the result. The whites had 16.0 and 16.3 per cent alcohol; the reds, 12.6, 16.5, and 18.6 per cent. The latter wine was the only sweet wine, with 6.0 per cent sugar. Although many components were analyzed over a period of two years, the only consistent differences were slightly better retention of sulfur dioxide in the corked wines. In two of the five wines, oxidation-reduction potential was much higher in the plastic-closed bottles, but in the other three there was little difference.

---

[1] Registered trademark.

The author stated that the corked wines had a better perfume ("il vino tappato con sughero ha accusato un maggior volume di profumo"). After one year the plastic-closed bottles also showed signs of oxidation (greater absorption at 525 mμ in four of the five wines). He concluded that plastics are not a substitute for cork for Sardinian wines. Bergeret and Feuillat (1967) compared normally corked Burgundy wines with screw caps of various kinds. The tests lasted nine to eighteen months. In general, there was a decrease in oxidation-reduction potential for six months followed by an increase. Cork closures gave no significant difference from screw caps with a polyvinylydene (PVDC) lining. Several other types of screw caps with other linings gave less desirable odors. We do not at present recommend polyethylene stoppers for sparkling or still wines that are to be aged for more than about one year.

Bottle-aging

The beneficial results of aging wine in the cask are generally recognized, but bottle-aging, particularly of red table wines (fig. 98), is frequently neglected. The main reasons for this are inadequate funds or space to hold bottled wines, failure of some wines to develop in the bottle, and fear that deposits will form during aging.

The main goals in bottle-aging of white table wines are reduction of the free sulfur dioxide content, recovery from the temporary "aerated" odor which develops in the wine after bottling and corking, and development of some aged odor or bouquet. The decrease in free sulfur dioxide may take from a few weeks to several months, depending on the composition of the wine. Some white wines have been bottled at so high a content of sulfur dioxide that even many months after bottling the quantity remained excessive. This is particularly true of white table wines of standard quality. Mercaptan formation very rarely takes place in bottled wines containing yeast cells and sulfur dioxide (Schanderl, 1965b). The prevention is clear: less sulfur dioxide and more careful filtration.

The time required for a wine to develop to full maturity and quality in the bottle is variable. Many white wines are at their best immediately after recovery from the bottling aeration, provided they do not have an excessive content of free sulfur dioxide. Most white table wines develop to their maximum quality within a year from the time of bottling. The heavier and more alcoholic types may require four or five years. A dry white table wine seldom improves after five years in the bottle. Pink wines, of course, have a low content of free sulfur dioxide. In other respects their bottle-aging requirements are similar to those of white table wines.

Figure 98. Storage cellar for bottle-aging.

Red wines are exceedingly variable in their bottle-aging requirements. Yet their response to aging is the greatest, and they show the greatest increase in quality. The lighter and lesser-quality red wines require no more aging than do ordinary white wines. With wine of ordinary quality the increase in quality after bottling is not likely to be enough to justify the bottle-aging. But there is a surprising improvement in the odor and taste of standard California red wines after one or two years in the bottle. The finer sorts, particularly Cabernet Sauvignon, may require five to ten years of bottle-aging to come to full maturity, and may continue to develop, albeit slowly, for five to ten years more (fig. 98).

Sweet table wines present a special problem. They are usually bottled with a fairly high total sulfur dioxide, and the free sulfur dioxide content just after bottling is normally high. However, their sugar content facilitates a rapid decrease in the latter. No studies seem to have been made of the changes in free sulfur dioxide in sweet table wines after five to ten years of bottling-aging. It seems, however, that the objectionable property of the sulfur dioxide disappears and that the quality may even be enhanced. We have seen some old French Sauternes of exceptional quality.

Bottle-aging of ordinary wines is impractical and probably unjustified because it brings little increase in quality. White and pink table wines should be kept at least three months before starting distribution. The best whites should be kept a year. Red table wines should be kept six months, and the better types for a minimum of eighteen months. Even then, the winery should retain some of its very finest red wines for a longer period in order to build up a supply of high-quality wine that will enhance the reputation of the winery. Wineries would do well to advise their customers also to lay aside some of the finer red table wines for five years in order to obtain the full development of quality. (For a more detailed discussion see Amerine, 1950b.)

LABELING

The label must show (1) the brand name, (2) the name and address of the bottler, or the name, registry number, and state where the premises of the bottler are located, (3) the kind of wine (class and type), (4) the alcohol content by volume (except that "Table" or "Light Wine" may be so designated in lieu of the alcohol content), and (5) the net contents of the bottle unless this is legibly blown into the bottle (p. 569). If there is no brand name the name of the bottler may take the place of a brand name. A certificate of label approval (Form 1649 of the Alcohol and Tobacco Tax Division of the Internal Revenue Service) or a certificate of exemption from label approval (Form 1650) must be obtained for all bottled wines for distribution.

The brand name either alone or in association with other printed or graphic matter, must not create any impression or inference as to age, origin, or identity. The class and type need not include "table" ("light") or "dessert." For champagne, "sparkling wine" need not be added.

The regulations of the U.S. Internal Revenue Service (1961b) define generic, semigeneric, and nongeneric designation of geographical origin and grape-type (varietal) appellations. Sake and vermouth are classified as generic. Semigeneric names of geographical origin used for table wines in this country include burgundy, claret, chablis, champagne, chianti, moselle, rhine (syn. hock), sauterne, and haut sauterne. In using any of these for wines produced in this country an appropriate appellation of origin disclosing the true place of origin of the wine must be given alongside the semigeneric appellation. The wine must also conform to the standard of identity, if any, for such wine contained in the regulations, or, if there is no such standard, to the trade understanding of such class or type. This has been very loosely interpreted.

Examples of nongeneric appellation which are distinctive designations of specific natural wines include Bordeaux, Médoc, Saint Julien, Margaux, Graves, Barsac, Pomerol, Saint Emilion, Bourgogne, Grand Chablis, Beaune, Pommard, Volnay, Beaujolais, Hermitage, Loire, Vouvray, Alsace or Alsatian, Mosel-Saar-Ruwer, and Mosel.

A wine is entitled to a geographical appellation of origin if it meets the following conditions: (1) at least 75 per cent of its volume is derived from fruit or other agricultural products both grown and fermented in the place or region indicated by such appellation, (2) it has been fully manufactured and finished within such place or region, and (3) it conforms to the requirements of the laws and regulations of such place or region governing composition, method of manufacture, and designation of wines for home consumption.

However, a further provision of the regulations permits filtering, fining, pasteurization, refrigeration, and certain treatments or corrections for deficiencies or abnormalities outside the region of origin without losing the geographical appellation. They also permit blending of wines of the same origin outside the region without losing the geographical appellation.

Grape type or varietal designations have been increasingly used for the finer California table wines. The variety of grape "may be employed as the type designation if the wine derives its predominant taste, aroma, and characteristics, and at least 51 per cent of its volume, from that variety of grape. If such type designation is not known to the consumer as the name of a grape variety, there shall appear in direct conjunction therewith an explanatory statement as to the significance thereof."

The name and address requirement is complicated:

(1) If the bottler or packer is also the person who made not less than 75 per cent of such wine by crushing the grapes or other materials, fermenting the must and clarifying the resulting wine, or if such person treated the wine in such manner as to change the class thereof, there may be stated, in lieu of the words "Bottled by" or "Packed by", the words "Produced and bottled by" or "Produced and packed by".[2]

(2) If the bottler or packer has also either made or treated the wine, otherwise than as described in (1) above, there may be stated, in lieu of the words "Bottled by" or "Packed by", the phrases "Blended and bottled (packed) by", "Rectified and bottled (packed) by", "Prepared and bottled (packed) by", "Made and bottled (packed) by", as the case may be, or, in the case of imitation wine only, "Manufactured and bottled (packed) by".[3]

(3) In addition to the name of the bottler or packer and the place where bottled or packed (but not in lieu thereof) there may be stated the name and address of any other person for whom such wine is bottled or packed, immediately preceded by the words "Bottled for" or "Packed for" or "Distributed by" or other similar statement; or the

---

[2] As amended by Treasury Decision 5618, approved May 28, 1948, effective June 5, 1948.

[3] As amended by Treasury Decision 5618, approved May 28, 1948, effective July 1, 1949.

name and principal place of business of the rectifier, blender, or maker, immediately preceded by the words "Rectified by", "Blended by", or "Made by", respectively, or, in the case of imitation wine only "Manufactured by".[4]

For imported wines the label requirements are also complicated:

(b) Imported wine.—On labels of container of imported wine, there shall be stated the words "Imported by" or a similar appropriate phrase, and immedately thereafter the name of the permittee who is the importer, agent, sole distributor, or other person responsible for the importation, together with the principal place of business in the United States of such person. In addition, but not in lieu thereof, there may be stated the name and principal place of business of the foreign producer, blender, rectifier, maker, bottler, packer, or shipper, preceded by the phrases "Produced by", "Blended by", "Rectified by", "Made by", "Bottled by", "Packed by", "Shipped by", respectively, or, in the case of imitation wine only, "Manufactured by".[4]

(1) If the wine is bottled or packed in the United States, there shall be stated, in addition, the name of the bottler or packer and the place where bottled or packed immediately preceded by the words "Bottled by", or "Packed by". If, however, the wine is bottled or packed in the United States by the person responsible for the importation there may be stated, in lieu of the above required statements, the name and principal place óf business in the United States of such person, immediately preceded by the phrase "Imported and bottled (packed) by" or a similar appropriate phrase.[2]

(2) If the wine is blended, bottled, or packed in a foreign country other than the country of origin and the country of origin is stated or otherwise indicated on the label, there shall also be stated the name of the bottler, packer, or blender, and the place where bottled, packed, or blended, immediately preceded by the words "Bottled by", "Packed by", "Blended by", or other appropriate statement.

The alcohol content of table wines (not over 14 per cent alcohol by volume) may be stated as "Alcohol . . . % by volume." A tolerance of ± 1.5 per cent is provided. Thus a wine labeled "Alcohol 12 per cent by volume" may actually contain as little as 10.5 per cent or as much as 13.5 per cent alcohol. The alcohol content may also be given for table wines as "Alcohol . . . % to . . . % by volume." The range may not be more than 3 per cent, and no tolerance above or below the range is permitted.

Labels must be designed so that all the required statements are readily legible under ordinary conditions, printed on a contrasting background in type not smaller than 8-point Gothic caps. All mandatory label information must be in English. (An exception is made for imported wines if "Product of" immediately precedes the name of the country of origin. The labels may not be placed over the mouth of the bottle, must be firmly affixed, and cannot be obscured by stamps.

There are also a number of prohibited statements: false, untrue, misleading, disparaging to a competitor, obscene or indecent; misleading analyses, standards, or tests; coined names that give the impression of entitling the wine to a class or type to which it is not entitled, statement of age (except see below): any implication of guaranty that might mislead;

---

[4] As amended by Treasury Decision 5618, approved May 28, 1948, effective July 1, 1949.

curative or therapeutic effects if untrue or if they tend to create a misleading impression; name of living individual of public prominence or of an existing private or public organization that might imply endorsement, or production by or supervision by the person or organization; designs simulating United States or foreign stamps, flags, seals, coats of arms, crests, or other insignia that imply a misleading relation to or supervision by the armed forces of the United States; use of the American flag, or a government, organization, family, or individual.

For domestic wine the vintage may be given if the wine was "bottled or packed in containers by the permittee who crushed the grapes, fermented the must, and clarified such wine." The date can be given only in direct conjuction with the name of the viticultural area in which the grapes were grown and the wine fermented.

Statements of bottling dates are permitted, but must be printed no larger than 8-point Gothic caps. The use of the word "old" is not considered to be a representation relative to age. If wine is rebottled in containers of one gallon or less by a person other than the producer, the year may be given if the original container had that date. This restricts vintage labeling in this country and prevents minor blending which would be useful (i.e., for topping, etc.).

Truthful statements of wine production methods are permitted, such as: "This wine has been mellowed in oak casks," "Stored in small barrels," "Matured at regulated temperatures in our cellars." These, however, must appear inconspicuously on the back label or other matter accompanying the container.

The California regulations are very similar to those of the federal government. They also provide that coined names may not imply mixture with (or presumably resemblance to) such standard types as burgundy, claret, sauterne, and so forth.

Labeling is done by automatic equipment except by very small operators, and even they use special equipment for gluing to speed up operations. Sparkling wine bottles, which are cold and often wet when labeled, may be passed through a heated tunnel to dry (p. 699).

CAPSULING

To prevent insect infestation of the top of the exposed cork and to make the seal airtight, it was once customary to seal it with beeswax. Imported and domestic metal caps, called foils, are placed over the neck of the bottle and crimped into place by a special machine. Some wineries continue to use these metal capsules, but plastic foils are becoming increasingly popular because of their lower cost and ease of application. These usually come in a moist condition and are placed on the bottle while they are still damp. When the plastic material dries, it shrinks and forms a smooth, almost airtight seal over the end of the bottle (fig. 99). Automatic cap crimpers are also widely used (fig. 100). The possibility that traces of lead may reach the wine from the lead capsules has been noted by Ferré and Jaulmes (1948).

Figure 99.   Automatic capsuling machine. (Courtesy Almadén Vineyards, Inc.)

INSPECTING AND STORAGE TESTS
All wines should be inspected after bottling.
If possible, special storage tests should be made.

Each bottle should be carefully inspected in a strong light for undesirable haze, cloud, or sediment. High-quality freshly bottled wine should be stored for at least several weeks before the final inspection in order to give it time to recover, show disease, or throw down sediment. Changes can be hastened by storing some bottles at 32.2°C (90°F) in an incubator and some at −3.9°C (25°F) in a refrigerator. Large bottlers will therefore find it very profitable to install such testing equipment and to incubate

Figure 100. Automatic capsule crimper. (Courtesy Budde & Westermann Corp.)

and refrigerate samples of each lot of bottled wine before shipment. If the test bottling has been properly made, this final check is largely perfunctory. Even where the test cannot be made before shipment, as with large bottlers, it is worth while; if the wine shows a tendency to become cloudy, the bottler can notify the wholesaler and can facilitate rapid distribution and consumption before cloudiness occurs.

SHIPPING . . .
. . . is now mainly done in cardboard boxes.

Before the war most bottle shipments were made in wooden cases. During the war fiberboard cases were widely used and have now largely replaced the wooden ones. When properly construced they are satisfactory for all except the roughest usage. To prevent excessive breakage the cases should be well stacked in the railroad car or truck.

Some wine is still shipped by barrel, although at only a fraction of the prewar volume. The barrels should be thoroughly cleaned, inside and out, before being filled. Unless the shipping distance is very short, it is better to use properly seasoned barrels rather than new ones.

An increasing volume of wine is shipped in railroad tank cars, by ship, and in tank trucks. The tanks should be lined with an inert material. Before they are filled a qualified person should inspect them, checking the inner walls for pitting or corrosion. Wines should not be shipped in defective tanks. Thorough washing and sterilizing of the tank are the minimum treatments.

PRODUCTION OF WINES

CHAPTER THIRTEEN

RED WINES

The greater flavor and aroma of red wines arise from the contact of the fermenting juice with the skins. The most important problem in red wine production is proper manipulation of the cap to obtain maximum color and flavor.

GRAPES FOR DRY RED WINES

Selection of varieties depends upon the region and the type and quality of wine desired.

Varieties

Only a few of the large number of red wine grape varieties that have been tested in California can be recommended for planting or purchase. Of these few, the Cabernet Sauvignon is still the most valuable. It grows well in regions I, and II, and III (see p. 72), and when properly pruned bears moderate crops. For detailed recommendations see Amerine and Winkler (1963a, 1963b) and Ough and Alley (1966). By use of clonal selections, yields of about 4.7 tons per acre are reported. It should be used to produce only Cabernet Sauvignon wines, for it loses some of its characteristics when blended. Its only disadvantage is that it requires both cask- and bottle-aging to bring it to its highest quality. For this reason, and because of the limited planting available in California, it is sometimes blended with about 25 per cent of a lighter variety, such as Refosco, or a relatively light Petite Sirah, to hasten maturity. It should never be blended with Zinfandel, with whose flavor it is particularly incompatible. Usually the free-run Cabernet wine is drawn off and fermented and aged separately from the press wine. The addition of some press wine, however, improves the body of California Cabernet and appears to give it some desirable characteristics.

Ruby Cabernet was recommended by Ough and Alley (1966) especially for regions IV and V. They noted the need of using virus-free material for optimum yield and composition. Yields of up to eight tons may then be expected. (See also Olmo, 1948.) Merlot was another cabernet-flavored variety recommended by Ough and Alley (1966). It is a good producer and has a mild flavor and normal tannin. It may be useful for an early-maturing wine or for blending with high-tannin Cabernet Sauvignon wines.

Cabernet-flavored varieties that are not recommended for planting in California include Malbec, Cabernet franc, and Pfeffer. Malbec (also called Côt) is undesirable because of the high pH of its musts. Another clone may give better results (Ough and Alley, 1966). Cabernet franc gives an early-maturing wine but of low color and acid. A better clone should be sought. Pfeffer does not mature well in the coolest region, and lacks color wherever planted. It does have a spicy cabernet-type aroma.

Pinot noir is recommended for planting only in the coolest locations in California, normally in region I. It should not be planted, however, by those who are unfamiliar with its early-ripening habit; and the wine maker should recognize that its fermentation and aging are difficult. For special precautions on its handling see Amerine (1949a). These include close control of harvesting, moderate fermentation temperatures, and prevention of excessive malo-lactic fermentation. One aspect of Pinot noir fermentations in Burgundy seems to have escaped our California wine makers (and writers): the fermentation starts at a relatively low temperature and lasts for seven or more days before pressing (Ferré, 1958; Feuillat and Bergeret, 1966, 1967b).

Gamay Beaujolais is a useful variety, although it too must be harvested early and the wine handled carefully to prevent spoilage. It is known to do well in region I and and will probably be useful in region II also. The grape called Gamay (see footnote, p. 77), which is now widely planted in region II, is a late-ripening variety of somewhat different fruit and vine characteristics and wine-making potentialities. It does, however, make fruity, early-maturing wines and is an excellent producer. Since it ripens late it may sometimes be too low in sugar.

Grenache is a useful variety in regions I and II. It is a good producer, is more disease-resistant than Carignane, and its wines are fruity and well balanced. In some years it may be deficient in color, but since it is acceptable in making pink wines also, it can be diverted to that use in such years. It should not be planted in regions IV or V with the intention of making pink wines from the fruit; when grown there its wines are too orange in color and too low in total acidity.

The Zinfandel is our most widely planted red wine grape variety. Nevertheless it has certain deficiencies. It ripens unevenly, and some berries become raisins before the others are ripe. The clusters are tight, and bunch rot is not uncommon in irrigated vineyards or in seasons of high humidity or early rainfall. The popularity of Zinfandel is probably due to its regular production and to the fact that it produces a wine of recognizable aroma and flavor. It should be harvested early, at about 21° or 22° Brix, so that the raisined fruit will not add enough sugar to the must during fermentation to make the alcohol too high. Amerine (1949a) has discussed other precautions which should be taken to obtain the best results with this variety, such as early pressing and separation of press wine. The use of Zinfandel grapes for making rosé wines seems appropriate when raisined grapes are present and it is desirable to press early.

Barbera is a high-acid variety, but is unfortunately of only moderate vigor and productivity. Its fruity flavor should be protected by preventing excessive malo-lactic fermentation, and blending is not recommended. It may be grown in regions III and IV for producing wines of above-average quality.

The Carignane, Petite Sirah, Pinot Saint George, and Refosco are varieties which produce sound but undistinguished wines. The last three may be grown in regions II and III for the production of standard wines. Carignane is best adapted to regions III and IV. It is very subject to bunch rot when grown in hot humid areas. Great care should be taken in proper harvesting of these varieties when they are grown in regions IV and V. We have seen wines in which the must pH was over 4, and slow and incomplete fermentation resulted. Petite Sirah wines of high color and tannin require considerable cask-aging before bottling. Under the most favorable conditions the variety produced a wine of moderately distinctive flavor.

Maturity

Grapes for red table wines should be picked when they have reached a Brix of not over 23° or at most 24°. The acid content of the grapes is rarely too high for making dry red table wines in California if the Brix exceeds 20°. But if it is over 24°, especially with certain varieties, the musts often have too low a total acid content and too high a pH for making satisfactory and stable dry red table wines. In warm years, for varieties such as Zinfandel and Petite Sirah, the picking should be started when the Brix reaches 20°; at higher sugar contents there will be numerous dried berries in the clusters. These markedly increase the apparent alcohol yield of these varieties and give the wines a raisin flavor.

For crushing and stemming operations, see pages 301 to 304.

Léglise (1964) disapproved of crushing Pinot noir grapes in Burgundy in the field rather than at the winery. He believed that contact of the stems and the juice has an unfavorable effect and that cooling the grapes before crushing was important. He strongly recommended destemming when the stems are still green and had no objection to some stems passing to the fermenter when the grapes were ripe, low in tannin, and had dry stems.

FERMENTATION

*The method selected depends on the physical and chemical
composition of the varieties used, the temperature,
and the quality of wine to be produced.*

Size of vats

Figure 28 shows a pre-World War II fermenting room with red must
in redwood vats. The size of the fermenters used for dry red wines in
California varies widely, ranging from vats of less than 1,000 gallons to
concrete or steel or lined-iron containers of 50,000-gallon capacity or more.
The larger the tank, the greater the difficulty of controlling the temperature and managing the cap. However, the greater the surface area to the
height the less thick the cap and the easier the control of temperature.
For medium-quality wines a 10,000-gallon vat is ordinarily large enough;
for high-quality wines, 100 to 5,000. At present most of the large-scale
fermenting tanks in California are either stainless steel, lined iron, or
concrete. Wood, either redwood or oak, is still used by the older wineries.
Stainless steel is probably the material of preference for fermenters.

Metal pressure tanks were introduced into the German wine industry
after World War II. Klenk (1958) has reviewed this trend. He considers
them to have been particularly valuable for red table wines. The main
advantages of the pressure tank are easy control of temperature, reduced
danger of bacterial contamination, and simplicity of cleaning. This is
probably correct for grapes with more or less fungal infection. Ough and
Amerine (1961a) found that pressure fermentation of Pinot noir in
California gave less satisfactory wines than with normal atmospheric
fermentations.

Adding sulfur dioxide

Usually, before addition of sulfur dioxide, several samples should be
removed from various parts of the fermenter to make a composite sample
of the must for testing its Brix degree, the total acidity, and the pH.
Such a sample is more representative than one or two small samples taken
during crushing. These determinations are helpful in deciding how much
sulfur dioxide to add and what other corrections should be made on the
must. It is important, however, that the sulfur dioxide be added as soon as
possible after crushing. This, of course, will be impossible where the
sulfur dioxide is added continuously to the must line during crushing.

The reasons for adding sulfur dioxide and the amounts to use under
different conditions have been discussed (pp. 380–408). Usually 15 ounces
of liquid sulfur dioxide per 1,000 gallons of must will be sufficient. It may

be added during the crushing if the sump, the must pump, and the must line are made of corrosion-resistant material; but ordinarily it is introduced directly into the vats during filling or immediately thereafter. By whatever means it is added, the sulfur dioxide may have to be distributed properly throughout the must by pumping over (p. 407). Léglise (1964) recommended 80 mg per liter for Pinot noir in Burgundy. When low-sugar high-acid musts are fermented, care should be taken not to add too much sulfur dioxide in order to ensure a prompt and complete malolactic fermentation (Audidier, 1965). Ascorbic acid should not be used, since it leads to oxidation of anthocyanins, according to Valuĭko (1966). Kushida et al. (1965) reported the optimum amount of antioxidants for Japanese red wine to be 60 mg per liter of sulfur dioxide and 60 mg per liter of 1-ascorbic or erythorbic acid. Their data did show that higher levels of antioxidant reduced the color after eight months' storage, but it also reduced the tendency to browning.

Several hours (2 to 4) after sulfiting, 1 to 3 per cent of a pure yeast culture (p. 340) should be added. After it has been in the fermenter for about 6 hours it too should be thoroughly mixed with the must. The short delay in mixing is to allow the yeast culture to start growing on the sulfited must. The practice of using material from an actively fermenting tank as a starter for a new vat is undesirable. Correction of the acidity is recommended when necessary. Color retention during aging was best improved by adding sulfuric acid (illegal) to musts in the experiments of Ough et al. (1967). Tartaric acid was best from the quality point of view. Fumaric was undesirable because of the formation of undesirable off-odors and its biological instability.

Testing during fermentation

The Brix degree and the temperature should be taken at least twice daily until the time for pressing. It is customary to make the temperature reading, after punching down the cap, by means of a long-stemmed, metal-encased thermometer (see fig. 45) inserted directly into and then through the cap. Temperature sensors can be permanently installed, but they do not extend into the cap.

Punching the cap

In conducting red wine fermentations the control of the cap is the major problem. Unless this is properly managed the extraction of color and tannin will be inadequate and the cap may reach high temperature levels, become a source of spoilage, and result in wines of off-flavor and brown color. In dry, hot climates the cap becomes oxygenated rapidly, and acetification takes place within a short time. Since the color is mainly in the skins, there will be a color deficiency in the wine unless the cap is punched down *frequently*. In the early stages of fermentation, punching the cap down also has the beneficial effect of aerating the must, but this is undesirable in the later stages. Consequently, very active punching down or pumping over should be restricted mainly to the period before the peak of fermentation or until about 6 per cent alcohol has been produced.

At least twice a day during the early stages of fermentation the cap should be punched down, or liquid from the bottom of the vat should be drawn off and pumped over it. Of course, this operation must be properly managed. Excessive pumping over may overaerate the must, will lead to loss of anthocyans, and is believed to result in slightly greater losses of alcohol. Stirring was the best method for increasing color extraction compared to pumping over, punching down, or fermentation under pressure in the tests of Ough and Amerine (1960), but this is not feasible for large containers. They also showed (1961b) that pumping over was slightly superior to punching down for color extraction, and in practice this seems to be the most feasible procedure. Later (1962) they found some advantage in color extraction by preferment blending of red and white grapes compared to postfermentation blending. The amount of color extracted is dependent on the amount of pigment present.

Cooling and heating

During fermentation there will be a steady rise in temperature. Where adequate cooling equipment is available, the cooling should be started well before the critical temperature limit is reached. The equipment for this and the amount of cooling needed have been previously discussed (pp. 376–380). Normal temperatures in red wine fermentation for Ruby Cabernet, 18.3° to 26.6°C (65° to 80°F), and early pressing, 8° to 14° Brix, are recommended by Ough and Alley (1966), especially in cooler regions, where the variety has excessive tannin and color if fermented too long on the skins. In the San Joaquin Valley, in contrast, it may be necessary to delay pressing to secure adequate color. For Cabernet Sauvignon, Ough and Alley (1966) recommended fermentation temperatures of 18.3° to 29.4°C (65° to 85°F), normal but not excessive agitation during fermentation, and pressing at 5° to 10° Brix. Ough and Amerine (1961b) reported that fermentation of Pinot noir at 21.1° to 26.6°C (70° to 80°F) gave better wines than at 15°C (53°F); with Cabernet Sauvignon, 21.1°C (70°F) was better than 26.6°C (80°F) or 15°C (53°F). Ásvány (1967b) recommended optimum fermentation temperatures of 28° to 30°C (72° to 76°F) for red wine production in Hungary. No optimum fermentation temperatures have been obtained for Zinfandel. Tentatively we favor slightly lower temperatures than for other varieties.

According to Biol and Siegrist (1966), Pinot noir is less resistant to mold attack than Gamay. Therefore, fermentation procedures involving partial crushing are not applicable to Pinot noir. Under European conditions of high mold growth, heating the must has the additional advantage of inactivating oxidizing enzymes that can be derived from the mold growth. This objective can be obtained by short-time high-temperature treatment. For Pinot noir in Burgundy, Biol and Siegrist (1966) favored separation of the free-run, heating the rest of the musts, combining the two, cooling, and fermenting at normal temperature. Some of the wines of heated musts were difficult to clarify. This system seemed better than heating all the grapes. They emphasized that the most favorable results occurred when the musts contained oxidizing enzymes. For musts in good condition we fail to see the advantage or necessity of heating. They note that separation of the stems is advantageous in years of considerable mold. Heating of the musts in Burgundy has one great advantage, according to Léglise (1964, 1966)—it destroys the polyphenoloxidases.

Color extraction is favored by longer periods of contact of the skins with the wine, higher temperatures, and high enough alcohol content. Great differences in cap-liquid temperature were observed in red wine fermentations between 18° and 10° Brix by Ough and Amerine (1961b). For red wines the research at Davis has favored temperatures about 28°C (80°F). For Pinot noir, Amerine and Ough (1957) recommended fermentation temperatures of 21.1°C (70°F) to 26.5°C (80°F), with cap temperatures held to 32.2°C (90°F). They did not recommend fermenting red musts at temperatures of 10.0°C (50°F) to 15.5°C (60°F), which resulted in poor flavor and low color. Prehoda (1963) has reported similar results in commercial wineries in Hungary.

Unfortunately, a high fermentation temperature leads to a shorter period of contact with the skins before pressing. A relatively cool fermentation, not over 26.5°C (80°F), and three to five days on the skins prior to pressing are usually desirable.

Toward the latter part of the season the grapes may occasionally be so cool that in order to start the fermentation the must may need to be warmed by pumping the free-run juice through a heat exchanger and over the cap.

Period of fermentation

The fermenation need not continue for a long period in contact with the skins if adequate color extraction has been obtained. As soon as the color extraction is complete or sufficient the wines may properly be drawn off. Leaving the grapes on the skins longer than is necessary may slightly increase the yield, but the quality often suffers, particularly from increased volatile acidity and excessive tannin with certain varieties. A visual inspection of successive samples of filtered must, taken during the fermentation, will show when color extraction has been adequate. The earlier the drawing off, the less the danger of overaerating the wine or of spoilage in the cap. In normal seasons in California the color extraction is completed within three to five days, but the drawing off may take place at any time between the second and tenth day, according to the amount of color present, the condition of the grapes, the temperature of the fermentation, and the degree of astringency and depth of color desired. The degree Brix is usually from 1° to 10° when color extraction has ceased, but the decrease in degree Brix cannot be used as an indication of color extraction, for there is no direct relation between the degree Brix of the fermenting must and its color content that is applicable to all varieties of grapes at all stages of maturity.

In Burgundy the fermentation on the skins may last four to six days, but in Beaujolais, where a fruity early-maturing wine is desired, it lasts only three days. Leaving the skins in contact with the must for two to three weeks is practicable only in very cool climates and when the grapes are in good condition. However, some believe that this practice is desirable. Laborde (1907) reported its general use in the Médoc region near Bordeaux. In that area closed tanks are sometimes used for red wine fermentation. Even so, long fermentation on the skins is now used by only a few Bordeaux producers. In the Rioja district of northern Spain, closed fermenters are used also, and there can be a rather prolonged period of contact between the fermenting wine and the skins. Ribéreau-Gayon and Peynaud (1950), in addition, have shown that, when closed tanks are used, aeration of the fermenting must should be accomplished at an early stage if all the sugar is to be fermented out.

SPECIAL FERMENTATION METHODS...
... may be of value in special cases.

The abnormal chemical composition of the must and the difficult temperature conditions during fermentation have led to the introduction of numerous special methods of fermentation for making red table wines in hot climates such as are found in much of California. Most of these methods are of questionable value, to be used only with caution and after trial and comparison with the standard procedures.

Submerged-cap fermentation

Among the most promising of these methods is fermentation in closed containers with or without a submerged cap. Submerged-cap fermentation, although used for many years in open containers in California, has not ordinarily been made in closed concrete tanks such as are now available in this state.

In a simple submerged-cap fermentation, a lattice of wood is fixed in the upper part of the open fermentation vat. The crushed grapes are put into the vat below the wooden framework and extend nearly up to it. When the fermentation begins, the cap rises against the lattice but cannot push through, while the fermenting liquid rises and covers the cap. Since the cap is the chief source of spoilage in red wine fermentation, this method has the obvious advantage of removing the cap from contact with the air and from possible contamination. Keeping the frames clean and preventing too much pressure against them, with consequent breakage, are important details of operation. Use of a metal framework has been suggested, but is probably too expensive and the danger of metal contamination may be too great. The large, free surface of liquid exposed may lead to oxidation or contamination. Some wine makers complain of too rapid a fermentation with submerged-cap fermentations. Ough and Amerine (1960) reported up to 5.6°C (10°F) higher temperature in the submerged cap than in the surrounding liquid. This may be due to the trapping of carbon dioxide and heat in the tightly packed submerged cap.

Closed concrete tanks

Two types of closed-tank fermentation have been used in California. In the simpler method the fermentation tank is an ordinary concrete vat with a concrete top that has a manhole for release of gas. As the tank is only about two-thirds filled, the cap does not rise sufficiently to press against the top. Cooling coils are placed inside the tank, since there will be reduced surface loss of heat. The chief advantage of this system is that an atmosphere of carbon dioxide is maintained over the surface to restrict spoilage in the cap. But punching down the cap is difficult, and color extraction will naturally be somewhat lowered unless vigorous pumping over is practiced. The must is sometimes sprayed on the cap while being pumped over. We do not recommend this, for it aerates the fermenting must too much. Occasionally a semipermanent framework is built inside the tank to keep the cap submerged, but this is very inconvenient to handle. Tanks of this nature can be used for storing bulk wines after the vintage season.

The second type of closed-tank system calls for a concrete tank with manholes and a raised edge on the top. The tank is filled fuller than in the first method, and the cap rises to the top. Juice and gas may rise through the manholes, but the cap sticks together sufficiently to prevent much of it from breaking through. The raised edge around the top prevents the loss of any liquid. Sometimes a small wooden latticework is placed just under the manholes.

Algerian lessivage system

Another submerged cap method is the Algerian *lessivage* system described by Winkler (1935), Sifnéos and Laurent (1937), and Fabre (1938). In this system a central tube is believed to permit the gas and liquid to rise faster than the gas can penetrate the cap, thus supposedly creating a circulation. This is doubtful but the submerging of the cap does give a better extraction of coloring material and prevents acetification. An objection has been that this system involves excessive aeration of the wine, with consequent overmultiplication of the yeasts. The must may be cooled, however, by cooling coils in the main tank; and the tanks can be used for storage throughout the year.

Ducellier system

The Ducellier system (Sifnéos and Laurent, 1937) now used in North Africa appears to be a modification of the *lessivage* process. In this system (fig. 101) there are three tanks, a large lower tank (A) and two small upper ones (B and C). The crushed grapes are introduced into the lower tank and some juice is pumped into tank B. As the fermentation starts, a pressure of carbon dioxide is built up over the must and liquid is forced up around the cooling coils, D, to tank B. When sufficient carbon dioxide pressure develops, the liquid trap allows carbon dioxide to escape through tank C. Liquid then flows back from tank B and the process is repeated. The returning liquid falls onto the cap, helps to keep it moist, and aids in color extraction (fig. 101). A very rapid fermentation is reported. Such procedures appear to be useful only for ordinary-quality wines.

Argentinian continuous process

Continuous-feed systems of fermentation have been tried successfully by many of the fermentation industries. They have been recommended in Argentina (Willig, 1950). The must is introduced continuously at a point slightly above the bottom of the tall, 100,000-gallon concrete fermenters. Partially fermented wine is drawn off through valves about halfway from the bottom, and the fermented pomace is scraped off the

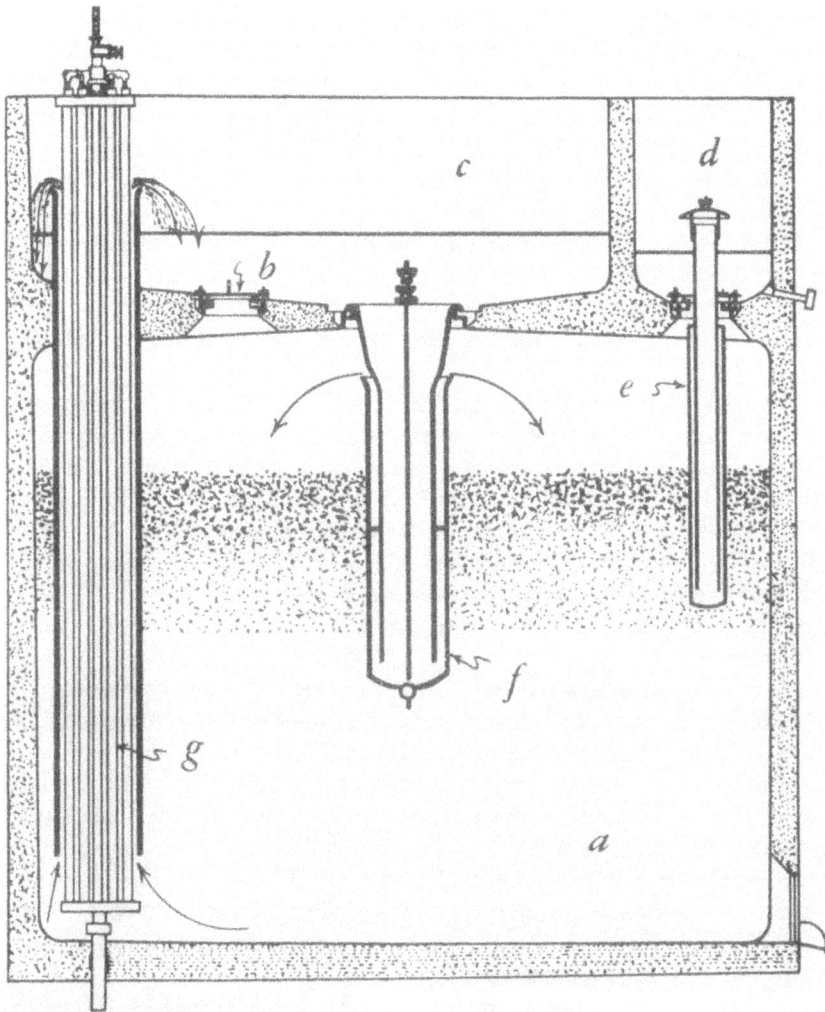

Figure 101. Ducellier system: *a*, fermenting must, also tank A; *b*, manhole for introducing must, also tank B; *c*, open fermenter space; *d*, space for water, also tank C; *e*, water valve; *f*, liquid valve for return of fermenting must; *g*, cooling coils.

top of the fermenter by a mechanically operated device. Advantages claimed for this system are that a uniform product is obtained with little labor and that, since the must is being introduced into a partially fermented wine, little sulfur dioxide is required.

Tarantola and Gandini (1966) reported that systems of fermentation such as the Cremaschi (Argentina) or Defranceschi (Italy) are really *superquatre* systems. That is, the fresh must is continuously introduced into a dilute alcohol solution which is unfavorable to the growth of *Pichia* sp., *Kloeckera* sp., and others. The fermentation is thus carried on mainly by *Saccharomyces cervisiae* var. *ellipsoideus*, with less frequent isolations of *Sacch. italicus, Sacch. oviformis, Sacch. chevalieri*, and *Sacch. bayanus*. No comparative chemical or sensory data relative to discontinuous systems are presented.

The probable disadvantages are the danger of contamination during operation, the difficulty of controlling the temperature, and the necessity of continuously introducing fresh must. There is no evidence that these systems of fermentation can be used for any except the most ordinary wines.

The Cremaschi, Ladousse, and Vico continuous systems for preparing red wines are believed to save as much as 45 per cent of fermenting room space. Minor differences in alcohol and color have been reported between wines produced in continuous and in discontinuous procedures. The absolute necessity of preventing unfavorable bacterial action by use of sulfur dioxide is rightly emphasized. To reëmphasize this, Nègre (1947) notes the critical importance of using high-quality musts. He concluded that "a notre avis, la fermentation continue est une des orientations à étudier."

Rémy (1967a) believes that continuous fermenters can be made small so as to require only one type of grape, but we find this concept difficult to understand. Since the period of optimum maturity for each variety is very short, it is unlikely that a continuous system can be brought to equilibrium in a short period of time. According to Rémy, little sulfur dioxide should be used (since maximum reducing conditions occur in continuous fermenters which may lead to sulfide formation). However, he also states that the system provides against biological and physico-chemical accidents. We should like to see considerable more analytical, microbiological, and sensory data.

Although Rémy (1967b) maintains that his continuous fermenter represents a new concept, it appears to us as another modification of the Sémichon *superquatre* concept: start the fermentation at 4 per cent alcohol and no undesirable microbial activity will occur. Proof of this is needed.

Color extraction by heat, etc.

Other procedures suggested for making red wines involve the extraction of color and tannin from the seeds and skins by the use of massive doses of sulfur dioxide (Barbet, 1912), by heat extraction (Bioletti, 1905, 1906b, Mathieu, 1913, Ferré, 1926, 1928b, 1947, 1958, Joslyn et al., 1929, Konlechner and Haushofer, 1956b, Marsh, 1948, Berg and Akiyoshi,

1957, Coffelt and Berg, 1965a, Prehoda, 1966), or by storage of the grapes under carbon dioxide (Flanzy, 1935, Garino-Canina, 1948). The resulting must is pressed and fermented in closed containers. Of these procedures, heat extraction alone has been used commercially in California. In the commercial process the grapes, after crushing, are heated to about 62.8°C (145°F) and then pressed. The mixture of skins, seeds, and juice may be heated together, or the juice may be separated, heated alone, and then mixed with the crushed skins. After pressing off, the juice is cooled, a pure yeast culture is added, and the fermentation is conducted as for white wines.

Much of the red table wine made in New York is produced by heating the red grapes, usually Concord, before pressing. In the old heat-extraction process the crushed grapes and juice are heated in steam-jacketed kettles to the required temperature (about 62.8°C, 145°F), the free-run juice is drawn off, and the remaining juice is pressed out of the pomace in rack-and-cloth hydraulic presses. Such a procedure is time-consuming, and exposes the grapes to excessive oxidation. However, with *Vitis labrusca* grapes the effect of the heat is much less deleterious to the flavor than with *V. vinifera* grapes.

Ferré (1958) noted that heating of Pinot noir grapes in Burgundy was proposed as early as 1772. He (1928b) tested several methods. Léglise (1966) has accepted the desirability of heating. Ferré's procedures were to dip whole grapes in boiling water for ten minutes to bring their temperature to 80°C (176°F), then cooling, sulfiting, and fermenting on the skins, or to heat to above 80°C (176°F) and, after six to seven hours, cooling to 63° to 64° C (146° to 147°F), sulfiting and fermenting. The fermentations remained on the skins seven days in the first procedures and two days in the latter. Much more color and polyphenolic compounds were extracted in the first than in the latter, and more polyphenol relative to color was extracted. Heating he thought not harmful, but no satisfactory sensory data are presented. Ferré (1958) reported the wines were slow to clear and at first had an odd (banana) odor.

Feuillat and Bergeret (1967b) indicated that heating to 72°C (161.6°F) for seventeen hours without the stems was preferable to heating to 80°C (176°F) for the same period with the stems. Heating resulted in a marked increase in total nitrogen, amine nitrogen, and color intensity both in the must and in the wine, particularly when the stems were removed. Among the amino acids which showed marked increase in the must and the wine were α-alanine, valine, isoleucine and leucine, threonine, serine, and γ-aminobutyric acid. Arginine increased only when the must was heated without the stems. Lysine increased in the heated musts, then fell to a low level, but about double the normal level.

Continuous, closed, tubular preheaters have been developed for improved extraction of color and flavor from grapes in the manufacture of grape juice. So far they have not been applied extensively in making red table wines. Marsh (1948) has suggested, for red table wines, color extraction in closed heat exchangers at temperatures high enough to coagulate the gums and mucilages, for periods short enough to avoid undesirable changes in flavor, followed by rapid separation of skins either by vibrating screens or properly designed and operated screw presses, and then prompt cooling in closed heat exchangers. When these conditions are achieved, it will be possible to ferment red wines with the same ease as white wines. The data available at present (Joslyn et al., 1929) indicate that heating at 73.9°C (165°F) for one minute is sufficient for extraction of color and flavor from Carignane and possibly from most red wine grapes.

Berg (1950b) and Berg and Marsh (1950) have reviewed the work on heat treatment of musts and have conducted extensive experiments of their own in commercial wineries. They used a heat exchanger, a mechanical agitator in the fermenters (preferably round), and a holding tank. They heated to 65.5°C (150°F), 76.7°C (170°F), and 90°C (194°F), with an average holding time of two minutes. The heated wines were of deeper color and lower dominant wave length, but scored slightly less in sensory tests and cleared much less rapidly than did the untreated wine. The color was fairly stable, in contrast to Bioletti's results (1905, 1906b). More data are needed on how much time may elapse during color and flavor extraction without injuring the flavor or producing wines that are difficult to clarify.

The oxidation that may occur in this procedure might be minimized or avoided by heating whole grapes. Amerine and De Mattei (1940) have shown that complete release of color from the skin cells may be obtained by dipping whole grapes into boiling water for less than one minute. Coffelt and Berg (1965a) also found this with super-heated steam.

Most of the heating procedures used thus far give musts and wines which are rather difficult to clear. The tint of color in the resulting wine is sometimes undesirable. It is frequently easy to detect a loss of fruity flavor in such wines. The development of a procedure for readily extracting all the desired constituents from the red grape skins before fermentation would be very useful if it did not adversely affect quality or prove too costly. The Coffelt-Berg process seems to be the most promising.

Valuĭko et al. (1964) suggested the use of infrared heating for five to ten minutes to facilitate extraction of color (ten to fifteen minutes for tannins). They got better results by heating whole grapes with super-heated steam to 150°C (302°F) for fifteen to twenty seconds or even to 170°C (338°F) with hot air. The pulp temperature did not exceed 25° to 30°C (77° to 86°F).

Biol and Siegrist (1966) suggested that the failure to adopt color extraction by heat is due to the technical difficulties of heating and cooling. Unless it is carefully done the process results in a "heated" odor. They recommended heating at 60° to 65°C (140° to 149°F). The settled juice, which is cool, is then placed on the heated crushed grapes and the fermentation is carred on as usual. This procedure obviously facilitates cooling the heated pomace, but it is unduly complicated and does not permit continuous operation. It does deactivate much of the polyphenoloxidase, but retains sufficient pectinase to facilitate clarification. They found that four hours of heating was insufficient to extract adequate color.

In the Côte d'Or, heating of only the crushed drained grapes gave wines of greater color, volatile acidity, residual sugar, and pectins, but of less clarity and some retardation of the malo-lactic fermentation.

One important difference between grapes in California and those of some other countries is that here grapes are seldom attacked by molds prior to harvesting, and so heat inactivation of oxidases from molds is unnecessary.

Complete anerobic fermentation

The ancient practice of filling the fermentation tank with intact grapes and allowing the fermentation to proceed spontaneously (with or without covering the tank) is still used in many regions: the Rioja district of Spain, in Italy, Switzerland, the south of France, on the Rhone, and particularly in Beaujolais (Chauvet et al., 1963; See also pp. 7–8). The persistence of the practice is apparently due to favorable effects on the quality of the wines: finer bouquet, earlier maturity, slightly more alcohol, and softer taste. Two processes seem to take place in this procedure: an intracellular fermentation of malic acid (and possibly of some tartaric) and an improved malo-lactic fermentation. Because of the longer period of fermentation, the high amount of press wine, and the danger of excessive bacterial activity, the process is not widely used.

Flanzy et al. (1967) studied the effect of temperature on the changes in berries held under anaerobic conditions. At 15°C (59°F) for thirty-one days shikimic acid plus quinic acid, fumaric acid, and succinic acid increased; citric acid showed little change; malic acid and tartaric acid decreased markedly. The percentage decrease in malic acid is much greater. They believe that it is the main source of succinic acid and ethyl alcohol. The maximum rate of increase in shikimic acid plus quinic acid lasts for only a few days and then the rate decreases. The changes were more marked at 25°C (77°F) and 35°C (95°F). The large amount of succinic acid formed and the marked decrease in tartaric acid are particularly noteworthy.

Flanzy et al. (1967) believe that the special odors produced by macération carbonique are not due to yeast activity under anaerobic conditions, as proposed by Peynaud and Guimberteau (1962a). This odor is greater with low temperature fermentation or storage, but the wines are poorly colored. They attribute this to less loss of volatile fermentation products owing to intracellular fermentation and the slow over-all fermentation. Flanzy et al. noted that a reduction in tartaric acid in warm regions where the total acidity is low would create difficult problems of fermentation and stability. For previous studies see Flanzy (1935), Gallay and Vuichoud (1938), Garino-Canina (1948), Peynaud and Guimberteau (1962a), Bénard and Jouret (1963), Marteau and Olivieri (1966), and Laszlo et al. (1966).

In Beaujolais, Bréchot et al. (1966b) reported only a small amount of malic acid metabolized by yeasts (0 to 13 per cent of the total in four studies). About 20 per cent was lost by intracellular respiration before alcoholic fermentation.

When grapes are not crushed but allowed to stand under carbon dioxide, anaerobic cellular processes result in formation of alcohol, production of carbon dioxide, loss of weight, death of the cells, and release of color. According to Flanzy and André (1965), the death of the cells is due to this intracellular formation of alcohol. The process is best at 35°C (95°F), but alcohol yield was independent of temperature. Further information on the effect of anaerobic fermentation of uncrushed grapes (as in the Rioja, the south of France, and Beaujolais) have been given by Chauvet et al. (1966). They found, surprisingly, that the yeast population of anaerobic fermentation was higher than with the usual crushing procedure. In both procedures the yeasts appeared to be without cytochromes. They claim that the efficiency of alcoholic fermentation is greater with fermentations of uncrushed grapes (i.e., more alcohol is produced per gram of sugar fermented), but the results are not altogether convincing.

The most recent defense of the system of macération carbonique is that of André et al. (1967). They admit that the amount of free-run juice is less than when the grapes are crushed. Even so, owing to the weight of the fruit, from 17 to 20 per cent of the grapes are crushed and thus are not submitted to macération carbonique. Pumping over should be avoided, for it reduces the percentage of grapes undergoing the anaerobic intracellular fermentation.

The decrease in redox potential during fermentation is much more rapid for the traditional red wine making procedure than it is by the *macération carbonique* method according to Mourgues *et al.* (1967). However, after pressing the redox potential of the latter drops to a lower value—mainly because there is more sugar still to ferment and the malolactic fermentation starts sooner. The potential of the macération carbonique wines remained lower than the regular wines for up to two years. They were not able to establish a correlation between redox potential and wine quality—probably because the redox potential is not itself *per se* a quality parameter. For young wines, however, the lower potential seemed to give better quality. This was not necessarily true after two years of aging.

The amount of alcohol produced by this procedure is essentially the same as that from normal fermentation, but the fermentation is much slower and the temperature rise is less. André *et al.* (1967) argue: "Cette différence de vitesse est extrêmement importante car elle nous explique une des raisons de succès de la méthode Macération Carbonique: la vitesse de fermentation est plus régulière". While this is true, it is certainly not a unique feature. Controlled fermentation by pressure or temperature are equally successful in controlling the rate of fermentation.

Laszlo *et al.* (1966), fermenting whole grapes by classic techniques, obtained the usual results: less yield of wine, lower tannin, volatile and total acidity, and a softer taste.

For alcohol extraction of red color for use in blending (where legal), Maruyama and Kushida (1965) found 30 per cent alcohol best. Moreover, the rate of browning of the extract was less when higher alcohol levels were used for color extraction.

DRAWING OFF AND PRESSING . . .
. . . separate the skins and seeds from the fermenting wine.

Many different systems for handling the pomace and wine have been devised; the choice depends on the arrangement and facilities of the winery. In all systems the wine is removed from the bottom of the vat, formerly often through a rough screen made of wooden slats.

If the color extraction has not been completed when most of the sugar has fermented, the pomace may be placed directly in a press. If the pomace still retains sugar, distilling material may be produced by adding water to the pomace and completing the fermentation. The pomace of two or three tanks may be thrown together for the final fermentation, but this involves additional handling. When all the sugar has been converted to alcohol, the pomace is pressed; or the distilling material may be drawn off, and the remaining alcohol extracted from the pomace by washing or other means. In a winery that has no still, the pomace is pressed as soon as the wine is drawn off, even though it may still contain some sugar. Pomace wines for beverage purposes are never made commercially in California. We do not favor adding the pomace from one tank (or two) into newly crushed grapes. This usually results in excessive tannin and may spread contamination. Nevertheless, it has been used with low-color varieties such as Pinot noir. We have not been impressed with the quality of the resulting wines.

Extraction of the alcohol remaining in the pomace with water and distilling it, or selling the wash for distillation, appears to be economically feasible. Some wineries wash the pomace more than once, and even press between washings. Direct distillation of the pomace in special stills is also quite common in this state.

Type of press

The hydraulic basket press is preferred when the press wine is to be used in the cellar instead of for distilling. This type of press, though somewhat more expensive to operate, extracts less of the finely divided pulp than does the continuous press. Laborde (1907) reports 5 to 6 per cent solids in the wine from a basket press and 10 per cent in that from a continuous press, and it is doubtful whether the difference is any less today than it was sixty years ago.

The horizontally or vertically mounted continuous press is the obvious choice where the press wine is used for distilling material or where refermented pomace is being pressed. It operates very well with partially or completely fermented red pomace. The new Serpentine press (p. 201) of Coffelt and Berg (1965b) appears useful for pressing red pomace. The Willmes (p. 199) or Vaslin press is now commonly used.

The press wine should be kept separate from the free-run because it ages less rapidly, may be difficult to clarify, and is usually of lower quality. The press wine amounts to from 10 to 15 per cent of the total wine according to the variety and its maturity, the length of time of extraction on the skins and of draining and pressing, the type of press, the amount of pressure, and other factors.

Diffusion batteries

Another system suggested for the removal of juice or wine from the pomace is the use of diffusion batteries. Water is introduced slowly at the bottom of a closed tank three-fourths filled with crushed grapes. The grape juice is forced up and flows out through a pipe to the bottom of the second tank and so on for eight or ten tanks. As the grapes in the first tank are exhausted of their sugar or wine, the tank is replaced at the end of the line with another. It is claimed that good yields of fairly clear juice or wine are obtained by this system, with the skins acting as a sort of filter. It has not, however, always met with favor by government agencies because unscrupulous operators might operate the battery so that water is allowed to get into the must or wine. Moreover, the cost of construction of a large-scale unit may be excessive. Coffelt and Berg (1965a) constructed a countercurrent extractor on this principle, but the product was intended only for distillation.

COMPLETING FERMENTATION

Prevention of oxidation and removal of
fermentable sugar are important.

The young wine usually contains a smaller or larger amount of unfermented sugar after drawing off or pressing. It should not, therefore, be stored in a cold place, but should be kept at a fairly warm temperature, 21.1°C (70°F), until the fermentation is complete. For this purpose closed tanks with a fermentation bung are desirable. Any simple bung that prevents free access of air to the wine and maintains a slight pressure of carbon dioxide in the container is satisfactory. The fermentation should always be completed at once. It is not safe to trust that it will restart and be finished in the spring, since the wines with residual sugar often become contaminated and spoil during the winter.

While the wine is completing fermentation, the container should be filled as full as possible and loosely bunged to allow escape of gases. As soon as the gross sediment has settled, usually not until after the fermentation is completed, the wine should be racked into its storage container in the cellar.

If the fermentation is especially difficult to complete, and if aeration and warming will not restart it, the product may have to be treated as a stuck wine and refermented by one of the procedures already mentioned (p. 311)—usually by adding the wine slowly to active fermenters.

Testing for sugar

The Brix reading is of little value in determining the dryness of the wine. At 0° on the Brix hydrometer there will still be 1 per cent or more of sugar present, because the alcohol formed has a lower specific gravity than water and tends to lower the reading. If the Brix reading is made on dealcoholized wine and is above 3.0, there is a strong suspicion that the wine contains unfermented sugar; hence a chemical test (see p. 746) is the only safe method of proving its presence. The wine is usually dry enough to prevent refermentation or bacterial activity if the sugar content is below 0.2 per cent. If a Brix hydrometer, scaled to register from −5 to +5, is used, *dry* red wines will usually have a reading of −1° to −2°.

CLARIFYING AND AGING . . .
. . . consist of storage and periodical rackings.

Clearing after fermentation

The appearance of the young wine changes rapidly during and after fermentation. During the active process it is murky and disturbed, but the bulkier sediment, mainly grape tissue and yeast cells, drops rapidly at the conclusion of the tumultuous fermentation. Within a month of the time of pressing, most of the remaining suspended yeast particles should settle out, leaving the wine reasonably clear. In this condition, if dry, well stored, and properly treated, it is fairly safe from spoilage.

If the wine fails to clear in this fashion, especially after an early racking, some undesirable condition in the wine is indicated. This may be because of a slow fermentation of residual sugar, bacterial contamination, colloidal cloudiness from excess metals or tannins, enzymes from moldy or rotten grapes, and so forth. In red wines an appreciable portion of the colloidal material is tannins, according to Nilov and Skurikhin (1967); the main part is due to polymerization of tannins and pigments. The cause of persistent cloudiness should be determined at once by appropriate chemical and microbiological analysis; it is undesirable for any wine to remain in this condition for even a short period of time. (See pp. 726–729 and 772 ff).

Racking

Young red wines made from nonmoldy grapes may usually be racked in contact with the air without fear of overoxidation. However, this is not true if the grapes were moldy, since the new wine may contain considerable polyphenoloxidase. Naturally, no red table wine should be overaerated. By December the wine should be free of the fermentable sugar, ready to be racked off the sediment (called lees). Care should be taken in racking to avoid stirring up the sediment; a good practice is to leave a generous layer of wine over the deposit. The lees (yeast, seeds, tartrates, and the like) and remaining wine from several tanks may then be pumped into a single tank; after a further but brief settling, an additional amount of clear wine can be racked or, better, filtered off. The collected deposits of lees must not stand for long in contact with the supernatant wine lest undesirable chemical and microbiological changes occur, leading to high- or low-pH wines of poor flavor and color. The remaining lees are generally used for distilling material or sold to a by-products plant for recovery of alcohol and cream of tartar. For a small container the racking may be done by siphoning, using a rubber hose. Preferably the clear wine may be drawn or pumped off through a hose connection on the side of the cask or tank. This connection should be several inches above the upper level of the lees. Sometimes a hose is lowered into the tank to the desired depth and the liquid pumped out.

If the first racking takes place in November or December, the next may be made in February or March and a third in August before the next vintage. The earlier and more frequent the rackings, the less the danger of a malo-lactic fermentation (pp. 496–505, and 778). In the experiments of Ough and Amerine (1961a) a delayed malo-lactic fermentation on Pinot noir wines gave better quality. Even in wines of only moderate acidity Brugirard et al. (1965) consider the malo-lactic fermentation to be necessary for the stability of red wines. Whether this is true in California red wines appears doubtful.

After the first year the wines will continue to deposit small amounts of sediment. They should be racked about twice annually—less if the wine is sensitive to oxidation—until they are ready for bottling. At each racking care should be taken to prevent excess aeration.

Sulfuring of red wines during aging has always been practiced, although perhaps not to the extent that some careless wine makers do nowadays. In our opinion the total sulfur dioxide should never be over 100 mg per liter. There is some danger that too much will retard the rate of aging and reduce the sensory quality which can be achieved. Kushida et al. (1965) reported 60 mg per liter each of sulfur dioxide and ascorbic (or erythorbic) acid gave the best results with a red table wine stored for eight months. For precautions on the use of ascorbic acid see p. 416.

Testing during aging

About the first of the year a general checkup of the wines in the cellar should be made. This includes an analysis of the wine (p. 727) as well as careful sensory examination. The wines should then receive a preliminary classification and, according to their soundness, varietal composition, and quality, a decision should be made regarding the proper future treatment. By this means the lesser-quality wines can be segregated for early clarification and sale, and the wines for aging can be properly stored. The mid-year check is also the occasion for detecting off-flavored and diseased wines before they have a chance to contaminate other tanks or equipment during racking.

Refilling

Between rackings the containers must be regularly filled. The chief source of danger during the aging of red table wines is spoilage through failure to keep containers full. This is also the chief source of spoilage where large air spaces are permitted. Maintaining a low carbon dioxide pressure over such wine has been suggested as a means of preventing spoilage.

In the late spring, when the cellar temperature rises too rapidly, a considerable expansion in volume occurs in large cooperage, and some wine may have to be removed. The contraction in volume, when the temperature drops, is greatest in large cooperage.

Small cooperage, especially, must be filled regularly. In fact, small cooperage, because of the greater ratio of surface to volume, should be watched most carefully. The lower the humidity the greater the circulation of air, and the warmer the air of the cellar the more frequent the filling will have to be. The type of container will also be a factor. During the first months the casks or tanks should be filled as often as every other week, and the barrels even more frequently; later a monthly filling is usually sufficient. When red wines are aged in barrels, it is customary to turn the barrel during the second and third years so that the bung is on the side and covered with wine. This reduces the amount of air entering the barrel from around the bung. The barrel is returned to the normal position for the periodic fillings that are necessary.

Period of aging

Light, fruity wines may and usually should be drunk when comparatively young. Wines of no particular character, with a light body, fall into this class, as do wines whose composition makes them unsuitable for aging (such as those moderately high in volatile acid or low in alcohol). Wines of this quality should be brought to a stable condition and sold within a year or eighteen months. By removing them from the cellar as soon as possible, the cost of storage is reduced; and, since they can never be sold for a high price, the margin of profit is increased. Such wines may be filtered, fined, and otherwise stabilized as necessary.

The wines of heavier body and richer flavor, which should be aged for some time, must be handled differently. These must be low in volatile acid when aging begins. They need a moderate alcohol content—12 to

13 per cent—and acidity of at least 0.6 per cent. To increase in value through aging they must be free of foreign odors and possess a well-balanced and distinct flavor. Any necessary blends not made at the time of crushing should be made as soon as possible after fermentation. If several fermentations of the same variety or blend of grapes have been made, they should be tasted and analyzed. If they are of equal quality they may be blended into one lot or, preferably, the least desirable can be kept separate for lesser-quality wine.

There is no necessity for complete finishing of the wine by fining or filtering at this time, since mellowing occurs naturally during the aging. The too early removal of all tart, astringent, or rough constituents by these operations or by undue lowering of the temperature sometimes seems to defeat the purpose of aging. At this time, also, the temperature should be reduced, naturally or artificially, to aid in precipitation of the tartrates and in general clearing of the wine, and to prevent undue losses by evaporation. (See p. 517.)

The wines of Pinot noir seem especially difficult to handle. There is a prejudice against their filtration. Feuillat and Bergeret (1966) recommended fining with kieselguhr rather than bentonite before filtration in order to reduce color loss and to prevent what they call *maladie de filtration* (excessive aeration). Filter aid also holds a fantastic amount of oxygen. When a very small volume is filtered the results may be quite different from those obtained when a large volume is processed. This may account for the flavor of wines prepared solely for a judging or special market in comparison with the regular large lot.

Aging changes are chemical reactions with temperature coefficients of 1 or 2. The larger the container and the lower the temperature, the slower the rate of aging; in small containers at higher temperatures, more rapid aging is obtained. Wines aged too rapidly by heating and oxidation do not acquire the desired bouquet and may taste flat. Excessively prolonged aging is expensive and may bring undesirable changes caused by overoxidation, bacterial contamination, or excessive extraction of woody flavors from containers. When too rapid, proper interaction of wine and container cannot take place. Long experience with aging procedures indicates that smaller casks, of 50 to 1,000 gallons, stored at low temperatures (10° to 15.5°C, 50° to 60°F), are the best.

Ullage and topping

During aging, even at low temperatures, there is a constant loss of wine through the wood. This is called "ullage". If nothing is done to keep the container full, the air space over the wine will be sufficient to cause acetification. The accepted practice is to refill the containers periodically, the smaller the container the more frequent the filling must be. Filling the containers is known as "topping" in California wineries. Some concept of the magnitude of the "topping" problem is indicated in figure 102. The wine required for topping is usually stored in smaller casks and demijohns.

During the aging of red table wines, the volatile acid content should be checked periodically. If it tends to rise unduly, small amounts of sulfur dioxide should be added, or some other method of controlling micro-organisms, such as filtration or pasteurization, should be followed.

Clarification

The question of whether gelatin or bentonite is preferred for fining red and rosé wines has been studied by Bergeret (1963). He finds that it depends on the type and age of the wine. Bentonite removes more color from young wines than gelatin; the opposite is true for old wines. Bergeret considers this to be due to the relatively greater activity of bentonite on the colloidal colored material of young wines. Bentonite is especially effective (often too much so) in removing color from rosé wines. Tannins are usually removed better by gelatin fining, but this may promote instability. See chapter 12 for further details on finishing.

CHAPTER FOURTEEN

ROSÉ AND PINK
TABLE WINES

Pink or rosé wines are intermediate in character between red and white wines. Since World War II many have been made in California. They are also widely produced and appreciated in France and Italy.

Rosé wines have long been produced abroad, particularly in France, Algeria, Austria, Hungary, Rumania, Cyprus, and Israel. They are now produced in Canada, California, Australia, and South Africa. The production of rosé wines was recommended by Hilgard many years ago, and before World War II by Amerine and Winkler (1941b). It was commercially introduced in California about twenty-five years ago and immediately became popular (Gallo, 1958). Dry pink, sweet pink, and sparkling rosé wines are now produced and marketed.

Pink wines are produced from low-colored red grapes, from normally colored red grapes left on the skins for only one or two days, from a mixture of red and white grapes, or from a blend of wines of different colors and ages. The first two procedures are considered best for the rosé character.

Pink wines are seldom high in alcohol and are meant for early consumption. A total acid content of 0.70 per cent and 11 per cent or less of alcohol are common. These wines should have a fresh, fruity flavor. Amerine (1949a) and Amerine and Winkler (1941b, 1963a, 1963b) have called attention to their desirable characteristics and discussed methods of producing them.

GRAPES FOR PINK WINES
Either pink or red grapes may be used.

Pink varieties
The pink variety most commonly used in France for very light-bodied, low-alcohol pink wines is the Aramon. It is a very heavy producer and at its best yields a pleasant, early-maturing, fruity product. However, it fails to ripen adequately in region I, and is too low in acidity in regions

III and IV. It is doubtful whether plantings of this variety are justified. Grenache, used in the Rhône Valley of France for production of the pink Tavel, is subject to oxidation.

The Grignolino has a characteristic orange-red color and is frequently used in Italy for producing a varietal wine. Its wines are better balanced than those of Aramon, but may be somewhat too rich in tannin unless aged for one or two years or blended. The Grignolino does not ripen well in regions I and II, and plantings are indicated only for region III and perhaps for IV. A variety of clones of varying color and tannin content (or possibly even different varieties) are grown as Grignolino both in Italy and California. Unfortunately, the trade demand for Grignolino in California is for a light red wine. This involves considerable blending, and one sometimes finds inadequate Grignolino character in wines so labeled.

Although several varieties widely planted in California—Mission, Mataro, Ribier, Tokay, Black Hamburg—yield light-colored wines, their musts are all low in total acid. Usually these grapes are planted in the warmer districts, with the result that their sugar content is too high and their quality too low. The quality of their wines does not warrant planting them for use in table wines.

Red wine varieties

Except for the few varieties which yield red juice on pressing, any red wine grape can be used to make pink wines by varying the time between crushing and pressing. The following varieties have been particularly useful: both the Gamay Beaujolais and the variety called Gamay in the Napa Valley yield fruity, early-maturing pink wines. Pinot noir produces a fairly heavy-bodied pink wine, but under warmer California conditions it may often be better adapted for producing pink than red wine. It should be grown for this purpose only in region I and possibly II. Pinot noir and Cabernet Sauvignon are best used for red wines, but the former, if overripe, may be used for a rosé.

Roson (1967) studied the effect of time on the skins, sulfur dioxide, and temperature on color extraction from Cabernet (franc ?) grapes. He noted that the less the amount of red color the more yellow it appeared and this was not related to oxidation. The best temperature for the most desirable color was 20°C (68°F). He recommended a short period on the skins—about 24 hours.

Zinfandel yields a tart wine of berrylike flavor; the early pressing reduces the extraction of sugar from the dried fruit and gives better-balanced wine.

Grenache, when grown in regions I and II and picked sufficiently early, yields a fruity-flavored pink wine of good character; some of the best pink wines of California and France have been produced from this variety. In warm seasons it is likely to become overripe and must, therefore, be harvested particularly early. Grenache wine grapes now are grown mainly in the Central Valley region, the warm climatic area classified as region V. In this region Grenache was recommended for use in white dessert wines, but was not recommended for rosé wines by Amerine and Winkler (1944). There its wines are often too orange-colored, low in acidity, and high in pH.

Carignane probably produces a better pink than a red table wine, but is rather neutral in flavor. It does best for rosés in regions II and III. Petite Sirah, in warm seasons, is probably used more rationally for pink than for red wine because fermentation of the raisined berries gives a wine of excessive alcohol content. Unfortunately, the color of a pink Petite Sirah wine is a little too orange for consumer acceptance.

Mixing white and red grapes

White and red grapes are sometimes mixed to make pink wines. The length of time on the skins depends on the variables previously listed and upon the percentage of red and white grapes. The wines made by this method are sometimes very pleasant, for they do not have the high tannin content which results when light-colored varieties, such as the Grignolino, are fermented for several days on the skins.

The stability of the anthocyanin pigments extracted by a short fermentation on the skins from red wine grapes is lower than the stability of anthocyans in pink grapes fermented for longer periods of time. Nègre (1942–43) reported that the musts drawn off from red grapes after a short period of fermentation on skins are appreciably lower in tannins than those prepared from pink wine grapes. In France there is a recognized difference between true rosés and the early-pressed rosés. The latter are actually classified as whites. The true pink wines are characterized by a lower extract and a higher alcohol-extract ratio than red wines and are more similar to white wines in this respect. The pink wines prepared from early pressed red wines, referred to as "24-hour wines" by Nègre, are low in extract content in relation to alcohol. Red wines of low color may be marketed as early-pressed "vin de café". Nègre proposed that the ratio of tannin to nontannin polyphenols be used to differentiate the true rosés from "vin de café". For Aramon grapes he reported this ratio to increase from 0.71 to 1.8 as the period of fermentation increased from one to seven days. The nontannin phenols reached a maximum level at two or three days, but tannins did not increase appreciably until five or six days. It is unfortunate that more information is not available on the polyphenols in rosé wines. Joslyn and Little (1967) reported that catechins appeared to be more reactive than other flavonoid compounds and were, therefore, a good index of storage deterioration. The change in dominant wave length of wines stored at various temperatures was closely related to the loss of flavonols and cyanins, that is, the destruction of anthocyanin pigments paralleled the loss of flavonols.

The total quantity of wine grapes suitable for rosé wine production at present in not enough to meet the demands, and this has led to blending. Grenache grapes from the cooler Napa Valley are usually fermented on the skins longer than those from Modesto and Lodi, and are then used for blending. Rosé wines may be prepared by blending Grenache base wine with Zinfandel and white free-run wines made from a neutral-flavored variety, such as Thompson grapes. The blended rosé may be marketed at a total acidity of 0.6, a reducing sugar content as high as 2.3 per cent, and and alcohol content of about 12 per cent.

FERMENTING AND PRESSING

Close attention to color extraction is required.

Pink-colored grapes are fermented in much the same way as are grapes for red wines. The pressing need not be delayed so long, since the necessary color extraction occurs rather early. The wine maker should make frequent tests of the rate of color extraction during fermentation, keeping in mind that losses in color occur during aging.

Grapes in good condition, especially if harvested before becoming overripe and if pressed *immediately* after crushing, yield white or very slightly tinted musts. The amount of color in the juice and the rate of extraction from the skins depend on the variety of grape, the region and season, the time of picking, the physical condition of the grapes, the temperature of the must, and other factors. Usually, sufficient color is extracted in 24 to 36 hours, with a little longer time required when the temperature is low.

In Algeria, large amounts of sulfur dioxide (400 mg per liter) are sometimes added to red grapes at the time of crushing to check fermentation; after about 24 hours the must is separated from the skins. This juice is relatively colorless, but contains a considerable quantity of the anthocyanins. It is allowed to settle, and the clear liquid is then drawn off and fermented by the procedures used for white wine. Much of the sulfur dioxide is lost during fermentation, and the pink color of the anthocyanin pigments returns. The pink wine so obtained is said to mature early and to be free from stemmy or earthy flavors. Color standardization is difficult, however, because the sulfur dioxide discolors the must so that the proper length of time on the skins cannot be determined accurately. Since the resulting wine may retain too much free sulfur dioxide, the process is not recommended.

CLARIFYING AND AGING...

... are not difficult, and require little time.

The clarification and handling of these wines closely resemble the methods for white wines (Chap. 15) except that the clarification is usually easier for pink wines because of their greater tannin content. Where color is restricted, as in rosés produced by limited periods on the skins, Feuillat and Bergeret (1966) strongly advise against fining with bentonite. They are seldom aged more than six months in the cask or tank.

PREPARATION FOR MARKET...

... requires greater attention.

A wide range in tint and depth of color is observable in California rosé wines. Even in wines from the same winery there is considerable variation. Very little information is available on the preferred hue and depth of color for California rosés. Based on a study with experienced and inexperienced personnel, Ough and Amerine (1967b) reported similar rosé wine color preferences for the two groups. The most popular color for rosé wines was that having brightness of 35 to 50 per cent, dominant wave length of 595 to 620 mμ, and purity of 25 to 40 per cent. The two groups were relatively stable in their color preferences over a four-month period, especially the experienced group. Further and more extensive data of this type are needed. This is emphasized by the finding of Ough and Berg (1959) that extremely small differences in color, as determined by tristimulus values, could be distinguished by their panel. It may also be significant that their panel especially disliked orange tints under fluorescent light. For methods of measuring color differences see Amerine *et al.* (1959a) and Berg *et al.* (1964). The color plate of Amerine *et al.* (1959a) may also be useful

There is a wide range in the sweetness of California rosés. This is partially based on Filipello and Berg's (1959) observation at California fairs that sweet rosé wines were preferred. This result appears logical, since many of the participants probably seldom drank wine. Amerine and Ough (1967) reported that of twelve experienced judges six always preferred the drier rosé wine, whereas five always preferred a rosé with 1 to 2 per cent sugar. Only one judge had a broad range of preference. The pH of the wine had little effect on preference. It would appear that the industry should reinvestigate the problem of sweetness preference of the consumer of rosé wines. If, indeed, there are two groups with different dryness or sweetness preference it would seem desirable to produce rosé wines of varying degrees of sweetness so that they could be identified by the potential consumer.

CHAPTER FIFTEEN

WHITE WINES

Using proper varieties, preventing handling injuries, rapid crushing and pressing, settling, controlling oxidation, avoiding metal contamination, restricting use of sulfur dioxide, and early bottling are important factors in white wine quality.

GRAPES FOR DRY WHITE WINES
Dry white wines call for high-acid grapes, harvested early and handled carefully.

Varieties

The best varieties of grapes for high-quality dry white table wines are those which give a product of distinctive flavor. Among these are the Chardonnay, White Riesling, and Sauvignon blanc.

The Chardonnay is, unfortunately, a poor producer unless it is cane-pruned and grown in fertile soils. It ripens early and is recommended for planting in regions I and II. Chardonnay wines should have a full, rich flavor and ripe-grape aroma. They are normally distributed with a varietal label. Chardonnay and especially Pinot blanc musts have a strong tendency to brown. Browning may occur during crushing, owing to polyphenoloxidase activity, and this may be difficult to control by sulfiting.

White Riesling is likewise recommended only for the cooler regions. It is a better producer than the Chardonnay and ripens considerably later. Its wines should be fruity and light in color, with a characteristic aroma. If produced with over 51 per cent White Riesling grapes and possessing the characteristic aroma of this variety, these wines should be labeled White Riesling (better than Johannisberger Riesling) rather than simply Riesling. Amerine (1949a) has discussed the production problems of the variety.

Sauvignon blanc is rather a small producer when young, but with time its vigorous growth provides a large bearing area and it bears well. It ripens in midseason, can attain a high sugar content, and is useful in regions I to III. Its wines have a very characteristic fragrance which is probably best described as aromatic. In California it is almost always bottled under its varietal name.

The Traminer is not extensively grown in California. It produces a very aromatic and characteristic wine which would seem to be of considerable interest to consumers. Recently its spicy clone, Gewürztraminer, has been widely planted in the state. Traminers probably should be grown only in regions I and II in view of their early ripening. The variety called Traminer in the Napa Valley is frequently Red Veltliner.[1]

Sémillon is one of the most popular white wine grapes of above-average quality grown in the state. It produces and bears well in regions I, II, and III, and may be used for either dry or sweet table wine. It can also be blended with other varieties to improve their character. Amerine (1949a) has discussed some of its enological properties.

The Chenin blanc and Pinot blanc are frequently confused. This confusion arose because the former is known in the Loire region of France as the Pineau de la Loire. The Chenin blanc produces well, and is adapted particularly to regions II and III, where it ripens nicely. Its wines, if not distinctive, are above average in quality. The Pinot blanc produces slightly more characteristic wines and is perhaps more resistant to disease. It has done well in regions I to III, but only a small acreage is now grown. It is to be hoped that both these wines will eventually become known to the trade under their correct names.

Red Veltliner resembles these two somewhat in the quality of its wines and will perhaps compete with them in the trade. It is a good producer and should be planted in regions II and III. One advantage is that its wines mature rapidly.

Sylvaner (sometimes known as Franken Riesling) produces very well, but is somewhat subject to rot and should be planted mainly in regions II and III. It ripens well and produces rather soft, early-maturing wines with only a slight characteristic aroma.

Folle blanche is of considerable value to the wine maker, but has definite viticultural handicaps (see Amerine, 1949a). These include only moderate production and a tendency to rot in humid locations or wet seasons. Its high acidity, however, makes it useful for blending, and it has a place in regions II and III, and possibly in IV.

French Colombard is an excellent producer, ripens in early midseason, and retains good total acidity. It is the variety of choice for standard white table wines in regions III and IV. Sufficient tests have not been made to established the utility of Peverella, a variety of similar characteristics, but it appears useful in producing standard white table wines in regions III

---

[1] "Red" in "Red Veltliner" (and sometimes with Traminer) simply indicates that the grape skins have a very slight pink blush when fully ripe. They produce only white wines.

and IV. Emerald Riesling, a variety with a slight muscat flavor, has excellent acidity and is best planted in regions III and IV. Its wines must be handled carefully because of their tendency to brown.

Frequently, however, the wine maker does not have a choice of varieties; he has to use those that are available. It is best to avoid low-acid grapes, whether of table varieties such as Malaga, or wine types such as Palomino (Napa Golden Chasselas), Sauvignon vert, and the like. Where there is a choice, other factors being equal, grapes from the cooler regions and from mature rather than young vines should be selected.

Harvesting

Grapes for white table wines should be picked with special care (see p. 297) and at a slightly lower degree Brix than for red wines.

Dry white wines should be made as light and fruity as possible. For this result an early picking is essential, particularly if July and August temperatures are high. The degree Brix should be from 20° to 23°, and the total acid as high as possible within these limits of sugar.

The vinification of White Riesling should, if possible, be made before the grapes lose their green color—usually at a Brix of about 20°. The preferred temperature of fermentation is below 60°F. Every effort should be made to prevent oxidation and to stabilize the wine for early bottling.

The white wine grape varieties generally have thinner and more delicate skins than the red and hence are more easily bruised in picking. Since white wines reflect off-colors and flavors more readily than do red, moldy and spoiled clusters should be carefully discarded. Injured berries brown rapidly, which makes the maintenance of a light color in the resulting wines more difficult. Léglise (1964) especially cautioned against crushing Chardonnay grapes in Burgundy in the field. Browning and loss of character resulted. Moldy fruit contains polyphenoloxidase that causes darkening of the must and may lead to oxidasic casse (see p. 280) in the wine. The picking boxes or gondolas used should be clean and free from remnants of red grapes.

Transporting

White wine grapes, which must not be carried over long distances, should be transported to the winery in boxes or small gondolas and crushed promptly. The wine maker should try to obtain from such grapes a must as free from an objectionable brown oxidized color as possible. Therefore he must take all possible precautions against breaking the skins before crushing and against subsequent oxidation.

Red grapes for white wines

Occasionally the wine maker must use red grapes to produce white wines. The varieties of red grapes to be used for making white wine should not yield a colored juice when pressed. Varieties which are naturally low in color, such as Pinot noir and Grenache, give the best results. Late in the season, even the sound berries of these varieties give some colored juice, and the rotten berries or those with broken skins always give it. Red grapes which are to be used for making white wine should therefore be picked early in the season, handled carefully to prevent injury to the berries, crushed without tearing, and pressed without delay. A white or nearly white juice may be obtained by a gentle crushing with a roller crusher and by drawing off only the free-run juice. The Garolla-type crusher tears the skins and always produces colored musts. The remaining pomace may be used for producing red wines or as distilling material. Or, as in the Champagne district in France, the uncrushed grapes could be pressed in special flat presses. Crushing directly into Willmes presses and using only the first pressing is possible. It is necessary to add sulfur dioxide (100 to 200 parts per million, depending on the condition and temperature) to the free-run juice and allow any remnants of skins, seeds, and other waste, which would add color to the wine during fermentation, to settle out. Some of the pink color, which is extracted by crushing or pressing, will disappear during fermentation, racking, and aging. Vigorous aeration of the must prior to fermentation removes much of the pink color. White wines from red grapes have a heavier body than do wines from comparable white grapes.

PREFERMENTATION PRACTICES
Careful pressing and settling are required.

Crushing and stemming

Crushers like those described for red wines (p. 301) are commonly used for white. The combination stemmer and crusher is usually satisfactory, but all the berries must be macerated. The seeds should not be crushed. In order to secure greater yields of free-run juice, some use a roller-crusher preceding the rotary crusher-stemmer; others use a roller-type macerator to squeeze the skins after crushing. Although roller-type macerators do increase the yield, the juice is very cloudy; and producers usually find the wines somewhat more difficult to clarify. The quantity of lees is also high.

Occasionally the stems are left on white grapes to facilitate handling of the mass in the press basket; but the extraction of bitter, stemmy flavors during pressing, or even by the contact of must and stems, has led most wineries to abandon this system.

Free-run and press juice

The free-run juice that drains off from the bottom of a vat containing freshly crushed grapes is nearly always the most desirable for making high-quality white table wines. For such wines it should not be mixed with even the first-press juice. The press juices from the second and third pressings, being often bitter, high in tannin, and very cloudy, must always be kept separate.

The free-run juice is usually the sweetest (see p. 238). The later pressings have from 0.5 to 2.0 per cent less sugar, although there may be some variation in this with the degree of maturity. In soluble-solids content or degree Brix, the later pressings decrease less because they are high in nonsugar materials, particularly minerals. The alkalinity of the ash increases by as much as 30 per cent in the third- and fourth-pressed juices. These differences are associated with the differences in composition of the outer and inner layers of the fruit. (See pp. 236–238.)

When to press

If the whole mass of freshly crushed grapes is left for a short period in a vat, more free-run juice will drain off, the skins will become slightly less slimy and thus easier to press, and, with some varieties, more flavor will be extracted. This period should not be extended too long, however, lest too much tannin, color, or off-flavors be extracted. There is danger, also, of extracting excessive amounts of pectins, gums, and other colloidal matters which contribute to subsequent clouding and hazing. Treatment with pectolytic enzymes may be necessary to increase yield, clarity, and acceptability of white must (see p. 308). Furthermore, the grapes may darken too rapidly in the open tank unless the sulfur dioxide content is fairly high. Drawing off in two to six hours is recommended.

With grapes of the warmer districts, somewhat better-balanced ordinary white table wines may be produced by a short fermentation on the skins because, in such localities, the musts are very low in tannin. By fermenting such white grapes on the skins for a limited time, more stable wines that are higher in tannin and extract content are produced. Musts handled in this manner, however, should be free of spoiled grapes; otherwise the wines may not be sound or palatable. If too much tannin is extracted, the amount can be reduced by a heavy fining. The press wine should, as usual, be kept separate, since it is harder to clarify and may be very high in tannin. This method, primarily applicable to ordinary

629

white table wines (Dugast, 1929; Fabre, 1929), not only gives stable wines but also increases the yield per ton of grapes.

Modifications of these two systems are used by other producers. In one, the grapes are left on the skins for about twelve hours, the free-run juice drawn off, and the residue loosely pressed in a continuous press followed by a basket press. Although yields of 180 gallons per ton are obtained, the volume of free-run is not much greater than when it is taken off immediately. The quality of the press juice remains about the same.

Where the grapes are not expensive and the winery needs distilling material, the grapes are not pressed at all. Water is introduced after the free-run juice has been drawn off, and the remaining sugar is then fermented to produce distilling material. (See Joslyn and Amerine, 1964.)

Pressing

Basket-type hydraulic presses were widely used in California until the introduction of the Willmes and Vaslin presses. Basket presses yield clear juice, but are slow in operation and of limited capacity. Horizontal basket presses are more efficient than the usual type. The Willmes press, particularly when used with press aid such as specially selected cellulose fiber, gives the highest yield of clear juice and is easier to operate. Continuous-screw expeller presses have the tendency to macerate the skins and seeds and yield cloudy wines which are difficult to clarify. With the basket press, the pressure should be very light so that the skins will not slip between the slats of the basket. The pressure is then increased very slowly. The rate of flow of the juice, very rapid at first, will soon decrease. The press is then opened, the pomace turned over, mixed, or transferred to another basket, and the whole re-pressed to produce an additional amount of juice. With the horizontal and Willmes presses the turning of the pomace is done mechanically, saving hand labor.

The pomace, instead of being turned in the press after the first pressing, may be moved to a second press and combined with other pomace before re-pressing. Mechanical hoists, elevators, and other methods of reducing manual labor and handling costs in pressing are used in modern wineries.

Continuous press

The continuous expeller press is not desirable for producing choice white table wine and is difficult to operate for some musts. The juice from a continuous press is frequently cloudy and the resulting wine too astringent. The more rational use of a continuous press is in pressing fermented pomace for distilling material. Continuous presses are used also for the last pressing of pomace from the basket-type press. The press wine is kept separate because of its high content of colloidal material.

Settling and first racking

To facilitate the earlier clearing of white wines, much of the coarser sediment should be eliminated before fermentation. This is especially true of free-run juice low in tannin, which yields a wine hard to clear.

The ready fermentation of warm musts hinders settling, but prompt cooling and the use of moderate amounts (60 to 120 mg per liter; see p. 403) of sulfur dioxide will prevent this difficulty. The coarser particles, bits of skins, seeds, wild yeasts, and the like gradually settle out; and, usually within 24 hours, the clear supernatant juice can be racked off. Fining of the must before racking has seldom been practiced in California. Settling removes not only the fragments of grape tissue but also many of the microflora, and thus facilitates control of fermentation.

The must may be centrifuged to eliminate the coarser particles, but this method of clearing has not found favor in California. Neither pasteurization of the musts (except the spoiled ones) nor filtration is ordinarily a necessary or desirable aid to settling.

Some wineries in California and Australia use pectolytic enzymes, either prior to pressing or during settling (Hickenbotham and Williams, 1940; Besone and Cruess, 1941; Kilbuck et al., 1949; Berg, 1959b). Properly used, these enzymes increase the juice yields when added before pressing, and increase the rate of settling and clarification of the juice when added either before or after pressing (pp. 308–309 and 546–547).

Too much clarification of white musts is believed by Benvegnin et al. in Switzerland (1951) and by Nègre and Françot in France (1965) to be undesirable because it eliminates too many of the pectins and mucilaginous materials, which they consider contribute to the mellowness and softness of the wine. Wines made from overclarified musts also do not have so satisfactory a malo-lactic fermentation. In Switzerland, at least, this fermentation is a necessary adjunct of wine making, but it is seldom desirable in low-acid California white wines (see p. 497).

Balancing the must

Use of high-acid varieties and green grapes is probably the cheapest and best method of raising the acidity. White musts lacking acid should be balanced before the addition of pure yeast. Adding acid before fermentation, although more expensive because of the amount lost in the lees, gives smoother wines which usually have cleaner fermentations. The amount of acid added will depend upon the original acidity of the must and upon the type of wine (p. 306). California white musts should usually be brought to at least 0.7 to 0.8 per cent acidity before fermentation, preferably with tartaric acid. The proper amount of acid should be dissolved in a small volume of must and added to the tank, the contents of which should then be thoroughly mixed. Citric acid should not be added before the fermentation, but may be used after the first racking. Other acids, such as fumaric and succinic, have not given satisfactory results, either because of cost or off-flavors from bacterial fermentation.

Adding tannin to white musts may be advantageous under some conditions here; but to get the full benefit of the tannin it should be added after the first racking. According to Nègre and Françot (1965), much of the tannin added before fermentation is adsorbed by the yeasts and thus lost.

To reduce the extent of the malo-lactic fermentation, Wejnar (1966) recommended adding 2.5 per cent sugar and 0.3 to 0.4 per cent tartaric acid to the wine. The malic-acid fermentation then proceeded to 60 per cent on the 10th to the 60th day after the start of the fermentation and to 80 per cent between the 10th and 100th day, compared to 100 per cent by the 60th day in the controls.

Prefermentation treatment of musts by ion exchange (H cycle) to reduce the pH from 3.83 (for Riesling) to 3.22, 3.14, and 2.87, and from 3.73 (for Steen) to 3.34, 3.00, and 2.95 was conducted by C. S. Du Plessis (1964). There was a reduction in bouquet at the two lower pH's for the Riesling and at the lowest pH for the Steen. He believes that it was caused by removal of amino acids, flavor precursors, and thiamine. While "soluble" protein content was reduced and browning and bacterial growth were retarded, the treatments, even the less extreme, were not recommended.

Sulfur dioxide is always added to white musts, usually at the time of crushing. (See pp. 380–409.)

FERMENTING

. . . requires closed containers and temperature control.

Adding the yeast

The must should be inoculated with pure yeasts immediately after being racked off the sediment from the settling tanks. For white wines a yeast which gives a rapidly settling, compact sediment is especially desirable. Usually from 1 to 3 per cent of a yeast culture is added. In California a Burgundy strain was used widely at first, but has now been supplanted by a Montrachet strain.

If any difficulty is experienced in getting the fermentation started, a racking with aeration, or addition of more yeast, or both, is indicated.

Filling the tanks

The fermentation of white wines should take place in closed containers about 70 to 90 per cent full. Some wineries favor filling the fermentation container so that the foam, containing yeasts and other matter, will rise to the surface and overflow through the bunghole during the violent fermentation. If this is done, special curved, corrosion-resistant metal tubes should be fitted into the bunghole and reach over the side of the container to a tube or drain so that the yeasts and other materials in the foam will not overflow onto the sides of the containers. Under no circumstances should the foam be allowed to run over the sides, where it will acetify, draw flies, and create an unsanitary condition. Although wines which have been allowed to overflow are said to clarify more easily, it is doubtful whether the practice is of general utility.

After the active period of fermentation, the tanks should be filled and kept full. Since the volume decreases rapidly at this period, they may need filling once a week.

Ordinarily, fermentation bungs may be used to prevent the free access of air to the fermenting must; but in the first stages, when the fermentation may be rather violent, there is often sufficient surface foam to plug the bungs. Although bungs are less necessary for large, closed fermenters, since the volume of gas liberated in such containers is very great, much unnecessary oxidation takes place when they are not filled during the later and slower stages of fermentation. It would be advisable to use fermentation bungs in all cases. What is usually forgotten is that the headspace tends to equilibrate to the partial pressures of all the gaseous components of air, and oxygen can be absorbed by the wine from such headspace.

White table wines fermented in open containers are not only subject to undue oxidation, especially in the later stages of fermentation, but may also become contaminated.

Temperature for fermentation

The fermentation should be conducted at a temperature lower than that used for red wines. The color, flavor, and aroma of dry white wines are injured at 29.4°C (85°F) or above. The most desirable temperature is apparently about 15.5°C (60°F), although many producers of fine dry white wines prefer a somewhat lower temperature. At 15.5°C (60°F) the rate of fermentation is slower than at higher temperatures, but the fermentation is cleaner and the temperature is not likely to get out of control during the most active stages. To maintain temperatures such as these, wineries use small cooperage and provide some means of cooling. Cooling during fermentation is usually necessary, especially with large tanks, and results in sound wines of improved quality (p. 376).

633

Those who prefer temperatures as low as 4.4° to 10°C (40° to 50°F) apparently do so because they have found it difficult to keep the fermentation temperatures from going above 15.5° or 21.1°C (60° to 70°F), even with cooling facilities, when the fermentation takes place in very large containers. Ough and Amerine (1966), on the basis of their own and other experiments, recommended 10° to 18°C (50° to 65°F). This agrees with the latest recommendation of 12° to 14°C (53.6° to 57.2°F) for Hungarian white wines in the experiments of Ásvány (1967b).

Because white wines are fermented at lower temperatures, and in closed containers, more or less in contact with the air, the period of fermentation is prolonged. If, however, the wines have not finished fermenting in four to six weeks, they usually should be aerated to invigorate the yeast; occasionally the addition of fresh, actively fermenting yeast is necessary. Toward the later part of the season, as the temperature drops, it may become difficult to finish the fermentation of white wines without one or two aerations or even warming of the musts. But the aeration of white musts and wines should not be excessive if darkening of color and spoilage of the products are to be avoided.

Special methods

Other methods of fermenting white wine grapes, such as the Semichon process of controlling fermentation by maintaining the alcohol content in the fermenting must above 4 per cent, the Mestre system (Mestre, 1947) of continuous fermentation of muté, and the Barbet and North African systems which also utilize muté (Fabre, 1929), have not been adequately tested in California, but they do not appear promising under present conditions, at least for high-quality wines.

Fermentation under pressure in specially constructed stainless steel tanks was introduced in Germany and has been used also in Austria, Australia, Switzerland, and South Africa. Pressure fermentation was introduced originally to improve alcohol yield and flavor, and to expedite production of wines with residual sugar (Geiss, 1952). The tanks used are of limited capacity and require particular care in operation.

Böhringer et al. (1956) reported that wines fermented in pressure tanks were relatively high in volatile acidity. They preferred cooled or cold fermentation of white musts. Similar conclusions were reached by Amerine and Ough (1957); they also found it difficult to speed up or slow down fermentation rates of white California musts at 15.5°C (60°F) with either oxygen, nitrogen, air, or carbon dioxide. This is in contrast to the retarding influence of anaerobic conditions reported from Bordeaux. Ough and Amerine (1960) reported pressure-fermented white table wines to be higher in pH and volatile acidity and to have lower quality ratings compared to wines from the same grapes fermented under atmospheric conditions. They could find no advantage in stirring under low carbon dioxide pressure (20 psi). Fermentation at 15.5°C (60°F) with 100 psi pressure required nineteen days to ferment dry compared to ten days at atmospheric pressure at 11.7°C (53°F).

For stability of bottled white table wines, Zinchenko (1964b, 1965) recommended fermenting under carbon dioxide pressure, a low nitrogen content (less than 350 mg per liter of total nitrogen and under 200 mg per liter of amine nitrogen), one or two rackings, and early sterile filtration and bottling.

True continuous fermentations in California will, in our opinion, probably be developed first for the production of distilling material. Suitable musts are available in California for producing distilling material for four to five months of the year (packing-house culls, rain-damaged grapes, late-harvested surplus table grapes, etc.) Sparkling wines (p. 691) and submerged yeast culture of flor sherries (Joslyn and Amerine, 1964) may also be produced by continuous processes. When suitable fruit is available for a long enough period, dry white wines will probably be produced by continuous fermentation systems. Ègamberdiev (1967) reported Soviet experiments using ten fermenters which lasted three months. However, the sugar content of some musts was only 16.3 per cent. They recommended a standard must of 19.6 per cent sugar and six to 8 fermenters. A fermentation time of five to eight days was contemplated. No economic study of the amortization of the cost of construction and operation of continuous systems for a three-month period as against the batch system is given nor do we find any comparative sensory evaluation of wines produced by continuous fermentation compared to these produced by the usual batch system.

Ribéreau-Gayon et al. (1963) reported increased quality of white wines prepared from musts heated to 65° to 75°C (149° to 167°F) before fermentation. This may be true of musts infected with Botrytis and other molds. We believe that this result should be accepted with caution for California conditions—if for no other reason than cost, but also because mold-infected grapes are seldom a problem here.

Browning

Recent investigations indicate that the browning of white wines is due mainly to the oxidation of phenolic compounds such as condensed tannins and leucoanthocyanins. The velocity of the change in color depends on the amount of phenolic compounds, on the temperature, and on the quantity of dissolved oxygen in the wine. To reduce the leucoanthocyan content light pressing is recommended. Earlier harvesting increases

635

the amount of natural reducing substances present but does not lower the tannin content. Use of soluble synthetic polymers, such as PVP, which selectively remove undesirable phenolic compounds, is recommended. Casein, charcoal, sulfur dioxide, ion exchange, and ascorbic acid all have their place in the handling of such wines.

Peterson and Caputi, Jr. (1967), confirmed that variety is an important factor in browning. Surprisingly, they found Palomino wines to brown less than those made from French Colombard. Browning of Palomino and Thompson Seedless wines was predominately oxidative, whereas browning of French Colombard wines occurred in the absence as well as in the presence of oxygen. They suggest two types of browning of white wines: (1) oxidative polymerization involving polyphenols and (2) nonoxidative reactions possibly involving nitrogenous compounds such as amino acids condensing with carbonyls. Peterson and Caputi, Jr., found that ion-exchanged wines generally browned less than untreated wines. Hydrogen exchange was more effective than sodium exchange in inhibiting browning. When hydrogen exchange was followed by hydroxide exchange the resulting wines resisted browning, but they were water white and retained little wine character.

CLARIFYING AND AGING
Early clarification and bottling are recommended.

Racking

White wines should be removed from the lees promptly after the fermentation is over. Otherwise, especially in warm cellars, unpleasant odors and flavors, caused by decomposition of yeasts and the like, may be added to the wine. At this stage, also, conditions favoring the formation of hydrogen sulfide are set up in the lees. Indeed, there is no objection to racking white wines off the sediment as soon as the fermentation is complete. Frequently, if only a trace of sugar remains, the aeration during racking will be beneficial in finishing the fermentation as well as in aiding the removal of undesirable metals by oxidation. However, the very lightest-colored white wines should be racked without aeration or even under carbon dioxide. This gas is used to displace the air in the tank to be filled. (See pp. 495–496 and 507–509.)

In California, early racking and cooling are especially important to maintain a high total acidity. The initial racking should always take place before the first of the year. These conditions are different from those in Europe, where free sulfur is used only occasionally in the vineyard and where bacterial decomposition of excess acidity is desired. Use of fresh, clean lees to treat new wines which have developed an off-flavor or odor is seldom practiced in California because of the danger of contaminating the wine with undesirable microörganisms. The number of additional rackings should be based on the amount of sediment being deposited and upon the clarity of the wine. Special aeration of the young white wine is seldom desirable except in the rare case where hydrogen sulfide has developed. Often, where very delicately colored and flavored wines are produced abroad, the racking are carried on under slight pressure of carbon dioxide to lessen access of air to the wine. This procedure and, in general, the handling of white wines under a protective blanket of carbon dioxide would help prevent some of the overoxidation of white wines which is common in California. (Pure dry nitrogen could also be used, but since nitrogen is lighter than air it is more difficult to avoid absorption of oxygen during its use.)

The use of carbon dioxide gas in protecting white wines against oxidation is apparently not prohibited by present federal regulations. More general use of carbon dioxide would be a boon to the California wine industry in preparing better white table wines. It would not only help prevent oxidation but would also help prevent escape of the carbon dioxide already present. Many of the finer light white table wines of Europe, particularly those of Germany, Switzerland, and Alsace, are prepared, stored, and marketed with a slight natural carbon dioxide content which improves their taste. Wine makers contemplating such use of carbon dioxide should ascertain the latest regulation of the appropriate government agencies before proceeding.

Adding sulfur dioxide

The use of sulfur dioxide is practically essential in handling white table wines. Its rational use protects these wines against a too dark color, and oxidized, sherry-like flavor and aroma, and bacterial disease. It is valuable also in removing small amounts of hydrogen sulfide. When improperly used (too large quantities) it is undesirable.

Since the sulfur dioxide is constantly being dissipated by oxidation, volatilization, and the like, it must be renewed at regular intervals, though in lesser quantities as the wine becomes biologically more stable with age. Naturally, the cooler the cellar, the cleaner, drier, and more acid the wine; and the larger the container, the less the danger of contamination or oxidation and the smaller the amount of sulfur dioxide necessary. But larger amounts must be used in warm cellars (or in the summer), and for low-acid, slightly sweet wines in small cooperage. Under the most favorable conditions, about 50 parts per million of total sulfur dioxide will suffice. Under the usual California conditions, 75 to 100 mg per liter or more may be necessary. (See pp. 403–404.)

Clarifying

Properly made and balanced white table wines should not be cloudy in the spring after the vintage. If cloudiness persists, the wine should be examined for bacterial and nonbacterial disease (chap. 19). The main problem in white wines is to secure, as early as possible, wines that will stay brilliantly clear in the bottle. Certain high-acid, low-pH, light wines of Europe reach this condition, without undue treatment, within a year after the vintage. Unfortunately, the ordinary white table wines of California seldom show any natural tendency to early clarification. This is probably because of their high content of various colloidal matters, their high pH values and consequent more populous microflora, and the prevailing high-temperature conditions of the cellars. Such wines often require more strenuous inducements to clearing. In Russian white wines the colloidal material is primarily arabans and galactans with small amounts of pectins, according to Nilov and Skurikhin (1967). Excess protein is more common in California. (See pp. 542–544.)

White wines should be stored at a low temperature, below 0°C (32°F), after the first or second racking, to facilitate the precipitation of excess tartrates and other slightly soluble substances. If cooled in a separate room they may be easily racked thereafter. At this time, care must be taken not to aerate the wine too much lest it take up excessive oxygen. It should not be kept at the low temperature any longer than is necessary to obtain the desired precipitation. Usually the wine is rough-filtered off the precipitate. The filtration should be done at a low temperature to avoid redissolving some of the precipitated materials. Wineries in the cooler regions can secure considerable precipitation of tartrates by opening the cellar doors and windows during the nights of January and February.

Fining usually takes place in the spring. For white table wines it should be completed before the warm weather. The usual tests (p. 530) should be conducted for each container, and the correct amount of the proper agent should be added. The fining agent should not be left in contact with the wine too long, or decomposition of the colloidal material in the precipitate may impart an undesirable flavor. The usual procedure is to filter the wine off the sediment—safer than simply pumping it away—although young white table wine should not be filtered too closely.

If the clarification has been successful, the wine should be nearly brilliant without undue darkening of the color. Early clarification of all wines, regardless of their quality and the length of time they are to spend in wood, results in less danger of bacterial spoilage; however, excessive manipulation—fining, filtration, and the like—harms quality.

Storing

White wines bottled young (one to two years old) are lighter in color, higher in fermentation aroma, less in bouquet and barrel flavor, and fresher and fruitier in taste. In some, the presence of small quantities of carbon dioxide improves the flavor and appeal. Judging from Laborde's test (Ribéreau-Gayon, 1937) with the white wines of Bordeaux, wines bottled at one year of age were always of better quality than those left in the cask for more than a year.

The best white table wines in California are usually kept in casks, placed in the coolest part of the cellar, from one to three years for additional aging and clarification. It is questionable whether keeping them in a cask after three years is ever advisable. Most California white wines have too much oak flavor and become excessively dark when kept in small containers in the usual warm cellars for over two years. Bottling after only about a year in cask is recommended for the lighter (less alcoholic) types. The rate of aging in large hard oak casks is slower than in small soft oak casks (figs. 31 and 102).

Figure 102. Oak barrels for table wines in California winery. (Courtesy Paul Masson Vineyards.)

CHAPTER SIXTEEN

SWEET TABLE WINES[1]

Varieties

The best sweet table wines thus far produced in California have been made from ripe Sémillon and Sauvignon blanc grapes. These varieties produce well and attain a sufficiently high degree of sugar for this type of wine in regions II, III, and IV. In years of high August temperatures, particularly in unirrigated vineyards in region III, some sunburning of Sémillon may occur, and when the vines are overcropped the fruit does not ripen properly. The aromatic flavor of wines of Sauvignon blanc may be unduly pronounced. Because its leaf surface is usually luxuriant, its fruit is seldom sunburned.

A sweet table wine of high fruitiness is produced also from Chenin blanc grapes. While the completely fermented Chenin blanc wines lack characteristic flavor, those produced by partial fermentation of ripe Chenin blanc grapes are quite pleasing. Malvasia bianca grapes are used for the production of sweet dessert wines, but may be useful also in modified sweet table wine production. Sweet muscat-flavored table wines of good quality with a stable flavor are difficult to produce. However, several wineries have succeeded using Muscat of Alexandria grapes. Muscat blanc (also called Muscat Canelli or Muscat Frontignan) and Orange Muscat can also be used.

For production of sweet table wines from Sauvignon blanc, Butănescu (1966) recommended late harvesting, leaving on the skins sixteen hours at 10° to 12°C (50° to 53.6°F) and settling the pressed juice by fining with bentonite and adding sulfur dioxide. To stop the fermentation he recommended cooling to 8° to 10°C (46.4° to 50°F), addition of sulfur dioxide, racking, addition of 200 ml per liter of sorbic acid, and filtration with germproof filters. If the free sulfur dioxide is kept to a minimum of 30 mg per liter he indicates that it should be possible to bottle the wine within six months.

Many other varieties may attain a sufficiently high sugar content to produce a sweet table wine. Among these are Muscat blanc, Gewürz-

----

[1] It should be noted that *vins doux naturels* as produced in France (Flanzy, 1959) are fortified wines, whereas the sweet table wines discussed here are not. However, since federal regulations permit blending of fortified and unfortified wines the distinction is not so clear.

traminer, Sylvaner, and Pinot blanc. While Pinot blanc and Sylvaner are known to ripen sufficiently in region IV, their utility for producing sweet table wines has not been tested. At higher sugar contents their acidity would have to be corrected. When the composition of the must permits, a number of varieties may be made as a sweet table wine. In general they should be fermented at a low temperature (not over 60°F) and be stabilized with 0.5 to 3.0 per cent sugar at not over 12 per cent alcohol.

Maturity

The process of making natural sweet wines in California differs markedly from that used in France, Hungary, and Germany. In those countries the growth of the fungus *Botrytis cinerea* Pers. on the surface of the grapes removes considerable water and thus increases the sugar concentration to 30 per cent or more. Increased content of glycerin and oxidizing enzymes, and decreased nitrogen and total acid also result.

Moser (1967) noted that the higher the sugar content of the berries at the time of botrytis infection the less the sugar loss owing to growth of the fungus. He found that high sugar occurs only under dry climatic conditions. The most damaging climatic conditions are periods of drizzle and fog. Under these conditions sugar losses occur. Large-berried, tight-clustered varieties should not be used for botrytis growth. For Austria Moser recommended White Riesling, Traminer, Weisser Burgunder (Pinot blanc ?) and Muscat-Sylvaner (Müller-Thurgau ?).

The presence of glycerin in musts whose grapes have been attacked by *B. cinerea* has long been known. In one study no glycerin was found in musts of sound grapes, 1.6 per cent in *Auslese*[2] musts and 2.0 per cent in *Trockenbeerenauslese* musts. The glycerin seemed to have its source in the saprophytic activity of the fungus even when growing on berries still on the vine. The favorable and unfavorable effects of *B. cinerea* on the quality of musts have been emphasized by Charpentié (1954). Malic and tartaric acids are attacked about equally at low pH's by the fungus. At higher pH's citric acid is formed, but *in vivo* this was not observed. The presence of glycerin in botrytised musts was again confirmed. Charpentié views botrytis as a biological means of deacidifying musts or wines. It has long been known that *B. cinerea* metabolizes organic acids. Musts of botrytised grapes also have a high dextran content. When grapes are attacked by *B. cinerea*, galacturonic acid may be converted to saccharic acid. Kielhöfer and Würdig (1961a, 1961b) found calcium saccharate as the crystal deposit in 1949 *Auslese* wines.

---

[2] *Auslese* is the term used to denote the selection of only fully ripe botrytised berries. *Trockenbeerenauslese* refers to harvesting grapes that are partially dried after attack by *B. cinerea*.

Lafourcade (1955) showed that musts from grapes attacked by B. cinerea fermented much slower than those of nonbotrytised grapes. This was believed to be due to an antibiotic, botryticine, produced by the fungus. It was suggested that sulfur dioxide destroyed part of the retarding effect of botryticine, either by reducing it or by combining with it. The experiments of Dittrich (1964b) do not substantiate the claim that botrytised musts ferment slowly because of the presence of an antibiotic, botryticine. Addition of peptone and growth factors stimulated fermentation, and botryticine did not retard fermentation. The slow fermentation of botrytised musts appeared to be due to their higher sugar content.

The excessive loss of acidity following inoculation of grapes by B. cinerea was verified by Lepădatu and Bellu (1959). They also reported that botrytis attack was especially favorable for Sauvignon blanc and Muscat Ottonel. With Wälschriesling and Fetească albă there was excessive loss of fruitiness, and with Traminer and Neuburger insufficient increase in sugar.

Recently Chaudhary et al. (1968) reported larger amounts of diethyl succinate, ethyl acid succinate, succinic acid, 2,3-butylene glycol, and less of n-hexanol and γ-butyrolactone in wines made from botrytised Sauvignon blanc grapes compared to those made from normal grapes.

The destructive effect of B. cinerea on flavonol and anthocyanin pigments was studied by Bolcato et al. (1965). Based on Rf values, there were some flavonols and anthocyanins that were not found in wines made of sound grapes but which were present in wines of botrytised grapes.

When no botrytis is present, those who wish to make natural sweet wines are restricted mainly to picking the grapes at the proper stage of maturity. The alternative is sweetening of the finished dry wine by blending with sweet musts or dessert wines.

The California climate, although it prevents the regular growth of this mold under vineyard conditions, makes possible a high sugar content in the musts. Since in most years the picking can come fairly late, it is not difficult to obtain properly matured grapes for making natural sweet wines. For the moderately sweet types, grapes should be 24° to 26° Brix. The sweeter "château" types should, if possible, be made from grapes with even higher Brix readings. The grapes should be kept from raisining unduly as a result of staying on the vines too long. Wines made from such grapes have undesirable flavors and are too dark in color. In extremely hot harvest periods, excessive drying may occur very quickly in certain varieties and much of the crop may be lost through raisining on the vines. In other years, early rains may damage the crop. Nevertheless, to obtain a sufficiently sweet must, as much delay in harvesting as possible is recommended.

Musts high in total acid (above o.8 per cent) should not be used for natural sweet wines, even if their sugar content is satisfactory, since they do not make a sweet table wine of the best sensory balance, particularly in flavor.

Most varieties, at maturity, reflect crop and soil conditions. The vineyards with poorer soils and the smallest crops will ordinarily give grapes with the highest sugar content. These are more suited to making the sweeter types of natural sweet wines. However, there is more likelihood of sunburning where the foliage is scanty.

It would therefore be desirable for the pickers to separate the grapes in harvesting—into riper bunches and greener bunches, and into sweeter fruit, from vines having a small crop, and less sweet fruit from heavily laden vines. By such sorting, with some varieties, one can obtain a good percentage of high-quality grapes with a Brix of 26° or more—well suited to making the sweeter types. The remaining fruit can be used for dry-wine production.

The precautions outlined for the picking and transportation of the grapes for dry red and white table wines apply also to those intended for making sweet table wines. Use of shears to remove raisined or rotten berries is recommended.

Crushing and pressing

The grapes are crushed and stemmed, the free-run juice is ordinarily separated, and the pomace is pressed in basket presses as for dry white wines. The grapes for sweet table wines are rarely fermented on the skins, although a brief period on the skins would appear to be a rational procedure in securing a higher extraction of sugar from shriveled fruit.

Crushing semidried grapes for sweet wines is much more difficult than crushing the plump, juicy grapes used for dry table wines. Rollers, if used, must be set closer together, but not so close as to break the seeds. If a rotating cylinder is used, the propellers must be operated more rapidly. Rollers, preceding the usual crusher-stemmer, are frequently desirable for these types of grapes. The first pressing should be very slow. Thereafter the pomace should be well stirred and re-pressed. Since it is particularly sweet, it is usually saved for producing distilling material.

In France and Algeria, special equipment is often used to disintegrate the pomace of the first pressing before making a second pressing. This usually consists of rotating cylinders leading from one press to another. Greater yields are obtained in this manner, but the cost of handling is greater and the musts from the second and later pressings often contain an excessive amount of suspended material.

Settling and racking

The very sweet, sometimes viscous, musts used for these types of wines are often high in colloidal matter and may yield wines that are very difficult to clarify. It possible, therefore, these musts should receive a preliminary settling before fermentation. Settling gives a clearer wine and also, by removing nutrients and microörganisms, permits better control of the fermentation, especially by using a selected yeast.

The musts should be cool at time of settling; they may be cooled immediately after pressing. A generous amount of sulfur dioxide is then added (100 to 150 mg per liter). To reduce the acetaldehyde content of sweet table wines, Lafourcade (1955) recommended reducing the amount of sulfur dioxide added to musts, eliminating the yeasts quickly after fermentation before furthur addition of sulfur dioxide, and storing new wines at 1° to 2°C (34° to 36°F) to prevent refermentation as much as possible.

In the absence of fermentation, the musts settle in about 24 hours and the clear supernatant liquid can then be racked off the sediment. Use of pectin-splitting enzymes during settling is probably more rational for such musts than for any others. (See pp. 308–309 and 546–547.)

Controlling fermentation

Musts with the highest sugar content should be used for the sweeter types; those with a lower sugar content, for the less sweet types. Those with Brix readings below 24° should not be used for making sweet table wines, since they will be too low in alcohol and in extract when properly finished. A rough calculation will show how much of the sugar present is necessary to produce a 12 to 14 per cent alcohol wine; the balance of the sugar will remain in the wine if the fermentation is properly conducted. As some sugar will be lost during the year or two which the wine must spend in wood, the fermentation should be stopped somewhat above the percentage of sugar desired in the finished wine.

The chief problem in making sweet table wines is to halt the fermentation at the proper stage. This involves complete control of the rate of fermentation throughout the vinification. Because this can be more satisfactorily achieved in small than in large cooperage, most sweet table wines are made in tanks of less than 1,000-gallon capacity. However, with modern equipment there is no reason why larger-scale operations should not be successful.

Adding sulfur dioxide

If there has been no preliminary settling and if no sulfur dioxide has been previously added, it should be added after pressing in amounts of about 50 to 150 mg per liter, according to the quality of the grapes and the temperature (p. 404). The must should be cooled to below 21.1°C (70°F).

Adding yeast culture

A pure yeast is added from one to several hours after the sulfur dioxide, or following the racking after settling. It should never be delayed until a "wild" fermentation has started. Greater attention should be given to the use of selected strains of yeast for producing sweet table wines. The Sauternes strain, which is apparently native to the Sauternes district of France, appears to ferment fructose faster than glucose. This type of yeast is helpful in controlling the rate and extent of fermentation because it ferments glucose rather slowly. To obtain stable sweet table wines (less than 14 per cent alcohol) Şeptilici et al. (1962–63) isolated several slow-growing strains of yeast. They recommended settling the musts and adding only 0.5 per cent of the yeast culture to the clear must. (See pp. 335–341 for a discussion of the value of special yeast strains.)

Slowing and stopping fermentation

The fermentation should be slow. If it tends to become too rapid, the fermenting must should be cooled; if necessary, more sulfur dioxide should be added. The practice of letting the temperature rise and having the fermentation stick with residual sugar is very undesirable (see also p. 311), for it is difficult to control sugar content by this method.

As the sugar content of the must gradually drops to the desired level, the rate of fermentation should also be further slowed down. The Brix reading is not an accurate measure of sugar concentration at this time because the varying amounts of alcohol which are present influence the specific gravity. For an accurate picture the alcohol, extract, and reducing sugar should be determined. If fermentation has not been too rapid, the yeasts will be mainly in the lees, and the process can be slowed down considerably by racking the must off the yeast one or more times during the fermentation. If the racking is properly done, only a small proportion of the yeast cells will be transferred, and they will have to multiply before appreciable alcoholic fermentation can take place. The fermentation is slowed down not only because of reduction in yeast population but also because the nitrogen and phosphate content of the must is gradually reduced so that the yeasts multiply much less rapidly. Recent experiments indicate, however, that other factors than the depletion of total nitrogen and phosphate are active in slowing down the rate of fermentation. If such a system is to be successful, the fermentation must not be violent enough to stir up the lees continuously. To achieve this purpose, cool fermentation is essential. (See also pp. 376–379.)

The fermentation may be stopped by addition of massive doses of sulfur dioxide and by racking. The objections to this practice are that the resulting wines contain excessive amounts of free and fixed sulfur dioxide and are often unpalatable.

Another method of stopping the fermentation is by filtration. Various types of rough filters will remove large amounts of yeast. By an appropriate setup of filters, an actively fermenting wine can be made yeast-free, but the procedure is slow and expensive. Where large volumes are to be filtered this method is not practicable; it requires many changes of the filter pads and might injure the flavor of the wine. Self-cleaning rotary filters might be used. Filtration is useful only where the musts have been settled and the rate of fermentation has been so slow that most of the yeasts are at the bottom of the fermenter.

Ribéreau-Gayon (1944–1945a) has shown that more sugar is retained in wines fermented under fermentation bungs, but that more sulfur dioxide is required. Further experiments of this nature should be conducted.

The most rational procedure for making sweet table wines is to control the rate, extent, and character of fermentation by cooling and by frequent careful rackings followed by the addition of small doses of sulfur dioxide. The fermentation is thus controlled by a gradual depletion of nutrients required for yeast growth and by the use of sulfur dioxide under optimum conditions (smaller number of microörganisms and lower temperatures). Such practices also result in a higher accumulation of desirable by-products of fermentation, particularly of glycerin.

Other methods

Other methods of making sweet table wines are being increasingly used in California, and perhaps predominate at present. These include the addition to dry white table wines of grape concentrate, or addition of white sweet dessert wines, or use of musts preserved with a high concentration of sulfur dioxide (*muté*). These methods are not considered favorable in the production of the highest-quality sweet table wines, but they are easier to follow and are generally cheaper. The clarification of wines prepared by addition of concentrate may, however, prove difficult. The use of a partially fortified wine is probably the best of these procedures. The fortified wine should be of the same variety as the dry wine, preferably Sémillon. Fortification to 16 or 17 per cent is usually sufficient to prevent refermentation and to hold the wine until it is used.

When grape concentrate, dessert wine, or *muté* are used in the production of sweet table wines, the following precautions should be noted. All the ingredients of the mixture should be as brilliant as possible in order to prevent contamination of the mixture and to facilitate subsequent clarification. The *muté* should have at least ½ of 1 per cent alcohol to meet the legal restrictions on the blending of wines. The required amount of the sweetening agent necessary to obtain the desired sugar concentration in the finished wine can be determined by the use of a blending formula. See p. 565. After blending, the wine should be stabilized by cooling, pasteurizing, adjusting the sulfur dioxide content, and filtering.

The possible improvement in fruitiness of wines produced from grape concentrate by adding back the essence stripped from the juice before vacuum evaporation has been considered. In limited experiments, no significant improvement in fruitiness was observed when grape essence was added to the concentrate after dilution and before fermentation. Addition of essence after fermentation was not found by Webb and Ough (1962) to improve quality. Both technological limitations and Internal Revenue service restrictions would have to be overcome.

Amerine and Ough (1960) and Ough and Amerine (1963a) have demonstrated that high-quality sweet table wines can be produced using grape concentrate. However, their concentrate was produced in a Howard "Lo-temp" vacuum concentrator in which the must temperature during concentration is kept between 15.6° and 21.1°C (60° and 70°F). The resulting concentrate (70° Brix) had a greenish color and no caramel odor. The concentrate gave no test for hydroxymethylfurfural. They preferred to add the concentrate to finished dry table wines. One possible disadvantage of adding the concentrate after the fermentation is that the glucose/fructose ratio is higher. Since fructose is sweeter than glucose, this should result in wines with a less sweet taste for wines of the same total reducing sugar content. Their sensory tests did not show this to be so. They reported that DEPC (100 to 200 mg per liter) plus 100 to 125 ml per liter of sulfur dioxide was effective in stabilizing the wines. Sterile filtration is also available. Matalas et al. (1965a, 1965b) have studied the use of concentrate for table wine production. They especially recommend holding the concentrate at low temperatures to prevent deterioration.

To differentiate sweet table wines produced by fermentation from fortified musts, Dimotaki-Kourakou (1964b) suggest determining the citramalic acid content, since this acid is not present in fortified musts.

In a pilot plant experiment Schenk and Orth (1965) stuck wines with 1.2 per cent sugar by adding sulfur dioxide. The same wine was allowed to ferment dry and was sweetened to 1.2 per cent sugar with slightly fermented must. All the treatments for the two wines were the same. Eleven tasters using the duo-trio procedure (p. 714) found the wines from the first procedure superior. The results were statistically highly significant. The difference is believed to be due to the fact that in the first procedure 80 per cent of the sugar was fructose, which is appreciably sweeter than glucose.

In areas of high humidity during grape maturation, infection with *B. cinerea* often occurs. In later warm dry periods the fruit dries and high sugar musts suitable for sweet table wines are produced. Charpentié (1954) has made a thorough study of the biochemistry of the mold's action. For data on the increase of glycerin see Dittrich (1964*a*). Because of the low humidity in California, *B. cinerea* seldom infects all or even a

Figure 103.   Trays for drying botrytised grapes.

majority of the grapes in a vineyard. Even when infection does take place, it is rare that the fruit is dehydrated enough by dry windy days to raise the sugar content appreciably.

Nelson and Amerine (1956, 1957a, 1957b) showed that *B. cinerea* could be used under pilot plant conditions to produce sweet table wines in California. Their procedure was to harvest sound ripe grapes. The grapes were arranged one layer thick on stainless steel trays (fig. 103). Spores of the fungus, in the form of an aerosol, were sprayed on them. The trays were then covered with polyethylene sheeting and the humidity was kept near saturation. After 24 hours the cover was removed and the grapes were maintained at below 75 per cent relative humidity in the temperature range 20° to 22°C (68° to 78°F) for ten or more days. The low relative humidity during drying is critical. The Brix degree of the resulting musts was usually over 30°. The volatile acidity of the resulting wines increased somewhat, owing to the long slow fermentation, but it was not considered excessive. The process has been used commercially in one California winery to produce very high-quality sweet table wines with a true *Botrytis* flavor. This would seem to be one quality product which could be produced by a number of wineries. Details of commercial operation are given by Nelson and Nightingale (1959). The nature of the drying is shown in figure 104.

Figure 104. Nonbotrytised and botrytised Sémillon grapes.

649

In order to facilitate commercial operation, Nelson *et al.* (1963) developed a technique for the large-scale production of spores of *B. cinerea*. The mold was grown in 1-liter Roux culture bottles at 20° to 22°C (68° to 72°F) in indirect light. The spores could be stored for at least ten months without appreciable reduction in viability.

De Soto *et al.* (1966) give the following analysis of four natural sweet table wines produced commercially in California by application of *B. cinerea* to Sémillon grapes:

	Minimum	Maximum	Average
Brix	9.4	11.1	10.3
Alcohol (per cent vol.)	12.0	12.7	12.4
Total acid (gr per 100 ml)	0.60	0.80	0.67
Volatile acid (gr per 100 ml)	0.063	0.096	0.078
pH	3.70	3.89	3.80
Total SO_2 (mg per liter)	224	262	246
Acetaldehyde (mg per liter)	98	171	141
Fusel oil (mg per liter)	122	205	155
Glycerin	1.40	2.58	2.01

In the production of sweet table wines De Soto *et al.* (1966) emphasized the distinctive odor derived from the growth of the fungus, and reasoned that growing the fungus in sterile musts would produce the characteristic flavor. This had been attempted earlier by Popova and Puchkova (1947) and by Popova (1960) in the Soviet Union, by Georgiev *et al.* (1956) in Bulgaria, by Popper *et al.* (1964) in California, and by others elsewhere. Popper *et al.* succeeded in growing *B. cinerea* in aerated submerged cultures, but the product was considered abnormal by some tasters.

The major problem in these tests is to avoid sulfur dioxide, which inhibits the enzyme system of the fungus. De Soto *et al.* (1966) tried heat sterilization, DEPC at 300 mg per liter and sulfur dioxide, and settling. None of the procedures was satisfactory. They concluded that the analyses did not reveal significant differences between wines made from botrytis-treated and normal musts. However, sensory evaluation revealed a significant difference between the wines. When the wines were sweetened after fermentation, experienced tasters preferred the botrytis-based wine. They also reported that 59 of 100 subjects preferred this blend.

The importance of experiments of this type is that they introduce a new flavor into wines. Probably only a minority of the population will immediately appreciate the new flavor. It is significant that Popper *et al.* (1964) reported greatly reduced tartaric acid in these experiments. This should be investigated in more detail.

Amerine and Kunkee (1965) have noted the Russian reports (Delle, 1911) on the repressive effect of sugar on fermentation. A slide rule and a nomogram based on Delle units has been presented by Stanescu (1965).

It is predicated on the original postulated equivalence that 80 per cent sugar has the equivalent antiseptic value of 18 per cent (vol.) of alcohol, or that 4.5 per cent sugar has an equivalent antiseptic value to 1 per cent (vol.) of alcohol. As Amerine and Kunkee have shown, this varies with the sugar content and the variety and strain of yeast. Nevertheless, the slide rule and nomogram should give an approximation of Delle units for various sugar and alcohol contents, and hence of approximate stability to refermentation. This is not important in present practice in this country, but it could be in the future. However, for sweet table wines the sugar content would need to be over 20 per cent if the alcohol content is to keep below 14 per cent. This seems to be excessively sweet for the American palate.

Using sulfur dioxide

During the first year, three to five rackings should be made to eliminate yeasts and other microörganisms. Since a considerable amount of sulfur dioxide is lost during racking, it must be renewed after each racking. The amounts added should be based upon an actual analysis of the wine (total and volatile acidity, reducing sugar, total and free sulfur dioxide) and upon the condition of the wine. A total of about 250 mg per liter should be maintained in the sweeter types. An unnecessarily high amount of sulfur dioxide is to be avoided. Proper fermentation to deplete the nutrient material and storage at low temperatures will reduce the amount needed.

The sooner the wine can be brought to stability and brilliancy after fermentation, the less the danger of bacterial spoilage, yeast contamination, and loss of sugar by a secondary fermentation. As Casale (1938) notes, wines low in acidity and rich in sugar must have a higher sulfur dioxide content, to prevent fermentation, than similar wines of higher acidity. During aging, the total amount of sulfur dioxide required in the wine gradually decreases. When the wine is properly matured, an objectionable amount of sulfur dioxide will not be required for stability.

Storing

Sweet table wines, like white table wines, are fined before the end of the first year and are bottled as soon as matured. They should be stabilized to fairly cold temperatures, since they are commonly cooled before serving. Though the aging period for sweet table wines may last from two to four or more years, the tendency is to free these wines of sediment as early as possible and to bottle them fairly young. Before bottling, the free and total sulfur dioxide should be determined very carefully. The wines should not exceed the legal limits for total sulfur dioxide and, in addition, must not have the objectionable smell of free sulfur dioxide (not over 20 to 30 mg per liter free). The color of a mature, natural sweet wine must not be bleached by high sulfur dioxide, but neither should it be darkened by overoxidation. The preferred color is a moderate gold, but with no traces of amber. It is especially useful, with sweet table wines, to make sample bottlings to see whether the wine will remain stable in the bottle (see p. 568). The wines should be germproof-filtered as they are bottled. Bottling hot to achieve a bottle pasteurization is less satisfactory, as it affects the quality.

These wines, because of their high sulfur dioxide content, must be kept from all contact with metals throughout their storage and handling. Pasteurization in copper-containing equipment must be particularly avoided. These wines should not be kept in concrete tanks.

Stabilizing

Stabilizing sweet wines by partial pasteurization has been used by several wineries. In one procedure the clear wine is racked into barrels and placed in a room at a temperature of about 48.9°C (120°F). The heating is continued for one or two weeks. Wines treated in this manner are said to be smoother and to show less tendency to re-ferment in the bottle. The danger of the method, in the hands of an unskilled operator, is that the wines may become brown in color, oxidized in flavor, and acquire a woody taste. The heating should certainly not be continued long enough for appreciable quantities of hydroxymethylfurfural to form as a result of the dehydration of fructose.

Singleton et al. (1964b) proposed heating white and red table wines in the complete absence of oxygen for 15 to 30 days at 53.4°C (128°F). They noted a definite improvement in bottle bouquet and increased complexity of flavor. They emphasized the necessity of heating for two or more weeks. There was some loss of grape aroma. They recommended the process not only for producing a new type of aged wine but also as a blending wine for young fruity wines. Similar, but less drastic, processes have been used in the Soviet Union. So far no commercial trials seem to have been made in California.

MISCELLANEOUS TYPES

Several types of sweet table wines other than those previously described have been produced in California. A pink-colored, natural sweet wine of muscat flavor, made from Aleatico grapes, was among the most promising. Unfortunately, little is produced at present. Slightly sweet white wines with a pleasing muscat flavor are now in production. Wines of these types are best when they have a fresh, fairly tart flavor. These types are produced in a fashion similar to that of other natural sweet wines. Fermentation at low temperature is especially desirable.

Many sweet table wines with a Concord flavor have been produced. Either fresh Concord grapes or Concord concentrates can be used. If concentrate is used, it should be carefully produced so that it has the least possible burnt or cooked flavor. Usually a dry red table wine is used as the base to which the grape concentrate is added. Some of these wines have 12 to 13 per cent sugar and 13 per cent alcohol. Sucrose is usually used to sweeten the blend. Sweet table wines without a Concord or muscat flavor are also made, usually with concentrate or sucrose. These frequently have as much sugar as alcohol—that is, 13 per cent—or even more.

A number of states limit the alcohol content of wines to 16 per cent. Several types of wines of moderate sugar and 15 to 16 per cent alcohol content are produced in California for sale in those states. The methods commonly used are the blending of dry white or red wine with dessert wine or grape concentrate or both. The blending formulas (see p. 565) may be used advantageously. These wines are very subject to refermentation, unless they are aged for some time, and to spoilage. To preserve them, over 250 parts per million of sulfur dioxide should be used, especially when they are shipped in bulk. (For use of other preservatives, see pp. 409–418.) When bottled by the winery, they should be pasteurized before and after bottling. Monochloracetic acid must not be used in their stabilization; its use is prohibited by state and federal agencies. Since these wines more nearly resemble dessert wines in usage, they should probably be classified as such.

CHAPTER SEVENTEEN

SPARKLING AND CARBONATED WINES

Preparation of high-quality sparkling wines involves use of a fine still wine, manipulation of the equipment required, and control of the fermentation, clarification, and aging.

DEFINITIONS

Champagnes and other sparkling wines are produced by secondary fermentation, carbonated wines by artificially charging with carbon dioxide.

Wines that retain a permanent excess of carbon dioxide of over about 0.5-pound pressure in the bottle (at about 10°C; 50°F) are said to be "sparkling." Gassy wines with less than this amount of pressure often but not always develop more by accident than by design. They may result from one or more of several factors: (1) the fermentation of a trace of residual sugar from the original fermentation, (2) vigorous malo-lactic fermentation in the bottle, (3) direct carbonation, or (4) bottle or tank fermentation of a wine of low sugar content. The *frizzante* red wines common in Italy are sometimes caused by the first or second factor; the *vinho verde* wines of northern Portugal, exclusively by the second; the "crackling" wines on the American market, by the third; some of the "Pearl" wines of Germany, South Africa, and Australia, by the fourth; and some of the slightly gassy wines of Switzerland, by the third. A number of French white wines are slightly gassy (*pétillant*), particularly those from the Loire and Alsace. These wines are not true sparkling wines, as their gassiness is neither uniform nor great, and it seems unreasonable that such wines should be taxed as sparkling wines. Imitation *vinho verde* or *frizzante* wines produced by artificial carbonation should, however, be considered as sparkling wines.

The various types of true sparkling wines produced in this country are distinguished from each other by the methods of their production. The largest volume of sparkling wine is a white type produced by a secondary fermentation of sugar added to a still white wine in a closed container.

It is called "champagne" in this country if the fermentation container is of 1-gallon capacity or less. Wines produced by a secondary fermentation in large, closed containers (Charmat and other processes) are called "champagne type" or "champagne style" or "American (or "California," etc.) champagne—bulk process" *if* they have the taste, aroma, and characteristics generally attributed to champagne. It is not clear what taste, aroma, and characteristics can be uniformly attributed to American (or California) bottle-fermented sparkling wines (champagnes). A wide range of raw material (varieties, acid content, etc.) is employed; and the length of time (varying from a few weeks to two or more years) the wine remains on the yeast before disgorging is an important factor in the character and quality of the finished wine. Unfortunately few data are available on the production and aging of various types of sparkling wines in this country.

White wines artificially charged with carbon dioxide, known to the American trade as carbonated moselle, or the like, may not be labeled "sparkling wine". However, sparkling burgundy, red champagne, and champagne rouge are red wines produced by a secondary fermentation in a closed container, with no restrictions on the size of the container. Various sparkling pink wines are also made, mainly by natural fermentation in closed containers.

A number of slightly gassy table wines are produced in California. They may not contain more than 0.277 gr per 100 ml of carbon dioxide if they are to escape the tax on carbonated wines. The presence of carbon dioxide improves their fruitiness. They resemble certain German, Swiss, and Italian wines produced by low carbonation.

The process of producing sparkling muscat wine, Asti Spumante, in northern Italy is unique. During the primary fermentation the wine is filtered several times. When the wine is bottled it contains only 4 to 8 per cent alcohol and 10 to 14 per cent sugar. Tarantola and Lovisolo (1955) showed that nearly 50 per cent of the total nitrogen was removed by five filtrations. Albuminoid nitrogen, particularly, decreased. The rate of refermentation decreased considerably after each filtration. The diminution of nitrogen may be so great that limited yeast growth occurs during the bottle fermentation. This, of course, is exactly what is desired, for, if all the sugar fermented a pressure of 20 atm or higher would be produced. This resembles the concept of Vandecaveye (1928), who proposed the production of stable wines by filtration of yeast after successive generations of yeast growth. He applied this procedure, however, only to cider, and reported stabilization of such ciders as a result of depletion of both phosphate and nitrogen. The amount of fermentation after final bottling was obviously small. Experience has shown that if muscat wines are fermented to dryness, they have a bitter taste (*sapore amaro*) and lack the muscat fragrance. The wines made from the various muscat varieties vary in their retention of muscat character and degree of bitter after-taste. Muscat of Alexandria wines retain a strong muscat aroma, but it is not as flowery and pungent as that retained in wines made from Muscat blanc. However, we have tasted some splendid California sparkling wines made from muscat wines. The Italian Asti Spumante is made from Muscat blanc and is a mild, sweet sparkling muscat-flavored wine. In spite of the depletion of nitrogen these wines are usually bottle-pasteurized at 45°C (113°F) for one hour.

Up to the time of bottling, the treatment of wine intended for bottle fermentation, for secondary fermentation in large tanks, and for carbonation is virtually the same. For all purposes, a brilliant, tart, and impeccably clean-flavored wine of low volatile acidity is required. Wines somewhat older and more alcoholic than would be suitable for bottle or bulk fermentation are normally used for carbonation.

The standard modern texts on sparkling wine production are those of Pacottet and Guittonneau (1930), Chappaz (1951), Agabal'iants (1954), Schanderl (1959), and De Rosa (1964b).

American production of sparkling wines now is nearly 10 million gallons a year—about 52 per cent in California and 43 per cent in New York. Production has increased every year since 1958.

THE STILL-WINE BASE

Brilliance, low volatile acidity, and high total acidity
are important.

Grapes for white sparkling wines

The details given for the picking, crushing, pressing, and fermentation of dry white table wines (pp. 627–634) are applicable to the making of a satisfactory still wine for use in producing white sparkling wines. Particular care should be taken in selecting proper varieties and harvesting at the recommended degree of maturity. The alcohol content of the finished wine should never exceed 12 per cent. The grapes should therefore be below 22° Brix. Attention should also be paid to obtaining grapes of a relatively high total acidity, to ensuring a thorough settling of the must following pressing, and to achieving a clean, cool fermentation.

The favorite varieties for these wines in the Champagne district of France are Chardonnay and Pinot noir, pressed before fermentation. The high acidity and mild Pinot flavor of wines made from these varieties in the cool Champagne region seem especially suitable for wines of this type. Under California conditions these varieties ripen before the usual wine grape varieties and, if they are to be used for making sparkling wines, must be picked very early, even when grown in regions II and III. Even so, they are sometimes too low in acid, and their musts or wines may have to be ameliorated by addition of acid or by blending with high-acid wines. However, the wines of properly matured and early-harvested Chardonnay have been used as the basis for some notable California sparkling wines.

Teodorescu et al. (1966) made a thorough study of the best varieties and regions for the production of sparkling wines in Rumania. Varieties which produced a good sparkling wine in one region, Wälschriesling, for example, did not do so in another. Some varieties need to be especially harvested when they are used for sparkling wines. Pinot gris must be harvested early, for example. The variety Iordană gave good sparkling wines, but needed to be aged a year and a half before bottling. Poorly balanced sparkling wines were produced in Rumania from Chardonnay, Traminer, and Sauvignon blanc. Several blends were recommended as suitable for sparkling wines. This indicates again the importance of climatic conditions on the composition of wines.

Other varieties have found favor in California, especially the Folle blanche and French Colombard. Although these varieties do not have very distinct flavors, they produce light wines with no particular disadvantages when grown under suitable conditions. Neutral-flavored varieties such as Burger, Chenin blanc, Green Hungarian, Saint Emilion, and Verdal, which seldom get overripe, have also been used, especially for blending. Unfortunately, these are often too low in acidity and too vapid in taste. This is true also of the wines normally made from Thompson Seedless grapes. However, a neutral-flavored but balanced wine can be produced from this variety when the grapes are harvested early in the season. Strong-flavored varieties such as Sémillon and Sauvignon vert should not be used, but, surprisingly, Amerine and Monaghan (1950) found that California wines of White Riesling were very desirable for sparkling wines.

In the eastern United States a number of *labrusca*-flavored varieties are used for sparkling wines: Catawba, Niagara, Delaware, Dutchess, and others. California wine may be blended in to make the cuvée.

It is seldom that the wine of a single crush will be perfectly balanced, in composition and flavor, for a high-quality sparkling wine. Blending to obtain the proper balance is thus very important for anyone wishing to produce a fine sparkling wine. In California the primary problem is to produce a wine that is not too alcoholic and heavy in body and color. It should likewise be neither too neutral nor too built-up in acidity. The importance of balancing the sparkling wine cuvée was emphasized by Popov and Semenenko (1964). They noted wide differences in the composition of sparkling wine stocks produced in one part of the Soviet Union compared to another. They recommend especially the use of wines of certain regions because of their higher quality. They make suggestions for desirable varieties and blends to improve the quality of Soviet sparkling wines. Similar studies have not been published in California. Except for the preliminary report of Amerine and Monoghan (1950) there are no published results on the relative value of the various varieties, either alone or in blends for producing sparkling wines.

Only a few muscat-flavored sparkling white wines have been produced in California. These have usually been blends of a good white wine and a highly flavored dry wine made from one of the muscat varieties (usually from Muscat of Alexandria), or a straight muscat wine with added citric acid. We believe that the production of this type of wine has been unduly neglected in California.

Grapes for red and pink sparkling wines

The Pinot noir makes a satisfactory sparkling red wine if it is not too low in acidity. The results in California, however, are often disappointing because of the low color and acid and the oxidized character and color of the Pinot noir wine. Blends containing Carignane, Refosco, or Petite Sirah are acceptable when the grapes are picked early enough. Wines made from Zinfandel grapes are useful as a base for red or rosé sparkling wine if they are distinctive enough to be recognized by the consumer, and if they are produced from grapes that are not too ripe; they could carry a varietal appellation. This is a type which deserves more attention from producers and consumers. Only unraisined grapes in the best of condition can be used. The red grapes should not be left on the skins too long or the wine will be too astringent for early bottling. Usually a fermentation of about 48 hours on the skins will give sufficient color, since sparkling red wines do not require a deep color.

Only a limited quantity of sparkling pink wines has been produced in California. These wines are generally easier to prepare than sparkling white and could well be promoted by the industry. The wine of an early-pressed, tart Zinfandel is useful. Grenache is also satisfactory when the grapes are harvested and pressed soon enough. The wine selected should have a pink and not a brown, orange, or violet tint. Therefore we do not recommend early-pressed Pinot noir for this type of wine in California. Furthermore, it often has too low an acidity and too high an alcohol content.

TABLE 65

COMPOSITION OF MUST AND RESULTING WINE OF EIGHT SUCCESSIVE FRACTIONS
(From same grapes in a champagne press)

Press no.	Amount (hl)	Sugar (gr/100 cc must)	Alcohol by volume in wine (per cent)	Titratable acidity (gr tar./100 cc must)	pH		Tartaric acid		Potassium acid tartrate		Calcium	
					Must	Wine	Must (gm/100 cc)	Wine (gm/100 cc)	Must (gr KHTa/100 cc)	Wine (gr KHTa/100 cc)	Must (gr/liter)	Wine (gr/liter)
1	2	19.3	11.8	0.79	2.98	3.01	0.612	0.471	0.471	0.358	0.220	0.094
2	2	19.2	11.75	0.85	2.94	2.98	0.728	0.542	0.575	0.363	0.212	0.094
3	6	19.3	11.8	0.96	2.87	2.85	0.810	0.584	0.598	0.373	0.236	0.094
4	6	19.1	11.7	0.93	2.94	2.94	0.777	0.565	0.650	0.387	0.196	0.094
5	4	19.3	11.8	0.82	2.96	2.94	0.687	0.462	0.678	0.368	0.204	—
6	4	19.2	11.8	0.66	3.12	3.16	0.517	0.415	0.603	0.462	0.252	0.086
7	2.7	19.1	11.7	0.51	3.43	3.43	0.410	0.288	0.655	0.537	0.212	0.062
8	2	18.3	11.4	0.45	3.69	3.84	0.349	0.152	0.874	0.725	0.180	0.082

Source of data: Françot (1950).

659

Crushing and pressing

Where white grapes are used, the regular procedure in the crushers and presses is satisfactory if the contact of the skins with the juice is of brief duration and the pressing not too prolonged. However, when red grapes are used to produce white wines special precautions must be taken. All rotten or broken berries should be removed. The crushing should be as gentle as possible and the free-run juice must be drained off at once. Even so, in California, the wines are usually slightly pink. To avoid this tint a different technique has been developed in the Champagne district of France.

The square presses used in the Champagne district of France have been described by Françot (1950). These presses hold about 4,000 kg (2.2 tons) and are approximately 8.5 feet square and 2.5 feet deep. About four pressings of the whole grapes are made. The first juice expressed is considered the most desirable for sparkling wine production, being lighter in color and fresher in taste. Françot has studied the composition of eight successive fractions taken from such a press. The grapes were turned three times—after the second, fourth, and sixth fractions were obtained. From the data summarized in table 65 it will be seen that, whereas there is little fluctuation in the sugar and alcohol contents, there is a significant increase in pH and potassium content and a decrease in titratable acidity and tartaric acid from the first to the last press. Françot found that the must of the last pressing was higher in ash, iron, total nitrogen, ammonia, tannin, pectins, and gums. It clarified less easily and spoiled more quickly.

The Willmes press may also be used for producing musts for sparkling wine production. Françot et al. (1957) compared it with the traditional flat press of Champagne. The successive juice fractions from the Willmes press had a composition similar to those from the usual press. The first fractions from the Willmes press had more color than those from the regular press, but later fractions had less color. The sensory evaluation showed no differences between the two. The cost of operation of the Willmes press is much less (one operator compared to four or five), but the initial cost is higher. The fermentation is carried out as for dry white wines (p. 632). For fermentation of Champagne musts, Geoffroy (1965) strongly advised a *maximum* temperature of no more than 20° to 22°C (68° to 71.6°F).

Composition of the base

The finished still wine should be dry, containing 11 to 12 per cent alcohol and at least 0.70 per cent total acid with a pH of less than 3.4. Wines below 10.5 per cent alcohol have poor capacity for holding carbon dioxide, whereas those above 12.5 per cent are frequently difficult to re-ferment. Before the secondary fermentation it is usually considered desirable to adjust the acidity to 0.75 per cent or higher, since there will be a decrease during aging and finishing. The pH of one New York *tirage* bottling was less than 3.0, according to Goldman (1963*a*). Most California producers would favor a pH of 3.2 to 3.4. For Austrian sparkling wines, Paul (1960) favored a pH of only 3.0 to 3.2 in the finished wine.

The volatile acid content must be low—below 0.050 per cent as acetic acid. To secure wine of low volatile acid content, the wine makers customarily bottle their product in the spring after the vintage. This is very desirable because the wines are then fruitier in flavor and lighter in color than they would be if kept longer in the cask or tank.

Sparkling wine fermentation of wines high in sulfates or sulfites leads to fat deposition, according to Schanderl (1965*b*). Reduction of sulfates or sulfites results in free sulfur, which is stored with the fat. In bottle fermentation with excess oxygen, sufficient lipoid is produced to give oval or egg-shaped masks on the sides of the bottle. To remove the mask the bottles are turned at a high speed in a special machine before being placed on the riddling racks. In tank-fermented wines, propeller-type agitators are often used. The yeast cells or plasma plaques erode, releasing particles of fat. These are not removed in the usual disgorging process in bottle-fermented wine nor by filtration in the transfer system or with tank-fermented wine.

The wine must be comparatively free of bacteria, since the secondary fermentation should be as clean and free of secondary odors as possible. Geoffroy (1963) recommends that if a malo-lactic fermentation is desired, it should take place before the secondary fermentation in the bottle. This is also true for cuvées for tank fermentation. A malo-lactic fermentation in the bottle or tank almost always results in wines which are difficult to riddle or filter. A noticeable amount of sulfur dioxide is undesirable because of its objectionable odor, and also because it may hamper the progress of the fermentation and produce compounds with objectionable odors during the secondary fermentation in closed containers. The free sulfur dioxide content should not exceed 10 parts per million. It is difficult to measure accurately such small quantities, but sensory tests and trial bottlings may be used to establish the maximum. The presence of free sulfur dioxide in amounts of 5 to 40 mg per liter delayed the tirage fermentation for four to twelve days, according to Goldman (1963*a*).

Soviet investigators apparently feel that an acid reduction is necessary in cuvées for sparkling wine production. Rodopulo and Sarishvili (1965) tried three procedures. They favored adding 0.3 per cent sugar and 2 per cent of a yeast slurry, holding for four to five days, heating, to 36° to 40°C (96.8° to 104°F) for two days and cooling to 15°C (59°F). The process was continuous. If the heating is done in the presence of air, peroxides may be formed which oxidize tartaric to oxalic acid. In the absence of oxygen, tartaric acid is partially converted to dioxyfumaric and thence to dioxysuccinic acid. No comparative sensory data of comparable wines were presented. We do not recommend acid reduction of wines intended for sparkling wines in California, at least not at the present state of knowledge. In fact, as previously noted (pp. 631), addition of acid is often needed.

Schanderl (1965c) recommended removal of excess polyphenols from sparkling wine cuvées. He used polyvinylpyrrolidone (PVP), which removed half or more of the leucoanthocyans but little of the catechin. He recommended 10 gr of PVP per 100 liters. He calculated that this would cost about $3.00 per 100 gallons.

To remove traces of red color from the cuvée before bottling, the traditional Champagne procedure is to use a small amount (25 gr per hectoliter) of animal carbon. This should be done after the first racking before addition of sulfur dioxide. It takes only 24 hours. Geoffroy and Perin (1966) showed that there is no advantage in adding the carbon in two fractions. As they say, preliminary tests on the proper quantities to use should always be made. They also recommended the use of 15 to 20 per cent by volume of clean fresh yeast from a white wine fermentation as a method of reducing color.

Schanderl (1965d) reported that sparkling wine cuvées that had been treated by cation exchange often fermented very slowly if at all. Addition of 40 gr per 100 liters of ammonium phosphate speeded up the fermentation more than did adding 1 gr per 100 liters of magnesium sulfate. However, neither addition resulted in as rapid a fermentation as occurred with wine that was not ion-exchanged.

Paul (1960) analyzed sparkling wines that had been awarded gold, silver, and bronze medals. The total sulfur dioxide content of the six gold-medal wines ranged from 53 to 92 mg per liter (average 70), in the twelve silver-medal wines from 18 to 220 (average 112), and in the five bronze-medal wines 51 to 201 (average 145). In seven French Champagnes the total varied from 23 to 72 mg per liter (average 55). Only one of the gold-medal wines had free sulfur dioxide and none of the Champagnes had any. Paul therefore recommends as small an amount of sulfur dioxide as possible in the fermentation of the musts and the preparation of the cuvée.

Schanderl (1938) has found that too high an iron or copper content or a deficiency of phosphate may inhibit the secondary fermentation of German sparkling wines. (See also Abeijon, 1942.) Phosphate should be added cautiously, however, lest the wine be subject to ferric phosphate cloudiness (p. 794). Schanderl (1942, 1943) questions the value of adding nitrogen in any form, and in some instances finds it dangerous. He reports that excess free sulfur may lead to masking (p. 670).

The importance of the cuvée was emphasized by E. A. Rossi, Jr. (1965): "a sparkling wine can be no better than the base wine used and the base wine in turn can be no better than the grapes originally crushed". He recommended good varieties and harvesting at optimum maturity. Bo (1965) strongly concurred. He also stated that "cold" fermentations were desirable. Rossi noted that where only a portion of the base wine was ion-exchanged there was no adverse effect on rate of fermentation. He preferred yeasts acclimated to sulfur dioxide (25 mg per liter higher than in the prospective cuvée) rather than frozen yeast cakes. No comparative data are presented. Rossi indicated that blending is a desirable method for adjusting the total acidity, but that in practice citric acid is the most important acidulant.

Clarifying the base

One important problem with the sparkling wine base is how to get it brilliant as soon as possible after the primary fermentation without an excessive number of unduly severe treatments. Early clarification can be aided by making the original fermentation on as clear a must as possible. After pressing, the must should be placed in a settling tank. If the conditions are cold enough (either artificially or naturally) much of the extraneous matter will be precipitated. Fermentation of the clear liquid after separation from the precipitate will give cleaner and more easily clarified wines. As soon as the fermentation is completed, the wine maker must take steps to facilitate the natural clarification.

Early and careful racking and lowering of the temperature are the first steps. Since the finished wine must be stable to temperatures as low as 0°C (32°F) (sparkling wines are always chilled before serving), the excess tartrates should be removed from the bulk wine before bottling. This is best done by chilling. Ion-exchange treatment (p. 524) may result in wines that ferment slowly. While it is possible to chill the bottled wine (0°C, 32°F) and thus precipitate excess tartrates and remove them from the bottled wine at the time of disgorging (p. 673), this is not recommended. Since a light color is essential for white sparkling wines, the racking must be done with the least possible aeration. Usually two or three rackings take place during winter and spring, the last one preceded by a fining. To tartrate-stabilize sparkling wines Filippov (1965) recommended addition of 100 mg per liter of metatartaric acid. We do not believe that this gives sufficiently lasting tartrate stability. Any necessary blending should be done before the fining and final racking.

Isinglass, gelatin, and gelatin plus tannin have been the favored fining agents for champagne material. According to Pacottet and Guittonneau (1930), filtration of the light, delicate wine of Champagne, France, is less desirable than fining. However, all wines intended for bottle fermentation should be close-filtered. Geoffroy (1963) recommended filtration after fining with the usual agents but not after fining with alginates. He noted that new fining agents should be developed for sparkling wine cuvée clarification to reduce the uncertainties of riddling.

Just before the secondary fermentation, the metal content should be accurately determined. If the iron or copper contents approach their critical limits of about 10 and 0.4 mg per liter, respectively, the excess should be removed or the wine blended with low-metal wines.

Many have observed that certain wines subjected to secondary fermentation in the bottle ferment slowly if at all. Schanderl (1962) reports that this is often due to the inhibitory effect of various polyphenolics and tannins on alcoholic fermentation when alcohol content is above 8 per cent. In order to detect wines which contain excessively high amounts of these compounds, Schanderl (1962) proposed a simple test. He adjusted the wine to pH 7 and added a drop of a 1 to 2 per cent solution of aqueous ferric ammonium sulfate. Wines which give only yellow or clear, pale, red-violet colors are relatively low in phenolic compounds and should be suitable for sparkling wine production. Darker colors indicate unsuitability. Black or dark violet indicates gallic acid; red or red-brown color, catechins; and red-brown precipitates, ellagic acid. Wines to be carbonated are used directly after clarification (see p. 693).

SECONDARY FERMENTATION IN GLASS . . .

. . . should be carefully controlled by laboratory tests.

Without adding sugar

A number of sparkling wines are prepared by the expensive process of bottling partially fermented wine and transferring (or filtering) it successively off the yeast so that there is a progressive decrease in nitrogen content and reduction in the tendency to referment. One of us (M. A. A.) has seen the process in Istria, in northern Italy, and in the south of France. Bizeau (1963) believes that depletion of nitrogenous constituents and phosphates by this process is most important for stabilization of sparkling wines of Clairette and Muscat blanc. Even though there were marked differences in the amino acids present in the two varieties, this did not seem to affect their biological stability. He also showed that it is very difficult to "exhaust" the nitrogenous constituents by successive filtrations and fermentations. It is unlikely that this type of sparkling wine production will be adopted in California because of labor costs and the generally high nitrogen content of California wines.

Adding sugar

The chief purpose of the secondary fermentation in the bottle or in bulk is to produce sufficient carbon dioxide to give an internal pressure of 3 to 4 atm (45 to 60 pounds per square inch) at about 10°C (50°F) in the final wine as it goes to the consumer. To secure this pressure the secondary fermentation is conducted in a closed system from which the carbon dioxide cannot escape. After the wine has been put in large mixing vats, the desired amounts of sugar, yeast, and, rarely, other substances such as tannin and charcoal are added. About 4.0 to 4.3 gr of sugar per liter will yield 1 atm of pressure in the bottle (4.3 gr of sugar per liter is equal to 1 pound of sugar in 27.3 gallons). Wines of fairly high alcohol content will require slightly more sugar. The sugar content should be determined, and the proper deduction made for residual sugar if the wine contains over 0.2 gr of reducing sugar per 100 ml. More accurate formulas, in which the absorption capacity of the wine is taken into account for calculating the exact amount of sugar needed, are given by Pacottet and Guittonneau (1930) and Weinmann and Telle (1929), but the quantity given above is usually satisfactory. Table 66 gives the pressure at 10°C (50°F) when measured at other temperatures.

The sugar was formerly added as a 50 per cent solution in wine. This is conveniently prepared by mixing 11 pounds of cane sugar, 1.81 gallons of wine, and 0.25 pound of citric acid. (See next section.) The mixture is heated nearly to boiling to invert the sugar. When cool, there should be 2.64 gallons of 50 per cent invert sirup. This is called the tirage liqueur. High-quality, commercially prepared invert-sugar preparations are now available in this country and are commonly used. A 77° Brix mixture of 40° Brix invert sugar and 37° Brix sucrose is often used. The sucrose is easily and rapidly inverted at the relatively low pH of the wine.

The higher the sugar content of the cuvée the higher the acetaldehyde and acetal content of the finished sparkling wine, and the lower the quality of the sparkling wine, according to Dzhurikiants et al. (1966). The sugar content of the cuvée thus appears to be important in tank fermentations. The base wine should be fermented dry and the necessary dosage (see p. 675) added after removal of the yeast. The base cuvée should have as low an acetaldehyde content as possible.

A test bottling of the proposed wine containing the calculated amounts of sugar and acid is advisable to check on the actual pressure obtainable. Several types of gauges are available for measuring the bottle pressure.

TABLE 66

CORRECTION OF PRESSURE TO 10°C (50°F) WHEN MEASURED AT OTHER TEMPERATURES

Temperature		Pressure as measured (in atmospheres)																	
°C	°F	3.0	3.2	3.4	3.6	3.8	4.0	4.2	4.4	4.6	4.8	5.0	5.2	5.4	5.6	5.8	6.0	6.2	6.4
5	41.0	3.7	4.0	4.2	4.5	4.7	4.9	5.2	5.4	5.7	5.9	6.2	6.4	6.7	6.9	7.2	7.4	7.7	7.9
6	42.8	3.6	3.8	4.0	4.3	4.5	4.7	5.0	5.2	5.4	5.7	5.9	6.2	6.4	6.6	6.9	7.1	7.4	7.6
7	44.6	3.4	3.6	3.9	4.1	4.3	4.5	4.8	5.0	5.2	5.5	5.7	5.9	6.1	6.4	6.6	6.8	7.1	7.3
8	46.4	3.3	3.5	3.7	3.9	4.1	4.3	4.5	4.7	4.9	5.2	5.4	5.6	5.8	6.1	6.3	6.5	6.8	7.0
9	48.2	3.1	3.3	3.5	3.7	3.9	4.1	4.3	4.5	4.7	5.0	5.2	5.4	5.6	5.8	6.0	6.2	6.5	6.7
10	50.0	3.0	3.2	3.4	3.6	3.8	4.0	4.2	4.4	4.6	4.8	5.0	5.2	5.4	5.6	5.8	6.0	6.2	6.4
11	51.8	2.9	3.1	3.3	3.5	3.7	3.8	4.0	4.2	4.4	4.6	4.8	5.0	5.2	5.4	5.6	5.8	6.0	6.2
12	53.6	2.7	2.9	3.1	3.3	3.5	3.7	3.8	4.0	4.2	4.4	4.6	4.8	5.0	5.2	5.4	5.5	5.7	5.9
13	55.4	2.6	2.8	3.0	3.2	3.4	3.5	3.7	3.9	4.0	4.2	4.4	4.6	4.8	5.0	5.1	5.3	5.5	5.7
14	57.2	2.5	2.7	2.8	3.0	3.2	3.4	3.6	3.6	3.9	4.1	4.3	4.5	4.6	4.8	5.0	5.1	5.2	5.4
15	59.0	2.4	2.5	2.7	2.9	3.0	3.2	3.3	3.5	3.7	3.9	4.0	4.2	4.3	4.5	4.5	4.7	4.9	5.1

Example: Pressure as measured at 14°C (57.2°F) is 5.8 atm., pressure at 10°C (50°F) is 5.0 atm.
Source of data: Pacottet and Guittonneau (1930).

One type of gauge has a pointed stem that pushes through the cork; the point drops off once the hollow tube has penetrated the cork, and the pressure is registered on a scale. (See Pacottet and Guittonneau, 1930.) The solubility of carbon dioxide in wines as influenced by temperature and pressure has been summarized by Rentschler (1965a). However, generally applicable formulas for different types of wines are still not available.

The problem of securing adequate carbon dioxide absorption of normal California wines was not noted by Amerine and Monaghan (1950), but they fermented all their experimental wines at relatively low temperatures (11.7°C; 53°F). According to Merzhanian (1950), the coefficient of carbon dioxide absorption depends only on the alcohol and sugar content of the wine. There is other evidence to indicate that rather wide variation in the extract content and in various components of the extract does not change the absorption coefficient appreciably. Rapid fermentation at high temperature, too low an alcohol content, a rough interior in the bottle, and serving the wine at too high a temperature are probably the main causes of very rapid loss of carbon dioxide when a bottle of sparkling wine is opened.

Merzhanian (1963a) believes that the amount of carbon dioxide chemically and physically bound increases if the bottle fermentation is carried out below 54°F (12°C). Heating of the cuvée before the secondary fermentation also increased the amount of bound carbon dioxide. The amount is related to the protein content of the wine. However, his method needs further investigation. The definition of bound carbon dioxide is not as precise as one would like. Substances formed during the process of champagnization and yeast autolysis display an important influence on the quality of wine, according to Merzhanian.

Adding yeast and other substances

A pure culture of a champagne-type yeast, characterized by its ability to flocculate readily, is required to produce the firm (agglutinating) sediment. (The initial fermentation may also be conducted satisfactorily with this yeast.) Apparently a variety of types of yeast are being successfully used both in this country and abroad. Various aids to promote flocculation have been proposed for addition to the sparkling wine cuvée. Vegetable carbon does this, but it also removes some color and odor; a trial bottling should be made to determine if it is necessary or desirable.

Goldman (1963a, 1963b) found wet compressed yeast more satisfactory for sparkling wines than traditional cultures as far as the time of fermentation and ability to grow at cooler temperatures are concerned. Geoffroy (1963) reported wide variations in the ease of riddling when different yeasts were used. More research is needed on this aspect. About 2 to 3 per cent by volume of a pure culture (p. 340) is usually added.

If too few yeasts are used in the secondary fermentation in the bottle the fermentation may be slow to develop and will take longer for completion. However, if too much yeast is used, one runs the risk of producing wines with a fermentation (sulfide) odor. Geoffroy and Perin (1965) recommended 1 to 2 million cells per ml. They used 5 per cent of an active culture containing 30 million cells per ml for best results. Citric acid may be used to bring the wine to above 0.75 per cent total acid. This also helps prevent iron casse (p. 794). Occasionally small amounts of tannin are added to facilitate the later "working down" of sediment; however, recent investigations (p. 424) cast considerable doubt on the desirability of any addition of tannin. In a number of tests bentonite has been added to the tirage bottling to improve riddling, but none has been completely successful, including our own trials.

The three most important yeast strains (65 per cent) isolated from sparkling wines by Cases (1959) were: *Sacch. carlsbergensis*, *Sacch. uvarum*, and *Sacch. florentinus*. He also reported *Sacch. bailii*, *Sacch. mellis*, *Sacch. elegans*, *Sacch. chevalieri*, *Sacch. lactis*, *Sacch. cerevisiae*, and *Sacch. willianus*.

A very slight off-color (as in white wines made from red grapes) may be removed by adding decolorizing carbons. Very small amounts should be used, or the wines may acquire an off- or oxidized flavor. The carbons are removed at the time of disgorging. As suggested earlier, it is better to remove the color before bottling.

Addition of nitrogen or phosphate compounds is seldom necessary unless the wines are very low in these compounds. When needed, the minimum quantity required should be used. Cases (1959) noted that most yeasts tested fermented to a higher pressure when the total nitrogen was 95 mg per liter than when it was 58 mg per liter. Excess nitrogen and phosphate in the finished wine may encourage the growth of spoilage organisms, and with high phosphate content there is danger of iron clouds (ferric phosphate *casse*; p. 794). The chelating effect of citric acid (p. 794) usually prevents this.

Addition of various amino acids (glycine, methionine, tryptophan, histidine, cysteine, glutamic acid, leucine, and phenylalanine) to the sparkling wine cuvée greatly increased the 2-methyl-1-propanol and 3-methyl-1-butanol contents in the study of Sisakian *et al.* (1963). The increase was especially large for leucine and glutamic acid. Storage of sparkling wine for one year on the yeast with phenylalanine resulted in greater maturity (equal to three year's aging without the amino acid). The statistical significance of the sensory results is not known. Rodopulo (1964) obtained improved sensory quality by adding tyrosine, tryptophan, or (especially) phenylalanine to sparkling wine cuvées and storing six months. The latter increased the 2-phenethyl alcohol content. Addition of 25 to 50 γ per liter of phenethyl alcohol also improved quality, but addition of 100 to 200 γ per liter of this alcohol and of leucine, methionine, and threonine had an adverse influence on sensory quality. The use of such additives needs further experimentation.

When C^{14}-labeled alanine was added to sparkling wine cuvées after three months in the bottle it was recovered as follows: as alanine, 33 per cent; as other amino acids, 26; in organic acids, 5; in keto acids, 1; in alcohols, 1; in aldehydes, 0.4; in the yeast, 7; leaving more than 26 per cent unaccounted for, according to Glonina and Dubinchuk (1967). When C^{14}-labeled aspartic acid was added, after three months, 42 per cent was recovered as aspartic acid, 17 in other amino acids, 23 in organic acids, 1.0 in keto acids, 0.4 as alcohols, 0.3 as aldehydes, 1.3 in the yeast, and 14 per cent was unaccounted for.

The wine, yeast, sugar, and other ingredients added are thoroughly mixed in large tanks before and during the bottling. Sometimes the mixture is allowed to stand overnight to permit the yeasts to multiply while the wine is still in contact with the air. The whole mass should be vigorously stirred throughout the period of actual bottling.

Bottling

The next step, the tirage bottling, requires special bottles capable of withstanding pressures up to about 9 atm. They should be carefully examined before use, either visually or by tapping them against one another (cracked bottles will not ring like sound ones), or preferably by polarized light (see p. 573). Defective bottles should be eliminated. Manufacturers normally make pressure tests to ensure that the bottles are capable of withstanding the required pressures.

The bottles are not filled as full as for dry wines. Special corks, called tirage corks, are used. These are one-, two-, or three-piece paraffined corks, usually without a "mirror" (see p. 697). They are thus of lesser quality and cheaper than the finishing cork used later. The tirage cork must not be too hard or it will be difficult to remove at the time of disgorging. It is tapped only about halfway into the bottle by a special machine. Another machine places a steel clamp, called an *agrafe*, over the cork so that it catches on the rim around the bottle neck. The clamp holds the cork in the bottle when the pressure develops. The bottled wine is stacked so that no air remains in contact with the cork.

Recently most American and many European wineries have employed crown caps for the tirage bottling. A special bottle is used, on which the crown cap fits. If such caps are resistant to the pressure during the long aging period and if no metal contamination or off-flavor development occurs, they will materially reduce costs. Disgorging is reported to be simpler when crown caps are used, and loss of pressure during aging is reduced to a minimum.

Fermenting in the bottle

For the secondary fermentation the bottles are stacked or placed in wooden bins. Too rapid fermentation in warm rooms may cause breakage and also leakage around the cork. A relatively slow fermentation at cool temperatures yields a wine with a more satisfactory aroma and a better absorption of carbon dioxide. The important Champagne caves in Reims, France, have a maximum temperature of 12.2°C (54°F) and the yeasts used are acclimated to these cool temperatures. Temperatures up to 21.1°C (70°F) or more are sometimes used in California to complete the fermentation as rapidly as possible. However, Tarantola (1937) found that the aroma, flavor, and persistence of the foam were better with a fermentation in the bottle at 5°C (41°F) than at about 15.5°C (60°F). Schanderl (1943) reported that a slow fermentation at 8.9°C (48°F) to 12.2°C (54°F) gave a more intimate union of the wine and carbon dioxide, produced less breakage, reduced the activity of harmful bacteria, and resulted in more complete precipitation of the cream of tartar than did fermentation above those temperatures. If new bottles are used and the fermentations are conducted at moderate temperatures, usually less than 1 per cent of the bottles break. Amerine and Monaghan (1950) especially favored fermentation at a low temperature. Goldman (1963a) indicated that at one sparkling winery in New York the secondary fermentation occurred at 20°C (68°F) in a *labrusca*-flavored wine.

It has been observed by many enologists that pink and red sparkling wines have a longer secondary fermentation period in the bottle than white (Goldman, 1963b).

After about six months, when the fermentation is practically over, the stacks are torn down, the broken bottles removed, and the bottles shaken well to break up the sediment. The bottles, when restacked, are so arranged that the lees will be deposited in the same place as before. During maturation in contact with the lees, the bottles may be shaken and restacked about once a year. By restacking the bottles so that the sediment always collects at the same place, sticking can occur only at one point and not on all sides. This sticking, or "masking," usually occurs when the wine has not been properly clarified, but it can take place also when free sulfur is present. Masking probably occurs from other causes as well. A chalk mark is placed on the bottom of the bottles on the side where the sediment first collected. This facilitates handling and returning bottles to their proper position.

In the early stages of fermentation there is danger of flying glass from the breakage of faulty bottles. A wire mask should be worn to protect the face, and gloves to protect the hands. This danger is not entirely over until the wine is disgorged, and workers should always wear a protective face mask in the sparkling wine department.

Figure 105.   Sparkling wines aging. Note use of fork lift. (Courtesy Almadén Vineyards, Inc.)

Aging on the sediment is very important and accelerated procedures may result in wines of lesser quality (fig. 105). The reason for this is that when the wine remains in contact with the sediment for two to five years the viable yeast population is greatly reduced. Despite various claims that diethylpyrocarbonate (DEPC) is present in sparkling wines, Kielhöfer (1963a) was unable to detect its presence. In fact, DEPC hydrolizes rapidly—normally in a few hours (see p. 412). After disgorging there is less danger of refermentation of the added sugar. When the fermentation is greatly speeded up (hot rooms, etc.) and disgorging (or bottling, for bulk-fermented wines) follows the fermentation by a very brief interval, a number of viable yeast cells may remain. This is one reason why sulfur dioxide is usually added to such wines with the final dosage (p. 674). It also explains why some bottle-fermented wines which have a high dosage of added sugar are sold with little aging: the wine maker is unwilling to risk a further fermentation in the bottle. Vogt (1945) recommended aging on the sediment for half a year to one and a half years or longer, and the best French sparkling wines are not disgorged until after two to four years. In fact, wines cannot be sold as Champagne in France if they have not been kept on the yeast for at least nine months. Although no definitive experiments have been made in California, there is general agreement that the best sparkling wines should be aged on the sediment to improve their flavor and stability, and this appears to be one of the critical factors in producing high-quality sparkling wine. Schanderl (1943) noted the favorable effect of a long aging on the yeast resulting in a better bouquet and an increase in the vitamin content of the wine. Amerine and Monaghan (1950) also believed that aging on the yeast was of value. They quote Russian experiments showing that enzymes from yeast autolysis accounted for the desirable changes that occurred during bottle-aging. Rabinovich (1960) also recognized the advantage of yeast autolysis. Unless bottle-fermented wine is aged on the yeast for at least one year, there seems to be little difference between the quality of the product from this process and that from tank fermentations or from bottle-fermented wines which are transferred (p. 677).

Yeast autolysis occurs also in the continuous process of sparkling wine production, according to Agabal'iants and Avakiants (1966), but no detailed tests comparing the quality of wines so produced with the quality of bottle- or tank-fermented wines are known to us.

The decrease and then increase in total nitrogen and in the amounts of the amino acids in bottle-fermented sparkling wines are well illustrated by the data of Bergner and Wagner (1965), in mg per liter:

			Days from addition of yeast				
	Cuvée	1	3	21	21[a]	180[b]	395[b]
Total N	855.2	785.9	766.3	768.3	792.2	805.5	837.7
Alanine	27.8	9.7	7.6	17.6	12.8	14.0	21.6
Arginine	38.2	24.3	23.5	23.7	27.2	30.4	33.6
Aspargine	28.2	4.8	3.4	3.8	4.6	8.0	18.8
Cystine	6.6	5.9	5.5	6.0	5.9	5.1	5.5
Glutamic acid	65.6	15.5	10.3	10.9	11.3	16.0	25.7
Glycine	23.0	11.3	11.5	11.6	14.1	19.4	21.9
Histidine	19.4	18.1	17.1	18.9	20.4	22.6	25.4
Leucine + isoleucine	23.2	6.7	4.8	4.9	6.6	10.9	14.4
Lysine	6.0	5.3	4.3	4.8	5.2	10.3	40.5
Phenylalanine	11.8	5.3	4.8	5.1	5.8	9.0	9.7
Proline	296.0	286.0	282.0	280.0	306.0	350.0	420.0
Serine	15.0	3.1	3.4	3.7	4.0	6.5	9.8
Threonine	7.4	5.4	6.0	6.2	5.6	5.6	5.7
Tryptophan	14.6	12.8	14.1	16.5	14.0	14.4	14.8
Valine	31.2	5.9	4.5	4.8	6.6	8.3	14.2

[a] Bottled tank wine after fining, filtering, and liqueuring.
[b] Bottle-fermented and aged on the yeast.

The increases in proline, lysine, glutamic acid, leucine and isoleucine, phenylalanine, serine, and valine are especially notable in the bottle-aged wine. Thus it is not the fermentation in the bottle which differentiates tank- and bottle-fermented wines, but the period of time the bottle-fermented wine remains in contact with the yeast and the secondary reactions therefrom. Cases (1959) emphasized that a newly finished sparkling wine had less of all of the amino acids (except cystine) than the base wine.

Avakiants (1965) emphasized the importance of the enzyme β-fructo-furanosidase in the development of the bouquet and taste of sparkling wines. This was due not only to its hydrolytic property but also to its synthetic property. He stressed that the inversion of sugar occurred not only in the tirage but in the final dosage. It is involved in the synthesis of β-ethyl fructoside also. This alcohol forms an acetal which he believes is important in the "champagne" character.

Disgorging

The secondary fermentation in the bottle produces considerable yeast sediment and often of tartrates. The simplest method of completing this precipitation is to lower the temperature to about $-3.9°C$ ($25°F$) for one to two weeks. The bottles are then ready for clarification, which is accomplished by gradually moving the sediment from the side of the bottle onto the cork. The bottles are placed on end in special racks (fig. 106), twirled rapidly to the right and left, and dropped back into the rack in a position slightly to the right or left of their original position. In this way the sediment is gradually worked onto the cork. Usually the bottles are turned only slightly to the left or right at first. About two to six weeks are required for complete clarification. The sediments in different wine lots vary greatly in ease of movement, depending on the clarity of the original wine, the length of aging on the sediment, the composition of the wine, the yeast strain employed, and other factors. When the sediment starts to stick to the sides of the bottle, disgorging should take place at once. If the sediment cannot be worked loose on the racks, it may be necessary to tap the bottles with rubber hammers or use shaking machines.

If there is not too much pressure in the bottles, the actual disgorging is comparatively simple. The sediment and a small portion of the liquid in the bottle neck are frozen solid by placing the neck in a liquid cooled to below $-15°C$ ($5°F$) (fig. 107). It is best to cool the wine itself to below $10°C$ ($50°F$) for the disgorging, to reduce the loss in pressure. The agrafe, or crown cap, is then removed, and the frozen plug of yeast, tartrates, and the like is forced out by the gas pressure in the bottle. Only

Figure 106.   Riddling racks. Note face mask. (Courtesy Almadén Vineyards, Inc.)

a little wine is lost by this procedure unless the pressure is very high, and only about 1 to 2 atm of pressure should be lost during the disgorging. The disgorger not only removes the cork and makes sure that all sediment is ejected, but also smells each bottle as it is opened and eliminates any bad bottles. A good disgorger, by working very rapidly, loses little wine. Some wine makers prefer not to freeze the neck, but instead ferment the wine to a higher pressure and blow the sediment out. More pressure and wine are lost by this system and it is not recommended. Even with the most careful disgorging all the viable yeast cells are rarely removed.

Final dosage

An empirical equation for determining the sugar content of sirups made of sucrose intended for the final dosage of sparkling wines was devised by Drboglav (1958), where C (as invert sugar in gr per 100 ml) = R-A-P/0.95 d. R is the refractometer reading at 20°C (68°F), A a correc-

Figure 107. Equipment for freezing the neck of sparkling wine bottles before disgorging. (Courtesy Beaulieu Vineyard.)

tion for the alcohol in the sirup, P the nonsugar nonalcohol constituents of the sirup, d the specific gravity, and 0.95 the correction for inversion. A is 1.30 for 4 per cent alcohol; 1.66 for 5; 2.13 for 6; 2.41 for 7; 2.79 for 8; 3.18 for 9; 3.57 for 10; 3.95 for 11; 4.36 for 12; and 4.73 for 13. A final dosage (*liqueur d'expedition*) with a refractometer reading of 58.5 and a specific gravity of 1.245 then has [(58.5 − 5.50) × 1.245]/0.95 or 70.1 per cent sugar. Data to confirm this formula would be welcome. It simply converts the sugar sirup to its invert sugar equivalent.

The wine should be free of fermentable sugar after disgorging, and will be very dry in taste. To satisfy the customer's palate, 1 to 3 per cent or more of sugar is usually added to the disgorged wine. The better-quality champagnes are left the driest, whereas the wines with a less desirable flavor are more generously sugared (up to about 5 per cent in this country). (See fig. 108 for a filling and dosage machine.)

Each firm has a slightly different formula for preparing the sweetening liqueur (called the *dosage*). This is composed of sucrose in aged wine. In Europe it usually includes 5 to 6 per cent of good-quality brandy. (It averages somewhat sweeter than the 50 per cent tirage liqueur previously mentioned; see p. 665.) The mixture must be filtered until absolutely

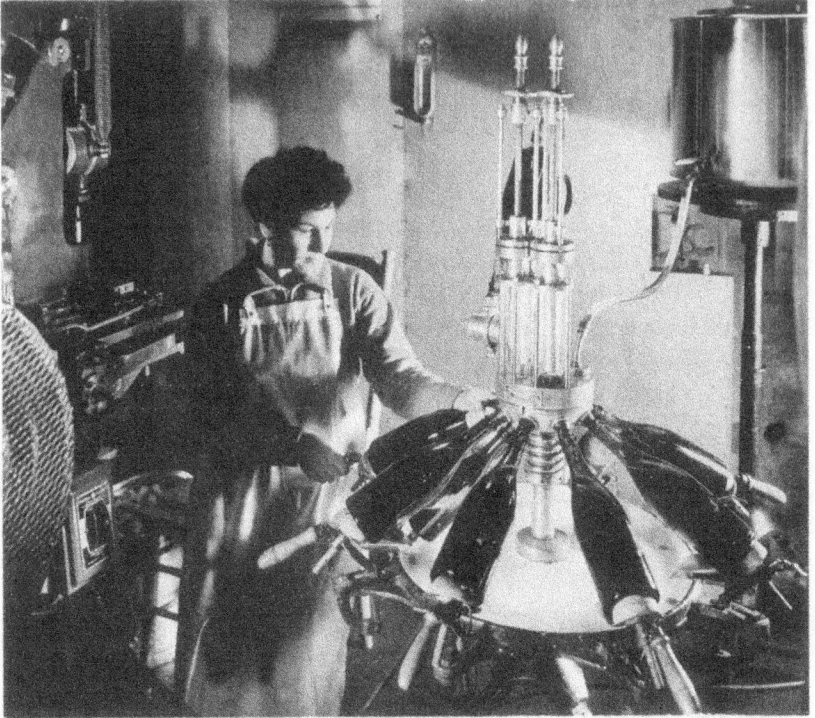

Figure 108. Filling and dosage machine for sparkling wines. (Courtesy Beaulieu Vineyard.)

brilliant, and is added immediately after disgorging. Liqueuring machines automatically add a measured amount of the liqueur—usually 10 to 50 cc per bottle, depending on the amount of sweetness desired—and, if necessary, wine of the same type to replace that lost during disgorging. American producers use invert sugar and seldom add brandy. E. A. Rossi, Jr. (1965), reported that in California the brandy dosage "is only infrequently used today." He believes that this reflects industry satisfaction with "natural grape flavors as developed with sound fermentation and aging techniques." Excessive use of brandy dosage is, of course, to be deplored. However, discreet use of a dosage containing some brandy may be useful, especially for some of the more neutral-flavored California sparkling wines.

De Rosa (1955) reported that all the sucrose added in the dosage was hydrolyzed in about 45 days. Only 0.01 to 0.03 per cent sucrose remained.

Gushing of red sparkling wines often occurs when they are opened.

To reduce gushing the tannin content should be low. Gelatin fining of the cuvée is recommended. The presence of crystals of potassium acid tartrate also favors gushing, as does an alcohol content of over 12 per cent. Schanderl (1965a) reported increases in leucoanthocyan and catechin in finished dosaged sparkling wines compared to pre-dosaged wine. He attributed this to the presence of polyphenols in the dosage liqueur. He suggested that this might be due to storage of the liqueur in new oak casks.

The bottle is kept closed as much as possible during these operations to prevent loss of pressure. (Corking and capping are described on pp. 697–698.)

Paul (1960) recommended that the sugar content of the different types of sparkling wines fall in the following ranges: brut (*Extra Trocken*), below 1.5 per cent; extra sec (*Trocken*), 1.5 to 3.0; sec (*Halbsüss, polosladké*), 3.0 to 4.5; and demi-sec (*Süss, sladké*), 4.5 to 6.0. The recommended sugar content for sparkling wines was given by Ribéreau-Gayon and Peynaud (1960–1961) as follows: brut, 0.2 to 1.0 per cent; extra-sec, 1 to 2; sec, 2 to 4; demi-sec, 4 to 6; and doux 8 to 10. We approve the recommendation that brut be low in sugar. We are not sure that the range 3.0 to 4.5 is wide enough for differentiation of the demi-sec from the sec. For the American market only three ranges would seem necessary: below 1.5, 1.5 to 3.0, and 3.0 to 6.0, for brut, sec, and demi-sec, respectively.

After disgorging and addition of the final dosage, bottle-fermented wines seldom re-ferment even though viable yeasts may be present. The reason for this is not known. The presence of DEPC appears unlikely (see p. 412).

Transfer system

The transfer system was introduced into California from Germany in the early 1950's. Its primary advantage is the elimination of the riddling process, since the bottle-fermented wine is transferred to a tank (under counterpressure) and from there filtered into a bottle (fig. 109). It has the legal advantage over the tank process that the wines can be labeled "bottle-fermented". On the American market one can sometimes distinguish bottle-fermented and regularly disgorged wines from transferred wines in that the former can state "naturally fermented in this bottle," which the latter cannot. Berti (1961) noted that the transfer system results in greater uniformity of dosage, more uniform pressure in the final wine, and predictable daily production of wines for processing. As disadvantages he mentions the cost of equipment, susceptibility to oxidation, and, if young, the possibility of refermentation. He

Figure 109. Transferring bottle-fermented sparkling wine into tanks. (Courtesy Paul Masson Vineyards.)

believed that the quality of transfer-system wines was better than that made with regularly disgorged bottles! But he added, correctly we believe, that it is superior to the bulk processed wine only when the transfer wine is aged in the bottle prior to transfer. Presumably, then, disgorging does more harm to the quality of the sparkling wine than does transferring. We should like to see more experimental evidence on this.

Bo (1965) stated that the secondary fermentation and proper bottle-aging before disgorging are critical quality factors in producing transfer system wines. He thus implies that all transfer-system wines are "properly aged". He recommends that the receiving tank be pressurized "with nitrogen or air . . . to a slightly higher pressure than exists in the tank". Filtration of wines containing carbon dioxide, whether produced by the transfer system or by tank fermentation, must be done under pressure on the exit side of the filter in order to prevent foaming.

Turbulent-flow transfer of sparkling wines in nonwettable or rough pipes results in large losses of carbon dioxide compared to transfer in smooth wettable pipes, according to Merzhanian (1963b). He also reported 20 per cent loss of gas during filtration.

In the transfer system the final dosage liqueur may be added to the empty tank before it is pressurized, or by means of a pressure cylinder after it is full. If the latter, the contents of the full tank are circulated with a pump to ensure thorough mixing. The sparkling wine is then filtered into another pressured tank under refrigeration and isobarometric conditions. Bo (1965) felt that the transfer system provided "quality control," but he presented no statistical sensory data.

Figure 110. Traditional champagne bottling room: right to left, disgorged bottles, machine for adding dosage, corking machine, and wire netting machine.

Schanderl (1964) is particularly critical of the transfer system. The loss of pressure (which should not be excessive if properly operated) and the oxygen pickup are given as reasons. Likewise, he is opposed to the Carsten process (where carbon dioxide from a tank fermentation of grape juice is collected, cooled, purified, and compressed; after the wine is clarified the compressed carbon dioxide is added back). He correctly notes that this is really just carbonation.

Hemphill (1965) makes the point that bottle-fermented sparkling wines constitute a decreasing percentage of the total American sparkling wine production. He notes, correctly we believe, that this is due to increasing labor costs rather than to any quality advantage of tank-fermented or transfer-system wines. To reduce costs Hemphill recommends plywood bins holding about 500 bottles, the use of a forklift, and automated riddling. He maintains that an electric vibrator eliminated most of the riddling. He used portable electric vibrators with permanently mounted air vibrators connected to solenoid valves and timed to work at night. He claims that no manual riddling is necessary, and that new combinations of vibration time and frequency will reduce the clarification time. We would be more easily convinced if time and economic studies were presented. Most important, comparative sensory data are not presented. Nevertheless, the traditional system requires much hand labor (fig. 110).

FERMENTATION IN TANKS...

... is more rapid, saves space and labor, and is recommended for standard-quality wine.

Fermentation in large-sized containers (500 to 1,500 gallons or larger) makes it possible to produce a large volume of uniform sparkling wine at a lower cost than is possible by the bottle-fermentation method just described. Less capital is tied up in bottles and wine during this process than during the long aging period in glass, and the clarification is more certain and requires less labor than does fermenting in the bottle. The effects on quality may differ somewhat because of the reduced time the wines are left in contact with the lees in the large containers and because of other factors (p. 686).

Comparison with bottle process

Schanderl (1943) reported that the character of bottle-fermented French Champagne was due in part to the long aging on the lees after the secondary fermentation in the bottle and before disgorging. He pointed out that the ratio of wine to yeast in bottles and in tanks is very different. Normally, wines fermented in tanks are not aged in tanks on the yeast because of the desire for a rapid turnover of wine and because the depth of yeast in the tank may lead to autolysis of the yeasts and subsequent formation of hydrogen sulfide and mercaptans. However, palatable wine can be produced by fermentation in large tanks provided good-quality wine is used, a reasonable length of time at low temperature is taken for the fermentation, the wine is bottled with minimum aeration, and the finished wine does not contain excessive amounts of sulfur dioxide. (See also Amerine and Monaghan, 1950.)

The difference between bottle-fermented and tank-fermented wines was studied by Janke and Röhr (1960). They developed two objective tests: (1) a carbon dioxide release test (the ratio of the amount of carbon dioxide released at 20°C (68°F) and 35°C (95°F) (expressed as percentage) and (2) the amount of nitrogen separated from 1 liter of wine by ultrafiltration. The first test gives a measure of the relative stability of carbon dioxide in wines, and the second may measure the colloidal nitrogen released by yeast autolysis. In fifty-three commercial wines the carbon dioxide release ratio was 21 to 31 (average 27.8) for twenty-four bottle-fermented wines, and 30 to 45 (average 35.5) for twenty-nine tank-fermented wines. Obviously, tank-fermented wines lose their carbon dioxide more rapidly than do bottle-fermented wines. The nitrogen content of the material separated by ultrafiltration was 9.6 mg for the bottle-fermented and 6.2 for the tank-fermented.

When the same cuvée was used for tank and bottle fermentation (Janke and Röhr, 1960) there was a significant difference between the wines after 42 days as measured by the carbon dioxide release test. This test measures the rate of loss of carbon dioxide. This was 29 to 45 (average 34) for twenty-eight tank-fermented wines and 23 to 35 (average 30.6) for thirty-four bottle-fermented wines. The average carbon dioxide release test value for tank-fermented bottles stored for eighteen months decreased to 31.0; that for bottle-fermented bottles disgorged after eighteen months was 26.0. After 250 to 380 days the nitrogen by the ultrafiltration test was 6.2 mg per liter for the tank-fermented wine and 8.2 for the bottle-fermented. Janke and Röhr (1960) concluded that the difference between tank- and bottle-fermented wines lay in the influence of yeast autolysis. They asked the question whether the quality of tank-fermented wines could be improved by influencing yeast autolysis.

The oxygen pickup during handling of tank-fermented sparkling wines contributed to subsequent oxidation during aging, according to Kielhöfer and Würdig (1963b, 1963d). Carbon dioxide as the counterpressure gas did not contribute significantly to reducing oxygen pickup. It is this oxygen pickup which requires addition of sulfur dioxide. Using a maximum pickup of 20 mg per liter would require 80 to 85 mg per liter of free sulfur dioxide, which would injure the quality. They therefore recommend the use of ascorbic acid.

They also write: "Der O_2-Gehalt ist das einzige bis jetzt bekannte eindeutige Unterscheidungsmerkmal des Tankgärsektes gegenüber dem nach der 'Methode Champenoise' hergestellten, völlig O_2-freien Flaschengärsekt. Da auch der mittels des Transvasierverfahrens hergestellte Sekt im O_2-Gehalt völlig dem Tankgärsekt gleicht, entspricht die Bezeichnung 'Flaschengärung' hierbei Verfahren hinweisenden Zusatz verwendet werde."[1] There have also been reports that tank-fermented sparkling wines are lower in certain amino acids than bottle-fermented. Use of enzyme concentrates for improving the quality of tank-fermented sparkling wines was discussed by Filippov (1963).

Bergner (1968) compared the amino acid content of tank-fermented sparkling wine with that of bottle fermented—both made from the same base. The latter aged 13 months on the lees had much higher proline, lysine, glutamic acid, and alanine contents than wines aged 6 months on the lees of tank-fermented wine. Tank-fermented wines could be increased in amino acids by pumping over twice a day for one-half hour.

Procedure

The tanks vary in size from 500 to 1,500 or more gallons, and are ordinarily made of steel lined with glass or with stainless steel; they must be capable of withstanding high pressures. Sparkling wine tanks are normally insulated with cork or, more recently, with urethane foam (which can be sprayed on, providing a seamless finish). A thin vaporproof outer coat is applied to prevent water absorption. Most German tanks for sparkling wine production are equipped with agitators to stir the yeast during fermentation. This is said to result in wines with a lower redox potential and improved quality. Stirring may also increase aldehyde content.

Means of cooling or heating the fermenting wine are provided within the tanks, or between the walls if they are jacketed. The fermentation can thus be closely regulated. Safety valves and pressure gauges are provided.

---

[1] "The oxygen content is one characteristic by which tank-fermented sparkling wine differs from that produced with the 'Methode Champenoise', which is completely free from oxygen. Since the sparkling wine produced by means of the transfer process resembles fully tank-fermented sparkling wine as regards oxygen content, the designation 'bottle-fermented' does not comply with the expectations of the consumer and should therefore be qualified by words indicative of this process."

The same type of wine, yeast, and sugar solution used for bottle fermentation should be employed (pp. 660, 665, 667). The sugar content need not be controlled so accurately, since excess pressure is easily allowed to escape. The wine may be pasteurized in the tank before adding the yeast and sugar. E. A. Rossi, Jr. (1965), reported that some producers of tank-fermented sparkling wines preferred to add all the sugar at the start and manipulate the temperature so as to retain the required sugar concentration in the finished wines. Others ferment "dry" and then sweeten. Biologically, the first probably makes sense, but the second seems to be more practical. Tank fermentation should not be carried out at too high a temperature: 10° to 15°C (50° to 59°F) is probably best, but more data would be welcome.

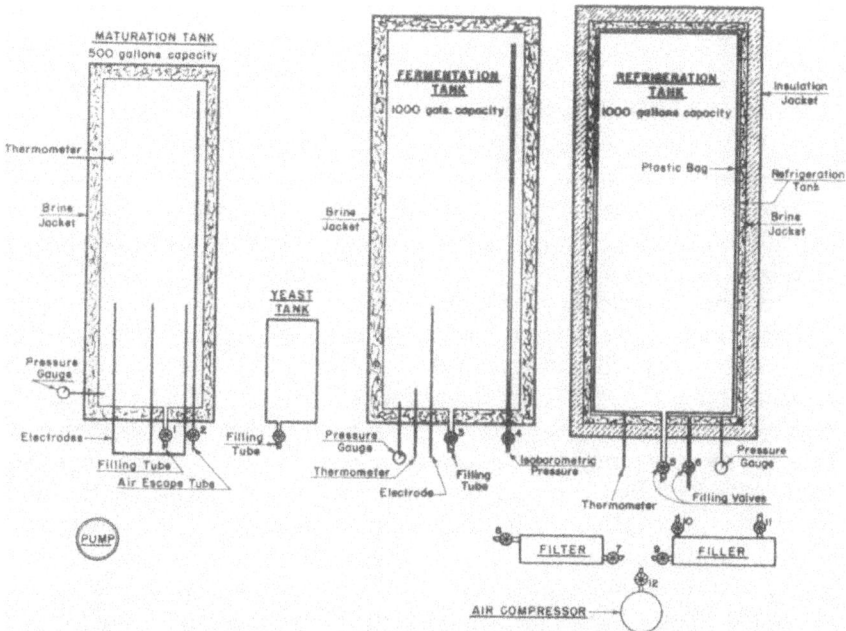

Figure 111.   Schematic diagram of original Charmat process.

Charmat process

Bulk-process champagne is usually produced by the closed cuvée process developed by Eugene Charmat in 1910. In the original process, as described by Tschenn (1934), a battery of four was used. The first, a heating or maturation tank, is a strong, 500-gallon, jacketed tank (fig. 111). Wine is pumped into this tank through valve 1, leaving valve 2 open.

The wine is heated to between 55°C (131°F) and 65°C (149°F), with predetermined amounts of air (obtained by closing valve 2 earlier or later during filling), to a pressure of 140 to 160 pounds per square inch. This is said to speed esterification. Chapidze (1960) recommended heating to 50°C (122°F). The wine is then brought to the required temperature, in eight to ten hours, by electrode heaters, maintained for four to six hours at the pasteurizing temperature, and then cooled in six hours to 25.5°C (78°F) by circulating refrigerated brine in the jacket. The cooled wine is next transferred to a 1,000-gallon fermentation tank by connecting valve 1 to valve 3, leaving valve 4 open during transfer. Prepared yeast starter and sugar are added. The sweetened wine is allowed to ferment ten to fifteen days to produce a wine with 5 to 6 atm carbon dioxide pressure. E. A. Rossi, Jr. (1965), indicated that the normal tank fermentation time in California is generally less than two weeks, unless the fermentation temperature is 10°C (50°F) or less.

The refrigeration tank developed by Charmat had an outer jacket and an inner leakproof plastic bag of 1,000-gallon capacity. When this bag is inflated, it adapts itself to the walls of the tank. The charged wine, with the final sweetening solution, is transferred from valve 3 of the fermentation tank to valve 6 of the refrigeration tank, in which a counter pressure is automatically established to the required degree. The transfer is made under isobarometric conditions so that no carbon dioxide gas escapes.

The wine goes into the tank itself, not into the inner bag, which remains deflated. The charged wine is cooled from 21.1° to 5.5°C (70° to 22°F) in twenty hours by circulating brine through the jacket of the tank, and remains stored at − 5.5°C (22°F) for four to five days. After this clarification by refrigeration, the sparkling wine is filtered under isobarometric pressure by connecting valve 6 through a pump to valve 7 of the filter and from valve 8 of the filter to valve 5 of the refrigeration tank. As the tank is emptied, the plastic bag lining, which permits carbon dioxide to diffuse through, expands and finally occupies the whole tank. The filtered wine is returned continuously to the refrigeration tank, this time inside the expanded plastic bag. This arrangement keeps the wine from coming into contact with air during the filtering and filling processes, and also saves a second tank.

When a new charge of wine is pumped from the fermentation tank into the refrigeration tank, between the tank wall and the full plastic bag, it automatically pushes the clarified and filtered wine out of the bag into the bottling machine. To bottle: valve 5 is connected to valve 9, valve 10 of the filler is connected to valve 4, and valve 11 is connected to valve 12 of the air compressor, and the pump is started.

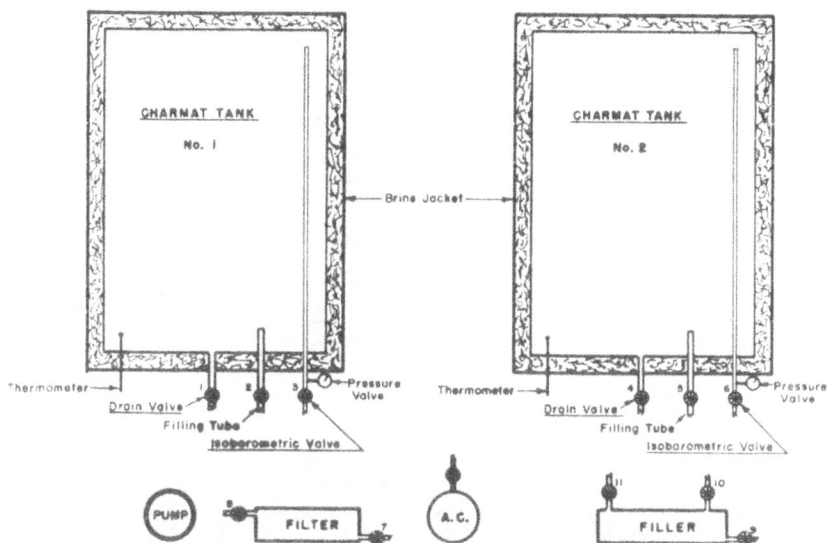

Figure 112.   Schematic diagram of two-tank bulk-fermentation process.

Two-tank process

The Charmat process has been modified under California conditions into a two-tank process (fig. 112) with a yeast-starter tank. (See fig. 113 for actual tanks used in California.) The yeast starter is prepared by growing champagne yeast in grape concentrate diluted to 18° Brix with water, in the presence of 100 parts per million of sulfur dioxide. The latter is added during fermentation in increments of 25, 25, and 50 parts per million. The yeast starter is built up from 1.5 liters to 19 and finally to 49 liters. When the 49 liters of starter are fermenting vigorously, they are added to 32 gallons of a blend of wine, invert sugar, and citric acid containing 100 mg per liter of sulfur dioxide. This wine has been previously pasteurized at 82.2°C (180°F) for fifteen minutes. Sufficient sugar is added to the wine base to produce 6 atm of pressure (fig. 114), and sufficient citric acid to bring the acidity to 0.75 per cent as tartaric acid. The total volume of yeast culture is about 45 gallons.

The 45 gallons of yeast culture is then added to 855 gallons of wine base in a jacketed, 1,000-gallon fermenting tank through valve 1 of tank 1, leaving valve 3 open during filling. The components of the 900-gallon tirage are prepared so as to produce after fermentation 6 atm pressure, 12.5 per cent alcohol, 0.75 gr acid as tartaric per 100 ml, 2 per cent reducing sugar, and about 75 mg per liter of sulfur dioxide, by addition of invert sirup, citric acid, and sulfur dioxide.

685

Figure 113.   Tanks for sparkling wine production. (Courtesy Valley Foundry and Machine Works, Inc., Fresno.)

Dzhurikiants *et al.* (1966) note the high aldehyde and acetal content of some tank-fermented Soviet sparkling wines. This they attribute to fermenting with too high a sugar content. When the fermentation is stopped by cold, not all the aldehyde is transformed into alcohol. With six commercial wines the brut had 2.1 to 3.0 per cent sugar and the aldehydes ranged from 32.5 to 61.7 mg per liter (average 49.6). In six demi-sec and doux the sugar content was 7.2 to 12.3; the aldehydes ranged from 44.9 to 64.2 mg per liter (average 54.4).

In the tank process, whether Charmat or two-tank, sulfur dioxide must be added to control the possible harmful effects of oxygen introduced during transfer, filtration, and filling. Kielhöfer and Würdig (1963*b*) recommend judicious use of ascorbic acid to reduce the amount of

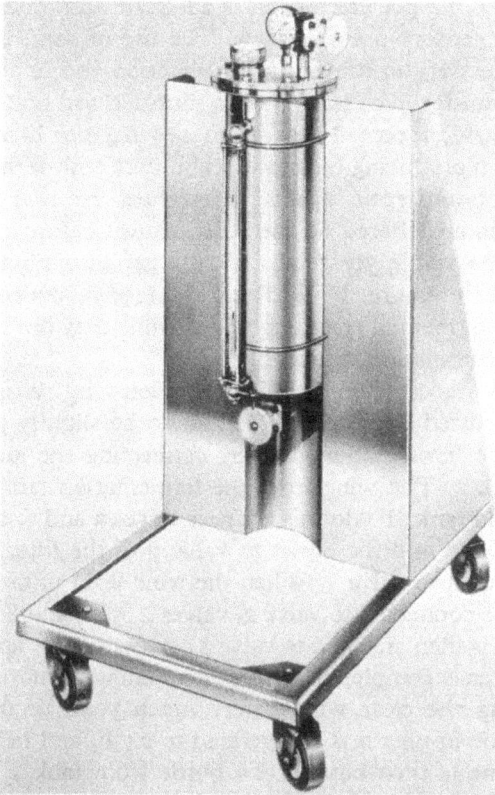

Figure 114. Dosage machine for adding sugar to tanks in bulk process. (Courtesy Valley Foundry and Machine Works, Inc., Fresno.)

sulfur dioxide. Oxygen is also introduced in the transfer process. Kielhöfer and Würdig therefore recommended labeling such wines "transfer wines" to distinguish them from bottle-fermented wines (see p. 682).

This wine is then fermented at over 15.5°C (60°F) by pumping refrigerated brine into the cooling jacket of the tank. At the end of the fermentation, or when the pressure in the tank has reached 90 pounds, a sample of the wine is withdrawn from the tank and analyzed for alcohol, reducing sugar, acidity, and sulfur dioxide. Should the sugar content not be high enough, it is possible to add more either to this tank or to tank 2. Fermentation is then stopped by circulating brine through the cooling jacket until the wine is brought to −4.4°C (24°F) to −1.1°C (30°F). The wine is now adjusted to the required sulfur dioxide content (which some operators feel is as high as 50 mg per liter) and is stabilized by refrigeration at $-4.4°C$ ($24°F$) to $-2.2°C$ ($28°F$) for ten days. We believe that 100 mg per liter or less is adequate and much less objectionable from the sensory point of view. The use of some sulfur dioxide is considered necessary to prevent refermentation and to absorb the excess air from the transfer from tank to tank, filtration and bottling.

Edelényi (1966) reported that 75 to 115 mg per liter of DEPC was very effective in producing biological stability of tank-fermented sparkling wines. Use of antiseptic agents is essential because tank-fermented sparkling wines are filtered off the yeast deposit before all the yeast cells are dead. Some viable yeast cells usually get through the filter during filtration under pressure. E. A. Rossi, Jr. (1965), does not recommend fining of tank-fermented sparkling wines while they are in the tank. The greater lees accumulation reduces net yield.

The second 1,000-gallon tank is then cooled by refrigeration and its pressure is adjusted with compressed air to be slightly greater than the pressure in the fermentation tank by connecting the air compressor to valve 5 of tank 2. The wine from the fermentation tank is now filtered into this second tank. To do this connect valves 3 and 6, attach the pump suction to valve 2 and the outlet to valve 7 of the filter. Valve 8 of the filter is connected to valve 5. When the wine level in tank 1 reaches the top of the tube connected to valve 2, valves 2, 5, 7, and 8 are closed. The pump suction is then attached to valve 1; valves 1, 5, 7, and 8 are opened; and the transfer is completed. This system allows a more rapid filtration by transferring the clear wine before much yeast accumulates on the filter. The wine in tank 2 is refrigerated to $24°F$, and held there for two days. The wine is then bottled. To bottle from tank 2 connect the air compressor to valve 10 of the filter. Valves 11 and 6 are connected and valves 5 and 9 are connected. The air pressure on valve 10 is adjusted to slightly greater than that in tank 2. Valves 10, 6, 11, and 5 are opened, in that order, and bottling is started. When the wine level reaches the top of the tube above valve 5, the hose from valve 5 is shifted to valve 4.

A packaging line for tank-fermented or transfer-system sparkling wine is given in figure 115.

One of the major problems in filtering and bottling from tanks is to prevent loss of pressure in the space above the wine as the tank is being emptied. Use of a counterpressure of compressed air is common practice, but some of the air is dissolved in the wine as the tank is emptied, and the bottled wine, especially that bottled last, suffers from overoxidation unless sulfur dioxide is employed. The logical procedure, used in France and elsewhere, is to use compressed carbon dioxide to maintain the pressure during emptying. (See also Ventre, 1930.) However, some government agency might question the legality of this practice, for an

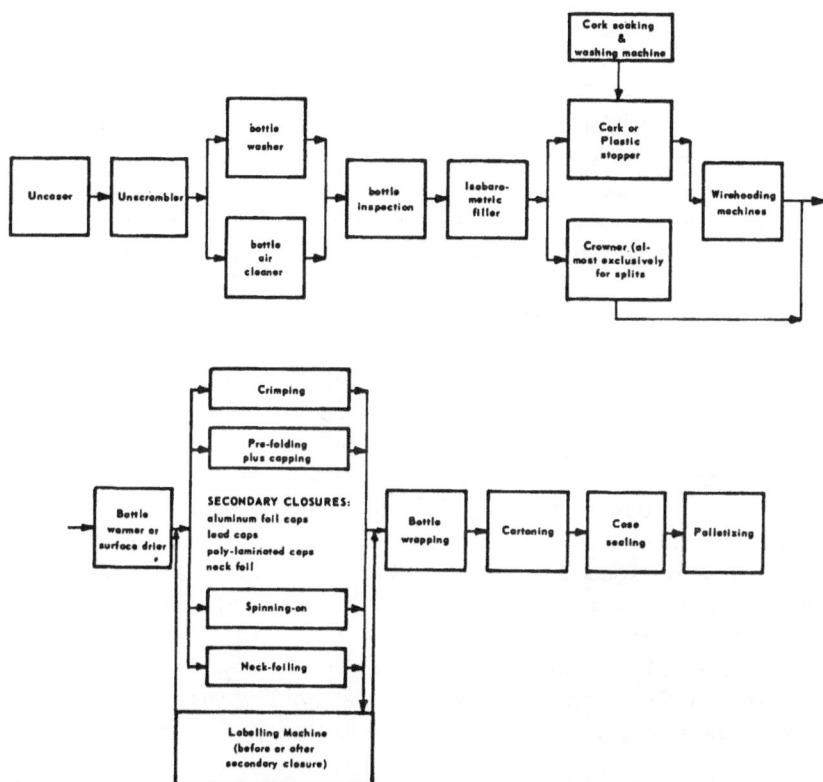

Figure 115. Packaging line for tank-fermented or transfer-system sparkling wines. (Courtesy Budde & Westermann Machinery Corp.)

unscrupulous operator could derive pressure from the compressed carbon dioxide gas and not from the natural fermentation in the tank. Use of compressed nitrogen might eliminate some of the objections of government agencies and it would also, like carbon dioxide, prevent excessive oxidation. Carpenè (1962) recommended it in Italy. It would not have the acidifying effect of dissolved carbon dioxide.

Drboglav (1951) and others have shown that the low quality of some tank-fermented sparkling wine is due to the introduction of oxygen during finishing, which increases the aldehyde content and darkens the color. Much better wines (Deibner, 1957a, 1957c, 1957d) were produced when the wine was handled under carbon dioxide, and the best under carbon dioxide with a small amount of sulfur dioxide. Some reduction of the oxidation-reduction potential is obtained by adding ascorbic acid at the time of the tirage bottling. During aging in the bottle there was usually a decrease in potential. Frolov-Bagreev (1948) reported that the oxidation-reduction potential of Champagne was markedly lower than that of Soviet sparkling wines. The technique for measuring the potential was criticized by Deibner (1956a, 1966c).

Kielhöfer and Würdig (1963b) found 5 to 200 mg per liter of oxygen in newly bottled tank-fermented wines, but none in bottle-fermented wine. To reduce the oxygen content in the first case, they recommended sulfur dioxide and ascorbic acid. They found that transfer-system wines must be treated the same as tank-fermented. In German tank-fermented sparkling wines which have a relatively high sulfur dioxide base, acetaldehyde is produced in considerable amounts during the fermentation in the tank. It is thus necessary to add more sulfur dioxide. Schanderl and Staudenmayer (1964) recommended adding fresh yeast to the wine and agitating. This reduces the acetaldehyde content and may remove some metals and tannins by adsorption. However, agitation of the wine with yeast must be of short duration or more acetaldehyde will be formed.

It is clear from the report of Dittrich and Staudenmayer (1968) that high accumulations of acetaldelyde can occur in slow pressure-tank fermentations, especially when using highly clarified musts. The problem of yeast growth in tank-fermented sparkling wines is considered by De Rosa (1964b). He noted the widespread bottle pasteurization of Italian tank-fermented sparkling wines (Carpenè, 1944; Paronetto, 1952; De Rosa, 1964a). Paronetto reported that holding the bottled wine for two hours or longer at 40°C (104°F) inhibited yeast growth.

Bottling

The isobarometric bottling of tank-fermented or transfer-system sparkling wines is carried out with special equipment. Occasionally foaming creates a problem. Use of approved anti-foam agents may be necessary. The bottling should be conducted at the lowest practicable temperature in order to reduce loss of pressure.

One of the persistent problems in the bottling of tank-fermentation production of sparkling wines is their "sweating" after bottling and before labeling. E. A. Rossi, Jr. (1965), used infrared lamps in the far infrared region. Automatic timer control and various arrangements of the circuits helped to standardize the process and prevent overheating of the bottles.

Composition

The extreme variation in the composition of sparkling wines is illustrated by the following data of Edelényi (1964) on twenty-eight sparkling wines: alcohol, 7.2–13.0 per cent; titratable acidity, 0.40–0.78 gr per 100 ml (the best, 0.53–0.68); free sulfur dioxide, 0–44 mg per liter;

total sulfur dioxide, 38–245 mg per liter; acetaldehyde, 21–163 mg per liter; ratio of extract to ash, 7–9; and total nitrogen, 140–200 mg per liter. This indicates to us that the process of sparkling wine production is not as standardized as is commonly believed.

Using GLC, Rodopulo and Pisarnitskiĭ (1966) reported the following compounds in the aroma of three Soviet (tank-fermented?) sparkling wines (as percentage of total):

	Moscow	Leningrad	Yalta
Pentane[a]	34.93	16.00	28.12
Unidentified	—	1.86	2.05
Unidentified	—	1.55	—
Ethyl formate	3.31	2.4	1.68
Ethyl acetate	23.15	9.54	4.68
Unidentified	—	6.7	0.66
Ethyl propionate	4.95	3.67	1.8
Ethyl butyrate	5.8	3.1	2.7
Ethyl valinate	3.99	4.35	1.8
Isoamyl acetate	3.49	1.68	4.5
Ethyl caproate	6.4	5.18	4.95
Unidentified	—	1.24	—
Isoamyl butyrate	—	0.93	2.1
Unidentified	—	1.65	1.5
Ethyl lactate	8.15	31.0	35.0
Unidentified	—	1.65	4.5
Isobutyl caproate	4.96	—	2.23
Unidentified	0.86	—	—

[a] Used for extraction.

The concern of Soviet investigators with sparkling wine standards is indicated only briefly in the summary. Nevertheless, it is apparently of no less concern to Soviet technicians than to American. They reported more 1-propanol and 2-methyl-1-propanol and more 1-ethyl-1-propanol and high-boiling esters in French than in Soviet sparkling wines. There was also more acetaldehyde, isovaleraldehyde, and 1-heptanal in the French wines. All samples contained farnesol, linaloöl, and phenethyl alcohol.

Continuous systems

The first descriptions of continuous tank fermentation of sparkling wines were in the Soviet Union. (See Amerine, 1959, 1963a, 1963b; Broussilovskiĭ, 1959; Agabal'iants and Avakiants, 1966.) It is quite common in the Soviet Union to give the wine a preliminary heat treatment, either in the barrel or in tanks. This may last for two months at 50°C (122°F).

Agabal'iants and Avakiants (1966) studied the composition of the yeast sediment in the final stages of a continuous sparkling wine fermentation. They found higher amounts of total nitrogen, amino acids, enzymes, and other compounds in the sediment than in the supernatant wine. They believe these to be important to quality, but no comparative sensory data on the products were presented. All the yeast cells were dead at the end of the fermentation, which is surprising. Yeast strain, age of culture, cells adapted to wine, and use of added nutrients did not materially improve growth or fermentation. Use of higher inocula of adapted cells did give adequate fermentation rates. Agabal'iants and Avakiants (1965) noted that sterilization of flow lines and pasteurization of the cuvée was essential to eliminate undesirable activity of lactic acid bacteria. Avakiants and Belousova (1965) reported that continuous fermentations provided adequate β-fructofuranosidase for sucrose inversion.

To prevent dehydration of β-fructofuranosidase and other enzymes, Avakiants (1965) recommended that the preliminary heat treatment be not over 40°C (104°F). The main inversion of sucrose is enzymatic; acids do not give complete hydrolysis. Hydrolysis of sucrose in the finished wine, even when held at a low temperatue, is complete in 6 to 9 days, according to Avakiants. The procedure calls for a series of six to eight closed tanks. Wine, plus a heavy yeast inoculum, is introduced into the first tank. Wine is pumped from tank to tank under pressure. The number of viable yeast cells decreases markedly in the last two tanks, 5 or 6 (7 or 8), and the Soviet enologists indicate that considerable yeast autolysis takes place in these tanks. The Russian continuous system calls for a 0.5 to 1.5 mm layer of vaseline or paraffin oil on the wine to prevent adsorption of air. Yeast is added ($\geqq$ 30,000,000 cells per ml) to the first of six 4,750,—5,000-liter tanks. The fermentation starts at 13° to 14°C (45.4° to 47.2°F) and finishes in the last tank 9° to 10°C (48.2° to 50°F). The wine is cooled to 0–2°C (52°–35.6°F) and the dosage added; finally it is cooled to −5°C (23°F). In both of the last two tanks the wine passes through layers of ring-shaped caps made of polyethylene to catch yeasts. The product is thereby enriched with yeast autolyzate and its flavor improved, or so it is claimed. The process is covered by U.S. patent 3.062.656 (Agabalianz and Merzhanian, 1962). Agabal'iants and Avakiants (1965) reported a slight increase in volatile acids (0.02 per cent) and in pH (0.1 unit), and a decrease in total acidity, (0.01 per cent). They emphasize that the lines must be sterilized to prevent growth of lactic acid bacteria.

Kunkee and Ough (1966) were unable to induce adequate yeast growth when the inoculation was made under pressure. They attribute this to the low sugar concentrations they used compared to the high sugar concentrations used by the Soviet experimenters.

De Rosa (1964b) noted correctly the effect of pressure on yeast development (based on earlier data of Schmitthenner):

Pressure (atm)	Increase in pressure from fermentation	Yeast cells (millions per ml)
0	normal	104.3
2	13.5	14.9
3	11.6	1.0
4	9.1	6.3
5	6.6	3.2
6	2.8	0.8
7	1.6	0.2
8	1.3	0.09
9	0.0	0.03

The success of the Russian process is apparently due to the large yeast inoculum and possibly to the use of young wines. Further experiments would seem to be needed. Where many tanks are being used for the tank process, as in California, bottling takes place almost daily, and the process has some of the features of a continuous system.

CARBONATION . . .

. . . has been neglected as a method of producing low-priced sparkling wine.

Artificially charged wines are made from brilliant wines sweetened with the proper amount of sugar and then charged with carbon dioxide gas in a suitable carbonating apparatus. This usually consists of a chamber in which the gas is introduced, as a fine stream, into cold wine which is agitated. Or the wine may be charged by flowing from one bottle to another over a series of balls or baffles through which carbon dioxide is rising under pressure. For best results the wine must be cold (about 0.0°C; 32°F) and the apparatus properly constructed and operated. The primary problem in carbonating wines is to obtain a wine which has small bubbles of carbon dioxide and remains brilliant. Even when aged wines that have been pasteurized and refrigerated are used, the wine may throw a deposit after carbonation. This is one reason why carbonated wines are not aged. Special care should be directed toward preventing metal pickup during carbonation. Correia and Parro (1967) reported fermented sparkling wines to contain 115 to 140 mg per liter of aspartic acids, while carbonated wines had only 27 to 57. The respective glutamic acid contents were 96 to 102 and 35 to 49.5. These differences would not hold if yeast autolyzate were added to the carbonated wine.

Properly carbonated wines will not lose their gas immediately after opening. However, Liotta (1956) left carbonated and bottle-fermented wines open for seven days at 4.4°C (40°F). He then determined their carbon dioxide content. Carbonated wines had less than 0.23 gr per 100 ml of carbon dioxide (0.15 to 0.22 in eight samples); bottle-fermented wines had more than 0.26 gr per 100 ml (0.27 to 0.54 in thirteen samples).

Miller (1964, 1966) has shown that excellent carbonation can be achieved in about 24 hours be removing dissolved gases from the wine before carbonation. Miller (1966) views carbonation not so much as a dissolving of carbon dioxide in the wine and subsequent formation of carbonic acid as a fine emulsion of carbon dioxide in wine. He reports that the rate of escape of gas from the carbonated wine varies directly with the level of disturbance of the wine during carbonation. To obtain minimum escape rates Miller recommends degassing the wine before carbonation. He uses a vacuum for degassing, but notes that carbon dioxide (but not nitrogen) may be used. Degassing not only removes oxygen but also makes subsequent pressure measurements more reliable. In Miller's review (1964) the composition of the wine did not influence the solubility of carbon dioxide so much as the rate of approach to equilibrium during carbonation. Therefore, other things being equal, a high-sugar wine will take longer to reach a given gas content, and the gas will escape at a slower rate when the system is exposed to the atmosphere than a wine of low-sugar content.

The practical problems of low-level carbonation of wines were considered by E. A. Rossi, Jr., and Thoukis (1960). They observed that a wine carbonated to a given pressure at a high temperature contained less carbon dioxide than if carbonated at low temperatures. For this type of carbonation they found higher carbon dioxide at low soluble-solids content (0.236 gr per 100 ml at 3.5° Brix versus 0.219 at 11.2, 0.208 at 16.9, 0.192 at 22.7, and 0.179 at 27.80). Alcohol had only a small effect, with a decreasing solubility from 11.9 per cent alcohol to 19.5 (0.257 gr per 100 ml versus 0.241). They recommended the use of gauge pressure as a means of controlling the carbon dioxide content of the wine. The fact that their carbonator was not 100 per cent efficient accounts for their values being lower than those given by Etienne and Mathers (1956), which are for equilibrium head pressures. One must allow for the influence of various factors on carbon dioxide solubility. Nevertheless, they found it possible to carbonate two wines to contain the same amount of carbon dioxide, even though their gauge pressure was slightly different. The relation is apparently a straight-line function in this range:

Gauge pressure (lb per in².)	Carbon dioxide (gr per 100 ml)
8	0.174
12	0.212
16	0.249
20	0.286

The U.S. Internal Revenue Service (1956) experimentally determined values for wine (at equilibrium) as follows, at 15.56°C (60°F).

Pounds per sq. in.	Volume of carbon dioxide	Carbon dioxide (gr per liter)
0	0.95	1.866
1	1.00	1.964
2	1.08	2.121
3	1.15	2.259
4	1.23	2.416
5	1.30	2.560
6	1.35	2.652
7	1.40	2.750
8	1.45	2.848
9	1.52	2.986
10	1.65	3.241

TABLE 67
SOLUBILITY OF CARBON DIOXIDE AT VARIOUS PRESSURES
AND TEMPERATURES IN 10 PER CENT ALCOHOL AND WINE

Pressure (ATM)	Temperature (°C)	Wine (Bachmann, 1967) (gr/liter CO_2)	10 per cent Alcohol (Deinhardt, 1961) (gr/liter CO_2)	Wine (Deinhardt, 1965) (gr/liter CO_2)
1	0	2.7	4.0	1.8
2	0	5.2	8.0	4.1
3	0	7.5	12.0	6.7
4	0	9.6	16.0	9.3
5	0	—	—	12.0
6	0	—	—	—
1	5	2.3	3.3	1.0
2	5	4.4	6.6	3.3
3	5	6.4	9.9	5.6
4	5	8.3	13.2	7.8
5	5	—	16.5	10.1
6	5	—	—	—
1	10	2.0	2.7	0.7
2	10	4.0	5.5	2.6
3	10	5.8	8.3	4.7
4	10	7.5	11.0	6.6
5	10	9.2	13.8	8.5
6	10	—	16.5	10.5
1	15	1.8	2.3	0.5
2	15	3.6	4.8	2.2
3	15	5.2	7.1	3.9
4	15	6.8	9.5	5.7
5	15	8.2	11.8	7.4
6	15	9.7	14.2	9.0
1	20	1.6	2.1	0.3
2	20	3.0	4.2	1.8
3	20	4.5	6.4	3.4
4	20	5.8	8.5	4.9
5	20	7.3	10.7	6.5
6	20	8.6	12.8	8.0
7	20	9.8	—	9.5

The work of Deinhardt (1961) indicates that no accurate data on the solubility of carbon dioxide in wine are available. Subtraction of about 1.5 atm from the solubility of carbon dioxide in 10 per cent alcohol gave approximately correct values. The solubility of carbon dioxide at various temperatures and pressures is summarized in table 67. The data of Bachmann (1967) should be used, as they take into account the solubility of carbon dioxide in the base wine, whereas Deinhardt's (1965) data do not. The value at 15.56°C (60°F) of 1.866 corresponds closely to Bachmann's value of 1.80 at 15°C (59°F). He found no greater solubility of carbon dioxide in wine than in distilled water; in fact, his value of 1.65 gr of carbon dioxide for 20°C (68°F) at 760 mm compares with the text book value of 1.675 for water. Increasing the sugar content from 2.3 to 2.5 per cent lowered the solubility from 1.652 to 1.640. Increasing the alcohol content 2.8 per cent lowered the solubility from 1.652 to 1.590. Doubling the acid (from 0.68 to 1.22 per cent) had no measurable effect, but an increase in the sugar-free extract lowered the solubility by 0.1 gr of carbon dioxide per liter.

Miller (1958) has published a useful test table for determining the carbon dioxide equilibrium pressure at a chosen temperature. The actual

Temperature °C	°F	Base pressure	Temperature °C	°F	Base pressure	Temperature °C	°F	Base pressure
−3.9	25	0.870	10.6	51	1.393	24.4	76	1.948
−3.3	26	0.889	11.1	52	1.415	25.0	77	1.971
−2.8	27	0.907	11.7	53	1.437	26.6	78	1.993
−2.2	28	0.926	12.2	54	1.459	26.1	79	2.015
−1.7	29	0.944	12.8	55	1.481	26.7	80	2.037
−1.1	30	0.963	13.3	56	1.504	27.2	81	2.059
−0.6	31	0.981	13.9	57	1.526	27.8	82	2.082
0.0	32	1.000	14.4	58	1.548	28.3	83	2.104
0.6	33	1.018	15.0	59	1.570	28.9	84	2.126
1.1	34	1.037	15.6	60	1.593	29.4	85	2.148
1.7	35	1.056	16.1	61	1.615	30.0	86	2.171
2.2	36	1.074	16.7	62	1.637	30.6	87	2.193
2.8	37	1.092	17.2	63	1.659	31.1	88	2.215
3.3	38	1.111	17.8	64	1.682	31.7	89	2.237
3.9	39	1.129	18.3	65	1.704	32.2	90	2.260
4.4	40	1.148	18.9	66	1.726	32.8	91	2.282
5.0	41	1.170	19.4	67	1.748	33.3	92	2.304
5.6	42	1.192	20.0	68	1.770	33.9	93	2.326
6.1	43	1.215	20.6	69	1.793	34.4	94	2.348
6.7	44	1.237	21.1	70	1.815	35.0	95	2.371
7.2	45	1.259	21.7	71	1.837	35.6	96	2.393
7.8	46	1.281	22.2	72	1.859	36.1	97	2.415
8.3	47	1.304	22.8	73	1.882	36.7	98	2.437
8.9	48	1.326	23.3	74	1.904	37.2	99	2.460
9.4	49	1.348	23.9	75	1.926	37.8	100	2.482
10.0	50	1.370						

gauge pressure and temperature are determined. Add 14.7 to the gauge pressure. Divide by the base pressure at the temperature of observation. The result is the equilibrium pressure in pounds per square inch on the absolute scale at 0°C (32°F). To secure the equilibrium pressure at another temperature multiply this by the base pressure at that temperature from the tabulation on page 696.

To find the approximate carbon dioxide content in gr per 100 ml divide the equilibrium pressure in pounds per square inch on the absolute scale at 0.0°C (32°F) by 49.

CORKING AND CAPPING

High-quality, branded corks are favored.

The bottle containing sparkling or carbonated wine was traditionally closed with a cork of high quality, usually a four-piece cork with a "mirror." The mirror is a thin piece of cork of the finest quality, glued to the end of the cork that is in contact with the wine. The corks are rinsed in water containing 25 mg per liter of sulfur dioxide, washed in cold water (but not in hot), and drained before use. The cork is placed only about halfway into the bottle. These corks are often branded with the producer's name. Both hand-operated and power-driven corking machines are available.

Polyethylene closures have been used by many producers. These are satisfactory if the wine is not to be aged. They are not recommended for the highest-quality wine because of their tendency to allow loss of pressure. Consumers generally liked polyethylene stoppers for sparkling wines because of their ease of removal and their cleanness. Stephan (1964), however, reported that an appreciable number of consumers thought them untraditional. Sharf and Lyon (1958) found sparkling wines with polyethylene stoppers to mature faster than those with cork stoppers, probably because of the more rapid entry of oxygen. Ough and Amerine (1961c) found that some tasters preferred cork closure to plastic and others ʳhe reverse, but the over-all preference was for cork. For sparkling wines that are sold and consumed while young, they believed that plastic closures offered an economic advantage.

Geoffroy (1963) noted that sparkling wines with plastic closures were better after aging if they had had 40 mg per liter of ascorbic acid added before and after the bottle fermentation and 20 mg per liter of sulfur dioxide in the latter case, compared to those without these additions. This, we take it, is another indication of the penetration of oxygen through plastic closures. He also recommends addition of 15 mg per liter of sulfur dioxide in the final dosage when the wine after disgorging shows any tendency to darkening of color. This is also advisable when the wine is to be subject to high temperatures during transportation.

Dubois (1955), in contrast, showed that the potential pressure loss of polyethylene stoppers in sparkling wine bottles was not excessive—about 1 atm in six years. The question of exchange of oxygen, across the pressure gradient, remains. One advantage of polyethylene stoppers for sparkling wines appears to be that there are fewer leakers. Another is the possibility of reinsertion of the polyethylene stopper.

To hold in the cork or polyethylene closure, a wire cap (muselet) is placed on a metal plaque over the cork. A large lead, plastic, or paper foil capsule is put over the entire end of the bottle when the wine is labeled, so that the wire and cork do not show. A strap-type wire hood is often used for plastic corks (fig. 116). After corking and wiring, the bottle should be shaken so that the sugar and wine are well mixed.

Figure 116.   Strap-type wire hood used for plastic closures. (Courtesy Budde & Westermann Machinery Corp.)

AGING AND STORING
Aging is less important after disgorging.

Champagnes and similar effervescent wines undergo aging in two stages, before and after disgorging. Practice in this respect varies. Champagnes are preferably allowed to age for one to three years in contact with the lees in the bottle. Little aging is then required after disgorging, although very high-quality dry sparkling wines may show marked improvement for several years (Amerine and Monaghan, 1950). In some California wineries the champagnes are disgorged young and are then aged in the bottle, although this is theoretically less desirable (see p. 671). Carbonated wines usually are not aged after carbonation. Bulk-fermented wines can be aged in the bottle only after removal from the tank. They improve during aging as the free sulfur dioxide decreases markedly.

Champagnes and other effervescent wines stored on their side should,

Figure 117.   Surface drier for sparkling wine bottles. Used in preparation for labeling. (Courtesy Budde & Westermann Machinery Corp.)

if properly cooled, remain in good condition for at least ten years. The limiting factor is the loss of pressure caused by leakage around the corks.

Moisture condensation on sparkling wine bottles, particularly those from the tank- or bottle-fermented process, makes label application difficult. Surface driers are available to insure a dry surface (fig. 117).

TASTING, ANALYSES, AND WINE DISORDERS

CHAPTER EIGHTEEN

TASTING AND ANALYZING

Sensory and laboratory examinations must be made before bottling or shipping in bulk.

TASTING . . .

. . . is neglected in some wineries, but will yield valuable results if properly conducted.

Regular examination of the wines in the cellar by qualified wine tasters should be standard winery practice. By this means the wine maker can follow the development of the wine, detect incipient spoilage, establish the type and quality of the wine, prepare blends, and eventually decide upon the necessary treatment and blends as well as upon the time for bottling.

European enologists have generally been reluctant to use modern panel tests. (Ribéreau-Gayon and Peynaud, 1960–1961, 1964–1966; Flanzy, 1965). Nevertheless, many French regulations call for or imply some kind of sensory examination. Flanzy would have the sensory examination precede the chemical analysis and be a sort of guide to the type of analysis needed. He also noted, correctly, that in judging ordinary wines one must evaluate them as they are, but for high-quality wines one needs to take into account what they may become. The current interest in sensory evaluation of foods in France is indicated by a symposium on this subject held in Dijon, sponsored by the Institut Technique du Vin (1967). Some of the papers given at the symposium recognized the need for statistical analysis of the results, but others did not. One, at least, likened the taster presented with a fine wine as being like an explorer in a dark cave! (For a perceptive French symposium on the same subject see Anon., 1966c.) Amerine (1961) has emphasized the practical and legal aspects of sensory evaluation of alcoholic beverages. He believes that sensory tests made by standard procedures and using proper statistical analyses would be accepted by the courts. Even without this acceptance, they are useful where the quantities are small and the effect on quality is large. He indicated the need for better tests to distinguish the quality of wines from various regions, varieties, and processes. There is also a need for determination of the sensory threshold for many compounds and for evaluation of their interrelationships.

Troost (1965) stresses three objectives of sensory evaluation of wines: (1) the care of the wines at the winery, (2) quality evaluation, and (3), judging by governmental food agencies. For the first objective, chemical analysis should be used to supplement the sensory data. If qualified personnel and adequate laboratory data (particularly stability tests) are available, no difficult problems should arise. In the second objective, Troost (1965) noted the difficulty of comparing wines of different regions or seasons. He emphasizes the psychological factors which influence

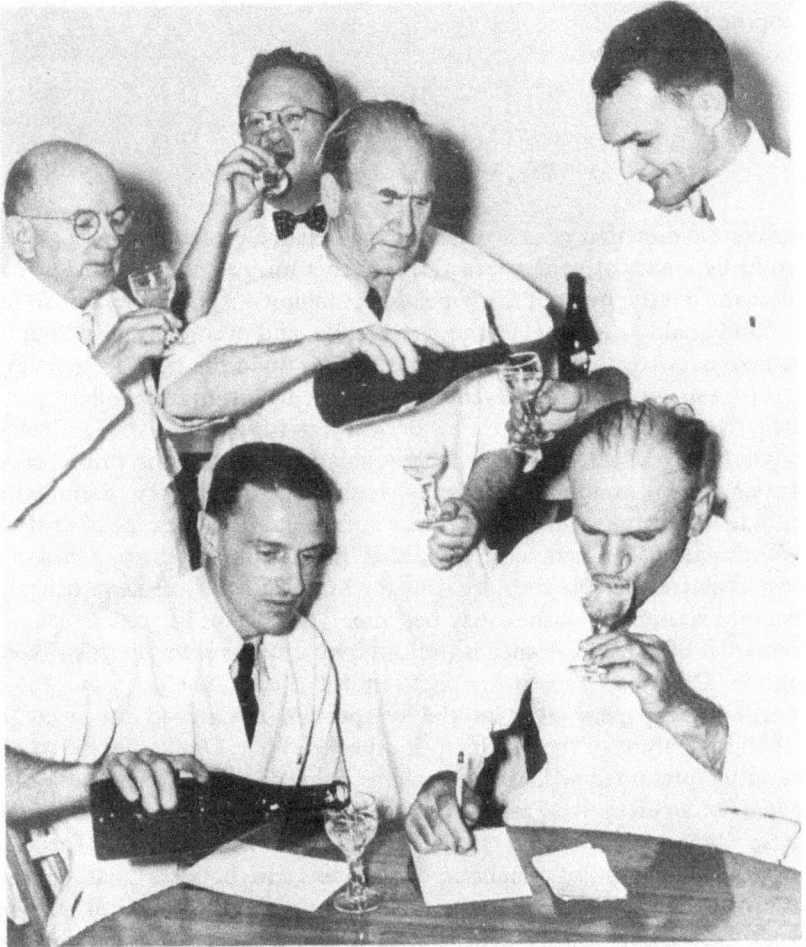

Figure 118. Tasters at the California State Fair about 1937. From left to right: Charles Wetmore, Professors George L. Marsh, Maynard A. Joslyn, William V. Cruess, A. J. Winkler, and Maynard A. Amerine.

sensory panels, particularly information on source or price, where analytical data are of little value. Wines with the same analysis may have very different tastes and odors. But, as Troost said, quality and market success are not necessarily related. In the third objective (not often used in this country) the judges are asked to assess the relative merit of a group of wines in relation to normal conditions. Is the wine deserving of a *Spätlese*? Is the year a bad, good, or excellent one? Analytical data are useful in reaching a decision, particularly for abnormal wines. (See fig. 118 for an early post-Prohibition photograph of wine evaluation at the California State Fair.)

Jakob and Schrodt (1966) note that in storage experiments it is difficult to obtain absolute judgments of wine quality because of the changing quality of the samples during storage. They also found scoring to give more information than paired comparison tests. This is evidence of European consideration of the techniques recommended by Amerine *et al.* (1959b, 1965a).

Not all European enologists are averse to modern procedures of sensory evaluation of wine. Daepp (1966–1967) gives a critical and perceptive survey of the value of panel testing and a statistical evaluation of the results. For a list of terms see Vedel (1966). See also Joslyn and Amerine (1964) and Paul (1964). For a general survey of the sensory evaluation of foods see Amerine *et al.* (1965a) and A. D. Little (1958).

Ribéreau-Gayon (1964) and Peynaud (1965) had called attention to the effects of one component on the perceived sensations of others. Sugars weaken the acid taste, acids accentuate the bitter taste, and increasing volatile acidity (above 0.06 to 0.08 gr per 100 ml) decreases smoothness, as does ethyl acetate (above 150 mg per liter). For a more detailed study see Berg *et al.* (1955a, 1955b) and Hinreiner *et al.* (1955a, 1955b).

The influence of the environment on the results was studied by Foulonneau (1967). He investigated the influence of light, sound, interfering odors, and temperature. Since the results are discussed in subjective terms, without statistical analysis, it is difficult to evaluate them. However, some of the general results seem plausible or at least worth detailed experimentation. Reduction in light (i.e., cloudy), he believes, results in more agreeable odors in white and red wines compared to those under ordinary light. One finds it difficult to believe that pleasant music could change the reaction to odor if the judges were at all sophisticated. One would also like to have detailed information on the deformation of taste allegedly caused by smoking two cigarettes. The presence of butyric acid could, of course, have disagreeable effects on the judges. Foulonneau preferred 26°–27°C (78.8°–80.6°F) to 14°–15°C (57.2°–59°F) for tasting, and this seems reasonable.

Chemical analyses should be available when examining the tasting results, as they may confirm those results with regard to the soundness of the wine. Analyses may or may not confirm the tasting results.

Appearance

The first step in inspecting a wine is to examine its appearance (figs. 119, 120). This frequently establishes the condition of the wine. Often the nature of the sediment or cloudiness will indicate the specific, immediate treatment necessary. Various diseases have characteristic forms of clouding which the taster soon learns to recognize, particularly if working with a limited number of wines from a single source: for example, floating white particles indicate a film yeast infection; certain wines infected with *Lactobacillus* spp. have a silky cloudiness which is entirely distinct from ferric phosphate casse (see chap. 19).

Figure 119. Glasses (left to right): first two for sensory evaluation of odor, for sparkling wines, last two for red or white wines.

The general appearance is usually classified as brilliant, clear, dull, or cloudy, according to increasing amounts of suspended material. Old wines that have been aged in the bottle have normal deposits of color and tannins, but the wine can be poured off in a brilliant condition, particularly if the bottle is not shaken when opened or is set on end a day or two before the cork is removed. This deposit is not, of course, a defect and is almost a guarantee of the age of the wine. However, on the highly competitive American wine market such a wine would encounter consumer resistance except from a limited clientele. Nephelometers are used to detect changes in appearance during filtration.

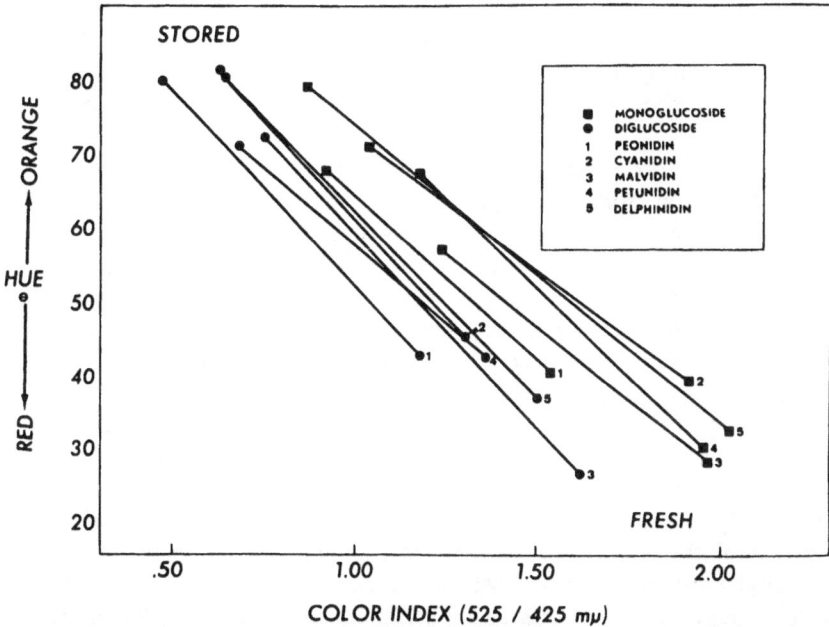

Figure 120. Relation of hue to color index (525–425 mμ ratio) for anthocyanin monoglucosides and diglucosides in wine before and after storage at 48.9°C (120°F). (From Robinson *et al.*, 1966.)

From the color, the taster can gain additional information. White wines that have become brown have usually been fermented on the skins too long, overaerated during cellar operations, or possibly overaged. Old white wines of good quality turn darker with age, but the color is more golden than brown. Old red table wines show a slight browning, quite characteristic and unobjectionable if not too pronounced. High-acid (or rather low-pH) wines always have an especially bright-red color not found in low-acid, high-pH wines. The first indications of the type of wine are given by the color. Light- and dark-colored types are quickly differentiated. Certain wines have distinctive tints, easily recognized. Where dyes have been added, these can sometimes be identified. Since many people "buy with their eyes," it is important to have a wine with an impeccable appearance. Each type should have a uniform and proper depth and shade of color.

White wines are usually graded from light yellow to gold, with green and amber as modifying tints. Red wines are marked as pink to deep red, with violet and brown the commonest modifying shades.

Odor

The difference thresholds for a number of odorous materials present in beers, as measured in beer, have been summarized by Harrison (1963). Difference thresholds in wine may be higher or lower, and research on olfactory thresholds of the volatile constituents of wine is urgently needed.

In order of difference sensitivity for a small laboratory panel these were: hydrogen sulfide (0.05), diacetyl (0.005), octaldehyde (0.001), ethyl isobutyrate (0.01), indole (0.1), linaloöl (0.1), methylnonyl ketone (0.1), caprylic acid (0.5), butyric acid (1), isobutyric acid (1), valeric acid (1), isovaleric acid (1), phenylacetic acid (1), butyraldehyde (1), isobutyraldehyde (1), ethyl propionate (1), isobutyl acetate (1), isoamyl acetate (1), acetal (1), 2-butanol (5), caproic acid (5), propionaldehyde (5), ethyl acetate (5), ethyl lactate (10), and acetaldehyde (25). The figures in parentheses are parts per 10^6. Other compounds, such as other acids and alcohols, had higher thresholds. Thresholds for 3-methyl-1-butanol in dry white wine varied from 100 to 900 mg per liter (average 300) for seven judges, according to Rankine (1967a). 2-Methyl-1-propanol and 1-propanol thresholds were over 500 mg per liter. The threshold for 3-methyl-1-butanol in water was 4 mg per liter. There is sufficient for certain judges to detect this alcohol. Differences in 2-methyl-1-propanol and 1-propanol in the range present in wines were not detected by seven panelists used. Kendall and Neilson (1966) showed that the odor of binary and quaternary mixtures cannot be predicted from those of the original components; synergistic, blending, and masking effects may occur.

Difference thresholds, mainly for taste components, in wine have been measured by Hinreiner et al. (1955a, 1955b) and Berg et al. (1955a, 1955b).

After careful visual inspection, the wine is smelled. Many diseases impart a characteristic odor or taste which is quickly recognized by the experienced judge. Wines with high volatile acid, hydrogen sulfide, sulfur dioxide, and the like can all be distinguished by smell alone. Other odors or tastes that indicate diseases or defects are mousy, moldy, corked, metallic, and bitter. If the odor indicates serious disease or defects, further testing is usually unnecessary.

Rankine (1966b) reported an off-flavor in Australian dry white table wine. The flavor was reported to be "yeasty, aldehydic". It was associated with growth of Pichia membranaefaciens. It grew in wine up to 11 per cent alcohol and did not tolerate any free sulfur dioxide. Acetaldehyde, ethyl acetate, isoamyl acetate, and an unidentified (but quantitatively important) compound were reported.

In California the "rubbery" off-odor described by Brown (1950) is associated with the fermentation of high pH musts. The origin of the "earth" odor is not known, but it is definitely associated with grapes grown in certain regions (e.g., Davis, California).

The wine should not be sniffed continuously. Olfactory fatigue or insensitivity to the odor—even to a pronouncedly bad one—occurs very quickly. Thus a lesser quantity of the same spoilage product in the next wine tested may pass unnoticed.

Aroma refers to odors which are derived wholly or in part from the grape, such as Concord or muscat, and when recognized should be noted by the taster. A well-trained judge should be able to identify the aroma of the common varieties, such as Cabernet, Zinfandel, Sauvignon blanc, and Sémillon. Wines without a distinct aroma are marked as vinous. Bouquet is the aged odor characteristic of fine wines that have been aged in the bottle for one or more years.

Flavor

The actual tasting should be made with a very small quantity of liquid and only after the visual and olfactory examinations have been completed. By drawing air in and over this liquid in the mouth, the more elusive aromas are brought out. Sweetness, astringency, metallic flavors, and other taste qualities can be identified. The degree of smoothness of a red wine is frequently an indication of how long it must be aged before bottling. On the basis of the appearance, odor, and flavor, the taster must be able to evaluate the quality of the wine. Naturally this decision must conform to the type of wine being judged. The odor and flavor appropriate for one type of wine may not be so for another.

Münz (1963) notes that the acid taste of wines is due primarily to the acid salts, since most of the acids are partially neutralized. Wines in which an appreciable part of the acids are not bound to minerals are generally too acid in taste. Soluble salts seem to reduce the acid taste. The buffering capacity of musts and wines is expressed in terms of potassium equivalent: 1 gr per liter of potassium versus 0.31 gr of magnesium and 0.59 gr of sodium. Münz favors adjusting the acidity of musts in preference to that of wines. If the acid taste of wines needs to be adjusted (as it frequently does in Germany) he prefers to do this by blending.

As Münz (1965a) has indicated, the amount of must potassium is of critical importance to the acid taste of the resulting wine. This is due, obviously, to the reduction in acidity which occurs when large amounts of acid tartrate precipitate. He also notes that the mineral acids, sulfurous and phosphoric, have a hard acid taste. The free organic acids have a sharp acid taste, while their acid salts have a soft and delicate acid taste. Both the amounts of acid and the buffer capacity of the wine influence the degree of acid taste. Münz added sulfuric acid and sulfuric and phosphoric acids together to wine. Sulfuric alone reduced the potassium content and increased the potassium sulfate content. Addition of both acids reduced the organic magnesium content and the potassium content, and increased the phosphate and sulfate contents. The original wine was slightly flat. Adding the acids decreased the buffer capacity and increased the hard acid taste. Münz considers that the organic bound potassium has only a buffer value, whereas the change of organic potassium from the organic to the inorganic form improves the fruity acid taste without changing the amounts of tartaric or malic acid. Munz (1967) has emphasized that the acid taste depends not only on the concentration of the acid but on its buffer capacity and on the effect of sugar and other compounds. The most important buffering is done by potassium, with lesser effects of calcium, magnesium, and sodium.

The relative sourness of malic, citric, tartaric, and lactic acids in wines was determined in two ways by Amerine et al. (1965b). At the same titratable acidity the order of sourness was malic, tartaric, citric, and lactic. At the same pH the order was malic, lactic, citric, and tartaric. A noteworthy feature of this report was the relatively small differences in pH (0.05 pH unit) and titratable acidity (0.02 to 0.05 per cent) which could be detected by the panel. They concluded that both pH and titratable acidity were important in determining sensory response to sourness. In the study of Ough and Kunkee (1967) acid taste correlated well with the logarithm of the titratable acidity. However, in these studies quality and total acidity or pH were not well correlated—possibly because all the wines underwent a spontaneous malo-lactic fermentation.

The sweet taste has been extensively investigated in wines (see Amerine et al., 1965a). Most California wineries now carefully control the sugar content of all the types of wines they produce. Amerine and Ough (1967) studied the sweetness preference for rosé wines of a group of twelve experienced judges. They found that six preferred drier types; five, sweeter types; and one had a broad range of acceptability. This bimodal type of distribution indicates that wineries may have to produce rosé wines with two levels of sugar in order to win maximum consumer acceptance. For interrelationships of taste see Pangborn et al. (1964) and Amerine et al. (1965a).

Comparing results with analyses

The tasting results should be compared with a chemical analysis of the wine. It is psychologically important that the taster not look at the analytical record before tasting. Accurate tasting may be used as a guide to the extent of chemical analysis necessary, and the two together act as the rational basis for any treatments necessary for the wine. To gain confidence and skill, the taster should record his results and check them by a later, independent tasting.

Records and scoring

In spite of the wealth of information on how to conduct sensory examination of wines, Ough (1959a) reported very haphazard practices in the California wine industry. The fact that only 25 per cent of the California wineries answered his questionnaire would indicate widespread indifference to critical sensory examination of wine in the industry. Our experience is that this is true in Europe also.

Many tasters find it useful to keep a permanent record of their successive tastings for each wine. This is helpful in determining the duplicability of the taster's results and also in detecting desirable and undesirable changes in the wine during aging. To facilitate this record it is desirable to have some form of tasting card and to systematize the scoring. The suggested score cards shown below provide for a complete sensory examination of the wine. A more detailed subdivision of the score card may be made if desired. In using a score card it is important that the previously recorded data on a given wine should not be seen by the taster immediately before or during the scoring. This can be accomplished by covering the used portions of the card or by having the taster dictate his remarks to a secretary.

	Suggested number of points			
	Amerine et al.	*Blaha*	*Cruess (1947)*	
Characters judged	*(1959b)*	*(1951)*	*Dry white*	*Dry red*
Appearance	2	12	20	15
Color	2	12	10	10
Aroma and bouquet	4	12	10	15
Volatile acidity	2	—	15	15
Total acidity	2	—	10	10
Sugar	1	—	10	10
Body	1	12	—	—
Flavor	2	20	20	—
Tannin and astringency	2	—	10	10
General quality	2	12	—	15
Varietal character	—	20	—	—

Since many food products are judged on a percentage basis, it has been suggested that wines be so judged (Cruess, 1947). Although scoring systems from 0 to 100 are frequently proposed, it is difficult for inexperienced judges to distinguish a range of more than 10 to 15 grades, while 20 to 25 is about the usable range possible for experienced judges. Ough and Baker (1961) recommended a scale of 9 steps: average quality with defects, 2–3 points; average, 4–6; above average 7–8; and superior, 9. (See Amerine, 1948a; 1948b; and Amerine et al., 1959b, 1965a, for further information.) If a 100-point system is used, most wines will be found to fall in the range of 65 to 85.

The first scale of André et al. (1963) included 200 points! Later they presented three scales: one for red wines, one for rosé and white wines, and one for high-quality wines. The red wine scale was color 5, appearance 5, bouquet 10, bouquet (persistence) 10, taste-structure (balance?) 8, taste-softness 6, mouth-odor (after-taste?) 6, for a total of 50. The rosé and white wine scale was color 5, appearance 5, aroma 10, aroma (persistence) 10, fruitiness 8, freshness (acidity) 6, and elegance and harmony 6. For fine wines with expert panels they recommended the following scale: odor 30 (10 each for nature and number of components, richness and finesse of aroma, and identity of origin), taste 30 (10 for the basic tastes and 20 for the form of the wine, to include form of the structure, amplitude and mobility of taste sensations), mouth odor 20 (10 for the nature, finesse, and richness of the mouth odor and 10 for the persistence-taste plus odor), and total impression 20 (the relation between the sensations) for a total of 100.

We feel that panels would need considerable training to understand the proper use of these terms. We also question scales of 50 to 100 points. The scale of the Office International de la Vigne et du Vin (1961) of 20 points for table wines is acceptable, but the 5-point range for special wines seems too narrow.

Use of an anchor or reference sample improved performance in comparison with a straight score card in some tests but not in others, in the experiments of Baker et al. (1965b).

For determining the significance of differences in scores, Amerine et al. (1959b) recommended use of analysis of variance. The F-distribution provides an over-all test of significance among the different means. For determining the significance of the differences between means the multiple-range test is used. For specific instructions on these procedures see Amerine et al. (1959b, 1965a). An application of statistical analysis to sensory data has been made by Marie et al. (1962). Both complete and incomplete blocks were used.

Systems based on degree of like or dislike, called hedonic, have recently been used. A typical scale would be:

Like very much _____   Dislike slightly _____
Like much _____   Dislike much _____
Like slightly _____   Dislike very much _____
Neither like nor dislike _____

An excellent example of the use of the hedonic system is the study of Huglin and Schwartz (1960) on grape juice. One advantage of hedonic systems is that they can be used with untrained judges. The hedonic scores are converted to figures and analyzed in the same way as the data from score cards.

Ranking methods are often useful for arranging a group of wines in the order of a specific characteristic. A series of wines might be arranged in the order of color, acidity, sweetness, or over-all quality. For tests to be meaningful, the judges must all be ranking on the same basis, that is, color in order of commercial acceptability, or quality based on experienced judges. Ranking gives no information on the degree of difference between any two wines. Paul (1967) praised rank-order as a method of distinguishing between wines, but had the wines separately ranked according to specific characteristics. Amerine and Ough (1967) found ranking and hedonic scoring gave essentially the same results in preference for sweetness in rosé wines. However, hedonic scoring appeared to give a larger scatter in results and a higher sugar preference. Ranking data can be evaluated statistically: either for the agreement in ranking between judges or as to how well a judge has ranked in comparison with some standard. For details and examples see Amerine *et al.* (1959*a*, 1965*b*).

Often the enologist is interested only in determining whether there is a difference between two wines. The difference may be due to processing, blending, aging, origin of grapes, fermentation procedures, or other factors. A number of techniques have been devised for such tests. Where the parameter of difference is known (sweetness, sourness, etc.) the paired test is favored. In this test the judge is given two samples and is asked, for example, which is sweeter, more sour, more muscaty. On the basis of the null hypothesis (that there is no difference between the wines), the probability of a taster identifying the sweeter wine by chance in each of several trials is 1 to 2. To determine the significance of the difference the χ^2 distribution is used. If N is the total number of tests (and N should be 7 or more) and X_1 the number of tests favorable to wine No. 1 and X_2 the number of tests favorable to wine No. 2, then

$$\chi^2 = \frac{([X_1 - X_2] - 1)^2}{N}$$

When the calculated value of χ^2 equals or exceeds 2.71, this indicates significance at the 5 per cent level ($p = 0.05$). In other words, this result would occur only once in twenty times by chance. If χ^2 is 5.40 or more, significance at the 1 per cent level, $p = 0.01$, is indicated; 9.55

or more, at the 0.1 per cent level, $p = 0.001$. These values are for identification of a known difference (a one-tailed test). The probabilities are also given in table 68.

TABLE 68

Significance of Paired Tests ($p = \frac{1}{2}$)

No. of judges or judgments	Minimum correct judgments to establish significant differentiation (one-tailed test)			Minimum agreeing judgments necessary to establish significant preference (two-tailed test)		
	Probability level[a]					
	0.05	0.01	0.001	0.05	0.01	0.001
7	7	7	7	7	—	—
8	7	8	—	8	8	—
9	8	9	—	8	9	—
10	9	10	10	9	10	—
11	9	10	11	10	11	11
12	10	11	12	10	11	12
13	10	12	13	11	12	13
14	11	12	13	12	13	14
15	12	13	14	12	13	14
16	12	14	15	13	14	15
17	13	14	16	13	15	16
18	13	15	16	14	15	17
19	14	15	17	15	16	17
20	15	16	18	15	17	18
21	15	17	18	16	17	19
22	16	17	19	17	18	19
23	16	18	20	17	19	20
24	17	19	20	18	19	21
25	18	19	21	18	20	21
30	20	22	24	21	23	25
35	23	25	27	24	26	28
40	26	28	31	27	29	31
45	29	31	34	30	32	34
50	32	34	37	33	35	37
60	37	40	43	39	41	44
70	43	46	49	44	47	50
80	48	51	55	50	52	56
90	54	57	61	55	58	61
100	59	63	66	61	64	67

[a] $p = 0.05$ indicates that the odds are only 1 in 20 that this result is due to chance; $p = 0.01$ indicates a chance of only 1 in 100; and $p = 0.001$, 1 in 1000.
Source of data: Roessler et al. (1956).

If the nature of the difference between the two samples is not known, the duo-trio or triangular test is favored. In a duo-trio test three samples are presented. One is labeled as the reference sample and the other two are coded. One is the same as the reference and the other is different. The judge's decision then is concerned with which of the two coded samples is different from the reference sample. The analysis is exactly as for the paired test, or table 68 may be used.

If, in addition to identifying the different sample, the judge is asked which of the two he prefers, the test becomes a two-tailed procedure and the right-hand columns in table 68 are used for determining the significance.

In the triangular system three glasses are presented, two of which, however, contain the same wine and the other a different wine. The judge's problem is to identify which of the three glasses contains wine which is different from the other two. If the wines are A and B the wines could be poured in any of six orders: AAB, ABA, BAA, BBA, BAB, or ABB. The chance of identifying the odd sample correctly is $\frac{1}{3}$ ($p = \frac{1}{3}$) and the χ^2 value is

$$\frac{([4X_1 - 2X_2] - 3)^2}{8N}$$

N should be at least 5. The same values for significance apply as for the paired test. It is simpler to determine the significance by use of table 69.

It is difficult to state concisely the factors which constitute quality in table wines. Experienced enologists acquire a knowledge of those factors by constant testing of a wide variety of wines and by observing other tasters. All wines offered for sale should be free of such obvious defects as high volatile acidity, excessive sulfur dioxide, cloudiness, and inappropriate color. Wines of quality should also have a color and aroma appropriate to the type, a balanced and desirable flavor, and a pleasing aftertaste. The most common defect of California wines is their lack of a distinctive aroma or bouquet rather than the presence of any particular disease or defect.

Although the relation between chemical composition and quality cannot be completely stated because of our lack of information about the nature and quantity of many chemical substances which contribute to quality, some relations may be quantitatively stated. Volatile acidities of over 0.10 per cent, particularly in new wines, are very objectionable. (See p. 440.) A volatile neutral ester content of over 150 to 200 mg per liter as ethyl acetate is likewise undesirable. Red wines with a tannin content of over 0.20 per cent, or white wines with over 0.05 per cent, usually taste objectionably astringent. Table wines, particularly when new, will taste too alcoholic if the alcohol content is over 13 per cent, especially if the glycerin content is low—below 0.5 per cent.

The results of Filipello (1955, 1957a, 1957b), as summarized by Filipello and Berg (1959), demonstrated that experienced judges were significantly better than inexperienced judges in evaluating wine quality. However, use of a reference sample was essential for both groups. Filipello also preferred attitude rating scales to numerical scoring scales.

TABLE 69

SIGNIFICANCE OF TRIANGULAR TESTS ($p = \frac{1}{3}$)

No. of judges or judgments	Minimum correct judgments to establish significant differentiation			No. of judges or judgments	Minimum correct judgments to establish significant differentiation		
	$p = 0.05$	$p = 0.01$	$p = 0.001$		$p = 0.05$	$p = 0.01$	$p = 0.001$
5	4	5		56	25	28	31
6	5	6	6	57	26	28	31
7	5	6	7	58	26	29	31
8	6	7	8	59	27	29	32
9	6	7	8	60	27	30	32
10	7	8	9	61	27	30	33
11	7	8	10	62	28	30	33
12	8	9	10	63	28	31	34
13	8	9	11	64	29	31	34
14	9	10	11	65	29	32	34
15	9	10	12	66	29	32	35
16	9	11	12	67	30	32	35
17	10	11	13	68	30	33	36
18	10	12	13	69	30	33	36
19	11	12	14	70	31	34	37
20	11	13	14	71	31	34	37
21	12	13	15	72	32	34	37
22	12	14	15	73	32	35	38
23	12	14	16	74	32	35	38
24	13	14	16	75	33	36	39
25	13	15	17	76	33	36	39
26	14	15	17	77	33	36	39
27	14	16	18	78	34	37	40
28	14	16	18	79	34	37	40
29	15	17	19	80	35	38	41
30	15	17	19	81	35	38	41
31	16	17	20	82	35	38	42
32	16	18	20	83	36	39	42
33	16	18	21	84	36	39	42
34	17	19	21	85	36	39	43
35	17	19	21	86	37	40	43
36	18	20	22	87	37	40	44
37	18	20	22	88	38	41	44
38	18	20	23	89	38	41	44
39	19	21	23	90	38	41	45
40	19	21	24	91	39	42	45
41	20	22	24	92	39	42	46
42	20	22	25	93	40	43	46
43	20	23	25	94	40	43	46
44	21	23	25	95	40	43	47
45	21	23	26	96	41	44	47
46	22	24	26	97	41	44	48
47	22	24	27	98	41	45	48
48	22	25	27	99	42	45	49
49	23	25	28	100	42	45	49
50	23	25	28	200	79	83	88
51	24	26	28	300	114	120	126
52	24	26	29	400	150	156	164
53	24	27	29	500	185	192	200
54	25	27	30	1,000	359	369	381
55	25	27	30	2,000	702	717	733

Source of data: Roessler et al. (1948). However, a number of minor corrections have been made in the data by Roessler.

In tests at the California State Fair, Filipello (1957a) showed consumer preference for sweetened red table wines, for rosés of 1 to 3 per cent sugar over dry rosés, low-alcohol (7 to 8 per cent) muscatel, and Concord wines at a carbon dioxide pressure of 20 psi over still. Filipello and Berg (1959) found no significant preference for wines sweetened with sucrose, invert sugar, or grape concentrate. Even in a very large test 53 per cent preferred sucrose-sweetened wine as against 47 per cent otherwise sweetened. The State Fair audience was 56 per cent males, with 47 per cent in the

21–35 age group. As far as drinking wine was concerned, 34 per cent drank some several times a week, 41 per cent a few times a month, 21 per cent a few times a year, and 2 per cent never. Older participants preferred wines more than did younger participants. Filipello and Berg were able to show that the "infrequent" users had wine available in their homes. They implied that these did not become regular users of wine because they did not like the taste of conventional wine types. Therefore, they believed that wine sales could increase only by development of new types. Many attempts to do this since 1959 have not yet caused a significant rise in per capita wine consumption in this country.

COLOR AND PIGMENT EVALUATION...

... could be improved by the use of better equipment and more specific procedures.

The color of a wine is one of its most important attributes. Standardization of the color of each nonvintage wine is an important duty of the careful wine maker. Unfortunately, color specification is not simple. The perception of color involves a source of energy, an object, and an observer. With wines, it is the observer's concept of color which interests us. To specify a color completely it is necessary to measure three of its attributes: (1) dominant wave length, which corresponds to the hue; (2) purity, which refers to the degree of saturation of the color and is the attribute of a chromatic color which determines its degree of difference from the achromatic (gray) color of the same luminance; and (3) luminance (or brightness), which is related to the ratio of the amount of light leaving an object compared with that which is incident upon it. More specifically, luminance is the attribute of a color which classifies it as a member of a series of achromatic colors (that is, of gray colors ranging from black to white).

Methods available

To determine the dominant wave length, purity, and luminance of a wine, the spectrophotometer is commonly used, and a transmission curve is prepared for the visible spectrum. From this transmission curve the tristimulus values for red and green are then determined for a standard illuminant, using the methods of Hardy (1936). The luminance (brightness), obtained directly from the tristimulus values for yellow, is equal to the ratio of the sum of the tristimulus values for yellow actually determined divided by the sum of the theoretical values for 100 per cent brightness over the same range for the same standard illuminant. This is possible because the tristimulus values for yellow have been made to fit the visibility curve of the eye. The dominant wave length and purity are then determined by graphic methods. The procedure is rather tedious, and most wineries do not have a spectrophotometer of sufficient sensitivity to obtain a detailed transmission curve. It is the only instrument, however, which makes possible a complete and accurate specification of the color of a wine, and for certain research problems—for example, determination of the influence of a process or treatment on the color of a wine—it is indispensable. Since the color of wines is due to two or more constituents, each with a different transmission curve, spectrophotometers are particularly useful in specifying color.

Ough and Berg (1959) showed that extremely small differences in color, as determined from tristimulus data, could be differentiated by their panel. A fluorescent light (Sylvania, 40-watt, cool white standard with white reflector) was significantly better for identifying small differences than daylight or incandescent light (a special 500-watt incandescent globe with a heavy blue filter). There was a one-blend preference shift under fluorescent light compared to the other two. Orange tints were especially disliked under fluorescent light.

Sudraud (1958) proposed using the ratio of the optical density at 420 mμ to that at 520 mμ as a measure of the tint or hue, and the sum of these optical densities (420 + 520 mμ) as a measure of luminance (or intensity of color). The former has been accepted by the Office International de la Vigne et du Vin (1962). However, for intensity they use only the optical density at 520 mμ.

The ratio of the optical density at 520 mμ to that at 420 mμ has been used by many investigators as a measure of the rate of browning of red wines. Robinson *et al.* (1966) caution against using such a ratio for comparing different wines. The reason is that the ten or more anthocyan pigments have optical metamers; so it is possible for two wines of varying pigment composition to have the same hue but widely different optical density ratios. The changes in hue compared to optical density ratio during storage are shown in figure 120.

In an attempt to determine luminance and dominant wave length of red wines without the laborious calculations of Hardy (1936), Ough *et al.* (1962) found that the sum of the optical densities at 420 and 520 mμ usually gave values within 3 per cent. The dominant wave length could be estimated to ± 3 mμ from a plot of the ratio of the optical densities (420 to 520 mμ) versus the dominant wave length between 590 and 630 mμ. Color matches of blends could be predicted only when both wines had similar dominant wave lengths.

Alexiu (1967) compared the abbreviated procedures for colors as

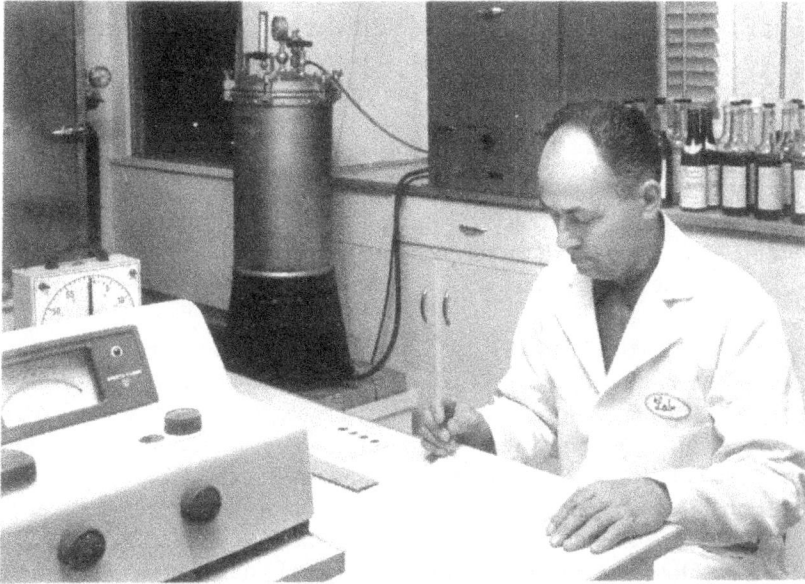

Figure 121. Applying wine to paper for chromatographic determination of acids. Laboratory autoclave in rear. (Courtesy Beaulieu Vineyard.)

against the regular trichromatic data. The method of the Office International de la Vigne et du Vin (1962) compared well with the results of Mareca Cortés (1964) or Sudraud (1958). Both compared favorably for the intensity of color, but only Mareca Cortés method approximated the dominant wave length. Pataky (1963) found that the procedure of the Office International de la Vigne et du Vin was inadequate for determining the dominant wave length, but was adequate for luminance characterization. It is encouraging that objective color standards are being developed abroad also, similar to those of Crawford et al. (1958) and Amerine et al. (1959a). Cuisa and Barbiroli (1967) published full data on the color characteristics of 49 Italian white wines and 102 reds.

Photometers (photoelectric colorimeters of the Klett-Summerson type) have been used by a number of California wineries. Actually these instruments measure the approximate luminance of the sample and give no information concerning the dominant wave length or the purity. Their accuracy can be improved by using a filter which transmits light in the region of the absorption band of the sample. For red wines, a blue-green filter would be indicated; for white wines, a blue filter.

Crawford et al. (1958) recommended specifying the color of white wines by using a Klett-Summerson photoelectric colorimeter with the Klett No.

719

42 blue filter. They found color specification with a Lovibond Tintometer or at 430 mμ in a 0.5-inch cuvette in the spectrophotometer unsatisfactory for white wines of similar color characteristics. For color differentiation of all types of wines, tristimulus values should be determined as suggested by Amerine *et al.* (1959a).

A series of three to eight spectral filters have been used to convert the photoelectric colorimeter into an abridged spectrophotometer. The tristimulus values are then determined as described above. The results obtained by this procedure are less satisfactory because the bands are of wider wave-length ranges, and a less accurate transmission curve is obtained.

Another system (Hunter, 1942) is that of photoelectric tristimulus colorimetry. This also requires filters, "but it does not primarily measure the variation of any property of samples with respect to wave length. Instead samples are measured with source-filter photocell combinations which spectrally duplicate, as nearly as possible, the three distribution functions characterizing the standard observer combined with some standard illuminant." Special filters designed to duplicate the I.C.I. Standard Observer are needed for this type of colorimetry. The results are very similar to those from the transmission curve obtained from a spectrophotometer. Worthington *et al.* (1949) have described a reflection meter based on this principle which appears useful.

Dujardin-Salleron Vino-Colorimeter has been widely used by enologists, particularly in Europe, but is seldom used here at present. It is not a precision instrument, but it does give some measure of hue and luminance. Furthermore, it is crudely constricted and the color of the disks fades when exposed to light. It can, of course, be used only for red wines.

Visual comparison of the color of the wine with that of color standards is also made. Mixtures of potassium permanganate and dichromate have been recommended. Winkler and Amerine (1938) modified Vogt's method using alizarin astroviolet and crocein scarlet. These do make a permanent color standard, but it is most useful for wines having the same color characteristics as the mixture of dyes. For white wines a color standard made of Eastman ABC dyes has been used. The Duboscq color comparator was used as a means of matching the standard and the wine. The disadvantage of color standards is that it is very difficult to match a wine with a standard if their dominant wave lengths are not similar. The values obtained are roughly proportional to the luminance.

Lovibond slides designed particularly for wines have been manufactured for many years, but, so far as known, none are available in this country and, although the data can be calculated in terms of the basic color parameters, this seems unnecessary.

In all colorimetric determinations the wine must be brilliant, that is, completely free of suspended material. Filtration through several thicknesses of filter paper, or prolonged centrifuging and filtration are sometimes necessary to accomplish this. The only exception may be where the color characteristics are being obtained by a reflection procedure. At greater depths there is an apparent change in color (known as dichroism). The effect of this on the various procedures is not known.

Little (1964) noted that with translucent liquids, such as wine, reflectance measurement is complicated by internal light transmission and the consequent light loss through trapping. Her thin-layer procedure was applied to wines by Little and Mackinney (1966). The advantage of reflectometers over spectrophotometers is the rapidity of measurement and calculation of results. Joslyn and Little (1967) reported close agreement between the results obtained by the conventional CIE tristimulus values and those from Little's (1964) thin-layer procedure. They also reported that color measurements were sensitive indicators of anthocyan destruction or of browning.

Procedure

Follow the directions in Hardy (1936) for the spectrophotometer, or Hunter (1942) for the photoelectric tristimulus colorimetry methods, to determine the tristimulus values and to convert them to dominant wave length, purity, and luminance (brightness).

For the direct approximate determination of luminance in a photoelectric colorimeter, follow the directions for the instrument used. A blue-green filter should be employed for red wines and a blue filter for white wines. Red wines require small cells, usually 2 to 5 mm; with white wines a 20- or even 40-mm cell may be used. The depth used for measurement should always be stated in conjuction with the result reported. Various instruments have different scales. So long as the name of the instrument is known, results obtained on the same instrument are comparable (if the equipment is properly calibrated and operated). But some of the disadvantages of the photoelectric colorimeter, vino-colorimeter, and color comparator are that the results are not interchangeable and they cannot be converted to the primary color specifications of dominant wave length, purity, or luminance.

To use the color standards in the Duboscq colorimeter, clean the cups and plungers with cleaning solution, rinse carefully, and dry with 95 per cent alcohol. Put the empty cups in place and check the scale readings to see that they read exactly zero with the glass plungers just touching the bottom of each cup. Using the slide adjustment, adjust the light so that equal amounts go through the two cups to the eyepiece. Now remove the screen and fill the left cup approximately two-thirds full of the color standard and fill the right cup with wine. Turn the plunger on the left until it is in the wine and reads 10.0 mm for the red color standard or 3.0 or 6.0 mm for the white color standard. Place the screen around the cups and, while looking through the eyepiece, turn the right plunger adjustment until the amount of light reaching the eyepiece is the same on both sides. Read the mm on the right by using the vernier and the magnifying lens in front of the colorimeter. Repeat the measurement at least three times by moving the right plunger up or down and readjusting until the amount of light reaching the eye is again the same for both sides. Average the readings obtained, which should not differ by more than about 1.5 mm. The color of the standard is taken as 100. The color value of the wine is then obtained as follows: $Cx/Cs = Ds/Dx$, where Cx is the color value of the wine, Cs that of the standard (taken as 100), and Dx and Ds the depths of the wine and the standards (3, 6, or 10). For red wines, this becomes $Cx = 1,000/Dx$, and for white wines, $Cx = 300/Dx$ or $600/Dx$, depending on whether the standard is set at 3.0 or 6.0 mm. Using this procedure, color values of red wines range from 100 to 400; of white wines, from 10 to 50.

The red color standard is prepared by mixing 35 mg of crocein scarlet and 65 mg of alizarin astroviolet and making to a liter with water. The white color standard is made by preparing 0.5 per cent solutions of Eastman ABC dyes. Pipette 10 ml of the red and 5 ml each of the blue and yellow stock solutions into a flask and make to 300 ml with water.

The photoelectric colorimeter most widely used now is the Bausch and Lomb Spectromic 20 with adjustable wavelength selector.

The problem of the color of white wines has recently been reviewed by P. Ribéreau-Gayon (1965). He envisages a primary light yellow and a secondary yellow-brown. The nature of the primary color is not known. The secondary color is attributed to oxidation of phenolic compounds. In musts and young wines, especially those from botrytised grapes, this oxidation is catalyzed by polyphenoloxidase. The leucoanthocyanins are indicated as the probable compounds involved, although no correlation between total leucoanthocyanins and tendency to brown has been observed. The presence of iron is known to favor darkening in white wines, probably by forming colored complexes with phenolic compounds. Citric acid, by complexing iron, would tend to reduce the effect of iron. P. Ribéreau-Gayon notes that compounds of iron and tartaric, malic, and citric acids are known to have a light yellow-green color. This would probably occur only at higher concentrations than observed in wine. Occasionally, in young wines of certain varieties, a slight green color which has been attributed to chlorophyll is observed. The oxidation products of the catechin fractions apparently accounted for the color of most white wines in the work of J. A. Rossi, Jr., and Singleton (1966a).

In white wines containing sugar there are other mechanisms for formation of brownish colors, especially when the wine is heated for some time (as in California sherry). A complex series of dehydrations and condensations, known as caramelization, then occurs. The Maillard reaction involving amino acids and sugars may also be involved.

Anthocyanins

Pataky (1965) has emphasized the importance of pH in the spectrophotometric determination of anthocyanins. He recommended a pH of 0.8 and establishment of the pH by pH meter rather than with buffers.

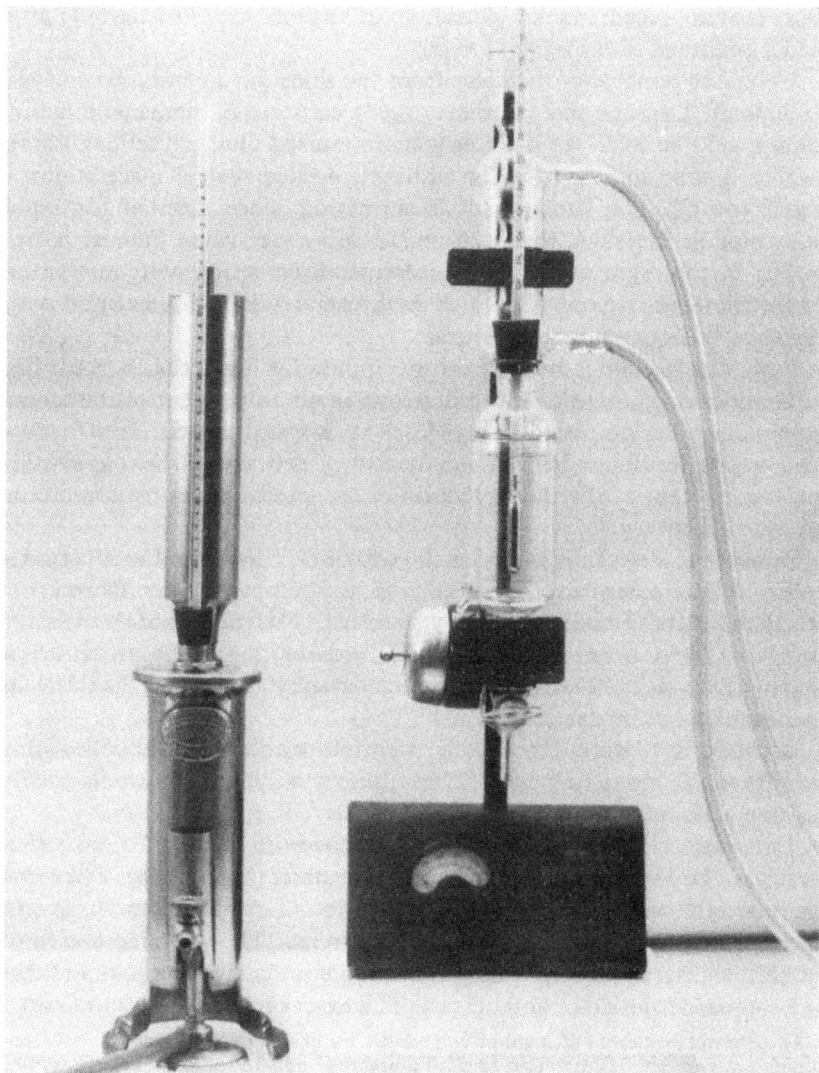

Figure 122. Two ebulliometers. Note micro-gas burner on left and rheostat to control heating on right.

Pataky doubts that there is a simple equilibrium of the type oxonium (colored) $\rightleftarrows$ leucobase (colorless) because there is no evidence of an isobestic point. Some third colored product may be formed. Because of the high optical density in alkaline media and the fact that malvidin appears to be the only pigment that fluoresces, Pataky believes that it is possible to develop a purely physical procedure for its determination. For recent procedures for detection of malvin see Burkhardt (1965b) and Eisenbrand et al. (1965a, 1965b).

Complete removal of the color from the skins for quantitative analyses is difficult. Deibner and Bourzeix (1966) used 0.1 N methanolic hydrochloric acid at 68°F (20°C). Carignane required four extractions lasting twenty hours; and Aramon, in addition, needed several macerations at 140°F (60°C). The latter result is surprising, since Aramon has much less color in the skins than Carignane. However, when interest is primarily in the diglucosides they recommend extraction with very dilute hydrochloric acid (0.001 N), which easily extracts diglucosides but does so only poorly for the monoglucosides.

Basic lead acetate is not a good precipitate for diglucosides, according to Deibner et al. (1966). It also destroys about 10 per cent of the monoglucosides. The destruction at pH 10–11 is even greater. Even neutral lead acetate results in deficiencies at 12 to 15 per cent of the diglucosides and 12 per cent of the monoglucosides. Ion-exchange resins also do not give good recovery.

In order to detect diglucosides in red wines, Eisenbrand et al. (1965a, 1965b) diluted them 1:100 in acetic acid and compared their fluorescence at 546 mμ, using Rhodamine B as a standard. (Malvin cannot be used as a standard, for it is not stable in acetic acid solution.) The procedure is much more rapid than paper chromatography (fig. 121). As little as 50 mμ/ml could be detected.

Bourzeit and Baniol (1966) call attention to the difficulty of isolating the flavonols from red wines. They present a procedure which clearly separates the anthocyanins from the flavonols.

Two methods for determining the anthocyanin content of red wines were used by P. Ribéreau-Gayon and Stonestreet (1965, 1966). They first measured the difference in absorbancy at 520 mμ at pH 0.6 and 3.5 compared to those of solutions of grape anthocyanins. The second method compared the absorbancy with and without sodium bisulfite at 520 mμ to that of a standard solution of anthocyans. The exact procedure is as follows:

To each of two tubes add 1 ml of wine and 1 ml of 95 per cent ethyl alcohol (containing 0.1 per cent hydrochloric acid). To one tube add 10 ml of 2 per cent hydrochloric acid (pH 0.6) and to the other 10 ml buffer (303.5 ml of 0.2 M sodium hydrogen phosphate [Na_2HPO_4] and 606.5 ml of 0.1 M citric acid; pH 3.5). Measure the absorbancy of each in a 1-cm cell at 520 mμ versus water. Calculate the difference in absorbancy at pH 0.6 and 3.5. Compare with the difference in absorbancy of a pure anthocyanin solution. Ribéreau-Gayon and Stonestreet prepared a pure anthocyanin by repeated crystallization with picric acid. To prepare the standard curve, 1 ml of 10 per cent ethyl alcohol (containing 5 mg of ⅓ neutralized tartaric acid) and 1 ml of anthocyanin solution (37.5 to 375 mg of crystal pigment in 95 per cent ethyl alcohol containing 0.1 per cent hydrochloric acid). The absorbancy was measured at 520 mμ as described. The difference is linear.

In the bisulfite procedure, mix 1 ml of wine, 1 ml of 95 per cent ethyl alcohol (containing 0.1 per cent hydrochloric acid), and 20 ml of 2 per cent hydrochloric acid. Place 10 ml of the solution in each of two tubes. Add 4 ml of 15 per cent sodium bisulfite to one tube and 4 ml of water to the other. After twenty minutes, measure the absorbancy at 520 mμ versus water. Calculate the difference. A standard wine is prepared by mixing 1 ml of pure pigment (prepared as above in 95 per cent alcohol and 0.1 per cent hydrochloric acid), 1 ml of 10 per cent ethyl alcohol (containing 5 mg of ⅓ neutralized tartaric acid), and 20 ml of 2 per cent hydrochloric acid. Place 10 ml in each of two tubes. Measure the absorbancy at 520 mμ as above. The differences (and the absorbancy) are approximately linear.

Comparable results were obtained by both procedures: 60 to 385 mg per liter in new (1959–1964) wines and 16 to 50 in old (1921–1957) wines. In the same wines total tannins were 1.50 to 2.09 and 2.00 to 4.40 gr per liter, respectively. P. Ribéreau-Gayon and Nedeltchev (1965) noted that colorimetric determination of anthocyanins is more accurate in young wines, where the color is due primarily to anthocyans. In old wines an appreciable amount of the color is due to condensed and oxidized tannins.

Because the official French procedure (Jaulmes and Ney, 1960) for detecting glucosides in wines is rather long, Dorier and Verelle (1966) proposed treating wines with sodium nitrite, adding ammoniacal ethanol, and observing the fluorescence under ultraviolet light. Diglucosides give a vivid green fluorescence, while normal anthocyanins result in insoluble brown deposits. Using Dorier and Verelle's (1966) procedure for malvin, Bieber (1967) reported malvin contents in pure *Vitis labrusca* hybrids of up to 275 mg per liter (in a 1964 Oberlin). Fermentation reduced the malvin content by 50 per cent or more.

Jouret (1967) found the Dorier and Verelle procedure (as modified by him) simple and specific. They preferred it to the official procedure since it uses a spectrofluorimeter to measure intensity of fluorescence rather than visual observation. The standard method of the Office International de la Vigne et du Vin (1962) for diglucosides is that of Dorier and Verelle (1966). For several examples of diglucosides in *V. vinifera* see Anon. (1967d). A circular chromatography procedure of Gombkőtő (1965) for separating diglucosides is also available. For early data on the partition chromatography of anthocyanidins see Spaeth and Rosenblatt (1950).

The procedure for detection of diglucosides proposed by Van Wyk and Venter (1964) is a modification of that developed by Bieber (1960).

The sensitivity of the procedure seems to have been enhanced by adding pure malvin to wines known not to contain malvin. Fluorescence of reference and test samples could thus be studied against comparable background. Swain and Hillis (1959) found that the quantitative estimation of anthocyanins by measuring the difference in red color at pH 2.0 to pH 5.4 and by bleaching with sodium sulfite was not as satisfactory, at least for methanol extracts of fruits, as decolorization with hydrogen peroxide. Joslyn and Little (1967) applied the latter procedure to rosé and red wines and found it to compare well with visible color changes and with thin-layer reflectance measurement of color.

Deibner et al. (1964) separated the anthocyanin diglucosides from the monoglucosides by thin-layer chromatography. Spectrophotometric determination (at 530 mμ) revealed 0.096 to 1.90 gr per liter of diglucosides (as malvidin) in wines made of several Seyve-Villard hybrids. As little as 5 mg per liter of diglucosides could be detected by their procedure. Bourzeix (1967b) used a new technique to identify diglucosides in wines which eliminated some of the uncertainties of previous procedures. It was specific for malvin. For a recent summary of methods for the determination of anthocyans see P. Ribéreau-Gayon and Nedeltchev (1965) and P. Ribéreau-Gayon (1968).

MICROSCOPIC EXAMINATION . . .

. . . requires a trained operator and incubation
of suspended samples.

Microscopic examination of fermenting musts and wines is often useful in detecting the presence of undesirable microörganisms. It is of limited diagnostic value by itself, because mere inspection will not always differentiate between living and dead cells or between desirable and undesirable ones. Changes in the microflora (both in type and in number) during fermentation or storage and the presence of abnormal forms are more significant. Incubation of samples of suspected wines, at 23.9°–29.4°C (75°–85°F) for 1–3 months, with periodic inspection and microscopic examination during the incubation, will usually reveal the presence of viable forms capable of developing in the wine. These observations should be confirmed by the usual bacteriological technique of isolating and identifying the organisms present. Analysis of the samples for total and volatile acidity and pH during storage will also be useful in diagnosing the nature of the spoilage.

Centrifuging of the wine to concentrate the microörganisms is usually desirable before examination. To make the examination, place a drop of the centrifuged sediment on a slide, fix, stain, and examine under a microscope. For information on the size and shape of the organisms, see pages 559, 775, and 784 and figs. 124 and 126.

For microbiological examination, membrane filters are now commonly used. Jakubowska (1963) has recommended them for wines. They are rapid and simple to use, and make identification of colonies easy. Geiss (1962a) gives a complete description of how to determine the biological sterility of a wine by means of the membrane filter. Special attention must be paid to sterilizing the equipment and the filter. (See fig. 48.)

ANALYSES

A knowledge of the composition of wine is of value in blending, in determining soundness and keeping quality, and ensuring that the wine meets government standards.

General surveys of analytical procedures suitable for wines are given in Jaulmes (1951), Ribéreau-Gayon and Peynaud (1958), Hennig (1962), Office International de la Vigne et du Vin (1962), Amerine (1965b), and Association of Official Agricultural Chemists (1965). For practical and theoretical information on volumetric analysis see Kolthoff and Stenger (1942, 1947).

Jaulmes (1965a, 1965b), correctly we believe, emphasizes the great complexity of wine and the importance of precise methods of analysis to prevent sale of sophisticated or spoiled wines. He reviewed the history of official French methods of analysis of wines since 1907. Since 1950 the Office International du Vin (now officially the Office International de la Vigne et du Vin) has worked to produce methods that would have international acceptance. These methods, first published as their "Cahier Vert," are continuously being revised and appear as the Recueil des Méthodes Internationales d'Analyse des Vins, Office International de la Vigne et du Vin (1962). (For recent additions see Anon., 1966a, 1967b, 1968). The recommended procedures of this organization are divided into two classes: those for precise "reference" work, and more rapid though less accurate methods for commercial control purposes. In Europe these procedures have a quasi-legal status. In this country the procedures of the A.O.A.C. have a similar status. Jaulmes (1965a, 1965b) reviewed the international attempts to achieve some sort of unification in methods of analysis. He also notes the efforts being made in France to modernize methods of wine analysis. Garoglio (1963) gives a useful summary of enzymatic methods of analysis as applied to wines.

Bober and Haddaway (1963) believed that it might be possible to identify various alcoholic beverages from significant peaks on their gas chromatograms. They were even hopeful that with aging new significant peaks would develop and that the chromatogram might be used as a means of dating the beverage. So far no conclusive evidence has been presented.

Although accurate enough for control purposes, the methods described below have been chosen, as far as possible, for simplicity and speed rather than for extreme accuracy. They have been written for persons familiar with, but not necessarily skilled in, laboratory technique. It is stressed, however, that the analyst should have sufficient knowledge of analytical procedures to be able to comprehend the principle of the procedures and the sources of error. Naturally, errors due to faulty technique and poor equipment must be eliminated. Many procedures, although similar, are not equivalent under all conditions and for all wines.

Hydrometry

The hydrometer is based on the Archimedean principle that the same body displaces an equal weight of liquids in which it floats. This reading can often be related to some characteristic constituent of the liquid. Hydrometers (fig. 46, p. 211) are widely used to determine the specific gravity of liquids when speed is more important than accuracy. A hydrometer consists of a closed tube filled with shot or metal at the lower end to keep the tube upright, a bulb or expanded portion which causes the greater part of the displacement, and a thin stem calibrated to read specific gravity, some arbitrary function of specific gravity, or to indicate the percentage composition of the solution tested.

The sensitivity of the hydrometer depends upon the ratio between the total volume displaced and the volume displaced by a unit length of the stem. Increasing the size of the immersed bulb and decreasing the diameter of the stem increases the instrument's sensitivity. The Brix (or Balling) hydrometer is calibrated to indicate the percentage by weight of sucrose when immersed in solutions of sucrose and water. It is calibrated for use at a given temperature and reads correctly only when the instrument and liquid are equilibrated to this temperature. In grape juice or wine the hydrometer merely indicates the percentage composition of a sucrose solution which has the same specific gravity as the solution being tested. This reading has become a rough measure of the soluble-solids content of the liquid under test. The specific gravity concentration relations of glucose and fructose are sufficiently close to that of sucrose to give approximately the same readings. The Brix degree of unfermented grape juice is higher than the true sugar content, whereas that of partially fermented must or wine is lower.

Hydrometers should be calibrated against a sucrose solution of known concentration before use. To reduce errors to a minimum, the hydrometer should be clean and dry before being immersed. The juice or must to be tested should be brought to the temperature at which the hydrometer is calibrated, or a temperature correction applied. The most convenient hydrometers have an enclosed thermometer and correction scale. However, temperature-correction tables are available for pure sugar solutions (table 70), and these apply approximately to grape juice.

TABLE 70

CORRECTIONS FOR BRIX OR BALLING HYDROMETERS CALIBRATED AT 20°C (68°F)

Temperature of solution	Observed percentage of sugar						
	0	5	10	15	20	25	30
Below calibration	SUBTRACT (PER CENT)						
°C °F							
15 59.0	0.20	0.22	0.24	0.26	0.28	0.30	0.32
15.56 60.0	0.18	0.20	0.22	0.24	0.26	0.28	0.29
16 60.8	0.17	0.18	0.20	0.22	0.23	0.25	0.26
17 62.6	0.13	0.14	0.15	0.16	0.18	0.19	0.20
18 64.4	0.09	0.10	0.11	0.12	0.13	0.13	0.14
19 66.2	0.05	0.05	0.06	0.06	0.06	0.07	0.07
Above calibration	ADD (PER CENT)						
°C °F							
21 69.8	0.04	0.05	0.06	0.06	0.07	0.07	0.07
22 71.6	0.10	0.10	0.11	0.12	0.13	0.14	0.14
23 73.4	0.16	0.16	0.17	0.17	0.20	0.21	0.21
24 75.2	0.21	0.22	0.23	0.24	0.27	0.28	0.29
25 77.0	0.27	0.28	0.30	0.31	0.34	0.35	0.36
26 78.8	0.33	0.34	0.36	0.37	0.40	0.42	0.44
27 80.6	0.40	0.41	0.42	0.44	0.48	0.52	0.52
28 82.4	0.46	0.47	0.49	0.51	0.56	0.58	0.60
29 84.2	0.54	0.55	0.56	0.59	0.63	0.66	0.68
30 86.0	0.61	0.62	0.63	0.66	0.71	0.73	0.76
35 95.0	0.99	1.01	1.02	1.06	1.13	1.16	1.18

Source of data: Association of Official Agricultural Chemists (1965).

For a quick, approximate correction, 0.033° Brix is added for each degree Fahrenheit (0.06° per degree Centigrade) above that at which the hydrometer is calibrated. Brix hydrometers calibrated in 0.1° are very useful. Specific gravity hydrometers calibrated at 0.001° or 0.0005° are also available. A set of five hydrometers to read 1.030–1.060, 1.060–1.090, 1.090–1.120, and 1.120–1.150 are employed for musts and sweet wines. A fifth hydrometer for 0.980 to 1.005 is used for dry table wines.

The hydrometer jar should be of clear glass, of a diameter large enough to allow the hydrometer to float freely (about 1 inch greater in diameter than the hydrometer bulb), and of sufficient height (greater than length of hydrometer) to enable full reading. The sample of the must (juice), wine, or extract to be tested should be carefully poured down the side of the inclined jar to avoid trapping bubbles of gas, and care should be taken to avoid air bubbles in the liquid or clinging to the sides of the vessel or hydrometer. The air bubbles, when present, will buoy up the hydrometer and cause a low reading. For accurate results freshly pressed grape juice should be deaerated before testing.

It has long been recognized that soluble-solids measurements on grape musts are influenced by the presence of solid materials. Cooke (1964) showed that the hydrometer readings of unfiltered musts were 0.4° to 0.7° Brix higher than those of the refractometer. He emphasizes that an empirical correction of the refractometric readings is needed, owing to the fact that the refractive index of glucose and fructose is less than that of sucrose, upon which the refractometer scale is based. This correction amounts to +0.3 for Brix readings of 15.0 to 15.5, +0.4 for Brix readings of 15.6 to 20.2, +0.5 for Brix readings of 20.3 to 24.7, and +0.6 for Brix readings of 24.8 to 29.3.

The hydrometer should be slowly immersed in the liquid to a point slightly beyond where it begins to float, and then allowed to float freely. The reading should not be made until the liquid and hydrometer are free from air bubbles and at rest. In reading the hydrometer scale, place the eye slightly below the level of the surface of the liquid. Then raise the eye until the surface, seen as an ellipse, becomes a straight line. The point where this line cuts the hydrometer scale, disregarding the film of liquid drawn up around the stem by capillarity, is taken as the reading of the hydrometer.

Surface tension, viscous drag, and parallax errors are inherent in using hydrometers. Because the surface tension exerted by the liquid on the stems tends to increase its displacement, it is particularly necessary to avoid handling the hydrometer except at the top of the stem. Otherwise, variable amounts of body oils left on the hydrometer would influence the results. The hydrometer should be washed immediately after use so that no undesirable residues remain on it.

The hydrometer should be checked against values obtained by an accurate pycnometer determination on the same solution. In fact, the pycnometer is the reference procedure of the Office International de la Vigne et du Vin (1962) for determining the specific gravity of wines. They give special correction tables for both Pyrex and ordinary glass pycnometers. The Westphal balance is also used for getting an approximate measure of the specific gravity.

The Brix or Balling reading is in grams per 100 gr. To convert to grams per 100 ml, multiply by the specific gravity of the solution. (See table 74, p. 745.)

Alcohol by ebullioscope

The ebulliometric procedure is based on the regular variation in the boiling point of mixtures of alcohol and water with alcohol content. For ordinary wines with low to medium extract, 50 ml of the wine is introduced directly into the ebullioscope (fig. 122), the thermometer is inserted, the condenser is filled with cold water, and heat is applied. The flame and ebullioscope must be protected from drafts if the boiling temperatures is to be determined accurately. The boiling point is observed on the special thermometer when the mercury remains at the same level for a minute or two.

For table wines with a sugar content of 5 to 10 per cent, dilute the wine accurately with an equal volume of water and use 50 ml of the diluted wine. Multiply the result obtained with this diluted wine by two. Correction tables for the effect of sugar are given by Love (1939) and Churchward (1940). It is also possible to distill off the alcohol into a volumetric flask and make to volume. This liquid can then be used directly in the ebullioscope.

The boiling point of water varies with atmospheric pressure, which must be ascertained at the time each series of determinations is made. The boiling point of water is usually determined with 50 ml of distilled water in the boiling chamber, and with the condenser empty. A sliding scale, on which the boiling point may be adjusted, accompanies the instrument and is used for calculating the alcohol content. The results are accurate to about ±0.25 per cent, although Jaulmes (1965b) and others noted that the results are usually better than this. They should be reported only to the nearest 0.1 per cent.

Alcohol by hydrometer

To determine the alcohol content by means of a special alcohol hydrometer, scaled for this purpose, it is necessary to separate the alcohol from the wine. The temperature of the wine should be approximately the same as the calibration temperature of the volumetric equipment—usually 20°C (68°F). A still composed of a Kjeldahl flask and trap, short-coupled to a water-cooled condenser, is required. Pipette 100 ml of wine into a Kjeldahl flask and add about 50 ml of water. Or use a 100-ml volumetric flask to measure the wine into the Kjeldahl flask, rinse it with about

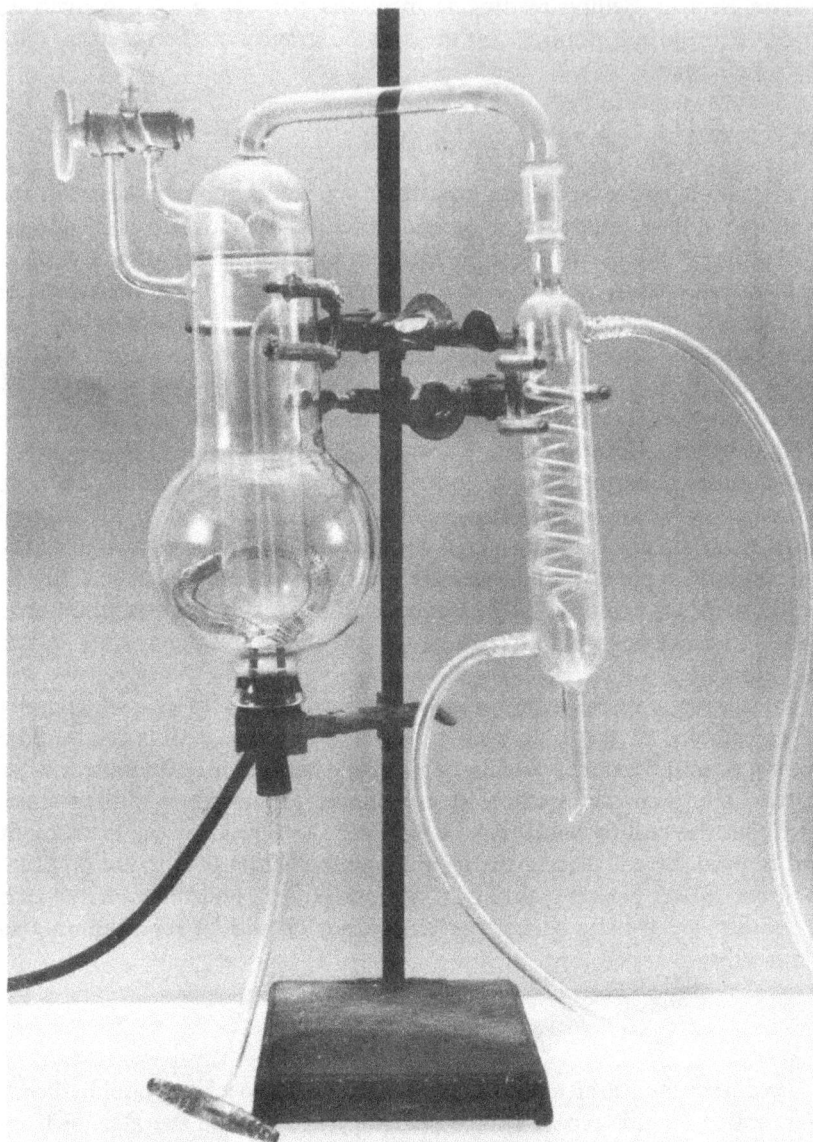

Figure. 123.   Cash volatile acid assembly showing coil heater.

50 ml of water, and add the rinsings to the Kjeldahl. Use the volumetric flask to collect the distillate. If the sample is high in volatile acidity—above 0.10 per cent—add a volume of normal sodium hydroxide sufficient

to neutralize the total acidity. This can be calculated from the total acid determination (p. 740). Join the Kjeldahl to a connecting tube and condenser, and distill about 95 ml into the volumetric flask. Rinse the residue in the Kjeldahl flask into another 100-ml volumetric flask and save for extract, tannin, and other determinations.

The alcohol distillate is then brought to volume at the calibration temperature of the volumetric flask, mixed, and again brought to volume if necessary, or allowed to equilibrate to room temperature if the second procedure was used. If possible, the flasks are then brought to the calibration temperature of the alcohol hydrometers—usually 15.56°C (60°F). The distillate is carefully poured into a clean, dry hydrometer cylinder. (A hydrometer cylinder in which the usual alcohol hydrometer will float in 100 ml of liquid may be made by cutting a glass tube of the proper size and making a base of a large rubber stopper.) Some prefer to use 200 ml of wine and collect the distillate in a 200-cc volumetric flask. A larger hydrometer cylinder may be used in this case. A clean, dry hydrometer in the appropriate alcohol range is gently lowered into the distillate to about the level at which it is at equilibrium. When nearly at equilibrium, it is pushed a little farther into the liquid and again allowed to come to equilibrium. The percentage of alcohol is read directly off the hydrometer stem. If the temperature is not at that for which the hydrometers are calibrated, it will be necessary to apply a correction. These corrections are given in table 71. (p. 734). Corrections for hydrometers calibrated at 20°C (68°F) are given in table 72. The apparent specific gravity of alcohol solutions from 0 to 22.5 per cent at 15.56°C (60°F) and 20°C (68°F) are given in table 73.

Jaulmes and Brun (1963*b*) point out that 20°C (68°F) is now the recommended reference temperature for specific gravity, density, and alcohol percentage. Should this be generally adopted, the present tables based on 15°C (56°F) as the reference temperature would need to be replaced. Among the related problems are the type of pycnometers used (Pyrex or ordinary glass). Because of the many errors inherent in this method, the results should only be reported to the nearest 0.1 per cent.

Alcohol by chemical analyses

The chemical method commonly used for alcohol determination is that developed by Semichon and Flanzy (1929) or modifications of it (Joslyn and Amerine, 1941*a*; Fessler, 1941; Zimmerman and Fessler, 1949). These methods are based on the fact that, in acid solution, dichromate oxidizes alcohol to acetic acid as follows:

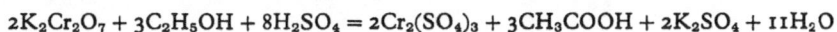

$$2K_2Cr_2O_7 + 3C_2H_5OH + 8H_2SO_4 = 2Cr_2(SO_4)_3 + 3CH_3COOH + 2K_2SO_4 + 11H_2O$$

TABLE 71

CORRECTIONS OF ALCOHOL HYDROMETERS CALIBRATED AT 15.56°C (60°F) IN PERCENTAGE BY VOLUME OF ALCOHOL WHEN USED AT TEMPERATURES ABOVE OR BELOW 15.56°C (60°F)

To or from the observed

Observed alcohol content	Add			Subtract														
	At 57°F	At 58°F	At 59°F	At 61°F	At 62°F	At 63°F	At 64°F	At 65°F	At 66°F	At 67°F	At 68°F	At 69°F	At 70°F	At 72°F	At 74°F	At 76°F	At 78°F	At 80°F
1	0.14	0.10	0.05	0.05	0.10	0.16	0.22	0.28	0.34	0.41	0.48	0.55	0.62	0.77	0.93	—	—	—
2	0.14	0.10	0.05	0.05	0.11	0.17	0.23	0.29	0.35	0.42	0.48	0.56	0.63	0.78	0.94	1.10	1.28	1.46
3	0.14	0.10	0.05	0.06	0.12	0.18	0.24	0.30	0.36	0.43	0.50	0.57	0.64	0.80	0.96	1.13	1.31	1.50
4	0.14	0.10	0.05	0.06	0.12	0.19	0.25	0.32	0.38	0.45	0.52	0.59	0.67	0.83	1.00	1.17	1.35	1.54
5	0.15	0.10	0.05	0.07	0.13	0.20	0.26	0.33	0.40	0.47	0.54	0.62	0.70	0.86	1.03	1.21	1.40	1.60
6	0.17	0.11	0.06	0.07	0.14	0.20	0.27	0.34	0.42	0.50	0.57	0.66	0.74	0.90	1.09	1.27	1.46	1.66
7	0.18	0.12	0.06	0.07	0.14	0.21	0.29	0.36	0.44	0.52	0.60	0.68	0.77	0.94	1.13	1.32	1.52	1.73
8	0.19	0.13	0.06	0.08	0.16	0.23	0.31	0.39	0.47	0.55	0.64	0.73	0.81	0.99	1.18	1.38	1.59	1.80
9	0.21	0.14	0.07	0.08	0.16	0.24	0.32	0.41	0.50	0.58	0.67	0.76	0.86	1.04	1.25	1.46	1.67	1.89
10	0.23	0.16	0.08	0.08	0.17	0.25	0.34	0.43	0.52	0.61	0.71	0.80	0.90	1.10	1.32	1.54	1.76	1.99
11	0.25	0.16	0.08	0.09	0.18	0.27	0.37	0.46	0.56	0.65	0.75	0.85	0.96	1.16	1.39	1.61	1.84	2.09
12	0.27	0.18	0.09	0.10	0.20	0.29	0.39	0.49	0.59	0.70	0.80	0.91	1.02	1.23	1.46	1.70	1.94	2.20
13	0.29	0.19	0.10	0.10	0.21	0.31	0.42	0.52	0.63	0.74	0.85	0.97	1.08	1.31	1.55	1.80	2.05	2.31
14	0.32	0.21	0.11	0.11	0.22	0.32	0.44	0.55	0.66	0.78	0.91	1.02	1.14	1.39	1.65	1.91	2.17	2.44
15	0.35	0.23	0.12	0.12	0.24	0.35	0.48	0.60	0.71	0.84	0.97	1.10	1.23	1.50	1.76	2.03	2.30	2.58
16	0.37	0.24	0.12	0.13	0.26	0.38	0.52	0.65	0.77	0.90	1.03	1.17	1.31	1.60	1.88	2.16	2.44	2.72
17	0.40	0.26	0.13	0.14	0.27	0.41	0.54	0.68	0.82	0.96	1.10	1.25	1.40	1.70	1.99	2.28	2.58	2.87
18	0.44	0.29	0.14	0.14	0.29	0.44	0.58	0.73	0.88	1.03	1.18	1.33	1.49	1.80	2.10	2.41	2.72	3.02
19	0.47	0.32	0.16	0.15	0.30	0.46	0.62	0.78	0.94	1.10	1.26	1.42	1.58	1.90	2.22	2.54	2.86	3.17
20	0.51	0.34	0.17	0.16	0.32	0.49	0.66	0.82	0.98	1.15	1.33	1.48	1.65	2.00	2.32	2.65	2.98	3.33
21	0.53	0.35	0.18	0.17	0.34	0.51	0.68	0.85	1.02	1.20	1.38	1.54	1.72	2.06	2.41	2.76	3.10	3.45
22	0.56	0.38	0.19	0.17	0.36	0.53	0.71	0.90	1.07	1.25	1.44	1.61	1.78	2.13	2.48	2.84	3.20	3.56
23	0.58	0.40	0.20	0.18	0.37	0.55	0.74	0.92	1.11	1.30	1.49	1.66	1.84	2.20	2.56	2.93	3.30	3.67
24	0.60	0.40	0.20	0.18	0.38	0.56	0.77	0.96	1.16	1.35	1.54	1.72	1.91	2.27	2.65	3.03	3.40	3.78

Source of data: United States Bureau of Internal Revenue (1945).

734

TABLE 72

CORRECTIONS OF ALCOHOL HYDROMETERS CALIBRATED AT 20°C (68°F) IN PERCENTAGE BY VOLUME OF ALCOHOL WHEN USED AT TEMPERATURES ABOVE OR BELOW 20°C (68°F)

To or from the observed

	Add								Subtract							
Observed alcohol content	12°C 53.6°F	13°C 55.4°F	14°C 57.2°F	15°C 59.0°F	16°C 60.8°F	17°C 62.6°F	18°C 64.4°F	19°C 66.2°F	21°C 69.8°F	22°C 71.6°F	23°C 73.4°F	24°C 75.2°F	25°C 77.0°F	26°C 78.8°F	27°C 80.6°F	28°C 82.4°F
0	0.74	0.67	0.60	0.51	0.42	0.32	0.23	0.11	—	—	—	—	—	—	—	—
1	0.76	0.69	0.61	0.52	0.43	0.33	0.24	0.12	0.13	0.26	0.41	0.56	0.71	0.86	—	—
2	0.78	0.70	0.62	0.54	0.45	0.34	0.24	0.12	0.13	0.27	0.42	0.57	0.72	0.88	1.05	1.22
3	0.80	0.73	0.65	0.56	0.46	0.35	0.25	0.13	0.13	0.28	0.43	0.58	0.73	0.91	1.08	1.25
4	0.84	0.76	0.68	0.58	0.48	0.37	0.26	0.13	0.14	0.29	0.44	0.59	0.75	0.93	1.10	1.28
5	0.88	0.80	0.71	0.61	0.50	0.39	0.27	0.13	0.14	0.29	0.45	0.61	0.78	0.96	1.14	1.33
6	0.94	0.85	0.75	0.64	0.53	0.41	0.28	0.14	0.15	0.30	0.47	0.64	0.81	1.00	1.18	1.38
7	1.00	0.90	0.79	0.68	0.55	0.43	0.30	0.15	0.15	0.32	0.49	0.67	0.84	1.04	1.23	1.44
8	1.07	0.95	0.85	0.72	0.58	0.46	0.31	0.16	0.16	0.34	0.51	0.70	0.89	1.09	1.28	1.50
9	1.15	1.02	0.90	0.77	0.62	0.48	0.33	0.16	0.17	0.35	0.54	0.73	0.93	1.14	1.34	1.56
10	1.23	1.10	0.96	0.82	0.66	0.51	0.34	0.17	0.18	0.37	0.56	0.77	0.98	1.19	1.40	1.62
11	1.32	1.18	1.03	0.87	0.71	0.55	0.36	0.18	0.19	0.40	0.60	0.81	1.02	1.24	1.46	1.68
12	1.43	1.28	1.11	0.94	0.76	0.58	0.39	0.20	0.20	0.42	0.63	0.85	1.07	1.30	1.53	1.75
13	1.55	1.38	1.19	1.01	0.82	0.63	0.42	0.21	0.21	0.44	0.67	0.89	1.12	1.36	1.60	1.83
14	1.68	1.48	1.28	1.09	0.88	0.67	0.45	0.23	0.23	0.46	0.71	0.94	1.18	1.43	1.68	1.92
15	1.82	1.60	1.39	1.17	0.94	0.71	0.47	0.24	0.24	0.48	0.74	0.99	1.24	1.51	1.76	2.02
16	1.97	1.72	1.49	1.26	1.01	0.76	0.50	0.26	0.26	0.51	0.78	1.04	1.32	1.57	1.85	2.11
17	2.13	1.85	1.59	1.34	1.08	0.81	0.53	0.28	0.28	0.54	0.82	1.09	1.38	1.65	1.93	2.21
18	2.27	1.99	1.70	1.44	1.15	0.85	0.56	0.29	0.30	0.57	0.86	1.14	1.44	1.73	2.02	2.31
19	2.40	2.11	1.81	1.51	1.20	0.90	0.60	0.30	0.31	0.60	0.90	1.21	1.51	1.82	2.12	2.43
20	2.51	2.20	1.89	1.58	1.26	0.95	0.63	0.31	0.32	0.62	0.94	1.27	1.60	1.90	2.22	2.54
21	2.63	2.29	1.96	1.64	1.31	0.99	0.65	0.32	0.33	0.65	0.98	1.31	1.65	1.97	2.30	2.63
22	2.73	2.38	2.03	1.71	1.36	1.02	0.68	0.33	0.34	0.67	1.00	1.35	1.69	2.02	2.36	2.70
23	2.84	2.47	2.11	1.77	1.40	1.06	0.70	0.34	0.35	0.69	1.03	1.38	1.72	2.06	2.42	2.76
24	2.93	2.57	2.19	1.82	1.45	1.09	0.72	0.35	0.36	0.71	1.06	1.42	1.76	2.12	2.47	2.83

Source of data: Office International de la Vigne et du Vin (1962).

TABLE 73

SPECIFIC GRAVITY[a] OF ALCOHOL SOLUTIONS AT 15.56°C (60°F) AND 20°C (68°F)

Alcohol per cent)	15.56°C	20°C	Alcohol (per cent)	15.56°C	20°C
0.0	1.00000	0.99821	11.5	0.98487	0.98294
0.5	0.99925	0.99747	12.0	0.98430	0.98235
1.0	0.99850	0.99672	12.5	0.98374	0.98177
1.5	0.99776	0.99599	13.0	0.98319	0.98119
2.0	0.99703	0.99525	13.5	0.98264	0.98062
2.5	0.99630	0.99453	14.0	0.98210	0.98006
3.0	0.99559	0.99381	14.5	0.98157	0.97950
3.5	0.99488	0.99311	15.0	0.98104	0.97894
4.0	0.99419	0.99241	15.5	0.98051	0.97839
4.5	0.99350	0.99172	16.0	0.97998	0.97784
5.0	0.99282	0.99104	16.5	0.97946	0.97729
5.5	0.99215	0.99037	17.0	0.97895	0.97675
6.0	0.99150	0.98971	17.5	0.97844	0.97621
6.5	0.99085	0.98907	18.0	0.97794	0.97567
7.0	0.99022	0.98842	18.5	0.97744	0.97514
7.5	0.98960	0.98779	19.0	0.97694	0.97461
8.0	0.98899	0.98717	19.5	0.97645	0.97409
8.5	0.98838	0.98656	20.0	0.97596	0.97357
9.0	0.98779	0.98594	20.5	0.97546	0.97503
9.5	0.98720	0.98533	21.0	0.97496	0.97250
10.0	0.98661	0.98473	21.5	0.97446	0.97196
10.5	0.98602	0.98413	22.0	0.97395	0.97142
11.0	0.98544	0.98353	22.5	0.97344	0.97087

Source of data: For 15.56°C, U.S. Treasury Department, Bureau of Internal Revenue (1945); for 20°C, Office International de la Vigne et du Vin (1962). The values are corrected for the buoyancy of air and were obtained with Pyrex pycnometers.

[a] It should be recalled that our specific gravity is what is called *densité* in France and what we call density is their *poids spécifique*. In order to prevent the obvious confusion, Jaulmes (1965a) proposed to substitute *masse volumique* for *poids spécifique*. *Densité* is thus the ratio of the mass of a certain volume of the liquid (or solid) to the mass of the same volume of water (both referred to the same temperature).

The completeness of the oxidation depends on the concentrations of the oxidizing agent, of the sulfuric acid, and of the alcohol, as well as on the time. The accuracy of the results depends on a number of factors: reliability of standardizations, draining time of pipettes, and so on. This procedure should therefore be attempted only by those who have some skill in quantitative chemical analysis. Jaulmes (1965a) notes that the chief objection to the chemical procedure is that there are no tables accurately correlating alcohol content as determined by the usual alcohol tables, and alcohol content as determined by density and chemical analyses.

Zimmerman (1963) reported that dichromate results were 0.24 per cent higher in alcohol than the results of densimetric procedures, while dichromate and pycnometric results were identical. This result he showed to be fortuitous because of a compensation of errors (owing to a highly volatile component other than alcohol which is eliminated by aerating the fresh distillates). The volatile acidity and sulfur dioxide content of wines depress the pycnometer and hydrometer results, but have little influence on the dichromate results. About 40 per cent of the remaining difference between pycnometer and dichromate results appears to be due to a substance which depresses the pycnometer results but does not effect those of the dichromate procedure. Under these circumstances Zimmerman argued that the dichromate results more accurately reflect the true alcohol content of the wine. This may be true under very precise laboratory controls.

The dichromate procedure was adapted for semiautomatic operation by Morrison and Edwards (1963). They point out particularly the need to use primary-standard potassium dichromate, high-quality distilled water, accurately calibrated volumetric glassware, and a standard ethanol-water solution as a daily check on the analyst and the apparatus. They indicate that the 95 per cent confidence limit for the standard deviation is ±0.12.

Accurately pipette 10 ml of wine into a distillation apparatus and steam-distill 80 ml of distillate into a 500-ml volumetric flask. Make to volume with water at the calibrated temperature. Pipette 10 ml of the diluted alcohol distillate into a 100-ml volumetric flask. Accurately pipette in 25 ml of potassium dichromate solution[1] and 10 ml of sulfuric acid[2]. Stopper the flask, shake, and allow the oxidation to proceed for ten minutes at 40° to 50°C. Dilute the oxidation mixture to 100 ml at 20°C with distilled water. Pipette 25 ml of this mixture and 10 ml of diluted sulfuric acid[3] into a 250-ml Erlenmeyer flask and titrate to a brownish-red end point with standardized ferrous ammonium sulfate,[4] using one

---

[1] Dissolve 42.576 gr of the CP salt in distilled water in a 1,000-ml volumetric flask and make to volume.

[2] Add 700 ml of concentrated sulfuric acid to 300 ml of distilled water and make to 1 liter at 20°C.

[3] Add 200 ml of concentrated sulfuric acid to 800 ml of distilled water and bring to 1 liter at 20°C. This solution should be roughly standardized by diluting 10 ml to 500 ml. A 10-ml aliquot of this rediluted solution should require about 15.40 ml of 0.1 N sodium hydroxide for titration.

[4] Dissolve 45 gr of CP $FeSO_4 \cdot (NH_4)_2SO_4 \cdot 6H_2O$ in about 800 ml of distilled water in a 1,000-ml volumetric flask. Add 20 ml of CP concentrated sulfuric acid and make to temperature and volume. To protect this solution from oxidation, keep it stored under hydrogen or illuminating gas.

drop of o-phenanthroline[5] as an internal indicator. The volume of ferrous ammonium sulfate required is designated as B.

To determine the volume of ferrous sulfate equivalent to the dichromate, carefully pipette 25 ml of the dichromate solution and 10 ml of the sulfuric acid solution into a 100-ml volumetric flask and bring to temperature and volume. Pipette 25 ml of this diluted dichromate solution into a 250-ml Erlenmeyer flask, add 10 ml of diluted sulfuric acid, and titrate with the ferrous ammonium sulfate to the same end point as above. The volume required is designated as A.

The calculation is based on the fact that the dichromate solution is made up so that 1 ml is equivalent to 10 mg of ethyl alcohol. When freshly prepared, 25 ml of the ferrous ammonium sulfate solution is equivalent to 25 ml of the diluted dichromate solution. The milligrams of alcohol in the aliquot are then equal to $(A - B)/A \times 250$. In the original sample of wine, the alcohol in milligrams $500/10 \times (A - B)/A$, or, in grams, $12.5 (A - B)/A$. The weight of alcohol in grams per 100 ml is $100/10 \times 12.5 (A - B)/A$, or $125 (A - B)/A$.

The alcohol percentage by volume at 20°C (68°F) can be calculated from the density of ethyl alcohol, which is 0.78934 at 20°C. The alcohol percentage is sometimes expressed at 15.56°C (60°F), at which temperature the specific gravity of ethyl alcohol is 0.79389.

In the Zimmerman and Fessler (1949) modification, the concentrations and aliquots are such that the milliliters of dichromate solution required are equivalent to the alcohol percentage of the sample. They also use diphenylamine as the indicator. The procedure of the Office International de la Vigne et du Vin (1962) also adjusts the dichromate concentration to give alcohol percentage at 15.56°C (60°F) or at 20°C (68°F).

Other methods for alcohol

Some wine makers prefer to use the pycnometer for determining alcohol. This procedure, described in the official methods of the Association of Official Agricultural Chemists (1965) should be attempted only by a skilled analytical chemist, but in his hands it gives excellent results and is frequently used when accurate results are desired. A rapid technique for use of the pycnometer has been given by Jaulmes and Brun (1963a). The procedure is based on accurately measuring the temperature of the pycnometer (made from specified glass) rather than bringing the pycnometer and its contents to a specified temperature. They prefer a pycnometer of Pyrex glass and a thermometer calibrated to read to 0.02°C.

---

[5] Prepare a 0.025 M solution by dissolving 0.695 gr of ferrous sulfate (FeSO$_4 \cdot$7H$_2$O) in water, adding 1.485 gr of o-phenanthroline monohydrate, and bring to 100 ml.

Jaulmes (1965*a*, 1965*b*) believes the pycnometer procedure to be the most accurate—to ± 0.02 at its best.

The alcohol corrections for pycnometers made of Pyrex at temperatures of 10° to 30°C and 1 to 30 per cent alcohol are given by Jaulmes and Brun (1966). These are, however, for pycnometers calibrated at 15°C and for the French alcohol tables of 1884. Jaulmes (1965*a*) states that the pycnometric procedure is still the standard in France and that a good analyst may obtain results to ± 0.1 per cent. Ebulliometry is no longer recommended in France even as a control procedure.

An article by Etienne and Breyer (1951) gives a modification of the Williams' procedure for alcohol. A special reagent is used to extract the alcohol from the wine, leaving the water as a lower layer. A specially calibrated tube reads the percentage of alcohol directly. The procedure is rapid and requires only 10 ml of sample, but has an accuracy of only ± 0.5 per cent.

The immersion refractometer gives rapid and accurate values. Correction factors for calculating the alcohol content of the distillate by immersion refractometer at temperatures other than 15.56°C (60°F) are given by Reeves (1965). A portable recording refractometer for the temperature range 10° to 30°C was tested by Vasben *et al.* (1966).

Jaulmes *et al.* (1965) noted that a new French differential refractometer had renewed interest in the refractometric determination of alcohol. This is really a double refractometer and the temperature effect is negligible. Higher alcohols have nearly twice the effect on the refractive index as ethanol. However, the effect at 500 mg per liter of higher alcohol is only 0.07 per cent. If the operator reads the instrument properly and if the temperature remains constant during the determinations, Jaulmes and his co-workers state that the method gives good results. A table of factors for computing the alcohol content of distilled spirits using the immersion refractometer was presented by Reeves (1965). This table gives factors at 20°, 25°, 30°, and 35°C; so the sample need not be at 15.56°C. Both the Office International de la Vigne et du Vin (1962) and the Association of Official Agricultural Chemists (1965) include the refractometric method as an official procedure.

The primary objection to refractometry is that it is quite sensitive to the influence of higher alcohols. Nevertheless, Jaulmes (1965*a*) reports that when well used it is capable of ± 0.05 per cent accuracy.

There has been some use of gas chromatography for alcohol determination (see Morrison, 1961). Besides the cost of the equipment it must be carefully calibrated and operated for accurate results. In contrast to earlier results, Garschagen (1967) reported only approximately ± 0.5 per cent results when gas chromatography was used to determine ethanol.

739

Total acid

Remove the carbon dioxide either by placing 25 ml in an Erlenmeyer flask, connecting to a water aspirator, and agitating for one minute under vacuum, or heat the wine to incipient boiling for thirty seconds, swirl, and cool.

Add 1 ml of 1 per cent alcoholic phenolphthalein solution to 200 ml of hot boiled distilled water in a 500-ml wide-mouth Erlenmeyer flask. Neutralize with dilute sodium hydroxide to a very faint pink color. Add 5.00 ml of the degassed must or wine and titrate with 0.1 N (or 0.0667 N) standardized sodium hydroxide to the same end point, using a well-lighted background. The total acidity, expressed as grams of tartaric acid per 100 ml, is obtained by multiplying the number of milliliters of 0.1 N sodium hydroxide used in the titration by 0.075 or, in general, N sodium hydroxide × 100/ml of wine used × 0.075 × ml NaOH used = grams of tartaric acid per 100 ml. If 0.0667 N alkali is used, grams of tartaric acid per 100 = ml 0.1 N sodium hydroxide. It would be useful for many purposes if the results were reported in milliequivalents of base used.

In Austrian grape juices Schaller (1958) reported end points of 7.4 to 8.0. He recommended 7.8. Wong and Caputi, Jr. (1966) present evidence that the true end point of the total acid titration curve in California wines is close to pH 7.7. Cresol Red (o-cresol-sulphonephthalein) has a distinctive color change at this pH and they recommend it for this determination. These values can be approximately converted to phenolphthalein or pH 8.4 end-point titrations by mulitplying by 1.05. The reference procedure of the Office International de la Vigne et du Vin (1962) is to titrate to pH 7 at 20°C (68°F). Their usual procedure is to use bromthymol blue as an indicator. This changes color at about 7. This will give much lower results than the official procedure using phenophthalein given above. On purely theoretical grounds, titration to 7.7 or 8.4 is more correct than titration to 7.0. Jaulmes (1965a) indicated that he regretted the new definition. We recommend 8.0 and the use of an automatic titrimeter.

There are numerous procedures for the determination of malic acid in musts and wines. Peynaud and Blouin (1965) have compared five procedures (chemical, chromatographic, enzymatic, manometric, and microbiological). For routine work they recommend the microbiological procedure of Peynaud and Lafon-Lafourcade (1965). For research laboratories the enzymatic method of Mayer and Busch (1963) appears more precise. A relatively rapid colorimetric procedure for malic acid was developed by Tarantola and Castino (1962, 1966). The wine is passed through both an anion and a cation exchanger, is treated with activated carbon, and a characteristic color is developed with 2,7-naph-thalene disulfonic acid in a sulfuric acid solution. Good recovery of added malic acid was obtained. For a rapid procedure for qualitative detection of malic and lactic acids see Kunkee (1968). The standard method for malic acid of the Office International de la Vigne et du Vin (1962) is based on the colorimetric chromotropic acid reaction. The most recent procedure of the Office International de la Vigne et du Vin (Anon., 1968) corrects for interference by tartaric and lactic acid. To follow the malo-lactic fermentation paper chromatography (fig. 121) is used.

Guimberteau and Peynaud (1966) compared five procedures for determination of lactic acid in wines. They favor the procedures of Peynaud and Charpentié (1950) and Dimotaki-Kourakou (1960). Colori-metric methods such as those of Rebelein (1961, 1964) and of Schaelderle and Hasselmann (1961) gave high results.

The determination of the tartrate content by precipitation of potassium bitartrate, first used by Pasteur, is known as the Pasteur-Reboul procedure. It has been modified by many workers. Nègre et al. (1958) proposed two bitartrate procedures which resulted in excellent recovery: (1) reducing 50 ml of must (or wine) to 7 ml and adding a 3.5 pH buffer and filtering after 4 days; and (2) using 20 ml of must (or wine), adding alcohol to 16.5 per cent and a 3.8 pH buffer, and proceeding as before. The first pro-cedure was recommended. They showed that the presence of malic acid reduced the speed of precipitation of potassium bitartrate. Some entrain-ment of potassium malate occurs, but this is partially compensated by the solubility of potassium bitartrate. Sulfates and phosphates did not interfere unduly. Their procedure was at least comparable in accuracy to those based on precipitation of calcium racemate, and is much more rapid. (See also Amerine, 1965b, for the similar procedures of Hennig.)

The usual European procedure is precipitation as calcium racemate, but the necessary reagents are not easily or cheaply available in this country and the racemate is slightly soluble. The vanadate procedure frequently used in this country is described by Amerine (1965b). Maurer (1967) also recommended the ammonium vanadate method for the rapid deter-mination of tartrates. He claimed an error of no more than ±0.02 gr of tartaric acid per 100 ml. It is now the standard procedure of the Office International de la Vigne et du Vin (1962), but the racemate method remains the reference procedure (Anon., 1968).

Pato and Beck (1963) proposed determining the tartaric and malic acids from titration data. Staudenmayer and Becker (1966) obtained good results on model solutions. They also noted that the Pato and Beck procedure gave higher results than the Hennig procedure on wines when ion exchange was used to separate the tartaric acid—due to the failure of the Hennig procedure to precipitate all the tartrate at the low pH of the ion-exchanged material.

An enzymatic procedure for determination of citric acid was presented by Mayer and Pause (1965a). They used citrate lyase to produce oxalacetic acid and acetic acid from citric acid. The oxalacetic acid is reduced by malate dehydrogenase using DPNH. The method is specific and rapid, does not require preliminary clarification, and uses only 0.2 ml of sample. They reported 0.0018, 0.0236, and 0.0165 per cent in a white wine, a red wine, and a white grape juice, respectively.

The most recent recommendation (Anon., 1968) for citric acid is to use the Henning and Lay (1965) procedure.

Rebelein (1967a) separated the citric acid as the barium salt and determined it with a colorimetric procedure. The determination takes only one hour. A gravimetric procedure for oxalic acid was developed by Hennig and Lay (1965). Castino (1967) has given a sensitive colorimetric procedure for the determination of citramalic acid.

Peynaud (1963), Mayer and Busch (1963), and Mayer and Pause (1965a) used photometric enzymatic procedures to determine organic acids in wines. Amerine (1965b) and Peynaud et al. (1966a) emphasized the rapidity, sensitivity, and specificity of these methods. Egorov and Boresova (1956, 1957) give methods for determining keto acids in wines.

Volatile acid

This method is based on the fact that the volatile acids which cause spoilage of wine are distillable by steam at atmospheric pressure. With a 10-ml pipette, introduce 10 ml of degassed (p. 740) wine into the central

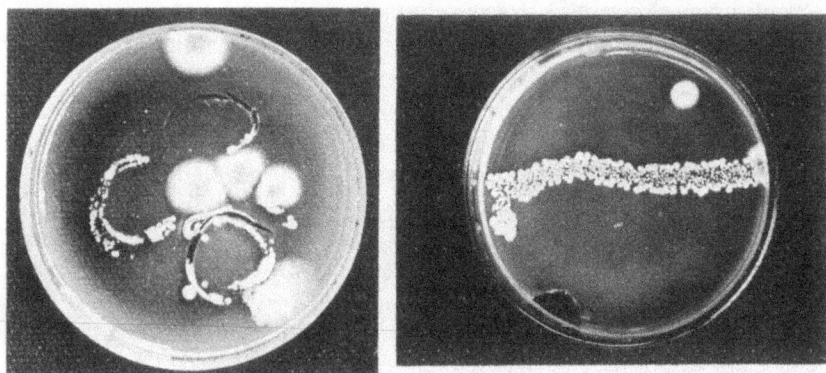

Figure 124. Microörganisms growing on agar: (above) from the top of a bottle and (below) from a drop from a cork. (Courtesy Seitz-Werke GmbH, Bad Kreuznach, Germany. See p. 13, Informationen No. 29, 1967.)

tube of a volatile acid distillation apparatus or (preferably) of a Cash electric still, fig. 123. Add 150 to 200 ml of recently boiled, hot distilled water to the outer flask, or 500 ml if the Cash electric still is used. Tightly connect the apparatus as directed by the manufacturer. Start cold water running through the condenser, and apply heat to the outer flask. When the water has boiled for a moment, close the pinchcock on the outlet of the outer flask and distill until 100 ml have been collected in a 300-ml flask. Heat the distillate with 0.1 N sodium hydroxide to a pink end point which persists fifteen seconds. The volatile-acid content, as grams of acetic acid per 100 ml of wine, is obtained by multiplying the titration in milliliters by 0.06. Somewhat more accurate results may be obtained for the titration by using a more dilute sodium hydroxide solution. Gowans (1965), comparing the usual steam distillation procedure for volatile acidity with the Cash electric still, found the results were very similar. In the present official procedure of the Association of Official Agricultural Chemists (1965), 25 ml of wine are used instead of 10, and 300 ml instead of 100 ml of distillate are collected. The reference procedure of the Office International de la Vigne et du Vin (1962) suggests a distilling apparatus with a rectification column to limit distillation of lactic acid. G. J. Pilone (1967), however, showed that even with very high lactic acid contents, the interference is minimal.

Owades and Dono (1967, 1968) used microdiffusion in a modified Conway cell to separate the volatile material. The volatiles, including sulfur dioxide, or the volatile acidity and total sulfur dioxide were determined. Recovery of added acetic acid varied from 92 to 104 per cent. Recovery of sulfur dioxide added to beer ranged from 90.0 to 107 per cent. No data on recovery from wine were given. The determination takes six hours (or more), but up to 50 microdiffusion cells could be done at once.

Federal and state limits in volatile acidity are exclusive of sulfur dioxide. If the volatile acidity percentage is near the legal limits, the determination should be made so that sulfur dioxide is not included in the calculation. One procedure for this is given in the official methods of the Association of Official Agricultural Chemists (1965). The wine (50 ml) is titrated with saturated barium hydroxide to a phenolphthalein end point or to pH 8. It is kept at this end point or pH (by adding more alkali if necessary) for thirty minutes. The wine is diluted to 100 ml and is filtered through fluted rapid paper (e.g., Whatman No. 2). Pipette 50 ml of the solution into the Cash electric still, add 1 ml of 1 + 3 sulfuric acid, and distill as before. Poirier (1966) points out that one cannot, without error, calculate the correction for sulfur dioxide in the volatile acidity determination from the total sulfur dioxide value.

Another source of error is due to the distillation of sorbic acid (p. 769). No provision for this is made in the regulations. However, Jaulmes *et al.* (1962) noted that the presence of 200 mg per liter of sorbic acid could result in a correction of 0.012 gr per 100 ml of acetic acid. Procedures for determining sorbic acid are given by Jacquin (1960) and Jaulmes *et al.* (1962).

Fixed acid

The total acidity less the volatile acidity is known as the nonvolatile or fixed acidity. Since, however, the total acidity is expressed as tartaric acid and the volatile acidity as acetic acid, the value for the fixed acidity cannot be obtained by direct subtraction (another reason for expressing results in milliequivalents). If the volatile-acid content as acetic is multiplied by 1.25, it will then be in terms of tartaric acid and can be subtracted from the total acid to give the fixed acid as tartaric acid per 100 ml. For a gas-liquid chromatographic procedure for quantitatively determining the fixed acids see Brunelle *et al.* (1967). See also Lehmann (1966).

pH

The best procedure for pH determination is by use of the glass electrode pH meter. A number of commercial models are available that vary in ruggedness and sensitivity. The most recent procedures of the Office International de la Vigne et du Vin (Anon., 1968) recommended use of a pH meter for determining the pH.

Total soluble solids

The total soluble solids or extract include nonvolatile material such as sugars, acids, and salts.

Pipette 100 ml of wine into a 250-ml beaker, add 50 ml of distilled water, place over the source of heat, and evaporate to approximately 50 ml. Remove from heat and cool to room temperature. Dilute to the original volume in a 100-ml volumetric flask by adding distilled water. After adjusting the temperature to exactly 20°C (68°F), place a portion of the liquid in a glass hydrometer cylinder. Read off the approximate extract with a special Brix or Balling hydrometer. A small hydrometer with a short neck and enclosed thermometer is very useful. When the dealcoholized wine has been adjusted to the same volume as that of the original sample, the Brix reading gives the approximate extract direct as grams per 100 gr of wine. To express the result in grams of extract per 100 ml, multiply the Brix reading by the corresponding specific gravity. (See table 74.) To convert from specific gravity to grams per 100 ml use table 75.

744

TABLE 74

SPECIFIC GRAVITY[a] AT 15.56°C (60°F) AND 20°C (68°F) CORRESPONDING
TO READINGS OF THE BALLING OR BRIX HYDROMETER

Brix or Balling degrees	Specific gravity		Brix or Balling degrees	Specific gravity	
	15.56/15.56	20/20		15.56/15.56	20/20
0.50	1.0019	1.00194	17.50	1.0701	1.07185
1.50	1.00575	1.00584	18.50	1.07456	1.07624
2.50	1.00966	1.00975	19.50	1.07926	1.08065
3.50	1.01356	1.0169	20.50	1.0836	1.08509
4.50	1.0174	1.01766	21.50	1.0881	1.08958
5.50	1.0212	1.02165	22.50	1.0926	1.09409
6.50	1.0252	1.02567	23.50	1.0971	1.09863
7.50	1.02906	1.02972	24.50	1.1017	1.102341
8.50	1.03336	1.03381	25.50	1.1063	1.10782
9.50	1.0372	1.03790	26.50	1.1109	1.11246
10.50	1.0414	1.04205	27.50	1.1155	1.11794
11.50	1.0454	1.04622	28.00	1.1180	1.11949
12.50	1.0495	1.05041	28.50	1.1203	1.12185
13.50	1.05356	1.05464	29.00	1.1227	1.12422
14.50	1.05753	1.05889	29.50	1.1251	1.12659
15.50	1.06163	1.06319	30.00	1.1274	1.12898
16.50	1.0660	1.06740			

[a] Assuming specific gravity of water at 15.56°C (60°F) and 20°C (68°F) as unity, respectively.
Sources of data: U.S. Bureau of Internal Revenue (1945) and Association of Official Agricultural Chemists (1965).

TABLE 75

EXTRACT CONTENT FROM SPECIFIC GRAVITY
(gr per 100 ml)

Specific gravity + 0.2 decimal	3d decimal of density									
	0	1	2	3	4	5	6	7	8	9
1.00	0	0.26	0.51	0.77	1.03	1.29	1.54	1.80	2.06	2.32
1.01	2.58	2.84	3.10	3.36	3.62	3.88	4.13	4.39	4.65	4.91
1.02	5.17	5.43	5.69	5.95	6.21	6.47	6.73	6.99	7.25	7.51
1.03	7.77	8.03	8.29	8.55	8.81	9.07	9.33	9.59	9.85	10.11
1.04	10.37	10.63	10.90	11.16	11.42	11.68	11.94	12.20	12.46	12.72
1.05	12.98	13.24	13.50	13.76	14.03	14.29	14.55	14.81	15.07	15.33
1.06	15.59	15.86	16.12	16.38	16.64	16.90	17.16	17.43	17.69	17.95
1.07	18.21	18.48	18.74	19.00	19.26	19.52	19.78	20.05	20.31	20.58
1.08	20.84	21.10	21.36	21.62	21.89	22.15	22.41	22.68	22.94	23.20
1.09	23.47	23.73	23.99	24.25	24.52	24.78	25.04	25.31	25.57	25.84
1.10	26.10	26.36	26.63	26.89	27.15	27.42	27.68	27.95	28.21	28.48

Source of data: Office International de la Vigne et du Vin (1962).

The residue from the determination of alcohol by distillation may be saved and used for this determination, provided the wine was not neutralized before distillation. The dealcoholized residue is poured into a 100-ml volumetric flask, the Kjeldahl flask is well rinsed into the volumetric flask, and the contents are brought to temperature and volume. The actual determination is made as described above.

The accurate determination of the extract is of more importance in Europe, where minimum standards for extract are often a part of the regulations, directly or indirectly. Hamelle's (1965) thesis deals entirely with the precise determination of the extract. She recommends evaporation of 10 ml of wine under vacuum at 70°C (158°F) in a current of dry air. The tared evaporating dish should contain 4 gr of cellulose. This is essentially the reference procedure of the Office International de la Vigne et du Vin (1962).

This figure divided by the extract determined at 100°C (212°F) gave a figure of 1.21 for dry red wines and 1.32 for dry white wines and dry rosés. The destruction of fructose at 100°C (212°F) makes calculation of such ratios for sweet wines very hazardous, but at 70°C (158°F) there is very little destruction in the time required to reach dryness.

Of the indirect methods for determining the extract, the formula of Tabarié has had the most general acceptance, and Usseglio-Tommasset (1965) has shown that it gives good results with Italian sweet wines. This formula is $d = d_w - d_a + 1000$, where d is the density of extract, d_w that of the wine and d_a that of an alcohol-water solution of the same alcohol percentage as that of the wine. To convert the density thus determined to grams per 100 ml, table 74 may be used.

To calculate the extract-less-sugar (p), Hamelle (1965) recommended the formula $p = 2,510 \ (D_{20}^{20} - 1) - 22.2$. Values obtained by this formula and density are reported by Hamelle to be very near those reported by extract by vacuum or at 70°C (158°F). Hamelle feels, however, that the extract calculated from the density is of less general utility since it will not give a correct value for abnormal wines. The older coefficient, 2,667, is applicable only to solutions containing more than 40 per cent sugar. For the extract-less-sugar content of musts Hamelle recommends $2,560 \ (D_{20}^{20} - 1) - 22.2$.

When used as an enological ratio, alcohol by weight to extract-less-sugar should not exceed 4.1 for red wines, 4.9 for rosé wines, and 5.4 for white wines.

Reducing sugars

The volumetric method given below is based on the Lane and Eynon titration method for determining substances capable of reducing copper in alkaline tartrate solution. Although most of the copper-reducing substances in wine are sugars, other reducing matter is present also. Most of the nonsugar reducing substances can be removed by clarifying with lead acetate and charcoal and then removing the excess lead with phosphate or oxalate. The usual lead acetate clarification does not remove galacturonic acid, according to Dimotaki-Kourakou and Sotiropoulos (1964).

Since this acid reduces Fehling's solution, one obtains high results for reducing sugar. They recommend passing the wine through a Dowex 3 column (20 to 50 mesh in the acetate form).

Jaulmes (1965b) notes the large number of procedures for reducing sugar that are used in different countries. In a comparison of three procedures for reducing sugar Bayonove (1966) favored a modified Bertrand procedure using ferrous orthophenanthroline as the indicator in the manganimetric titration of ferrous salts. This is now the official French procedure.

Jacob and Anghelide (1967) used a Zeiss interferometer for determining ethanol and glucose in wines. The error claimed was 0.001 per cent! However, a number of correction factors are used and one would want to see determinations on wines of widely variant alcohol, sugar, and acidity.

Dealcoholize the wine (as for extract), cool, and make to the original volume by adding distilled water (or use the dealcoholized sample of extract from the alcohol distillation). Pipette 50 ml of either of these solutions into a 100-ml volumetric flask. Add 10 ml of saturated neutral lead acetate solution and about 2 gr of an acid-washed, decolorizing carbon. Shake well, bring to volume, and filter onto anhydrous disodium acid phosphate. Shake the solution and either allow it to settle clear or filter it. Use 20 ml of this solution for the reducing sugar determination. For sweet wines, clarify only 5 or 10 ml of wine.

Immediately before use, prepare the Soxhlet reagent by mixing 50 ml each of Fehling's A and B solutions in a clean, dry flask. Pipette accurately 25 ml of the mixed Soxhlet reagent into a clean, 300-ml Erlenmeyer flask. To standardize the method, fill a burette with a standardized 0.5 per cent glucose solution (or better, 0.5 per cent invert sugar solution) and also pipette 20 ml of this solution into the flask. Set the burette over the flask on a wire gauze, heat this cold mixture to boiling, and maintain a moderate boiling for about fifteen seconds; lower the flame enough to avoid bumping. Add 5 drops of a 1 per cent solution of methylene blue. Rapidly add further quantities of the sugar solution, and titrate drop by drop until the indicator is completely decolorized. The total volume of the standard solution necessary should be about 24 ml.

To determine the sugar content of the wine, proceed as above, adding 20 ml of the clarified wine solution to 25 ml of the mixed Soxhlet reagent. After boiling the solution for fifteen seconds, finish the titration by adding standard glucose solution from the burette. If the 20-ml wine solution completely decolorizes the Soxhlet reagent, dilute the wine solution further and repeat the determination. For example, take 20 ml; dilute to 100 ml; take 20 ml of the diluted liquid and multiply the result by five in this case.

The sugar content of the wine solution can be calculated from the amount of the standard glucose solution necessary to reduce the copper completely and from the amount necessary to finish the reduction after the addition of the wine solution, as follows: percentage of copper-reducing matter as glucose in the wine solution = (ml of glucose for direct titration − ml of glucose for back titration) × 0.005 × 100 ÷ ml of wine in the 20-ml aliquot.

The recommended procedures of the Association of Official Agricultural Chemists (1965) and the Office International de la Vigne et du Vin (1962) involve separation of the precipitated cuprous oxide and either weighing it or by dissolving and titrating or by determining the excess copper.

Ough and Cooke (1966) and Cooke and Ough (1966) used the Ames Company Reducing Sugar Test Kit for semiquantitative determination of reducing sugars in wines of 1 per cent or less reducing sugar. This test consists of boiling 10 drops of wine with a prepared tablet for fifteen seconds. The color is then compared with a chart and readings to 0.0, 0.05, 0.1, 0.2, 0.4, 0.6, and 1.0 per cent reducing sugar made (sucrose does not react). Comparisons with the Lane and Eynon procedure in this range were good for both red and white wines. This appears to be a convenient method for determining the completeness of fermentation of dry table wines, shermat, and distilling material. Red wine needs to be treated with charcoal (such as Darco G60), heated, and filtered before testing.

Tannin and coloring matter

The method depends on the determination of the permanganate reducing matter of the wine before and after decolorization with carbon. This Neubauer-Loewenthal procedure has had great utility. However, as P. Ribéreau-Gayon and Nedeltchev (1965) noted, no completely satisfactory method based on permanganate oxidation of tannins is possible. For a comparison of the Neubauer-Loewenthal and Pro procedures see Smit et al. (1955).

Dealcoholize the wine, cool, and make to the original volume (or use the extract solution from the alcohol distillation). Transfer 5 ml to an 800-ml beaker. Add about 500 ml of water and exactly 5 ml of indigo carmine solution.[6] Titrate with standardized potassium permanganate solution (about 0.1 N), 1 ml at a time, until the blue color changes to green; then add a few drops at a time until the color becomes golden yellow. Thoroughly stir the solution with a glass rod after each addition. Designate the milliliters of potassium permanganate solution required to

---

[6] Prepare by dissolving 6 gr of indigo (free of indigo blue) in 500 ml of water and 50 ml of concentrated sulfuric acid. When cool, make to a liter.

reach the golden-yellow color, using the dealcoholized sample by shaking 25 ml with 10 gr of carbon, as a. Filter, pipette the same volume as used previously into the 800-ml beaker, add water and 5 ml of indigo carmine solution, and titrate. Designate the number of milliliters of potassium permanganate used for the dealcoholized, decolorized, and detanninized samples as b.

Then $a - b = c$, the number of milliliters of potassium permanganate solution required for oxidizing the tannin and coloring matter in 5 ml of the wine. The amount of tannin and coloring matter as grams of tannin per 100 ml of wine is equal to c × normality of $KMnO_4$ × 0.0416 × 100/volume of wine. Since b has almost a constant value for wines of the same composition, it is customarily determined only when starting analyses on a new type of wine.

To determine only the tannins, Nègre (1942–1943) recommended determining the ml of permanganate required to oxidize 10 ml of indigo carmine in 500 ml of water, d. To determine the permanganate required to oxidize only the tannin, pipette 10 ml of dealcoholized wine into a 25-ml centrifuge tube, add 10 ml of ammoniacal zinc acetate,[7] stir, centrifuge three minutes, and decant the supernatant liquid. Wash the precipitated tannin by stirring with 20 ml of 1 per cent ammonium hydroxide, and centrifuge and decant the supernatant liquid. Dissolve the tannin precipitate in 10 ml of dilute sulfuric acid and transfer to a 1-liter beaker. Add water to 500 ml and 10 ml of indigo carmine and titrate as previously: ml permanganate = e. Then $e - d = f$ or the ml of permanganate to titrate only tannins. The oxidizable nontannin material = $c - f$, which Nègre reports as nontannin polyphenols.

Tannins produce a blue color by reduction of phosphomolybdic-phosphotungstic acid; this is the basis of the latest official method of the Association of Official Agricultural Chemists (1965).

Pipette 1 ml of wine into a 100-ml volumetric flask containing 75 ml of water. Add 3 ml of the Folin-Denis reagent (to 750 ml of water add 100 gr of sodium tungstate ($Na_2WO_4 \cdot 2H_2O$), 20 gr phosphomolybdic acid, and 50 ml of phosphoric acid, reflux for two hours, cool, and dilute to 1 liter). Add 10 ml of 20 per cent sodium carbonate and dilute to volume. Shake and let stand for thirty minutes. Determine the percentage of transmittance at 760 mμ in a spectrophotometer or in a photoelectric colorimeter with a suitable filter. If the optical density is over 0.9 dilute the wine 1:1 or 1:2 (always necessary with red wines). A standard curve is prepared using tannic acid solutions in the range 0.01 to 0.3 per cent. Improvements on Burkhardt's (1963) procedure for catechin were given

---

[7] Prepared by dissolving 40 gr of zinc acetate in distilled water; add 110 ml of concentrated ammonium hydroxide, stir, make to a liter. Store in a glass-stoppered bottle.

by Rebelein (1965a). These have especially to do with the amount and concentration of hydrochloric acid used. The vanillin procedure used by Joslyn and Little (1967) for catechin and related compounds gave better correlations with the color stability than did the Folin-Denis procedures.

J. A. Rossi, Jr., and Singleton (1966a) reported the catechin fraction (at the levels found in wine) to be bitter but not astringent. The leuco-anthocyanin and condensed tannin fractions were bitter and astringent. The thresholds for the three groups in water were 0.002, 0.003, and 0.0035 gr per 100 ml, respectively. All three were effective in reducing the apparent sourness of acids. Recently Diemair and Polster (1967) separated brown and blue tannin fractions from red wines. Their molecular weights were 5,000 to 10,000 and 1,000 to 5,000, respectively.

P. Ribéreau-Gayon and Stonestreet (1966) also reported the vanillin colorimetric procedure to be more sensitive to oxidative changes in wine phenolics and more specific for astringency.

The colorimetric procedure based on Folin-Denis reagent is recognized to be influenced by the occasional formation of a white, dense, crystalline material, by erratic results due to inadequate mixture of samples and reagents and addition of reactants in wrong sequence, and by the use of an ill-defined standard of comparison. Singleton and J. A. Rossi, Jr. (1965) investigated several of the factors influencing the determination of phenolics by this procedure and developed an improved method with better precision and accuracy. In their modification 1 ml samples of wine (white wines usually undiluted, red wines diluted 1:10) are added to 60 ml water in a 100 ml volumetric flask and, after mixing, 5.0 ml of Folin-Ciocalteu reagent are added and mixed; then after 30 seconds and before 8 minutes 3.0 gr anhydrous sodium carbonate in aqueous solution (e.g., 15 ml of 20 per cent solution). After 2 hours at 23.9°C (75°F) absorbance is measured in 1 cm cells at 765 mμ. Gallic acid standard is used instead of tannic acid. Folin-Ciocalteu reagent is more stable than Folin-Denis reagent and does not produce sediment.

Joslyn et al. (1968) also report data on the factors influencing accuracy and reproducibility of the Folin-Denis phenol procedure. In the micro method the proposed Folin-Ciocalteu reagent is used in preference to Folin-Denis reagent but anhydrous D-catechin is used as the standard of comparison since this more closely resembles the actual phenolics present in wine. Interference due to sulfite, ascorbic acid, and ferrous ion was investigated. In their method 1 ml of diluted wine or catechin solution containing 1–20γ of phenolic, 1 ml of Folin-Ciocalteu reagent (diluted 1:4 before use) and, after 3 minutes, 1 ml of 10 per cent sodium carbonate solution are added, mixed, and after 60 minutes absorbance at 760 mμ is read.

Sulfur dioxide

The method of the Office International de la Vigne et du Vin (1962) for total sulfur dioxide is the iodometric procedure of Deibner and Bénard (1955), later modified by Deibner and Sabin (1962) and by Deibner (1965). See also Jaulmes (1965b). In this procedure the distillation is made into 50 ml of 2 N sodium hydroxide. This is sufficiently alkaline to decompose the acetaldehyde sulfurous acid complex (α-hydroxyalcoyl sulfuric acid), which distills over, and the titration is done rapidly enough to prevent the recombining of acetaldehyde and sulfurous acid. Deibner (1965) notes that the distillation of sweet wines or of grape juices should not reduce the volume below one-half. Apparently, heating sugars will produce compounds which react with iodine.

The total sulfur dioxide is best determined by distillation. Pipette 50 or 100 ml of wine into an 800-ml Kjeldhal flask. Add 200 ml of water, 10 ml of concentrated hydrochloric acid, and 10 ml of saturated sodium bicarbonate solution. Connect to a condenser and attach a calcium chloride tube as an adapter. Then distill into 50 ml of a freshly standardized 0.02 N solution of iodine in an Erlenmeyer flask, cooled with ice water. After 150 to 175 ml of distillate are collected, determine the residual iodine by titration with standard 0.02 N sodium thiosulfate, using a few drops of freshly prepared starch solution as the indicator. In the same way, determine the volume of sodium thiosulfate solution initially required to reduce 50 ml of the iodine. Then the difference between the initial and final sodium thiosulfate titration values of the iodine solution × 32 × the normality of the thiosulfate solution × 1,000 ÷ ml of wine used gives the sulfur dioxide content in mg per liter or parts per million.

Deibner (1966a), in a specially designed apparatus, distilled the sulfur dioxide from an acidified solution under nitrogen in a vacuum into an alkaline solution. The iodometric titration was made under alkaline conditions. Tanner (1962, 1963) also distilled under nitrogen. Deibner reported results ±0.3 per cent for the total. The official procedure of the Association of Official Agricultural Chemists (1965) is the Monier-Williams procedure, which also distills under nitrogen. When this method is used the modification of the apparatus proposed by Thrasher (1966) should be considered.

The free sulfur dioxide content of white wine can be determined approximately by pipetting 50 ml of wine into a 250-ml Erlenmeyer flask, adding 5 ml of sulfuric acid (1 to 3; that is, 1 volume of concentrated sulfuric diluted with 3 volumes of water) and 0.5 gr of sodium carbonate, to expel air, and titrating the sulfurous acid thus formed with 0.02 N iodine using a starch indicator. The volume of iodine solution required × its normality × 32 × 1,000 ÷ ml of wine used gives the free sulfur dioxide content in mg per liter or parts per million. For more accurate determinations in white wines, and for dark-red wines whose pigment interferes with the starch end point, the electrometric titration developed by Ingram (1947a, 1947b) is preferable. (See also Monier-Williams, 1927; Benvegnin and Capt, 1931; Nichols and Reed, 1932; Archinard, 1937; Ponting and Johnson, 1945; and Ripper, 1892.)

In the usual procedure for determining free sulfur dioxide the pH is brought to 1–1.5. This reduces the rate of decomposition of the bound sulfur dioxide and makes it more volatile. When anthocyanins are present, the sulfur dioxide with which they are combined will be released at these low pHs and the apparent free sulfur dioxide will be too high. Timberlake and Bridle (1967b) note that the error is considerable with products, such as black currant juice, which have a high anthocyanin content. They suggested that the strong combination of flavylium salts unsubstituted in the A ring, such as 4'-hydroxyflavylium chloride, might make them useful for the quantitative determination of free sulfur dioxide.

The procedure of the Office International de la Vigne et du Vin (1962) uses 50 ml of wine, 3 ml of 0.1 N sulfuric acid, 5 ml of 0.5 per cent of starch solution, and 30 mg of sodium versenate. Titrate with 0.05 N iodine to the usual end point—n. Add 8 ml of 4 N sodium hydroxide, stir, and leave five minutes. Transfer quickly to a beaker containing 10 ml of 0.1 N sulfuric acid and titrate to a blue end point with 0.05 N iodine—n'. Add 20 ml of 4 N sodium hydroxide, shake, leave five minutes, and dilute with 200 ml of ice-cold water. Shake well and transfer rapidly to a beaker containing 30 ml of 0.1 N sulfuric acid. Titrate again with 0.05 N iodine—n''. Since substances other than sulfurous acid are oxidized by iodine, these should be determined. To 50 ml of wine add 5 ml of acetaldehyde (6.9 gr per liter) and 5 ml of propionaldehyde (10 gr per liter). Stopper and leave at least thirty minutes. Add 3 ml of 0.1 N sulfuric acid and titrate with 0.05 N iodine = n'''. Then $n - n'''$ equals the free sulfur dioxide. The total sulfur dioxide is $n + n' + n'' - n'''$. The directions note that n''' is low, 0.2 to 0.3 ml of 0.05 N iodine. However, when ascorbic acid has been used it is higher, and approximately indicates the ascorbic acid content. Note that 1 ml of 0.05 N iodine is equivalent to 4.4 mg of ascorbic acid.

It is important that the free sulfur dioxide be determined at the same temperature at which the wine is sampled. A rise in temperature increases the free sulfur dioxide. For further information on the precise determination of sulfur dioxide in wines see Jones (1964).

In the iodometric determination of ascorbic and sulfurous acids, Burkhardt and Lay (1966) recommend additon of glycolaldehyde to bind the sulfur dioxide in the second titration. They report less fading of color with glycolaldehyde than with acetaldehyde. Ponting and Johnson (1945) previously reported formaldehyde superior to acetaldehyde.

The iodimetric determination of free sulfur dioxide by titration in acid solution is subject to considerable error, according to Joslyn and Braverman (1954), Joslyn (1955), and others. More accurate determination is possible by removing free sulfur dioxide by a stream of air from acidified wine, absorbing it in neutralized hydrogen peroxide, and titrating the sulfuric acid so formed by 0.01 N sodium hydroxide. Burroughs and Sparks (1964) modified this by removing the free sulfur dioxide in an air stream under reduced pressure. Lloyd and Cowle (1963) introduced a modification of the para-rosaniline colorimetric method previously developed for determination of total sulfur dioxide by Beetch and Oetzel (1957). Beetch and Oetzel (1957) distilled the sulfur dioxide in 300 gr of beer after acidification with 20 ml of concentrated hydrochloric acid in a stream of nitrogen into 100 ml of sodium tetrachloromercurate (II). The disulfitemercurate (II) ion which forms in the absorbing medium is stable and does not require immediate analysis. The sulfur dioxide so trapped is determined colorimetrically by addition of the absorbing medium to hydrochloric acid bleached para-rosaniline hydrochloride solution. This method was found to be rapid, sensitive, and specific for sulfur dioxide in brewing materials and should be investigated for wine. Lloyd and Cowle (1963) used this procedure as the basis of a procedure for determining free sulfur dioxide in soft drinks. At pH 1 to 2 the desorption of free sulfur dioxide is much faster than the dissociation of combined sulfur dioxide and the free sulfur dioxide may be desorbed at room temperature by a stream of oxygen-free nitrogen. The free sulfur dioxide may be trapped in alkaline glycerol and the determination completed iodimetrically for beverages containing 50 mg per liter or more of free sulfur dioxide. Lower concentrations are trapped in sodium tetrachloromercurate reagent and the determination is completed colorimetrically.

Iron

Total iron in wines is easily and accurately determined by the procedure developed by Saywell and Cunningham (1937). The method involves wet-ashing and the development of a colored complex which ferrous iron forms with o-phenanthroline. The danger of using laboratory-fermented samples for metal analysis was demonstrated by Corrao and Gattuso (1964a). In sixteen laboratory-fermented wines the iron content varied from 0 to 3.3 mg per liter (average 1.5), whereas in eighty-five commercial wines from the same region it ranged from 0.5 to 36.3 (average 12.3).

Corrao (1963*a*) recommended ashing of samples before determining the total iron content. Failure to do so generally gave very high results.

Pipette 2 ml of wine into 25 × 150 mm Pyrex test tubes previously marked at 10 ml with a file or wax pencil. Evaporate just to dryness, cool, and add 1 ml of concentrated sulfuric acid. Heat with care over a flame, under a hood, until the contents of the tube are completely liquefied. Cool partially, then carefully add 2 to 4 drops of the perchloric acid. Continue the digestion and addition of perchloric acid until the sample is clear *and until all the excess perchloric acid has been evaporated off.* Usually about 10 drops are required. At this state, set the tubes aside to cool. *Caution:* The entire digestion must be conducted behind a shatter-proof glass because perchloric acid in the presence of organic material occasionally explodes during heating. The hand behind the glass should be protected by an asbestos glove.

Add 2 ml of distilled water and a small piece (0.5 cm square) of Congo red or wide-range pH paper. Then add 1 ml of a 10 per cent aqueous solution of hydroxylamine hydrochloride and 1 ml of a 0.1 per cent solution of *o*-phenanthroline in 50 per cent alcohol. Titrate to the color change (blue to a light red) of the Congo red paper or to a pH of 5 on the wide-range paper, with concentrated ammonium hydroxide, and set aside to cool. Then make to 10 ml with distilled water and transfer to standardized test tubes for comparison with a series of standards, or to colorimeter tubes for comparison in a visual or photoelectric colorimeter at about 490 mμ (a blue-green filter will do.)

Prepare a standard stock solution of iron to contain 1 mg of iron per ml. To prepare a standard solution, carefully dissolve 1 gr of clean, rust-free iron wire (special for standarization) in a few ml of hot, concentrated hydrochloric acid and dilute to 1 liter with distilled water. From this standard stock solution, prepare a series of solutions containing known concentrations (in parts per million of iron). One ml of this stock solution diluted to a liter gives a standard containing one part per million. Run 2 ml of these solutions through the procedure given above for the unknown, using all the reagents and following the directions closely. Transfer to standard-diameter test tubes, stopper tightly, and save for use as a series of standards. These standards are stable for long periods, and the procedure outlined automatically corrects for the iron in the reagents. In like manner, the solutions can be used to establish a curve for use with a photoelectric colorimeter. In this case, however, a blank containing water instead of an iron solution, but containing all the reagents, must be prepared. Use this blank to set the instrument to zero reading.

The recommended procedure of the Office International de la Vigne et du Vin (1962) is essentially the same, but the transmission is determined at 474 mμ. Jaulmes (1965b) also recommends this procedure. Using an atomic absorption spectrophotometer, Caputi, Jr., and Ueda (1967) successfully determined copper and iron in wines by aspiration of the undiluted sample. Corrections for the effect of alcohol and solids need to be made. The determination took only twelve seconds.

Rapid method for iron

The following method measures the inorganic iron in solution. It is more rapid than the digestion procedure described above, and is sufficient for tests on cloudy wines where high iron is suspected. The method depends on the red color developed by ferric and thiocyanate ions in acid solutions.

Pipette 10 ml of the standard solutions containing 0, 1, 5, and 10 parts per million of iron into large test tubes. Add 2 ml of dilute hydrochloric acid to each, then 1 ml of potassium thiocyanate (50 per cent solution), and 5 drops of hydrogen peroxide. (The hydrogen peroxide is to make certain that the iron present is in the ferric state.) Shake the solution. Pipette 10 ml of white wine into each of two large test tubes. Add 2 ml of dilute hydrochloric acid and 1 ml of the thiocyanate solution to both tubes. To one of the tubes (A), add 5 drops of hydrogen peroxide and to the other (B) add 5 drops of water. Shake and compare the color. The tube to which hydrogen peroxide has been added should indicate the total inorganic iron content. The other tube will indicate the iron in the ferric state in the wine. $A - B$ is the ferrous-iron content. Compare the colors by matching with the standards to indicate the range of concentration; for a more exact determination, make a colorimeter comparison with the standard having the same approximate concentration. Report results as mg per liter.

For red wines, proceed through the last step, then add 10 ml of a 1:1 mixture of ethyl ether plus ethylene glycol mono-n-butyl ether, and shake. Draw off the colored ether solution and compare with the color of an ether solution extracted from a properly prepared standard. Jaulmes (1965b) believes that a better solvent should be developed. We agree. A number of organic solvents may be used, providing the comparison standards are prepared with the same solvents, but none are perfect.

Copper

Copper in wine is accurately determined by a modified Coulson (1937)–Drabkin (1939) procedure. It is a colorimetric test using sodium diethyldithiocarbamate as the reagent for color production. When used under the conditions of the test, the reagent is specific for copper, and forms a reasonably stable color complex which can be compared against a series of standards, by visual examination or in colorimeters of the Duboscq type, or against standard curves in photoelectric photometers.

Pipette 25 ml of wine into a 100-ml Kjeldahl flask. Place the flask in a hot-air oven at 100°C and heat until the sample is dry. Remove from the oven and, when *cool*, add 5 ml of concentrated sulfuric acid and 5 ml of concentrated nitric acid. Heat gently until red nitrous oxide fumes appear, and set aside overnight. Then heat, gently at first and more strongly as frothing ceases, to the complete disappearance of nitrous oxide fumes or to the appearance of sulfur trioxide fumes. Set aside to cool and, when cool, add 1 ml of 70 per cent perchloric acid. Again heat until the fumes disappear and there is complete clearing of the solution. Cool.

Rapidly add about 55 ml of distilled water and transfer the contents of the Kjeldahl flask to a 150-ml beaker. Wash out the flask and transfer the washings to the beaker, using the minimum amount of water necessary. Add 10 ml of hydrochloric–citric acid reagent.[8] Then neutralize to litmus with concentrated ammonium hydroxide and add 0.2 ml excess. Add 1 ml of 1 per cent aqueous solution of sodium diethyldithiocarbamate and transfer the content of the beaker, by careful washing, to a 125-ml, pear-shaped, separatory funnel which has been previously marked at 50 ml. Make up to this volume with distilled water. Then accurately pipette 10 ml of isoamyl acetate over the solution, stopper the funnel, and shake with reasonable vigor for one minute. Allow the two liquid phases to separate, draw off the aqueous phase, dry the stem of the separatory funnel with a pipe cleaner, and transfer the organic phase of the solution to the colorimetric comparator tubes.

Standard copper solution is best prepared from reagent-quality copper metal so that it will contain 1.25 mg copper per ml. The solution for the preparation of the series of standards is prepared by pipetting 5 ml of the above solution into a 500-ml volumetric flask and making up to volume with distilled water. Pipette 1-, 2-, 3-, 4-, 5-, and 6-ml aliquots of the latter solution into 100-ml Kjeldahl flasks. Add 5 ml of concentrated sulfuric acid and 5 ml of concentrated nitric acid to each flask and boil until all the nitric acid has been dispelled. Cool, add 1 ml of 70 per cent perchloric acid, and again boil to dispel this latter acid. Again allow to cool and then quickly add 5 ml of distilled water. Transfer the contents of the Kjeldahl flasks to 150-ml beakers, and proceed as directed above for the unknown. Draw off the extracted colored solution into tubes of

---

[8] Prepared by dissolving 75 gr of citric acid in 300 ml of distilled water in a 500-ml volumetric flask, adding 50 ml of concentrated hydrochloric acid, and making to volume with distilled water.

the same size and diameter as those used for the unknown, and stopper tightly with corks. Compare the unknowns against the standards in a Walpole block or other convenient comparator. Standards prepared as described above automatically correct for copper in the reagents and are stable for at least one month.

These standards contain 0.0125, 0.025, 0.0375, 0.050, 0.0625, and 0.075 mg of copper per 10 ml of colored solution, respectively, which, under the conditions of the test, correspond to 0.5, 1.0, 1.5, 2.0, 2.5, and 3.0 mg per liter of copper in the wine, respectively. When the unknown contains more than 3.0 mg per liter of copper, repeat the determination, using a smaller aliquot as sample, but make due allowance for the new volume of the aliquot when calculating the copper concentration.

Where a Duboscq colorimeter is available, only two standards at the most need be prepared, one using 1.5 ml of the diluted standard solution, the other using 3.0 ml. At these concentrations of copper the color intensity is proportional to the concentration, and balancing methods yield reliable values. A photoelectric colorimeter may also be used.

Marsh copper method

The Coulson-Drabkin procedure is time-consuming, and careful attention to details is required for accurate results. Nearly as accurate values can be obtained by the following method devised by Marsh (1950) and in a very much shorter period of time. The Marsh method is better suited to the average winery laboratory. Wet-ashing is avoided and the procedure is simpler.

Pipette 10 ml of wine into a 25 × 150-mm Pyrex test tube. Add 1 ml of the hydrochloric–citric acid reagent, shake, and then add 2 ml of ammonium hydroxide (333 ml concentrated NH_4OH per liter), and again shake. Then add 1 ml of a 1 per cent aqueous solution of sodium diethyldithiocarbamate, shake, and set aside for a minute or so before adding 10 ml of amyl acetate. Follow the addition of amyl acetate with 5 ml of absolute methyl alcohol.

Place the palm of the hand over the top of the test tube and shake vigorously for at least thirty seconds. Set aside and allow the two phases to separate. When separation is complete, draw off the aqueous phase by inserting a length of glass tubing to the bottom of the test tube and applying suction. The organic phase is dried by adding powdered anhydrous sodium sulfate from the tip of a spatula, and shaking. Add just enough to accomplish the purpose. It should be added while holding the tube at an angle and rotating it to dry the moisture film adhering to the walls.

Transfer the dried organic phase to clean, dry test tubes for color comparison against a set of standards, or to the colorimeter tubes of the Duboscq color comparator, or to a photoelectric colorimeter. If the organic phase is turbid, as it frequently is, it should be filtered.

The absolute methyl alcohol which is added to the reaction mixture serves two purposes. It markedly reduces the tendency of the two phases to emulsify and thereby aids in a quick and clean separation of the aqueous and organic phases. More important, however, is its second purpose. Coulson showed that the intensity of the color which is extractable from the aqueous phase by amyl acetate is dependent upon the pH value of the aqueous phase. At pH 8.0 to 8.5, maximum color intensity is developed; great care must be exercised in adjusting the pH value if accurate values are to be obtained. Marsh, in developing the procedure outlined above, found that when methyl alcohol was added to tubes containing samples adjusted to varying pH values, no difference in color intensity of the extracted colored solution could be detected. Without methyl alcohol the color intensity of the extracted colored solution was dependent upon the pH value of the aqueous phase.

Standard copper solution for this procedure is best prepared from Merck copper metal (reagent quality) so that it will contain 0.50 mg of copper per ml. Solutions for the preparation of the series of standards are prepared by pipetting 1, 2, 3, 4, 5, 6, 8, and 10 ml of the above stock solution into separate 1,000-ml volumetric flasks, adding 150 ml of 95 per cent ethyl alcohol to each, and making them up to volume with distilled water. These solutions then contain 0.5, 1.0, 2.0, 2.5, 3.0, 4.0, and 5.0 mg per liter of copper, respectively. Pipette 10 ml of each solution into separate 25 × 150-mm test tubes and proceed as directed for the unknown. The solutions so obtained can be used as a series of standards or can be used to establish standard curves for use with photo-electric photometers.

There is no copper procedure of the Office International de la Vigne et du Vin (1962) at present. Copper forms a colored compound with zinc dibenzyldithiocarbamate (ZDBT) which can be extracted with carbon tetrachloride. Strunk and Andreasen (1967a) obtained good results on sherry and vermouth using this procedure. They (1967b) found equal precision using atomic absorption spectrophotometry. See also Caputi, Jr. and Ueda (1967). This procedure is now preferred.

Other cations

For further information on the colorimetric determination of traces of metals, see Sandell (1944). Semimicro methods for the metals in wines are given by Bonastre (1959). The early methods of Heide and Hennig

(1933) should also be consulted for arsenic, zinc, and manganese. Méranger and Somers (1968) used atomic absorption spectrophotometry for determination of copper, zinc, nickel, chromium, lead, cadmium, and cobalt in wines. The most recent corrections to the Office International de la Vigne et du Vin's Recueil (Anon., 1968) give a procedure for determination of arsenic. We know of no reason for making this determination on normal American wines.

For the determination of ash, 20 ml of wine are ashed in a tared platinum dish. The wine is carefully evaporated to dryness to avoid splattering. The residue should be further dried at 100°C (212°F) for an hour or more and then ashed in a muffle furnace at 525°C (977°F), cooled, 5 ml of distilled water added, again evaporated and ashed. If the ash is not white or grayish after fifteen minutes the process is repeated and ashed at 550°C (1022°F). Cool in a desiccator and weigh.

The determination of total cations can be made by the procedure of Bonastre (1959). Place 5 or 6 gr of Dowex 50 in the acid form in a burette, wash and pass 10 ml of wine through (one drop every two seconds). Rinse the column three times with 10 ml of distilled water. Determine the acidity, n', as ml 0.1 N alkali. Determine the acidity, n, of 10 ml of wine. Then $n' - n = \Delta n$ and 10 Δn is the total cation content in milliequivalents per liter.

The alkalinity of the ash represents the carbonates derived from organic anions during ashing. If the alkalinity of the ash is expressed in milliequivalents as A, then 10 $\Delta n - A$ represents the cations combined to mineral anions. To determine the alkalinity of the ash place the platinum dish containing the ash in a beaker and add 10 ml of 0.10 N sulfuric acid. Heat to incipient boiling using a glass rod to be sure all the ash is dissolved. Add 1 ml of 0.01 per cent methyl orange and titrate the excess sulfuric acid with 0.1 N sodium hydroxide. Net ml of sulfuric acid used = n. The alkalinity of the ash = $5(10 - n)$ in milliequivalents per liter. Expressed in potassium carbonate per liter = $0.345(10 - n)$.

In using flame photometry for potassium, sodium, and calcium in fruit juices, Ditz (1965) emphasized the importance of preparing the standard curves containing normal amounts of the other cations as well as sugar and citric acid. Jouret and Poux (1961a) reported that simply diluting wines 1:25 or 1:50 made it possible to determine potassium directly with an error of ± 3 per cent in the usual wine and ± 5 per cent in the most unfavorable case. They especially did not recommend ashing, as there was often a loss, apparently due to volatilization, during ashing. The chief interference is from sulfates and phosphates. For precise details for the flame photometer procedure for potassium and sodium see Association of Official Agricultural Chemists (1965). The reference procedure of the Office International de la Vigne et du Vin (1962) is gravimetric, using sodium tetraphenylborate or precipitation of potassium acid tartrate from an ashed sample. For determining sodium by flame photometry, Jouret and Poux (1961b) recommend diluting 1:10 or 1:25. Since sulfates and phosphates both cause low sodium results, they recommend adding 0.01 to 0.3 milliequivalent of sodium per liter to the diluted samples. Of course, sulfate and phosphate could be removed by use of ion exchange. The procedure of the Office International de la Vigne et du Vin (1962) also is a flame photometric method using a reference solution. For information on flame photometry see Dean (1960).

A flame photometer and complexometric (EDTA) procedure were favorably compared for calcium determination by Farey and Macpherson (1966). However, the sample for flame photometry was prepared by oxalate precipitation. They found that oxalate ion interfered, and added 0.1 N perchloric to compensate for the oxalate. The procedure of the Office International de la Vigne et du Vin (1962) uses an ashed sample and complexometry. Iwano and Sawanobori (1962) used EDTA for calcium and magnesium.

Bălănescu (1963–1964a) diluted wines 1:100 or 1:50 for potassium and 1:10 or 1:5 for sodium or calcium, using flame photometry and acetylene-air. Reducing sugar up to 5 per cent did not influence the results.

To determine manganese directly in wines Ramos and Gomes (1966) used formaldoxime following defecation with cadmium sulfate and charcoal. The color was measured at 450 mμ. The results were as good as those obtained using the more laborious permanganate procedure. A new colorimetric procedure for zinc in wines is given by Jouret and Bénard (1967).

Polarographic procedures for the determination of antimony and nickel were given by Eschnauer (1965a, 1965b, 1966a). Many of these determinations can be more rapidly made with an atomic absorption spectrometer.

Other anions

A simple potentiometric procedure for determination of chloride ion in wines was presented by Hubach (1966). The reference procedure of the Office International de la Vigne et du Vin (1962) uses barium hydroxide precipitation and the usual silver nitrate method. The ordinary procedure uses silver nitrate directly. See Amerine (1965b) for a satisfactory procedure. The official method of the Association of Official Agricultural Chemists (1965) ashes an alkaline solution, extracts with hot water, dissolves in nitric acid, and determines gravimetrically as silver chloride or volumetrically by titrating excess silver with thiocyanate. Stella's procedure (1967) for detecting organic chloride compounds in musts and wines should detect residual organic herbicides containing chlorine. A colorimetric procedure for total bromide was given by Guyot and Balatre (1966). Their data and that of others indicated less than 1 mg per liter.

Sulfate is most conveniently determined by barium precipitation (Amerine, 1965b). For a more accurate method involving removal of sulfur dioxide see Office International de la Vigne et du Vin (1962). The official procedure of the A.O.A.C. (1965) is to acidify and heat the wine and add hot barium chloride by drops. The precipitated barium sulfate is separated, ashed, and weighed. Results are usually reported as potassium sulfate.

Phosphate should be determined by the official procedure of the A.O.A.C. (1965). The sample is either wet- or dry-ashed and the color is developed with a molybdovanadate reagent. The transmittance is measured at 400 mμ. A standard curve is prepared. The rapid procedure of Joslyn and Luckton (1953) usually gives good results. In a colorimetric procedure for nitrates developed by Rebelein (1967b) recovery was good. In about 500 musts and wines the nitrate content (as N_2O_5) was 6 to 8 mg per liter. The ratio of N_2O_5 to ash for undiluted wines should not exceed 0.008, according to Rebelein.

Esters

Numerous esters are found in wines: neutral ethyl esters of acetic, lactic, malic, citric, tartaric, and probably other organic acids; and ethyl acid esters of dibasic acids. The total esters are thus classified as neutral esters and acid esters. The determination of neutral esters is more important because the volatile neutral ester, ethyl acetate, is an important spoilage product. The method outlined here for neutral esters is essentially that of Peynaud (1937) as used by Amerine (1944). A continuous liquid-liquid extraction apparatus is required for the determination.

Pipette 25 ml of wine into a beaker and neutralize to a pH of 7.5 in a pH meter, using N and 0.1 N sodium hydroxide. Carefully wash the neutralized wine into the extraction chamber of a liquid-liquid extraction apparatus, using carbon-dioxide-free distilled water. Pipette 10 ml of 0.05 N carbonate-free sodium hydroxide into the receiving flask and pour freshly distilled, neutral petrol ether through the condenser until about 40 or 50 ml overflow into the receiving flask. Adjust the heater so that about 500 ml of ether condense per hour, and continue the extraction for 24 hours. Disconnect the receiving flask and carefully titrate the residual sodium hydroxide with 0.05 N sulfuric acid to a phenolphthalein end point. The total neutral esters as mg of ethyl acetate per liter = (ml NaOH × N NaOH) − (ml H_2SO_4 × N H_2SO_4) × (1,000/ ml wine used) × 88. A blank extraction should be made, and the ml of sodium hydroxide used up during extraction should be subtracted.

Should one wish to know the approximate ethyl acetate content of the sample as distinguished from the ethyl acetate plus ethyl tartrate, etc., determined above, transfer the neutralized total neutral ester solution to a separatory funnel, using carbon-dioxide-free distilled water. Remove the aqueous layer into a 100-ml volumetric flask. Wash the petrol ether several times with water, and add the washings to the volumetric flask. Bring the flask to volume. Transfer the contents of the volumetric flask into the boiling flask of a distillation assembly (similar to that used in the alcohol determination). Rinse the flask with 2 ml of N sulfuric acid and 8 ml of water and add the rinsings to the boiling flask. Distill 100 ml of the 110 ml into a 250-ml Erlenmeyer flask. About 90 per cent of the volatile acids are distilled. Heat the distillate to boiling and titrate to a phenolphthalein end point, using 0.05 N sodium hydroxide. The blank extraction should also be distilled. Subtract the number of ml required for the blank from that for the sample. The volatile neutral esters as mg of ethyl acetate per liter = net ml NaOH × N NaOH (1,000/ml wine used) × 88.

Direct procedures for the colorimetric determination of esters have been developed by Hestrin (1949) and Thompson (1950). These depend on the reaction of esters with hydroxylamine in alkaline solution to form a hydroxamate complex. The reaction is very sensitive and only 1 ml of wine is required. At present the following procedure of Libraty (1961) is used at Davis. The reactions are:

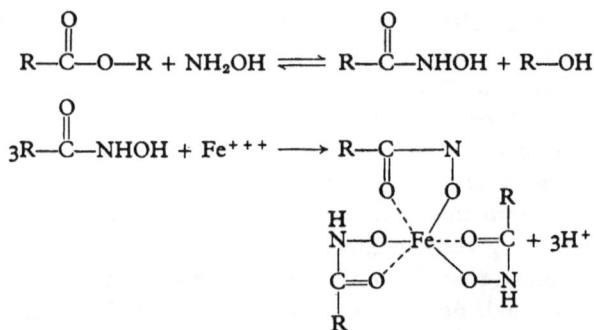

Prepare pure ethyl acetate by washing ethyl acetate (vol. to vol.) with 5 per cent sodium carbonate and then with a saturated solution of sodium chloride, separating and drying over calcium chloride, followed by fractional distilling. From a stock solution of 1 gr ethyl acetate per liter make up solutions of 25, 50, 75, and 100 mg per liter. (For the alternate method when using alcohol distillates make up the solution in 10 per cent ethyl alcohol solution saturated with potassium acid tartrate; dilute each from 100 ml to 150 ml with water, distill and collect 90 ml of distillate, and bring to 100 ml volume.)

Pipette 3 ml of each of the standard solutions into a standard Klett-Summerson colorimeter tube; add 2 ml of hydroxylamine solution and 2 ml of 3.5 N sodium hydroxide. Stopper with a neoprene stopper, shake, and allow to stand for ten minutes. Then add 2 ml 4 N hydrochloric acid and 2 ml of ferric chloride reagent (10 per cent $FeCl \cdot 6H_2O$ in 0.1 N hydrochloric acid). Shake and read directly in colorimeter at 510 mμ or with the Klett-Summerson colorimeter use the No. 54 filter. From these colorimeter values plot a standard curve—colorimeter reading versus standard solution ethyl acetate concentration.

For the sample to be tested, pipette 50 ml wine into a Kjeldahl flask plus 4 ml water and distill 48 ml into a 50-ml flask immersed in ice. The flask contains 2 ml of water. Or, alternately, use the distillate of 100 ml wine diluted with 50 ml water and distill 90 ml into a 100-ml flask. Bring to volume. Take 3 ml of the distillate and treat as the standard solutions to develop the color and read in colorimeter. Determine the concentration of volatile esters from the standard curve and report as ethyl acetate. P. Ribéreau-Gayon (1964a) used gas chromatography for determining ethyl acetate in wines.

Aldehyde

Acetaldehyde is a normal by-product of alcoholic fermentation. However, unless considerable sulfur dioxide has been added during fermentation, there is little accumulation of the aldehyde. New wines thus contain only 20 or 30 mg of aldehyde (calculated as acetaldehyde) per liter. But wines fermented in the presence of sulfur dioxide may contain 100 mg per liter or more. During aging, some aldehyde is produced by microörganisms and by direct oxidation, particularly if the temperature is raised—as in sherry production—or if the wine is stored in contact with air over a long period of time. Finally, during cellar handling of wines—racking, filtering, bottling, etc.—variable quantities of oxygen are introduced into the wine. Following such aeration there is a rapid increase in the free-aldehyde content, and the wine acquires a distinct "flat" odor. This "aerated" odor is transitory (if the aeration is not continued over a long period) and disappears ofter a few weeks in the absence of air. The following method has been studied by Guymon and Crowell (1963). In the tests of Guymon and Wright (1967) it gave standard deviations of 2.7 to 4.5 per cent of the mean.

763

Pipette 50 ml of wine (containing not over 0.03 gram of aldehyde as acetaldehyde) into a 500-ml Kjeldahl flask. Add 50 ml of saturated borax solution (or a phosphate buffer when using wines of over 4 per cent sugar) and boiling chips (to prevent bumping). Connect the Kjeldahl flask to the condenser and place a calcium chloride tube on the end of the condenser as an adapter. Mark a 750-ml Erlenmeyer flask at 370 ml, add 300 ml of water and 10 ml each of metabisulfite[9] and phosphate-EDTA[10] solutions. The pH should be 7.0 to 7.2. If necessary, adjust by adding hydrochloric acid or sodium hydroxide to the metabisulfite solution. Place this flask so that the calcium chloride tube just dips under the surface of the liquid. Distill about 50 ml.

Lower the receiving flask and rinse off the calcium chloride tube into the flask with a small amount of distilled water. In the neutral buffer solution the reaction of aldehyde and bisulfite to form the aldehyde-bisulfite complex is rapid and complete. Add 10 ml of fresh 0.2 per cent starch indicator and 10 ml of 1 + 3 hydrochloric acid (to bring the pH to below 1.0). Swirl to mix. Oxidize the excess bisulfite by titrating rapidly with 0.1 N iodine (unstandardized) to a faint blue end point. Wash the sides of the flask down with water. It is best to approach the end point with 0.1 N iodine and then the complete the titration with 0.01 N iodine. Should excess iodine be added, use dilute thiosulfate to remove the blue color and again titrate to the faint blue end point, using 0.01 N iodine. The reaction of hydrogen ion and the aldehyde-bisulfite complex at a pH of 0 to 1 is slow. But at the same pH, the oxidation of bisulfite by iodine is rapid.

Add 10 ml of sodium borate solution[11] to the contents of the Erlenmeyer flask to hydrolyze the aldehyde-bisulfite complex. At the pH of 8.5 to 9.0, the hydrolysis is rapid. Carefully titrate the released sulfite with 0.02 N (standardized) iodine to a faint blue end point. The oxidation reaction at this pH is: $SO_3^= + I_3^- + 2OH^- \rightarrow 3I^- + SO_4^= + H_2O$. The number of milliequivalents of sulfite used in this titration is equal to the number of equivalents of aldehyde originally present, but from the last equation it can be seen that 1 millimol of sulfite is equal to 2

---

[9] Dissolve 15 gr of potassium metabisulfite in water, add 70 ml of hydrochloric acid and dilute to a liter with water. The titer of 10 ml of this solution should not be less than 24 ml of 0.1 N iodine.

[10] Dissolve 200 gr of sodium phosphate ($Na_3PO_4 \cdot 12H_2O$) and 4.5 gr of EDTA (disodium salt) in water and dilute to a liter. Instead of the sodium phosphate use 188 gr sodium acid phosphate ($Na_2HPO_4 \cdot 12H_2O$) and 21 gr sodium hydroxide, or 72.6 gr sodium acid phosphate ($NaH_2PO_4 \cdot H_2O$) and 42 gr sodium hydroxide, or 71.7 gr sodium acid phosphate (NaH_2PO_4) and 42 gr of sodium hydroxide.

[11] Prepared by dissolving 100 gr of boric acid and 170 gr of sodium hydroxide in water and making to a liter.

milliequivalents of iodine. The milliequivalent weight of the aldehyde as aldehyde is thus $\frac{44}{2}$, or 22. The milligrams of aldehyde per liter (as acetaldehyde) = ml of iodine × N iodine × 22 × (1,000/ml of wine used). The recovery of acetaldehyde by this procedure is approximately 90 per cent. In warm weather or when the condenser water is warm, cool the neutral buffer-bisulfite solution during the distillation.

Glycerin

Glycerin (or glycerol) is an important by-product of fermentation. On the average, about 4 per cent of the sugar fermented is converted to glycerin, but many factors influence the percentage conversion: temperature, percentage of sugar, amount of sulfur dioxide, acidity, yeast strain, and so on.

The determination of glycerin is rarely made in winery practice. It has some value in distinguishing between sweet table wines which have been produced by fermentation and those which have been produced by adding fortified wines or grape juice to a dry wine. In some instances of spoilage the decrease in glycerin is of diagnostic value. It is useful also in experiments to determine the influence of temperature, aeration, and other factors on the quality of the wines produced.

The older and more popular methods for glycerin determination are based on the principle of destroying the sugars by heating with calcium hydroxide, and extracting the residual glycerin with an alcohol-ether mixture. This is the basis of the official A.O.A.C. procedure (Association of Official Agricultural Chemists, 1965). The separated glycerin is either weighed or oxidized with dichromate. While the method seems satisfactory for wines of low sugar content (less than 5 per cent), it is difficult to determine how much calcium hydroxide to use for clarifying sweeter wines.

If there is less than 2 per cent reducing sugar, proceed as indicated in the next paragraph. For other wines a preliminary clarification is necessary. Pipette 50 ml of wine into a 200-ml volumetric flask and place the flask in a boiling-water bath. While it is heating, add small amounts of freshly prepared calcium hydroxide solution (15 gr of CaO per 100 ml of water). The color should first turn dark brown and, after heating, a lighter color. Discontinue the additions of the base when this second color change occurs. Cool and bring to volume with 95 per cent ethyl alcohol. Shake well and allow to settle.

Pipette 100 ml of the clarified solution described in the paragraph above (25 ml of the original wine) or 100 ml of wine (of less than 2 per cent reducing sugar) into an evaporating dish. Place the dish on a hot water (not boiling) bath and evaporate to 20 ml. In this and other evaporations the area of the dish exposed to the bath should at no time be greater than the area in the dish covered by liquid. This is to prevent loss of glycerin by volatilization because of local overheating.

While stirring, treat the residue with 5 gr of fine sand and about 5 ml of the freshly prepared calcium hydroxide solution per gram of extract (usually 10 to 15 ml). Continue heating and stirring the solution until it is nearly dry; then add 50 ml of 90 per cent ethyl alcohol. Use a rubber policeman to rub the precipitated material from the sides of the dish. Continue heating and stirring for a few minutes and then decant off through filter paper into another evaporating dish. Add 10-ml portions of hot 90 per cent alcohol, stir, and decant through the filter paper. Continue this until there is about 150 ml of filtrate. The success of this determination depends on extracting all the glycerin from the gummy mass.

Evaporate the filtrate at low heat, not over $32.2°C$ $(90°F)$, to a sirup. Carefully transfer to a glass-stoppered, 50-ml volumetric flask, using about 20 ml of absolute alcohol and 30 ml of anhydrous ether. Shake and let stand until clear. Mannite, a spoilage product in wines, is insoluble in a $2:3$ alcohol-ether mixture, and is removed by this step. Pour the contents of the 50-ml volumetric flask through a filter into a small porcelain evaporating dish (previously heated in a muffle furnace to $700°C$ $(1292°F)$. Wash the flask and filter with a mixture of two parts absolute alcohol and three parts anhydrous ether. Evaporate the filtrate to a sirup at room temperature on a hot-water bath (not over $80°C$; $176°F$). Dry in an oven at $89°$ to $100°C$ $(192°$ to $212°F)$ for one hour, cool in a desiccator, and weigh. Place the evaporating dish on a muffle furnace and heat at $600°C$ $(1112°F)$, cool in a desiccator, and reweigh. Loss in weight equals weight of glycerin in 25 or 100 ml of wine. Results are reported as grams of glycerin per 100 ml of wine.

The determination of glycerin by periodic acid oxidation is the basis of several procedures (Amerine and Dietrich, 1943; Elving et al. 1948; Peynaud, 1948). Excellent results with relatively rapid operation are obtained on wines containing less than 0.5 per cent sugar. However, separation of glycerin from larger quantities of sugars is often difficult and time-consuming. Peynaud (1948) has shown that 2,3-butylene glycol causes a 4 to 7 per cent positive error in the glycerin determination by periodic acid, but he gives methods for avoiding this.

The Office International de la Vigne et du Vin (1962) suggests a reference procedure of converting glycerin to acrolein and then to quinoline using sulfuric acid, aniline, and nitrobenzene. The precipitated quinoline is weighed. For ordinary work they suggest oxidation with periodic acid. The formaldehyde produced by periodic acid oxidation of glycerol forms a color with phloroglucinol which is measured at 480 mμ. The acetaldehyde produced by oxidation of 2,3-butylene glycol forms a color with piperidine and sodium nitroprussiate which is measured at 570 mμ. (See also Jaulmes, 1965a.)

Other compounds

For precise determination of methanol see improvements in the chromotropic acid procedure by Rebelein (1965b). One important change is the more precise control of the sulfuric acid content. For recent discussions of the determination of higher alcohols see Webb and Kepner (1961), Maurel et al. (1965), and Brunelle (1967). The latter, using gas-liquid chromatography, obtained results as good as by the A.O.A.C. procedure.

Methods for the determination of the total nitrogen and nitrogen fractions were reviewed by Cordonnier (1966). A comparison of methods for determining total nitrogen and the various nitrogen fractions was given by Kichkovsky and Mekhouzla (1967). Methods for determining proteins in wine were studied by Avakiants et al. (1967). The latter recommended two precipitations with trichloracetic acid. See also Diemair et al. (1961a, b). For methods for the determination of amino acids see Lafon-Lafourcade and Peynaud (1959) and Ough and Bustos (1969).

France authorized addition of up to 100 mg per liter of ascorbic acid in 1962. Maurel et al. (1964) give a procedure accurate to 10 per cent for its determination. For a more precise determination of ascorbic acid Ournac (1966) recommends clarification with mercuric acetate followed by precipitation as the dinitrophenylhydrazone and thin-layer paper chromatography. The R_f of the hydrazone is 0.75. No dioxyfumaric or dioxytartaric acids were detected. The author concludes that the hydrazone precipitated under these conditions is suitable for the determination of ascorbic acid. By this procedure wines appear to have 3 to 9 mg per liter and musts 20 to 60 (see Ournac, 1965).

Fruit wines contain a high sorbite content, 5 to 10 mg per liter, according to Rentschler and Tanner (1966). Using their sensitive test, addition of 5 per cent fruit wine to grape wine can be detected. Tanner (1967) has given thin-layer paper chromatographic procedures for identifying sorbite, mannite, and other polyhydroxy compounds in wines. The R_f values were as follows: sorbite, 0.25; mannite, 0.33; sucrose, 0.35; glucose-fructose, 0.43; glycerin, 0.57; ethyleneglycol, 0.63; 1,2-propylene glycol, 0.70; 2,3-butylene glycol, 0.74; and divinyl glycol, 0.82.

The small amount of sucrose in grapes disappears during fermentation. A paper chromatographic procedure for detecting small amounts of sucrose in wines was given by Guimberteau and Peynaud (1965). Sucrose added to wines hydrolyzes very rapidly—in less than six days in two wines and in twenty days in another. A small but detectable amount of sucrose remained in the other two. Therefore, while the presence of sucrose is an indication of sugaring, the determination must be made soon after addition of the sucrose.

The problem of detecting added sucrose is important in countries where such addition is against the law or needs to be controlled. Navara et al. (1966b) demonstrated that sucrose hydrolysis proceeds very rapidly in musts and wines. During fermentation there is considerable difference in the speed of inversion using various yeasts. The speed of hydrolysis was greater at higher sugar concentrations. Tartaric acid catalyzed the hydrolysis better than malic acid. The products of sucrose hydrolysis, besides glucose and fructose, include raffinose, melibiose, lactose, and maltose. In the relatively high-acid, low-pH wines of Czechoslovakia they reported that sucrose hydrolysis was complete in 36 hours for 1.8 per cent sucrose. The Office International de la Vigne et du Vin (1962) suggests several provisionary procedures including paper chromatography, reaction with diphenylamine hydrochloride or increase in reducing ability following hydrochloric acid hydrolysis. (See also Jaulmes, 1965a.)

As of 1965 the enzymatic, volumetric, and manometric procedures for carbon dioxide in wine have been accepted by the Association of Official Agricultural Chemists, but only the enzymatic procedure is to be retained. For the latest information on the enzymatic procedure see Morrison (1965). For carbon dioxide, preference for the Hennig and Lay procedure is expressed by the Office International de la Vigne et du Vin (Anon., 1968).

For the enzymatic determination cool the sample to 0°C (32°F) or lower in order to pipette the sample with minimum loss of carbon dioxide. With an automatic 25- or 30-ml pipette with a Teflon stopcock transfer an aliquot of 0.1 N sodium hydroxide into a beaker. Rinse a 20-ml pipette with the cold sample and pipette 20 ml of wine. Transfer to beaker with tip of pipette in the alkali. Add 3 to 4 drops of carbonic anhydrase solution (an aqueous solution containing 1 mg per ml which is stable for about two weeks if kept in a refrigerator). Titrate to pH 8.45 with 0.1 N sulfuric acid (from a burette calibrated to 0.01 ml in a pH meter). To correct for presence of acids other than carbonic, place 50 ml of wine in a heavy-walled flask at room temperature and agitate one minute under 27 inches vacuum. Titrate 20 ml to pH 7.75 with 0.1 N sodium hydroxide. Net ml sodium hydroxide × normality × 100/20 × ·44 = mg carbon dioxide per 100 ml wine.

A rapid polarographic procedure for determining carbonyl compounds in wines was given by Petrova et al. (1966). The results checked well with those obtained by the iodometric procedure.

Amati and Formaglini (1965) used thin-layer chromatography for detection of various possible preservatives in wines, including benzoic, sorbic, cinnamic, p-chlorobenzoic and dehydroacetic acids, vitamin K_3, various hydroxybenzoic acid esters and o-hydroxyquinoline.

Excellent recovery of sorbic and benzoic acids was obtained by Würdig (1966) using gas chromatography. More than 90 per cent of the sorbic acid was recoverable after four years in wine! Losses reported elsewhere should be checked. The Office International de la Vigne et du Vin (1962) suggests determining sorbic acid on an aliquot of the volatile acid distillate, since practically all the sorbic acid is distilled with the acetic acid. A 0.5 ml aliquot (of 250 to 300 ml distillate from 20 ml of wine in their procedure) is placed in a 1-cm silica cuvette to which are added 1.5 ml of a solution made as follows: 0.5 gr sodium bicarbonate, 0.001 gr copper sulfate (Cu $SO_4 \cdot 5H_2O$) and water to a liter. Leave the cuvette a few minutes and then determine the optical density at 256 mμ compared to water. A standard curve should be prepared with pure sorbic acid (using potassium sorbate). The procedure is good from 10 to 300 mg per liter of sorbic acid. One gram of sorbic acid corresponds to 8.92 ml N base. If a wine has 200 mg per liter of sorbic acid the interference with the volatile acid determination amounts to 1.7 milliequivalent, or 0.0107 gr per 100 ml of acetic acid.

Benzoic acid and its derivatives (o-hydroxybenzoic, p-hydroxybenzoic, and their halogen derivatives and sodium salts) may not be used in wines in most countries. Data for their detection and semiquantitative determination were given by Maurel and Touyé (1963). For a general discussion of the detection of preservatives in foods, see Diemair and Postel (1967) and Petrischek and Schillinger (1967).

The use of ultraviolet light to detect sophistication of wines with various fruits was carried out by Gentilini (1966).

The natural occurrence of cyanides in musts and wines (Albonico, 1958; Jaulmes and Mestres, 1960; Tarantola and Castino, 1961; Miconi, 1962a) has led to a variety of methods for their determination (Jaulmes and Mestres, 1962; Tarantola and Castino, 1964; Deibner, 1966b). The procedure of Deibner seems to avoid interference best. Using this technique he found no cyanide in musts and 0–16.5 μg per liter of free cyanide in twelve wines and 0 to 15 combined. More sensitive procedures for the determination of free cyanide, cyanohidrin, and ferrocyanide in musts and wines are given by Albonico (1958), Jaulmes and Mestres (1960, 1962), Tarantola and Castino (1961, 1964), Mestres (1961), Deibner and Mourgues (1963), and Deibner (1966b). The Office International de la Vigne et du Vin (1962) gives distillation procedures for total and free hydrogen cyanide. They also give a rapid qualitative test based on absence of formation of ferric ferrocyanide on addition of ferric ion to an acidified wine. See also Habach (1948).

A rather complicated procedure for the quantitative determination of hydrogen sulfide in wines was developed by Eschnauer and Tölg (1966). Results were accurate to ±3 per cent with a hydrogen sulfide content of 1 mg per liter. With 0.1 mg per liter the error was ±20 per cent.

Pauli and Genth (1966) give a simple procedure for determination of DEPC. They extract 100 ml of wine three or four times with 20 ml of a 105 to 45 mixture of pentane and ether. To the extracts 50 mg of 4-aminoazobenzol is added and the liquid dried over anhydrous sodium sulfate. Add 0.1 ml of acetic acid and take up the residue with 30 ml of the pentane-ether mixture. Treat with n-hydrochloric acid and 40 per cent acetic acid until the water phase is colorless. Determine the O.D. at 450 mμ. If some color remains it can be destroyed. This procedure detects 0.5 gr of DEPC, and up to 2 mg it gives ±10 to 15 per cent recovery. By gas chromatography 1 mg or less can be detected. To determine whether DEPC has reacted with amino acids to form N-carbethoxy amino acids, check the IR spectrum for the characteristic N-C-O group with the band at 1500 cm^{-1}. They can be detected also by thin-layer paper chromatography. Alkyl reactions of DEPC would be very undesirable from the toxicology point of view. Pauli and Genth (1966) do not believe that they occur. Garschagen (1967) recommended gas chromatography for determination of diethyl carbonate (in DEPC-treated wines).

Boidron and P. Ribéreau-Gayon (1967) have outlined the procedures for isolating and identifying the odorous materials in wines. For data on sensory and chromatographic analyses of odors see Sihto et al. (1962, 1964) and Kendall and Neilson (1966).

The problem of the precise determination of the redox potential of wines has been extensively studied by Deibner (1953, 1956b, 1957b). See Anon (1968). Among the many variables which must be controlled are preparation of electrodes, number of electrodes (Deibner recommends four), and time to reach a limiting value (measurements should be made at ten- to fifteen-minute intervals and continued, often for one to two and a half hours, until the change is less than 0.2 mV per minute.) Deibner and Mourgues (1964a) showed a decreased potential as the temperature was increased. They recommended measuring the potential at 20°C (68°F) in millivolts with a simultaneous determination of the pH. The redox potential, rH$_2$, may also be calculated. They recommended no agitation or circulation of the wine during the determination (in contrast to Deibner, 1956a) even in the absence of aeration.

Hennig and Lay (1963) recommended the use of the Beckman oxygen electrode for determination of dissolved oxygen in wines. See Ough and Amerine (1959) for the original procedure. See also Kolthoff and Lingane (1952), Gualandi *et al.* (1959), Sawyer and Interrante (1961), Strohm and Dale (1961), and Letey and Stolzy (1964) for the theoretical discussions of oxygen determination.

A new procedure using Besthorn's hydrazone (3-methyl-benzthiazolon -(2)-hydrazone) was used by Voigt and Noske (1966) to determine the polyphenoloxidase activity.

CHAPTER NINETEEN

WINE DISORDERS

Wines are subject to various types of clouding and spoilage resulting primarily from activity of microörganisms and from metallic contamination.

APPEARANCE
*Changes in the appearance of wines
are usually danger signals.*

Haziness or cloudiness and the presence of sediment indicate a defective product to many consumers, although the wine may be fundamentally sound and, except for the cloudiness, of excellent quality. When consumers learn how to store wine, it need not be stabilized to withstand extremes of temperature and exposure to light and air.

Under certain conditions, however, cloudiness or the presence of a sediment is a good indication of spoilage. Clouding may result from microbial spoilage, metallic deposits, precipitation of certain oxidized products, or coagulation or precipitation of certain colloids or other substances during aging or clarification by heat or cold.

MICROBIAL SPOILAGE
*The bacteria responsible for the spoilage of wine are widely
distributed in nature and are present in practically all wineries.*

The development of bacteria and other microörganisms in wines is restricted by the composition of the must and of the wine, and by the method of fermenting, storing, and handling. The high acidity of dry table wines, their alcohol, their low content of fermentable sugar, of nitrogenous matters, and of other nutrient elements, and the presence of free sulfur dioxide all restrict the development of bacteria. Immunity to spoilage can be increased by balancing a must deficient in acid, by cool and complete fermentation, and by prompt and efficient racking. When reasonable care is taken with such wines, when general cleanliness is practiced, and when contamination with diseased wines is avoided, most of the microörganisms remain dormant or are so poorly adapted to the medium that they develop very slowly.

The acid- and alcohol-tolerant organisms that can develop in wine differ in their requirements for oxygen. Some, like the *Acetobacter*, will not grow in the absence of oxygen; others, like the *Lactobacilli*, while they will grow in the presence of oxygen, develop best under conditions of reduced oxygen content (microaerophiles). The most serious spoilage is caused by the latter.

Engel (1950) investigated the microflora of wine cellars to determine the source of microbial infection of wine. Wine cellars of good and of poor sanitation were selected for the study of the occurrence of molds, bacteria, and yeasts on the walls, floors, and ceilings of fermenting, storage, and bottling rooms, as well as on the equipment used in wine making. Samples were taken under sterile conditions and plated out after dilution with water. The types of molds, yeasts, and bacteria isolated at two wineries are shown in table 76; the number indicates the average frequency of occurrence of a particular species. Some of the strains of *Mucor* isolated had a marked ability to form aromatic esters. The number of bacteria isolated was relatively small, but Engel did not consider them without significance in potential wine spoilage.

TABLE 76

Occurrence of Microörganisms in Swiss Wineries

	Winery 1	Winery 2
Wine yeast	41	41
Saccharomyces	2	–
Wild yeast	9	16
Endomyces	9	–
Torulopsis	49	2
Pichia	17	–
Aspergillus	7	19
Penicillium	109	59
Mucor	18	4
Oïdium	16	–
Citromyces	44	10
Oospora	39	3
Trichoderma	4	–
Rhizopus	–	3
Absidia	2	–
Coccus forms	3	16
Diplococcus	3	12
Streptococcus	–	5

Source of data: Engel (1950).

Similar surveys of incidence of microörganisms in California wineries are not available, although limited bacteriological surveys have been made (Vaughn, 1955). Such surveys are needed to determine the actual sources of infection in the plant. The studies of Minárik (1967) on Czechoslovakia and of Ribéreau-Gayon and Peynaud (1960–1961) revealed a rich flora in various parts of the winery, particularly from equipment. Yeasts were most common. They concluded—and we concur —that the winery is the major source of microbial contamination. This emphasizes again the importance of sanitation in winery operation.

The principal investigations of wine bacteria were the early studies of Pasteur (1866), Gayon and Dubourg (1894, 1901), and Müller-Thurgau and Osterwalder (1913, 1917, 1919). Unfortunately, the bacteria encountered were not described sufficiently to permit complete identification, and some of the conclusions have had to be revised in the light of present knowledge. More recently, Arena (1936), Ribéreau-Gayon (1937–38, 1938, 1941, 1942, 1944–1945b), Fornachon (1938, 1943), and Vaughn et al. (1949) have reported on the bacterial spoilage of wine. The Office International de la Vigne et du Vin has a special technical commission on wine microbiology. Subjects under investigation (Anon., 1966b) included procedures for detecting added antiseptic agents, and a study of the physiology and growth factors of the bacteria involved in the malo-lactic fermentation. Currently they are preparing a report on the taxonomy of such bacteria.

Vaughn (1955), in his review of bacterial spoilage of wines in California, emphasized the unsatisfactory description and taxonomy of the cocci which decompose malic acid. He pointed out that, while the tetrad-forming cocci of wines are more closely related to *Micrococcus* than to *Streptococcus*, more study was necessary to justify this. Pederson (1949) proposed that similar bacteria should be placed in the genus *Pediococcus*, and Vaughn pointed out that five cultures isolated by him did have characteristics which would place then in the genus *Pediococcus*. Subsequently, Gini and Vaughn (1962) classified the *Lactobacillus* associated with the spoilage of California dessert wines as species of *Pediococcus*. Recently Sauvard et al. (1968) isolated three heterofermentative bacteria from wine. They resembled *Pediococcus* sp. except that they contained catalase and have an active aerobic metabolism.

Ribéreau-Gayon (1938, 1941), in reviewing the available literature on the subject, concluded that wine bacteria may be divided into two large groups: those capable of decomposing tartrates, and those unable to do so, whatever the acidity of the wine. Vaughn (1949) believed that *Acetobacter*, some varieties of which can decompose tartrates, may have been confused with the *Lactobacillus* by the European investigators. However, later work indicates that tartrate-decomposing bacteria are present in California wines. Practically all lactic acid bacteria growing in wine can decompose malic acid to form lactic acid and carbon dioxide, thus producing a decrease in total acidity. All are able, in varying degree, to decompose sugar to form lactic acid and acetic acids.

Ribéreau-Gayon (1938, 1942) pointed out that the nature of the bacteria present and the composition of the medium (malic acid content, sugar content, pH) determine bacterial development. The same bacteria can produce different transformations, and different bacteria can produce the same or similar changes in wine, depending on environmental conditions; hence the gross chemical changes encountered cannot be used as an indication of the type of spoilage. Morphological as well as biochemical criteria are used for species differentiation in modern bacteriology (Breed et al., 1957).

The gas-producing types, both *Leuconostoc* and *Lactobacillus* species, are often called "heterofermentative"; the non-gas-producing types, "homofermentative." The former have been found in wines more often than have the latter. Many of the lactic acid bacteria may act upon glycerin and fixed acids (malic, tartaric) as well as upon glucose and fructose.

The presence of microörganisms in wine can be detected by direct microscopic observation with or without staining. Viable microörganisms are detected by incubation, by plating on nutrient agar, and more recently by separation on and growth in membrane filters. The microbiological analysis of wine by membrane filter was described by Jakubowska (1963). This is capable of detecting one organism per 5 ml in comparison with 10^6 per ml by direct count or 20 per ml by plate count. A rapid staining procedure for distinguishing between living microörganisms and protein cloudiness was developed by Staudenmayer (1966). A membrane filter was used to separate the solid material. Examination of the fluorescence under ultraviolet light also may distinguish live from dead yeasts.

Nobile (1967) developed serial filtration with two membrane filters for the simultaneous viable counts of yeast and bacteria in fermented beer and wines. The yeast cells were successfully retained and cultivated on 1.2 μ membrane filter. All bacteria filtered avoided entrapment on the 1.2 μ membrane filter, but were successfully retained and cultivated on a 0.22 μ membrane. The final filtrates from these serial filtrates were free from all yeast cells and bacteria when tested with fluid thioglycollate medium.

Lactic acid bacteria

The most widely distributed acid-tolerant organisms which account for the spoilage of wine are the Gram-positive bacteria, both rod and spherical forms, which produce lactic acid. The presence of lactic acid bacteria in suspension often causes a silky, cloudy appearance and streaming, as the wine is shaken. This phenomenon is due to the alignment of the bacteria in chains, and is caused by any rod-shaped bacterium that is longer than it is wide. The growth of these organisms is accompanied by the production of an objectionable "mousy" flavor, clouding or hazing of the wine, and formation of abnormal flocculents and deposits.

Two main groups of rod-shaped *Lactobacillus* are recognized by Breed *et al.* (1957). One produces only traces of by-products—for example, carbon dioxide gas, alcohol, acetic acid, and mannitol—from fructose; the other produces considerable amounts of these products. The spherical, or coccus, form of lactic acid bacteria (classified as *Leuconostoc*) produces carbon dioxide gas, lactic acid, acetic acid, and ethanol from glucose, and mannitol from fructose.

A heterofermentative *Lactobacillus* which differed from the better-known species in metabolizing fewer organic compounds was isolated from California table wine by Douglas and Cruess (1936) and designated *Lact. hilgardii.* Vaughn *et al.* (1949) isolated ten cultures of this organism from California table wines and described it more completely. Since then the species has been isolated from Australian wines (Fornachon, 1963*b*), French wines (Barre, 1966*a*), and South African wines (Du Plessis and Van Zyl, 1963). In South African wines L. de W. Du Plessis (1964) isolated 64 strains of lactic acid bacteria representing *Lact. leichmannii, Lact. hilgardii, Lact. brevis, Lact. buchneri, Pediococcus cerevisiae,* and one unidentified *Lactobacillus* species.

On the basis of a study of 750 strains of lactic acid bacteria, Peynaud and Domercq (1967*a*) classified 253 as heterolactic bacilli, 398 as heterolactic cocci, 34 as homolactic cocci, and 23 as a homolactic bacteria. These were identified as follows: nine of *Lact. plantarum,* two of *Lact. casei* var. *casei,* four of *Lact. casei* var. *alactosus,* and eight of unidentified *Streptobacterium* species. Peynaud and Domercq (1967*a*) note that the classification is of little value for enologists because it does not allow identification of strains or of the constituents which the strains are able to metabolize. They suggest an enological classification: fermentation of pentoses and citric acid versus no fermentation of pentoses with subgroups fermenting citric acid. They also noted strains which attack tartaric acid or glycerin. They (1967*b*) classified the 34 strains of homolactic cocci as *Pediococcus cerevisiae.* Some of these were innocuous, but most were undesirable (or at least were present in spoiled wines).

Peynaud *et al.* (1967*b*) distinguished three types of bacteria in wines: those forming L(+) lactic acid (mainly *Lact. casei*), those producing D(−) lactic acid (with 2 to 5 per cent L(+)), and those forming D(−) and L(+) (such as *Lact. plantarum*). The fact that malo-lactic fermentation gives only L(+) lactic acid makes it difficult to consider pyruvic acid as an intermediary.

The lactic acid bacteria isolated from wine may utilize the residual sugars or decompose organic acids as a source of carbon for growth and for energy. Malic, citric, and tartaric acids may be metabolized, depending on conditions. While the microbial degradation of tartaric acid is now well established, it is interesting that Wejnar (1966), very recently considered tartaric acid to be only organic acid resistant to microbial degradation. See, however, p. 774.

From 450 bottled dry wines Du Plessis and Van Zyl (1963) recovered a variety of Gram-positive, catalase-negative, micro-aerophilic, non-motile, noncapsulated, nonspore-forming, nonnitrate-reducing, and nondextran-producing bacteria. Twenty-one were of two types of *Lact. leichmannii*, sixteen of *Pediococcus cerivisiae*, two of *Lact. hilgardii*, three of *Lact. brevis*, and four were unidentified isolates of *Lactobacillus*.

From 3,000 table wines of the south of France, Barre (1966b) isolated fifty-eight bacteria: the homofermentative *Lactobacillus plantarum* (4) and *Lact. casei* (5), and the heterofermentative *Lact. hilgardii* (12), *Lact. buchneri* (3), *Lactobacillus* sp. (1), and *Leuconostoc citrovorum* (33). All strains fermented malic acid, forty-two fermented citric, and two attacked tartaric.

Nonomura *et al.* (1965) studied nine isolates of lactic acid bacteria from Japanese table wines. They were classified as *Lactobacillus plantarum*, *Lact. yamanashiensis* nov. sp., *Leuconostoc lactosum* nov. sp., *Leuc. lactosum* var. *intermedium* nov. var., *Leuc. infrequens* nov. sp., *Leuc. vinarum* nov. sp., and *Leuc. vinarum* var. *debile* nov. var.

Iustratova (1967) systematically studied the lactic acid bacteria in sound and unsound Moldavian wines. Of a hundred cultures very few were identified. The principal ones were *Lactobacillus* (*Lact. plantarum* and *Lact. brevis*) and *Leuconostoc* (*Leuc. mesenteroides* and *Leuc. dextranicum.*)

After the massive and widespread attack of *Lactobacillus* in California dessert wines of the vintages of 1936 and 1937, there have been few serious reports of the so-called cottony or Fresno "mold". Malan (1966), however, reported the typically cottony formation in Italian vermouths and identified it as due to *Lact. trichodes*, Fornachon, Douglas and Vaughn. The optimum pH (4.2 to 4.4), temperature, aeration, and alcohol content were the same as the strain isolated in California. The control measures used in California—75 to 100 mg per liter of sulfur dioxide or a pH of 3.5 or lower—also proved effective.

Krumperman and Vaughn (1966) isolated 64 tartrate-fermenting lactobacilli from 184 cultures. Only twelve were from wine, wine lees, or brandy stillage. The rest were from fermenting cucumbers or olives, sour orange juice, sour milk, or sour tomato juice. Of these 24 were heterofermentative and 40 homofermentative. Of the twelve isolated from vinous sources, six were *Lactobacillus plantarum*, two *Lact. brevis*, two *Lact. buchneri*, and two *Lact. pastorianus*. (See also Berry and Vaughn, 1952.)

Successful culture or growth of these organisms depended on the presence of yeast autolyzate and exclusion of oxygen. The products found included succinic and acetic acids and carbon dioxide, but not, surprisingly, propionic acid.

In a laboratory-scale continuous fermentation of musts of high pH (3.5–3.6) Flanzy *et al.* (1966) reported growth of lactic acid bacteria and formation of volatile acidity. If sulfur dioxide was periodically added, growth of bacteria and volatile acid formation were controlled, but the fermentation rate was reduced and acetaldehyde formation increased.

Jakubowska and Piatkiewicz (1963) made paper chromatographic studies of the fixed organic acids in sound and microbially contaminated wines. In most of these wines there were more or less profound losses of one or more organic acids, which suggested that the technique could be used to identify spoiled wines. The procedure appears to be useful, although microscopic studies would probably be needed also.

The most recent taxonomic classification of the malo-lactic bacteria is that of Nonomura and Ohara (1967). The genus *Lactobacillus* was divided into homofermentative *Lact. plantarum*, *Lact. plantarum* var. *arabinosus*, *Lact. yamanashiensis*, *Lact. casei*, and *Lact. casei* var. *pseudo-plantarum;* and heterofermentative *Lact. buchneri*, *Lact. buchneri* var. *parvas*, *Lact. brevis*, *Lact. brevis* var. *gravesensis*, and *Lact. brevis* var. *otakiensis*. *Leuconostoc* and *Pediococcus* were classified as follows: hetero-fermentative *Leuc. mesenteroides*, *Leuc. mesenteroides* var. *lactosum*, *Leuc. infrequens*, *Leuc. blayaisense*, *Leuc. dextranicum*, *Leuc. dextranicum* var. *kefiri*, *Leuc. dextranicum* var. *citrovorum*, and *Leuc. dextranicum* var. *debile;* and homofermentative *Pediococcus parvelus*. Of these the strains which induced a malo-lactic fermentation vigorously during alcoholic fermentation were: *Leuc. dextranicum* var. *vinarium*, *Leuc. infrequens*, *Leuc. blayaisense*, and *Leuc. buchneri*. It is of interest that α-ϵ-diaminopimelic acid was detected in the cell hydrolyzates of *Lact. yamanashiensis* and of *Lact. plantarum*.

It is now clear that a variety of bacterial phenomena are associated with the decarboxylation of malic acid. In many instances this is a desirable enological practice, but the bacterial activities which cause degradation of tartaric acid or glycerin (usually associated with *tourne* and propionic acid) are harmful to the quality of wines.

The mechanism of the conversion of malic acid into lactic acid was discussed previously (pp. 496-505).

The products of the bacterial degradation of citric acid have been studied by the Bordeaux enologists, and are summarized by Peynaud (1956a). The first products are oxalacetic acid and acetic acid. The former is reduced to pyruvic acid and carbon dioxide. Pyruvic acid may form (1) acetylmethylcarbinol (acetoin) and thence 2,3-butylene glycol, (2) acetic acid, or (3) lactic acid. The Bordeaux workers report that degradation of 5 milliequivalent of citric acid could result in the formation of as much as 45 mg of 2,3-butylene glycol.

Tartaric acid appears to give rise to oxalic acid via diketosuccinic acid as follows:

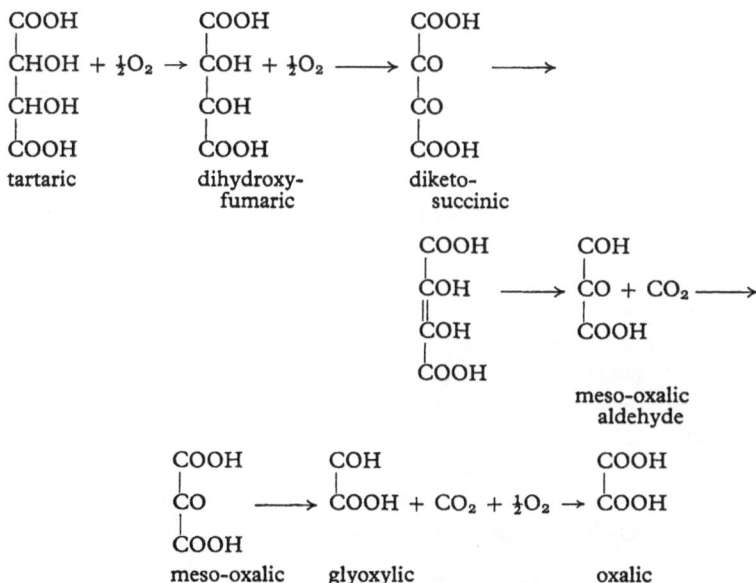

Apparently little dihydroxyfumaric or diketosuccinic acid accumulates, according to Dupuy (1960).

Fernández and Ruiz-Amil (1965) reported that one strain of *Rhodotorula glutinis* metabolized L(+)-tartrate (but *not* D(−)-tartrate). An enzyme catalyzing decarboxylation of tartrate to glycerate and the presence of glycerate kinase were also noted. The pathway postulated is decarboxylation of tartrate to glyceric acid and phosphorylation to phosphoglyceric acid.

Dittrich and Kerner (1964) clearly established that the lactic sour odor of spoiled German wines was due to high diacetyl: a typical odor at 0.9 mg per liter, with some badly spoiled wines containing 4.3 mg per liter. Normal wines had 0.2 to 0.4 mg per liter. To remove diacetyl a brief refermentation (with 1 per cent sugar) is recommended. The reduction products are acetoin and finally 2,3-butylene glycol. Some reduction of volatile acidity also occurs if the refermentation is carefully made.

Ribéreau-Gayon and Peynaud (1964–1966) claimed that small amounts of diacetyl enhance the aroma of some wines. Fornachon and Lloyd (1965) believe that this needs to be confirmed. However, 4 mg per liter of diacetyl appears to be excessive. They showed that wines which have undergone a malo-lactic fermentation are significantly higher in diacetyl and acetoin than wines which have not. It is strange that Peynaud and Lafon (1951) and Charpentié et al. (1951) did not duplicate this result. Radler (1962a) found more acetoin plus diacetyl in wines that had undergone a malolactic fermentation. Acetoin probably arises from a reaction of the type 2 pyruvate → acetoin + $2CO_2$. Fornachon and Lloyd (1965) conclude that more than 1 mg per liter of diacetyl indicates bacterial action. *Leuconostoc mesenteroides* is the chief organism involved, and pyruvic and citric acids are the substrate—not malic acid. The presence of glucose in the media interfered with the formation of both acetoin and diacetyl.

The determination of the optical rotation of the lactic acid produced by *Lactobacillus* is of taxonomic importance. Bréchot et al. (1966a), Barre (1966a), and Peynaud et al. (1966b) presented simple and precise procedures for determining the proportions of L(+) and D(−) lactic acid present to replace the usually tedious procedures: L(+) by lactic dehydrogenase, total lactic acid by chemical procedures, and D(−) lactic acid by difference. In Barre's report, strains yielding from 47 to 53 per cent of one of the isomers are considered to produce racemic lactic acid. With 59 strains his results were:

	Racemic	R^a + L(+)	L(+)	R^a + D(−)	D(−)
Lactobacillus					
Homofermentative	4	3	2	0	0
Heterofermentative	8	0	0	7	1
Leuconostoc	0	0	0	3	31

^{a}R = racemic.

In wines lactic acid can come from two sources: alcoholic or malolactic fermentation. The former produces D(−) lactic acid up to about 6 or 7 milliequivalents per liter. Peynaud et al. (1966b) indicate that *Saccharomyces veronae* is an exception as far as solely producing D(−) lactic acid is concerned. It produced L(+) lactic acid. Most malo-lactic acid bacteria produce L(+). Bréchot et al. (1966a) reported that some wines after the malo-lactic fermentation contained no D(−) lactic acid.

Peynaud *et al.* (1966*b*) consider the method of difference to be insufficiently sensitive to establish that the disappearance is due to isomerization.

Lockwood *et al.* (1965) have summarized our knowledge of the chemistry of lactic acid. Apparently the L(+) isomer is levorotatory, but its apparent dextrorotation is due to the formation of an ethylene oxide bridge between atoms 1 and 2 by tautomeric shift of the hydrogen atom of the hydroxyl group on carbon atom 2 to the carboxyl group of the carboxyl radical, thus:

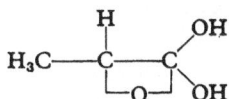

$$H_3C-\underset{\underset{O}{\overset{H}{|}}}{C}-C\underset{OH}{\overset{OH}{<}}$$

The strongly dextrorotatory ethylene oxide form is in equilibrium with the straight chain form. Therefore the (+) and (−) signs should be reversed in relation to structure but retained on the basis of observed values. D(−) and L(+) isomers are produced by various bacteria. They show that lactic acid is a stronger acid than is generally believed (pK_a = 3.73 at 26°C). For further details on the chemistry of lactic acid their paper should be consulted.

Malo-lactic fermentation

The desirable conversion of malic acid into lactic acid and some of the conditions governing this transformation have been described (see p. 496) by Ribéreau-Gayon (1936, 1946, 1947) and Ribéreau-Gayon and Peynaud (1938*a*, 1938*b*, 1964–1966). Malo-lactic fermentation occurs spontaneously under some conditions in Beaujolais: Bréchot *et al.* (1966*a*) noted that the malo-lactic fermentation was completed within twelve days after pressing. Gendron (1967*a*) considers that an important factor in favoring the malo-lactic fermentation is the increase in arginine and alanine caused by pressing the skins and stems (particularly in reds). Constituents from the stems may, however, inhibit use of arginine and alanine by yeasts. Absence of arginine seems to inhibit the malo-lactic fermentation completely. Kielhöfer (1966) specifically warns against an excessive malo-lactic fermentation which, he says, leads to a foreign flavor (called the sauerkraut flavor) caused by diacetyl. To remove diacetyl he recommends refermentation (or addition of fresh yeast), which reduces diacetyl to 2,3-butylene glycol.

As L. de W. Du Plessis (1964) writes, "It is clear . . . that there is no clearcut differentiation between spoilage bacteria and malo-lactic bacteria. An organism may . . . cause either spoilage or malo-lactic fermentation, depending on the conditions existing in the wine in which it occurs." When a marked increase in volatile acidity occurs during a malo-lactic fermentation it is probably due to *Acetobacter* sp. rather than to lactic acid bacteria. Pilone (1967) showed that with the usual procedures for volatile acidity too little lactic acid is distilled to account for the rise in volatile acidity.

A strain of *Lactobacillus plantarum* isolated by Barre and Galzy (1962) produced little volatile acidity. They showed that one can determine whether the lactic acid produced is L(+), D(−), or racemic by using the lactic-dehydrogenase properties of yeasts. The strain isolated produced both L(+) and D(−) lactic acid.

Fornachon (1963a) reported that *Lact. hilgardii* and *Leuc. mesenteroides* rapidly attack acetaldehyde. In a media containing sulfur dioxide and acetaldehyde this resulted in the release of sufficient sulfur dioxide (from the bound sulfur dioxide) to inhibit growth. *Lact. arabinosus* consumed much less acetaldehyde. Thus it may be difficult to start a malo-lactic fermentation in wines containing a high level of bound sulfur dioxide. Fornachon argues that bound sulfur dioxide must therefore be considered to be inhibitory to bacterial growth.

The velocity of the malo-lactic acid fermentation varied markedly between strains of the same species. Gomez and Galzy (1966) reported two types of bacteria in French cellars, both capable of carrying on a malo-lactic fermentation. One was heterofermentative, fermenting glucose and other sugars to give lactic and acetic acids plus some alcohol, mannitol, and carbon dioxide. The other was homofermentative, fermenting glucose to lactic acid.

Nutrient requirements

The amino acids affecting 64 lactic acid bacteria were studied by Peynaud *et al.* (1965). The amino acids found to be indispensible for all the rods studied included glutamic acid, isoleucine, leucine, and valine. About three-fourths required α-alanine, aspartic acid, methionine, and phenylalanine. Those required by few bacteria were proline, lysine, glycine, and histidine.

For the cocci studied, in addition to the four found indispensable for the rods, arginine, histidine, methionine, phenylalanine, serine, tryptophan, and tyrosine were essential. Also, isoleucine and leucine were only stimulating, and α-alanine and aspartic acid had less effect than with the rods. Thus the cocci clearly have less capacity to synthesize amino acids. This probably explains why cocci are rarely found in aged wines, where the necessary amino acids for their growth are rarely found.

The amino acids which are present in the largest amounts, proline and threonine, are not needed for bacterial growth. The vitamins affecting bacterial growth were studied by Peynaud *et al.* (1965). Pantothenic and nicotinic acids were found to be indispensable for all the strains studied. Thiamine and riboflavin were more necessary for the rods than for the cocci, but folic acid was more important for the cocci than for the rods. Biotin and cobalamine were not stimulating. Purine bases such as uracil were also stimulating. DL-aminobutyric acid was a strong inhibitor of the cocci, and orotic acid stimulated rod growth. It should be noted that the yeasts can synthesize much of their amino acid requirements, but are stimulated by thiamine and biotin. Additional information on their biotin requirement is given by Potter and Elvhjem (1948).

Off-odor and -flavor development

Bacteria capable of decomposing tartrates have been described in French, Swiss, and Argentinian wines, and more recently they have been encountered in California wines. The term *tourne* or *pousse* was originally used by Pasteur to describe red table wine infected with such organisms. Subsequently it was shown that these organisms attack tartaric acid, glycerin, or both. Arena (1936) reported that the tartrate-decomposing bacteria isolated from Argentinian wine would not grow above 12 per cent of alcohol. The alcohol tolerance of the European species is not known, but presumably it is low.

The mousy flavor produced by these organisms has been emphasized by Osterwalder (1948). He finds either *Bacterium mannitopoeum* or *B. tartarophthorum* to be associated with the production of a mousy flavor and that it does not always accompany glycerin destruction. Vaughn (1949) reports mousiness also in *Acetobacter* spoilage.

Recently Schanderl (1948*a*) has questioned the microbiological origin of the "mousy" odor. He finds that mousiness occurs in German wines when the oxidation-reduction potential is very high. Villforth (1950–51) believes that the mousy flavor is formed by a polymerization of formaldehyde (from formic acid) and acetaldehyde. There is some evidence in California that it occurs in high pH wines under oxidizing conditions. Since the mousy flavor is associated with oxidized conditions, the use of charcoal to remove the mousiness is not recommended. Sulfur dioxide is the preferred treatment.

The *tourne* disease of wines is characterized by loss of tartrates, increase in volatile acidity, decrease in fixed acidity, and increase in pH value. Other types of spoilage are characterized by loss of sugar, increase in both fixed and volatile acidity, and decrease in pH value. Diseased wines

TABLE 77

Bacterial Diseases of Wine and Their Causative Agents

Disease	Appearance of wine	Changes in flavor and composition of wine	Causative organisms	Physiological properties
Wine flower	Chalky, fragile, white film on surface if exposed to air, slight haziness, and considerable deposit	Decrease in alcohol, and total acid content; increase in acetaldehyde, giving the wine an objectionable vapid oxidized flavor	Aerobic yeasts, particularly *Candida mycoderma*, elongated branching asporogenous cells, 3–10 × 2–4μ	Oxidize alcohol to carbon dioxide and water, do not attack tartaric very much and citric not at all, but destroy acetic acid and glycerin
Acetic tinge (acescence)	Translucent, adhesive film, or uniform clouding with deposit, according to type of organism	Decrease in alcohol and sugar; increase in acetic acid, acetaldehyde, ethyl acetate, and fixed acids, such as gluconic and ketogluconic, giving wine a characteristic acescent, or vinegary, odor	*Acetobacter ascendens, A. xylinum, Bacterium vini acetati, B. xylinoides, B. orleanense*; short rods occurring singly, in pairs, or in chains, either free or surrounded by a zoögleal sheath: nonmotile; usually 0.5 × 1μ	Oxidize alcohol to acetic acid, and ferment sugars to acids and alcohols
Tartaric fermentation (*tourne*)	Haziness with slow evolution of carbon dioxide gas; on slight agitation there appear silky streamers with iridescent reflection which move slowly through the liquid	Decrease in tartrate and sugar content; increase in acetic acid, lactic acid, propionic acid, carbon dioxide, and ammonia, giving wine a sour odor and insipid and "mousy" taste	*Bacterium tartarophthorum*; long thin rods in early stages and curved rods in older cultures; 0.5–2 × 2–5μ	One strain acts on both tartrates and glycerin, another only on glycerin (apparently these are heterofermentative lactic bacteria)
Mannitic fermentation	Uniform clouding with deposition of mannite and bacterial cells; very little gas formation but wine becomes appreciably viscous	Increase in extract content owing to formation of mannite and glycerin, giving a sweet-sour taste	*Bacterium mannitopoeum*, rod-shaped, 0.7 × 1.3μ	Ferment fructose to mannite, lactic acid, acetic acid, and carbon dioxide; ferment glucose to lactic acid, acetic acid, and carbon dioxide; attack xylose, arabinose, and citrate, but not lactose

784

TABLE 77 cont.

Haziness or bitter fermentation	Haziness; loss in color and precipitation of tannins and coloring matters	Decrease in reducing sugar, tartrates, and glycerin; increase in total and volatile acidity, and ammonia; butyric acid and acrolein are claimed to occur; bitter and disagreeably acid taste	Similar to that involved in tartaric fermentation	Convert tartrate and glycerin in solution into lactic and acetic acids and some carbon dioxide
"Oiliness or fattiness"	Oily appearance, wine pouring in a viscous stream, slow evolution of gas, uniform turbidity, deposits	Decrease in sugar; increase in total and volatile acidity; mannite, acetic acid, lactic acid, glycerin, alcohol, and gum (dextran?) formed	*Bacterium gracile*; small rod-shaped organisms, $0.4 \times 0.6\mu$	Ferment fructose to mannite, lactic acid, acetic acid, and carbon dioxide: do not decompose xylose; convert malic acid into lactic acid and carbon dioxide
Lactic	Uniform turbidity or deposit with growth at bottom only; slight gas evolution	Decrease in sugar; increase in volatile and fixed acids; "mousy," or lactic, taste	*Micrococcus variococcus, M. acidovorax*, and *M. malolacticus*, coccus forms, $0.5-0.7 \times 0.7-1.5\mu$, singly, as diplococci, or tetracocci; *Bacterium gayoni* and *B. intermedium*, rod-shaped bacteria $0.7-1.3\mu$, singly or in chains	Ferment fructose to lactic acid, acetic acid, and carbon dioxide, (and mannite in the case of *B. gayoni* and *B. intermedium*); ferment glucose to lactic acid, acetic acid, and carbon dioxide; some species (*M. malolacticus* particularly) convert malic acid to lactic acid and carbon dioxide

Source of data: Description of organisms taken from Müller-Thurgau and Osterwalder, 1913, 1917, 1919: Ribéreau-Gayon, 1938; Arena, 1936; and Ventre, 1931. For more recent descriptions and classification of the causative organisms see Breed *et al.* (1957); and Vaughn (1949).

have been differentiated into various classes according to the relative amounts of lactic acid and volatile acid produced, since these constituents often serve to distinguish the types of organisms involved. Some of the classifications are: little lactic acid and little volatile acid; much lactic acid and little volatile acid; little lactic acid and much volatile acid; much lactic acid and much volatile acid. The bacterial diseases which have been recognized in European dry wines are described briefly in table 77. Mannite, mannose, acetoin (acetylmethylcarbinol), butyric acid, propionic acid, and so on have been reported as spoilage products. Acetamide has been reported as one of the spoilage products of "mousy" diseased California wines, but this has not been verified.

The origin of the musty odor of some wines is still not known. There have been many suggestions that it is caused by actinomycetes growing in empty tanks. Dougherty et al. (1966) isolated a compound produced by actinomycete cultures that had a strong musty odor.

While Marinova-Boiadzhieva (1966) did isolate bacteria from bentonites, they did not grow well in wine and caused no undesirable changes in it.

The development of ropiness in certain wines is facilitated by two microörganisms growing in symbiosis, according to Lüthi (1957a, 1957b). *Streptococcus mucilaginosus* and *Acetobacter rancens* produced extreme ropiness in three weeks, but *Strep. mucilaginosus* alone only slightly increased the viscosity in the same period.

Sulfur dioxide (70 mg per liter) in the presence of an excess of acetaldehyde does not inhibit growth of the homofermentative *Lactobacillus plantarum*, but completely inhibits the heterofermentative *Lact. hilgardii* and *Leuc. mesenteroides*. In the latter two the bacteria reduce the aldehyde content so that the resulting wine contains sufficient free sulfur dioxide to inhibit growth.

Acetic acid bacteria

Several species of *Acetobacter* that oxidize the alcohol to acetic acid often occur in wine. Vogt (1945) reports *A. aceti, A. ascendens, A. pasteurianum*, and *A. xylinum* as having been found in wines. Wines high in alcohol are less liable to acetic fermentation than are those low in alcohol content. The growth of these bacteria may be checked by the addition of about an ounce of sulfur dioxide per 100 gallons or by pasteurization. Their growth is greatly favored, however, by temperatures of 30° to 40°C (86° to 104°F). Since they are quite aerobic, they normally develop only in casks which have not been kept filled.

The acetic acid bacteria are unique among bacteria in oxidizing ethanol to acetic acid under acid conditions. Some, however, oxidize acetic acid to carbon dioxide and water. These have been called "overoxidizers" and the former "nonoveroxidizers".

The acetic acid bacteria are classified in two genera. Rainbow (1966) classified those which oxidize acetate to carbon dioxide and water, operate the tricarboxylic acid (TCA) cycle, grow well on lactate, and when motile possess peritrichous flagella as *Acetobacter*. Members which do not oxidize acetate probably do not operate the TCA cycle, require certain sugars alcohols for growth, oxidize glucose to gluconate, do not grow on lactate, and when motile possess one or more polar flagella he classified as *Acetomonas*.

In the Midi of France Dupuy (1957*a*) reported that 32 per cent of the *Acetobacter* of wines were *A. paradoxum* Frateur, 24 per cent *A. rancens* Beijerinck, 19 per cent *A. ascendens* (Hennesberg) Bergey *et al.*, and 19 per cent *A. mesoxydans* Frateur. *A. suboxydans* Kluyver and de Leouw was seldom found. Dupuy makes the special point that the acetic bacterial flora of wines is different from that of vinegar. Apparently the *A. mesoxydans* Frateur is the same as *A. aceti* (Pasteur) Beijerinck. The classification used by Dupuy was that of Frateur (1950).

Aside from oxygen, Dupuy (1957*b*) reported that pH and alcohol percentage were the most important factors influencing the growth of *Acetobacter*. He found little growth at pH 3.2 or 13 per cent alcohol. Sulfur dioxide was most effective as an antiseptic. Other factors had little influence on growth of *Acetobacter:* sugar, glycerin, tannin, bluefining, or addition or elimination of minor elements.

When musts containing many acetic acid bacteria and few yeasts are fermented under conditions unfavorable to the rapid multiplication of yeasts, there occurs a rapid acetification that retards alcoholic fermentation. Vaughn (1938) found the bacteria causing this reaction to be *Acetobacter*, but not the usual strains (see table 78). All strains of *Acetobacter* examined by him formed a "mousy" off-flavor in grape juice. The rate of acetification was influenced by the types of yeasts grown in association with the bacteria, by the strains of the bacteria, and by the temperature. The effect of temperature is particularly striking (table 78). The use of sulfur dioxide to check the growth of acetic acid bacteria, cool fermentation, and pure yeast will prevent this condition.

It has long been known that more acetic acid is formed in fermenting musts of high sugar content. Sudraud (1967) reports that in the 1966 vintage a part of the high volatile acidity was formed in the musts by bacteria and yeasts. Both *Saccharomyces acidifaciens* and *Sacch. elegans*, which produce considerable volatile acidity, were present in fermentations of botrytised grapes. Unfortunately, no reference was made to Vaughn's (1938) pioneer work which showed that the high volatile acidity resulted from the association (symbiosis) of certain bacteria and yeasts. Sudraud noted that this high volatile acidity made it difficult or impossible for French Sauternes to conform to the French legal limit of 0.125 per cent volatile acid (as acetic).

TABLE 78

EFFECT OF TEMPERATURE OF INCUBATION[a] AND STRAIN OF ACETOBACTER ON ACETIFICATION OF MUST

	Volatile acid as acetic			
Yeast 66 with Acetobacter *number*	At 25°C (77°F) (gr per 100 cc)	At 31°C (87.8°F) (gr per 100 cc)	At 37°C (98.6°F) (gr per 100 cc)	At 42°C (107.6°F) (gr per 100 cc)
68	0.360	0.465	0.741	0.239
69	0.288	0.429	0.771	0.192
73	0.279	0.252	0.571	0.114
83	0.562	0.321	0.546	0.184
84	0.233	0.498	0.551	0.110
85	0.258	0.465	0.609	0.140
86	0.145	0.270	0.402	0.158
88	0.203	0.543	1.164	0.176
90	0.204	0.498	0.648	0.098
92	0.180	0.186	0.300	0.260
98	0.189	0.477	0.903	0.286
99	0.165	0.228	0.468	0.256
100	0.621	0.588	0.780	0.238
101	0.276	0.183	0.633	0.246
102	0.147	0.213	0.537	0.180
108	0.180	0.282	0.405	0.036
109	0.102	0.306	0.477	0.268
110	0.136	0.117	0.171	0.175
111	0.099	0.303	0.591	0.162
112	0.153	0.330	0.618	0.138
113	0.153	0.149	0.606	0.535
114	0.183	0.333	0.483	0.480
115	0.150	0.172	0.378	0.220
116	0.297	0.177	0.516	0.225
117	0.165	0.306	0.510	0.326
118	0.168	0.168	0.450	0.381
119	0.105	0.180	0.324	0.305
A.T.C.C.[b] 4920 alone[c]	—	0.180	0.068	—
Yeast 66 alone	0.050	0.054	0.069	0.021
Blank	0.012	0.012	0.012	0.012

[a] Incubation period of three days.
[b] American Type Culture Collection No.
[c] Volatile acid not produced by other typical *Acetobacter* strains in absence of yeast.
Source of data: Vaughn (1938).

Yeasts

Aerobic, oxidative, alcohol-destroying yeasts often grow as films on wines of low alcoholic content that are exposed to air. These organisms, usually strains of *Pichia, Debaryomyces,* or *Candida,* are not very

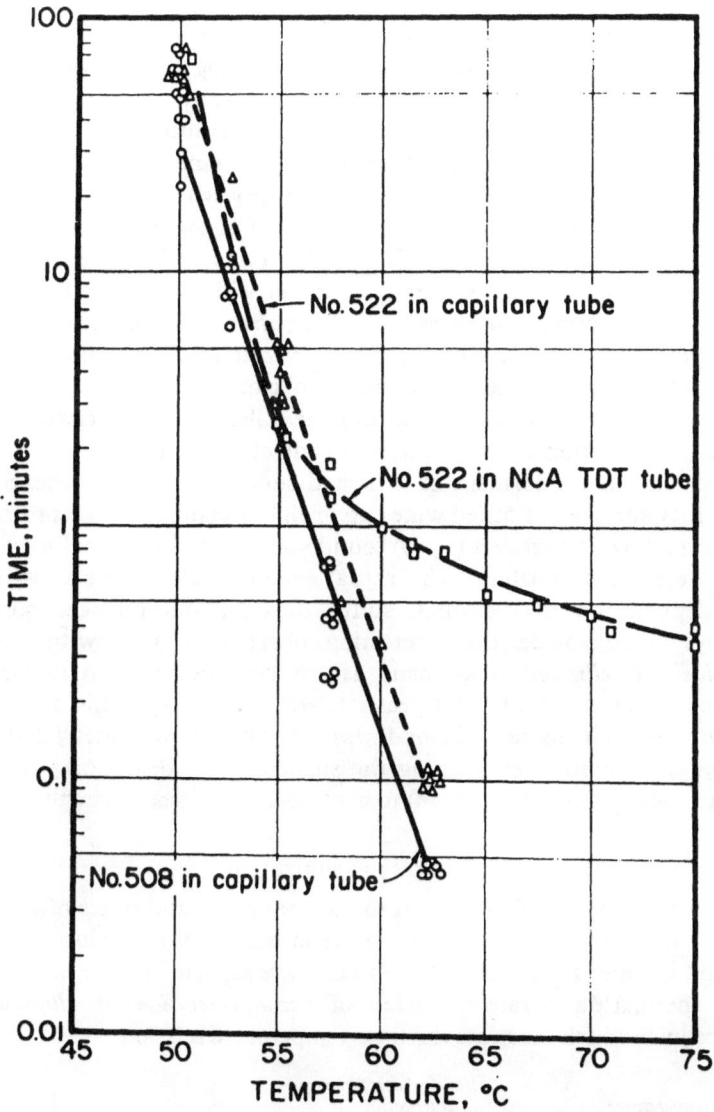

Figure 125. Thermal death-time results with capillary tubes and with NAC tubes. At 60°C (140°F) and above, NAC tubes gave unreliable results because of time required for heat to penetrate tubes. This gave non-uniform temperature distribution. Source of data: Jacob et al., 1964.

tolerant of alcohol, and develop only on the surface in contact with air. For a review of yeasts as spoilage organisms see Castor (1952). They are, moreover, often present in bottles and corks, fig. 124.

Spoilage of California table wines by film-forming yeasts is rare because of (1) their relatively high alcohol content (usually 11.5–12.5), (2) the wide use of sulfur dioxide to which these organisms are particularly sensitive, and (3) the use of germ-proof pad or membrane filters. Certain yeasts produce coarsely granular deposits in dry white wines. Species of *Pichia* and *Saccharomyces* may cause clouding of sweet table wines low in sulfur dioxide content. Vogt (1945) also reports clouding caused by species of *Torula* and *Torulopsis* in German white wines. Scheffer and Mrak (1951) found *Saccharomyces chevalieri*, *Sacch. carlsbergensis* var. *monacensis*, *Sacch. oviformis*, *Sacch. cerevisiae*, *Pichia alcoholophila*, and *Candida rugosa* as common causes of cloudiness in California dry white wines. As they are sulfur dioxide tolerant, pasteurization and sterilization filtration are the preferred control methods.

Yeast can remain dormant in wine and survive for a long time. Some strains may survive in bottled wines up to eighty-eight years, according to Dubourg (1897). Schanderl (1959) found viable yeasts in sparkling wines fifteen years after bottling. The actual extent of cloudiness caused by yeast depends on pH, alcohol, sulfur dioxide, and nitrogen source. Edelényi (1966) made the interesting observation that with *Sacch. cerevisiae* var. *ellipsoideus* as many as 10^2 cells per ml did not cause cloudiness, but that with wild yeasts (*Brettanomyces* sp.) this or even a lesser number of cells caused cloudiness. For white table wines Jakubowska (1963) recommended keeping the sugar content below 0.23 to 0.67 per cent (depending on the wine) to avoid danger of yeast growth.

Control by pasteurization

There are relatively few data on the temperatures and times of heating required to destroy the yeasts and bacteria which may develop in wine. Aref and Cruess (1934) and Tracy (1932) have reported considerable data on the thermal death rate of a strain of *Sacch. cerevisiae* var. *ellipsoideus* in grape juice which illustrate the time-temperature relation:

Temperature °C	°F	Time for complete destruction (minutes)	Source
62.0	143.6	1	
59.0	138.2	5	Tracy (1932)
56.0	132.8	15	
57.5	135.5	10	
56.4	133.5	20	
55.3	131.5	40	Aref and Cruess (1934)
54.7	130.5	60	
53.8	128.8	120	

Beamer and Tanner (1939) reported that pH had little effect on thermal death rates of yeast; the time required to kill several yeasts was only slightly reduced when the pH was lowered. Aref and Cruess (1934) also noted that the thermal death rates for yeasts, unlike bacteria, were reduced only slightly with reduction in pH.

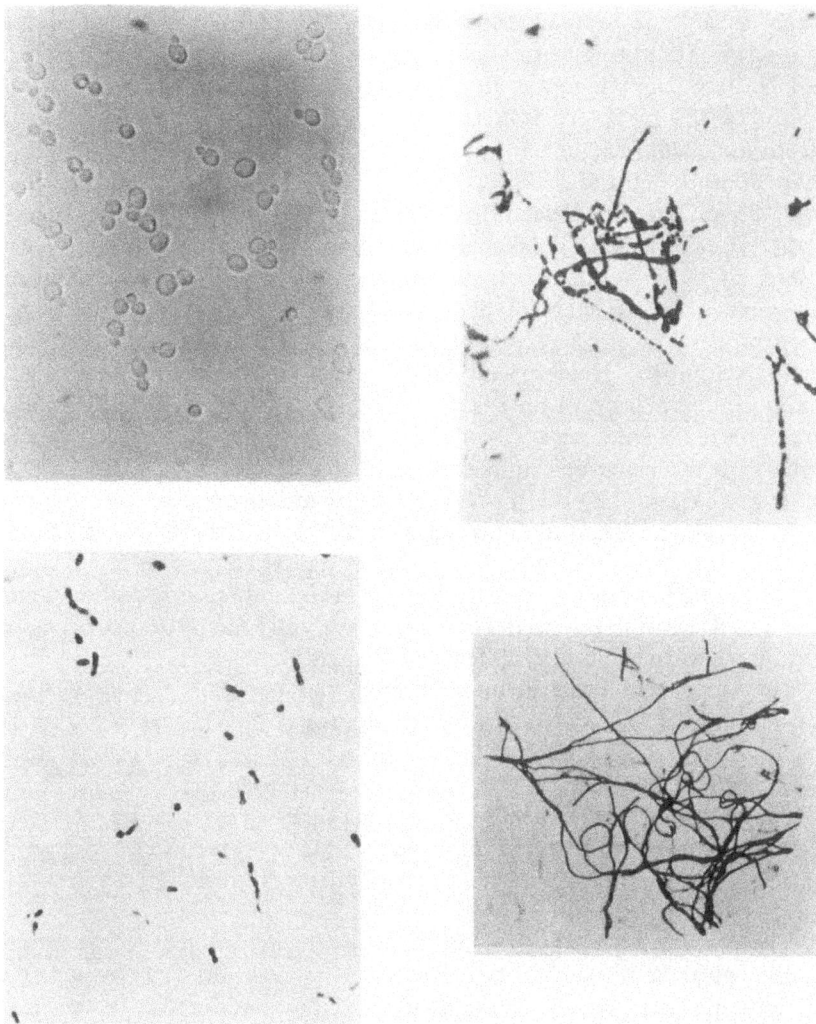

Figure 126. Photomicrographs of some wine microorganisms: *A*, growing wine yeast, *Saccharomyces cerevisiae* var. *ellipsoideus*—note the budding method of reproduction; *B*, vinegar bacteria, *Acetobacter aceti*—involution forms; *C*, pure culture of *Leuconostoc citrovorum* used for the malo-lactic fermentation; *D*, pure culture of *Lactobacillus trichodes*. (*B* from Prof. R. H. Vaughn.)

The most recent detailed study of the thermal death rate of yeast is that of Jacob et al. (1964). They developed capillary tubes which they preferred to the National Canners Association thermal death-point tubes, especially above 60°C (140°F). The test organism was *Sacch. cerevisiae* var. *ellipsoideus*, strains No. 522 (Montrachet) and No. 508 (Burgundy). Their data (fig. 125) corroborated that of earlier investigators except for Yang et al. (1947). They concluded that the thermal death point for No. 522 is 1.0 minute at 57.5°C (135°F) or 0.1 minute at 62°C (143.6°F). For No. 508 it is 1°C lower. In practice they recommended 0.1 minute at 70°C (158°F) for a dry sauterne wine. Yang and his co-workers used electronic (radio-frequency) pasteurization of wine. They obtained a thermal death point of 4 seconds at 55°C (131°F). Jacob et al. consider this to be unverified, and indeed their data so indicate, even when the synergistic effect of alcohol is taken into account. They considered that yeast killing by radio frequency (short electromagnetic waves) is due to heat. Although more expensive, it is a satisfactory method of pasteurization.

In modern thermal process determination (Baumgartner, 1943; Ball and Olson, 1957; Hersom and Hulland, 1964; Stumbo 1965; Frazier, 1967) the thermal death time is plotted as log of time versus temperature to define the z value or slope of the resulting straight line. When this is done with the value determined by Aref and Cruess (1934), the z value in grape juice at pH 3.8 is 3.6°C (6.5°F) per minute, and at pH 1.45 it is 3.7°C (6.6°F) per minute. This high z value for the thermal rate of destruction of yeast indicates that a rise in temperature is of more value in destroying the yeast than is an increase in time. By raising the temperature and decreasing the period of heating, the yeast can be destroyed more completely with less undesirable effect on quality.

The lactic acid bacteria in table wine can be destroyed by heating for 1 minute at 62.8°C (145°F) for 10 minutes at 57.2°C (135°F), according to Douglas and Cruess (1936). More heat-resistant strains were isolated from Australian dessert wines by Fornachon (1943). These required heating for 1 minute at 70°C (158°F) or 15 minutes at 65°C (149°F) for their destruction. Heating to 82.2°C (180°F) for 1 minute is usually sufficient to kill most injurious microörganisms found in California wine.

The destruction by heat of the microörganisms in wine is influenced by biological factors such as type and extent of infection, and composition of the wine (pH, alcohol content, sugar content, sulfur dioxide content, and so on), and by factors determining the rate of heat transfer. In general, the larger the number of microörganisms present, the more difficult it will be to destroy them. The death rate is logarithmic, but the probability of the occurrence of more heat-resistant strains increases with population. The greater the population, the greater the heat required to kill them.

The relative importance of all the biological factors has not been evaluated for wine. The rate of heat transfer from the heating medium through the container walls to the coldest spot in the container, for "in the bottle" pasteurization, or from the heating medium through plates or tubes to the coldest point in the fluid, in continuous pasteurization, must also be known, since it is necessary to raise all points to the required temperature and hold it for the necessary period of time. When wine is filled hot into the bottle, consideration must be given to the more rapid cooling which occurs in the neck of the bottle, where the ratio of glass volume to wine is larger than in the body of the bottle. The sterilization of the closure (cap or cork) and its tightness are additional factors to consider. The evaluation of these factors under the conditions selected must precede actual pasteurization of wine.

Ribéreau-Gayon (1937) has shown that marked changes occur in wine colloids on heating. Both coagulation of some colloids and formation of protective colloids by heating may occur. In the latter condition, permanent hazes may develop.

Control of microbial growth by sterilization filtration has been discussed (p. 558). Geiss (1962a) gives credit to Schmitthener for developing the basic concepts of germproof filtration in 1913.

Use of ionizing radiation has been frequently attempted on wines, with varying claims of success. Paunovic (1963) reported negative results when lethal activity was attempted. He recommended further experimentation.

E. A. Rossi, Jr. (1963), reported that Westinghouse Sterilamps, which have about 80 per cent of their radiation at 253 mμ, were effective in preventing acetification of low-alcohol wines or of distilling material held in large partially filled containers. Where exposure was not excessive, less than six weeks, no harmful effects in wine quality were noted.

A summary of wine disorders caused by microörganisms is given in table 77. See also figure 126.

NONMICROBIAL DISORDERS
. . . are preventable, although still too common.

Iron clouds

Clouding of wine owing to the formation of colloidal complexes of metals is referred to as "casse." The presence of iron in white wine often causes clouding after aeration, although iron alone is by no means the only cause. The formation of iron clouds depends upon a number of factors: the concentration of iron, the nature of the predominating acid and its concentration, the pH, the oxidation-reduction potential, and the concentration of phosphates or tannins. Ribéreau-Gayon (1933*b*) made the most complete study of the conditions under which iron cloud will form. Unless the other conditions are proper for their formation, iron clouds will not occur even in the presence of fairly high concentrations of iron.

Dissolved iron in wine can exist in several different forms, according to the physical and chemical constitution of the wine (Marsh, 1940). Under normal conditions of vinification, most of the iron is present in the ferrous (Fe^{++}) state, the porportion depending upon the amount of oxygen dissolved by the wine. When the quantity of dissolved oxygen is small in comparison with the potential content of reducing matters, a reduction of the ferric iron to the ferrous form will usually occur in storage. These two forms of iron exist both as free ions and as soluble complexes with constituents, such as citrates. Under certain conditions, however, such as low acidity and the presence of considerable tannin or phosphate, the ferric ions formed during exposure of the wine to air may combine with the tannins or the phosphates to form insoluble, usually colloidal, ferric complexes.

In white wines the milky ferric phosphate casse occurs only in the range of pH 2.9 to 3.6. Although many California wines are of a higher pH, and thus not affected, this casse is still the one most commonly found in California. Ferric tannate casse is rare here.[1] It is found in white wines only after tannin has been added. In red wines it forms an inky-blue cloud and later a blue deposit. If the concentrations of dissolved iron and tannin or phosphate are large enough to form an appreciable amount of these ferric complexes when the wine is exposed to oxidative conditions, clouding will eventually occur. The cloud thus formed results from the agglomeration of the colloidal ferric complexes to particles of visible size. Ribéreau-Gayon (1933*b*) has shown that ferric phosphate casse occurs in some wines at much lower concentrations of iron than normal, when traces of copper are present, and that sulfur dioxide tends to inhibit its formation. But the opposite effects are observed in other wines.

The ferric phosphate casse can usually be controlled by adding citric acid at the rate of one pound per 1,000 gallons of wine. However, some

---

[1] The casse due to high ferric ion content in red wine has been incorrectly called ferric tannate. The phenolic substances involved are leucoanthocyanins or similar condensed tannins. Gallotannic acid does not occur in wines unless they have been stored in oak casks or are fermented for too long a time with grape stems and seeds. The characteristic blue-black ferric tannate precipitate, however, may occur in white wines of high iron content when they are treated with added gallotannic acid as part of the gelatin-tannin fining.

susceptible wines do not respond to this treatment. In some white wines, aeration, which leads to the formation of ferric phosphate, followed by fining and filtration to remove the colloidal material, gives a brilliant wine. Many California wines, however, become brown in color and vapid in taste after such treatment. Virtually no California white table wines can stand the use of oxygen or peroxides. The threshold for iron clouding was placed at 3 mg per liter by Eschnauer (1966b), but many wines remain stable at 10 mg per liter.

The more iron a wine contains the more it will oxidize in the presence of oxygen. Blue-fining to remove iron is thus a help in preventing undue oxidation.

Copper deposits

In sulfited white wines containing over 0.3 to 0.5 part per million of copper, and stored in sealed containers, a reddish-brown deposit often forms. This deposit occurs only in the absence of oxygen and ferric iron, and redissolves readily upon exposure to oxygen. Its formation is accelerated by light and heat. It is believed to consist of colloidal cupric sulfide (Ribéreau-Gayon, 1933b, 1935b). Vogt (1945) reports that the copper compounds in wines have a very bitter taste. A comparison of the behavior of wines susceptible to white casse and to copper casse is given in table 79.

The formation of undesirable clouds and sediments in California wines containing excessive amount of copper and iron ions constitutes a serious problem in the industry. The presence of these metallic ions in excess has resulted in undesirable changes not only in appearance but also in flavor brought about by the catalytic powers of these ions. This condition has existed in the California wine industry since Repeal, but is common to all wine-producing regions of the world. It was aggravated soon after Repeal by (1) bottling wine at the winery and shipping to consuming centers in the bottle rather than in bulk; (2) developing better closure for wine bottles which, in preventing losses in quality by leakage and oxidation, results in the production of reducing conditions favorable to the formation of copper casse; (3) storing wines in the bottle at the winery or the bottling plant, thus increasing the time between bottling and consumption during which the wine is expected to remain bottle-bright; (4) using equipment and fittings of brass, bronze, or other copper alloys to minimize contamination of the wine with iron; and (5) using sulfites which, while necessary to control growth and activity of undesirable microörganisms and prevent overoxidation under the prevailing winery conditions, render the wine more susceptible to copper pickup and copper casse.

TABLE 79

BEHAVIOUR OF WINES SUSCEPTIBLE TO CLOUDING BY IRON CASSE AND BY
COPPER CASSE

Points compared	Type of turbidity and its behavior	
	White iron casse	Copper casse
Essential nature: Appearance	A uniform white turbidity appearing after exposure of wine to air	A white haze appearing in wines stored in the absence of air; later a reddish-brown sediment
Causes	Colloidal ferric phosphate; oxidation of free ferrous ions, in presence of excess phosphate, and formation of suspended aggregates of ferric phosphate which eventually settle out and form a white deposit	Colloidal cupric sulfide; a complex reduction of cupric ions and sulfite, involving reduction of other metals; does not happen in presence of free ferric ions; cupric sulfide forms first as a white haze which later becomes reddish in color and finally deposits as a reddish-brown sediment
Action of various physical and chemical agents: Air	Turbidity appears after exposure to air; wines that become turbid on oxidation often clear when stored in absence of air, owing to deposition of ferric phosphate	Turbidity disappears on exposure to air and appears in its absence; the disappearance of turbidity after oxidation may be accompanied by slight sediment formation
Light	Light hinders the appearance of turbidity, and decreases its intensity	Light hastens the appearance of turbidity and increases its intensity
Heat	Hinders turbidity	Hastens turbidity
Hydrogen peroxide and similar strong oxidizing agents	Turbidity appears immediately after addition of H_2O_2	Turbidity, if present, disappears immediately
Sulfur dioxide and similar reducing agents	Prevent turbidity	Hasten and increase turbidity
pH	Turbidity appears only in range of pH 2.9 to 3.6	——
Citric acid	Specific inhibitor and preventative agent even in small amounts	No effect unless added in excessive quantities
Precautionary treatments	Avoid contamination of wine with phosphates, iron, and copper salts, and oxidation	Avoid contamination of wine with iron and copper salts, and use of excessive amounts of sulfur dioxide

The mechanism of copper casse and the factors affecting its development were investigated in detail by Joslyn and Lukton (1956), Lukton and Joslyn (1956), Kean and Marsh (1956a), and Peterson et al. (1958). Copper casse may occur in the presence of a large amount of copper and sulfur dioxide (i.e., copper sulfide), but more commonly it is due to reactions between copper and sulfur-containing amino acids or with proteins. Kean and Marsh (1956a) concluded that the various types of clouds were formed independently. The catalytic action of sunlight was emphasized. If copper but not sulfur dioxide was present, the protein cloud was partially reversible. If both were present, at first an almost completely reversible cloud formed, but became progressively more irreversible with time.

Although protein or polypeptide constitute a major part of the haze material and sediment in copper casse (Ribéreau-Gayon, 1933a, 1935b;

Kean and Marsh, 1956*a*; Lukton and Joslyn, 1956), the nature and the source of the polypeptide constituent are not known. Although the polypeptide constituent is probably derived from grapes, certain identification of it has not been made and progress has been slow in devising methods of controlling haze by suitable selection of grapes of desirable variety and maturity, and in devising methods of fermentation and aging to reduce the concentration of this substance. Polypeptides are known to constitute the major part of the haze material in beer. Surveys have shown that beer hazes from different sources yield, on acid hydrolysis, amino acids in approximately the same relative amounts (Pollock, 1962). Whether this is true of wine is not known, as data on the amino acid composition of the polypeptide component of copper casse are very limited.

Both cuprous sulfide and cupric sulfide clouds were shown to occur by reduction of sulfite to sulfide, but reduction of sulfur-containing proteins was observed also. More recently Eschnauer (1966*b*) distinguishes copper sulfite from copper-protein and copper-leucoanthocyan clouding with excessive copper. He placed the threshold for clouding at 2 mg per liter, which is four to ten times as high as reported in California. It is known that under existing conditions the copper tolerance of California wines is fairly low (0.3)—lower than the 0.5 part per million tolerance of Bordeaux established by Lherme (1931–32).

With replacement of copper-bearing alloys by stainless steel, glass, or other inert material and the better elimination of proteins and other colloids contributing to clouding, thus increasing copper tolerance, copper casse is no longer the problem it was twenty years ago.

The relative contribution of the naturally occurring heavy-metal ions of the grape and of those picked up during the prevailing cellar operations is not known, but Amerine's data (see p. 274) indicate that little copper is derived from the grapes. It is not known whether the complete elimination of iron- and copper-bearing metals and fittings would prevent the formation of iron and copper casses. (Both the white phosphate casse and the blue tannate casse have occurred in California wines, although the former predominates.) The available data indicate that the iron and copper salts introduced during crushing and pumping are precipitated during fermentation. European enologists have frequently implicated the use of copper-bearing fungicides or grapes as a source of copper in wine. Gasiuk *et al.* (1965) investigated this in detail. They report that treatment of grapes with Bordeaux mixture increased the copper content in the fruit and that washing eliminated only 14 to 40 per cent of the copper added. Most of the added copper was eliminated during pressing and settling; the juice, whether from washed or unwashed grapes, contained 5 mg of copper per liter.

Other metals

Aluminum precipitates as the hydroxide, according to Eschnauer (1966*b*). He places the threshold for clouding at 10 mg per liter, which is higher than the amounts reported in normal wines. In eighteen German wines, Eschnauer (1964) reported 0.51 to 0.93 mg per liter of aluminum (average 0.8). When the aluminum content is high (about 10 mg per liter), owing to contamination in storage in aluminum containers, undesirable changes in clarity, color, odor, and taste occur. Aluminum haze is rare in California, since equipment containing this metal is seldom used. In Australia, Rankine (1962) showed that a maximum of 5 mg per liter could be tolerated in dry white table wine and white dessert wine. The maximum haze occurred at pH 3.8.

Tin clouding is due to tin-protein compounds, often with traces of other metals and with sulfur. Eschnauer (1963*c*, 1966*b*) set the threshold for clouding at 1 mg per liter—higher than the amounts reported in normal wines. Nickel is very seldom present in sufficient amounts to cause clouding, since the normal amount found is less than 1 mg per liter and the threshold for clouding is 100 mg per liter, according to Eschnauer (1966*b*). Nickel hydroxide precipitates, usually in the presence of high protein and with true amounts of other metals. Zinc clouding is seldom described. However, Eschnauer (1966*b*) believes that it must be kept in mind when zinc contamination is possible.

In German wines from 1959 to 1964, calcium contents of 50 to 180 mg per liter are reported (Anon., 1966*a*). The suggested limit was 80 mg per liter. The solubility of calcium tartrate is particularly sensitive to the alcohol content at alcohol concentrations of less than 10 per cent. Consequently, even slight increases in alcohol content in this region may result in calcium tartrate precipitation. The calcium content could not be correlated to the pH. Treatment with different types of bentonite resulted in varying degrees of calcium pickup. In wines approaching the suggested limit, care should be taken to use sodium rather than calcium bentonite. Treatment with cation ion-exchangers should reduce the calcium content sufficiently to prevent haze formation. Cambitzi (1947) claimed racemization of tartaric acids in wines after long standing (or heating) was responsible for the formation of the insoluble calcium racemate which he reported to occur. This should be confirmed.

Blue-fining

At present the most efficient method known of reducing both the iron and copper content is by blue-fining, i.e., precipitation by the addition of potassium ferrocyanide under carefully controlled conditions. Blue-fining for removal of excess iron and copper was developed by

Moslinger in Germany during the period 1902–1905. It has been authorized in Germany since 1923, in Luxembourg since 1927, Austria 1928, Jugoslavia 1932, Russia 1936, Hungary and Bulgaria 1953, France 1962, Italy 1965, and Argentina 1965. In the United States it is not allowed, but a proprietary mixture containing ferrocyanide is permitted (see p. 546).

Tarantola and Castino (1964) emphasized that prolonged contact of wine with ferrocyanide residues would lead to formation of appreciable amounts of hydrogen cyanide. At 25°C (77°F) this occurs in eight to fifteen days. The pH of the wine did not influence the rate of decomposition, but the fixation of hydrogen cyanide in organic form is greater at higher pH values. They note that even under the most adverse conditions the amount of hydrogen cyanide formed is only 1/60th to 1/120th of the lethal dose. It is sufficient, however, to give the wine a characteristic bitter-almond smell.

Deibner (1966a) confirmed the results of Tarantola and Castino (1961) that cyanide, both free and combined as cyanohydrin, occurs in wines that have not been treated with ferrocyanide. Free cyanide is present in very small amounts—4 to 5 mg per liter. Wines treated with ferrocyanide had much higher amounts (Tarantola and Castino, 1964).

Ionescu (1966) reported that Rumanian wines may contain up to 40 mg per liter of iron as a result of accidental contact with iron-bearing metals (filters, fillers, metal tanks). The average iron content is in the range of 9 to 15 mg per liter. Ionescu proposed two procedures for using blue-fining in Rumania. When properly used, under conditions that eliminate every trace of excess ferrocyanide, blue-fining will effectively remove both copper and iron without markedly affecting the other constituents of the wine. The reluctance to use it, in both Europe and the United States, is based upon the possible health hazards of cyanogenetic compounds. There are few reliable data on the toxicity of ferrocyanide itself, or on the conditions under which it can be hydrolyzed into the free cyanides which are toxic.

While the Food and Drug Administration has not expressly prohibited the use of potassium ferrocyanide in the manufacture of wine, it does not approve such practice, and will hold the wine maker responsible if the use of the material has a deleterious effect on the consumer.

OXIDATION DEFECTS . . .
. . . are primarily a problem in white wines.

Overoxidation causes the oxidized products of tannin and of coloring matter to be precipitated. Aldehyde formation also occurs; it gives table wines unpleasant flavors and may cause the development of insoluble complexes of aldehyde and coloring matter, which settle out.

Oxidasic casse

Oxidasic casse is characterized by clouding and changes of color on exposure to air. Red wines turn brown, and white wines become yellow. A "cooked" or a rancio (aldehydic) odor also usually develops. These changes are caused by an enzyme known as phenolase or polyphenoloxidase, normally present in small quantities in sound grapes, and in abnormally high quantities in moldy ones (p. 280). A wine may be protected from the action of this enzyme by the use of sufficient sulfur dioxide or by destroying the oxidase by flash pasteurization at 70° to 85°C (158° to 185°F), according to the condition of the wine. This form of casse, though less common in America than in Europe, may be expected in any wine made from moldy grapes, particularly whites. (See Hussein and Cruess, 1940b.) It has been observed on grapes in the coast counties of California in seasons of early fall rain and high humidity which result in excessive mold growth on the fruit.

Young wines which contain polyphenoloxidase must be especially protected against contact with oxygen. Addition of sulfur dioxide *before* the first racking is thus justified in such cases.

Glucose oxidase has been recommended as an agent for removal of oxygen from dry white table wines. Ough (1960a) found treated wines to be darker than untreated on exposure to air. He did not recommend it.

Color changes in red wines can take place as a result of changes in the amounts present or of shifts in the equilibrium between colored (anthocyanin) and colorless (anthocyanogen) forms of the pigments. Decrease of color, for example, can occur because of oxidation or polymerization of colored forms, or might take place as a result of a shift in equilibrium from the colored to the colorless form. An example of the shift from colorless to colored forms occurs, according to Berg and Akiyoshi (1962), during early aging of the wine. During longer aging of red wines the color loss was less in wines containing sulfur dioxide or higher alcohol contents. Owing to the shift from colorless to colored forms, old wines have a much higher percentage of their total anthocyans present in the colored form.

Harborne (1967) cites the Ribéreau-Gayon (1964) report that the major anthocyanins of grapes are delphinidin, petunidin, and malvidin 3-glucosides, and malvidin-3-p-coumaroylglucoside. The major grape flavonol glucosides are kaempferol, quercetin, and myrecitin 3-glucosides and quercetin 3-glucuronide. See p. 255 for further information.

The anthocyanin pigments are particularly sensitive to photochemical decomposition. When protected from light by storage in dark-colored bottles in cellars in complete darkness, the color is retained almost indefinitely. Changes in pigment tint, however, do occur; these are reflected in a change in absorption maximum from 530 mμ in new red wines to 470 mμ in old vintage wines. The fading of color on exposure to light occurs in aqueous acid solutions of anthocyanins both in the presence and the absence of oxygen. The rate and extent of this fading were investigated by Spaeth and Rosenblatt (1950) and Nordström (1956). Oxidation, however, is involved in decrease in pigment concentration of anthocyan solutions. This is accelerated by addition of ascorbic acid or hydrogen peroxide. Karrer and de Meuron (1932) and Lukton et al. (1956) found that the pseudobase of the anthocyanin was more susceptible to oxidation than the colored form itself. In the presence of oxygen this decomposition was pH-dependent.

Berg (1963) was not able to find a correlation between pH or tannin content and the loss of colored pigments from oxidation. The fact that certain tannins may have a protective effect on the colored pigments may partially account for this result. Berg was able to show that the response of pure pigments to pH was different from that of the pigments in wine. The degree of pigment association with other polyphenols is suggested as a factor.

The chemical transformations of the anthocyanin pigments during aging are poorly understood. At least three processes are involved: (1) hydrolysis of the glucoside releasing the unstable aglycones, (2) condensations between molecules so that the pigments gradually pass to a colloidal condition, and (3) demethoxylation.

The precipitation of coloring matter in a red wine (Weger, 1965a), was prevented by bentonite-fining at above 50°C (122°F).

Berg (1953c) showed a clear relation of grape variety to susceptibility to browning. This was confirmed by Berg and Akiyoshi (1956b). Although iron and copper usually increased the tendency to browning, some browning occurred even in the absence of these minerals or of sulfur dioxide. Enzymatic action did not seem to be a factor. Caputi, Jr., and Peterson (1965) showed that at room temperatures Palomino wines darkened readily in the presence of dissolved oxygen, but French Colombard darkened in the absence of oxygen. Even with dissolved oxygen the darkening was little influenced by the presence of oxygen unless the temperature was raised. To reduce browning they found that nylon (Nylasent 66 of National Polymer Products) was helpful in some cases but harmful in others. PVP (Polyclar AT of Antara Chemicals), at the rate of 4 to 6 pounds per 1,000 gallons, was more generally beneficial. Even so, the effect was small in some tests. They concluded that browning is strongly influenced by temperature but that the effect is not predictable in the light of our present knowledge.

The main factors increasing the browning rate of white wines are temperature, oxygen content, and increasing pH, according to Berg and Akiyoshi (1956b). Citric acid decreased the rate; increasing amounts of copper or iron increased it (iron more than copper in the presence of oxygen; the effect of the two is additive). The most important factor, however, was the variety of grape, as shown by Berg (1953c), particularly the region where grown and the state of maturity.

The tendency of the white table wines of Verdicchio to darken is strongly associated with their leucoanthocyanin content, according to Pallotta et al. (1966). They reported a direct correlation between leucoanthocyanin content and the optical density at 410 mμ. The leucoanthocyanin content of these samples varied from 0.048 to 0.696 gr per liter.

Chapon and Urion (1960) in their work on beer have provided the probable explanation for the action of ascorbic acid in wine. Investigating rate of disappearance of added ascorbic acid, they found that the oxidation of ascorbic acid was accompanied by the oxidation of an equivalent amount of various organic substances, and that the rate was directly proportional to the sum of the copper ions and iron complexes. This is the phenomenon known as coupled oxidation, in which the system acts as a powerful oxidant, oxidizing an equal amount of other substances present which molecular oxygen cannot touch. As long as free sulfur dioxide is present the reaction is greatly inhibited, but in its absence it is greatly accelerated.

PROTEIN CLOUDINESS ...

. . . is poorly understood.

The appearance of cloudiness or haze when wines are heated to 60°–65.5°C (140°–150°F) in the presence of air may be caused by proteins in the wine. Kielhöfer (1951) found that the precipitated material is alkali-soluble, contains about 60 per cent protein, and is negatively charged. Pasteurization and either addition of tannin or ultracentrifuging removed most of the proteins. It is not known how extensive such cloudiness may be in California, but many white wines have been observed to throw down a voluminous precipitate when heated to 65.5°C (150°F).

Farkas (1966) showed that the efficiency of bentonite in the reduction of protein depended on the way the bentonite was added, the amount of wine used for activating the bentonite, the time of activation, the way in which the bentonite was mixed with the wine, the pH values, and the temperature. The closer the pH value to the isoelectric point of wine protein, the more efficient the elimination of protein. High temperature also increased the rate of reaction. Farkas recommended ten times as much wine as bentonite for its hydration. In these studies the color and flavor were improved, but sensory data were not statistically analyzed.

Koch and Sajak (1961) observed that some wines eventually clouded in storage even though they had not given a positive test by the usual heating test (24 hours at 55°C; 131°F). They recommended mixing 95 ml of wine and 5 ml of a saturated solution of ammonium sulfate, holding nine hours at 45°C (113°F), cooling to 0°C (32°F), and observing at fifteen-minute intervals for the appearance of cloudiness. They consider this a much more severe and sure test for the "soluble" protein which may eventually cloud the wine. The Koch-Sajak test gave different results depending on the temperature at which the test was made. The presence of yeasts increased the amount of turbidity.

MERCAPTAN AND HYDROGEN SULFIDE

Wines occasionally have a hydrogen sulfide taste owing to formation of hydrogen sulfide and mercaptans during fermentation or storage. Tanner and Rentschler (1965) investigated the factors accounting for this, and its control. They report that these substances arise from reduction of sulfur dripping into the wine vat from a sulfur wick, but it is known to result also from reductions of sulfites and sulfates by yeast in the brewing industry (Ricketts and Coutts, 1951). The formation of mercaptans as well as hydrogen sulfide by yeast is discussed by Brenner et al. (1955) and in Colloque sur la Biochimie du Soufre (1956).

While activated carbon will remove mercaptans, it reduces the quality of the treated wine. Schneyder (1965) recommended treatment of the wine with silver chloride (2 gr mixed with 100 gr of kieselguhr). The process is authorized in Austria providing no silver remains in the wine. It is *not* permitted in this country. Aeration, sulfur dioxide, and close filtration are used here.

De Rosa (1965a) and Capelleri et al. (1965a) recommended treatment with palladium chloride to remove mercaptans from wines. A Swiss recommendation that ascorbic acid be used to reduce the mercaptan to sulfide which can be removed with carbon dioxide does not work, according to De Rosa (1965a), nor did it in tests in this laboratory.

CRYSTALLINE DEPOSITS

Crystalline deposits in wine may be formed by precipitation of excess cream of tartar (potassium bitartrate), calcium tartrate, saccharic acid, or calcium oxalate. It is reported that acid tartrate is more soluble in the presence of malic and lactic acids. This is one reason for precipitation of tartrate after a malo-lactic fermentation. However, Nègre et al. (1960)

showed that the specific effect of malic acid is small, and that tartrate precipitation is slow, reaching equilibrium only after four to six months. The values for the solubility of potassium acid tartrate calculated by Berg and Keefer (1958) are similar to those obtained by Nègre et al. (1960).

For Italian wines Usseglio-Tomasset (1961) gave the following maximum potassium values which will not lead to bitartrate precipitation at the tartaric acid (per cent), alcohol (per cent) and pH values shown:

Tartaric acid	pH 3 Alcohol			pH 3.2 Alcohol		
	8	12	16	8	12	16
0.09	0.204	0.160	0.135	0.167	0.128	0.105
0.15	0.122	0.096	0.081	0.100	0.077	0.063
0.21	0.087	0.068	0.057	0.071	0.055	0.045
0.27	0.068	0.053	0.045	0.055	0.042	0.035
0.33	0.055	0.043	0.036	0.045	0.035	0.029
0.39	0.047	0.037	0.031	0.038	0.029	0.024

Tartaric acid	pH 3.4 Alcohol			pH 3.6 Alcohol		
	8	12	16	8	12	16
0.09	0.147	0.110	0.088	0.140	0.103	0.079
0.15	0.088	0.066	0.053	0.084	0.062	0.047
0.21	0.063	0.047	0.038	0.060	0.044	0.033
0.27	0.049	0.037	0.029	0.046	0.034	0.026
0.33	0.040	0.030	0.024	0.038	0.028	0.021
0.39	0.034	0.025	0.020	0.032	0.023	0.018

Kielhöfer and Wurdig (1961a) first called attention to precipitates of mucic (galactaric) acid crystals in wine (as the calcium salt). The precipitates were either white and gravelly or had quadratic cross-section crystals. The precipitates often did not occur until long after (months or years) the wine had been bottled.

The precipitates were noted mainly in years of considerable botrytis growth. Schormüller et al. (1966) reported that mucic acid was found also in musts produced from grapes which had been attacked by Botrytis cinerea. It occurs only when botrytis attacks intact grapes. Würdig and Clauss (1966) state that it is produced by oxidation of galacturonic acid. This oxidation reaction is catalyzed by an NAD-specific dehydrogenase. The authors state that there are several possible mechanisms for the oxidation. The solubility product of calcium mucate at 20°C (68°F) is 1.3×10^{-7} (mol per liter)^2. Wines should not contain more than 0.1 gr per liter of mucic acid. This amount usually occurs when more than

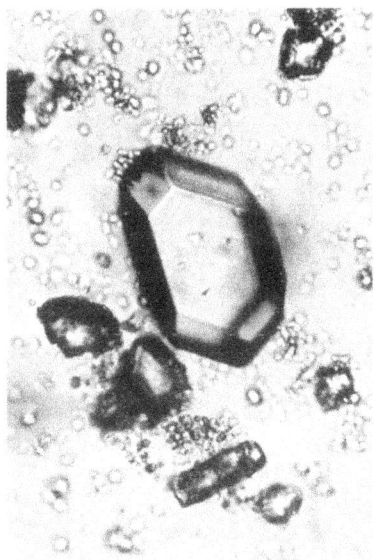

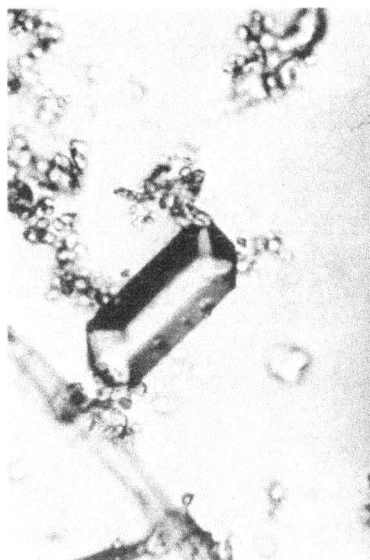

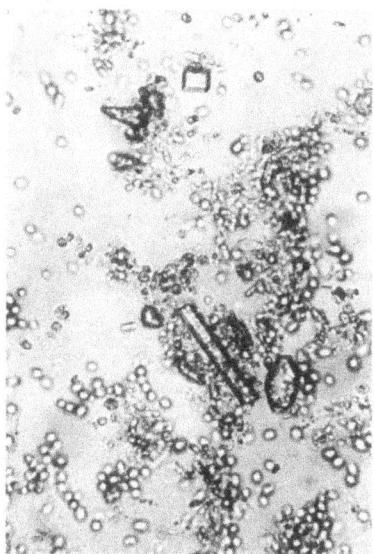

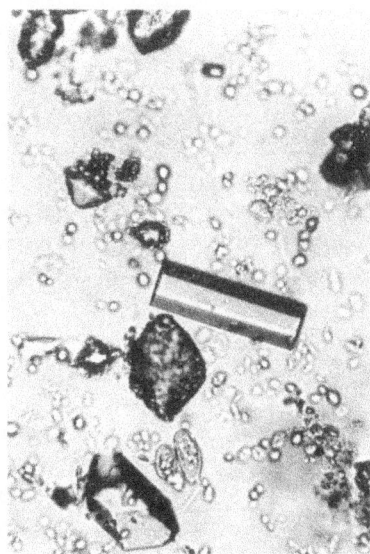

Figure 127.   Various types of crystals found in wine. Upper left, calcium tartrate; upper right, calcium tartrate in deposit from a calcium bentonite fining; lower left, calcium oxalate; lower right, potassium acid tartrate (lemon-shaped). (Courtesy Dr. I. Tuzson, Budapest.)

10 to 25 per cent of the berries have been attacked by botrytis. Würdig and Clauss (1966) recommend determining the calcium and mucic acid contents before bottling. Addition of a small amount of calcium carbonate removes mucic acid, but has the disadvantage of reducing the acidity.

In the lees of Hungarian wines Tuzson (1966) reported normal crystals of calcium tartrate ($CaTa \cdot 4H_2O$). The crystals of potassium acid tartrate, however, were less normal in shape. Calcium oxalate crystals were occasionally found. Wines fined with sodium bentonite contained only potassium acid tartrate crystals; wines fined with calcium bentonite contained less of the acid tartrate crystals and considerable calcium tartrate crystals. (See figs. 127 and 128.)

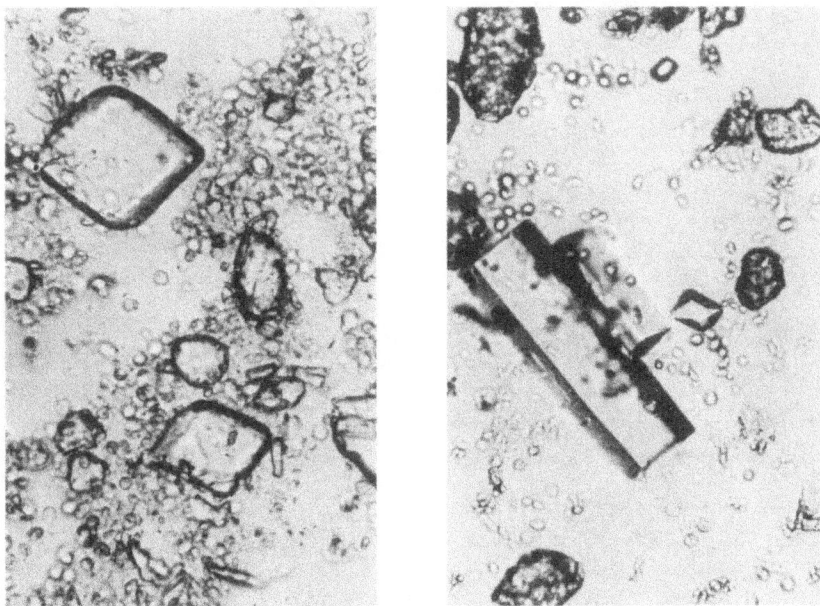

Figure 128. Various types of crystals found in wine. Left, calcium tartrate (flat form); right, calcium tartrate (half-formed crystal). (Courtesy Dr. I. Tuzson, Budapest.)

Jakob (1965) compared the effect of acids on six sodium bentonites and three calcium bentonites as far as solutions of sodium and calcium were concerned. Both sodium and calcium were dissolved by acids from sodium bentonite. Negligible amounts of sodium were dissolved from calcium bentonite, but more calcium than from sodium bentonite. The sodium bentonite was a better ion-exchange media for potassium than calcium bentonite. Sodium bentonite has a higher pH and 2.4 to 2.9 times as much swelling property in wine than calcium bentonite.

Berg (1962) gives the following directions for identifying crystalline deposits in wines:

1. Collect the deposit on a sintered glass funnel and wash three times with water and once with alcohol. Dissolve a few crystals in dilute hydrochloric acid and make a flame test. A violet color indicates potassium bitartrate. If calcium is present then

2. Crush about 20 mg in 1 ml of water and dissolve by adding 2 drops of concentrated hydrochloric acid. If crystals form on rubbing the side of the dish with a glass rod, calcium saccharate is present and tartrates and oxalates are absent. The melting point of calcium saccharate is 213°C. Saccharic acid can be identified from the pyrrole reaction: wash the crystalline deposit with water and alcohol and dry by suction. Put in 1 ml of water and add a little ammonia to dissolve. Evaporate a few drops to dryness. Place pine shavings moistened with hydrochloric acid over the deposit. Now heat. The rising fumes color the pine shavings an intense red-violet if saccharic acid is present. Tartaric acid colors the pine shavings only a faint red.

3. If heat must be applied dissolve 20 mg in 1 ml of dilute hydrochloric acid; if long needles form on cooling, calcium sulfate is present.

4. If 20 mg of the deposit are soluble in dilute hydrochloric acid and no deposit forms on rubbing with a glass rod, potassium bitartrate, calcium oxalate, or calcium tartrate is present. The flame test can eliminate the first. To distinguish the other two dissolve in dilute hydrochloric acid, add a few drops of methyl red, and neutralize with 20 per cent ammonia. Calcium oxalate precipitates on slight heating; calcium tartrate does not.

The prevention and control of tartar deposition may be carried out by removal of excess potassium bitartrate by refrigeration and cold storage (see p. 518); by reduction of potassium content by treating with cation exchange resin in the H-form or Na-form; by reduction of tartrate ion by treating with an ion-exchange resin in the hydroxyl form; or by addition of inhibitors such as metatartaric acid (Scazzola, 1956; Miconi, 1962b).

Appendix A
SELECTED REFERENCES

A SELECTED, CLASSIFIED, AND ANNOTATED LIST OF REFERENCES[1]

GENERAL

ADAMS, L. D.
1964. The commonsense book of wine. New York, David McKay Co., Inc. xiv + 178 pp. Popular.

ALLEN, H. W.
1964. The wines of Portugal. New York, McGraw-Hill Book Co. 192 pp. Popular.

AMBROSI, H.
1959. Die Weine Südafrikas. Deut. Wein-Ztg. **95**:482, 484, 498, 500, 502. Popular.

AMERINE, M. A.
1963. Viticulture and enology in the Soviet Union. Wines & Vines **44**(10):29–34, 36; (11):57–62, 64; (12):25–26, 28–30. Popular.
1964. Der Weinbau in Japan. Wein-Wissen. **19**:225–231. Popular

AMERINE, M. A., H. W. BERG, AND W. V. CRUESS
1967. The technology of wine making. 2d ed. Westport, Conn., The Avi Publishing Co., Inc. ix + 799 pp. Technical.

AMERINE, M. A., AND V. L. SINGLETON
1965. Wine: an introduction for Americans. Berkeley and Los Angeles, University of California Press. viii + 357 pp. Popular. Also in a revised paperback edition, 1967.

ANON.
1953. The wine industry of Argentina. Buenos Aires. 56 pp. Popular.
1957. Wines of Italy. Rome, Istituto Nazionale per il Commercio Estero. 36 pp. Popular.
1961. Wine-producing in Yugoslavia. Belgrade, Federal Chamber of Foreign Trade. 64 pp. Popular.

AUSTIN, C.
1968. The science of wine. London, University of London Press. 216 pp. Also a U.S. edition. Technical.

BABO, A. W. VON, AND E. MACH
1922–1927. Handbuch des Weinbaues und der Kellerwirtschaft. Berlin, P. Parey. 2 vols. The great German classic on grapes and wines.

BERIDZE, G. I.
1956. Tekhnologiia i ėnokhimicheskaia kharakteristika vin Gruzii. Tbilisi, Izd. Akademiia Nauk Gruzinskoi S.S.R. 405 pp. Mainly technical.

---

[1] This list is not complete and is primarily devoted to publications since 1955, with exceptions for especially notable or unique earlier material. Only books are included unless the periodical reference cited contains material not available elsewhere.

1965. Vino i kon'iake Gruzii; les vins et les cognacs de la Georgie. Tbilisi, Izda-tel'stvo "Sabchota Sakartvelo." 360 pp. Popular.

BIRON, M.
1950. Vignes et vins de Turquie, Istanbul, Trace-Marmara. 64 pp. Semipopular.

BLAHA, J.
1961. Réva vinná. Prague, Nakladatelství Československé Akademie Věd. 462 + 43 pp. Technical on grape varieties for wine.

BODE, C.
1956. Wines of Italy. New York, The McBride Co., Inc. 135 pp. Popular.

BOSDARI, C. de
1966. Wines of the Cape. 3d ed. Cape Town, Amsterdam, A. A. Balkema. 95 pp. Popular.

BRUNI, B.
1964. Vini italiani portanti una denominazione di origine. Bologna, Edizioni Calderini. 255 pp. Popular.

CAPONE, R.
1963. Vini tipici e pregiati d'Italia. n.p. Editoriale Olimpia. 282 pp. Popular.

CHAPPAZ, G.
1951. Le vignoble et le vin de Champagne. Paris, L. Larmat. xvii + 414 pp. Popular and technical; authentic.

COCKS, C., AND E. FÉRET
1949. Bordeaux et ses vins, classés par ordre de merité. 11th ed. Bordeaux, Féret et Fils. 1129 pp. Popular and useful for many vineyards; out of date for others.

CORNELSSEN, P. A.
1954. Das Buch vom Deutschen Wein. Mainz, Deutscher Weinverlag. 262 pp. Popular.

COSMO, I.
1966. Guida viticola d'Italia. Treviso, Longo and Zoppelli. 244 pp. Popular, with useful production data.

COX, H.
1967. The wines of Australia. London, Hodder and Stoughton. 192 pp. Popular.

FORBES, P.
1967. Champagne: The wine, the land and the people. London, Victor Gollancz Ltd. 492 pp. Popular, interesting.

GOLDSCHMIDT, E.
1951. Deutschlands Weinbauorte und Weinbergslagen. 6th ed. Mainz, Verlag der Deutschen Wein-Zeitung. 263 pp. German vineyards and their wines.

GROSSMAN, H. J.
1964. Grossman's guide to wines, spirits and beers. 4th ed. New York, Charles Scribner's Sons. xviii + 508 pp. Popular.

HALÁSZ, Z.
1962. Hungarian wines through the ages. Budapest, Corvina. 185 pp. Popular. Also published in French, German, and Hungarian.

HALLGARTEN, S. F.
1957. Alsace and its wine gardens. London, André Deutsch. 187 pp. Popular.
1965. Rhineland; wineland. Manchester, Withy Grove Press. xiv + 210 pp. Popular.

HEALY, M.
1963. Stay me with flagons. London, Michael Joseph Ltd. 262 pp. Popular. Mainly on French wines.

HEDRICK, U. P.
1945. Grapes and wines from home vineyards. New York, Oxford University Press. 326 pp. Popular.

ISNARD, H.
1951–1954. La vigne en Algérie. Gap, Ophyrs. 2 vols. Popular and historical.

JAQUELIN, L., AND R. POULAIN
1965. The wines and vineyards of France. New York, G. P. Putnam's Sons. 418 pp. Too popular. Also in a French edition.

JAMES, W.
1962. Wine in Australia: a handbook. 3d ed. London, Phoenix House. 148 pp. Popular; in the form of a dictionary; includes non-Australian wines.

KITTEL, J. B., AND H. BREIDER
1958. Das Buch vom Frankenweine. Würzburg, Universitätsdruckerei H. Stürz. 207 pp. Popular and complete.

LAFFORGUE, G.
1947. Le vignoble Girondin. Paris, Louis Larmat. xi + 319 pp. Technical and authentic.

LANGENBACH, A.
1951. The wines of Germany. London, Harper and Co. xii + 168 pp. Technical and authentic.
1962. German wines and vines. London, Vista Books. 190 pp. Popular and practical. (Not a revision of preceding entry.)

LAYTON, T. A.
1961. Wines of Italy. London, Harper Trade Journals Ltd. xi + 221 pp. Popular.

LEIPOLDT, C. L.
1952. Three hundred years of Cape wine. Cape Town, Stewart. 230 pp. Popular.

LÉON, V. F.
1947. Uvas y vinos de Chile. Santiago de Chile, Sindicato Nacional Vitivinícola. 287 pp. Too popular.

LEONHARDT, G.
1962. Das Weinbuch: Werden des Weines von der Rebe bis Glase. Leipzig, VEB Fachbuchverlag. 418 pp. Popular, on many countries.

LICHINE, A.
1964. Wines of France. 6th ed. London, Cassell. xiv + 379 + xxvii pp. Popular and authentic.

MARECA CORTÉS, I.
1969. Enologia. Enfoques cientificos y tecnicos sobre la vid y el vino. Madrid, Editorial Alhambra, S.A. xiv + 308 pp.

MARRISON, L. W.
1958. Wines and spirits. London, Penguin Books. 320 pp. Popular.

PESTEL, H.
1959. Les vins et eaux-de-vie à appellations d'origine contrôlées en France. Mâcon, Imprimerie Buguet-Comptour. 44 pp. Popular.

POUPON, P., AND P. FORGEOT
1959. Les vins de Bourgogne. 2d ed. Paris, Presses Universitaires de France. 175 pp. Very popular. 4th French edition was published in 1967 in Paris by Presses Universitares de France. 196 pp.

PRISNEA, C.
1964. Bacchus in Rumania. Bucharest, Meridiane Publishing House. 245 pp. Popular.

PUBLIC RELATIONS DEPARTMENT OF THE KWV
1967. A survey of wine growing in South Africa. Paarl, KWV. 55 pp. Popular; updated annually.

RADENKOVIĆ, D. G.
1962. Vino, savremeni problemi vinarstva u svetu i kod nas. Belgrade. 136 pp. National and world problems of wine industry.

ROGER, J. R.
1960. The wines of Bordeaux. New York, Dutton. 166 pp. Popular and personal.

SCHOONMAKER, F., AND T. MARVEL
1934. The complete wine book. New York, Simon and Schuster. xi + 315 pp. Popular and authentic.

SHAND, P. M.
1960. A book of French wines. 2d ed. New York, Alfred A. Knopf. 415 pp. Popular; useful data on geographical names for wines.

SIMON, A. L.
1957. The noble grapes and great wines of France. New York, McGraw-Hill Book Co., Inc. xi + 180 pp. Popular.
1962a. Champagne; with a chapter on American champagne by Robert J. Misch. New York, McGraw-Hill Book Co., Inc. 224 pp. Popular.
1962b. The history of Champagne. London, Ebury Press. 192 pp. Popular but authentic; good statistics.
1967. The wines, vineyards and vignerons of Australia. London, Paul Hamlyn. xiii + 194 pp. Popular.

SIMON, A. L., AND S. F. HALLGARTEN
1963. The great wines of Germany and its famed vineyards. New York, McGraw-Hill Book Co., Inc. 191 pp. Popular.

THUDICHUM, J. L. W., and A. DUPRÉ
1872. A treatise on the origin, nature and varieties of wine. London, Macmillan and Co. xvi + 760 pp. Popular; many shrewd and still applicable observations.

WAGNER, P. M.
1965. A wine-grower's guide. Rev. ed. New York, Alfred A. Knopf. 11 + 224 + xii pp. Popular and authentic.

WINKLER, A. J.
1962. General viticulture. Berkeley and Los Angeles, University of California Press. 612 pp. Technical and practical.

HISTORY

ALLEN, H. W.
1932. The romance of wine. New York, E. P. Dutton & Co. 264 pp. A classic in praise of old wines.
1962. A history of wine: great vintage wines from the Homeric Age to the present day. New York, Horizon Press. 304 pp. Published in London in 1961. Popular.

BASSERMANN-JORDAN, F. VON
1923. Die Geschichte des Weinbaus. 2d ed. Frankfurt am Main, Frankfurter Verlags-Anstalt. 3 vols. Notable history of German vineyards and wines.

BILLIARD, R.
1913. La vigne dans l'antiquité. Lyon, H. Lardanchet. viii + 560 pp. A fine history.

Carosso, V. P.
1951. The California wine industry, 1830–1895: a study of the formative years. Berkeley and Los Angeles, University of California Press. ix + 241 pp. Thorough on California history.

Dalmasso, G.
1937. Le vicende techniche ed economiche della viticoltura e dell'enologia in Italia. Milan, Presso Arti Grafiche Enrico Gualdoni. ix + 481 pp. History of Italian wines. (From A. Marescalchi and G. Dalmasso's *Storia della vite e del vino in Italia*, same publisher, in 3 vols.)

Dion, R.
1959. Histoire de la vigne et du vin en France des origines au XIXᵉ siècle. Paris. xii + 768 pp. Much original work.

Emerson, E. R.
1902. The story of the vine. New York and London, G. P. Putnam's Sons. ix + 252 pp. General.

Galtier, G.
1960. Le vignoble du Languedoc méditerranéan et du Roussillon: étude comparative d'vignoble de masse. Montpellier, Causse, Graille & Castelnau. 3 vols. Scholarly.

Hymans, E. S.
1965. Dionysus: a social history of the wine vine. New York, The Macmillan Co. 381 pp. Popular, with some new ideas.

Kerdéland, J. de
1964. Histoire des vins de France. Paris, Hachette. 317 pp. Popular.

Lucia, S. P.
1963. A history of wine as therapy. Philadelphia, J. B. Lippincott Co. xiii + 234 pp. History of use of wine as medicine.

Peninou, E., and S. Greenleaf
1954. Winemaking in California. San Francisco, Peregrine Press. 2 vols. Original data.
1967. A directory of California wine growers and wine makers in 1860 ... Berkeley, Tamalpais Press. vii + 84 pp. Much rare information.

Schoonmaker, F., and T. Marvel
1941. American wines. New York, Duell, Sloan and Pearce. ix + 312 pp. Contains useful historical material.

Seltman, C. T.
1957. Wine in the ancient world. London, Routledge & Paul. 196 pp. Popular.

Vizetelly, H.
1882. A history of Champagne. London, H. Sotheran & Co. xii + 263 pp. Classic text.

Warner, C. K.
1960. The winegrowers of France and the government since 1875. New York, Columbia University Press. 303 pp. Scholarly.

Younger, W. A.
1966. Gods, men, and wine. Cleveland, World Publishing Co. 516 pp. Popular but scholarly.

Zaragoza, C. L.
1964. Historia y mitología del vino; con una antología báquica y un diccionario del vino. Buenos Aires, Editorial Mundi. 322. pp. Popular.

815

ANALYSIS AND COMPOSITION

AMERINE, M. A.
1954. Composition of wines. I. Organic constituents. Adv. Food Research 5:354–510. Technical.
1958. Ibid. II. Inorganic constituents. Ibid. 8:133–224.
1965. Laboratory procedures for enology. 6th rev. Davis, Calif., Associated Students Bookstore. 109 pp. Technical.

AMERINE, M. A., E. B. ROESSLER, AND F. FILIPELLO
1959. Modern sensory methods for evaluating wines. Hilgardia 28:477–567. Technical.

AMERINE, M. A., AND A. J. WINKLER
1963. California wine grapes: composition and quality of their musts and wines. Calif. Agr. Exper. Stat. Bull. 794:1–83. Technical.

ANON.
1955. Convention internationale pour l'unification des méthodes d'analyse des vins dans le commerce internationale. Bull. Office Inter. Vin 28 (291):22–34. (See also Ann. ferment. 1:310–319.) Technical.
1956. Analisis de vinos. Métodos oficiales para los laboratorios dependientes del Ministerio de Agricultura. Madrid, Ministerio de Agricultura. 240 pp. Official Spanish methods.
1958. Methodi ufficiali analisi per materie che interessano l'agricoltura. II (Part I). Mosti, vini, birre, aceti, sostanze tartriche, materie tanniche. Rome, Librerie dello Stato. 161 pp. Official Italian methods.
1961. Méthodes d'analyses et éléments constitutifs des vins. Bull. Office Inter. Vin 34 (367):57–90. Technical.
1962. Recueil des méthodes internationales d'analyse des vins. Paris, Office International de la Vigne et du Vin. Unpaged. Technical. New sections are added from time to time. A German edition is available.
1963a. Méthodos oficiais para a análise de vinhos, vinagres e azeites. Lisbon, Ministério de Comércio, Indústia e Agricultura, Direcção Geral da Acção Social Agrária. Official Portuguese methods.
1963b. Textes d'intérêt général, répression des fraudes. Méthodes officielles d'analyses des vins et des moûts (Arrêté du 24 juin 1963). Paris, Journaux Officiels. 37 pp. Official French methods.

ASSOCIATION OF OFFICIAL AGRICULTURAL CHEMISTS
1965. Official methods of analysis. 10th ed. Washington D.C., Association of Official Agricultural Chemists. 957 pp. (See pp. 171–179.) The American reference source for analytical procedures.

BEYTHIEN, A., AND W. DIEMAIR
1963. Laboratoriumsbuch für den Lebensmittelchemcher, mit Mitwirkung von L. Acker . . . 8th ed. Dresden and Leipzig, Verlag von Theodor Steinkopff. xxiii + 804 pp. (See pp. 479–551.) The German A.O.A.C.

DUJARDIN, J., L. DUJARDIN, AND R. DUJARDIN
1928. Notice sur les instruments de prècision appliqués à l'oenologie. Paris, Dujardin-Salleron. xii + 1096 pp. Now largely of historical interest.

FROLOV-BAGREEV, A. M., AND G. G. AGABAL'IANTS
1951. Khimia vins. Moscow, Pishchepromizdat. 392 pp. Technical.

HEIDE, C. VON DER, AND F. SCHMITTHENNER
1922. Der Wein. Braunschweig, F. Vieweg und Sohn. 350 pp. Technical; historically important.

SELECTED REFERENCES

HENNIG, K.
1962. Chemische Untersuchungsmethoden für Weinbereiter und Süssmosthersteller. 5th ed. Stuttgart, Verlagsbuchhandlung Eugen Ulmer. 126 pp. Technical and practical.

HESS, D., AND F. KOPPE
1968. Weinanalytik. *In* Handbuch der Lebensmittelchemie. VII. Alkoholische genusmittel, W. Diemair, ed. Berlin, Springer-Verlag. pp. 312–495. Technical; excellent.

JAULMES, P.
1951. Analyse des vins. 2d ed. Montpellier, Librairie Coulet, Dubois et Poulain. 547 pp. Technical and standard.

JUNTA NACIONAL DO VINHO
1942. Contribuição para o cadastro dos vinhos Portugueses de área de influència da J.N.V. Lisbon, Tipografia Ramos. 2 vols. Technical; many analyses of Portuguese wines.

KÖNIG, J.
1903. Chemische Zusammensetzung der menschlichen Nahrungs- und Genussmittel. Berlin, Julius Springer. Vol. 1:1–1935. Historically important analytical data.

LAHO, L., AND E. MINÁRIK
1959. Vinárstvo. II. Chémia, mikrobiológia, analytika vína. Bratislava, Slovenské Vydavatel'stvo Technickej Literatúry. 310 pp. Technical and practical.

LEAKE, C. E., AND M. SILVERMAN
1966. Alcoholic beverages in clinical medicine. Chicago, Year Book Medical Publishers, Inc. 160 pp. Useful analytical tables; wine in the diet.

MACKINNEY, G., AND A. C. LITTLE
1962. Color of foods. Westport, Conn., Avi Publishing Co., Inc. 308 pp. Technical.

NILOV, V. I., AND I. M. SKURIKHIN
1967. Khimiia vinodeliia i kon'iachnogo proizvodstva. 2d ed. Moscow, Pishchepromizdat. 439 pp. Technical and complete.

PARONETTO, L., AND G. DAL CIN
1954. I prodotti chimici nella tecnica enologia. Verona, Scuola d'Arte Tipografica "D. Bosco." 458 pp. Composition of chemicals used in wine making and their use.

PEYNAUD, E.
1947. Contribution à l'étude biochimique de la maturation du raisin et de la composition du vin. Lille, Imp. G. Sautai & Fils. 93 pp. Technical.

PROSTOSERDOV, N. N.
1952. Osnovy degustatsii vina. Moscow, Pishchepromizdat. 81 pp. Technical and practical.

RIBÉREAU-GAYON, J., AND E. PEYNAUD
1958. Analyse et contrôle des vins. 2d ed. Paris, Librairie Polytechnique Ch. Béranger. xxxii + 557 pp. Technical; contains much data on composition.
1966. Traité d'oenologie. II. Composition, transformations et traitements des vins. Paris, Librairie Polytechnique Ch. Béranger. 1065 pp. Technical; practical and theoretical.

VALAER, P.
1950. Wines of the world. New York, Abelard Press. 576 pp. Popular but useful; some unique analyses.

WILEY, H. W.
1903. American wines at the Paris exposition of 1900. U.S. Dept. Agr. Bur. Chem. Bull. 72:1–40. Historically interesting; data on composition.

817

MICROBIOLOGY AND SANITATION

AMERINE, M. A., AND R. E. KUNKEE
1968. Microbiology of wine making. Ann. Rev. Microbiol. 22:323–358. Technical.

ASSOCIATION OF FOOD INDUSTRY SANITARIANS, INC.
1952. Sanitation for the food-preservation industries. New York, McGraw-Hill Book Co. 284 pp. Practical.

CASTELLI, T.
1959. Introduzione alla microbiologia enologica. Milan, Tipografia Setti e Figlio. 75 + 5 pp. Technical.
1960. Lieviti e fermentazioni in enologia. Rome, Luigi Scialpi Editore. 63 pp. Technical and practical.

DAVIDSON, A. D.
1963. Wine Institute sanitation guide. San Francisco, Wine Institute. iv + 68 pp. Practical directions.

DOMERCQ, S.
1957. Etude et classification des levures de la Gironde. Ann. Technol. Agr. 6:5–58, 139–183. Technical.

LODDER, J., AND N. J. W. KREGER-VAN RIJ
1952. The yeasts: a taxonomic study. New York, Interscience Publishers, Inc. 713 pp. The standard text on yeast taxonomy.

MAESTRO PALO, F.
1959. Defectos y enfermedades de los vinos. 2d ed. Zaragoza, Tip. "La Academica." 527 pp. Descriptive.

MRAK, E. M., AND H. J. PHAFF
1948. Yeasts. Ann. Rev. Microbiol. 2:1–46. Technical.

MÜLLER-THURGAU, H., AND A. OSTERWALDER
1912. Die Bakterien im Wein und Obstwein und die dadurch veruchten Veränderungen. Centr. Bakt. Abt. II, 36: 129–339. Classic.
1918. Weitere Beiträge zur Kenntnis der Mannitbakterien im Wein. Ibid. 48:1–36. Classic.

PEYNAUD, E., AND S. DOMERCQ
1959. A review of microbiological problems in wine-making in France. Am. J. Enol. Viticult. 10:69–77. Technical.

PRILLINGER, F.
1966. Der wissenschaftliche Weinkenner. Vienna, Austria-Press. Pp. 131–150. Technical.

SCHANDERL, H.
1959. Die Mikrobiologie des Mostes und Weines. 2d ed. Stuttgart, E. Ulmer Verlag. 321 pp. Technical; useful practical observations.

SKOFIS, E. C. (ed.)
1960. Symposium on the sanitation problems of the wine industry and their solution. Am. J. Enol. Viticult. 11:80–101. Practical suggestions for California wineries.

VAN KERKEN, A. E.
1963. Contribution to the ecology of yeasts occurring in wine. Pretoria, University of the Orange Free State. ix + 119 pp. Technical.

VAUGHN, R. H.
1955. Bacterial spoilage of wines with special reference to California conditions. Adv. Food Research 6:67–108. Technical and perceptive.

VERONA, O., AND G. FLORENZANO
1956. Microbiologia applicata all'industria enologica. Bologna, Edizioni Agricole. 191 pp. Technical.

BIBLIOGRAPHY[1]

ALAN-BAKER, E.
1958. Bibliography of food. A select international bibliography ... New York, Academic Press, Inc.; London, Butterworths Scientific Publications. xii + 331 pp. (See pp. 271–283.) Limited.

AMERINE, M. A.
1951. The educated enologist. Proc. Am. Soc. Enol. **1951**:1–30. A selected list.
1954. Composition of wines. I. Organic constituents. Adv. Food Research **5**: 353–510. From 1930.
1958. *Ibid.* II. Inorganic constituents. *Ibid.* **8**: 133–224. From 1930.
1959. A short check list of books and pamphlets in English on grapes, wines and related subjects 1949–1959 ... n.p. 61 pp. Useful for period.
1969. A check list on grapes and wine. 1960–68. With a supplement for 1949–1959. Davis, Calif. 92 pp.

AMERINE, M. A., AND L. B. WHEELER
1951. A check list of books and pamphlets on grapes and wine and related subjects, 1938–1948. Berkeley and Los Angeles, University of California Press. 240 pp. Very incomplete.

ARAMIAN, N. G.
1957. Bibliograficheskiĭ ukazatel' otechestvennoĭ literatury po tekhnologii vina 1948–1956. Erevan, Izdatel'stvo Glavnogo Upravleniia Sel'khoz. 223 pp. 1957–1958. 1959. 119 pp. 1959–1962. 1965. 216 pp. Includes periodical references.

DICEY, P.
1951. Wine in South Africa: a selected bibliography. Rondebosch, University of Cape Town, School of Librarianship. 29 pp.

DUMBACHER, E.
1966. Internationale Weinbibliographie 1955–1965 ... Klosterneuburg, Mitteilungen Rebe und Wein. n.p. Classified; very useful.

KOROTKEVICH, A. V.
1948. Bibliograficheskiĭ ukazatel' knig i zhurnal'nykh statei nei Russkom iazyke po voprosu ispol'zovaniia produktsii vinogradarstva za 1887–1947. Chast. I. Vinodelie. Yalta, Magarach. 77 pp. Useful.

MEYER, H. M.
1966. Wein-Bibliographie. Deutschsprachiges Schrifttum 1966 in Auswahl und Nachträge aus früheren Jahren. Weinberg-Keller **15** (1):1–39. Published annually in this journal since vol. 4, 1958, and in Der Deutsche Weinbau, 1954. 25 pp.

MOLON, G.
1927. Bibliografía orticola ... Milan, Tip. Terragni & Calegari, viii + 428 pp. Much on grapes.

MUÑOZ PEREZ, J., AND J. BENITO ARRANZ
1961. Guía bibliográfica para una geografía agraria de España. Madrid, Instituto "Juan Sebastian Elcano" de Geografia. 887 pp. Mainly on Spain.

PÉREZ MARTINEZ, M.
1963. Bibliográfia de los vinagres. Havana. 120 pp. Surprisingly complete.

---

[1] The Department of Viticulture and Enology of the University of California at Davis maintains a card file of books and pamphlets on grapes and wines. This is fairly complete for the period 1900 to date.

SCHNEIDER, J.
1961. Der Wein im Leben der Völker. Eine empirische Bibliographie deutschsprachiger Hochschulschriften . . . Neustadt-Weinstrasse, Verlag D. Meininger. 64 pp. (*Also in* Weinblatt 55:19 *et seq.* 1961.) Wine in daily life.

SIMÕES, C., AND B. SIMÕES
1938. A vinha e o vinho (bibliografia). Catálogo dos volumes e folhetos em portugues e de autores portugueses e alguns estrangeiros. Lisbon. 81 pp. Good listings for Portugal.

SIMON, A. L.
1913. Bibliotheca vinaria: a bibliography of books and pamphlets dealing with viticulture, wine-making . . . London, Grant Richards, Ltd. vii + 399 pp. Incomplete and difficult to verify some entries.

1927–1932. Bibliotheca bacchica: bibliographie raisonnée des ouvrages imprimés avant 1800 . . . London and Paris, Maggs Bros. 2 vols. Essential.

TAIROV, V. E.
1891. Bibliograficheskiĭ ukazatel' knig', borshiur' i zhurnalnykh' stateĭ po vinogradarstva i vinodieliiu napechatannykh' s" 1750 po 1890 god" vkliuchitel'no. St. Petersburg, Tipo-lit Vineke. vii + 196 pp. A classic.

VICAIRE, G.
1954. Bibliographie gastronomique: a bibliography of books appertaining to food and drink and related subjects . . . London, D. Verschagle. xviii pp. + 927 cols. Facsimile of first edition of 1890; introduction by André L. Simon. Indispensable.

DICTIONARIES AND ENCYCLOPEDIAS

BRUNET, R.
1946. Dictionnaire d'oenologie et de viticulture. Paris, M. Ponsot. 534 pp. Popular.

ENGEL, R.
1959. Vade-mecum de l'oenologue et du buveur "Très prétieux." Paris, M. Ponsot. 466 pp. Popular.

JAMES, W.
1960. Wine: A brief encyclopedia. New York, Alfred A. Knopf. vi + 208 pp. Popular.

KRAMER, O.
1962. Kellerwirtschaftliches Lexikon. 2d ed. Neustadt-Weinstrasse, D. Meininger. 276 pp. Popular.

LICHINE, A.
1967. Alexis Lichine's encyclopedia of wines and spirits. New York, Alfred A. Knopf. xii + 713 pp. Popular and original.

MARCUS, I.
1969. Dictionary of wine terms. 14th ed. San Francisco, Wines & Vines. 64 pp. Popular.

MENDELSOHN, O. A.
1965. The dictionary of drink and drinking. New York, Hawthorn Books. ix + 382 pp. General.

MÜLLER, K.
1930. Weinbau-Lexikon. Berlin, Paul Parey. vii + 1015 pp. Contains historically valuable material.

OFFICE INTERNATIONAL DE LA VIGNE ET DU VIN
1963. Lexique de la vigne et du vin: Français, Italiano, Español, Deutsch, Português, English, Russkii. Paris. 674 pp. Thorough and very useful.

PÉREZ, J., AND R. ALSINA
1966. Diccionario de vinos españoles. Barcelona, Editorial Teide. 237 pp.
RENOUIL, Y.
1962. Dictionnaire du vin. Bordeaux, Féret et Fils. 1374 pp. Valuable and original.
RÖSSLER, O.
1958. Getränkekundliches Lexikon von Absinth bis Zythos. Neustadt-Wein-strasse, D. Meininger. 280 pp. Popular.
SCHOONMAKER, F.
1964. Frank Schoonmaker's encyclopedia of wine. New York, Hastings House. vi + 410 pp. Useful descriptions.
SIMON, A. L.
1935. A dictionary of wine. London, Cassell & Co., Ltd. xi + 266 pp. Brief descriptions, but some unique; still useful.
1958. A dictionary of wine, spirits and liqueurs. London, Jenkins. 167 pp. General.
VOEGELE, M. C., AND G. H. WOOLEY
1961. Drink dictionary. New York, Ahrens Publishing Co. 192 pp. General.

LEGAL AND STATISTICAL

ANON.
1967. Das Oesterreichische Weingesetz 1961, mit Erläuterungen für die Praxis; Bundesgesetz vom 6. Juli 1961 über den Verkehr mit Wein und Obstwein.... 2d ed. Vienna, Oesterreichischer Agrarverlag. 88 pp. Austrian wine law.
BAMES, E.
1938. Auslandische Gesetzgebung über alkoholische Genussmittel. In B. Bleyer, Alkoholische Genussmittel. Berlin, Julius Springer. (See pp. 760–797.) Non-German laws.
BLANCHET, B.
1962. Code du vin et textes viti-vinicoles. New ed. Montpellier, Edition de "La Journée Vinicole." 195 pp. French laws and regulations.
CALIFORNIA. CROP AND LIVESTOCK REPORTING SERVICE
1967. California grape acreage ... as of 1966. Sacramento, Crop and Livestock Reporting Service. 23 pp. Bearing and nonbearing by variety and county.
CALIFORNIA. DEPARTMENT OF PUBLIC HEALTH
1954. Regulations establishing standards of identity, quality, purity, sanitation, labeling and advertising of wine. Sacramento, California State Printing Office. 243 pp. (California regulations from California Administrative Code, Title 17, Chap. 5, Art. 14, Secs. 17,000–17, 105.)
CALIFORNIA. LAWS AND STATUTES
1967. Alcoholic beverage control act; Business and Professions Code, Division 9 and related statutes. Sacramento, California State Printing Office. 243 pp. 1968 Supplement. 32 pp.
COMBES, D.
1957. La fraude et le code du vin. Montpellier, Causse, Graille & Castelnau. 197 pp. Technical.
DEAGE, P., AND M. MAGNET
1959. Le vin et le droit. 2d ed. Montpellier, Editions de "La Journée Vinicole." 6 + 297 pp. French wine laws interpreted.
DISTILLED SPIRITS INSTITUTE
1966. Public revenues from alcoholic beverages. Washington, D.C., Distilled Spirits Institute. 63 pp. Useful statistics; new editions issued annually.

FARRELL, K. R.
1961. Statistics relating to California grape and wine industries. Berkeley, Giannini Foundation of Agricultural Economics. Miscellaneous Unnumbered Publication. 5 + 28 pp. Data on production, etc.

GALET, P.
1964. Cépages et vignobles de France. Vol. IV. Le raisins de table. La production viticole française. Montpellier, Paul Déhan. pp. 2901–3500. Important historical statistics on French vineyards and wines included.

HIERONIMI, H.
1958. Weingesetz. 2d ed. Munich and Berlin, C. H. Beck'sche Verlagsbuchhandlung. 551 pp. Ergänzungsband. 1967. 144 pp. German wine laws.

HOLTHÖFER, H., AND K.-H. NÜSE
1959. Das Weingesetz. 2d ed. Berlin and Cologne, Carl Heymanns Verlag. xv + 325 pp. German wine laws.

KOCH, H. J.
1958. Das Weingesetz. Neustadt-Weinstrasse, Verlag Daniel Meininger. 280 pp. German wine laws.

LEYTE MARRERO, J.
1960. El estatuto del vino y legislación complementaria posterior; su interpretación práctica. La Coruña, Litografía e Imprenta Roel. 269 pp. Technical.

MEHREN, G. L., AND S. W. SHEAR
1950. Economic situation and market organization in the California grape industries. Berkeley, Giannini Foundation of Agricultural Economics Rept. 107: 1–92, 35, 19, 93, 18. (Mimeo.) Valuable data on production and distribution.

OFFICE INTERNATIONAL DE LA VIGNE ET DU VIN
1964. Codex eonologique international. Paris, Office International de la Vigne et du Vin. Vol. 1:1–75. Standards of identity and purity for chemicals permitted in wine.
1965. Mémento de l'O.I.V. Paris, Office International de la Vigne et du Vin. 1196 pp. Statistical and legal data.

PARONETTO, L.
1963. Ausiliari fisici chimici biologici in enologia. Verona, Enostampa Editrice. xix + 819 pp. Chemicals and practices permitted in wine.

PROTIN, R.
1967. Situation de la viticulture dans le monde en 1966. Bull. Office Intern. Vin 40: 1091–1134. Technical. Published annually.

QUITTANSON, C., A. CIAIS, AND R. VANHOUTTE
1949–1965. La protection des appellations d'origine des vins et eaux-de-vie et le commerce des vins; législation et jurisprudence suivies des documents officiels et de tableaux analysant toutes les appellations contrôlées reglementées. Montpellier, "La Journée Vinicole." 3 vols. All the laws and regulations on protected geographical names for wines in France.

SCIALPI, L.
1960. Codice del vino. Legislazione, giurisprudenza, circolari e resoluzioni Ministeriali. Rome, Luigi Scialpi Editore. xx + 287 pp. Italian wine laws. (See also latest edition of his Annual agenda vinicola and his Codice degli alcoli, delle acqueviti e dei liquori. 1962. 291 pp.)

SERRA, E., AND M. RODRIQUES
1938. Legislação sôbre vinhos. Lisbon, Emprêsa Jurídica Editora. 472 pp. Portuguese laws and regulations.

SELECTED REFERENCES

UNITED STATES. BUREAU OF INTERNAL REVENUE
1945.   Regulations No. 7 relative to the production, fortification, and tax payment,
etc., of wine. Washington, D.C., Government Printing Office. xxvii + 289 pp.
UNITED STATES. FOOD, DRUG, AND COSMETIC ADMINISTRATION
1958.   General regulations for the enforcement of the Federal Food, Drug and
Cosmetic Act, Title 21, Part 1. June, 1958, revision FDC Regs., Part 1.
UNITED STATES. INTERNAL REVENUE SERVICE
1960.   Liquor dealers regulations. Part 194 of Title 26, Code of Federal Regulations.
Washington, D.C., U. S. Government Printing Office. (U. S. Treasury Dept.,
IRS Publ. No. 195, rev. pp. 9–60.)
1961a. FAA Regulations No. 4 relating to the labeling and advertising of wine.
Part 4 of Title 27, Code of Federal Regulations. Washington, D.C., U. S.
Government Printing Office. (U.S. Treasury Dept., IRS Publ. No. 449,
pp. 1–70.)
1961b. FAA Regulation No. 6 relating to inducements furnished to retailers under
the provision of the Federal Alcohol Administration Act, Part 6 of Title 27,
Code of Federal Regulations. Washington, D.C., U.S. Government
Printing Office. (U.S. Treasury Dept., IRS Publ. No. 449, pp. 2–6).
1961c. Wine. Part 240 of Title 26, Code of Federal Regulations. Washington,
D.C., U.S. Government Printing Office. (U.S. Treasury Dept., IRS
Publ. No. 146, rev. pp. 1–114.)
1962.   Gauging manual embracing instructions and tables for determining the
quantity of distilled spirits by proof and weight. Washington, D.C., U.S.
Government Printing Office. (U.S. Treasury Dept., IRS Publ. No. 455.)
1965.   Excise Tax Reduction Act of 1965. Public Law 89–44. Title VIII. Miscellaneous structural changes. P. 26. (Amends Internal Revenue Code of
1954, par. 23, 370A.)
UNITED STATES. LAWS AND STATUTES
1954.   Internal Revenue Code of 1954, as amended. Sub-title E, chap. 51. Title 26,
United States Code 1. Washington, D.C.
1955.   Federal wine regulations. 26 CFR (1954), Part 240, effective January 1,
1955. Chicago, Commerce Clearing House, Inc. pp. 16075–16240.
1958.   Excise Tax Technical Changes Act of 1958, Public Law 85–859, 85th
Congress, H. R. 7125, September 2. Washington, D.C.
1959.   Food additives, Part 121, Title 21, Code of Federal Regulations. (See
Food, Drug, and Cosmetic Law Reports, Commerce Clearing House,
Inc.)
WINE INSTITUTE
1967a. Annual industry statistical surveys. San Francisco, Wine Institute. Published
since 1937 in its Bulletin; valuable and up to date.
1967b. California wine type specifications, recommended as desirable for California
wine and brandy types for guidance of judges at fairs. San Francisco, Wine
Institute. 7 pp. Published at intervals for last several years.
WINES & VINES
1967.   Statistical issue. San Francisco, Wines & Vines. 67 pp. Useful statistical
summaries; published annually.

WINE MAKING

BENVEGNIN, L. E., E. CAPT, AND G. PIGUET
1951.   Traité de vinification. 2d ed. Lausanne, Librairie Payot. 583 pp. Technical;
excellent on white wines.
BERNAZ, D., I. DUMITRESCU, GH. BERNAZ, AND M. MARTIN
1962.   Tekhnologia vinului. Bucharest, Editura Agro-Silvică. 392 pp. Technical.

BREMOND, E.
1965. Techniques modernes de vinification et de conservation des vins en pays méditerranéens. 2d ed. Paris, Maison Rustique. 296 pp. Practical.

COSMO, I., AND T. DE ROSA
1960. Manuale di enologia, guida del buon cantiniere. Bologna, Edizioni Agricole. xi + 164 pp. Popular and practical.

DE ROSA, T.
1964. Tecnica dei vini spumanti. Conegliano, Rivista di Viticoltura e di Enologia. xxiv + 48 pp. Sparkling wines; often descriptive.

FABRE, J. H.
1946–1947. Traité encyclopédique des vins. Algiers, Chez l'Auteur. Vol. I. Procédés modernes de vinification. 5th ed. 1946. 352 pp. Vol. III. Maladies des vins, vinifications spéciales. 4th ed. 1947. 304 pp. Mainly as applied to Algerian conditions and often out of date.

FARKAS, J.
1960. Vinárstvo. I. Technológia vína. 2d ed. Bratislava, Slovenské Vydavatel'stvo Technickej Literatúry. 394 pp. Technical and practical.

FERRARESE, M.
1951. Enologia practica moderna. 3d ed. Milan, Editore Ulrico Hoepli. xv + 348 pp. Popular and practical.

FERRÉ, L.
1958. Traité d'oenologie bourguignonne. Paris, Institut National des Appellations d'Origine des Vins et Eaux-de-Vie. viii + 303 pp. Technical; authentic.

GAROGLIO, P. G.
1965. La nuova enologia. 2d ed. Florence, Libreria LI.COSA. viii + 503 pp. Technical.

GIESS, W.
1957. Kaltsterile Abfüllung von Wein. Mainz, J. Diemer Verlag der Deutschen Wein-Zeitung. 116 pp. Technical on sterile filtration.
1959. Technische Fortschritte bei der Sektherstellung. Deut. Wein-Ztg. 95:616, 618, 620, 622, 624. Technical and practical.
1960. Lehrbuch für Weinbereitung und Kellerwirtschaft. Bad Kreuznach, Im Selbstverlag. 290 pp. Practical, especially on stabilization.

GEORGREV, IV.
1949. Vinarstvo. Sofia, Zemizdat. xvi + 456 pp. Wine making in Bulgaria.

GERASIMOV, M. A.
1964. Tekhnologiia vina. Moscow, Izd-vo "Pishchevaia Promyshlennost'," 639 pp. Technical; the standard Soviet text.

GIANFORMAGGIO, F.
1955. Manuale pratico di enologia moderna. 3d ed. Milan, U. Hoepli. 443 pp. Popular.

HULAČ, V.
1949. Príručka sklepniho hospodárství. Brno, Nákladem Ústredního Snazu Československých Vinařú. 227 pp. Practical cellar management.

JOSLYN, M. A., AND M. A. AMERINE
1964. Dessert, appetizer and related flavored wines. Berkeley, University of California, Division of Agricultural Sciences. xii + 483 pp. Technical; practical and theoretical.

JOSLYN, M. A., AND M. W. TURBOVSKY
1954. Commercial production of table and dessert wines. In L. A. Underkofler and R. J. Hickey (eds.), Industrial fermentations. New York, Chemical Publishing Co., Inc. Vol. 1:196–251.

SELECTED REFERENCES

Laborde, J.
1907. Cours d'oenologie. Paris, L. Mulo. 1:1–344. Still valuable.

Laho, L.
1962. Vinohradnictvo. Bratislava, Slovenské Vydavatel'stvo Pôdohospodárskej Literatúry. 555 pp. Practical and theoretical.

Larrea, A.
1965. Tratado práctico de viticultura y enología; manual para capataces bodegueros. Barcelona, Editorial Aedos. 333 pp. Practical for beginners.

Magistocchi, G.
1955. Tratado de enología adaptado a la República Argentina. Buenos Aires, Ed. El Ateneo. 765 pp. Descriptive.

Marcilla Arrazola, J.
1946. Tratado práctico de viticultura y enología españoles. 2d ed. Madrid, SAETA. 2 vols. Technical and still useful.

Marescalchi, C.
1965. Manuale dell'enologo. 13th ed. Casale Monferrato, Casa Editrice Fratelli Marescalchi. 700 pp. Semipopular.

Mayer-Oberplan, M.
1956. Das Schönen und Stabilisieren von Wein, Schaumwein und Süssmost. Frankfurt, Verlag Sigurd Horn. 152 pp. Technical on stabilizing wines.

Nègre, E., and P. Françot
1965. Manual pratique de vinification et de conservation des vins. Paris, Flammarion. 456 pp. Practical.

Oreglia, F.
1964. Enología téorico-práctica. Mendoza, Rodeo del Medio. xxxviii + 864 pp. General wine making.

Ough, C. S., and M. A. Amerine
1966. Effects of temperature on wine making. Calif. Agr. Exper. Stat. Bull. 827:1–36. Technical.

Prostoserdov, N. N.
1955. Osnovy vinodeliia. Moscow, Pishchepromizdat. 244 pp. Technical principles of enology.

Ribéreau-Gayon, J., and E. Peynaud
1964. Traité d'oenologie. I. Maturation du raisin; fermentation alcoolique, vinification. Paris, Librairie Polytechnique Ch. Béranger. 753. pp. Technical; practical and theoretical.

Rodopulo, A. K.
1962. O biokhimicheskikh protsessakh v vinodelii. Moscow, Pishchepromizdat. 178 pp. Technical; theoretical principles.

Saller, W.
1955. Die Qualitätsverbesserung der Weine und Süssmoste durch Kälte. Frankfurt, Verlag Sigurd Horn. 98 pp. Technical; value of cool fermentations for white wines.

Sannino, F. A.
1948. Tratado de enología. Buenos Aires, Ediciones G. Gili. vii + 920 pp. Technical.

Theron, C. J., and C. J. G. Niehaus
1947. Wine making. Union of South Africa Dept. Agr. Bull. 191:1–94. Practical and still useful.

Troost, G.
1961. Die Technologie des Weines. 3d ed. Stuttgart. Eugen Ulmer. 702 pp. Technical, practical; especially useful for white table wines.

825

VEGA, L. A. DE
 1958. Guía vinícola de España. Madrid, Editora Nacional. 318 pp. Popular.

VENTRE, J.
 1930. Traité de vinification pratique et rationnelle. Montpellier, Librairie Coulet. 3 vols. Technical; practical and still useful.

VOGT, E.
 1963. Der Wein, seine Bereitung, Behandlung und Untersuchung. 4th ed. Stuttgart, Verlag Eugen Ulmer. 267 pp. Technical.

 1968. Weinbau und Weinbereitung. *In* Handbuch der Lebensmittelchemie. VII. Alkoholische genusmittel. W. Diemair, ed. Berlin, Springer-Verlag. pp. 173–311. Technical and up-to-date.

Appendix B
JOURNALS

JOURNALS WITH MATERIAL ON GRAPES AND WINE[1]

- -

Algeria
[2] Revue Agricole Hebdomadaire. Paris. No. 231, 1965. (T; I; G, W, O).[3]
Ceased publication ?

Argentina
[2,4] Boletín Técnico, Estación Experimental Agropecuaria. Mendoza. No. 2, 1963.
(T; I; G, W, O). Ceased publication ?
[4] Oeste. (Centro Universitario de Estudiantes Mendocinos). Mendoza. Vol. 6,
1967. (SP, T; 2 × M; G, W, O).
[2,4] Revista de Investigaciones Agropecuarias. Serie 2, Biología y Producción
Vegetal. Buenos Aires. Vol. 3, 1966. (T; I; G, O) *Also* Serie 6, Economía
y Administración Rural. Vol. 2, 1966. (T; I; W, O)
El Viñatero. San Juan. Vol. 8, 1967. (P, SP; W ?; G, W, O)
[2,4] Vinos y Vinas; Revista mensual ilustrada de agricultura . . . Buenos Aires.
Vol. 62, 1967. (SP, T; M; G, W, F) Formerly Vinos, Vinas y Frutas.

Australia
[2,4] Wine Board. Annual Report. Adelaide. (T; A; W, F)
[2] The Australian Grapegrower. Adelaide. No. 48, December, 1967. (SP, T; M ?;
G, W, F)
[2,4] Australian Journal of Agricultural Research (Commonwealth Scientific and
Industrial Research Organization). Mildura. Vol. 18, 1967. (T; M; G, W,
F. O)
[2] Australian Journal of Applied Science. Melbourne. Vol. 19. 1967. (T; M; G,
W, O)
[2,4] Australian Journal of Biological Sciences (Commonwealth Scientific and Industrial Research Organization). Melbourne. Vol. 20, 1967. (T; M; G, W,
F. O)
[2] Australian Wine, Brewing and Spirit Review. Melbourne. Vol. 76, 1967. (P,
SP; M; G, W, F) (until 1959 Australian Brewing and Wine Review)

---

[1] This is by no means a definitive list. Most of the Journals listed deal primarily
with the grape and wine industry, but a number have been included because they
frequently contain significant articles in this field. See also the Food Science and
Technology Abstracts, monthly, Vol. 1, 1969, Shinfield (Reading), England, International Food Information Service.

[2] In library of University of California, usually at Davis. Some Davis holdings are
in A. J. Winkler Library, Department of Viticulture and Enology.

[3] P for popular, SP for semipopular, T for technical; W for weekly, M for monthly,
Q for quarterly, A for annual, I for irregular; G for grapes or grape growing, W for
wines or wine making, F for fermentation or other fermentation industries, O for
related subjects, particularly horticulture.

[4] In National Agricultural Library, Washington, D.C.

[2,4] Federal Wine and Brandy Producers' Council of Australia. Annual Report. Kensington, Victoria. 1967. (T; A; W, F)
[2,4] Journal of the Department of Agriculture of South Australia. Adelaide. Vol. 70, 1966. (SP, T; M ?; G, W, F, O)

Austria
[2,4] Mitteilungen Rebe und Wein, Obstbau und Früchteverwertung. Klosterneuburg. Vol. 17, 1967. (T; 6 × A; G, W, F, O) (French, English, and Spanish summaries)
[2,4] Mitteilungen der Versuchsstation für das Gärungsgewerbe und des Institutes für angewandte Microbiologie der Hochschule für Bodenkultur. Vienna. Vol. 21, 1967. (T; 2 × M; W, F, O)
Neuer Wein-Kurier. Vienna. No. 21, 1966. (SP; 2 × M; G, W)
Obst- und Weinbau (Landes-Obst- und Weinbauverein für Steiermark). Graz. Vol. 36, 1967. (SP, T; M; G, W, O)
Oesterreichischer Weinbaukalender. Vienna. 1965. (SP, T; A; G, W)
[4] Oesterreichische Wein-Zeitung. Vienna. Vol. 21, 1966. (SP, T; 2 × M; G, W, F)
Der Winzer; Fachblatt de österreichischen Weinbaues. Baden bei Wien. Vol. 23, 1967. (SP, T; M; G, W)

Belgium
[2] Rapport de la Station Provinciale de Recherches Scientifiques Agricoles et Viticoles. La Hulpe. Vol. 16, 1966. (T; A; G, O)
[2,4] Revue des Fermentations et des Industries Alimentaires. Brussels. Vol. 22, 1967. (T; M; W, F, O)
Revue Belge des Vins et Spiriteux. Bruxelles. Vol. 21, 1965. (P, SP; M; W, O)

Bulgaria
[2] Gradinarska i Lozarska Nauka (Bulgarska Akademiia na Naukite). Sofia. Vol. 4, 1967. (SP, T; M; G, W, F, O) (English, French, and Russian summaries)
[2,4] Lozarstvo i Vinarstvo. Sofia. Vol. 16, 1967. (T; 6 × A; G, W, F, O)
[2] Nautschni Trudove (Nauchno-izsledovatelski Tekhnologicheski Institut po Vinarska i Pivovarna Promishlenost). Sofia. Vol. 7, 1966. (T; A; G, W, O) (German, French, and Russian summaries)

Canada
[2,4] Report of the Horticultural Experiment Station. Vineland, Ontario. 1967. (T; A; G, W, F, O)

Chile
[2] Boletin de la Vitivinicultura Nacional. Santiago de Chile. Vol. 6, 1963. (P; 6 × A; G, W, O)
[2] Boletin Informativo de la Asociación Nacional de Viticultores. Santiago de Chile. Vol. 6, 1968. (SP, T; M; G, W, F)

Czechoslovakia
[2] Kvasný promýsl. Prague. Vol. 13, 1967. (T; M; G, W, F, O)
[2] Pokroky vo Vinohradníkom a Vinárskom Výskume. Bratislava. Vol. 4, 1966. (T; A; G, W) (French, Russian, and German summaries)
Sborník Vysoké Škole Chemicko-Technologické v Praze. Potravinárski Technologie. Prague. Vol. 11, 1967. (T; M ?; G. W, F, O)
[2,4] Vinohrad. Bratislava. Vol. 5, 1967. (SP, T; M; G, W, F) (followed journal Vinarstvi in 1961. Tables of content in French, German, and Russian)

France
[4] Alimentation et la Vie. Bulletin de la Société Scientifique d'Hygiène Alimentaire de l'Alimentation Rationnelle de l'Homme. Paris. Vol. 55, 1967. (T; M ?; W, F)
[2,4] Annales de l'Amélioration des Plantes. Paris. Vol. 17, 1967. (T; Q; G, O)
[2,4] Annales de l'Ecole National d'Agriculture. Montpellier. Vol. 31, 1966. (T; I; G, W, F, O)

[2,4] Annales des Falsifications et de l'Expertise-Chimique (Société des Experts-Chimistes de France). Paris. Vol. 60, 1967. (T; M; G, W, F, O)

[2,4] Annales de Technologie Agricole. Paris. Vol. 16, 1967. (T; Q; G, W, F, O)

Le Bourguignon Viticole. Organe Mensuel de la Vigne et du Vin de la Bourgogne. Mâcon. Vol. 23, 1966. (P, SP; M; G, W)

[2,4] Bulletin de l'Institut National des Appellations d'Origine des Vins et Eaux-de-Vie. Paris. No. 100, 1967. (T; Q; G, W, F)

[2] Bulletin Officiel de la Fédération Nationale de la Viticulture Nouvelle et du Syndicat des Hybrideurs et Métisseurs Viticoles Fénavino. Poitiers-Bordeaux. Vol. 20, 1967. (SP, T; M; G, W)

[4] Bulletin Technique d'Information des Ingénieurs des Services Agricoles (Ministère de l'Agriculture). Paris. No. 206, 1967. (T; M; G, W, F, O)

La Champagne Viticole. Épernay. 1966. (P, SP; M; G, W)

[2,4] Comptes Rendus Hebdomadaires des Séances de l'Académie d'Agriculture de France. Paris. Vol. 54, 1967. (T; 2× M; G, W, F, O)

[2] Cuisine et Vins de France. Paris. Vol. 20, 1967. (P, SP; M; G, W, O)

[2,4] Industries Alimentaires et Agricoles (Association des Chimistes et Ingénieurs...) Paris. Vol. 84, 1967. (T; M; W, F, O)

[2] La Journée Vinicole; Journal Quotidien des Boissons. Montpellier. Vol. 41, 1967. (P, SP, T; 5× W; G, W, F)

[2] La Journée Vinicole Export; Journal Quotidien des Boissons. Montpellier. No. 246, October, 1967. (P; M; G, W, F) (in French, English, and German)

Le Midi Vinicole; Organe de la Production et du Commerce des Vins et Alcools. 1967. (P, SP; 24× A; G, W)

Le Moniteur Vinicole. Paris. Vol. 112, 1967. (P, SP; 2× W; G, W) (Published since 1856)

[2] Le Paysan. Cognac. Vol. 40, 1967. (P, SP; M; G, W, F, O)

[2,4] Le Progrès Agricole et Viticole. Montpellier. Vol. 84, 1967. (T; 2× M; G, W, O)

[2] Revue de l'Embouteillage et des Industries Connexes. Paris. No. 85, 1967. (SP, T; M; G, W, F, O)

[2] Revue Française d'Oenologie. Paris. Vol. 26, 1967. (T; Q; G, W, O) (formerly Bulletin de l'Union Nationale des Oenologues)

[2] La Revue du Vin de France. Paris. No. 214, 1967. (SP, T; Q; G, W) (some articles in English)

[2] La Revue Vinicole International; Revue Internationale des Vins de Liqueur et Apéritifs, Eaux-de-Vie et Liqueurs, Cidres et Boissons de Qualité. Paris. No. 135, 1967. (P, SP; 6× A; W) (in French, German, Italian, Spanish, and English) Not to be confused with Revue Vinicole, a popular monthly.

[2] Science et Vin. Organe de la Foire Internationale de la Vigne et du Vin. Montpellier. No. 13, 1966. (SP; A; G, W)

[2] Tastevin en Main; Gazette Périodique de la Confrérie des Chevaliers du Tastevin. Nuits-St.-Georges. No. 43, April, 1967. (P, SP; Q; W)

[2] Le Vigneron Champenois (Association Viticole Champenoise). Épernay. Vol. 88, 1967. (SP, T; M; G, W)

Le Vigneron des Côtes-du-Rhône et du Sud-Est. Mâcon. Vol. 22, 1968. (P; M; G, W) (formerly Le Vigneron du Sud-Est)

[2,4] Vignes et Vins. (Institut Technique du Vin) Paris. No. 161, July-August, 1967. (T; M; G, W, F)

[2] Le Vignoble Girondin (Ligue des Viticulteurs de la Gironde). Bordeaux. Vol. 70, 1967. (P; 2× M; G, W)

Les Vins d'Alsace; Revue Viticole et Vinicole. Colmar. Vol. 62, 1966. (SP, T; M; G, W) (in German and French)

Vins, Cidres, Spiritueux et Liqueurs de France. Paris. 1965. (SP, T; M; W, O)

Vins et Spiritueux Informations (Confédération Nationale des Industries et des Commerces en Gros des Vins, Jus de Fruits, Spiritueux et Liqueurs de France). Paris. Vol. 8, 1967. (SP, T; 2× M; W, O)

[2] Le Vrai Cognac (Fédération des Viticulteurs Charentais). Cognac. No. 215, July, 1967. (P, SP, T; M; G, W, F, O)

Germany

Die Alkohol-Industrie, Deutsche Spirituosen-Zeitung und Brennerei-Zeitung. Düsseldorf. Vol. 80, 1967. (T; 2 × M; W, F) (includes a supplement: Wissenschaftlichtechnische Brennereibeilage)

[2] Berichte der Hessische Lehr- und Forsuchungsanstalt für Wein, Obst- und Gartenbau. Geisenheim. Jahresbericht 1966. (T; A; G, W, F, O)

[2,4] Die Branntweinwirtschaft; Zeitschrift für Spiritusindustrie; Institut für Gärungsgewerbe . . . Berlin. Vol. 108, 1968. (SP, T; M; W, F) (includes a supplement: Der Destillateur-Lehring. Vol. 18, 1968)

[2,4] Deutsche Lebensmittel-Rundschau: Zeitschrift für Lebensmittelkunde und Lebensmittelrecht. Stuttgart. Vol. 63, 1967. (T; M; G, W, F, O)

[2] Deutsche Weinbau-Jahrbuch (*earlier* Deutsche Weinbaukalender). Waldkirch i. Br. Vol. 17, 1966. (T; A; G, W)

[2,4] Deutsche Weinbau; Organ des Deutscher Weinbauverbandes. Wiesbaden. Vol. 22, 1967. (SP, T; 2 × M; G, W, F)

[2] Deutsche Wein-Zeitung; Illustrierte Wein-Zeitung; Wein und Rebe. Mainz am Rhein. Vol. 103, 1967. (SP, T; 2 × M; G, W) Published since 1864.

Flüssiges Obst; Zeitschrift für gärunglose Früchteverwertung. Obererlenbach, Bad Homburg. Vol. 34, 1967. (T; M; G, O)

Forschung-Schule-Praxis; Mitteilungsblatt d. Vereins ehem. Schüler d. Landes-, Lehr- und Forschungsanstalt für Wein- und Gartenbau. Neustadt an der Weinstrasse. Vol. 13, 1965. (T; I; G, W) Ceased publication?

[2] Fruchtsaft-Industrie. Internationale Zeitschrift für Herstellung und Untersuchung von Fruchtsaften und Aromakonzentraten. Frankfurt am Main. Vol. 12, 1967. (T; 6 × A; G, W, O)

Getränke-Industrie; Unabhängiges Fachblatt für die gesamte Getränke-Industrie. Munich. Vol. 20, 1966. (T; 2 × M; G, W, F, O)

[2] Rebe und Wein. Weinsberg (Heilbronn). Vol. 20, 1967. (SP, T; M; G, W, F, O)

[2] Seitz Informationen. Praxis, Forschung, Technik. Bad Kreuznach. Vol. 29, 1967. (T; I; G, W, F, O)

[2,4] Vitis, Berichte über Rebenforschung. Geilweilerhof (Landau). Vol. 7, 1968. (T; Q; G, W, F, O)

[2,4] Weinberg und Keller; Monatshefte für Weinbau- und Kellerwirtschaft . . . Traben-Trarbach (published in Frankfurt am Main). Vol. 14, 1967. (T; M; G, W)

Das Weinblatt; Allgemeine Deutsche Weinfachzeitung. Neustadt an der Weinstrasse. Vol. 62, 1967. (SP, T; W; G, W)

[2] Weinfach-Kalender. Jahrbuch des deutschen Weinfaches. Mainz am Rhein. Vol. 77, 1966. (T; A; G, W)

[2] Die Wein-Wissenschaft. Beilage zur Fachzeitschrift der Deutsche Weinbau. Mainz am Rhein. Vol. 22, 1967. (T; M; G, W)

[2,4] Zeitschrift für Lebensmittel-Untersuchung und-Forschung. Munich. Vol. 134, 1967. (T; I [2–3 vols. per year]; G, W, F, O)

Great Britain

Harper's Directory and Manual of the Wine and Spirit Trade. London. 1968 edition. (P, SP; A; W, O)

Harper's Export Wine and Spirit Gazette. London. Vol. 15 ?, 1967. (P, SP; 3 × A; W, O)

Harper's Wine and Spirit Gazette. London. No. 4330, 1967. (P, SP; 3 × A; W, O) Started in 1880.

[2,4] Journal of the Science of Food and Agriculture. London. Vol. 18, 1967. (T; M; W, F, O)

[2] Process Biochemistry. London. Vol. 2, 1967. (T; M; F, O)
Ridley's Wine and Spirit Trade Circular. London. No. 1442, 1967. (P, SP; M; W, O) Established in 1848.
Ridley's Wine and Spirit Handbook. London. 1967 edition. (SP, T; A; W, O)
[2] Vintage (Harvey's of Bristol, Ltd.). Bristol. Vol. 4, 1966. (P, SP; Q; G, W) Ceased publication?
The Wine-Butler. London. No. 145, January, 1967. (P; M; W, O)
[2] Wine and Food. London. No. 134, 1967. (P, SP, T; Q; G, W, O)
[2] Wine Magazine. London. No. 56, January–February, 1968. (P, SP; 2× M; G, W)
Wine and Spirit Trade Record. London. Vol 94, 1967. (P, SP; M; W, O)
Wine and Spirit Trade Review. London. Vol. 105, 1967. (P; W; W, O) Published since 1862.

Greece
[4] Deltion Agrotikes Trapezes. Athens. Vol. 153, 1966. (SP; 6× A; G, W, O)

Hungary
Acta Agronomica Academiae Scientiarum Hungaricae. Budapest. Vol. 18, 1967. (T; G, W, F, O) (German, English, and Russian summaries)
[2,4] Borgazdaság (Wine Industry). Budapest. Vol. 15, 1967. (T; M; G, W) (summaries in French)
[2,4] Budapest. Nemzetközi Borverseny, Katalógus; Concours International des Vins à Budapest. Budapest. Vol. 4, 1966. (T; I; G, W)
Élelmiszerviszgalati Közlemények (Food Research). Budapest. Vol. 12, 1966. (T; M?; W, F, O)
[2,4] A Kertészeti és Szőlészeti Főiskola Közlemények; Annales Academiae Horti- et Viticulturae. Budapest. Vol. 30, 1966. (T; I; G, W, F, O) (German and Russian summaries; title varies)
Kertészeti Kutató Intézet Évkönyve (Yearbook of Horticulture Research Institute). Budapest. Vol. 18, 1967. (T; A; G, O)
[4] Kisérletügyi Közlémenyek (Experimental Publications). (A Földmüvelézügyi Minisztérium Kiadvanya). Budapest. Series C, Kertészet (Horticulture and Viticulture), Vol. 56, 1964. (T; 3× A; G, W, O) (German, French, and Russian summaries)
[4] Magyar Tudományos Akadémia Agrártudományok Osztálya Kőlzemenyei (Hungarian Scientific Academy of Agric. Scien., Publicat. series). Vol. 25, 1966. (T; Q?; G, W, O)
[4] Növénytermelés (Plant Science). Budapest. Vol. 16, 1966. (T; Q; G, F) (Russian and English summaries)
[2,4] Szölészeti Kutató Intézet Évkönyve; Annales Instituti Investigandum Viticulturae. Budapest. Vol. 12, 1958–1962. (T; I; G, W, F, O) (summaries in French, English, German, and Russian)
[2] Szőlő-és Gyümölcstermesztés (Grape and Fruit Production). Budapest. Vol. 3, 1967. (T; A; G, W, F, O)

International
[2,4] Bulletin O. I. V. (Office International de la Vigne et du Vin). Paris. Vol. 39, 1966. (T; M; G, W, F)
[2] Feuillets Oenologiques (Office International de la Vigne et du Vin.) Paris. No. 277, 1968. (T; I; G, W, F)
[2] Lebensmittel-Wissenschaft Technologie. Zurich. Vol. I, 1968. (T; M; W, F, O) Articles in English, French, and German.

Israel
[4] Alon. (Israel Wine Institute), Rehovot. Vol. 1, 1966. (T; M; G, W)

TABLE WINES

Italy

[2,4] Annali, Facoltà di Agraria, Università. Bari. Vol. 20, 1966. (T; A; G, W, F, O)
[2,4] Annali, Facoltà di Agraria, Università degli Studio. Perugia. Vol. 21, 1966. (T; A; G, F, O)
[2,4] Annali, Facoltà di Agraria, Università. Pisa. Vol. 26, 1965. (T; A; G, W, F, O) English summaries)
[2,4] Annali della Sperimentazione Agraria. Rome. Vol. 21, 1967. (T; 3 × A; G, W, F, O)
[2,4] Atti del'Accademia Italiana della Vite e del Vino. Florence and Siena. Vol. 18, 1966. (T; A; G, W)
[2] Bollettino di Ricerche Informazioni, Centro Reg. Spec. Industria Enologica. Marsala. Vol. 1, 1962. (T; Q; G, W) Title has varied.
[2,4] Il Coltivatore e Giornale Vinicolo Italiano. Casale Monferrato. Vol. 113, 1967. (P, SP; M; G, W, O)
[2,4] Il Corriere Vinicolo; Organo delle Industrie del Vini, Liquori e Prodotti Affini. Milan. Vol. 40, 1967. (P, SP; W; G, W)
[2] Enotria; Rivista Vinicola d'Italia. Milan. Vol. 36, 1967. (P, SP; Q; G, W)
[2,4] Frutticoltura; Rivista di Frutticoltura, Viticoltura, Orticoltura, Floricoltura. Bologna. Vol. 28, 1967. (T; 2 × M; G, F)
[2] Industrie Agrarie. Florence. Vol. 4, 1966. (T; M; G, W, F, O)
[2,4] L'Italia Agricola; Giornale di Agricoltura. Rome. Vol. 104, 1967. (SP, T; M; G, W, O)
[2] Italia Vinicola ed Agraria. Casale Monferrato. Vol. 57, 1967. (SP, T; M; G, W, F, O)
[2] Rivista di Viticoltura e di Enologia. Conegliano. Vol. 20, 1967. (T; M; G, W, F, O)
[2,4] Il Torchio; Giornale Indipendente dei Viti-Vinicola Italiani. Rome. Vol. 19, 1967. (SP; M; G, W, O)
[2] Vini d'Italia; Rivista Internazionale di Enotecnica. Rome. Vol. 9, 1967. (P, SP; 6 × A; G, W) (English, French, and German summaries)

Japan

Agricultural and Biological Chemistry. Tokyo. Vol. 32, 1968 (T; M; F) In English.
[2,4] Hakkô Kôgaku Zasshi (Journal of Fermentation Technology). Osaka. Vol. 45, 1967. (T; M; W, F, O)
[2,4] Journal of the Japanese Society for Horticultural Science. Tokyo. Vol. 36, 1967. (T; Q; G, O) (also in English)
[2,4] Kyoto Daigaku, Schokuryo Kagaku Kenkyusho (Memoirs of the Research Institute for Food Science). Kyoto-fu. No. 28, 1967. (T; A; W, O).
[2] Nippon Jôzô Kyokai Zasshi (Journal of the Society of Brewing, Japan). Vol. 62, 1967. (T; M; W, F)
[2] Yamanashi Daigaku, Hakko Kenkyusho Kenkyu Hokoku (Bulletin of the Research Institute of Fermentation, Yamanashi University). Kofu. No. 13, 1966. (T; A; W, F, O)

Jugoslavia

[4] Arhiv za Poljoprivredne Nauke i Tehnicu. Belgrade, Vol. 22, 1967. (T; M?; G, W, O)
[4] Glasnik. Belgrade. Vol. 16, 1967. (T; M; W, O)
Sadjarstvo, Vinogradništvo, Vrtnarstvo. Ljubljana. Vol. 48, 1961. (P; 6 × A; G, W, F, O)
[4] Savremena Poljoprivredna (Savez Zemljoredničkih Zadruga A. P. Vojvodine). Sremski Karlovci. Vol. 14, 1966. (T; M; W, O)
Socialisticno Kmetijstvo in Gozdarstvo. Ljubljana. Vol. 18, 1967. (T; I; G, F, O)

Mexico
[2] Mexico Vitivinícola. Mexico City. Vol. 2, 1967. (SP; M; G, W)
[4] Revista Latinoamericana de Microbiología. Mexico City. Vol. 9, 1967. (T; M?; W, O)

Poland
[4] Przemysl Spożywczy. Warsaw. Vol. 20, 1966. (T; M; W, F, O)
[4] Wyzsza Szkola Rolnicza. Poznán. Vol. 21, 1964. (T; M; W, F, O)

Portugal
[2] Anais do Instituto do Vinho do Porto. Porto. Vol. 19, 1963–64. (T; A; G, W)
[2] Anais da Junta Nacional do Vinho. Lisbon. Vol. 13, 1961. (T; A; G, W) (German, English, and French summaries)
[2,4] Cadernos, Instituto do Vinho do Porto. Porto. Nos. 332–333, 1967. (SP, T; M; G, W)
[4] Casa do Douro. Régua. Vol. 20, 1965. (T; M; G, W)
[2,4] De Vinea et Vino Portugaliae Documenta. Lisbon. Vol. 3, 1966. (T; I; G, W) (French summary)

Rumania
[4] Analele Institului de Cercetări Agronomice, Series C, Fiziologie ... Anexa: Viticultura, Pomicultura şi Legumicultura. Bucharest. Vol. 33, 1966. (T; M?; G, W, F, O) (English and Russian summaries)
[2] Buletin Ştiinţific, Secţia de Biologie şi Ştunte Agricole, Academia Republici. Populare Romîne. Bucharest. Vol. 17?, 1968. (T; M?; G, W, O)
[2,4] Grădina Via şi Livada; Revista Lunara de Ştiinta şi Practica Hortiviticola. Bucharest. No. 12, 1966. (SP, T; M; G, W, O)
[2,4] Industria Alimentara; Produce Vegetale. Bucharest. Vol. 8, 1965. (T; M; G, W, O)
[2,4] Lucrări Ştiinţifice, Institutul de Cercetări Horti-Viticole. Bucharest. Vol. 9, 1966. (T; A; G, W, F, O) (French or German and Russian summaries)
[2] Revista de Horticultură şi Viticultură. Bucharest. Vol. 16, 1967. (T; M; G, O)

South Africa
[2,4] Wine, Spirit and Malt. Stellenbosch. Vol. 36, 1967. (P; M; G, W)
[2,4] Die Wynboer. Stellenbosch. No. 432, September, 1967. (SP; M; G, W) (official organ of the KWV; includes its Jaarverslag)

Soviet Union[5]
[2] Biokhimiia Vinodeliia; Akademiia Nauk SSSR Institut Biokhimii. Moscow. Vol. 7, 1963 (T; I; G, W, F) Ceased publication?
[4] Izvestiia Biologicheski i Sel'skokhoziaistvennye, Akademiia Nauk Arminskoi SSR Erevan. Vol. 19, 1966. (T; M?; G, W, F)
[2,4] Prikladnaia Biokhimiia Mikrobiologiia. Moscow. Vol. 3, 1967. (T; 6 × A; W, F, O) Also published in English.
[2,4] Sadovodstvo, Vinogradarstvo i Vinodelie Moldavii. Kishinev. No. 165, December, 1966. (T; M; G, W, F, O) Formerly published in Moscow.
Sbornik Vinodelie i Vinogradarstvo (Gosudarstvennyi Nauchno-Issledovatel'-skii, Institut Nauchnoi i Tekhnicheskoi Informatsii. Moscow. Vol. 5, 1966. (T; I; G, W)
[4] Soobshcheniia Akademiia Nauk Gruzinoskoi SSR Tbilisi (Tiflis). Vol. 45, 1967. (T; M?; G, W, F, O)
Trudy Institut Sadovodstva, Vinogradarstva i Vinodeliia; Akademiia Sel'sko-khozyaĭstvenniukh Nauk Gruzinskoi SSR Tbilisi. 1967. (T; I; G, W, F, O)

---

[5] This is a very small representation of the Soviet journals on viticulture and enology. However, it probably includes most of the important journals available in this country. Not included are journals published in Alma-Ata, Baku, Krasnodar, Novocherkassk, and Odessa.

[4] Trudy Moldavskii Nauchno-Issledovatel'skiĭ Institut Sadovodstva, Vinogradarstva i Vinodeliia. Kishinev. Vol. 12, 1966. (T; A; G, W)
Trudy Nauchno-Issledovatel'skii Vinogradarstva, Vinodeliia i Plodovodstva. Minist. Sel'skogo Khozyaistva Armianskoi SSR. Erevan. 1961 (T; I; G, W, O) Also publishes a bulletin.
Trudy Nauchno-Issledovatel'skii Institut Sadovodstva, Vinogradarstva i Vinodeliia; Minist. Sel'skogo Khoz. Uzbek. SSR. Samarkand (or Tashkent?); Vol. 29, 1964. (T; I; G, W, O)
[2,4] Trudy Vsesoiuznyĭ Nauchno-Issledovatel'skiĭ Institut Vinodeliia Vinogradarstva "Magarach." Moscow. Vol. 16, 1967 (T; G, W) (summaries in French) Issues usually alternate for wine making and grape growing. This Institute also publishes a Biulleten' Nauchno-Tekhnicheskoi Informatsii and a Referaty Nauchnykh Rabat (Abstracts).
[2,4] Vinodelie i Vinogradarstvo SSSR. Moscow. Vol. 25, 1967. (T; 8 × A; G, W, F)
Vinogradarstvo i Sadovodstvo Kryma. Simferopol. Vol. 10, 1967. (T; M;· G, W, O)
[4] Visnyk Sil'skohospodars'koyi Nauky. Kiev. Vol. 2, 1966. (SP, T; M; G) (in Ukrainian)

Spain
[4] Anales Bromatología; Sociedad Española de Bromatología. Madrid. Vol. 19, 1967. (T; M; G, W, F, O)
[2,4] Boletín del Instituto Nacional de Investigaciónes Agronómicas. Madrid. Vol. 25, 1965. (T; I; G, W, O) (summaries in English and French)
[2,4] Cuadernos, Consejo Superior de Investigaciónes Científicas, Patronato "Juan de la Cierva" de Investigación Técnica, Sección de Fermentaciónes Industriales. Madrid. No. 24, 1966. (T; G, W, O)
[2] Dionysos en la Viticultura, las Artes y las Letras. Villafranca del Panadés. No. 121, 1967. (P, SP; M; G, W)
[2] Microbiología Española. Madrid. Vol. 19, 1966. (T; M; F) (English summaries)
[2,4] Revista de Agroquímica y Tecnología de Alimentos. Valencia. Vol. 7, 1967. (T; Q; W, F, O)
[2] La Revista Vinicola y de Agricultura. Zaragoza. No. 1984, 1967. (P, SP; M; G, W, O)
[2,4] Semana Vitivinícola. Valencia. No. 1039, 1966. (SP, T; W?; G, W)
Vid; Sindicato Nacional de la Vid, Cerevezas y Bebidas. Madrid. Vol. 1, 1962. (T; A?; G, W, F, O)

Switzerland
[2] L'Ami du Vin; der Weinfreund; l'Amico del Vino; Association Nationale des Amis du Vin. Neuchâtel. Vol. 16, 1967. (P; M; G, W)
[2,4] Annuaire Agricole de la Suisse; Landwirtschaftliches Jahrbuch der Schweiz. Bern. Vol. 81, 1967. (T; A; G, W, F, O) (in German and French)
[2,4] Mitteilungen aus dem Gebiete Lebensmittel Untersuchung und Hygiene. Bern. Vol. 58, 1967. (T; 2 × M; G, W, F, O) (in German and French, summaries in English)
[2] Publications, Station Fédérale d' Essais Viticoles et de Chimie Agricole. Lausanne. No. 650, 1961. (T; I; G, W, O) (mainly reprints)
[2,4] Revue Romande d'Agriculture, de Viticulture et d' Arboriculture. Lausanne. Vol. 6, 1967. (SP, T; M; G, W)
[2,4] Schweizerische Zeitschrift für Obst- und Weinbau. Wädenswil. Vol. 76, 1967. (T; 2 × M; G, W, O)
Schweizerische Weinzeitung; Journal Vinicole de la Suisse. Zürich. Vol. 75, 1967. (SP, T; M?; G, W) (in German and French)

Turkey

Ankara Üniversitesi, Ziraat Fakültesi Yilliği. Ankara. Vol. 13, 1963. (T; I; G, W. O)

Ege Üniversitesi, Ziraat Fakültesi Dergesi. Izmir. Vol. 1, 1964. (T; I; G, W, O)

United States

[2,4] American Journal of Enology and Viticulture. Davis, Calif. Vol. 19, 1968. (T; Q; G, W, F)

[2,4] Applied Microbiology. Bethesda, Maryland. Vol. 16, 1968. (T; M; W, F, O)

[2] Bottles and Bins. St. Helena, Calif. Vol. 20, 1967. (P; Q; W)

[2] Distilled Spirits Institute, Inc., Summary of State Laws and Regulations Relating to Distilled Spirits. Washington, D.C. Vol. 18, 1966. (T; A; W, F)

[2,4] Food Technology. Chicago. Vol. 21, 1967. (T; M; G, W, F, O)

[2,4] Gourmet. New York. Vol. 27, 1967. (P; M; W, F, O)

[2,4] Journal of Agricultural and Food Chemistry. Washington (Easton, Pa.). Vol. 14, 1966. (T; M; W, F, O)

[2,4] Journal of Association of Official Analytical Chemists. Washington, D.C. Vol. 50, 1967. (T; Q; G, W, F, O) (formerly Journal of Association of Official Agricultural Chemists)

[2,4] Proceedings of the American Society of Horticultural Science. St. Joseph, Michigan. Vol. 91, 1967. (T; 2 × A; G, O)

[2,4] Quarterly Journal of Studies on Alcohol. Rutgers, New Jersey. Vol. 28, 1967. (T; Q; W, F, O)

[2] The Raisin Industry News (Raisin Administrative Committee). Fresno. Vol. 13, 1967–68. (SP; I; G)

[2] Raisinland News. (California Raisin Advisory Board). Fresno. No. 12, June, 1967. (P; Q; G)

[2,4] Western Fruit Grower. San Francisco. Vol. 21, 1967. (P, SP; M; G, O)

[2,4] Wine Institute Bulletin. San Francisco. No. 1451, December 27, 1967. (SP, T; W; G, W)

[2] Wine and Spirits Wholesalers of America. Annual Operations Survey. St. Louis. 1966. (T; A; W, F) (also publishes or has published a Monthly Round-up Report, a Quarterly Market Roundup Report, a Warehouse Bulletin, and an Office Operations and Proceedings Bulletin)

[2,4] Wines and Vines. San Francisco. Vol. 48, 1967. (P, SP, T; M; G, W)

Appendix C
EXPERIMENT STATIONS

EXPERIMENT STATIONS WORKING PRIMARILY ON VITICULTURAL AND ENOLOGICAL SUBJECTS[1]

Algeria
1. Alger-Maison Carée. Ecole Nationale d'Agriculture (includes Laboratoire de Recherches de la Chaire de Viticulture and Laboratoire de Chimie et d'Oenologie). (V,[2] E[3])

Argentina
1. Buenos Aires, Rivadavia 1439. Instituto Nacional de Tecnología Agropecuaria. (V, E)
2. Luján de Cuyo (Mendoza), Casilla Correo No. 3. Centro Regional Andino, Estación Experimental Agropecuaria. (V, E)
3. Mendoza, Calle Rivadavia 65. Universidad Nacional de Cuyo, Liceo Agrícola y Enológico "Domingo F. Sarmiento." (V, E)

Australia
1. Adelaide, Private Mail Bag No. 1, G.P.O. The Australian Wine Research Institute. V, E)
2. Griffith (New South Wales). Viticultural Research Station. (V)
3. Hartley Grove, Glen Osmond (South Australia). C.S.I.R.O. Horticultural Research Section. (V)
4. Merbein (Victoria), C.S.I.R.O. (Commonwealth Research Station). Horticultural Research Section. (V)
5. Perth (West Australia), St. George's Terrace. Swan Research Station (Department of Agriculture). (V, E)
6. Roseworthy (South Australia). Roseworthy Agricultural College. (V, E)

Austria
1. Klosterneuburg A-3400, Wienerstrasse 74. Höhere Bundeslehr- und Versuchsanstalt für Wein und Obstbau. (V, E)
2. Vienna, II, Trunnerstrasse 5. Landwirtschaftlich-Chemische Bundesversuchsanstalt, Abteilung für Weinuntersuchung. (V, E)

----

[1] This list was taken primarily from "Répertoire des Stations de Viticulture et Laboratoires d'Oenologie," published by the Office International de la Vigne et du Vin, Paris, 1961; 327 pp. This publication also gives information on the personnel and subjects under study at each of the stations. Because of limitations of space many important experiment stations have had to be omitted. The location and address precedes the name of the station.
[2] Viticultural research and/or teaching.
[3] Enological research and/or teaching.

Belgium
1. Brussels 7, 1, avenue E.-Cryson, Anderlecht. Centre d'Enseignement et de Recherches pour les Industries Alimentaires et Chimiques. (E)
2. La Hulpe, 17, rue Saint-Nicolas. Station Provinciale de Recherches Scientifiques Agricoles et Viticoles. (V)

Brazil
1. Campinas (Estado de São Paulo), Caixa Postal 28. Instituto Agronômico. (V, E)

Bulgaria
1. Plovdiv. Institut Supérieur d'Agronomie "V. Kolarev" and Institut Technologie Supérieur des Industries Alimentaires (V, E)
2. Sofia, 134, boulevard du "9-Septembre." Nauchnoizsledovatelski Tekhnologicheski Institut po Vinarska i Pivovarna Promishlenost. (V, E)
3. Sofia. Institut Supérieur d'Agriculture "G. Dimitrov." (V)

Canada
1. Vineland Station (Ontario). Horticultural Experiment Station and Products Laboratory, Department of Agriculture. (V, E)

Chile
1. Cauquenes, Casilla 1655. Estación Viti-Vinícola Experimental. (V)
2. Cauquenes, Casilla 5560. Bodega y Estación Viti-Vinícola Experimental. (E)
3. Maipu. Escuela de Agronomía and Estación Experimental Agronómica, Universidad de Chile. (V, E)
4. Santiago de Chile, Casilla 5427. Instituto de Investigaciones Agropecuarias. (V, E)

Cyprus
1. Athalassa (Nicosia). Agricultural Research Institute. (V, E)
2. Limassol. Government Wine Section. (V, E)

Czechoslovakia
1. Bratislava, Matúškova ulica 97. Výskumný Ustavu Vinársky a Vinohradnícky (Research Institute for Enology and Viticulture). (V, E)
2. Karlstjen (Prague).

France
1. Beaune (Côte-d'Or), 10, boulevard Bretonnière. Station Oenologique de Bourgogne. (E)
2. Bordeaux (Gironde), 351, cours de la Libération, Talence. Station Agronomique et Oenologique. (V, E)
3. Cognac (Charente), 45, rue de Bellefonds. Station Viticole du Bureau National Interprofessionnel du Cognac. (V, E)
4. Colmar 68 (Haut-Rhin), 8, rue Kléber. Station de Recherches Viticoles et Oenologiques. (V, E)
5. Dijon (Côte-d'Or), boulevard Gabriel. Centre d'Etudes de Viticulture et d'Oenologie, Faculté des Sciences, Université de Dijon. (V, E)
6. Épernay (Marne), 5, rue Henri-Martin. Services Techniques du Comité Interprofessionnel du Vin de Champagne. (V, E)
7. Montpellier (Hérault). Ecole Nationale Supérieure Agronomique (including Station de Recherche Viticoles and Station de Technologie). (V, E)
8. Narbonne (Aude), Boulevard du General de Gaulle. Station Centrale de Technologie des Produits Végétaux. (V, E)
9. Pont-de-la-Maye (Gironde), Domaine de la Grande-Ferrade. Station de Recherches Viticoles et d'Arboriculture Fruitière du Sud-Ouest. (V)

Germany

1. Ahrweiler (Rheinland/Pfalz), Walporzheimerstrasse 48. Landes- Lehr- und Versuchsanstalt für Weinbau, Gartenbau und Landwirtschaft. (V, E)
2. Bad Kreuznach, Rudesheimerstrasse 60/68. Landes- Lehr- und Versuchsanstalt für Weinbau, Gartenbau und Landwirtschaft and Höherer Landbauschule und Höherer Weinbauschule. (V, E)
3. Bernkastel-Kues. Biologische Bundesanstalt für Land- und Forstwirtschaft für Rebenkrankheiten. (V, E)
4. Freiburg, Stefan Meier Strasse, 21. Staatlisches Weinbauinstitut, Versuchs- und Forschungsanstalt für Weinbau und Weinbehandlung. (V, E)
5. Geilweilerhof, Siebeldingen bei Landau (Pfalz). Bundes Forschungsanstalt für Rebenzüchtung. (V, E)
6. Geisenheim 6222 (Rheingau). Hessische Lehr- und Forschungsanstalt für Wein-, Obst- und Gartenbau. (V, E)
7. Neustadt/Weinstrasse (Rheinland/Pfalz), Maximilianstrasse 43/45. Landes- Lehr- und Forschungsanstalt für Wein- und Gartenbau. (V, E)
8. Oppenheim 6504 (Rheinland/Pfalz), Zuckerberg 19. Landes- Lehr- und Versuchsanstalt für Wein- und Gartenbau. (V, E)
9. Trier (Rheinland/Pfalz), Egbertstrasse 18/19. Landes- Lehr- und Versuchsanstalt für Weinbau, Gartenbau und Landwirtschaft. (V, E)
10. Trier (Rheinland/Pfalz), Egbertstrasse 18. Weinforschungsinstitut der Landeslehr und Versuchsanstalt für Weinbau, Gartenbau und Landwirtschaft. (E)
11. Veitshochheim 8702 (Bayern), Postfach 49. Lehr- und Versuchsanstalt, Bayerischen Landesanstalt für Wein-, Obst- und Gartenbau. (V, E)
12. Weinsberg (Heilbronn), Hallerstrasse 6. Staatliche Lehr- und Versuchsanstalt für Wein- und Obstbau. (V, E)
13. Würzburg (Bayern), Residenzplatz 2. Institut für Zuchtungsforschung, Bayerische Landesanstalt für Wein-, Obst- und Gartenbau. (V, E)

Greece

1. Athens, Votanikos T3. Ecole des Hautes Etudes Agronomique. (V, E)
2. Lykovrissi-Kifissia (Athens). Institut du Vin (Ministère de l'Agriculture). (V, E)
3. Pyrgos. Georgikon Institouton Stafidos (Corinth Raisin Institut). (V)
4. Salonika (Macedonia), rue Aristotelous, No. 3. Laboratoire de Viticulture and Laboratoire de Technologie Agricole, Université de Thessaloniki. (V, E)

Hungary

1. Budapest II, Hermann Ottó út, 15. Szőlészti Kutato Intézet (Ampelologiai Intézet) (Institut de Recherches Viticoles et Institut Ampélologique). (V, E)
2. Budapest XI, Menesi út, 44. Kertészeti és Szőlészeti Föiskola (École Supérieure Horticole et Viticole). (V, E)
3. Modra. Výskumného Procoviska Vinárskych Závodov. (V, E)

Israel

1. Haifa. Department of Food and Bio-Technology, Israel Institute of Technology. (E)
2. Rehovot-Bet-Dagan. Division of Pomology and Viticulture and Division of Food Technology, Agricultural Research Station (Ministry of Agriculture). (V, E)
3. Rehovot. Israel Wine Institute. (V, E)

Italy

1. Alba (Cuneo). Istituto Tecnico Agrario Specializzato per la Viticoltura e la Enologia. (V, E)
2. Asti, via Pietro Micca, 35. Stazione Enologica Sperimentale. (V, E)
3. Catania (Sicily). Istituto Siciliano della Vite e del Vino. (V, E)
4. Conegliano (Treviso). Stazione Sperimentale di Viticoltura e di Enologia. (V, E)

5. Florence, via Agnolo Poliziano, 2. Istituto di Industrie Agrarie and Istituto di Microbiologia Agraria e Tecnica, Facoltà di Scienze Agrarie e Forestali. (E)
6. Marsala (Sicily). Istituto Tecnico Agrario Specializzato per la Viticoltura e l'Enologia. (V, E)
7. Pavia, P.O. Box 165. Istituto Botanico e Laboratorio Crittogamico, Università di Pavia. (V)
8. Perugia (Umbria), Piazza d'Anti. Istituto di Microbiologia Agraria e Tecnica and Istituto Industrie Agrarie, Facoltà di Scienze Agrarie, Università di Perugia. (E)
9. Pisa, via San Michele degli Scalzi, 2. Istituto di Patologia Vegetale e Microbiologia, Istituto Chemica Agraria, and Istituto de Coltivazioni Arboree, Università de Pisa. (V, E)
10. Sassari (Sardegna), Piazza Conte di Moriana, 8. Istituto Industrie Agrarie, Università Sassari. (E)
11. Torino, via Pietro Giuria, 15. Istituto Industrie Agrarie, Istituto di Microbiologia Agraria e Tecnica and Istituto di Coltivazioni Arboree, Facoltà di Scienze Agrarie, Università di Torino and Stazione Chemico-Agraria Sperimentale. (V, E)

Japan
1. Kitakoma-gun (Yamanashi), Onta, Futaba-cho. Suntory Research Institute of Crops for Brewing. (V)
2. Kofu, Kitashin-machi. Research Institute of Fermentation, Yamanashi University. (E)
3. Tokyo, Bunkyo-ku. Department of Agricultural Chemistry, University of Tokyo. (E)

Jugoslavia
1. Belgrade, Studentski Trg 1. Faculté Agronomique, Université de Belgrade (V,E)
2. Ljubljana, Livarska 9. Faculté de l'Agriculture et l'Oenologie, Université de Ljubljana. (V, E)
3. Novi Sad. Faculté de Technologie, Université de Novi Sad. (E)
4. Skopje (Macédoine). Institut de Recherches Viticoles et Vinicoles. (V, E)
5. Split (Croatia). Institut de Cultures de la Côte Adriatique. (V, E)
6. Sremski Karlovci (Serbie-Voivodina). Institut of Recherches Viticoles et Arboricoles. (V, E)
7. Zagreb (Croatia), Kačićeva ul. 9. Institut za Vocarstvo, Vinogradarstvo, Vinarstvo i Vrtlarstvo (Institut d'Arboriculture Fruitière, de Viticulture, d'Oenologie et d'Horticulture). At Trg Maršala Tita 14: Faculté d'Agriculture, Université de Zagreb, Section de Viticulture et d'Oenologie. (V, E)

Lebanon
1. Tel Amara-Rayak. Section d'Arboriculture Fruitière et de Viticulture, Institut de Recherches Agronomiques. (V)

Luxembourg
1. Remich, Station Viticole. (V, E)

Morocco
1. Ellouizia (formerly St-Jean-de-Fédala) (Casablanca). Station de Viticulture de l'Ecole d'Agriculture Xavier Bernard. (V)

New Zealand
1. Te Kauwhata, P.O. Box 19, Private Mail Bag. Horticultural Research Station. (Department of Agriculture). (V, E)

Peru
1. Ica. Stación Vitivinícola Nacional del SIPA. (V, E)

Poland

1. Łodz, Zwirki 36. Politechnika Łódzka (Faculty of Chemistry of Foodstuffs, Technical Microbiology). (E)
2. Warsaw, Ul. Rakowiecka 36. Instytut Przemystu Fermentacyjnego (Institute for the Fermentation Industries). (E)

Portugal

1. Anadia. Estação Vitivinícola da Beira-Litoral. (V, E)
2. Lisbon, Tapada de Ajuda. Istituto Superior de Agronomia. (V, E)
3. Lisbon, Rua Capitão Renato Baptista. Centro Nacional de Estudos Vitivinícolas. (V, E)
4. Lisbon, Rua Mousinho da Silveira 5. Junta Nacional do Vinho. (E)
5. Oeiras. Estação Agronomica Nacional. (V, E)
6. Peso da Regua (Douro). Estação Vitivinícola do Douro. (V, E)
7. Porto, Rua de Ferreira Borges. Istituto do Vinho do Porto. (V, E)
8. Porto. Laboratorio da Comissão de Viticultura da Região dos Vinhos Verdes. (E)

Roumania

1. Băneasa (Bucharest), Bd. N. Bălcescu 4–6. Institutul de Cercetări Horti-Viticole. (branches, usually called Statiunea Experimentala Viticola, at Cluj, Istrița, Iași, Drăgășani, Valea Călugărească, Odobești, Miniș, Greaca-Bucureşti and Ostrov- Constanța). (V)
2. Bucharest, Str. Lt. Lemnea 16. Colectivul Ampelografie. (V)
3. Cluj, Str. Mănăştur 3. Agronomic Institute "Dr. Petru Groza." (V, E)

South Africa

1. Stellenbosch (Cape Province), Private Bag 26. Viticultural and Enological Research Institute. (V, E)
2. Suider-Paarl. Ko-operatieve Wijnbouwers Vereniging. (E)

Soviet Union

1. Alma-Ata (Kazakstan). Nauchno-Issledovatel'skiĭ Institut Plodovodstva i Vinogradarstva. (V, E)
2. Baku, Zavokzal'naia, 20. Azerbaĭdzhanskii Nauchno-Issledovatel'skii Institut Sadovodstva, Vinogradarstva i Subtropicheskikh Kul'tur. (V, E)
3. Erevan (Armenia). Armianskiĭ Nauchno-Issledovatel'skiĭ Institut Vinogradarstva, Vinodeliia i Plodovodstva. (V, E)
4. Kishinev (Moldavia), Kostiuzhenskoye shosse 9. Nauchno-Issledovatel'skiĭ Institut Sadovodstva, Vinogradarstva i Vinodeliia. (V, E)
5. Krasnodar. Severo-Kavkazskii Zonal'nyĭ Nauchno-Issledovatel'skiĭ Institut Vinogradarstva. (V)
6. Moscow. Institut Biokhimii im. A. N. Bakha, Akademiia Nauk SSSR. (E)
7. Novocherkassk (Rostov), Arsenal'naia 15. Nauchno-Issledovatel'skiĭ Institut Vinogradarstva i Vinodeliia RSFSR. (V, E)
8. Odessa. Ukrainskiĭ Nauchno-Issledovatel'skiĭ Institut Vinogradarstva i Vinodeliia im. V. E. Tairova (also has a branch in Kiev). (V, E)
9. Tashkent (Uzbekistan). Sredneaziatskaia Stantsiia Vsesoiuznogo Instituta Rastenievodstva and Uzbekskaia Institut Sadovodstva, Vinogradarstva i Vinodeliia im. R. R. Shredera (also has a branch in Samarkand). (V, E)
10. Tbilisi (Georgia). Gruzinskii Nauchno-Issledovatel'skiĭ Institut Sadovodstva, Vinogradarstva i Vinodeliia. (V, E)
11. Yalta, ul. Kirova 25. Vsesoiuznyĭ Nauchno-Issledovatel'skiĭ Institut Vinodeliia i Vinogradarstvo "Magarach." (V, E)

Spain

1. Haro. Estación de Viticultura y Enología. (V, E)
2. Jerez de la Frontera. Estación de Viticultura y Enología. (V, E)

845

3. Madrid 3, Avenida Puerta de Hierro. Instituto Nacional de Investigaciones Agronómicas, Centro de Ampelografía y Viticultura. (V, E)
4. Madrid, Serrano, 150. Patronato Juan de la Cierva de Investigación Técnica, Consejo Superior de Investigaciónes Científicas. (E)
5. Requena (Valencia), Alza Garcia Morato, 1. Estación de Viticultura y Enología. (V, E)
6. Reus (Tarragona). Estación de Viticultura y Enología. (V, E)
7. Villafranca del Panadés, Calle de la Fuente 57. Estación de Viticultura y Enología. (V, E)

Switzerland
1. Lausanne. Stations Fédérales d'Essais Agricoles. (V, E)
2. Wädenswil. Eidgenossische Versuchsanstalt für Obst-, Wein- und Gartenbau. (V, E)

Tunisia
1. Tunis. Ecole Supérieure d'Agriculture. (V)
2. Tunis. Office du Vin de Tunisie. (E)

Turkey
1. Ankara. College of Agriculture, University of Ankara. (V, E)
2. Bornova-Izmir. Institute of Agricultural Research, University of Izmir. (V)
3. Istanbul. Enological Laboratory, State Institute of Monopolys. (E)

United States
1. Albany 94710 (Calif.), 800 Buchanan St. Western Regional Research Laboratory, Western Utilization Research and Development Division, U.S. Dept. Agr. (V, E)
2. Berkeley 94720 (Calif.). University of California, College of Agriculture. (E)
3. Davis 95616 (Calif.). University of California, College of Agriculture. (V, E)
4. Fresno 93726 (Calif.). Fresno State College, College of Agriculture. (V, E)
5. Fresno 93702 (Calif.), 2021 South Peach Avenue. Western Fruit and Vegetable Investigations Laboratory, Horticultural Crops Branch, Agricultural Research Service, U.S. Dept. Agr. (V)
6. Geneva (New York). New York State Agricultural Experiment Station. (V, E)
7. Prosser (Washington). Irrigation Station, Washington Agricultural Experiment Station. (V)

Uruguay
1. Montevideo. Laboratoire de Fermentations et Oenologie, Faculté de Chimie, Université de Montevideo. (E)

LITERATURE CITED

LITERATURE CITED[1]

ABEIJON, J.
1942. Difficultés de fermentation des vins mousseux. Bull. O.I.V. (Office Intern. Vigne Vin) 15(151):34–38.

ADAMS, A. M.
1964. Malo-lactic fermentations in Ontario wines. Rept. Hort. Exper. Stat. and Prod. Lab., Vineland Station, Ontario, 1964:108–111.
1965a. The diethyl ester of pyrocarbonic acid as an antimicrobial agent. *Ibid.* 1965:138–145.
1965b. Taste and aroma of wines treated with diethyl ester of pyrocarbonic acid. *Ibid.* 1965:133–138.
1966. Studies on storage of yeast. III. Effect of low temperatures (−29°C) on viability of stored starter. Rept. Hort. Res. Inst. Ontario 1966:114–117. (*See also* Rept. Hort. Exper. Stat. and Prod. Lab. 1953–1954:94–97, 98–101.)

AGABAL'IANTS, G. G.
1954. Khimiko-tekhnologicheskii kontrol' proizvodstva sovetskogo shampanskogo. Moscow, Pishchepromizdat. 383 pp.

AGABAL'IANTS, G. G., and S. P. AVAKIANTS
1965. Issledovanie preobrashchenii organicheskikh kislot pri nepreryvnoi shampanizatsii. Izv. Vysshikh Uchebn. Zavedeniĭ, Pishchevaia Tekhnol. 1965(6): 34–38.
1966. Issledovanie protsessa avtoliza drozhzheĭ pri nepreryvnoĭ shampanizatsii. Vinodelie i Vinogradarstvo SSSR 26(1):17–20.

AGABAL'IANTS, G. G., and N. N. GLONINA
1966. Aminokisloty i okislennost' vina. Vinodelie i Vinogradarstvo SSSR 26(8): 9–16.

AGABALIANZ, G. G., and A. A. MARZHANIAN
1962. Method of champagnizing wine in a continuous stream and installation for same. U.S. Patent 3,062,656, November 6, 1962. 2 pp.

AIROLA, A.
1941. Ueber die Bestimmung des saures und alkalischen Gase in Gasmischungen mit Hilfe der Ermittlung des pH. Svensk Kem. Tidsk. 53:123–125.

AIR REDUCTION CO.
1961. Nitrogen in the wine industry. Part I. Sparging process. Bull. Air Reduction Co., Inc., Customer Serv. Lab., Madison, Wisconsin.

ALBACH, R. F., R. E. KEPNER, and A. D. WEBB
1959. Comparison of anthocyan pigments of red *Vinifera* grapes. II. Am. J. Enol. Viticult. 10(4):164–172.

ALBERTSON, N. F., and J. P. McREYNOLDS
1943. Mechanism of the reaction between hydrogen sulfide and sulfur dioxide in liquid media. J. Am. Chem. Soc. 65:1690–1691.

------

[1] The abbreviations used for journals, with few exceptions, follow those of Chemical Abstracts.

ALBONICO, F.
1958. Richerche sul trattamento dei vini con ferrocianuro di potassio. Riv. Viticolt. Enol. 11:352–362, 387–392.

ALDERTON, G., W. H. WARD, and H. L. FEVOLD
1945. Isolation of lysozyme from egg white. J. Biol. Chem. 157:43–58.

ALEXIU, A.
1967. Studiul comparativ al unor metode pentru caracterizarea cromatică a vinurilor roşii. Inst. Cercetări Horti-Viticole, Lucrări Ştiinţ. 9:499–511.

ALEXIU, A., G. ENĂCHESCU, A. MIHALCA, G. SANDU-VILLE, E. POPA, GH. BUTĂNESCU, and C. BASAMAC
1967. Prezenţa acidului malic în strugurii şi vinurile unor podgorii din republica socialistă România. Inst. Cercetări Horti-Viticole, Lucrări Ştiinţ. 9:471–490.

ALLEN, H. W.
1963. The wines of Portugal. New York, Toronto, and London, McGraw-Hill Book Company, Inc. 192 pp.

ALLEWELDT, G.
1965. Der Rebenanbau in der Türkei. Wein-Wissen. 20:109–126.

ALLMENDINGER, W.
1965. Some production and marketing trends and prospects. Fresno, Fermenting Material Processors Advisory Board. (See table 2.)
1967. California grape industry—economic situation and outlook. San Francisco. 19 pp. (Processed.)
1968. Personal communication.

ALWOOD, W. B.
1914. Crystallization of cream of tartar in the fruit of the grape. J. Agr. Res. 1:153–154.

AMANO, Y., and M. KAGAMI
1966. On the characteristics of rate of ion exchange treatment of wine. Bull. Res. Inst. Ferm. Yamanashi Univ. 13:25–31.

AMATI, A., and A. FORMAGLINI
1965. Reconoscimento di antifermentativi nei vini mediante cromatografia su strato sottile. Riv. Viticolt. Enol. 18:387–395.

AMATI, A., and R. RASTELLI
1967. Sul contenuto in litio di vini italiani. Ind. Agr. 5:233–237.

AMBROSI, H., and J. FLOCKEMANN
1961. Die Berechnung von Süssweinverschnitten. Weinberg Keller 8:9–19.

AMERICAN PETROLEUM INSTITUTE
1948. API toxicological review: sulfur dioxide. New York, American Petroleum Institute. 5 pp.

AMERINE, M. A.
1944. Determination of esters in wines—liquid-liquid extraction. Food Res. 9(5):392–395.
1947. The composition of California wines at exhibitions. Wines & Vines 28(1): 21–23, 42–43, 45; (2):24–26; (3):23–25, 42–46.
1948a. An application of "triangular" taste testing to wines. Wine Rev. 16(5):10–12.
1948b. Organoleptic examination of wines. Wine Technol. Conf., University of California, Col. Agr., Davis, August 11–13, 1948. Pp. 15–26. (Mimeo.)
1949a. Wine production problems of California grapes. Wine Rev. 17(1):18–22, (2):10–11, 22; (3):10–12; (5):10–12; (6):11–12, 14–15; (7):6–7, 19–20.
1949b. Some factors influencing the yield of juice and wine. Wine Institute, Technical Advisory Committee, San Francisco, March 9, 1949. 3 pp.

1950a. The acids of California grapes and wines. I. Lactic acid. Food Technol. 4:117–181.
1950b. The response of wine to aging. Wines & Vines 31(3):19–22; (4):71–74; (5):28–31.
1951. The acids of California grapes and wines. II. Malic acid. Food Technol. 5(1):13–16.
1954. Composition of wines. I. Organic constituents. Advan. Food Res. 5:354–510.
1955. Further studies with controlled fermentations. Am. J. Enol. 6(1):1–16.
1956. The maturation of wine grapes. Wines & Vines 37(10):27–30, 32, 34–36; (11):53–55.
1958. Composition of wines. II. Inorganic constituents. Advan. Food Res. 8:133–224.
1959. Continuous flow production of still and sparkling wine. Wines & Vines 40(6):41–42.
1961. Legal and practical aspects of the sensory examination of wines. J. Assoc. Offic. Agr. Chemists 44:380–383.
1962a. Hilgard and California viticulture. Hilgardia 33 (1):1–23.
1962b. Physical and chemical changes in grapes during maturation and after full maturity. Proc. XVI Intern. Hort. Cong. 3:479–483. (Published 1963.)
1963a. Continuous fermentation of wines. Wines & Vines 40(6):41–42.
1963b. Viticulture and enology in the Soviet Union. Ibid. 44(10):29–34, 36; (11):57–62, 64; (12):25–26, 28–30.
1964. Der Weinbau in Japan. Wein-Wissen. 19(5):225–231.
1965a. The fermentation industries after Pasteur. Food Technol. 19:75–80, 82.
1965b. Laboratory procedures for enology. University of California, Dept. Viticulture and Enology. Davis, Calif., Associated Students Bookstore. 100[9] pp. (Reprinted with corrections, 1968.)
1969. An introduction to the pre-Repeal history of grapes and wines in California. Agr. Hist. 43 (2):259–268.

AMERINE, M. A., and C. B. BAILEY
1959. Carbohydrate content of various parts of the grape cluster. Am. J. Enol. Viticult. 10:196–198.

AMERINE, M. A., H. W. Berg, and W. V. CRUESS
1967. The technology of wine making. 2d ed. Westport, Conn., Avi Publishing Co., Inc. ix + 799 pp.

AMERINE, M. A., and W. DE MATTEI
1940. Color in California wines. III. Methods of removing color from the skins. Food Res. 5(5):509–519. (See also Wines & Vines 22[4]:19–20. 1941.)

AMERINE, M. A., and W. C. DIETRICH
1943. Glycerol in wines. J. Assoc. Offic. Agr. Chemists 26:408–413.

AMERINE, M. A., and M. A. JOSLYN
1940. Commercial production of table wines. Calif. Agr. Exper. Stat. Bull. 639: 1–143.
1951. Table wines; the technology of their production. Berkeley and Los Angeles, University of California Press. xv + 397 pp.

AMERINE, M. A., AND T. T. KISHABA
1952. Use of the flame photometer for determining the sodium, potassium and calcium content of wine. Proc. Am. Soc. Enol. 1952:77–86.

AMERINE, M. A., and R. E. KUNKEE
1965. Yeast stability tests on dessert wines. Vitis 5:187–194.
1968. Microbiology of winemaking. Ann. Rev. Microbiol. 22: 323—358.

AMERINE, M. A., L. P. MARTINI, and W. DE MATTEI
1942. Foaming properties of wine. Ind. Eng. Chem. 34:152–157.

AMERINE, M. A., and M. W. MONAGHAN
1950. California sparkling wines. Wines & Vines 31(8):25–27; (9):52–54.

AMERINE M. A., and C. S. OUGH
1957. Studies on controlled fermentations. III. Am. J. Enol. 8:18–30.
1960. Methods of producing sweet table wines. Wines & Vines 41(12):23–39.
1967. Sweetness preference in rosé wines. Ibid. 18:121–125.

AMERINE, M. A., C. S. OUGH, and C. B. BAILEY
1959a. Suggested color standards for wines. Food Technol. 13:170–175.

AMERINE, M. A., R. M. PANGBORN, and E. B. ROESSLER
1965a. Principles of sensory evaluation of food. New York and London, Academic
Press. x + 602 pp.

AMERINE, M. A., and E. B. ROESSLER
1958a. Field testing of grape maturity. Hilgardia 28(4):93–114.
1958b. Methods of determining field maturity of grapes. Am. J. Enol. 9:37–40.
1963. Further studies on field sampling of wine grapes. Am. J. Enol. Viticult.
14:144–147.

AMERINE, M. A., E. B. ROESSLER, and F. FILIPELLO
1959b. Modern sensory methods of evaluating wine. Hilgardia 28:447–567.

AMERINE, M. A., E. B. ROESSLER, and C. S. OUGH
1965b. Acids and the acid taste. I. The effect of pH and titratable acidity. Am. J.
Enol. Viticult. 16:29–37.

AMERINE, M. A., and G. A. ROOT
1960. Carbohydrate content of various parts of the grape cluster. II. Am. J. Enol.
Viticult. 11:137–139.

AMERINE, M. A., and G. THOUKIS
1958. The glucose-fructose ratio of California grapes. Vitis 1:224–229.

AMERINE, M. A., and A. D. WEBB
1943. Alcohol-glycerol ratio of California wines. Food Res. 8(4):280–285.

AMERINE, M. A., and A. J. WINKLER
1941a. Maturity studies with California grapes. I. The Balling-acid ratio of wine
grapes. Proc. Am. Soc. Hort. Sci. 38:379–387.
1941b. Color in California wines. IV. The production of pink wines. Food Res.
6(1):1–14.
1942. Maturity studies with California grapes. II. The titratable acidity, pH, and
organic acid content. Proc. Am. Soc. Hort. Sci. 40:313–324.
1943. Grape varieties for wine production. Calif. Agr. Exper. Stat. Cir. 356:1–15.
1944. Composition and quality of musts and wines of California grapes. Hilgardia
15(6):493–674.
1947. The relative color stability of the wines of certain grape varieties. Proc. Am.
Soc. Hort. Sci. 49:183–185.
1963a. California wine grapes: composition and quality of their musts and wines.
Calif. Agr. Exper. Stat. Bull. 794:1–83.
1963b. Grape varieties for wine production. Berkeley. Agr. Ext. Serv. Leaflet.
154:1–2.

ANDERSON, A. B.
1961. The influence of extractives on tree properties. 1. California redwood. J.
Inst. Wood Science No. 8:14–34.

ANDRÉ, L.
1966. Caractérisation des composés carbonyles et dosage de l'acétaldéhyde dans
certains vins par chromatographie en phase gazeuse. Ann. Technol. Agr.
15:159–171.

ANDRÉ, P.
1966. Influence du foulage de la vendange sur la qualité des vins du Beaujolais. Vignes et Vins, numéro spécial, **1966**:122–124.

ANDRÉ, P., P. BÉNARD, Y. CHAMBROY, C. HANZY, and C. JOURET
1967. Méthode de vinification par macération carbonique. I. Production de jus de goutte en vinification par macération carbonique. II. La production d'alcool en vinification par macération carbonique. Ann. Technol. Agr. **16**:109–116, 117–123.

ANDRÉ, P., P. CHARNEY, and R. VIOT
1963. Recherche d'une méthode de dégustation rationelle applicable aux vins à appellation d'origine. Bull. Inst. Nat. Appel. Orig. No. **86**:1–18.

ANON.
1927. Partie officielle: loi tendant à completer la loi du 6 mai 1919 relative à la protection des appellations d'origine. J. Agr. Pratique **48**:239–241.

1932. Per la tutela del vino Chianti e degli altri vini tipici Toscani. Bologna, Tipografia Antonio Brunelli. 537 pp.

1933. Loi tendant à completer et à modifier la loi du 4 juillet sur la viticulture et la commerce du vins. J. Agr. Pratique **60**:123–125, 140–142.

1965a. System blends two liquids in-line. Food Eng. 37(8):96.

1965b. Panel discussion on methods of pressing and juice separation of white wines. Wines & Vines 46(4):65–66, 68–70.

1966a. Méthodes d'analyse et éléments constitutifs des vins. Bull. O.I.V. (Office Intern. Vigne Vin) **39**:1035–1067. (See pp. 1056–1059.)

1966b. Microbiologie du vin. *Ibid.* **39**:1068–1090.

1966c. Méthodes objectives et subjectives d'appréciation des caractères organoleptiques des denrées alimentaires. Journées scientifiques du Centre National de Coordination des Études et Recherches sur la Nutrition de l'Alimentation. Paris, 23–27 novembre 1964. Paris, C.N.R.S. 606 pp.

1966d. Fruit production expands in the Soviet Union. Foreign Agriculture Circular, FDAP **3-66**:1–18.

1966e. Grape and apple processing. Food Technol. **20**:49.

1966f. Thirtieth annual wine industry statistical survey. Part 1. Wine Institute Bull. **1377**:15.

1966g. Untersuchungen von Kristallausscheidungen in Wein. Jahresbericht Staatl. Lehr- u. Versuchsanstalt. Wein- u. Obstbau, Weinsberg **1965**:112–113.

1966h. Vini d'origine controllata signora riconosciuti della legge. Ital. Vinicola Agrar. **56**:299–300.

1967a. California vintners extend filtration by metering filter aid. Food Eng. 39(10):135.

1967b. Méthodes d'analyse et éléments constitutifs des vins. Bull. O.I.V. (Office Intern. Vigne Vin) **40**:934–972.

1967c. A survey of wine growing in South Africa 1966–1967. Suider Paarl, Public Relations Dept. of KWV. 55 pp.

1967d. Qualitative Sortimentspolitik im Einzelhandel mit Wein. Deut. Wein-Ztg. **103**:137–138.

1967e. Kaltsterilfüllung—Grundlagen, Erfahrungen und Praxis. Seitz Informationen **29**:3–15.

1968. Methodes d'analyse et elements constitutifs des vins. Bull. O.I.V. (Office Intern. Vigne Vin) **41**:951–978.

ARCHER, T. E., and J. G. B. CASTOR
1956. Phosphate changes in fermenting must in relation to yeast growth and ethanol production. Am. J. Enol. **7**:62–68.

ARCHINARD, P.
1937. Sur le dosage de l'anhydride sulfureux et des acidités fixe et volatile dans les vins sulfités. Rev. Viticult. **87**:257–263.

AREF, H., and W. V. CRUESS
1934. An investigation of the thermal death point of *Saccharomyces ellipsoideus*. J. Bacteriol. 27:443–452.

ARENA, A.
1936. Alteraciones bacterianas de vinos Argentinos. Rev. Facultad de Agron. y Vet. (Buenos Aires) 8:155–315.

ARNOLD, A.
1957. Beiträge sur refraktometrischen Methode der Mostgewichtsbestimmung. Vitis 1:109–120.

ARONSON, H.
1967. Cold sterilization by filtration. San Francisco, Wine Institute, Technical Advisory Committee, June 8, 1967. 3 pp.

ASSOCIATION OF OFFICIAL AGRICULTURAL CHEMISTS
1965. Official methods of analysis. 10th ed. Washington, D.C., Association of Official Agricultural Chemists.

ÁSVÁNY, Á.
1963. Adatok és tabasztalatok, a szorbinsavas bortartósítás köréből. Szőlészeti Kutató Intézet Évkönyve 12:289–299.
1965. Bentonit alkalmazása a must erjedése közben. Borgazdaság 13:146–151.
1967a. Personal communication.
1967b. Influence des températures de fermentation et de conservation du vin et des vins spéciaux sur leurs caractères chimiques, microbiologiques et organoleptiques. Bull. O.I.V. (Office Intern. Vigne Vin) 40:611–621.

AUDIDIER, L.
1965. Des méthodes chimiques ou biologiques pour la vinification des vins fins. Compt. Rend. Acad. Agr. France 51:1163–1166.

AVAKIANTS, S. P.
1965. Fermentational changes in champagne with participation of β-fructofurano-side. Dokl. Akad. Nauk SSSR 165(1):221–223. (English edition.)
1967. Kolorimetricheskiĭ metod opredileniia belkov vina (Determination of proteins in wine). Vinodelie i Vinogradarstvo SSSR 27(5):29–31.

AVAKIANTS, S. P., and I. D. BELOUSOVA
1965. Activity of β-fructofuranosidase in continuous flow conversion of wine to champagne. Prik. Biokh. Mikrobiol. 1:57–65. (English edition.)

AXELROD, B.
1967. Glycolysis. *In* D. M. Greenberg (ed.), Metabolic pathways. 3d ed. New York and London, Academic Press. Vol. 1, 460 pp. *See* pp. 112–145.)

ÄYRÄPÄÄ, T.
1962. Phenethyl alcohol in wines. Nature 194:472–473.
1967a. Formation of higher alcohols from amino acids derived from yeast proteins. J. Inst. Brew. 73:30–33.
1967b. Formation of higher alcohols from ^{14}C-labeled valine and leucine. *Ibid.* 73:17–30.

AZIZ, P. M., and H. P. GODARD
1952. Pitting corrosion characteristics of aluminum. Ind. Eng. Chem. 44:1791–1795.

BABO, A. W., AND E. MACU
1922–27. Handbuch des Weinbaues and der Kellerwirtschaft. 5th and 6th eds. Berlin, P. Parey. 2 vols in 4.

BACHMANN, O.
1967. Die manometrische Bestimmung der Kohlendioxide. Wein-Wissen. 22: 154–159.

BACQUE, J.
1966.   Les vins et la cryoconcentration. Bull. Inst. Intern. Froid 46, Annexe 1966(3):149–156.

BADRAN, A. M., and D. E. JONES
1965.   Polyethylene glycols-tannins interaction in extracting enzymes. Nature 206:622–624.

BÄCKSTRÖM, H. L. J.
1934.   Kettenmechanism bei der Autoxydation von Natrium sulfit Lösungen. Z. physik. Chem. B25:122–138.

BAGLIONI, S., L. CASALE, and C. TARANTOLA
1935–37.   L'attività proteolitica del succo d'uva. Ann. R. Staz. Enol. Sper. Asti (II) 2:204–219.

BAKER, G. A., and M. A. AMERINE
1953.   Organoleptic ratings of wines estimated from analytical data. Food Res. 18:381–389.

BAKER, G. A., M. A. AMERINE, and E. B. ROESSLER
1965a.   Characteristics of sequential measurements on grape juice and must. Am. J. Enol. Viticult. 16:21–28.

BAKER, G. A., C. S. OUGH, and M. A. AMERINE
1965b.   Scoring vs. comparative rating of sensory quality of wines. J. Food Sci. 30:1055–1062.

BALAKIAN, S., and H. W. BERG
1968.   The role of polyphenols in the behavior of potassium bitartrate in red wines. Am. J. Enol. Viticult. 19:91–100.

BĂLĂNESCU, G.
1963–1964a.   Na, K, and Ca determination in wine by flame photometry (transl.). Lucrările Inst. Cercetări Aliment. 7:469–492. (Chem. Abst. 65:625h. 1966).
1963–1964b.   Contributii la determinarea potenţialului de oxido-reducere al unor vinuri şi rachiuri româneşti. Lucrările Inst. Cercetăre Aliment. 7:375–391.

BALDWIN, E.
1967.   Dynamic aspects of biochemistry. 5th ed. London, Cambridge University Press. 525 pp.

BALL, C. O., and F. C. W. OLSON
1957.   Sterilization in food technology: theory, practice, and calculations. New York, McGraw-Hill Book Co. xxvii + 654 pp. (See pp. 133–192, 291–312.)

BANOLAS, E.
1948.   On the new apparatus for the rational equipment of wine-making cellars. Bull. Inst. Intern. Froid Annexe 2:9–23. (See also Refrig. [Sydney] 3:372, 374, 378, 380–382, 384–385. 1950.)

BARAUD, J.
1951.   Une nouvelle méthode de dosage de l'acide ascorbique. Bull. Soc. Chim. France 1951:837–834.
1953.   Dosages simultanés de l'acide ascorbique, de l'acide dehydroxymaléique et de la réductone. Ibid. 1953:521–525.

BARAUD, J., L. GENEVOIS, and C. HEBRE
1966.   Recherches sur les dérivés carbonyliques des vins et eaux-de-vie. Bull. O.I.V. (Office Intern. Vigne Vin) 39:586–593.

BARBET, E. A.
1912.   La vinerie. 2d ed. Paris, H. Dunod et E. Pinot. vii + 190 pp.

BARRE, P.
- 1966a. Détermination rapide de la nature optique de l'acide lactique produit dans les fermentations bactériennes. Les applications dans la classification des *Lactobacillaceae* Winslow *et al.* Ann. Technol. Agr. **15**:203–209.
  1966b. Recherches sur les bactéries lactiques des vins. *Ibid.* **15**:173–180.

BARRE, P., and P. GALZY
  1960. Étude d'une nouvelle bactérie malolactique. Ann. Technol. Agr. **9**:331–343.
  1962. Étude d'un lactobacille homofermentatif isolé du vin. *Ibid.* **11**:121–130.

BARRET, A., P. BIDAN, and L. ANDRÉ
  1955. Sur quelques accidents de vinification dus à des levures à voile. Compt. Rend. Acad. Agr. France **41**:426–430.

BARTON-WRIGHT, E. C.
  1949. Some nitrogenous constituents of wort and their fate during fermentation by top and bottom fermentation yeasts. European Brewing Conv. Proc. Lucerne **1**:19–31.

BASSERMANN-JORDAN, F. VON
  1923. Geschichte des Weinbaues. 2d ed. Frankfurt am Main, Frankfurter Verlags-Anstalt AG. 3 vols.

BAUMGARTNER, J. G.
  1943. Canned foods: an introduction to their microbiology. London, J. A. Churchill Ltd. viii + 157 pp. (*See* pp. 76–95.)

BAYER, E.
  1957. Aromastoffe des Weines. II. Aliphatische Aldehyde des Weines und der Trauben. Vitis **1**:93–95.
  1966. Quality and flavor by gas chromatography. J. Gas Chromat. **4**:67–73.

BAYES, A. L.
  1950. Investigations on the use of nitrogen for the stabilization of perishable food products. Food Technol. **4**(4):151–157.

BAYLY, F. C., and H. W. BERG
  1967. Grape and wine proteins of white wine varietals. Am. J. Enol. Viticult. **17**:18–32.

BAYONOVE, C.
  1966. Étude de trois méthodes de dosage des sucres réducteurs. Ann. Technol. Agr. **15**:139–147.

BEAMER, P. R., and F. W. TANNER
  1939. Heat resistance studies on selected yeasts. Zentr. Bakteriol. Parasitenk. Abt. II, **100**:202–211.

BECKER, W., and H. LORENZ
  1966. Rapport allemand. Bull. O.I.V. (Office Intern. Vigne Vin) **39**:1149–1179.

BEDFORD, C. L.
  1942. A taxonomic study of the genus *Hansenula*. Mycologica **34**:628–649.

BEETCH, E. G., AND E. I. OETZEL
  1957. Sulfur dioxide in malt and beer. Colorimetric determination of sulfur dioxide from malt and beer by complexing with sodium tetrachloromercurate (II). J. Agr. Food Chem. **5**:951–952.

BEGUNSOVA, P. D., and A. E. LINETSKAIA
  1967. Stravitel'naia otsenka ranznykh metodov opredelnia belkov v vine. Vinodelie i Vinogradarstvo SSSR **27**(5):31–34.

BELAVOINE, P.
  1950. Saveur de fer dans le vin. Mitt. Gebiete Lebensm. Hyg. **41**(1/2):56–57.

BELITZER, W. A.
1934. Ueber die Beeinflussung der Selbstgärung der Hefezellen. Protoplasma 22:17–21.

BELL, J. W.
1967. Winery inspections 1966. Wine Institute Bull. 1420–C:3.

BÉNARD, P., and C. JOURET
1963. Essais comparatifs de vinification en rouge. Ann. Technol. Agr. 12:85–102.

BÉNARD, P., C. JOURET, and M. FLANZY
1963. Influence des porte-greffes sur la composition minérale des vins. Ann. Technol. Agr. 12:277–285.

BENDA, I.
1962. Torulopsis burgeffiana nov. spec., eine von Weinbeeren isolierte, neue Hefeart. Antonie von Leeuwenhoek J. 28:208–213.

BENDA, I., and A. SCHMITT
1966. Oenologische Untersuchungen zum biologischen Säureabbau in Most durch Schizosaccaromyces pombe. Weinberg Keller 13:239–254.

BENDA, I., and E. WOLF
1965. Versuche einer Rassendifferenzierung bei Saccharomyces cerevisiae var. ellipsoideus. Mitt. Rebe u. Wein, Serie A (Klosterneuburg) 15:300–316.

BENES, V., and V. KRUMPHANZL
1964. Stabilizace vín kyselinou metavinnou a její vyroba. Kvasný Průmysl 10(11):258–261.

BENVEGNIN, L., and E. CAPT
1931. Du dosage de l'acide sulfureux libre et total dans les vins rouges. Mitt. Gebiete Lebensm. Hyg. 22:257–263, 365–368.

BENVEGNIN, L., and J. MICHAEL
1952. Doses d'acide sulfureux à ajouter à un vin pour porter sa teneur en SO₂ à une valeur donnée. Ann. Agr. Suisse (N.S.) 1:1107–1112.

BENVEGNIN, L., E. CAPT, and G. PIGUET
1951. Traité de vinification. 2d ed. Lausanne, Librairie Payot. 584 pp.

BERAND, P., and J. MILLET
1949. Observations sur le pouvoir alcoogéne des levures cultivées à basse température. Ann. Inst. Pasteur 77:581–587.

BERG, H. W.
1948. Cooperage handling. Wine Technol. Conf., University of California, Col. Agr., Davis, Calif., August 11–13, 1948. Pp. 38–45. (Mimeo.)
1949. Wrinkles in winery operations. Proc. Wine Technol. Conf., University of California, Davis, Calif., August 10–12, 1949:122–128.
1950a. Unpublished results on nitrogen stripping of wines. Division of Viticulture, Davis, Calif.
1950b. Heat treatment of musts. Wines & Vines 31(6):24–26.
1951. Stabilization practices in California wineries. Proc. Am. Soc. Enol. 1951: 90–147.
1953a. Present practices in California wine industry. J. Agr. Food Chem. 1:152–157.
1953b. Wine stabilization factors. Proc. Am. Soc. Enol. 4:91–111.
1953c. Varietal susceptibility of white wines to browning. I. Ultraviolet absorption of wines. II. Accelerated storage test. Food Res. 18:399–406, 407–410.
1959a. Investigation of defects in grapes delivered to California wineries: 1958. Am. J. Enol. Viticult. 10:61–68.
1959b. The effects of several fungal pectic enzyme preparations on grape musts and wines. Ibid. 10:130–134.
1962. Personal communication.

1963. Stabilisation des anthocyannes. Comportement de la couleur dans les vins rouges. Ann. Technol. Agr. 12(numéro hors série 1):247–257.

1967. Continuous wine stabilizing in South Africa. Wines & Vines 48(10):27.

BERG, H. W., and M. AKIYOSHI

1956a. The effect of contact time of juice with pomace on the color and tannin content of red wines. Am. J. Enol. 7:84–90.

1956b. Some factors involved in browning of white wines. Ibid. 7:1–7.

1957. The effect of various must treatments on the color and tannin content of red grape juices. Food Res. 22:373–383.

1961. Determination of protein stability in wine. Am. J. Enol. Viticult. 12:107–110.

1962. Color behavior during fermentation and aging of wines. Ibid. 13:126–132.

BERG, H. W., R. J. COFFELT, and G. M. COOKE

1968. Countercurrent sugar extraction from pomace on a commercial scale. Am. J. Enol. Viticult. 19:108–115.

BERG, H. W., and J. F. GUYMON

1951. Countercurrent extraction of alcohol from grape pomace. Wines & Vines 32(10):27–31.

BERG, H. W., and R. M. KEEFER

1958. Analytical determination of tartrate stability in wine. I. Potassium bitartrate. Am. J. Enol. 9:180–183.

1959. Analytical determination of tartrate stability in wine. II. Calcium tartrate. Ibid 10:105–109.

BERG, H. W., and G. L. MARSH

1950. Heat treatment of musts. Wines & Vines 31(7):23–24; (8):29–30.

BERG, H. W., and A. D. WEBB

1955. California wine types. University of California, Davis, Calif. 20 pp. (Mimeo.)

BERG, H. W., C. J. ALLEY, and A. J. WINKLER

1958. Investigations of defects in grapes delivered in California wineries. Am. J. Enol. 9:24–31.

BERG, H. W., F. FILIPELLO, E. HINREINER, and A. D. WEBB

1955a. Evaluation of thresholds and minimum difference concentrations for various constituents of wines. I. Water solutions of pure substances. Food Technol. 9:23–26.

1955b. Ibid. II. Sweetness: the effect of ethyl alcohol, organic acids, and tannin. Ibid:138–140.

BERG, H. W., C. S. OUGH, and C. O. CHICHESTER

1964. The prediction of perceptibility of luminous-transmittance and dominant wave-length differences among red wines by spectrophotometric measurements. J. Food Sci. 29:661–667.

BERGERET, J.

1963. Action de la gélatine et de la bentonite sur la couleur et l'astringence de quelques vins. Ann. Technol. Agr. 12:15–25.

BERGERET, J., and M. FEUILLAT

1967. Étude de la maturation en bouteilles de vins fins de Bourgogne. Ind. Aliment. Agr. (Paris) 84:1599–1605.

BERGNER, K. G.

1968. Les amino-acides dans les vins mousseux et leurs variations en fonction des procédés de fabrication. Bull. O.I.V. (Office Intern. Vigne Vin) 41:460–467.

BERGNER, K. G., and H. WAGNER

1965. Die freien Aminosäuren während der Flaschen- und Tankgärung von Sekt. Mitt. Rebe u. Wein, Serie A (Klosterneuburg) 15:181–198.

LITERATURE CITED

BERIDZE, G. I.
1965. Vino i kon'iaki Gruzii; les vins et les cognacs de la Georgie (The wines and brandies of Georgia). Tbilisi, Gos. Izdatel'stvo "Sabchota Sakartvelo." 262 pp.

BERIDZE, G. I., and M. G. SIR'ILADZE
1963. Sostov azotistykh veshchestv v gruzinskikh vinakh. Biokh. Vinodeliia 7:102–118.

BERRY, J. M., and R. H. VAUGHN
1952. Decomposition of tartrates by lactobacilli. Proc. Am. Soc. Enol. 1952: 135–138.

BERTI, L.
1949. New table evaluates wine. Wines & Vines 30(5):20–21.
1961. A review of the transfer system of champagne production. Am. J. Enol. Viticult. 12:67–68.
1965. Preparation of juice for white wine production. Wines & Vines 46(4):65–66.

BERTRAND, G., and D. BERTRAND
1949. Recherches sur la teneur des vins en rubidium. Ann. Inst. Pasteur 77: 541–543.

BERTRAND, G., and L. SILVERSTEIN
1950. La fermentation du sucre par la levure produit-elle normalement du méthanol? Compt. Rend. 230:800–803.

BESONE, J.
1940. Pectin constituents in certain varieties of grapes. Unpublished manuscript. University of California, Div. Viticulture, Davis, Calif.

BESONE, J., and W. V. CRUESS
1941. Observations on the use of pectic enzymes in wine making. Fruit Prod. J. 20(12):365–367.

BEST, C. H., W. S. HARTROFT, C. C. LUCAS, and J. H. RIDEOUT
1949. Liver damage produced by feeding alcohol or sugar and its prevention by choline. Brit. Med. J. 2:1001–1006.

BIDAN, P.
1956. Sur quelques bactéries isolées de vins en fermentation malolactique. Ann. Technol. Agr. 5:597–617.
1966. Étude préliminaire de l'influence de certains pesticides sur l'action de quelques microörganismes. Vignes et Vins, numéro spécial 1966:27–31, 33–37.

BIDAN, P., and L. ANDRÉ
1958. Sur la composition en acides amines de quelques vins. Ann. Technol. Agr. 7:403–432.

BIEBER, H.
1960. Der papierchromatographische Nachweis von rotem Hybridenfarbstoff. Deut. Wein-Ztg. 96:104–106.
1967. Die fluorimetrische Bestimmung von Malvin in Traubenmost und Wein. Deut. Lebensm.-Rundschau 63:44–46.

BIELIG, H.-J.
1966. Ein Beitrag zur Frage der Sauerstoffaufnahme bei der Abfüllung von Getränken. Weinberg Keller 13:65–78.

BIGELOW, W. C.
1900. The composition of American wines. Bull. U.S. Dept. Agr., Div. Chem. 59:1–76.

859

BIOL, H., and C. FOULONNEAU
1961. Le paeonidol 3.5 diglucoside dans le genre *Vitis*. Ann. Technol. Agr. 10:345–350.

BIOL, H., and A. MICHEL
1961. Étude chromatographique des vins rouges de cépages réglementés. Ann. Technol. Agr. 10:339–344.
1962. Étude chromatographique des vins rouges issus de cépages réglementés. *Ibid.* 11:245–247.

BIOL, H., and J. SIEGRIST
1966. Comparison d'un type de vinification en rouge à d'autres types: influence du chauffage. Vignes et Vins, numéro spécial 1966:60–77.

BIOLETTI, F. T.
1905. The manufacture of dry wines in hot countries. Calif. Agr. Exper. Stat. Bull. 167:1–66. (Out of print.)
1906a. A new wine-cooling machine. *Ibid.* Bull. 174:1–27. (Out of print.)
1906b. A new method of making dry red wine. *Ibid.* Bull. 177:1–36. (Out of print.)
1912. Sulfurous acid in wine making. Eighth Intern. Cong. Applied Chemistry 14:31–59.
1914. Winery directions. Calif. Agr. Exper. Stat. Cir. 119:1–8. (Out of print.)
1915. The wine-making industry of California. Intern. Inst. Agr., Agr. Intelligence and Plant Dis. Mo. Bull. 6(2):1–13.
1938. Outline of ampelography for the vinifera grapes in California. Hilgardia 11:227–293.

BIOLETTI, F. T., and W. V. CRUESS
1912. Enological investigations. Calif. Agr. Exper. Stat. Bull. 230:1–118. (Out of print.)

BIZEAU, C.
1963. Étude des facteurs limitant la croissance des levures dans les moûts de "Clairette" at de "Muscat blanc à petits grains." Ann. Technol. Agr. 12:247–276.

BLAHA, J.
1951. Personal communication.

BLOUIN, J.
1963. Constituants du vin combinant de l'acide sulfureux. Ann. Technol. Agr. 12 (numéro hors série 1):97–98.
1966. Contribution à l'étude des combinations de l'anhydride sulfureux dans les moûts et les vins. *Ibid.* 15:223–287, 360–401.

BLOUIN, J., and E. PEYNAUD
1963a. Présence constante des acides glucuronique et galacturonique dans les moûts de raisins et les vins. Compt. Rend. 256:4774–4775.
1963b. Présence constante des acides pyruvique et α-cétoglutarique dans les moûts de raisin et les vins. *Ibid.* 256:4521–4522.

BO, M. J.
1965. The transfer methods. San Francisco, Wine Institute, Technical Advisory Committee, June 7, 1965. 2 pp.

BOBADILLA, G. F. DE, and E. NAVARRO
1949. Vinos de Jerez. Estudio de sus ácidos, desde el período de madurez de la uva hasta el envejecimiento del vino. Bol. Inst. Nac. Invest. Agron. (Madrid) 9(21):473–519.

BOBER, A., and L. W. HADDAWAY
1963. Gas chromatographic identification of alcoholic beverage. J. Gas Chromat. 1(12):8–13.

BOCKIAN, A. H., R. E. KEPNER, and A. D. WEBB
1955. Skin pigments of the Cabernet Sauvignon grape and related progeny. J. Agr. Food Chem. 3:695–699.

BOCKS, S. M.
1961. Nutritional requirements for the induced formation of malic enzyme in *Lactobacillus arabinosus.* Nature 192:89–90.

BOEHM, E. W.
1966. Rapport australien. Bull. O.I.V. (Office Intern. Vigne Vin) 31:985–993.

BÖHRINGER, P., and H. DÖLLE
1959. Ueber die Eiweisstrubungen hervorrufenden Eiweissarten des Weines. Z. Lebensm.-Untersuch. -Forsch. 111:121–136.

BÖHRINGER, P., A. STÜHRK, and B. BERGDOLT
1956. Der Einfluss verschiedner Gärverfahren auf die Physiologie der Hefen und die Zusammensetzung des Weines. Weinberg Keller 3:513–526. (*See also* Mitt. Rebe u. Wein, Serie A [Klosterneuburg] 6:309–317. 1955.)

BOIDRON, J.-N., and P. RIBÉREAU-GAYON
1967. Les techniques de laboratoire appliquées à l'identification des arômes des vins. Ind. Aliment. Agr. (Paris) 84:883–893.

BOLCATO, V., F. LAMPARELLI, and F. LOSITO
1964. Azione della *B. cinerea* a del lievito sulle sostanze coloranti dei mosti d'uva. Riv. Viticolt. Enol. 17:415–421.

BOLCATO, V., C. PALLAVICINI, and F. LAMPARELLI
1965. Separazione con Sephadex degli enzimi dai mosti d'uva e dai vin. Riv. Viticolt. Enol. 18:42–48.

BOLIN, H. R., F. S. NURY, and F. BLOCH
1964. Effect of light on processed dried fruits. Food Technol. 18:1975–1976.

BOLOTOV, M. P.
1939. The content and sources of lead and copper in wines (transl.). Voprosy Pitan. 8:100–108. (Chem. Abst. 33:6520. 1939.)

BONASTRE, J.
1959. Contribution à l'étude des matières minérales dans les produits végétaux. Application au vin. Ann. Technol. Agr. 8:377–446.

BONETTI, W.
1965. Problems encountered in storage of wine under nitrogen. San Francisco, Wine Institute, Technical Advisory Committee, December 10, 1965. 2 pp.

BONNET, A.
1903. Recherches sur la structure du grain de raisin. Ann. École Nat. Agric. Montpellier 3:58–102.

BOSTICCO, A.
1966. I pigmenti antocianici dell'uva, quali fattori di accrescimento nel pollo (contributo sperimentale). Ann. Facoltà Sci. Agr. Univ. Studi Torino 1:41–51.

BOUFFARD, A.
1895. Détermination de la chaleur dégagée dans la fermentation alcoolique. Prog. Agr. Vitic. 24:345–347.

BOULARD, M.
1926. Sur un procédé permettant d'arrêter à volonté les fermentations à n'importe quel moment. Compt. Rend. Acad. Agr. France 12:615–620.

BOURDET, A., and J. HÉRARD
1958. Influence de l'autolyses des levures sur la composition phosphorée et azotée des vins. Ann. Technol. Agr. 7:177–202.

BOURZEIX, M.

1967a. L'isolement et le dosage des faibles quantités d'anthocyanes diglucosides dans les vins et les jus de raisin. Ann. Technol. Agr. 16:357–364.

1967b. Le dosage des flavonols du vin et des extraits de raisin. Ibid. 16:349–355.

BOURZEIX, M., and P. BANIOL

1966. L'isolement des flavonols du vin par chromatographie sur couche mince de cellulose. Ann. Technol. Agr. 15:211–217.

BRANAS, J.

1967. L'irrigation dans le Midi de la France. Prog. Agr. Vitic. 168:585–597.

BRANDT, K. M.

1945. The metabolic effect and the binding of carbon dioxide in baker's yeast. Acta Physiol. Scand. 10, suppl. XXX:1–206.

BRAVERMAN, J. B. S.

1953. Le mécanisme de l'action de l'anhydride sulfureux sur certains sucres. Conf. Comm. Sci. Féderation Internationale Producteurs Jus de Fruits. (See also Mechanism of the interaction of sulfur dioxide and certain sugars. J. Sci. Food Agr. 4:540–547. 1953.)

BRAVERMAN, J. B. S., and J. KOPELMAN

1961. Sugar sulfonates and their behavior. J. Food Sci. 26:248–252.

BRÉCHOT, P., J. CHAUVET, and H. GIRARD

1962. Identification des levures d'un moût de Beaujolais au cours de sa fermentation. Ann. Technol. Agr. 11:235–244.

BRÉCHOT, P., J. CHAUVET, M. CROSON, and R. IRRMANN

1966a. Configuration optique de l'acide lactique apparu au cours de la fermentation malolactique pendant la vinification. Compt. Rend., Ser. C 262:1605–1607.

1966b. Disparition de l'acide malique pendant la vinification de vendanges entières en Beaujolais (année 1965): rôle du raisin, des levures et des bactéries. Compt. Rend. Acad. Agr. France 52:582–587.

BREED, R. S., E. G. D. MURRAY, and N. R. SMITH

1957. Bergey's manual of determinative bacteriology. Baltimore, The Williams and Wilkins Co. 1094 pp. (See Acetobacter sp., pp. 183–189; Leuconostoc sp., pp. 531–533; Lactobacillus sp., pp. 541–552; Pediococcus sp., pp. 529–531.)

BREIDER, H., and E. WOLF

1966. Qualität und Resistenz. V. Ueber das Vorkommen von Biostatica in der Gattung Vitis und ihren Bastarden. Züchter 36:366–379.

BREIDER, H., E. WOLF, and A. SCHMITT

1965. Embryonalschäden nach Genuss von Hybridenweinen. Weinberg Keller 12:165–182.

BRÉMOND, E.

1937a. Contribution à l'étude analytique et physico-chimique de l'acidité des vins. Algiers, Imprimeries La Typo-Litho et Jules Carbonel Reunies. 139 pp.

1937b. L'anhydride sulfureux en oenologie. Rev. Agr. Afrique Nord 37:518–523, 533–537, 548–552.

1957. Techniques modernes de vinification et de conservation des vins en pays chauds. Paris, Edit. Maison Rustique. 296 pp.

BRÉMOND, E., and J. ROUBERT

1958. Le plomb dans les moûts et les vins. Ann. Ecole Natl. Agr. Alger. 1(1):1–11.

BRENNER, M. W., J. L. OWADES, and R. GOLYZNIAK

1955. Determination of volatile sulfur compounds. II. Further notes on hydrogen sulfide in beer. Am. Brewer 88(1):43–46. (See also Am. Soc. Brewing Chemists Proc. 1954, 81–87.)

LITERATURE CITED

BRIZA, K.
1955. Physical resistance of berries of the common types of table grapes. Personal communication.

BROCKMANN, M. C., and T. J. B. STIER
1948. Influence of temperature on the production of glycerol during alcoholic fermentation. J. Am. Chem. Soc. 70:413–414.

BROUSSILOVSKIĬ, S.
1959. Usovershenstvovanie metoda nepreryvonĭ shampanizatsii. Vinodelie i Vinogradarstvo SSSR 19(3):12–26. (See also ibid. 15(4):15–22. 1955.)

BROWN, E. M.
1950. A new off-odor in sweet wines. Proc. Am. Soc. Enol. 1950:110–112.

BROWN, G. (ed.)
1961. The X-ray identification and crystal structure of clay minerals. London, Mineralogical Society, Clay Minerals Group. 544 pp.

BROWN, W. L.
1940. The anthocyanin pigment of the Hunt Muscadine grape. J. Am. Chem. Soc. 62:2808–2810.

BRUGIRARD, A., J. ROQUES, and E. DIXONNE
1965. L'acide malique du raisin et du vin. Bull. Tech. Chamb. Agr. Pyrénées-Orientale 34:14–27.

BRUNELLE, R. L.
1967. Evaluation of gas-liquid chromatography for the quantitative determination of fusel oil in distilled spirits. J. Assoc. Offic. Anal. Chemists 50:322–329.

BRUNELLE, R. L., G. E. MARTIN, and V. G. OHANESIAN
1965. Comparative study of the anthranilic acid esters by ultraviolet spectrophotometry, colorimetry, gravimetry, and gas-liquid chromatography. J. Assoc. Offic. Agr. Chemists 48:341–343.

BRUNELLE, R. L., R. L. SCHOENEMAN, and G. E. MARTIN
1967. Quantitative determination of fixed acids in wines by gas-liquid chromatographic separation of trimethylsilylated derivatives. J. Assoc. Offic. Anal. Chemists 50(2):329–334.

BRUNI, B.
1964. Vini italiani portanti una denominazione di origine. Bologna, Edizioni Calderini. 255 pp.

BRYAN, J. M.
1948. Aluminum and aluminum alloys in the food industry. (Gt. Brit.) Dept of Scientific and Industrial Research. Food Invest. Spec. Rep. 50:1–143. London, H.M. Stationery Office.

BUĬKO, M. A.
1966. O priamougol'nykh rezervuarakh dlia vina. Vinodelie i Vinogradarstvo SSSR 26(1):35–39.

BUNDESVERBAND DER WEINBAUTREIBENDEN ÖSTERREICHS
1963. Das österreichische Weinbuch. Vienna, Verlag Austria Press. 499 pp.

BUR'IAN N. I., G. D. VODOREZ, and I. G. MAKSIMOVA
1964. O soderzhanii vitaminov gruppy B v krasnykh vinogradnykh vinakh. Trudy Vses. Nauchno-Issledov. Institut Vinodeliia i Vinogradarstva "Magarach" 13:80–83.

BURKHARDT, R.
1963. Einfache und schnelle quantitative Bestimmung der kondensierbaren Gerbstoffe in weissen Weinen and Tresterweinen. Weinberg Keller 10:274–285.

1965*a*. Nachweis der *p*-Cumarylchinasäure in Weinen und das Verhalten der Depside bei der Kellerbehandlung. Mitt. Rebe u. Wein, Serie A (Klosterneuburg) **15**:80–86.

1965*b*. Zum Hybridennachweis. Mitt. Bl. GDCh, Fachgr. Lebensmittelchemie **19**:87–88.

BURKHARDT, R., and A. LAY
1966. Bestimmung der Ascorbinsäure mit Glykolaldehyd in Most und Weissweinen neben schwefliger Säure. Mitt. Rebe u. Wein, Obstbau u. Früchteverwertung (Klosterneuburg) **16**:457–462.

BURROUGHS, L., and G. WHITING
1960. The sulphur dioxide combining power of cider. Ann. Rept. Agr. Hort. Exper. Stat. Long Ashton **1960**:144–147.

BURROUGHS, L. F., AND A. H. SPARKS
1964. Determination of free sulphur dioxide content of ciders. Analyst **89**:55–60.

BUTĂNESCU, GH.
1966. Studiul cîtorva aspecte privind tehnologia obținerii tipului de vin "Sauvignon de Drăgășani." Inst. Cercetări Horti-Viticole, Lucrări Științ. **7**:845–857.

BYRNE, J., L. G. SAYWELL, and W. V. CRUESS
1937. The iron content of grapes and wine. Ind. Eng. Chem., Anal. Ed. **9**(2):83–84.

CABEZUDO, D., C. LLAGUNO, and J. M. GARRIDO
1963. Contentido en biotina y otros componentes fundamentales en vinos de las principales zonas vinícolas de España. Agroquim. Tecnol. Alim. **3**:369–375.

CABINIS, J. E.
1962. Le brome dans les vins. Montpellier. 160 pp.

CADDELL, J. R. (ed.)
1959. Fluid flow in practice. London, Chapman & Hall, Ltd. vi + 119 pp.

CAHN, R. D.
1967. Detergents in membrane filters. Science **155**:195–196.

CAILLEAU, R., and L. CHEVILLARD
1949. Teneur de quelques vins français en aneurine, riboflavine, acide nicotinique et acide pantothénique. Ann. Agron. **19**:277–281.

CALDERON, P., J. V. BUREN, and W. B. ROBINSON
1968. Factors influencing the formation of precipitate and hazes by gelatin and condensed and hydrolyzable tannins. J. Agr. Food Chem. **16**:479–482.

CALIFORNIA CROP AND LIVESTOCK REPORTING SERVICE
1967. California fruit, grape crushing price. Sacramento. 1 p. February 15, 1967 (table 1). See also February 19, 1968 (table 1).

CALIFORNIA STATE DEPARTMENT OF PUBLIC HEALTH, BUREAU OF FOOD DRUG INSPECTION
1946. Regulations establishing standards of identity, quality, purity, sanitation, labeling and advertising of wine. Adopted May 23, 1942, and amended as of May 8, 1946. Excerpt from California Administrative Code, Title 17, Public Health, pp. 384–396. Department of Public Health, San Francisco.

CAMBITZI, A.
1947. The formation of racemic calcium tartrate in wines. Analyst **72**:542–543.

CAMERON, A. T.
1944. The relative sweetness of certain sugars, mixtures of sugar and glycerol. Can. J. Res. **22E**:45–63.

CAMPBELL, W. B., and O. MAASS
1930. Equilibrium in sulfur dioxide solutions. Can. J. Res. **2**:42–64.

LITERATURE CITED

CANO MAROTTA, C. R., and D. BRACHO DE KALAMAR
1962–1964. I lieviti della fermentazione vinaria in Uruguay: I. I lieviti della zone di El Colorado; 2. I lieviti della zona "Rincón de la Gallinas." Atti Accad. Ital. Vite Vino 14:275–285; 16:85–94.

CANT, R. R.
1960. The effect of nitrogen and carbon dioxide treatment of wines on dissolved oxygen levels. Am. J. Enol. Viticult. 11:164–169.

CANTARELLI, C.
1957. On the activation of alcoholic fermentation in wine making. Am. J. Enol. Viticult. 8:113–120, 167–175.
1958. The increase of the fermentation speed in wine making. Rev. Ferm. Ind. Aliment. 13:59–71.
1962. I trattamenti con resine poliammidiche in enologia. Atti Accad. Ital. Vite Vino 14:219–249.
1963. Prévention des précipitations tartriques. Ann. Technol. Agr. 12 (numéro hors série 1), 343–357.
1964. Il difetto da idrogeno solforato dei vini, la natura, le cause, i trattamenti preventivi e di risanamento. Atti Accad. Ital. Vite Vino 16:163–175.
1966. Il colore dei vini bianchi. Corso Nazionale di Aggiornamento per Enotecnici "G. Battista Cerletti" 2:261–283.

CANTARELLI, C., and C. PERI
1964. The leucoanthocyanins in white grapes: their distribution, amount, fate during fermentation. Am. J. Enol. Viticult. 15:146–153.

CANTARELLI, C., F. TAFURI, and A. MARTINI
1964. Chemical and microbiological surveys on the effects of dithiocarbamate fungicides on wine-making. J. Sci. Food Agr. 15:186–196.

CAPONE, R.
1963. Vini tipici e pregiate d'Italia. Rome, Editoriale Olimpia. 282 pp.

CAPPELLERI, G.
1965a. La ricerca della malvina nei vini di Vitis vinifera. Riv. Viticolt. Enol. 18:350–356.
1965b. Risultati di un'indiagine sulla ricerca della malvina in una serie di vini di Vitis vinifera. Atti Accad. Ital. Vite Vino 17:153–159.
1968. Contenuto in ferro dei vini del Veneto e Friuli-Venezia Giulia, cause di arricchimento e conseguenti fenomeni d'instabilità della limpidezza. Riv. Viticolt. Enol. 21:307–322.

CAPPELLERI, G., C. S. LIUNI, and A. CALÒ
1966. Esame dei principali vitigni da vino coltivati in Italia per mezzo della carto-cromatografia dei loro componenti antocianici. Atti Accad. Ital. Vite Vino 18:425–428.
1968. Sulla composizione dei principali vini prodotti nel Veneto. Riv. Viticult. Enol. 21:189–216.

CAPT, E., and G. HAMMEL
1953. Le traitement des vins par l'acide carbonique. Rev. romande agr., viticult., arboricult. 9:41–43, 55–57, 96–97. (See also Bull. O.I.V. (Office Intern. Vigne Vin) 26[269]:71–74; [270]:97–101. 1953.)

CAPUS, J.
1935. Protection des appellations d'origine viticoles. Compt. Rend. Acad. Agr. France 21:522–533.
1947. L'évolution de la législation sur les appellations d'origine; genèse des appellations contrôlées. In G. Lafforgue, Le vignoble Girondin. Paris, Louis Larmat. 84 + 317 pp. (See pp. 3–72.)

CAPUTI, A., JR.
1967. Personal communication.

CAPUTI, A., JR., and R. G. PETERSON
1965. The browning problem of wines. Am. J. Enol. Viticult. 16:9–13.

CAPUTI, A., JR., and M. UEDA
1967. The determination of copper and iron in wine by atomic absorption spectrophotometry. Am. J. Enol. Vitic. 18:66–70.

CARLES, J., A. ALQUIER-BOUFFARD, and J. MAGNY
1963. De quelques variations apparaissant dans le jus de raisin au cours du pressurage. Comp. Rend. Acad. Agr. France 48:773–780.

CARLES, J., M. LAMAZOU-BETBEDER, and R. PECH
1958. Les acides aminés libres du vin. Compt. Rend. 246:1254.

CARLES, J., J. LAYOLE, and A. LATTES
1966. Un nouvel acid organique du vin, l'acide diméthylglycérique. Compt. Rend. Ser. D 262:2788–2790.

CARON, P.-A.
1964. L'embouteillage, techniques et matériels. Paris, Compagnie Française d'Editions. xx + 438 pp.

CAROSSO, V. P.
1951. The California wine industry; a study of the formative years. Berkeley and Los Angeles, University of California Press. ix + 241 pp.

CARPENÈ, A.
1944. La pastorizzazione del vino spumante in bottiglia. Ital. Vinicola Agrar. 34:171–175.
1962. L'impiego dell'azota nella tecnica dello spumante. Riv. Viticolt. Enol. 15:306–320.

CARPENTER, D. C., C. S. PEDERSON, and W. F. WALSH
1932. Sterilization of fruit juices by filtration. Ind. Eng. Chem. 24:1218–1222.

CARRANTE, V., and M. PERNIOLA
1967. Contenuto in manganese di alcuni vini pugliesi. Riv. Viticolt. Enol. 20: 509–514.

CASALE, L.
1930. Ricerche fisico-chimiche sulle materie coloranti delle uve e dei vini rossi. Ann. Chim. Appl. 20:559–566.
1938. L'acide sulfureux en vinification. Vème Congrès Internatl. de la Vigne et du Vin, Lisbonne, Rapps., Tome II, Oenologie. Pp. 86–94.

CASALE, L., and E. GARINA-CANINA
1935–1937. Ricerche sugli enzimi del vino e del mosto. Ann. R. Staz. Enol. Sper. Asti (II) 2:239–250.

CASES, P.-Y.
1959. Essai de selection de levures pour un vin mousseux naturel. Ann. École Nat. Supér. Agron., Faculté Sciences Toulouse 7:47–125.

CASSIGNARD, R.
1966. La polyphénol oxydase et la vinification des moûts de raisins blancs en Bordelais. Vignes et Vins, numéro spécial 1966:13–17, 24–25.

CASTAN, P.
1927. Contribution à l'étude des levures de vin. Ann. Agr. Suisse 1927:311–319.

CASTELLI, T.
1938. Nuovi blastomiceti isolati da mosti del Chianti e zone limitrofe. Arch. Mikrobiol. 9:449–467.
1939a. Ancora sui lieviti della fermentazione vinaria nel Chianti classico. Nuovi Ann. Agr. 19(1):85–90.

LITERATURE CITED

1939b. I lieviti della fermentazione vinaria nel Chianti classico e zone limitrofe. *Ibid.* 19(1):47–84.
1941. Temperatura e chimismo dei blastomiceti. Ann. Microbiol. 2(1):8–22.
1942. Nella vinificazione con fermenti selezionati e bene attenersi sempre a culture di *Sacch. ellipsoideus*? *Ibid.* 2(4):131–134.
1947a. I lieviti della fermentazione vinaria del picemo. Ann. Facoltà di Agrar. Univ. Perugia 4:45–73.
1947b. Recenti progressi nelle ricerche zimotecniche. Congresso Nazionale Viti-vinicolo, Siena-Roma, vol. 2, Sessione di Roma, 26–30 nov. 1946. Pp. 432–437. Rome, Tipografia Operaia Romana. (*Reprinted and expanded as* La fermentazione vinarie. Rome, Stab. Tipogr. Ramo Editoriale degli Agricoltori. 8 pp.
1947c. I lieviti della fermentazione vinaria in Sicilia. Ital. Agr. 84(9):521–524.
1948a. I lieviti della fermentazione vinaria nella regione Pugliese. Ricera Sci. 18: 66–94.
1948b. Gli agenti della fermentazione vinaria in diverse regioni Italiane. Riv. Viticolt. Enol. 1(8):258–264.
1954. Fermentazione e rifermentazione nei paesi caldi. Intern. Congr. Agr. Indust. (Madrid) 5:1891–1911.
1965. Ruolo della microbiologia nell'enologia di oggi ed in quella di domani. Atti Accad. Ital. Vite Vino 17:3–13.

CASTELLI, T., and E. DEL GUIDICE
1955. Gli agenti della fermentazione vinaria nella regione Etnea. Riv. Viticolt. Enol. 8:127–141, 167–173.

CASTINO, M.
1967. Microdeterminazione colorimetrica dell'acido α-metilmalico nei vini. Riv. Viticolt Enol. 20:247–257.
1968. La presenza nei vini dell'acido 2,3-diidrossiisovalerianico. *Ibid.* 21:177–188.

CASTOR, J. G. B.
1952. Yeast spoilage of wines. Proc. Amer. Soc. Enol. 1952:139–159.
1953a. The B complex vitamins of musts and wines as microbial growth factors. Appl. Microbiol. 1:97–102.
1953b. Experimental development of compressed yeasts as fermentation starters. Wines & Vines 34(8):27–29; (9):33–34.
1956. Bacteriological test of the sterility of factory-closed, new wine bottles. Am. J. Enol. 7:137–141.

CASTOR, J. G. B., and T. E. ARCHER
1956. Amino acids in musts and wines, proline, serine and threonine. Am. J. Enol. 7:19–25.

CASTOR, J. G. B., and J. F. GUYMON
1952. On the mechanism of formation of higher alcohols during alcoholic fermenta-tion. Science 115:147–149.

CASTOR, J. G. B., K. E. NELSON, and J. M. HARVEY
1957. Effect of captan residues on fermentation of grapes. Am. J. Enol. 8:50–57.

CAUSERT, J., D. HUGOT, M. THUISSIER, E. BIETTE, and J. LECLERC
1964. L'utilisation des sulfites en technologie alimentaire: quelques aspects toxi-cologiques et nutritionnels. IV Congrès d'Expertise Chimique, Athènes spécial no. 4:215–224.

CAVANAUGH, G. C., E. J. CECIL, and K. ROBE
1961. Inert gas generator protects vegetable oil, saves $1500 per month. Food Processing 22(11):46–48.

867

CELMER, R. F.
1961. Continuous fruit juice production. *In* D. K. Tressler and M. A. Joslyn (eds.), Fruit and vegetable juice processing technology. Westport, Conn., Avi Publishing Co. (See pp. 254–277.)

CERUTTI, G.
1963. Manuale degli additivi alimentari. Milan, Et/As Kompass. 540 pp.

CHAPIDZE, E. E.
1960. O nekotory vaprosakh proizvodstva shampanskogo reservuarynym metodom. Vinodelie i Vinogradarstvo SSSR **20**(5):11–14.

CHAPON, L., and E. URION
1960. Ascorbic acid and beer. Wallerstein Lab. Commun. **23**:38–44.

CHAPPAZ, G.
1951. Le vignoble et le vin de Champagne. Paris, Louis Larmat. xiii + 414 pp.

CHARPENTIÉ, Y.
1954. Contribution à l'étude biochimique des facteurs de l'acidité des vins. Ann. Technol. Agr. **3**:89–167.

CHARPENTIÉ, Y., J. RIBÉREAU-GAYON, and E. PEYNAUD
1951. Sur la fermentation de l'acide citrique par les bactéries malolactiques. Bull. Soc. Chim. Biol. **33**:1369–1378.

CHAUDHARY, S. S., R. E. KEPNER, and A. D. WEBB
1964. Identification of some volatile compounds in an extract of the grape, *Vitis vinifera* var. Sauvignon blanc. Am. J. Enol. Viticult. **15**:190–198.

CHAUDHARY, S. S., A. D. WEBB, and R. E. KEPNER
1968. GLC investigation of the volatile compounds in extracts of Sauvignon blanc wines from normal and botrytised grapes. Am. J. Enol. Viticult. **19**:6–12.

CHAUVET, J., P. BRÉCHOT, M. CROSON, and R. IRRMANN
1966. Étude de la croissance anaérobie des levures au cours de vinifications par macération de raisin entiers. Ann. Technol. Agr. **15**:99–111.

CHAUVET, J., P. BRÉCHOT, P. DUPUY, M. CROSON, and R. IRRMANN
1963. Évolution des acides malique et lactique dans la vinification par macération carbonique de la vendange. Ann. Technol. Agr. **12**:237–246.

CHEN, L. F., and B. S. LUH
1967. Anthocyanins of Royalty grapes. J. Food Sci. **32**:66–74.

CHENARD, P. F.
1963. Règles d'utilisation de l'acide metatartrique. Ann. Technol. Agr. (numéro hors série 1) **12**:362–363.

CHINN, C. H.
1950. Effect of oxidation on the cytochrome systems of the resting cells of brewers' yeast. Nature **165**:926–927.

CHURCHWARD, C. R.
1940. Dujardin-Salleron ebulliometer. Australian Chem. Inst. J. and Proc. **7**:18–30.

CIFERRI, R., and O. VERONA
1941. Descrizione dei lieviti delle uve, dei mosti e dei vini. *In*: P. G. Garoglio, Trattato di enologia. Firenze, Stamperia Fratelli Parenti di G. Vol. 2, (*See* pp. 275–309.)

CIOFFI, R. M.
1948. Sul contenuto in alcoli superiori dei vini Italiani. Riv. Viticolt. Enol. **1**: 341–343.

CLARK, W. M.
1960. Oxidation-reduction potentials of organic systems. Baltimore, The Williams & Wilkins Co. 7 + 584 pp.

CLAUSS, W., G. WÜRDIG, and J. SCHORMUELLER
1966. Untersuchungen über das Vorkommen und die Entstehung der Schleimsäure in Traubenmosten und Weinen. I. Nachweis und Bestimmung der Schlcimsäure in Traubenmost. II. Bestimmung der Schleimsäure in Wein. Z. Lebensm.-Untersuch. -Forsch. 131:274–278, 278–280.

CLEMENS, R. A., and A. J. MARTINELLI
1958. PVP in clarification of wines and juices. Wines & Vines 39(4):55–58.

COFFELT, R. J.
1965. A continuous-crush press for the grape industry—the Serpentine fruit press. Calif. Agr. 19(6):8–9.

COFFELT, R. J., and H. W. BERG
1965a. Color extraction by heating whole grapes. Am. J. Enol. Viticult. 16:117–128.
1965b. New type of press—the Serpentine. Wines & Vines 46(4):69.

COFFELT, R. J., H. W. BERG, P. FREI, and E. A. ROSSI, JR.
1965. Sugar extraction from grape pomace with a 3-stage countercurrent system. Am. J. Enol. Viticult. 16:14–20.

COLAGRANDE, O., and G. GRANDI
1960. Contributo allo studio dei pigmenti antocianci dell'uva. Ann. Sper. Agrar. (Rome) 14(3):325–337.

COLBY, G. E.
1896. On the quantities of nitrogenous matters contained in California musts and wines. University of California, Rept. Vitic. Work during Seasons 1887–1893. Part II, pp. 422–446.

COLLOQUE SUR LA BIOCHIMIE DU SOUFRE
1956. Colloque sur la biochimie du soufre, Roscoff 14–18 Mai 1956. Paris, Editions du Centre National de la Recherche Scientifique. 244 pp.

COMMITTEE ON FOODS. ADVISORY BOARD ON QUARTERMASTER RESEARCH AND DEVELOPMENT
1957. Chemistry of natural food flavors—a symposium. Natick, Mass., Quartermaster Research and Engineering Center. 200 pp.

COMMONWEALTH ECONOMIC COMMITTEE
1961. Fruit. London, Her Majesty's Stat. Office. 219 pp. (See pp. 60–61, 173–181, 191.)

CONSTANTINESCU, G. H., I. POENARU, and V. LĂZĂRESCU
1959–1967. Ampelografia republicii populare romîne. Bucharest, Editura Academiei Republicii Populare Romîne. 8 vols.

COOK, A. H. (ed.).
1958. The chemistry and biology of yeasts. New York, Academic Press. xii + 763 pp.

COOKE, G. M.
1964. Effect of grape pulp upon soluble solids determinations. Am. J. Enol. Viticult. 15:11–16.

COOKE, G. M., and C. S. OUGH
1966. A rapid semi-quantitative reducing sugar test for dry wines. San Francisco, Wine Institute, Technical Advisory Committee, June 9, 1966. 3 pp.

TABLE WINES

CORDONNIER, R.
1953. Le fer et ses origines dans le vin. Ann. Technol. Agr. 2:1–14.
1956. Recherches sur l'aromatisation et le parfum des vins doux naturels et des vins de liqueur. *Ibid.* 5:75–110.
1961. Conséquences oenologiques du traitements des vins par le ferrocyanure de potassium. Bull. O.I.V. (Office Intern. Vigne Vin) 34(366):39–46.
1966. Etude des protéines et des substances azotées. Leur évolution au cours des traitements oenologiques. Conditions de la stabilité protéique des vins. *Ibid.* 39:1475–1489.

CORIGLIANO, F., and S. DI PASQUALE
1965. Contenuto in fluoro di vini delle provincie di Messina e Reggio Calabria. Atti Primi Convegno Regionale Aliment. e Quarto Convegno Naz. Qualità, Trieste 1965:1–11.

CORRAO, A.
1957. Sulla presenza dello xilosio fra i prodotti di idrolisi delle gomme del vino. Riv. Viticolt. Enol. 10:241–244.
1963a. Sulla determinazione colorimetrica del ferro totale nei vini. *Ibid.* 16:165–170.
1963b. Sul contenuto in manganese dei vini siciliani. *Ibid.* 16:343–349.

CORRAO, A., and A. M. GATTUSO
1964a. Contenuto in ferro e fosforo e stabilità colloidale di vini da pasto di produzione siciliana. Riv. Viticolt. Enol. 17:203–219.
1964b. L'indice de Rebelein nei vini Siciliani. I. Vini bianchi. *Ibid.* 17:531–534.

CORREIA, E. M., and A. DA C. PARRO
1967. Chromatographie et électrophorèse dans la différenciation des mousseux de provenances diverses et leur distinction des vins gazéifiés. Bull. O.I.V. (Office Intern. Vigne Vin.) 40:607–610.

CORSE, J., R. E. LUNDIN, and A. C. WAISS, JR.
1965. Identification of several components of isochlorogenic acid. Phytochem. 4(3):527–529.

CORSETTI, M.
1966. I vasi vinari in calcestruzzo di cemento armato (sidero-cemento). Ann. Fac. Sci. Agr. Univ. Studi Torino 1:389–404.

COSMO, I.
1957. Si coltiva la vite in Cina? Venezia, Fantoni. 44 pp. (*From* Agricoltura delle Venezie, February, 1957.)
1964. Romania: probabile nuovo grande paese viti-vinicolo. Atti Accad. Vite Vino 16:179–220.
1966. Guida viticola d'Italia. Treviso, Longo & Zoppelli. 244 pp.

COSTA, E. N.
1959. Investigations into measuring redox potential in wines. Am. J. Enol. Viticult. 10:56–60.
1968. A simple analog computer for blending calculations. *Ibid.* 19:84–90.

COULSON, E. J.
1937. Report on copper. J. Assoc. Offic. Agr. Chemists 20:178–188.

CRAWFORD, C. M., R. J. BOUTHILET, and A. CAPUTI, JR.
1958. A review and study of color standards for white wines. Am. J. Enol. Viticult. 9:194–201.

CROWELL, E. A., and J. F. GUYMON
1963. Influence of aeration and suspended material on higher alcohol, acetoin, and diacetyl during fermentation. Am. J. Enol. Viticult. 14:214–222.

870

CROWTHER, R. F.
1951–52. Flavours and odours from yeasts. Rept. Hort. Prod. Lab., Vineland, Ontario, Canada. Pp. 80–83.

CRUESS, W. V.
1911. The effect of sulfurous acid on fermentation organisms. J. Ind. Eng. Chem. 4:581–585.
1912. The effect of sulfurous acid on fermentation organisms. Ibid. 4:581–585.
1918. The fermentation organisms of California grapes. Univ. Calif. Publ. Agr. Sci. 4(1):1–66.
1935a. Notes on producing and keeping wines low in volatile acidity. Fruit Prod. J. 15:76–77, 108–109.
1935b. Further data on the effect of SO₂ in preventing high volatile acidity in wines. Ibid. 15:324–327, 345.
1947. The principles and practice of wine making. 2d ed. New York, Avi Publishing Co., Inc. 476 pp.
1948. Fermentation of wines at lower temperatures. Wines & Vines 29(9):19–21. (See also Wine Technol. Conf., University of California, Col. Agr., Davis, Calif., August 11–13, 1948. Pp. 90–97.) (Mimeo.)

CRUESS, W. V., and C. R. HAVIGHORST
1948. How California wines are made. Food Ind. 20:523–530.

CRUESS, W. V., and J. H. IRISH
1932. Further observations on the relation of pH to toxicity of preservatives to microorganisms. J. Bacteriol. 23:163–166.

CRUESS, W. V., P. H. RICHERT, and J. H. IRISH
1931. The effect of hydrogen ion concentration on the toxicity of several preservatives to microorganisms. Hilgardia 6(10):295–314.

CRUESS, W. V., T. SCOTT, H. B. SMITH, and L. M. CASH
1937. A comparison of various treatments of cement and steel wine tank surfaces. Food Res. 2(5):385–396.

CUISA, W., and G. BARBIROLI
1967. La misura delle caratteristiche del colore nei vini. Riv. Viticolt. Enol. 20:433–458.

CUNNINGHAM, R. C.
1947. Proper specification for usage of stainless steel to achieve sanitation and corrosion resistance in food processing. Food Technol. 1:470–477.

CURTEL, M.
1912. L'extrait sec des vins rouge de Bourgogne. Ann. Fals. Fraudes 5:33–35.

CZECHOSLOVAK COLLECTIONS OF MICROORGANISMS
1964. Catalogue of cultures. Brno. 191 pp.

DAEPP, H. U.
1966–1967. Grundlagen und Methodik der Sinnenprüfung. Schweiz. Z. Obst-Weinbau 102:611–618, 695–706; 103:12–20.

DAGHETTA, A., and G. AMELOTTI
1965. Sugli accorgimenti da osservarsi nell'applicazione del metodo Goldenberg-Geromont per la determinazione dell'acido tartarico presente nelle materie tartariche. Riv. Viticolt. Enol. 18:411–414.

DANA, E. S., and W. E. FORD
1932. A textbook of mineralogy. 4th ed. New York, John Wiley & Sons. 851 pp.

DANILATOS, N., and S. SOTIROPOULOS
1968. Étude des protéines et des substances azotées; leur évolution au cours des traitements oenologiques; conditions de la stabilité protéique des vins. Bull. O.I.V. (Office Intern. Vigne Vin) 41:468–479.

DATUNASHVILI, E. N.
1964. Vliianie razlichnykh faktorov na vydelenie drozhzhami azotistykh veshchestv. Trudy Vses. Nauchno-Issledov. Institut. Vinodeliia i Vinogradarstva "Magarach" **13**:68–77.

DATUNASHVILI, E. N., Iu. NURMEDOV, and A. I. SEĪDER
1967. Vliianie pektoliticheskikh fermentnykh prepartov na gidroliz vinogradnogo pektina. Vinodelie i Vinogradarstvo SSSR **27**(7):26–29.

DAUWE, F.
1905. Ueber die Absorption der Fermente durch Kolloide. Beit. Chem. Physiol. Path. **6**:426–453.

DAVIDIS, U. X.
1956. Rôle démographique et social de la vigne et du vin en Grèce. Athens. 15 pp. (*See also* Bull. O.I.V. [Office Intern. Vigne Vin] **31**[333]:70–79. 1958.)

DAVIES, R., E. A. FALKINER, J. F. WILKINSON, and J. L. PEEL
1951. Ester formation by yeasts. I. Ethyl acetate formation by *Hansenula* species. Biochem. J. **49**:58–61.

DAVIS, D. S.
1932. Behavior of sulfur dioxide toward water. Chem. Metal. Eng. **39**:615–616.

DAVISON, A. D.
1959. *Drosophila* control with air currents. San Francisco, Wine Institute, Technical Advisory Committee, December 4, 1959. 1 p.
1961. Review of wine industries' sanitation program. Am. J. Enol. Viticult. **12**:31–36.
1963. Wine Institute sanitation guide for wineries. San Francisco, Calif. 68 pp.

DE ALMEIDA, H.
1947. Investigação acerca des causas da possível presença do chumbo no vinho do porto. Anais Inst. Vinho Porto **8**:11–28.

DEAN, J. A.
1960. Flame photometry. New York, McGraw-Hill Book Co., Inc. 354 pp.

DECAU, J., and M. LAMAZOU-BETBEDER
1964. Étude des effets de la fertilisation boratée des vignes carencées en bore sur la vinification et sur la composition minérale des vins. Ann. Technol. Agr. **13**:19–29.

DE CORES, M., E. DE POITTEVIN, and A. CARRASCO
1966. Estudio sobre la fermentación malolàctica en vinos del Uruguay. V. Estudio del metabolismo de *Lactobacillus plantarum* (*pentosus y arabinosus*) y del *Lactobacillus buchneri* aislados de vinos y su applicación en enología. Rev. Latinoam. Microbiol. Parasitol. **8**:33–37.

DE EDS, FLOYD
1949. Vitamin P properties in grapes and grape residue. Wine Technol. Conf., University of California, Col. Agr., Davis, Calif. August 10–12, 1949. Pp. 48–50. (Mimeo.)

DE FRANCESCO, F.
1967. Il vino, la chimica, la tecnica. Vini d'Italia **9**:219–228.

DEHNER, J.
1965. Die Weinsteinstabilisierung von Schaumweinen mit Hilfe von Kationenaustauschern. Weinberg Keller **12**:403–424.

DEHORE, M., and M. DELLENBACH
1966. Nouveaux matériaux pour la fabrication et le revêtement des récipientes vinaires. Bull. O.I.V. (Office Intern. Vigne Vin) **39**:900–928.

LITERATURE CITED

DEIBNER, L.

1953. Quelques dispositifs spéciaux facilitant les mesures potentiométriques dans les vins à l'abri de l'air. Ann. Fals. Fraudes **46**:1–2. See also *Ibid.* **43**: 283–246. 1950.

1956a. Recherches sur les techniques de mesure du potentiel d'oxydo-réduction dans les jus de raisin et les vins. Détermination de ce potentiel dans quelques jus de raisin et quelques vins. Ann. Technol. Agr. **5**:31–67.

1956b. Potentiel d'oxydo-réduction et ses relations avec les propriétés organoleptiques des vins doux naturelles. *Ibid.* **3**:399–415.

1957a. Factors regulating the maturation and aging of alcoholic beverages. Am. J. J. Enol. **8**:94–104.

1957b. Potentiel oxydoréducteur des vins: son importance, sa signification; tendances actuelles de la technique de sa mesure. Ind. Aliment. Agr. (Paris) **74**:273–283.

1957c. Modifications du potentiel oxydoréducteur au cours de l'élaboration des vins de différents types. Ann. Technol. Agr. **6**:313–345.

1957d. Évolution du potentiel oxydoréducteur au cours de la maturation des vins. *Ibid.* **6**:347–362.

1957e. Effet de différents traitements sur le potentiel oxydoréducteur des vins au cours de leur conservation. *Ibid.* **6**:363–373.

1964a. Évolution de quelques éléments constitutifs dans le grain de raisin au cours de la maturation. I. Hexosamines libres et alpha-aminoacides libres. II. Acides cétoniques. Rev. Ferm. Ind. Aliment. **19**:141–146.

1964b. Évolution de quelques éléments constitutifs dans le moût de raisin rouge au cours de la fermentation alcoolique ainsi que dans le vin obtenu laissé sur lies. I. Hexosamines libres et alpha-aminoacides libres. II. Acides cétoniques. *Ibid.* **19**:201–209.

1965. Étude comparative des méthodes de dosage iodométrique et acido-complexométrique de l'anhydride sulfureux total dans les jus de raisin et les vins après séparation par distillation. Ind. Aliment. Agr. (Paris) **82**:391–399.

1966a. Dosage iodométrique de l'anhydride sulfureux libre et combiné dans les vins et les jus de raisin après séparation par entraînement à l'azote et fixation dans la solution alcaline. Chim. Anal. (Paris) **48**:66–67, 143–147.

1966b. Dosage spectrophotométrique de l'acide cyanhydrique libre et combiné dans les vins et les jus de raisin au moyen du réactif pyridinobarbiturique. *Ibid.* **48**:278–289.

1966c. Sur les particularités de mesure du potentiel oxydoréducteur de jus de raisin et des vins. Bull. O.I.V. (Office Intern. Vigne Vin) **39**:312–326.

DEIBNER, L., and C. BAYONOVE

1965. Dosage gazovolumétrique des hexosamines libres et des alpha-aminoacides libres dans les jus de raisin et les vins après leur séparation au moyen des échangeurs d'ions. Chim. Anal. (Paris) **47**:512–523.

DEIBNER, L., and P. BÉNARD

1955. Dosage de l'anhydride sulfureux total et l'ion sulfurique dans les liquides organiques. Application aux vins, aux jus de raisin et aux solutions de sucre cristallisé. Ind. Aliment. Agr. (Paris) **72**:565–573, 673–676.

1958. Emploi des alginates pour la déferrisation des vins. Ann. Technol. Agr. **7**:103–109.

DEIBNER, L., and M. BOURZEIX

1960. Sur les incertitudes dans la différenciation des cépages *Vitis vinifera* et hybrides rouges par chromatographie sur papier de leurs substances colorantes. Compt. Rend. Acad. Agr. France **46**:968.

1964. Recherches sur la détection des anthocyannes diglucosides dans les vins et les jus de raisin (par chromatographie sur papier et fluoriscopie de taches obtenues). Ann. Technol. Agr. **13**:263–282.

1965a. État actuel de la détection des anthocyannes diglucosides dans les vins et les jus de raisin. Ann. Fals. Expertise Chim. **58**:149–159.
1965b. Ueber die Extraktion der Anthocyane der Rebe. Mitt. Rebe u. Wein, Serie A (Klosterneuburg) **15**:165–177.
1966. Ueber die Gesamtextraktion der Anthocyane aus den Schalen von roten Weintrauben. Mitt. Rebe u. Wein, Obstbau u. Früchteverwertung (Klosterneuburg) **16**:200–206.

DEIBNER, L., M. BOURZEIX, and M. CABIBEL-HUGUES
1964. Séparation des anthocyannes diglucosides par chromatographie sur mince et leur dosage spectrophotométrique. Ann. Technol. Agr. **13**:359–375.
1966. Sur la valeur analytique de l'isolement des anthocyannes des vins et des jus de raisin au moyen des acétates de plomb et des résines échangeuses d'ions. Ann. Fals. Expertise Chim. **59**:39–47.

DEIBNER, L., and M. CABIBEL-HUGUES
1965. Modification de quelques éléments constitutifs des vins doux naturels au cours de leur traitement thermique prolongé à l'abri de l'air. II. Acides alpha-cetoniques. Ind. Aliment. Agr. (Paris) **82**:15–16. (See also *Ibid.* **81**:1195–1198. 1964.)
1966. Dosage électrophotométrique des acides pyruvique et alpha-cétoglutarique dans les jus de raisin et les vins après séparation de leurs dinitrophenyl-hydrazones au moyen de la chromatographie sur couche mince de cellulose. Ann. Technol. Agr. **15**:127–134.

DEIBNER, L., and J. MOURGUES
1963. Détection des ferrocyanures et de l'acide cyanhydrique dans les jus de raisin et les vins au moyen de la réaction de Moir. Ann. Technol. Agr. **12**:177–202.
1964a. Influence de quelques facteurs sur les résultats de mesures du potentiel oxydoréducteur des vins. *Ibid.* **13**:31–43.
1964b. Potentiel oxydoréducteur de quelques vins. Ind. Aliment. Agr. (Paris) **81**:1075–1080.

DEIBNER, L., J. MOURGUES, and M. CABIBEL-HUGUES
1965a. Évolution de l'indice des substances aromatiques volatiles des raisins de deux cépages rouges aux cours de leur maturation. Ann. Technol. Agr. **14**:5–14.
1965b. Sur la maturation des raisins de quatre cépages méridionaux. Ind. Aliment. Agr. (Paris) **82**:85–88, 207–211.

DEIBNER, L., and H. RIFAÏ
1958. Variations de l'indice des substances volatiles des jus de raisin au cours de leur conservation. Ann. Technol. Agr. **7**:21–29.
1963. Die Polyphenoloxydase in Traubensaft. I, II. Mitt. Rebe u. Wein, Serie A. (Klosterneuberg) **13**:56–70, 113–119.

DEIBNER, L., and G. SABIN
1962. Confrontation de quelques procédés de dosage iodométrique de l'anhydride sulfureux dans les vins. Ind. Aliment. Agr. (Paris) **79**:1057–1069.

DEINHARDT, H.
1961. Löslichkeit von Kohlensäure im Wein. Deut. Wein-Ztg. **97**:68, 70, 72.
1965. Die Löslichkeit von Kohlensäure im Wein und Sekt. Weinberg Keller **12**:428–434.

DELAUNAY, H.
1925. Hygiène des ouvriers du vin. Chim. Indus. (Paris) Spec. No., pp. 615–618.

DE LEON, J.
1959. La viticoltura in Palestina. Atti Accad. Ital. Vite Vino **11**:231–306.

LITERATURE CITED

DELISLE, A. L., and H. J. PHAFF
1961. The release of nitrogenous substances by brewers' yeast. Am. Soc. Brewing Chemists Proc. **1961**:103–118.

DELLE, P. N.
1911. The influence of must concentration on the fermentation and composition of wine and its stability. Odessa, Otchet' vinodeiel'cheskoi stantsii russkikh' vinogradarei i vinodielov' za 1908 i 1909g. Pp. 118–160.

DEMEAUX, M., and P. BIDAN
1967. Étude de l'inactivation par la chaleur de la polyphénoloxydase du jus de raisin. Ann. Technol. Agr. **16**:75–79.

DE ROSA, T.
1955. Sulla velocità di inversione del saccarosio in un vino reso spumante mediante rifermentazione in autoclave. Riv. Viticolt. Enol. 8:123–126.
1964a. Esperienze sulle sovrapressioni originantise in bottiglia in sede di pastorizzazione di un vino spumante. *Ibid.* **17**:125–130.
1964b. Tecnica dei vini spumanti. Conegliano, Tipografia Editrice F. Scarpis. xxiv + 481 pp.
1965a. Contributo alla soluzione del problema dell'eliminazione dei mercaptani dal vino. Riv. Viticolt. Enol. **18**:537–543.
1965b. Materiali per la costruzione ed il vivestimento dei recipienti vinari. *Ibid.* **18**:415–425.

DE ROSA, T., and V. BIONDO
1959. Indagine sull'impiego del PVP come chiarificante in un vino bianco. Riv. Viticolt. Enol. **12**:381–394.

DESCOMBLES, M., and J. CRESPY
1956. Emploi des matières plastique dans l'industrie vinicole. Vignes et Vins, Feuillets Techniques No. **53**:12–18.

DE SOTO, R. T.
1955. Integrating yeast propagation with winery operation. Am. J. Enol. **6**(3): 26–29.

DE SOTO, R. T., and H. YAMADA
1963. Relationship of solubility products to long range tartrate stability. Am. J. Enol. Viticult. **14**:43–51.

DE SOTO, R. T., M. S. NIGHTINGALE, and R. HUBER
1966. Production of natural sweet table wines with submerged cultures of *Botrytis cinerea* Pers. Am. J. Enol. Viticult. **17**:191–202.

DE VILLIERS, F. J.
1926. Physiological studies of the grape. The anatomy of the grape berry and the distribution of its primary chemical constituents. Union S. Africa Sci. Bull. **45**:6–15.

DE VILLIERS, J. P.
1961. The control of browning of white table wines. Am. J. Enol. Viticult. **12**: 25–30.

DICENTY, D., G. REQUINYI, S. PALINKAS, E. SZABÔ, I. SOÔS, L. RAKCSÁNYI, and E. WETTSTEIN
1935. Le point de vue hongrois. IVe Cong. Intern. Vigne Vin, Lausanne, **1**:179–192. (See pp. 189–190.)

DICKEY, G. D.
1961. Filtration. New York, Reinhold Publishing Co. 353 pp.

DICKEY, G. D., and C. L. BRYDEN
1946. Theory and practice of filtration. New York, Reinhold Publishing Corp. 346 pp.

DIDDENS, H. A., and J. LODDER
1942.   Die Hefesammlung des "Centraalbureau voor Schimmelcultures." Beiträge zu einer Monographie der Hefearten. II Teil. Die Anaskosporogenen Hefen. Zweite Hälfte. Amsterdam, Holland, N. V. Noord-Hollandsche Uitgevers-Maatschappij. 7 + 511 pp.

DIEMAIR, W., and S. DIEMAIR
1966.   Ueber das Histamin. Deut. Wein-Ztg. 102:1146–1150.

DIEMAIR, W., J. KOCH, and D. HESS
1960a.   Ueber den Einfluss der schwefligen Säure und L-Ascorbinsäure bei der Weinbereitung. I. Ueber die Bindung der schwefligen Säure an Acetaldehyde und Glucose. Z. Lebensm.-Untersuch -Forsch. 113:277–289.
1960b.   Ibid. II. Inaktivierung der Polyphenoloxydase. Ibid. 113:381–387.
1960c.   Ibid. III. Antioxydative Wirkung der schwefligen Säure und der L-Ascorbinsäure. Ibid. 114:26–38.
1961d.   Zur Bestimmung des gesamten und freien schwefligen Säure im Wein. Z. anal. Chem. 178:321–330.

DIEMAIR, W., J. KOCH, and E. SAJAK
1961a.   Zur Bestimmung des "löslichen" Proteins in Most und Wein. Z. Lebensm.-Untersuch. -Forsch. 116:5–7.
1961b.   Zur Kenntnis der Eiweissstoffe des Weines. V. Allgemeine Eigenschaften des löslichen Traubenproteins. Ibid. 116:7–13.
1961c.   Ibid. VIII. Die Eiweisstrübung. Ibid. 116:327–335.

DIEMAIR, W., and G. MAIER
1962.   Bestimmung des Eiweissgehaltes. Z. Lebensm.-Untersuch. -Forsch. 118:148–152.
1963.   Die Polypholoxydase im Traubensaft. Mitt. Rebe u. Wein, Serie A (Klosterneuburg) 13:113–119.

DIEMAIR, W., and A. POLSTER
1967.   Ueber Gerbstoffe im Rotwein. II. Eigenschaften und Isolierung. Z. Lebensm.-Untersuch. -Forsch. 134:80–86, 345–352.

DIEMAIR, W., and W. POSTEL
1967.   Nachweis and Bestimmung von Konservierungsstoffen in Lebensmitteln. Stuttgart, Wissenschaftliche Verlagsgesellschaft M.B.H. XI + 255 pp.

DIETRICH, K. R.
1954.   Die Vermeidung von Schwundlusten an Alkohol bei der Gärung. Deut. Wein-Ztg. 90:448.

DIETRICH, V.
1966.   Influence des phases préfermentatives de la vinification sur la qualité du vin. Vignes et Vins, numéro spécial, 1966:113–115, 117–119.

DILLON, C. L.
1958.   Current trends in glass technology. Am. J. Enol. Viticult. 9:59–63.

DIMOTAKI-KOURKAKOU, V.
1960.   Dosage de l'acide lactique à l'acide des échangeurs d'ions. Ann. Fals. Expertise Chim. 53:569–580.
1962.   Le présence de l'acide α-methyl-malique dans les vins. Ibid. 55:149–158.
1964a.   Absence d'acide glycuronique dans les vins. Ann. Technol. Agr. 13:301–308.
1964b.   Differenciation des mistelles d'avec les vins doux. Congrès Assoc. Intern. Expertise Chim. 4:355–359.

DIMOTAKI-KOURKAKOU, V., and S. SOTIROPOULOS
1964.   Un nouveau procédé de défécation en vue du dosage des sucres. Chim. Chron. (Athens) 29A:245–248.
1967.   Reported in literature.

LITERATURE CITED

DIMOTAKIS, P. A.
1956. Acides aminés libres dans les vins grecs. Bull. O.I.V. (Office Intern. Vigne Vin) 29(299):81–82.
1958. Determination of free amino acids in Greek wines by paper chromatography. Am. J. Enol. Viticult. 9:79–85. (See also Chim. Chron. [Athens] 20A:3–8. 1955.)

DITTRICH, H. H.
1963a. Versuche zum Aepfelsäureabbau mit einer Hefe der Gattung Schizosaccharomyces. Wein-Wissen. 18:392–405.
1963b. Zum Chemismus des Aepfelsäureabbaues mit einer Hefe der Gattung Schizosaccharomyces. Ibid. 18:406–410.
1964a. Ueber die Glycerinbildung von Botrytis cinerea auf Traubenbeeren und Traubenmosten sowie über der Glyceringehalt von Beeren- und Trockenbeerenausleseweinen. Ibid. 19:12–20.
1964b. Zur Vergärung edelfauler und hochkonzentrierter Moste. Ibid. 19:169–182.

DITTRICH, H. H., and E. KERNER
1964. Diacetyl als Weinfehler; Ursache und Beseitingung des "Milchsäuretones." Wein-Wissen. 19:528–535.
1966. Ueber die gärhemmende Wirkung der Alkylester der Gallussäure. Z. Lebensm.-Untersuch. -Forsch. 129:364–369.

DITTRICH, H. H., and T. STAUDENMAYER
1968. Die Acetaldehydbildung bei der Mostgärung und bei der Sussreservebereitung. Wein-Wissen. 23:1–7.

DITZ, E.
1965. Contribution au dosage des jus de fruits. Détermination du phosphore, du potassium, du sodium et du calcium. Ann. Technol. Agr. 14:67–78.

DODDS, D. D.
1963. Electricity in food processing operations. In M. A. Joslyn and J. L. Heid (eds.), Food processing operations, Vol. I. Westport, Conn., Avi Publishing Co. 644 pp. (See pp. 246–310.)

DOMERCQ, S.
1956. Étude et classification des levures de vin de la Gironde. Paris, Institut National de la Recherche Agronomique. 114 pp.
1957. Étude et classification des levures de vin de la Gironde. Ann. Technol. Agr. 6:5–58, 139–183.

DORIER, P., and L.-P. VERELLE
1966. Nouvelle méthode de recherche des glucosides anthocyaniques dans les vins. Ann. Fals. Expertise Chim. 59:1–10. (See also Rev. Franç. Oenol. No. 24:5–8. 1966.)

DOUGHERTY, J. D., R. D. CAMPBELL, and R. L. MORRIS
1966. Actinomycete: isolation and identification of agent responsible for musty odors. Science 152:1372–1373.

DOUGLAS, H. C., and W. V. CRUESS
1936. A Lactobacillus from California wine: Lactobacillus hilgardii. Food Res. 1:113–119.

DRABKIN, D. L.
1939. Report on copper. J. Assoc. Offic. Agr. Chemists 22:320–333.

DRAWERT, F.
1965. Ueber Inhaltsstoffe von Mosten und Weinen. V. Nachweis von biogenen Aminen im Wein und deren Bedeutung. Vitis 5:127–130.

DRAWERT, F., and A. RAPP
1966. Ueber Inhaltsstoffe von Mosten und Weinen. VI. Gaschromatographische Untersuchung der Aromastoffe des Weines und ihrer Biogenese. Vitis 5:351–376.

DRAWERT, F., and H. STEFFAN
1966. Biochemisch-physiologische Untersuchungen an Traubenbeeren. III. Stoffwechsel von zugeführten C^{14} Verbindungen und die Bedeutung des Säure-Zucker-Metabolismus für die Reifung von Traubenbeeren. Vitis 5:377–384.

DRAWERT, F., A. RAPP, and H. ULLEMEYER
1967. Radio-gaschromatographische Untersuchung der Stoffwechselleistungen von Hefen (Saccharomyces und Schizosaccharomyces) in der Bildung von Aromastoffen. Vitis 6:177–197.

DRAWERT, F., A. RAPP, and W. ULRICH
1965a. Bildung von Apfelsäure, Weinsäure und Bernsteinsäure durch verschiedene Hefen. Naturwissenschaften 52:306.

1965b. Ueber Inhaltsstoffe von Mosten und Weinen. VI. Bildung von Hexanol als Stoffwechselprodukt der Weinhefen sowie durch Reduktion von Hexen-2-al-1 während der Hefegärung. Vitis 5:195–198.

1965c. Ueber die Bildung von organischen Säure durch Weinhefen. II. Quantitative Beziehungen zwischen Stickstoffquelle und Weinsäurebildung in Modellgärversuchen. Ibid. 5:199–200.

DRBOGLAV, E. S.
1958. Uskorennoe opredelenie soderzhaniia sakhara v shampanskikh likerakh. Vinodelie i Vinogradarstvo SSSR 18(2):4–6.

DRBOGLAV, N. I.
1951. O srokakh vyderzhki shampanskikh materiolov i shampanskogo. Vinodelie i Vinogradarstvo SSSR 11(2):19–22.

DREWS, B.
1936. Ueber die Autolyse einiger Kulturhefen. Biochem. Z. 288:207–237.

DRUCKER, C.
1927. Versuche mit Diffusionselektroden am Palladium. Z. Elektrochem. 33:504–507.

DRUCKER, K.
1904. Messungen und Berechnungen von Gleichgewichten stark dissociierter Säuren. Z. physik. Chem. 9:563–589. (See also Ibid. 46:827–852. 1904.)

DUBAQUIÉ, J., and G. DÉBORDES
1935. Antiseptiques et fermentations électives. Ann. Ferment. 1(1):33–40. (See also Proc.-Verb. Soc. Sci. Phys. Nat. Bordeaux 1934–35:27–33. 1935.)

DUBOIS, P.
1955. Quelques remarques au sujet du polyéthylène et des plastiques employés dans le bouchage du vin ou des autres boissons. Vigneron Champenois 76:250–255.

1964a. Autoxydation des huiles de pépins de raisin. I. Rôle des tocophérols. Ann. Technol. Agr. 13:97–103.

1964b. Ibid. II. Présence d'un synergiste des tocophérols. Ibid. 13:105–108.

DUBOIS, P., and C. JOURET
1965. Note sur la composition quantitative de l'acidité volatile des vins de Carignan (Vitis vinifera). Comp. Rend. Acad. Agr. France 51:595–599.

DUBOURG, E.
1897. Contribution à l'étude des levures de vin. Rev. Viticult. 8:467–472.

DUBRUNFAUT, M.
1947. Sur une propriété analytique des fermentations alcoolique et lactique, et sur leur application à l'étude des sucres. Ann. Chim. Phys. (3d ser.) 21:169–178.

DUDEN, R., and I. R. SIDDIQUI
1966. Rolle der Ascorbinsäure als Inhibitor enzymatischer Bräunungsreaktionen. Z. Lebensm.-Untersuch. -Forsch. 132:1–4.

DUGAST, J.
1929. La vinification dans les pays chauds. Algers, Editions Aumeran. 396 pp.

DUHM, B., W. MAUL, H. MEDENWALD, K. PATZSCHKE, and L. A. WEGNER
1966. Zur Zenntnis des Pyrokohlensäurediäthylesters. II. Radioaktive Untersuchungen zur Klärung von Reaktionen mit Getränkesbestandteilen. Z. Lebensm.-Untersuch. -Forsch. 132:200–216.

DU PLESSIS, C. S.
1964. The ion exchange treatment (H cycle) of white grape juice prior to fermentation. II. The effect upon wine quality. S. African J. Agr. Sci. 7:3–16.

DU PLESSIS, L. DE W.
1964. The microbiology of South African wine-making. VII. Degradation of citric acid and L-malic acid by lactic acid bacteria from dry wines. S. African J. Agr. Sci. 7:31–42.

DU PLESSIS, L. DE W., and J. A. VAN ZYL
1963. The microbiology of South African winemaking. IV. The taxonomy and incidence of lactic acid bacteria from dry wines. S. African J. Agr. Sci. 6:261–273.

DUPUY, P.
1957a. Les Acetobacter du vin. Identification de quelques souches. Ann. Technol. Agr. 6:217–233.
1957b. Les facteurs du développement de l'acescence dans le vin. Ibid. 6:391–407.
1957c. Une nouvelle altération bactérienne dans les vins de liqueurs. Ibid. 6:93–102.
1960. Le métabolisme de l'acide tartrique. Ibid. 9:139–184.

DUPUY, P., and M. FLANZY
1954. Influence de la thiamine sur la fermentation du jus de raisin. Comp. Rend. Acad. Agr. France 40:273–277.

DUPUY, P., M. NORTZ, and J. PUISAIS
1955. Le vin et quelques causes de son enrichissement en fer. Ann. Technol. Agr. 4:101–112.

DURMISHIDZE, S. V.
1955. Dubil'nye veshchestva i antotsiany vinogradnoi lozy i vina. Moscow, Izdatel'stva Akademii Nauk. 324 pp.
1959. Tannins and anthocyans in the grape vine and wine. Am. J. Enol. Viticult. 10:20–28.
1965. Puti prevrashcheniia oksikislot i aminokislot vinograda v protsesse ego pererabotki. Prikl. Biokhim. Mikrobiol. 1:129–138.
1966. Contribution à l'étude de la formation et de l'évolution chimique des produits secondaires de la fermentation alcoolique. Bull. O.I.V. (Office Intern. Vigne Vin) 39:465–481.
1967. Voprosy biokhimii pererabotki vinograda; questions de biochimie du traitement du raisin. Tbilisi. 54, [3] p.

DZHURIKIANTS, N. G., V. I. VINICHENKO, and V. I. KHALINA
1966. Vliianie sakharistosti prodil'noĭ smesi na kachestvo shampanskogo. Vinodelie i Vinogradarstvo SSSR 26(2):47–50.

EDELÉNYI, M.
1964. Pezsgöborok összehasonlitó vizsgàlata (Comparative studies of sparkling wines). Borgazdasàg 12:30–32.
1966. Pezsgöstabilitási vizsgálatok (Study on the stabilisation of sparkling wines). Ibid. 14:123–129.

EDWARDS, J. W., JR., and M. A. JOSLYN
1952. Inhibition and activation of orange pectinesterase. Archiv. Biochem. Biophys. 39:51–55.

ÈGAMBERDIEV, N. B.
1967. Izuchenie dinamiki shrazhivaniia vinogradnogo susla drozhzhomi Saccharomyces vini (rasa Parkentskaia-1) v usloviiakh nepreryvnogo kul'tivirovaniia. Prikl. Biokhim. Mikrobiol. 3:458–463.

EGOROV, I. A., and N. B. BORESOVA
1955. Razdelenie i kolichestvennoe opredelenie ketokislot v vine; metodom raspredelitel'noǐ khromatografii na bumage. Doklady Akad. Nauk RSSR 104:433–435.
1956. Opredelenie ketokislot v vine (Determination of keto acids in wines). Vinodelie i Vinogradarstva SSSR 16(2):23–25.
1957. Razdelenie i kolichestvennoe opredelenie ketokislot v vine; metodom raspredelitel'noǐ khromatograffii na bumage (Separation and quantitative determination of keto acids in wines; partition paper chromatography method). Biokhim. Vinodeliia 5:253–258.

EHLERS, S.
1961. The selection of filter fabrics re-examined. Ind. Eng. Chem. 53:552–556.

EHRLICH, F.
1907. Ueber die Bedingungen der Fuselölbildung und über ihren Zusammenhang mit dem Eiweissaufbau der Hefe. Ber. Deut. Chem. Ges. 40:1027–1047.
1912. Ueber Tryptophol (B-Indolyläthylalkohol), ein neues Garprodukt der Hefe aus Aminosäuren. Ibid. 45:883–889.

EISENBRAND, J., O. HETT, and G. BECKER
1965a. Ueber den direkten Nachweis von Malvin in Lösungen mit Hilfe seiner Fluoreszenz und die Anwendung auf Rotweinverdünnungen. Deut. Lebensm.- Rundschau 61:8–11. (See also Z. anal. Chem. 125:385. 1964.)
1965b. Ueber die Rotfluoreszenz von Rotweinberdünnungen. Ibid. 61:177–181.

ELVING, P. J., B. WARSHOWSKY, E. SHOEMAKER, and J. MARGOLIT
1948. Determination of glycerol in fermentation residues. Anal. Chem. 20:25–29.

EMBS, R. J., and P. MARKAKIS
1965. The mechanism of sulfite inhibition of browning caused by polyphenol oxidase. J. Food Sci. 30:753–758.

EMEL'IANOV, V. D., G. A. ZHDANOVICH, and P. V. EREMII
1966. Ob udarnoǐ prochnosti iagod vinograda. Izv. Vysshikh Uchebn. Zavedenǐ, Pishchevaia Tekhol. 1966(6):59–64.

ENGEL, F.
1950. Beitrag zur Topographie der Microorganismen im Weingarten und im Keller unter besonderer Berücksichtigung der Infektionen im Keller. Mitt. Versuchsanstalt Gärungsgewerbe 1(5/6):63–68; (7/8):98–100; (9/10):117–118.

EPHRUSSI, B., and P. P. SLONIMSKI
1950. Effet de l'oxygène sur la formation des enzymes respiratoires chez la levure de boulangerie. Compt. Rend. 230:685–686.

LITERATURE CITED

ERCZHEGYI, L., and Á. MERCZ
1966. A szőlőfeldolgozás és sajtolás hatása a must minőségére. Borgazdaság 14: 84–88.
ESAU, P.
1967. Pentoses in wines. I. Survey of possible sources. Am. J. Enol. Viticult. 18:210–216.
ESAU, P., and M. A. AMERINE
1964. Residual sugar in wine. Am. J. Enol. Viticult. 15:187–189.
1966. Quantitative estimation of residual sugars in wine. Ibid. 17:265–267.
ESCHNAUER, H.
1961. Spurenelemente im Weinbau. Mitt. Rebe u. Wein, Serie A (Klosterneuburg) 11:123–130.
1962. Topographie der Mikronährstoffe und der entbehrlichen Spurenelemente der Weinbaugemarkung von Ober-Ingelheim (Rheinhessen). Ibid. 12: 293–314.
1963a. Aluminiumtrübungen im Wein. Vitis 4:57–61.
1963b. Bewährung der Aluminium-Herbstbütten. Wein-Wissen. 18:613–619.
1963c. Zinntrübungen im Wein. Weinberg Keller 10:523–528.
1964. Aluminium im Wein. Aluminium 40:700–703.
1965a. Beiträge zur analytischen Chemie des Weines. XIII. Bestimmung von Cadmium im Wein. Z. Lebensm.-Forsch. 127:4–10.
1965b. Ibid. XIV. Bestimmung von Nickel im Wein. Ibid. 127:268–271.
1966a. Ibid. XV. Bestimmung von Antimon im Wein. Ibid. 128:337–340.
1966b. Metallische Trübungen im Wein. Deut. Wein-Ztg. 102:450, 452.
1967. Ueber Spurenelemente im Wein. Z. Lebensm.-Untersuch. -Forsch. 134: 13–17.
ESCHNAUER, H., and G. TÖLG
1966. Beiträge zur analytischen Chemie des Weines. XVI. Bestimmung von Schwefelwasserstoff in Most und Wein. Z. Lebensm.-Untersuch. -Forsch. 128:337–340.
ETIENNE, A. D., and G. F. BREYER
1951. Determination of alcohol in wines and liqueurs. Wines & Vines 32(4):63–64.
ETIENNE, A. D., and A. P. MATHERS
1956. Laboratory carbonation of wine. J. Assoc. Offic. Agr. Chemists 39:844–848.
EYLAUD, J.-M.
1960. Vin et santé; vertus hygiéniques et thérapeutiques du vin. Soissons, La Diffusion Nouvelle du Livre. 247 pp.
FABRE, J.-H.
1929. Traité encyclopédique du vins. I. Procédés modernes de vinification. 4th ed. Paris, Dujardin Fils. 336 pp.
1938. La fermentation des moûts dans les pays chauds. Véme Cong. Intern. Vigne Vin, Lisbon. Rapports, Tome II, Œnologie. (See pp. 17–33.)
FAREY, M. G., and C. C. H. MACPHERSON
1966. A rapid flame photometric procedure for determining calcium in wine. Am. J. Enol. Viticult. 17:203–205.
FARKAS, JAN
1966. Eliminácia bielkovinných zákalov vo víne (Elimination of protein turbidities from wine), Kvasný Průmysl 12:87–89, 108–112.
FARRELL, K. R.
1966. The California grape industries: economic situation and outlook. Berkeley, University of California. 37 l.

FARRELL, K. R., and O. P. BLAICH
1964. World trade and the impacts of tariff adjustments upon the United States wine industry. Univ. Calif., Giannini Found. Agr. Econ. Res. Rept. 271:1–114.

FEDUCHY MARIÑO, E., J. A. SANDOVAL PUERTA, and T. HIDALGO ZABALLOS
1964. Contribución al estudio analítico de las dosis de metanol existentes en productos procedentes de la fermentación vínica. Bol. Inst. Nac. Invest. Agron. 51:453–484.

FEHRMANN, K., and M. SONNTAG
1962. Mechanische Technologie der Brauerei; Einrichtung für die Herstellung des Bieres vom Rohstoff bis zum Versand. Berlin, P. Parey. 538 pp.

FEJES, A.
1964. Situation actuelle et développement de la viticulture et de l'oenologie en grand. Budapesti Nemzetközi Borverseny 4:7.

FELL, G.
1961. Étude sur la fermentation malolactique du vin et les possibilités de la provoquer par ensemencement. Landwirt. Jahrb. Schweiz. 75:249–264.

FELLENBERG, TH. VON
1914. Ueber des Ursprung der Methylalkohols in Trinkbranntweinen. Mitt. Gebiete Lebensm. Untersuch. Hyg. 5:172–178.

FERENCZI, S.
1966a. A magyar borok nitrogén- es fehérjetartalmáról. Borgazdaság 14:110–116.
1966b. Étude des protéines et des substances azotées. Leur évolution au cours des traitements oenologiques. Conditions de la stabilité protéique des vins. Bull. O.I.V. (Office Intern. Vigne Vin) 39:1311–1336.
1966c. Studio comparativo dei alcune bentoniti nei riguardi della stabilizzazione proteica dei vini. Ind. Agr. 4:469–474.

FERNÁNDEZ, M. J., and M. RUIZ-AMIL
1965. Tartaric acid metabolism in Rhodotorula glutinis. Biochim. Biophys. Acta 107:383–385.

FERRÉ, L.
1925. L'analyse des moûts dans l'état actuel de nos connaissances et les renseignements d'ordre pratique qui puevent en découler. Chim. et Ind. Special No., pp. 582–588.
1926. Autolyse de la matière colorante dans les raisins entiers soumis à l'action de la chaleur humide—application à la vinification des vins rouges. Compt. Rend. Acad. Agr. France 12:370–375.
1928a. Indices oenologiques et rétrogradation de l'acide malique. Ann. Fals. Fraudes 21(230):75–84.
1928b. Vinification des vins de Bourgogne par chauffage préalable des raisins. Rev. Viticult. 69:5–11, 21–29.
1947. Sur maturation des raisins par la chaleur artificielle. Bull. O.I.V. (Office Intern. Vigne Vin) 20(196):30–37.
1958. Traité d'oenologie bourguignonne. Paris, Institut National des Appellations d'Origine des Vins et Eaux-de-Vie. VIII, 303 pp. (Also in Bulletins 57, 58, 59, 60, and 61, Institut National des Appellations d'Origine des Vins et Eaux-de-Vie.)

FERRÉ, L., and P. JAULMES
1948. Les capsules en étain plombifère causé de la présence de plomb dans les vins. Compt. Rend. Acad. Agr. France 34:864–865.

FERREIRA, J. D.
1959. The growth and fermentation characteristics of six yeast strains at several temperatures. Am. J. Enol. Viticult. 10:1–7.

FERRY, J. D.
1936. Ultra-filter membranes and ultra filtration. Chem. Rev. **18**:373-455.

FESSLER, J. H.
1941. Alcohol determination by dichromate method. Wines & Vines **22**(4):17-18.
1949. Sterile filtration of wines. *Ibid.* **30**(6):23. (*See also* Wine Rev. **17**[5]:14-15; [6]:9-10.)
1961. Erythorbic acid and ascorbic acid as antioxidants in bottled wines. Am. J. Enol. Viticult. **12**:20-24.
1952. Development of the Fessler compound. Wines & Vines **33**(7):15.
1953. Centrifugal clarification of wines versus filtration of wines. Proc. Am. Soc. Enol. Davis, August 12-14 **4**:151-154.
1966. Review and some studies of practical methods of filtering wines. Am. J. Enol. Viticult. **17**:277-282.

FESSLER, J. H., and S. S. NASLEDOV
1951. The clarification of wines with the centrifuge. Proc. Am. Soc. Enol. **1951**: 201-206.

FESSLER, J. H., J. PARSONS, and S. NASLEDOV
1949. Sterile filtration. Wine Technol. Conf., University of California, Col. Agr., Davis, Calif. August 10-12, 1949. Pp. 52-68. (Mimeo.)

FETTER, K.
1966a. Weinpumpen und ihre Anwendung in der Kellerwirtschaft. Deut. Weinbau-Jahrb. **1966**:125-146.
1966b. Dampf und seine Anwendung in Kellereien. Das Weinblatt **60**(35/36):1-3.
1968. Edelstähle in der Kellerwirtschaft Weinberg u Keller **15**:189-203. *Also in* Fluss. Obst **4**:1-7.

FEUILLAT, M., and J. BERGERET
1966. Étude de quelques facteurs conditionnant la filtration des vins nouveaux. Ann. Technol. Agr. **15**:79-97.
1967a. Identification et dosage des aminoacides dans les moûts et les vins de Bourgogne. Compt. Rend. **264**:1757-1759.
1967b. Influence des traitements thermiques de la vendange sur l'extraction des constituants azotes. *Ibid.* **264**:2520-2523.

FILAUDEAU, G.
1912a. Les vins de la récolte 1911. Ann. Fals. Fraudes **5**:36-44.
1912b. Les vins de l'Yonne de la récolte 1911. *Ibid.* **5**:230-236.
1916. Résumé de l'enquête annuelle sur la composition des vins de consommation courante. *Ibid.* **9**:347-413.

FILIPELLO, F.
1955. Small panel taste testing of wine. Am. J. Enol. **6**(4):26-32.
1957a. Organoleptic wine-quality evaluation. I. Standards of quality and scoring vs. rating scales. Food Technol. **11**:47-51.
1957b. *Ibid.* II. Performance of judges. *Ibid.* **11**:51-53.

FILIPELLO, F., and H. W. BERG
1959. The present status of consumer tests on wine. Am. J. Enol. Viticult. **10**:8-12.

FILIPPOV, B. A.
1963. Poluchenie fermentnykh kontsentratov i ikh primenenie (Production of enzyme concentrates and their use). Vinodelie i Vinogradarstvo SSSR **23**(2):11-14.
1965. Primenenie metavinnoï kisloty v proizvodstve shampanskogo. *Ibid.* **25**(8): 39-40.

FITELSON, J.
1967. Paper chromatographic detection of adulteration in Concord grape juice. J. Assoc. Offic. Anal. Chemists **50**:293-299.

FLANZY, C., and P. ANDRÉ
1965. Note sur le métabolisme des raisins en situation d'anaérobiose. Ann. Technol. Agr. 14:173–178.

FLANZY, C., P. ANDRÉ, M. FLANZY, and Y. CHAMBROY
1967. Variations quantitatives des acides organiques stables, non–cétoniques, non-volatils, dans les baies de raisin placées en anaérobiose carbonique. I. Influence de la température. II. Influence de la durée d'anaérobiose. Ann. Technol. Agr. 16:27–34, 89–107.

FLANZY, C., and C. POUX
1965a. Les levures alcooliques dans les vins. Protéolyse, protéogenèse (III). Ann. Technol. Agr. 14:35–48.
1965b. Note sur la teneur en acides aminés du moût de raisin et du vin en fonction des conditions de l'année (maturation et fermentation). Ibid. 14:87–91.

FLANZY, C., C. POUX, and M. FLANZY
1964. Les levures alcooliques dans les vins. Protéolyse et protéogenèse. Ann. Technol. Agr. 13:283–300.

FLANZY, J., and M. FLANZY
1959. Note sur la valeur de l'huile de pépins de raisin en acides gras essentiels. Ann. Technol. Agr. 1:107–111.

FLANZY, M.
1935. Nouvelle méthode de vinification. Compt. Rend. Acad. Agr. France 21: 935–938. (See also Rev. Viticult. 83:315–319, 325–329, 341–347.)
1959. Recherches sur la vinification des vins doux naturels. Ann. Technol. Agr. 8:285–320.
1961. Action physiologique sur l'homme des vins préparés ou conservés suivant divers procédés. Bull. O.I.V. (Office Intern. Vigne Vin) 34(367):35–56.
1965. Comment définir et mesurer les caractères organoleptiques des boissons. Ann. Nutr. Aliment. 19:A235–A264.

FLANZY, M., and L. BOUZIGUES
1959. Pectines et méthanol dans les moûts de raisin et les vins. Ann. Technol. Agr. 8:59–67.

FLANZY, M., J. CAUSERET, D. HUGOT, and J. GUERELLOT
1955. Contribution à l'étude des boissons alcooliques. II. Étude comparée de différents vins et de diverses techniques oenologiques. Ann. Technol. Agr. 4:359–380.

FLANZY, M., and Y. COLLON
1962a. Sur la présence de l'hydroxyméthylfurfural dans certains jus de raisin. Ann. Technol. Agr. 11:227–233.
1962b. Sur l'origine de l'hydroxyméthylfurfural de certains jus de raisin. Ibid. 11:271–273.

FLANZY, M., and L. DEIBNER
1956. Sur la variation des teneurs en fer dans les vins, obtenus en présence ou en absence d'une terre ferrugineuse. Ann. Technol. Agr. 5:69–73.

FLANZY, M., and P. DUBOIS
1964. Étude du dosage des tocophérols totaux. Application à l'huile de pépins de raisin. Ann. Technol. Agr. 13:67–75.

FLANZY, M., P. DUPUY, C. POUX, and P. ANDRÉ
1966. Fermentation du jus de raisin en continu. Ann. Technol. Agr. 15:311–320.

FLANZY, M., and Y. LOESEL
1958. Évolution des pectines dans les boissons et production du méthanol. Ann. Technol. Agr. 7:311–321.

FLANZY, M., and A. OURNAC
1963. Fermentation des jus de raisins frais et désulfités. Influence d'additions de levures et de thiamine. Ann. Technol. Agr. 12:65–84.

FLANZY, M., and L. THÉROND
1939. Le manganèse dans les vins de *Vitis vinifera* et les vins d'hybrides. Riv. Viticult. 90:433–437, 454–459.

FLESCH, P., and B. HOLBACH
1965. Zum Abbau der L-Aepfelsäure durch Milchsäurebakterien. I. Ueber die Malat-abbauenden Enzyme des Bakterium "L" unter besonderer Berücksichtigung der Oxalessigsäure-Decarboxylase. Arch. Mikrobiol. 51:401–413.

FLORENZANO, G.
1949. La microflora blastomiceta dei mosti e dei vini di alcune zone Toscane. Ann. Sper. Agrar. n.s. 3(4):887–918.

FLOYD, W. W., and G. S. FRAPS
1939. Vitamin C content of some Texas fruits and vegetables. Food Res. 4:87–91.

FORBES, P.
1967. Champagne, the wine, the land and the people. London, Victor Gollancz. 492 pp.

FORNACHON, J. C. M.
1938. Bacterial fermentations in fortified wines. Adelaide, Australian Wine Board. 19 pp. (Mimeo.)
1943. Bacterial spoilage of fortified wines. Adelaide, Australian Wine Board. ix + 126 pp.
1950. Yeast cultures. Aust. Brew. and Wine J. 69(3):32.
1953. The accumulation of acetaldehyde by suspensions of yeasts. Australian J. Biol. Sci. 6:222–233.
1957. The occurrence of malo-lactic fermentation in Australian wines. Australian J. Appl. Sci. 8:120–129.
1963a. Inhibition of certain lactic acid bacteria by free and bound sulfur dioxide. J. Sci. Food Agr. 14:857–862.
1963b. Travaux récents sur la fermentation malolactique. Ann. Technol. Agr. 12(numéro hors série 1):45–53.
1964. A *Leuconostoc* causing malo-lactic fermentation in Australia wines. Am. J. Enol. Viticult. 15:184–186.
1968. Influence of different yeasts on the growth of lactic acid bacteria in wine. J. Sci. Food Agr. 19:374–378.

FORNACHON, J. C. M., and B. LLOYD
1965. Bacterial production of diacetyl and acetoin in wine. J. Sci. Food Agr. 16:710–716.

FOULONNEAU, M. C.
1967. De l'influence de certaines conditions d'ambiance sur la dégustation des vins. Vignes et Vins, numéro spécial 1967:65–75.

FRANÇOT, P.
1945. Acide total et acidité fixé réele des moûts et des vins de Champagne. Bull. O.I.V. (Office Intern. Vigne Vin) 18(167–170):114–118.
1950. Champagne et qualité par le pressurage. Différenciations organoleptiques, biologiques et chimiques des divers moûts et vins provenant de la fragmentation du pressurage champenois. Vigneron Champenois 71:250–255, 273–283, 342–351, 371–382, 406–416.

FRANÇOT, P., and P. GEOFFROY
1951. Les pectines et les gommes dans les moûts et les vins de Champagne. Vigneron Champenois 72(2):54–59.

FRANÇOT, P., P. GEOFFROY, and J. PERIN
1957. Résultats au premier essai de pressurage destiné à comparer le pressoir horizontal "Pneumabilpress" au pressoir classique champenois entrepris au cours des vendanges 1956. Vigneron Champenois 78:268–274.

FRANZEN, K.
1968. SO₂-Bildung und Schwefelbedarf des 1967er. Deut. Wein-Ztg. 104:123–124.

FRATEUR, J.
1950. Essai sur la systématique des *Acetobacter*. La Cellule 53:287–392.

FRAZIER, W. C.
1967. Food microbiology. 2d ed. New York, McGraw-Hill Book Co. xii + 537 pp. (*See* pp. 82–108.)

FROLOV-BAGREEV, A. M.
1948. Sovetskoe shampanskoe, tekhnologiia proizvodstva shampanskikh (igristykh) vin (Soviet sparkling wine; technology of production of sparkling wines). 2d ed. Moscow, Pishchepromizdat. 271 pp.
1952. O formakh uglekisloty v shampanskom. Vinodelie i Vinogradarstvo SSSR 12(6):20–21.

FROLOV-BAGREEV, A. M., and G. G. AGABAL'IANTS
1951. Khimiia vina. Moscow. Pishchepromizdat. 392 pp.

FROLOV-BAGREEV, A. M., and E. G. ANDREEVSKAIA
1950. Oroli mikroelementov v vinodelii. Vinodelie i Vinogradarstvo SSSR 10(6): 38–40.

FÜGER, A.
1957. Oesterreichische Spätleseweine. Mitt. Rebe u. Wein, Serie A (Klosterneuburg) 7:67–70.

FULLER, W. L., and H. W. BERG
1965. Treatment of white wine with Nylon 66. Am. J. Enol. Viticult. 16:212–218.

FUNK, L. P., and J. H. PRESCOTT
1967. Study shows wine aids patients. Modern Hospital 108:182–184.

GALET, P.
1956–1964. Cépages et vignobles de France. I. Les vignes américaines. II. Les cépages de cuve. III. Les cépages de table. IV. Les raisins de table; la production viticole française. Montpellier, Paul Déhan. 4 vols.

GALLAY, R.
1966. Rapport Suisse. Bull. O.I.V. (Office Intern. Vigne Vin) 39:993–1002.

GALLAY, R., and A. VUICHOUD
1938. Premiers essais de vinification en rouge d'après la méthode Flanzy. Rev. Viticult. 88:238–242.

GALLO, E.
1958. Outlook for a mature industry. Wines & Vines 39(6):27–28, 30.

GALZY, P.
1956. Nomenclature des levures du vin. Ann. Technol. Agr. 5:473–491.

GALZY, P., and J.-A. RIOUX
1955. Observations sur quelques vins atteints par la maladie "la fleur" dans le "Midi" de France. Prog. Agr. Vitic. 144:365–370.

GANCEDO, C., J. M. GANCEDO, and A. SOLS
1968. Glycerol metabolism in yeasts. Pathways of utilization and production. Eur. J. Biochem. 5:165–172.

GANDINI, A., and A. TARDITI
1966. Vinificazione di mosti piemontesi con lieviti in associazione controllata e scalare. Ind. Agr. 4:411–420.

GARINO-CANINA, E.
1928. Le sostanze pectiche e la tecnica enologica. Ann. R. Accad. Agr. Torino 71:49–68.
1935. Il potenziale di ossidoriduzione e la tecnica enologica. Ann. Chim. Appl. 25:209–217.
1948. Fermentation "vinaire" avec des détails biochimiques du processus de la fermentation. Bull. O.I.V. (Office Intern. Vigne Vin) 21(204):55–61.
1953. Ricerche chemico-biologiche sulla Vernaccia di Sardegna. Atti Accad. Ital. Vite Vino 5:1–22.

GAROGLIO, P. G.
1963. Proposta di una diagnostica enzimatica per alcuni componenti principali e secondari dei mosti, vini e succhi di frutta. Riv. Viticolt. Enol. 16:155–157.

GAROGLIO, P. G., and G. BODDI-GIANNARDI
1967. Contributo critici sperimentali sulla formazione e la ricerca dei prodotti furilici risultanti dal riscaldamento dei mosti e dei vini e valutati come caramello. Proposta di un indice (numero) di caramello. Riv. Viticolt. Enol. 20:147–164.

GAROGLIO, P. G., and V. SALATI
1967. Primo contributo allo studio sperimentale della composizione in potassio e sodio su vini di varie regioni d'Italia ed allo studio della natura ed entità delle variazioni derivanti dai trattamenti in cantina. Riv. Viticolt. Enol. 20:70–78.

GAROGLIO, P. G., and C. STELLA
1964. Ricerche sull'impiego in enologia dell'estere dietilico dell'acido pirocarbonico (DEPC) allo stato puro. Riv. Viticolt. Enol. 17:380–394, 422–453.
1965. Cloro-organico componente naturale dei mosti d'uva botritizzati. Ibid. 18:373–378.

GARSCHAGEN, H.
1967. Ueber die Bestimmung von Diäthylcarbonat in mit Pyrokohlensäure-diäthylester behandelten Weinen. Weinberg Keller 14:131–136.

GÄRTEL, W.
1967. Action sur la vigne des produits chimiques utilisés dans la lutte contre les parasites et ses conséquences. Bull. O.I.V. (Office Intern. Vigne Vin) 40:1342–1351.

GASIUK, G. N., M. I. ZELENSKAIA, and A. E. BELOKON
1965. The influence of washing grapes on the content of copper in grape juice (transl.). Trudy Moldavsk. Nauchn.-Issled. Inst. Pischevoi Prom. 6:42–44. (Chem. Abst. 65:1294. 1966.)
1966. O soderzhanii medi v vinogradnom soke. Vinodelie i Vinogradarstvo SSSR 26(4):13–14.

GATET, L., and L. GENEVOIS
1941. Sur le pouvoir réducteur des vins. Bull. Soc. Chim. France 8(5):485–487.

GAYON, U., and E. DUBOURG
1890. Sur la fermentation alcoolique du sucre interverti. Compt. Rend. 110:865–868.
1894. Sur les vins mannités. Ann. Inst. Pasteur 8:108–116.
1901. Nouvelles recherches sur le ferment mannitique. Ibid. 15:527–569.

GEHMAN, H., and E. M. OSMAN
1954. The chemistry of the sugar sulfite-reaction and its relationship to food problems. Advan. Food Res. 5:53–96.

GEISS, W.
1952. Gezügelte Gärung. Frankfurt, Sigurd Horn Verlag. 69 pp.
1962a. Biologische Kontrolle kaltsteril abgefüllter Weine (nach Prof. Zsigmondy). Bad Kreuznach, Seitz-Werke. 20 pp.
1962b. Flaschenverschlüsse für Wein, Perlwein, Sekt, Spirituosen, Trauben- und Obstsäfte. Deut. Wein-Ztg. 98:296–302.
1963a. Kohlensäure im Stillwein. Ibid. 98:314–315.
1963b. La filtration des vins. Filtration avec alluvionnage continu et filtration sur plaques stérilisantes. Ann. Technol. Agr. 12(numéro hors série 1):205–214.
1965. Die Bedeutung des Luftsäuerstoffes bei der Abfüllung von Wein. Deut. Wein-Ztg. 101:238, 240, 242.
1966. L'importanza dell'ossigeno per l'imbottigliamento del vino. Vini d'Italia 44:269–273.

GELOSO, J.
1930–31. Relation entre le vieillissement des vins et leur potentiel d'oxydo-réduction. Ann. Brass. Distill. 29:177–181, 193–197, 257–261, 273–277. (See also Chim. Indus. [Paris] 27:430–431. 1932.)

GENDRON, C.
1967a. Recherches sur la fermentation malolactique, Thèse, Faculté des Sciences, Université de Nantes. 279 pp.
1967b. Le pressurage des raisins à la vendange. Rev. Franç. Oenol. 7(27):26–34.

GENEVOIS, L.
1936. L'énergétique des fermentations. Ann. Ferment. 2:65–78.
1950. Essais de bilans de la fermentation alcoolique dû aux cellules de levures. Biochim. Biophys. Acta 4:179–192. (Reprinted in Metabolism and function, ed. D. Nachmansohn. New York, Elsevier Publishing Co., Inc., 1950. 348 pp.)
1961. Die Sekündarprodukte des alkoholischen Gärung. Brauwissen. 14:52–55.

GENEVOIS, L., and H. FLAVIER
1939. Dosage des vitamines B_1 et B_2 dans quelques vins de la Gironde. Proc.-Verb. Soc. Sci. Phys. Nat. Bordeaux 1938–39:72–73.

GENEVOIS, L., E. PEYNAUD, and J. RIBÉREAU-GAYON
1946. Sur un bilan des produits secondaires de la fermentation alcoolique. Compt. Rend. 223:693–695.
1948a. Action du milieu sur les produits secondaires de la fermentation alcoolique des levures elliptiques. Ibid. 226:126–128.
1948b. Bilans des produits secondaires de la fermentation alcoolique dans les vins rouges de la Gironde. Ibid. 226:439–440.

GENEVOIS, L., and J. RIBÉREAU-GAYON
1936. Sur les substances azotées des moûts et des vins. Ann. Ferment. 1:541–546.
1947. Le vin. Paris, Hermann & Cie. 150 pp.

GENTH, H.
1964. On the action of diethylpyrocarbonate on micro-organisms. In International Symposium on Microbial Inhibitors in Food, 4th Göteborg, Sweden. Stockholm, Almqvist and Wiksell. 402 pp. (See pp. 77–85.)
1965. Baycovin zur Kaltentkeimung von Flaschenwein. Weinberg Keller 12:499–504.

GENTILINI, L.
1966. Applicazioni della luce di Wood in campo enologico. Riv. Viticolt. Enol. 19:275–306.

GENTILINI, L., and G. CAPPELLERI
1966. A proposito della sofisticazione con vini di I.P. Riv. Viticolt. Enol. 19:438–453.

GEOFFROY, P.
1963. Clarification et stabilisation des vins de Champagne. Ann. Technol. Agr. 12(numéro hors série 1): 366–375.
1965. Les applications du froid en oenologie champenoise. Vigneron Champenois 86: 257–265.

GEOFFROY, P., and J. PERIN
1960. Les alginates en tant qu'adjuvants de remuage. Vigneron Champenois 81: 169–175.
1965. Influence de la concentration en levures sur la prise de mousse. Ibid. 86: 303–308.
1966. Le décoloration des moûts et des vins tachés (Action du noir et levures). Ibid. 87: 416–426.

GEORGEAKOPOULOS, G., V. DIMOTAKI-KOURKAKOU, and N. LYDAKI-KRIESE
1963. La teneur des vins en acides uroniques. Chim. Chronika (Athens) 28A: 161–168.

GÉORGIEV, I., I. POPOV, T. A. TONTCHEV, and S. MANTCHEV
1956. De l'effet d'un produit enzymatique de Botrytis cinerea sur l'accelération de la maturation du vin. Prog. Agr. Vitic. 146: 211–220, 237–244.

GEORGIEV, L., L. POPOV, and V. LITCHEV
1950–51. Esterification and oxidation reduction potential in native wines in relation to their maturation and their aging (transl.). Ann. Inst. Super. Agr. V. Kolarov 6: 1–31

GERASIMOV, M. A., and E. S. KULESHOVA
1965. Change in the content of B group vitamins by the treatment of grape wines with absorbents. Prikl. Biokh. Mikrobiol (English ed.) 1: 697–706.

GERASIMOV, M. A., and E. S. KULESHOVA
1965. Deistvie ionitov na vitaminy vina. Vinodelie i Vinogradarstvo SSSR 25(8): 4–8.

GETOW, G., and G. PETKOW
1966. Nachweis der Anwesenheit von Malvin in Vitis-Vinifera-Sorten. Mitt. Rebe u. Wein, Obstbau u. Früchteverwertung (Klosterneuburg) 16: 207–210.

GILROY, P. E., and F. A. CHAMPION
1948. Aluminum and fruit juices. J. Soc. Chem. Ind. 67: 407–410.

GINI, B., and R. H. VAUGHN
1962. Characteristics of some bacteria associated with spoilage of California dessert wines. Am. J. Enol. Viticult. 13: 20–31.

GIRARD, A., and B. LINDET
1898. Sur le phlobaphène du raisin. Bull. Soc. Chim. France, sér. 3, 19: 583–586. (See also Bull. Min. Agr. France 14: 694–782. 1895.)

GIUDICE, E. DEL
1955. Intorno all'azione dell'aneurina sul processo di fermentazione vinaria. Riv. Viticolt. Enol. 8: 311–315.

GLONINA, N. N., and S. P. AVAKIANTS
1966. Kolichestvennyĭ analiz aminokislot metodom khromatografii na ionoobmen-nykh smolakh. Vinodelie i Vinogradarstvo SSSR 26(4): 4–6.

GLONINA, N. N., and L. V. DUBINCHUK
1967. Izuchenie prevrashcheniĭ alanina-C¹⁴ i asparaginovoĭ kisloty-C¹⁴ pri shampanizatsii vina. Prikl. Biokh. Mikrobiol. 3: 78–85.

GMELIN INSTITUT FÜR ANORGANISCHE CHEMIE UND GRENZGEBIETE IN DER MAX-PLANK GESSELSCHAFT ZUR FORDNUNG DER WISSENSCHAFTEN
1960. Gmelins Handbuch der Anorganischen Chemie. 8th ed. Schwefel. Teil B, Lieferuna 2, System Numm. 9. Berlin, Weinheim, Verlag Chemie. (See pp. 463–468.)

GOETZ, A., and N. TSUNEISHI
1951. Application of molecular filter membranes to the bacteriological analysis of water. J. Am. Water Works Assoc. 43:943–969.

GÖTZ, B.
1966. Mikrolepidopteren als Korkschädlinge. Wein-Wissen. 21:263–274.

GOLDMAN, M.
1963a. Factors influencing the rate of carbon dioxide formation in fermented-in-the-bottle champagne. Am. J. Enol. Viticult. 14:36–42.
1963b. Rate of carbon dioxide formation at low temperatures in bottle-fermented champagne. Ibid. 14:155–160.

GOLDSTEIN, J. L., and T. SWAIN
1965. The inhibition of enzymes by tannins. Phytochem. 4:185–192.

GOLODRIGA, P. IA., and KH. PU-CHAO
1963. Rannespelost' vinograda i nekotorye biokhimicheskie pokazateli. Trudy Vses. Nauchno-Issled. Inst. Vinodeliia i Vinogradarstva "Magarach" 12:74–83.

GOMBKÖTÖ, G.
1965. Anthocyaninfarbstoffe der Weintrauben. III. Einige neue Verfahren zur chromatographischen Trennung von Anthocyaninen. Kerteszet. Szőlész. Főiskola Közlemények 29:171–182.

GOMEZ, R., and P. GALZY
1966. Observations sur la flore lactique d'un cellier. Ann. École Nat. Sup. Agr. Montpellier 31(1):117–131.

GOOR, A.
1966. The history of the grape-vine in the holy land. Econ. Bot. 20:46–64.

GORANOW, N., Z. GANEWA, and W. LITSCHEW
1965. Veränderung des Redoxpotentials der Weine bei Stabiliesierung mit Kationenaustauscherharzen. Mitt. Rebe u. Wein Serie A (Klosterneuburg) 15:135–139.

GOTTLIEB, L. S., L. A. BROITMAN, J. J. VITALE, and N. ZAMECHECK
1959. The influence of alcohol and dietary magnesium upon cholesterolemia and atherogenesis in the rat. J. Lab. Clin. Med. 53:433–441.

GOTTSCHALK, A.
1946. Mechanism of selective fermentation of d-fructose from invert sugars by Sauternes yeast. Biochem. J. 40:621–626.

GOWANS, W. J.
1965. Total volatile acidity in wine. J. Assoc. Offic. Agr. Chemists 48:473–474.

GRAY, P. P., and I. STONE
1939a. Oxidation in beers. I. A simplified method for measurement. Wallerstein Lab. Commun. 2(1):5–16.
1939b. Ibid. II. Oxidation stability of the finished beer. Ibid. 2(6):25–34. Also in J. Inst. Brewing 45:443–452. 1939.

GRAY, W. D.
1941. Studies on the alcohol tolerance of yeasts. J. Bacteriol. 42:561–574.
1945. The sugar tolerance of four strains of distillers yeast. Ibid. 49:445–452.
1946. The acclimatization of yeast to high concentrations of glucose; the subsequent effect upon alcohol tolerance. Ibid. 52:703–709.
1949. Initial studies on the metabolism of Hansenula anomala (Hansen) Sydow. Am. J. Botany 36:475–480.

GREENFIELD, J. R.
1965. Microbiological controls in the brewery: use of chlorine, bromine, and alkaline compounds. Am. Soc. Brewing Chemists Proc. 1965:72–83.
1967. Modern sanitation for wineries. Wines & Vines 48(4):61–62.

GUALANDI, G., G. MORISI, G. UGOLINI, and E. B. CHAIN
1959. Continuous measure of dissolved oxygen in fermentations in the presence of microorganisms. Selected Scientific Papers, Instituto Superiore di Sanità 2(1):4–49.

GUIMARÃES, A. F.
1965. O álcool metílico no vinho do porto e noutras bebidas alcoólicas. De Vinea et Vino, Portugaliae Documenta, série II 2(5):1–12.

GUIMBERTEAU, G., and E. PEYNAUD
1965. Recherche et estimation du saccharose ajouté aux moûts et aux vins à l'aide de la chromatographie sur papier. Ann. Fals. Expertise Chim. 58:32–38.
1966. Comparison de quelques méthodes de dosage de l'acide lactique dans les vins. Ann. Technol. Agr. 15:303–309.

GUINOT, Y., and Y. MÉNORET
1965. Le chauffage continu du raisin avant pressurage. Compt. Rend. Acad. Agr. France 51:866–872.

GUSTAVSON, K. H.
1949. Some protein-chemical aspects of the tanning process. Advan. Protein Chem. 5:353–421.
1954. Interaction of vegetable tannins with polyamides as proof of the dominant function of the peptide bond of collagen for its binding of tannins. J. Polymer Sci. 12:317–324.
1956. The chemistry of tanning processes. New York, Academic Press. 403 pp. (See pp. 142–201.)

GUYMON, J. F.
1964. Studies of higher alcohol formation by yeasts through gas chromatography. Qualitas Plant. Mater. Végétabiles 11:194–201.

GUYMON, J. F., and E. A. CROWELL
1963. Determination of aldehydes in wines and spirits by the direct bisulfite method. J. Assoc. Offic. Agr. Chemists 46:276–284.
1965. The formation of acetoin and diacetyl during fermentation, and the levels found in wines. Am. J. Enol. Viticult. 16:85–91.

GUYMON, J. F., and J. E. HEITZ
1952. The fusel oil content of California wines. Food Technol. 6:359–362.

GUYMON, J. F., J. L. INGRAHAM, and E. A. CROWELL
1961. Influence of aeration upon the formation of higher alcohols by yeasts. Am. J. Enol. Viticult. 12:60–66.

GUYMON, J. F., and D. L. WRIGHT
1967. Determination of aldehydes in wines by the direct bisulfite method: American Society of Enologists—AOAC joint report. J. Assoc. Offic. Anal. Chemists 50:305–307.

GUYOT, A. M., and P. BALATRE
1966. Le brome total dans les boissons. Ann. Fals. Expertise Chim. 59:329–335.

GVALADZE, V.
1936. Relation between the products of alcoholic fermentation (transl.). Moscow Lenin Agr. Acad. USSR. 76 pp.

HAAGEN-SMIT, A. J., F. N. HIROSAWA, and T. H. WANG
1949. Chemical studies on grapes and wines. I. Volatile constituents of Zinfandel grapes (Vitis vinifera var. Zinfandel). Food Res. 14:472–480.

HABALA, I., and V. ŠVEJCAR
1965. Vztah různých druhà kvasinek v moště a ve víně. Vinohrad 58:155.

HAILER, E.
1911. Versuche über die Entwicklungshemmenden und Keimtötenden Eigenschaften der freien schwelfigen Säure, der schwefligsäuren Salze und einiger komplexer Verbindungen der schwefligen Säure. Arb. Gesundheitsamte **36**:297–340.

HALL, A. P., L. BRINNER, M. A. AMERINE, and A. F. MORGAN
1956. The B vitamin content of grapes, musts and wines. Food Res. **21**:362–371.

HALLGARTEN, S. F.
1965. Rhineland wineland. Manchester, Withy Grove Press Limited. 319 pp.

HALTER, P.
1959. Biologische Untersuchungen fabrikneuer Weinflaschen. Schweiz. Z. Obst-Weinbau **68**:211–213, 241–245.
1960. Biological investigations of new wine bottles. Am. J. Enol. Viticult. **11**:15–18.

HAMELLE, G.
1965. L'extrait sec des vins et des moûts de raisin; sa mesure, son intérêt pour la recherche des fraudes. Montpellier. 170 pp.

HARBORNE, J. B.
1967. The comparative biochemistry of the flavonoids. New York, Academic Press. viii + 383 pp.

HARDEN, A.
1932. Alcoholic fermentation. 4th ed. London, Longmans, Green and Co. 243 pp. (*See esp.* pp. 192–194.)

HARDY, A. C.
1936. Handbook of colorimetry. Cambridge, The Technology Press. 87 pp.

HARRIS, G.
1958. Nitrogen metabolism. *In* A. H. Cook (ed.), The chemistry and biology of yeasts. New York, Academic Press. (*See* pp. 437–533.)

HARRIS, M. B.
1964. Grape juice clarification by filtration. Am. J. Enol. Viticult. **15**:54–62.

HARRISON, G. A. F.
1963. Investigations on beer flavour and aroma by gas chromatography. Proc. European Brewery Conv. **1963**:247–256.

HARTONG, B. D.
1929. Gerbstoffbestimmungen. Woch. Brauerei **46**:11–15.

HARVALIA, A.
1965. La couleur des vins rouges. Chim. Chronika (Athens) **20**(9):155–159.

HASLAM, E.
1966. Chemistry of vegetable tannins. London, Academic Press. 179 pp.
1967. Gallotannins. XIV. Structure of the gallotannin. Jour. Chem. Soc. (C) **1967**:1734–1738.

HAUBS, H.
1966. Die Förderung des Weines zur Füllmaschine. Deut. Wein-Ztg. **102**:400, 402, 404, 406.

HAUSHOFER, H.
1966. Kritische Betrachtungen über die Art der Aufstellung von Weinlagertanks. Mitt. Rebe u. Wein, Obstbau u. Früchteverwertung (Klosterneuburg) **16**:349–356.

HAUSHOFER, H., and A. RETHALLER
1964. Die Heissabfüllung von Wein unter besonderer Berücksichtigung des Kohlensäuregehaltes und der Reduktone. Mitt. Rebe u. Wein, Serie A (Klosterneuburg) **14**:1–20.

1965. Ueber die Auswirkungen von l-Askorbinsäure aud die wichtigsten österreichischen Weinsorten. *Ibid.* **15**:230–240.

1966. Ergebnis einer Eignungsprüfung des Holzimprägnierungsmittels "Wolmanit M" zur Aussenkonservierung von Holzfässern. Mitt. Rebe u. Wein, Obstbau u. Früchteverwertung (Klosterneuburg) **16**:35–39.

HEID, J. L., and A. M. SHIPLEY
1961. Plant location and design. *In* D. K. Tressler and M. A. Joslyn (eds.), Fruit and vegetable juice processing technology. Westport, Conn., Avi Publishing Co. (*See* pp. 216–232.)

HEIDE, C. VON DER
1933. Die Blauschönung. Wein u. Rebe **14**:325–335, 348–359, 400–408; **15**:35–44.

HEIDE, C. VON DER, and K. HENNIG
1933. Bestimmung des Arsens und der Phosphosäure, des Kupfers, Zinks, Eisens, und Mangans im Most und Wein. Z. Untersuch. Lebensm. **66**:341–348.

HEIDE, C. VON DER, and F. SCHMITTHENNER
1922. Der Wein. Braunschweig, Germany, F. Vieweg und Sohn. 350 pp.

HEIMANN, W.
1965. Zur Biogenese der Aromastoffe des Traubenmostes und des Weines. Münch. medizin. Wochensch. **107**:1726–1729.

HEMPHILL, A. J.
1965. The traditional method of champagne production. San Francisco, Wine Institute, Technical Advisory Committee, June 7, 1965. 3 pp.

HENDERSON, W. W., J. M. KITTERMAN, and F. H. VANCE
1965. California grape acreage by varieties and principal counties as of 1964. Sacramento, California Crop and Livestock Reporting Service. [2] + 5 pp.

HENDRICKSON, A. H., and F. J. VEIHMEYER
1950. Irrigation experiments with grapes. Calif. Agr. Exper. Stat. Bull. **728**:1–31.

HENGST, G.
1963. Die moderne Kellerwirtschaft und ihre Enrichtung. Deut. Wein.-Ztg. **99**:308–313.

HENNIG, K.
1943. Bilans de l'azote dans les moûts et les vins nouveaux en fermentation. Bull. O.I.V. (Office Intern. Vigne Vin) **16**(159):82–86. (*See also* K. Hennig and P. Oshke, Bilanz der Stickstoffverbindungen in einen gärenden Most und Jungwein. Vorratspflege u. Lebensmtlforsch. **5**:408–424, 1942.)

1950. Die Schönung eiweisstrüber Weissweine mit Bentoniten (Montmorilloniten). Deut. Wein-Ztg. **86**:341.

1955. Der Einfluss der Eiweiss- und Stickstoffbestandteile auf den Wein. *Ibid.* **91**:377–378, 380, 394–397.

1956. Ueber die Vergärung von Mosten unter Zusatz verschiedener Enzyme. Weinberg Keller **3**:435.441.

1957. Die Most- und Weinbehandlung für Klein-, Mittel- und Grossbetriebe. *Ibid.* **4**:81–90.

1959. Pyrokohlensäurediäthylester, ein neues gärhemmendes Mittel. Deut. Lebensm.-Rundschau **55**:297–298.

1962. Chemische Untersuchungsmethoden für Weinbereiter und Süssmosthersteller. 5th ed. Stuttgart, Verlag Eugen Ulmer. 126 pp.

1963. Emploi du pyrocarbonate d'éthyle dans le traitement des vins. Ann. Technol. Agr. **12**(numéro hors série 1):115–124.

HENNIG, K., and R. BURKHARDT
1957. Ueber die Gerbstoffe und Polyphenole der Weine. Naturwissenschaften **11**:328–329.

1958. Die Nachweis phenolartiger Verbindungen und hydroaromatischer Oxy-carbonsäuren in Traubenbestandteilen Wein und Weinähnlichen Getränken. Weinberg Keller 5:542–552, 593–600.

1960a. Detection of phenolic compounds and hydroxy acids in grapes, wines, and similar beverages. Am. J. Enol. Viticult. 11:64–79.

1960b. Vorkommen und Nachweis von Quercitrin und Myricitrin in Trauben und Wein. Weinberg Keller 7:1–3.

HENNIG, K., and A. LAY

1963. Die Bestimmung des im Wein molekular gelösten Sauerstoffs mit der Beckmann-Sauerstoffelektrode, Adapter und Zeromatic. Weinberg Keller 10:165–169.

1965. Die gewichtsanalytische Bestimmung der Oxalsäure im Most und Wein. Ibid. 12:425–427.

HENNIG, K., and F. VILLFORTH

1938. Spurenelemente in Most und Wein. Vorratspflege u. Lebensmtlforsch. 1:563–592. (See also F. Villforth, Die quantitative Bestimmung von Zink in Most und Wein. Wein Rebe 22[12]:271–279, 1940.)

1942. Die Aromastoffe der Weine. Ibid. 5:181–199, 313–333. (See also K. Hennig, Les substances de l'arôme des vins. Nouveaux procédés de détermination de la qualité des vins. Bull. O.I.V. (Office Intern. Vigne Vin) 16(158):8–16, 1943; and H. Müller-Thurgau, Weine im Lichte der neuzeitlichen Forschung. Wein u. Rebe 1950–51:15–24.)

HENWOOD, D., JR.

1957. Savings through engineering design. San Francisco, Wine Institute, Technical Advisory Committee Report, March 8, 1957. 3 pp.

HERRMANN, K.

1959. Ueber Katechine und Katechin-Gerbstoffe und ihre Bedeutung in Lebensmitteln. Z. Lebensm.-Untersuch. -Forsch. 199:487–507.

1963. Ueber die phenolischen Inhaltsstoffe der Trauben und des Weines (Flavonoide, Phenolkarbonsäure, Farbstoffe, Gerbstoffe). Weinberg Keller 10: 154–164, 208–220.

HERSOM, A. C., and E. D. HULLAND

1964. Canned foods. An introduction to their microbiology. New York, Chemical Publishing Co. 291 pp.

HERZ, K. O.

1967. The literature of brewing. Wallerstein Lab. Commun. 30(101):17–29.

HESTRIN, S.

1949. The reaction of acetylcholine and other carboxylic acid derivatives with hydroxylamine and its analytical application. J. Biol. Chem. 180:249–261.

HEWITT, J. T.

1928. The chemistry of wine making. A report on oenological research. Great Britain Empire Marketing Board Publ. No. 7. London, H.M. Stationery Office. 57 pp. (See p. 24.) (See also Science Progress 33:625–644, 1939.)

HEYDENREICH, G. A.

1967. Die schweflige Säure und ihre Salze in der Lebensmittelverarbeitung und -lagerung. Z. Ernährungswiss. 8:44–65.

HICKINBOTHAM, A. R.

1948. Glycerol in wine. Australian Chem. Inst. J. and Proc. 15:89–100.

HICKINBOTHAM, A. R., and J. L. WILLIAMS

1940. The application of enzymic clarification to commercial wine making. J. Agr. South Austral. 43:491–495, 596–602.

LITERATURE CITED

HIDALGO, L.
1967. Personal communication.

HIDVÉGHY, S.
1950. Szőlőtalajvizsgálatok Tokajhegyalján. Szőlészeti Kutató Intézet Évkönyve 10:175–191.

HILLIS, W. E.
1962. Wood extractives and their significance to the pulp and paper industries. New York, Academic Press. 450 pp.

HINREINER, E., F. FILIPELLO, H. W. BERG, and A. D. WEBB
1955a. Evaluation of thresholds and a minimum difference concentration for various constituents of wines. IV. Detectable differences in wines. Food Technol. 9:489–490.

HINREINER, E., F. FILIPELLO, A. D. WEBB, and H. W. BERG
1955b. Ibid. III. Ethyl alcohol, glycerol and acidity in aqueous solution. Ibid. 9:351–353.

HITIER, J.
1923. Les appellations d'origine dans le domaine international. Compt. Rend. Acad. Agr. France 9:883–886.

HOCH, F. L., and B. L. VALLIE
1953. Gravimetric estimation of protein precipitated by trichloroacetic acid. Anal. Chem. 25:317–320.

HÖTZEL, D., E. MUSBAT, and H. D. CREMER
1966. Toxicität von schwefliger Säure in Abhängigkeit von Bindungsform und Thiaminversorgung. Z. Lebensm.-Untersuch. -Forsch. 130:25–31.

HOFFMAN, J., and L. J. BERG
1967. Vacuum filtration in the winery. Am. J. Enol. Viticult. 18:97–99.

HOHL, L. A.
1938. Further observations on production of alcohol by Saccharomyces ellipsoideus in syruped fermentations. Food Res. 3:453–465.

HOHL, L. A., and W. V. CRUESS
1936. Effect of temperature, variety of juice, and method of increasing sugar content on maximum alcohol production by Saccharomyces ellipsoideus. Food Res. 1:405–411.
1939. Observations on certain film forming yeasts. Zentr. Bakteriol. Parasitenk. Abt. II, 101:65–78.

HOHL, L. A., and M. A. JOSLYN
1941a. The production of lactic acid in alcoholic fermentation. Plant Physiol. 16:343–360.
1941b. The production of formic acid in alcoholic fermentation. Ibid. 16:755–769.

HOLDEN, C.
1953. Studies on polish filtration of wine with plate and frame filters. Proc. Am. Soc. Enol., Davis, August 12–14, 4:133–138.
1955. Combined method for the heat and cold stabilization of wine. Am. J. Enol. 6:47–49.

HOLLMANN, S.
1964. Non-glycolytic pathways of metabolism of glucose. Transl. and rev. by O. Touster. New York, Academic Press. ix + 276 pp.

HOLM, HANS C.
1908. A study of yeasts from California grapes. Calif. Agr. Exper. Stat. Bull. 197:169–175. (Out of print.)

895

HOLZBERG, I., R. K. FINN, and K. H. STEINKRAUS
1967. A kinetic study of the alcoholic fermentation of grape juice. Biotechnol. Bioeng. **9**:413–427.

HOPKINS, R. H., and R. H. ROBERTS
1935. Kinetics of alcoholic fermentation of sugar by brewer's yeast. II. The relative rates of fermentation of glucose and fructose. Biochem. J. **29**: 931–936.

HORECKER, B. L.
1963. Pentose metabolism in bacteria. New York and London, John Wiley & Sons, Inc. ix + 100 pp.

HORNE, F. L.
1942. Filtration and beer clarification. Some technical aspects. Wallerstein Lab. Commun. **5**(16):171–180.

HORNSTEIN, I. (ed.)
1966. Flavor chemistry. Advances in Chemistry. Series 56. Washington, D.C. American Chemical Society. ix + 278 pp.

HOWARD, B. J.
1905. Microscopic examination of fruits and fruit products. U.S. Dept. Agr. Bur. Chem. Bull. **66**:103–107.

HUBACH, C. E.
1948. Detection of cyanides and ferrocyanides in wines. Anal. Chem. **20**:1115–1116.
1966. Potentiometric determination of chlorides in wine, distilled spirits, and wine vinegar. J. Assoc. Offic. Anal. Chemists **49**:498–501.

HUBER, P.
1967. What line of action for California. Wines & Vines **48**(7):16–18.

HUGLIN, P., and SCHWARTZ, J.
1960. Essai d'obtention et de dégustation de jus de raisin à partir d'hybrides-producteurs. Ann. Technol. Agr. **9**(1):53–65.

HUMEAU, G.
1966. Pressure tests (transl.). Chim. Chronika (Athens) **31**(3):23–37. (Chem. Abst. **65**(2):2965h. 1966.)

HUNTER, R. S.
1942. Photoelectric tristimulus colorimetry with three filters. Natl. Bur. Standards Cir. **C429**:1–46.

HUSSEIN, A. A., and W. V. CRUESS
1940a. Properties of the oxidizing enzymes of certain *vinifera* grapes. Food Res. **5**:637–648.
1940b. A note on the enzymatic darkening of wine. Fruit Prod. J. **19**:271.

HYMANS, E.
1965. Dionysis: a social history of the wine vine. New York, The Macmillan Co. 381 pp.

IAKIVCHUK, A.
1966. Antotsiany kozhitsy iagod vinograda. Vinodelie i Vinogradarstvo SSSR **26**(7):29–30.

IBARRA, M., and W. V. CRUESS
1948. Observations on removal of excess color from wine. Wine Rev. **16**(6):14–15.

ILIESCU, L. V., F. MUJDABA, G. SANDU-VILLE, and GH. BUTĂNESCU
1965. Studiul unor conditi de prelucrare a mustului la vinificarea în alb. Inst. Cercetări Horti-Viticole, Lucrări Ştiinţ. **9**:451–469.

INGALSKE, D. W., A. M. NEWBERT, and G. H. CARTER
1963. Concord grape pigments. J. Agr. Food Chem. 11:263–268.

INGRAHAM, J. L.
1966. Personal communication.

INGRAHAM, J. L., and G. M. COOKE
1960. A survey of the incidence of the malo-lactic fermentation in California table wines. Am. J. Enol. Viticult. 11:160–163.

INGRAHAM, J. L., and J. F. GUYMON
1960. The formation of higher aliphatic alcohols by mutant strains of Saccharomyces cerevisiae. Arch. Bioch. Biophys. 88:157–166.

INGRAHAM, J. L., R. G. VAUGHN, and G. M. COOKE
1960. Studies on the malo-lactic organisms isolated from California wines. Am. J. Enol. Viticult. 11:1–4.

INGRAM, M.
1947a. An electometric indicator to replace starch in iodine titrations of sulfurous acid in fruit juices. J. Soc. Chem. Ind. 66:50–55.
1947b. Investigation of errors arising in iodometric determination of free sulfurous acid by the acetone procedure. Ibid. 66:105–115.
1948. Germicidal effects of free and combined sulfur dioxide. Ibid. 67:18–21.

IÑIGO LEAL, B., and F. BRAVO ABAD
1963. Acidez y levaduras vínicas. II. Evolución de la acidez fija del mosto por la acción de distintas especies de levaduras vínicas. III. Curso cinético de la acidez volátil en fermentados de mosto originados por distintas especies de levaduras vínicas. Rev. Cien. Apl. (Madrid) 17:40–43, 132–135.

IÑIGO LEAL, B., D. VÁZQUEZ MARTÍNEZ, and V. ARROYO VARELA
1963. Los agentes de fermentación vínica en la zona de Jerez. Rev. Cien. Apl. (Madrid) 17:296–305.

INSTITUT TECHNIQUE DU VIN
1967. La dégustation. Vignes et Vins, numéro spécial 1967:1–137 [5].

IONESCU, A. I.
1966. Consideratti în legatură cu reglementarea aplicării cleirii albastre (Some considerations on the laws regulating the application of blue fining). Grădina Via Livada 15(2):34–43.

IONESCU, T. D.
1966. Schimbul ionic în chimia şi tehnologia alimentară. Bucharest, Editura Tehnica. 446 pp.

IRMAY, S.
1958. On the theoretical derivation of Darcy and Forchheimer formulas. Trans. Am. Geophys. Union 39:702–707.

IUSTRATOVA, L. S.
1967. Vidovoĭ sostav molochnokislykh bakterii vin Moldavii. Vinodelie i Vinogradarstvo SSSR 27(2):21–24.

IVANOV, T. P.
1966. Activity of polyphenoloxidase during the ripening of Cherven muscat, Dimyat, Riesling, and Aligoté grapes (transl.). Lozarstvo Vinarstvo (Sofia) 15(2):24–27.
1967. Sur l'oxydation du moût de raisin. I. Activité de la polyphénoloxydase du raisin des cépages "Muscat rouge", "Dimiat", "Riesling" et "Aligoté". II. Étude comparée de l'anhydride sulfureux et de la bentonite en tant qu'inactivateurs de la polyphénoloxydase du moût de raisin. Ann. Technol. Agr. 16:35–39, 81–88.

IVANOV, T., T. TONTSCHEV, and G. BAMBALOV
  1966. Der Pyrokohlensäurediäthylester (PKE) als Konservierungsmittel. Weinberg
  Keller 14:167–171.

IWANO, S., and SAWANOBORI, H.
  1962. Titration of calcium and magnesium in wine with ethylenediamine tetra-
  acetate. Am. J. Enol. Viticult. 13:54–57.

IWANOFF, N. N.
  1913. Ueber die flüchtigen Basen der Hefeautolyse. Biochem. Z. 58:217–224.
  1921a. Ueber die Verwandlung stickstoffhaltiger Substanzen bei den Endphasen
  der Hefenautolyse. Ibid. 120:1–24.
  1921b. Ueber Eiweissspaltung in Hefen während der Gärung. Ibid. 120:25–61.
  1921c. Ueber den Einfluss der Gärungsprodukte auf den Zerfall der Eisweisstoffe
  in den Hefen. Ibid. 120:62–80.

JACOB, F. S., T. E. ARCHER, and J. G. B. CASTOR
  1964. Thermal death time of yeast. Am. J. Enol. Viticult. 15:69–74.

JACOB, M., and N. ANGHELIDE
  1967. O metodă rapidă pentru determinarea alcoolului şi glucozei din vin. Rev.
  Hort. Viticult. 11:29–35.

JACOBS, M. B.
  1951. Sulfur dioxide. In R. E. Kirk and D. F. Othmer (eds.), Encyl. Chem.
  Technol. 6:835–848.

JACOBS, P. B., and H. P. NEWTON
  1938. Motor fuels from farm products. U.S. Dept Agr., Misc. Publ. 327:1–129.

JACQUIN, P.
  1960. Dosage de l'acide sorbique dans les boissons. Ann. Technol. Agr. 9:393–408.

JAKOB, L.
  1965. Bedeutung und Verhaltensweise der sogenannten Natrium- und Calcium-
  Bentonite in der Kellerwirtschaft. Weinblatt 59:613–619.

JAKOB, L., and O. BACHMANN
  1964. Isolierung von geschmackswirksamen Komponenten aus Wein mittels
  Dünnschicht-Chromatographie. Mitt. Rebe u. Wein, Serie A (Klosterneuburg) 14:187–192.

JAKOB, L., and W. SCHRODT
  1966. Qualitätsbeeinflussung von Weissweinen durch Abfüllverfahren und
  Lagerung. Weinberg Keller 13:397–418.

JAKUBOWSKA, J.
  1963. Analyse microbiologique des vins à l'acide des filtres à membrane. Bull.
  O.I.V. (Office Intern. Vigne Vin) 36:812–824.

JAKUBOWSKA, J., and A. PIATKIEWICZ
  1963. Quelques données sur les modifications chimiques des vins sous l'influence
  de la flore microbienne. Ann. Technol. Agr. 12 (numéro hors série 1):
  74–75.

JANKE, A., and M. RÖHR
  1960. Ueber Schaumweine und deren Untersuchung. I. Objektive Teste zur
  Beurteilung von Schaumweinen. II. Ueber einem kontrollierten Vergleichs-
  versuch Tankgärverfahren/Flaschengärverfahren. Mitt. Rebe u. Wein,
  Serie A (Klosterneuburg) 10:111–123, 210–217.

JAULMES, P.
  1951. Analyse des vins. 2d ed. Montpellier, Librairie Coulet, Dubois et Poulain.
  574 pp.
  1965a. Les méthodes internationales et les nouvelles méthodes officielles d'analyse
  des vins et des moûts. Mise au Point de Chimie Anal. 14:182–215.

1965*b*. Les nouvelles méthodes officielles d'analyse des vins et des moûts. Bull. Techn. Infor. Ing. Serv. Agr. No. **196**:3–15.

1967. Utilisation du gaz carbonique et de l'azote en oenologie et dans l'industrie du jus de raisin. Bull. O.I.V. (Office Intern. Vigne Vin) **40**:147–165.

JAULMES, P., C. BESSIÈRE, S. FOURCADE, and C. CHAMPEAU
1963–64. Dosage microbiologique des vitamines et des acides aminés dans les vins après différents traitement. Trav. Société Pharm. Montpellier **23**:361–369; **24**:36–41.

JAULMES, P., and S. BRUN
1963*a*. La mesure pycnométrique de la masse volumique de la densité et du degré alcoolique des vins. Ann. Fals. Expertise Chim. **56**:129–142.

1963*b*. Les nouvelles tables de corrections de température pour le masse volumétrique des moûts, vins, vins doux, etc. *Ibid.* **56**:143–178.

1966. Table des corrections de température pour la détermination du titre alcoométrique à l'aide d'un pycnomètre en verre pyrex. *Ibid.* **59**:35–38.

JAULMES, P., S. BRUN, and J. P. LAVAL
1965. Titre alcoométrique par réfractométrie. Ann. Fals. Expertise Chim. **58**:304–310.

JAULMES, P., G. HAMELLE, and J. ROQUES
1960. Le plomb dans les moûts et les vins. Ann. Technol. Agr. **9**:189–245.

JAULMES, P., and R. MESTRES
1960. Recherche et dosage de l'acide cyanhydrique dans vins traités au ferrocyanure. Ann. Fals. Expertise Chim. **53**:455–475.

1962. Dosage de l'acide cyanhydrique dans les vins. Ann. Technol. Agr. **11**:249–269.

JAULMES, P., R. MESTRES, and B. MANDROU
1962. Influence de l'acide sorbique de l'acidité volatile sur le dosage. Trav. Société Pharm. Montpellier **21**:25–27.

JAULMES, P., and J. M. NEY
1960. Recherche des vins d'hybrides producteurs directs par chromatographie. Ann. Fals. Fraudes **53**:180–183.

JELLINEK, K.
1911. Ueber die Leitfähigkeit und Dissoziation von Natriumhydrosulfit und hydroschwefliger Säure im Vergleich zu analogen Schwefelsauerstoffverbindungen. Z. Physik. Chem. **76**:257–354.

JENNINGS, W. G.
1961. How to select and use detergents and sanitizers in the winery. Wines & Vines **42**(9):29–31. (*See also* J. Am. Oil Chemists Soc. **40**:17–20. 1963.)

JOHNSTONE, H. F., and P. W. LEPPLA
1934. The solubility of sulfur dioxide at low partial pressures. The ionization constant and heat of ionization of sulfurous acid. J. Am. Chem. Soc. **38**:901–936.

JONES, G. T.
1964. The determination of sulphur dioxide in beers and wines. Analyst **89**:678–679.

JOPPIEN, P. H.
1960. Zur Kenntnis der Weine Morokkos. Deut. Wein-Ztg. **96**:518, 520, 522, 524.

JORDAN, JR., R.
1911. Quality in dry wines through adequate fermentation by means of defecation, aeration, pure yeast, cooling and heating. San Francisco, California. Privately published. 146 pp.

JOSLYN, M. A.
1938. Electrolytic production of rancio flavor in sherries. Ind. Eng. Chem., Ind. Ed. **30**:568–577.
1940. The by-products of alcoholic fermentation. Wallerstein Lab. Commun. **3**(8):30–43.
1941. The mineral metabolism of yeast. *Ibid.* **4**(11):49–65.
1949. California wines. Oxidation reduction potentials at various stages of production and aging. Ind. Eng. Chem. **41**:587–592.
1950. Hard chrome plating and avoidance of metal pick-up. Wines & Vines **31**(4):67–69.
1951. Nutrient requirements of yeast. Mycopath. Mycol. Appl. **5**(2/3):260–276.
1953. The theoretical aspects of the clarification of wine by gelatin fining. Proc. Am. Soc. Enol. **4**:39–68.
1954a. How to cut down on SO₂ content. San Francisco, Wine Institute, Technical Advisory Committee, July 30, 1954. 5 pp.
1954b. How to cut down on SO₂ content. Wines & Vines **35**(12):31–34.
1955a. Yeast autolysis. I. Chemical and cytological changes involved in autolysis. Wallerstein Lab. Commun. **18**(61):107–122.
1955b. Determination of free and total sulfur dioxide in white table wines. J. Agr. Food Chem. **3**(8):686–695.
1956. Changes in fermenting wort and beer produced by yeast autolysis. Techn. Proc. 48th Annual Conv. Master Brewers Assoc. **1955**:49–56.
1961a. Blending formulas and syrup algebra. *In* D. K. Tressler and M. A. Joslyn (eds.), Fruit and vegetable juice processing technology. Westport, Conn., Avi Publishing Co. 1028 pp. (*See* pp. 586–619.)
1961b. Luncheon address at meeting of Am. Soc. Enologists, June 24, 1961. (*Referred to in* Wines & Vines, **42**(7):27–28. 1961.)
1963a. Water in food processing. *In* M. A. Joslyn and J. L. Heid (eds.), Food processing operations. Vol. I. Westport, Conn., Avi Publishing Co. 664 pp. (*See* pp. 311–339.)
1963b. Steam in food processing. *In* Joslyn and Heid, *Ibid.*, pp. 340–361.

JOSLYN, M. A., and M. A. AMERINE
1941a. Commercial production of dessert wines. Calif. Agr. Exper. Stat. Bull. **651**:1–186. (Out of print.)
1941b. Commercial production of brandy. *Ibid.* **652**:1–80. (Out of print.)
1964. Dessert, appetizer and related flavored wines; the technology of their production. Berkeley, Division of Agricultural Sciences, University of California. xii + 482 pp.

JOSLYN, M. A., and J. B. S. BRAVERMAN
1954. The chemistry and technology of the pretreatment and preservation of fruit and vegetable products with sulfur dioxide and sulfites. Advan. Food Res. **5**:97–160.

JOSLYN, M. A., and C. L. COMAR
1938. Determination of acetaldehyde in wines. Ind. Eng. Chem., Anal. Ed. **10**:364–366.
1941. The role of aldehydes in red wine. Ind. Eng. Chem., Ind. Ed. **33**:919–928.

JOSYLN, M. A., and W. V. CRUESS
1934. Elements of wine making. Calif. Agric. Exten. Cir. **88**:1–64. (Out of print.)

JOSLYN, M. A., and H. F. K. DITTMAR
1967a. The proanthocyanidins of Pinot blanc grapes. Am. J. Enol. Viticult. **17**:1–10.
1967b. Die Proanthocyanide der Trauben. I. Die Proanthocyanide der Schalen und Kerne von Trauben der Sorte Pinot blanc. Mitt. Rebe u. Wein, Obstbau u. Früchteverwertung (Klosterneuburg) **17**(2):92–108.

JOSLYN, M. A., and R. DUNN
1938. Acid metabolism of wine yeast. I. The relation of volatile acid formation to alcoholic fermentation. J. Am. Chem. Soc. 60:1137–1141.

JOSLYN, M. A., H. B. FARLEY, and H. M. REED
1929. Effect of temperature and time of heating on extraction of color from red juice grapes. Ind. Eng. Chem. 21:1135–1137.

JOSLYN, M. A., and J. L. GOLDSTEIN
1964. Astringency of fruit and fruit products in relation to phenolic content. Advan. Food Res. 13:179–217.

JOSLYN, M. A., and A. LITTLE
1967. Relation of type and concentration of phenolics to the color and stability of rosé wines. Am. J. Enol. Viticult. 18:138–148.

JOSLYN, M. A., and A. LUKTON
1953. Prevention of copper and iron turbidities in wine. Hilgardia 22:451–533.
1956. Mechanism of copper casse formation in white table wine. I. Relation of changes in redox potential to copper casse. Food Res. 21:384–396.

JOSLYN, M. A., A. LUKTON, and A. CANE
1953. The removal of excess copper and iron ions from wine. Food Technol. 7:20–29.

JOSLYN, M. A., S. MIST, and E. LAMBERT
1952. The clarification of apple juice by fungal pectic enzyme preparations. Food Technol. 6:133–139.

JOSLYN, M. A., M. MORRIS and G. HUGENBERG
1968. Die Bestimmung der Gerbsäure und verwandter Phenolsubstanzen mittels des Phosphomolybdat- Phosphowolframat-Reagenz. Mitt. Klosterneuburg Rebe u. Wein, Obstbau u. Früchteverwertung. 18:17–34.

JOSYLN, M. A., and M. W. TURBOVSKY
1954. Commercial production of table and dessert wine. In L. A. Underkofler and R. J. Hickey (eds.), Industrial fermentations. Vol. I. New York, Chemical Publishing Co. (See pp. 196–251.)

JOSLYN, M. A., and D. C. VOSTI
1955. Yeast autolysis. II. Factors influencing the rate and extent of autolysis. Wallerstein Lab. Commun. 18(62):191–205.

JOURET, C.
1967. Détection des diglucosides anthocyaniques dans les vins suivant la technique de Dorier et Verelle. Ann. Technol. Agr. 16:373–377.

JOURET, C., and P. BÉNARD
1965. Influence des porte-greffes sur la composition minérale des vins des vignes de terrains salés. Ann. Technol. Agr. 14:349–355.
1967. Dosage colorimétrique du zinc par le "zincon" résultats sur quelques vins. Ann. Fals. Expertise Chim. 60:182–187.

JOURET, C., and C. POUX
1961a. Étude d'une technique de dosage du potassium par spectrophotométrie de flamme dans les vins et les moûts. Ann. Technol. Agr. 10:351–359.
1961b. Étude d'une technique de dosage du sodium dans les vins et les moûts par spectrophotométrie de flamme. Ibid. 10:361–368.
1961c. Note sur les teneurs en potassium, sodium et chlore des vignes de terrains salés. Ibid. 10:369–374.

JURD, L.
1964. Reactions involved in sulfite bleaching of anthocyanins. J. Food Sci. 29:16–19.

JURD, L., and S. ASEN
1966. The formation of metal and "co-pigment" complexes of cyanidin-3-glucoside. Phytochem. 5:1263–1271.

JURICS, É. W.
1967. Papierchromatographische Bestimmung von Catechin und Epicatechim in Früchten. Z. Lebensm.-Untersuch. -Forsch. 135:269–275.

KAIN, W.
1967a. Untersuchung und Beurteilung von Gelatine zum Weinbehandlung. Mitt. Rebe u. Wein, Obstbau u. Früchteverwertung (Klosterneuburg) 17:109–128.
1967b. Untersuchung und Beurteilung von Bentoniten zur Weinbehandlung. I. Der Reinheitsgrad von Handelprodukten. II. Ueber die Bestimmung der Werksamkeit auf Wein. Ibid. 17:10–24, 201–222.

KARRER, P., and G. DE MEURON
1932. Pflanzenfarbstoffe. XL. Zur Kenntnis des oxydativen Abbaus der Anthocyane. Konstitution des Malvons. Helv. Chim. Acta 15:507–512.

KASIMATIS, A. M., J. J. KISSLER, J. V. LIDER, C. D. LYNN, B. B. BURLINGAME, and E. A. YEARY
1967. Economic review of the California grape industry. Berkeley, Calif. Agr. Ext. Serv. (Irregularly paged.)

KATONA, J.
1966. Rapport hongrois. Bull. O.I.V. (Office Intern. Vigne Vin) 39:1141–1149.

KEAN, C. E., and J. L. MARSH
1956a. Investigation of copper complexes causing cloudiness in wines. I. Chemical composition. Food Res. 21:441–447. (See also Am. J. Enol. 8:80–85.)
1956b. Investigation of copper complexes causing cloudiness in wine. II. Bentonite treatment of wines. Food Technol. 10:355–359.

KEHAT, E., A. LIN, and A. KAPLAN
1967. Clogging of filter media. Ind. Eng. Chem., Proc. Develop. 6:48–55.

KENDALL, D. A., and A. J. NEILSON
1966. Sensory and chromatographic analysis of mixtures formulated from pure odorants. J. Food Sci. 31:268–274.

KEPNER, R. E., and A. D. WEBB
1956. Volatile aroma constituents of Vitis rotundifolia grapes. Am. J. Enol. 7:8–18.

KERN, A.
1964. Die Bedeutung des Hydroxymethylfurfurols als Qualitätsmerkmal für Fruchtsäfte und Konzentrate. International Fruchtsaft-Union, Berichte wissensch.-techn. Kommission 5:203–214.

KERP, W.
1903. Ueber organisch gebundene schweflige Säure in Nahrungsmitteln. Z. Untersuch. Lebensm. 6:66–68.
1904a. Ueber die schweflige Säure im Wein. I. Allgemeines über die schweflige Säure im Wein. Arb. Gesundheitsamte 21(2):1–15.
1904b. Ueber die schweflige Säure im Wein. II. Ueber die aldehydschweflige Säure im Wein. Ibid. 21(2):16–40.
1904c. Die schweflige Säure und ihre Verbindungen mit Aldehyden und Ketonen. 1 Teil. Berlin, Julius Springer. 236 pp. (From Arb. Gesundheitsamte.)

KERP, W., and E. BAUER
1904. Zur Kenntnis der gebundenen schwefligen Säuren. I. Arb. Gesundheitsamte. 21:180–185.
1907a. Ibid. II. Ibid. 26:231–248.
1907b. Ibid. III. Ibid. 26:269–279.
1907c. Die elektrolytische Dissoziations Konstante der schwefligen Säure. Ibid. 26:297–300.

KICHKOVSKY, Z., and N. MEKHOUZLA
1967. Étude des protéines et des substances azotées. Leur évolution au cours des traitements oenologiques; conditions de la stabilité protéique des vins. Bull. O.I.V. (Office Intern. Vigne Vin) 4:926–933.

KIEFFER, M.
1949. Säureumlagerungen in der Traube und im Wein. Oesterreich. Weinztg. 4:41–42. (Rev. in Bull. O.I.V. [Office Intern. Vigne Vin]) 22(217):69–70. 1949.)

KIELHÖFER, E.
1942. Troubles albuminoïdes du vin. Bull. O.I.V. (Office Intern. Vigne Vin) 16(152):7–10.
1948a. Die Wärmetrübung der 1947er Weine. Deut. Wein-Ztg. 84:293.
1948b. Die Eiweisstrübung der 1947er Weine. Der Weinbau, wissen. Beih. 2:233–240.
1949a. Die Eiweisstrübung der 1947er Weine. Ibid. 3:10–17, 33–39.
1949b. Erfahrungen bei der Behandlung von Weinen, die zu Eiweisstrübungen neigen. Weinblatt 29:1–3.
1951. Die Eiweisstrübung des Weines. IV. Ueber die Beschaffenheit des Trubes und die Verminderung des N-Gehaltes des Weines durch die Trubabscheidung. Z. Lebensm.-Untersuch. -Forsch. 92(1):1–9.
1954. Die Eiweisstrübung des Weines. Weinberg Keller 1:292–298.
1956. Die Vermendung von Ascorbinsäure (Vitamin C) in des Weinbehandlung an Stelle der schwefligen Säure. Ibid. 3:324–331.
1957. Die Weinentsäuerung mittels Ionenaustauscher im Vergleich zu der Entsäuerung mit kohlensaurem Kalk. Ibid. 4:136–145.
1958. Die Bindung der schwefligen Säure an Weinbestanteile. Ibid. 5:461–476.
1959. Neue Erkenntnisse über die Wirking der schwefligen Säure im Wein und die Moglichkeit ihres Ersatzes durch Ascorbinsäure. Frankfurt/M., Verlag. Sigurd Horn. 78 pp. (Reprint of articles appearing in Weinberg und Keller.)
1960a. Neue Erkenntnisse über die schweflige Säure im Wein und ihrem Ersatz durch Ascorbinsäure. Deut. Wein-Ztg. 96:14, 16, 18, 20, 22, 24.
1960b. Bekanntes und Neues über die schweflige Säure im Wein. Deut. Weinbau 15:659–669.
1963a. Emploi du pyrocarbonate d'éthyl dans le traitement des vins. Ann. Technol. Agr. 12(numéro hors série 1):125–126.
1963b. État et action de l'acide sulfureux dans les vins; règles de son emploi. Ibid. 77–89.
1966. Zwölf Jahre önologische Forschung im Forschungsring des Deutschen Weinbaues bei der DLG. Deut. Wein-Ztg. 102:200, 202, 204, 206.

KIELHÖFER, E., and H. AUMANN
1955. Das Verhalten von Zinn gegenüber Wein. Mitt. Rebe u. Wein, Serie A (Klosterneuburg) 5:127–135.

KIELHÖFER, E., and G. WÜRDIG
1959. Der Verbrauch von schwefliger Säure im Most und Wein aus gesunden und faulen Trauben. Weinberg Keller 6:364–379.
1960a. Die Bestimmung von Acetoin und Diacetyl im Wein und der Gehalt deutscher Weine an diesen Substanzen. Wein-Wissen. 15:135–146.
1960b. Untersuchungen über die gedoppelte Oxydation von Ascorbinsäure und schwefliger Säure im Wein. Ibid. 15:103–117. (See also Wein u. Rebe 42:1–6. 1960.)
1960c. Die an Aldehyd gebundene schweflige Säure im Wein. I. Acetaldehydbildung durch enzymatische und nicht enzymatische Alkohol-Oxydation. Weinberg Keller 7:16–22.
1960d. Ibid. II. Acetaldehydbildung bei der Gärung. Ibid. 7:50–61.

1960e. Die an unbekannte Weinbestandteile gebundene schweflige Säure (Rest-SO_2) und ihre Bedeutung für den Wein. *Ibid.* 7:313–328, 361–372.

1960f. Die Verminderung der schwefligen Säure im Wein. Weinblatt 54:154–158.

1961a. Ueber Vorkommen, Nachweis und Bestimmung der Schleimsäure im Wein und Weintrub. Z. Lebensm.-Untersuch. -Forsch. 115:418–428.

1961b. Kristalltrübungen in Wein durch das Kalsalz einer bisher unbekannten Säure des Weines. Deut. Wein-Ztg. 97:478–480.

1961c. Die von Aldehyd gebundene schweflige Säure im Wein. Weinberg Keller 7:16–22, 50–61.

1963a. Die Entsäuerung sehr saurer Traubenmoste durch Ausfällung der Weinsäure und Apfelsäure als Kalkdoppelsalz. Deut. Wein-Ztg. 99:1022, 1024, 1026, 1028.

1963b. Die Oxydationsvorgänge im Wein. 6. Die Sauerstoffaufnahme durch den Sekt bei der Sektbereitung nach dem Grossraumgärverfahren. Mitt. Rebe u. Wein, Serie A (Klosterneuburg) 13:18–35.

1963c. Nachweis und Bestimmung von Diäthylacarbonat und Pyrokohlensäure-diäthylester im Wein und Schaumwein. Deut. Lebensm.-Rundschau. 59:197–200, 224–228.

1963d. Ueber das Vorkommen von Kohlensäureestern im Wein und Schaumwein. Weinberg Keller 10:201–207.

1964. Einige bei der Anwendung von Pyrokohlensäurediäthylester zu biologischen Weinstabilisierung auftretende Probleme. I. *Ibid.* 11:495–504.

KILBUCK, J. H., F. NUSSENBAUM, and W. V. CRUESS
1949. Pectic enzymes. Investigations on their use in making wines. Wines & Vines 30(8):23–24.

KLEINZELLER, A.
1941. The formation of succinic acid in yeast. Biochem. J. 35:495–501.

KLENK, E.
1958. Erfahrungen mit Anwendung von Metalltanks zur Rot- und Weisswein-bereitung. Deut. Wein-Ztg. 94:398–406.

KLENK, E., and R. MAURER
1967. Ueber die Störung der Weinsteinkristallisation durch phenolische Sub-stanzen im Wein. Weinberg Keller 14:674–676.

KLIEWE, H., and H. ESCHNAUER
1967. Der gesundheitliche Wert der Mineralien, Spurenelemente und Vitamine in Spätburgunderweinen. Wein-Wissen. 23:30–44.

KLIEWER, W. M.
1964. Influence of environment on metabolism of organic acids and carbohydrates in *Vitis vinifera*. I. Temperature. Plant Physiol. 39:869–880.

1965. The sugars of grapevines. II. Identification and seasonal changes in the concentration of several trace sugars in *Vitis vinifera*. Am. J. Enol. Viticult. 16:168–178.

1966. Sugars and organic acids of *Vitis vinifera*. Plant Physiol. 41:923–931.

1967a. Concentration of tartrates, malates, glucose and fructose in the fruits of the genus *Vitis*. Am. J. Enol. Viticult. 18:87–96.

1967b. The glucose-fructose ratio of *Vitis vinifera* grapes. *Ibid.* 17:33–41.

KLIEWER, W. M., and H. B. SCHULTZ
1964. Influence of environment on metabolism of organic acids and carbohydrates in *Vitis vinifera*. II. Light. Am. J. Enol. Viticult. 15:119–129.

KLIEWER, W. M., L. HOWARTH, and M. OMORI
1967a. Concentrations of tartaric and malic acids and their salts in *Vitis vinifera* grapes. Am. J. Enol. Viticult. 18:42–54.

LITERATURE CITED

KLIEWER, W. M., L. A. LIDER, and H. B. SCHULTZ
1967b. Influence of artificial shading of vineyards on the concentration of sugar and organic acid in grapes. Am. J. Enol. Viticult. **18**:78–86.

KLIEWER, W. M., A. R. NASSAR, and H. P. OLMO
1966. A general survey of the free amino acids in the genus *Vitis*. Am. J. Enol. Viticult. **17**:112–117.

KNUCHEL, F., L. DESHUSSES, and J. CORBAZ
1954. Influence de la vitamine B_1 et de quelques sels minéraux sur la fermentation alcoolique du jus de raisin et la rétrogradation malolactique. Rev. Ferm. Ind. Aliment. **9**:39–43.

KOBLET, W., and P. ZWICKY
1965. Der Einfluss von Ertrag, Temperatur und Sonnenstunden auf die Qualität der Trauben. Wein-Wissen. **20**:237–244.

KOCH, J.
1955. The determination of ITT value in wine. Am. J. Enol. **6**(3):23–25.
1956a. Kellertechnische Erfahrungen mit der neuzeitlichen Most- und Weinbehandlung. Weinberg Keller **3**:49–59.
1956b. Neuere Ergebnisse auf dem Gebiet der Traubensaftherstellung. Fruchtsaft-Ind. **1**:7–15.
1957. Die Eiweissstoffe des Weines und ihre Veränderungen bei verschiedenen kellertechnischen Behandlungsmethoden. Weinberg Keller **4**:521–526.
1963. Protéines des vins blancs. Traitements des précipitations protéiques par chauffage et à l'aide de la bentonite. Ann. Technol. Agr. **12** (numéro hors série 1):297–311.

KOCH, J., and E. BREKER
1953. Zur Bestimmung und Bedeutung der Aromazahl bei naturreinen Apfel- und Traubensäften. Z. Lebensm.-Untersuch. -Forsch. **96**:329–335.

KOCH, J., and G. BRETTHAUER
1957. Zur Kenntnis der Eiweissstoffe des Weines. I. Chemische Zusammensetzung des Wärmetrubes kurzzeiterhitzter Weissweine und seine Beziehung zur Eiweisstrübung und zum Weineiweiss. II. Einfluss der Mosterhitzung auf die Eiweissstabilität der Weissweine. Z. Untersuch.-Lebensm. -Forsch. **106**:272–280, 361–367.
1960. Das Glucose-Fructose Verhältnis der Konsumweine in Abhängigkeit von verschiedenen kellertechnischen Massnahmen. *Ibid.* **112**:97–105.

KOCH, J., G. BRETTHAUER, and H. SCHWAHN
1956. Ueber die chemische Zusammensetzung des Wärmetrubes kurzzeiterhitzter Weine und seine Beziehung zu der sogenannten "Eiweisstrübung." Naturwissenschaften **43**:421–422.

KOCH, J., and E. GEISS
1955. Der Einfluss verschiedener kellertechnischer Massnahmen auf den Aufbau der Weissweine. II. Die Kurzzeiterhitzung der Weine. Z. Lebensm.-Untersuch. -Forsch. **100**:15–24.

KOCH, J., and E. SAJAK
1959. A review and some studies on grape protein. Am. J. Enol. Viticult. **10**:114–123.
1961. Ein neuer Wärmtest für bentonitgeschönte Weine. Weinberg Keller **8**:152–155.
1963. Zur Frage der Eiweisstrübungen der Weine. *Ibid.* **10**:35–51.

KOCH, J., and H. SCHILLER
1964. Kinetik der Kristallisation von Weinstein. Z. Lebensm.-Untersuch. -Forsch. **124**:180–183.

905

KOCH, J., and H. SCHWAHN
1958. Zur Kenntnis der Eiweissstoffe des Weines. III. Papierelektrophoretische-Untersuchungen der "löslichen" Traubenproteine. IV. Einfluss der Bentonitschönung auf das Traubenprotein. Z. Lebensm.-Untersuch. -Forsch. 107:20–24, 413–415.

KOCSIS, P.
1958–1962. Ú jabb szőlőfajtak és szőlőfajta-jelöltek. Szőlészeti Kutató Intézet Évkönyve 12:125–132.

KÖNIG, J.
1903. Chemische Zusammensetzung der menschlichen Nahrungs- und Genussmittel. Vol. I. Berlin, Julius Springer. 1535 pp.

KOEPPEN, B. H., and D. S. BASSON
1966. The anthocyanin pigments of Barlinka grapes. Phytochem. 5:183–187.

KOLTHOFF, I. M.
1920. Die Neutralisationskurve der schwefligen Säure. Chem. Weekblad. 16: 1154–1163.

KOLTHOFF, I. M., and J. L. LINGANE
1952. Polarography. Vol. I. Theoretical principles, instrumentation and technique. 2d ed. New York, Interscience Publishers. xvi + 420 pp.

KOLTHOFF, I. M., and V. A. STENGER
1942. Volumetric analysis. Vol. I. Theoretical fundamentals. New York, Interscience Publishers, Inc. 215 pp.
1947. Volumetric analysis. Vol. II. Titration methods: neutralization, precipitation, and complex formation reactions. New York, Interscience Publishers, Inc. 400 pp.

KONDAREV, M.
1965. Die Entwicklung des Weinbaues in Bulgarien. Wein-Wissen. 20:428–432.

KONLECHNER, H., and H. HAUSHOFER
1956a. Beurteilung eines Maische-Vorensafters. Mitt. Rebe u. Wein, Serie A (Klosterneuburg) 6:64–65.
1956b. Versuche mit Rotwein Bereitungsverfahren. Ibid. 6:158–173.
1962. Ergebnisse von Arbeitsversuchen mit der Horizontalpresse "Garnier." Ibid. 12:19–23.

KORKES, S., A. DEL CAMPELLO, and S. OCHOA
1950. Biosynthesis of dicarboxylic acids by carbon dioxide fixation. IV. Isolation and properties of an adaptive "malic" enzyme from Lactobacillus arabinosus. J. Biol. Chem. 187:891–905.

KOWALA, C.
1965. Apparatus for pressure filtration in inert atmosphere. Chem. Ind. (London) 1965(43):1784.

KOZENKO, E. M.
1952. O soderzhanii v vine stoïkikh éfirov ugol'noï kisloty. Vinodelie i Vinogradarstvo SSSR 12(4):25–28.
1957. Content of stable ethyl carbonate in wine (transl.). Trudy Krasnodar Pishevoi Prom. 1:47–50. (Chem. Abst. 54:21629. 1960.)
1962. Dinamika tituremoï kislotnosti v protsesse sozrevaniia vinograda. Vinodelie i Vinogradarstvo SSSR 22(7):7–10.

KOZENKO, E., and N. M. ROMANOVA
1962. Kislotno-solevoï balans soka i vina. Vinodelie i Vinogradarstvo SSSR 22(4):19–22.

KRASINSKII, N. P., and E. A. PRYAKHINA
1946. Theory of aging (transl.). Vinodelie i Vinogradarstvo SSSR 6(2):7–11.

KRAUS, V.
1966. Beitrag zur characteristik der Rebsorten in Hinsicht auf das Korrelationsmass zwischen Reifesumme und Qualität der Ernte in nordlichen Weinbaugebieten. Wein-Wissen. 21(2):53–60.

KREGER-VAN RIJ, N. W.
1964. Endomycopsis vini and Pichia etchellsii, spp. n. Antonie van Leeuwenhoek J. 30:428–432.

KRÖMER, K., and F. HEINRICH
1922. Ueber eine in überschwefelten Mosten auftretende Hafe der Gattung Saccharomyces. Geisenheimer Festschr. Deut. Wein-Ztg. Mainz. 258–295.

KRUG, K.
1967. Die Eiweissnachtrübungen im Wein und ihre Verhütung. Deut. Wein-Ztg. 103:1029–1035.
1968. Die Ursachen der Eiweissnachtrübungen bei Weinen sowie des Ausfall der Eiweisschönungen. Wein-Wissen. 23:8–29.

KRUMPERMAN, P. H., and R. H. VAUGHN
1966. Some lactobacilli associated with decomposition of tartaric acid in wine. Am. J. Enol. Viticult. 17:185–190.

KUDRIAVTSEV, V. I.
1960. Die Systematik der Hefen. Berlin, Akademie Verlag. 324 pp.

KUEHNER, R. L. (ed.)
1964. Recent advances in odor: theory, measurement and control. Ann. N.Y. Acad. Sci. 116:357–746.

KUFFERATH, A.
1954. Filtration und Filters. 3d ed. Berlin, G. Bodenbender. 512 pp.

KUHNHOLTZ-LORDAT, G.
1963. La genèse des appellations d'origine des vins. Mâcon, Imprimerie Buguet-Comptour. 148 pp.

KUNKEE, R. E.
1967a. Control of malo-lactic fermentation induced by Leuconostoc citrovorum. Am. J. Enol. Viticult. 18:71–77.
1967b. Malo-lactic fermentation. Adv. Appl. Microbiol. 9:235–279.
1968. Simplified chromatographic procedure for detection of malo-lactic fermentation. Wines & Vines 49(3):23–24.

KUNKEE, R. E., and M. A. AMERINE
1965. Personal communication.

KUNKEE, R. E., J. F. GUYMON, and E. A. CROWELL
1965a. Formation of a fusel oil component by cell-free extracts of yeast. San Francisco, Wine Institute, Technical Advisory Committee, June 7, 1965. 3 + [2] pp.

KUNKEE, R. E., and C. S. OUGH
1966. Multiplication and fermentation of Saccharomyces cerevisiae under carbon dioxide pressure in wine. Appl. Microbiol. 14:643–648.

KUNKEE, R. E., C. S. OUGH, and M. A. AMERINE
1964. Induction of malo-lactic fermentation by inoculation of must and wine with bacteria. Am. J. Enol. Viticult. 15:178–183.

KUNKEE, R. E., G. J. PILONE, and R. E. COMBS
1965b. The occurrence of malo-lactic fermentation in southern California wines. Am. J. Enol. Viticult. 16:219–223.

KUSHIDA, T., C. MARUYAMA, and K. ITO
1965. Use of antioxidants in wine making. III. Effects of the mixtures of ascorbic, erythorbic, and sulfurous acid on the quality of red wines (transl.). Yamanashi Daigaku, Hakkô Kenkyusho Kenkyu Hokoku No. 12:35–39.

KUSHIDA, T., C. MARUYAMA, and K. YOSHIDA
1964. Antioxidants in wine making. II. Effect of mixtures of ascorbic, erythorbic, and sulfurous acid on the quality of wines during storage. Yamanashi Daigaku, Hakkô Kenkyusho Kenkyu Hokoku. No. 11:65–69.

KUTTER, F.
1959. Klärung und Filtration. Schweizer Brauerei-Rundschau 70(5):75–81.

KUTTER, F., and W. HIRT
1959. Die technische Hilfsmittel für die Separation Fest/Flüssig. Schweizer Brauerei-Rundschau 70(5):82–158.

LABORDE, J.
1907. Cours d'oenologie. Vol. I. Paris, L. Mulo; Bordeaux, Feret & Fils. 344 pp.

LAFON-LAFOURCADE, S., J. BLOUIN, P. SUDRAUD, and E. PEYNAUD
1967. Essais d'utilisation de la thiamine au cours de vinifications expérimentales en vins blancs liquoreux. Compt. Rend. Acad. Agr. France 60:1046–1051.

LAFON-LAFOURCADE, S., and G. GUIMBERTEAU
1962. Évolution des aminoacides au cours de la maturation des raisins. Vitis 3:130–135.

LAFON-LAFOURCADE, S., and E. PEYNAUD
1958. L'acide p-aminobenzoïque, l'acide ptérolglutamique et la choline (vitamines du group B) dans les vins. Ann. Technol. Agr. 7:303–309.

1959. Dosage microbiologique des acides aminés des moûts de raisin et des vins. Vitis 2:45–46.

1961. Composition azotée des vins en fonction des conditions de vinification. Ann. Technol. Agr. 10:143–160.

1965. Sur l'évolution des acides pyruvique et α-cétoglutamique au cours de la fermentation alcoolique. Compt. Rend. 261:1778–1780.

1966. Sur les taux des acides cétoniques formés au cours de la fermentation alcoolique. Ann. Inst. Pasteur 110:766–778.

LAFOURCADE, S.
1955. Contribution à l'étude des activeurs et des inhibiteurs de la fermentation alcoolique des moûts de raisin. Ann. Technol. Agr. 4:5–66.

LAMBION, R.
1963. Contaminations microbiennes des vins; mesures préventives-hygiène. Ann. Technol. Agr. 12(numéro hors série 1):27–40.

LAMBION, R., and A. MEKHI
1957. Les bactéries de la fermentation malolactique. Rev. Ferm. Ind. Aliment. 12:131–144.

LAMOURIA, L. H., A. J. WINKLER, and A. N. KASIMATIS
1961. Mechanical harvesting of grapes; an engineering and economic appraisal. Wines & Vines 42(5):29–31.

LANG, K., M. FINGERHUT, E. KRUG, and W. REIMOLD
1967. Zur Kenntnis des Pyrokohlensäurediäthylesters. IV. Ueber enzymatische Spaltung und Stoffwechsel von Umsetzungsprodukten des Pyrokohlensäurediäthylesters mit Bestandteilen von Lebensmitteln. Z. Lebensm.-Untersuch. -Forsch. 132:333–341.

LANTEAUME, M.-T., P. RAMEL, P. GIRARD, P. JAULMES, M. GASQ, and J. NANAU
1965. Effets physiologiques à long terme de l'anhydride sulfureux ou des sulfites utilisés pour le traitement des vins rouges. Ann. Fals. Expertise Chim. 58:16–31.

LA ROSA, W. V.
1955. Maturity of grapes as related to pH at harvest. Am. J. Enol. 6(2):42–46.
1958. Observations with the use of PVP on wines. San Francisco, Wine Institute, Technical Advisory Committee, March 7, 1958. 1 p.
1963. Enological problems of large-scale wine production in the San Joaquin Valley. Am. J. Enol. Viticult. 14:75–79.

LA ROSA, W. V., and U. NIELSON
1956. Effect of delay in harvesting on the composition of grapes. Am. J. Enol. 7:105–111.

LASSERRE, A.
1933. Sur les quantités de silicium, de calcium et de magnésium contenus dans les vins et leurs moûts. Proc.-Verb. Soc. Sci. Phys. Nat. Bordeaux 1932–33: 66–72.

LASZLO, I., V. LEPĂDATU, T. GIOSANU, M. MICICI, and S. TARAŞ
1966. Prepararea vinurilor roşii prin macerarea carbonica. Inst. Cercetări Horti-Viticole, Lucrări Ştiinţ. 7:859–868.

LATHROP, R. E., and H. S. PAINE
1931. Some properties of honey colloids and the removal of colloids from honey by bentonite. Ind. Eng. Chem. 23:328–332.

LATIMER, W. M.
1952. The oxidation states of the elements and their potentials in aqueous solutions. 2d ed. New York, Prentice-Hall. 408 pp.

LAUREMA, S., and J. ERKAMA
1968. Formation of ethyl acetate in Hansenula anomala. Acta Chem. Scand. 22:1482–1486.

LAVOLLAY, J., and J. SEVESTRE
1944. Le vin, considéré come un aliment riche en vitamin P. Compt. Rend. Acad. Agr. France 30:259–261.

LAYTON, T. A.
1961. Wines of Italy. London, Harper Trade Journals, Ltd. xi + 221 pp.

LEAKE, C. D., and M. SILVERMAN
1966. Alcoholic beverages in clinical medicine. Chicago, Ill., Year Book Medical Publishers. 160 pp.
1967. The clinical use of wine in geriatrics. Geriatrics 22:G3–G6.

LEE, J. A.
1950. Modern materials solve food processing equipment problems. Food Eng. 22(5):101–112.

LEE, T. C., and M. J. LEWIS
1968a. Identifying nucleotidic materials released by fermenting brewer's yeast. J. Food Sci. 33:119–123.
1968b. Mechanisms of release of nucleotidic material by fermenting brewer's yeast. Ibid. 33:124–128.

LEGENT-FOURNÈS, P.
1944. Le bouquet des vins. Bull. O.I.V. (Office Intern. Vigne Vin) 17(163/166): 54–61.

LEGGETT, H. B.
1939. The early history of the wine industry in California. Master's thesis, University of California, Berkeley. 124 l. (Typewritten.)

LÉGLISE, M.
1964. Particularités de la vinification bourguignonne. Rev. Ferm. Ind. Aliment. 19:87–93.

1966. L'emploi des hautes températures dans la phase préfermentative de la vinification en rouge comme moyen d'extraction rapide et de stabilisation des matières phénoliques (tanins et matières colorantes). Vignes et Vins, numéro spécial **1966**:39–46, 55–59.

LÉGLISE, M., and A. MICHEL
1958. Le déferrage partiel des vins blancs par le phytate de calcium. Ann. Technol. Agr. **7**:433–439.

LEHMAN, A. J.
1950. Some toxicological reasons why certain chemicals may not be permitted as food additives. Quart. Bull. Assoc. Food Drug Off. **14**(3):82–93.

LEHMANN, G.
1966. Trennung organischer Säuren auf Cellulose-Dünnschichtplatten. Z. Lebensm.-Untersuch. -Forsch. **130**:269–273.

LE MAGNEN, J.
1966. Les bases physiologiques de l'analyse et de l'appréciation des qualités organoleptiques. Colloque Oenologique de l'Inst. Tech. Vin. **11**:19–21, 23. (*See also* Vignes et Vins No. **149**:13.)

LEMPERLE, E., and R. MECKE
1965. Gas chromatographische Analyse der flüchtigen Inhaltsstoffe von Weinen, Mosten und Spirituosen. Z. analyt. Chem. **212**:18–30.

LEONHARDT, G.
1963. Das Weinbuch; Werden des Weines von der Rebe bis zum Glase. Leipzig, VEB Fachbuchverlag. 418 pp.

LEPĂDATU, V., and O. BELLU
1959. Modificarea caracteristicilor fizico-chemice ale strugurilor provocată de *Botrytis cinerea* prin infecţii artificiale. Comm. Acad. Repub. Pop. Romîne **9**:467–472.

LEPĂDATU, V., O. BELLU, T. GIOSAN, and E. POPA
1963. Prevenirea oxidării mustului în timpul prelucrării strugurilor prin administrarea de SO_2. Inst. Cercetări Horti-Vinicole, Lucrări Ştiinţ. **6**:687–696.

LETEY, J., and L. H. STOLZY
1964. Measurement of oxygen diffusion rates with platinum microelectrodes. I. Theory and equipment. Hilgardia **35**:545–554.

LEWIS, M. J.
1964. Aspects of the nitrogen metabolism of brewers' yeast. Wallerstein Lab. Commun. **27**:29–39.

LEWIS, M. J., and H. J. PHAFF
1963. Release of nitrogenous substances by brewers' yeast. 2. Effect of environmental conditions. Am. Soc. Brew. Chem. Proc. **1963**:114–123.

LHERME, G.
1931–32. La teneur en cuivre des vins de la Gironde (récolte 1931). Proc. Verb. Soc. Sci. Phys. Nat. Bordeaux **1931–32**:119–121.

LIBRATY, V.
1961. Ester determinations and their application to wine. Master's thesis, University of California, Davis. 76 pp.

LICHEV, V., N. GORANOV, and T. GANEVA
1966. V'rkhu okisitelno-reduktsionnite protesi pri proizvodsvoto na vino. Lozarstvo Vinarstvo (Sofia) **15**(3):24–30.

LICHINE, ALEXIS
1963. Wines of France. 4th ed. New York. Alfred A. Knopf, Inc. ix + 388 + xxvii pp.
1967. Alexis Lichine's encyclopaedia of wines and spirits. New York, Alfred A. Knopf, Inc. xii + 713 pp.

LITERATURE CITED

LICHTBLAU, S.
1965. Food technology in Israel. *In* M. S. Peterson and D. K. Tressler (eds.), Food technology the world over. Vol. 2. South America, Africa, and the Middle East, Asia. Westport, Conn., Avi Publishing Co. 414 pp. (*See* pp. 195–241.)

LIEBERT, H. P., and H. WARTENBERG
1965. Untersuchungen des Wein- und Apfelsäuregehaltes reifender Weinbeeren. Ber. Deut. Botan. Ges. 78:115–121.

LINDNER, J.
1912. Die elektrolytische Dissoziation der schwefligen Säure. Monatsh. Chem. 33:613–672.

LIOTTA, C.
1956. Interim report concerning experiments on naturally fermented and artificially carbonated wines. Washington, D.C., Internal Revenue Service No. 21174. 2 pp.

LITSCHEV, W., S. GANEV, and M. GANEVA
1966a. Ueber den Mechanismus des Zusammenwirkens zwischen Bentonittonen und Wein. Weinberg Keller 13:445–451.

LITTLE, A. C.
1964. Color measurement of translucent food samples. J. Food Sci. 29:782–789.

LITTLE, A. C., and G. MACKINNEY
1966. A proposal for rapid measurement of color and color differences in wine samples. San Francisco, Wine Institute, Technical Advisory Committee, June 9, 1966. 4 pp.

LITTLE, ARTHUR D. (Inc.)
1958. Flavor research and food acceptance. A survey of the scope of flavor and associated research. New York, Reinhold Publishing Corp. 6 + 391 pp.

LIUNI, C. S., A. CALÓ, and G. CAPPELLERI
1965. Contributo allo studio sui pigmenti antocianici di alcune specie del genere *Vitis* e di loro ibridi. Atti Accad. Ital. Vite Vino 17:161–167.

LLOYD, W. J. W., and B. C. COWLE
1963. Determination of free sulphur dioxide in soft drinks by a desorption and trapping method. Analyst 88:394–398.

LOCKWOOD, L. B., D. E. YODER, and M. ZIENTY
1965. Lactic acid. Ann. N.Y. Acad. Sci. 119:854–865.

LODDER, J.
1934. Die Hefesammlung des "Centraalbureau voor Schimmelcutures." Beiträge zu einer Monographie der Hefearten. II Teil. Die Anaskosporogenen Hefen. Erste Hälfte. Amsterdam, Holland, N.V. Noord-Hollandsche Uitgevers-Maatscjapprij. ix + 256 pp. (*See also* K. Akad. van Wetensch. te Amsterdam Verhandel. [Tweed Sec.] 32:1–256.)

LODDER, J., and N. J. W. KREGER-VAN RIJ
1952. The yeasts. A taxonomic study. Amsterdam, North Holland Publishing Co. xi + 713 pp.

LOGOTHÉTIS, B.
1965. Le Malvasie. Atti Accad. Ital. Vite Vino 17:69–122.

LOOMIS, W. D., and J. BATTAILE
1966. Plant phenolic compounds and the isolation of plant enzymes. Phytochem. 5:423–438.

LOTT, R. V., and H. C. BARRETT
1967. The dextrose, levulose, sucrose, and acid content of the juice from 39 grape clones. Vitis 6:257–268.

LOVE, R. F.
1939. A table for ebulliometers. Ind. Eng. Chem., Anal. Ed. 11:548–550.

LUCIA, S. P.
1954. Wine as food and medicine. Toronto and New York, The Blakiston Co., Inc. 149 pp.
1963a. A history of wine as therapy. Philadelphia and Montreal, J. B. Lippincott Co. xviii + 234 pp.
1963b. Alcohol and civilization. New York, McGraw-Hill Book Co. 416 pp.

LUCIA, S. P., and M. L. HUNT
1957. Dietary sodium and potassium in California wines. Am. J. Digest. Diseases 2:26–30.

LÜCK, E., and H. NEU
1965. Verhütung des Nachgarens von restsüssen Weinen durch Sorbinsäure. Z. Lebensm.-Untersuch. -Forsch. 126:325–335.

LÜTHI, H.
1957a. La rétrogradation malolactique dans les vins et les cidres. Rev. Ferm. Ind. Aliment. 12:15–21.
1957b. Symbiotic problems relating to the bacterial deterioration of wines. Am. J. Enol. 8:176–181.
1958. Die Bedentung des kontinuierlichen Pressen von obst- und Traubensäften. Flüssiges Obst. 25(4):16–19.
1959a. Fortschritte in der Fruchtsaftherstellung. Proc. Fifth Int. Fruit Juice Congress in Vienna. 15 pp.
1959b. The progress realized in the production of fruit juices. Fruit 14:447–457.

LÜTHI, H., and R. HOCHSTRASSER
1950. Zur Behandlung der Schläuche in den Wein- und Obstweinkeltereien. Schweiz. Z. Obst- Weinbau 59(18):321–325.

LÜTHI, H., and U. VETSCH
1952. Papierchromatographische Bestimmung von Aminosäuren in Wein. Schweiz. Z. Obst- Weinbau 61:390–394, 405–408.
1953. Papierchromatographische Trennung und Bestimmung von Aminosäuren in Traubenmost und Wein. Deut. Weinbau, wissen. Beih. 7:3–6, 33–54.

LUKTON, A., and M. A. JOSLYN
1956. Mechanism of copper casse formation in white table wine. II. Turbidimetric and other physico-chemical considerations. Food Res. 21: 456–476.

LUKTON, A., C. O. CHICHESTER, and G. MACKINNEY
1956. The breakdown of strawberry anthocyanin pigment. Food Technol. 10: 427–432.

McCOLLY, D. W.
1967. Production and marketing of wine and brandy in the United States. Fresno, Wine Institute. 13 pp. plus 21 pp. statistical appendix. Processed.

McGARVEY, F. X., R. W. PERCEVAL, and A. L. SMITH
1958. Ion exchange develops as a process in the wine industry. Am. J. Enol. Viticult. 9:168–179.

MACHER, L.
1952. Hefegärung und Schwefelwasserstoffbildung. Deut. Lebensm.-Rundschau 48:183–189.

McLAREN, A. D.
1954. The adsorption and reactions of enzymes and proteins on kaolinite. I. J. Phys. Chem. 58:129–137.

LITERATURE CITED

McLaren, A. D., G. H. Peterson, and I. Barshad
1958. The adsorption and reactions of enzymes and proteins on clay minerals. IV. Kaolinite and montmorillonite. Soil Science Soc. America Proc. 22: 239–244.

Mahler, H. R., and E. H. Cordes
1966. Biological chemistry. New York, Harper and Row. 872 pp. See pp. 404–474.

Malan, C. E.
1966. Sul deposito cotonaceo prodotto dallo sviluppo di *Lactobacillus trichodes* Fornachon, Douglas e Vaughn in vermut piemontesi. Ann. Facoltà Sci. Agr. Univ. Studi Torino 1:27–40.

Malan, C. E., and C. Cano Marotta
1959. I lieviti della fermentazione vinaria in Piemonte. V. I lievite dei mosti del Cortese. Atti Accad. Ital. Vite Vino 11:405–420.

Maltabar, V. M.
1959. Sorta vinograda i sroki sbora dlia proizvodstva kon'iakov Moldavii. Trudy Mold. Nautchn. -Issled. Inst. Sad., Vinogr. Vinod. 4:187–211.

Malya, E.
1965. A reduktontartalom alakulása a szőlő érése és a bor fejlődése során. Borgazdaság 13:141–145.

Manfredi, E., and G. C. Ropa
1962. Orientamenti nella meccanizzazione della viticoltura. Progressi tecnici della meccanizzazione et organizzazione della cantina. Pavia, Camera di Commercio, Industria, Agricoltura. 95 pp.

Mannelli, G., and P. Mancini
1966. Les constituants mineurs dans les milieux hydro-alcooliques. II. Le méthanol dans les vins italiens. Ann. Fals. Expertise Chim. 59:225–229.

Marcilla Arrazola, J.
1946. Tratado práctico de viticultura y enología españolas. Vol. II. Enología. 2d ed. Madrid, Sociedad Anónima Española de Traductores y Autores. 501 pp.

Mareca Cortés, I.
1954. Estado actual de los conocimientos sobre los fenómenos de oxidación y reducción en los vinos. Rev. Cienc. Appl. (Madrid) 8:224–230.
1964. Mesure de l'intensité et de la tonalité de la couleur d'un vin quelconque. Paris, Office International de la Vigne et du Vin. Cahier Vert No. 162–168.

Mareca Cortés, I., and M. de Campos Salcedo
1957. Sur la combinaison de l'éthanal et des polyphénols dans les vin rouges. Ind. Aliment. Agr. (Paris) 74:103–106.

Mareca Cortés, I., C. Diez de Bethencourt, and A. M. Plasencia Plasencia
1959. La vitamina C, nuevo antioxidante para elaborar mostos y vinos. Revista Agricultura (Madrid) 28:512–516.

Mareca Cortés, I., and A. Gonzalez
1964. Sur la composition de la matière colorante des vins rouges. Ind. Agr. Aliment. (Paris) 81:391–397.
1965a. Rapports à l'étude de l'évolution de la matière colorante des vins. Compt. Rend. Acad. Agr. France 51:636–643.
1965b. Contribución al conocimiento de la evolución de la materia colorante desde la uva al vino anejo. Vitis 5:201–211.

Mareca Cortés, I., and M. de Miguel
1965. Contribution à l'étude polarographique des flavonoïdes. Compt. Rend. Acad. Agr. France 51:123–128.

Marescalchi, A., and G. Dalmasso
1931–1937. Storia della vite et del vino in Italia. Milan, Presso Arti Grafiche E. Gualdoni. 3 vols.

913

MARESCALCHI, C.
1966. Ancora perplessità sull'alcole metilico nei vini. Italia Vinicola Agrar. 56:413–420.

MARGHERI, G., and M. RIGOTTI
1964. Le sostanze minerali dei vini della provincia de Trento. Riv. Viticolt. Enol. 17:405–414.

MARIE, R., D. BOUBALS, and P. GALZY
1962. Sur la dégustation rationnelle des boissons. Bull. O.I.V. (Office Intern. Vigne Vin) 35:756–787.

MARINOVA-BOIADZHIEVA, B.
1966. Po v'prosa za mikroflorata na k'rdzhaliiskiia i rum'nskiia bentonit i vliianieto i v'rkhu mifroflorata na vinoto. Nauchni Trudove (Nauchnoizsledovatelski Institut po Vinarska i Pivovarna Promishlenost, Sofiia) 7:121–126.

MARKAKIS, P., and R. J. EMBS
1966. Effect of sulfite and ascorbic acid on mushroom phenol oxidase. J. Food Sci. 31:807–811.

MARKLEY, K. S., C. F. SANDO, and S. B. HENDRICKS
1938. Petroleum ether-soluble and ether-soluble constituents of grape pomace. J. Biol. Chem. 123:641–654.

MARKMAN, A. L., and W. M. EL'GORT
1962. Polarographic determination of pyruvic acid in fermenting musts and wines (transl.). Izv. Vysshikh Uchebn. Zavedenii, Pishchevaia Tekhnol. 1962(6): 128–131. (Chem. Abst. 58:11,924. 1963.)

MARQUARDT, P., and H. W. J. WERRINGLOER
1965. Toxicity of wine. Food Cosmet. Toxicol. 3:803–810.

MARQUES GOMES, J. V., M. F. DA SILVA BABO, and A. F. GUIMARAIS
1956. L'emploi des bactéries selectionnées dans la fermentation malolactique du vin. Bull. O.I.V. (Office Intern. Vigne Vin) 29(299):349–357.

MARSAIS, P.
1941. Les huiles de pépins de raisin. Bull. O.I.V. (Office Intern. Vigne Vin) 14(143):82–116; (144):61–108.

MARSH, G. L.
1940. Metals in wine. Wine Rev. 8(9):12–14, 24; (10):24–26, 28–29.
1948. Automatic temperature control recommended for improved wine-growing. Calif. Agr. 2(4):5, 8.
1949. Proper filtration technique. Wines & Vines 30(6):21–22.
1950. Personal communication.
1952. New compound ends metal clouding. A report on the Fessler compound. Wines & Vines 33(6):19–21.
1958. Alcohol yield: factors and methods. Am. J. Enol. 9:53–58.

MARSH, G. L., and J. F. GUYMON
1964. Refrigeration in wine making. Am. Soc. Refr. Eng. Data Book, Vol. I, chap. 10.

MARSH, G. L., and M. A. JOSLYN
1935. Precipitation rate of cream of tartar from wine. Ind. Eng. Chem. 27:1252–56.

MARSH, G. L., and K. NOBUSADA
1938. Iron determination methods. Wine Rev. 6(9):20–21.

MARSH, G. L., and G. A. PITMAN
1930. Pectin content of grapes. Fruit Prod. J. 9:187–188.

MARTAKOV, A. A., and V. A. KOLESNIKOV
1967. Adsorbtsiia i usloviia vydeleniia β-fruktofuranozidazy drozhzhevymi kletkami. Vinodelie i Vinogradarstvo SSSR 27(1):13–16.

MARTEAU, G.
1963. Propriétes clarificantes des enzymes pectolytiques. Ann. Technol. Agr. 12 (numéro hors série 1):218–230.
MARTEAU, G., and C. OLIVIERI
1966. Perspectives et données actuelles de la vinification en rouge par macération à chaud. Prog. Agr. Vitic. 166:133–136, 150–163, 191–195, 215–219.
MARTEAU, G., J. SCHEUR, and C. OLIVIERI
1961. Cinétique de la libération enzymatique du méthanol au cours des transformations pectolytiques du raisin. Ann. Technol. Agr. 10:161–183.
1963. Le rôle des enzymes pectolytiques du raisin ou de préparations commerciales dans le processus de la clarification des jus. Ibid. 12:155–176.
MARTINI, L. P.
1966. The mold complex of Napa Valley grapes. Am. J. Enol. Viticult. 17:87–94.
MARTINI, M.
1965. L'acido tartarico racemico come decalcifiante nei vini. Riv. Viticolt. Enol. 18:379–386.
MARUYAMA, C., and T. KUSHIDA
1965. Enological studies on the color of red wines. 6. Extraction and utilization of red color from grape skins. Yamanashi Daigaku, Hakkô Kenkyusho Kenkyo Hokoku. No. 12:41–46.
1966. Effects of EDTA and ascorbic acid on the quality of white wines. Ibid. No. 13:43–49.
MASCOLO, A.
1966a. Contenuto in alcole metilico dei vini prodotti nelle provincie di Modena e Reggio Emilia. I. Vendimia 1965. Riv. Viticolt. Enol. 19:336–359.
1966b. Osservazioni sul metodo ufficiale di analisi per la determinazione dell'alcole metilico nei vini e nei mosti. Ibid. 19:213–219.
MASQUELIER, J.
1956. Identification et dosage des facteurs vitaminiques P dans diverses boissons fermentées. Bull. Soc. Chim. Biol. 38:65–70.
1959. The bactericidal action of certain phenolics of grapes and wine. In J. W. Fairbairn (ed.), The pharmacology of plant phenolics. New York and London, Academic Press. 151 pp. (See pp. 123–131.)
MASQUELIER, J., and D. DELAUNAY
1965. Action bactéricide des acides-phénols du vin. Bull. Soc. Pharm. Bordeaux 104:152–156.
MASQUELIER, J., and H. JENSEN
1953. Bactericidal action of red wines (transl.). Bull. Soc. Pharm. Bordeaux 91:24–29, 105–109. (Chem. Abst. 48:325. 1954.)
MASQUELIER, J., and J. LAPARRA
1967. Action cholérétique des constituents cinnamiques du vin. Rev. Franç. Oenol. No. 26:33.
MASQUELIER, J., and R. RICCI
1964. Chromatographie des dérivés cinnamiques du vin. Qual. Plant. Mater. Végétabiles 11:244–248.
MASUDA, H., and H. MURAKI
1966. Utilization of yeast cells in wine making. III. Amino acids discharged by yeast cells into wine. Yamanashi Daigaku, Hakkô Kenkyusho Kenkyo Hokoku No. 13:1–9.
MASUDA, H., N. SHIJO, and H. MURAKI
1964. Fermentative processing of apple fruit. V. Use of complex yeast cultures in cider making. Hakkô Kôgak. Zasshi 42:383–387.

MATALAS, L., G. L. MARSH, and C. S. OUGH
1965a. The effect of concentration conditions and storage temperatures on grape juice concentrate. Am. J. Enol. Viticult. 16:129–135.
1965b. The use of reconstituted grape concentrate for dry table wine production. Ibid. 16:136–143.

MATHEWS, J.
1958. The vitamin B complex content of bottled Swiss grape juice. Vitis 2:57–64.

MATHIEU, L.
1913. Influence de la temperature du cuvaison sur les qualités du vin rouge. Prog. Agr. Vitic. 58:340–346.

MATHIEU, L., and R. MATHIEU
1938. Problèmes pratiques sur la vinification. Cannes, France, Institut Œnotechnique de France. 62 pp.

MATTA, M., and V. ASTEGIANO
1967. Ricerca gascromatografica dell'essenza di Salvia sclarea nel Moscato d'Asti. Riv. Viticolt. Enol. 20:165–172.

MATTICK, L. R., W. B. ROBINSON, L. D. WEIRS, and D. L. BARRY
1963. Grape juice flavor: determination of methyl anthranilate in grape juice by electron affinity–gas chromatography. J. Agr. Food Chem. 11:334–336.

MATZ, S. A.
1965. Texture in foods. Westport, Conn., Avi Publishing Co. 275 pp.

MAUREL, A., and S. TOUYÉ
1963. Détection et dosage des dérivés de l'acide benzoïque dans les vins. Compt. Rend. Acad. Agr. France 49:150–157.

MAUREL, A., S. REY, and M. REY
1964. Séparation par chromatographie en couche mince de l'acide ascorbique du glucose. Dosage de l'acide ascorbique dans le vin. Compt. Rend. Acad. Agr. France 50:1081–1083.

MAUREL, A., O. SANSOULET, and Y. GEFFARD
1965. Étude de la détermination des alcools supérieurs dans les eaux-de-vie. Ann. Fals. Expertise Chim. 58:219–227.

MAURER, R.
1967. Eine einfache kolorimetrische Schnellmethode zur Weinsäurebestimmung für das Betriebslabor. Weinberg Keller 14:323–328.

MAYER, K.
1965. Biologischer Säureabbau mit Spalthefen. Schweiz. Z. Obst-Weinbau 101: 368–370.

MAYER, K., and I. BUSCH
1963. Ueber eine enzymatische Apfelsäurebestimmung in Wein und Traubensaft. Mitt. Gebiete. Lebensm. Hyg. 45:60–65.

MAYER, K., I. BUSCH, and G. PAUSE
1964. Ueber die Bernsteinsäurebildung während der Weingärung. Z. Lebensm.-Untersuch. -Forsch. 125:375–381.

MAYER, K., and H. LÜTHI
1960. Versuche mit Pyrokohlensäurediäthylester, einem neuen Getränkekonservierungsmittel. Mitt. Gebiete Lebensm. Hyg. 51:132–137.

MAYER, K., and G. PAUSE
1965a. Eine enzymatische Citronensäure-Bestimmung. Mitt. Gebiete Lebensm. Hyg. 56:454–458.
1965b. Ueberprüfung einiger aus der Praxis isolierter Weinhefen. Schweiz. Z. Obst- Weinbau 101:235–240.
1966. Indole Verbindungen in Wein. Mitt. Gebiete Lebensm. Hyg. 57:147–160.

LITERATURE CITED

MECKE, R., R. SCHINDLER, and M. DE VRIES
1960. Gaschromatographische Untersuchungen an Weinen. Wein-Wissen. 15: 183–191.

MEHLITZ, A., and H. TREPTOW
1966. Ueber die Einwirkung von Pyrokohlensaüre-diäthylester (PKE) auf Enzyme. Ind. Obst- Gemüseverwert. 51(3):65–66. (Chem. Abst. 64:17938. 1966.)

MEHLITZ, A., H. TREPTOW, and K. GIERSCHNER
1967. Ueber die Einwirkung von Pyrokohlensäurediäthylester (PKE) auf Alkohol-dehydrogenase und auf Trypsin. Flüssiges. Obst 34:20–24.

MEHREN, G.
1950. Economic situation and market organization in the California grape industries. Univ. Calif., Giannini Found., Agr. Econ. Rept. 107:1–92 + suppls. (Mimeo.)

MEHREN, G., and S. W. SHEAR
1950. Trends and outlook in the California grape industries. Calif. Agr. Exper. Stat. Cir. 397:1–24.

MELAMED, N.
1962. Détermination des sucres résiduels des vins, leur relation avec la fermentation malolactique. Ann. Technol. Agr. 11:5–31.

MERANGER, J. C., and E. SOMERS
1968. Determination of heavy metals in wines by atomic absorption spectrophotometry. J. Assoc. Off. Anal. Chem. 51:922–925.

MERCER, W. A., and I. I. SOMERS
1957. Chlorine in food plant sanitation. Advan. Food Res. 7:129–169.

MERCZ, A., L. ÉRCZHEGYI, and S. ROHRSETZER
1963. A borban szuszpendált anyagok—a gravitáció és a centrafugális erő hatására beálló—változásainak bortechnológiai értékelése. Szőlészeti Kutató Intézet Évkönyve 7:251–271.

MERZHANIAN, A. A.
1950. Carbon dioxide absorption coefficient of wine (transl.). Vinodelie i Vinogradarstvo SSSR 10(5):34–37. (Chem. Abst. 44(20):9620. 1950.)
1952. O povedenii diétilovogo éfira pirougol'noĭ kisloty v igristykh i gazirovannykh vinakh. Vinodelie i Vinogradarstvo SSSR 11(3):19–22.
1963a. Faktory nakopleniia v shampanskom sviazannoĭ uglekisloty. Biokh. Vinodeliia 7:148–163.
1963b. Nekotorye fizicheskie usloviia rozliva shampanskogo (Some physical conditions in transfer of sparkling wines). Vinodelie i Vinogradarstvo SSSR 23(8):3–8.

MESTRE ARTIGAS, C.
1947. Las superfermentaciones en enología. R. Acad. Ciénc. Artes Barcelona Mem. 28(14):509–556.

MESTRE ARTIGAS, C., and I. CAMPLLONCH Y ROMEN
1942. La producción de aldehidos en la fermentación de mostos sulfitados y su influencia en los vinos. Minist. Agr., Inst. Nac. Invest. Agron., Estac. Vitic. Enol., Villafranca del Panadés, Cuaderno 14:1–16.

MESTRE ARTIGAS, C., and A. JANÉ
1946. Fermentaciones comparativas con diferentes levaduras. Minist. Agr., Inst. Nac. Invest. Agron., Estac. Vitic. Enol., Villafranca del Panadés, Cuaderno 68:1–28.

917

MESTRES, R.
1961. Note sur la recherche du traitement des vins au ferrocyanure de potassium. Extraction d'acide cyanhydrique à partir de vins normaux. Ann. Fraud. Expertise Chim. 54:284–298.
MESTRES, R., and C. MARTIN
1964. Teneur en cobalt des vins. Trav. Soc. Pharm. Montpellier 24:42–46.
MEYERHOF, O.
1945. The origin of the reaction of Harden and Young in cell-free alcoholic fermentation. J. Biol. Chem. 157:105–119.
MEYRATH, J., and H. R. LÜTHI
1969. On the metabolism of hexoses and pentoses by *Leuconostoc* isolated from wines and juices. Lebensm.-Wiss. u. Technol. 2:22–27.
MICONI, C.
1962a. A proposito dell'acido cianidrico nei vini. Riv. Viticolt. Enol. 15:33–38.
1962b. Valutazione degli acidi metatartarici. *Ibid.* 15:134–137.
MILLER, J. F.
1958. Carbon dioxide in water, in wine, in beer and in other beverages. Oakland, Calif. 49 pp.
1964. Carbon dioxide stability in beverages. Food Technol. 18:60–63.
1966. Viewpoint: quality carbonation of wine. Wines & Vines 47(6):49–50.
MILLER, S. A.
1947–1949. Filtration. Ind. Eng. Chem. 39:5–7; 40:25–27; 41:38–41.
MILLS, D. R., and E. H. WIEGAND
1942. Effect of storage on sulfur dioxide in wine. Fruit Prod. J. 22(1):5–9.
MINÁRIK, E.
1962. Doterajšie poznatky so stabilizáciou vín kyselinou sorbovou. Kvasný Průmysl 9:253–257.
1964a. Beitrag zur Hefeflora gärender Rotweinmaischen. Vitis 4:368–372.
1964b. Die Hefeflora von Jungweinen in der Tschechoslowake. Mitt. Rebe u. Wein, Serie A (Klosterneuburg) 14:306–315.
1965. Vplyv čistej kultúry a zmesi cistých kultúr kvasiniek na kvasenie hroznového muštu (Influence of pure yeast and mixed yeast cultures on the fermentation of grape musts). Kvasný Průmysl 11:82–85.
1966. Ekológia prírodných druhov vínnych kvasiniek v Československii. Biol. Práce 12(4):1–107.
1967. Zum Vorkommen von kontaminierenden Hefen und hefeartigen Mikroorganismen im Wein bei der Abfüllung. Wein-Wissen. 22:67–74.
MINÁRIK, E., and A. KOCKOVÁ-KRATOCHVÍLOVÁ
1966. Príspevok k taxonomickej príslušnosti niektorých vinárskych kvasiniek a príbuzných druhov. Kvasný Průmysl 12:13–15.
MINÁRIK, E., and L. LAHO
1962. Štabilizácia sladkastých vín dietylesterom kyseliny pyrouhličitej. Kvasný Průmysl 8:86–89.
MINÁRIK, E., L. LAHO, and A. NAVARA
1960. Beitrag zur Kenntnis der Hefeflora von Trauben, Mosten und Weinen. Mitt. Rebe u. Wein, Serie A (Klosterneuburg) 10:218–223.
MINÁRIK, E., and M. NAGYOVÁ
1966a. Mikroflóra mustov a vín nitrianskej a podunajskej vinohradníckej oblasti. *In* Pokroky vo Vinohradnickom a Vinárskom Výskume. Bratislava, Vydavatel'stvo Slovenskej Akadémie Vied. **1966.** (Pp. 277–305.)
1966b. Poznatzy o stabilizách sladkých vín voči kvasinkovým zákalom. *In Ibid.* (*See* pp. 259–276.)

LITERATURE CITED

Minárik, E., and P. Rágala
1966. Einfluss einiger Fungizide auf die Hefeflora bei der spotanen Mostgärung. Mitt. Rebe u. Wein, Obstbau Früchteverwertung (Klosterneuburg) **16**:107–114. (*See also* Agrochemia, **7**:196–207. 1967.)

Moeller, W. M.
1957. The use of gelatin as a precipitant in American cellaring operation. Brewers Digest **32**(2):52–53, 57. (See also *Ibid.* **39**(1):60–70, 1964; **41**(1):50–54, 1966.)

Molnar, A., and A. Mercz
1964. A specifikus nyomás alakulása a szőlő sajtolásakor. Borgazdaság **12**:1–4.

Mondavi, P., and K. Robe
1961. Two inert gases hold peak quality, improve wine production scheduling. Food Processing **22**(5):45–47.

Monier-Williams, G. W.
1927. The determination of sulfur dioxide. [Great Britain] Min. Health, Repts. Pub. Health and Med. Subjs. **43**:1–56.

Moreau, L., and E. Vinet
1928. L'acide sulfureux en vinification. Son réglage. Doses antiseptiques. Ier Cong. Internatl. de la Vigne et du Vin. Bordeaux [France]. (*See* J. Ribéreau-Gayon, 1947.)
1937. Sur la détermination du pouvoir antiseptique réel de l'acide sulfureux dans les moûts et les vins par la méthode de l'index iodé. Compt. Rend. Acad. Agr. France **23**:570–576. (*See also* Rev. Viticult. **87**:25–28, 61–67.)

Moretti, R. H., and H. W. Berg
1965. Variability among wines to protein clouding. Am. J. Enol. Viticult. **16**:69–78.

Morgan, A. F., L. Brinner, C. B. Plaa, and M. M. Stone
1957. Utilization of calories from alcohol and wines and their effects on cholesterol metabolism. Am. J. Physiol. **189**:290–296.

Morgan, A. F., H. L. Nobles, A. Wienes, G. L. Marsh, and A. J. Winkler
1939. The B vitamins of California grape juices and wines. Food Res. **4**:217–219.

Morgan, K.
1965. Fusel oil in beer. Quantitative analysis by gas-liquid chromatography. J. Inst. Brewing **71**:167–171.

Morgan, O. M., and O. Maass
1931. An investigation of the equilibria existing in gas-water systems forming electrolytes. Can. J. Research **5**:162–199.

Morrison, R. L.
1961. Determination of ethanol in wine by gas-liquid partition chromatography. Am. J. Enol. Viticult. **12**:101–106.
1965. Determination of carbon dioxide in lightly carbonated wine. J. Assoc. Offic. Agr. Chemists **48**:471–472.

Morrison, R. L., and T. E. Edwards
1963. Semi-automatic determination of ethanol in wine by the micro-dichromate method. Am. J. Enol. Viticult. **14**:185–193.

Mortimer, R. K., and D. C. Hawthorne
1966. Yeast genetics. Ann. Rev. Microbiol. **20**:151–168.

Moser, L.
1967. Die Abhängigkeit des Zuckergehaltes edelfauler Trauben von den Witterungsverhältnissen. Mitt. Rebe u. Wein, Obstbau u. Früchteverwertung (Klosterneuburg) **17**:173–179.

919

Moțoc, D., and C. Dimitriu
1963–64. Acțiunea unor antibiotice asupra drojdiilor și bacteriilor acetice (The action of some antibiotics on yeasts and acetic bacteria). Lucrările Inst. Cercetări Aliment. 7:393–417.
1966. Acțiunea unor esteri asupra fermentației alcoolice. Inst. Cercetări Horti-Vinicole, Lucrări Științ. 11:303–313.

Moțoc, D., B. Segal, and R. Segal
1966. Studiul factorilor care influentează formarea alcoolului metilic la tratamentul cu enzime pectolitice. Inst. Cercetări Horti-Vinicole, Lucrări Științ. 11: 117–124.

Mourgues, J., P. Bénard, C. Flanzy, and C. Jouret
1967. Techniques de vinification en rouge et potentiel oxydoréducteur. Ann. Technol. Agr. 16:333–347.

Mrak, E. M., L. Cash, and D. C. Caudron
1937a. Effects of certain metals and alloys on claret- and sauterne-type wines made from vinifera grapes. Food Res. 2:539–547.

Mrak, E. M., D. C. Caudron, and L. M. Cash
1937b. Corrosion of metals by musts and wines. Food Res. 2:439–455.

Mrak, E. M., and J. F. Fessler
1938. Changes in iron content of musts and wines during vinification. Food Res. 3:307–309.

Mrak, E. M., and L. S. McClung
1938. Concerning the genera of yeasts occurring on grapes and grape products in California. J. Bacteriol. 36:74–75.

Mrak, E. M., and H. J. Phaff
1948. Yeasts. Ann. Rev. Microbiol. 2:1–46.

Müller-Thurgau, H.
1884. Ueber den Einfluss der Temperatur auf Verlauf und Produkt der Wein-gährung. Generalversamm. des Deutschen Weinbau-Ver. Ber. Pp. 50–57. Geisenheim, Germany.

Müller-Thurgau, H., and A. Osterwalder
1913. Die Bakterien im Wein und Obstwein und die dadurch verursachten Veränderungen. Zentr. Bakteriol. Parasitenk., Abt. II 36:192–339.
1917. Weitere Beiträge zur Kenntnis der Mannitbakterien im Wein. Ibid. 48:1–35.
1919. Ueber die durch Bakterien verursachten Zersetzung von Weinsäure und Glyzerin im Wein. Landw. Jahrb. Schweiz 33:313–361.

Münz, T.
1963. Die Kalium-Pufferung im Most und Wein. Wein-Wissen. 18:496–502.
1965a. Die Bedeutung der organischen und anorganischen Kalium-Bindung für die Ausbildung des säuren Geschmacksbildes im Wein. Ibid. 20: 560–562.
1965b. Die Veränderung der Säure von Most zum Wein und ihre natürliche Geschmacks Ausbildung. Mitt. Rebe u. Wein, Serie A (Klosterneuburg) 15:67–71.
1967. Die geschmackliche Wandlungsfähigkeit der Aepfelsäure. Wein-Wissen. 22:266–272.

Mulvany, J.
1966. Filtration-sterilization of beverages. Process Biochem. 1(9):470–473.

Muradov, A. G.
1966. Vitaminnyĭ sostav vinogradnogo soka Azerbaĭdzhana. Konserv. Ovosh-chesushil. Prom. 21(8):28–29.

Murolo, G.
1967. Azoto totale ed assimilabile in alcuni mosti meridionali. Ind. Agr. 4:585–588.

MUTH, FR., and L. MALSCH
1934. Versuche zur Aufstellung einer Stickstoffbilanz in Traubenmosten und -weinen. Z. Untersuch. Lebensm. 68:487–500.

NAITO, R.
1966. Studies on the coloration of grapes. VII. Behavior of anthocyanins and leucoanthocyanins in the skin of some black and red grapes as affected by light intensity. J. Jap. Soc. Hort. Sci. 35:225–232.

NAKAGAWA, S., and Y. NANJO
1965. A morphological study of Delaware grape berries. J. Jap. Soc. Hort. Sci. 34:85–95.
1966. Comparative morphology of the grape berry in three cultivars. Ibid. 35: 117–126.

NAKAYAMA, O., N. YANOSHI, and T. NAGATA
1966. Studies on the application of honey to alcoholic beverages. 7. On the browning reaction and the application of "solid antioxidants." Yamanashi Daigaku, Hakkô Kenkyusho Kenkyo Hokoku No. 13:19–24.

NASLEDOV, S. N.
1963. Heat instability of wines. San Francisco, Wine Institute, Technical Advisory Committee, June 3, 1963. 3 pp.

NASSAR, A. R., and W. M. KLIEWER
1966. Free amino acids in various parts of Vitis vinifera at different stages of development. Proc. Am. Soc. Hort. Sci. 89:281–294.

NAVARA, A., K. DOBROVODA, and P. BAJČI
1966a. Obsah vol'ných aminokyselín a cukrov vo vínach rôznych odrôd lokalít Modra a Nitra. In Pokroky vo Vinohradníckom a Vinárskom Výskume. Bratislava, Vydavatel'stvo Slovenskej Akadémie Vied. 1966 (pp. 209–228).

NAVARA, A., L. LAHO, and A. PEŠKO
1966b. Výsledky sledovania inverzie sacharózy v muštoch a vo vinach. In Pokroky vo Vinohradníckom a Vinárskom Výskume. Bratislava, Vydavatel'stvo Sloveneskej Akadémie Vied. 1966 (pp. 229–258).

NAVASSART, E.
1910. Ueber den Einfluss der Alkalien und Säuren auf die Autolyse der Hefe. Z. Physiol. Chem. 70:189–197.
1911. Ueber den Einfluss der Antiseptica bei der Hefeautolyse. Ibid. 72:151–157.

NÈGRE, E.
1939. Sur le collage des vins. Ann. École Nat. Agr. Montpellier. n.s. 25:279–294.
1942–1943. Les matières tannoïdes et la composition des vins. Bull. O.I.V. (Office Intern. Vigne Vin) 15(154):20–52; 16(155):25–56.
1967. Le point actuel sur la vinification continue. Prog. Agr. Vitic. 167:511–524.

NÈGRE, E., and R. CORDONNIER
1953. Les origines du fer des vins. Compt. Rend. Acad. Agr. France 39:52–56.

NÈGRE, E., and A. DUGAL
1967. Les variations d'acidité fixe lors de l'élaboration des vins du Midi méditerranéen. Rev. Franç. Oenol. 7(27):4–12.

NÈGRE, E., A. DUGAL, and J. M. EVESQUE
1958. Dosage de l'acide tartrique sous forme de bitartrate de potassium dans les moûts et les vins. Ann. Technol. Agr. 7:31–101.
1960. La prévision de l'acidité fixe des vins est-elle possible? Ibid. 9:247–321.

NÈGRE, E., and P. FRANÇOT
1965. Manuel pratique de vinification et de conservation des vins. Paris, Flammarion. 456 pp.

NEILSON, N. E., and M. A. JOSLYN
1951. Effect of oxygen on growth, respiration and fermentative activity of yeast. Personal communication.

NELSON, E. K., and D. H. WHEELER
1939. Natural aging of wine. Ind. Eng. Chem. 31:1279–1281.

NELSON, K. E., and M. A. AMERINE
1956. Use of Botrytis cinerea for the production of sweet table wines. Am. J. Enol. 7:131–136.
1957a. The use of Botrytis cinerea Pers. in the production of sweet table wines. Hilgardia 26:521–563.
1957b. Further studies on the production of natural, sweet table wines from botrytised grapes. Am. J. Enol. 8:127–134.

NELSON, K. E., T. KOSUGE, and A. NIGHTINGALE
1963. Large-scale production of spores to botrytise grapes for commercial natural sweet wine production. Am. J. Enol. Viticult. 14:118–128.

NELSON, K. E., and M. S. NIGHTINGALE
1959. Studies in the commercial production of natural sweet wines from botrytised grapes. Am. J. Enol. 9:123–125.

NEMEC, P., and L. DROBNICA
1963. Energy metabolism of yeast and its importance in industrial fermentation technology. Proc. Intern. Cong. Biochem, 5th Moscow, 1961, 8:241–256.

NÉMETH, L.
1964. La culture de la consommation du vin. Budapesti Nemzetközi Borverseny 4:42–43.

NEUBECK, C. E.
1959. Pectic enzymes in fruit juice technology. J. Assoc. Offic. Agr. Chemists 42:374–382.

NEUBERG, C.
1946. The biochemistry of yeast. Ann. Rev. Biochem. 15:435–474.

NEUBERG, C., and E. REINFURTH
1919. Natürliche und erzwungene Glycerinbildung bei der alkoholischen Gärung. Biochem. Z. 92:234–266. (Also in Ber. Deut. Chem. Ges. 52B:1677–1703. 1919.)

NEW YORK CROP REPORTING SERVICE
1966. New York fruit tree and vineyard survey, 1966. Albany. 34 pp. (AMA Release No. 98.)

NEY, M.
1965. Le lithium naturel dans les vins. Ann. Fals. Expertise Chim. 58:263–266.

NICHOLS, P. F.
1934. Public health aspects of dried foods. Amer. J. Pub. Health 24:1129–1134.

NICHOLS, P. F., and W. V. CRUESS
1932. Sulfur dioxide as a dried fruit preservative. Ind. Eng. Chem. 24:649–650.

NICHOLS, P. F., and H. M. REED
1932. Distillation methods for determination of sulfur dioxide. Ind. Eng. Chem., Anal. Ed. 4:79–84.

NIEHAUS, C. J. G.
1937. Sugar-alcohol ratios in South African musts and wines. S. Africa Dept. Agr. Sci. Bull. 161:1–11.
1938. Studies on the nitrogen content of South African musts and wines. Ibid. 172:1–15.

NIERENSTEIN, M.
1934. The natural organic tannins. History, chemistry and distribution. London J. & A. Churchill, Ltd. 319 pp.

NIGHTINGALE, M.
1961. The technology of wine making. Wines & Vines 42(1):31.
1965. Some aspects of in-line sparging of dry wine. San Francisco, Wine Institute, Technical Advisory Committee, December 10, 1965. 2 pp.

NIKANDROVA, V. N.
1959. O sortakh vinograda i technologii pervichnogo vinodeliia desertnykh vin tsentral'noi zony Moldavii. Trudy Mold. Nautch.-Issled. Inst. Sad., Vinogra., Vinod. 4:97–122.

NILOV, V. I., and S. T. OGORODNIK
1965. Vzaimodeĭstvie aminokislot s sakharami. Prikl. Biokhim. Mikrobiol. 1:139–143.

NILOV, V. I., and I. M. SKURIKHIN
1967. Khimiia vinodeliia. 2-e izd. Moskva, Izdatel'stvo "Pishchevaia Promyshlennost'." 442 pp. (See pp. 274, 276–277, 280–281.)

NILSSON, R., and F. ALM
1936. Ueber die Loslösung des Zymasesystems aus der Hefezelle durch Autolyse. Z. Physiol. Chem. 239:179–187.

NOBILE, J.
1967. Use of membrane filter technique in the microbiological control for the brewing industry. Appl. Microbiol. 15(4):736–737.

NONOMURA, H., and Y. OHARA
1967. Die Klassifikation der Apfelsäure-Milchsäure-Bakterien. Mitt. Rebe u. Wein, Obstbau u. Früchteverwertung 17:449–465.

NONOMURA, H., T. YAMAZAKI, and Y. OHARA
1965. Die Apfelsäure-Milchsäure-Bakterien welche aus japanischen Weinen isoliert werden. Mitt. Rebe u. Wein, Serie A (Klosterneuburg) 15:241–254.

NORDSTRÖM, C. G.
1956. Flavonoid glycosides of Dahlia variabilis. IV. 3-Glucosido-5-arabinosidocyanidin from the variety Dandy. Acta Chim. Scand. 10:1491–1496.

NORDSTRÖM, K.
1961. Formation of ethyl acetate in fermentation with brewer's yeast. J. Inst. Brewing 67:173–181.
1962a. Formation of ethyl acetate in fermentation with brewer's yeast. II. Kinetics of formation from ethanol and influence of acetaldehyde. Ibid. 68:188–196.
1962b. Ibid. III. Participation of coenzyme A. Ibid. 68:388–407.
1963a. Ibid. IV. Metabolism of acetyl-coenzyme A. Ibid. 69:142–153.
1963b. Formation of esters from acids by brewer's yeast. I. Kinetic theory and basic experiments. Ibid. 69:310–322.
1963c. Formation of esters, acids and alcohols from α-keto acids by brewer's yeast. Ibid. 69:483–495.
1964a. Formation of esters from acids by brewer's yeast. II. Formation from lower fatty acids. Ibid. 70:42–55.
1964b. Ibid. III. Formation by various strains. Ibid. 70:226–233.
1964c. Ibid. IV. Effect of higher fatty acids and toxicity of lower fatty acids. Ibid. 70:233–242.
1964d. Formation of ethyl acetate in fermentations with brewer's yeast. V. Effect of some vitamins and mineral nutrients. Ibid. 70:209–221.
1964e. Formation of esters from alcohols by brewer's yeast. Ibid. 70:328–336.
1965a. Formation of esters from lower fatty acids by various yeast species. Ibid. 72:38–40.

1965*b*. Possible control of volatile ester formation in brewing. Proc. European Brewing Conv., Stockholm 10:195–208.
1965*c*. Formation of volatile esters by brewer's yeast. Brewers Digest 40(11):60–67.
1966. Yeast growth and glycerol formation. Acta Chem. Scand. 20:1016–1025.

NORDSTRÖM, K., and B.-O. CARLSSON
1965. Yeast growth and formation of fusel alcohols. J. Inst. Brewing 71:171–174.

NUZUBIDSE, N. N., and D. J. GULBANI
1964. Flavonols in grapes (transl.). Soobshcheniia Akad. Nauk Gruz. SSR 36: 345–352.

NYKÄNEN, L., E. PUPUTTI, and H. SUOMALAINEN
1966. Gas chromatographic determination of tyrosol and tryptophol in wines and beers. J. Inst. Brew. 72:24–28.

O'BRIEN, D. F.
1960. New concept of cunilate-bearing tank treatments. San Francisco, Wine Institute, Technical Advisory Committee, March 23, 1960. 1 p.

OESPER, P.
1968. Error and trial. The story of the oxidative reactions of glycosis. J. Chem. Ed. 45:607–610.

OFFICE INTERNATIONAL DE LA VIGNE ET DU VIN
1961. Concours internationaux de vins; modèle de règlement. Paris, O.I.V. 19 pp.
1962. Recueil des méthodes internationales d'analyse des vins. Paris. n.p. (Sections of corrections and new procedures have been added since 1962.)
1964. Codex oenologique international. Vol. 1. Paris, 1964. 75 pp. (Also published in German, Vienna, 1965.)
1965. Mémento de l'O.I.V. éd. 1965. Paris. 1196 pp.

OLMO, H. P.
1948. Ruby Cabernet and Emerald Riesling. Calif. Agr. Exper. Stat. Bull. 704:1–12.

OLMO, H. P., and H. E. STUDER
1967. Mechanical harvesting of Thompson Seedless grapes. Wines & Vines 48(2):25–27.

OLMO, H. P., H. E. STUDER, A. N. KASIMATIS, P. P. BARANEK, L. P. CHRISTENSEN, J. J. KESSLER, D. A. LUVISI, and C. D. LYNN
1968. Training and trellising grape vines for mechanical harvest. Berkeley, University of California Agricultural Extension Service AXT 274:1–16.

OLPHEN, H. VON
1963. An introduction to clay colloid chemistry. New York, Interscience Publishers. 301 pp.

O'NEAL, R., L. MEIS, and W. V. CRUESS
1951. Observations on the fining of wines with casein. Food Technol. 5(2):64–68.

OREGLIA, F.
1964. Enología téorico-práctica. Mendoza, Rodeo del Medio. 37 + 834 pp.

OSTERAS, P.
1966. The Gallo vacuum grape harvester. Wines & Vines 47(4):49–50.

OSTERWALDER, A.
1924. *Schizosaccharomyces liquefaciens* n. sp., eine gegen freie schweflige Säure widerstandsfähige Gärhefe. Mitt. Gebiete Lebensm. Hyg. 15:5–28.
1934*a*. Die Vergärung überschwefelter Traubenmoste. Eine gegen freie schweflige Säure widerstandsfähige Saccharomyces Gärhefe. Landwirtsch. Jahrb. Schweiz 48:1101–1132.
1934*b*. Von Kaltgärhefen und Kaltgärung. Zentr. Bakeriol. Parasitenk., Abt. II 90:226–249.
1941. Die verkannten Kaltgärhefen. Schweiz. Z. Obst- Weinbau 50:487–490.

1948.   Vom Mäuselgeschmack der Weine, Obst- und Beerenweine; eine Erwiderung. *Ibid.* 57:420–421.

OUGH, C. S.

1959a. A survey of commercial practices in sensory examination of wines. Am. J. Enol. Vitic. 10:191–198.

1959b. Personal communication.

1960a. Die Verwendung von Glukose Oxydase in trockenem Weisswein. Mitt. Rebe u. Wein, Serie A (Klosterneuburg) 10A:14–23.

1960b. Gelatin and polyvinylpyrrolidone compared for fining red wines. Am. J. Enol. Viticult. 11:170–173.

1964a. California commercial wine analyses—1963. Wines & Vines 45(5):29–30.

1964b. Fermentation rates of grape juice. I. Effect of temperature and composition on white juice fermentation rates. Am. J. Enol. Viticult. 15:167–177.

1965.   Wine production and development of the research winery. Report to the government of Israel. FAO Rept. 2025:1–73.

1966a. Fermentation rates of grape juices. II. Effect of initial °Brix, pH and fermentation temperature. Am. J. Enol. Viticult. 17:20–26.

1966b. *Ibid.* III. Effects of initial ethyl alcohol, pH, and fermentation temperature. *Ibid.* 17:74–81.

1966c. The analyses of 1963-bottled California commercial wines. Wines & Vines 47(1):17–18.

1968.   Proline content of grapes and wines. Vitis 7:321–331.

OUGH, C. S., and C. J. ALLEY

1966.   An evaluation of some Cabernet varieties. Wines & Vines 47(5):23–25.

OUGH, C. S., and M. A. AMERINE

1959.   Dissolved oxygen determination in wine. Food Res. 24:744–748.

1960.   Experiments with controlled fermentation. IV. Am. J. Enol. Viticult. 11:5–14.

1961a. Studies on controlled fermentation. V. Effects on color, composition, and quality of red wines. *Ibid.* 12:9–19.

1961b. Studies with controlled fermentation. VI. Effects of temperature and handling on rates, composition, and quality of wines. *Ibid.* 12:117–128.

1961c. Polyethylene and cork closure and the fermentation temperature for sparkling wines. Wines & Vines 42(10):28–29.

1962.   Studies with controlled fermentation. VII. Effect of ante-fermentation blending of red must and white juice on color, tannins, and quality of Cabernet Sauvignon wine. Am. J. Enol. Viticult. 13:181–188.

1963a. Use of grape concentrate to produce sweet table wines. *Ibid.* 14:194–204.

1963b. The production of table wines in regions IV and V. Wines & Vines 45(6):56–58, 60–62.

1965.   Studies with controlled fermentations. IX. Bentonite treatment of grape juice prior to wine fermentation. Am. J. Enol. Viticult. 16:185–194.

1966.   Fermentation rates of grape juice. IV. Compositional changes affecting prediction equations. *Ibid.* 17:163–173.

1967a. Studies with controlled fermentation. X. Effect of fermentation temperature on some volatile compounds in wine. *Ibid.* 18:157–164.

1967b. Rosé wine color preference and preference stability by an experienced and an inexperienced panel. J. Food Sci. 32:706–711.

OUGH, C. S., and G. A. BAKER

1961.   Small panel sensory evaluations of wines by scoring. Hilgardia 30:587–619.

OUGH, C. S., and H. W. BERG

1959.   Studies of various light sources concerning the evaluation and differentiation of red wine color. I. Am. J. Enol. Viticult. 10:159–163.

OUGH, C. S., H. W. BERG, and C. O. CHICHESTER
1962. Approximation of per cent brightness and dominant wave length and some blending application with red wines. Am. J. Enol. Viticult. 13:32–39.

OUGH, C. S., H. W. BERG, and C. LOINGER
1967. Acid treatment of red table wine musts for color retention. Am. J. Enol. Viticult. 18:182–189.

OUGH, C. S., and O. BUSTOS
1969. A review of amino acid analytical methods and their application to grapes and wine. Wines and Vines 50(4):50–51, 53, 55–58.

OUGH, C. S., and G. M. COOKE
1966. A rapid semi-quantitative reducing sugar test for dry wines. Wines & Vines 47(8):27, 29.

OUGH, C. S., J. F. GUYMON, and E. A. CROWELL
1966. Formation of higher alcohols during grape juice fermentations at various temperatures. J. Food Sci. 31:620–625.

OUGH, C. S., and J. L. INGRAHAM
1960. Use of sorbic acid and sulfur dioxide in sweet table wines. Am. J. Enol. Viticult. 16:117–122.
1961. The diethyl ester of pyrocarbonic acid as a bottle wine sterilizing agent. Ibid. 12:149–151.

OUGH, C. S., and R. E. KUNKEE
1967. Effects of acid additions to grape juice on fermentation rates and wine qualities. Am. J. Enol. Viticult. 18:11–17.
1968. Fermentation rates of grape juice. V. Biotin content of juice and its effect on alcoholic fermentation rate. Appl. Microbiol. 16:572–576.

OUGH, C. S., E. B. ROESSLER, and M. A. AMERINE
1960. Effects of sulfur dioxide, temperature, time and closures on the quality of bottled dry white wines. Food Technol. 14(7):352–356.

OURNAC, A.
1953. Recherches sur la variation de la teneur en acide ascorbique et en facteur P du jus de raisin en fonction de son mode de préparation. Ann. Technol. Agr. 2:99–111.
1965. Étude du dosage de l'acide ascorbique dans les vins et dans les jus fortement colorés. Ibid. 14:341–347.
1966. Corps interférant dans le dosage de l'acide ascorbique par la dinitrophényl-hydrazine dans les jus de raisin et les vins. Ibid. 15:113–125.

OURNAC, A., and M. FLANZY
1957. Localisation et évolution de la vitamine B₁ dans le raisin, au cours de la maturation. Ann. Technol. Agr. 6:257–292.
1967. Enrichissement en vitamine B₁ des vins conservés au contact des lies. Ibid. 16:41–54.

OURNAC, A., and C. POUX
1966. Acide ascorbique dans le raisin au cours de son développement. Ann. Technol. Agr. 15:193–202.

OVERBY, D. D.
1959. Drosophila control with air currents. San Francisco, Wine Institute, Technical Advisory Committee, December 4, 1959. 1 p.

OWADES, J. L., and J. M. DONO
1967. Simultaneous determination of volatile acids and sulfur dioxide in alcoholic beverages by microdiffusion. J. Assoc. Offic. Anal. Chemists 50:307–311.
1968. Determination of volatile acids in wine by microdiffusion. Am. J. Enol. Viticult. 19:47–51.

LITERATURE CITED

Ozino, O. I.
1967. Gli schizomiceti predominanti accanto ai batteri lattici, nel corso della fermentazione malo-lattica di alcuni vini del Piemonte. Atti Accad. Ital. Vite Vino 19:99–109.

Pacottet, P., and L. Guittonneau
1930. Vins de Champagne et vins mousseux. Paris, J.-B. Baillière et Fils. 412 pp.

Page, R. O.
1942. The tannin-gelatin reaction and the molecular weight of tannins. J. Intern. Soc. Leather Trades Chemists 26:71–84.

Palieri, G.
1938. Studio comparativo tecnico-economico su alcuni sistemi di torchiatura (presse idrauliche e torchi continui). Ital. Vinicola Agrar. 28:308–311, 324–327.

Pallavicini, C.
1966. Determinazione degli enzimi fenolasi, fosfatasi, proteinasi e saccarasi in 20 vini di diversa natura. Riv. Viticolt. Enol. 19:492–496.

1967. Distribuzione degli enzimi fenolasi, fosfatasi, proteinasi e saccarasi nelle parti constituente gli acini di 4 uve e nei vini provenienti dalle une stesse. Ind. Agr. 5:603–606.

Pallotta, U., and A. M. Donati
1965. Sulla composizione chimica e caratteristiche chimico-fisiche del vino Verdicchio. Nota 1. Ind. Agr. 3:275–284.

Pallotta, U., A. M. Donati, and P. Spallacci
1966. Sulla composizione chimica e caratteristiche chimico-fisiche del vino Verdicchio. Nota 3. Ind. Agr. 4(1):1–4.

Pancsev, T.
1964. La viticulture et l'oenologie en Bulgarie. Budapesti Nemzetközi Borverseny 4:66–67.

Pangborn, R. M., C. S. Ough, and R. B. Chrisp
1964. Taste interrelationship of sucrose, tartaric acid, and caffeine in white table wine. Am. J. Enol. Viticult. 15:154–161.

Pantanelli, E.
1912. Ein proteolytisches Enzym in Most überreifer Trauben. Zentr. Bakteriol. Parasitenk., Abt. II 31:545–559.

1915. Weitere Untersuchungen über die Mostprotease. Ibid. 42:480–502.

Parfent'ev, L. N., and V. I. Kovalenko
1951. Ovogmozhnoĭ roli pirougol'nykh éfirov v formirovanii shampanskikh kachestv igrstykl vin. Vindodelie i Vinogradarstvo SSSR. 11(3):16–19. (Chem. Abst. 45:6795.)

1952. K voprosu teorii shampanizatsii (The theory of champagnization). Ibid. 12(4):28–29.

Parle, J. N., and M. E. di Menna
1966. The source of yeasts in New Zealand wines. New Zealand J. Agr. Res. 9:98–107.

Paronetto, L.
1952. Indagni e considerazioni sull'inattivazione dei lieviti con mezzi termia nei vini spumanti. Riv. Viticolt. Enol. 5:357–364.

1966. Ruólo dell'ossigeno in enólogia: dalla vinificazione all'imbottigliamento. Corso Nazionale di Aggiornamento per Enotecnici "G. Gattista Cerletti." 2:285–313.

Pasteur, L.
1866. Études sur le vin. Paris, Imprimerie Impériale. 264 pp.

TABLE WINES

PASTOR, I. G.
1967. La viticoltura peruviana. Riv. Viticolt. Enol. **20**:367–370.

PATAKY, M. B.
1963. Az O.I.V. borszínmérési ajálásának vizsgálata vörösborok esetére. Szőlészeti Kutató Intézet Évkönyve **12**:301–310.

1965. Sur les particularités du dosage spectrophotométrique des anthocyannes. Ann. Technol. Agr. **14**:79–85.

PATO, C. M.
1959. Effect of pH on the removal of iron and copper from wine with ion exchange resins. Am. J. Enol. Vitic. **10**(2):51–55.

PATO, C. DE M., and M. DE S. E H. BECK
1963. Método para determinação simultânea dos ácidos tartárico e málico e da alcalinidade dos mostos por electrotilulação. De Vinea et Vino, Portugaliae Documenta, Série II **1**(2):1–35.

PATO, M. A. DA S.
1967. O ácido tartárico na correcção ácida dos mostos e dos vinhos. De Vinea et Vino, Portugaliae Documenta, Série II, **3**(4):1–25.

PATO, M. DOS SANTOS
1932. Química-física aplicada aos mostos e aos vinhos. Bol. Estação Vitivinícola da Beira Litoral (Bairrada) **1**:1–57.

PATSCHKY, A.
1967. Zur Feststellung eins Alkoholzusatzes bei Tischweinen. Deut. Lebensm.-Rundschau **63**:197–200.

PAUL, F.
1958. Ueber den Acetaldehyd im Wein. Seine Enstehung während der alkoholischen Gärung. Mitt. Rebe u. Wein, Serie A (Klosterneuburg) **8**:123–134.

1960. Chemische Untersuchungen an Schaumweinen I. II. Der Gehalt an schwefeliger Säure. *Ibid.* **10**:138–155, 238–247.

1963. Sur la teneur des vins en réductones naturelles et leur influence. Ann. Technol. Agr. **12**(numéro hors série 1):171–176.

1964. Die technische Durchführung der organoleptischen Beurteilung von Weinen. Mitt. Rebe u. Wein, Serie A (Klosterneuburg) **14**:197–209.

1967. Die "Rangziffern-Methode," eine einfache Möglichkeit für den organoleptischen Vergleich zweier oder mehrerer Proben. Mitt. Rebe u. Wein, Obstbau u. Früchteverwertung (Klosterneuburg) **17**:280–288.

PAULI, O.
1967. Le pyrocarbonate d'éthyle en oenologie. Bull. O.I.V. (Office Intern. Vigne Vin) **40**:764–772.

PAULI, O., and H. GENTH
1966. Zur Kenntnis des Pyrokohlensäurediäthylesters. I. Eigenschaften, Wirkungsweise und Analytik. Z. Lebensm.-Untersuch. -Forsch. **133**:216–227.

PAULUS, W., and D. LORKE
1967. Zur Kenntnis des Pyrokohlensäurediäthylesters. III. Herstellung und toxikologische Prüfung repräsentativer Urenestzungsprodukte des Pyrokohlensäurediäthylesters. Z. Lebensm.-Untersuch. -Forsch. **132**:325–333.

PAUNOVIC, R.
1963. Possibilité d'utilisation des radiations dans la conservation des vins. Ann. Technol. Agr. **12**(numéro hors série 1):143–153.

PEDERSON, C. S.
1949. The genus *Pediococcus*. Bact. Reviews **13**:225–232.

PEEL, J. L.
1951. Ester formation by yeasts. II. Formation of ethyl acetate by washed suspensions of *Hansenula anomala*. Biochem. J. **49**:62–67.

928

PENINOU, E., and S. GREENLEAF
1954. Winemaking in California. San Francisco, The Peregrine Press. 2 vols.
1967. A directory of California wine growers and wine makers in 1860. Berkeley, Tamalpais Press. 7 + 84 pp.

PERI, C.
1967. Acid degradation of leucoanthocyanidins. Am. J. Enol. Viticult. 18:168–174.

PERINI, D.
1966. Rapport italien. Bull. O.I.V. (Office Intern. Vigne Vin) 38:801–825.

PERLMAN, L., and A. F. MORGAN
1945. Stability of B vitamins in grape juices and wines. Food Res. 10:334–341.

PETERS, L.
1965. Stainless steel tanks in the wine industry. Wines & Vines 46(3):26–26.

PETERS, P.
1953. Selection, arrangement and use of filtration equipment. Proc. Am. Soc. Enol. Davis, August 13–14. 4:139–142.

PETERSON, C. G.
1964. Pumps and pumping. In M. A. Joslyn and J. L. Heide (ed.), Food processing operations. Vol. III. Westport, Conn., Avi Publishing Co. (See pp. 394–416.)

PETERSON, R. G., and A. CAPUTI, JR.
1967. The browning problem in wines. II. Ion exchange effects. Am. J. Enol. Viticult. 18:105–112.

PETERSON, R. G., M. A. JOSLYN, and P. W. DURBIN
1958. Mechanism of copper casse formation in white table wine. III. Source of the sulfur in the sediment. Food Res. 23:518–524.

PETERSON, W. H., and M. S. PETERSON
1945. Relation of bacteria to vitamins and other growth factors. Bact. Reviews 9:49–109.

PETRITSCHEK-SCHILLINGER, A.
1967. Zulässige Fremdstoffe und Zusatzstoffe für Lebensmittel in der Bundesrepublik Deutschland. Z. Lebensm.-Untersuch. -Forsch. 135 (Gesetze): 1–36.

PETRÓ, I.
1965. Korai fehérjestabilizáció a must bentonitos kezelésével. Borgazdaság 13: 151–157.

PETROVA, O., G. JANICEK, and G. DAVIDEK
1966. Stanovení karbonylových látek ve vine. Kvasný Prumýsl 12:40–41.

PETTIGIANI, A. E.
1943. Los alcoholes superiores en los vinos argentinos. La Plata, Univ. Nac., Rev. Quím. Facultad de Cien. 18:95–104.

PEYER, E.
1964. Économie du vin en Suisse. Budapesti Nemzetközi Borverseny 4:100–101.

PEYNAUD, E.
1936. L'acétate d'éthyle dans les vins atteints d'acescence. Ann. Ferm. 2:367–384. (See also Rev. Viticult. 90:321–327. 1939.)
1937. Étude sur les phénomènes d'estérification dans les vins. Rev. Viticult. 86:209–215, 227–231, 248–253, 299–301, 394–396, 420–423, 440–444, 472–475; 87:49–52, 113–116, 185–188, 242–229, 278–285, 297–301, 344–350, 362–364, 383–385. (Summary in Ann. Ferm. 3:242–252. 1937.)
1939a. L'acide malique dans les moûts et les vins de Bordeaux. Rev. Viticult. 90(2323):3–12; (2324):25–30.

1939*b*. L'azote amine et l'azote amide dans les vins de Bordeaux. Ann. Fals. Fraudes **32**:228–243.

1939–40. Sur la formation et la diminution des acides volatils pendant la fermentation alcoolique en anaérobiose. Ann. Ferm. **5**:321–327, 385–401.

1948. Dosage du glycérol dans les vins par oxydation périodique. Ann. Fals. Fraudes **41**:384–402.

1956*a*. New information concerning biological degradation of acids. Am. J. Enol. **7**:150–156. (*See also* Mitt. Rebe u. Wein, Serie A [Klosterneuburg] **5**:183–191. 1955.)

1956*b*. Sur la formation d'acétate d'éthyle par les levures de vin. Ind. Aliment. Agr. (Paris) **73**:253–257.

1957. Charakterisierung der verschiedenen Spezies von Weinhefe. Mitt. Rebe u. Wein, Serie A (Klosterneuburg) **7**:1–15.

1963. Emploi de l'acide sorbique dans la conservation des vins. Ann. Technol. Agr. **12**(numéro hors série 1):99–114.

1965. Le goût et l'odeur du vin. Aliment. Vie **53**:249–260.

PEYNAUD, E., and J. BLOUIN
1965. Comparaison de quelques méthodes de dosage de l'acide L-malique. Ann. Technol. Agr. **14**:61–66.

PEYNAUD, E., J. BLOUIN, and S. LAFON-LAFOURCADE
1966*a*. Review of applications of enzymatic methods to the determination of some organic acids in wines. Am. J. Enol. Viticult. **17**:218–224.

PEYNAUD, E., and Y. CHARPENTIÉ
1950. Note sur le dosage de l'acide lactique dans les boissons fermentées. Ann. Fals. Fraudes **43**:246–252.

PEYNAUD, E., and S. DOMERCQ
1953. Étude des levures de la Gironde. Ann. Technol. Agr. **2**:265–300.

1955. Étude de la microflore des moûts et des vins de Bordeaux. Compt. Rend. Acad. Agr. France **41**:103–106.

1959*a*. A review of microbiological problems in wine-making in France. Am. J. Enol. Viticult. **10**:69–77.

1959*b*. Possibilité de provoquer la fermentation malolactique en vinification à l'acide de bactéries cultivées. Compt. Rend. Acad. Agr. France **45**:355–358.

1961. Études sur les bactéries lactiques. Ann. Technol. Agr. **10**:43–60.

1967*a*. Étude de quelques bacilles homolactiques isolés de vins. Arch. Mikrobiol. **57**:255–270.

1967*b*. Étude de quelques coques homolactiques isolés de vins. Rev. Ferm. Ind. Aliment. **22**:133–140.

PEYNAUD, E., and G. GUIMBERTEAU
1958. Sur la teneur des vins en alcools supérieurs. Estimation séparée de alcools isobutylique et isoamylique. Ann. Fals. Fraudes **51**:70–80.

1961. Recherches sur la constitution et l'efficacité anticristallisante de l'acide métatartarique. Ind. Aliment. Agr. (Paris) **78**:131–135, 413–418.

1962*a*. Modification de la composition des raisins au cours de leur fermentation propre en anaérobiose. Ann. Physiol. Veg. **4**:161–167.

1962*b*. Sur la formation des alcools supérieurs par les levures de vinification. Ann. Technol. Agr. **11**:85–105.

PEYNAUD, E., G. GUIMBERTEAU, and J. BLOUIN
1964. Die Löslichkeitsgleichgewichte von Kalzium und Kalium in Wein. Mitt. Rebe und Wein, Serie A (Klosterneuburg) **14A**:176–186.

PEYNAUD, E., and S. LAFON
1951. Présence et signification du diacétyle, de l'acétoine et du 2,3-butanediol dans les eaux-de-vie. Ann. Fals. Fraudes **44**:263–283.

PEYNAUD, E., and S. LAFON-LAFOURCADE

1962. Sur la nutrition azotée des levures de vin. Rev. Ferm. Ind. Aliment. **17**: 11–21.

1965. Étude d'un dosage simple de l'acide malique appliqué aux vins à l'aide de *Schizosaccharomyces pombe.* Ann. Technol. Agr. **14**:49–59.

1966. Facteurs de la formation des acides pyruvique et α-cétoglutarique au cours de la fermentation alcoolique; conséquences pratiques sur les combinaisons sulfitiques des vins. Ind. Aliment. Agr. (Paris) **83**:119–126.

PEYNAUD, E., S. LAFON-LAFOURCADE, and S. DOMERCQ

1965. Besoins nutritionnels de soixante-quatre souches de bactéries lactiques isolées de vins. Bull. O.I.V. (Office Intern. Vigne Vin) **38**:945–958.

PEYNAUD, E., S. LAFON-LAFOURCADE, and G. GUIMBERTEAU

1966b. L(+) lactic acid and D(−) lactic acid in wines. Am. J. Enol. Viticult. **17**: 302–307. (*See also* Compt. Rend. **263**:634–635. 1966.)

1967a. Nature de l'acide lactique formé par les levures—un caractère spécifique de *Saccharomyces veronae* Lodder et van Rij. Antonie van Leeuwenhoek J. **33**:49–55. (*See also* Rev. Franç. Oenol. 7[27]:17–20. 1967.)

1967b. Sur la nature de l'acide lactique formé par les bactéries lactiques isolées de vins. Rev. Ferm. Ind. Aliment. **22**:61–66.

1968. Ueber den Mechanismus der Aepfelsäure-Milchsäure-Gärung. Mitt. Rebe u. Wein, Obstbau u. Fruchteverwertung (Klosterneuburg). **18**:343–348.

PEYNAUD, E., and S. LAFOURCADE

1952. Sur les conditions d'emploi de l'anhydride sulfureux dans les vins liquoreux. Bull. O.I.V. (Office Intern. Vigne Vin) **25**(252):110–120 (*See also* Bull. Inst. Nat. Appl. Orig. Vins et Eaux-de-Vie **42**:20–31. 1952.)

1955a. L'acide pantothénique dans les raisins et dans les vins de Bordeaux. Ind. Aliment. Agr. (Paris) **72**:575–580, 665–670.

1955b. L'inositol dans les raisins et dans les vins: son dosage microbiologique. Ann. Technol. Agr. **4**:381–396.

1957. Teneurs en pyridoxine des vins de Bordeaux. *Ibid.* **3**:301–302.

1958. Évolution des vitamines B dans le raisin. Qual. Plant. Mater. Végétabiles **34**:405–414.

PEYNAUD, E., and A. MAURIÉ

1953a. Évolution des acides organiques dans le grain de raisin au cours de la maturation en 1951. Ann. Technol. Agr. **2**:83–94.

1953b. Sur l'évolution de l'azote dans les différentes parties du raisin au cours de la maturation. *Ibid.* **2**:15–26.

1958. Synthesis of tartaric and malic acids by grape vines. Am. J. Enol. Viticult. **9**:32–36. (*See also* Ann. Technol. Agr. **4**:111–139. 1956.)

PEYNAUD, E., and J. RIBÉREAU-GAYON

1947. Sur les divers types de fermentation alcoolique déterminés par diverses races de levures elliptiques. Compt. Rend. **224**:1388–1390.

PEYNAUD, E., and P. SUDRAUD

1964. Utilisation de l'effet désacidifiant des *Schizosaccharomyces* en vinification de raisins acides. Ann. Technol. Agr. **13**:309–328.

PEYROT, E.

1934. Sulla probabile quantità e qualità delle principali sostanze coloranti nei vini bianchi. Ann. Chim. Appl. **24**:512–519.

PHAFF, H. J., and H. C. DOUGLAS

1944. A note on yeasts occurring in dessert wines. Fruit Prod. J. **23**(11):332–333.

PIFFERI, P. G.

1966. Studi sui pigmenti naturali. V. Ricerche sui flavonoidi e sui tannini estraibili dai vini con acetato d'etile. Ind. Agr. **4**:475–479.

PIFFERI, P. G., and A. ZAMORANI
1964. Contributo alla conoscenza della sostanza colorante dei vini. 2. Azione dell'anidride solforosa e della rifermentazione sugli antociani del vino. Riv. Viticolt. Enol. 17:115–121.

PILNIK, W.
1964. Ueber den säuren Geschmack von Früchtesäuren. Report Scientific Technical Commission, International Federation of Fruit Juice Producers, Vienna, Fruchtsaftforschung und -Technologie. Pp. 149–157.

PILNIK, W., and M. FADDEGON
1967. Ueber den Nachweis eines Zusatzes von DL-Aepfelsäure zu Aepfelsaft. Mitt. Gebiete Lebensm. Hyg. 58:151–154.

PILONE, B. F., and H. W. BERG
1965. Some factors affecting tartrate stability in wine. Am. J. Enol. Viticult. 16:195–211.

PILONE, F. J.
1953. The role of pasteurization in the sterilization and clarification of wines. Proc. Am. Soc. Enol. Davis, August 12–14, 4:77–83.

PILONE, G. J.
1967. Effect of lactic acid on volatile acid determination of wine. Am. J. Enol. Viticult. 18:149–156.

PILONE, G. J., and R. E. KUNKEE
1965. Sensory characterization of wines fermented with several malolactic strains of bacteria. Am. J. Enol. Viticult. 16:224–230.
1966. Chemical characterization of wines fermented with various malolactic bacteria. Appl. Microbiol. 14:608–615.

PLODERL, F., and W. E. WEYMAN
1957. Corrosion resistant coatings—properties and uses. Am. J. Enol. 8:135–138.

POIRIER, L. M.
1966. Calcul de l'acidité volatile corrigée de la présence d'anhydride sulfureux. Ann. Fals. Expertise Chim. 59:443–446.

POITTEVIN, M. A., A. CARRASCO, and M. N. GIOIA
1963. Estudios sobre la fermentación malolactica en vinos del Uruguay. IV. Estudio de las bacterias lácticas en vinificaciones experimentales. Rev. Latinoamer. Microbiol. 6:147–158.

POLAKOVIČ, F., and A. VEREŠ
1966. Príodné podmienky Tokajskej vinohradníckej oblasti CSSR. Pokrovky vo Vinohradníkrom a Vinárskom Výskume, Bratislava, Vydavatel'stvo Slovenskej Akádémie Vied 1966:41–84.

POLLOCK, J. R. A.
1962. Anthocyanogens in malt and their enzymatic degradation. In Food Group Symposium, Soc. Chem. Ind., Recent Advances in Processing Cereals. London, Society Chemical Industry. 199 pp. (See pp. 70–77.)

PONTING, J. D., and G. JOHNSON
1945. Determination of sulfur dioxide in fruits. Ind. Eng. Chem., Anal. Ed. 17:682–686.

PONTING, J. D., and M. A. JOSLYN
1948. Ascorbic acid oxidation and browning in apple tissue extracts. Arch. Biochem. 19:47–63.

POPOV, K. S., and G. F. SEMENENKO
1964. Raïony proizvodstva i sorta vinograda dlia shampanskikh vinomaterialov. Trudy Vses. Nauchno-Issledov. Institut Vinodeliia i Vinogradarstva "Magarach" 13:84–107.

POPOVA, E. M.
1960. Biokhimicheskoe issledovanie kul'tury *Botrytis cinerea* Biokh. Vinodeliia 6:31–52.

POPOVA, E. M., and M. G. PUCHKOVA
1947. Znachenie fermentativnogo kompleksa preparata iz *Botrytis cinerea* dlia vinogradnykh susel. Biokh. Vinodeliia 1:71–76.

POPPER, K., and F. S. NURY
1964. Recoverable static regenerant ion exchange treatment of Thompson Seedless grape juice. Am. J. Enol. Viticult. 15:82–86.

POPPER, K., F. S. NURY, W. M. CAMIRAND, and W. L. STANLEY
1964. Development of botrytis character in must by aerated submerged culture. San Francisco, Wine Institute, Technical Advisory Committee, December 11, 1964. 3 pp.

PORCHET, B.
1931. Contribution à l'étude de l'adaptation des levures à l'acide sulfureux. Ann. Agr. Suisse 32(2):135–154.
1935. Influence de l'acide acétique sur la fermentation du sucre par les levures, en présence d'alcool. Mitt. Gebiete Lebensm. Hyg. 26:19–28.
1938. Biologie des levures provoquant la fermentation alcoolique à basse temperature. Ann. Ferm. 4:578–600.

POTTER, R. L., and C. A. ELVHJEM
1948. Biotin and the metabolism of *Lactobacillus arabinosus*. J. Biol. Chem. 172: 531–537.

POUX, C.
1966a. Polyphénoloxydase dans le raisin. Ann. Technol. Agr. 15:149–158.
1966b. Polyphénoloxydase dans le raisin. Vignes et Vins, numéro spécial 1966: 18–25.
1967. Enzymes oxydatifs dans le raisin. Ann. Nutr. Aliment. 21(5):B205–222.

POUX, C., C. FLANZY, and M. FLANZY
1964. Les levures alcooliques dans les vins; protéolyse et protéogenèse. Ann. Technol. Agr. 13:5–18.

POWER, F. B.
1921. The detection of methyl anthranilate in fruit juice. J. Am. Chem. Soc. 43:377–381.

POWER, F. B., and V. K. CHESTNUT
1921. The occurrence of methyl anthranilate in grape juice. J. Am. Chem. Soc. 43:1741–1742.
1923. Examination of authentic grape juices for methyl anthranilate. J. Agr. Res. 23:47–53.

POWERS, J. J., D. SAMAATMADJA, D. E. PRATT, and M. K. HAMDY
1960. Anthocyanins. II. Action of anthocyanin pigments and related compounds on the growth of certain microörganisms. Food Technol. 14:626–632.

PRAKKE, F., and E. STIASNY
1933. Ueber die Einwirkung von Thiosulfat auf verdünnte Säurelösungen. Rec. Trav. Chim. 52:615–639.

PREHODA, J.
1963. Hömérsékletszabályozás a vörös borok erjesztésénél (Temperature regulation of the fermentation of red wines). Borgazdaság 11:15–23.
1966. Rotweinbereitung mit und ohne Maischegärung. Mitt. Rebe u. Wein, Obstbau u. Früchteverwertung. (Klosterneuburg) 16:463–468.

PRESCOTT, J. A.
1965. The climatology of the vine (*Vitis vinifera* L.). The cool limits of cultivation. Trans. Roy. Soc. South Australia 89:5–23.

PRÉVOT, A., and F. CABEZA
1962. Note sur la composition de quelques corps gras peu communs par chromatographie en phase gazeuse. Rev. Franç. Corps Gras 9, 3:149–152.

PRILLINGER, F.
1957. Ueber die stickstoffhältigen Substanzen im Wein. Mitt. Rebe u. Wein, Serie A (Klosterneuburg) 7:138–147.
1963. Protection des vins de l'oxydation par l'emploi de l'acide sulfureux et de l'acide ascorbique. Ann. Technol. Agr. 12(numéro hors série 1):159–169.
1964. Ueber den Nachweis und die Bestimmung von Diäthylkarbonat in mit Baycovin behandelten Weinen. Mitt. Rebe u. Wein, Serie A (Klosterneuburg) 14:29–32.
1965. Uso dell'acido solforoso e dell'acido ascorbico per impedire l'ossidazione dei vini. Riv. Viticolt. Enol. 18:99–108.

PRILLINGER, F., and H. HORWATITSCH
1964. Ueber ein rasches Verfahren zum Nachweis des Pyrokohlensäurediäthylesters in Getränken und zum natürlichen Vorkommen von Kohlensäureestern in Gärprodukten. Mitt. Rebe u. Wein, Serie A (Klosterneuburg) 14:251–257.

PRILLINGER, F., H. HORWATITSCH, and A. MADNER
1967. Versuche zur Charakterisierung von Weinen auf Grund ihrer flüchtigen Inhaltsstoffe. Mitt. Rebe u. Wein, Obstbau u. Früchteverwertung (Klosterneuburg) 17:271–279.

PRILLINGER, F., and A. MADNER
1968. Versuche zur charakterisierung von Weinen auf Grund ihrer flüchtigen Inhaltsstoffe. Mitt. Rebe u. Wein, Obstbau u Früchteverwertung (Klosterneuberg) 18:1–9.

PRILLINGER, F., A. MADNER, and J. KOVACS
1968. Die flüchtigen Inhaltsstoffe des Aepfel- und Traubensaftes. Mitt. Rebe u. Wein, Obstbau u. Früchteverwertung. (Klosterneuburg) 18:98–105.

PRISNEA, C.
1964. Bacchus in Rumania. Bucharest, Meidiane Publishing House. 245 pp.

PRONI, G., and G. PALLAVICINI
1962. Introduzione allo studio della viticoltura nell'economia dell'azienda agraria in Piemonte. Ann. Facoltà Sci. Agrar. Univ. Torino 1:253–360. (Published 1966.)

PROTIN, R.
1966. Situation de la viticulture dans le monde in 1965. Bull. O.I.V. (Office Intern. Vigne Vin) 39:1206–1247.
1967. Situation de la viticulture dans le monde en 1966. Ibid. 40:1019–1134.
1968. Situation de la viticulture dans le monde en 1967. Ibid. 41:1081–1123. (The figures for Lebanon and Israel have been corrected.)

PUISSANT, A.
1960. Étude sur la variation de la teneur en NH$_4$, K, Ca, Mg de quelques raisins au cours de la maturation. Ann. Technol. Agr. 9:321–330.

QUEUILLE, H.
1927. Partie officielle: circulaire relative à la protection des appellations d'origine. J. Agr. Pratique 48:338–340.

QUINN, D. G.
1940. Sulfur dioxide, its uses in the wine industry. J. Dept. Agr. Victoria 38:200–204.

QUITTANSON, C.
1965. L'appellation d'origine, facteur permanent de recherche de la qualité. Rev. Ferm. Ind. Aliment. 20:49–61.

RABINOVICH, Z. D.
1960. Vliianie dissotsiastsii drozhzheĭ na proizvodstvennĭ protsess butylochnoĭ shampanizatsii. Vinodelie i Vinogradarstvo SSSR 20(3):14–17.

RADENKOVIČ, D.
1962. Réglementation et encouragement du marché du vin yougoslave. Katalog Mednarodni Sejem Vin, Žganih Pijač, Sadnih Sokov in Opreme, Ljubljana 8:88–92.

RADER, R. D.
1946. Designing the proper concrete floors for the food plant. Food Ind. 18 (7):1026–1030, 1156, 1158.

RADLER, F.
1957. Untersuchungen über den Gehalt der Moste einiger Rebensorten und -arten an den Vitaminen Pyridoxin, Pantothensäure, Nicotinsäure und Biotin. Vitis 1:96–108.
1958. Der Nähr- und Wuchsstoffbedarf der Apfelsäureabbauenden Bakterien. Arch. Mikrobiol. 32:1–15.
1962a. Die Bildung von Acetoin und Diacetyl durch die Bakterien des biologischen Säureabbaus. Vitis 3:136–143.
1962b. Ueber die Milchsäurebakterien des Weines und den biologischen Säureabbau. I. Systematik und chemische Grundlagen. Ibid. 3:144–176.
1965a. The main constituents of the surface waxes of varieties and species of the genus Vitis. Am. J. Enol. Viticult. 16:159–167.
1965b. The surface waxes of the Sultana vine (Vitis vinifera cv. Thompson Seedless). Australian J. Biol. Sci. 18:1045–1056.
1966. Die mikrobiologischen Grundlagen des Säureabbaus im Wein. Zent. Bakteriol. Parsitenk., Abt. 2, 120:237–287.
1968. La structure et la composition chimique de la cire cuticulaire de la baie de raisin. Bull. O.I.V. (Office Intern. Vigne Vin) 41:403–415.

RAHN, O.
1932. Physiology of bacteria. Philadelphia, Penn., P. Blakiston's Son & Co., Inc. 14 + 438 pp. (See pp. 27, 42.)

RAHN, O., and J. E. CONN
1944. Effect of increase in acidity on antiseptic efficiency. Ind. Eng. Chem. 36:185–187.

RAINBIRD, G.
1966. Sherry and the wines of Spain. London, Michael Joseph. 224 pp.

RAINBOW, C.
1966. Nutrition and metabolism of acetic acid bacteria. Wallerstein Lab. Commun. 29:5–14.

RAINBOW, C., and A. H. ROSE
1963. Biochemistry of industrial micro-organisms. New York, Academic Press. 19 + 708 pp.

RAKCSÁNYI, L.
1964. A szölöhj cserzöangagtartalmának kitermelése. Borgazdaság 12:5–15.

RAMOS, M. DE CUNHA, and L. GUEDES GOMES
1966. Um método colorimétrico para a determinação do manganês no vinho. De Vinea et Vino Portugaliae Documenta, Série 2, 3(3):1–8.

RANKINE, B. C.
1953. Quantitative differences in products of fermentation by different strains of wine yeasts. Australian J. Appl. Sci. 4:590–602. (See also Am. J. Enol. 6[1]:1–9. 1955.)
1955. The lead content of some Australian wines. J. Sci. Food Agr. 10:576–579.
1957. Factors influencing the lead content of wines. Ibid. 12:458–460.

1962. Aluminum haze in wine. Australian Wine, Brewing, Spirit Rev. 80(9):14, 16.

1963. Nature, origin, and prevention of hydrogen sulfide aroma in wine. J. Sci. Food Agr. 14:79–91.

1965a. Factors influencing the pyruvic acid content of wines. *Ibid.* 16:394–398.

1965b. Ion-exchange treatment of wine. Australian Wine, Brewing, Spirit Rev. 85:59–62.

1966a. Decomposition of L-malic acid by wine yeasts. J. Sci. Food Agr. 17:312–316.

1966b. *Pichia membranaefaciens,* a yeast causing film formation and off-flavor in table wine. Am. J. Enol. Viticult. 17:82–86.

1966c. Sulphur dioxide in wines. Food Technol. Australia 18:134–135, 137, 139, 141.

1967a. Formation of higher alcohols by wine yeasts, and relation to taste thresholds. J. Sci. Food Agr. 18:583–589.

1967b. Influence of yeast strain and pH on pyruvic acid content of wines. *Ibid.* 18:41–44.

1968a. The importance of yeasts in determining the composition of quality wines. Vitis 7:22–49.

1968b. Formation of α-ketoglutaric acid by wine yeasts and its oenological significance. J. Sci. Food Agr. 19:624–627.

RANKINE, B. C., K. M. CELLIER, and E. W. BOEHM
1962. Studies on grape variability and field sampling. Am. J. Enol. Viticult. 13:58–72.

RANKINE, B. C., and W. W. EMERSON
1963. Wine clarification and protein removal by bentonite. J. Sci. Food Agr. 14:685–689.

RANKINE, B. C., R. E. KEPNER, and A. D. WEBB
1958. Comparison of anthocyan pigments of *Vinifera* grapes. Am. J. Enol. 9(3):105–110.

RANKINE, B. C., and B. LLOYD
1963. Quantitative assessment of dominance of added yeast in wine fermentations. J. Sci. Food Agr. 14:793–798.

RAPPAPORT, S.
1958. Wine production in Israel and its future prospects. Israel Economic Forum IX(1–2):79–82.

RASPINO, G.
1966. I fabbricati per l'imbottigliamento dei vini da pasto presso le cantine sociali del Piemonte. Ann. Facoltà Sci. Agr. Univ. Studi Torino 1:175–198.

REBELEIN, H.
1957a. Unterscheidung naturreiner von gezuckerter Weinen und Bestimmung des natürlichen Alkoholgehaltes. Z. Lebensm.-Untersuch. -Forsch. 105:293–311, 403–420.

1957b. Zur chemisch-analytischen Erkennung gezuckerter und gespriteter Weine. Deut. Wein-Ztg. 93:291–292.

1958. Zur Erkennung naturreiner Weine mittels des K-Wertes. Deut. Lebensm.-Rundschau 54:297–307.

1961. Kolorimetrisches Verfahren zur gleichzeitigen Bestimmung der Weinsäure and Milchsäure in Wein und Most. *Ibid.* 57(2):36–41.

1962. Untersuchungen über gesetzmässige Beziehungen zwischen Weininhaltsstoffen. Mitt. Rebe u. Wein, Serie A (Klosterneuburg) 12:227–258.

1964. Möglichkeiten der Erkennung gezuckerter, gespriteter und gestreckter Weine. Mitt.-Bl. GDCH., Fachgr. Lebensmitt. -Chem. 18:8–15.

1965a. Beitrag zur Bestimmung des Catechingehaltes in Wein. Deut. Lebensm.-Rundschau 61:182–183.

1965b. Beitrag zur Bestimmung des Methanolgehaltes in Weinen. *Ibid.* 61:211–212.

1965c. Beitrag zum Catechin- und Methanolgehalt von Weinen. *Ibid.* **61**:239–240.

1965d. Die analytische Bedeutung von Kationen-Anionenbilanzen bei Traubenmosten und Weinen. *Ibid.* **61**:304–308.

1967a. Vereinfachtes Verfahren zur kolorimetrischen Bestimmung der Citronensäure in Wein und Traubenmost. *Ibid.* **63**:337–340.

1967b. Beitrag zur Bestimmung und Beurteilung des Nitratgehaltes von Traubenmosten und Wein. *Ibid.* **63**:233–239.

REED, G.
1966. Enzymes in food processing. New York, Academic Press. 483 pp. (*See* pp. 73–88, 313–318.)

REEVES, P. A.
1965. Temperature corrections for dilutions in the refractometer method of determining alcohol. J. Assoc. Offic. Agr. Chemists **48**:476–478.

REHM, H. J.
1964. The antimicrobial action of sulphurous acid. *In* International Symposium on Microbial Inhibitors in Food, 4th Göteborg, Sweden. Stockholm, Almquist and Wiksell. 402 pp. (*See* pp. 105–115.)

REHM, H. -J., E. SENING, H. WITTMANN, and P. WALLNÖFER
1964. Beitrag zur Kenntnis der antimikrobiellen Wirkung der schwefligen Säure. III. Aufhebung der antimikrobiellen Wirkung durch Bildung von Sulfonaten. Z. Lebensm.-Untersuch. -Forsch. **123**:425–432.

REHM, H. -J., P. WALLNÖFER, and H. KESKIN
1965. Beitrag zur Kenntnis der antimikrobiellen Wirkung der schwefligen Säure. IV. Dissoziation und antimikrobielle Wirkung einiger Sulfonate. Z. Lebensm.-Untersuch. -Forsch. **127**:72–85.

REHM, H. -J., and H. WITTMANN
1962. Beitrag zur Kenntnis der antimikrobiellen Wirkung der schwefligen Säure. I. Uebersicht über einflussnehmende Faktoren auf die antimikrobielle Wirkung der schwefligen Säure. Z. Lebensm.-Untersuch. -Forsch. **118**: 413–429.

1963. *Ibid.* II. Die Wirkung der dissozierten und undissozierten Säure auf verschiedene Microorganismen. *Ibid.* **120**:465–478.

REICHARD, O.
1936. Nachweis und Bestimmung von Natrium in Wein und sein Gehalt bei Pfälzer Weinen. Z. Untersuch. Lebensm. **71**(6):501–515.

REIFF, F. R., R. KAUTZMANN, H. LÜERS, and H. LINDEMANN
1960–1962. Die Hefen. Bd. 1. Die Hefen in der Wissenschaft. Bd. 2. Technologie der Hefen. Nuremberg, Hans Carl. 2 vols.

RÉMY, R. H.
1967a. Introduction à l'étude de la vinification continue. Vignes et Vins No. **160**: 49–53.

1967b. Fermentation continue et vinificateur continu à multiple effet. Ind. Aliment. Agr. (Paris) **84**:1265–1285.

RENAUD, J.
1939–1940. La microflore des levures du vin. Son rôle dans la vinification. Ann. Ferm. **5**:410–417.

RENTSCHLER, H.
1950. Ueber das Braunwerden der 1950er Weine. Schweiz. Z. Obst- Weinbau **59**(25):455–458.

1960. Weinbehandlung mit Ascorbinsäure? *Ibid.* **69**:148–152.

1965a. Die Löslichkeit von Kohlensäure in Wein in Abhängigkeit von Temperatur und Druck. *Ibid.* **74**:662–663.

1965b. Ueber den Nachweis der chemischen Konservierung von Weinen mit Pyrokohlensäure-Diäthylester. Mitt. Gebiete Lebensm. Hyg. **56**:265–269.

RENTSCHLER, H., and H. TANNER
1955. Ueber den Nachweis von Gluconsäure in Weinen aus edelfaulen Trauben. Mitt. Gebiete Lebensm. Hyg. **46**:200–208.
1958. Ascorbinsäure (Vitamin C) als Ersatz für schweflige Säure in der Kellerwirtschaft. Schweiz. Z. Obst- Weinbau **67**:3–6.
1966. Zum Nachweis von Sorbit. *Ibid.* **75**:151–153.

REYNOLDS, T. M.
1963. Chemistry of nonenzymic browning. I. The reaction between aldoses and amines. Advan. Food Res. **12**:1–52.
1965. Chemistry of nonenzymic browning. II. *Ibid.* **14**:167–283.

RIBÉREAU-GAYON, G.
1968. Étude des mechanismes de synthese et de transformation de l'acide malique, de l'acide tartrique et de l'acide citrique chez *Vitis vinifera* L. Phytochem. **7**:1471–1482.

RIBÉREAU-GAYON, G., and A. LEFEBVRE
1967. Relations entre le métabolisme respiratoire et la synthèse des acides organiques, en particulier de l'acide tartarique, dans les baies de *Vitis vinifera*. Compt. Rend **264**:1112–1115.

RIBÉREAU-GAYON, J.
1932. Sur les matières albuminoïdes des vins blancs. Ann. Fals. Fraudes **25**: 518–524.
1933a. Contribution à l'étude des oxydations et réductions dans les vins.... 2d ed. Bordeaux, Delmas. 7 + 213 pp. (*See esp.* pp. 52–55, 92–97.)
1933b. États, réactions, équilibres et précipitations du fer dans les vins. Casses ferriques. Bordeaux, Delmas. 102 pp.
1935a. Collage bleu; traitement des vins par le ferrocyanure. Complexes du fer dans les vins. Bordeaux, Delmas. 42 pp.
1935b. Le cuivre des moûts et des vins. Ann. Fals. Fraudes **28**:349–360.
1936. Sur la "désacidification biologique" des vins. Proc.-Verb. Soc. Sci. Phys. Nat. Bordeaux **1936–37**:23–25.
1937. Les effets protecteurs dans les vins chauffés. Ann. Ferm. **3**:382–383.
1937–1938. Note sur l'action des bactéries dans les vins. Proc.-Verb. Soc. Sci. Phys. Nat. Bordeaux **1937–38**:73–74.
1938. Les bactéries du vin et les transformations qu'elles provoquent. Bull. Assoc. Chim. Sucr. Distill. de France et Colon. **55**:601–656. (*See also* Progr. Agr. Vitic. **111**, Suppl.:1–32. 1939.)
1941. Note sur l'action des bactéries dans le vin. Ann. Ferm. **6**:228–240. (*See also* Bull. O.I.V. [Office Intern. Vigne Vin] **15**[150]:95–99. 1942.)
1942. Etudes sur les transformations du vin par les bactéries. *Ibid.* **7**(1–4):21–31; (5–7):74–92; (8–10):142–156. (*See also* Bull. O.I.V. [Office Intern. Vigne Vin] **16**[156]:88–92. 1943.)
1943. Les diastases en oenologie. Bull. O.I.V. (Office Intern. Vigne Vin) **16**(160): 43–53.
1944–1945a. Sur la possibilité de préparer les vins doux stables par fermentation en anaérobiose absolue. Proc.-Verb. Soc. Sci. Phys. Nat. Bordeaux **1944–1945**:60–63.
1944–1945b. Sur certaines actions bactériennes dans les vins. *Ibid.* **1944–1945**: 40–45.
1946. Sur la fermentation de l'acide malique dans les grands vins rouges. Bull. O.I.V. (Office Intern. Vigne Vin) **19**(182):26–29.
1947. Traité d'oenologie. Paris and Liège, Librairie Polytechnique Ch. Béranger. 546 pp.
1952. Le potentiel d'oxydo-réduction en oenologie. Chim. Ind. (Paris) **68**:1–7.
1963. Phenomena of oxidation and reduction of wines and applications. Am. J. Enol. Viticult. **14**:139–143.

1964. Analyse chimique et qualité des denrées alimentaires. Qual. Plant. Mater. Végétabiles 11:202–206.

RIBÉREAU-GAYON, J., R. CASSIGNARD, P. SUDRAUD, J. BLOUIN, and J. C. BARTHE
1963. Sur la vinification en blanc sec. Compt. Rend. Acad. Agr. France 49:509–512.

RIBÉREAU-GAYON, J., and J. GARDRAT
1957. Application du titrage potentiométrique à l'étude du vin. Ann. Technol. Agr. 6:185–216.

RIBÉREAU-GAYON, J., and A. MAURIÉ
1938. Arrêts de fermentation. Bull. O.I.V. (Office Intern. Vigne Vin) 11(118): 46–48.

RIBÉREAU-GAYON, J., and E. PEYNAUD
1934–1935. Études sur le collage des vins. Rev. Viticult. 81:5–11, 37–44, 53–59, 117–124, 165–171, 201–205, 310–315, 341–346, 361–365, 389–397, 405–411; 82:3–13.
1936. Estérification chimique et biologique des acides organiques du vin. Bull. Soc. Chim. France 3(5):2325–2330.
1938a. La désacidification des vins par les bactéries. Compt. Rend. Acad. Agr. France 24:600–605.
1938b. Bilan de la fermentation malolactique. Ann. Ferm. 4:559–569.
1946. Sur la formation des acides acétiques, lactique et citrique au cours de la fermentation alcoolique. Compt. Rend. 222:457–458.
1950. Perfectionnements techniques en vinification. Vignes et Vins 6(12):21–23.
1952. Sur l'emploi en vinification de quelques activeurs vitaminiques de la fermentation. Compt. Rend. Acad. Agr. France 38:444–448.
1958. Analyse et contrôle des vins. 2d ed. Paris, Librairie Polytechnique Ch. Béranger. xxxii + 557 pp.
1960–1961. Traité d'oenologie. Paris, Librairie Polytechnique Ch. Béranger. 2 vols.
1963. Application à la vinification de quelques métabolisant l'acide malique. Compt. Rend. Acad. Agr. France 48:555–560.
1964–1966. Traité d'oenologie. New ed. Paris, Librairie Polytechnique Ch. Béranger. 2 vols. (See vol. 2, chap. 14, pp. 607–687.)

RIBÉREAU-GAYON, J., E. PEYNAUD, and S. LAFOURCADE
1951. Sur l'influence de l'aération au cours de la fermentation. Ind. Aliment. Agr. (Paris) 68(3/4):141–150.

RIBÉREAU-GAYON, J., and P. RIBÉREAU-GAYON
1958. The anthocyans and leucoanthocyans of grapes and wines. Am. J. Enol. Viticult. 9:1–9.

RIBÉREAU-GAYON, P.
1959. Recherches sur les anthocyanes des végétaux. Application au genre *Vitis*. Paris, Librairie Générale de l'Enseignement. 114 pp.
1963. La différenciation des vins par l'analyse chromatographique de leur matière colorante. Ind. Aliment. Agr. (Paris) 80:1079–1084.
1964a. L'acétate d'éthyle dans les vins, son dosage par chromatographie en phase gazeuse. Qual. Plant. Mater. Végétabiles 11:249–255.
1964b. Les composés phénoliques du raisin et du vin. I. II. III. Ann. Physiol. Végétale 6:119–147, 211–242, 259–282.
1964c. Les composés phénoliques du raisin et du vin. Paris, Institut National de la Recherche Agronomique. 83 pp.
1964d. Identification d'esters des acides cinnamiques et de l'acide tartarique dans les limbes et les baies de *V. vinifera*. Compt. Rend. 260:341–343.
1965. La couleur des vins. Bull. Soc. Sci. Hyg. Aliment. 53:232–248.
1968. Les composés phenoliques des végétaux. Paris, Dunod. 10 + 254 pp.

RIBÉREAU-GAYON, P., and N. NEDELTCHEV
 1965. Discussion et application des méthodes modernes de dosage des anthocyanes et des tannins dans les vins. Ann. Technol. Agr. 14:321–330. (*Also in* Rev. Franç. Oenol. No. 26:28–32. 1967.)

RIBÉREAU-GAYON, P., and J. C. SAPIS
 1965. Sur la présence dans le vin de tyrosol, de tryptophol, d'alcool phényl-éthylique et de γ-butyrolactone, produits secondaires de la fermentation alcoolique. Compt. Rend. 261:1915–1916.

RIBÉREAU-GAYON, P., and E. STONESTREET
 1964. La constitution du raisin et du vin. Compt. Rend. Acad. Agr. France 50:662–670.
 1965. Le dosage des anthocyanes dans le vin rouge. Bull. Soc. Chim. France 1965:2649–2652. (*See also* Chim. Anal. [Paris] 48:188–195. 1966.)
 1966. Vorkommen und Bedeutung der Catechine, Leukoanthocyanidine und der Gerbstoffe in Rotweinen. Deut. Lebensm.-Rundschau 62:1–5.

RICE, A. C.
 1965a. Identification of grape varieties. J. Assoc. Offic. Agr. Chemists 48:525–530.
 1965b. The malo-lactic fermentation in New York State wines. Am. J. Enol. Viticult. 16:62–68.
 1968. Effects of certain p-hydroxybenzoic acid esters in two American wines. Ibid. 19:101–107.

RICE, A. C., J. W. FERGUSON, and R. S. BELSCHER
 1968. Residual sugars in New York state wines Amer. Jour. Enol. Vit. 19:1–5.

RICKETTS, J., and M. W. COUTTS
 1951. Hydrogen sulfide in fermentation gas. Amer. Brewer 84(8):27–30; (9):27–30, 74–75; (10):33–36, 100–101.

RIDEAL, E. K., and W. THOMAS
 1922. Adsorption and catalysis in fuller's earth. J. Chem. Soc. 121:2119–2123.

RIMSKIĬ, L. S.
 1966. Promyshlennaia éstetika vinodel'cheskikh predpriiatiĭ. Vinodelie i Vino-gradarstvo SSSR 26(2):56–58.

RIPPER, M.
 1892. Die schweflige Säure im Weine und deren Bestimmung. J. Prakt. Chem. (N. F.) 45:428.

ROBINSON, W. B., L. D. WEIRS, J. J. BERTINO, and L. R. MATTICK
 1966. The relation of anthocyanin composition to color stability of New York State wines. Am. J. Enol. Viticult. 17:178–184.

ROCQUES, X.
 1899. Le bouquet des vins. Rev. Viticult. 12:95–99.

RODOPULO, A. K.
 1964. O roli produktov prevrashcheniia aminokislot v obrazovanii buketa sham-panskogo (Role of the conversion products of amino acids in the formation of the bouquet of sparkling wines). Vinodelie i Vinogradarstvo SSSR 24(1):6–9.
 1965. Phénomènes d'oxydo-réduction au cours de la maturation et du vieillisse-ment du vin et procédés pour les modifier. Bull. O.I.V. (Office Intern. Vigne Vin) 38(415):959–969.

RODOPULO, A. K., and I. A. EGOROV
 1965. Karbonil'nye soedineniia kheresa. Vinodelie i Vinogradarstvo SSSR 25(1):6–9.

LITERATURE CITED

RODOPULO, A. K., and A. F. PISARNITSKIĬ
1966. K izucheniiu buketoobrazuiushchikh veshchestv shampanskogo. Vinodelie i Vinogradarstvo SSSR 26(6):5–9. (See also Prikl. Biokhim. Mikrobiol. 2:452–460. 1960.)

RODOPULO, A. K., and N. G. SARISHVILI
1965. O biokhimicheskikh protsessak pri biologicheski potochnom metode obrabotki vinomaterialov pered shampanizatsieĭ. Prikl. Biokhim. Mikrobiol. 1:669–674.

ROESSLER, E. B., and M. A. AMERINE
1958. Studies on grape sampling. Am. J. Enol. 9:139–145.

ROESSLER, E. B., G. A. BAKER, and M. A. AMERINE
1956. One-tailed and two-tailed tests in organoleptic comparisons. Food Res. 21:117–121.

ROESSLER, E. B., J. WARREN, and J. F. GUYMON
1948. Significance in the triangle taste tests. Food Res. 13:503–505.

ROGERS, S. S., H. B. STAFFORD, and A. E. MAHONEY
1940. The wine grape resting law. Calif. Dept. Agr. Bull. 29(1):3–11.

RONKAINEN, P., S. BRUMMER, and H. SUOMALAINEN
1967. Chromatographic identification of carbonyl compounds. VIII. The carbonyl compounds in a fermented glucose solution. J. Chromatog. 27:443–445.

ROOS, L., and F. CHABERT
1897. Contribution à l'étude des fermentations viniques. Rev. Viticult. 8:33–37, 69–74, 89–94, 122–125.

ROSE, A. H.
1961. Industrial microbiology. London, Butterworths. 8 + 286 pp.
1965. Chemical microbiology. London, Butterworths. 8 + 247 pp.

ROSENSTEIN, E.
1964. Wine production in countries with hot climates. Wines & Vines 45(8): 29–32.

ROSON, J.-P.
1967. Recherches sur la vinification des vins rosés. Ann. Technol. Agr. 16:227–240.

ROSSI, JR., E. A.
1951. Report on chrome plating. Proc. Am. Soc. Enol. 2:192–200.
1963. Ultraviolet light as an aid in winemaking. Am. J. Enol. Viticult. 14:178–184.
1965. Sparkling wine production by Charmat process. San Francisco, Wine Institute, Technical Advisory Committee, June 7, 1965. 4 pp.

ROSSI, JR., E. A., and G. THOUKIS
1960. Low-level carbonation of still wines. Am. J. Enol. Viticult. 11:35–45.

ROSSI, JR., J. A., and V. L. SINGLETON
1966a. Contributions of grape phenols to oxygen absorption and browning of wines. Am. J. Enol. Viticult. 17:231–239.
1966b. Flavor effects and adsorptive properties of purified fractions of grape-seed phenols. Ibid. 17:240–246.

ROSSI, JR., R.
1962. United Vintners expands its Madera premises. Wines & Vines 43(9):25–27

RUBNER, M.
1904. Die Umsetzungswärme bei der alkoholischen Gärung. Arch. Hyg. 49: 355–418.
1913. Die Ernährungsphysiologie der Hefezelle bei der alkoholischen Gärung. Arch. Anat. Physiol., Physiol. Abt. 1912:1–396.

RUDIGER, M., and E. MAYR
1929. Die Weinschönung. Kolloid-Z. 47:141–145.

941

Rumpf, P.
1933. Sur le titrage électrométrique des acides sulfureux, selenieux et α-oxyalcoyl-sulfoniques. Compt. Rend. 197:686–689.

Russell, A.
1935. The natural tannins. Chem. Rev. 17:155–186.

Russell, J. P., F. R. Ingram, and E. W. Dakan
1939. Industrial hygiene survey of California wineries. Calif. Dept. Pub. Health, Indus. Hyg. Serv., Invest. Rept. 2:1–36.

Rzędowski, W., and H. Rzędowsak
1960. The biological decomposition of fruit musts (transl.). Prace Inst. Lab. Badawczych Przemystu Rolnego i Spożywczego 10(3):1–24. (Chem. Abst. 55:9775. 1961.)

Sadjadi, A.
1936. Recherches sur les matières pectiques dans le vin. Bordeaux, Imprimerie Brusau Frères. 45 pp.

Sai, T., and M. Amaka
1967. Studies on (−)-citramalic acid formation by respiration-deficient yeast multants: citramalic formation and a variety of respiration-deficient strains of brewers', bakers', saké and wine yeasts. J. Gen. Appl. Microbiol. (Tokyo) 13:15–23.

Sakthivadivel, R., and S. Irmay
1966. A review of filtration theories. Univ. Calif., Hydraulic Engineering Lab. Publ. HEL 15–4:1–65.

Sale, J. W., and J. B. Wilson
1926. Distribution of volatile flavor in grapes and grape juices. J. Agr. Res. 33: 301–310.

Saller, W.
1957. Die spontane Sprosspilzflora frisch gepresster Traubensäfte und die Rein-hefegärung. Mitt. Rebe u. Wein, Serie A (Klosterneuburg) 1:130–138.
1958. Control of cold fermentation. Am. J. Enol. 9:41–48.

Saller, W., and C. de Stefani
1962. Untersuchungen über die Brauchbarkeit von Presshefe zur Umgärung von Wein. Mitt. Rebe u. Wein, Serie A (Klosterneuburg) 12:11–18.

Sandell, E. B.
1944. Colorimetric determination of traces of metals. New York, Interscience Publishers, Inc. xvi + 487 pp.

Santagostino, E. A., and C. Sapetti
1963. Ossidazione dell'acido ascorbico in presenza e in assenza di anidride sol-forosa. Ann. Sperim. Agr. (Rome) 17:561–581.

Sauvard, D., L. L. Dorange, and P. Galzy
1968. Etude de trois bactéries isolees du vin. Rev. Ferm. Ind. Aliment. 23:17–24.

Sawyer, D. T., and L. V. Interrante
1961. Electrochemistry of dissolved gases. II. Reduction of oxygen at platinum, palladium, nickel and other metal electrodes. J. Electroanal. Chem. 2: 310–327.

Saywell, L. G.
1934a. Clarification of vinegar. Ind. Eng. Chem. 26:379–385.
1934b. The clarification of wine. Ibid. 26:981–982. (Also in Wine Rev. 2[5]:16–17. 1934.)
1935. Effects of filter aids and filter materials on wine composition. Ibid. 27: 1245–1250.

LITERATURE CITED

SAYWELL, L. G., and B. B. CUNNINGHAM
1937. Determination of iron. Colorimetric o-phenanthroline method. Ind. Eng. Chem., Anal. Ed. 9:67–69. (*See also* F. C. Hummel and H. H. Willard. *Ibid.* 10:13. 1938.)

SCARDOVIC, V.
1951. Studi sulla resistenza dei lieviti all'anidride solforsa. I. Prime indagini sul meccanismo della resistenza all'anidride solforosa. Ann. Microbiol. 4:131–172.
1952. *Ibid.* II. Resistenza e contenuto in glutatione delle discendenze ottenute per isolamento monosporiale dal ceppo assuefatto. *Ibid.* 5:5–16.
1953. *Ibid.* Influenza del NaHSO₃ su alcune funzioni metaboliche del ceppo originario e del ceppo assuefatto. *Ibid.* 5:140–161.

SCAZZOLA, E.
1956. Sur un produit inhibiteur de la cristallisation. Ann. Fals. Fraudes 49:159–163.

SCHAELDERLE, D., and M. HASSELMANN
1961. Application de la réaction d'Eegrive au dosage de l'acide lactique dans le vin. Ann. Falsif. Expertise Chim. 54(633/634):421–428.

SCHALLER, A.
1958. Die Lage des Aequivalenzpunktes bei potentiometrischer Bestimmung der Titrationsacidität von Fruchtsäften. Fruchtsaft-Ind. 4:160–163.

SCHALLIS, A., and P. MACULOSO
1954. Sulfur dioxide. *In* R. E. Kirk and D. F. Othmer (eds.), Encyclopedia of chemical technology. New York, The Interscience Encyclopedia, Inc. 13:417–426.

SCHANDERL, H.
1938. Kellerwirtschaftliche Fragen zur Schaumweinbereitung. Wein u. Rebe 20(1):1–8.
1942. Le "bleuissement biologique" des vins mousseux. Bull. O.I.V. (Office Intern. Vigne Vin) 15(150):45–50.
1943. Eine vergleichende Studie über Champagner- und Schaumweinbereitung. Wein u. Rebe 25(5/6):74–82.
1948a. Oxidation-reduction potentials of wine. Wines & Vines 29(11):27–28. (*See also* Schweiz. Z. Obst- Weinbau 57[18]:301–305. 1948.)
1948b. Die Reduktions-Oxydations-Potentiale während der Entwicklungsphasen des Weines. Der Weinbau, wissen. Beih. 2(7):191–198; (8):209–229.
1950–1951. Ueber den Einfluss des Entsäuerns, verschiedener Schönungen und Lichtes auf das rH und pH der Weine. Wein u. Rebe 1950–1951:118–128.
1952. Ueber die Wirkung der schwefligen Säure auf Schimmelpilze, Hefe und Bakterien. Rhein. Weinztg. 2:181–183.
1957. Ueber den Keimgehalt direkt am Kühkanal der Glashütte verpackter neuer Weinflaschen. Deut. Wein-Ztg. 93(10)155–160.
1959. Die Mikrobiologie des Mostes und Weines. 2d ed. Stuttgart, Verlag E. Ulmer. 321 pp.
1962. Der Einfluss von Polyphenolen und Gerbstoffen auf die Physiologie der Weinhefe und der Wert des pH-7-Tests für des Auswahl von Sektgrundweinen. Mitt. Rebe u. Wein, Serie A (Klosterneuburg) 12A:265–274.
1964. Vergangenheit, Gegenwart und Zukunft der Schaumweine. Deut. Wein-Ztg. 100:188, 190, 192.
1965a. Beispiele aus der Mikrobiologischen Praxisberatung der letzten Jahre. Weinberg Keller 12:490–491.
1965b. Ueber die Entstehung von Hefefett bei der Schaumweingärung. Mitt. Rebe u. Wein, Serie A (Klosterneuburg) 15:13–20.

1965c. Möglichkeiten der Verringerung hohen Polyphenolgehaltes von Weinen, insbesondere von Sektgrundweinen. Jahresbericht Hessische Lehr- u. Forschungsanstalt Wein-, Obst- und Gartenbau, Geisenheim 1965:17–18.

1965d. Der Einfluss von Kationenaustausch des Grundweines auf Sektgärungen. Ibid. 1965:18–19.

SCHANDERL, H., and M. DRACZNYSKI
1952. Brettanomyces, eine lästige Hefegattung im flaschenvergorenem Schaumwein. Deut. Wein-Ztg. 88:462–464.

SCHANDERL, H., and T. STAUDENMAYER
1964. Ueber den Einfluss der Schwefligen Säure auf die Acetaldehydbildung verschiedener Hefen bei Most- und Schaumweiningärungen. Mitt. Rebe u. Wein, Serie A (Klosterneuburg) 14:267–281.

SCHATZLEIN, CH., and E. FOX-TIMMLING
1940. Untersuchungen über den Vitamin C-Gehalt von Gemüse und Obst. Z. Untersuch. Lebensm. 79:167–174.

SCHEFFER, W. R., and E. M. MRAK
1951. Characteristics of yeast causing clouding of dry white wine. Mycopath. Mycol. Appl. 5(2/3):236–249.

SCHELHORN, M. VON
1951a. Control of microörganisms causing spoilage in fruit and vegetable products. Advan. Food Res. 3:429–482.

1951b. Efficacy and specificity of sulfurous acid (transl.). Deut. Lebensm. Rundschau 47:170–173.

1953. Efficacy and specificity of chemical food preservatives. Food Technol. 7:97–101. (See also Z. Lebensm.-Untersuch. -Forsch. 92:256–266. 1957.)

SCHENK, W., and H. ORTH
1965. Restzuckerhaltige Weine durch individuelle Gärungsunterbrechung oder durch Zusatz von Süssreserve. Weinberg Keller 12:525–531.

SCHEURER
1944. Les vitamines du vin. La terre vaudoise. Bull. O.I.V. (Office Intern. Vigne Vin) 17(163/66):159. (See Méd. et Hyg., June 4, 1943.)

SCHMIDT, H. C.
1965. Der Weinbau in Oesterreich. Wein-Wissen. 20:525–536.

SCHNEYDER, J.
1965. Die Behebung der durch Schwefelwasserstoff und Merkaptane versursachten Geruchsfehler der Weine mit Silberchlorid. Mitt. Rebe u. Wein, Serie A (Klosterneuburg) 15:63–65.

SCHÖN, K. A., J. A. DE GOUVELA, and F. P. COELHO
1939. Determination of vitamin A, ergosterol, lactoflavin, and vitamin C by physicochemical methods. Investigation of the red wines of Bairrada (transl.). Coimbra Univ., Rev. Faculdade de Cien. 8:130–147. (Chem. Abst. 38: 1317. 1944.)

SCHOONMAKER, F.
1964. Frank Schoonmaker's encyclopedia of wine. New York, Hastings House. vi + 410 pp.

SCHORMÜLLER, J., and W. CLAUSS
1966. Untersuchungen über das Vorkommen und die Enstehung der Schleimsäure in Traubenmosten und Weinen. III. Säulen- und Papierchromatographische Untersuchungen über die im Traubenmost enthaltenen Säuren. Z. Lebensm. -Untersuch. -Forsch. 133:65–72.

SCHORMÜLLER, J., W. CLAUSS, and G. WÜRDIG
1966. *Ibid.* IV. Parasitäre Bildung von Schleimsäure in den Beeren von *Vitis vinifera* aus Galakturonsäure nach dem Befall durch *Botrytis cinerea*. *Ibid.* **132**:270–276.

SCHRATT, H.
1966. Rapport Autrichien. Bull. O.I.V. (Office Intern. Vigne Vin) **39**:1003–1008.

SCHREFFLER, C.
1952. Heat transfer in winery refrigeration. Proc. Am. Soc. Enol. **3**:211–217.

SCHROETER, L. C.
1966. Sulfur dioxide. Applications in foods, beverages, and pharmaceuticals. New York, Pergamon Press. 14 + 342 pp.

SCHULTZ, A. S., L. ATKIN, and C. N. FREY
1940. The effect of bios on the nitrogen metabolism of yeast. I. Ammonia and carbamide. J. Biol. Chem. **135**:267–271.

SCHULTZ, H. B., and L. A. LIDER
1964. Modification of the light factor and heat load in vineyards. Am. J. Enol. Viticult. **15**:87–92.

SCHULTZ, H. W., E. A. DAY, and L. M. LIBBEY (eds.)
1967. Symposium on foods: The chemistry and physiology of flavors. Westport, Conn, Avi Publishing Co. 9 + 552 pp.

SCOTT, R. D.
1923. Methyl anthranilate in grape beverages and flavors. Ind. Eng. Chem. **15**:732–733.

SCOTT, R. S.
1967. Clarification—the better half of filtration. Wines & Vines **48**(10):29–30.
1969. New approaches to clarification. *Ibid.* **50**(3):22–23.

SCULL, R. S.
1949. Observations on use of stainless steel and chromium plating for the food industry. Fruit Prod. J. **28**:179–181.

SEARLE, H. E., F. L. LAQUE, and R. H. DOHROW
1934. Metals and wines. Ind. Eng. Chem. **26**:617–627.

SEIFERT, W.
1901. Ueber die Säureabnahme im Wein und den dabei sich vollziehenden Gährungsprocess. Mitt. gärungsphysiol. Lab. Klosterneuburg, **1901**:1–17.
1903. Ueber die Säureabnahme im Wein und dem dabei stattfindenen Gärungsprozess. Z. land. Versuch. Oesterreich. **6**:567–585. (*See alse ibid.* **4**:980–992.)

SEMICHON, L.
1930. La loi du 1er janvier 1930 et la renaissance des vins d'origine. Compt. Rend. Acad. Agr. France **16**:461–467. (*See also* J. Agr. Pratique **53**:491–492.)

SEMICHON, L., and M. FLANZY
1926. Sur les pectines des raisins et le moelleux des vins. Compt. Rend. **183**(6):394–396.
1929. Dosage de l'alcool dans les vins et spiritueux par l'emploi du mélange sulfi-chromique. Ann. Fals. Fraudes **22**:139–153. (*See also ibid.* **22**:414–415; **23**:347–349.)

ŞEPTILICI, G., M. GHERMAN, F. MUJDABA, E. POPA, O. BELLIC, and M. MACICI
1962–1963. Suşe di drojdii cu activitate lentă folosite la obţinera vinurilor demiseci. Inst. Cercetări Horti-Viticole, Lucrări Ştiinţ. **6**:697–712.

SEQUIN, G.
1966. Sols du Médoc et qualité des vins. Compt. Rend. Acad. Agr. France **51**:1197–1209. (*See also* Rev. Franç. Oenol. **7**[27]:21–25, 1967.)

SHARF, J. M., and C. A. LYON
1958. Historical development of stoppers for sparkling wines. Am. J. Enol. Viticult. 9:74–78.

SHAULIS N. J., and W. B. ROBINSON
1953. The effect of season, pruning severity, and trellising on some chemical characteristics of Concord and Fredonia grape juice. Proc. Am. Soc. Hort. Sci. 62:214–220.

SHAULIS, N. J., E. S. SHEPARDSON, and J. C. MOYER
1960. Grape harvesting research at Cornell. I. Proc. New York State Hort. Soc. 105:250–254.
1964. Grape harvesting research at Cornell. VI. Pruning, training and trellising Concord grapes for mechanical harvesting in New York. Proc. New York State Hort. Soc. 109:234–241.

SHEAR, S. W., and G. G. PEARCE
1934. Supply and price trends in the California wine-grape industry. Part 2. Univ. Calif. Giannini Found. Agr. Econ. (mimeo.) Rept. 34:1–63.

SHEPARDSON, E. S., H. E. STUDER, N. J. SHAULIS, and J. C. MEYER
1962. Mechanical grape harvesting, research progress and developments at Cornell. Agr. Engin. 43:66–71.

SHEPPARD, W. L., JR.
1947. Better floors for food plants with acid-proof brick or tile. Food Indus. 19(12):1656–1659, 1757.

SHERRILL, M. S., and A. A. NOYES
1926. The inter-ionic attraction theory of ionized solutes. VI. The ionization and ionization constants of moderately ionized acids. J. Am. Chem. Soc. 48: 1861–1873.

SHIJO, N., and K. TANAKA
1965. Biosynthesis of iron-containing substances by yeasts. I. Fate of iron during fermentation of grape musts (transl.). Yamanashi Daigaku, Hakkô Kenkyusho Kenkyu Hokoku No. 12:47–54.

SHIMATANI, Y., and Y. NAGATA
1967. Studies on microflora related to wine making. I. Mould and yeast flora in a vineyard. II. Distribution of film yeasts in vineyards and wine making processes in five grape and wine producing districts of Japan. Hakkô Kôgaku Zasshi (J. Ferm. Technol.) 45:179–184, 185–190.

SIEGEL, A., R. G. ROTTER, and L. SCHMID
1965. Zur Frage der Weinbeurteilung. Z. Lebensm.-Untersuch. -Forsch. 126: 321–324.

SIFNÉOS, C., and P. LAURENT
1937. Différentes méthodes de macération des vins rouges. Rev. Viticult. 87:81–84, 116–129.

SIGALIN, A.
1965. Usefulness of epoxy resins produced in Poland as protective coat for steel and concrete containers for wine and must (transl.). Prace Inst. Lab. Badawczych Przemyslu Rolnego Spozyw czego 15(4):51–68. (Chem. Abst. 64[13]: 20588–29589. 1966.)

SIHTO, E., L. NYKÄNEN, and H. SUOMALAINEN
1962. Gas chromatography of the aroma compounds of alcoholic beverages. Teknillinen Kemian Aikakausilehti 19:753–762.
1964. Gas chromatography of the aroma compounds of alcoholic beverages. Qual. Plant. Mater.Végétabiles 11:211–228.

ŠIKOVEC, S.
1966. Der Einfluss einiger Polyphenole auf die Physiologie von Weinhefen. I. Der Einfluss von Polyphenolen auf den Verlauf der alkoholischen Gärung, insbesondere von Umgärungen. II. Der Einfluss von Polyphenolen auf die Vermehrung und Atmung von Hefen. Mitt. Rebe u. Wein, Obstbau u. Früchteverwertung (Klosterneuburg) 15:127–138, 272–281.

SILICA PRODUCTS CO., KANSAS CITY, MO.
1930. Bentonite, properties, sources, geology, production. Bull. 107.

SILOX, H. E., and L. B. LEE
1948. Fermentation. Ind. Eng. Chem. 40:1602–1608.

SILVA, F., P. FIGUEIREDOE, and O. SALES PETINGA
1962. Identificação de aminoácidos em mostos de algumas castas das regiões do Dão a do Oeste. Agros (Lisbon) 45:195–199.

SILVA BABO, M.
1963. Essais d'application des bactéries malolactiques aux vins verts. Ann. Technol. Agr. 12(numéro hors série 1):57–58.

SILVESTRE, J.
1953. Contribution à l'étude des tocophérols du pépin de raisin. Thèse, Faculté de Pharmacie, Toulouse. 79 pp.

SIMON, A. L.
1962. Champagne, with a chapter on American champagne by R. J. Misch. New York, McGraw-Hill Book Co., Inc. 224 pp. (There is an English edition without the Misch chapter. London, Ebury Press, 1962. 192 pp.)
1967. The wines, vineyards, and vignerons of Australia. London, Paul Hamlyn. 13 + 194 pp.

SINGLETON, V. L.
1962. Aging of wines and other spirituous products: acceleration by physical treatments. Hilgardia 32:319–392.
1963. Changes in quality and composition produced in wine by cobalt-60 gamma irradiation. Food Technol. 18(6):112–115.
1966a. Fining-phenolic relationships. San Francisco, Wine Institute, Technical Advisory Committee, December 9, 1966. 6 pp.
1966b. The total phenolic content of grape berries during maturation of several varieties. Am. J. Enol. Viticult. 17:126–134.
1967. Fining-phenolic relationships. Wines & Vines 48(2):23–26.

SINGLETON, V. L., H. W. BERG, and J. F. GUYMON
1964. Anthocyanin color level in port-type wines as affected by the use of wine spirits containing aldehydes. Am. J. Enol. Viticult. 15:75–81.

SINGLETON, V. L., and D. E. DRAPER
1961. Wood chips and wine treatment; the nature of aqueous alcohol extracts. Am. J. Enol. Viticult. 12:152–158.
1963. Ultrasonic treatment with gas purging as a quick aging treatment for wine. Ibid. 14:23–35.
1964. The transfer of polyphenolic compounds from grape seeds into wines. Ibid. 15:34–40.

SINGLETON, V. L., D. E. DRAPER, and J. A. ROSSI, JR.
1966. Paper chromatography of phenolic compounds from grapes, particularly seeds, and some variety-ripeness relationships. Am. J. Enol. Viticult. 17:206–217.

SINGLETON, V. L., and C. S. OUGH
1962. Complexity of flavor and blending of wines. J. Food Sci. 27:189–196.

SINGLETON, V. L., C. S. OUGH, and M. A. AMERINE
1964. Chemical and sensory effects of heating wines under different gases. Am. J. Enol. Viticult. 15:134–145.

SINGLETON, V. L., and J. A. ROSSI, JR.
1965. Colorimetry of total phenolics with phosphomolybdic-phosphotungstic acid reagents. Am. J. Enol. Vitic. 16:144–158.

SISAKIAN, N. M., and E. H. BESINGER
1953. Izmenenie aminokislotnogo sostava vina pri ego pervichnoĭ tekhnologii. Biokhim. (USSR) 18:412–422.

SISAKIAN, N. M., I. A. EGOROV, N. G. SARISHIVILI, A. K. RODOPULO, and V. V. AGAPOV
1961. Primenie fermentnykh preparatov pri shampanizatsii vina rezervuarnym sposobom. Vinodelie i Vinogradarstvo SSSR 21(7):15–20.

SISAKIAN, N. M., A. K. RODOPULO, I. A. EGOROV, and N. G. SARISHVILLI
1963. Produkty prevrashcheniia aminokislot drozhzhami i ikh vliianie na kachestvo shampanskogo. Biokh. Vinodeliia 7:131–147.

SISANI, L.
1948. Indagini chimico microbiologiche sul vino santo. Riv. Viticolt. Enol. 1:142–147.

SKAVYSH, V. I.
1965. Lechenie vin, infitsirovannykh molochnokislymi bakterriiami. Vinodelie i Vinogradarstvo SSSR 25(8):41–44.

SKOFIS, E.
1953. The role of refrigeration in the stabilization and clarification of wines. Proc. Am. Soc. Enol. 4:69–76.
1960. In-plant production problems. Wines & Vines 41(11):45–46.

SMIT, C. J. B., M. A. JOSLYN, and A. LUKTON
1955. Determination of tannins and related polyphenols in foods; comparison of Loewenthal and Pro methods. Anal. Chem. 27:1159–1162.

SMITH, J. C.
1961. Applications of centrifugation equipment. Ind. Eng. Chem. 53:439–444.

SMITH, M. B., and H. P. OLMO
1944. The pantothenic acid and riboflavin in the fresh juice of diploid and tetraploid grapes. Am. J. Bot. 31:240–241.

SMITH, R. M., and B. S. LUH
1965. Anthocyanin pigments of the hybrid grape variety Rubired. J. Food Sci. 30:995–1005.

SMITH, W. C., and R. C. GIESSE
1961. Filtration equipment. Design. Ind. Eng. Chem. 53:439–444.

SOMAATMADJA, D., and J. J. POWERS
1963. Anthocyanins. IV. Anthocyanin pigments of Cabernet Sauvignon grapes. J. Food Sci. 28:617–622.

SOMAATMADJA, D., J. J. POWERS, and R. WHEELER
1965. Action of leucoanthocyanins of Cabernet grapes on reproduction and respiration of certain bacteria. Am. J. Enol. Viticult. 16:54–61.

SOMERS, T. C.
1966a. Grape phenolics: the anthocyanins of Vitis vinifera, variety Shiraz. J. Sci. Food Agr. 17:215–219.
1966b. Wine tannins—isolation of condensed flavonoid pigments by gel-filtration. Nature 209:368–370.
1967. Resolution and analysis of total phenolic constituents of grape pigment. J. Sci. Food Agr. 18:193–196.

1968. Pigment profiles of grapes and of wines. Vitis 7:303–320.

SONG, PILL-SOON and C. O. CHICHESTER
1967a. Kinetic behavior and mechanism of inhibition in the Maillard reaction. III. Kinetic behavior of the inhibition in the reaction between D-glucose and glycine. J. Food Sci. 32:98–106.
1967b. Ibid. IV. Mechanism of the inhibition. Ibid. 32:107–115.

SPAETH, E. C., and D. H. ROSENBLATT
1950. Partition chromatography of synthetic anthocyanidin mixtures. Anal. Chem. 22:1321–1326.

SPATZ, G.
1967. Improving winery filtration. San Francisco, Wine Institute, Technical Advisory Committee, June 8, 1967. 5 pp.

STANESCU, H.
1965. Disque pour le calcul de l'indice de stabilisation biologique des vins. Bull. O.I.V. (Office Intern. Vigne Vin) 38:1428–1429.

STAUDENMAYER, T.
1966. Eine verbesserte Methode zur Unterscheidung von Lebenden Mikroorganismen und Eiweisstrub im Wein. Wein-Wissen. 21(3):127–129.

STAUDENMAYER, T., and N. J. BECKER
1966. Bestimmung von Weinsäure und Aepfelsäure in Mosten nach dem stufentitrimetrischen Verfahren von Carlos de Miranda Pato und Manuel de Souza e Holstein Beck. Wein.-Wissen. 21:28–34.

STELLA, C.
1967. Ricerca degli acidi alogeno-carbossilici nei mosti e nei vini. II. Il clororganico nei mosti e nei vini. Riv. Viticolt. Enol. 20:393–404.

STELLING-DEKKER, N. M.
1931. Die Hefesammlung des "Centraalbureau voor Schimmelcultures." Beiträge zu einer Monographie der Hefearten. I Teil. Die Sporogenen Hefen. 7 + 547 pp. K. Akad. van Wetensch. te Amsterdam. Amsterdam, Holland. (See also K. Akad. von Wetensch. te Amsterdam Verhandel. Afdeeling Natuurkunde [Tweede Sec.] 28[1]:1–547.)

STENGER, V. A., and I. M. KOLTHOFF
1947. Volumetric analysis. Vol. II. Titration methods: neutralization, precipitation, and complex formation reactions. New York, Interscience Publishers. 400 pp.

STEPHAN, E.
1964. Polyäthytenstopfen für Wein- und Sektflaschen, Weinberg Keller 11:447–450.

STERN, D. J., A. LEE, W. H. MCFADDEN, and K. L. STEVENS
1967. Voltatiles from grapes. Identification of volatiles from Concord essence. J. Agr. Food Chem. 15:1100–1103.

STEVENS, K. L., J. L. BOMBEN, A. LEE, and W. H. MCFADDEN
1966. Volatiles from grapes. Muscat of Alexandria. J. Agr. Food Chem. 14:249–252.

STEVENS, K. L., J. L. BOMBEN, and W. H. MCFADDEN
1967. Volatiles from grapes, Vitis vinifera (Linn.) cultivar Grenache. J. Agr. Food Chem. 15:378–380.

STEVENS, K. L., A. LEE, W. H. MCFADDEN, and R. TERANISHI
1965. Volatiles from grapes. I. Some volatiles from Concord grapes. J. Food Sci. 30:1006–1007.

STEVIĆ, B.
1963. The significance of bees (*Apis* sp.) and wasps (*Vespa* sp.) as carriers of yeasts, for the microflora of grapes and the quality of wine (transl.). Arhiv Poljoprivredne Nauke 15:80–94.

STOCKER, H. R.
1967. Zur Chemie der Flavanoidgerbstoffe. Synthese und Analytik des Leucocyanidins. Schweiz. Brau.-Rundschau 78:33–56.

STORER, T. I.
1949. Control of field rodents in California. Calif. Agr. Ext. Cir. 148:1–50.

STOWE, C.
1957. Plastic pipe and the wine industry. Am. J. Enol. Vitic. 8:146–148.

STRADELLI, A.
1951. Evaporazione di alcool durante la fermentazione dei mosti. Riv. Viticolt. Enol. 4:50–53.

STROHM, J. A., and R. F. DALE
1961. Dissolved oxygen measurement in yeast propagation. Ind. Eng. Chem. 53:760–764.

STRUNK, D. H., and A. A. ANDREASEN
1967a. Collaborative study using ZDBT colorimetric method for the determination of copper in alcoholic beverages. J. Assoc. Offic. Agr. Chemists 50:334–338.
1967b. Collaborative study using atomic absorption spectrophotometry for the determination of copper in alcoholic products. *Ibid.* 50:338–339.

STUDER, H. E., and H. P. OLMO
1967. Mechanical harvesting of wine and raisin grapes. Am. Soc. Agr. Eng. Pap. 143:1–32.

STUMBO, C. R.
1965. Thermobacteriology in food processing. New York, Academic Press. 16 + 236 pp.

SUDARIO, E.
1966. Vini in scatola di banda stagnata. Ind. Agr. 4:211–215.

SUDRAUD, P.
1958. Interprétation des courbes d'absorption des vins rouges. Ann. Technol. Agr. 7:203–208.
1963. Stabilisation biologique des vins par chauffage. Ann. Technol. Agr. 12 (numéro hors série 1):131–140.
1967. L'acidité volatile des vins de vendanges botrytisées. Compt. Rend. Acad. Agr. France 53:339–342.

SUOMALAINEN, H.
1963. Changes in the cell constitution of baker's yeast in changing growth conditions. Pure Appl. Chem. 7:639–654.
1965. Flüchtige Begleitstoffe in Gärunglösungen insbesondere in alkoholischen Getranken. Branntweinwirtschaft 105:1–11.

SUOMALAINEN, H., O. KAUPPELA, L. NYKÄNEN, and R. J. PELTONEN
1968. Branntweine. *In* W. Diemair. Alkoholische Genussmittel, Handbook der Lebensmittelchemie 7:495–653. Berlin, Springer Verlag.

SUOMALAINEN, H., and A. J. A. KERÄNEN
1967. Keto acids formed by baker's yeast. J. Inst. Brewing 73:477–484.

SUOMALAINEN, H., and L. NYKÄNEN
1964. The formation of aroma compounds by yeast in sugar fermentation. Suomen Kemistilehti B37:230–232.

SUOMALAINEN, H., and P. RONKAINEN
1963. Keto acids in baker's yeast and in fermentation solution. J. Inst. Brewing 69:478–483.

LITERATURE CITED

SUTER, M. A.
1944. The organic chemistry of sulfur. Tetracovalent sulfur compounds. New York, John Wiley & Sons, Inc. 5 + 858 pp. (*See especially* pp. 126–130, 136–139.)

SUVERKROP, B., and A. TCHELISTCHEFF
1949. Malo-lactic fermentation in California wines. Wines & Vines 30(7):19–23.

SWAIN, T., and W. E. HILLIS
1959. Phenolic constituents of *Prunus domestica*. I. Quantitative analysis of phenolic constituents. J. Sci. Food Agr. 10:63–68.

SWINDELLS, R., and R. H. ROBBINS
1966. Extraction of fruit juices. Process Biochem. 1(9):457–460.

SZABÓ, J., and L. RAKCSÁNYI
1937. Das Mengenverhältnis der Dextrose und der Lävulose in Weintrauben, im Mosten und im Wein. Ve Cong. Intern. Techn. Chim. Agr. Compt. Rend. 1:936–949.

TABACHNIK, J., and M. A. JOSLYN
1953a. Formation of esters by yeast. I. The production of ethyl acetate by standing surface cultures of *Hansenula anomala*. J. Bacteriol. 65:1–9.
1953b. Formation of esters by yeast. II. Investigations with cellular suspensions of *Hansenula anomala*. Plant Physiol. 28:681–692.

TAGUNKOV, IU. D.
1966. Snizhenie aktivnosti katekoloksidazy susla adsorbentami. Vinodelie Vinogradarstvo SSSR. 26(6):20–23.

TAMMANN, G.
1925. The states of aggregation. New York, D. Van Nostrand Company. 297 pp. (*See* pp. 226–282.)

TANNER, H.
1962. Neues über die schweflige Säure und deren Bestimmung in Getränken und Konzentraten. Schweiz. Z. Obst- Weinbau 71:22–226, 245–248.
1963. Die Bestimmung der gesamten schwefligen Säure in Getränken. Mitt. Gebiete Lebensm. Hyg. 54:158–174.
1967. Ueber den dünnschichtchromatographischen Nachweis von Sorbit, Mannit und anderen Polyhydroxyverbindungen. Schweiz. Z. Obst- Weinbau 103: 610–617.

TANNER, H., and A. H. RENTSCHLER
1956. Ueber Polyphenole der Kernobst- und Traubensäfte. Fruchsaft-Ind. 1: 231–245.
1965. Weine mit Böckser; Ursachen des Böcksers und Wiederherstellung davon befallener Weine. Schweiz. Z. Obst- Weinbau 74(1):11–14.

TARANTOLA, C.
1937. La preparazione dell' "Asti Spumante" con fermentazione a bassa temperatura. Ann. R. Staz. Enol. Sper. Asti. 2(II):315–321.
1946. Nouvelle contribution à l'étude des levures apiculées. Ann. Acad. Agr. Turino 1945–1946. (*Reviewed in* Bull. O.I.V. [Office Intern. Vigne Vin] 21[208]:70–72, 1948. *See also* p. 74.)
1948. Le sostanze riduttrici infermentescibili dei vini. Riv. Viticolt. Enol. 1:44–47.
1950. Pentosi e pentosani nei vini e bilancio dei carboidrati fermentescibili e infermentescibili. *Ibid.* 3:287–292.
1954. Separazione e identificazione cromatografica degli aminoacidi nei vini. Atti Accad. Ital. Vite Vino 6:146–157.
1955. Ulteriori indagini sul tenore in azoto assimilabile dei vini. Riv. Viticolt. Enol. 8:10–20.
1963. Traitement des vins par le ferrocyanure de potassium. Ann. Technol. Agr. 12(numéro hors série 1):279–289.

1966. Studio delle proteine e delle sostanze: loro evoluzione nel corso dei trattamenti enologici; condizioni della stabilità proteica dei vini. Ind. Agr. 4: 535–538.

TARANTOLA, C., and M. CASTINO
1961. Sulla naturale presenza di microquantità di acido cianidrico nei vini. Ann. Sperim. Agrar. (Rome) 15:3–15.
1962. Determinazione colorimetrica dell'acido malico nei vini. Annali Facoltà Sci. Agrar. Univ. Studi Torino 1:137–150.
1964. La formazione di acido cianidrico nei vini trattati con ferrocianuro potassico. Riv. Viticult. Enol. 17:483–493.
1966. Determinazione colorimetrica dell'acido malico nei vini. Annali Facoltà Sci. Agrar. Univ. Studi Torino 1:137–150.

TARANTOLA, C., C. CAMPISI, E. BOTTINI, and F. EMANUELE
1954. Industrie agrarie. Turino, Unione Tipografico-Editrice Torinese. 31 + 972 pp.

TARANTOLA, C., and A. GANDINI
1966. La microflora lievitiforme nella vinificazione continua. Atti Accad. Ital. Vite Vino 18:43–62.

TARANTOLA C., and A. LIBERO
1958. Microelementi nei vini. II. Il piombo. Riv Viticolt. Enol. 11:47–60.

TARANTOLA, C., and R. LOVISOLO
1955. Evoluzione delle sostanze azotate durante la lavorazione del Moscato di Canelli ed influenza di alcuni fattori vitaminici sulla sua rifermentazione. Ann. Sperim. Agrar. (Rome) 9:167–188.

TARANTOLA, C., and L. USSEGLIO-TOMASSET
1963. I colloidi delle uve nei vini. Riv. Viticolt. Enol. 16:449–463.

TARTAR, W. V., and H. H. GARRETSON
1941. The thermodynamic ionization constants of sulfurous acid at 25°. J. Am. Chem. Soc. 63:808–816.

TCHELISTCHEFF, A.
1948. Comments on cold fermentation. Wine Technol. Conf., University of California, Col. Agr., Davis, Calif., August 11–13, 1948. Pp. 98–101. (Mimeo.)

TEODORESCU, S. C.
1966. Personal communication.

TEODORESCU, ŞT., C. MATRAN, J. JURUBIŢĂ, G. ŞEPTILICI, V. LEPĂDATU, E. POPA, F. MUJDABA, C. BASAMAC, T. GIOSANU, GH. BUTĂNESCU, G. SANDU-VILLE, C. VODARICI, AL. MIHALCA, and G. BRĂGUTA
1966. Studiul comportării la şampanizare a vinurilor din principalele podgorii şi soiuri din Republica Socialistă România. Inst. Cercetări Horti-Viticole, Lucrări Ştiinţ. 7:881–894.

TERČELJ, D.
1965. Etude des composés azotés du vin. Ann. Technol. Agr. 14:307–319.
1966. La diminution des composés azotés dans le vin par différents collages. Ibid. 15:73–77.

TERČELJ, D., and J. ADAMIČ
1965. Biologische Stabilisierung süsser Weine mit Sorbinsäure und Pyrokohlensäurediäthylester (PKE). Mitt. Rebe u. Wein, Serie A (Klosterneuburg) 15:279–290.

TERRILE, A. P.
1965. Proper maintenance and cleansing procedures of stainless steel in the wine industry. San Francisco, Wine Institute, Technical Advisory Committee, June 7, 1965. 1 p.

THEDDEN, H.
1965. Fragen der Reinigung und Desinfektion in der Kellerwirtschaft. Weinberg Keller **12**:492.

THERON, D. J., and C. J. G. NIEHAUS
1938. Wine-making. Union S. Africa Dept. Agr. Bull. **191**:1–98.

THOMPSON, A. R.
1950. A colorimetric method for the determination of esters. Australian J. Sci. Res. Ser. A **3**(1):128–135.

THORNE, R. S. W.
1946. The nitrogen nutrition of yeast. Wallerstein Lab. Commun. **9**(27):97–114.
1950. Mechanisms of nitrogen assimilation by yeast and their relation to the problem of yeast growth in wort. *Ibid.* **13**(43):319–340.

THOUKIS, G.
1958. The mechanism of isoamyl alcohol formation using tracer techniques. Am. J. Enol. **9**:161–167.

THOUKIS, G., and M. A. AMERINE
1956. The fate of copper and iron during fermentation of grape musts. Am. J. Enol. **7**:45–52.

THOUKIS, G., R. J. BOUTHILET, M. UEDA, and A. CAPUTI, JR.
1962. Fate of diethyl pyrocarbonate in wine. Am. J. Enol. Viticult. **13**:105–113.

THOUKIS, G., G. REED, and R. J. BOUTHILET
1963. Production and use of compressed yeast for winery fermentation. Am. J. Enol. Viticult. **14**:148–154. (*See also* Wines & Vines **44**[1]:25–26.)

THOUKIS, G., and L. A. STERN
1962. A review and some studies of the effect of sulfur on the formation of off-odors in wine. Am. J. Enol. Viticult. **13**:133–140.

THOUKIS, G., M. UEDA, and D. WRIGHT
1965. The formation of succinic acid during alcoholic fermentation. Am. J. Enol. Viticult. **16**:1–8.

THRASHER, J. J.
1966. Sulfur dioxide determination by modified Monier-Williams method. J. Assoc. Offic. Anal. Chemists **49**:834–836.

TILLER, F. M., and C. J. HUANG
1961. Filtration equipment. Theory. Ind. Eng. Chem. **53**:529–537.

TIMBERLAKE, C. F., and P. BRIDLE
1967a. Flavylium salts, anthocyanidins and anthocyanins. I. Structural transformations in acid solutions. J. Sci. Food Agr. **18**:473–478.
1967b. *Ibid.* II. Reactions with sulphur dioxide. *Ibid.* **18**:479–485.

ŢÎRDEA, C.
1964. Dinamica acizilor tartric și malic în timpul perioadei de maturare a strugurilor la principalele soiuri din Podgoria Copou-Iași. Ind. Aliment. Produse Vegetale **15**:294–302.

TIURIN, S. T., A. I. BAZANOVA, and B. N. IL'CHENKO
1964. Rezul'taty issledovaniia sanitarno-gigienicheskikh i fizikomekhanicheskikh svoĭstv otechestvennykh plastmass, prednaznachaemykh dlia primeneniia v vinodel'cheskoĭ promyshlennosti. Trudy Vses. Nauch.-Issledov. Inst. Vinodeliia Vinogradarstva "Magarach" **13**:149–163.

TIURINA, L. V.
1960. O vykhode spirta po khodu brozheniia vinogradnogo susla. Trudy Vses. Nauch.-Issledov. Inst. Vinodeliia Vinogradarstva "Magarach" **9**:96–106.

TIURINA, L. V., N. I. BUR'IAN, and I. G. MAKSIMOVA
1967. Soderzhanie vitaminov gruppy B v krasnom igristom vine tipa tsimlianskogo. Prikl. Biokhim. Mikrobiol. 3:371–374.

TONCHEV, T., T. IVANOV, and G. BANIBALOV
1967. Deistvie na dietilovira ester na perov'lgenata kislima v'rkhu niakoi mikroogranizmi v grozdov sok. Lozarstvo Vinarstvo (Sofia) 16(3):36–39.

TRABERT, R.
1967. Automated pressure leaf filtration. San Francisco, Wine Institute, Technical Advisory Committee, June 8, 1967. 4 pp.

TRACY, R. L.
1932. Lethal effect of alternating current on yeast cells. J. Bacteriol. 24:423–428.

TRAUTH, F., and K. BÄSSLER
1936. Ein Beitrag zur Frage der Beziehung zwischen Mostgewicht und Alkoholgehalt und deren Nutzanwendung bei der Verbesserung der Moste. Z. Untersuch. Lebensm. 72:476–498.

TREPTOW, H., and K. GIERSCHENER
1968. Pyrokolensaürediäthylester (PKE). Seine Wirkung auf die Mikroorganismen und die Inhaltsstoffe in fruchthaltigen Getränken. Flüss. Obst. 35:292–298.

TRESSLER, D. K., R. F. CELMER, and E. A. BEAVENS
1941. Bulk fermentation process for sparkling cider. Ind. Eng. Chem., Ind. Ed. 33:1027–1033.

TROOST, G.
1961. Die Technologie des Weines. 3d ed. Stuttgart, Eugen Ulmer. 702 pp.
1965. Unter welchen Voraussetzungen lässt eine organoleptische Weinprüfung im Zusammenspiel mit einer Weinanalyse ein Höchstmass an Zuverlässigkeit erwarten? Deut. Weinbau 20:262–264.
1966a. Entwicklungstendenzen in der Kellerwirtschaft. Deut. Fass- Weinküfer 17:280–282.
1966b. Nouveaux matériaux pour la fabrication et le revêtement des récipients vinaires. Bull. O.I.V. (Office Intern. Vigne Vin) 39:598–617.
1966c. Nouveaux matériaux pour la fabrication et le revêtement des récipients vinaires: cuves d'élaboration et de stockage, réservoirs de transport, bouteilles et emballages divers. Ibid. 39:737–756.
1967a. Personal communication.
1967b. Weinbehälter und ihre Werkstoffe. Deut. Fass- Weinküfer 20 (No. 20): 289–292.
1968. Die Kaltsterilabfüllung von Getränken bei Zusatz von Baycovin mit Hilfe des Dosiergerätes "Burdomat." Deut. Wein-Ztg. 104:191–193.

TROOST, G., and K. FETTER
1960. Zur Praxis der Eiweissstabilisierung der Weine durch Bentonite. Weinberg Keller 7:444–459.
1966a. Nichtrostends Stähle in der Kellerwirtschaft unter besonderer Berücksichtigung ihrer Oberflächenbeschaftenheit. Weinblatt 60(35/36):797–801.
1966b. Erfahrungen mit der Säureminderung beim Jahrgang 1965. Deut. Wein-Ztg. 102:637–638, 810, 812, 814, 816, 818, 820, 822.

TRUNKEL, H.
1910. Ueber Leim und Tannin. Biochem. Z. 26:458–492.

TSCHENN, C.
1934. Champagnization by the Charmat process. Fruit Prod. J. 13(11):334–336.

TUFANOV, D. G., S. T. TIURIN, and V. V. CHERVANEVA
1966. Isledovanie nerzhaveiu schikh stalei posle dlitel'nogo kontakta s vinom. Vinodelie i Vinogradartsvo SSSR 26(3):52–55.

LITERATURE CITED

TURBOVSKY, M. W.
1949. Practical value of sterile filtration. Wines & Vines 30(6):22.
TURBOVSKY, M. W., F. FILIPELLO, W. V. CRUESS, and P. ESAU
1934. Observations on the use of tannin in wine making. Fruit Prod. J. 14:106–107.
TUZSON, I.
1964. Magyar borok mangántartalma. Borgazdásag 12:106–113.
1966. A magyar borok üledékében előforduló kristályos kiválások. Ibid. 14:116–122.
1967. Kristallausscheidungen in ungarischen Weinen. Mitt. Rebe u. Wein, Obstbau u. Früchteverwertung (Klosterneuburg) 17:129–134.
UCHIMOTO, D., and W. V. CRUESS
1952. Effect of temperature on certain products of vinous fermentation. Food Res. 17:361–366.
UNGARIAN, P. N.
1953. Raïony i sorta vinograda dlia proizvodstva sovetskogo shampanskogo v Moldavskoï SSR. Trudy Vses. Nautch.-Issled. Inst. Vinodeliia Vinogradarstva "Magarach" 4:61–130.
UNITED STATES, BUREAU OF INTERNAL REVENUE
1945. Regulations No. 7, wine. Washington, D.C., U.S. Government Printing Office. 289 pp.
UNITED STATES, DEPARTMENT OF AGRICULTURE, FOOD AND DRUG ADMINISTRATION
1939. Federal food, drug and cosmetic act and general regulations for its enforcement. Serv. and Regulat. Announc., Food, Drug and Cosmetic No. 1:1–50. Washington, D.C., U.S. Government Printing Office
UNITED STATES, INTERNAL REVENUE SERVICE
1956. The determination of carbon dioxide in wine. Washington, D.C., Internal Revenue Service, Alcohol and Tobacco Tax Division Laboratory (IRS-14719). 2 pp.
1961a. Wine. Part 240 of Title 26, Code of Federal Regulations. Washington, D.C., U.S. Govt. Print. Office. 114 pp. (IRS Publ. No. 146. Revised 1–61.)
1961b. FAA Regulations No. 4 relating to the labeling and advertising of wine. Part 4 of Title 27, Code of Federal Regulations. Washington, D.C., U.S. Govt. Print. Office. (U.S. Treasury Dept.) IRS Publ. No. 449:1–70. (Revised 2–61.)
UNITED STATES, LAWS
1963. Production of volatile flavor concentrates, 26 CFR (1954). Part 198. Chicago, Commerce Clearing House.
1965. Excise tax reduction act of 1965. Public Law 89–44. Title VIII. Miscellaneous structural changes. P. 26. (Amends Internal Revenue Code of 1954, par. 23,370A.)
UNITED STATES, TARIFF COMMISSION
1939. Grapes, raisins and wine. U.S. Tariff Comm. Rept. 2d ser. 134:1–408.
UNIVERSITY OF CALIFORNIA, SCHOOL OF PUBLIC HEALTH
1946. Preliminary draft of a sanitation manual for food industries. Berkeley and Los Angeles, University of California Press. 84 pp.
UPPERTON, A. M.
1965. Some observations on detergents and sterilizing agents in British breweries. Wallerstein Lab. Commun. 28:137–142.
USSEGLIO-TOMASSET, L.
1961. La stabilità del vino nei reguardi delle precipitazioni tartariche. Vini d'Italia 3:323–325.
1963. Travaux sur les colloïdes des vins en relation avec les problèmes de la limpidité. Ann. Technol. Agr. 12(numéro hors série 1):195–202.
1964. Gli alcoli superiori nei fermentati alcolici. Riv. Viticolt. Enol. 17:497–530.

1965. L'estratto densimetrico nei vini dolci. liquorosi e verm ut. *Ibid.* **18**:314–320.
1966. Osservazioni sul pigmento ceduto al substrato fermentativo da uno stipite di *Schizosaccharomyces pombe. Ibid.* **19**:454–458.
1967a. L'alcol β-feniletilico nei vini. *Ibid.* **20**:10–34.
1967b. L'aroma di moscato nelle uve e nei vini. Ind. Agr. **4**:216–227.
1967c. Gli acidi volatili omologhi dell'acetico nei fermentati con specie diverse di lieviti. Atti Accad. Ital. Vite Vino **19**:165–188.
1967d. L'acido formico nei vini. *Ibid.* **19**:229–242.

Usseglio-Tomasset, L., V. Astegiano, and M. Matta
1966. Il linalool composto responsible dell'aroma delle uve a dei vini aromatico. Ind. Agr. **4**:583–584.

Valaer, P.
1950. Wines of the world. New York, Abelard Press. 576 pp.

Valdner, R. Ch.
1967. Development of the wine industry of Czechoslovakia. Medzinárodrý Konkurz Vín, Bratislava **2**:78–82.

Valuĭko, G. G.
1965a. Binding of tannin substances and pigments of wine by aldehydes (transl.). Prikl. Biokhim. Mikrobiol. **1**(2):242–243. (Chem. Abstr. **63**:7622d. 1965.)
1965b. Primenenie sernistogo angidrida pri vinodelii po krasnomu sposobu. Sadovodstvo, Vinogradarstvo Vinodelie Moldavii **1965**(4):31–33.
1966. Biochemical and chemical changes in pigments during production of red wines (transl.). Tecknol. Pishch. Prod. Rast. Proiskhozhd. **1966**:72–78. (Chem. Abst. **67**:8426. 1967.)

Valuĭko, G. G., W. P. Aldeev, and K. G. Godin
1967. Izolechenie susla iz mezgi s pomoshch'iu vakuuma. Trudy Vses. Nauch.-Issledov. Inst. Vinodeliia Vinogradarstva "Magarach" **15**:27–35.

Valuĭko, G. G., K. G. Godin, and M. N. Pozmanskaia
1964. Rezhimy termicheskoĭ obrabotki vinograda. Trudy Vses. Nauch.-Issledov. Inst. Vinodeliia Vinogradarstva "Magarach" **13**:44–56.

Vandecaveye, S. C.
1928. The effect of successive generations of yeast on the alcoholic fermentation of cider. J. Agr. Res. **37**:43–54.

Van der Walt, J. P., and A. E. Van Kerken
1958. The wine yeasts of the Cape. II. The occurrence of *Brettanomyces inter-medius* and *Brettanomyces schanderlii* in South African table wines. Antonie van Leeuwenhoek J. **24**:240–251.
1960. *Ibid.* IV. Ascospore formation in the genus *Brettanomyces. Ibid.* **26**:292–296.

Van Note, R. H., and F. T. Weems
1961. Filtration equipment. Application. Ind. Eng. Chem. **53**:546–551.

Van Wyk, C. J., and P. J. Venter
1964. An investigation into the occurrence of malvine in South African dry red wines. S. African J. Agr. Sci. **7**:731–738.

Van Wyk, C. J., A. D. Webb, and R. E. Kepner
1967. Some volatile components of *Vitis vinifera* variety White Riesling. 1, 2, 3. J. Food Sci. **32**:660–664, 664–668, 669–674.

Van Zyl, J. A.
1962. Turbidity in South African dry wines caused by development of *Brettano-myces* yeasts. S. Africa Sci. Bull. **381**:1–42 (Viticultural series No. 1).

Van Zyl, J. A., and L. de W. Du Plessis
1961. The microbiology of South African wine making. I. The yeasts occurring in vineyards, musts and wines. S. African J. Agr. Sci. **4**:393–401.

LITERATURE CITED

VAN ZYL, J. A., M. J. DE VRIES, and A. S. ZEEMANN
1963. The microbiology of South African winemaking. III. The effect of different yeasts on the composition of fermented musts. S. African J. Agr. Sci. 6:165–180

VAS, K., and M. INGRAM
1949. Preservation of fruit juices with less SO₂. Food Manuf. 24:414–416.

VAS, K., M. NEDBALEK, H. SCHEFFER, and G. KOVÁCS-PROSZT
1967. Methodological investigations on the determination of some pectic enzymes. Fruchtsaft-Ind. 12:164, 166–183.

VASBEN, V. Z., B. M. MEDINETS, R. I. MOSKALENKO, and D. T. NATSVLISHVILI
1966. Avtomaticheskie refraktometry dlia opredeleniia sakharistosti susla. Vinodeli i Vinogradarstvo SSSR 26(5):15–17.

VAUGHN, R. H.
1938. Some effects of association and competition on Acetobacter. J. Bacteriol. 36:357–367.
1949. The bacteriology of wines—a critical review. Wine Technol. Conf., Univ. Calif., Col. Agr. August 10–12, 1949, Davis, Calif. Pp. 33–36. (Mimeo.)
1955. Bacterial spoilage of wines with special reference to California conditions. Advan. Food Res. 6:67–108.

VAUGHN, R. H., H. C. DOUGLAS, and J. C. M. FORNACHON
1949. The taxonomy of Lactobacillus hilgardii and related heterofermentative Lactobacilli. Hilgardia 19(4):133–139.

VAUGHN, R. H., and A. TCHELISTCHEFF
1957. Studies on the malic acid fermentation of California table wines. I. An introduction to the problem. Am. J. Enol. 8:74–79.

VEDEL, A.
1966. Terminologie gustative oenologique. Vignes et Vins No. 149:15.

VENEZIA, M.
1938. Sull'acido ascorbico (vitamin C) nell'uva e nel vino. Ann. R. Staz. Sper. Viticolt. Enol. Conegliano 8:67–82.
1944. Propriétés alimentaires et charactéristiques du raisin. Bull. O.I.V. (Office Intern. Vigne Vin) 17(161):50–59.

VENTRE, J.
1930. Traité de vinification pratique et rationnelle. Vol. I. Le raisin. Les vinifications. Montpellier, France, Librairie Coulet. 490 pp.
1931. Traité de vinification pratique et rationnelle. Vol. II. Le vin. Sa composition, ses maladies, sa conservation. Montpellier, France, Librairie Coulet. 487 pp.
1935. Les levures en vinification. Montpellier, France, Coulet et Fils. 59 pp. (See also Prog. Agr. Vitic. 106:111–115, 135–140, 153–155, 183–187. 1936.)

VEREŠ, A.
1966. Rapport Tchécoslovaque. Bull. O.I.V. (Office Intern. Vigne Vin) 39:1429–1441.

VERONA, O., and G. FLORENZANO
1956. Microbiologia applicata all'industria enologica. Bologna, Edizione Agricole. 191 pp.

VESELOV, K. IA., and I. M. GRACHEVA
1963. Intensity of metabolism in brewers' yeast in various conditions of fermentation. Proc. Intern. Congr. Biochem. 5th, Moscow, 1961 8:257–271.

VETSCHE, U., and H. LÜTHI
1964. Farbstoffverluste während des biologischen Säureabbaues. Schweiz. Z. Obst- Weinbau 73:124–126.

VETTORI, M.
1966. I componenti del vino in rapporto alla sua genuinità. Corso Nazionale di Aggiornamento per Enotecnici "G. Battista Cerletti" 2:179–219.

VILLFORTH, F.
1950–1951. Die Bestimmung der Ameisensäure im Wein. Wein u. Rebe 1950–1951:144–179.
1954. Säureverminderung in Wein durch Austauscher-Kunstharze. Mitt. Rebe u. Wein, Serie A (Klosterneuburg) 4:212–225.

VILLFORTH, F., and W. SCHMID
1954. Ueber hohere Alkohole im Wein. II. Wein-Wissen. 8:107–121.

VIRTANEN, O. E.
1955. Autolyysista ja sen käytöstä biokemiallisessa teollisvvsessa. Suomen Kemistilehti A28:103–111.

VISINTINI-ROMANIN, M.
1967. Ricerche sui pigmenti antocianici nella vite e nei vini. IV. L'ossidazione della malvina. Riv. Viticolt. Enol. 20:79–88.

VITAGLIANO, M., and G. MURA
1966. L'influenza de alcuni fattori technologici sul tenore di acido metil-malico nel vino. Ital. Vinicola Agrar. 56:163–166, 195–200.

VITELES, H.
1928. Cultivation of grapes in Palestine. Bull. Palestine Econ. Soc. 3(3):1–90.

VODRET, A.
1965. La tappatura con sughero o con plastica sulle caratteristiche del vino in bottiglia. Ital. Vinicola Agrar. 55:49–61.

VOGT, E.
1945. Weinchemie und Weinbereitung. Revision of W. Seifert, Die Chemie des Mostes und Weines. 2d ed. Mainz, Verlag d. Deutschen Weinzeitung, J. Diemer; Wiesbaden, R. Bechtold & Co. 361 pp.

VOIGT, J., and R. NOSKE
1966. Zur Bestimmung der Polyphenoloxydaseaktivität. Z. Lebensm.-Untersuch. -Forsch. 129:359–364.

VOSTI, D. C., and M. A. JOSLYN
1954a. Autolysis of baker's yeast. Appl. Microbiol. 2:70–78.
1954b. Autolysis of several pure culture yeasts. Ibid. 2:79–84.

WAGNER, P. M.
1955. The French hybrids. Am. J. Enol. Viticult. 6(1):10–17.

WAHAB, A., W. WITZKE, and W. V. CRUESS
1949. Experiments with ester forming yeasts. Fruit Prod. J. 28(7):198–200, 202, 219.

WALDSCHMIDT-LEITZ, E.
1929. Enzyme actions and properties. (Transl. by R. P. Walton.) New York, John Wiley & Sons, Inc. 255 pp. (See pp. 92–100.)

WALLNÖFER, P., and H.-J. REHM
1965. Beitrag zur Kenntnis der antimikrobiellen Wirkung der schwefligen Säure. V. Die Wirkung der schwefligen Säure auf den Stoffwechsel atmender und gärender Hefe- und Colizellen. Z. Lebensm.-Untersuch. -Forsch. 127:195–206.

WALTON, A. G.
1967. The formation and properties of precipitates. New York, John Wiley & Sons, Inc. 232 pp.

WANNER, E.
1938. Beiträge zur Frage der Ueberschwefelung von Wein. Wein u. Rebe 20(9/10):267–292.

WARKENTIN, H.
1950. Wine stability tests. San Francisco, Wine Institute, Technical Advisory Committee, August 11, 1950. 4 pp.
1951. Wine stability tests. San Francisco, Wine Institute, Technical Advisory Committee, March 6, 1951. 3 pp.

WARKENTIN, H., and M. S. NURY
1963. Alcohol losses during fermentation of grape juice in closed containers. Am. J. Enol. Viticult. 14:68–74.

WATANABE, M., Y. AMANO, and M. KAGAMI
1966. On the adjustment of sulfur dioxide contents in wine and must. Yamanashi Daigaku, Hakkô Kenkyusho Kenkyu Hokoku No. 13:33–41.

WATT, B. K., and A. L. MERRILL
1950. Composition of foods—raw, processed, prepared. U.S. Dept. Agr., Agriculture Handbook 8:1–147.

WEBB, A. D.
1959. The Australian wine industry. Wines & Vines. 40(7):29–30.
1964. Anthocyanins of grapes. In V. C. Runeckles, Phenolics in normal and diseased fruits and vegetables. Norwood, Mass., Plant Phenolics group, North America. (See pp. 21–39.)
1965. Personal communication.
1967. Some aroma compounds produced by vinous fermentation. Biotech. Bioeng. 9:305–319.
1968a. L'arôme du muscat. Connaissance de la Vigne et du Vin 2(1):12–23, 25.
1968b. Lactones in wine. San Francisco, Wine Institute, Technical Advisory Committee, June 6, 1968. 3 pp.

WEBB, A. D., and R. E. KEPNER
1957. Some volatile aroma constituents of Vitis vinifera var. Muscat of Alexandria. Food Res. 22:384–395.
1961. Fusel oil analysis by means of gas-liquid partition chromatography. Am. J. Enol. Viticult. 12:51–59.
1962. The aroma of flor sherry. Am. J. Enol. Viticult. 13:1–14.

WEBB, A. D., R. E. KEPNER, and W. G. GALETTO
1964a. Comparison of the aromas of flor sherry, baked sherry, and submerged-culture sherry. Am. J. Enol. Viticult. 15:1–10.

WEBB, A. D., R. E. KEPNER, and L. MAGGIORA
1966. Gas chromatographic comparison of volatile aroma materials extracted from eight different muscat-flavored varieties of Vitis vinifera. Am. J. Enol. Viticult. 17:247–254
1967. Identification of ethyl acid tartrate and one isomer of ethyl acid malate in California flor sherry. J. Agr. Food Chem. 15:334–339. (See also Wines & Vines 46[7]:25. 1965.)

WEBB, A. D., and C. S. OUGH
1962. Comparison of wines from aroma-stripped musts with wines to which aroma was re-added. Am. J. Enol. Viticult. 13:51–53.

WEBB, A. D., P. RIBÉREAU-GAYON, and J. N. BOIDRON
1964b. Composition d'une essence extraite d'un vin de V. Vinifera (variété Cabernet Sauvignon). Bull. Soc. Chim. France 1964:1415–1420.

WEBB, R. B.
1962. Laboratory studies of the malo-lactic fermentation. Am. J. Enol. Viticult. 13:189–195.

WEBB, R. B., and J. L. INGRAHAM
1960. Induced malo-lactic fermentations. Am. J. Enol. Viticult. 11:59–63.

WEEKS, C.
1969. Production of sulfur dioxide-bending compounds and of sulfur dioxide by two *Saccharomyces* yeasts. Am J. Enol. Viticult. **20**:32–39.

WEGER, B.
1963a. Analisi e prove di chiarificazione con la gelatina liquida. Riv. Viticolt. Enol. **16**:446–448.
1963b. Ueber einen Vergleich von Preis und Qualität verschiedener Handelsbentonite. Weinberg Keller **10**:533–536.
1965a. Ueber ein abnormales verhalten eines Rotweines beim Wärmetest. *Ibid.* **12**:481–484.
1965b. Calcium- und Natriumbentonite. Wein-Wissen. **20**:545–559.
1965c. Ueber das Glukose: Fruktose-Verhältnis in Südtiroler Trauben zum Zeitpunkt der Lese. Mitt. Rebe u. Wein, Serie A (Klosterneuburg) **15**:291–294.
1966. Sui rapporti estratto dedotto zuccheri: ceneri e alcalinità delle ceneri: ceneri. Atti Accad. Ital. Vite Vino **18**:209–223.
1967. Il contenuto in sodio dei vini ed il trattamento con bentoniti. Riv. Viticolt. Enol. **19**:73–78.
1968. Können Analysenwerte von Laborproben zur Beurteilung eines Weines herangezogen werden? Mitt. Rebe u. Wein, Obstbau u. Früchteverwertung (Klosterneuburg) **18**:441–442.

WEIMARN, P. P. VON
1925. Die Allgemeinheit des Kolloidsaustandes, Kolloides und Kristalloides Lösen Niederschlagen. 2d ed. Dresden, T. Steinkopff. 1 vol.

WEINGES, K.
1964. The occurrence of catechins in fruits. Phytochem. **3**:263–266.

WEINMANN, J., and L.-F. TELLE
1929. Manuel de travail des vins mousseux. 5th ed. Épernay Institute de Oenologique de Champagne. viii + 335 pp.

WEIS, H., A. ROUSETT, M. MAILLARD, and R. BONNET
1960. L'évolution qualitative et semi-quantitative des acides aminés au cours de la fermentation des moûts de raisin provenant de cépages d'Alsace. Compt. Rend. **250**:1322–1324.

WEJNAR, R.
1965. Der Einfluss der Temperatur auf die Bildung von Zucker, Aepfelsäure und Weinsäure in Weintrauben. Ber. Deut. Botan. Ges. **78**:314–321.
1966. Der biologische Säureabbau im Wein. I. Zur Frage der Stabilität organischer Säuren gegenüber Mikroorganismen im Wein. II. Die Regulierung des Aepfelsäureabbaues in Traubenwein durch Zusatz von Weinsäure und Zucker. Zentr. Bakteriol., Parasitenk., Abt. 2 **120**:123–131, 132–140.
1967. Weitere Untersuchungen zum Einfluss der Temperature auf die Bildung von Aepfelsaure in Weinbeeren. Ber. Deut. Bot. Ges. **80**:447–450.
1968. Untersuchungen über die Bedeutung der Weinsäure fur die Wasserstoffionenkonzentration des Traubenweines. Mitt. Rebe u. Wein, Obstbau u. Früchteverwertung (Klosterneuberg) **18**:349–358.

WERNER, H., and G. HARTMANN
1966. Ueber die Möglichkeit der kolorimetrischen Bestimmung des "löslichen" Traubenproteins im Hinblick auf die Frage der Eiweissstabilität von Wein. Weinberg Keller **13**:5–36.

WERRINGLOER, H. W. J.
1966. Quantitative Bestimmung von Histamin in Wein. Arzneimittel-Forsch. **16**:1654–1656.

WEURMAN, C., and C. DE ROOIJ
1958. Chlorogenic acid isomers in "Black Alicante" grapes. Chem. Ind. **1958**:72.

WHITNEY, R. P., S. T. HAN, and J. L. DAVIS
1953. The mechanism of sulfur dioxide absorption in aqueous media. Tappi 36:172–175.

WHITNEY, R. P., and J. E. VIVIAN
1949. Absorption of sulfur dioxide in water. Chem. Eng. Progr. 45:323–337.

WIENHAUS, H.
1967. Untersuchungen über Zucker- und Aepfelsäurevergärung durch Schizosaccharomyces pombe var. liquefaciens. Wein-Wissen. 22(1):25–39.

WILEY, H. W.
1903. American wines at the Paris Exposition of 1900: their composition and character. U.S. Dept. Agr., Bur. Chem. Bull. 72:1–40.
1919. Beverages and their adulteration. Philadelphia, P. Blakiston's Son and Co. 421 pp. (See chapter on wines, especially pp. 192, 231–233.)

WILKINSON, W. P.
1918 (i.e., 1919). The nomenclature of Australian wines. Victorian Geog. J. 34: 18–70.

WILLAMAN, J., and Z. I. KERTEZ
1931. The enzymatic clarification of grape juice. New York (Geneva) Agr. Exper. Stat. Tech. Bull. 178:1–15.

WILLIAMS, B. L., and S. H. WENDER
1952. The isolation and identification of quercetin and isoquercitrin from grapes (Vitis vinifera). J. Am. Chem. Soc. 74:4372–4373.

WILLIAMS, R. J.
1941. Growth-promoting nutrilites for yeasts. Biol. Rev. 16:49–80.

WILLIG, R.
1950. Continuous fermentation of wine. Wynboer 19(223):14–15.

WILLSON, K. S., W. O. WALKER, C. V. MARS, and W. R. RINELLI
1943a. Liquid sulfur dioxide in the fruit industries. Fruit Prod. J. 23(2):72–82.

WILLSON, K. S., W. O. WALKER, W. R. RINELLI, and C. V. MARS
1943b. Liquid sulfur dioxide. Chem. Ind. 53:176–186.

WILLSTÄTTER, R.
1927. Problems and methods in enzyme research. George Fisher Baker Non-Resident Lecture in Chemistry, Cornell University, Ithaca, New York. 62 pp.

WILLSTÄTTER, R., and E. H. ZOLLINGER
1915. Untersuchungen über die Anthocyans. VI. Ueber die Farbstoffe der Weintrauben und der Heidelbeere. Ann. Chem. 408:83–109.
1916. Ibid. XVI. Ueber die Farbstoffe der Weintraube und der Heidelbeere. Ibid. 412:195–216.

WINDISCH, K.
1906. Die chemischen Vorgänge beim Werden des Weines. Stuttgart, Eugen Ulmer. 122 pp. (See especially p. 37.)

WINE INSTITUTE
1946. Winery sanitation guide. San Francisco, Wine Inst. Bull. 331-A:1–16.
1957. Bulletin No. 871. San Francisco. 14 pp. (See p. 12.)
1965. Twenty-ninth annual wine industry statistical survey. San Francisco, Wine Inst. Bull. 1318:1–15 (see p. 7). (See also no. 1380, 1966; no. 1420, 1967.)
1966. Bulletin No. 1380. San Francisco. 14 pp. (See p. 5.)
1967. Bulletin No. 1420. San Francisco. 13 pp. (See p. 7.)
1968a. California wine type specifications. Rev. ed. June 8, 1968. San Francisco, Calif. 7 pp.

1968b. Thirty-first annual wine industry statistical survey. San Francisco, Wine Inst. Bull. Economic Research Report 2:1–23. Thirty-second survey. 3:1–22.

WINES & VINES
1967. Wines & Vines annual directory. San Francisco, Wines & Vines. 167 pp.

WINKLER, A. J.
1932. Maturity tests for table grapes. Calif. Agr. Exper. Stat. Bull. 529:1–35.
1935. Making red wines in Algeria. Wine Rev. 3(7):14–15, 40.
1958. The relation of leaf and climate to vine performance and grape quality. Am. J. Enol. Viticult. 9:10–23.
1962. General viticulture. Berkeley and Los Angeles, University of California Press. 663 pp. (See pp. 504–516.)
1964. Varietal wine grapes in the central coast counties of California. Am. J. Enol. Viticult. 15:204–205.

WINKLER, A. J., and M. A. AMERINE
1937. What climate does—the relation of weather to the composition of grapes and wines. Wine Rev. 5(6):9–11; (7):9–11, 16.
1938. Color in California wines. I. Methods for measurement of color. II. Preliminary comparisons of certain factors influencing color. Food Res. 3(4): 429–447.

WINKLER, A. J., L. H. LAMOURIA, and G. H. ABERNATHY
1957. Mechanical grape harvesting problems and progress. Am. J. Enol. 8:182–187.

WINKLER, A. J., and W. O. WILLIAMS
1936. Effect of seed development on the growth of grapes. Proc. Am. Soc. Hort. Sci. 33:430–434.

WINTER, F. A.
1967. Personal communication.

WINTON, A. L., and K. B. WINTON
1935. The structure and composition of foods. New York, John Wiley & Sons, Inc. 2 vols. (See 2:745–768.)

WINZLER, R. J., and J. P. BAUMBERGER
1938. The degradation of energy in the metabolism of yeast cells. J. Cellular Compar. Physiol. 12:183–211.

WINZLER, R. J., D. BURK, and V. DU VIGNEAUD
1944. Biotin in fermentation, respiration, growth and nitrogen assimilation by yeast. Arch. Biochem. 5:25–47.

WOLF, E., and I. BENDA
1965. Qualität und Resistenz. III. Das Futterwahlvermögen von Drosophila melanogaster gegenüber natürlichen Weinhefe-Arten und -Rassen. Biolog. Zbl. 84:1–8.
1966. Zur Differenzierung von Heferassen durch Drosophila melanogaster bei Vertreten der Gattung Schizosaccharomyces. Weinberg Keller 14:163–166.

WOLF, E., and G. REUTHER
1959. Qualität und Resistenz. II. Das Futterwahlvermögen von Drosophila im Hinblick auf ökologisch verschiedene Weinhefe-Populationen. Biolog. Zbl. 78:813–821.

WONG, G., and A. CAPUTI, JR.
1966. A new indicator for total acid determination in wines. Am. J. Enol. Viticult. 17:174–177.

WORTHINGTON, O. J., R. F. CAIN, and E. H. WIEGAND
1949. Determination of color of unclarified juices by reflectometer. Food Technol. 3(8):274–277.

LITERATURE CITED

WUCHERPFENNIG, K., and G. BRETTHAUER
1962. Versuche zur Stabilisierung von Wein gegen oxydative Einflusse durch Behandlung mit Polyamidpulver. Weinberg Keller 9:37–55. (*See also* Fruchtsaft-Ind. 7:40–54. 1962.)

WUCHERPFENNIG, K., and I. FRANKE
1964. Beitrag zur Bestimmung einer Kennzahl für Polyphenole in Weinen durch Gel-Filtration. Wein-Wissen. 19:362–369.
1967. Zur Frage der Eiweissstabilisierung von Wein durch eine Bentonitbehandlung des Mostes. *Ibid.* 22:213–226.

WUCHERPFENNIG, K., and E. M. KLEINKNECHT
1965a. Beitrag zur Veränderung der Farbe und der Polyphenole bei Abfüllung von Weisswein durch Einwirkung von Säuerstoff und Warme. Wein-Wissen. 20:489–514.
1965b. Betrag über die Sauerstoffaufname der abzufüllenden Flüssigkeit in verschiedenen Füllsystem. Brauwissenschaft 18(8):296–296.
1965c. Ueber das Verhalten des Kohlendioxyds bei der Abfüllung von Wein mit Hilfe von verschiedenen Füllsystemen und- verfahren und über seine Bestimmung. Weinberg Keller 12:547–556.
1966a. Beitrag zur Sauerstoffaufnahme von Wein unter besonderer Berücksichtigung der Veränderungen seines Gehaltes an freier Schwefliger Säure und seines ITT-Wertes. Mitt. Rebe u. Wein, Obstbau u. Früchteverwertung (Klosterneuburg) 16:19–30.
1966b. Organoleptische Prüfung kalt- und warmgefüllter Weine. Deut. Weinbau 21:6, 8, 10–12.

WUCHERPFENNIG, K., and A. LAY
1967. Zur Bildung und zum Vorkommen von Hydroxymethylfurfurol in Weinen. Weinberg Keller 14:209–216.

WUCHERPFENNIG, K., and D. RATZKA
1967. Ueber die Verzögerung der Weinsteinausscheidung durch polymere Substanzen des Weines. Weinberg Keller 14:499–509.

WÜRDIG, G.
1966. Die gaschromatographische Bestimmung von Sorbinsäure und Benzoesäure im Wein. Deut. Lebensm.-Rundschau 62:147–149.

WÜRDIG, G., and W. CLAUSS
1966. Herkunft und Entstehung von Schleimsäure; Ursache häufiger Kristalltrübungen im Wein. Weinberg Keller 13:513–517.

WÜRDIG, G., and H. A. SCHLOTTER
1967. SO$_2$-Bildung in Gärenden Traubenmosten. Z. Lebensm.-Untersuch. -Forsch. 134:7–13.

WURZIGER, J.
1954. Beitrag zur Kenntnis des Mangangehaltes im Wein. Deut. Lebensm.-Rundschau 50:49–51.

WYSS, O.
1948. Microbial inhibition by food preservatives. Advan. Food Res. 1:373–393.

YAMADA, T.
1959. Economic considerations in selecting storage and processing tanks. Am. J. Enol. Viticult. 10:13–16.

YANG, H. Y., and R. E. ORSER
1962. Preservative effect of vitamin K$_5$ and sulfur dioxide on sweet table wines. Am. J. Enol. Viticult. 13:152–158.

YANG, H. Y., J. H. JOHNSON, and E. H. WIEGAND
1947. Electronic pasteurization of wine. Fruit Prod. J. 26:295–299.

YERINGTON, A. P.
1958. What wineries can do about drosophila. San Francisco, Wine Institute, Technical Advisory Committee. May 26, 1958. 3 pp.
1963. Studies with aerosol insecticides for winery use. San Francisco, Wine Institute, Technical Advisory Committee. March 15, 1963. 2 pp.
1964. The use of dichlorvos (DDVP) in wineries for Drosophila control. San Francisco, Wine Institute, Technical Advisory Committee. December 11, 1964. 2 pp.
1967. The use of dichlorvos in wineries. San Francisco, Wine Institute, Technical Advisory Committee. December 8, 1967. 2 pp.

YOSHIZAWA, K.
1965. The formation of higher alcohols in the fermentation of amino acids by yeasts. The formation of isobutanol and isoamyl alcohol from pyruvic acid by washed yeast cells. Agr. Biol. Chem. (Tokyo) 29:672–677.
1966. On various factors affecting formation of isobutanol and isoamyl alcohol during alcoholic fermentation. Agr. Biol. Chem. (Tokyo) 30:634–641.

YOUNGER, W.
1966. Gods, men, and wine. Cleveland, Ohio, The Wine and Food Society. 516 pp.

YUI, N.
1940. Dissociation constants of sulfurous acid. Bull. Inst. Phys. Chem. Research (Tokyo) 19:1229–1236.

ZAMBONELLI, C.
1964a. Ricerche biometriche sulla produzione di idrogeno solforato da solfati e solfiti in Saccharomyces var. ellipsoideus. Ann. Microbiol. Enzimol. 14:129–141.
1964b. Ricerche genetiche sulla produzione di idrogeno solforato in Saccharomyces cerevisiae var. ellipsoideus. Ibid. 14:143–153.
1965a. Ibid. II. Ereditarietà del carattere dal punto di vista quantitativo. Ibid. 15:89–97.
1965b. Ibid. III. Ulteriori studi sulla ereditarietà del carattere quantitativo. Ibid. 15:99–106.
1965c. Stabilità ed ereditarietà della variazione nella produzione di H_2S indotte dal nitrato fenil-mercurico in Saccharomyces cerevisiae var. ellipsoideus. Ibid. 15:181–195.

ZAMORANI, A., and P. G. PIFFERI
1964. Contributo alla conoscenza della sostanza colorante dei vini. 1. Identificazione e valutazione quantitativa degli antociani vini da Vitis vinifera (Merlot) e da ibridi (Clinton e Baco). Riv. Viticolt. Enol. 17:85–93.

ZANG, K.
1963. Schweflige Säure im ungeschwefelten Most? Deut. Wein-Ztg. 99:214.

ZANG, K., and K. FRANZEN
1966. Schweflige-Säure-Bildung im Verlauf der Traubenmost-Gärung. Deut. Wein-Ztg. 102:128, 130.
1967. Bildung von schwefligen Säure bei der Weinbereitung und ihre möglichen Ursachen. Ibid. 103:88–89.

ZAVARIN, E., and K. SNAJBERK
1965. The chemistry of natural phlobaphenes. III. Pyrolysis of the phlobaphenes from five representative softwood species. Tappi 48:612–616.

ZAVARIN, E., K. SNAJBERK, and R. M. SMITH
1963. The chemistry of natural phlobaphenes from redwood (Sequoia sempervirens Endl) Tappi 46:320–323.
1965. The chemistry of natural phlobaphenes. II. Further pyrolysis studies of the phlobaphenes from redwood (Sequoia sempervirens Endl) Tappi 48:574–577.

LITERATURE CITED

ZHDANOVICH, G. A., V. D. EMEL'IANOV, and L. L. GEL'GAR
1967a. K metodike sravnitel'nykh ispytaniia vinodel'cheskikh drobil'nogrebneot-deliaiushchikh mashin. Trudy Vses. Nauch.-Issledov. Inst. Vinodeliia Vinogradarstva "Magarach" 15:13–20.

ZHDANOVICH, G. A., V. P. NECHAEV, and P. M. IAKOVLEV
1967b. Issledovanie protsessa izvlecheniia susla pervoĭ fraktsii iz vinogradnoĭmezgi. Trudy Vses. Nauch.-Issledov. Inst. Vinodeliia Vinogradarstva "Magarach" 15:21–26.

ZIEMBA, J. V.
1954. Production streamlined on semicircular line. Food Eng. 26(8):74–76, 170.

ZIMMERMAN, H. W.
1963. Studies on the dichromate method of alcohol determination. Am. J. Enol. Viticult. 14:205–213.

ZIMMERMAN, H. W., and J. FESSLER
1949. Chemical method for alcohol. Wines & Vines 30(9):65.

ZIMMERMAN, H. W., E. A. ROSSI, JR., and E. WICK
1964. Alcohol losses from entrainment in carbon dioxide evolved during fermentation. Am. J. Enol. Viticult. 15:63–68.

ZIMMERMAN, W., and L. MALSCH
1938. Der Gehalt an Vitamin C und A in Süssmosten. Vorratspflege u. Lebensmittelforsch. 1:311–314.

ZIMMERMAN, W., L. MALSCH, and R. WEBER
1940. Die Veränderung des Vitamin-C-gehaltes von Süssmosten. Vorratspflege u. Lebensmittelforsch. 3(1/2):1–7.

ZINCHENKO, V. I.
1964a. Sortovye razlichiia vinograda, sviazannye s azotonakopleniem (v Zakar-patskoĭ oblasti USSR). Trudy Vses. Nauch.-Issledov. Inst. Vinodeliia Vinogradarstva "Magarach" 13:11–23.

1964b. Predel'noe soderzhanie azotistykh veshchestv v susle i vine, predotvrash-chaiushchee pereokislennost'. Ibid. 13:24–43.

1965. O faktorakh, vliiaiushchikh na pereokislennost' belykh stolovykh vin. Vino-delie i Vinogradarstvo SSSR 25(1):10–12.

ZUBECKIS, E.
1964. Ascorbic acid in Veèrport grape during ripening and processing. Rept. Hort. Exper. Stat. and Prod. Lab. Ontario 1964:114–116.

INDEX

INDEX

Includes grape varieties and chemical compounds, but districts within countries, unless extensively discussed in the text, are not separately listed.

977

TABLE WINES